MW00710622

# Handbuch der Pflanzenphysiologie · Encyclopedia of plant physiology

# ENCYCLOPEDIA
# OF PLANT PHYSIOLOGY

EDITED BY

## W. RUHLAND

COEDITORS

E. ASHBY · J. BONNER · M. GEIGER-HUBER · W. O. JAMES
A. LANG · D. MÜLLER · M. G. STÅLFELT

VOLUME I

# GENETIC CONTROL OF
# PHYSIOLOGICAL PROCESSES ·
# THE CONSTITUTION OF THE PLANT CELL

CONTRIBUTORS

D. G. CATCHESIDE · M. COHEN · H. DRAWERT · K. EGLE
L. VON ERICHSEN · L. GEITLER · S. GRANICK · C. HARTE
P. J. KRAMER · A. PISEK · R. D. PRESTON · M. M. RHOADES
W. SEIFRIZ · J. A. SERRA · K. STEFFEN · E. TREIBER
F. W. WENT · S. G. WILDMAN

SUBEDITORS

H. ULLRICH AND H. J. BOGEN

WITH 283 FIGURES

Springer-Verlag Berlin Heidelberg GmbH
1955

# HANDBUCH DER PFLANZENPHYSIOLOGIE

HERAUSGEGEBEN VON

## W. RUHLAND

IN GEMEINSCHAFT MIT

E. ASHBY · J. BONNER · M. GEIGER-HUBER · W. O. JAMES
A. LANG · D. MÜLLER · M. G. STÅLFELT

BAND I

## GENETISCHE GRUNDLAGEN PHYSIOLOGISCHER VORGÄNGE · KONSTITUTION DER PFLANZENZELLE

BEARBEITET VON

D. G. CATCHESIDE · M. COHEN · H. DRAWERT · K. EGLE
L. VON ERICHSEN · L. GEITLER · S. GRANICK · C. HARTE
P. J. KRAMER · A. PISEK · R. D. PRESTON · M. M. RHOADES
W. SEIFRIZ · J. A. SERRA · K. STEFFEN · E. TREIBER
F. W. WENT · S. G. WILDMAN

REDIGIERT VON

## H. ULLRICH und H. J. BOGEN

MIT 283 ABBILDUNGEN

Springer-Verlag Berlin Heidelberg GmbH
1955

ISBN 978-3-642-49071-2 ISBN 978-3-642-94653-0 (eBook)
DOI 10.1007/978-3-642-94653-0

ALLE RECHTE,
INSBESONDERE DAS DER ÜBERSETZUNG IN FREMDE SPRACHEN,
VORBEHALTEN

OHNE AUSDRÜCKLICHE GENEHMIGUNG DES VERLAGES
IST ES AUCH NICHT GESTATTET, DIESES BUCH ODER TEILE DARAUS
AUF PHOTOMECHANISCHEM WEGE (PHOTOKOPIE, MIKROKOPIE) ZU VERVIELFÄLTIGEN

ⓒ BY SPRINGER-VERLAG BERLIN HEIDELBERG 1955
ORIGINALLY PUBLISHED BY SPRINGER-VERLAG OHG. BERLIN · GÖTTINGEN · HEIDELBERG 1955
SOFTCOVER REPRINT OF THE HARDCOVER 1ST EDITION 1955

# Vorwort.

Dem Handbuch der gesamten Pflanzenphysiologie, das mit dem vorliegenden ersten Band zu erscheinen beginnt, ist als einziger Vorläufer die klassische „Pflanzenphysiologie" Wilhelm Pfeffers vorausgegangen. In deren zwei Bänden „Stoffwechsel" und „Kraftwechsel", die 1897 und 1904 in Leipzig erschienen, bewältigte der Altmeister auf 1600 Seiten die gesamte Literatur, welche schon damals in großer Fülle vorlag. Diese Leistung muß selbst dann noch bewundernswert erscheinen, wenn man absehen will von dem hohen Niveau kritischer Durchdringung und der Vielfalt eigener, neuer und fruchtbarer Gedanken, von welchen jede Seite des Werkes zeugt.

Seither hat die Physiologie alle anderen Zweige der Botanik in immer wachsendem Tempo überflügelt und immer größere Anziehungskraft auf die Forscher ausgeübt. Ihr Streben, hinter die äußere, symptomatische Erscheinung der Dinge vorzudringen und die bewegenden Zusammenhänge zu suchen, hat ebensosehr unser allgemeines Verständnis für das pflanzliche Lebewesen erstaunlich vergrößert, wie es die Grundlage für erfolgreiches Voranschreiten der Land- und Forstwirtschaft, Pharmakologie, Pharmakognosie, Pathologie und anderer Zweige der angewandten Botanik gefestigt hat. Zur Entwicklung der physiologischen Forschung haben freilich auch Chemie und Physik entscheidend beigetragen, indem sie neue Gesichtspunkte und Methoden zur vertieften Behandlung alter Probleme und zum Angriff auf immer neue Ziele darboten.

Als zwangsläufige Folge dieser stürmischen Entwicklung mußte indessen auch eine ständig wachsende Spezialisierung der physiologischen Arbeitsbereiche in Kauf genommen werden. Heute vermag der einzelne Forscher kaum noch die Nachbargebiete des eigenen Betätigungsfeldes zu überschauen, niemand aber mehr das Ganze.

So war denn schon in den Jahren nach dem ersten Weltkriege das Bedürfnis nach einer groß angelegten, durch Zusammenarbeit vieler Forscher zu gewinnenden Darstellung des bisher Erreichten gewachsen und vielfach laut geworden. Dieses starke Bedürfnis veranlaßte Herrn Dr. Ferdinand Springer, im Jahre 1937 an den Unterzeichner mit dem Vorschlage zur Organisierung und Herausgabe eines solchen Werkes heranzutreten. Nach längerem Zaudern ob der Größe des Unternehmens und der hohen Verantwortung, besonders auch für eine glückliche Auswahl und Gewinnung der besten Autoren, nahm der Befragte die ehrenvolle Aufgabe schließlich an, nicht zuletzt auch auf das Drängen seiner alten Leipziger Mitarbeiter, auf deren Hilfsbereitschaft er bei diesem Werke vertrauen durfte. Die alsbald erfolgenden Zusagen der Kollegen des In- und Auslandes zur Bearbeitung der ihnen angetragenen Kapitel waren über Erwarten zahlreich und ermutigend. Als der Kriegsausbruch allen Plänen ein jähes Ende bereitete, waren diese immerhin schon so weit verwirklicht, daß für die Mehrzahl der Kapitel die Kontrakte aus den wesentlichen Kulturländern abgeschlossen waren und sogar zum Teil die Manuskripte zum Druck bereit lagen.

Bald nach Kriegsende trafen auch aus dem Auslande eine ganze Reihe von Anfragen nach dem weiteren Schicksal des Planes ein, wobei immer wieder hervorgehoben wurde, daß das Bedürfnis nach einem ausführlichen Handbuch in den Jahren unterbrochener wissenschaftlicher Kommunikation, aber erheblicher Erkenntnisfortschritte noch gewachsen sei. Für den Springer-Verlag hätte es solcher ermutigender Stimmen kaum bedurft, um mit gewohntem Wagemut das alte Ziel anzusteuern; den Herausgeber haben sie veranlaßt, die alten Pläne doch wieder aufzunehmen und sie Gestalt gewinnen zu lassen.

Dabei erforderte natürlich das inzwischen vermehrte Wissen eine radikale Umarbeitung der Disposition; an ihr haben Coeditoren und Bandherausgeber hervorragenden Anteil. Die Frage, wie weit man bei dieser Neugliederung und Aufteilung in Kapitel gehen solle, ist allgemeingültig kaum zu beantworten. So bevorzugten denn manche Stimmen umfassendere Kapitel, innerhalb derer dem sachkundigen Autor freierer Spielraum für Anordnung, Auslese und Bewertung des Stoffes bleibt. Zwischen dieser beachtenswerten Auffassung und dem Grundsatz einer strengen Spezialisierung kam es nach mancherlei Erwägungen schließlich zu einem Kompromiß, der sich allerdings der zweiten Alternative zuneigte. Herausgeber und Verleger glaubten, nur mit solcher Entscheidung dem durch die Spezialisierung enorm angewachsenen Wissensstoff gerecht werden zu können und mußten dafür freilich ein erhebliches Steigen der Zahl mitarbeitender Autoren in Kauf nehmen, deren Individualität unvermeidlich eine entsprechende Ungleichheit in Darstellung und Diktion bringen mußte. Doch verblieb so der Nachdruck auf der Authentizität des gebotenen Inhalts. Praktisch bewährte sich dieses Verfahren insofern, als die Autoren fast immer um so schwerer für das Handbuch zu gewinnen waren, je weiter das ihnen angebotene Kapitel abgesteckt war. Oft mußte ein Kapitel sogar noch stärker als vorgesehen unterteilt werden, um die Zusage von kompetenten Autoren zu sichern. In seiner weit ausgedehnten Korrespondenz wurde dem Unterzeichner dagegen nur in zwei Fällen der Vorwurf einer übertriebenen thematischen Gliederung gemacht.

Um diese detaillierte Aufgliederung nach Möglichkeit zu kompensieren, wurde jedem Bande eine „Einführung und Übersicht" vorangeschickt, welche dem weniger unterrichteten Leser, namentlich auch manchem Praktiker aus den angewandten Gebieten der Physiologie willkommen sein wird. Sie wurde vom jeweiligen Bandherausgeber erst nach Kenntnis der von den Autoren des Bandes eingegangenen Manuskripte abgefaßt.

Häufig erschien es geboten, die Hinweise auf verwandte Beiträge anderer Bände zu geben oder eine kurze Zusammenfassung solcher Beiträge (meist aus der Feder des gleichen Autors) in den betreffenden Band einzufügen. Dadurch mag zunächst dem Benutzer des Handbuches das Verständnis erleichtert werden. Mindestens ebenso bedeutungsvoll erschien es dem Herausgeber jedoch, daß durch solche Hinweise die vielfältige Verschlungenheit der Probleme als Hauptmerkmal der Physiologie zum Ausdruck kommt.

Es bedarf wohl keiner Begründung dafür, daß die Untersuchungsmethodik überall nur kurz und in grundsätzlicher Hinsicht berührt wurde, während wir uns für die Details und die praktische Anwendung mit literarischen Hinweisen begnügten. Eine neue ausführliche Darstellung wird in den „Modernen Methoden der Pflanzenanalyse", herausgegeben von PAECH und TRACEY, des gleichen Verlages geboten.

So glauben wir hoffen zu dürfen, den Wissensstand, wie er zum Zeitpunkt der Abfassung dieses Werkes vorlag, annähernd vollständig wiedergegeben zu haben. Der Verlag beabsichtigt, nach dem vollständigen Erscheinen des Handbuches Supplemente herauszugeben, damit es nicht streng zeitgebunden bleibe, sondern mit der Forschung Schritt halte.

Zum Schluß sei allen Mitarbeitern, die das Zustandekommen des Werkes ermöglichten, der gebührende Dank gesagt. Daß sie sich aus den deutschen, englischen, französischen und anderen Sprachgebieten so bald nach der Kriegskatastrophe zu gemeinsamer Arbeit zusammenfanden, möge auch als tröstliches Zeichen wachsender Völkerversöhnung nicht ganz übersehen werden.

Schloß Unterdeufstetten, Oktober 1955.                    WILHELM RUHLAND.

# Preface.

The Encyclopedia of Plant Physiology, which is inaugurated by the present volume, has only one comparable, comprehensive forerunner, namely, WILHELM PFEFFER's "Plant Physiology", published in 1897–1904 in Leipzig and soon after translated into English (1900–1906) and quite a number of other languages. In the 1600 pages of this classic, PFEFFER summarized the entire plant physiological literature which existed at that time and which had already reached an impressive volume. Even more significant than this achievement, however, was the author's thorough, critical evaluation of the entire subject matter, and the multitude of new and stimulating ideas which he contributed and which are evident on almost every page of the work.

Since PFEFFER's days, the development of plant physiology has increasingly overtaken all other branches of botany and has attracted a steadily growing number of investigators to this field. Their efforts to penetrate through the external aspects of phenomena to the underlying, causative processes have greatly enhanced our insight into the plant as an organism. Their contributions have also served as the foundation for numerous advances in agriculture, forestry, pharmacology, plant pathology and other fields of applied botany. An important feature in the development of modern plant physiology has been the growing role of physics and chemistry in furnishing new ideas and new techniques, both for the solution of old problems and the recognition of new ones.

An inevitable effect of this rapid development, however, has been an increasing specialization in plant physiological research. Today, the individual plant physiologist has a hard time keeping up with the progress in fields immediately bordering on his own; he cannot hope to follow the progress of his science as a whole.

Because of this growing specialization, the need for a comprehensive and authoritative review of all achievements of plant physiology became apparent as early as the years following World War I. It was evident, however, that such a review could no longer be accomplished by a single author, but would have to be the joint effort of a team of many workers. This situation caused Dr. Ferdinand Springer in 1937 to suggest to the present editor the organization and preparation of an encyclopedia of plant physiology. The size of the undertaking and the responsibility which it carried, particularly in selecting and engaging the best contributors, caused the undersigned to hesitate for some time. However, he ultimately decided to accept the challenge and to assume the responsibility, not the least inducement in this decision being the encouragement which he received from his old students in the Botany Department of the University of Leipzig, and the assurance that he could count on their enthusiastic support. The response from plant physiologists at home as well as abroad, who were approached as contributors, was also favorable and encouraging beyond hopes. When the outbreak of World War II brought an abrupt end to all plans, authors of many countries had been signed for the majority of chapters, and some manuscripts were ready to go to press.

Soon after the end of the war, numerous inquiries were received about the project, both from Germany and from abroad. It was emphasized that the need for the encyclopedia had, if anything, further increased because of the long years of interrupted exchange, but continued growth in information. The Springer Verlag hardly needed such an encouragement to tackle the old plan again;

and, under such opportune circumstances, the editor, in turn, felt he should not hesitate to resume the project.

It was obvious that the advances in our knowledge called for radical alterations in the organization of the encyclopedia; coeditors and subeditors enthusiastically shared in this task with the editor-in-chief. The most difficult problem, and one that can hardly be settled in a general way, was the question of how far to carry the subdivision of the subject matter. Preference was voiced all the way from comprehensive chapters, which would leave the final word on arrangement, selection and evaluation of the material with the author, to a great number of strictly specialized chapters, each extremely limited in scope. In the end, a compromise was reached between the two extremes, which, however, tends to the second rather than the first alternative. Both editor and publisher felt this to be the best possible solution in view of the high degree of specialization and the enormously increased volume of research. They appreciated the fact that the number of authors and, thus, the heterogeneity in treatment and presentation would by necessity increase, but they believed that the whole work would benefit by maximum authenticity of the content. It soon became evident, too, that it was much easier to enlist authors for shorter chapters than for longer ones. In order to secure the cooperation of competent authors, it proved in some cases necessary to subdivide a chapter even further than originally planned. On the other hand, in all of his very extensive negotiations with authors the editor was taken to task not more than twice for excessive fragmentation of the chapters.

To compensate for the high degree of specialization, each volume commences with an "Introduction and Survey", which will aid the nonspecialist reader in finding all the information he desires and which should be particularly welcome to readers coming from applied fields of plant physiology. These introductory chapters have been written by the editors of the respective volumes after all contributions have been available in their final form. In many instances, cross-references are given to related chapters in other volumes; in some, brief summaries of such chapters—written, as a rule, by the same author—have been incorporated. This not only will aid the reader in obtaining as comprehensive a grasp of the subject as possible, but will at the same time serve to emphasize the close and manifold interrelations of the individual problems which have become a major characteristic of modern physiology.

It will be appreciated, without further explanation, that methods could be treated only briefly and with emphasis on principles rather than technical details. The latter can be found either in the original publications, as cited in the literature lists, or in the new and comprehensive publication "Modern Methods of Plant Analysis", edited by PAECH and TRACEY and also published by Springer Verlag.

It is our earnest hope that the encyclopedia will present a genuine and well balanced picture of our knowledge in plant physiology such as is stood when the work was published. However, to keep it abreast of development, it is planned to publish, upon completion of the basic volumes, supplementary volumes summarizing the advances made in various fields.

In conclusion, the editor wishes to express due thanks to all coworkers and contributors who have joined in a truly international accomplishment of this enterprise. That this has been possible so soon after the last, disastrous war, should be taken as an encouraging sign of the growing goodwill between the nations of the world.

Schloss Unterdeufstetten, October 1955.                    WILHELM RUHLAND.

# Einführende Bemerkungen zur Gliederung des Handbuches.

Das Handbuch zerfällt in 3 (Haupt-) Teile, von denen I die „Allgemeinen Grundlagen" in den Bänden 1 und 2 behandelt, während der weit umfangreichere Teil II (Bände 3—13) dem Stoff- und Energiewechsel gewidmet ist. Teil III endlich bringt die Physiologie des Wachstums, der Entwicklung und der Bewegungen.

Teil I (Allgemeine Grundlagen) beginnt mit einem Kapitel über die Grundlagen genetischer Art, d. h. mit inneren, erblichen Faktoren, welche in erster Linie die Gestaltung, Differenzierung, Entwicklung und Korrelationen beherrschen und damit auch die besondere Reaktion einer Pflanze auf die Umwelt beeinflussen (CATCHESIDE). Die dafür verantwortlichen lebenden Elemente der Protoplasten können die Chromosomen als die Träger der Erbanlagen („Gene") sein oder, wie der 2. Artikel (RHOADES) zeigt, nicht-genische Erbfaktoren, welche nicht im Zellkern lokalisiert sind. Dahin gehören die Plastiden („das Plastom" oder „Plastidom"), andere partikuläre Bestandteile des Plasmas und das mikroskopisch homogene Cytoplasma mit seinen „Plasmonen".

Da sich unsere physiologische Erkenntnis auf die richtige Analyse von Ursache und Wirkung stützt, müssen wir für unsere Experimente auf möglichst erbeinheitliches Pflanzenmaterial bedacht sein, das gleichwohl individuell noch etwas verschieden auf die von uns gestellten experimentellen Bedingungen, also biologisch variabel reagieren kann, so daß unsere Schlüsse einer statistischen Sicherung bedürfen. Diese physiologische Variabilität wird in dem 3. Artikel (WENT), besonders am Beispiel entwicklungsphysiologischen Geschehens behandelt, während der darauffolgende Artikel sich mit der statistischen Methodik befaßt (HARTE).

In Band 1 werden an weiteren allgemeinen Grundlagen die morphologisch faßbaren Strukturen der Zelle dargestellt: das Cytoplasma, der Zellkern, die Plastiden, Chondriosomen und Mikrosomen, die Vacuole und die Zellwand, wobei ihre physikalischen und chemischen Eigenschaften und die Beziehungen zu den physiologischen Vorgängen besondere Berücksichtigung finden. Aber auch dem Wasser sind mehrere Kapitel gewidmet.

Von der Tatsache ausgehend, daß alle physiologischen Vorgänge sich zuerst in der Zelle abspielen, behandeln wir im Band 2 dann eine allgemeine Physiologie der Pflanzenzelle, womit der Teil I des Handbuchs abgeschlossen wird.

Die Teile II und III behandeln dann die Vielzahl der physiologischen Einzelvorgänge, deren Gruppierung ohne weiteres aus den Bandüberschriften verständlich wird. Teil II (Band 3—13) ist, wie schon bemerkt, dem Stoff- und Energiewechsel, Teil III (Band 14—18) der Physiologie von Wachstum, Entwicklung und Bewegungen gewidmet.

## Introductory remarks concerning the arrangement of the subject matter.

The Encyclopedia is divided into three main parts. Part I (Volumes 1 and 2) deals with "General Principles", Part II—by far the largest, comprising Volumes 3 to 13—with Metabolism, including energy relations, and Part III with the Physiology of Growth, Development, Reproduction, and Movements (Volumes 14 to 18).

The first chapter of the first of the two volumes on "General Principles" deals with genetical principles, that is, with the internal, hereditary factors which control metabolism, differentiation, development and correlations and also determine the response of the plant to its environment (CATCHESIDE). The living units of the protoplasm, which are responsible for this control, can be carried in the chromosomes (genes), or, as shown in the second article (RHOADES), they can be nongenic hereditary factors which are carried outside of the nucleus. This second group of hereditary factors is represented by the plastids ("plastome" or "plastidome"), by other particulate components of the protoplast, and by the microscopically homogeneous cytoplasm with its "plasmones".

Since our insight into physiological phenomena is based on the correct analysis of cause and effect, we should attempt to perform our experiments with as uniform plant material as possible. However, even plant material which is completely uniform in its genetic properties, will show some individual variability in its response to our experimental conditions. In view of this biological variability, our results and conclusions must be subjected to biostatistical analysis. The third article (WENT) deals with this biological variability, particularly in growth and development, while the following article (HARTE) is devoted to biostatistical methods.

Other general foundations which are treated in Volume 1 are the different morphological structures of the cell: cytoplasm, nucleus, plastids, mitochondria and microsomes, vacuole and cell wall. In every case, emphasis is placed on the physical and chemical properties and on the relation to physiological processes. Several chapters are devoted to water and the water relations of cells.

Since all physiological processes are, at least originally, localized within cells, we next proceed to a general treatment of the physiology of plant cells. This is given in Volume 2, which completes Part I of the Encyclopedia.

Parts II and III will then deal with the multitude of individual physiological processes which occur in plants. The arrangement of these parts can easily be seen from the titles of the separate volumes. As noted before, Part II (Volumes 3 to 13) deals with metabolism and energy changes, Part III (Volumes 14 to 18) with growth, development and movements.

# Inhaltsverzeichnis. — Contents.

## I. Genetische Grundlagen physiologischer Vorgänge.

# II. Die Strukturen der Zelle
## und ihre chemische und physikalische Konstitution.

# Mitarbeiter von Band I. — Contributors to volume I.

Dr. D. G. CATCHESIDE, Professor of Genetics, University of Adelaide (South Australia).

Dr. MORRIS COHEN, Assistant research botanist, Department of Botany, University of California, Los Angeles 24, California (USA.).

Professor Dr. HORST DRAWERT, Pflanzenphysiologisches Institut der Freien Universität Berlin, Berlin-Dahlem, Königin-Luise-Straße 1—3.

Dr. KARL EGLE, o. Professor für angewandte Botanik an der Universität Hamburg, Staatsinstitut für Angewandte Botanik, Hamburg 36, Bei den Kirchhöfen 14.

Dr.-Ing. LOTHAR VON ERICHSEN, o. Professor der Verfahrenstechnik und der Physik. Chemie, Valparaiso (Chile), Univ. Técnica/Bonn, Melbweg 28.

Univ. Professor Dr. LOTHAR GEITLER, Direktor des Botanischen Gartens der Universität Wien, Wien III, Botanisches Institut der Universität, Rennweg 14.

Professor Dr. S. GRANICK, Associate Member, Rockefeller Institute for Medical Research, New York City, N. Y. (USA.).

Professor Dr. CORNELIA HARTE, Entwicklungsphysiologisches Institut der Universität, Köln-Riehl, Amsterdamer Str. 36.

Dr. PAUL J. KRAMER, James B. Duke Professor of Botany, Department of Botany, Duke University, Durham, North Carolina (USA.).

Dr. ARTHUR PISEK, o. ö. Professor der Botanik und Direktor des Botanischen Institutes der Universität, Innsbruck (Österreich), Sternwartestr. 15.

Professor Dr. R. D. PRESTON, F. R. S., Department of Botany. The University, Leeds 2 (Great Britain).

Dr. MARCUS M. RHOADES, Professor of Botany, University of Illinois, Urbana, Illinois (USA.).

Dr. WILLIAM SEIFRIZ†, Botanical Laboratory. The University of Pennsylvania 38 th and Woodland Ave., Philadelphia Pa. (USA.).

Professor Dr. J. A. SERRA, Faculty of Science, University-Lisbon (Portugal).

Professor Dr. KURT STEFFEN, Marburg/Lahn, Botanisches Institut, Pilgrimstein 4.

Privatdozent Dr. ERICH TREIBER, Cellulosaindustriens Centrallaboratorium, Stockholm (Schweden).

Dr. F. W. WENT, Professor of Plant Physiology, Earhart Plant Research Laboratory, California Institute of Technology, Pasadena, California (USA.).

Dr. SAM G. WILDMAN, Associate Professor, Department of Botany, University of California, Los Angeles 24, California (USA.).

# I. Genetische Grundlagen physiologischer Vorgänge.

## The physiology of gene action.

By

### D. G. Catcheside.

With 5 figures.

### The nature of the gene.

The attributes of a gene are several. They include self-duplication, physiological action, crossing-over, breakage and rearrangement and lastly mutation. Before considering the action of the gene, we must consider its mode of organization with respect to these various properties, which need not be perfectly correlated. Primarily, the gene of genetic experiments is a unit of action, for unless two adjacent parts of a chromosome have separable actions their separate mutations and usually their recombinations cannot be observed. The gene is usually said to be recognised by producing a particular character in an organism. In fact, what is generally observed is a difference in character which is inherited in a Mendelian manner when one allelic gene is substituted for another. An allele (or allelomorph) is one of two or more dissimilar genes which on account of their corresponding position in corresponding chromosomes are subject to alternative, Mendelian, inheritance. It is usually assumed that alleles have similar or related physiological effects and that two factors producing unrelated physiological effects are not allelic.

By the property of self-duplication is meant simply that each chromosome promotes, once in each cell cycle, the formation of a daughter chromosome, identical to the smallest detail. Any supposition that the genetic chromosome functions as a whole with respect either to self-duplication, other than mechanical continuity, or to physiological action is contradicted by the high proportion of structural rearrangements that persist and cause at the most only trivial changes in physiology. Evidently the parts, at least down to a certain size and order of organization are each independently endowed with the property of self-duplication and are independent in physiological function to a considerable degree. At the other extreme is the assumption that the genetic chromosome is essentially a single enormously long molecule, perhaps a super polypeptide (or polynucleotide), in which self-duplication is a unit determination by the parent molecule of the order of different kinds of constituent residues in a daughter molecule. Crossing-over and chromosome breakage might then be expected to occur about equally frequently between any two residues and the genes as structural units would tend to be reduced to these residues. Small structural changes, such as short inversions, deficiencies or duplications, would then be expected to show frequent false allelism, that is mutations which appear to be strictly allelic to two, or more, genes that are not allelic to one another. This is contrary to experience.

The morphological appearance of chromosomes, with their numerous identifiable chromomeres in the thread stages, shows there is an organization of the chromosome into units, perhaps even a hierarchy of units variously integrated, of intermediate size. With respect to the property of self-duplication, it is possible that there are groupings larger than the ultimate building blocks. The separation of the components of such groupings by structural rearrangement could result in the loss of the property of self-duplication. This may explain why rearrangements so frequently have a lethal effect, at or very close to the breakage points and correlated with a small chromosome deficiency. If crossing-over occurred at the same time as duplication, it would not occur within the self-duplicating blocks, though it could if the events were separated in time.

Whatever the degree of integration with respect to self-duplication, it seems certain that physiological action depends on the pattern of the genic material over considerable lengths of the chromosome. There is abundant evidence that the blocks defined by physiological properties are also ones within which crossing-over is very infrequent, and ones within which breakage into components that retain the property of self-duplication is also rare. If the primary gene products are formed by a process essentially the same as duplication, a strong correlation between the units of self-duplication and of physiological action may be expected.

Complexity of pattern within blocks characterized by a particular type of physiological action seems usual. Series of multiple alleles all produce the same kind of effect, but with different efficiencies. Those at the white eye locus in Drosophila melanogaster are a good example. The normal, or wild type, has a red eye and the alleles range down through coral, which is very little lighter than red, to white, by a whole series (eosin, cherry, apricot, buff, ivory and so on) of intermediate shades. White is not quite colourless, for a deficiency-white is even lighter. This quantitative relationship is characteristic of multiple alleles, as regards their physiological effects. It is also the case that they usually involve pleiotropic effects, namely effects on apparently unrelated characters. The series of diminishing effectiveness of the genes is often different (Table 1) for these different phenotypic actions (Stern 1930).

Table 1. *Orders of pleiotropic effects of multiple alleles.*

Eye colour  $w^+ > w^b > w^e > w$

Viability    $w^+ > w^e > w^b > w$

Fertility    $w^+ > w \ > w^b > w^e$

The differences among the orders of effect on the different characters suggest these pleiotropic effects are primary. The simplest interpretation is that mutations occur at different points in a complex pattern rather than at a single one.

The gene or its product tends to function as a whole, despite complexity of pattern, for it is not usual for different recessive mutations, with a given type of effect to act in a complementary fashion with respect to one another. Usually the phenotype is intermediate, as in the case of eosin and apricot alleles of white eye in Drosophila. But exceptions occur in both directions, to show that correlation between the structural and physiological units is imperfect. There are cases in which recessive mutations with identical or closely similar effects and no crossing-over nevertheless reconstitute the normal wild type. The two yellow body alleles in Drosophila respectively producing grey body with yellow bristles and hairs ($y^n$) and yellow body with grey or brown bristles ($y^z$), together give the grey bodied, black bristled wild type. Similarly, the genes $R^g$ and $r^r$ in maize, which respectively give anthocyanin in endosperm and plant only, together give anthocyanin in both, like the allele $R^r$. In the latter case, no recombination has been observed in over 90,000 gametes tested (Stadler 1951).

Alternatively, there is an increasing number of cases in which similar mutations, which can be recombined by rare crossing-over, have physiological effects which suggest allelism. The work of Green and Green (1949) on the lozenge locus (*lz*) in *Drosophila* is illustrative. Many alleles have been described, all affecting the shape and pigmentation of the eye. They found, using marker genes on both sides of the lozenge locus, that crossing-over occurred between three different sites, which may be called *lz*-1, *lz*-2 and *lz*-3. These sites could be arranged in order with 0.09 and 0.06% recombination between them. This indicates that there are at least three sites of mutation separable by crossing-over, but not by physiological action. For, whereas flies of the genotype $\frac{+\ +}{lz\text{-}1\ lz\text{-}2}$ are phenotypically wild type as would be expected, flies of the genotype $\frac{lz\text{-}1\ +}{+\ lz\text{-}2}$ are mutant. Thus either there are not three units of physiological action or the units are coordinated into a compound unit, any defect in which leads to the same ultimate physiological disfunction.

Evidence for two closely linked genes forming a compound locus comes from recent studies of the white eye series of multiple alleles in *Drosophila*. It has long been known that these fall into two groups, the eosin series and the apricot series. Members of the eosin group produce an eye colour which is lighter in the male than in the female, and the presence of the "Pale" translocation reduces the intensity of the eye pigmentation. Members of the apricot series show the reverse situation, males having a darker eye colour than females. Crossing-over has been observed (McKENDRICK and PONTECORVO 1952, LEWIS 1952) between members of the two groups, namely between white and blood, white and coral and white and apricot, the recombination frequency being about 0.01% in each case. White belongs to the eosin series, the others all to the apricot group. The phenotypes of the coupling and repulsion heterozygotes of white and apricot are respectively wild type and mutant, thus agreeing with the pattern shown by lozenge.

There seems no compelling *a priori* reason for supposing that mutations within a single structural unit should never have complementary effects, or that physiological integration should not extend over two or more structural units, especially if the latter were evolutionarily duplicates or together formed a complex which was a physiological unity, with some novel integrated action, any defect in which could cause failure of the whole effect.

The gene, of genetic experiments, is a unit of action and experimentally a given physiological gene can be associated with a particular part of a cytological chromosome. At first, the separable physiological units were regarded as separate physical units. However, some genes, such as scute and yellow in *Drosophila*, are visibly composite in the giant salivary chromosomes, and have become regarded as composite physical units. Any structural change with one break located within six bands, on the salivary chromosome, of the presumed locus of yellow produces a yellow mutant effect. Indeed scute and yellow are even thought to overlap in their physical boundaries. An increasing number of genes, formerly thought to be allelic, have been found to show recombination at very low frequencies and therefore to be the consequence of alterations in the chromosome at separate though close positions. Perhaps, as GOLDSCHMIDT (1946) holds, the chromosome is a continuous pattern, with the physiological gene corresponding to variously overlapping segments. It is even possible that the fields of cooperation of the parts and therefore of the division of the chromosome into genes, may vary during the life cycle. It is, however, more probable that there is an organization into independent structural units, otherwise rearrangements might well cause more drastic defects than they do. While we cannot be certain what the structural nature of the gene may be, it is surely some segment of the chromosome. Moreover, the uncertainty is not a serious handicap in considering the physiological action, at our present state of knowledge of the latter.

1*

# Time and place of gene action: Differentiation.

The physiology of gene action may be said to include all aspects of the processes by which the presence of a given gene is translated into the physiological effect by which we recognise its presence. Before considering the manner of this process, it is convenient to consider where and when the action is manifested. Different genes are manifested at different stages of growth and differentiation, from the earliest phase of the zygote to the manner of death at the end of the cycle. The action is thus bound up with growth and development.

A cell may have properties which are independent of its nucleus (and genes), both that cells with unlike nuclei may behave similarly and that cells with similar nuclei may behave differently. The breeding systems of *Primula* and *Lythrum* provide instructive examples (MATHER 1948).

*Primula* species are generally distylic, with two kinds of plants, one long and the other short styled, the stamens being borne at corresponding antithetic positions. The short styled is heterozygous ($Ss$) and the long styled homozygous ($ss$) for genes determining the flower differences. The short style produces two kinds of pollen, $S$ and $s$, the long style only one, $s$, which is genetically like one of the kinds produced by the short styled parent. These plants display incompatibility of pollen and style, both kinds of pollen from the short style behaving alike, being large in size, and growing only on the long style, while the pollen from the long style behaves differently, being small in size, and growing only on the short style. The size and compatibility properties of the pollen grains cannot be understood by reference to their genotypes, but only by referring the behaviour to the cytoplasm. All pollen borne by one type of plant receives from it the same kind of cytoplasm; hence, to explain the difference in behaviour it is necessary to assume merely that the two kinds of plant endow their respective pollens with different kinds of cytoplasm which directly determine the compatibility reactions. While the phenotype of the parent plant determines the behaviour of the pollen through the cytoplasm, the phenotype of the plant itself is determined by the genotype ($Ss$ or $ss$) of the zygote. The genes in the nucleus are the ultimate agents of control; the cytoplasm is the immediate agent of gene action, and moreover an agent by which gene action may be extended beyond the immediate presence of a gene.

The tristylic species *Lythrum salicaria* has the reproductive organs borne at three, instead of two, levels in the flower. The stigma occurs at one level and the stamens at the other two in a given type. In different flower types, the stigma occupies a different level, the three types being known as long, mid and short styled. The type of plant is determined genetically, the system involving two independent factors and tetraploidy, so that it is genetically more complicated than *Primula*. The pollen from one plant may, and usually does, include grains of several genotypes. The pollen from stamens at the two levels in a flower includes the same genotypes, similarly distributed in frequency. Yet all the pollen grains from any anther borne at the same level, in different flower types, behave alike in compatibility reactions, while those from stamens from the two levels of one flower have entirely different compatibility reactions. Pollen from long stamens grows only on long styles, mid on mid and short on short. The behaviour is incomprehensible in terms of the genotype of the pollen, but is simple if referred to the cytoplasm. But in this case the cytoplasm depends not only upon the genotype of the mother plant, but also on the level at which the stamens are borne, all pollen in a stamen borne at a given level being endowed with exactly similarly differentiated cytoplasm. Two different kinds of genotype endow pollen borne at the same level with similar cytoplasm.

In these heterostyled plants, the immediate control of the reaction is cytoplasmic, but it is more remotely controlled by the nuclear genes. How soon does a gene exert its influence on the cytoplasm ? When does it do so ? How long may the effect persist after the disappearance of the gene ? These questions are perhaps somewhat simpler to answer than how the gene exerts its influence, which is the crux of the whole problem of its physiology. The gene takes some time to express itself through the production of its own characteristic modification of the cytoplasm. The lag may extend from less than one cell generation to more than one life cycle.

In *Nicotiana* and other plants, the compatibility reactions of the pollen with respect to the style depend only on the genotype of the pollen itself, and not on that of the plant that produced the pollen. In each plant, a given compatibility gene is always associated with a second different gene, allelic to it. The plants produce two kinds of pollen, which react each according to the properties conferred by its own compatibility gene, quite independent of the other with which it had been associated in the zygote. Here differentiation acts early, within the compass of a cell generation, as it does also in waxy pollen in maize (DEMEREC 1924) and other grasses and in the ascospore colour mutants of the fungus *Bombardia* (ZICKLER 1934). Similar situations are encountered in heterokaryotic fungi, such as *Aspergillus* (PONTECORVO 1947).

In heterostyled plants, the action is delayed, but how much is hard to determine except that in *Lythrum* it can hardly be earlier than the time when the two types of stamen are differentiated in the developing flower.

The delay can extend to a whole generation as, apparently, in the determination of direction of coiling in snails, as to whether it is sinistral or dextral, the former being recessive. The direction of coiling reflects, not the snail's own genotype, but that of its mother. The actual persistence in the cytoplasm may be much less than a generation even here, for the direction of coiling is determined early in the ontogeny of the embryo. The gene may have exerted its effect on the mother's cytoplasm only a short time before the egg was formed and the cytoplasmic determinant decayed shortly thereafter. Nearly a whole life cycle would have to elapse before a further change could occur.

In the case of "grandchildlessness" in *Drosophila subobscura* (SPURWAY 1948) the lapse of at least a whole generation is certain. This is a recessive gene whose effect is limited to females homozygous for it. They are normal and fertile, but all their offspring are sterile, even when normal males are used to father them, when the offspring can be only heterozygous for the gene.

Microorganisms provide particularly interesting evidence. In *Paramecium*, characters ultimately determined by genes may show cytoplasmic control extending to as many as 36 cell generations before fading out. In yeast, adaptive enzymes are under gene control, but are not formed in perceptible amounts in the absence of the substrate. In its presence, the enzymes take an appreciable time to appear in detectable concentration and take a considerable time to disappear on removal of the substrate. Almost more striking are the antigenic states in *Paramecium*. Their character and range are potentiated by the nucleus, but the particular state at a given time is activated by factors of the external environment such as food level and temperature. Once developed a particular antigenic type tends to persist until a substantial environmental change jolts it into a different state (SONNEBORN 1950). Sometimes the delay in the time at which the effects of a gene become apparent depends upon the store or degree of increase of the substance it produces in the cytoplasm. The disappearance of the old gene's products and the accumulation of the new gene's products may also be slow. In the moth *Ephestia*, KÜHN (1937) found that in the cross of pigmentless *aa* mothers with *Aa*, the unpigmented eyes of *aa* progeny were at once distinguishable from the pigmented *Aa* progeny. But in the young progeny from *Aa* mothers no distinction could be made. The pigment determinants in the cytoplasm of the egg provided full pigmentation until a late larval instar, when only did the segregation appear. The determinants, moreover, are diffusible, for transplantation of *Aa* testes caused pigmentation of the eyes of an *aa* host and of its young *aa* offspring.

We have then these facts about the agents of gene action. A gene can cause an alteration in the cytoplasm which may persist after the gene itself has gone from the cell. The effect persists for various numbers of cell generations. In microorganisms, where alone such tests are possible, the persistence of the genic effect on the cytoplasm can be controlled by adjusting other conditions affecting the cytoplasm.

Genes must act by altering the cytoplasm, which thereby becomes the agent of gene action. The change in the cytoplasm is plausibly understood as due to the release into it of some characteristic substance from the gene. The persistence of the effect in the cytoplasm implies the persistence of this substance or of some compound of it. The prolonged persistence, in some cases, after the gene has disappeared, through a number of cell cycles without apparent dilution, implies that the substance is not merely able to persist but also able to reproduce at least to some extent under certain conditions. The simplest assumption is perhaps that each gene releases its characteristic substance in all cells, perhaps once in each cell cycle, and that the substance shows impersistence or persistence, with or without reproduction, according to the immediate state of the cytoplasm. The behaviour of each gene product is thereby assumed to be conditional.

The characteristic substances, or their compounds, in the cytoplasm may be called plasmagenes, thought of as being capable of self-reproduction, that is capable of bringing about the manufacture of replicas of themselves from simpler compounds. Obviously, the self-reproduction of any substance in the cytoplasm, as also in the nucleus, is conditioned by the supply of raw materials. It is also conditioned by other similar substances, in the cytoplasm, with which it competes. Some cytoplasmic bodies can reproduce in a wide range of conditions, as can nuclear genes, and so are permanent in the same sense. These are truly plasmagenes, and being normal constituents of the cell they can be detected only when they themselves change and show cytoplasmic inheritance of the alteration, as in the "petite" cytoplasmic mutant of yeast, or when the nucleus changes in respect of a gene with which the plasmagene interacts, and the character change shows Mendelian inheritance, as in the "petite" genic mutant in yeast (EPHRUSSI 1951).

It is the less permanent plasmagenes which offer a solution of the process of differentiation and gene action. The fate of each plasmagene is dependent upon the state of the cytoplasm at a given time. Thus the action of a given gene will become effective only when the composition of the cytoplasm is within the range necessary for the gene's product to reproduce and multiply. This multiplication in turn would alter the cytoplasm in such a way as to affect the consequences of other genes. The cytoplasm in which a plasmagene is newly multiplying will become less favourable to some other plasmagenes previously common and more favourable to still others previously rare or completely suppressed. Thus this provides a model whereby a constant nucleus can be associated with a changing cytoplasm, which alters in a deterministic and progressive manner, and yet whose changes are under the control of a constant nucleus. An effect of a gene continued after it has disappeared can be understood either by persistence of its characteristic product or by a diversion in the path of change in the cytoplasm caused by the presence of a critical gene product at an early stage in development.

The process of differentiation can be understood as due to changes in the composition of the cytoplasm, through alterations in the kinds of plasmagene that persist and the proportions and balance they display. Some differentiation could perhaps follow with constant proportions of all plasmagenes in all cells, but arranged in different patterns in different cells. This might be too unstable and offer too narrow a range of variation.

## Interaction of genes.

**Dominance.** Allelic genes are characteristically related to one another as dominant and recessive. That is, one of the pair, the dominant, shows complete expression in the heterozygote, suppressing all signs of the recessive gene. This relationship is not a necessary one; indeed, all gradations are known between those cases in which the heterozygote shows the full effect of one gene (complete dominance) and those in which the heterozygote is intermediate and it is not possible to call either gene dominant. However, a considerable degree of dominance is much commoner than

Table 2.

| Genotype | Percent of amylose in carbohydrate | Genotype | Number of spots per kernel |
|---|---|---|---|
| + + + | 29.0 | Dt Dt Dt | 121.9 |
| + + wx | 27.5 | Dt Dt + | 22.2 |
| + wx wx | 21.2 | Dt + + | 7.2 |
| wx wx wx | 0 | + + + | 0 |

Data of SAGER (1951)     Data of RHOADES (1941)

would be expected purely by chance; this has been supposed to be a consequence of the evolution of dominance, either by the selection of modifiers which make the heterozygote more like the homozygote (FISHER 1931) or by the selection of more efficient dominants as wild type genes (HALDANE 1930). Complete dominance may be illusory, for example the large class of recessive lethals so fully studied in *Drosophila melanogaster* characteristically have a slightly depressive effect on viability in the heterozygote. Also, detailed chemical analysis in some cases, such as waxy in maize, has disclosed slight differences between the phenotypic effects of different dosages of the dominant gene, of which from none to three are possible in the endosperm (Table 2). The effects of the "dominant" gene, Dotted, are more clearly intermediate.

The effect of the rest of the genotype upon expression and dominance will be referred to later. The environment can also affect dominance. The recessive gene *cubitus interruptus* in *Drosophila melanogaster* breaks the cubitus vein of the wing. Homozygotes grown at 19° C all show the effect, but if grown at 25° C only half of them express it. Heterozygotes grown at 25° C do not express it at all, but if grown at 13° C it is expressed in about 10 % of the flies. But such environmental effects on dominance are rare. The reason is that the normal or wild type gene, which is usually the dominant, exists as part of a genotype which has become adapted to give a nearly constant phenotype in varying environments, i.e. it shows homeostasis. The heterozygote of a wild type dominant would be expected to give a more stable phenotype than would its new untested allele when homozygous. It does so because the genotype has been selected to give stable dominance, as FISHER (1931) has shown.

**Interaction.** The dominance-recessiveness relation depends on the interaction between two allelic genes, but similar interactions occur between non-allelic genes. Indeed, all the genes in the genotype probably react with one another during development. Probably these reactions are always between gene-products rather than between genes themselves. Where two genes which segregate independently do not interact in respect to the obvious characters they determine, they give the familiar Mendelian $F_2$ ratio of $9:3:3:1$. When they interact this ratio is simplified, by combination of the classes in most of the ways possible with complete dominance.

Interaction in the simplest case consists of anticipation, one gene acting before another, whose effect is cut out by it. The former is said to be *epistatic* to the other. Epistasy occurs at all stages of development. A lethal is epistatic to all other genes affecting the organism whose life it terminates. A gene suppressing, or failing to produce, a substance, is epistatic to one modifying the substance or using it for its own activity. Epistasy is expressed in two types of $F_2$ ratio according to whether the epistatic gene is itself dominant or recessive. Colourless aleurone in maize is generally due to a gene which is recessive, and also of course epistatic to the colour genes, purple and red. The 9 purple: 3 red: 4 white ratio is characteristic of such $F_2$'s. In maize there is also a dominant gene which inhibits aleurone colour and is also epistatic to the colour genes; in $F_2$ they give a 12 white: 3 purple: 1 red segregation ratio.

In epistasy the expression of one gene difference conditions that of another. With reciprocal relations other $F_2$ ratios are found, of three kinds according to whether the two dominants, the two recessives or one of each are needed for the combined effect. Only two classes are phenotypically distinguishable in $F_2$. When two dominants are needed, they are complementary genes. The first case of interaction ever analysed was of this type (Bateson 1909). It was the production of coloured sweet peas by a cross between two different whites, the coloured:white ratio in $F_2$ being 9:7. When two recessives are needed we have duplicate genes, as in the case of certain chlorophyll and albino genes in maize. The $F_1$ resembles both green parents, when the cross is $Ab \times aB$, but the $F_2$ gives a 15 green:1 albino segregation. The duplicate genes are so named because the alleles appear to be performing identical functions, whereas complementary genes, themselves dominant, appear to have different functions. But the 15:1 ratio could equally well indicate a complementary action of recessives.

The remaining case appears to show the complementary action of dominant and recessive giving the 13:3 ratio of white to coloured shown by a cross of the recessive and dominant white aleurone types in maize. The recessive allele of one gene is often here called a suppressor of the recessive allele of the other; equally we may regard the dominant allele of the first gene as suppressed by the dominant allele of the other. The same kinds of interactions may underly all the sorts of interaction, the terms applied to the different cases distinguishable by their segregation ratios being traditional and not implying full knowledge of the physiological basis.

The groupings can of course be divided into their genetic classes by appropriate breeding tests. They can also be subdivided by closer examination, by microscopic or chemical tests. Two strains of *Rudbeckia* each with yellow bud cones give an $F_1$ with purple bud cones and a ratio of 9 purple to 7 yellow in $F_2$. The yellow ones can be subdivided in a ratio of 4 which turn red to 3 which turn black upon treatment with caustic potash solution (Blakeslee 1921). Apparent complementary interaction is converted to recessive epistasy.

The expression of all genes is affected by modifying genes which are otherwise not readily detectable. Suppressor genes eliminate the expression of another gene, but have no other obvious effect. In *Drosophila melanogaster*, the recessive purple eye gene, *pr*, is suppressed by a recessive, *su*, situated in a different chromosome. Suppression is an extreme case of modification; usually the effect is less marked. Many modifying genes probably do not differ in any fundamental way from ordinary genes. They merely produce a slight effect which is concealed by the factor of safety characteristic of dominant wild type genes. Differences between such wild type isoalleles can often be revealed in the less balanced mutant forms. Thus scute in *Drosophila* has an effect on certain bristles which can be detected only if another bristle gene, Hairless, is also present; one can be regarded as a modifier of the other.

Indeed any gene has not only a main effect by which it is usually recognized, but also a diversity of smaller effects which may be difficult to detect in the normal organism but may be apparent as modifying effects upon other mutants. The white eye gene in *Drosophila* changes not only the eye colour, but also that of the testicular membrane and the Malphigian tubules, the shape of the spermatheca, length of life, and general viability and fertility. The vestigial wing gene reduces wing size, modifies the halteres, makes certain bristles erect instead of horizontal, changes the wing muscles, the shape of the spermatheca, the rate of growth, fertility and the length of life. Under favourable external conditions vestigial also reduces the number of ovarioles in the ovaries, while under unfavourable conditions it has the opposite effect. Teopod in maize induces strong tillering, narrow leaves, an increased number of nodes, many small ears with enlarged glumes covering some of the kernels, a simple staminate inflorescence with long bracts and poor pollen production; the homozygote is probably lethal or sterile. In some cases it is clear that the multiple effects, or some of them, are of the same nature, as with eye and testis colour in *Drosophila* and with flower and plant colour in numerous plants. But sometimes as in the cases cited the connection is far from obvious. In some cases different sets of phenotypic effects are produced under different kinds of environmental conditions. The generality of phenomena of these kinds is shown by the usual effect of genes on viability. Almost any gene substitution changes the general capacity of an organism to develop and maintain itself, the alteration probably being an expression of the influence of the gene on the whole complex of reactions constituting the life of the organism. Cells, and tissues, usually die unless they have a full complement of genes; even very small deficiencies are generally lethal to cells in which they are homozygous.

Any given character may be affected by several different genes, e.g. chlorophyll development and distribution in maize by well over a hundred; and any given gene may affect a number of the reactions in an organism. Therefore, there is no simple one-to-one relation between a gene and a character, but only between the phenotype and the genotype as a whole. It is incorrect to say that $w^+$ corresponds to red eyes and $w$ to white eyes in *Drosophila*, instead it is proper to say that, in the usual genotypes encountered, a substitution of $w^+$ for $w$ changes the eyes from white to red. The rest of the genotype, other than the particular gene under consideration, may be referred to as the genetic background, or genotypic milieu. The background will affect the degree of expression of a gene (its expressivity), the proportion of individuals which show the effect at all (its penetrance), and even the actual mode of expression (TIMOFÉEFF-RESSOVSKY 1934).

**Dosage Relations.** One of the most important variables is the quantity of genes present. MULLER (1932) has made an especial study in *Drosophila melano-*

*gaster* of the effect on expression of a gene when more of it is added to the same background, and when more of the background is added to the same quantity of the gene. He has proposed a fivefold classification of genes, on the basis of a comparison between the gene considered and some standard allele of it, preferably the wild type. The classes are hypomorph, amorph, hypermorph, antimorph and neomorph.

The mutant gene scute normally removes, or appears to remove, certain bristles. When an extra mutant scute gene is added the number of bristles is almost normal. With two extra scute genes the fly has more bristles than the normal. In fact scute helps to produce bristles, not remove them, but works less than half as effectively as wild type. It is a hypomorph. Amorphs are extreme cases of hypomorphs, being the limit of variation in having very little of the same effect as the standard.

Hypermorphs are more efficient than the standard, antimorphs oppose its action. Ebony is an antimorph, $ee$ being darker than $e^+ee$ and the latter darker than $e^+e$; that is, adding $e$ to $e^+e$ darkens the fly, while adding $e^+$ to $ee$ lightens it. Neomorphs are genes which are doing something quite different from anything done by the standard gene, the latter behaving, towards a neomorph, like an amorph. Bar eye and Hairy wing are examples. Clearly all the terms are comparative. Antimorphs and neomorphs frequently, if not invariably, involve structural changes in the chromosomes. It is possible that they depend on a position effect, namely an integrated action of more than one genic constituent brought together for the first time by the structural change.

Of the five types, only the hypomorph is common, the rest are rare. This indicates that mutation is generally a loss of efficiency, a breakdown in a complex mechanism, and broadly pathological. This is not unexpected, for in a complex and delicately balanced system, any change is more likely to break down the efficiency of the old mechanism rather than build up a new and more efficient one.

Genes which are related as hypo- and hypermorphs control the quantity of effect which occurs in the phenotype. They may be supposed to determine the production of greater or lesser amounts of some substance whose quantity controls the phenotypic effect. In most cases the nature of the substance concerned is quite unknown. Generally, at the best only the ultimate substance or character is capable of measurement. In general the curve of effect rises with increasing dosage of the relatively hypermorphic gene in an approximately hyperbolic form, rising rapidly for small doses, but flattening out and approaching an asymptote as the doses become larger.

## Gene products.

**Primary.** It has been argued that the quantitative control exerted by genes is a sign that different alleles themselves differ only in quantity. They may represent different amounts of an enzyme catalyzing the reaction producing the substance, which is finally responsible for the character effect. But it is more likely that the alleles differ in producing different enzymes which catalyze the same reaction to different degrees, or perhaps sometimes even catalyze different reactions.

In some cases there is no doubt that the two alleles cause the production of different chemical substances. Perhaps the clearest cases are the blood group genes in man. Others will be mentioned later. The two alleles $A$ and $B$ each

produce a specific antigen quite independently of the presence of the other allele
or of any other genes in the nucleus. A third allele at the locus produces no
antigen; it behaves as an amorph to the other two which are neomorphs to it
and to each other. In some species hybrids of doves new antigens appear to
arise which are found in neither parent. The cooperation of more than one gene
in their production is implied. However, so close is the connection, the distinction between gene and antigen is formally maintained only because of the necessity for distinguishing between determinant and product, the nuclear and
the cytoplasmic representatives. The antigens are complex proteins, which thus
appear to be the immediate products of gene activity.

There is similar evidence of the production of substances which are actually
enzymes. A well-known example is the recessive which causes the loss of the
enzyme enabling man to oxidize homogentisic acid, which is therefore excreted
unchanged in the urine, a condition known as alkaptonuria (GARROD 1923). A
case in *Trifolium repens* (ATWOOD and SULLIVAN 1943) is particularly complete.
Some strains of white clover contain HCN derived from cyanogenic glucosides
and the enzymes necessary to hydrolyse them. The glucosides are lotaustralin
and linamarin, which yield HCN, a glucose and a ketone (ethyl methyl ketone
for the former and acetone for the latter) under the hydrolytic action of the
enzyme, linamarase. Different strains contain the glucosides alone, the enzyme
alone, neither or both. The differences are determined by two independent
dominant genes, one determining ability to produce the glucosides, the other
determining presence of the enzyme. *In vitro* cyanogenesis can be produced by
the combination of extracts from acyanogenic plants, the one providing the substrates and the other the enzyme.

These and similar cases have given rise to the hypothesis that genes function
by producing or determining specific enzymes and that there is a one gene to
one enzyme relationship, a hypothesis put forward by BEADLE (1945a).

As stated by HOROWITZ and LEUPOLD (1951) "The concept is that of a gene
whose sole activity aside from self-duplication is that of functioning in the
synthesis of a particular enzyme or enzyme precursor. It is not thereby implied
that genes at other loci may not also function directly in the formation of the
enzyme ... but if two or more genes do, in fact, cooperate in the production of
a given enzyme, then their respective contributions must be different". A different
allele is presumed to produce an enzyme of different efficiency in catalyzing
the specific reaction. The altered enzyme may be changed in various ways,
so that it is differently susceptible to the environment and to the genetic
background.

**Secondary.** Variation in the activity of an enzyme would be expected to give
greater or lesser amounts of various products. Variations in the amounts of
these could give rise to character differences, and by providing greater or lesser
amounts of substrate for other enzymes, and in other ways, could give rise to
the familiar pleiotropic effects. Complete absence or inertness of an enzyme
may cause a block at a particular place in the metabolic fabric and the accumulation or diversion of precursors. Many instances of this are known, for example
the phenylalanine-tyrosine metabolism in man, the formation of eye pigment
from tryptophane in insects (KIKKAWA 1941) and flower pigmentation (LAWRENCE
and PRICE 1940). But, since the known suite of gene controlled reactions is far
more complete and concerns generally simpler compounds, the nutritional
mutants in *Neurospora* (HOROWITZ 1950) provide a more complete picture of
the kinds of behaviour encountered.

## Biochemical genetics of Neurospora.

Beadle and Tatum (1941) showed that simple nutritional mutants (auxotrophs) were unable to grow on a medium which would support the growth of the wild type, but that many of them would grow with a vigour equal to the wild type if a single organic compound such as an amino acid, a vitamin or a nucleotide, was added to the medium. It could be argued that such a mutant, e.g. one that requires an exogenous supply of arginine for its growth, exhibits this requirement because it is unable to synthesize sufficient of this essential amino acid for its needs, whereas the wild type could do so. It was soon shown that such differences in nutritional requirement segregated as though controlled by single genes. A number of mutants were found all with the same apparent nutritional requirement, but due to different non-allelic genes. It was surmised

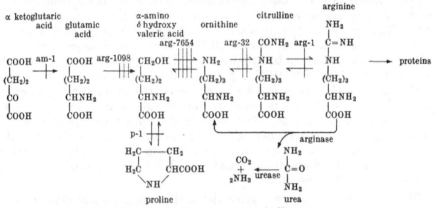

Fig. 1. Genetic control of arginine synthesis in *Neurospora crassa*.

that these different genes might control different steps in synthesis. In the case of arginine synthesis the surmise could be tested on the assumption that it would be similar to that in mammalian tissue. It was found that of seven different genetic types, one of the mutants would only grow on arginine, two others would grow on arginine or citrulline, and the remaining four on arginine or citrulline or ornithine (Srb and Horowitz 1944). Thus it could be argued that the order of synthesis is ornithine → citrulline → arginine, and that each step in the synthesis is controlled by a single gene. If so the ornithine-citrulline stage consists of at least two steps (Fig. 1).

If, as the evidence shows, metabolic products are characteristically formed in a stepwise manner it seems probable that only one step is generally blocked by any particular mutant. The steps before and after the block should be unaffected. Absence of affect after the block is shown, for example, by the capacity of certain of the arginine mutants for growth on ornithine. They are able to carry out all the steps from ornithine to arginine. Absence of affect before the block is shown by the accumulation of precursors, or shunt products of them, shortly before the block. The first case of this actually found, by Tatum, Bonner and Beadle (1944), was of anthranilic acid by a mutant requiring indole or tryptophane for growth (Fig. 3). Indeed this compound is accumulated by several tryptophane mutants blocked at different points, and it seems clear that the particular precursor accumulated must depend upon unknown reaction equilibria between the various steps.

In some cases an accumulated compound is readily used for growth by mutants blocked at earlier steps in the synthesis. Thus, in methionine synthesis (Fig. 2) cystathionine is accumulated by mutant *me-2* and utilised by mutant *me-3*. In other cases, the compound accumulated is clearly not the actual precursor, but something formed from it by secondary reactions, as in the cases of the purple pigment formed by certain adenine mutants or the quinolinic acid secreted by one nicotinic acid mutant, but not utilized by a mutant acting earlier in the synthesis (Fig. 3). The order of action of mutants, one of which accumulates a precursor or a derivative, may be deduced from the properties of double mutants, i.e. strains having both of the blocks. Thus the double mutant *me-1 me-2* accumulates cystathionine, while the double *me-3 me-2* does not (Fig. 2).

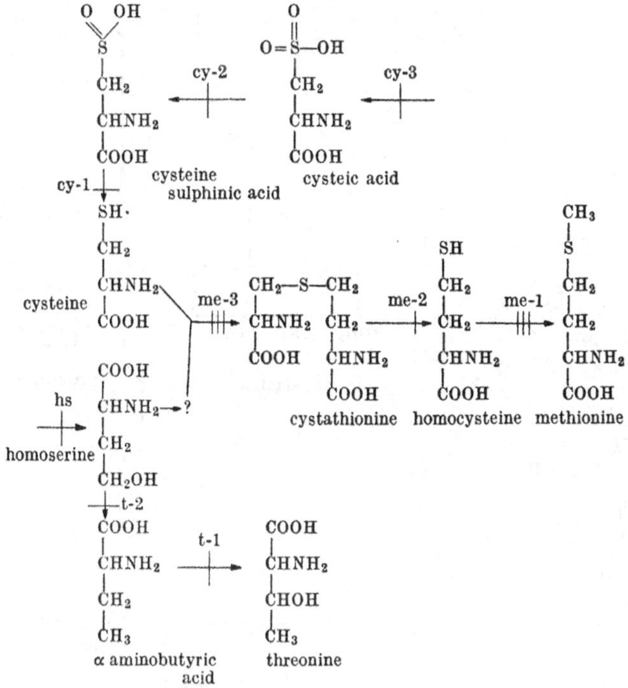

Fig. 2. Genetic control of methionine and threonine synthesis in *Neurospora crassa*.

In a few cases it has been found that a precursor is inhibitory towards another reaction in the organism, sometimes of the utilization of the growth factor into which it is normally converted. Thus ε-hydroxynorleucine, a precursor of lysine (Fig. 4), is inhibitory towards mutants blocked at later stages in lysine synthesis (GOOD, HEILBRONNER and MITCHELL 1950). A mutant which requires a balanced proportion of isoleucine and valine, appears to have a block in isoleucine synthesis (Fig. 5). This results in the accumulation of a precursor which in turn inhibits a reaction in valine synthesis which is similar to the blocked reaction in isoleucine synthesis. Precursors are accumulated in both syntheses (ADELBERG and TATUM 1950). There is now evidence that instead of inhibition being concerned, the dual requirement is the consequence of an absence of one enzyme which catalyses equivalent steps in both valine and isoleucine synthesis.

The genetically controlled reactions in *Neurospora* illustrate the various modes of gene action. The direct product of the gene, the enzyme, will be discussed in detail later. Through these enzymes, the genes control reactions which may be (1) successive, where one compound is converted into a second one and that in turn into a third; (2) cooperative, where two compounds are combined to make one; and (3) competitive, where one compound is the precursor of two or more reactions. In fact, nearly all successive reactions are cooperative since usually something is added to, or taken from, the precursor to form the product. In competitive reactions, as for example, the utilization of homoserine for both methionine and threonine synthesis (Fig. 2), a block in one usually leads to an over-production in the other, because the same amount of common precursor is still available. In a complex system, such as a developing organism, and indeed

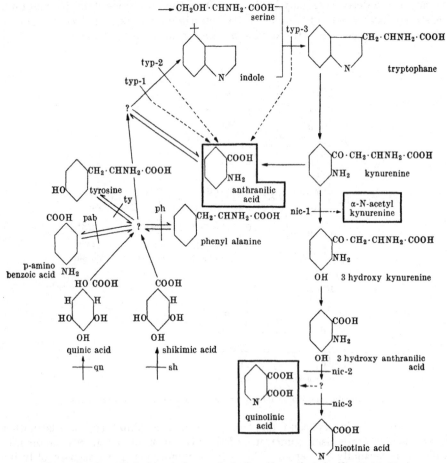

Fig. 3. Genetic control of tryptophane and nicotinamide synthesis in *Neurospora crassa*.

the intricate metabolism in each of its cells, there must be a large number of such competitive reactions. It is therefore easy to see how a breakdown at one point may affect a multitude of apparently unrelated events.

Evidence for the absence or disfunction of an enzyme at the metabolic step is available for two cases in *Neurospora*, namely tryptophane desmolase and glutamic dehydrogenase. Tryptophane desmolase couples indole and serine to form tryptophane. Two allelic mutants are known and Yanofsky (1952) has been unable to detect enzyme activity in them by methods which would respond to 1/250th of the wild type activity. The use of mixtures of wild type and mutant extracts gave no evidence of an inhibitor in the mutant or an activator in the wild type.

Unfortunately the enzyme systems involved in many of the nutritional blocks in *Neurospora* mutants are at present unknown. Nevertheless it is possible to show that a requirement does not necessarily mean an absence of the appropriate enzyme. WAGNER and HADDOX (1951) found that mutants requiring panthothenic acid are able to synthesize the vitamin from precursors under certain culture conditions, provided some pantothenic acid is initially present to start growth. The block is therefore incomplete and in fact absent under growing conditions. Similar partial blocks, presumably due to hypomorphic genes, have been found by BONNER (1951) in the synthesis of tryptophane and nicotinamide (Fig. 3). He used $N^{15}$ incorporated into various intermediates in the synthesis and studied the distribution of the isotope in the end products. Under

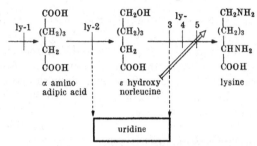

Fig. 4.  Genetic control of lysine synthesis in *Neurospora crassa*.

growing conditions anthranilic acid is utilized in the synthesis of tryptophane and nicotinamide by a mutant which is unable to commence growth on this compound. In fact, the only two tryptophane mutants with complete blocks are those in which it has been shown that tryptophane desmolase is absent.

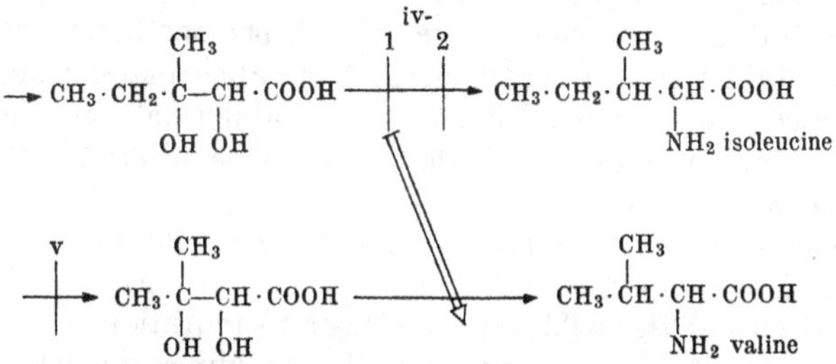

Fig. 5.  Isoleucine and valine mutants in *Neurospora crassa*.

However, HASKINS and MITCHELL (1952) and NEWMEYER and TATUM (1953) have shown that the apparent position of the block in the mutant studied by BONNER may be shifted by modifiers to a position so far back that it will grow on phenylalanine, considered to be several steps prior to anthranilic acid. The effect of the modifiers in causing an apparent, though incomplete, block is quite unknown.

If a single gene determines a single specific enzyme, how are all the different alleles related to the enzyme concerned ? May the enzyme be altered in a number of different ways, to account for the variety of alleles ? If so, what alterations are possible ?

In both *Neurospora* and *Escherichia coli* a considerable number of temperature sensitive mutants are known. They behave like the wild type at $25^0$ C but are unable to carry out a particular synthesis at a higher temperature, usually $35^0$ C. Apparently the enzyme functions normally at $25^0$ C, but at $35^0$ is somehow inactivated. MAAS and DAVIS (1952) have studied a mutant of *E. coli* which requires pantothenate when grown above $30^0$ C. It was derived from an auxotroph which requires pantothenate at all temperatures, and which lacks the enzyme, pantothenic desmolase, which catalyses the coupling of β-alanine and pantoyl lactone, the final step in pantothenate synthesis. The enzyme extracted from the temperature sensitive mutant and the wild type differ strikingly in heat lability. Whereas the wild type enzyme is completely stable for two hours at 35⁰ and is only slightly inactivated at 47⁰ during the same period, the mutant enzyme is almost completely inactivated at 30⁰ within an hour. The presence of either a heat activated inhibitor in the mutant or of a stabiliser in the wild type is excluded by experiments in which mixtures of the enzyme extracts show a rate of heat inactivation of each component unaffected by the presence of the other.

Thus the change in the gene, giving a strain with a temperature sensitive requirement, has also resulted in a qualitatively changed enzyme. In the strain of *E. coli* used, it is unfortunately not possible to determine whether the temperature sensitive mutant is allelic with the one having the absolute requirement, but in analogous cases in *Neurospora* they have been shown to be allelic. This would suggest that if the gene in the strain with the absolute requirement is also producing an enzymic protein, it is one that completely lacks the reaction characteristic of the wild type. This is rendered more probable by the evidence from the activity of suppressors, discussed later.

Pauling et al. (1949) have found evidence for the production of an altered protein as a result of gene change, in sickle cell anaemia in man. Erythrocytes of affected individuals contain a haemoglobin which differs in electrophoretic mobility from normal haemoglobin. The difference probably lies in the globin part of the molecule and is concerned with the number of ionisable groups, the net number of positive charges being 2 to 4 more than in normal haemoglobin. The difference is due to the single gene, and in persons heterozygous for it the erythrocytes contain a mixture of the normal and abnormal haemoglobin whereas the two homozygotes contain all normal or all abnormal haemoglobin. This is the most direct demonstration of the formation of an altered protein as a result of gene change, so far known.

Auxotrophic mutants sometimes revert to the wild phenotype, no longer having a nutritional requirement. This may be due either to back mutation of the mutant gene to the wild type allele, or to mutation at a different locus to a suppressor of the mutant phenotype. Is the suppressor now producing the enzyme or has it some indirect effect, for example so altering the cellular environment that the gene is now producing an active enzyme instead of an inert protein? In the case of the tryptophane desmolase mutants in *Neurospora*, a suppressor has been found (Yanofsky 1952) which enables one of them (S1952), but not the other (C83) to grow without added tryptophane. The growth rate was slower than wild type, but eventually the same dry weight was achieved. The mutant plus suppressor had tryptophane desmolase activity, though only to the extent of 5% of that in the wild type, and it agreed in all measurable properties with the enzyme from wild type. The suppressor reduces the growth rate of wild type and also depresses the amount of growth achieved by the mutant C-83, whose requirement it does not relieve, on tryptophan supplemented media.

It appears that the two mutants, seemingly allelic, have qualitatively different alterations to the same gene. It seems likely that the mutant alleles are continuing to give rise to products which have the same patterns as the respective genes and are inactive in the reaction concerned. If so, the action of the suppressor would be to alter the environment in such a way that the product of one strain is reactivated, but this method of reactivation is ineffective for the other and actually harmful to it, as it is to the wild type.

MITCHELL and MITCHELL (1952) have made an extensive study of a suppressor gene *s*, with radical and apparently unrelated effects. It was discovered by its action in suppressing the requirement of a pyrimidine mutant (*pyr*-3a) and those of three other different mutants. However, in combination with any one of six other mutants it acts as an inhibitor, i.e. prevents the mutants from growing with the supplement that normally enables them to grow. Indeed *pyr*-3a can act as an inhibitor of a citrulline mutant (*cit*-1) and three ornithine mutants act as partial suppressors of *pyr*-3a, so there is no reason to suppose that the suppressor *s* is in any different category. It is difficult to imagine that the gene *s* could be capable of duplicating the functions of three different genes. Rather can it be imagined that its product is concerned in interacting with a great many other genes and their products, in fact with the whole genotype in a complex series of interactions, and that its major or primary action is unknown.

One further point arises from these interactions. In the presence of the suppressor *s*, three different ornithine mutants are prevented from utilising ornithine, so that the apparent position of the block is shifted to come between ornithine and citrulline (Fig. 1). Similarly in the presence of *pyr*-3a, *cit*-1 appears to have its block shifted from between ornithine and citrulline to between citrulline and arginine. Similar effects of modifiers in altering nutritional requirements are known for tryptophan-nicotinamide mutants (HASKINS and MITCHELL 1952, NEWMEYER and TATUM 1953). It is no longer safe to conclude that a particular nutritional requirement indicates a block in synthesis immediately prior to it nor that a given mutant gene bears a direct relationship to the reaction which appears to be blocked.

*There is a considerable body of evidence that the gene acts as a unit of physiological action through its control of individual enzymes.* Enzymes differing from their wild type counterparts in physical properties, such as heat lability, affinity for substrate or complete loss of function, are all the result of gene alteration and have a direct relationship to the altered phenotype. Whether the gene acts as a template for the enzyme molecules, either directly or indirectly through plasmagenes, or whether it functions by controlling in some way the time and rate of enzyme formation is largely unknown. That effects may persist and multiply after the gene itself has gone from the cell seems certain, and this makes it more probable that the gene somehow produces the enzymes. Whether the enzymes are the plasmagenes, or whether the latter exist as independent entities without other functions than producing enzymes cannot be decided. However, it would appear more economical if the enzymes themselves, or organized aggregates of them, had autocatalytic properties, so reducing the hierarchy to gene-enzyme-reaction.

## Literature.

ADELBERG, E. A., and E. L. TATUM: Characterization of a valine analog accumulated by a mutant strain of *Neurospora crassa*. Arch. of Biochem. **29**, 235–236 (1950). — ATWOOD, S. S., and J. T. SULLIVAN: Inheritance of a cyanogenetic glucoside and its hydrolyzing enzyme in *Trifolium repens*. J. Hered. **34**, 311–320 (1943).

BATESON, W.: MENDEL's principles of heredity. London: Cambridge Univ. Press 1909. — BEADLE, G. W.: Biochemical genetics. Chem. Rev. **37**, 15–96 (1945). — BEADLE, G. W., and E. L. TATUM: Genetic control of biochemical reactions in *Neurospora*. Proc. Nat. Acad. Sci. U.S.A. **27**, 499–506 (1941). — BLAKESLEE, A. F.: A chemical method of distinguishing genetic types of yellow cones in *Rudbeckia*. Z. Abstammgslehre **25**, 211–220 (1921). — BONNER, D.: Gene-enzyme relationships in *Neurospora*. Cold Spring Harbor Symp. Quant. Biol. **16**, 143–157 (1951).

CATCHESIDE, D. G.: Gene action and mutation. Biochem. Soc. Symposia **4**, 32–39 (1950). DEMEREC, M.: A case of pollen dimorphism in maize. Amer. J. Bot. **2**, 461—464 (1924).

EPHRUSSI, B.: The interplay of heredity and environment in the synthesis of respiratory enzymes in yeast. Harvey Lect. **1951**.

FISHER, R. A.: The evolution of dominance. Biol. Rev. **6**, 345–368 (1931).

GARROD, A. E.: Inborn errors in metabolism, 2. Ed. London: Oxford Univ. Press 1923. — GOLDSCHMIDT, R.: Physiological genetics. New York: McGraw Hill 1938. — Position effect and the theory of the corpuscular gene. Experientia (Basel) **2**, 1–40 (1946). — GOOD, N., R. HEILBRONNER and H. K. MITCHELL: Hydroxynorleucine as a substitute for lysine for *Neurospora*. Arch. of Biochem. **28**, 464–465 (1950). — GREEN, M. M., and K. C. GREEN: Crossing-over between alleles at the lozenge locus in *Drosophila melanogaster*. Proc. Nat. Acad. Sci. U.S.A. **35**, 586–591 (1949).

HALDANE, J. B. S.: A note on FISHER's theory of the origin of dominance. Amer. Naturalist **64** (1930). — HASKINS, F. A., and H. K. MITCHELL: An example of the influence of modifying genes in *Neurospora*. Amer. Naturalist **86**, 231–238 (1952). — HOROWITZ, N. H.: Biochemical genetics of *Neurospora*. Adv. Genet. **3**, 33–71 (1950). — HOROWITZ, N. H., and U. LEUPOLD: Some recent studies bearing on the one gene-one enzyme hypothesis. Cold Spring Harbor Symp. Quant. Biol. **16**, 65–74 (1951).

KIKKAWA, H.: Mechanism of pigment formation in *Bombyx* and *Drosophila*. Genetics **26**, 587–607 (1941). — KÜHN, A.: Entwicklungsphysiologisch-genetische Ergebnisse an *Ephestia kuhniella*. Z. Abstammgslehre **73**, 419–455 (1937).

LAWRENCE, W. J. C., and J. R. PRICE: The genetics and chemistry of flower colour variation. Biol. Rev. **15**, 35–58 (1940). — LEWIS, E. B.: The pseudoallelism of white and apricot in *Drosophila melanogaster*. Proc. Nat. Acad. Sci. U.S.A. **38**, 953–961 (1952).

MAAS, W. K., and B. D. DAVIS: Production of an altered pantothenate-synthesizing enzyme by a temperature-sensitive mutant of *Escherichia coli*. Proc. Nat. Acad. Sci. U.S.A. **38**, 785–797 (1952). — MATHER, K.: Nucleus and cytoplasm in differentiation. Symposia Soc. Exper. Biol. **2**, 196–216 (1948). — McKENDRICK, M. E., and G. PONTECORVO: Crossing-over between alleles at the *w* locus in *Drosophila melanogaster*. Experientia (Basel) **8**, 390 (1952). — MITCHELL, M. B., and H. K. MITCHELL: Observations on the behaviour of suppressors in *Neurospora*. Proc. Nat. Acad. Sci. U.S.A. **38**, 205–214 (1952). — MULLER, H. J.: Further studies on the nature and causes of gene mutations. Proc. 6. Int. Congr. Genetics **1**, 213–255 (1932).

NEWMEYER, D., and E. L. TATUM: Gene expression in *Neurospora* mutants requiring nicotinic acid or tryptophane. Amer. J. Bot. **40**, 393–400 (1953).

PAULING, L., H. A. ITANO, S. J. SINGER and I. C. WELLS: Sickle cell anaemia, a molecular disease. Science (Lancaster, Pa.) **110**, 543–548 (1949). — PONTECORVO, G.: Genetic systems based on heterokaryosis. Cold Spring Harbor Symp. Quant. Biol. **11**, 193–201 (1947).

RHOADES, M. M.: The genetic control of mutability in maize. Cold Spring Harbor Symp. Quant. Biol. **9**, 138–144 (1941).

SAGER, R.: On the mutability of the waxy locus in maize. Genetics **36**, 510–540 (1951). — SONNEBORN, T. M.: The cytoplasm in heredity. Heredity (Lond.) **4**, 11–36 (1950). — SPIEGELMAN, S.: Differentiation as the controlled production of unique enzymatic patterns. Symposia Soc. Exper. Biol. **2**, 285–325 (1948). — SPURWAY, H.: Genetics and cytology of *Drosophila subobscura*. IV. An extreme example of delay in gene action, causing sterility. J. Genet. **49**, 126–140 (1948). — SRB, A. M., and N. H. HOROWITZ: The ornithine cycle in *Neurospora* and its genetic control. J. of Biol. Chem. **154**, 129–139 (1944). — STADLER, L. J.: Spontaneous mutation in maize. Cold Spring Harbor Symp. Quant. Biol. **16**, 49–63 (1951). — STERN, C.: Multiple Allelie. In Handbuch der Vererbungswissenschaft, Bd. 14. 1930.

TATUM, E. L., D. BONNER and G. W. BEADLE: Anthranilic acid and the biosynthesis of indole and tryptophane by *Neurospora*. Arch. of Biochem. **3**, 477–478 (1944). — TIMOFÉEFF-RESSOVSKY, N. W.: Über den Einfluß des genotypischen Milieus und der Außenbedingungen auf die Realisation des Genotyps. Nachr. Ges. Wiss. Göttingen, N. F. **1** (1934).

WAGNER, R. P., and C. H. HADDOX: A further analysis of the pantothenicless mutant in *Neurospora*. Amer. Naturalist **85**, 319–330 (1951). — WRIGHT, S.: The physiology of the gene. Physiologic. Rev. **21**, 487–527 (1941).

YANOFSKY, C.: The effects of gene change on tryptophane desmolase formation. Proc. Nat. Acad. Sci. U.S.A. **38**, 215–226 (1952).

# Interaction of genic and non-genic hereditary units and the physiology of non-genic inheritance.

By

## M. M. Rhoades.

## I. Introduction.

In the first mitosis of a fertilized egg cell the chromosomes divide equationally and the two identical halves of each chromosome pass to opposite poles of the spindle. Thus the nuclei in the two daughter cells are identical. Since the chromosomes divide equationally in all of the thousands of subsequent mitoses occurring during development and differentiation, it is believed that all nuclei of an individual, whether located in undifferentiated or specialized tissues, are alike—*i.e.*, they all have the same genotype. Yet these diverse somatic cells of presumed identical genotype come to possess during ontogeny certain differences and these differences have been shown to be permanent modifications since they are maintained indefinitely in tissue culture. The inherited difference between somatic cells of differentiated tissue, or cell heredity as it has been termed, must have its basis in the cytoplasm. It follows that the architecture of the cytoplasm is such as to permit the reproduction of its kind. Cytoplasmic differences first arise, so it is argued, from the action of nuclear genes; therefore the ultimate control of development would seem to lie with the gene and the kind of genes present in the nucleus determines in a given environmental milieu the course of development. We may grant that gene-induced cytoplasmic changes are concerned in somatic differentiation and that these modifications once acquired are part of the genetic endowment of the cell. But does this mean that the nucleus has a monopoly on the control of hereditary characteristics or does the cytoplasm possess a fundamental, stable, self-perpetuating constitution not subject to modification by nuclear genes which can be shown to play a vital, and in some instances a decisive, role in developmental physiology? If the latter be true then it is the interplay between nuclear and cytoplasmic systems which controls developmental processes. Normal development would result if the two genetic systems were able to interact in a harmonious manner while the combination of dissimilar systems could lead to abnormalities of various kinds.

Not until recent years has extra-nuclear or cytoplasmic inheritance been widely accepted. Only in relatively few of the thousands of characteristics in different organisms whose mode of transmission in sexual reproduction has been investigated does it appear that non-Mendelian heredity is involved. Inasmuch as the first examples all concerned maternally inherited plastid variegations in plants it seemed unlikely that the cytoplasm, with the exception of plastids, possessed specific and heritable properties whose fundamental nature remained unaltered by the action of nuclear genes. This heritable and constant quality of the cytoplasm was designated by v. WETTSTEIN as the plasmon in contradistinction to the genom or the genic determiners of heredity, and it is largely because of the careful and long-continued experiments of CORRENS, v. WETTSTEIN, RENNER, MICHAELIS, OEHLKERS, SCHWEMMLE and other German investigators that evidence of plasmatic heredity became so overwhelming as

2*

finally to bring widespread although not universal acceptance of its existence. It must be admitted that geneticists of the American school, which played so prominent a rôle in the erection of the chromosome theory of heredity, were among the skeptics. In his 1934 critique of plasmatic inheritance, EAST concluded that the case for plasmatic inheritance was not completely proven. According to him the cytoplasm and nucleus cooperate in development but the only ascertained agent of heredity is the nucleus. He conceded that cytoplasmic differentiation occurs and that this differentiation is an important part of the machinery of developmental physiology but was not convinced that more could be claimed for the cytoplasm. As late as 1949 BEADLE stated that there was no convincing evidence of the existence of extranuclear elements having a degree of autonomy comparable to that of the genes. He does not deny the existence of cytoplasmic heredity but maintains that except for a very few special cases the cytoplasmic factors responsible for extra-nuclear inheritance are not gene independent. Although BEADLE grants that plastids have a limited degree of autonomy he does not believe that they can be primary units of heredity. On the other hand SONNEBORN (1951a, b) argues eloquently for fuller recognition of the cytoplasm as a carrier of hereditary determinants. No doubt the reluctance to admit the cytoplasm to partnership with the gene as a vehicle for heredity is fully justified. It is, as has often been stated, not an easy matter to obtain conclusive evidence of the genetic constancy of the plasmon or of particulate cytoplasmic inclusions. So many instances are known where gene induced changes in the cytoplasm of the egg cell (predetermination) or environmentally produced modification (dauermodifications) are transmitted maternally for several generations before becoming slowly dissipated that it is felt by some that more persistent cases could also gradually disappear if the cytoplasm were exposed to nuclear genes for a longer period of time. And, too, in certain cases which were believed to demonstrate cytoplasmic inheritance it was later shown by more critical tests that the ultimate control was genic. For example, GOLDSCHMIDT long maintained that in the moth *Lymantria* the factor (or factors) for femaleness was in the cytoplasm but in 1942 stated that it is most probably carried in the Y chromosome.

In sexual reproduction of higher forms the genic contribution of the maternal and paternal parents is usually equal in terms of chromosome number but the egg cell contributes either all, or the greater mass, of the cytoplasm to the zygote. This unequal cytoplasmic contribution by the two parents offers a material basis for recognition of any rôle of the maternal cytoplasm in developmental physiology. In general, the occurrence of dissimilar hybrids from exact reciprocal crosses is suggestive of a cytoplasmic effect. However, before any part can be assigned to the cytoplasm it must first be established that the nuclei of the reciprocal hybrids are not merely equivalent but are identical. Nuclear identity cannot be assumed; it is well known from RENNER's studies in *Oenothera* where systems of genic complexes exist as a consequence of reciprocal translocations, that one or both parents may produce two types of gametes of which one is not transmissable through the ovules while the other is not male transmissable. In such cases the reciprocal hybrids would possess quite dissimilar sets of genes and would differ profoundly in appearance irrespective of any cytoplasmic effect. If, however, it can be shown that nuclei with identical genic constitutions are present in the reciprocal hybrids, then the differences between them must be ascribed to the maternal cytoplasm. But the question then arises as to the nature of this cytoplasmic effect. Are its properties due to changes impressed upon it by the action of nuclear genes during the course of development or does the cytoplasm possess stable and self-reproducing properties which are not subject to genic modification? Only when the latter has been demonstrated can it be said that the plasmon exists as an independent system in heredity.

Theoretically, it should be easy to make a distinction between maternally inherited effects due to predetermination or dauermodifications and those of true cytoplasmic inheritance. In the former where the cytoplasm of the unfertilized egg cell has certain properties impressed upon it by the maternal genes or by the environment, which influence the development of specific characteristics in the offspring, the effect should always be temporary while in true plasmatic inheritance there should be no diminution, over an indefinitely large number of generations, of the phenotypic alteration produced by the action of a stable plasmon. As OEHLKERS puts it, in cases of predetermination or dauermodification we are dealing with transient changes in the phenotype of the cytoplasm while true cytoplasmic inheritance involves the action of a stable structure with specific and enduring qualities in the same sense that the genotype or genom consists of a constellation of stable, self-reproducing genes.

The paucity of cases of cytoplasmic heredity in animals as compared with plants has also contributed to an unwillingness to concede that the cytoplasm plays a positive rather than a passive role in heredity. As we have said, the existence of cytoplasmic inheritance rests upon differences in reciprocal crosses where the egg cell contributes by far the larger mass of cytoplasm to the zygote. It is true that in many plants the male gamete brings little or no cytoplasm into the zygote so that in reciprocal crosses identical nuclei lie in the cytoplasm of the maternal parent. If the cytoplasms of the reciprocally crossed individuals are fundamentally different the interaction of the same kind of nucleus in two distinct types of cytoplasms may alter certain developmental processes with a consequent phenotypic effect. But in animals the spermatozoon is more than a naked nucleus; in many instances its midpiece is composed of mitochondria and there is a cytoplasmic layer enveloping its nuclear portion. The frequent occurrence in animals of polyspermy, where more than one spermatozoon penetrates the egg but only one succeeds in uniting with the egg nucleus, would also lead to an accumulation of sperm cytoplasm in the fertilized egg cell. It is wholly probable that the cytoplasm of the animal zygote contains cytoplasmic elements from both parents and certain of these may be present in not too unequal proportions. Reciprocal crosses in animals would not usually provide the situation found in higher plants where the cytoplasm of the zygote is largely, and in some cases exclusively, of maternal origin; therefore reciprocal differences due to a difference in cytoplasmic units in the two parents would be much more infrequent. Further, the fact that many instances of extra-nuclear inheritance in plants involve plastids, which are not found in animals, has led to doubt of plasmatic heredity as a general phenomenon.

In recent years there have been a number of excellent reviews on the relative rôles of the nucleus and cytoplasm in heredity. I refer particularly to the papers by RENNER (1934, 1936), CORRENS (1937), v. WETTSTEIN (1937), CASPARI (1948), SONNEBORN (1949, 1950, 1951b), OEHLKERS (1952, 1953), and EPHRUSSI (1953). The genetic data on extra-nuclear inheritance are well presented in these admirable papers and I suggest that the reader refer to them for details and further references which must necessarily be omitted from this dissertation. What shall be attempted here is the presentation and evaluation of certain of the diverse investigations where there is evidence of an independent set of genetic determiners in the cytoplasm.

Heuristically, instances of cytoplasmic heredity can be divided into two groups: (1) those involving visible particulate cytoplasmic entities and (2) those for which it has not yet been possible to ascribe the genetic effects to discrete bodies. These might be termed particulate and non-particulate cytoplasmic heredity. The term plasmon was used by Wettstein (see his 1937 review) for cases of non-particulate cytoplasmic inheritance. His conception of the plasmon was that of a homogeneous mass of cytoplasm lacking in individual elements comparable to genes or plastids. This reasoning was largely based on his results with interspecific and intergeneric moss hybrids where there was no indication of plasmatic segregation. However, extensive studies with other plants, notably *Epilobium* where plasmon alterations occur, have led to the concept that the genetic determiners of the plasmon are probably self-reproducing protein particles, not all of them identical, which may undergo segregation during somatic divisions so that some cells possess more and other cells fewer of certain of these genetic units. It is possible that the microsomes have the essential qualities of genetic determiners. They range in size from 0.06 to 0.2 micra, are the centers of enzyme localization, and contain ribosenucleic acid and phospholipids in definite proportions (Claude 1943).

## II. Particulate cytoplasmic inheritance.

Before particulate bodies in the cytoplasm can be conceded to constitute an independent set of hereditary determiners they must first be shown to possess the fundamental properties of nuclear genes—namely those of self-duplication and of mutability. Suppose, for example, that the plastids of plant cells are able to reproduce themselves and never arise save from pre-existing bodies. It will then be granted that they have genetic continuity. Further suppose that they undergo mutational changes, thus apparently satisfying the second criterion. However, before placing plastids on the same level with nuclear genes as primary units of heredity, one must next consider the cause of these mutations. Gene mutability has been shown to be controlled by other genes (see Rhoades 1941) but no convincing evidence has been presented of the influence of the cytoplasm on gene mutation. If plastid mutations are not induced by nuclear genes then plastids must be accorded the same rank with genes as hereditary units. On the other hand if plastid mutations can be shown to be gene-induced then they should be assigned a lesser rank even though the gene-induced plastid mutations maintain their mutant phenotypes in all genic environments. In evaluating cases of cytoplasmic inheritance these criteria should be kept in mind although in few instances will the data now available permit so rigorous a distinction. In my opinion the fact that the functioning of a cytoplasmic body is gene-directed is irrelevant. It could also be argued that plasmon-sensitive genes are cytoplasmically controlled. Developmental processes involve the interaction of nucleus and cytoplasm and the kind of interaction has no significance to the question of the autonomy of the two systems.

Among the visible particulate components of the cytoplasm for which there is evidence that they arise only from division of their kind are mitochondria, plastids of various types, centrioles, blepharoplasts, kinetosomes and kinetoplasts. Virus and virus-like particles such as the kappa bodies of *Paramecium* and the sigma genoid of *Drosophila* should also be included even though they may be of extrinsic origin. And then there are those intracellular bacteria-like bodies, the bacteroids, found in certain insects, which are not only self-duplicating but are essential for the viability of their hosts.

Since the original observations of non-Mendelian inheritance came from the early studies of CORRENS and BAUR on maternally transmitted plastid varie-gations, it is appropriate that we consider first the evidence for plastids as autonomous hereditary units. Plastids are restricted to plant cells, although some flagellates possess them, but this difference between plant and animal cells may not be significant if plastids represent a species of mitochondria as has been suggested by many workers. The morphological evidence of chloro-plast continuity is inconclusive in higher plants where the mature plastids deve-lop from minute bodies or proplastids, which cannot always be distinguished in meristematic cells from mitochondria, but it has been established in certain algae that chloroplasts are self-duplicating bodies. There is also cogent evidence for plastid mutations. Thus, being both self-duplicating and mutable, plastids could serve as a vehicle of extra-nuclear heredity.

## A. Plastid mutations.

Hundreds of examples are known of gene-controlled plastid characters in many different plants. The maize plant alone has more than one hundred mutant genes which produce a variety of phenotypic effects on the plastids. Some are albinos, totally devoid of all chlorophyll and carotenoid pigments, while other mutants are only partially deficient for certain of these pigments. In some instances the plants are variegated for green and white (or yellow) sectors. In the great majority of these cases the gene has produced a transient physiological condition in the cytoplasm unfavorable for the normal functioning and develop-ment of the plastids but has not induced a permanent alteration in their fun-damental nature. It must be granted in the above cases that plastid function is gene-controlled and in this sense they are not independent entities, but as we have indicated this cannot be taken to mean a lack of true autonomy.

On the other hand there are cases where the inheritance of plastid variega-tion is non-Mendelian. As the result of many investigations by CORRENS, BAUR, NOACK, IMAI and others (see CORRENS 1937) it appears that non-Mendelian in-heritance of plastid variegation can arise from direct plastid transmission through the cytoplasm or as the consequence of a variable cytoplasmic condition in-hibiting plastid development. In certain of these cases, plastid variegation is inherited through the maternal parent only—*i.e.*, it is uniparental. CORRENS was the first to discover this type of heredity. When green-white variegated plants of *Mirabilis jalapa* were used in experimental crosses, the following breed-ing behavior was found: (1) flowers on green sectors gave only green offspring irrespective of the constitution of the pollen parent, (2) flowers from white branches produced only white seedlings no matter what kind of pollen was applied to the stigmas while (3) flowers from variegated branches yielded a mixed progeny of green, white and variegated plants in widely varying ratios. Crosses of flowers on green branches by pollen from flowers on white sectors gave only green off-spring. In the reciprocal crosses there was no effect of the pollen parent on the type of offspring. This case in *Mirabilis* and others showing the same strictly maternal plastid inheritance were designated as "status albomaculatus" by CORRENS. The term paralbomaculatus has been used for those cases of non-Mendelian plastid variegation which are biparental. A number of names have been given to describe variegated forms according to the developmental pattern of green and mutant tissue. These distinctions are of importance morphogenetic-ally but there is no reason to confuse the point at issue here by introducing these terms.

The simplest and most widely accepted interpretation of these breeding data is Baur's hypothesis of plastid segregation. It is assumed that variegated plants possess two distinct kinds of plastids—normal and mutant. It is believed that the white plastids arose by spontaneous plastid mutation. If the cytoplasm of a fertilized egg cell has both types of plastids, contributed in albomaculatus types solely by the maternal parent, segregation of the two kinds of plastids could occur during somatic mitoses so that some cells would possess only mutant plastids while others would have only normal or mixed populations of normal and mutant plastids. Subsequent divisions of a meristematic cell with only mutant plastids would give rise to a pure white sector on the plant, cells with only (or predominantly) normal plastids would produce only green tissue while cells with a mixed population of plastids give rise to variegated sectors as continuing plastid segregation occurs. The hypothesis of somatic segregation of normal and mutant plastids demands not only a genetic autonomy of the two kinds of plastids but requires that both types of plastids be found in the cytoplasm of single cells. Gregory (1915) reported that in a variegated race of *Primula sinensis*, whose breeding behavior showed it belonged in the status albomaculatus class, the mature green portions had only normal plastids and the chlorotic sectors only smaller, nearly colorless plastids while both types were observed in meristematic cells. Correns (see his 1937 paper for earlier references) rejected the hypothesis of plastid segregation as the explanation for all cases of albomaculatus. He did not believe it tenable because in his extensive studies with diverse material he rarely found cells with more than one kind of plastid and these should be the most frequent class. Another objection was that, even beginning with small numbers of normal and mutant plastids, there was a serious discrepancy between the number of mitoses needed to produce cells with only mutant plastids, if there is a random assortment of the two types and both multiply at the same rate, and the number of cell divisions occurring during development. Randolph (1922) found no evidence of two distinct types of plastids in single cells in his studies with an albomaculata type of maternally inherited plastid variegation in *Zea*. He reported that embryonic cells of normal plants have plastids in various stages of development and suggested that the range in size was mistakenly interpreted to indicate the existence of distinct plastid types.

Correns concluded that the cytoplasm in embryonic cells of certain variegated plants exists in an indifferent labile state which changes fortuitously into a normal condition permitting plastid development or into a plastid-inhibiting state. That these cytoplasmic states, once realized, were stable and transmissable was demonstrated by breeding experiments since flowers from white sectors gave only colorless offspring even when pollen came from green plants or green sectors of variegated plants, while flowers from green sectors produced all green seedlings. On either hypothesis, cytoplasmic inheritance is involved although if Correns' hypothesis is correct, there is in many cases of "status albomaculatus" no evidence of plastid autonomy. It has been found in maize, sorghum and in a number of other plants (see Correns 1937) that when the seeds borne on a variegated inflorescence were planted according to their relative spatial positions, there was a definite grouping or clustering of each class—i.e., groups of each kind of ovule were found along the axis of the female inflorescence. For example, Anderson (1923) and Demerec (1927) working with a maternally inherited chlorophyll variegation in maize reported that the three classes of offspring (green, striped and yellowish) were in definite clusters on the ear and that the sector yielding striped seedlings was situated between green and yellow areas.

This non-random distribution has been taken to indicate the somatic segregation of some kind of cytoplasmic unit which, if not the plastids, is one controlling plastid functioning (RHOADES 1946) but this does not necessarily negate the labile cytoplasm hypothesis since it could also give a non-random grouping of similar classes.

Even though the hypothesis of the somatic segregation of mixed plastids is untenable for some of the cases of maternally inherited chlorophyll mosaicism (*Mirabilis, Humulus, Borago*) it is almost certainly the mechanism which is operating in *Nepeta cataria*. WOODS and DU BUY (see their 1951 paper for complete list of publications) reported that at least 15 different and distinct plastid mutations occurred in plants homozygous for the recessive *m* gene. The type of mutant plastid ranged from plastids only slightly different from normal through intermediate forms not only deficient in pigments but having gross morphological modifications to those which were indistinguishable from mitochondria. Since as many as three distinct kinds of mutant plastids as well as normal ones were found in a single mesophyll cell, it is evident that the gene *m* does not induce mutations or even the same mutation in all of the plastids of a cell. In crosses where pollen from variegated plants homozygous for *m* was used on stigmas of wholly green plants, 27% of the ensuing 292 plants were variegated, although as seedlings they were green and mosaic sectors did not appear until later. The simplest explanation is that the mutant plastids were brought into the egg cytoplasm by the male gamete so the cytoplasm of the zygote had both normal and mutant plastids and that the mutant plastids subsequently became sorted out. In the cross cited above the variegated plant used as the pollen parent possessed sectors with different kinds of mutant plastids and in making the cross onto green stigmas no attempt was made to determine the kind of plastids present in the sectors providing the source of pollen. However, in another cross (line 7 of table 8 in WOODS and DU BUY 1951) where stigmas of a stable green strain were pollinated with pollen from flowers borne on a sector containing "cream" mutant plastids, 24% of the $F_1$ plants were variegated and the colorless sectors contained the same kind of mutant plastids, a class morphologically recognizable, present in the pollen parent. Although WOODS and DU BUY do not emphasize this point it would seem that this is an *experimentum crucis* in establishing that, despite CORRENS' theoretical objections, variegation can arise by the somatic segregation of mixed plastids.

Transmission of plastids through the male gamete is by no means confined to *Nepeta*. BAUR (1909) in his study of variegation in *Pelargonium zonale*, where inheritance of plastid mosaicism is biparental though non-Mendelian, believed that the appearance of variegated $F_1$ plants in crosses of green ♀ × variegated ♂ was due to the inclusion in the egg cytoplasm of plastids from the sperm cell thus producing a zygote with mixed plastids in the cytoplasm. The two plastid types became, it was argued, separated by chance segregation during embryogenesis. BAUR's observations, those of MICHAELIS on *Epilobium*, and especially those of RENNER on variegation in *Oenothera* produced by the introduction of male plastids into the egg cytoplasm provide strong evidence that plastids are autonomous units and that variegation can arise from the somatic segregations of mixed plastids.

Convincing evidence that plastids constitute an autonomous genetic system comes from RENNER's (1934, 1936) magnificent studies with hybrids in *Oenothera*. Species of this genus not only differ in their genotypes but possess different plastid systems which are adapted to specific genotypes. Normal plastid development occurs in certain combinations of gene-plastid systems but in other combinations the plastids are adversely affected; *i.e.*, the plastid phenotype is determined by the interaction of plastids and genoms. Since many of the *Oenothera* species have distinct and specific plastid systems, it is believed that during the course of evolution of the genus they became differentiated by plastid mutation as did the genic complexes which characterize this genus by gene mutation. One example will suffice to illustrate the general conclusions reached from his extensive investigations which by 1936 included 120 different combinations of plastid systems and genoms. Genoms from *Oenothera muricata* and *Oe. Hookeri* were combined to give *hHookeri-curvans* hybrids. (*Hookeri* has only one type of gamete or genom while *muricata* has both the *rigens* and *curvans* gene complexes but we will consider only the *curvans* genom.) From the cross with *Hookeri* as the female parent the hybrids *hHookeri-curvans* were yellow and lethal. It was concluded that the plastids from *Hookeri* were not adapted to the hybrid nucleus and failed to develop normally. The reciprocal cross gave hybrids of the same genic constitution but had *muricata* plastids. The fact that these plants were green indicated that *muricata* plastids were able to develop normally in the presence of the same hybrid nucleus which impaired the *Hookeri* plastids. RENNER found that these green hybrids developed yellow sectors resembling in appearance the phenotype of the reciprocal cross. He assumed that the sperm cell from *Hookeri* brought some plastids from the paternal parent into the embryo sac and that these *Hookeri* plastids were included as a mixture with the more numerous *muricata* plastids from the female parent. The yellow sectors came, it was argued, from the somatic segregation of the two plastid types and contained plastids derived from the male sperm cell. In some instances flowers arose from sectors possessing the putative *Hookeri* plastids, thus permitting genetic tests. When flowers from yellow sectors were self-pollinated, two kinds of viable offspring were produced: (1) *hHookeri-curvans* which were yellow as expected since they carried *Hookeri* plastids and (2) *hHookeri-hHookeri* which were green since *Hookeri* plastids were combined with a *Hookeri* nucleus. Other combinations with different genoms were made and it was found that these plastids from the yellow sectors invariably behaved in a manner characteristic of *Hookeri* plastids. Although their development had been abnormal in the hybrid with a *hHookeri-curvans* nucleus, no modifications in their fundamental nature had taken place. An even more dramatic demonstration of the stability of plastid systems is that plastids from one species were combined with foreign nuclei for 14 generations and were then recovered. Appropriate tests disclosed that no perceptible change had been induced in their specific properties even though they had long been exposed to the action of alien genoms.

    In the above cited investigations of plastid variegation in *Oenothera* RENNER found no evidence of any effect of the plasmon. The phenotypes of the plastids were determined solely by the interaction of specific plastid and genom systems. It is, therefore, interesting to note that another type of variegation found by RENNER was the consequence of a different kind of mutation. With a frequency of about 5 in 10,000, variegated plants appeared with white (rarely yellow) variegation. At first the young plants had a haphazard distribution of white flecks but in later development sectorial and periclinal chimeras of green and white tissue arose. This type of variegation was not due to hybridization. All tested cases exhibited a typical paralbomaculatus type of breeding behavior. The colorless plastids remained colorless in all tested genom combinations; they were really defective, wholly different in nature from phenotypically modified plastids in an unfavorable genomic environment.

Mention has already been made of Woods and Du Buy's studies with *Nepeta* where a wide spectrum of permanent plastid mutations was produced in the presence of a recessive gene. Although this case is unique in that so many different kinds of induced plastid mutations were obtained, it is neither the first nor the only example of nuclear genes inducing irreversible changes in plastids. Other cases were plastid mutations have been induced by genic action are those in barley (Imai 1936, and Arnason, Harrington and Friesen 1946), rice (Pal 1941) and maize (Rhoades 1943).

The recessive iojap (*ij*) gene located on chromosome 7 in maize produces green and white striped plants when homozygous. When ears of iojap plants are crossed with pollen from unrelated green plants the offspring may consist only of normal green plants or may consist of green, white and green-white striped plants in widely varying ratios. Occasionally an $F_1$ progeny is comprised only of white offspring but this is relatively infrequent. The reciprocal cross of green females by iojap males yields only green plants. The pure white $F_1$ plants coming from the cross of iojap by green are lethal and cannot be subjected to genetic testing but the striped $F_1$ plants are viable since they possess green sectors capable of sustaining growth. When these $F_1$ striped plants, which are all heterozygous for the dominant allele of iojap, were pollinated by green males it was found that the ensuing progenies consisted either of all white, of all green, or of a mixture of green, white and occasionally a few striped individuals. In the all white progenies half of the plants were homozygous for the normal allele of iojap yet all the plastids were small and colorless. These all white populations arise when the ear bud is located in a white sector of the striped $F_1$ plants. All green populations are produced when the ear bud arises from a green sector and a mixed progeny results when the ear is a composite of green and white areas.

It was concluded that the iojap allele induces irreversible plastid mutations and that these mutant plastids retain their changed characteristics in outcrossed generations even when the inducing gene has been wholly replaced by the normal allele. Although I encountered in my experiments no indication of gene-dependence of the mutant plastids, Mazoti (1950) reported that when homozygous iojap plants were pollinated by strains carrying the dominant *R* alleles no colorless seedlings were obtained in the $F_1$ populations although they were found in progenies of similar crosses when the pollen parent had the $r^r$ allele. Mazoti believes that the *R* alleles cause the mutated plastids to revert to normal while this does not occur if the $r^r$ allele is present. If it be granted that the genic constitution of some lines controls the appearance of colorless offspring, there is no proof in his experiments that it is the *R* locus which is involved but this is relatively unimportant. The pertinent point in his work is the demonstration of nuclear control of the phenotype of gene-induced mutant plastids. These observations strengthen the concept of the primacy of nuclear genes as hereditary units.

While the genetic results obtained from the crossing of iojap plants are understandable on the assumption that sporogenous cells contain either normal, mutant, or both normal and mutant plastids (this class giving rise to the striped $F_1$ plants by somatic segregation of plastids), it is also possible that the breeding data could be interpreted on Correns' hypothesis of a labile cytoplasm. If this is the correct explanation, then it follows that the labile condition of the cytoplasm is produced by the action of the iojap gene. On either hypothesis a genically induced change in the cytoplasm has taken place. A decision between the two alternative hypotheses could be reached if the presence of normal and mutant plastids in single cells could be conclusively demonstrated in the striped $F_1$ plants. This is difficult to do because of the minute size of the mutant plastids, but Woods and Du Buy's finding that individual cells of *Nepeta* have mixed populations of normal and mutant plastids provides the desired evidence for plastid rather than plasmon mutation.

The variegations and plastid inhibitions which Renner studied so extensively resulted from combining disharmonious plastid and genom systems. As has been indicated above the differentiation of specific plastid systems is believed to have come about by plastid mutations which occurred in the past history of the species. Therefore the evidence of plastid mutability is in a sense inferential but none-the-less a reality since plastids of different species are demonstrably unlike in their properties. Inasmuch as there was no indication of heritable gene-induced plastid changes in Renner's experiments, it might be concluded that the postulated plastid mutations leading to the formation of species-specific plastid systems arose from spontaneous mutations. By spontaneous, one means that they were apparently not gene-induced. [Imai (1937) uses the term automutation for spontaneous plastid mutation and exomutations for those which are gene-induced.] Since, however, it has been unequivocally established in other plants that permanent plastid mutations can be induced with a high frequency by nuclear genes, I am unable to see by what reasoning the less frequently occurring plastid mutations can be said with certainty to be of spontaneous origin. Indeed they may be, but since plastids and genes are invariably associated in the cell how can it be shown for any given plastid mutation that nuclear genes were not responsible? Once it has been demonstrated that some plastid mutations are gene-induced, all plastid mutations become suspect of arising in a similar way. Since these gene-induced plastid mutations are, in general, irreversible modifications it is clear that plastids have a high degree of autonomy. I am inclined to agree with Beadle's (1949) statement that, though they possess the power of self-duplication and can mutate, they cannot be said to be primary units of heredity comparable to genes; they do not have the same degree of autonomy. Plastid mutation can be caused by specific genes but there is as yet little or no evidence that the cytoplasm or any of its constituents controls the mutability of genes. To be sure, as Sonneborn says, genes and cytoplasm are mutually and specifically dependent upon one another for their maintenance, but this does not confer equal rank upon them in the cell's hierarchy.

No definite conclusions can be drawn of the genetic significance of kinetoplasts, kinetosomes, blepharoplasts and centrioles which are all visibly self-duplicating. The latter three are concerned with the production of fibers or fibrillae of one kind or another. Centrioles are found in many animal cells, in certain algae and fungi and in the spermatogenous cells of bryophytes and those vascular plants with motile male gametes but rarely, if ever, occur in cells of Angiosperms. The blepharoplasts of bryophytes and vascular plants are undoubtedly homologous to centrioles and in any event are restricted to certain plant groups. The finding by Pollister and Pollister (1943) that, during meiosis in viviparid snails, centromeres (kinetochores) become detached from their chromosomes, pass into the cytoplasm and there function as centrioles casts serious doubt upon the status of centrioles as autonomous structures. Lwoff (1950) has presented impressive evidence in the ciliated protozoon *Gymnodinioides inkystans* that kinetosomes arise only by division, and Werbitzki (1910) found that, if division of the kinetoplast of Trypanosomes was inhibited by acridine dyes so one of the daughter cells did not receive this body, its loss was irreplaceable. In all of the above cases, however, there is no evidence of the mutability of these cytoplasmic elements and a final judgment of their genetic continuity should be held in abeyance. This is also true for bacteroids and other symbiotic bodies. The reader is referred to LEDERBERG's (1952) lucid and thoughtful review of the significance of hereditary symbiosis.

## B. Cytoplasmic inheritance in Paramecium.

The protozoan *Paramecium aurelia* offers many advantages for the study of cytoplasmic inheritance and it is not surprising that a number of significant studies have been made with the organism by SONNEBORN, PREER, BEALE, DIPPEL and others. This protozoan is a single-celled individual which can multiply asexually by simple fission thus giving rise to a clone with all animals of the same genetic constitution, although not necessarily homozygous since this depends upon the past history of the culture. In each individual there is one large (polyploid?) macronucleus and two diploid micronuclei of identical genic constitutions. The macronucleus is derived from micronuclei and plays no significant part in sexual reproduction. During autogamy (uniparental sexual reproduction) the macronucleus degenerates while each of the two micronuclei undergo meiosis to produce 8 haploid nuclei. Seven of these degenerate and the remaining one divides to form two daughter haploid nuclei which fuse to produce a single homozygous diploid nucleus. If the animal undergoing autogamy is heterozygous for a gene pair, for example *Aa*, the probability that the diploid nucleus arising at the end of autogamy will be homozygous *aa* or *AA* is equal. Two mitotic divisions occur giving rise to four diploid nuclei of which two become micronuclei and the other two are concerned in the formation of the two new macronuclei. Prior to fission the two micronuclei divide again so when fission takes place each daughter cell receives one macronucleus and two micronuclei.

In biparental sexual reproduction, which involves individuals of different mating types, nuclear behavior in each animal is the same as in autogamy but instead of the fusion of the two haploid nuclei, formed by the division of the one functional nucleus produced at the end of meiosis, an exchange of haploid micronuclei takes place between the two conjugating cells. This is followed by fusion of the two haploid nuclei, one from each parent, to restore the diploid number of chromosomes. When two animals of opposite mating type conjugate, gametic nuclei are exchanged between the two cells, but ordinarily no exchange of cytoplasm takes place. Since the two exconjugants have an identical nuclear constitution but cytoplasms of different origin, any heritable difference between the two exconjugants can be provisionally ascribed to the diverse cytoplasms. There is the further advantage that an exchange of cytoplasm can exceptionally occur during conjugation thus providing animals alike in both cytoplasmic and genic components. That SONNEBORN and his associates have fully exploited the technical advantages of *Paramecium* will become apparent when we consider the inheritance of the killer character and of antigen specificity.

Certain strains of *Paramecium* produce and liberate into the culture fluid a toxic substance, paramecin, which kills animals of certain other strains but which does not affect paramecin-producing animals. Strains able to form this poison are called killers and those which are killed are known as sensitives. Killer strains have kappa particles in the cytoplasm and the dominant gene *K* in the nucleus. The cytoplasmic factor kappa responsible for the killer trait is a minute particulate entity about 0.4 micra long. It possesses desoxyribose-nucleic acid and is similar in many respects to the *Rickettsia* viruses (PREER

1950). Paramecin is produced by these kappa particles. Sensitives have no kappa bodies, they may carry either gene $K$ or the recessive gene $k$. An animal which possesses kappa and gene $k$ soon reverts to a sensitive because kappa cannot be maintained in $k$ animals; they are incapable of being transformed into killers. When a killer strain with kappa and homozygous for $K$ is mated with a sensitive strain with no kappa but possessing gene $K$ the two exconjugant lines derived from this mating are genically identical but one is a killer and the other is a sensitive clone. The difference between these clones resides in the cytoplasm; the clone with the kappa bodies is a killer and the one without kappa is a sensitive. Here we have an unquestioned case of cytoplasmic inheritance determined by the presence or absence of a cytoplasmic particle. However when the two conjugating cells are exceptionally in prolonged contact thus allowing cytoplasm from each cell to flow into the other the two exconjugant clones are both killers since each has kappa bodies in its cytoplasm.

When sensitive cells carrying the gene $K$ are placed in a concentrated suspension of disintegrated killer animals some are converted into killers, that is, kappa is infectious. If kappa particles are lost or destroyed by exposure to high temperature (SONNEBORN 1946), X-rays (PREER 1948), nitrogen mustard (GECKLER 1949), or chloromycetin (BROWN 1950) a killer is transformed into a sensitive simply because kappa is no longer present to produce paramecin. Although the tempo of reproduction of kappa is usually correlated with fission rate, the two processes respond differently to temperature changes. At higher temperatures fission rate is accelerated more than reduplication of kappa with the end result that some animals maintained at these temperatures will eventually lack kappa and become transformed to sensitives. The opposite effect is found at lower temperatures where cell division is slower than kappa multiplication. If a single kappa particle is present it can, under favorable environmental conditions, multiply sufficiently over a period of cell divisions to produce a strong killer phenotype (something in the neighborhood of 250 particles is needed). Not only is kappa a self-reproducing cytoplasmic entity containing DNA but it is subject to mutation. DIPPEL (1948, 1950) has found a number of mutant kappas which differ in the amounts and kinds of paramecin. One animal can carry two kinds of kappa particles, each forming its specific kind of paramecin. They are not mutually exclusive. Kappa has all of the essential characteristics of an autonomous cytoplasmic component, genetic continuity and mutability. The killer phenotype is produced only in the presence of kappa but the fact that kappa is not maintained in $kk$ animals shows that an interaction of nuclear and cytoplasmic factors is involved.

Although it has been established beyond doubt that the killer character depends upon a particulate cytoplasmic body, the criticism has been made that the causal agent is not a normal cellular constituent but is a symbiont or parasite of external origin. Kappa has many of the characteristics of a virus-like particle; it is infectious, possesses desoxyribosenucleic acid which is not true for other cytoplasmic constituents, and is not found in most strains of *Paramecium*. The same criticism has been leveled against the sigma particle which determines $CO_2$ sensitivity in *Drosophila* and it must be admitted that both may be of extrinsic origin but, as SONNEBORN states, the same charge has in the past been made against other cytoplasmic components such as plastids and mitochondria which are now generally accepted as normal cellular organelles.

The mate-killer trait described by SIEGEL (1953) and LEVINE (1953) is similar to the killer trait in that both depend upon cytoplasmic particles which are genically controlled. The mate-killer phenotype is manifested when the sensitive member of a conjugating pair dies. The killing action of mate-killers depends upon prolonged pellicular contact and is not produced by fluids in which mate-killers have lived. No substance comparable to paramecin is liberated. The cytoplasm of mate-killers contains Feulgen positive bodies similar to kappa. These have been designated mu particles which are maintained only in animals with the dominant $M$ gene and disappear in individuals with the recessive $m$ gene. It has not been determined if the $K$ and $M$ genes are at the same locus. Resistant animals (mu plus $M$) are transformed to sensitives by the loss of mu bodies.

Three different mate-killer stocks have been found which differ in their effects on sensitive cells. The mu particles of these different mate-killers are reported to be morphologically distinguishable. When the three mate-killer strains were crossed to one another, the exconjugant from one of the stocks died, depending upon the relative mate-killing strengths of the conjugating cells which proved to be serially arranged.

PREER, SIEGEL and STARK (1953) found three types of cytoplasmic bodies in *Paramecium aurelia:* kappa, mu and pi. Kappa particles are associated with the killer phenotype, mu particles with mate-killing but pi bodies have not been associated with any lethal effect. The pi bodies were first found cytologically in a strain which had once been a killer and probably represent a mutation of kappa. Certain kappa mutants unable to kill are similar to pi particles. It is possible that mu bodies are also kappa mutants although an independent origin has not been excluded. Phase contrast microscopy disclosed that kappa bodies in a killer strain are of two types: "brights" and "non-brights". The bright particles contain one or two refractile regions not present in non-brights. The bright particles were demonstrated to be the carriers of the poison, paramecin, which kills sensitive animals. Brights were not found in the mu or pi particles and correspondingly cells with these particles do not produce paramecin.

## C. Mitochondria.

Mitochondria or chondriosomes, which are universally present in the cyto-plasm, are minute bodies of varying size and appearance but commonly are seen as rod or thread-like structures. Although conflicting observations have been reported, there is good reason to believe that mitochondria are self-duplicat-ing; it has long been held that they play a rôle in heredity (MEVES 1908) but not until recently has there been experimental evidence in support of this con-tention. Mitochondria are known to be associated with many important respira-tory enzymes; all of the enzymes of the Krebs cycle are carried by mitochondria of both plant and animal cells. They undoubtedly have a vital function in cell metabolism. There may be distinct species of mitochondria which differ not only in their biochemical properties but also in their genetic potentialities. Some investigators believe, however, that there is a common pool of undifferen-tiated mitochondria, and that in the course of growth and development some of them acquire certain specific properties not conferred upon others. It has been argued that the large chloroplasts of plant cells are derived from mitochondria and that in the study of plastid mutations we are dealing with changes in the chondriome. Whether or not this hypothesis will be eventually confirmed is not germane to the present discussion, although it is my opinion that no convinc-ing evidence of the homology of plastids and mitochondria has been presented. It might be said in passing that one of the strongest arguments against the origin of chloroplasts from mitochondria is that in certain algae, *Euglaena* for example, loss of chloroplasts is irreversible. If plastids arise from mitochondria it is diffi-cult to understand why new plastids are not formed. However, several recent investigations with yeasts, which have no chloroplasts but do possess mito-chondria, are strongly suggestive of the genetic function of mitochondria in heredity.

That mitochondria play the decisive role in the loss of spore germinating ability following inbreeding in yeast is suggested by Winge and Lausten's (1940) studies with *Saccharomyces cerevisiae*. They found widely differing per-centages of spore germination in diploid races of diverse origin. In diploid strains arising by fusion of two haploid cells the percentage of spores able to germinate was much greater than in those diploids coming from single cells where no cell fusion is involved but the diploid number of chromosomes results from nuclear fusion of two haploid sister nuclei or possibly by the formation of a restitution nucleus (direct diploidization). The loss of germinating power in those lines arising by direct diploidization proved to be irreparable and was not restored in diploids arising subsequently by cell fusion. Although no cytological observa-tions were made they offered the hypothesis that in direct diploidization divi-sion of the mitochondria has not occurred at the time of nuclear fusion thus producing a cell which is diploid for chromosome number but is haploid with respect to number of mitochondria. Since mitochondria unquestionably have a vital function in cell metabolism, it would not be astonishing if cells unbalanced for cytoplasmic and genic determiners would be abnormal. On the other hand in the diploids arising from cell fusion, mitochondria are contributed by both haploid cells and the resulting diploid cell has a full quota of mitochondria. If their hypothesis is correct, and it certainly has not been completely established, it follows that the reduplication of mitochondria is closely correlated with nuclear and cell division since once a cell has an insufficient number of mitochondria it is not able to increase this number to a normal level by a more frequent redupli-cation of the mitochondria. Indeed the multiplication rates of all essential self-duplicating cell entities must be correlated for otherwise an unbalanced system would be created. There is some cytological evidence in higher forms that the number of mitochondria is fairly constant. Wilson (1916) reported that the primary spermatocytes of the scorpion *Opisthacanthus* had exactly 24 mitochondria which were divided equally among the two secondary spermato-cytes and the four spermatids.

Some remarkable work on cytoplasmic inheritance of enzyme systems has been done in recent years with yeast. Ephrussi, Hottinguer and Chimènes (1949) found that any population of baker's yeast gave rise to 'petite' (small or dwarf) colonies with a frequency of about 1 in 100 although different strains varied considerably in the frequency of these dwarf colonies. The slow growth of the petite colonies is due to the loss of the respiratory enzymes, cytochrome oxydase and succinic acid dehydrogenase, and utilization of glucose occurs only by the less efficient fermentative processes. Cells of the petite colonies contain cytochrome c but are lacking in cytochromes b and a (Tavlitzki 1949, Slonimski 1949, Slonimski and Ephrussi 1949). Slonimski (in Ephrussi 1953) found that the mutants were also totally deficient in cytochrome e, reduced coenzyme I-cytochrome c reductase and α-glycero-phosphate-dehydrogenase and contained lesser amounts of malic dehydrogenase linked to coenzyme I. Increased amounts of alcohol-dehydrogenase and of cytochrome c were found in mutant as compared to normal cells. Cytochrome $a_1$ and malic cytochrome c reductase, independent of coenzyme I, were found in mutant cells. These are not present in normal cells.

During vegetative reproduction, yeast cells lacking these respiratory enzymes give rise only to cells deficient for these enzymes—*i.e.*, the mutation is irreversible since no reversions were found in many thousands of cell generations. However, in crosses between normal and petite cells the characteristics of the mutant cells, designated as vegetative mutants, are lost in the diploid generation and do not appear in the haploid progeny nor even after repeated backcrosses to mutant cells. This clearly indicated that their inability to form these enzyme systems was not genically determined.

Not only do mutations from normal to petite arise spontaneously as noted above but they can be induced with a frequency approaching 100% by growing normal cells on a medium containing the acridine dyes, euflavine and acriflavine. By means of a study of cell lineage, where the successive buds produced by single mother cells grown in euflavine were individually isolated and tested, it was conclusively demonstrated that mutation of a normal to a mutant condition was induced by the dye. These induced mutants are identical to those arising spontaneously in that both lack cytochrome oxidase and succinic acid dehydrogenase. EPHRUSSI concluded that the most probable explanation of the action of the euflavine in inducing cells deficient for these enzymes was that it destroyed or rendered inactive by mutation those cytoplasmic particles, presumably mitochondria, necessary for the formation or functioning of these enzymes. That is, the mutant character results from the loss of an extranuclear, particulate and self-reproducing cytoplasmic particle. Consequently the daughter cells arising by budding from treated mother cells fail to have included in their cytoplasms those particles which were selectively acted upon by euflavine. Normal and mutant cells have identical nuclear constitutions but differ in their cytoplasmic constituents. If the character of the mutant cells is due to the lack of a specific class or classes of mitochondria it is evident that these missing mitochondria do not arise anew by genic action since mutant cells are irreversible.

A mutant type phenotypically similar to the vegetative mutants and deficient for the same respiratory enzymes was found in a French strain whose characteristics were due to a recessive mutant gene. Crosses of this new mutant with a normal strain gave asci with two ascospores producing normal cultures and two ascospores which yielded petite colonies. This new mutant was called a segregational mutant in contradistinction to the vegetative mutants of spontaneous or induced origin. Crosses of segregational and vegetative mutants gave diploid cells with normal growth, but when sporulation took place each ascus was found to contain two normal and two mutant spores. These unusual results were interpreted to indicate that the segregational mutant carried a recessive gene which inactivated (or perhaps was unable to confer specificity) but did not destroy the cytoplasmic particles while the vegetative mutant had the normal allele of this gene but lacked the cytoplasmic particles. The normal growth of the diploid cells arising from fusion of the two cell types was a consequence of combining the physiologically inactive but potentially normal particles from the segregational mutant with the dominant allele from the vegetative mutant. The inactive particles recovered their full activity in the presence of this allele. The synthesis of respiratory enzymes in yeast thus involves the interaction of a nuclear gene and cytoplasmic particles. These particles are apparently autonomous insofar as their reproduction is concerned but their functioning is gene-controlled for only in the presence of the dominant allele are they physiologically active. But, as EPHRUSSI emphasizes, their complete autonomy in reproduction is based on negative evidence. Indeed, there is evidence in certain populations that the rate of loss, perhaps due to a slower reproduction, is controlled by a recessive gene. It is also clear that the missing enzymes cannot be identified with the cytoplasmic particles for in the segregational mutant these particles are present but not the enzymes.

When sporulation of the diploid cells from the cross of segregational by vegetative mutant was induced immediately following their formation, a high frequency of ascospores was found which carried the recessive genes but whose cytoplasms were deficient for the cytoplasmic particles necessary for enzyme activity. These double mutants, however, were of infrequent occurrence when sporulation did not occur until after several serial transfers had been made. It was argued that the number of cytoplasmic particles in the newly arisen diploid cells was low since they were contributed only by the segregational mutant cell and that by chance some were not included in the cytoplasms of certain of the ascospores, while in cultures which had been maintained for many cell generations as diploids before sporulation, the number of these putative particles had increased sufficiently to insure that some would be in the cytoplasms of all or nearly all of the spores.

Later (1951) EPHRUSSI and HOTTINGUER reported evidence of an unstable state in freshly arisen petite colonies produced by euflavine treatment. When these colonies were suspended in saline and replated, there was a mixture of normal and mutant colonies. If this could not be ascribed to cell mixture it signified that mutation did not always involve the total loss of particles. It was further found that scalloped colonies appeared when normal cells were exposed to acriflavine on a solid medium. Upon suspending in saline and replating, these scalloped colonies gave rise to a mixture of normal and petite cultures. These observations suggest the persistence of high mutability. It was concluded that the existence of unstable cell states could occur if there was a reduction in number of particles while those remaining were unchanged or if the number remained unchanged but total inactivation was preceded by an intermediate and reversible unstable condition leading either to recovery or to irreversible inactivation (either inability to reproduce or impaired heterocatalytic function).

## D. Cytoplasmic inheritance in Neurospora.

Inheritance of mutant characteristics in *Neurospora* is ordinarily independent of the way the cross is made between mutant and wild type strains but MITCHELL and MITCHELL (1952) found that the "poky" characteristic, so designated because of the slow growth of the mycelia, was maternally inherited. The nature of the protoperithecial parent, which presumably contributes most of the cytoplasm to the diploid zygote and which may be considered to be the maternal parent, determined the growth rate of all the progeny. When poky was the protoperithecial parent in crosses with normal strains, all of the ascospores produced poky mycelia while in the reciprocal crosses only cultures with a normal growth rate were found. Since in the reciprocal crosses the nuclear constitutions of the zygotic nuclei were identical, it may be concluded that the poky character is controlled by cytoplasmic factors.

The possibility that the poky character was due to infective particles, such as a virus, was investigated but in mixed cultures of poky and normal strains there was no transference by contact of the poky growth habit to normal mycelia.

In a more recent study, HASKINS, TISSIERES, MITCHELL and MITCHELL (1953) found by spectroscopic analysis that the abundant red pigment produced by poky mycelia was due to a higher than normal concentration of cytochrome c. In young poky cultures little or no cytochromes a or b were present but slight amounts, about 2% of that in normal strains, later developed. Investigations of the enzyme systems in particle preparations obtained by centrifugation revealed that young poky strains were deficient in the respiratory enzymes, succinic acid oxidase and cytochrome oxidase, but with increasing age of the culture the succinic acid oxidase activity of poky became nearly that of wild type while the level of cytochrome oxidase activity, although it increased with age, remained far below that of normal. The level of succinic acid dehydrogenase was essentially the same in poky and normal cultures. Poky strains possessed an enzyme system, "cytochromase", not found in wild strains which was capable of destroying the cytochromes. The high concentration of cytochrome c in poky was possible only because of the simultaneous presence of another system which inhibited the "cytochromase" reaction.

Three additional strains with slow growth rates were investigated by MIT-CHELL, MITCHELL and TISSIÈRES (1953). Two of these, C115 and C117, were due to a single gene difference but the third, mi-3, was maternally inherited, resembling poky in this respect. Although both poky and mi-3 are similar in their mode of inheritance they differ phenotypically in growth rates. This difference was maintained in reciprocal crosses between poky and mi-3—*i.e.*, the dry weights of mycelial pads from poky × mi-3 were typical of those from poky × wild and the dry weights from mi-3 × poky were comparable to those from the cross of mi-3 × wild. Combinations of nuclear and cytoplasmic factors for slow growth were obtained by placing the C115 and C117 genes in poky and mi-3 cytoplasms. The C117 poky mycelia grew very slowly and usually disintegrated but could be maintained by varying the medium. The growth characteristics of the various strains studied are as follows:

| Strain | Dry weight, Mg. | |
|---|---|---|
| | minimal | yeast extract |
| wild | 75 | 95 |
| poky | 11 | 14 |
| mi-3 | 40 | 53 |
| C115 | 30 | 2 |
| C117 | 24 | 26 |
| C115 poky | 4 | < 0.5 |
| C117 poky | < 0.5 (6 days) | 14 (6 days) |
| C115 mi-3 | 6 | < 0.5 |
| C117 mi-3 | 23 | 24 |

Striking differences were found in the concentrations of the cytochromes. Poky, mi-3 and C115 all had above normal amounts of cytochrome c. The b band was normal in mi-3, weaker than normal in C115 and not visible in poky. No cytochrome c was found in C117 cultures but the b band was stronger than normal while a band which may be due to the presence of cytochrome $b_1$ was observed. Since cytochrome c is in higher concentrations in poky, mi-3, and C115, the double mutants had more cytochrome c than any of the single mutants. The b band was weak in the mi-3 C115 combination as was true in C115 and was non-existent in the poky C115 combination. The absorption bands in the mi-3 C117 combination were similar in those in C117 which is understandable if C117 inhibits the appearance of cytochrome c. The enzyme activities of mi-3, C115 and C117 or combinations of these with poky and among themselves have not yet been reported. No evidence was found from mixed cultures that the various mutants could interact to produce a normal growth.

The important feature of these results is that the cytoplasms of poky and mi-3 carry an altered cytochrome system which perpetuates itself from generation to generation without being modified in any way by nuclear genes. Since these respiratory enzymes are associated with particulate cytoplasmic granules, mitochondria or microsomes, the evidence suggests that the poky and mi-3 strains contain mutant mitochondria unable to form certain enzymes. It could be further argued that these mutant mitochondria have a genetic continuity, are capable of duplication, and have been irreversibly modified since they are not influenced by nuclear genes. In the C115 and C117 strains, the upset in the cytochrome system is gene-controlled but evanescent since no permanent modification is induced in the cytoplasm. These results parallel in a striking manner those found with plastids.

The petite strains of yeast and young poky strains are similar in that both have deficiencies of the respiratory enzymes but they differ in that poky has a normal level of succinic acid dehydrogenase activity while the petite yeast colonies are deficient for this enzyme.

## E. Inheritance of $CO_2$ sensitivity in Drosophila.

In 1937 L'Héritier and Tessier reported that certain strains of Drosophila melanogaster either did not recover or remained adversely affected by exposure to an atmosphere rich in carbon dioxide while other strains soon recovered from the narcotisation induced by this gas. Sensitivity to $CO_2$ proved to be due to the presence of a cytoplasmic self-reproducing particle called sigma. Females from a pure-breeding sensitive line gave only sensitive offspring in crosses with resistant males while the reciprocal cross produced both resistant and sensitive flies of both sexes. It is not surprising that a cytoplasmic constituent should be so readily transmitted by the sperm of Drosophila. During spermiogenesis in Drosophila, mitochondria and other cytoplasmic bodies are believed to be incorporated into the mature spermatozoon (Cooper 1950) and polyspermy is of frequent occurrence in Drosophila (Sonnenblick 1950). That sigma is infectious was demonstrated by the fact that resistant flies could be transformed into sensitives by the transplantation of an organ or by the injection of an extract from sensitive flies (L'Héritier and Hugon de Scoeux 1947[1]). Sigma has no known effect other than rendering flies susceptible to irreversible $CO_2$ narcosis.

When flies acquire sigma through organ transplantation or injection, inheritance of $CO_2$ sensitivity is only through the females. Varying proportions of the offspring from the cross of induced sensitive females by resistant males are sensitives but acquired sensitivity is never transmitted by the male gametes since the reciprocal cross of resistant females by induced sensitive males yields only resistant offspring. Evidently none of the male germ cells become infected and only part of the oocytes. From the sensitive daughters, coming from the cross of induced sensitive females by resistant males, a pure breeding sensitive strain can be obtained. Females of such strains give only sensitive offspring in crosses with resistant males while their sensitive brothers when mated with resistant females yield a mixed progeny of sensitive and resistant offspring. Flies which inherit sensitivity from their fathers transmit sensitivity to their offspring in a manner similar to that of flies with acquired sensitivity—i.e., the females produce a mixed progeny while the males give only resistant offspring in crosses with resistent females.

---

[1] Oehlkers (1952) is unwilling to accept the infectivity of sigma because Kalmus and Mitchison (1946) were not able to obtain infection by transplantation. It is difficult to understand why he chooses to ignore evidence to the contrary (L'Héritier 1951).

Females from a true breeding sensitive strain transmit sensitivity to all of their offspring irrespective of the male parent and since this performance is repeated over many generations and since the basis for $CO_2$ sensitivity rests on a particulate cytoplasmic entity, sigma, cytoplasmic inheritance is clearly involved. Replacement of the chromosomes in a sensitive strain with those from resistant stocks did not result in the loss of sensitivity. The maintenance of sigma at a *high* concentration depends upon several factors including temperature and to a limited extent on the genotypic constitution of the fly. Sigma is therefore not wholly independent of nuclear genes but they do not play a decisive role as does the gene $K$ for kappa in *Paramecium*. Indeed, sigma can multiply and be inherited in several species of *Drosophila* so it can adapt itself to a variety of genotypes. Some genes, however, provide so distinctly an unfavorable environment that sensitivity is maintained only by rigid selection. SIGOT (1951) found that, when meiosis in the male germ line occurs at a temperature of 30° C for six days, none of the spermatozoa transmit sigma and MERCIER (1951) reported that egg laying females kept at a temperature of 30° C for six days produced only resistant offspring. Thus sigma is inactivated or destroyed at high temperatures.

Since sigma is a self-reproducing body with genetic continuity it is not surprising that it undergoes mutation. One such mutation (GOLDSTEIN 1949, 1951) was to the omega mutant which differs from sigma only in that it is rarely transmitted through spermatozoa. Females carrying omega breed true for sensitivity. Although sigma and omega particles can be recovered, presumably from different cells, from one individual which has been infected with both, they are mutually exclusive in the oocytes inasmuch as the offspring of a female with both kinds of particles produces offspring with one or the other of the two but never with both sigma and omega particles. A second mutant of sigma, called iota, was studied by BRUN (in L'HÉRITIER 1951). Flies with the iota particle had some immunity to infection by sigma. It was found that iota particles are not infectious. There is some evidence that iota is unstable and frequently mutates to sigma.

Neither sigma nor its mutant forms have been observed cytologically by the light microscope, but studies of inactivation of sigma by X-irradiation give a target diameter of 420 Å units. Although sigma is thus too minute to be visible under the ordinary microscope it should be detectable by electron microscopy. Sigma has all of the essential characteristics of a plasmagene but it is by no means certain that it is not a virus-like symbiont of external origin. However, as we have seen for kappa, it is difficult to decide what is and what is not a normal cell constituent.

# III. Non-particulate cytoplasmic heredity.

From the work of v. WETTSTEIN (1924, 1926, 1928, 1937) with interspecific and intergeneric moss hybrids came the first convincing evidence of the plasmon. Since these extensive studies have been so frequently reviewed in the past, nothing more than a short summary of these exemplary investigations will be presented here.

Reciprocal crosses between varieties or races of *Funaria hygrometrica* revealed no differences which could be attributed to unlike plasmons, but crosses of *F. hygrometrica* with *F. mediterranea* gave dissimilar reciprocal hybrids. The two species differ in many ways. The sporophyte of *F. mediterranea* is small, has tall and acute opercula, paraphyses of spirally arranged oval cells and gametophytes with filamentous leaves whose midribs terminate abruptly below the apex. *Funaria hygrometrica* has a larger sporophyte with broad and flat opercula; the gametophyte has leaves with blunt tips whose midribs extend to the apex. Since the two species had numerous gene differences, many new combinations of characters were found in the gametophytes arising from spores produced by the $F_1$ hybrid sporophytes, but the plasmon inherited from the maternal parent played a decisive part in determining the formation of certain characters. Many characters appeared to be primarily gene-controlled but leaf shape resembled that of the female parent and length of the midrib of all the segregants was identical to that of the female parent. These maternal effects persisted in later generations without abatement. Evidently different races of *F. hygrometrica* have similar or identical plasmons while unlike plasmons are present in *hygrometrica* and *mediterranea*. It might be anticipated that more distantly related mosses would possess even more dissimilar plasmons and this was found to be true in intergeneric crosses involving *Physcomitrium piriforme* and *Funaria hygrometrica*. The $F_1$ gametophytes from the cross of *P. piriforme* ♀ × *F. hygrometrica* ♂ had predominantly maternal characteristics. The failure to find gametophytes resembling the paternal parent, *F. hygrometrica*, was taken to indicate that genoms consisting chiefly of *hygrometrica* genes were inviable in *piriforme* cytoplasm. In agreement was the fact that many aborted spores were found; presumably these were the ones with a majority of genes from *hygrometrica*. The $F_1$ gametophytes from the reciprocal cross, which was difficult to make, resembled the gametophytes of *hygrometrica*. These maternal effects were maintained for 15 years during which time further crosses to the male parents were made but there was no indication of a phenotypic shift toward the male parent.

More critical evidence of plasmon constancy came from regeneration experiments in which diploid, triploid, and tetraploid gametophytes were obtained directly from sporophytic tissue without the intervention of meiosis. The original cross of *P. piriforme* ♀ × *F. hygrometrica* ♂ produced a diploid $F_1$ hybrid sporophyte with one genom from each parent and a *piriforme* plasmon. Diploid gametophytes derived by regeneration were backcrossed to *F. hygrometrica* to produce a triploid sporophyte with one genom from *piriforme* and two from *hygrometrica*. The triploid gametophytes obtained by regeneration from sporophytic tissue were again crossed to *hygrometrica* thus producing a tetraploid sporophyte from which tetraploid gametophytes were produced by regeneration. These tetraploid gametophytes had three genoms from *hygrometrica*, one from *piriforme* and a *piriforme* plasmon. The *piriforme* plasmon was exposed to three sets of genes from *hygrometrica* over a period of six years but there was no increase in spore fertility and the gametophytes still resembled the maternal parent.

One of the most convincing examples of plasmon differences is that reported in *Streptocarpus* by Oehlkers (see his 1952 paper for references to earlier publications). *Streptocarpus rexii* and *S. wendlandii* are fertile hermaphroditic species. The $F_1$ hybrids from the cross with *rexii* as the female parent were wholly male fertile but only partially female fertile since there was a considerable reduction in seed set. When these $F_1$ plants were pollinated by *wendlandii*, the backcross populations consisted of two classes. One of these was similar to the $F_1$ hybrids in having hermaphroditic flowers with a lowered female fertility while the other class was comprised of plants with normal appearing flowers which, although male fertile, had completely sterile ovules—*i.e.*, they were functionally male plants. The reciprocal cross with *wendlandii* as the egg parent produced $F_1$ plants which were male sterile owing to the fact that the stamens were transformed into sterile staminodia. These male sterile $F_1$ hybrids were pollinated by *rexii* and the ensuing backcross progenies were found to consist of male sterile plants with staminodia instead of stamens, similar to the $F_1$ hybrids, and of plants in which the anthers had been transformed into functional ovules. These results, as well as comparable ones from other interspecific crosses in *Streptocarpus*, were interpreted by OEHLKERS to mean that *rexii* and *wendlandii* contained plasmons and genoms with opposed sex potentialities. For the sake of convenience let us speak of a plasmon tending towards maleness as a male plasmon and one with female tendencies as a female plasmon. Similarly, a genom favoring the development of female reproductive organs will be designated as a female genom and one with predominantly male factors as a male genom. Since both *rexii* and *wendlandii* are fertile hermaphrodites, a harmonious balance must exist within each species between the opposed sex tendencies of plasmon and genom but the results obtained from the reciprocal crosses and from the backcrosses to the recurrent male parents show that the two species differ in the relative sex strengths of plasmons and genoms. *Rexii* has a female genom and a male plasmon while *wendlandii* has a male genom and a female plasmon. Thus in both species a balance or equilibrium between two opposing systems for sexual development is established and the plants are both male and female fertile. The evidence for the differing potencies of plasmon and genom are revealed by the reciprocal crosses between the two species. For example, the cross of *wendlandii* as the egg parent gives an $F_1$ with a female plasmon and a nucleus composed equally of a female genom from *rexii* and a male genom from *wendlandii*. The combined tendencies for femaleness are thus so strong that staminodia are found instead of stamens. The backcross to *rexii* produces plants with the female plasmon from *wendlandii* and a greater concentration of the *rexii* female genom. Consequently the shift to femaleness is more marked in approximately half the plants than was found in the $F_1$ hybrids. The antithetical sexual tendencies of the plasmons and genoms found within each species must be emphasized since this constitutes critical evidence for the existence of the plasmon as an independent genetic system. These species of *Streptocarpus* are well established ones whose plasmons and genoms have been combined for many generations yet in *wendlandii*, for example, the male genom has not impressed its sex-determining qualities upon the plasmon which has retained its female tendencies.

It is not within the scope of this paper to discuss all of the cases where a plasmatic effect has been reported but SIRKS' (see his 1938 paper for citations) investigations with *Vicia* merit mention. In crosses between two subspecies of *Vicia Faba*, *major* and *minor*, he found lethal zygotes were produced when a series of linked genes from the *major* parent became homozygous in *minor* cytoplasm. The *minor* plasmon completely inhibited the action of these genes although they obviously functioned normally with a *major* plasmon. SIRKS also reported other instances of supposed plasmatic heredity but a critical analysis was not made.

WARIS (1950) found that enucleated cells, obtained by centrifugation, of both normal and a defective form of the desmid *Micrasterias* were able to form new semicells which became unilaterally lobed in a manner characteristic of the parent species. The control of the pattern of bilateral symmetry is believed by WARIS to lie in the structural organization of the cytoplasm, and this is assumed to be a permanent and heritable property which is not under nuclear control. WARIS' conclusions should be only tentatively accepted since his experiments do not negate the possibility of cytoplasmic predetermination by nuclear genes.

That *Aegilops ovata* and *Triticum durum* have different plasmons was demonstrated by Fukasawa (1953) who successively backcrossed the amphidiploid *Aegilotricum*, derived from the cross *Aegilops ovata* ♀ × *Triticum durum* ♂, to *Ae. ovata* until all of the *Triticum* chromosomes had been eliminated. The resulting plants with *ovata* chromosomes and a plasmon originally obtained from *ovata* were morphologically similar to *Ae. ovata* and were fertile. In a similar manner the amphidiploid was backcrossed to *Triticum durum* until the *Aegilops* chromosomes were lost and only *Triticum* chromosomes remained. These plants with *Triticum* chromosomes and an *Aegilops* plasmon were completely male sterile. When crossed with *durum* pollen all of the offspring were again male sterile.

Further evidence that different species have unlike and stable plasmons comes from Clayton's (1950) investigations with interspecific crosses in the genus *Nicotiana*. Clayton reported that plants with a plasmon from *N. debneyi* and chromosomes from *N. tabacum* were male sterile. Starting with plants possessing the *debneyi* plasmon and a full set of chromosomes from both parents, ten successive backcrosses to *tabacum* were made. Since no pairing took place between *debneyi* and *tabacum* chromosomes the *debneyi* genes were soon eliminated during the backcross experiments so that eventually plants with a *debneyi* plasmon and *tabacum* genoms were produced. The complete male sterility which was due to the incompatibility between unlike plasmon and genoms was not diminished by continued exposure to genes from *tabacum*. Similar results were obtained in the cross of *N. megalosiphon* to *N. tabacum*.

The classical example of a male sterile condition resulting from the interaction of a specific cytoplasm and a nuclear gene is that in flax. In 1921 Bateson and Gairdner reported that although crosses between two normally hermaphroditic races, normal tall and procumbent, gave fertile $F_1$ plants, male sterile individuals appeared in the $F_2$ generation in a 3:1 ratio when the procumbent strain was the maternal parent. The reciprocal cross yielded only fertile $F_2$ plants. Wettstein (1924) and Chittenden and Pellew (1927) interpreted these results as indicating that the tall and procumbent races differed in their cytoplasms and also by a single gene. The tall race carried the recessive allele *m* which was ineffective in producing male sterility in tall cytoplasm and the procumbent race carried the dominant allele *M* but possessed a cytoplasm which when combined with the recessive *m* allele from the tall strain produced male sterility. Purely cytoplasmic inheritance is not involved here since the appearance of the unexpected male sterile phenotype came from the interaction of a specific gene from the tall parent with the cytoplasm contributed by the procumbent race. The constancy of the procumbent plasmon was established by Wettstein (1946) who, beginning with $F_1$ plants from the cross of procumbent female by tall male, made eight successive backcrosses to the tall race. In all backcrossed generations he selected female plants for further crossing which more nearly resembled the tall parent. This continued backcrossing produced plants with genoms from the tall race and a plasmon from the procumbent strain. Yet all plants were completely male sterile; there was no indication that the genes from the tall parent had in any perceptible way modified the procumbent cytoplasm.

Other examples where male sterility results from the interaction of a specific cytoplasm with nuclear genes have been reported in *Nicotiana* (East 1932), *Allium* (Jones and Clark 1943), *Beta* (Owen 1942, 1945), and *Dactylis* (Myers 1946). The *Dactylis* case is of some interest since this involves a tetraploid species and the gene producing the male sterile condition is a dominant rather than a recessive thus making it possible to determine the effect of varying doses.

Plants with two, three or four doses of the dominant allele and the sterile cytoplasm were completely male sterile, plants with one dominant and three recessive alleles had varying degrees of sterility while those homozygous for the recessive allele were male fertile.

Cases have been reported where a new phenotype other than pollen abortion arises by combining a plasmon from one race or species with a nuclear gene from another strain. MICHAELIS (1940a, b, and c) found that although the Kew race of *Epilobium hirsutum* had normal fertile flowers it was homozygous for a gene which when placed in the cytoplasm of three German *hirsutum* races caused the formation of deformed and sterile flowers. Another example where a gene with no discernible effect with its own plasmon produces an unexpected phenotype when combined with a different plasmon was reported by OEHLKERS (1934) in *Streptocarpus*. Both *S. rexii* and *wendlandii* have trumpet-shaped corollas. The $F_1$ hybrid from the cross of *rexii* ♀ × *wendlandii* ♂ has trumpet-shaped corollas as do the $F_2$ plants but the mating of *wendlandii* ♀ × *rexii* ♂ gives an $F_2$ progeny in which one-fourth of the plants have a slit corolla. *S. rexii* carried the recessive *s* gene for slit corollas but this gene was not expressed until it was combined with the *wendlandii* plasmon. The dominant *S* allele was homozygous in *wendlandii* and slit corollas would not appear in this species until a mutation of *S* to *s* occurred. One such mutation did in fact arise, but no plasmon mutations were found in the *rexii* plasmon which would permit the expression of the recessive *s* gene which is homozygous in *rexii*. In these experiments the stability of the plasmon was fully equal to that of the *S* gene.

## A. Cytoplasmic inheritance of male sterility in maize.

A case of cytoplasmic inheritance of male sterility in maize was reported by RHOADES (1933) where the expression of the male sterile phenotype was not markedly influenced by the genic constitutions of the pollen parents. It was possible to determine the effect of nuclear genes by a systematic replacement, using chromosomes marked with mutant genes, of the chromosomes present in the original male sterile plants and when this was accomplished, many of the plants were still male sterile. It may be argued, as OEHLKERS (1952) has done, that more extensive crosses with diverse races of maize would reveal the existence of nuclear genes capable of suppressing the male sterility factors of the cytoplasm. Unfortunately, viable seed of the strain studied by RHOADES is no longer available so this conjecture remains untestable. However, an interaction with nuclear genes has been found in other cytoplasmic male sterile lines of independent origin in maize.

If the male sterile condition is due to the combination of a sterile cytoplasm and a specific gene (or genes) and if in the outcrosses the pollen parents carry the same genes for male sterility, all of the offspring will be male sterile and inheritance is apparently purely cytoplasmic. However, if the male parent contributes fertility-restoring genes then gene-cytoplasm interaction becomes apparent. It will be recalled that in SONNEBORN's studies with the killer character in *Paramecium* a purely cytoplasmic inheritance of the killer phenotype was obtained in matings of a killer stock having kappa and gene *K* with sensitive races no kappa but gene *K*, if no cytoplasmic exchange occurred between the paired conjugants. Only when sensitive races possessing the recessive *k* allele were used did it become apparent that the killer phenotype depends upon both gene and cytoplasm.

With the possible exceptions of certain of the plastid mutations and of $CO_2$ sensitivity in *Drosophila*, where it is probable that a cytoplasmically borne virus-like particle of foreign origin is the hereditary unit, it is doubtful if other cases have been reported where a hereditary characteristic is determined solely by the cytoplasm. In many examples the interaction of nuclear genes and the cytoplasm is clearly indicated. Even in CORRENS' (1928) studies with *Cirsium* and *Satureia*, where successive crosses of female plants by hermaphrodites produced only female offspring, it is not necessary to assume a purely cytoplasmic determination. It has been well established, chiefly by the work of LEHMANN, MICHAELIS, SCHWEMMLE and others in *Epilobium*, that the action of specific genes is inhibited to varying degree by the nature of the cytoplasm. In *Cirsium* and *Satureia* it may be that the plasmon of the female race completely inhibits the action of the male sex-determining genes by raising the threshold for their activity. The plasmon may be nothing more than a substrate through which gene products mediate developmental processes, but the enduring quality of this substrate is in some instances the decisive factor determining the effectiveness of genic action.

JOSEPHSON and JENKINS (1948) found an inbred line of maize (33–16) which, although male fertile, possessed a type of cytoplasm producing pollen abortion but only when combined with specific genes. Some inbred lines carried genes which almost completely inhibited the cytoplasmic effect, while other inbred lines lacked these fertility-restoring genes. It is evident that an interaction of nucleus and cytoplasm controls the expression of male sterility. They believed that a minimum of two dominant genes was involved in the suppression of the cytoplasmic sterility factors of inbred 33–16. It was also found that environmental conditions significantly affected the degree of pollen abortion, a statement which applies to all reported cases of cytoplasmic male sterility. JONES (1950, 1951) confirmed the earlier findings of JOSEPHSON and JENKINS that certain strains were able to restore pollen fertility. His data suggested that only one gene was involved rather than a minimum of two. In the $F_2$ obtained by the self pollination of an $F_1$ from the cross of a male sterile plant by a fertility-restoring line, normal and male sterile plants occurred in a 3:1 ratio. The recovery of sterile plants in the $F_2$ indicated that the fertility-restoring genes had not affected the nature of the cytoplasmic factor or factors for pollen abortion. ROGERS (1950) working with the Texas source of male sterile cytoplasm reported that a limited number of inbred lines carried fertility-restoring genes. SCHWARTZ (1951) analyzed an unusual type of male sterility which involved interaction between a specific sterile cytoplasm and two genic factors. Male sterile individuals have the sterile cytoplasm, a dominant gene for male sterility and the recessive allele of a suppressor locus while plants lacking any one of these are male fertile.

The dozen or so cytoplasmic male sterile lines which have been found represent independent occurrences of some type of heritable change in the cytoplasm. There is no evidence that a virus-like particle of external origin is responsible but this has not been rigorously excluded. While the breeding behavior of those which have been studied is similar, with the exception of that described by SCHWARTZ, there is no reason to assume that the different sterile strains are identical insofar as the nature or degree of the cytoplasmic change is concerned. For example, in the first studied cytoplasmic sterile strain (RHOADES 1933) the effect of fertility-restoring genes was less obvious than in the sterile races described by JOSEPHSON and JENKINS and JONES and ROGERS. Further, one inbred line has fertility-restoring genes which are very effective on one type of sterile cytoplasm but less so on another. The nature or cause of the cytoplasmic mutations in the various sterile strains is wholly conjectural. They appeared sporadically in lines where no such sterility had been noticed previously. Whether particulate or not these cytoplasmic mutations represent a change in some cytoplasmic component which is self-reproducing; the mutant component perpetuates itself indefinitely, apparently unaltered by nuclear genes. It is known in maize that the plastids can be induced to mutate by the gene iojap and in 1950 RHOADES reported that this same gene induced a change in the cytoplasm which led to a male sterile condition. The breeding behavior of these induced cytoplasmic male steriles was similar to that of the spontaneously arisen cases. It may be that these induced cytoplasmic changes are due to modifications in the plastids rather than in some other cytoplasmic component such as mitochondria; the data on this point do not permit a decision. BRIGGLE (1953) reported that a cytoplasmic condition leading to male sterility in maize was found in the offspring of plants carrying the gene for vestigial glumes. Similar observations had been made by R. R. ST. JOHN (unpublished). There is thus some evidence that cytoplasmic mutations in maize can be gene induced. But it is possible that the vestigial glumes gene has not induced cytoplasmic mutations but has merely revealed the presence of a potentially sterile cytoplasm whose action could only be expressed in combination with certain genotypes.

GABELMAN (1949) in his study of JENKINS' strain of cytoplasmic male sterility in maize concluded that pollen abortion was due to a particulate cytoplasmic element and that the presence of one or more of these particles in the microspores resulted in their failure to develop into mature, functional pollen. However, his conclusions were only tentative and the particulate basis, although probable, cannot be said to be rigidly established.

## B. The plasmon of Oenothera.

Before discussing the extensive work on cytoplasmic inheritance in *Oenothera* a brief statement is necessary about the genetic and cytological peculiarities of this genus. All species have a diploid number of 14 chromosomes. It would be expected, therefore, that there would be 7 pairs of chromosomes at meiosis and that there would be 7 independent groups of linked genes, each corresponding to one of the 7 chromosomes of the haploid complement. However, the great majority of races behave as though they had but one pair of chromosomes and one linkage group. With minor exceptions all of the chromosomes contributed by the paternal parent pass to the same pole during meiosis while all of the chromosomes from the maternal parent go to the opposite pole, thus producing but two kinds of gametes. One of these has only paternal and the other only maternal genes except for exchange of segments due to crossing over. These sets of paternal and maternal genes were designated by RENNER as "complexes". Another remarkable fact is that although two dissimilar kinds of eggs and sperms are formed, and the plants are highly heterozygous, a uniform progeny may be produced upon self-pollination. This was shown to be due to the presence of balanced lethals which prevented the formation of viable zygotes homozygous for either the maternal or paternal chromosomes. Thanks to the masterful genetic analyses by RENNER and his students and to the cytological studies by CLELAND, the unusual behavior of *Oenothera* became understandable. It was found that, instead of 7 pairs at meiosis, many races of *Oenothera* had all 14 chromosomes arranged end to end in a circle or ring as a consequence of a series of reciprocal translocations between non-homologous chromosomes. The distribution of the chromosomes in the ring is such that chromosomes of paternal origin alternate with those of maternal origin and the orientation of the chromosomes on the metaphase I spindle leads to alternate members of the ring passing to the same pole. Consequently all chromosomes of paternal origin go to one pole and all seven of maternal origin pass to the opposite pole, and thus the Renner complexes are produced.

Renner found no evidence of a plasmon in his *Oenothera* studies; phenotypic disturbances of diverse sorts other than plastid inhibition were attributed to the faulty interaction of plastid systems (plastidoms) and genoms. However, in the investigations of Schwemmle, Haustein, Sturm and Binder (1938) on *Oenothera*, it was reported that a distinction could be made between the morphogenetic effects of plasmon and plastidom. This conclusion was based upon the assumption that only the maternal parent contributes cytoplasm to the zygote while plastids come from both parents, although the number of maternal plastids is in excess of the paternal. They found that some morphological changes other than variegation or plastid inhibition were due to plastids, thus confirming Renner's observations, but reported that the reduction in hypanthia length and petal size was plasmon controlled.

*Oe. Berteriana* has a *Berteriana* plasmon, *Berteriana* plastids and the *B* and *l* gene complexes while *Oe. odorata* has an *odorata* plasmon, *odorata* plastids and the *v* and *I* complexes. Crosses between these two species made possible a number of combinations of plasmons, plastidoms and gene complexes. For example, a comparison was made between plants with *odorata* plasmon, *odorata* plastids and the *BI* complexes and those with *odorata* plasmon, *Berteriana* plastids and the *BI* complexes; the two differ only in the nature of the plastid system since plasmons and genoms are identical. The former class with *odorata* plastids consisted of pale yellowish plants while the class with *Berteriana* plastids contained green plants. Leaf shape also differed in the two classes with dissimilar plastid systems; both the chlorophyll effect and that of leaf shape were due to plastid differences. The plasmon effect became evident when plants with identical complexes and the same plastid system but differing in plasmons were obtained. The *odorata* plasmon produced longer hypanthia and larger petals than did the *Berteriana* plasmon. The validity of this claim of a plasmon effect rests upon the unproved assumption that plastids, but no other part of the paternal cytoplasm, are brought in with the sperm nucleus. The *odorata* plasmon, *Berteriana* plastid combination is possible only if the pollen parent brings in *Berteriana* plastids but no *Berteriana* plasmon, and the *Berteriana* plastids are sorted out during subsequent mitoses to produce a sector or branch pure for *Berteriana* plastids but possessing an *odorata* plasmon. Flowers produced on this sector can be used in crosses to obtain the desired combinations of plasmon, plastidom and gene complexes for the critical comparisons.

Schwemmle found that the *BI*, *lv*, *Bv* and *vI* complexes in *Berteriana* cytoplasm and the *BI* and *Bv* complexes in *odorata* cytoplasm had hypanthia whose lengths became reduced in successive generations of selfing. The same change was found for petal size. When the *v* complex was maintained in *lv* hybrids for several generations with *Berteriana* cytoplasm and was then reintroduced into *Oe. odorata* by crosses where the female parent brought in the *odorata* cytoplasm and the *I* genom while the pollen parent contributed the *v* complex, it was found that hypanthia length was reduced. This heritable change in *v* was attributed to the *Berteriana* cytoplasm. Differences of this sort were consistently obtained—*i.e.*, complexes kept for a number of generations in a foreign cytoplasm became changed but this did not occur when the same complexes were maintained with their own cytoplasm. In every case there was a shortening of hypanthia length and a reduction in petal size. The nature of these changes remains obscure but SCHWEMMLE does not favor the view that mutations in the *v* genom have been induced by *Berteriana* cytoplasm (or plastids) although this is one possibility.

When the *lv* and *BI* complexes are first combined with *odorata* cytoplasm the plants are yellowish in color rather than green and have a reduced fertility. Selfed seed, however, can be obtained and when successive generations of selfed plants were grown it was found that the *lv* and *BI* plants gradually lost their yellow color and became green and fertile. By the fifth selfed generation the plants were normal green and highly fertile. The abnormalities observed in the original cross are due to the poor functioning of *odorata* plastids with *lv* and *BI* genoms. That recovery of green color and of fertility in later generations was not due to plastid mutation was demonstrated by pollinating a recovered *BI* plant with *odorata* cytoplasm by pollen from a plant with the *l* complex to produce zygotes of *Bl* constitution (= *Berteriana*) and of *Il* constitution in *odorata* cytoplasm. No seedlings were obtained from these crosses. Since it was known from previous studies that combinations of *Bl* and *Il* with *odorata* plastids were lethal, the fact that no offspring were obtained was taken to prove that the plastids in the recovered *BI* plants were typical *odorata* plastids. This does not necessarily follow since they could be somewhat different from *odorata* plastids and still give the same kind of lethal interaction with specific genoms. It must be admitted, however, that there was no positive indication of any change in their properties.

Although it may be questioned if a plasmon effect has been definitely separated from a plastid effect in *Oenothera*, the important point in SCHWEMMLE's studies is that the cytoplasm (plasmon and plastidom) was found to be at least as stable as the nuclear complexes.

## C. The plasmon of Epilobium.

In no other plant has plasmatic heredity been so assiduously studied as in *Epilobium*. It is unfortunate that, despite the enormous and painstaking experiments, the constancy of its plasmon remains in doubt. All students of *Epilobium* are agreed that plasmatic heredity is responsible for the often striking differences in reciprocal crosses but there is a divergence of opinion as to the nature of the plasmon. MICHAELIS (see his 1951 paper for full references to his work) RENNER, v. WETTSTEIN and others believe that the plasmon of *Epilobium* is a gene-independent system while LEHMANN and SCHWEMMLE feel that the nucleus is able to modify the plasmon and that each nuclear combination brings about a specific kind of plasmon by the release of gene products into the cytoplasm. This view is essentially that of predetermination.

No attempt will be made here to review the *Epilobium* work in detail; for a fuller account the reader is referred to MICHAELIS' paper mentioned above and to CASPARI's (1948) able summary. A brief statement of the pertinent observations is however necessary. Reciprocal crosses between *E. hirsutum* and *E. roseum*, *E. hirsutum* and *E. luteum*, *E. parviflorum* and *E. roseum*, and other races or species give dissimilar reciprocal hybrids. Many characteristics such as habit of growth, abnormal development of leaves, stem and flowers, sterility and lethality are found. All of these modifications are ascribed to the inhibiting effect of the plasmon on the expression of nuclear genes. In one kind of plasmon certain genes, sometimes only one and sometimes many, may be affected while other genes are affected in a different plasmon. Genes whose activity is adversely controlled by the plasmon have been termed plasmon-sensitive genes. As Oehlkers has pointed out, all genes are plasmon-sensitive in that they exert their influence in development only by interacting with the cytoplasm but the term is useful in a descriptive sense of relative inhibition. Commonly, it is the paternal genes which are inhibited upon being brought into a foreign cytoplasm but maternal genes also may be inhibited in hybrid combinations. Changes of dominance relations may even occur, as Michaelis has found.

The stability of the plasmon in v. Wettstein's moss experiments was established by eliminating all maternal chromosomes through successive backcrosses to the male parent and finding that certain maternal characteristics remained constant in their expression. Michaelis, accordingly, attempted a similar demonstration using *E. hirsutum* and *E. luteum* which give dissimilar reciprocal hybrids. The $F_1$ hybrids from the cross of *hirsutum* ♀ × *luteum* ♂ are completely male sterile and after two backcrosses to *luteum* no viable offspring were obtained. The reciprocal hybrids from *luteum* ♀ × *hirsutum* ♂ had varying degrees of pollen abortion. An increasing amount of pollen sterility was observed during the 25 successive backcrosses which were made with *hirsutum* as the pollen parent. After long continued backcrossing it can be assumed that all *luteum* genes have been replaced and that plants with a *luteum* plasmon and only *hirsutum* genes have been produced. Yet these plants resembled the *luteum* parent in certain physiological characteristics although in most respects they were identical to *hirsutum*. Reciprocal crosses were made between plants with a *hirsutum* plasmon and *hirsutum* genes and those with a *luteum* plasmon and *hirsutum* genes. Those plants from the cross where pure *hirsutum* was the female parent were all like typical male fertile *hirsutum* but the reciprocal cross with a *luteum* plasmon gave male sterile and partially sterile plants, thus resembling the parental type with *hirsutum* genes and a *luteum* plasmon. Further evidence of the persisting action of the *luteum* plasmon came from reciprocal crosses of plants with *hirsutum* genes and a *luteum* plasmon to pure *luteum*. The reciprocal hybrids would have identical hybrid nuclei and identical plasmons if the *luteum* plasmon in the back-crossed strain had not been modified by exposure to *hirsutum* genes. The reciprocal classes were alike in consisting of plants with varying degrees of pollen sterility as was true for the original cross of *luteum* ♀ × *hirsutum* ♂. There was, however, a difference in the average percentage of pollen abortion. Those plants coming from the cross where the egg parent possessed a *luteum* plasmon and *hirsutum* genes had a significantly higher amount of pollen abortion than did the plants from the reciprocal cross. Evidently the *luteum* plasmon had been modified somewhat so that it resembled the *hirsutum* plasmon in the amount of sterility produced in plants with hybrid genoms. Can this change in the *luteum* plasmon be ascribed to the action of *hirsutum* genes or is it due to the transmission of small amounts of the male plasmon during the backcrosses to *hirsutum*? It is on this point that sharp disagreement exists. Michaelis has evidence from other experiments indicating the constancy of the plasmon but this issue has not been fully resolved.

East (1934) in considering the evidence of a plasmon from the *Epilobium* investigations raised the objection that in species other than *hirsutum* and *luteum* there was no indication of a maternal plasmatic effect in the reciprocal crosses. He believed it unlikely that plasmon heredity would be confined to only one of several interspecific combinations. This criticism is no longer valid, however, since plasmon differences have been found in other interspecific *Epilobium* hybrids. In general, it appears probable that, inasmuch as related species differ in the nature of their plasmons, these differences must have arisen during the course of speciation. Plasmon mutations have occurred as well as gene mutations but the antithetical natures of plasmon and genom make it unlikely that plasmon modifications have come from nuclear conditioning of the cytoplasm. Closely related races have similar though not identical genoms and plasmons, and as a rule greater differences in both genetic systems are found in more distantly forms. The extent of differentiation of genom and plasmon within one race is not always of the same degree; the Jena strain of *E. hirsutum* possesses a plasmon with marked inhibitory effects when combined with certain *hirsutum* races while the reciprocal crosses gave normal vigorous plants. They have similar genoms but dissimilar plasmons. Some of the interracial hybrids with the Jena strain were normal or only slightly abnormal in both reciprocal crosses, thus plasmons and genoms were similar, while two South African races gave inviable hybrids in Jena cytoplasm indicating a marked divergence in the plasmons of the South African and Jena races of *hirsutum*.

The real or assumed independence of the maternal plasmon is not easily demonstrated if paternal cytoplasm is brought in with the male nucleus at fertilization. In those examples where the plasmatic effect persists unabated during many generations of backcrossing to the male parent, it would appear that there was no appreciable mixing of maternal and paternal cytoplasms. There is, however, good evidence in a number of cases that male cytoplasm is included in the fertilized egg cell. Not only has direct cytological observation revealed the presence of plastids in the male cell of a number of plants but transmission of paternal plastids has been shown by breeding tests to be of frequent occurrence in *Oenothera*, *Pelargonium* and *Nepeta*. It is much less frequent in *Epilobium* since only 2 in 1,000 plants from the cross of green ♀ × variegated ♂ have sectors with paternal plastids, and male transmission of plastids apparently does not occur at all in the Gramineae. The inheritance of $CO_2$ sensitivity through the sperm of *Drosophila* and the effect of the male plasmon on the size of thoracic spots in *Lymantria* are clear illustrations of the transmission of male cytoplasm in animals. Indeed, as we have mentioned elsewhere, there is good reason to believe that in animals transmission of sperm cytoplasm is of regular occurrence and may account for the relatively few instances of differences in reciprocal crosses.

It is obvious that when the male gamete makes a cytoplasmic contribution to the zygote, evidence of plasmatic heredity can only be obtained under exceptional circumstances. Persistence of an unchanged plasmon is possible only when no male cytoplasm enters the egg cell. The conflicting claims of the constancy of the plasmon in *Epilobium*, where a diminishing effect of the maternal plasmon has been reported following successive backcrosses to the recurrent male parent, may well be due to a contamination of the egg cytoplasm by that from the male.

## D. Plasmon heterozygosity.

HARDER (1927), working with different strains of the basidiomycete *Pholiota mutabilis*, obtained by micromanipulation mycelia with nuclei from only one parent in a mixed cytoplasm. He was able to isolate haploid strains with distinct types of growth behavior which proved to be fairly constant. Presumably, the differences between the various isolates is due to the interaction of the nuclei from one strain with cytoplasms comprised of varying proportions of cytoplasm from the two parents. Evidently a heterozygous plasmon is produced when cytoplasmic exchange occurs between the two conjugating cells; this is followed by segregation so that descendant cells differ in their plasmons. It would appear that the plasmons eventually become stabilized since the different growth patterns, once established, remained more or less constant. Although in *Pholiota* we are dealing with cytoplasmic components other than plastids, the results are comparable to the variegation produced by the somatic segregation of mixed plastids, especially in those cases where the male gamete has introduced plastids into the egg cytoplasm.

Plasmon segregation has also been found in *Epilobium* by Brücher (1940) in certain hybrids between *E. hirsutum* and *E. parviflorum*. The $F_1$ hybrids with *hirsutum* as the female parent were dwarf, stunted plants owing to the inhibiting effect of the *hirsutum* plasmon. The reciprocal cross with the *parviflorum* plasmon gave vigorous hybrids. Occasionally side branches or shoots were produced by the stunted hybrids which were normal appearing and which flowered profusely. Crosses onto flowers of these normal branches with *parviflorum* pollen yielded normal offspring free from the growth inhibitions usually produced by the *hirsutum* plasmon. Brücher believed that the normal shoots on stunted hybrids from the *hirsutum* ♀ × *parviflorum* ♂ cross had a *parviflorum* plasmon introduced by the male gamete and that this male plasmon underwent segregation during ontogeny.

Michaelis (1949) in more extensive studies found that environmental conditions under which the stunted hybrids were grown influenced the number of normal appearing branches. All of the normal shoots, however, did not have a plasmon different from that in the dwarfed portions of the plants, but constant and different plasmon changes had occurred in some normal branches. Michaelis is not inclined to attribute these plasmon changes to the introduction of male plasmon, thus producing plasmon heterozygosity followed by somatic segregation, but suggests that selective multiplication of unlike plasmatic units takes place in slowly growing forms and that this will lead to the formation of new kinds of plasmons. On either hypothesis, and an unequivocal distinction between them is not possible at this writing, plasmon heterozygosity and segregation are involved.

If the male plasmon possesses discrete elements it is not difficult to visualize how plasmon segregation could occur since the mechanism would be comparable to that of plastid segregation, but plasmon segregation would seem unlikely if the male cytoplasm were a homogeneous mass which became completely and uniformly distributed within the maternal cytoplasm thereby losing its individuality.

## E. Physiological differences in plasmons.

Certain physical and chemical properties of the plasmon have been reported which may be responsible for its role in character determination. Dellingshausen (1935, 1936) found that the plasmon of *Epilobium luteum* was more permeable, as indicated by plasmolysis experiments, to potassium chloride, glycerine and succinimide than was the *hirsutum* plasmon. Differences in viscosity of the two cytoplasms were found by centrifugation, and a differential permeability to urea was obtained after treatment with chloral hydrate, which increased the viscosity of *hirsutum* cytoplasm but decreased that of *luteum*. These physical properties in which the *luteum* and *hirsutum* plasmons differed were maintained for many generations in plants with a *luteum* plasmon but homozygous for *hirsutum* genes.

Michaelis (1951) in a comparison of plants with a *luteum* plasmon and *hirsutum* genes and those with a *hirsutum* plasmon and *hirsutum* genes reported that they differed in sensitivity to poisons, infection by mildew, enzymatic activities and photoperiodic response. LEHMANN (1936) and HINDERER (1936) using the avena coleoptile test for auxins found that inhibited hybrids with a *hirsutum* plasmon contained smaller amounts of growth hormones than did more normal appearing hybrids. Ross (1941) found that the application of indole acetic acid to the growing points of stunted hybrids resulted in more normal growth. Peroxidase activity in inhibited hybrids was higher than in the un-inhibited reciprocal hybrids and this difference in activity was reported to be directly proportional to the degree of inhibition. Possibly the greater amount of enzyme was responsible for the reduction in carbohydrate reserves and this led to abnormal development.

SCHLÖSSER (1935) reported that two wild races of *Lycopersicum esculentum* giving unlike reciprocal hybrids differed in osmotic pressure values as determined cryoscopically from expressed juice. In both the $F_1$ and $F_2$ generations the osmotic pressure level corresponded to that of the maternal parent, apparently unin-fluenced by paternal genes. However, despite these demonstrated physiological differences between plasmons, the kind or the nature of the interaction between plasmons and genoms leading to modified physiological reactions remains wholly speculative.

GOLDSCHMIDT (1938) suggested that the effect of the cytoplasm on develop-ment might be as follows. The genes determine hereditary traits by their control of chains of reaction of definite velocity. The speed of these reactions is dependent upon the substratum (*i.e.*, the cytoplasm) in which they take place. The speci-ficity of the cytoplasm may be of a physical nature such as viscosity or perme-ability or it may be chemical such as $p_H$. It is the physical or chemical properties of the cytoplasm which may decrease, or in some cases increase, the velocities of gene-controlled developmental reactions and in this way determine the pheno-type of those characteristics whose development is influenced by the relative rates of differentiation.

A further elucidation of the physiological nature of plasmon-genom inter-actions may provide an effective means of studying cellular dynamics and could lead to a solution of the most basic of all biological problems, namely that of growth and differentiation.

# IV. Dauermodifications.

The term Dauermodification describes environmentally-induced changes in the cytoplasm which though transient in nature may persist in the offspring for several, sometimes many, generations before disappearing (JOLLOS 1921, 1924). Dauermodification is similar to predetermination in that a cytoplasmic change is produced but in the former the external environment is the causal agent while in the latter it is the action of nuclear genes which modifies the cyto-plasm. Inasmuch as phenotypic changes due to Dauermodifications are trans-mitted through the maternal cytoplasm to the offspring for a number of genera-tions after the removal of the inducing agent, it may legitimately be asked if cases of plasmon inheritance do not represent extremely persistent Dauermodifi-cations. If a large number of generations is needed for the genes to restore the cytoplasm to its normal condition, a distinction between a plasmon effect and a Dauermodification would be difficult, were it not that the latter effect slowly diminishes while that of the plasmon, for example in WETTSTEIN'S mosses, persists undiminished. It should not be denied that both phenomena exist or may even co-exist but the one is characterized by stability and the other by in-stability. In fact certain of the *Epilobium* data, where there was at first a measur-able reversion of the phenotype towards that of the paternal parent later followed by relative stability, may be due to a combination of Dauermodification and plasmon action. Whether or not the quality of the cytoplasm comes from its own inherent properties or from a change impressed upon it, character determination is the result of interaction between the nucleus and cytoplasm.

A number of students, JOLLOS and KNAPP among others, have taken the position that there is no fundamental difference between plasmon heredity and that due to Dauermodification and they would rule out extranuclear heredity in the sense that it is autonomous. This would appear to be an untenable stand in view of many experiments where the constancy of the plasmon over long periods has been demonstrated.

There is an abundant literature dealing with Dauermodifications. HÄMMER-LING (1929) gives a full account of the earlier work. Many of the experiments were done with Protista, particularly with *Paramecium* and *Arcella*, although the phenomenon has been reported in higher plants, notably in *Antirrhinum* by STEIN, and in the Metazoa. In general, the methodology has been to grow the organisms in an unfavorable environment (sub-lethal doses of poisons, high or low temperatures, an excess of certain ions such as $Ca^{++}$, partial starvation, antisera, etc.) which induced phenotypic changes. After varying periods of exposure the phenotypically modified animals were placed in a normal environment and the persistence of the modified character followed during subsequent generations. It is, however, not within the purview of this chapter to describe in detail these cases. We wish to point out the existence of the phenomenon and to emphasize its bearing on the question of the plasmon as a gene-independent system.

## A. Inheritance of serotype specificity in Paramecium.

We have seen that in the killer system the basis of the cytoplasmically inherited trait rests upon the kappa bodies, but in the inheritance of antigen specificity in *Paramecium* we find that the role of the cytoplasm is more complex and less well understood. Although the data have been interpreted in terms of cytoplasmic genes they can be accounted for by other mechanisms. The essential facts are as follows (SONNEBORN 1950): On the thread-like cilia of *Paramecium* are found antigens which invoke the formation of a specific antiserum when animals of one strain are injected into the blood of a rabbit. Within a given stock there exist a number of different serological strains. Each strain produces one and only one kind of antiserum which is specific in that it immobilizes only animals of the same serological type and has no effect on individuals of different serotypes. The role of the cytoplasm in the determination of antigen specificity may be illustrated by a single example. Among the strains of stock 51 of variety 4 of *P. aurelia* are the antigenic types A and B. When individuals of these two strains conjugated in pairs the two exconjugant cells had identical nuclei, but the one with cytoplasm from strain A gave rise to a type A culture while the other with cytoplasm from strain B produced a type B culture. Since the two cultures differed only in their cytoplasm it would appear that antigen type was cytoplasmically transmitted. Further analysis, however, showed the situation was not so simple. Changes in serotype may occur in a strain derived from a single homozygous individual. For example, serotypes A, B, C, D, E, G, H and J have arisen in stock 51 and types A, B, C, D, F, H and J in stock 29 of variety 4. These changes or transformations from one serotype to another are obtained by exposure to specific antisera, by partial starvation and by modifying the temperature at which the culture is raised. Unlike the irreversible change of killer to sensitive, these changes in serotype are not only reversible but can be directed in that a specified treatment will transform one type to a certain other type. Thus serotype D of stock 29 is transformed to serotype B by the action of antiserum D if the animals are grown at a temperature of 32° C but to serotype H at a temperature of 20° C. Under specified and standard cultural conditions each of these new serotypes remained stable but could be transformed back to a type D or to other serotypes by varying the environmental conditions. Inheritance of antigen type was apparently through the cytoplasm but different cytoplasmic states could be induced; these were stable under specified conditions.

That nuclear genes do play a role in the control of serotype formation became evident when serotype A of stock 51 was mated with serotype A of stock 29. Both form antisera A but these antisera, though similar, are not identical. Crosses between these two stocks showed that the difference between their A antisera was due to a pair of allelic genes with the gene in stock 51 dominant to that in stock 29. All of the $F_1$ animals were of serotype 51A including those with cytoplasm from stock 29, but in the $F_2$ those animals homozygous for gene 29A became serotype 29A. Similar allelic differences were demonstrated at other serotype-determining loci in the two stocks. SONNEBORN believes that stocks 51 and 29 carry at different loci a series of genes for serotype specificity. Since at least 8 antigen types are known in stock 51 there are 8 different loci concerned with antigen specificity and in stock 29 with 7 antigen types there is a minimum of 7 such loci. It follows that in a strain of serotype A in stock 51 which has 8 serotype genes, only the $A$ gene is effective, the others being inhibited, while in a serotype B strain only the $B$ gene is acting, etc. Which of the various genes is in effective control of antigen formation depends upon the treatment of the animals since transformations from one serotype to another are readily accomplished.

BEALE's (1951, 1952) studies on the inheritance of antigen systems in variety 1 of *P. aurelia*, which in some ways are more detailed than those in variety 4, show no essential difference between varieties 1 and 4. As both SONNEBORN and BEALE state, there is no evidence that the determination of the serotype systems is due to gene-initiated plasmagenes capable of surviving and reduplicating in the absence of the initiating gene. It is possible, though not at present subject to experimental proof, that the genes for serological types confer specificity upon non-specific plasmagene precursors. Another possibility is that the activity of a given serotype gene depends upon the concentration in the cytoplasm of the products of this gene. If there were more of gene-product A than of other gene-products, it would lead to the increased efficiency of gene $A$ while the others would be less productive. Changes in serotype could result when environmental conditions altered the proportion of the different gene products. However, it is difficult to reconcile this hypothesis with the fact that only one kind of antiserum is produced; others should be present although in lesser concentrations.

Clearly antigen specificity is gene-controlled and the nature of the cytoplasm favors the expression of a particular locus but little can be concluded about the nature of these different cytoplasmic states. DELBRUCK (1949) suggested an explanation in which autonomous cytoplasmic particles play no part. He suggested that the data could be satisfactorily accounted for by assuming the operation of a flux equilibrium system with a number of alternative, mutually exclusive and competitive reactions. Changes in the environment could result in a shift from one alternative "steady state" to another.

Similarities between the killer trait and serological specificity are that both are cytoplasmically inherited and environmental changes bring about inherited

differences. They differ (1) in that kappa is able to mutate and different kinds
of kappa have characteristic effects while the specific types of antisera depend
upon nuclear genes, (2) serotype changes are reversible while the change from
killer to sensitive is irreversible unless kappa particles are introduced into the
cytoplasm, (3) only one kind of serological specificity can exist (*i.e.*, they are
mutually exclusive) while two kinds of kappa may co-exist in the cytoplasm of
one individual and (4) the killer phenotype has a particulate cytoplasmic basis
while this has not been demonstrated for serological specificity.

Changes in antigen specificity due to altered cytoplasmic states induced by
environmental conditions can be regarded as Dauermodifications. Sonneborn (1947, 1951b) so interpreted the earlier data of Jollos, Sonneborn,
Harrison and Fowler on the acquisition of immunity to specific antisera as
well as his more recent work in this field. Although it is quite likely that the
basic mechanisms are similar in all cases of Dauermodification, the role of genes
was never demonstrated in the earlier work as has been done in the later studies
on antigen specificity (see Sonneborn 1951b). Other mechanisms, such as a
reduction in the concentration of self-reproducing cytoplasmic particles, may
produce results analagous to Dauermodifications. For example, a killer strain
can be transformed into a non-killer by growing the animals at a temperature
where kappa multiplication fails to keep pace with fission rate. This change in
phenotype remains constant upon transferring the animals to a normal temperature if kappa particles have been completely lost while those animals possessing
one or more kappa particles can revert to killers when the concentration of kappa
becomes sufficiently high.

## B. The barrage phenomenon in Podospora.

In Rizet's (1952) interesting studies of the barrage phenomenon in *Podospora anserina* there is evidence of some kind of a cytoplasmic effect although
its precise nature is unknown. There are two kinds of mycelia in *Podospora*
with respect to their behavior when the frontiers of two or more strains make
contact, or are confronted, by growing them on the same medium. These are
designated as the *S* and *s* mycelia and a single pair of genes is involved. When
two *S* or two *s* mycelia meet, growth does not cease but there is an anastomosing
of the hyphae for a short distance on either side of the boundary. If the two
strains, both *S* or *s*, are of the same mating type no perithecia are formed at
the region of anastomosis which appears macroscopically as a thin transparent
line. This region is, however, marked by a narrow row of perithecia if the two
strains are of opposite mating type. A different result occurs when the two
confronted strains are *S* and *s*. Instead of an anastomosing of hyphae at the
meeting line there is a mutual inhibition of growth; the transition zone appears
as a broad transparent region or barrage. If the two strains are of opposite
mating type a double line of perithecia are formed, one on either side of the barrage,
while no perithecia are produced by similar mating types.

Mycelia arising from spores of the cross *S* × *s* were isolated and subsequently
tested for their behavior when confronted by known *S* and *s* strains. Although
other characteristics due to nuclear genes underwent normal Mendelian segregation, this was not true for the *s* character. Instead, a modified-*s* phenotype was
recovered in which no barrage is produced in the presence of either *s* or *S* strains.
Although the modified-*s* strains were stable when maintained in pure culture,
Rizet found that reversion of modified-*s* to *s* occurred if the two strains were
in contact. He was properly concerned with the constitution of these modified-*s*

spores, and why it was that the effect of the *s* gene was not normally manifested in those spores coming from the cross of $S \times s$. Several possibilities were suggested. Predetermination of the hybrid cytoplasm by the $S$ allele seems unlikely since the modified-*s* characteristic does not disappear during prolonged vegetative reproduction. It also appears improbable that the $S$ gene induces the *s* allele to mutate to a new allele producing the modified-*s* barrage reaction. Moreover, the kind of spores found in reciprocal crosses between *s* and modified-*s* strains depends upon which strain contributes the greater mass of cytoplasm. If the *s* parent is the female parent then all or a great majority of the spores are *s* while the reverse is true when a modified-*s* strain is the female parent. It appears that the modified-*s* class contains the *s* allele but has a cytoplasm of altered state. The fact that this cytoplasmic state is reversible, and in this respect resembles antigen specificity in *Paramecium*, led to the suggestion (EPHRUSSI 1953) that flux equilibria transformations may be involved.

## C. Adaptive enzyme formation in yeast.

The constitution of the substrate on which microorganisms are cultured has a profound effect on the enzyme systems of these cultures. In many instances the addition of a certain compound elicits the formation of a specific enzyme, capable of utilizing this substance, whose presence was not detectable prior to the change in substrate—*i.e.*, the chemical nature of the substrate determines the phenotype of the cell in terms of its enzyme activity. Many instances of enzyme adaptation have been reported in bacteria and yeast but let us consider only the yeast studies since the genetics of yeast has been more fully elucidated. Some yeast species, or strains within species, are able to break down rapidly certain sugars upon being placed in a medium containing these compounds while other strains lack this ability. This difference has been shown to be gene controlled (WINGE and ROBERTS 1948, WINGE 1949). For example, *Saccharomyces cerevisiae* hydrolyzes galactose while *S. Chevalieri* is unable to do so when first grown on a galactose-containing substrate but requires a period of adaptation of approximately six days before its level of enzyme activity equals that of *cerevisiae*. However, once fully adapted this modification is transmitted indefinitely from one cell generation to another in the presence of substrate. Genetic analysis of crosses between these two species disclosed that *cerevisiae* had a dominant gene ($G$) for fast adaptation and *Chevalieri* a recessive gene ($g_s$) for slow (long term) adaptation. Both SPIEGELMAN (1951) and WINGE (1952) agree that synthesis of galactozymase is gene-controlled but they have differing views of the mechanism by which this is achieved. According to WINGE, all $g_s$ cells respond uniformly to galactose. Their capacity to form galactozymase is at first very low but the amount of enzyme increases autocatalytically with time. If this is true it follows that the rate of enzyme production is controlled by the amount of enzyme present in the cytoplasm. This is possible if in an equilibrium reaction: precursor $\rightleftharpoons$ enzyme, the enzyme is removed from the system (EPHRUSSI 1952). SPIEGELMAN, on the other hand, finds that only a few of the $g_s$ cells form the enzyme upon contact with the substrate, the remaining ones being unable to do so. Since mutation of $g_s$ to $G$ is not involved here, although in other cases in yeast long term adaptation has been reported to be due to selection following mutation (MUNDKUR and LINDEGREN 1949, MUNDKUR 1951), he argues that the ability of certain of the $g_s$ cells to form galactozymase comes from the circumstance that they possess one or more of the cytoplasmic particles needed for enzyme synthesis while the majority of the cells lack the essential particle. Spiegelman suggests that a cell with the dominant gene $G$ possesses a mechanism of high efficiency for producing these cytoplasmic particles in the absence of substrate so that the cytoplasm of every cell contains at least one particle. In cells with the $g_s$ gene the production of these postulated particles is so inefficient that from only one of about 1,000 divisions is a cell formed which has the cytoplasmic particle.

Under the conditions of their experiments, Spiegelman, de Lorenzo and Campbell (1951) calculated, on the basis of de-adaptation experiments where fully adapted single $g_s$ cells were placed in a galactose-free medium and the successive buds isolated and tested for galactozymase activity, that a fully adapted $g_s$ cell contained approximately 100 particles. Originally positive mother cells began to produce negative progeny at the fifth division and the proportion of negative cells increased with each succeeding cell generation. These data were taken to indicate that the particulate elements responsible for enzyme formation are randomly assorted between mother and daughter cells; negative buds do not arise until the fifth cell generation since the number of particles initially present does not until then become sufficiently diluted to permit the origin of a bud with a cytoplasm devoid of these particles. Although Winge (1952) remains unconvinced of the validity of Spiegelman's argument, a decision between his interpretation and that of Spiegelman apparently rests upon the heterogeneity, or lack of it, in the original population of $g_s$ cells exposed to substrate. Evidence of this heterogeneity has been repeatedly found in Spiegelman's laboratory but as Spiegelman (1951) states it must be admitted that genetic situations other than one involving cytoplasmic particles may prove to be correct, although the available data can be logically accounted for by a particulate mechanism.

Spiegelman, de Lorenzo and Campbell found that when galactose-adapted cells were treated with euflavine, vegetative mutants similar to those found by Ephrussi et al. were produced. These vegetative mutants, however, contain galactozymase. Evidently the particulate cytoplasmic units responsible for galactozymase formation were not affected by euflavine although those concerned with the deficient enzymes of petite cells were.

Not only is the mechanism bringing about the formation of adaptive enzymes uncertain but there is in these experiments no evidence of the functioning of autonomous cytoplasmic elements. Since the environment (presence of sugar in substrate) induces the formation of a specific response (production of enzyme) which gradually disappears in the course of a few generations upon removal of the sugar from the media these cases might properly be considered as examples of Dauermodifications.

The investigations on antigen specificity and adaptive enzymes differ notably from previous studies on Dauermodifications in that the underlying mechanisms have been elucidated. The puzzle of Dauermodifications, as Sonneborn says, has at least been solved in principle.

## Literature.

Anderson, E. G.: Maternal inheritance of chlorophyll in maize. Bot. Gaz. **76**, 411 to 418 (1923). — Arnason, T. J., J. B. Harrington and H. A. Friesen: Inheritance of variegation in barley. Canad. J. Res. 24, 145–157 (1946).

Bateson, W., and A. F. Gairdner: Male sterility in flax, subject to two types of segregation. J. Genet. 11, 269–276 (1921). — Baur, E.: Das Wesen und die Erblichkeitsverhältnisse der "varietates albomarginatae" von Pelargonium zonale. Z. Vererbungsl. 1, 330 to 351 (1909). — Beadle, G. W.: Genes and biological enigmas. Chapt. 9 from Science in Progress, 6th Series. New Haven: Yale Univ. Press 1949. — Beale, G. H.: Nuclear and cytoplasmic determinants of hereditary characters in Paramecium aurelia. Nature (Lond.) **167**, 256–260 (1951). — Antigen variation in Paramecium aurelia, variety 1. Genetics **37**, 62–74 (1952). — BRIGGLE, L. W.: A comparison of cytoplasmic-genotypic interactions in a group of cytoplasmic male sterile corn types. Ph. D. Thesis Iowa State College 1953. — BROWN, C. H.: Elimination of kappa particles from "Killer" strains of Paramecium aurelia by treatment with chloromycetin. Nature (Lond.) **166**, 527 (1950). — BRÜCHER, H.: Spontanes Verschwinden der Entwicklungshemmungen eines Artbastardes. Flora (Jena) N. F. **34**, 215–228 (1940).

CASPARI, E.: Cytoplasmic inheritance. Adv. Genet. **2**, 2–66 (1948). — CHITTENDEN, R. J., and C. PELLEW: A suggested interpretation in certain cases of anisogeny. Nature (Lond.) **119**, 10–11 (1927). — CLAUDE, A.: Distribution of nucleic acids in the cell and the morphological constitution of cytoplasm. Biol. Symposia **10**, 111–130 (1943). — CLAYTON, E. E.: Male sterile tobacco. J. Hered. **41**, 171–175 (1950). — COOPER, K. W.: Normal spermatogenesis in Drosophila. Chapt. 1 from Biology of Drosophila. New York: John Wiley & Sons 1950. — CORRENS, C.: Nichtmendelnde Vererbung. In Handbuch der Vererbungswissenschaft, Bd. 2, S. 1–159. 1937.

DELBRUCK, M.: Unités biologiques douées de continuité génétique. Colloques Internationaux du C.N.R.S., Bd. VIII, S. 33–35. 1949. — DELLINGSHAUSEN, M. v.: Entwicklungsgeschichtlich-genetische Untersuchungen an Epilobium. V. Permeabilitätsstudien an zwei genetisch verschiedenen Plasmen. Planta (Berl.) **23**, 604–622 (1935). — Entwicklungsgeschichtlich-genetische Untersuchungen an Epilobium. VIII. Über vergleichende Viskositätsmessungen an zwei genetisch verschiedenen Plasmen. Planta (Berl.) **25**, 282–301 (1936). — DEMEREC, M.: A second case of maternal inheritance of chlorophyll in maize. Bot. Gaz. **84**, 139–155 (1927). — DIPPEL, R. V.: Mutations of the killer plasmagene, kappa, in variety 4 of Paramecium aurelia. Amer. Naturalist **82**, 43–58 (1948). — Mutation of the killer cytoplasmic factor in Paramecium aurelia. Heredity (Lond.) **4**, 165–187 (1950).

EAST, E. M.: Studies on self-sterility. IX. The behavior of crosses between self-sterile and self-fertile plants. Genetics **17**, 175–202 (1932). — The nucleus-plasma problem. Amer. Naturalist **68**, 289–303 and 402–439 (1934). — EPHRUSSI, B.: The interplay of heredity and environment in the synthesis of respiratory enzymes in yeast. The Harvey Lectures, Ser. XLVI. Springfield, Illinois: Ch. C. Thomas 1952. — Nucleo-cytoplasmic relations in micro-organisms. Oxford: Clarendon Press 1953. — EPHRUSSI, B., and H. HOTTINGUER: On an unstable cell state in yeast. Cold Spring Harbor Symp. Quant. Biol. **16**, 75–84 (1951). — EPHRUSSI, B., H. HOTTINGUER et A. M. CHIMÈNES: Action de l'acriflavine sur les levures. I. La mutation "petite colonie". Ann. Inst. Pasteur **76**, 351–364 (1949).

FUKASAWA, H.: Studies on restoration and substitution of nucleus in Aegilotricum. I. Appearance of male sterile durum in substitution crosses. Cytologia **18**, 167–175 (1953).

GABELMAN, W. H.: Reproduction and distribution of the cytoplasmic factor for male sterility in maize. Proc. Nat. Acad. Sci. U.S.A. **35**, 634–639 (1949). — GECKLER, R. P.: Nitrogen mustard inactivation of the cytoplasmic factor, kappa, in Paramecium. Science (Lancaster, Pa.) **110**, 89–90 (1949). — GOLDSCHMIDT, R.: Physiological genetics. New York: McGraw-Hill Book Comp. 1938. — Sex-determination in Melandrium and Lymantria. Science (Lancaster, Pa) **95**, 120–121 (1942). — GOLDSTEIN, L.: Contribution à l'étude de la sensibilité héréditaire au gaz carbonique chez la Drosophile. Mise en évidence d'une forme nouvelle du génoide. Bull. biol. France et Belg. **83**, 177–188 (1949). — Action de facteurs transmissibles sur certaines modalités de la transmission de la sensibilité héréditaire à l'anhydride carbonique chez la Drosophile. Thèse Paris 1951. — GREGORY, R. P.: On variegation in Primula sinensis. J. Genet. **4**, 305–321 (1915).

HÄMMERLING, J.: Dauermodifikation. In Handbuch der Vererbungswissenschaft, Bd. I E, S. 1–69. 1929. — HARDER, R.: Zur Frage nach der Rolle von Kern und Protoplasma im Zellgeschehen und bei der Übertragung von Eigenschaften. Z. Bot. **19**, 337–407 (1927). — HASKINS, F. A., A. TISSIERES, H. K. MITCHELL and M. B. MITCHELL: Cytochromes and the succinic acid oxidase system of poky strains of Neurospora. J. of Biol. Chem. **200**, 819–826 (1953). — HINDERER, G.: Versuche zur Klärung der reziproken Verschiedenheiten von Epilobium-Bastarden. II. Wuchsstoff und Wachstum bei reziprok verschiedenen Epilobium-Bastarden. Jb. wiss. Bot. **82**, 669–686 (1936).

IMAI, Y.: Recurrent auto- and exomutation of plastids resulting in tricolored variegation of Hordeum vulgare. Genetics **21**, 752–757 (1936). — The behaviour of plastid as a hereditary unit. Cytologia (Fujii Jubilee Volume) **1937**, 934–947.

JOHNSON, T., and M. NEWTON: Crossing and selfing studies with physiologic races of oat stem rust. Canad. J. Res. **18**, 54–67 (1940). — JOLLOS, V.: Experimentelle Protistenstudien I. Arch. Protistenkde **43**, 1–22 (1921). — Untersuchung über Variabilität und Vererbung bei Arcellen. Arch. Protistenkde **49**, 307–374 (1924). — JONES, D. F.: The interrelation of plasmagenes and chromogenes in pollen production in maize. Genetics **35**, 507 to 512 (1950). — The cytoplasmic separation of species. Proc. Nat. Acad. Sci. U.S.A. **37**, 408 to

410 (1951). — JONES, H. A., and A. E. CLARK: Inheritance of male sterility in the onion and the production of hybrid seed. Proc. Amer. Soc. Horticult. Sci. **43**, 189–194 (1943). — JOSEPHSON, L. M., and M. T. JENKINS: Male sterility in corn hybrids. J. Amer. Soc. Agron. **40**, 267–274 (1948).

KALMUS, H., and A. N. MITCHISON: Transplantation of larval ovaries in Drosophila from and to individuals susceptible to carbon dioxide. Nature (Lond.) **157**, 230–231 (1946).

LEDERBERG, J.: Cell genetics and hereditary symbiosis. Physiologic. Rev. **32**, 403–430 (1952). — LEHMANN, E.: Versuche zur Klärung der reziproken Verschiedenheiten von Epilobium-Bastarden. I. Der Tatbestand und die Möglichkeit seiner Klärung durch differente Wuchsstoffbildung. Jb. Bot. **82**, 657–668 (1936). — LEVINE, M.: The diverse mate-killers of Paramecium aurelia, variety 8: their interrelations and genetic basis. Genetics **38**, 561 to 578 (1953). — L'HERITIER, PH.: The $CO_2$ sensitivity problem in Drosophila. Cold Spring Harbor Symp. Quant. Biol. **16**, 99–112 (1951). — L'HERITIER, PH., and F. HUGON DE SCOEUX: Transmission par greffe et injection de la sensibilité héréditaire au gaz carbonique chez la Drosophile. Bull. biol. France et Belg. **81**, 70–91 (1947). — L'HERITIER, PH., and G. TESSIER: Une anomalie physiologique héréditaire chez la Drosophile. C. r. Acad. Sci. Paris **205**, 1099–1101 (1937). — LWOFF, A.: Problems of morphogenesis in ciliates. New York: John Wiley & Sons 1950.

MAZOTI, L. R.: Nuevos hallazgos acerca del comportamiento de las unidades de la herencia, genes y plasmonio. Rev. argent. Agronomia **17**, 145–162 (1950). — MERCIER, L.: Action de la température sur la transmission par les Drosophiles femelles de la sensibilité héréditaire au gaz carbonique. Bull. biol. France et Belg. **85**, 226–236 (1951). — MEVES, F.: Die Chondriosomen als Träger erblicher Anlagen. Cytologische Studien an Hühnerembryo. Arch. mikrosk. Anat. **72**, 816–867 (1908). — MICHAELIS, P.: Über die reziprok verschiedenen Sippenbastarde bei Epilobium hirsutum. I. Die reziprok verschiedenen Bastarde der Epilobium hirsutum-Sippe Jena. Z. Vererbungslehre **78**, 187–222 (1940a). — Über die reziprok verschiedenen Sippenbastarde bei Epilobium hirsutum. II. Über die Konstanz des Plasmons der Sippe Jena. Z. Vererbungslehre **78**, 223–237 (1940b). — Über die reziprok verschiedenen Sippenbastarde bei Epilobium hirsutum. III. Über die genischen Grundlagen der im Jena-Plasma auftretenden Hemmungsreihe. Z. Vererbungslehre **78**, 295–337 (1940c). — Prinzipielles und Problematisches zur Plasmavererbung. Biol. Zbl. **68**, 173–195 (1949). — Interactions between genes and cytoplasm in Epilobium. Cold Spring Harbor Symp. Quant. Biol. **16**, 121–129 (1951). — MITCHELL, M. B., and H. K. MITCHELL: A case of "maternal" inheritance in Neurospora crassa. Proc. Nat. Acad. Sci. U.S.A. **38**, 442–449 (1952). — MITCHELL, M. B., H. K. MITCHELL and A. TISSIERES: Mendelian and non-Mendelian factors affecting the cytochrome system in Neurospora crassa. Proc. Nat. Acad. Sci. U.S.A. **39**, 606–613 (1953). — MUNDKUR, B.: Adaptation to galactose fermentation in Saccharomyces species. Bacter. Proc. 67 (1951). — MUNDKUR, B., and C. LINDEGREN: An analysis of the phenomenon of long-term adaptation to galactose by Saccharomyces. Amer. J. Bot. **36**, 722–726 (1949). — MYERS, W. M.: Effects of cytoplasm and gene dosage on expression of male sterility in Dactylis glomerata. Rec. Gen. Soc. Amer. **14**, 55–56 (1946).

OEHLKERS, F.: Bastardierungsversuche in der Gattung Streptocarpus Lindley. I. Z. Bot. **32**, 305–393 (1938). — Neue Überlegungen zum Problem der Außerkaryotischen Vererbung. Z. Vererbungslehre 84, 213–250 (1952). — Außerkaryotische Vererbung. Naturwiss. **3**, 78–85 (1953). — OWEN, F. V.: Male sterility in sugar beets produced by complementary effects of cytoplasmic and Mendelian inheritance. Amer. J. Bot. **29**, 692 (1942). — Cytoplasmically inherited male-sterility in sugar beets. J. Agricult. Res. **71**, 423–440 (1945).

PAL, B. P.: A new type of variegation in rice. Indian J. Agricult. Sci. **11**, 170–176 (1941). — POLLISTER, A. W., and P. F. POLLISTER: The relation between centriole and centromere in atypical spermatogenesis of viviparid snails. Ann. New York Acad. Sci. **45**, 1–48 (1943). — PREER, J. R.: The killer cytoplasmic factor kappa. Amer. Naturalist **82**, 35–42 (1948). — Microscopically visible bodies in the cytoplasm of the "Killer" strains of Paramecium aurelia. Genetics **35**, 344–362 (1950). — PREER, J. R., R. W. SIEGEL and P. S. STARK: The relationship between kappa and paramecin in Paramecium aurelia. Proc. Nat. Acad. Sci. U.S.A. **39**, 1228–1233 (1953).

RANDOLPH, L. F.: Cytology of chlorophyll types of maize. Bot. Gaz. **73**, 337–375 (1922). — RENNER, O.: Die pflanzlichen Plastiden als selbständige Elemente der genetischen Konstitution. Ber. math.-physik. Kl. Sächs. Akad. Wiss. Leipzig **86**, 241–266 (1934). — Zur Kenntnis der nichtmendelnden Buntheit der Laubblätter. Flora (Jena), N. F. **30**, 218–290 (1936). — RHOADES, M. M.: The cytoplasmic inheritance of male sterility in Zea mays. J. Genet. **27**, 71–93 (1933). — The genetic control of mutability in maize. Cold Spring Harbor Symp. Quant. Biol. **9**, 138–144 (1941). — Genic induction of an inherited cytoplasmic difference. Proc. Nat. Acad. Sci. U.S.A. **29**, 327–329 (1943). — Plastid mutations. Cold Spring Harbor Symp. Quant. Biol. **11**, 202–207 (1946). — Gene induced mutation of a heritable cytoplasmic factor producing male sterility in maize. Proc. Nat. Acad. Sci. U.S.A. **36**,

634–635 (1950). — RIZET, G.: Les phénomènes de barrage chez Podospora anserina. I. Analyse génétique des barrages entre souches $S$ and $s$. Rev. Cytol. et Biol. végét. **13**, 51–92 (1952). — ROGERS, J. S.: The utilization of male-sterile inbred lines in the production of corn hybrids. Mimeographed report of the 9. Southern Corn Improvement Conference. 1950. — ROSS, H.: Über die Verschiedenheiten des dissimilatorischen Stoffwechsels in reziproken Epilobium-Bastarden und die physiologisch-genetische Ursache der reziproken Unterschiede. I. Die Aktivität der Peroxydase in reziproken Epilobium-Bastarden der Sippe Jena. Z. Vererbungslehre **79**, 503–529 (1941).

SCHLÖSSER, L. A.: Beitrag zu einer physiologischen Theorie der plasmatischen Vererbung. Z. Vererbungslehre **69**, 159–192 (1935). — SCHWARTZ, D.: The interaction of nuclear and cytoplasmic factors in the inheritance of male sterility in maize. Genetics **36**, 676–696 (1951). SCHWEMMLE, J., E. HAUSTEIN, J. STURM u. M. BINDER: Genetische und zytologische Untersuchungen an Eu-Oenotheren. I.–VI. Z. Vererbungslehre **75**, 358–800 (1938). — SIEGEL, R. W.: A genetic analysis of the mate-killer trait in Paramecium aurelia, variety 8. Genetics **38**, 550–560 (1953). — SIGOT, A.: Contribution à l'étude de la sensibilité héréditaire au gaz carbonique chez la Drosophile. Influence de divers facteurs sur la transmission de la sensibilité par les mâles. Thèse Strasbourg 1951. — SIRKS, M. J.: Plasmatic inheritance. Bot. Review **4**, 113–131 (1938). — SLONIMSKI, P.: Action de l'acriflavine sur les levures. IV. Mode d'utilisation du glucose par les mutants "petite colonie". Ann. Inst. Pasteur **76**, 510–530 (1949). — SLONIMSKI, P., et B. EPHRUSSI: Action de l'acriflavine sur les levures. V. Le système des cytochromes des mutants "petite colonie". Ann. Inst. Pasteur **77**, 47–64 (1949).— SONNEBORN, T. M.: Experimental control of the concentration of cytoplasmic genetic factors in Paramecium. Cold Spring Harbor Symp. Quant. Biol. **11**, 236–255 (1946). — Recent advances in the genetics of Paramecium and Euplotes. Adv. Genet. **1**, 264–358 (1947). — Beyond the gene. Amer. Scientist **37**, 33–59 (1949). — The cytoplasm in heredity. Heredity (Lond.) **4**, 11–36 (1950). — Beyond the gene—two years later. Science in Progress, 7th Series. New Haven: Yale Univ. Press 1951a. — The role of the genes in cytoplasmic inheritance. Genetics in the 20th Century. New York: Macmillan & Co. 1951b. — SONNENBLICK, B.P.: The early embryology of Drosophila melanogaster. Biology of Drosophila. New York: John Wiley & Sons 1952. — SPIEGELMAN, S.: The particulate transmission of enzyme-forming capacity in yeast. Cold Spring Harbor Symp. Quant. Biol. **16**, 87–98 (1951). — SPIEGELMAN, S., W. F. DE LORENZO and A. M. CAMPBELL: A single-cell analysis of the transmission of enzyme-forming capacity in yeast. Proc. Nat. Acad. Sci. U.S.A. **37**, 513–524 (1951).

TAVLITZKI, J.: Action de l'acriflavine sur les levures. III. Étude de la croissance des mutants "petites colonies". Ann. Inst. Pasteur **76**, 497–509 (1949).

WARIS, HARRY: Cytophysiological studies on micrasterias. II. The cytoplasmic framework and its mutation. Physiol. Plantarum **3**, 236–246 (1950). — WERBITZKI, F. W.: Über blepharoblastlose Trypanosomen. Zbl. Bakter. **53**, 303–315 (1910). — WETTSTEIN, F. v.: Gattungskreuzungen bei Moosen. Z. Vererbungslehre **33**, 253 (1924). — Über plasmatische Vererbung sowie Plasma- und Genwirkung. Nachr. Ges. Wiss. Göttingen, Math.-physik. Kl. **1926**, 250–281. — Über plasmatische Vererbung und über das Zusammenwirken von Genen und Plasma. Ber. dtsch. bot. Ges. **46**, 32–49 (1928). — Die genetische und entwicklungsphysiologische Bedeutung des Cytoplasmas. Z. Vererbungslehre **73**, 349–366 (1937). — Untersuchungen zur plasmatischen Vererbung. I. Linum. Biol. Zbl. **65**, 149–166 (1946). — WILSON, E. B.: The distribution of the chondriosomes to the spermatozoa in scorpions. Proc. Nat. Acad. Sci. U.S.A. **2**, 321–324 (1916). — WINGE, Ö.: Inheritance of enzymatic characters in yeasts. Proc. 8th Int. Congr. Genet., Hereditas (Lund) Suppl. 520–529 (1949). — The basis for the present position of yeast genetics. Wallerstein Laboratories Communications **15**, 21–44 (1952). — WINGE, Ö., and O. LAUSTEN: On a cytoplasmatic effect of inbreeding in homozygous yeast. C. r. Labor. Carlsberg **23**, 17–38 (1940). — WINGE, Ö., and C. ROBERTS: Inheritance of enzymatic characters in yeasts, and the phenomenon of long term adaptation. C. r. Labor. Carlsberg **24**, 263–315 (1948). — WOODS, M. W., and H. G. DU BUY: Hereditary and pathogenic nature of mutant mitochondria in Nepeta. J. Nat. Canc. Inst. **11**, 1105–1151 (1951).

# Physiological variability in connection with experimental procedures and reproducibility.

By

## F. W. Went.

With 4 figures.

With the improvements in instrumentation and the application of more and more precise chemical and physical methods in the physiological analysis of plants, the uniformity of the experimental material usually becomes the limiting factor in experiments with plants. Therefore an analysis of the basis for variability among plants of genetic homogeneity becomes necessary. In this way the limits to which variability can be reduced become known and methods can be devised to obtain the most uniform plants. This aspect of experimental procedure has not received the attention of Botanists it deserves. It has been shortly discussed in KOSTYTSCHEW-WENT (1931) on pp. 301–303.

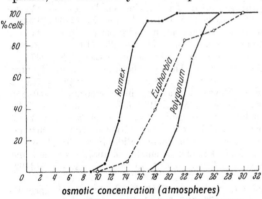

Fig. 1. Percentage of plasmolyzed cells (ordinate) at different osmotic concentrations (abscissa), for *Rumex*, *Euphorbia* and *Polygonum* leaves (data from LAMBRECHT 1929).

In the 19th century physiological uniformity of plant material was not so important as yet, since Botanists first had to investigate the basic responses of plants, which were at least partly of a qualitative rather than quantitative character and thus did not require such careful matching of plants. Yet, already very precise quantitative studies were made; as an example can be mentioned "Eine Methode zur Analyse der Turgorkraft" by DE VRIES (1884). He used cells from the same leaf of *Rhoeo discolor* to compare quantitatively the osmotic activity of different substances. He found that not all cells in the leaf have the same osmotic concentration of their cell sap, but that the epidermal cells under the midrib are most uniform, whereas the other epidermal cells are much more variable. In spite of the uniformity of adjacent cells, there is a gradient in osmotic concentration over the length of the leaf; in the middle of the leaf the osmotic concentration of the cell sap increased 0.01–0.02 mol. $KNO_3$ per 2–3 cm. closer to the insertion of the leaf. LAMBRECHT has found similar differences in the osmotic value of the leaf cells of a moss: low values near the midrib, high near the edge. Some of LAMBRECHT's (1929) data are plotted in Fig. 1, in which the percentage plasmolyzed cells at each osmotic concentration is shown for leaf cells of *Rumex*, *Euphorbia* and *Polygonum*. Not only are there absolute differences in osmotic concentration between the cells of different plants, but the variability is *e.g.* twice as great in *Euphorbia* as it is in *Rumex* or *Polygonum*.

Using the experimental plant of DE VRIES, *Rhoeo discolor*, we find that the variability in osmotic value between cells is less than for the plants which

LAMBRECHT investigated. Fig. 2 shows that the epidermal cells under the midrib are far more uniform than those below the rest of the leaf. Furthermore we see that the temperature conditions under which the plants grow determine the osmotic concentration of the cell sap: at higher temperatures it decreases, and is lowest in an ordinary green house after the hot summer months; under these uncontrolled conditions the variability between cells is also twice as great as when the plants are grown in controlled conditions. The plasmolytic difference between epidermal cells under the midrib and under the rest of the leaf blade as described by DE VRIES, can also be seen in Fig. 2. Close observation, however,

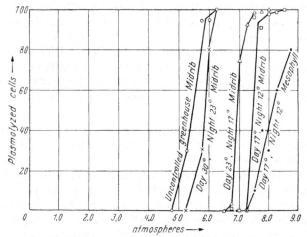

Fig. 2. Percentage plasmolyzed epidermal cells (ordinate) at different osmotic concentrations (abscissa, in atmospheres), of the leaf of *Rhoeo discolor*. All curves refer to the elongated cells under the midrib except the right-hand line, which gives values for the wider epidermal cells between the stomata, on both sides of the midrib. From left to right the curves show plasmolysis of cells from the longest leaves of plants, grown in

| | | | | | |
|---|---|---|---|---|---|
| Non-controlled greenhouse; | | | osmotic value 5.43 ± 0.18 atm. | | |
| Day temperature 30⁰ night temperature 23⁰; | | | ,, | ,, | 5.85 ± 0.09 ,, |
| ,, | ,, 23⁰ ,, | ,, 17⁰; | ,, | ,, | 6.90 ± 0.09 ,, |
| ,, | ,, 17⁰ ,, | ,, 12⁰; | ,, | ,, | 7.45 ± 0.09 ,, |
| ,, | ,, 17⁰ ,, | ,, 12⁰; | ,, | ,, | 8.0 ± 0.33 ,, |

seems to indicate that this difference is due to the type of plasmolysis rather than to differences in osmotic concentration. In the elongated cells under the midrib the protoplasm detaches itself first from the two ends, which is easily measured under the microscope. In the more isodiametric cells under the mesophyll the protoplast frequently detaches itself first from the upper or lower cell wall in the plane of observation, and thus plasmolysis does not show up. This decreases the percentage of observably plasmolyzed cells and increases the apparent variability.

The previously discussed cases give a good demonstration of physiological differences between cells within one organism or even within one organ. Such osmotic and protoplasmatic differences between cells have been described and discussed by a number of investigators (for instance WEBER 1925 and 1929, and HÖFLER 1937). In addition, RUHLAND et al. (1938) have shown that the plasmolytic behavior of cells is strongly influenced by their previous treatment. With these restrictions in mind a population of cells within one organism can be used to obtain physiologically uniform material for experiments. In general one cell type will have the least variability, whereas different cell types or cells from different organs or tissues of the same plant will usually differ much more.

Still further precautions are in order when cells or organs of different age from the same individual are compared. For many changes occur in an organ as it becomes older: its chemical composition changes, cell walls become thicker, auxin production and content decreases etc. Therefore only leaves or other organs of the same age can be compared and used in comparative experiments. This is independent of the growing conditions and no degree of uniformity in the external environment can make organs of different age more uniform. Only meristems remain similar throughout their life time provided they do not initiate floral primordia, go into dormancy or become separated from their roots by too long a stem. There are some curious exceptions to this rule, however.

In *Araucaria, Coffea, Hydnocarpus* and a few other plants the apical meristem differs from lateral meristems in that it can produce orthotropic shoots, whereas the lateral meristems can produce only plagiotropic shoots (reversals are very rare and cannot be induced experimentally). Other modifications of the growing point or apical meristem are less permanent. In *Hedera Helix* two types of meristem exist, one giving rise to vegetative shoots with palmately lobed leaves, and the other to flowering shoots with entire leaves. By grafting the latter form can be changed back to the form with palmate leaves (DOORENBOS 1954).

Root meristems also pass through one or more permanent changes. In peas *(Pisum sativum)* the primary root meristem can be cultivated indefinitely in culture by weekly or bi-weekly transfers. The meristems of lateral roots, however, stop growth after a few transfers. In tomatoes *(Lycopersicum esculentum)* the opposite case holds; if one wishes to maintain a clone of a root in culture, lateral root meristems have to be transferred, because the apical meristem decreases its growth potentiality after a few transfers. In monocotyledons, such as *Palmae* and *Gramineae*, many modifications of root meristems occur, each one maintaining its characteristic morphological potentialities, expressed *e.g.* in the number of xylem rays formed.

In the previous paragraphs the variability of cells or small cell complexes was discussed. For larger samples different methods are required. It is generally accepted that the leaf punch method, initiated by SACHS (1884), is the most reliable to obtain physiologically comparable leaf samples. For this method one leaf sample is punched with a cork borer or a special punch out of one side of a leaf, and later another similar sample of exactly the same surface is punched from the leaf blade opposite the midrib. It has been pointed out that in spite of their equal surface at punching time, a possible change in turgor of leaf cells between the two harvesting periods may result in dissimilar samples being collected with the punch method. It is therefore recommended that at the beginning of an experiment with a rubber stamp equal areas be printed on the two sides of the midrib, and that these areas be cut out at the two harvesting times. A variant of this method uses opposite leaves or opposite leaflets, one of each pair being harvested as a control for the other.

Both methods of leaf sampling tacitly assume that the wound made in punching the first sample or removing the first leaf or leaflet does not influence the other half. That this is not generally true was shown by WENT and CARTER (1945). They found that removal of leaves near the path of translocation seriously inhibited sugar translocation through stems, and that this effect of wounding persisted for several days.

When many plants are required to perform a single experiment, it becomes desirable if not imperative to use genetically uniform material. When the experiment involves perennial plants, it often will be possible to make clones of the material by dividing root stocks, or making cuttings or grafts. This does not substitute for uniform growing conditions, but it insures genetic identity of the experimental plants. This method was used for instance by BONNIER, by TURESSON and by CLAUSEN, KECK and HIESEY in their transplant experiments, to establish the existence of ecotypes. The ease with which strawberries *(Fragaria)*, potatoes *(Solanum tuberosum)*, *Saintpaulia* or carnations *(Dianthus)* can be propagated as runners, stem tubers, leaf cuttings or stem cuttings, makes them a very uniform experimental material.

In the case of annual plants it becomes imperative to use selected strains, in which the genetic variability is reduced to a minimum. Apomictic plants, such as *Draba verna*, or self pollinators, such as *Pisum sativum*, usually are genetically very uniform when planted from seed collected from individual plants. But in other cases selected seed material is seldom available, at least in the case of wild plants. Exceptions to this rule are 1. wild plants investigated by geneticists (such as *Arabidopsis*, *Crepis*) and 2. cultivated plants, most of which are available in selected strains of considerable genetic homogeneity. Some cultivated plants are deliberately kept heterozygous because heterosis increases yield (such as *Zea Mays* and *Beta vulgaris*), but in tomatoes, most cereals like wheat, barley, oats and rice, leafy vegetables like lettuce, spinach and endive, and particularly legumes like beans, soy beans and peas, the commercially available varieties are of remarkable genetic uniformity when the seed is bought from a reputable grower, and when well-established varieties are used. Therefore the use of crop plants for physiological experiments is highly recommended, not only from the standpoint of reduced variability, but also because many of these plants have been investigated by hundreds of scientists and thus are well-known culturally and physiologically. Therefore much information is available on the mineral nutrition, growth responses and biochemical composition of these plants, which greatly simplifies further experimentation. That any results obtained in experiments with cultivated plants might have practical significance is unimportant from a purely scientific standpoint, but may attract considerable financial assistance and also may give special satisfaction to the investigator inclined to take his social responsibilities seriously.

Other advantages of using cultivated plants as experimental material are: 1. it is easier to obtain additional seed material of the same strain when the original seed supply is exhausted; 2. it is possible to receive the same material other investigators have used in their experiments, and 3. usually many different strains with differences in morphological and physiological characters are available (differences in rate of stem elongation, leaf form, relative development of leaves, flowers and stems; vitamin, protein or carbohydrate content, etc.). On the other hand it has to be remembered that selection of these plants is continuous, and that in the course of years a gradual genetic change occurs in well-established varieties even though the name remains the same. This also explains why the same variety when obtained from different seed firms may be different, such as the Alaska variety of *Pisum sativum*.

The ultimate size, form and chemical composition of a plant is the result of the interaction between its genetic composition and the environment in which it grew up. Therefore differences in phenotype will occur in plants of genotypic uniformity, if they were not raised under strictly similar conditions. The greatest difficulty in growing plants under controlled conditions is the high light intensity required for normal growth, which is hard to obtain when at the same time the temperature has to be controlled. Since seedlings normally start their development in darkness, it is easy to control the conditions under which they are grown in physiological dark rooms, and therefore they have been used in many investigations where uniform material was essential. For this reason the extent to which the environment influences the phenotype of plants and their physiological variability can easily be demonstrated with seedlings.

Oat seedlings have been used for the quantitative determination of auxin (the so-called *Avena* test), after having been raised under strictly constant temperatures (usually 25⁰ in warm and 20⁰ in cool climates), constant humidity (about 90%), and constant weak red light. At 25⁰ it takes 3 and at 20⁰ it takes 4 days for the coleoptiles to reach a size of 3–4 cm. at which they are used for the test. When the red light is omitted, the hypocotyl elongates and the length of the coleoptile is reduced. Under the rigidly controlled conditions mentioned above there is still considerable variability in the *Avena* (Victory variety) seedlings. This is due to 1. differences in size of the seeds according to their position in the inflorescence, 2. a certain amount of dormancy which occurs in freshly harvested oat seeds, which results in 3. differences in rate of germination so that the seedlings do not all have the same physiological age. By rigidly selecting seeds of equal size and only transplanting seedlings of equal degree of germination the uniformity of the test plants can be increased; yet, considerable individual variability between seedlings remains. Part of the variability between plants in the *Avena* test is due to the variability between seedlings just discussed, but another part is due to insufficient contact of the plant with the auxin source, or too low humidity in the test room, causing differential drying of the agar blocks. In properly carried out experiments the standard error of one auxin determination is about 5% of the average curvature of 12 plants; this is the variability inherent in the *Avena* seedlings when grown under standard conditions, and is not reduced by further practice in carrying out the test. Even slight deviations from the proper test conditions will double or even quadruple the error. No systematic study has been made to reduce the inherent variability of the *Avena* seedlings.

In spite of the apparent uniformity of growing conditions for the *Avena* seedlings, daily and seasonal variations in sensitivity of the test plants to auxin have been reported by many investigators. In general, during the afternoon and early evening, and in summer and autumn, the same auxin concentration gives smaller curvatures than morning or winter tests. Only recently has it been found that these fluctuations in sensitivity disappear when the air in the growing and test rooms is filtered through activated carbon filters. These remove gaseous impurities from the air, which tend to lower the sensitivity of *Avena* seedlings (and pea seedlings) to auxin and also lower their growth rate (HULL et al. 1954). These gaseous impurities, identified with smog as found in the Los Angeles area, are now found in all large cities and they introduce a new variable factor in the environment of plants which has to be controlled in future experiments with plants. In Pasadena we have made the observation, that air which is in contact with activated charcoal for 0.1 second has lost 95% of its toxic components. Therefore air renewal in rooms in which plants are grown for physiological experiments has to occur with air filtered through activated carbon (like Norit C) at the rate of 1 liter per dm². per second at a bed depth of 1 cm.

As shown in the *Avena* test, the variability in the responses of plants can be due to many causes, and it is impossible to determine by statistical analysis of the experiment which percentage of this variability is due to differences in the environment (phenotypic variability) and how much is inherent in the plants (genotypic variability). If we could eliminate all variables in the environment, we would end up with pure genotypic variability, but as shown for the *Avena* test, we may not be aware of all variables affecting the development of plants. The elimination of variables is essential in obtaining uniform plant material.

This fact cannot be over-stressed. Not only does reproducibility become greater, but also the significance of the individual experiment is increased several-fold. Alternatively the number of experimental plants can be reduced, which simplifies experiments.

Frequently plants respond differently to an experimental variable according to the other growing conditions. Therefore fluctuations in these growing conditions will cause differences in response in an experiment, and consequently increases variability. As an example: strawberries *(Fragaria)* will initiate inflorescences on short days only, but at lower temperatures they become photoperiodically indifferent. Therefore, when a temperature gradient occurs in a greenhouse where strawberries are grown near the critical daylength, the plants in the cooler location may flower, whereas those in the center of the house might remain vegetative and form runners. This type of variability should be completely eliminated with proper control over the growing conditions, for it can produce major discrepancies between parallel experiments, which cannot be eliminated, or accounted for, or warned against by statistical treatment of the individual experiment. The following two examples illustrate this. VIRTANEN in Helsinki and WILSON in Madison, Wisconsin, had both measured excretion of nitrogenous compounds by leguminous roots. The former obtained positive, the latter negative results, apparently using the same experimental plants and technique. It was later found (WILSON 1940) that this N-excretion occurs only at low temperatures and in long days at high light intensity. The difference in results could be explained because these conditions occurred in Helsinki, but not in Madison; by changing the conditions in Madison N-excretion was observed there too. The second example refers to the response of plants to thiamin. In statistically unimpeachable experiments BONNER and GREENE (1939) had found that thiamin application to the root medium increased growth in a number of plants. Other investigators (*e.g.* HAMNER 1941) in experiments with equal statistical significance, obtained no effects of thiamin application. This controversy was resolved when BONNER (1943) showed that these differences in response were due to differences in temperature during execution of the experiments: at lower temperatures (20⁰) thiamin increased growth of *Cosmos*, at higher temperatures (26⁰) it did not.

With the improvements in air-conditioning it is now possible to grow plants throughout their life cycle under strictly controlled conditions of light, temperature, humidity, wind, gas content of the air, etc. This has been achieved in so-called phytotrons, of which the Earhart Plant Research Laboratory is the prototype. Plants are grown there from germination to maturity in controlled temperatures, obtained by passing large volumes of conditioned air evenly over plants. In this way no local differences in temperature or humidity or $CO_2$ content of the air surrounding the plant can occur, and this results in greatly reduced phenotypic variability. In *e.g.* tomato plants the use of an air-conditioned instead of a standard greenhouse reduced variability in length or weight to one half, and the further use of artificial light caused another reduction (about twofold) in variability of the experimental plants (WENT 1953). This same reduction in variability is found in the individual cells of a plant grown under controlled conditions, as can be seen in Fig. 2. The variability in osmotic concentration of the epidermal cells of plants grown in controlled temperatures was about half of that of plants grown in an ordinary greenhouse. This emphasizes the fundamental nature of the problem of organismal phenotypic variability: it exists not only on the macroscopical but also on the cellular level. Therefore sister plants which look different are presumably differing in their cellular composition as well. This is comparable to SACHS' dictum (1880): to produce different structures the composition of cells from which these structures arise must of necessity have been different too.

There are many reasons why plants grown in conventional greenhouses show great phenotypic variability. The ventilation of the greenhouse benches is left to convection. Whereas a rising air current along the edges of a bench will reduce the air temperature around plants along the edge, in the center of the bench air will start to rise only after its temperature has been raised several degrees. This is one of the main reasons why plants in the center of a greenhouse bench are taller and more spindly than those along the edge, even though they did not shade each other unduly. Mutual shading and impaired $CO_2$ supply enhance these differences. In greenhouses with forced ventilation all plants are surrounded by air of the same temperature, humidity and $CO_2$ content and consequently are more uniform. When also the illumination is made uniform by using artificial light only (if possible of 10,000–15,000 Lux intensity), phenotypic variability is reduced to the lowest level reached as yet. There is no reason to assume that the ultimate degree of control over plant growth has been reached already in phytotrons. Since it is theoretically highly important to know, how close phenotypic variability can ever approach genotypic variability, much more work should be carried out on further control of the plant environment. And from the standpoint of experimental technique the problem is equally important, because any reduction in variability will increase the significance of results.

The potential or genotypic uniformity of genetically identical plants is of an order of magnitude, seldom equalled in our physical world. Let us assume for a moment that through inbreeding we have developed a strain of homozygous peas, and that the selfed progeny of a single seed is collected after the third growing season, each plant producing 22 seeds. We will then have a bag with 10,000 peas, genetically identical except for an occasional mutation. If we assume a high mutation rate, 0.1% per generation, then we would expect at most 30 mutated genes in this bag, giving rise to not more than a few visible mutants, because they do not show up in the heterozygous condition in which they are formed. Since each pea has presumably over 3000 genes, there are 30 million genes represented in that bag of peas (each being present in thousands of cells per pea), of which 30 are mutated. This equals a purity of one part per million, a purity seldom realized in chemically pure compounds. To what extent is it now possible to translate this potential genetic homogeneity into phenotypic homogeneity among the 10,000 plants grown from this seed?

If gene reactions which initiate development were mass reactions, then we could expect absolute uniformity in the structures derived from identical sets of genes. Genetic evidence, however, indicates that genes are single or double units, as indicated by their purely statistical distribution over the progeny. This made JORDAN (1936) suggest that developmental reactions depending on genes were essentially microphysical reactions resulting in great variability due to the few controlling molecules (genes). This would explain the great phenotypic variability which was assumed to be typical for living organisms, and would make growth and development statistical phenomena like gene distribution. Actually the situation is just the reverse: under uniform growing conditions there exists perhaps nothing in nature which is as reproducible as an organism, in spite of its great complexity. As an example Fig. 3 shows a series of young pea plants, grown at $17^0$ C. in 12,000 Lux light intensity of "Daylight" fluorescent lamps augmented with 10% incandescent light. The plants in the upper row had grown from the heaviest seeds, those in the lower row from the lightest seeds (about half the weight of the former), and in the two middle rows the seeds had intermediate weight. The plants in the figure were not selected from a larger number of plants but show the actual variability which obviously is very small. Therefore instead of producing great variability, the gene mechanism of growth control is entirely stable and quantitatively reproducible. In addition to finding reasons to explain variability in plants as we did in previous paragraphs, we therefore must find mechanisms by which in a controlled environment a homogeneous growth rate is insured regardless of seed weight or gene control.

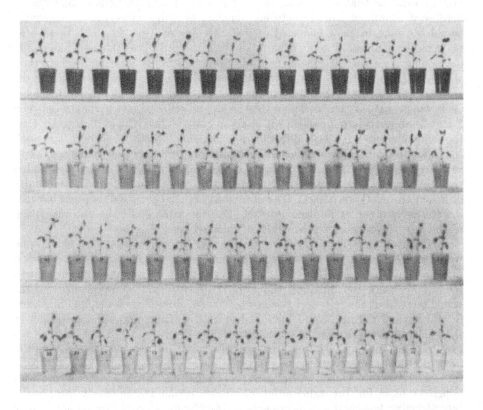

Fig. 3. Pea plants, sown individually in cups in vermiculite, all of the same age. Upper row: plants developed from the heaviest seeds, lower row, plants growing from the lightest seeds, intermediate rows: plants from seeds of intermediate weight.

Before discussing such mechanisms another phenomenon should be mentioned. When peas (or other plants) are germinated, the seedlings will exhibit a certain degree of phenotypic variability. If the differences in length were due to random fluctuations in the growth controlling processes, one would expect a gradual increase in the absolute difference in length between the tallest and the shortest plants. The coefficient of variability would remain approximately constant under those conditions. Actually a continued *decrease* in the coefficient of variability was found in growing peas, and the absolute length differences in a given group of plants remained approximately constant throughout their growth up to the flowering stage (WENT 1953). The same thing can be observed in wheat or corn fields derived from rigidly selected seed. In spite of differences between the plants soon after emergence, the mature plants are of remarkable uniformity in length. Therefore, the growth process is not autocatalytic, and is not a mass reaction, in which a greater initial rate would engender a greater subsequent elongation.

When we analyse this situation in more detail, we find that in peas a negative correlation exists between instantaneous growth rate and subsequent growth. The peas growing slowest in one period of a few days grow fastest in the next period and vice versa. This means that a mechanism exists which in a controlled environment keeps the growth rate constant regardless of the previous conditions. It is hard to consider such control as chemical, for one would expect that as a plant became larger it would produce more growth factors and thus increase its growth rate, or that the plant growing fastest at one particular time did so because it contained more growth hormones or growth factors and consequently would continue to grow faster. But just the opposite was found in these peas (as it has been found in many other plants): in the early growth stages the plants developing from the biggest seeds grow slowest and only after many days they start to catch up with the faster growing plants developing from the smaller seeds.

There are several other indications that growth in peas is not limited by a chemical process. Whereas below $15^0$ the $Q_{10}$ of cell elongation ranges between 2 and 3, above this temperature it is about unity. This $Q_{10}$ of 1 suggests that the controlling process is a diffusion phenomenon, which is proportional with the absolute temperature. On completely different grounds (e.g. the impossibility to increase the growth rate of plants beyond certain limits) it was concluded that the growth process at higher temperatures was limited by an intercellular diffusion process (WENT 1954). Therefore many crucial growth data point towards a diffusion process as limiting growth. On this same basis the autoregulation of the growth process of peas can be explained too.

If a diffusion process between nucleus and cytoplasm controls growth, one would expect that as the cell enlarges the time required to complete the diffusion process would increase. This would be equivalent to a slowing down of the growth rate. Conversely the diffusion time would decrease and consequently growth would increase as the cells decrease in size. On this basis cells which had increased in size beyond normal would have a decreased diffusion rate and grow slower than cells of average size. If at the same time the rate of cell division had remained constant, such slower growing cells would decrease in size during the next divisions and soon would have attained the standard size, diffusion rate and growth. Thus the diffusion control of the growth process provides a perfect basis for the explanation of the autonomous control of the growth rate at higher temperatures in peas and other plants. It equally explains 1. why increased ploidy (with concomitant increase in cell size) usually does not increase the growth rate of plants, 2. why most fast-growing plants have small cells, and 3. why the largest nuclei are found in plants with slow growth rates. On this same basis we can understand why the growth rates of most fast-growing plants tend to attain the same upper limit (2 mm./hour in *Zea Mays*, tomato and peas), and why the growth rates of bacteria with very small cells surpass those of other organisms with more differentiated and larger cells.

Such a diffusion mechanism does not only account for a low $Q_{10}$ of the growth rate and the compensating nature of the growth process, but explains another curious phenomenon observed concerning the variability of plants. When both growth rate and variability of pea plants are plotted against temperature (Fig. 4), then we find that over the range where there is no change in growth rate with temperature, the variability is lowest, but that it increases several-fold at lower temperatures, where presumably growth is controlled by chemical processes. And it is only where growth is controlled by a diffusion process that the compensating action could come into play and could decrease variability. At the lower temperatures, where growth is limited by chemical processes, no compensation mechanism would be expected on the basis of the above hypothesis, and variability would be high. From Fig. 4 it also follows that variability is lowest under optimal growing conditions, when the growth rate could not be increased any further. This same conclusion was reached already by PAAL in 1914, and was restated by WEBER in 1925.

The mechanism which reduces variability in the development of organisms actually makes ideal experimental material of them since it becomes possible to obtain as many objects as desired with a degree of similarity unsurpassed in our physical world, provided at least that the material was raised under absolutely uniform conditions. This statement is very different from the usual idea of extreme variability in biological materials, current among Biologists as well as protagonists of the Physical sciences. Only a few scientists have stressed the remarkable reproducibility of organisms, such as BOHR and SCHRÖDINGER (1945).

Because the phenotypic variability is reduced in plants as the control over the growing conditions is increased, experiments can be simplified. With a 2–5 fold decrease in the coefficient of variability of plants in the Earhart Plant Research Laboratory, the differences between plants in a single treatment usually are smaller than the dif-

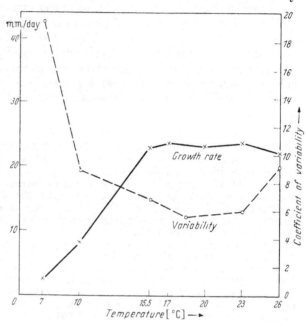

Fig. 4. Crosses and solid line: growth rate of pea plants developing at different constant temperatures (abscissa). For each group of plants also the coefficient of variability is given (circles and broken line) (from WENT 1953).

ferences between treatments. This obviates the necessity for statistical treatment, and makes it possible to use only a few plants per treatment, enough to discover the exceptional behavior of an occasional plant. For some reason the growing point of a plant may be injured, which makes its growth rate lag behind that of the others. Such a plant can be eliminated from the calculation of the average, if 3 or 4 plants are used per treatment. This is being done regularly in tomato experiments in the Earhart Laboratory, where usually 4 plants each time are subjected to the same conditions, like a chemist who carries out his analyses in triplicate.

By using the averages of plants which are all very much alike another error common in biological experiments can be avoided. In many quantitative experiments we try to establish the relationships between factors in the development of an organism. In only the minority of cases do we find arithmetical progression between these factors, but usually logarithmic or more complicated progressions are found. Yet, when we average plant size or weight within one treatment, we do this arithmetically. This introduces an error, which becomes more serious as the variability of the plants increases. Furthermore, in cases where we are dealing with periodic functions or discrete series, averages are meaningless. E.g. in the case of peas the first flower appears either on one or another node and never in between; yet in averaging nodal insertion of the first flower, we arrive at such physical impossibilities as the 10.35th node. In another case, investigated by J. F. Rijswijk (1954), vessel width in successive year rings of tropical leguminous trees seemed to increase gradually when averages were calculated. In reality however, all vessels belonged to a few classes of vessel width, and the proportion of the wider classes increased in the outer year rings. Therefore, if we can dispense with averages or if we take only averages of closely similar objects as obtained in rigidly controlled environments, many quantitative errors or impossible averages can be avoided.

In conclusion we can say, that with proper experimental precautions, phenotypic variability in experimental plants can be reduced to such an extent that it becomes only a minor factor in the interpretation of experiments. This is partly due to the remarkable genotypic uniformity of properly selected organisms, and partly to an innate compensation mechanism, which tends to even out individual differences within growing points. This mechanism, which is of the nature of a diffusion process, was shown to be involved in the control of growth in general.

## Literature.

Bohr, N.: Licht und Leben. Naturwiss. 21, 245–250 (1933). — Bonner, J.: Effects of application of thiamine to Cosmos. Bot. Gaz. 104, 475–479 (1943). — Bonner, J., and J. Greene: Further experiments on the relation of vitamin $B_1$ to the growth of green plants Bot. Gaz. 101, 491–500 (1940). — Bonnier, G.: Recherches experimentales sur l'adaptation des plantes au climat alpin. Ann. Sci. natur., Bot. Ser. VII 20 (1895).

Clausen, J., D. D. Keck and W. M. Hiesey: Experimental studies on the nature of species. I. Effect of varied environments on Western North American Plants. Carnegie Instn. Publ. No 520 1940, 1–452.

Doorenbos, J.: "Rejuvenation" of Hedera Helix in graft combinations. Proc. Kon. Ned. Akad. v. Wetensch. C. 57, 99–102.

Hamner, C. L.: Effects of vitamin $B_1$ upon the development of some flowering plants. Bot. Gaz. 102, 156–168 (1941). — Höfler, K.: Spezifische Permeabilitätsreihen verschiedener Zellsorten derselben Pflanze. Ber. dtsch. bot. Ges. 40, 133–148 (1937). — Hull, H., F. W. Went and Noboru Yamada: Fluctuations in sensitivity of the Avena test due to air pollutants. Plant Physiol. 29, 182–187 (1954).

Jordan, P.: Anschauliche Quantentheorie. Berlin: Springer 1936.

Kostytschew, S., and F. A. F. C. Went: Lehrbuch der Pflanzenphysiologie. II. Berlin: Springer 1931.

Lambrecht, E.: Beitrag zur Kenntnis der osmotischen Zustandsgrößen einiger Pflanzen des Flachlandes. Beitr. Biol. Pflanz. 17, 87–136 (1929).

Paal, A. V.: Individuelle Abweichungen in physiologischen Reaktionen. Math.-naturw. Ber. Ungarn 30 (1914).

Rijswijk, J. T.: Thesis, Wageningen, Holland. 1954. — Ruhland, W., H. Ullrich and S. Endo: Untersuchungen zu den "spezifischen Permeabilitätsreihen" Höflers. I. Zur Frage der Alkoholpermeabilität von Pflanzenzellen unter verschiedenen Versuchsbedingungen. Planta (Berl.) 27, 650–688 (1938).

Sachs, J.: Stoff und Form der Pflanzenorgane. I. Arb. bot. Inst. Würzburg 2, 452 (1880). — Ein Beitrag zur Kenntnis der Ernährungstätigkeit der Blätter. Arb. bot. Inst. Würzburg 3, 1–33 (1884). — Schrödinger, E.: What is life? The physical aspect of the living cell. Cambridge: Univ. Press 1945.

Turesson, G.: The genotypical response of the plant species to the habitat. Hereditas (Lund) 3, 211–350 (1922).

Vries, H. de: Eine Methode zur Analyse der Turgorkraft. Jb. wiss. Bot. 14, 427–601 (1884).

Weber, F.: Physiologische Ungleichheit bei morphologischer Gleichheit. Österr. bot. Z. 1925, 256–261. — Protoplasmatische Pflanzenanatomie. Protoplasma (Berl.) 8, 291–306 (1929). — Went, F. W.: The Earhart Plant Research Laboratory. Chronica Bot. (Waltham, Mass.) 12, 1–19 (1950). — Gene action in relation to growth and development. 1. Phenotypic variability. Proc. Nat. Acad. Sci. U.S.A. 39, 839–848 (1953). — Physical factors affecting growth in plants. Dynamics of growth processes, edited by E. J. Boell, Princeton Univ. Press 1954, p. 130–147. — Went, F. W., and M. Carter: Wounding and sugar translocation. Plant Physiol. 20, 457–460 (1945). — Wilson, P. W.: The biochemistry of symbiotic nitrogen fixation. Madison: Univ. of Wisc. Press 1940.

# Variabilität und statistische Behandlung physiologischer Experimente.

Von

**Cornelia Harte.**

Mit 7 Textabbildungen.

## A. Einleitung.
## Die Variabilität physiologischer Beobachtungen.

Die Pflanzenphysiologie gehört zu den Wissenschaften, die mit induktiver Methode arbeiten; sie versucht, aus den einzelnen Beobachtungen die allgemeineren Gesetzmäßigkeiten zu erkennen. Ihre Arbeitsmethode ist dabei das Experiment. In planmäßig durchgeführten Versuchen werden die Bedingungen, die für den Ablauf des untersuchten Vorgangs von Bedeutung sind, erforscht. Das Objekt der Untersuchung sind dabei Pflanzen oder Teile von Pflanzen. Hieraus ergibt sich, daß alle Besonderheiten, die das biologische Material von nichtlebenden Substanzen unterscheiden, berücksichtigt werden müssen. Das in diesem Zusammenhang wichtigste Kennzeichen der biologischen Reaktionen ist ihre Variabilität. Keine zwei Pflanzen einer Art, Blätter einer Pflanze oder Zellen eines Blattes sind in allen ihren Maßen und allen Einzelheiten ihres Aufbaues genau gleich. Diese Variabilität, die in morphologischer Hinsicht überall feststellbar ist, zeigt sich auch an den physiologischen Reaktionen. Als Folge davon sind Bestimmungen, die an verschiedenen Zellen oder verschiedenen Pflanzen vorgenommen wurden, nie ganz übereinstimmend. Weil die lebende Substanz dauernden Veränderungen unterworfen ist, können auch Messungen einer Reaktion, die nacheinander an einem Objekt vorgenommen wurden, nie das gleiche Ergebnis bringen.

Diese Abweichungen sind kein Ausdruck eines fehlerhaften Versuchs, genügende Genauigkeit des Meßinstruments und der Arbeitsweise vorausgesetzt, sondern sind darin begründet, daß wirkliche Unterschiede in der Reaktion bei den einzelnen Bestimmungen vorliegen. Aus dieser Variabilität ergibt sich, daß in allen derartigen Fällen eine einzelne Bestimmung zwar eine genaue Aussage über die zu diesem Zeitpunkt, unter diesen bestimmten Bedingungen bei diesem einen Objekt vorhandene physiologische Situation erlaubt, aber keine Verallgemeinerung zuläßt außer der einen, daß dieser gefundene Wert im Bereich der physiologischen Möglichkeiten des Objektes liegt. Bei der induktiven Arbeitsmethode ist es aber das Ziel, eine Verallgemeinerung zu ermöglichen und für die Aussage eine allgemeingültige Formulierung zu finden.

Da die Bestimmungen alle mehr oder weniger voneinander abweichen, kann eine einzige Beobachtung hierzu nicht ausreichen. Es muß vielmehr eine Vielzahl von Beobachtungen durchgeführt werden, die so groß ist, daß bei der Verallgemeinerung der Schlußfolgerungen die Abweichungen der einzelnen Beobachtungen nicht mehr ins Gewicht fallen. Für eine solche allgemeine Aussage ist es daher notwendig, aus dieser Vielfalt heraus den „wahren" Wert, der die betreffende Reaktion charakterisiert, zu finden. Dieses Ziel kann angestrebt werden mit Hilfe der Variationsstatistik, wobei versucht wird, durch eine Analyse der Verteilung der gefundenen Beobachtungsergebnisse, die in ihrer Gesamtheit die zu untersuchende statistische Masse bilden, eine Verallgemeinerung auf Grund dieser variierenden Beobachtungen durchzuführen. Darüber hinaus ist der Variationsstatistik noch die andere Aufgabe gestellt, nämlich Angaben darüber zu machen, mit welcher Sicherheit im gegebenen Fall eine solche Verallgemeinerung möglich ist.

In der Pflanzenphysiologie wurden bisher nur in relativ geringem Umfang statistische Methoden für die Auswertung der Versuchsergebnisse herangezogen. Soweit aber Berechnungen durchgeführt wurden, handelte es sich zum Teil um Methoden, die unzulänglich oder für den speziellen Fall nicht geeignet und der Struktur des Materials nicht angemessen sind. Für die Auswertung der Versuche ergeben sich hieraus zwei Fehlerquellen. Einmal besteht die Gefahr, daß durch Überschätzung der Bedeutung der Differenzen aus den Versuchen Schlußfolgerungen gezogen werden, die durch die empirischen Befunde nicht gestützt werden. Zum anderen gehen aber ebenso oft wertvolle Informationen verloren, weil die Sicherheit, mit der aus den Versuchsergebnissen Aussagen gemacht werden können, nicht erkannt wird. Beide Gefahren, sowohl die Überschätzung wie die Unterbewertung des Versuchsergebnisses, können vermieden werden, wenn das Material einer geeigneten statistischen Analyse unterworfen wird.

Die Methoden, mit deren Hilfe derartige Berechnungen durchgeführt werden können, sind von der Mathematik in der Wahrscheinlichkeitsrechnung, der Fehlertheorie und der mathematischen Statistik entwickelt. Da es sich hier nicht um ein Handbuch der Statistik, sondern um ein solches der Physiologie handelt, soll im folgenden nicht von der Systematik der mathematischen Methoden ausgegangen werden, sondern von den Eigenschaften der biologischen Objekte, um zu zeigen, in welcher Weise physiologische Untersuchungen einer variationsstatistischen Untersuchung zugänglich sind.

Für Formeln und ihre Ableitung, die in fast allen Lehrbüchern der biologischen Statistik zu finden sind, wurde keine Literatur angegeben. Wenn dagegen ein bestimmtes Problem in einem Buch besonders berücksichtigt wird, ist dies eigens erwähnt. Im folgenden wird soweit wie möglich auf Hand- und Lehrbücher verwiesen, in denen jeweils Hinweise auf die zuständigen Einzelarbeiten zu finden sind. Zeitschriftenartikel werden nur angeführt, wenn es sich um Probleme handelt, die in den angeführten Lehrbüchern nicht behandelt sind, aber doch allgemeineres Interesse für die Durchführung physiologischer Versuche haben. Dafür sei auf die Zeitschriften hingewiesen, in denen laufend über neue statistische Methoden, Verbesserung von Versuchsplänen usw. berichtet wird. Es handelt sich um Biometrics, Journal of the Biometric Society, USA.; Biometrika, Great Britain; Sankhya, India.

# B. Voraussetzungen der statistischen Bearbeitung physiologischer Versuche.

## I. Voraussetzungen von seiten der Statistik.

### a) Statistische Maßzahlen.

Zunächst sind die Voraussetzungen zu betrachten, die die Statistik bietet, und wodurch sie die geeigneten Methoden für die Auswertung physiologischer Versuche liefert. Die beiden wichtigen Hilfsmittel sind die statistischen Maßzahlen und die Prüffunktionen, die darauf aufbauen. Von einer Auswertung eines Versuchs wird eine Formulierung der Schlußfolgerungen erwartet, die es erlaubt, die Ergebnisse in wenigen statistischen Maßzahlen zusammenzufassen. Diese Maßzahlen müssen von der Art sein, daß sie nach allgemeingültigen Regeln berechnet werden und dadurch einen Vergleich zwischen verschiedenen Versuchen erlauben; außerdem müssen sie erkennen lassen, mit welcher Sicherheit bei einer Wiederholung des Versuchs die gleichen Ergebnisse erwartet werden können, da es hiervon abhängt, inwieweit sie als Grundlage für die zu ziehenden allgemeinen Schlußfolgerungen dienen können.

Die Regeln, nach denen die statistischen Maßzahlen als richtige Beschreibung einer statistischen Masse berechnet werden, werden von der mathematischen Statistik aufgestellt und fallen damit außerhalb des Rahmens dieses Handbuches. Die statistischen Maßzahlen und die Formeln zu ihrer Berechnung werden als gegeben hingenommen, und es ist hier zu prüfen, welche dieser Zahlen zur Beschreibung bestimmter physiologischer Erscheinungen und zur Lösung der sich daraus ergebenden Probleme herangezogen werden können.

Die statistischen Maßzahlen dienen der Beschreibung einer statistischen Masse; sie sollen einen Eindruck davon vermitteln, wie die Verteilung in dem zu untersuchenden Kollektiv aufgebaut ist. Als erstes sind zu nennen die Mittelwerte, die angeben, um welchen Wert der Meßskala sich die einzelnen Meßwerte gruppieren. Es gibt eine Reihe verschiedener Mittelwertmaße, von denen einige mathematisch exakt definiert werden können, andere nur beschreibenden Wert haben. Es ist danach zu streben, möglichst nur die exakt definierten Mittelwerte, das arithmetische und das geometrische Mittel zu verwenden, aber für bestimmte Fragen kann die Struktur des Materials dies verbieten, nämlich immer dann, wenn die Berechnung eines solchen Mittels nur möglich ist auf Kosten der Genauigkeit der Versuchsanstellung, so daß dann der „exakte" Mittelwert in Wirklichkeit weniger exakt ist als einer der anderen, etwa Zentralwert oder Dichtemittel. Sowohl Zentralwert wie Dichtemittel können nicht immer genau bestimmt werden, und vor allem kann für beide nicht angegeben werden, mit welcher Sicherheit bei einer Wiederholung des Versuchs ein ähnliches Ergebnis zu erwarten ist. In der englischen Literatur wird praktisch nur das arithmetische Mittel verwendet, während in den deutschen Lehrbüchern der Theorie der Mittelwerte meist ein größerer Raum gewährt wird (GEBELEIN, WEBER).

Für die Charakterisierung einer Verteilung ist es weiter wichtig, zu wissen, in welcher Weise sich die einzelnen Meßwerte um den Mittelwert gruppieren. Von den Maßen, mit denen diese Anordnung gemessen werden kann, ist praktisch nur die Streuung $\sigma$, definiert als die Wurzel aus der mittleren quadratischen Abweichung, von Bedeutung, die als Streuungsmaß dem arithmetischen Mittel zugeordnet ist. Die mittlere quadratische Abweichung $\sigma^2$ wird als die Varianz bezeichnet.

Manchmal ist es nötig, ein Maß dafür zu haben, ob sich die Werte symmetrisch um den Mittelwert gruppieren, oder ob die Werte um diesen Punkt asymmetrisch angeordnet sind. Für die Beschreibung der Schiefe einer Verteilung gibt es fast soviele Maßzahlen, wie es schiefe Kurven gibt, von denen aber kaum eine noch verwendet wird, ebenso wie die weiteren Maßzahlen, die der näheren Beschreibung der Überhöhung oder anderer Eigenschaften der Verteilungskurve dienen.

Sehr wichtig ist dagegen eine andere Maßzahl, die dazu verwendet wird, die Zuverlässigkeit anzugeben, die einem Mittelwert zuerkannt werden kann. Diese hängt ab von zwei Faktoren, der Streuung der Einzelwerte um den Mittelwert und der Anzahl von Beobachtungen, aus denen das Mittel berechnet wurde. Die Formel für den „mittleren Fehler" des arithmetischen Mittels, das im folgenden immer gemeint ist, wenn von einem Mittelwert gesprochen wird, in die die beiden genannten Faktoren eingehen, wurde der Fehlerrechnung der physikalischen Messungen entnommen. Es ist aber zu beachten, daß in der Biologie dieser Zahl eine ganz andere Bedeutung zukommt als bei dem Problem, für das sie zuerst geschaffen wurde. In der Physik bezeichnet der mittlere Fehler einer

Messungsreihe das Ausmaß der echten Meßfehler des Instruments, in der Biologie ist er aber ein Maß für die Variabilität des Materials. Wenn die Messungen, die den Berechnungen zugrunde liegen, so ungenau sind, daß in der Streuung und damit im mittleren Fehler des Mittelwertes tatsächliche Messungsfehler enthalten sind, sind sie für eine weitere statistische Bearbeitung unbrauchbar. Um diese Mißverständnisse zu vermeiden, wird daher manchmal der Ausdruck „Streuung des Mittelwertes" verwendet, wobei aber wieder eine Möglichkeit der Verwechslung mit der Streuung $\sigma$ der Einzelbeobachtungen gegeben ist. Auf weitere Maßzahlen, die der Beschreibung des Zusammenhangs zwischen mehreren Messungsreihen dienen, wird später eingegangen.

Wenn bestimmte statistische Maßzahlen bekannt sind, mit denen eine gegebene Verteilung beschrieben werden kann, vor allem Mittelwert und Streuung, dann kann man noch einen Schritt weiter gehen und die theoretische Verteilung berechnen, die sich mit den gleichen Maßen beschreiben läßt und durch bekannte Funktionen charakterisiert ist. Die Methoden, nach denen sich aus den gefundenen Maßzahlen die zugehörigen theoretischen Verteilungen berechnen lassen, sind in allen Lehrbüchern beschrieben, so daß hier darauf verwiesen werden kann. Zuerst wird meist die Prüfung mit einer Normalverteilung in Frage kommen, aber häufig wird sich ergeben, daß die gefundene Kurve eine bessere Übereinstimmung mit den Verteilungen nach Poisson, Lexis, einer Binominalverteilung oder anderen zeigt. Die Identifizierung einer empirisch gefundenen Verteilung mit einer Kurve, deren Funktion bekannt ist, kann für die Lösung mancher Fragen bedeutungsvoll sein. Da für die theoretischen Kurven die mathematischen Grundlagen bekannt sind, ist es möglich, festzustellen, in welcher Weise die Ursachen zusammenwirken müssen, damit eine derartige Verteilung zustande kommen kann, woraus wiederum vielfach Rückschlüsse auf die Wirkungsweise bekannter biologischer Faktoren, die auf das Material Einfluß hatten, möglich sind.

### b) Die statistischen Prüffunktionen.

Eine empirisch gefundene Verteilung wird nie mit einer auf Grund der verschiedenen Funktionen berechneten völlig übereinstimmen. Die Ursache hierfür ist darin zu suchen, daß die theoretischen Verteilungen Gültigkeit haben für unendlich große statistische Massen, während das empirische Material, auch wenn es einen sehr großen Umfang hat, immer nur eine Stichprobe aus einem derartigen Kollektiv darstellt, bei deren Entnahme mit Zufallseinflüssen gerechnet werden muß. Die Theorie der Stichproben ist mit Hilfe der Wahrscheinlichkeitsrechnung durchgearbeitet. Da keine vollständige Übereinstimmung zwischen der Stichprobe und der theoretischen Verteilung erwartet werden kann, ergibt sich die Frage, ob im vorliegenden Fall die zwischen beiden Verteilungen gefundene Differenz noch durch die Zufälligkeit der Probenentnahme zustande kommen kann, oder ob sie als echte Abweichung gewertet werden muß. Es geht also um die Feststellung, ob die der theoretischen Verteilung zugrunde liegenden Ursachen als zureichende Erklärung für die gefundene empirische Verteilung angesehen werden können, oder ob auf Grund der Differenz die dem Vergleich zugrunde gelegte theoretische Verteilung als Erklärung abgelehnt werden und nach einer anderen Erklärung für das Zustandekommen der Verteilung gesucht werden muß.

Der für eine solche Entscheidung notwendige Vergleich der beiden Verteilungen wird mit Hilfe der statistischen Prüffunktionen durchgeführt. Das Ziel bei der Anwendung dieser Funktionen ist die Feststellung, mit welcher Wahrscheinlichkeit die gefundenen Abweichungen als zufällige Varianten bei der Entnahme einer Stichprobe vom gegebenen Umfang aus einer unendlich großen statistischen Masse auftreten können. Die Prüffunktionen wurden von verschiedenen Mathematikern für verschiedene Zwecke ausgearbeitet, so daß ihr Zusammenhang nicht sofort ersichtlich ist, obwohl alle auf den gleichen Grundsätzen aufbauen. Allen gemeinsam ist, daß ein Quotient aus zwei Varianzen (oder aus zwei Streuungen) gebildet und mit einem Grenzwert verglichen wird, der angibt, mit welcher Wahrscheinlichkeit dieser Quotient unter den gegebenen Umständen diese bestimmte Größe erreicht. Die Grundlage hierfür bildet immer die Normalverteilung, und es wird eine Aussage darüber gemacht, mit welcher Wahrscheinlichkeit das gefundene Ergebnis gerade noch als zufällige Abweichung auf Grund der Normalverteilung der Differenzen angesehen werden kann. Die Grenzwerte für die einzelnen Prüffunktionen sind in Tabellen zusammengefaßt und ausführlich in Tafelwerken, in abgekürzter Form in fast allen Lehrbüchern der Biometrie zu finden (FISHER u. YATES 1938, KOLLER 1942).

Um im Einzelfall festzustellen, welche der vier verschiedenen Prüffunktionen anzuwenden ist, ist es nötig, den Quotienten, mit dem gearbeitet wird, näher zu betrachten. Der entscheidende Gesichtspunkt für die Wahl einer bestimmten Prüffunktion ist die Anzahl der Freiheitsgrade, die für Zähler und Nenner zur Verfügung stehen. Wenn eine dieser Zahlen einer theoretischen Verteilung entnommen ist, stehen hierfür unendlich viele Freiheitsgrade zur Verfügung, stammt sie dagegen aus einer empirischen Beobachtungsreihe, so ist die Anzahl der Freiheitsgrade gleich der Anzahl der Beobachtungen, vermindert um die Anzahl der aus der Verteilung bereits berechneten statistischen Maßzahlen, die zur Bestimmung der Streuung notwendig waren.

Die Zusammenhänge zwischen den vier Prüffunktionen $c$, $t$, $\chi^2$ und $z$ sind in anschaulicher Form bei MATHER (1951) dargestellt. Die allgemeinste Form stellt die $z$-Funktion dar. Sie findet Anwendung, wenn zwei aus empirischem Material stammende Varianzen als Varianzverhältnis geprüft werden sollen. Infolgedessen weisen sowohl Zähler wie Nenner eine von Fall zu Fall wechselnde Anzahl von Freiheitsgraden auf. Die $z$-Tafel ist demnach dreidimensional, da zwei Dimensionen von den Freiheitsgraden der beiden Zahlen gebildet werden, während die dritte die Wahrscheinlichkeitswerte $P$ umfaßt. Die einzelnen Fächer der Tafel enthalten die Grenzwerte, die $z$ bei der gegebenen Anzahl von Freiheitsgraden in Zähler und Nenner mit einer bestimmten Wahrscheinlichkeit erreicht.

Zwei Sonderfälle der $z$-Verteilung stellen die Funktionen für $t$ und $\chi^2$ dar. Für die $t$-Funktion ist es charakteristisch, daß im Quotienten der Zähler einer Differenz aus zwei Werten entspricht und daher mit nur einem Freiheitsgrad behaftet ist, während der Nenner eine wechselnde Anzahl von Freiheitsgraden besitzt, die zwischen 1 und $\infty$ liegen kann. Bei der $\chi^2$-Funktion ist umgekehrt die Anzahl der Freiheitsgrade im Nenner festgelegt auf $\infty$, da hier die Streuung der Normalverteilung auftritt, während der Zähler aus einer wechselnden Anzahl von Beobachtungen gebildet wird und 1—$\infty$ Freiheitsgrade aufweisen kann. Die Grenzwerte lassen sich in zweidimensionalen Tafeln darstellen, deren Koordinaten von der Anzahl der Freiheitsgrade und den $P$-Werten gebildet werden. Der $t$-Test wird verwendet, wenn eine Differenz mit Hilfe ihrer aus empirischem Material bestimmten Streuung auf ihre Signifikanz hin geprüft werden soll. Einer der beiden Werte, aus denen die Differenz gebildet wird, kann dabei durch eine Hypothese festgelegt sein, etwa auf Null, es können aber auch zwei empirisch gefundene Werte verglichen werden. So können z.B. zwei Mittelwerte aus verschiedenen Bestimmungen untersucht werden, um festzustellen, ob die beiden Kollektive, aus denen sie bestimmt wurden, als Stichproben aus einer einheitlichen statistischen Masse angesehen werden können, die nur zufällig verschieden ausgefallen sind. Es kann aber auch eine statistische Maßzahl mit Null oder einem anderen theoretischen Wert verglichen werden, um festzustellen, mit welcher Wahrscheinlichkeit der empirisch gefundene Wert von dem theoretisch angenommenen Wert abweicht. Wenn diese Art der Prüfung etwa zur „Sicherung" von Mittelwerten, die aus Messungen errechnet wurden, verwendet wird, ist der Test wenig sinnvoll, er erhält aber eine Bedeutung, wenn etwa ein Mittelwert aus Differenzen geprüft werden soll, zur Feststellung darüber, ob die mittlere Differenz zwischen je zwei zusammengehörenden Versuchen real, d.h. größer als Null ist (vgl. S. 104, Paarbildung). Ebenso ist die Prüfung der Differenz gegen Null wichtig für die Beurteilung von Regressions- und Korrelationskoeffizienten (vgl. S. 90). Der $\chi^2$-Test wird angewendet, wenn nicht nur eine, sondern eine Reihe von Differenzen gleichzeitig geprüft werden soll.

Die $c$-Funktion stellt wiederum einen Sonderfall sowohl der $t$- wie der $\chi^2$-Verteilung dar, indem hier für beide Anteile des Quotienten die Anzahl der Freiheitsgrade festgelegt ist, den für Zähler auf 1, für den Nenner auf $\infty$, weil hier eine Differenz mit der Streuung der zugehörigen Normalverteilung verglichen wird.

Tabelle 1.

| Prüffunktion | Anzahl der Freiheitsgrade des geprüften Quotienten | |
|---|---|---|
| | im Zähler | im Nenner |
| $c$ | 1 | $\infty$ |
| $t$ | 1 | $1—\infty$ |
| $\chi^2$ | $1—\infty$ | $\infty$ |
| $z$ | $1—\infty$ | $1—\infty$ |

Theoretisch ist es also durchaus möglich, in allen Fällen die $z$-Verteilung als Prüffunktion zu verwenden, es bedeutet aber eine Erleichterung und Vereinfachung der Arbeit, wenn jeweils die dem zu lösenden Problem angepaßte Funktion verwendet wird. Die Tabelle 1 gibt noch einmal die Zusammenhänge zwischen den vier Prüffunktionen nach Mather (1951) wieder.

## II. Voraussetzungen von seiten des biologischen Materials.

### a) Grundfragen.

Damit eine variationsstatistische Auswertung eines Versuchs möglich ist, muß das Material und die Durchführung des Versuchs bestimmten Bedingungen genügen. Nicht jeder Versuch und jede Fragestellung eignen sich für eine statistische Bearbeitung. Es ist daher zunächst festzustellen, welche Voraussetzungen das Material erfüllen muß, um eine statistische Auswertung zuzulassen. In den meisten Büchern, die von einer Darstellung der mathematischen Methoden ausgehen, sind diese Bedingungen stillschweigend vorausgesetzt; sie bilden aber die Grundlagen für die Auswertung, und es entstehen viele Fehlschlüsse aus der Vernachlässigung dieser Bedingungen.

Die erste Forderung ist eine klare Fragestellung und eine Versuchsplanung, die eine eindeutige Beantwortung des Problems ermöglicht. Das biologische Problem, das gelöst werden soll, muß so formuliert werden, daß völlig eindeutig zu erkennen ist, auf welche Fragen eine Antwort gesucht wird. Dann muß die Versuchsanstellung technisch so durchgearbeitet werden, daß tatsächlich die gewünschte Auskunft aus den Versuchsergebnissen erwartet werden kann. Die Berechnungen aus einem Versuch, der seiner Planung nach vieldeutig ist, können keine eindeutige statistische Aussage ergeben.

Das Material, mit dem gearbeitet wird, muß einheitlich, aber unselektioniert sein und es muß eine Variabilität der zu untersuchenden Reaktion aufweisen, weil die Verschiedenheit des Versuchsmaterials untersucht wird und die Grundlage für die Berechnung bildet. Ein Versuch, der seiner Natur nach nur eine Alternative für die Reaktion ergibt und in dem mit Sicherheit vorauszusehen ist, daß alle Einzelbeobachtungen, die unter gleichen Bedingungen durchgeführt werden, ohne Ausnahme eine gleichartige Reaktion zeigen müssen in der Weise, daß innerhalb des Versuchs keine individuellen Unterschiede auftreten, ergibt auch ohne statistische Berechnungen eindeutige Aussagen über den Verlauf des Versuchs und die Wirkung der Versuchsbedingungen. Andererseits können über Größe oder Ertrag einer einzelnen Pflanze, den Wert der Permeationskonstanten einer bestimmten Zelle oder die Assimilationsleistung eines Blattes, genügend genaue Messungen vorausgesetzt, bereits aus einer Bestimmung sichere Aussagen gemacht werden. Auch Untersuchungen, die den Vergleich von Individuen zum Ziel haben und zu einer Aussage führen sollen von der Art: ,,Der Meßwert für dieses Objekt ist um den Betrag $x$ größer als für das andere Objekt'' benötigen keine statistischen Berechnungen.

Sobald aber eine solche Bestimmung durchgeführt wird mit dem Ziel, eine Verallgemeinerung zu treffen von der Art: ,,Die Meßwerte werden für Zellen des einen Gewebes größer sein als für Zellen des anderen Gewebes'', wird die Lage anders. Aus dem Vergleich von zwei Individuen, die verschiedenen Gruppen angehören, können keine Rückschlüsse auf die Gruppenunterschiede gezogen werden, denn die beiden Meßwerte sind in diesem Fall von zwei Variablen abhängig, einmal ,,verschiedene Objekte'', zum anderen ,,verschiedene Gruppen'', und es kann nicht ohne weitere Prüfung die gefundene, tatsächlich vorhandene Differenz der Meßwerte zwischen den zwei Individuen auf nur einen dieser Faktoren zurückgeführt werden unter Vernachlässigung des Einflusses, den der andere Faktor auf das Ergebnis gehabt haben kann. In einem derartigen Fall ist es also nötig, zu bestimmen, in welchem Ausmaß die beiden im Vergleich enthaltenen Variabilitätsfaktoren auf den gemessenen Wert einen Einfluß haben. Anders ausgedrückt: Wenn aus einem Versuch auf die Wirkung eines Faktors geschlossen werden soll, so muß die Durchführung des ganzen Versuchs derart sein, daß für eine aufgefundene Differenz wirklich nur der Unterschied in diesem einen Faktor als Erklärung in Frage kommt.

## b) Ursachen der Variabilität und die Forderung nach Einheitlichkeit des Materials.

Von Bedeutung wird die statistische Bearbeitung also dann, wenn im Versuchsmaterial die Einzelbeobachtungen nicht völlig gleiche Ergebnisse bringen, sondern eine gewisse Variabilität zeigen. Vor jeder statistischen Berechnung ist nun soweit wie möglich den Ursachen dieser Variabilität nachzugehen. Das Ziel dieser Inspektion ist es, zuerst alle Abweichungen, die durch erkennbare Versuchsfehler zustande gekommen sind, von vornherein auszumerzen.

Hierher gehören z. B. Verunreinigungen oder Infektionen, Mängel einzelner Kulturgefäße, Beschädigungen durch Insekten, Tierfraß, Verlust von Versuchsmaterial, Meßfehler durch Verwendung eines mit wechselnder Genauigkeit arbeitenden Meßgerätes, aber auch Abweichungen, die durch die Ausführung der Messungen durch zwei Beobachter entstehen, durch Verwendung von verschiedenen Meßgeräten für gleichartige Ablesungen und ähnliche Ursachen.

Bei dieser Vorsortierung ist streng darauf zu achten, daß nur die Meßwerte verworfen werden, bei denen die Ursache der extremen Abweichung deutlich feststellbar ist und aus ihr eindeutig hervorgeht, daß es sich um einen abnormen

Fall handelt, der durch das Auftreten einer im Versuchsplan nicht vorgesehenen, einmaligen Veränderung der Versuchsbedingungen zustande gekommen ist. Alle anderen Abweichungen, auch wenn sie scheinbar sehr groß sind, müssen als Ergebnisse der zufälligen Einflüsse behandelt und ohne weitere Selektion in die Auswertung einbezogen werden.

Diese Bedingung wird oft als die Forderung nach Einheitlichkeit des Materials formuliert. Hierunter ist zu verstehen, daß im Material keine Einflüsse vorhanden sein dürfen, durch die zusätzliche Variabilitätsfaktoren geschaffen werden. Diese Forderung des einheitlichen, unselektionierten Materials muß aber nach zwei Richtungen hin eingeschränkt werden. Einmal darf das Bestreben, die Einheitlichkeit zu wahren oder zu erreichen, nicht dazu führen, eine Vorselektion der Beobachtungswerte oder des Materials so zu treffen, daß davon die Meßwerte berührt werden können. Die Wirkung der Zufallsfaktoren bringt es mit sich, daß immer auch scheinbar extreme Plus- und Minusabweicher auftreten müssen. Wenn diese entfernt werden, werden zugleich alle statistischen Maßzahlen, die aus dem Material berechnet werden können, davon beeinflußt und verfälscht. Es gibt zwar verschiedene Methoden, durch die festgestellt werden kann, ob ein einzelner abweichender Wert noch als zufällige Abweichung vom Durchschnitt gewertet werden kann; alle setzen jedoch ein großes Versuchsmaterial voraus, in dem aber ein einzelner Abweicher nicht besonders störend wirkt. Bei kleinem Versuchsumfang kann durch die Verwerfung von mehreren Extremwerten eine solche Selektion zustande kommen, daß dadurch die Versuchsergebnisse erheblich verfälscht und damit wertlos werden können. Ein unselektioniertes Material ist die Voraussetzung jeder weiteren statistischen Bearbeitung.

Zum anderen liegt es aber im Ziel der experimentellen Arbeit begründet, in das Untersuchungsmaterial mit Absicht neue Faktoren einzuführen, von denen gerade geprüft werden soll, ob sie als zusätzliche Variabilitätsfaktoren wirken. Solange es nur um die Beschreibung der natürlichen Variabilität geht, bleibt die Forderung des einheitlichen Materials uneingeschränkt bestehen, aber sobald die Zielsetzung der Untersuchung darauf eingestellt ist, die Ursachen dieser Variabilität aufzudecken, ist es unerläßlich, den Grundsatz der Einheitlichkeit zu durchbrechen und die durch die absichtlich eingeführten variierenden Faktoren hervorgebrachte Veränderung der Variabilität in den Mittelpunkt der Untersuchung zu stellen.

Wie hieraus hervorgeht, ist das Ziel der Berechnungen ein mehrfaches: Die Erfassung und Beschreibung der Variabilität des biologischen Materials, die Aufdeckung der Ursachen dieser Variabilität und der Bedeutung, die den einzelnen Variationsursachen im gegebenen Fall zuzuschreiben ist. Als nächstes sind also diese Variabilitätsfaktoren zu betrachten.

## III. Die Variabilitätsfaktoren.

### a) Zufällige und systematische Einflüsse.

Die biologischen Objekte werden durch eine Unzahl von äußeren und inneren Bedingungen beeinflußt, die nie für zwei Pflanzen während ihres ganzen Lebens genau gleich sind und nie für mehrere Objekte zum Zeitpunkt des Versuchs genau gleich gehalten werden können. Da nun der Zustand einer Pflanze bestimmt ist durch die vorangegangenen Einwirkungen und die im Augenblick der Beobachtung herrschenden Bedingungen, geht daraus hervor, daß bei allen Versuchen, die einen physiologischen Zustand oder eine Reaktion bei mehreren Individuen feststellen, nie genau gleiche Ergebnisse erzielt werden können.

Das Ausmaß der Schwankungen, die festgestellt werden, hängt dabei ab von der Beeinflußbarkeit des untersuchten Vorgangs und der Genauigkeit der verwendeten Meßgeräte. Die Gesamtheit dieser unbeeinflußbaren und unkontrollierbaren Einwirkungen auf die Pflanze werden als die zufälligen Einflüsse bezeichnet, ihr Ergebnis, das sich in der natürlichen Variabilität des Objektes ausdrückt, als die Zufallsschwankungen, denen die geprüfte Reaktion unterworfen ist.

Diese Variabilität des biologischen Materials ist, wie bereits gesagt, etwas anderes als die Variabilität physikalischer Messungen. Bei mehrfacher Wiederholung der gleichen Messung mit demselben Meßgerät an demselben Gegenstand werden immer geringe Differenzen der Bestimmungen auftreten, die zurückzuführen sind auf geringe Ungenauigkeiten des Meßgeräts. Für jedes Meßgerät ist die Grenze der Zuverlässigkeit feststellbar; bei dem Versuch, noch genauere Bestimmungen mit demselben Gerät zu erhalten, treten Abweichungen auf, die als echte Fehler anzusehen sind. Diese Fehler haben eine bestimmte Eigenschaft: sie bleiben bei einem Gerät gleich, werden aber durch Verwendung eines genauer arbeitenden Gerätes geringer. Bei physikalischen Beobachtungen ist der zu untersuchende Gegenstand immer derselbe, aber das Gerät hat einen „Fehler", der sich in Differenzen der Messungen anzeigt. Der gemessene Vorgang ist exakt reproduzierbar, die Messung dagegen nicht.

Bei der Variabilität, die sich bei der Messung biologischer Vorgänge zeigt, liegt dagegen ein grundsätzlich anderer Tatbestand vor: Die Messung ist immer gleich, hinreichend exakte Geräte vorausgesetzt, aber der gemessene Vorgang ist jedesmal anders auf Grund der nie genau reproduzierbaren biologischen Voraussetzungen. Bei Verwendung von Meßgeräten mit größerer Genauigkeit werden diese Abweichungen nicht kleiner, sondern oft, im Gegensatz zum Verhalten der physikalischen Meßfehler, noch deutlicher sichtbar.

Es wird dabei oft gesagt, daß für diese Schwankungen der Meßwerte die zufälligen Abweichungen verantwortlich seien. Diese Zufallsfaktoren sind noch näher zu betrachten. Unter „Zufall" ist in diesem Zusammenhang, wie bereits gesagt, die Summe aller derjenigen Einflüsse zu verstehen, die unbeeinflußbar und unkontrollierbar auf die Versuchsobjekte eingewirkt haben und dadurch die Endreaktion veränderten. Falls es möglich wäre, für jedes einzelne Individuum alle derartigen Einflüsse zu kennen, wäre es auch möglich, damit eine Erklärung für die Richtung und Größe der Abweichung zu geben, die gerade dieses Individuum bei der Messung dieses bestimmten Vorgangs ergibt. Wenn von „Zufall" gesprochen wird, so heißt dies also nicht, daß die betreffenden Vorgänge keine Ursache haben, sondern nur, daß hier viele bekannte und unbekannte Ursachen zusammenwirken, wobei der Einfluß der einzelnen sehr klein und nicht von dem der anderen Faktoren zu trennen ist.

Neben diese beiden Variationsursachen, die Zufallseinflüsse, die nicht auszuschalten sind, und die Versuchsfehler, die soweit wie möglich vermieden werden müssen, treten bei jedem Experiment die Versuchsbedingungen. Diese systematischen Einflüsse sind bei jedem Experiment von besonderer Bedeutung, da der Versuch zu dem Zweck durchgeführt wird, eine Entscheidung darüber zu treffen, ob sie tatsächlich auf den untersuchten Vorgang einen Einfluß ausüben, sowie in welcher Richtung und in welchem Ausmaß sich ihre Wirkung erstreckt. Dies geschieht durch die Beurteilung der durch sie im Verhältnis zu den Zufallseinflüssen hervorgerufenen Variabilität.

Der Weg, dies zu erreichen, ist der, daß die Bestimmungen nicht nur an einem Individuum, sondern an mehreren vorgenommen werden, wobei für die einzelnen die äußeren Bedingungen gleich sein müssen, damit nicht durch die Vergrößerung des Materials zugleich neue Variabilitätsfaktoren eingeführt werden. Dann ist aus den Differenzen zwischen den Bestimmungen, die sich nur im Faktor „verschiedene Individuen" unterscheiden, zu erkennen, welches Ausmaß der Einfluß dieses Faktors erreicht. Nachdem dies bekannt ist, kann dann dieser Einfluß rechnerisch ausgeschaltet werden und durch den Vergleich der Gruppen der Einfluß des Faktors „verschiedene Gruppen" erkannt werden. Alle Methoden, die angewendet werden, gehen auf dieses einfache Prinzip zurück. So wie hier erst ein Faktor ausgeschaltet wurde und dann der Einfluß des anderen geprüft werden kann, so können auch mehrere Variabilitätsfaktoren nacheinander ausgeschaltet und ihr Einfluß auf die Meßwerte untersucht werden. Damit ist die Durchführung von Versuchen, in die mehrere Variabilitätsfaktoren einbezogen werden sollen, möglich gemacht. Zugleich wird dadurch aber deutlich, daß eine so weitgehende Auswertung der Versuchsdaten nur dann geschehen kann, wenn der Versuch vorher im Hinblick auf die Auswertungsmöglichkeiten geplant und durchgeführt wurde. Eine Trennung der Besprechung von Planung und Auswertung der Versuche ist daher nicht möglich, beide bestimmen sich gegenseitig.

### b) Die qualitativen und quantitativen Meßergebnisse.

Für die Planung des Versuchs und die Festlegung der Auswertungsmethoden sind neben anderem zwei Faktoren zu berücksichtigen: die Art der in einem Versuch enthaltenen Variablen und die sich daraus ergebenden Gesichtspunkte, die für die erste Ordnung der Versuchsergebnisse maßgebend sind. Die Versuchsergebnisse können in verschiedener Weise erhalten werden, je nachdem, ob es sich um die Feststellung qualitativer oder quantitativer Differenzen handelt.

Die quantitativen Merkmale werden auf einer Skala gemessen, die stetig oder unstetig sein kann, und die Ergebnisse werden erhalten durch Messen, Wägen, Abzählen, Ablesen einer Skala am Meßgerät oder ähnliche Methoden. Wichtig ist dabei, daß die Skala theoretisch wenigstens nach einer Richtung hin unendlich sein muß, so daß immer die Möglichkeit besteht, bei einer folgenden Messung einen noch größeren oder noch kleineren Wert zu finden als bereits im Material vorliegt.

Bei der zweiten Gruppe wird festgestellt, ob qualitative Differenzen zwischen den einzelnen Individuen bestehen. Dieser Fall ist immer dann gegeben, wenn eine exakte Messung des Individuums für die untersuchte Eigenschaft nicht möglich ist. Diese Gruppe umfaßt demnach alle Versuche, bei denen die Ausbildung eines bestimmten Merkmals oder einer bestimmten Reaktion festgestellt wird. Dabei können eine oder mehrere Klassen aufgestellt werden, für die aber immer charakteristisch ist, daß auch bei linearer Anordnung der Grundsatz der gleichen Abstände der Klassenmittelwerte nicht gewahrt werden kann, weil keine exakte Skala mit genau gleichen Entfernungen zwischen den einzelnen Meßpunkten gebildet werden kann. Ebenso gehören hierher aber alle Versuche, bei denen an sich eine exakte Messung des Ergebnisses möglich wäre, aber aus versuchstechnischen oder anderen wichtigen Gründen darauf verzichtet und nur eine gröbere Klasseneinteilung vorgenommen wird. Manchmal ist es bei diesen qualitativen Verschiedenheiten nicht möglich, eine lineare Anordnung der Einzelbeobachtungen durchzuführen, oft sind aber mehrdimensionale Anordnungen vorhanden, oder es läßt sich zwar eine Klassifizierung durchführen, bei der jede Beobachtung eindeutig einer bestimmten Klasse zuzuordnen ist, während für die Klassen selber eine eindeutige Reihenfolge nicht festzulegen ist, und sie beliebig gegeneinander vertauscht werden können.

Beispiele: Farb- und Formmuster von Blüten und Blättern, Qualitätsunterschiede, Eltern- und Austauschklassen in crossing-over-Versuchen, Klassifizierung von Reaktionen als stark — mittel — schwach oder größer als $x$ — kleiner als $x$, oder schnell — langsam.

Die Beurteilung einer solchen qualitativen Einzelbeobachtung kann in der Weise erfolgen, daß jede für sich mit der vorher festgelegten Skala verglichen und danach benotet bzw. einer Klasse zugewiesen wird, oder so, daß alle Beobachtungen zueinander in Beziehung gesetzt werden und ihre Reihenfolge bestimmt wird. Diese letztere Methode ist nur dann anzuwenden, wenn eine lineare Anordnung der Varianten möglich ist, und läßt sich meist nur bei relativ kleinen Messungsreihen durchführen. Die besondere Bedeutung dieser Aufstellung einer Rangordnung der Ergebnisse liegt darin, daß sie besonders bei kleinen Versuchsreihen in einigen Fällen doch noch eine Auswertung ermöglicht, wenn andere Methoden versagen oder nur mit Einschränkungen angewendet werden können (vgl. Rangkorrelationen, S. 88).

Wenn eine alternative Variabilität mit zwei qualitativ verschiedenen Klassen vorliegt, so wird häufig der Anteil einer der beiden Klassen an der Gesamtzahl als Versuchsergebnis angesehen. Hierbei ist es gleichgültig, welche der beiden Klassen gewertet wird, wenn nur für alle zu vergleichenden Gruppen dieselbe gewählt wird. Die Anzahl der Werte kann umgerechnet werden in relative Häufigkeiten (bezogen auf die Gesamtzahl 1) oder in Prozente (bezogen auf 100), wobei aber das letztere Bezugssystem nur verwendet werden sollte, wenn tatsächlich etwa 100 Beobachtungen vorliegen, um eine Verzerrung des Eindrucks, den die Ergebnisse machen, zu vermeiden. Die so gewonnenen Zahlen werden dann als „Beobachtungen" einer quantitativen Variablen gewertet und für alle weiteren Rechnungen verwendet (vgl. S. 93—94).

## c) Einteilung der Veränderlichen.

Für eine Beurteilung der Variabilität und ihrer Ursachen ist es als nächstes nötig, die Veränderlichen, die in einer statistischen Masse auftreten, näher zu betrachten. Die übliche Einteilung in abhängige und unabhängige Veränderliche kann zwar in großen Umrissen beibehalten werden, aber das biologische Material läßt eine feinere Begriffsbestimmung zu, und für die richtige Lösung der auftretenden Probleme ist eine solche sogar oft erforderlich. Für die Veränderlichen ist es, wie es ihrer Definition entspricht, wichtig, daß sie im Material in verschiedenen Stufen auftreten. Es kann sich dabei um eine kontinuierliche Variabilität handeln, wobei alle Zwischenstufen möglich sind, und nur durch die Messung mit unzureichenden Geräten eine Diskontinuität der Verteilung vorgetäuscht (z. B. Längenmessung, Gewicht usw.) oder eine solche durch eine Einteilung in Größenklassen künstlich geschaffen wird. Es ist auch möglich, daß sich von vornherein eine deutliche Klasseneinteilung durchführen läßt und für jeden Meßwert eindeutig zu bestimmen ist, zu welcher Klasse er gehört, weil Übergänge der Natur des Meßwertes nach unmöglich sind (z. B. die Anzahl der gezählten Einheiten auf einer bestimmten Grundfläche ergibt immer ganze Zahlen). Da sich jede kontinuierliche Variantenreihe durch Einteilung in Klassen in eine wenigstens scheinbar diskontinuierliche Reihe überführen läßt, ist dieser Unterschied aber für die Berechnungen, die auf den Messungen aufbauen, meist ohne Bedeutung.

Wichtiger sind einige andere Eigenschaften der Veränderlichen. Bei einigen sind die Stufen, in denen sie im Material auftreten, nicht zufällig verteilt, sondern vom Beobachter bestimmt. Diese hängen von keinem der übrigen Faktoren ab, sind also im eigentlichen Sinn als „unabhängig" zu bezeichnen, und zwar zur näheren Charakterisierung als „unabhängige, willkürliche Veränderliche". In diese Gruppe gehören alle Faktoren, die als vom Beobachter willkürlich festzusetzende Versuchsbedingungen anzusehen sind: Zeit, Intensität, Dosis, Temperatur. Entscheidend ist, daß alle diese Faktoren keine Zufallsverteilung ergeben, wenn die Stufen und die Häufigkeit ihres Auftretens im Material betrachtet werden. Sie können in verschiedener Weise im Material auftreten, und die Differenzen können sowohl quantitativ wie qualitativ sein.

Zeit: Untersuchung in verschiedenen Zeitabständen vom Versuchsbeginn.

Temperatur: Untersuchung bei verschiedener Temperatur, nach Vorbehandlung bei verschiedener Temperatur usw.

Intensität und Dosis: Versuche über den Einfluß der Lichtmenge oder anderer Strahlung, verschiedene andere quantitative Variable der Versuchsanordnung, z.B. Wellenlänge des Lichtes, Nährstoffkonzentrationen, verschiedene Menge eines Stoffes, verschiedene Abstufungen eines mechanischen Reizes u. a. Diese unabhängigen willkürlichen Veränderlichen können auch qualitativer Natur sein: vergleichende Prüfung verschiedener Stoffe, verschiedener Reize usw.

Demgegenüber stehen Veränderliche, die eine dem Material eigene Variabilität besitzen. Diese Faktoren können zwar eine echte Zufallsverteilung zeigen insofern, als sie nicht durch den Beobachter zu beeinflussen sind und infolgedessen zu den echten „unabhängigen Veränderlichen" gehören; ob sie aber in die Auswertung des Versuchs einbezogen werden müssen oder als echte „Zufallsfaktoren" betrachtet werden können, hängt davon ab, ob sie ihrerseits einen Einfluß auf die Versuchsergebnisse haben. Hierher gehören alle Messungen an unbeeinflußtem Material, die der Feststellung und Beschreibung der Variabilität dienen, weiter aber auch z. B. die Variabilität der Eltern in Vererbungsversuchen, die Variabilität in Größe, Gewicht, Zellenzahl usw. des Versuchsmaterials zu Beginn des Versuchs, wenn die Beeinflussung dieser Faktoren nicht Ziel des Versuchs ist. Dazu rechnen alle meßbaren, quantitativen Merkmale, aber auch qualitative Eigenschaften wie Farbe, Form. Diese alle sind als „zufällige, unabhängige Veränderliche" zu bezeichnen.

Die gleichen Faktoren können aber auch zu den Veränderlichen gehören, deren Abhängigkeit von den Versuchsbedingungen geprüft werden soll. Richtiger ist es aber, zu sagen, daß geprüft werden soll, ob sich die im Material vorhandene Variabilität des betreffenden Merkmals unter dem Einfluß der gewählten Versuchsanordnung verändert. Es sind also keine echten „abhängigen" Veränderlichen, sondern sie sind als „zufällige, beeinflußbare" Variable zu bezeichnen (z. B. Größe, Gewicht, Assimilationsintensität, Mutationsrate).

Die einzige vollständig abhängige Veränderliche ist die Häufigkeit der Beobachtungen in den durch die Kombinationen der übrigen Variationsfaktoren bestimmten Klassen.

### d) Erwünschte und unerwünschte Veränderliche und ihre Berücksichtigung bei der Planung.

In einer statistischen Masse aus biologischem Material sind immer mehrere dieser Variablen enthalten, und es fragt sich, welchen davon für die angestrebte Aussage eine Bedeutung zukommt, da es unmöglich ist, alle zu berücksichtigen. Willkürliche, unabhängige Veränderliche müssen nicht notwendig in einer Versuchsanstellung enthalten sein, andererseits steht es im Belieben des Beobachters, wie viele er davon einbeziehen will. Aus praktischen Gründen ist es zweckmäßig, nicht mehr als drei in einem Versuch zu kombinieren, aber theoretisch können beliebig viele in den Versuchsplan eingebaut werden. Sobald eine im voraus geplante Versuchsreihe vorliegt, wird meist mindestens ein Faktor aus dieser Gruppe darin enthalten sein.

Die zufälligen Veränderlichen sind in jedem Material in großer Menge gegeben, und es hängt vom Beobachter ab, welche davon er in die Versuchsanstellung aufnimmt, und welche Rolle er ihnen zuschreibt. Vor allem ist festzustellen, ob sie als unabhängige oder beeinflußbare Faktoren gewertet werden sollen. Zu unterscheiden ist zwischen den erwünschten Faktoren, deren Einfluß auf das Versuchsergebnis oder deren Beeinflussung interessiert, und den unerwünschten Faktoren, deren Vorhandensein als Störung der Einheitlichkeit des Versuchsmaterials oder des Versuchsplanes gewertet wird. Zu diesen unerwünschten, zufälligen, unabhängigen Veränderlichen gehören z. B. die Blattgröße in einem Assimilationsversuch, die Größe der Gewebestücke bei einem Versuch über die Stoffaufnahme, die Zellgröße bei zellphysiologischen Messungen, Anzahl der Zellen im Gewebe bei Wachstumsversuchen, Anzahl der eingebrachten (eingeimpften) Keime (Sporen, Samen, Zellen) bei Versuchen über die weitere Entwicklung dieser Organe usw.

Eine Ausschaltung des unerwünschten Einflusses ist natürlich dadurch möglich, daß dieser Faktor möglichst konstant gehalten wird, d. h. aber, daß nur die Messungen an den Objekten verwendet werden dürfen, bei denen er einen bestimmten Wert annimmt oder allenfalls in einem sehr eng begrenzten Bereich variiert. Dies wird sich nur dann durchführen lassen, wenn so viel Material vorhanden ist, daß eine Auswahl möglich ist; meist würde aber die dadurch hervorgerufene Einschränkung des Materials die Durchführung des ganzen Versuchs in Frage stellen. Für diesen Fall gibt es mehrere Möglichkeiten, trotz der unerwünschten, als störend empfundenen Variabilität doch das gesamte Material zu verwenden und zu einer exakten statistischen Behandlung der Frage zu gelangen.

Einmal können derartige Faktoren in die Auswertung als unabhängige, zufällige Veränderliche einbezogen werden, so daß ihr Einfluß auf das Versuchsergebnis zu erkennen ist und bei den Schlußfolgerungen ausgeschaltet werden kann. Daneben ist es aber auch möglich, sie als „Zufallsfaktoren" zu betrachten und die durch sie vergrößerte Variabilität als unvermeidliches Übel hinzunehmen. Da in jedem Material sehr viele derartige Faktoren enthalten sind, können diese theoretisch alle in die Berechnung einbezogen werden. Es ist aber zweckmäßiger und beeinträchtigt die Exaktheit des Versuchs in keiner Weise, wenn die wichtigsten konstant gehalten und weiter möglichst viele als Zufallsfaktoren behandelt werden und nicht ausdrücklich als Variable bei den Berechnungen in Erscheinung treten, weil sonst die Berechnungen unnötig kompliziert werden, während die Sicherheit der Schlußfolgerungen in keiner Weise gewinnt. Ihr Einfluß auf das Versuchsergebnis kann aber trotzdem ausgeschaltet werden, wenn der Beobachter darauf achtet, daß diese Faktoren nicht bei einer Versuchsgruppe eine andere Größe besitzen als bei der anderen, d. h. sie müssen tatsächlich dem Zufall nach auf die Einzelversuche verteilt sein. Der Einfluß auf die *Einzelbestimmungen* ist damit zwar nicht ausgeschaltet, aber es ist sicher, daß sie auf alle *Gruppen* von Messungen den gleichen Einfluß ausüben können und damit genau die Eigenschaft besitzen, die von den echten „Zufällen" erwartet wird.

Bei kleinen Messungsreihen, die nur ein geringes Zahlenmaterial liefern und bei denen der Einzelbeobachtung ein relativ großes Gewicht zukommt, wird die erste Methode vorzuziehen sein, da sie eine Erfassung der unerwünschten Einflüsse ermöglicht, aber überall, wo es sich um größere Messungsreihen an zahlreichen Individuen handelt, wird die Genauigkeit der Ergebnisse durch das zweite Verfahren in keiner Weise beeinflußt. Wenn der Weg gewählt wird, die Veränderlichen tatsächlich zu berücksichtigen, so ist es doch erwünscht, dies nur für einen oder zwei Faktoren aus dieser Gruppe zu tun, da sonst die notwendige Rechenarbeit sehr stark zunimmt und die Vermehrung der Arbeit in keinem Verhältnis zum Zuwachs an Exaktheit steht.

Da vielfach versucht wird, die unerwünschten Variationsursachen durch andere Rechenmanipulationen auszuschalten, soll auf diese später nochmals eingegangen werden (s. Abschnitt relative Werte, S. 101 f.).

Von den erwünschten, zufälligen, beeinflußbaren Variablen können dagegen so viele in die Auswertung einbezogen werden, wie für die Schlußfolgerungen notwendig sind. Wie viele und welche Faktoren hierzu ausgewählt werden, ist nicht eine Frage der Statistik, sondern bestimmt sich einmal aus der Struktur des biologischen Materials und der Art der Erkenntnisse, zu denen der Versuch führen soll, zum anderen daraus, ob die für eine zusätzliche Bestimmung an einem einmal vorhandenen Versuchsmaterial aufzuwendende Mehrarbeit einen entsprechenden Zuwachs an Informationen verspricht.

# C. Das Korrelationsproblem in der Biologie.

## I. Korrelationsrechnung.

### a) Die Grundlagen für ihre Anwendung.

In der Physiologie wird sich selten, etwa bei der Erschließung eines neuen Arbeitsgebietes, der Heranziehung eines neuen Objektes oder einer abweichenden Versuchsanstellung, die Aufgabe ergeben, die Variabilität eines Merkmals oder einer Reaktion unter gleichbleibenden äußeren Bedingungen zu beschreiben. Hierfür stehen die statistischen Maßzahlen und Verteilungen zur Verfügung, mit deren Hilfe die gestellte Aufgabe gelöst werden kann, aber alle diese Angaben lassen dann, unabhängig von dem Umfang der Versuchsserie, nur den einen Schluß zu, daß hiermit das Verhalten des Objektes unter den gegebenen Bedingungen charakterisiert ist. Ein Rückschluß auf die Wirkung der Bedingungen, unter denen der Versuch vorgenommen wurde, ist hierbei nicht möglich, weil neben den bekannten noch vielerlei unbekannte Faktoren mitspielen und keine Vergleichsserie über deren Wirken Auskunft gibt.

In den meisten Fällen ist dagegen eine scheinbar ganz andere Aufgabe gestellt, nämlich der Vergleich verschiedener Werte, etwa einer neuen Versuchsserie mit dem durchschnittlichen „Normal"verhalten, oder häufiger der Vergleich von „Versuch" und „Kontrolle" oder verschiedener Versuche untereinander. Während im ersten Fall die Variabilität des Materials erfaßt werden kann, ist im anderen Fall jetzt eine weiter gehende Aussage möglich. Wenn die Versuchsergebnisse sich unterscheiden, ist durch die Anwendung der Versuchsbedingungen eine zusätzliche Variabilität in das Material hineingebracht. Durch den Vergleich der in ihren Ursachen nicht kontrollierbaren Variabilität innerhalb der Serien mit derjenigen zwischen den Serien, die also einen zusätzlichen, aber kontrollierbaren Variationsfaktor enthält, eben die variierenden Versuchsbedingungen, ist eine Aussage darüber möglich, ob die Versuchsbedingungen tatsächlich eine vergrößerte Variabilität zustande gebracht, also „gewirkt" haben. Es ist außerdem eine Aussage über die Größe und Richtung dieser Veränderungen möglich, die ihrerseits wieder Rückschlüsse über die Wirkung der Versuchsbedingungen auf die untersuchte Reaktion zuläßt. Beide scheinbar so verschiedenen Fragestellungen unterscheiden sich also nur dadurch, daß im ersten Fall die Variabilität des Materials als gegeben hingenommen und beschrieben wird, während im zweiten Fall die Aufgabe gestellt ist, den Ursachen dieser Variabilität nachzugehen.

Das Grundproblem, um das es sich hierbei handelt, ist die Frage, ob zwischen den verschiedenen Versuchsbedingungen und der Reaktion der Objekte eine bestimmte Beziehung besteht. In vielen Fällen genügt dabei die Feststellung, ob eine Abhängigkeit zwischen den Versuchsbedingungen und dem Ergebnis vorliegt, aber für die Lösung anderer Fragen ist es notwendig, ein Maß für die

Stärke dieser Abhängigkeit zu gewinnen, um daraus Rückschlüsse auf die Wirkung des Versuchsfaktors zu ziehen. Für die Lösung beider Problemstellungen gibt es mehrere statistische Möglichkeiten, die mathematisch in engem Zusammenhang stehen, die aber so verschieden sind, daß sie bei der Bearbeitung ganz verschiedener biologischer Probleme zur Anwendung kommen. Die Grundlagen bilden in allen Fällen die Fragen der Regression und der Korrelation.

Bei der Beurteilung des Korrelationsproblems bei biologischen Untersuchungen werden oft die Begriffe Korrelation und Korrelationskoeffizient so wenig scharf getrennt, daß der Eindruck entsteht, sie seien fast identisch. Der Begriff der Korrelation ist so weit zu fassen, daß darunter alle Probleme fallen, die mit der gegenseitigen Abhängigkeit und Beeinflussung mehrerer Veränderlichen in Zusammenhang stehen. Der Korrelationskoeffizient ist dagegen nur eine der Möglichkeiten, das Bestehen einer Korrelation durch eine statistische Maßzahl zum Ausdruck zu bringen. Die Problemstellung der Regression ist gegeben, wenn es sich darum handelt, den Einfluß einer unabhängigen Veränderlichen auf den Wert, den eine andere Veränderliche annehmen kann, zu untersuchen. Dabei wird also nur eine einseitige Abhängigkeit geprüft, während sich die Korrelation mit der gegenseitigen Beeinflussung befaßt. Dieser Unterschied kommt sowohl bei der mathematischen Ableitung der Formeln wie bei der Festlegung des Anwendungsbereichs für biologische Untersuchungen zum Ausdruck. Die Regressionskoeffizienten können daher neben dem Korrelationskoeffizienten zur Feststellung einer Abhängigkeit verwendet werden, aber ebensogut sind noch eine ganze Reihe anderer Verfahren möglich. Eine Zusammenstellung der verschiedenen, bisher vorgeschlagenen Korrelationsmaße ist bei GEBELEIN (1943) und bei MITTENECKER (1952) zu finden, wo neben der theoretischen Ableitung der Formeln auch die statistischen Voraussetzungen für die Verwendung der einzelnen Maßzahlen besprochen werden. Der gemeinsame Nachteil fast aller Korrelationsmaße außer den beiden genannten Koeffizienten ist der, daß für sie bisher keine Methoden der Fehlerrechnung ausgearbeitet wurden, so daß eine Prüfung der Zuverlässigkeit der gegebenen Informationen und der Nullhypothese mit Hilfe eines Wahrscheinlichkeitswertes nicht möglich ist. Sie können daher sinnvoll für weitere Schlüsse nur dann verwendet werden, wenn das zugrunde liegende Material sehr umfangreich ist und die Beziehung zwischen den Variablen sehr deutlich hervortritt, so daß ein Zufallsergebnis ausgeschlossen werden kann. Gerade die für eine Entscheidung in biologischen Fragen meist besonders wichtigen Grenzfälle schwacher Korrelationen sind aber deshalb mit diesen statistischen Methoden, auch wenn sie theoretisch sehr gut begründet und dem gegebenen Fall angemessen sind, nicht zu bearbeiten. Man wird daher meist, auch wenn aus theoretischen Gründen einer der anderen Methoden der Vorzug gebührt, doch vorläufig zu einem der Korrelationsmaße greifen, die eine Beurteilung mit Hilfe der Fehlerrechnung ermöglichen.

Neben den Fällen, bei denen eine exakte Beurteilung des Grades der Abhängigkeit gesucht wird, gibt es aber viele Probleme in der Biologie, für deren Lösung es nicht notwendig ist, genaue Aussagen über diesen Punkt zu machen, sondern wo nur das Bestehen oder Fehlen einer Abhängigkeit interessiert. Auch wenn hierfür nicht die eigentliche Korrelationsrechnung, sondern andere, diesem Zweck besser angepaßte Methoden verwendet werden, darf doch nicht vergessen werden, daß es sich immer um das gleiche Korrelationsproblem handelt, also um die Frage, ob eine gegenseitige Abhängigkeit oder Beeinflussung zwischen zwei oder mehreren Veränderlichen besteht, auch wenn auf die Berechnung einer Maßzahl, die den *Grad* dieser Abhängigkeit zum Ausdruck bringt, wegen der Besonderheiten des biologischen Problems verzichtet wird.

## b) Der Regressionskoeffizient.

Eine Berechnung der Regression ist immer dann angezeigt, wenn es sich darum handelt, die Veränderungen einer abhängigen Variablen $y$ von einer unabhängigen, willkürlichen Veränderlichen $x$ so zu untersuchen, daß dabei Aufschlüsse über die Art und den Grad der Abhängigkeit erhalten werden. Zwei Sonderfälle dieser Fragestellung, die in der Physiologie häufig auftreten, sind die Untersuchung von Zeitreihen und von Dosis-Effektreaktionen. Im ersten Fall ist die Zeit die unabhängige Veränderliche, und es wird die Veränderung in der Größe eines Meßwertes bestimmt in Abhängigkeit von der Zeit, die seit Versuchsbeginn verstrichen ist. Zu berechnen ist also die Regression des Meßwertes gegenüber der Zeit. Im zweiten Fall wird die Dosis eines Stoffes, von dem eine Wirkung erwartet wird, in verschiedenen Stufen verändert, und es ist zu untersuchen, ob die Stärke der Reaktion sich in bestimmter Abhängigkeit von der Dosis verändert. Auch hier gilt es, die Regression Effekt/Dosis zu bestimmen.

Unter den verschiedenen statistischen Maßzahlen, die für derartige Fragen ausgearbeitet wurden, ist der Regressionskoeffizient der bekannteste und auch für die meisten Fälle der am besten geeignete Ausdruck der bestehenden Zusammenhänge. Immer dann, wenn eine Regression einer Veränderlichen $y$ in Abhängigkeit von einer anderen vorliegt, ist hierdurch eine Veränderung der Mittelwerte von $y$ mit steigendem $x$ bedingt, während die echte Zufallsstreuung eine Abweichung der einzelnen Meßwerte für $y$ von dem für das betreffende $x$ gültigen Mittelwert bewirkt. Die Gesamtstreuung von $y$ läßt sich daher in die beiden Komponenten zerlegen, nämlich den Anteil, der auf Kosten der Regression geht, und die eigentliche, zufallsbedingte Reststreuung. Eine besonders leicht zu handhabende Berechnungsmethode, die dabei zu sicheren Aussagen führt, ist die varianzanalytische Bearbeitung der Regression (Mather 1949).

Der aus den Meßwerten berechnete Regressionskoeffizient ist aber nur dann geeignet, die bestehende Abhängigkeit richtig wiederzugeben, wenn die Regressionslinie, die die Mittelwerte von $y$ bei wechselndem $x$ verbindet, gerade ist. Die Prüfung der Linearität muß also immer der Berechnung des Regressionskoeffizienten folgen. Bei gekrümmten Regressionslinien, die in biologischem Material viel häufiger sind als gerade, ist es für manche weiteren Berechnungen notwendig, eine Transformation durchzuführen. Hierfür kommt sowohl bei Zeitreihen wie bei Dosis-Effektkurven zunächst die logarithmische Transformation in Frage, und zwar ist jeweils für diejenige Achse des Koordinatensystems die Umformung zuerst durchzuführen, zu der die Regressionslinie sich hinkrümmt. Oft müssen beide Skalen transformiert werden, bis sich die Beziehungen durch eine gestreckte Regressionslinie darstellen lassen. Die Art der hierzu notwendigen Transformationen, wobei unter Umständen verschiedene Möglichkeiten durchprobiert werden müssen, gibt gleichzeitig wertvolle Hinweise auf die Natur der geprüften Abhängigkeit. Bei besonders komplizierten Abhängigkeiten, vor allem dann, wenn Faktoren, die im Versuch variieren, aber nicht berücksichtigt wurden, auf die geprüfte Reaktion einen wesentlichen Einfluß ausüben, ist eine lineare Regressionslinie manchmal nicht zu erreichen. Wenn bei Dosis-Effektkurven die Reaktion in Anteilen der reagierenden Individuen an der Gesamtzahl bestimmt wird, ist hierfür eine besondere Methode ausgearbeitet, die Probittransformation, wobei die damit erhaltene Regressionslinie bereits eine weitgehende Deutung der gesuchten Zusammenhänge ermöglicht (Durchführung s. Mather 1949).

Auswertungsplan für Regressionsbestimmungen:

| Rechenoperation | ergibt Aufschluß über: |
|---|---|
| 1. Zerlegung der Streuung von $y$ in ihre Komponenten „Regression" und „Zufall" | |
| 2. Berechnung des Regressionskoeffizienten | Grad und Art der Abhängigkeit des Faktors $y$ von $x$ |
| 3. Prüfung der Geradlinigkeit | |
| 4. Zeichnung der Regressionslinie | |

Wenn die Regressionslinie nicht geradlinig ist:

| | |
|---|---|
| 5. Transformation einer Skala | |
| 6. Erneute Regressionsrechnung und Prüfung der Linearität | genauere Bestimmung der Art der Abhängigkeit und der Wirkungsweise der unabhängigen Variablen |
| 7. Transformation der anderen Skala | |
| 8. Erneute Regressionsrechnung | |

Schwierig wird die Analyse, wenn der Zeitfaktor zwar in Rechnung gestellt werden muß, aber eigentlich für die Versuchsabsicht als störend empfunden wird und damit zu den unabhängigen, unerwünschten Variablen gehört. Dies ist immer dann gegeben, wenn bei mehrfacher Wiederholung einer Untersuchung eine gesetzmäßige Veränderung zwischen aufeinanderfolgenden Bestimmungen, die der Theorie nach gleich sein sollten, gefunden wird. Hierbei kann natürlich die 1., 2. usw. Bestimmung als zusätzliche Variable in Rechnung gestellt werden. Dies läßt sich aber nicht anwenden, wenn Versuchs- und Kontrollbedingungen abwechselnd nacheinander durchgeführt werden müssen und sich dabei aus Gründen, die im Material oder im technischen Aufbau der verwendeten Versuchsgeräte liegen, eine bestimmte Richtung der aufeinanderfolgenden Abweichungen ergibt. Hier sind je nach der Art des Versuchs verschiedene Auswertungsmethoden möglich, die auf Grund einer richtigen Versuchsplanung das gleiche Ziel, nämlich die Ausschaltung der Regression zwischen der gemessenen Reaktion und der Zeit, erreichen.

Besondere Methoden für die Untersuchung einer Abhängigkeit, die sich durch eine Potenzfunktion ($y = x^2$, $y = x^3$ oder andere) darstellen läßt, sind in der Mutationsforschung zur Feststellung der für eine Reaktion notwendigen Trefferzahlen ausgearbeitet, wodurch besonders die Entscheidung darüber, ob eine Funktion dieser Art vorliegt, und durch welche Potenz sich die Abhängigkeit am besten darstellen läßt, herbeigeführt wird (TIMOFÉEFF-RESSOVSKY 1947). Die dort verwendeten Formeln können auch in allen anderen Fällen, in denen gekrümmte Regressionslinien vorliegen, zur Prüfung der verschiedenen Hypothesen für die Deutung dieser Abhängigkeit verwendet werden.

Aus jeder zweidimensionalen Verteilungstafel lassen sich zwei Regressionskoeffizienten berechnen, von denen für die biologische Regressionsbestimmung immer nur einer von Bedeutung ist, nämlich die Regression der abhängigen gegenüber der unabhängigen Variablen. Der Regressionskoeffizient besagt, daß sich bei einem Fortschreiten auf der Abszisse um 1 der zugehörige Wert von $y$ um den Betrag des Regressionskoeffizienten vergrößert oder verkleinert, in Abhängigkeit vom Vorzeichen. Um einen Anhaltspunkt dafür zu haben, von welchem Wert diese Berechnung auszugehen hat, ist es also noch nötig, einen Punkt des Koordinatensystems anzugeben, durch den die Regressionslinie geht. Dies ist der Wert, den $y$ annimmt beim Mittelwert von $x$. Der andere Regressionskoeffizient, der angibt, um wieviel sich $x$ verändert, wenn man vom Mittelwert $y$ ausgehend auf der $y$-Achse um 1 fortschreitet, ist für die biologische Deutung sinnlos.

Beispiel: Bei einem Versuch über die Stoffaufnahme soll der Gehalt am betreffenden Stoff im untersuchten Gewebe in verschiedenen Zeitabständen vom Versuchsbeginn untersucht werden. Hierbei ist ohne Zweifel die Zeit die unabhängige Variable $x$, die gefundene

Stoffmenge die abhängige Variable $y$. Der Regressionskoeffizient $r_{yx}$ besagt, daß bei einer Verlängerung der Beobachtungszeit um 1 die Menge des im Gewebe vorhandenen Stoffes durchschnittlich um den Wert von $r_{yx}$ zunimmt. Der andere Regressionskoeffizient $r_{xy}$ gibt dann an, wieviel Zeit durchschnittlich vergeht, bis die Stoffmenge 1 aufgenommen ist.

Es ist ein Vorteil des Regressionskoeffizienten, daß die Versuchspunkte nicht gleichmäßig über die $x$-Achse verteilt sein müssen, d.h. daß bei Zeitreihen die Probeentnahme nicht unbedingt in immer gleichen Zeitabständen erfolgen muß, da für die verschiedenen Proben nicht nur der Meßwert, sondern auch die Beobachtungszeit mit in die Rechnung eingeht. Ebenso ist es möglich, die Abhängigkeit einer bestimmten Reaktion von der Größe der Objekte, ihrem Alter oder von anderen Faktoren zu untersuchen, ohne daß es nötig ist, eine völlig regelmäßige Verteilung der Punkte über die $x$-Achse anzustreben, wenn auch eine gleichmäßige Berücksichtigung aller Abschnitte erwünscht ist. Bei der Schlußfolgerung ist zu beachten, daß in den meisten Fällen ein solcher Versuch nur Aufschluß gibt über den Verlauf der Regression im untersuchten Bereich der $x$-Skala. Wenn dagegen aus theoretischen Überlegungen zu folgern ist, daß die Regressionslinie unbedingt durch einen bestimmten Punkt außerhalb dieses Bereichs gehen muß, etwa den Nullpunkt des Koordinatensystems, so kann diese Annahme als Prüfhypothese formuliert werden. Die zu prüfende Nullhypothese lautet dann: Der Punkt, in dem die berechnete Regressionslinie die $y$-Achse schneidet, ist nur zufällig von Null verschieden. Durch die Bestätigung oder Verwerfung dieser Hypothese wird dann festgestellt, ob die Regressionslinie im Bereich zwischen Null und dem ersten Prüfpunkt der Abszissenskala gerade verläuft, und gibt damit die Möglichkeit, über die Zusammenhänge zwischen den beiden Veränderlichen in diesem, im eigentlichen Versuch nicht geprüften Bereich etwas auszusagen (Beispiel bei Mather 1951).

Die Verwendung des Regressionskoeffizienten für Aussagen über den Zusammenhang zwischen zwei Veränderlichen ist also immer dann angebracht, wenn es sich um eine willkürliche, unabhängige Variable handelt, deren Einfluß auf eine quantitativ faßbare und exakt meßbare Reaktion geprüft werden soll.

### c) Der Korrelationskoeffizient.

Der Korrelationskoeffizient hängt mit dem Regressionskoeffizienten in der Weise zusammen, daß er definiert wird als das geometrische Mittel aus den beiden Regressionskoeffizienten, die aus einer Verteilungstafel berechnet werden können. Trotz dieser eindeutigen mathematischen Zusammenhänge beider Maßzahlen ist ihre Verwendung an ganz verschiedenartige Probleme geknüpft. Der Korrelationskoeffizient wird überall dort verwendet, wo es darum geht, den Zusammenhang zwischen zwei zufälligen Veränderlichen darzustellen, der sich in den Veränderungen einer dritten Variablen, nämlich der Häufigkeit in den durch die verschiedenen Kombinationen der beiden Veränderlichen gebildeten Klassen, zum Ausdruck bringt. Aus einer Tafel, die diese Beziehungen darstellt, können natürlich zuerst die Regressionskoeffizienten berechnet werden, und in manchen Fällen werden diese zur Charakterisierung der Verteilung und als Grundlage für die Schlußfolgerungen ausreichen, aber sie sind für diesen Zweck unzureichend, wenn unter den beiden Veränderlichen nicht eine eindeutig als „unabhängig", die andere als „abhängig" klassifiziert werden kann. Es wird in einem solchen Fall, in dem auf Ordinate und Abszisse zwei Veränderliche zu verteilen sind, deren Rollen ohne weiteres vertauscht werden könnten, die Verwendung der Regressionskoeffizienten eine unzulässige Bevorzugung jeweils einer dieser Veränderlichen bedeuten. Der Korrelationskoeffizient schaltet diese

Schwierigkeit dadurch aus, daß beide Variablen in gleicher Weise in die Berechnungen eingehen.

Ebenso wie aus jeder zweidimensionalen Verteilungstafel die Regressions-koeffizienten berechnet werden können, es aber nicht in allen Fällen möglich ist, die so gewonnenen Werte als zuverlässige Charakteristik der Verteilung anzusehen, so darf auch nicht in jedem Fall von den Regressionskoeffizienten aus zur Berechnung des Korrelationskoeffizienten übergegangen werden. Es müssen mehrere Voraussetzungen dafür erfüllt sein, wenn der Korrelations-koeffizient ein richtiges Bild der bestehenden Zusammenhänge geben soll. Zuerst muß für beide Veränderliche, wenn sie unabhängig voneinander betrachtet werden, die Verteilung der Häufigkeiten auf die einzelnen Klassen einer Normal-verteilung folgen; außerdem müssen die Regressionslinien im geprüften Bereich linear sein. Wenn gekrümmte Regressionslinien vorliegen, kann durch eine Transformation der Skala versucht werden, eine Beziehung zu finden, die sich durch lineare Regressionslinien darstellen läßt, aber in den meisten Fällen wird dies schwierig sein. Wenn für eine dieser beiden Voraussetzungen nur geringe Abweichungen vorliegen, müssen diese bei der Beurteilung des gefundenen Korrelationskoeffizienten berücksichtigt werden, da er hierdurch zu groß oder häufiger zu klein ausfallen kann und dann als Maß für die untersuchte Abhängig-keit nur bedingt brauchbar ist. Wenn die Abweichungen sehr groß sind, gibt der Korrelationskoeffizient unter Umständen ein völlig falsches Bild, wobei vor allem zu beachten ist, daß eine durch gekrümmte Regressionslinien charak-terisierte Abhängigkeit bei der Berechnung des Korrelationskoeffizienten völlig verschwinden kann, was seinen Grund darin hat, daß hierbei von einer nicht zutreffenden hypothetischen Voraussetzung, nämlich der Linearität der Re-gressionslinien, ausgegangen wurde. In derartigen Fällen ist es also besser, allen weiteren Schlüssen nicht die Regressionskoeffizienten, sondern die Regres-sionslinien zugrunde zu legen. Weil bei gekrümmten Regressionslinien die Re-gressionskoeffizienten kein genaues Bild der bestehenden gegenseitigen Be-ziehungen der beiden Veränderlichen geben, kann der aus ihnen abgeleitete Korrelationskoeffizient nicht als exaktes Maß zur Beurteilung der vorliegenden Verteilung angesehen werden. Daran ändert sich nichts, wenn bei der Berechnung die Regressionskoeffizienten umgangen wurden und direkt der Korrelations-koeffizient bestimmt wurde.

Ein wesentlicher Nachteil des Korrelationskoeffizienten ist es, daß mit seiner Hilfe sich nie eine Unabhängigkeit beider Merkmale nachweisen oder auch nur wahrscheinlich machen läßt, sondern nur das Bestehen einer Abhängig-keit. Bei völliger Unabhängigkeit beider Veränderlichen wird $r = 0$, aber dieser Schluß ist nicht umkehrbar, so daß aus dem Befund $r = 0$ nicht auf das Fehlen einer Abhängigkeit geschlossen werden darf. Auch bei einer deutlichen Abhängig-keit kann $r$ den Wert 0 annehmen, wenn nämlich in bestimmter Weise gekrümmte Regressionslinien vorliegen. Aus $r = 1$ kann dagegen auf völlige Abhängigkeit geschlossen werden, aber wieder ist keine Umkehrung möglich, weil auch bei Bestehen einer derartig engen Beziehung $r$ kleinere Werte annehmen kann, wenn nämlich Besonderheiten der Verteilung einer oder beider Veränderlichen gegeben sind.

Zu beachten ist ferner, daß die Werte für $x$ und $y$ in die Berechnung ein-gehen. Sowohl der Korrelations- wie die Regressionskoeffizienten dürfen daher streng genommen nur dann berechnet werden, wenn für beide Variablen eine Verteilung mit quantitativ faßbaren, zahlenmäßig auszudrückenden Varianten vorliegt. In Fällen mit qualitativen Merkmalen müssen den Varianten Zahlen-werte zugeordnet werden, was immer willkürlich ist, so daß der Korrelationskoeffizient dann fast jede beliebige Größe annehmen kann, je nach der Bewertung der einzelnen Varianten. Da in vielen Fällen die anderen Korrelationsmaße, die diese Schwierigkeit vermeiden, andere Nachteile haben, durch die sie für den gerade vorliegenden Fall noch ungeeigneter erscheinen, wird in solchen Fällen meist so vorgegangen, daß angenommen wird, alle Klassenabstände zwischen den Mittelpunkten der qualitativ verschiedenen, aber linear anzuordnenden Klassen seien gleich; den Differenzen wird der Wert 1 zuerkannt, und die mittlere Klasse jeder Veränderlichen wird als 0-Punkt der Skala angesehen. Wenn bei den Schlußfolgerungen beachtet wird, daß diese Annahmen alle hypothetisch sind und nur in den wenigsten Fällen ohne Einschränkung zutreffen werden, können die so gewonnenen Korrelations- und Regressionskoeffizienten zwar nicht als objektiv richtige Maße für die Abhängigkeit gelten, sind aber als relatives Maß für den Vergleich zwischen gleichartigen oder vergleichbaren Verteilungen zu verwenden. Es ist jedoch in jedem Fall zu überlegen, ob nicht andere statistische Methoden, die nicht auf unbewiesenen Hypothesen aufbauen müssen, für den jeweiligen Fall besser geeignet sind. Derartige Methoden für die Untersuchung von qualitativen Differenzen werden in einem späteren Abschnitt besprochen ($\chi^2$-Test, S. 100).

### d) Rangkorrelationen.

Einen Sonderfall der Korrelationen stellt die Berechnung der Rangkorrelationen dar. Die Methode geht aus von der bereits erwähnten Rangordnung der Beobachtungen und kann sowohl bei quantitativen wie bei qualitativen Veränderlichen verwendet werden, kommt aber vor allem für die letzteren in Frage, wenn an einem Material zwei oder mehr Eigenschaften begutachtet wurden und festgestellt werden soll, ob diese beiden in einer gegenseitigen Abhängigkeit stehen. Auf diese Weise sind auch Bestimmungen an Versuchsreihen relativ geringen Umfangs möglich, wenn auch die Sicherheit der Schlußfolgerungen mit wachsender Beobachtungszahl steigt (Kendall, Weber).

### e) Mehrfachkorrelationen.

Oft sind an einem Material mehrere Messungen durchgeführt zum Zweck der Bestimmung der gegenseitigen Abhängigkeit. Wenn dabei Korrelationen gefunden werden, besteht manchmal der Verdacht, daß eine enge Korrelation zwischen zwei Faktoren nur dadurch zustande kommt, daß sie untereinander zwar unabhängig sind, aber beide von einem dritten Faktor beeinflußt werden. Um die Wirkung eines solchen Einflusses auszuschalten, wäre es natürlich möglich, nur diejenigen Messungen zu verwenden, die einen bestimmten Wert dieses Faktors zeigen, und an diesen dann die Korrelation zwischen den beiden übrigen Faktoren zu berechnen. Dabei erfolgt aber immer eine starke Verkleinerung des Materials, und es wird zudem, falls die vermuteten Beziehungen tatsächlich bestehen, auch noch ein falsches Bild der Verhältnisse zwischen den beiden restlichen Faktoren erhalten, da eine Auswahl getroffen wurde. Es liegt also das gleiche Problem vor, wie am Anfang der Korrelationsrechnung: es soll das ganze Material zur Begutachtung herangezogen werden, aber der Einfluß einer Variablen auf die Ausbildung eines anderen Merkmals, hier der Korrelation zwischen zwei anderen Faktoren, soll ausgeschaltet werden.

Eine rechnerische Möglichkeit zur Erreichung dieses Ziels ist die Berechnung der Partialregressionen und der Partialkorrelationen. Zunächst wird dabei die Korrelation zwischen je zwei der Faktoren bestimmt, so als ob der andere nicht im Material enthalten wäre. Dann folgt die Ausschaltung jeweils des zunächst vernachlässigten Faktors. In jedem Fall ist zuerst zu prüfen, welche Partialkorrelationen sinnvoll sind. In entsprechender Weise können auch mehrere Veränderliche ausgeschaltet werden, indem die Werte für die Partialkorrelationen als neuer Ausgangspunkt für die Berechnungen verwendet werden. Bei der Durchführung dieser Berechnungen ist aber eins zu beachten. Wenn für den Korrelationskoeffizienten Werte erhalten wurden, die nicht sehr groß sind, ist eine statistische Sicherung meist nur mit einem sehr großen Material möglich. Diese hierdurch entstehende Unsicherheit der Schlußfolgerungen bedingt, daß meist über 100, oft noch mehr Beobachtungen nötig sind, um zu gesicherten Schlüssen zu kommen. Bei der Berechnung der Partialkorrelationen werden je drei mit einem Unsicherheitsfaktor behaftete Korrelationskoeffizienten zueinander in Beziehung gesetzt. Damit wird die Sicherheit der Aussage aber so stark beeinträchtigt, daß als Mindestwert etwa 500 Beobachtungen angesehen werden können. Wenn weniger Beobachtungen vorliegen und die Korrelationskoeffizienten nicht sehr groß sind, muß von einer Berechnung der Partialkorrelationen abgesehen und ein anderer Weg der Auswertung gesucht werden. Außerdem ist zu berücksichtigen, daß sich die Anzahl der Rechenoperationen mit jeder Ausschaltung eines Faktors sehr stark steigert und die Sicherheit der Bestimmung der Partialkorrelation und damit die Beweiskraft für die Schlußfolgerungen sehr stark abnimmt. Eine Berechnung dieser Art, vor allem wenn es um die Ausschaltung von mehr als einem Faktor geht, wird daher nur in Ausnahmefällen zu empfehlen sein. Meist wird sich mit einiger Überlegung wohl eine andere Versuchsanordnung finden lassen, die mit weniger Rechenarbeit und mehr Sicherheit zu den gewünschten Aussagen über die gegenseitigen Beziehungen der einzelnen Faktoren führt.

Außerdem ist bei der Berechnung der Partialkorrelationen noch besonders darauf zu achten, daß die Voraussetzungen für die Berechnung des Korrelationskoeffizienten mit Sicherheit erfüllt sind. Wenn bei der Bestimmung der Korrelationskoeffizienten noch geringe Abweichungen von der Forderung der Normalverteilung für alle Variablen und der linearen Regression gestattet werden konnten, so führt die Berechnung der Partialkorrelationen zu völlig falschen Ergebnissen, wenn diese Voraussetzungen nicht erfüllt sind. Es ist also in jedem einzelnen Fall zu prüfen, ob sie tatsächlich zutreffen. Da streng lineare Regressionen bei biologischem Material zu den Seltenheiten gehören, wird sich meist aus diesem Grunde eine Berechnung der Partialkorrelationen verbieten.

Eine größere Bedeutung haben dagegen die Partialregressionen, die vor allem dann zur Anwendung kommen, wenn es darum geht, den Einfluß von zwei Veränderlichen, die zur Klasse der unabhängigen, willkürlichen Variablen gehören, auf eine dritte, abhängige Veränderliche zu prüfen und es nicht möglich ist, alle möglichen Kombinationen zwischen den beiden unabhängigen Veränderlichen an einem größeren Material zu untersuchen, so daß nur Einzelbeobachtungen bei willkürlichen Kombinationen der beiden unabhängigen Veränderlichen zur Verfügung stehen (Beispiel bei MATHER 1949, FISHER 1949). Da hierbei von den einschränkenden Voraussetzungen, die für die Partialkorrelationen beachtet werden mußten, einige wegfallen, lassen sich viele Fragen hiermit bearbeiten, und durch die Kenntnis dieser Methode kann in manchen Fällen eine kompliziertere, umfangreichere Versuchsplanung unterbleiben. Vor allem wird sich die Berechnung der Partialregressionen (multiple regression) dann empfehlen, wenn ein vollständiger Kombinationsversuch durch die Besonderheiten des Materials auf technische Schwierigkeiten stößt und gleichzeitig besonderer Wert nicht nur auf den Nachweis einer Beeinflussung, sondern auch auf die Kenntnis des Grades der Abhängigkeit gelegt wird.

### f) Mittlerer Fehler von Regressions- und Korrelationskoeffizient.

Sowohl für die Regressionskoeffizienten wie für den Korrelationskoeffizienten lassen sich mittlere Fehler berechnen, aus denen mit Hilfe des $t$-Testes die Sicherheit bestimmt werden kann, mit der aus den berechneten Werten Schlüsse gezogen werden dürfen. Die zugrunde liegende Nullhypothese ist die, daß keine Abhängigkeit zwischen den untersuchten Faktoren besteht, die Abweichung der berechneten Werte von Null also nur zufällig ist. Beim Korrelationskoeffizienten ist aber zu beachten, daß dieser mittlere Fehler aus statistischen Gründen zur Beurteilung nicht geeignet ist. Vielmehr ist es zu empfehlen, mit Hilfe von Tabellen eine Umrechnung in die Größe $z$ vorzunehmen, deren mittlerer Fehler leicht zu bestimmen ist und eine exakte Beurteilung zuläßt. Weil die Sicherheitskoeffizienten für beide Regressions- und den Korrelationskoeffizienten einer Verteilung gleich sind, ist es nur nötig, für einen von ihnen die Prüfung durchzuführen, um zugleich für die anderen eine Aussage machen zu können. Andererseits ist die Berechnung des $t$-Wertes für beide Regressionskoeffizienten eine gute Kontrolle für die Richtigkeit der Berechnungen, da bei Abweichungen auf Rechenfehler geschlossen werden kann.

Mit Hilfe der mittleren Fehler können auch zwei Regressionskoeffizienten oder Korrelationskoeffizienten bzw. deren $z$-Werte miteinander verglichen werden, wenn es um die Feststellung geht, ob die Regression in zwei verschiedenen Versuchen gleich ist. Sinnvoll ist dieser Vergleich nur dann, wenn es sich um die Regressionskoeffizienten zwischen gleichen oder wenigstens vergleichbaren· Veränderlichen handelt und geprüft werden soll, ob ein dritter Faktor, durch den sich die zwei Versuche unterscheiden, einen Einfluß auf den Grad der geprüften Abhängigkeit hat, oder ob zwei Faktoren den gleichen Vorgang in verschieden starkem Maße beeinflussen (Berechnung bei Mather 1949).

Beispiel. Es wird der Einfluß der Temperatur ($x$) auf einen bestimmten Vorgang ($y$) untersucht und mit Hilfe des Regressionskoeffizienten $r_{yx}$ die Temperaturabhängigkeit ausgedrückt. Der Versuch wird mit genau gleicher Anordnung bei anderer Lichtintensität wiederholt, und es ist zu prüfen, ob das Licht einen Einfluß ausübt in der Weise, daß etwa der Grad der Veränderung des untersuchten Vorgangs durch die Temperaturerhöhung von der Belichtung abhängt. Der Vergleich der beiden Regressionskoeffizienten, die aus den Versuchen mit starker und schwacher Lichtintensität gewonnen wurden, kann hier Aufschluß geben.

### g) Diskriminanzanalyse.

Um den Mittelpunkt der Korrelationstafel, der durch die Mittelwerte von $x$ und $y$ bestimmt ist, lassen sich mit Hilfe der Regressionslinien bei linearer Regression sog. Korrelationsellipsen zeichnen in der Weise, daß innerhalb des durch die Ellipse abgegrenzten Teils der Verteilungstafel ein bestimmter Anteil der Beobachtungen liegt. Von Bedeutung sind hierbei nur die Ellipsen, die nach dem Anteil des Gesamtmaterials, den sie einschließen, der $1\sigma$, $2\sigma$ und $3\sigma$ Grenze entsprechen, also 66%, 95% und 99,7% des Materials umfassen. Wenn eine einzelne neue Beobachtung hinzukommt, von der die Größe der zusammengehörenden $x$- und $y$-Werte bekannt ist, kann nach ihrer Lage zu diesen Korrelationsellipsen bestimmt werden, mit welcher Wahrscheinlichkeit sie zu dem bereits bekannten Kollektiv gehört oder als diesem Kollektiv fremd anzusehen ist. Von Bedeutung für biologische Untersuchungen wird diese Möglichkeit praktisch nur dann werden, wenn es darum geht, für eine neue Beobachtung zu entscheiden, welcher von zwei durch die Verteilung von $x$, $y$ und dem Korrelationskoeffizienten charakterisierten statistischen Massen sie zuzuordnen ist.

Dies ist aber eigentlich ein Sonderfall, der sich in eine allgemeinere statistische Frage einordnen läßt. Dabei handelt es sich um folgendes Problem: Es sind zwei Kollektive gegeben, die durch die Verteilung einer Reihe von Veränderlichen $x$, $y$, $z$ usw. und deren Korrelationskoeffizienten charakterisiert sind. Für eine neu hinzukommende Beobachtung, für die die Werte für $x$, $y$, $z$ usw. gemessen wurden, soll bestimmt werden, welchem der beiden Kollektive sie mit der größeren Wahrscheinlichkeit zuzuordnen ist. Diese Frage ist mathematisch verwandt mit dem Problem der Partialregressionen und wird gelöst mit Hilfe der *Diskriminanzanalyse*. Der Unterschied zwischen beiden Methoden in der Bedeutung für biologische Probleme liegt im Anwendungsbereich, der sich durch die statistischen Möglichkeiten dieser neuen Methode bietet. In der Physiologie wurde die Methode bisher noch nicht angewendet. Da das Problem, für das sie ausgearbeitet wurde, aber auch hier vorkommt, dürfte ihre Anwendung immer dann angezeigt sein, wenn es darum geht, für eine einzelne neue Beobachtung die durch mehrere Meßwerte charakterisiert ist, zu entscheiden, welcher von zwei oder mehreren Reaktionsgruppen sie zuzuordnen ist, wenn die Gruppen durch eine bestimmte Variation dieser Merkmale und ihre Korrelationen beschrieben werden können (discriminant functions: Mather 1949, Fisher 1950).

### h) Anwendung der Korrelationsmethoden.

Für die Anwendung der verschiedenen statistischen Methoden ist es entscheidend, zuerst die Art der im Versuch enthaltenen Veränderlichen zu bestimmen. Der *Regressionskoeffizient* wird immer dann geeignet sein, wenn die Häufigkeit der Beobachtungen gemessen wird in Klassen, die durch die Kombination einer willkürlichen mit einer zufälligen, beeinflußbaren Veränderlichen entstehen, wobei es durchaus möglich ist, daß die Häufigkeit in allen Klassen nur 0 oder 1 ist. Wenn mehrere willkürliche Veränderliche vorhanden sind oder der Einfluß einer unabhängigen, zufälligen Veränderlichen ausgeschaltet werden soll, ist unter Umständen die Berechnung der partiellen Regressionen durchzuführen, wobei zu beachten ist, daß die Partialregressionen nur dann berechtigt sind und richtige Auskünfte geben, wenn im bearbeiteten Bereich alle einfachen Regressionen geradlinig sind. Die Linearität ist also zuerst für alle Regressionslinien zu prüfen. Der *Korrelationskoeffizient* ist dagegen das gegebene Maß, wenn beide Veränderliche zu den zufälligen, unabhängigen Variablen gehören und die Häufigkeit in den durch die Kombination entstehenden Klassen als abhängige Variable gemessen wird. Die Verwendung des Korrelationskoeffizienten für die sog. Verwandtenkorrelation (Eltern-Kinderkorrelation, Geschwisterkorrelationen usw.) ist dagegen zwar weit verbreitet, aber es liegen viele Beobachtungen vor, die die Gültigkeit derartiger Schlußfolgerungen auf Grund umfangreicher Versuche in Zweifel ziehen, so daß zumindest große Vorsicht bei der weiteren Verwendung geboten ist (Haskall 1952, Bateman u. Mather 1951).

## II. Auswertung von Korrelationen ohne Berechnung der Koeffizienten.

### a) Grundfragen.

Die dargestellten Methoden der Regressions- und Korrelationsrechnung beziehen sich alle auf Untersuchungen, bei denen es darauf ankommt, die Größe einer bestehenden Abhängigkeit zum Ausdruck zu bringen. Vielfach stellt sich aber in der Physiologie die Frage, festzustellen, ob ein bestimmter Faktor einen Einfluß auf eine Reaktion ausübt, ohne daß es interessiert, ein genaues Maß für die Größe der Veränderung zu haben, die bei einer Veränderung des untersuchten Faktors um 1 zu erwarten ist. In solchen Fällen ist es überflüssig, die

Regressionskoeffizienten zu berechnen, sondern es genügt eine Auswertungsmethode, die erkennen läßt, ob dem fraglichen Faktor eine Wirkung zukommt, und die außerdem noch eine Aussage ermöglicht, ob die Differenzen zwischen bestimmten Einzelversuchen der ganzen Serie als gesichert angesehen werden können. Es soll also nur geprüft werden, ob durch eine Veränderung der Größe dieses Faktors innerhalb einer Versuchsserie eine zusätzliche Variable eingeführt und die Gesamtvariabilität hierdurch vergrößert wird. Das zu lösende statistische Problem ist in diesem Fall das, die gefundene Gesamtvariabilität so zu zerlegen, daß erkannt werden kann, welcher Einfluß den beiden Komponenten „Zufallsstreuung" und „Einfluß des variierenden Faktors" zukommt.

Die Grundlage für die Auswertung bildet demnach die Streuung der Werte im Material. Die Methode, die im Einzelfall angewendet wird, hängt davon ab, ob es sich um qualitativ faßbare oder nur qualitativ zu beschreibende Merkmale handelt. Für Messungen an quantitativen Merkmalen wird der $z$-Test in der Form der Varianzanalyse verwendet, bei qualitativen Merkmalen dagegen der $\chi^2$-Test. Es wird also der Einfluß geprüft, den unabhängige, willkürliche Veränderliche ausüben, im ersten Fall auf eine zufällige, beeinflußbare Variable, deren Größe gemessen wird, in zweiten Fall auf eine abhängige Veränderliche, nämlich die Häufigkeit in den einzelnen qualitativ verschiedenen Klassen. Die Kenntnis der Natur der im Versuch enthaltenen Variablen ist für die Entscheidung über die Wahl einer der beiden Berechnungsweisen notwendig. Beide Auswertungsmethoden machen es möglich, nicht nur, wie beim $t$-Test, eine einzige Differenz zwischen „Versuch" und „Kontrolle" zu vergleichen, sondern der zu variierende Versuchsfaktor kann in mehreren Stufen angewendet werden, wobei alle Ergebnisse zusammen ausgewertet werden. Der Unterschied liegt aber nicht nur hierin. Bei Verwendung des $t$-Testes werden die beiden zu vergleichenden Gruppen zunächst getrennt behandelt, als verschiedene Stichproben, für die zu prüfen ist, ob sie noch aus einer einheitlichen statistischen Masse herstammen können, während bei den beiden genannten Methoden das gesamte Material als Einheit angesehen und geprüft wird, welchen Einfluß die verschiedenen Variabilitätsursachen hierauf gehabt haben.

## b) Varianzanalyse.
### α) Bei quantitativen Merkmalen.

Für die Auswertung von Versuchen mit quantitativen Meßergebnissen wird dabei so vorgegangen, daß zunächst die Varianz, das Quadrat der Streuung, für das ganze Material bestimmt und so die gesamte Variabilität des Versuchs erfaßt wird. Wichtig ist, daß mit Wiederholungen gearbeitet wird; jeder Teilversuch muß mehrmals im gesamten Material vorkommen. Die Grundlage der Auswertung ist dann die Zerlegung der Gesamtvarianz in die Variabilität zwischen den verschiedenen Gruppen und innerhalb der gleichartig behandelten Gruppen. Bei einer getrennten Auswertung jedes einzelnen Teilversuchs steht für die Bestimmung der Zufallsstreuung nur die Anzahl von Beobachtungen zur Verfügung, die diese einzelne Gruppe umfaßt. Im Rahmen der Varianzanalyse wird nun nicht für jeden Teilversuch die Zufallsstreuung getrennt bestimmt, sondern es wird eine mittlere Zufallsstreuung berechnet, für die die Summe der mittleren quadratischen Abweichungen aus allen Teilversuchen zur Verfügung steht, die also mit einer viel größeren Anzahl von Freiheitsgraden versehen ist und daher Anspruch auf eine größere Genauigkeit erheben kann.

Wenn zwischen den zu vergleichenden Versuchen keine zusätzliche Variabilitätsursache wirksam war, muß die Streuung zwischen ihnen ausschließlich auf der Wirkung der Zufallsfaktoren beruhen und darf auch nur im Rahmen des zufällig Möglichen von der Streuung innerhalb der Gruppen, die bestimmt ausschließlich von den Zufallseinwirkungen abhängig war, abweichen. Wenn dagegen der Faktor, durch den sich die Teilversuche unterscheiden, eine zusätzliche Variabilitätsursache darstellt, muß sich dies in einer gegenüber der Zufallsstreuung vergrößerten Varianz zwischen den Gruppen anzeigen. Verglichen werden demnach die Varianzen „zwischen den Gruppen" und „innerhalb der Gruppen". Die gegebene Prüffunktion hierfür ist das Varianzverhältnis, die z-Funktion. Diese ist in verschiedenartigen Tafeln tabelliert, die alle eine Aussage ergeben darüber, mit welcher Wahrscheinlichkeit gesagt werden kann, daß eine der beiden verglichenen Varianzen größer ist als die andere. Wenn die Varianz „zwischen den Gruppen" um so viel größer ist, daß eine zufällige Abweichung sehr unwahrscheinlich wird, kann angenommen werden, daß sich die Gruppen durch eine zusätzliche Variabilitätsursache unterscheiden. Wenn nun die Durchführung des Versuchs derart war, daß zwischen den Gruppen nur eine kontrollierte Differenz in den Versuchsbedingungen bestand, dann kann daraus geschlossen werden, daß dieser Unterschied in der Behandlung der einzelnen Gruppen die Ursache der aufgefundenen zusätzlichen Variabilität ist, also im Sinne der Versuchsabsicht „gewirkt" hat. Sind dagegen beide Varianzen im Rahmen des Zufalls gleich, dann ist die Folgerung daraus, daß die Versuchsbedingungen entweder unwirksam waren, oder daß der Versuchsumfang nicht ausreicht, um eine reale Differenz von der gefundenen Größe nachzuweisen.

Schwierig wird die Deutung, wenn die Binnenklassenvarianz die größere ist. Dann muß innerhalb der gleichartig behandelten Gruppen eine systematisch wirkende Variabilitätsursache vorhanden sein, die durch die Zusammenfassung der zusammengehörenden Teilversuche ausgeschaltet wurde. In einem solchen Fall ist also immer nach einem Versuchsfehler zu fahnden, der in einer solchen Weise gewirkt haben könnte. Wenn ein solcher nicht gefunden werden kann, so sind die Versuche nicht wertlos, aber es muß bei den weiteren Schlußfolgerungen damit gerechnet werden, daß dieser versteckte Faktor vorhanden ist.

### β) Varianzanalyse mit Verhältniszahlen.

Die Methode der Varianzanalyse ist immer dann anwendbar, wenn die Versuchsergebnisse exakt gemessen oder gezählt werden können, kann aber auch durchgeführt werden mit Zahlen, die den Anteil der reagierenden Individuen unter einer bestimmten Anzahl angeben, wenn diese Gesamtanzahl für die zusammenfassenden Einzelversuche immer gleich ist. Zum Beispiel von $n$ Zellen sind jeweils $x$ in Mitose, $n-x$ Ruhekerne, oder: von $n$ Individuen sind $x$ abgestorben, $n-x$ überleben. Auch hier können die Originalzahlen den Berechnungen zugrunde gelegt werden, aber völlig exakt ist dies nicht, da es für die Deutung der gefundenen Abweichungen zwischen den Einzelversuchen nicht gleichgültig sein kann, ob die gesuchte Reaktion beobachtet wurde etwa bei „8 Objekten von 10" oder bei „8 Objekten von 500". Es ist daher richtiger, derartige Werte nicht als absolute, sondern als relative Häufigkeiten zur Grundlage der weiteren Berechnungen zu nehmen, im gegebenen Fall also $^8/_{10}$ oder $^8/_{500}$. Die Berechnung der relativen Häufigkeit, ebenso wie die sehr oft verwendeten Prozentzahlen, hat aber den Nachteil, daß jeder Wert mit einer anderen Zufallsstreuung behaftet ist, auch wenn alle Bestimmungen mit der gleichen Individuenzahl ausgeführt wurden. Dadurch wird die Auswertung derartiger Versuchsergebnisse mit theoretischen Bedenken belastet, die eine exakte Ausdeutung der Versuche erschweren. Wenn die Werte der Einzelversuche sehr dicht zusammenliegen, sind die praktischen Auswirkungen dieser theoretischen Schwierigkeiten so gering, daß sie bei den Schlußfolgerungen vernachlässigt werden können; sie sind aber zu berücksichtigen, sobald größere Differenzen auftreten.

Alle diese Komplikationen können umgangen werden, wenn statt der Umrechnung in Prozent oder relative Häufigkeiten eine Winkeltransformation durchgeführt wird in der Weise, daß die relative Häufigkeit $p$ ersetzt wird durch einen Winkel $\Phi$, so daß $p = \sin^2 \Phi$ ist. Diese Zahlen sind zwar weniger anschaulich als die bekannten Prozentzahlen, haben aber den Vorteil, daß alle Werte, die sie annehmen können, mit der gleichen Zufallsstreuung behaftet sind, da bei der Berechnung des mittleren Fehlers dieser Zahlen nur noch die Individuenzahl des Versuchs und nicht mehr der absolute Wert der Messung verwendet wird. Sie können daher ohne Bedenken einer statistischen Analyse zugrunde gelegt werden.

Voraussetzung ist allerdings, daß alle Beobachtungswerte aus Gruppen mit gleicher Individuenzahl gewonnen wurden! Hieraus ergibt sich aber die Bedingung, bei allen derartigen Versuchen so zu planen, daß alle Beobachtungsgruppen (Einzelversuche), aus denen die relative Häufigkeit der „günstigen" Fälle bestimmt werden soll, die gleiche Individuenzahl aufweisen. Nur wenn die Individuenzahl der Gruppen sehr groß ist, können kleine Abweichungen vernachlässigt werden, bei größeren Differenzen der Individuenzahl ergeben sich aber für jeden Wert, der den weiteren Berechnungen zugrunde gelegt werden soll, verschiedene mittlere Fehler, wodurch eine gemeinsame Auswertung theoretisch unmöglich wird und zu falschen Schlüssen führen kann, wenn sie dennoch durchgeführt wird.

Beispiele. Bei Untersuchungen über die Mitosehäufigkeit in einem bestimmten Gewebe muß jede Probe die gleiche Zellenzahl enthalten, nicht einmal 10, die nächste Probe 500.

Bei einem Versuch über die Beeinflussung der Blütenbildung muß jede Gruppe die gleiche Individuenzahl enthalten, nicht die Kontrolle 5, die Versuchsgruppen 10—20 und eine „besonders wichtige" Gruppe 50.

### γ) Kombinationsversuche.

Die Methode der Varianzanalyse kann nun noch weiter ausgedehnt werden. Ebenso, wie es möglich ist, die Wirkung eines systematisch variierten Faktors von der Zufallswirkung zu trennen, ist es auch möglich, mehrere Variationsursachen rechnerisch zu erfassen, wenn ein Kombinationsversuch vorliegt, in dem mehrere systematische Faktoren enthalten sind.

Wenn in einem Versuch mehrere Faktoren einbezogen werden, so gibt es für die Wirkung zwei Möglichkeiten, nämlich daß diese Faktoren unabhängig voneinander ihren Einfluß auf den zu prüfenden Vorgang ausüben, und daß sie sich gegenseitig beeinflussen. Wenn das erste zutrifft, gibt der Versuch über die Wirkung der Einzelfaktoren denselben Aufschluß, als wenn er im gleichen Gesamtumfang für jeden dieser Faktoren getrennt ausgeführt wäre.

Der Kombinationsversuch bedeutet dann also eine besonders rationelle Ausnutzung des Materials und der aufgewendeten Arbeitskraft. Zudem kann die Unabhängigkeit der einzelnen Faktoren eindeutig nachgewiesen werden, was für die Rückschlüsse auf die Art der Wirkung dieser Faktoren und ihren Eingriff in die für die Ergebnisse verantwortlichen Reaktionen der Objekte oft von besonderer Bedeutung ist.

In Kombinationsversuchen kann nun neben der Wirkung der einzelnen Faktoren, die im Versuch enthalten sind, auch ihr Zusammenwirken gesondert erkannt werden. Wenn ein Faktor in Kombination mit verschiedenen Stufen eines anderen Faktors eine verschiedenartige Reaktion ergibt, so bedeutet diese Kombinationswirkung die Einführung einer weiteren Variabilitätsursache. Ihr Einfluß kann mit Hilfe der Wechselwirkungsvarianz errechnet werden. Wenn sich für die Wechselwirkungsvarianzen gesicherte Werte ergeben, muß daraus geschlossen werden, daß die Faktoren sich gegenseitig beeinflussen und die Versuchsobjekte bei einer Veränderung der Stufe des einen Faktors gegenüber den Einwirkungen eines der anderen Faktoren andersartig reagieren. Ist dies der Fall, dann ist es nicht mehr möglich, aus einem solchen Versuch eine allgemeingültige Aussage über die Wirkung der einzelnen Faktoren zu machen, wohl aber über ihre gegenseitige Abhängigkeit. Da eine solche Feststellung oft wichtiger ist als die Beobachtungen über eine Wirkung eines isoliert betrachteten Faktors, ist ein solcher Versuch keinesfalls als wertlos zu betrachten. Eine derartige Wechselwirkung wird immer dann festzustellen sein, wenn die Einzelfaktoren nicht additiv wirken, sondern in einer anderen Weise, z.B. multiplikativ. Allein die Aussage auf Grund des Nachweises der gesicherten Wechselwirkungsvarianzen, daß sicher keine additive Wirkung der untersuchten Faktoren auf den geprüften Vorgang vorliegt, ist für sich bereits in den meisten Fällen besonders aufschlußreich.

Wenn aber zusätzlich eine Aussage über die Wirkung der einzelnen Faktoren gemacht werden soll, gibt es zwei Möglichkeiten hierzu. Man kann den Versuch so zerlegen, daß diese Teilversuche keine wesentlichen Kombinationswirkungen mehr erkennen lassen. Die so gewonnenen Aussagen beziehen sich dann eindeutig nur auf die durch die bestimmte Kombination der anderen Faktoren bedingten Verhältnisse, und der Vorteil des großen Versuchsumfanges geht für die Aussage über die Einzelwirkungen verloren.

Es kann aber notwendig sein, über die Art der gegenseitigen Abhängigkeit nicht nur eine negative, sondern eine positive Aussage machen zu müssen, und darüber hinaus ist es erwünscht, auch für die Untersuchung der Einzelfaktoren das gesamte Versuchsmaterial verwenden zu können. Die Möglichkeit hierzu ist gegeben durch eine Übertragung der Meßwerte auf eine Skala, die die Wirkung der Einzelfaktoren additiv wiedergibt. Es ist dazu nicht nötig, alle Messungen mit einem neuen Maßstab zu wiederholen, sondern es kann mit den vorhandenen Meßwerten eine Transformation durchgeführt werden, so daß diese umgewandelten Werte jetzt die gesuchte additive Wirkung der Einzelfaktoren zeigen. Häufig führt eine logarithmische Transformation ($x' = \log x$) bereits zum Ziel; in anderen Fällen, bei denen kompliziertere Verhältnisse vorliegen, ist der Umformung eine andere Formel zugrunde zu legen. Die Schwierigkeit liegt immer darin, die richtige Skala zu finden.

Zu diesem Zweck sind die scaling-tests ausgearbeitet, die bisher nur in der quantitativen Genetik verwendet werden, wenn es darauf ankommt, den Einfluß verschiedener Gene auf ein Merkmal zu erfassen. Es handelt sich dabei um ein Problem, bei dem seiner Natur nach mit komplizierten Wechselwirkungen der einzelnen Faktoren (hier der Gene) gerechnet werden muß. Das Ziel ist dabei, eine Skala zu finden, auf der sich die Wirkung der einzelnen Einflüsse additiv darstellt. Wenn die weiteren Berechnungen von diesen Werten ausgehen, sind die Wechselwirkungsvarianzen alle unbedeutend, und es kann das gesamte Material ohne Zerlegung in Einzelversuche für die Untersuchung jedes einzelnen Faktors herangezogen werden. Daneben gibt die Art der notwendigen Transformationen bereits wichtige Aufschlüsse über das Zusammenwirken der einzelnen Faktoren beim Zustandekommen der geprüften Reaktion (MATHER, Biometrical genetics 1949).

Manchmal ist es unmöglich, eine Skala zu finden, auf der die Wirkung aller beteiligten Faktoren additiv ist, nämlich immer dann, wenn in einem Versuch sowohl additiv wie multiplikativ wirkende Faktoren vereinigt sind. Dann muß unter Umständen eine mehrmalige Transformation durchgeführt werden, wobei jedesmal nur Aufschlüsse über einen Einzelfaktor erhalten werden können. Der vollständige Gang der Auswertung ist in solchen Fällen:

| Rechenoperation | ermöglicht Schlußfolgerungen über: |
| --- | --- |
| 1. Varianzanalyse mit den Meßwerten, unter Berücksichtigung der Besonderheiten der vorliegenden Versuchsplanung. | Vorkommen von Wechselwirkungen. Wenn diese nicht vorhanden sind, dann Wirkung der Einzelfaktoren.<br>Wechselwirkungen gesichert: Wirkung der Einzelfaktoren bei bestimmten Kombinationen der anderen Faktoren. |

Nur wenn gesicherte Wechselwirkungen vorhanden sind:

| | |
| --- | --- |
| 2. Durchführung des scaling-tests. | Art der vorhandenen Kombinationswirkungen. |
| 3. Transformation der Meßwerte. | — |
| 4. Varianzanalyse mit den transformierten Meßwerten. | Wirkung der Einzelfaktoren bei Ausschaltung der störenden Wechselwirkungen. Nur für die Faktoren, die auf der gewählten Skala additiv wirken. |
| 5. Eventuell erneute Transformation und Wiederholung der Varianzanalyse, bis für alle Faktoren befriedigende Ergebnisse erhalten werden. | |

Wenn die Wechselwirkungsvarianzen alle unbedeutend sind oder nur vereinzelt gesicherte Kombinationswirkungen vorliegen, wird die Auswertung der Beobachtungszahlen nicht gestört, und die Schlußfolgerungen auf Grund dieser Daten sind zuverlässig. Wenn dagegen viele gesicherte Wechselwirkungen, unter Umständen auch höheren Grades, vorliegen, sind alle Schritte des gegebenen Schemas zur vollständigen Auswertung und richtigen Schlußfolgerung nötig. Nur die Verwendung aller Angaben zusammen kann zu einer richtigen biologischen Deutung des zu untersuchenden Vorganges und seiner Abhängigkeit von den geprüften Bedingungen führen.

### δ) Confounding.

Die bisher besprochenen Anwendungen der Varianzanalyse sind alle darauf eingestellt, daß der Kombinationsversuch vollständig ist, d.h. daß alle möglichen Kombinationen zwischen allen Stufen aller einbezogenen Faktoren tatsächlich in einer gleichen Anzahl von Wiederholungen geprüft werden.

Die Abwandlung der Varianzanalyse, die als Methode des „confounding" in der genetischen und landwirtschaftlichen Versuchstechnik bekannt ist, wurde bisher bei physiologischen Versuchen noch nicht verwendet, obwohl gerade hier sehr viele Möglichkeiten bestehen, diese Methode in die Versuchsplanung einzubeziehen: zum Teil sind sogar die speziellen physiologischen Versuchsprobleme derart, daß erst durch Zuhilfenahme dieses Mittels umfangreiche Versuche ermöglicht werden. Der Grundsatz bei dieser Methode ist, durch Verzicht auf einige weniger wichtige Informationen eine starke Einschränkung des Versuchsumfangs zu erreichen. Bereits hieraus ergibt sich, daß bei physiologischen Versuchen, deren Umfang sehr häufig aus arbeitstechnischen Gründen beschränkt werden muß, die Anwendungsmöglichkeiten um so größer sein werden.

Grundsätzlich ist die Verwendung dort angezeigt, wo durch Vergrößerung des Versuchs auf den eigentlich notwendigen Umfang zwangsläufig neue, unkontrollierbare Variable eingeführt werden und dadurch der Zuwachs an Genauigkeit bei vergrößertem Umfang durch die gleichzeitig vergrößerte Variabilität wieder verlorengeht oder sogar diese Verhältnisse sich derart auswirken, daß eine Vergrößerung des Versuchs nicht nur zwecklos, sondern ungünstig ist.

Einige der Faktoren, die in dieser Weise einen Versuchsplan beeinflussen können, sind: Begrenztes Fassungsvermögen der Thermostaten, WARBURG-Apparaturen und anderer Geräte, wodurch einer Ausweitung des Versuchsumfangs Grenzen gesetzt sind, da bei einer Wiederholung in der gleichen Apparatur oder gleichzeitiger Verwendung mehrerer gleichartiger Geräte immer eine zusätzliche Variable geschaffen wird. Weiter ist die Aufteilung der Bestimmungen auf mehrere Beobachter oder eine Begrenzung durch die Unmöglichkeit, arbeitsmäßig mehr als nur eine geringe Anzahl von Bestimmungen gleichzeitig durchzuführen, auch wenn die technischen Möglichkeiten dies an sich zulassen würden, in gleicher Weise als eine Einschränkung des Versuchsumfangs zu bewerten. Auch im Material kann eine solche Begrenzung begründet sein, z.B. können von einer Pflanze nur eine sehr begrenzte Anzahl von vergleichbaren Blättern entnommen werden, oder auf einem heterogenen Boden können nur wenig Pflanzen auf vergleichbaren Stellen herangezogen werden.

In allen diesen Fällen wurde bisher darauf verzichtet, Kombinationsversuche durchzuführen, und immer nur die Wirkung eines einzigen Faktors im Vergleich zur Kontrolle geprüft. Dabei verzichtet man von vornherein auf einen direkten Vergleich der Versuche untereinander und auch auf die Kenntnis der Wechselwirkungen zwischen den verschiedenen Faktoren, die unter Umständen zur Lösung des Problems viel mehr beitragen können als viele Einzelversuche. Trotz der technischen Begrenzung braucht man aber nicht auf die Vorteile der Kombinationsversuche und ihrer weitgehenden Informationen zu verzichten. Es können mehrere Faktoren in allen möglichen Kombinationen geprüft werden, obwohl nicht alle gleichzeitig durchgeführt werden können, und doch kann der Versuch so angesetzt sein, daß eine vollständige Auswertung möglich wird und alle gewünschten Informationen erhalten werden. Das confounding, das Zusammenfallen bestimmter Bedingungskonstellationen mit den „Wiederholungen", was bei einem vollständigen Versuch als verboten bezeichnet werden muß, führt hier bei systematischer Anwendung zu besonders instruktiven Versuchsplänen.

Beispiel. In einer Versuchsserie soll der Einfluß von drei Stoffen in je drei Stufen und allen möglichen Kombinationen auf einen bestimmten Vorgang geprüft werden. Der vollständige Versuch umfaßt also $3 \times 3 \times 3 = 27$ Bestimmungen, die mehrmals durchgeführt werden müssen, um den Einfluß der Zufallsschwankungen erfassen zu können. Mit Rücksicht auf die Genauigkeit der Ablesung der Meßergebnisse soll aber eine Person nur neun Bestimmungen gleichzeitig durchführen. Der Versuch kann mehrmals nacheinander (z.B. an verschiedenen Tagen) wiederholt werden, und es sollen zwei Mitarbeiter zur Verfügung stehen, so daß alle 27 Bestimmungen zwar gleichzeitig stattfinden können, aber auf Grund der Besonderheiten der verwendeten Apparaturen muß mit geringen, nicht zu vermeidenden, individuellen Schwankungen der Ablesung gerechnet werden. Es ist also zu erwarten, daß die Einzelbestimmungen, die von einer Person durchgeführt werden, eine größere Einheitlichkeit aufweisen, als diejenigen, die von verschiedenen Personen stammen.

Wenn in einem solchen Fall jede Person alle Bestimmungen durchführen kann, ist es möglich, zunächst in einem Blankoversuch von gleichem Umfang wie der geplante Hauptversuch die individuellen Differenzen zu bestimmen und dann diese in Form der Covarianzanalyse (s. S. 99) in die Berechnungen des Hauptversuchs eingehen zu lassen. Die Planung, Durchführung und Berechnung eines solchen Versuchs ist aber zeitraubend und schwierig, und oft ist es auf Grund der Beschaffenheit des Materials gar nicht möglich, eine vollständige Parallelisierung der Einzelbestimmungen von Blanko- und Hauptversuch vorzunehmen, so daß dieser Methode enge Grenzen gesetzt sind. Man kann auch die individuellen Unterschiede als Blockdifferenzen bei der Berechnung auffangen, erhält dann aber eine relativ zu große Zwischenklassenvarianz, was ebenfalls für die weiteren Schlußfolgerungen Nachteile hat. Bei der beschriebenen Konstellation der Bedingungen ist es nach den klassischen Vorschriften unmöglich, den Versuch im geplanten Umfang durchzuführen; es müßte eine Aufteilung in Einzelversuche durchgeführt werden, wobei alle Informationen über die Kombinationswirkung der Faktoren verlorengehen.

Bei einer solchen Lage ist es das gegebene, nicht auf alle Informationen in dieser Richtung zu verzichten, sondern einen Teil der Sicherheit für einige Schlußfolgerungen, d.h. aber der Freiheitsgrade, zu opfern, um dadurch die Durchführung des ganzen Versuchs zu ermöglichen. Die richtige, erfolgversprechende Planung eines solchen Versuchs, erfolgt nach dem Beispiel, das FISHER (Design of experiments, S. 152) gibt. Die Anwendung der Methode des confounding ist hier nicht freiwillig, sondern durch die technischen Bedingungen des Versuches erzwungen.

Die 27 Einzelbestimmungen werden so auf die drei Beobachter verteilt. daß jeder alle drei Stufen aller drei Faktoren prüft, aber nur an einer begrenzten Anzahl von Kombinationen, wobei sich die so gebildeten drei Gruppen zum Gesamtversuch mit 27 Bestimmungen ergänzen. Die Zusammenstellung der Gruppen kann bei den Wiederholungen gleichbleiben oder verändert werden, so daß jeder Beobachter in den verschiedenen Wiederholungen alle Kombinationen zur Untersuchung erhält.

In welcher Weise das confounding, das Zusammenfallen zweier Variationsursachen in einem solchen Versuch geplant wird, hängt weitgehend von den Einzelheiten der Problemstellung ab und muß für jeden Fall neu bestimmt werden. Eine Anzahl von Beispielen für die verschiedenartigsten Fälle sind bei FISHER, MATHER (1951) und COCHRAN und COX (1953) zu finden.

Ein zweites Beispiel soll die Anwendungsmöglichkeiten noch weiter illustrieren: Es sollen die Wirkungen einer sehr großen Anzahl verschiedener Stoffe auf einen Vorgang geprüft werden. Es stehen acht Thermostaten zur Verfügung, die je acht Proben aufnehmen können. Es muß also mit geringen Temperaturschwankungen zwischen den Thermostaten gerechnet werden, die wegen ihrer Kleinheit nicht zu erfassen sind, aber doch verursachen, daß die Proben in einem Apparat unter einheitlicheren Bedingungen stehen als die in verschiedenen Gruppen. Da die Bestimmungen zeitlich schnell ausgeführt werden sollen, sei es nicht möglich, jeden Stoff mehrmals in jedem Thermostaten zu prüfen, um auf diese Weise eine Schätzung des Einflusses, den die verschiedenen Gruppen erlitten haben, durchzuführen. Es ist nun nicht möglich, immer nur acht Stoffe gleichzeitig zu prüfen, wobei die gleichartigen Gruppen in den verschiedenen Thermostaten als „Wiederholungen" in die Rechnung eingehen. Hierbei wird auf jeden exakten Vergleich zwischen nicht gleichzeitig geprüften Stoffen verzichtet. Die Planung erfolgt vielmehr (nach FISHER 1949) derart, daß zunächst 64 Stoffe in Achtergruppen geprüft werden, dann werden neue Achtergruppen gebildet und geprüft. Dies wird so lange fortgesetzt, bis jeder Stoff in einer gleichen Serie von Versuchen geprüft wurde. Dabei ist nur darauf zu achten, daß die Gruppen immer eine andere Zusammensetzung zeigen und nicht zwei Stoffe mehrmals in einer Gruppe zusammen vorkommen. Es ist schließlich noch möglich, weniger wichtige Stoffe in einer geringeren Anzahl von Proben auftreten zu lassen als diejenigen, auf deren Kenntnis der größere Wert gelegt wird. Das Beispiel von FISHER zeigt auch die Auswertung derartiger Versuche.

Wichtig ist aber immer, daß der Versuchsplan in allen Einzelheiten der Verteilung der Bestimmungen auf die Serien und Gruppen vorher genau festgelegt wird. Eine nachträgliche Auswertung eines Versuchs, der nur mehr oder weniger zufällig nach einem ähnlichen Plan durchgeführt wurde, ist in keinem Fall möglich, weil nie die Bedingungen der Verteilung, die allein eine Auswertung ermöglichen, erfüllt sein werden.

Drittes Beispiel. Für einen Versuch werden mehrere Blätter gebraucht, die von verschiedenen Pflanzen genommen werden müssen und naturgemäß nicht alle gleich alt und gleich groß sein können. Es sollen andererseits so viele Faktoren und Kombinationen geprüft werden, daß es unmöglich ist, jedes Blatt so zu zerteilen, daß alle Versuchsstufen mit jedem Blatt durchgeführt werden, und außerdem ist dabei noch mit physiologischen Differenzen innerhalb des Blattes zu rechnen, so daß statt der größeren Einheitlichkeit eine zusätzliche, unkontrollierbare Variable eingeführt würde.

Diese Situation führte bisher mit Sicherheit zu einer Beschränkung des Versuchs mit allen Folgen, die sich daraus durch den Mangel an wichtigen Informationen für die theoretische Ausdeutung des Versuchs ergeben. Auch hierbei ist es nicht nötig, den geplanten Versuchsumfang einzuschränken oder von vornherein auf die Planung von Kombinationsversuchen zu verzichten, um innerhalb des Versuchs die Einheitlichkeit zu wahren, sondern mit einem diesen Gegebenheiten angepaßten System des confounding läßt sich auch dieser Versuch durchführen, wobei zwar im Vergleich mit dem zunächst theoretisch geforderten „vollständigen" Versuch einige Informationen verlorengehen oder mit verminderter Sicherheit gewonnen werden, aber im Vergleich mit der Zerlegung in unabhängige Einzelversuche werden doch bedeutende Informationen über die Wechselwirkungen zwischen den Faktoren erhalten, die meist nicht wesentlich geringer sind, als wenn tatsächlich der ganze Versuch mit jedem Blatt durchgeführt worden wäre. Außerdem muß bei einer richtigen Planung des confounding-Versuchs mit einer gegenüber einer umfangreicheren Serie von Einzelversuchen oft bedeutenden Arbeitsersparnis gerechnet werden, wodurch, wenn man die gewonnenen Erkenntnisse in Beziehung zur aufgewendeten Arbeit setzt, die Zeit, die für die kompliziertere Versuchsplanung aufgewendet wurde, sich vielfach gelohnt hat.

### ε) Covarianzanalyse.

Ein weiterer Sonderfall der Varianzanalyse ist gegeben, wenn in einem Versuch zwei zufällige, beeinflußbare Variablen vorhanden sind, die von den Versuchsbedingungen beeinflußt werden, und für die vermutet werden muß, daß sie voneinander abhängig sind. Es ist damit das gleiche Problem gestellt, wie es den Partialkorrelationen zugrunde lag. Die Auswertung geht hier in der Weise vor sich, daß zunächst für jeden dieser beiden Messungsreihen die Varianzanalyse durchgeführt wird, um den Grad der Veränderung durch Versuchsfaktoren zu erkennen. Dann lassen sich mit Hilfe der Korrelationsrechnung die Werte der einen Reihe mit denen der anderen so korrigieren, daß der Einfluß dieser zweiten Variablen auf die Messungsergebnisse der ersten Veränderlichen beseitigt wird. Diese Covarianzanalyse ermöglicht also die Ausschaltung einer unerwünschten, aber unvermeidlichen Veränderlichen. Die notwendige Rechenarbeit ist jedoch erheblich, so daß sich eine solche Korrektur nicht mehr durchführen läßt, wenn in einem Versuch bereits mehrere willkürliche Veränderliche kombiniert sind, die gesicherte Wechselwirkungen zeigen. Für manche Fragestellungen ist die Covarianzanalyse die einzige Möglichkeit, zu den gewünschten Informationen über die Wirkung und das Zusammenwirken der in einem Versuch enthaltenen Variabilitätsursachen zu gelangen, aber meist wird es zu empfehlen sein, durch eine veränderte Versuchsanstellung ohne diese Methode zu den gewünschten gesuchten Ergebnissen zu kommen, wenn dies irgendwie möglich ist.

Das wichtigste Anwendungsgebiet der Covarianzanalyse liegt bei einer anderen Problemstellung. Häufig ist es bekannt, daß in einem Versuch unerwünschte, aber bekannte und unvermeidliche Faktoren, die nicht zufällig, sondern systematisch wirken, die Variabilität vergrößern. Wenn nun gleichzeitig bekannt ist, daß diese Faktoren nicht mit den Versuchsbedingungen selber zusammenhängen, sondern auch außerhalb des Versuchs in gleicher Weise vorhanden sind, können sie bei der Auswertung in Rechnung gestellt werden. Dies ist aber nur dann möglich, wenn der Versuch als Ganzes so wiederholt werden kann, daß die Bestimmungen des ersten und zweiten Versuchs vollständig zu parallelisieren sind, wenn also für jede Messung des einen Versuchs mit Sicherheit eine Messung des anderen Versuchs angewiesen werden kann, die ihr im Hinblick auf den Einfluß der auszuschaltenden Faktoren genau entspricht. Wenn diese günstigen Umstände vorhanden sind, kann man vor oder nach dem Hauptversuch einen zweiten Versuch durchführen, der aus der gleichen Anzahl von Messungen besteht, aber ohne daß die willkürlichen Veränderlichen darin aufgenommen wurden. Der ganze Wiederholungs- oder Vorversuch besteht dann ausschließlich aus völlig einheitlich behandelten „Kontroll"gruppen. Dieser Blankoversuch wird so ausgewertet, als ob in ihm alle Faktoren des Hauptversuchs enthalten seien, und ermöglicht es so, die Wirkung der unerwünschten Faktoren auf die Einzelmessungen zu bestimmen. Es können dann die Werte des Hauptversuchs in Beziehung gesetzt werden zu den ihnen entsprechenden Kontrollmessungen. Mit Hilfe der Covarianzanalyse ist eine Korrektur der Bestimmungen möglich, wobei die Messungen des eigentlichen Versuchs so umgerechnet werden, daß der Einfluß des unerwünschten Faktors, der in den Differenzen des Blankoversuchs zu erkennen ist, ausgeschaltet wird.

Beispiel. Die Pflanzen für den Versuch müssen auf einem Versuchsgelände herangezogen werden, das in Bodenbeschaffenheit und Wasserführung nicht einheitlich ist. Es ist zu vermuten, daß der durch diese unterschiedliche Vorbehandlung bedingte verschiedene physiologische Zustand einen Einfluß auf das Ergebnis hat. Es wird nun zuerst ein Blankoversuch angepflanzt und ohne jede Behandlung durchgeführt, dann der Hauptversuch und danach die Korrektur vorgenommen.

Ein anderes Beispiel für einen Laborversuch ist folgendes: Es sollen Messungen an verschiedenen Versuchsgruppen gleichzeitig vorgenommen werden, aber für jede Bestimmung wird ein eigener Apparat benötigt. Wenn nun etwa Geräte verschiedener Bauart verwendet werden müssen, ist es möglich, daß eine systematische Differenz zwischen den Messungen der einzelnen Geräte auftritt. Die Ergebnisse sind also nicht ohne weiteres zu verwenden, weil der Einfluß des Meßgerätes berücksichtigt werden muß. Die Durchführung ist dann derart, daß vor oder nach dem eigentlichen Versuch der Blankoversuch durchgeführt wird und beide Versuche zusammengefaßt in einer Covarianzanalyse ausgewertet werden.

### c) $\chi^2$-Test bei qualitativen Variablen.

In den vorhergehenden Abschnitten wurden Auswertungsmethoden besprochen, die für einfache und Kombinationsversuche zur Verfügung stehen, wenn die Ergebnisse der Teilversuche an einem quantitativ faßbaren Merkmal gemessen werden. Für qualitative Differenzen sind diese Methoden nicht brauchbar. Die Erfassung erfolgt bei derartigen Fällen durch Zuteilung der einzelnen Individuen zu den qualitativ verschiedenen Klassen. Die Differenzen zwischen den Versuchsergebnissen werden gemessen an den Unterschieden in der Häufigkeit, mit der die verschiedenen Klassen in den Teilversuchen vertreten sind. Zur Diskussion steht dann die Frage, ob diese Reihe von Differenzen in der gefundenen Größe noch als zufälliges Ergebnis auftreten kann. Die Prüffunktion hierfür ist die $\chi^2$-Verteilung. Die Ausführung des Tests ist etwas verschieden, je nach der zugrunde liegenden Fragestellung. Es können dabei nicht nur zwei, sondern auch eine größere Anzahl von Versuchen gleichzeitig verglichen werden. Ebenso wie bei der Varianzanalyse ist auch hier eine Prüfung möglich, durch die der Einfluß verschiedener Variationsursachen auf die Gesamtstreuung einzeln erfaßt wird. Mit Hilfe einer solchen $\chi^2$-Zerlegung können also auch Kombinationsversuche durchgeführt und ausgewertet werden, deren Wirkung sich an qualitativen Merkmalen zeigt.

Die Verbindung zur Varianzanalyse ist dadurch gegeben, daß der $\chi^2$-Test ebenfalls auf einer Untersuchung der Streuung beruht. Die Größe $\chi^2$ wird definiert als die Summe der quadratischen Abweichungen, die im Material aufgetreten sind. Für die Beurteilung ist wieder die Anzahl der Freiheitsgrade, aus denen die Größe errechnet wurde, von Bedeutung, da sofort einzusehen ist, daß eine bestimmte Größe des $\chi^2$ verschieden beurteilt werden muß, je nach der Anzahl von Vergleichen, aus denen sie berechnet wurde. Ebenso, wie bei der Varianzanalyse die Division der Summe der quadratischen Abweichungen durch die zugehörige Anzahl von Freiheitsgraden die Varianz ergab, so ist auch aus $\chi^2$ und den Freiheitsgraden eine Varianz zu berechnen, die ihrerseits wieder zur

Bildung eines Varianzverhältnisses und damit zur Durchführung des $z$-Testes verwendet werden kann, wenn aus einem Material mehrere Varianzen für verschiedene Variabilitätsfaktoren berechnet werden können. Meist wird sich aber dieser Rechengang erübrigen, weil sich aus den Werten einer $\chi^2$-Zerlegung auch direkt die Wirksamkeit der einzelnen im Versuch kombinierten Faktoren im Verhältnis zur Zufallsstreuung ersehen läßt.

Es kann nicht nur ein Vergleich mehrerer empirisch gefundener Verteilungen durchgeführt werden, sondern es ist auch möglich, eine gefundene Verteilung zu vergleichen mit einer anderen, die durch theoretische Überlegungen gefordert wird. Dieser letztere Fall ist immer gegeben bei genetischen Untersuchungen, denen bestimmte Erwartungen über die auftretenden Spaltungen zugrunde liegen. Das gleiche liegt aber vor, wenn geprüft werden soll, ob eine empirische Verteilung mit einer bestimmten theoretischen Kurve, etwa einer Normalverteilung, übereinstimmt, oder ob eine berechnete Regressionslinie als linear angesehen werden kann. Die Tatsache, daß aus den Daten des Materials die Berechnung einer theoretischen Kurve oder eines Regressionskoeffizienten möglich ist, berechtigt noch nicht sofort zu der Annahme, daß diese Kurve oder Gerade die richtige Beschreibung des empirischen, statistischen Materials gibt. Erst wenn die Prüfung des aus dem $\chi^2$-Test sich ergebenden $P$-Wertes zeigt, daß die auftretenden Abweichungen von dieser theoretischen Kurve sich im Rahmen dessen halten, was durch die Einwirkungen der zufälligen Probeentnahme erklärt werden kann, darf die der Kurve zugrunde liegende theoretische Annahme als richtige Beschreibung der vorhandenen Verhältnisse angesehen werden. Wenn mehrere theoretische Deutungen möglich sind, kann in dieser Weise festgestellt werden, welcher von ihnen die größere Wahrscheinlichkeit zukommt. Es wurde bereits mehrfach erwähnt, daß zur Ausschaltung von unerwünschten Veränderlichen verschiedene Methoden angewendet werden können. Das Ziel ist immer, unerwünschten Faktoren, die sich versuchstechnisch nicht ausschalten lassen, keinen Einfluß auf das Versuchsergebnis einzuräumen.

## III. Relative Werte.
### a) Berechnung in Prozent der Kontrolle.

Neben den geeigneten Berechnungen wie multiplen Regressionen, der Covarianzanalyse und verwandten Verfahren werden in der physiologischen Literatur sehr oft Methoden angewendet, die statistisch nicht einwandfrei sind. Es handelt sich also um die Auswertung von Versuchen, in denen unerwünschte, systematisch oder zufällig verteilte Veränderliche enthalten sind.

Wenn es sich um die Wirkung von Zufallsfaktoren handelt, ist es ein häufig beschrittener Weg, in jeder Serie eine „Kontrolle" mitlaufen zu lassen und die Ergebnisse des „Versuchs" in „Prozent der Kontrolle" anzugeben. Angeblich sollen hierdurch die Differenzen zwischen den immer etwas variierenden unkontrollierbaren Faktoren, durch die sich nicht genau parallel geführte Versuche unterscheiden, ausgeschaltet werden. Bei diesem Vorgehen werden aber im günstigsten Fall unbewiesene, meist sogar nachweislich falsche Hypothesen zugrunde gelegt. Die erste dieser Annahmen ist, daß die Kontrollen immer genau den gleichen Wert ergeben müßten, die Differenzen zwischen den „Kontrollen" also auf realen Unterschieden beruhen und eine zuverlässige Schätzung des Einflusses der unkontrollierbaren Versuchsbedingungen erlauben. Die zweite Zusatzannahme ist, daß diese unkontrollierbaren, variierenden Versuchsbedingungen auf Kontroll- und Versuchsmaterial so einwirken, daß dieser Einfluß sich in einer proportional genau gleichen Weise bemerkbar macht und damit durch die erwähnte Berechnungsweise „in Prozent der Kontrolle" völlig zum Verschwinden gebracht wird. Es läßt sich aber beweisen, daß beide Annahmen nicht zutreffen. Zunächst ist auch für die Kontrollen eine Zufallsschwankung anzunehmen, so daß verschiedene Bestimmungen allein auf Grund der Eigenschaften des biologischen Materials immer voneinander abweichen müssen. Wenn nun einmal eine „Kontrolle" etwas mehr Minusabweicher erhalten hat, so ist damit noch nicht gesagt, daß die parallel laufende Versuchsgruppe unbedingt ebenfalls Minusabweicher enthalten muß, die nach ihrer Anzahl und der Größe der Abweichung vom Mittel genau mit der Kontrolle übereinstimmen. Bei der Begrenzung der beiden Stichproben auf einen meist relativ geringen Umfang ist dies sogar äußerst unwahrscheinlich. Auf diesen Zufallsschwankungen bei der Auswahl der Objekte beruht die ganze bisher betrachtete statistische Auswertungsmöglichkeit. Wenn nun bei der Durchführung des neuen Versuchs die Kontrollgruppe etwas mehr Plusabweicher erhalten hat, die parallel gehende Versuchsgruppe etwas mehr Minusabweicher, so können bei einer Berechnung „in Prozent der Kontrolle" diese reinen Zufallsschwankungen ganz erhebliche Differenzen zwischen der Wirkung der beiden Versuchsfaktoren vortäuschen und zu Fehlschlüssen führen, vor allem dann, wenn die auf diese Weise erhaltenen Werte die Grundlagen für weitere statistische Berechnungen bilden, in deren Verlauf die unsichere Herkunft dieser Zahlen meist nicht mehr beachtet wird und auf keinen Fall bei den statistischen Schlußfolgerungen berücksichtigt werden kann. Eine richtige Auswertung darf nicht den Versuch machen, diese Zufallsschwankungen vor Beginn der Berechnungen zu eliminieren, da sie gerade eine wichtige Rolle bei der Beurteilung der Ergebnisse spielen.

Der Fehler der zweiten Annahme ist ebenfalls bei genauerer Betrachtung deutlich: Wenn die variierenden Faktoren tatsächlich einen Einfluß haben, so kann dieser muliplikativ oder additiv sein, um nur die einfachsten der vielen möglichen Beziehungen zu nennen. Der Einfluß kann sich weiter auf die gesamte Größe des gemessenen Wertes oder nur auf seine Veränderung während der Dauer des Versuchs beziehen. Nur wenn die Beeinflussung multiplikativ ist, sich auf den ganzen gemessenen Wert bezieht und in genau gleicher Richtung und gleichem Ausmaß auf Kontrolle und Versuch wirkt, kann durch die genannte Rechenoperation ihre Ausschaltung erfolgen, in allen anderen Fällen wird das Material nicht besser und einheitlicher, sondern dadurch, daß die Variabilität des einen Faktors zu der des anderen hinzukommt, wird das Material uneinheitlicher und damit vom statistischen Standpunkt aus schlechter, ohne daß dann noch eine Möglichkeit besteht, diese zusätzliche Variabilität auszuschalten, während dies bei der Verwendung der Originalzahlen ohne weiteres möglich ist.

Aus diesen Überlegungen ergibt sich, daß man, bevor man zu einer Berechnung der Versuchsergebnisse „in Prozent der Kontrolle" berechtigt ist, zu beweisen hat: 1. daß die Differenzen zwischen den „Kontrollen" in vollem Umfang real und nicht durch Zufallsfaktoren bedingt sind, und 2. daß die Wirkung der unkontrollierbaren, variierenden Faktoren multiplikativ ist, den ganzen Meßwert beeinflußt und Versuch und Kontrolle gleichartig berührt. Der erste Punkt ist nie zu beweisen, weil er den Eigentümlichkeiten des biologischen Materials widerspricht, und zum Nachweis des zweiten sind so umfangreiche Versuche mit so komplizierter Planung, Durchführung und Ausrechnung nötig, daß dies außerhalb der praktischen Durchführbarkeit einer begrenzten Versuchsmöglichkeit liegt, von der ausgegangen wurde. Es ist daher in allen Fällen besser, von einer Berechnung „in Prozent der Kontrolle" abzusehen und eine andere Versuchsplanung zu suchen, bei der die Zufallsfaktoren in irgendeiner Weise legal in die Berechnung eingehen, und die Differenzen zwischen den Versuchsreihen als zusätzliche Variable zu betrachten, deren Einfluß bei einer geeigneten Planung genau erkannt werden kann.

Die Methode der Berechnung „in Prozent der Kontrolle" ist aber anwendbar, wenn Bezug genommen werden kann auf einen Standardwert, der keiner Zufallsstreuung unterworfen ist, sich aber durch Veränderungen der Außenbedingungen gesetzmäßig so verschieben kann, daß bei einer Abweichung des Standards mit Sicherheit auf die Anwesenheit und die Wirkungsstärke der unkontrollierbaren Variationen der Versuchsbedingungen geschlossen werden kann. Dies ist immer dann der Fall, wenn der Standardwert dem Ausgangswert des Ablesegerätes entspricht, und dieser jedesmal, etwa in Abhängigkeit von Temperaturschwankungen oder Luftfeuchtigkeit, neu eingestellt werden muß. Bei biologischen Versuchen wird die geforderte Konstanz nie vorhanden sein, wenn der Standardwert nicht durch das Meßgerät, sondern durch das Material, den parallelgeführten Kontrollversuch, bestimmt wird.

Die Berechnung der „Abweichung vom Standardwert" ist ebenfalls erlaubt und für manche Zwecke besonders geeignet, wenn dieser Standard als Mittelwert aus einer sehr großen Beobachtungsserie gewonnen wurde, so daß der Fehler dieser Bestimmung vernachlässigt werden kann. Aber auch wenn diese Bedingung erfüllt ist, muß vor der Anwendung dieses Bezugssystems noch das Zutreffen der zweiten Voraussetzung nachgewiesen werden.

Beispiel. Die Kontrolle ergibt in einer Serie 100, der Versuch 110, bei der Prüfung des nächsten Faktors sind die Werte Kontrolle 110, Versuch 100. Bei unvoreingenommener Betrachtung zeigt sich eine Schwankung der Kontrollwerte, die ebenso groß ist wie die Differenzen zwischen den Versuchen. Allein aus diesen Versuchen ergibt sich also, daß wahrscheinlich beide Faktoren unwirksam sind und sich weder untereinander noch von der Kontrolle unterscheiden. Die Berechnung in der üblichen Weise ergibt aber, daß der eine Faktor eine Wirkung von 110% der Kontrolle, der andere nur von 90% der Kontrolle hatte, wodurch bedeutende Differenzen in der Wirkung beider Faktoren vorgetäuscht werden, während der Versuch in Wirklichkeit als ergebnislos bezeichnet werden muß und höchstens Veranlassung zu sehr zahlreichen Wiederholungen geben könnte, bei denen nicht nur Versuch und Kontrolle, sondern ebensohäufig auch die beiden Versuchsbedingungen nebeneinander geprüft werden müßten.

Ein anderes Beispiel. In aufeinanderfolgenden Versuchen zur Prüfung verschiedener Faktoren ergibt die Kontrolle die Werte 100, 90, 110; die jeweils parallel durchgeführten Versuchsgruppen dagegen 90, 110, 100. Die Variabilität zwischen den verschiedenartigen Versuchen ist also nicht größer als zwischen den Kontrollen, während für die Variabilität desselben Versuchs keine Bestimmungen vorliegen. Die Schlußfolgerung ist also, daß eine Wirkung der Versuchsfaktoren nicht sehr wahrscheinlich ist, auf keinen Fall aber sehr groß sein kann, und weitere Parallelbestimmungen erforderlich sind. Die übliche Umrechnung ergibt aber 90%, 125% und 91% der Kontrolle, wodurch eine sehr große Differenz in der Wirkung von über 30%, einmal im Verhältnis zur Kontrolle, zum anderen in dem Verhältnis der einzelnen Faktoren, vorgetäuscht wird.

Wenn nur ein Faktor im Vergleich zur Kontrolle geprüft werden soll, besteht die Möglichkeit der Paarbildung, wobei stets Versuch und zugehörige Kontrolle zusammen betrachtet werden und ihre *Differenz* das eigentliche Versuchsergebnis ist. Die zu prüfende Nullhypothese besagt dann, daß keine Unterschiede zwischen Versuch und Kontrolle bestehen: Der Erwartungswert für die Differenz ist Null. Es ist dann nur zu prüfen, mit welcher Wahrscheinlichkeit der tatsächlich gefundene Mittelwert der Differenzen von Null abweicht, um eine Aussage darüber machen zu können, ob die Ergebnisse von Versuch und Kontrolle als gesichert verschieden anzusehen sind. Es gibt noch verschiedene andere Möglichkeiten, einen solchen Versuch auszuwerten, und je nach der Fragestellung ist die eine oder die andere Art vorzuziehen (grundsätzliche Fragen: FISHER 1949; ausführliche Darstellung der Berechnungen: MATHER 1949; Abwandlung: LINDER 1953).

Für den erwähnten Fall, daß zwei Faktoren neben der Kontrolle geprüft werden sollen und zwei Parallelbestimmungen unter vergleichbaren Bedingungen möglich sind, ergibt sich eine sehr gute Möglichkeit daraus, daß hier die „1." und „2." Bestimmung als Versuchsfaktoren betrachtet werden, für die je drei Stufen möglich sind, nämlich Kontrolle und die beiden zu prüfenden Faktoren. Es werden nacheinander alle Kombinationen zwischen diesen beiden Parallelen mehrfach geprüft, wobei natürlich in einem Drittel der Fälle zwei gleichartige Versuche parallel laufen. Die bei gleichzeitiger Durchführung der gleichen Bestimmung in beiden Parallelen auftretenden Differenzen zeigen die Größe des Zufallseinflusses an und die zwischen den Wiederholungen dieser Bestimmungen auftretenden Unterschiede die Seriendifferenzen. Auf Grund dieser beiden Bestimmungen ist dann eine Beurteilung derjenigen Differenzen möglich, die bei unterschiedlichen Parallelbestimmungen auftreten.

Eine andere Möglichkeit der Planung ist wieder in der Paarbildung gegeben. Es werden hierzu mehrfache Parallelbestimmungen von Versuch 1 und Kontrolle und von Versuch 2 und Kontrolle durchgeführt. Die Nullhypothese besagt jetzt, daß keine Differenz zwischen Versuch und Kontrolle und folglich auch nicht zwischen den verschiedenen Versuchen besteht. Es wird für jede Serie der Mittelwert der Differenzen zur Kontrolle berechnet und geprüft, mit welcher Wahrscheinlichkeit dieser von Null abweicht. Für die Feststellung, ob die beiden Versuchsbedingungen verschieden wirken, können die beiden Mittelwerte mit Hilfe ihrer mittleren Fehler miteinander verglichen werden. Wenn sich dabei ergibt, daß die mittlere Differenz zur Kontrolle für die Versuchsbedingung 1 einen gesichert anderen Wert ergibt als für Versuchsbedingung 2, so kann angenommen werden, daß die Reaktion des Objektes auf beide Bedingungen verschieden ist.

Durch eine angemessene Versuchsplanung ist es also durchaus möglich, die Anwendung des Bezugssystems „in Prozent der Kontrolle" zu vermeiden und trotzdem die Unterschiede zwischen Versuch und Kontrollbestimmung der Auswertung zugrunde zu legen.

### b) Berechnung durch Bezug auf eine andere Messungsreihe.

Die besprochene Methode der Ausschaltung einer unerwünschten Variabilität durch Berechnung „in Prozent der Kontrolle" (oder des Standards) ist eigentlich nur ein Sonderfall eines allgemeineren Problems der biologischen Statistik. Es soll dabei eine Variabilität, deren Ursache im einzelnen nicht bekannt ist, ausgeschaltet werden. Häufiger liegt aber der Fall vor, daß die Ursache der unerwünschten Variabilität bekannt ist, nämlich immer dann, wenn im Versuchsmaterial eine der unabhängigen, zufälligen, unerwünschten Veränderlichen enthalten ist und vermutet wird, daß hierdurch ein Einfluß auf das Versuchsergebnis ausgeübt wird, der für eine exakte Auswertung der Versuchsdaten entfernt werden muß. Die am häufigsten verwendete Methode, um eine solche Variable auszuschalten, ist die Berechnung der „relativen Werte", die nicht mit den „relativen Häufigkeiten" zu verwechseln sind. Nach Kenntnis der verschiedenen Möglichkeiten einer exakten rechnerischen Behandlung des Problems, die durch die Regressions- und Korrelationsrechnung mit ihren Abwandlungen der Partialregression und der Covarianzanalyse gegeben sind, können jetzt die Nachteile oder besser die Fehler der relativen Werte dargelegt werden. Die statistische Ursache dieser Fehlschlüsse liegt darin, daß hier die Variabilität zweier Werte in einer einzigen Zahl enthalten ist, ohne daß die Möglichkeit besteht, dies bei den weiteren Berechnungen und Schlußfolgerungen zu berücksichtigen.

Beispiele für derartige Umrechnungen sind: Bestimmungen (z.B. Assimilation, Transpiration, Atmung) an ganzen Organen, bei denen, um Verletzungen und damit Wundreize zu vermeiden, die Größe nicht konstant gemacht werden kann, und Berechnung „pro Flächeneinheit", Stoffaufnahme oder -abgabe von Organen, Pflanzen oder Gewebestückchen und Berechnung „pro Gramm Ausgangsgewicht" (oder Trockensubstanz).

Oft wird bereits bei der Durchführung des Versuchs versucht, die unerwünschten Variationen von vornherein auszuschalten, indem die zu prüfende Substanz bereits dosiert wird in Abhängigkeit vom Ausgangsgewicht (-größe) des einzelnen Objektes. Dabei wird ebenso wie bei der nachträglichen Umrechnung von der Vorstellung ausgegangen, daß hierdurch eine besondere Exaktheit des Versuchsergebnisses erreicht wird. Wie sehr beide Methoden, die im Grunde genommen gleich sind, die Versuchsergebnisse beeinflussen, läßt sich am anschaulichsten an Hand der graphischen Darstellung der Regressionslinien demonstrieren.

In allen Abbildungen bedeutet die Linie 1 die tatsächlich vorliegende Regressionslinie, 2 die nach der Umwandlung vorhandene Regressionslinie und 3 die fälschlicherweise für die weiteren Schlußfolgerungen zugrunde gelegte Regressionslinie, die nach der Berechnung der relativen Werte angeblich vorhanden sein soll. Die Pfeile geben die Richtung an, in der sich die Werte bei der Umrechnung verändern. Für die $x$- und $y$-Achse wurde eine arithmetische Skala angenommen und in das so entstehende Koordinatensystem die Mittelwerte für $y$ bei steigendem $x$ eingetragen.

Am schlimmsten sind die Auswirkungen, wenn die relativen Werte in Beziehung gesetzt werden zu dem Wert, der gerade ausgeschaltet werden sollte. Dieser Fehler kommt so auffallend häufig vor, daß hier auf die Folgen davon näher eingegangen werden soll, weil immer wieder die gleichen Fehlschlüsse aus derartigem Zahlenmaterial gezogen werden.

### α) Veränderungen des Mittelwertes.

Bei einer Umrechnung der Meßergebnisse in relative Werte wird davon ausgegangen, daß erstens eine lineare Regression zwischen der untersuchten und der unerwünschten Veränderlichen bestehe, und zweitens, daß eine Korrelation der Streuung von $y$ mit dem Mittelwert vorhanden sei von der Art, daß für jedes $x$ $\sigma_y$ in Prozent des zugehörigen Mittelwertes $\bar{y}$ ausgedrückt immer gleich sei. Wenn diese Voraussetzungen tatsächlich zutreffen, dann wird durch die Berechnung der relativen Werte wirklich dieser Einfluß der unerwünschten Variablen ausgeschaltet und eine größere Einheitlichkeit des Versuchsmaterials erreicht, das jetzt nur noch die Zufallsstreuung des Faktors $y$ enthält, die bestimmt werden sollte. Dieser Idealfall ist in Abb. 1 dargestellt. Diese Voraussetzungen treffen aber in den seltensten Fällen uneingeschränkt zu, so daß meist von unzutreffenden, zumindest aber unbewiesenen Hypothesen ausgegangen wird, wenn in einem Material, in dem zwei Veränderliche enthalten sind, eine davon durch die Berechnung der relativen Werte entfernt wird.

Wenn zwischen der Variablen, deren Einfluß ausgeschaltet werden soll, und derjenigen, deren Veränderungen gerade geprüft werden müssen, keine Beziehung besteht, ist die Ausschaltung dieses Faktors nicht nur eine überflüssige Arbeit, durch die auf keinen Fall eine größere Exaktheit erzielt wird, sondern im Gegenteil wird dadurch, daß die Berechnungen auf einer nicht zutreffenden Hypothese aufbauen, ein Fehler in die derartig umgewandelten Meßergebnisse hineingebracht, dessen Auswirkungen in Abb. 2 dargestellt sind. Statt der eigentlich vorhandenen Unabhängigkeit kommt eine deutliche negative Regression zustande, und die Streuung $\sigma_y$ ist so verändert, daß statt der ursprünglichen Normalverteilung von $x$ eine extrem schiefe Kurve auftritt mit starker Häufung bei den niedrigen $y$-Werten und einem sehr weit zu den hohen Werten hin ausgezogenen Schwanz.

Wenn eine lineare Regression besteht, werden durch die Berechnung der relativen Werte die Mittelwerte der einzelnen Klassen tatsächlich, wie es der

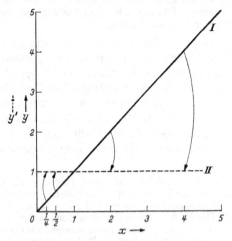

Abb. 1. Umwandlung einer linearen Regressionslinie $r_{yx}$ durch Berechnung der relativen Werte $y'$ aus $y$ zur Ausschaltung des Einflusses von $x$.

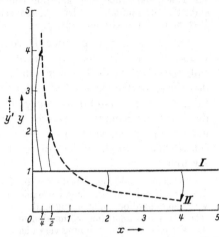

Abb. 2. Veränderung der Regressionslinie $r_{yx}$ durch Berechnung der relativen Werte $y'$ aus $y$ zur Ausschaltung des Einflusses von $x$ bei völliger Unabhängigkeit beider Merkmale.

Für alle Abbildungen gilt: Kurve I stellt die Funktion dar $y = f(x)$. Kurve II stellt die Funktion dar $y' = f\left(\dfrac{y}{x}\right)$. Kurve III stellt die Funktion dar $y = 1$, unabhängig von $x$.

Absicht entspricht, auf eine gerade Linie gebracht, die parallel der Abszisse verläuft, d.h. der Einfluß des betreffenden Faktors auf den Mittelwert wurde

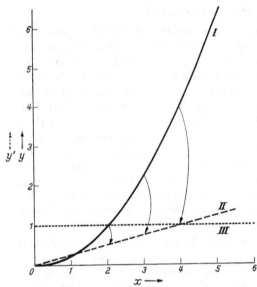

Abb. 3. Veränderung der Regressionslinie $r_{yx}$ durch Berechnung der relativen Werte $y'$ bei gekrümmter Regressionslinie.

ausgeschaltet, und das Material ist einheitlicher geworden, was sich in einer Verringerung der Streuung von $y$ ausprägt. Bei Vorliegen einer linearen Regression kommt die Verwendung der relativen Werte also im Prinzip auf das gleiche hinaus wie die Berechnung der Partialregressionen, nur sind diese wegen der Berücksichtigung beider Faktoren exakter und lassen viel weiter gehende Schlußfolgerungen zu, da beide Einflüsse in ihrer Wirkungsweise und Wirkungsstärke erkannt werden können. Liegt dagegen eine nicht-lineare Regression vor, dann kommt durch die Berechnung der relativen Werte wieder ein grober Fehler in die Berechnungen hinein, der um so gefährlicher ist, weil bei der Beurteilung eines derartig umgeformten Materials gerade davon ausgegangen wird, daß hierdurch eine besondere Exaktheit der Werte erreicht wird. Wenn die Regressionslinie sich von der Abszisse weg krümmt, so ist der Erfolg, daß durch die Umrechnung die großen Werte noch immer zu groß und die kleinen noch immer zu klein sind (Abb. 3). Die erstrebte Einheitlichkeit des Materials wird damit nicht erreicht, aber wohl wird bei den Schlußfolgerungen so verfahren,

als ob dies der Fall sei. Schließlich sei der Fall betrachtet, daß die Regressionslinie zur Abszisse hin gekrümmt ist. Die y-Werte, die kleinen x-Werten zugeordnet sind, werden dann stark vergrößert, diejenigen, die zu hohen x-Werten
gehören, dagegen übermäßig
verkleinert (Abb. 4). Die Mittelwerte für y der verschiedenen
x-Klassen liegen jetzt auf einer
Regressionslinie, die in ihrem
Verlauf der ursprünglichen entgegengesetzt ist, so daß aus
einer positiven Zuordnung der
einzelnen Werte eine negative
Abhängigkeit geworden ist. Die
gewünschte Vereinheitlichung
des Materials ist nicht eingetreten, dafür sind aber die
ursprünglichen Verhältnisse völlig umgedreht.

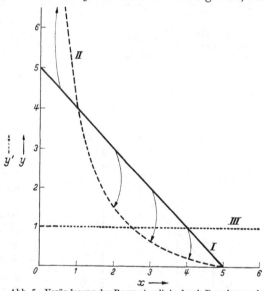

Abb. 4. Veränderung der Regressionslinie $r_{yx}$ durch Berechnung
der relativen Werte y' bei gekrümmter Regressionslinie.

In diesen Fällen wurde eine
positive Regression zwischen
den beiden Variablen voraus-
gesetzt. Wenn dagegen eine negative Regression vorliegt, also y mit wachsendem x kleiner wird, wird der große Wert für y bei kleinem x noch größer, der
kleine Wert bei großem x noch
stärker verkleinert, wodurch
eine eindeutige lineare, aber
negative Regressionslinie in eine
nur schwer zu deutende gekrümmte Regressionslinie verwandelt wird, die mit der ursprünglichen Linie nur das Vorzeichen gemeinsam hat. Statt
der stärkeren Zusammenziehung
der Werte ist aber der Erfolg
eingetreten, daß die Werte noch
weiter auseinander liegen als
vorher, d. h. die Variabilität
wurde vergrößert, während bei
den weiteren Folgerungen so
getan wird, als ob sie durch
die Ausschaltung eines Faktors
verkleinert wurde (Abb. 5).

Abb. 5. Veränderung der Regressionslinie durch Berechnung der
relativen Werte y' aus y zur Ausschaltung des Einflusses von x
bei linearer, negativer Regression.

Wenn gar gekrümmte Re-
gressionslinien mit negativem
Vorzeichen des Regressions-
koeffizienten vorliegen, werden die Verzerrungen, die die Verteilung für y durch
die Umrechnung in relative Werte erleidet, völlig grotesk.

Alle diese Fehler wirken sich natürlich besonders aus, wenn die Korrelation
mit dem eben ausgeschalteten Faktor für die weiteren Schlußfolgerungen wichtig
ist, aber werden noch verhängnisvoller, wenn z. B. anschließend die relativen Werte
in Beziehung gesetzt werden zu dem soeben ausgeschlossenen Faktor oder gar
zu einem anderen, der mit dem ausgeschlossenen korreliert ist, ohne daß dies

bekannt war. Es wird deutlich, daß dann die ursprünglich im Material vorhandenen Beziehungen völlig verkehrt dargestellt werden, was zu „statistisch gesicherten" Fehlschlüssen führen muß.

### β) Veränderungen der Streuung.

Diese Verzerrung der Mittelwerte ist aber nicht die einzige Verzerrung in den Meßwerten, die durch die Berechnung der relativen Werte eintritt. Neben der Veränderung der Mittelwerte ist auch zu beachten, welchen Einfluß die Berechnung der relativen Werte auf die Streuung von $y$ bei wachsendem $x$ hat, wenn zu jeder Stufe von $x$ mehrere $y$-Werte gehören, die naturgemäß etwas differieren, also eine bestimmte Streuung ergeben. Da bei gegebenem $x$ die größeren $y$-Werte relativ stärker reduziert werden als die kleineren, bleibt bei einer Umrechnung zwar die Richtung der Abweichung vom zugehörenden Mittelwert $\bar{y}$ erhalten, aber die Entfernung vom Mittel der relativen Werte, in Einheiten der $y$-Skala gemessen, hat sich verändert. Bei $x < 1$ wird die Streuung für $y'$ größer als für $y$, bei $x > 1$ dagegen kleiner, und zwar ist diese Veränderung der Streuung um so größer, je weiter die Entfernung von $x = 1$ ist. Dies bedeutet aber, daß durch die Berechnung der relativen Werte sich nicht nur die Mittelwerte von $y$ verändern, sondern daß sich auch die Streuung innerhalb der den verschiedenen $x$ zugeordneten Gruppen von $y$-Werten verändert. Am deutlichsten geht dies aus der graphischen Darstellung hervor, die nur für den Idealfall der linearen, positiven Regression gezeichnet wurde (Abb. 6). Wenn die ursprüngliche Streuung von $y$ unabhängig vom Mittelwert ist, entsteht eine Verzerrung der ganzen Verteilung dadurch, daß nach der Umrechnung bei kleinem $x$ die Streuung $\sigma_y$ zu groß, bei großem $x$ dagegen zu klein geworden ist. Wenn dagegen, wie in Abb. 7, die Streuung derart mit dem Mittelwert korreliert ist, daß $\sigma$ in Prozent des Mittelwertes immer gleich ist, also sich mit steigendem Mittelwert vergrößert, dann tritt durch die Umrechnung in relative Werte eine Angleichung der Streuung von $y$ in den zu verschiedenen $x$-Werten gehörenden Verteilungen ein.

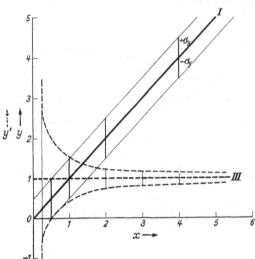

Abb. 6. Veränderung der Streuung von $y$ um die Regressionslinie durch die Berechnung der relativen Werte $y'$, wenn die Streuung $\sigma_y$ unabhängig von $x$ ist.

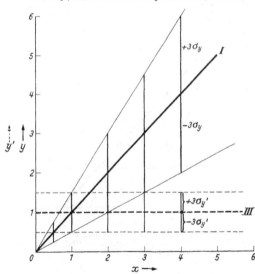

Abb. 7. Veränderung der Streuung um die Regressionslinie durch die Berechnung der relativen Werte $y'$ bei einer mit dem Mittelwert positiv korrelierten Streuung $\sigma_y$.

### γ) Folgen der Verwendung relativer Werte.

Um mit Hilfe der Umrechnung in relative Werte für $y$ unter Ausschaltung von $x$ zu der erstrebten Vereinheitlichung des Materials zu gelangen, müssen also zwei Voraussetzungen erfüllt sein: einmal muß eine streng lineare, positive Regression $r_{yx}$ vorliegen, zum andern muß aber die Streuung in der beschriebenen Weise mit dem Mittelwert korreliert sein. Wenn eine dieser beiden Voraussetzungen nicht erfüllt ist, führt die Umrechnung zu einer vollständigen Veränderung der Struktur des Materials. Immer dann, wenn dieses Material einer weiteren Analyse unterworfen wird, in der Annahme, daß ein unerwünschter Variabilitätsfaktor ausgeschaltet und es dadurch verbessert wurde, muß dies zu Fehlschlüssen führen. Eine richtige Bestimmung von Mittelwert und Streuung des angeblich berichtigten Faktors ist unmöglich geworden. Alle weiteren Überlegungen, die auf diesen Werten aufbauen, werden von diesem Fehler betroffen und sind damit für exakte Schlußfolgerungen unbrauchbar.

Ein völlig falsches Bild entsteht aber erst recht, wenn etwa die Regression zwischen den beiden Werten $x$ und $y$ sich durch die Versuchsbedingungen selber ändert. Durch eine Umrechnung in relative Werte nimmt man sich die Möglichkeit, über diesen Punkt überhaupt etwas auszusagen.

Wenn relative Werte von $y$ unter Ausschaltung von $x$ berechnet werden, so liegt aller weiteren Verwendung dieser Werte die Hypothese zugrunde, daß die Veränderlichkeit von $x$ als Variabilitätsursache von $y$ ausgeschaltet wurde und die verbleibende Streuung von $y$ nur noch durch die restlichen Variabilitätsfaktoren verursacht wird. Wie aber gezeigt wurde, trifft dies nur in einem einzigen Fall zu, der in der Biologie der Erfahrung nach praktisch nicht auftritt, nämlich bei positiver, linearer Regression und gleichzeitig mit dem Mittelwert korrelierter Streuung. In allen übrigen Fällen ist in der für $y$ vorhandenen Verteilung und damit bei der Bestimmung von Mittelwert und Streuung noch der Einfluß der Variabilität von $x$ wirksam, aber jetzt in einer Weise, die nicht mehr kontrollierbar und durch keine Rechenoperationen zu erkennen und zu entfernen ist. Alle weiteren Schlußfolgerungen, die aus derartig verzerrten Verteilungen gezogen werden, bauen auf einer nicht zutreffenden Hypothese auf und entbehren damit jeder Grundlage.

Vor einer Berechnung der relativen Werte muß also geprüft werden, ob die dieser Transformation der Meßwerte zugrunde liegenden Hypothesen im vorliegenden Fall zutreffen. Dazu sind folgende Berechnungen durchzuführen: Es ist nachzuweisen, 1. daß zwischen dem zu untersuchenden und dem auszuschaltenden Faktor eine positive, lineare Regression besteht; wenn eine gekrümmte Regressionslinie vorliegt, kann diese unter Umständen durch die Durchführung einer Skalentransformation gestreckt werden; 2. daß die Streuung um diese Regressionslinie in allen Abschnitten mit dem jeweiligen Mittelwert von $y$ so korreliert ist, daß $\sigma_y$ in Prozent des Mittelwertes konstant ist, 3. daß der auszuschaltende Faktor mit keinem derjenigen Faktoren, deren Beziehung zu dem zu untersuchenden Faktor geprüft werden soll, korreliert ist und selber bei den weiteren Berechnungen nicht mehr auftritt.

Wenn man sich die Mühe macht, tatsächlich alle diese Berechnungen durchzuführen, und das Glück hat, das Zutreffen der Hypothesen beweisen zu können, so ist es für die weitere Auswertung einfacher, alle diese Erkenntnisse bereits für die Schlußfolgerungen zu verwerten, statt erst die relativen Werte zu berechnen und dann mit der Auswertung von vorne anzufangen. Ohne diese Prüfung aber zur Berechnung der relativen Werte überzugehen, bedeutet, daß man sich durch eine völlige Verfälschung der Versuchsergebnisse der Gefahr von

Fehlschlüssen aussetzt, die ziemlich sicher zu einer vollkommen falschen Beurteilung des zu untersuchenden biologischen Problems führen und damit die erstrebten Auswirkungen einer statistischen Auswertung des Materials in ihr Gegenteil verkehren.

### c) Richtige Auswertungsmethoden.

Die richtige Auswertung von Versuchsdaten, in denen eine unerwünschte, aber unvermeidliche Veränderliche enthalten ist, kann ebenso wie im zuerst besprochenen Fall in verschiedener Weise geschehen. Alle Methoden zur Ausschaltung des unerwünschten Variabilitätsfaktors lassen sich aber meist mit Erfolg nur dann anwenden, wenn der Versuch von vornherein dafür geplant war und in entsprechender Weise durchgeführt wurde.

Einmal kann in einigen Fällen, wenn die Lage vor Versuchsbeginn erkannt wurde und es versuchstechnisch durchzuführen ist, der störende Faktor konstant gehalten werden. Wenn dies nicht möglich ist, kann seine Auswirkung dadurch ausgeschaltet werden, daß man bei der Versuchsplanung dafür sorgt, daß er dem Zufall nach auf alle Versuchsgruppen verteilt wird und so auf alle Teilversuche den gleichen Einfluß ausüben kann. Ist auch dies nicht möglich, was in einem biologischen Material durchaus vorkommen kann und immer eintreten wird, wenn der auszuschaltende Faktor selber durch die Versuchsbedingungen beeinflußt wird, so sind immer noch verschiedene einwandfreie rechnerische Möglichkeiten gegeben. Je nach den weiteren Bedingungen und der Natur der verschiedenen Veränderlichen, die im Versuch enthalten sind, kann die Berechnung der Partialregressionen oder der Partialkorrelationen angebracht sein, unter Umständen erst, nachdem die Meßwerte auf Skalen transformiert wurden, die lineare Regressionslinien ergeben. Wenn aber der auszuschaltende Faktor selber von den Versuchsbedingungen beeinflußt wird, ist die einzige Möglichkeit seiner Ausschaltung, seinen Einfluß durch eine Covarianzanalyse oder bei kleinem Material durch eine abgewandelte Regressionsrechnung (Mather 1951, Fisher 1949; concommitant observations) zu erfassen und dadurch sowohl den Grad der Abhängigkeit dieses Faktors von den Versuchsbedingungen wie seinen Einfluß auf die zu untersuchende Reaktion zu analysieren und so zu einer Einsicht in die Abhängigkeit der eigentlich interessierenden Veränderlichen von den Versuchsbedingungen unter Ausschaltung des störenden Einflusses zu gelangen. Die sich hierdurch ergebende Rechenarbeit ist in Kombinationsversuchen sehr groß, und es ist daher bei diesen zu empfehlen, den störenden Faktor durch eine andere Versuchsanordnung zu entfernen, wenn hierzu nur irgendwie die Möglichkeit besteht.

### d) Versuche mit Milligramm-Prozent (mg-%).

Oft werden die relativen Werte gewonnen in einer Form, die nicht ohne weiteres als hierher gehörig zu erkennen ist und meist sogar als besonders exakte Versuchsplanung gewertet wird. Es handelt sich darum, daß bei der Prüfung von Substanzen im Tierversuch, aber auch bei der Anwendung an Pflanzen, die Dosierung nicht erfolgt in „Einheiten pro Individuum" sondern in „Einheiten pro Gewichtseinheit des Individuums", wobei die Angabe der Dosis als Milligramm-Prozent erfolgt, d.h. es wurden $x$ mg des zu prüfenden Stoffes je 100 g Gewicht des Individuums verabreicht. Hierbei wird vorausgesetzt, daß zwischen dem Gewicht des Individuums und der für einen bestimmten Grenzwert der Reaktion notwendigen Menge des zu prüfenden Stoffes eine lineare Beziehung besteht, und daß außerdem die Streuung bei größeren (schwereren) Individuen zahlenmäßig größer ist als bei den kleineren, und daß infolgedessen die durch das verschiedene Ausgangsgewicht der Versuchsobjekte verursachte zusätzliche Streuung durch die Dosierung in Milligramm-Prozent ausgeschaltet werden kann. Dies sind aber die gleichen, zumindest unbewiesenen Voraussetzungen, die der Berechnung der relativen Werte zugrunde liegen, nur daß dort die Umrechnung nach Beendigung des Versuchs, hier aber bereits vorher erfolgt.

Das Ziel, das mit Hilfe der besonderen Dosierung nach dem Gewicht des Versuchsobjektes erreicht werden soll, ist dies, daß in allen Individuen eine gleichmäßige Konzentration des zu untersuchenden Stoffes hergestellt wird, unabhängig von der verschiedenen Größe des Objektes. Durch die übliche Methode wird dieses Ziel aber mit Sicherheit nicht erreicht, da der relative Anteil der Gewebe, in die der Stoff gelangen soll, bei Individuen verschiedenen Ausgangsgewichtes verschieden groß ist. Die Zufuhr des Stoffes in Prozent des Gesamtgewichts erreicht also mit Sicherheit eine verschiedene Konzentration in den zu untersuchenden Geweben. Durch den verschiedenen relativen Anteil des interessierenden Gewebes bei Individuen verschiedenen Gewichts wird bei einer Dosierung „pro Individuum" ebenfalls eine verschiedene Konzentration des zu prüfenden Stoffes geschaffen, die aber eine bestimmte, wenn auch nicht lineare Regression zum Gewicht zeigen wird. Wenn die Voraussetzung zutrifft, daß zwischen der Konzentration des Stoffes im Gewebe und der Reaktion darauf wenigstens in dem Bereich, in dem die Konzentration bei einer Dosierung „pro Individuum" variiert, eine durch eine Funktion darstellbare Beziehung besteht, dann ist es möglich, den Einfluß der Größe des Individuums in verschiedener Weise auszuschalten. Je nach dem Umfang des Materials und den weiteren Besonderheiten des Versuchs wird hierzu die Berechnung einer abgewandelten Regressionsmethode (s. MATHER 1951, FISHER 1949), der Partialregressionen, eine Covarianzanalyse oder eine Versuchsanordnung mit Paarbildung von gleichgroßen Individuen zum Ziel führen. Alle diese Methoden stellen eine einwandfreie Lösung des Problems dar, das sich aus der unterschiedlichen Größe der für den Versuch verwendeten Objekte ergibt, während die übliche Dosierung in Milligramm-Prozent auf nicht zutreffenden Hypothesen über die Korrelation der Wirkungsweise des zu untersuchenden Stoffes mit dem Gewicht der Versuchsobjekte aufbaut und damit zu einer Pseudoexaktheit führt, die in Wirklichkeit eine Verzerrung oder Verfälschung der Versuchsergebnisse zur Folge hat, wie sie für die relativen Werte geschrieben wurde.

# IV. Zeitreihen.

Eine besondere Behandlung erfordern die Versuche, in denen die Korrelation der Veränderungen eines bestimmten Vorgangs mit der Zeit untersucht werden soll, wenn die Untersuchung längere Zeiträume umfaßt. Solche Zeitreihen kommen in biologischem Material häufig vor; als Beispiele seien die Untersuchungen rhythmischer Vorgänge und Wachstumsmessungen genannt. Das Problem ist dabei, aus der sich ergebenden, stark gezackten Kurve die einzelnen Bestandteile herauszuholen. In einer solchen Kurve sind als Variabilitätsfaktoren enthalten: 1. die Zufallsschwankungen, 2. eine oder mehrere periodische Funktionen, z. B. Tagesschwankungen, jahreszeitliche Schwankungen, Jahresrhythmik, 3. eine zügige Funktion, die die allgemeine Richtung der Entwicklung angibt, der Trend, und 4. unstetige Funktionen, die einmalige, bleibende Veränderungen des Kurvenniveaus hervorrufen. Die Menge des für die einzelne Bestimmung herangezogenen Materials richtet sich dabei vor allem nach dem Ausmaß der Zufallsschwankungen, die Länge des Zeitraumes, über den die Untersuchungen ausgedehnt werden müssen, nach der Periodenlänge der unter 2. genannten Funktionen. Die

Möglichkeit solcher Zerlegung der Kurve ist durch die Periodogrammrechnung gegeben, die für wirtschaftsstatistische Untersuchungen und für meteorologische Untersuchungen ausgearbeitet wurde. Wegen der komplizierten Rechenoperationen wurde sie bisher in der Biologie nicht angewendet, sie soll aber doch erwähnt werden, weil es eine Reihe von Problemen in der Physiologie gibt, auf die sie grundsätzlich anwendbar ist. Einige dieser Fragen können auch mit einfacheren Methoden angegangen werden, wenn man auf einen Teil der Informationen über die Zusammensetzung der einzelnen Funktionen verzichtet. (Gebelein 1943).

Ein Vergleich mehrerer Zeitreihen erfordert besondere Methoden der Korrelationsrechnung. Für den Zusammenhang physiologischer Vorgänge, die alle eine Zeitkorrelation zeigen, sind mehrere Möglichkeiten gegeben. Beide untersuchten Phänomene können sich gleichsinnig und gleichzeitig verändern: Gleichbewegung. Der eine Vorgang kann eine Folge des anderen sein, wodurch seine Veränderungen sich zwar gleichsinnig, aber zeitlich immer etwas später vollziehen als beim ersten Vorgang: Folgebewegung. Die Untersuchung dieser Zusammenhänge führt zu Fragen, die in der Wirtschaftsstatistik als Lag-Problem bekannt sind. Die Bewegung der beiden Vorgänge kann weiter zwar gleichsinnig und gleichzeitig sein, aber mit verschieden starkem Ausschlag, was sich bei einer einfachen Regressionsrechnung an Teilabschnitten der Kurven in verschiedenen Größen der Regressionskoeffizienten bemerkbar machen würde: Strahlenbündel. Eine derartige Korrelation zweier Vorgänge ist sehr oft erst an sehr langen Beobachtungsreihen deutlich zu erkennen. Als letzte Möglichkeit kann ein Anstieg des einen Faktors mit einem Absinken des anderen verbunden sein: Gegenbewegung. Alle diese Fälle sind mit dem Problem der Partialregressionen und Partialkorrelationen verbunden, und ebenso wie dort ist immer dann, wenn ein Zusammenhang festgestellt wurde, zu prüfen, ob dieser ursächlich ist, oder ob eine gemeinsame Beeinflussung durch eine dritte Variable, die sich mit der Zeit verändert, vorliegt.

# D. Versuchsplanung.
## I. Grundlagen.

Wie bereits mehrfach betont, ist eine richtige Anwendung der Auswertungsmethoden für einen Versuch nur dann möglich, wenn dieser Versuch im Hinblick auf die geeignete statistische Methode geplant wurde. Für eine solche Planung ist nicht nur die Kenntnis der rechnerischen Möglichkeiten nötig, sondern es sind eine Reihe weiterer Faktoren zu berücksichtigen. Dies sind zuerst die Art des biologischen Problems, dann die Eigenschaften des Materials, die Natur der in den Versuch einzubeziehenden und darin notwendigerweise enthaltenen Variablen, die technischen Möglichkeiten der Versuchsanstellung und der erstrebte Grad der Sicherheit der Aussagen. Durch die Zusammenfassung der ersten drei dieser Bedingungen wird bestimmt, ob es sich um einen Versuch mit quantitativen oder qualitativen Variablen handelt, sowie ob ein einfacher oder ein Kombinationsversuch durchgeführt werden soll. Wenn diese Punkte geklärt sind, beginnt erst die eigentliche Planung des Versuchs. Diese setzt sich zusammen aus der Bestimmung des Umfangs der Einzelversuche und der Aufstellung des Auswertungsschemas, dem der ganze Versuch sich einfügen muß, und wird von den Faktoren bestimmt, die in den zwei letzten Punkten der obigen Aufstellung zum Ausdruck kommen.

Die Versuche und die Bestimmung der statistischen Maßzahlen aus dem Material werden durchgeführt, um daraus Rückschlüsse über das allgemeine

Verhalten des Objektes unter den bestimmten Bedingungen zu ziehen. Für eine völlig exakte Bestimmung der wahren Werte der statistischen Maßzahlen wäre dabei die Kenntnis der Reaktion aller Objekte nötig, also die Untersuchung eines unendlich großen Materials. Alle Bestimmungen an Stichproben aus dieser unendlich großen statistischen Masse können nur Schätzwerte der wahren statistischen Maßzahlen ergeben, die natürlich um so zuverlässiger sind, je größer die gewählte Stichprobe war. Die Größe des mittleren Fehlers der einzelnen Maßzahlen, der den Grad dieser Annäherung an den wahren Wert angibt, hängt außerdem aber noch von der Größe der Streuung ab. Die Streuung ihrerseits kann mit Hilfe der verschiedenen Methoden zerlegt werden in die Anteile, die den Zufallseinflüssen und den systematischen Faktoren zukommen. Ebenso, wie aus der Streuung und dem Umfang des Materials der Sicherheitsfaktor berechnet werden kann, ist es umgekehrt auch möglich, mit einer bekannten Streuung den zur Erreichung eines vorher festgelegten Sicherheitsgrades notwendigen Umfang des Versuchs zu berechnen. Für die Planung des Umfangs eines Versuchs müssen also drei Faktoren bekannt sein: 1. der Grad der erstrebten Sicherheit der Aussagen, 2. die durch die Zufallseinflüsse bedingte Streuung und 3. die durch die systematischen Einwirkungen zusätzlich in das Material hineingebrachte Variabilität.

### a) Bestimmung des Sicherheitsfaktors.

Von diesen drei Punkten kann der erste willkürlich festgesetzt werden. Meist wird angestrebt, die statistischen Schlüsse mit einer Wahrscheinlichkeit von 99% ($P = 0,01$) zu sichern. Für Versuche, die der ersten Orientierung dienen sollen, die sich auf weniger wichtige Fragen beziehen, oder bei denen aus einem möglichst kleinen Material einige Hinweise erhalten werden sollen, wird der zu erreichende Grenzwert oft auf $P = 0,05$ oder gar $0,1$ festgesetzt, was aber heißt, daß hierbei 5 bzw. 10% Fehlschlüsse in Kauf genommen werden. Bei Versuchen, die der Entscheidung über grundsätzliche Probleme dienen sollen, bei denen also aus den Ergebnissen sehr wichtige Schlüsse gezogen werden müssen, von denen unter Umständen die weitere Arbeitsrichtung eines ganzen Wissenschaftszweiges abhängt, wird man das Risiko von Fehlschlüssen viel stärker einschränken müssen und hierfür meist eine Sicherung mit $P = 0,001$ verlangen.

### b) Bestimmung der Zufallsstreuung, Blankoversuch.

Die Bestimmung des zweiten Faktors, der durch die Zufallseinflüsse bedingten Streuung und ihre Trennung von der Einwirkung unerwünschter Faktoren, erfolgt durch einen Blankoversuch mit dem Material, mit dem gearbeitet werden soll. Diese Untersuchung der natürlichen Variabilität muß immer dann vorgenommen werden, wenn ein neues Objekt eingeführt wird. Sie kann so angesetzt werden, daß gleichzeitig Aufschlüsse darüber erhalten werden, welche der natürlicherweise im Material vorhandenen Unterschiede als zusätzliche Variabilitätsfaktoren für die zu untersuchende Reaktion in Frage kommen. Hierher gehören Unterschiede zwischen Teilen von Individuen (z.B. verschiedene Blätter einer Pflanze), zwischen Individuen bei gleicher Vorbehandlung, zwischen verschieden alten Individuen. Eine derartige Untersuchung gibt also nicht nur Aufschluß über die natürliche Streuung, sondern auch darüber, auf welche Variabilitätsursachen bei den weiteren Versuchen geachtet werden muß. Damit ist z.B. eine Entscheidung möglich, ob für einen Versuch die zusätzliche Variabilität zwischen Individuen in Rechnung zu stellen oder zu vernachlässigen ist, oder ob bei der Auswahl der zu untersuchenden Teile etwa auf gleiches Alter oder gleiche Größe geachtet werden muß. Derartige Blankoversuche beanspruchen zunächst für die Durchführung der Messungen und die Auswertung der Daten genau soviel Zeit wie ein „echter" Versuch, aber mit dieser einmaligen Arbeit ist auch die Grundlage für alle weiteren Untersuchungen an dem betreffenden Objekt gelegt, und es wird weiterhin viele unnütze Arbeit gespart, wenn einmal bekannt ist, mit welchen Faktoren zu rechnen ist und welche keinen Einfluß auf die zu prüfende Reaktion haben.

Bei einer Vernachlässigung einer solchen Untersuchung mit dem Ziel der Arbeitsersparnis wird oft darauf verwiesen, daß die Kontrollgruppen eines jeden Versuchs doch Auskunft über diese Fragen geben. Es wird dabei aber übersehen, daß die Kenntnis der materialeigenen Streuung und derjenigen Faktoren, die entweder konstant zu halten, zufällig zu verteilen oder gesondert in Rechnung zu stellen sind, für die Planung eines Versuchs unumgänglich notwendig ist und in dieser Ausführlichkeit, wie sie durch einen Blankoversuch ermittelt wird, aus den oft zu kleinen Kontrollgruppen nie herausgelesen werden kann, abgesehen davon, daß solche erst nachträglich gewonnenen Informationen für die Planung des Versuchs nicht berücksichtigt werden können. Die Frage, welche von den möglichen Variabilitätsursachen tatsächlich auf die geprüfte Reaktion einen Einfluß haben und daher ausgeschaltet werden müssen, kann nicht durch theoretische Überlegungen oder Analogieschlüsse aus andersartigen Versuchen entschieden werden, sondern erfordert die Kenntnis der Ursachen der natürlichen Variabilität, die nur die beschriebene Voruntersuchung liefern kann.

Beispiel. Wenn es durch den Blankoversuch bekannt ist, daß das Alter der Individuen auf die geprüfte Reaktion keinen Einfluß hat, ist es sinnlos, zum Zweck einer „exakten" Versuchsanstellung großen Wert auf die gleichzeitige Anzucht zu legen und etwa den Versuch zu beschränken, weil nicht genug gleichaltriges Material zur Verfügung steht. Wenn in einem anderen Fall die individuelle Variabilität sehr groß ist, der Einfluß der Größe des untersuchten Stückes aber vernachlässigt werden kann, wird das Ergebnis nicht exakter dadurch, daß mit viel Mühe genau gleichgroße Teilstücke verschiedener Individuen herausgesucht werden.

Die vorbereitenden Untersuchungen müssen ebenso wie für die Bestimmung der Zufallsstreuung auch für die Prüfung des Einflusses vermuteter zusätzlicher Variationsursachen verwendet werden, etwa zu Untersuchungen über das Ausmaß der Differenzen zwischen verschiedenen Meßgeräten oder Versuchsapparaturen oder zur Prüfung auf das Vorkommen von individuellen Unterschieden bei der Klassifizierung von Versuchsergebnissen durch verschiedene Beobachter. Falls derartige Faktoren eine Rolle spielen, müssen sie bei der Versuchsplanung berücksichtigt werden, es ist aber eine überflüssige Vorsicht, wegen eines vermuteten Einflusses, dessen Größe unbekannt ist, den Versuchsplan kompliziert zu machen. Als Grundsatz muß gelten, in den Voruntersuchungen alle möglicherweise störend wirkenden Variabilitätsursachen einer einmaligen statistischen Prüfung auf das Ausmaß ihrer Wirkung hin zu unterziehen, aber in den endgültigen Versuchsplänen nur diejenigen zu berücksichtigen, deren Einfluß tatsächlich im Vergleich zu den von den systematischen Faktoren hervorgerufenen Differenzen bedeutend ist und wirklich als Störungsfaktor angesehen werden muß.

Beispiele. Wenn durch die Voruntersuchung bekannt ist, daß mehrere Beobachter gleiche Meßergebnisse erzielen, oder daß verschiedene Meßgeräte mit gleicher Genauigkeit arbeiten, ist es überflüssig, diese Differenz der Auswertungsbedingungen der Teilversuche im Plan zu berücksichtigen; andererseits darf nicht so verfahren werden, als ob mehrere Apparate einheitlich arbeiten würden, solange dies nicht nachgewiesen ist.

Bei Kenntnis dieser Daten ist es oft möglich, die weiteren Untersuchungen zu vereinfachen oder den Versuch in einem viel kleineren Umfang durchzuführen, als zuerst geplant war, so daß die einmal für die Blankobestimmung und Voruntersuchung aufgewendete Zeit in keinem Fall als verloren angesehen werden kann, auch wenn der Beginn der eigentlich geplanten Versuche dadurch etwas hinausgeschoben wird.

### c) Blindversuche.

Von den Blankoversuchen, die keine systematischen Faktoren enthalten dürfen und die manchmal als Blindversuche bezeichnet werden, sind die echten Blindversuche zu unterscheiden. Hierunter sind Versuche zu verstehen, bei denen Versuchsansteller und Beobachter nicht identisch sind, und derjenige, der die Messung der Ergebnisse vornimmt, nicht weiß, welche Behandlung die Teilversuche erfahren haben, oder ob es sich etwa um einen Blankoversuch ohne jede Behandlung handelt. Dies betrifft einen Teil des Versuchsplanes, der nichts mit der statistischen Planung und Auswertung zu tun hat, der aber immer dann eingeführt werden muß, wenn die Möglichkeit einer Beeinflussung der Messung durch das zu erwartende Ergebnis gegeben ist. Die Bestätigung für das Vorliegen eines derartigen Einflusses wird gefunden, indem ein echter Blankoversuch nach einem vorgegebenen Versuchsschema ausgemessen und statistisch ausgewertet wird. Eine Beeinflussung der Messung zeigt sich an „gesicherten" Differenzen im Blankoversuch an, wenn diese mit den angeblichen Versuchsunterschieden zusammenfallen, die dem Beobachter vorgelegen haben.

### d) Vorversuche.

Der letzte Faktor, der für die Planung bekannt sein muß, ist das Ausmaß der durch die Versuchsbedingungen zusätzlich in das Material hineingebrachten Streuung, also die Größenordnung der zu erwartenden Differenzen zwischen den Teilversuchen. Hierzu dienen die Vorversuche, die an kleinem Material mit den Extremwerten der durch die Versuchsbedingungen variierten Faktoren ohne Berücksichtigung der Zwischenwerte durchgeführt werden und nur zeigen sollen, mit welchen Unterschieden etwa gerechnet werden muß. Eine eigene statistische Sicherung der hierbei gefundenen Differenzen kann unterbleiben, weil der Umfang der Versuche meist zu klein ist.

## II. Versuchsumfang und Sicherheitsfaktor.

### a) Bestimmung des Umfangs der Teilversuche bei festgelegtem Sicherheitsfaktor.

Es ist jetzt der nächste Schritt, die im Blankoversuch bestimmte materialeigene Streuung zu den zu erwartenden Differenzen unter Berücksichtigung des gewünschten Sicherheitsgrades so in Beziehung zu setzen, daß sich daraus der notwendige Umfang der Teilversuche ergibt. Hierbei kann aus dem Blankoversuch, der Mittelwert und Streuung der Reaktion unter „natürlichen" Bedingungen ergeben hat, eine sog. Fehlerkurve aufgestellt werden (POST 1952). Unter der Voraussetzung, daß der Umfang der Bestimmungen nicht zu klein war, wird angenommen, daß der für die Streuung gefundene Schätzwert dem wahren Wert der zugrunde liegenden, unendlich großen statistischen Masse so nahe kommt, daß beide für die praktischen Zwecke als gleich angesehen werden können. Ausgehend von diesem Streuungswert $\sigma$ wird dann bestimmt, welche Größe der mittlere Fehler $m$ annimmt in einer Stichprobe bestimmten Umfangs. Zur Vereinfachung der Berechnungen wird der Umfang $N$ so festgesetzt, daß die zugehörige Anzahl der Freiheitsgrade $n$ eine Zahl ergibt, die rechnerisch einfach zu handhaben ist. Als Testwerte, für die $m$ bestimmt wird, werden meist gewählt $n = 5, 10, 20, 30, 50, 100, 150, 200, 500, 1000$. Die erhaltenen Werte werden in ein Koordinatensystem eingetragen, auf dessen Abszisse als unabhängige Veränderliche die Anzahl der Freiheitsgrade aufgetragen ist, während die Ordinate als abhängige Veränderliche die Größe des mittleren Fehlers angibt. Die so erhaltene Kurve zeigt, daß zunächst bei wachsender Individuenzahl die Größe des mittleren Fehlers ehr schnell abnimmt, während sie dann allmählich umbiegt und sich schließlich assymptotisch der $x$-Achse nähert, die bei $n = \infty$ erreicht wird. Im ersten steilen Abschnitt der Kurve wird durch eine Vergrößerung des Materials eine starke Verringerung des mittleren Fehlers, also ein großer Zuwachs an Genauigkeit erreicht, während nach der Biegung der Kurve eine Verdoppelung des Versuchsumfangs nur eine relativ geringe Verminderung des mittleren Fehlers zur Folge hat, der meist nicht mehr ins Gewicht fällt. Sobald bei einer Vergrößerung des Versuchsumfangs dieser Teil der Kurve erreicht wird, hat eine weitere Vermehrung zwar eine erhebliche Mehrarbeit zur Folge, die aber in keinem Verhältnis zum erzielten Zuwachs an Genauigkeit steht. In einem solchen Fall wird man also, wenn es technisch möglich ist, den Versuchsumfang zu verdoppeln, nicht den Umfang der Einzelversuche oder die Anzahl der Wiederholungen vermehren, sondern bei gleichbleibendem Umfang der Teilversuche neue Versuchsfaktoren einführen oder die vorhandenen weiter variieren, denn hierdurch verspricht die aufzuwendende Mehrarbeit tatsächlich neue Informationen, während sie im ersteren Fall das Versuchsergebnis nicht verbessern kann.

Durch die Größe des mittleren Fehlers wird andererseits gleichzeitig die Größe der in einem Versuch nachzuweisenden Differenzen bestimmt. Wenn der Vorversuch ergab, daß die zu erwartenden Unterschiede sehr klein sind, ist dies bei der Planung der Größe der Teilversuche zu berücksichtigen, während bei großen Differenzen oft die Versuche wesentlich eingeschränkt werden können, weil der Umfang nicht größer zu sein braucht, als zum Nachweis der Unterschiede mit der geforderten Sicherheit notwendig ist.

Während meist erst nach dem Versuch aus der Ausrechnung bestimmt wird, mit welcher Sicherheit die Aussagen gemacht werden können, ist es also bei einer Planung auf Grund von Blanko- und Vorversuchen möglich, den erstrebten Sicherungsgrad im voraus festzulegen, wodurch häufig wesentlich Arbeit eingespart werden kann, die dann für die Durchführung zusätzlicher Versuche zur Verfügung steht.

### b) Bestimmung des Sicherheitsfaktors bei festgelegtem Umfang der Teilversuche.

In einem biologischen Material ist oft der Versuchsumfang nicht bestimmt durch die Überlegungen über die für einen Nachweis notwendige Anzahl von Einzelbeobachtungen, sondern viel häufiger ist es unmöglich, das Material über einen bestimmten Umfang hinaus zu vergrößern, ohne zusätzliche Variabilitätsfaktoren einzuführen. In einem solchen Fall gibt die Fehlerkurve Aufschluß über die Größe der Differenzen, die mit einem Material derartig begrenzten Umfangs noch nachgewiesen werden können. Wenn der Vorversuch dann zeigt, daß die zu erwartenden Differenzen kleiner sind, kann der Hauptversuch unterbleiben, weil er mit Sicherheit ergebnislos verlaufen, d.h. zur Sicherung der auftretenden Unterschiede nicht ausreichen wird, oder es muß eine Versuchsplanung gesucht werden, die eine Berücksichtigung der zusätzlichen Variabilität ermöglicht.

## c) Sequenzanalyse.

Eine extreme Anwendung dieser rationellen Planung stellt die Sequenzanalyse dar, die bisher erst für einige Arbeitsgebiete verwendet wird, deren Anwendungsbereich sich aber auf alle Versuche erstreckt, bei denen die Durchschnittswerte der geprüften Reaktion bekannt sind, die Möglichkeit orientierender Vorversuche nicht gegeben ist, und das Versuchsmaterial nacheinander anfällt. Die Auswertung wird dabei vorweggenommen und der Versuch dann abgebrochen, wenn der gewünschte Sicherheitsgrad zum Nachweis einer Differenz in der vorher festgelegten Größe erreicht ist, oder wenn mit einer ebenfalls vorher bestimmten Sicherheit gesagt werden kann, daß die endgültige Differenz unter einer willkürlich festgelegten unteren Grenze liegen wird. Die Erfahrung hat gezeigt, daß hierbei wesentlich schneller Ergebnisse erhalten werden, als wenn erst der Versuch ohne Zwischenauswertung in einem großen Umfang durchgeführt wird, wobei oft nachträglich festgestellt wird, daß der erreichte Sicherheitsgrad weit über das notwendige Maß hinausgeht, oder die nachgewiesenen Differenzen in einer Größenordnung liegen, die praktisch nicht mehr interessiert (WALD 1947).

# III. Bestimmung der Anzahl von Einzelversuchen.

Wenn für ein Material die natürliche Streuung und daraus die Fehlerkurve bekannt sind, dazu die Größenordnung der zu erwartenden Differenzen und die Art der Variablen, so hängt die weitere Planung von den technischen Möglichkeiten der Versuchsanstellung ab. Diese entscheiden darüber, wie groß der Einzelversuch sein kann, wie viele Einzelversuche gleichzeitig durchgeführt werden können, und wie viele Wiederholungen möglich sind. Die technischen Beschränkungen machen sich darin bemerkbar, daß sie die Anzahl der unter gleichen Bedingungen durchzuführenden Bestimmungen beschränken und bei einer Überschreitung dieser Grenze zusätzliche Variabilitätsfaktoren schaffen.

Bei günstigsten, aber nur selten gegebenen Voraussetzungen liegen keinerlei Beschränkungen dieser Art vor; der Versuch könnte theoretisch auf ein unendlich großes Material ausgedehnt werden. In diesem Fall ist ein vollständiger Kombinationsversuch, in den mehrere willkürliche Variablen einbezogen und in allen Kombinationen geprüft werden, und die Auswertung je nach der Art der Messungen mit der Varianzanalyse oder dem $\chi^2$-Test das gegebene. Die Größe der Einzelversuche bestimmt sich dann ausschließlich nach der Größe der nachzuweisenden Differenzen und dem Grad der hierfür erstrebten Sicherheit. Durch die technischen Begrenzungen kann es nötig werden, in den Versuch zusätzliche Veränderliche aufzunehmen, die Anzahl der willkürlichen Veränderlichen einzuschränken oder bestimmte Versuchspläne und Auswertungsmethoden wie confounding, Berechnung von Partialregressionen u. a. anzuwenden. Auskunft hierüber geben jeweils die Voruntersuchungen.

# IV. Rationelle Planung.

Außer den technischen Begrenzungen eines Versuchsumfangs ist für eine richtige Versuchsplanung noch ein anderer Gesichtspunkt zu berücksichtigen, der bereits mehrfach erwähnt wurde. Das Ziel ist nicht nur, zu gesicherten wissenschaftlichen Befunden zu gelangen, sondern auch, diese so rationell wie möglich zu erhalten, also mit einem Arbeits- und Materialaufwand, der so gering ist, wie es mit der Forderung nach Sicherheit der Aussagen zu vereinbaren ist. Auch bei diesem Teil der Planung ist die statistische Voruntersuchung des

Materials unentbehrlich, um ein möglichst günstiges Verhältnis zwischen der aufgewendeten Arbeit und den erhaltenen Informationen herzustellen. Der Versuchsumfang, sowohl was die Größe der Teilversuche wie die Anzahl der variierenden Faktoren und ihrer Kombinationen betrifft, ist dann immer ein Kompromiß zwischen den statistischen Forderungen nach einem möglichst großen Material und den praktischen Möglichkeiten einer arbeitsmäßigen Bewältigung. „Richtig" ist der Umfang eines Versuchs dann, wenn aus der aufgewendeten Arbeit möglichst viel Ergebnisse gewonnen werden, also das Verhältnis zwischen Arbeit einerseits und Menge und Sicherheit der Aussagen andererseits ein Optimum erreicht. Dabei wird immer auf Sicherheit der Schlußfolgerungen verzichtet, um Arbeit zu sparen, aber der Versuchsansteller muß in jedem Fall wissen, wie groß dieser Verzicht ist, um sein Ausmaß richtig zu planen und die erhaltenen Ergebnisse richtig interpretieren zu können.

Eine richtig geführte Voruntersuchung des Materials gibt die Möglichkeit, zu vermeiden, den Hauptversuch zu klein oder zu umfangreich anzusetzen. Arbeitsverschwendung durch einen ergebnislosen Versuch, der die Differenzen gerade nicht sichern läßt, oder durch Aufwendung für einen geringen, oft gar nicht benötigten Zuwachs an Sicherheit der Aussagen wird verhindert. Es können in der gleichen Zeit oder mit den gleichen Mitteln oft mehrere Versuche gemacht werden, die dann viel weiter gehende Aussagen erlauben, oder es wird erreicht, daß an Stelle von mehreren zu kleinen Versuchen ein großer Kombinationsversuch gemacht wird, der dann mit fast dem gleichen Arbeitsaufwand viel weiter gehende Schlüsse erlaubt, als es ohne Planung möglich gewesen wäre.

Vor allem dann, wenn es darum geht, zu entscheiden, ob ein vollständiger Kombinationsversuch oder ein solcher mit einem Zusammenfallen (confounding) bestimmter Variablen anzusetzen ist, sind diese Überlegungen zu berücksichtigen. Wenn der Verlust an Sicherheit der Aussagen in einem Versuch mit „confounding" gering ist im Vergleich zu der Einsparung an Arbeit gegenüber dem vollständigen Versuch, wird immer der rationellere Versuchsplan vorzuziehen sein.

Ebenso, wie eine rationelle Planung oft zu einer Einschränkung der Versuchsarbeit führen kann, ist es in anderen Fällen nötig, den Versuch auszudehnen, etwa durch Einführung zusätzlicher Wiederholungen, Vermehrung der Stufen einiger Veränderlichen, Konstanthaltung mehrerer Bedingungen, Berücksichtigung der unerwünschten Variablen usw., wenn sich aus dem vorher festgelegten Auswertungsschema ergibt, daß die Vergrößerung des Versuchs eine erhebliche Vereinfachung der auszuführenden Berechnungen ergibt.

# V. Versuchsschema und Versuchsplan.

Es ist vielleicht nicht überflüssig, darauf hinzuweisen, daß das auf Grund der statistischen Vorarbeiten aufgestellte Versuchsschema nicht mit dem vollständigen Versuchsplan identisch ist. Das Versuchsschema enthält nur alle Einzelheiten des Versuchs, die für die spätere statistische Auswertung wichtig sind und berücksichtigt werden müssen, der Versuchsplan dagegen umfaßt auch alle sonstigen wichtigen Anweisungen über die Vorbehandlung der Objekte, Vorbereitung und Durchführung des Versuchs, Ausführung der Messungen und die Anleitung zur Auswertung, Zeitplan und Arbeitsverteilung, kurz alle Angaben, die für die vollständige Vorbereitung und Ausführung des Versuchs bis zur Beendigung der statistischen Auswertung und der Aufstellung von Anweisungen für die folgenden Versuche notwendig sind. Das Versuchsschema stellt einen wichtigen Teil dieses Planes dar. Am deutlichsten wird dieser Unterschied für Freilandversuche, wo der Versuchsfeldplan den Grundriß darstellt, nach dem das Feld aufgeteilt und der Versuch angepflanzt wird, und der später das Schema für die statistische Auswertung darstellt, während der Feldversuchsplan dazu alle weiteren Angaben enthalten muß, von den ersten Vorbehandlungen des Feldes oder Versuchsgartens und der Gewinnung des Saatgutes im Vorjahr bis zur Durchführung des Versuchs, Ausführung der Messungen und der folgenden statistischen Auswertung.

## VI. Versuche mit unvollständiger Planung.

Bisher wurden die Versuche besprochen, bei denen vorher die Aufstellung eines bestimmten Auswertungsschemas, das dem Versuchsplan zugrunde gelegt wird, möglich ist. Bei manchen Fragestellungen ist dagegen der Umfang des anfallenden Materials nicht im voraus zu bestimmen, weil jede Absicht, einen gleichen Umfang der Teilversuche zu erzielen, auf eine Selektion des anfallenden Materials hinauslaufen würde. Vor allem wenn hinzukommt, daß die Durchführung des Versuchs und die Vorbereitung der Auswertung viel Zeit in Anspruch nehmen im Vergleich zu der Aufwendung für die dann folgenden einzelnen Bestimmungen, ist es oft angebracht, die Auswertung nicht auf den Umfang des kleinsten Teilversuches zu begrenzen, sondern das gesamte Material zunächst auszuwerten und dann auch in die Berechnungen einzubeziehen. Dieser Fall liegt häufig bei cytologischen Untersuchungen vor, tritt aber auch auf anderen Gebieten ein. Hier läßt sich zwar der Versuch in großen Zügen planen, aber die endgültige Auswertung ist im einzelnen nicht vorher festzulegen, weil nicht feststeht, welche Teilversuche überhaupt verwertbares Material ergeben werden und welchen Umfang es bei den übrigen haben wird. Die Auswertungsmethode muß sich dann nach den Möglichkeiten richten, die durch die Befunde gegeben sind. Für derartige Versuche mit wechselndem Umfang der Teilgruppen sind besondere Abwandlungen der Varianzanalyse und des $\chi^2$-Testes ausgearbeitet (MATHER, FISHER, WEBER).

## E. Schluß.

### I. Arbeitshypothese und statistische Prüfhypothese.

Bei der statistischen Auswertung von Versuchen ist zu beachten, daß die dem Versuch zugrunde liegende Arbeitshypothese in den seltensten Fällen die für die statistische Prüfung angenommene Nullhypothese ist. Dies folgt einfach aus der Tatsache, daß nur in sehr seltenen Fällen ein direktes Interesse am Nachweis der Unwirksamkeit bestimmter Versuchsbedingungen besteht. Die Arbeitshypothese geht meist davon aus, daß bestimmte Bedingungen eine zu prüfende Reaktion in bestimmter Weise beeinflussen, und der Versuch wird durchgeführt mit der Absicht, hierfür eine Bestätigung zu finden.

Die Arbeitshypothese liegt meist in einer Form vor, die besagt, daß der Faktor $x$ einen Einfluß auf den Vorgang $y$ hat, wobei oft sogar bestimmte Vorstellungen über das Ergebnis dieser Beeinflussung und die ablaufende Reaktionskette bestehen. Der Versuch wird dann so angesetzt, daß der Faktor $x$ variiert und die Größe von $y$ gemessen wird, die unter diesen verschiedenen Bedingungen erreicht wird. Die Abhängigkeit der Größe $y$ vom Wert, den $x$ annehmen kann, soll festgestellt werden. Eine derartige Hypothese „es besteht ein Einfluß bestimmter Größe" kann zwar auch geprüft werden, aber wenn gefunden wird, daß das Ergebnis nicht damit übereinstimmt, ist nichts darüber ausgesagt, ob der geprüfte Einfluß nicht vorhanden, kleiner oder größer als die angenommene Abhängigkeit ist. Bei einem negativen Ausfall dieser Prüfung ist also eine Wiederholung mit einer anderen Ausgangshypothese notwendig, unter Umständen wären sogar ungezählte Rechnungen nötig, um endlich, mehr oder weniger zufällig, eine Prüfhypothese zu finden, bei der die Erwartung, die auf Grund dieser Hypothese aufgestellt wird, eine Übereinstimmung der gefundenen Werte aufweist. Die Formulierung läßt sich nämlich auch in die Form kleiden: „Es besteht zwischen den Größen $x$ und $y$ eine Abhängigkeit, die sich durch eine bestimmte Funktion ausdrücken läßt." Es leuchtet sofort ein, daß es äußerst unwahrscheinlich ist, durch bloßes Herumprobieren zufällig diese oft sehr komplizierte Funktion zu finden, da sie zunächst immer unbekannt ist.

Die statistische Prüfhypothese wird daher so formuliert, daß sie von allen hypothetischen Zusatzannahmen über die Art und Größe der Abhängigkeit zwischen beiden Variablen unabhängig ist. Die Formulierung, die hierfür in den meisten Fällen geeignet ist, lautet: Es besteht kein Unterschied zwischen den zu vergleichenden Versuchsreihen oder zwischen den empirischen Befunden und der theoretisch geforderten Verteilung; die Ergebnisse unterscheiden sich nicht mehr, als zufällig möglich ist; die zu erwartende Differenz ist gleich Null. Oder anders ausgedrückt: Die zu vergleichenden Versuche können als Stichproben aus einer homogenen, unendlich großen statistischen Masse angesehen werden. Es wird dann geprüft, mit welcher Wahrscheinlichkeit diese Nullhypothese als ausreichende Erklärung für die in den Versuchen aufgetretenen Differenzen der Ergebnisse angesehen werden kann, d.h. ob diese Unterschiede noch als zufällige Varianten bei der Entnahme von mehreren Proben aus einer einheitlichen statistischen Masse auftreten können. Wenn die gefundenen Differenzen so groß sind, daß die Wahrscheinlichkeit für das Zutreffen in einem bestimmten Versuch des gegebenen Umfangs über einem vorher bestimmten Grenzwert bleibt, wird angenommen, daß das Versuchsergebnis der zugrunde gelegten Nullhypothese nicht widerspricht, also eine Wirkung der angewendeten variierenden Faktoren der Versuchsbedingungen nicht nachgewiesen ist. Dies heißt nicht, daß mit Sicherheit keine Wirkung vorliegt, sondern nur, daß bei einem Versuch bei den gegebenen Bedingungen und dem gegebenen Umfang eine Differenz von der gefundenen Größenordnung auch noch zufällig auftreten kann. Es ist also durchaus möglich, daß bei einer Wiederholung des Versuchs, einer Vergrößerung des Versuchsumfangs, einer Veränderung der bisher konstant gehaltenen weiteren Bedingungen usw., aber auch bei einer weiteren Variation der untersuchten Faktoren sich doch Unterschiede in der Wirkung nachweisen lassen könnten. Die Berücksichtigung derartiger Möglichkeiten und die richtige Einschätzung des Versuchsergebnisses ist nicht mehr Sache der statistischen Auswertung des gerade vorliegenden Versuchs, sondern der Deutung der erhaltenen statistischen Maßzahlen. Sie kann wichtige Hinweise geben für die Planung der Bedingungen und des Umfangs weiterer Versuche und für die Deutung der biologischen Ergebnisse, vor allem aber für den Vergleich mit anderen, vielleicht abweichenden Befunden, die zum gleichen Thema bereits vorliegen.

## II. Auswertung und Deutung der statistischen Zahlen.

Bei jeder statistischen Bearbeitung eines biologischen Materials ist daher eine strenge Scheidung notwendig zwischen den statistischen Schlüssen und den biologischen Schlußfolgerungen, die daraus gezogen werden. Die statistische Bearbeitung allein kann nur entscheiden, ob eine bestimmte Hypothese, die logisch und biologisch möglich sein muß, als zureichende Erklärung für die beobachteten Differenzen im Versuchsmaterial angesehen werden kann. Da die zur Diskussion stehende Hypothese in den weitaus meisten Fällen eine sog. Nullhypothese ist, ist ihre Verwerfung von größerer Bedeutung als ihre Bestätigung, da im ersten Fall eine Wirkung der angewendeten Versuchsbedingungen wahrscheinlich ist. Darüber, was bei Nichtzutreffen der geprüften Nullhypothese als Erklärung für die gefundenen „gesicherten" Differenzen anzunehmen ist, und vor allem darüber, welche biologischen Vorgänge im einzelnen dafür verantwortlich zu machen sind, kann die Statistik keine Auskunft geben. Die „statistische Sicherung" gilt immer nur für den Nachweis der Differenzen, die unter dem Einfluß der Versuchsbedingungen aufgetreten sind, nicht aber für die Reaktionskette, die zwischen Wirkung und Ursache abgelaufen sein könnte, und die auf Grund anderer Voraussetzungen als Deutung für diese Wirkung der Versuchsbedingungen angenommen werden kann.

Wenn diese Trennung deutlich beachtet wird, kann nie die ursprünglich aufgestellte Arbeitshypothese, die selten mit der Nullhypothese der statistischen Prüfung identisch ist, bei einer begründeten Verwerfung der letzteren als „statistisch bewiesen" angesprochen werden, und bei einem Fehler in dem Teil der Schlußfolgerungen, der sich mit der Wahrscheinlichkeit des Zutreffens der Arbeitshypothese befaßt, bleiben doch die Versuchsergebnisse als solche unangetastet und sind einer neuen Deutung zugänglich. Falls dagegen eine solche Trennung bei der Darstellung der Versuche nicht eingehalten wird, ist es oft unmöglich, später nachzuprüfen, ob ein falscher Schluß seine Ursache in einem Versuchsfehler, einem Fehlschluß bei der statistischen Berechnung oder einem logischen Fehler bei der Deutung hat, wodurch eine erneute Verwendung des Materials ausgeschlossen ist. Bei Beachtung der Regel, daß die statistischen Schlußfolgerungen und die sich darauf aufbauenden biologischen Deutungen getrennt formuliert und deutlich gegeneinander abgesetzt werden müssen, ist eine Fehldeutung der statistischen Arbeitsweise nicht möglich. Viele biologisch falsche Schlüsse, die angeblich „statistisch bewiesen" wurden und dann Anlaß zu einer Verurteilung der statistischen Arbeitsweise gaben, sind in Wirklichkeit solche Fehler in der Ausdeutung der statistischen Maßzahlen.

Die Aufgabe der biologischen Statistik ist es nicht, die Versuchsergebnisse in Zahlen auszudrücken, sondern dazu zu führen, die Versuche rationell anzusetzen und auszuwerten und die erhaltenen Ergebnisse richtig zu deuten. Mit einer Reihe von statistischen Maßzahlen ist nichts gewonnen, wenn diese nicht interpretiert werden können. Die Zahlen an sich sind Beobachtungsergebnisse, die ebenso wie jede andere Beobachtung verschiedenen Deutungen zugänglich sind. Eine richtige Anwendung der statistischen Arbeitsmethoden liegt dann vor, wenn beachtet wird, daß die Rechnungen nicht nur richtig, sondern vor allem sinnvoll sein müssen. Die statistische Auswertung kann zu wesentlichen Einsichten in die untersuchten biologischen Zusammenhänge der geprüften Reaktion führen, aber sie setzt ihrerseits bereits Kenntnisse voraus, mit denen die Auswahl des Objektes und der zu variierenden Faktoren so geschehen kann, daß die erwarteten Befunde tatsächlich erzielt werden. Die biologisch richtige Deutung der in den statistischen Zahlen ausgedrückten Befunde zu finden, ist aber nicht mehr Sache der Statistik. Jede statistische Auswertung von Versuchen kann nur dann angebracht sein, wenn die zur Deutung der Ergebnisse notwendigen Kenntnisse der untersuchten biologischen Zusammenhänge vorhanden sind. Bei Mängeln an dieser Voraussetzung ist eine statistische Auswertung von Beobachtungsdaten Unsinn. Für die Richtigkeit der Formel, die angewendet wird, ist die mathematische Statistik verantwortlich, für die richtige und sinnvolle Anwendung dagegen die Gesamtheit der biologischen Kenntnisse und wissenschaftlichen Fähigkeiten dessen, der die Formel verwenden will. Die Statistik ist für die Biologie nicht Selbstzweck, sondern eine wissenschaftliche

Methode, ein Hilfsmittel, ebenso wie das Mikroskop oder die Waage, mit dessen Anwendung biologische Erkenntnisse gewonnen werden sollen. Ohne genaue Kenntnis dieses Hilfsmittels ist eine zu sicheren Erkenntnissen führende experimentelle, induktive Arbeitsweise nicht möglich, aber die Anwendung der Methode muß zu dem Ziel führen, eine immer weitere Kenntnis der Lebenserscheinungen zu gewinnen.

## Literatur.

Bateman, A. J., u. K. Mather: The progress of inbreeding in barley. Heredity (Lond.) 5, 321—348 (1951).

Cochran, W. G., and G. M. Cox: Experimental designs. New York: John Wiley & Sons u. London: Chapman & Hall 1950.

Fisher, R. A.: The design of experiments. Edinburgh u. London: Oliver a. Boyd 1949. — Statistical methods for research workers. Edinburgh u. London: Oliver a. Boyd 1950. — Fisher, R. A., and F. Yates: Statistical tables for biological, agricultural and medical research. Edinburgh u. London: Oliver a. Boyd 1948.

Gebelein, H.: Zahl und Wirklichkeit. Leipzig: Quelle & Meyer 1943. — Gebelein, H., u. H. J. Heite: Statistische Urteilsbildung. Berlin-Göttingen-Heidelberg: Springer 1951.

Haskell, G.: Genetics of cold tolerance in maize and sweet corn seed. Heredity (Lond.) 6, 377—386 (1952).

Kendall, M. G.: The advanced theory of statistics. London: Charles Griffin & Comp. 1948. — Rank correlation methods. London: Charles Griffin & Comp. 1948. — Koller, S.: Graphische Tafeln zur Beurteilung statistischer Zahlen. Darmstadt: Dr. Dietrich Steinkopff 1953.

Linder, A.: Statistische Methoden für Naturwissenschafter, Mediziner und Ingenieure. Basel: Birkhäuser 1945. — Planen und Auswerten von Versuchen. Basel: Birkhäuser 1953.

Mather, K.: Biometrical genetics. London: Methuen 1949. — Statistical analysis in biology. London: Methuen & Co. 1951. — Mittenecker, E.: Planung und statistische Auswertung von Experimenten. Wien: Franz Deuticke 1952. — Mudra, A.: Anleitung zur Durchführung und Auswertung von Feldversuchen nach neueren Methoden. Leipzig: S. Hirzel 1949.

Quantitative Inheritance. Herausgeg. von E. C. R. Reeve u. C. H. Waddington. London: Her Majesty's Stationery Office 1952. Agricultural Research Council.

Timoféeff-Ressovsky, N. W., u. K. G. Zimmer: Das Trefferprinzip in der Biologie. In Biophysik, Bd. I. Leipzig: S. Hirzel 1947.

Wald, A.: Sequential analysis. New York: Wiley u. London: Chapman & Hall 1947. — Weber, E.: Grundriß der biologischen Statistik für Naturwissenschaftler und Mediziner. Jena: Gustav Fischer 1948.

# II. Die Strukturen der Zelle und ihre chemische und physikalische Konstitution.

## Normale und pathologische Anatomie der Zelle[1].

Von

**Lothar Geitler.**

Mit 39 Textabbildungen.

## I. Übersicht.

Alle Organismen bestehen aus einer Zelle oder aus mehreren Zellen, oder aber sie besitzen eine Organisation, die sich aus der cellulären als Sonderfall, phylogenetisch betrachtet als abgeleitet, ergibt und die wenigstens in gewissen Stadien der Entwicklung in die celluläre übergeht. Eine Zelle ist eine bestimmt differenzierte Protoplasmamenge, die sich morphologisch und physiologisch als Einheit verhält. Die Zelle ist das Bauelement aller Organismen, so verschieden sie auch sein mögen; sie ist die Struktur, die grundsätzlich für lebende Systeme charakteristisch ist. Die letzte Behauptung bedürfte allerdings einer Einschränkung, wenn die Viren als Lebewesen aufgefaßt werden (HARTMANN 1953). Die echten, kristallisierbaren Viren sind aber keineswegs Organismen; die „großen" oder Pseudoviren (Cysticetes RUSKAS) besitzen dagegen offenbar cellulären Bau (vgl. die klare Auseinandersetzung bei WEIDEL; ausführlicher TROLL).

Die typische Zelle besteht aus einem Protoplasmakörper (Protoplast), der einen Zellkern enthält; sie besitzt also grundsätzlich eine Differenzierung in zwei Teile: in *Cytoplasma* (kurz Plasma) und *Karyoplasma*. Als Kern ist dabei verstanden ein hochkompliziertes Organell, das aus morphologisch exakt definierten, ihrerseits hochdifferenzierten Chromosomen besteht und das mitotisch teilungsfähig ist (oder, wie im Fall der Makronuclei der Ciliaten, aus einem solchen Kern entstanden ist). Ebenso ist das Cytoplasma durch bestimmte Teildifferenzierungen wie Mitochondrien, Plastiden u. a. wohl definiert. Diese Organisation zeigen sämtliche pflanzlichen und tierischen Zellen mit Ausnahme der Cyanophyceen, Bakterien und Spirochäten. Die Zellen dieser, oft mißverständlich als „Kernlose" bezeichneten Pflanzen besitzen eine *andere* Organisation, die im einzelnen noch nicht ganz aufgeklärt ist und eben wegen ihrer offenbar geringeren Differenziertheit schwerer enträtselbar ist, für die aber sichergestellt ist, daß eine analoge (nicht homologe) Gliederung in zwei Teile besteht, nämlich in einen dem Cytoplasma entsprechenden Bezirk und in Kernäquivalente. Letztere enthalten, wie sie auch im einzelnen ausgebildet sein mögen — sei es als mehr diffuser Chromidialapparat oder streng lokalisiert — die für die Kerne, genauer die Chromosomen, charakteristische Desoxyribose-Nucleinsäure, und sie oder in ihnen liegende Teilstrukturen müssen der identischen Reproduktion fähig sein, die für die Genonemen der Chromosomen wesentlich ist.

---

[1] Das Manuskript wurde im Jänner 1954 abgeschlossen. — Kleindruck im Text war nicht vorgesehen.

## II. Cyanophyceen, Bakterien, Spirochäten.

Bei den Cyanophyceen läßt sich klar erkennen, daß eine Organisation vorliegt, die keineswegs nur durch das „Fehlen" eines Zellkerns gekennzeichnet ist: der Protoplast ist überhaupt andersartig als Protoplasten mit Kern-Cytoplasma-Differenzierung organisiert. Die Assimilationspigmente sind nicht an Plastiden gebunden, sondern in einer analogen, undifferenzierteren Struktur, dem peripheren Chromatoplasma, enthalten; Plastiden fehlen überhaupt. Ebenso fehlen Mitochondrien und sind wahrscheinlich durch andere, noch unbekannte Bildungen ersetzt. Es gibt ferner kein Vacuom im Sinne anderer Pflanzenzellen: gelegentliche Vacuolisierung beruht auf degenerativer, irreversibler Verflüssigung; die Keritomie, d. h. Wabenbildung in Chromato- und Centroplasma, ist zwar reversibel und unpathologisch, jedoch nicht vergleichbar mit der Schaumbildung im Cytoplasma, wie dies Küster (S. 9) tut; die Wabenwände enthalten ja zum Teil die Kernäquivalente. Die Blaualgenzelle entspricht somit nicht einer typischen Zelle, aus der man den Kern gedanklich entfernt hat, sondern sie entspricht morphologisch wie physiologisch einem typischen Protoplasten samt Kern (vgl. im einzelnen Geitler 1936).

Als Kernäquivalent läßt sich im groben das Centroplasma oder sein innerer Teil ansehen. Über den feineren Bau herrscht aber noch keine Klarheit. Kernäquivalente im präziseren Sinn sind vielleicht die von Bringmann als „Karyoide" bezeichneten feulgenpositiven Gebilde, in die sich der „Chromidialapparat" mit entsprechender Methodik auflösen läßt. Sie enthalten aber auch Ribosenucleinsäure sowie Lipoide und Phosphatide, die sonst für Nucleolus und Cytoplasma bezeichnend sind. Bei *Lyngbya coerulea* liegen sie je Zelle in der Einzahl — bei anderen Arten mit größeren Zellen wohl in der Mehrzahl — und sind offenbar zusammengesetzte Bildungen. Ihre zu erwartende Zahlen- und Volumenkonstanz ist nicht nachgewiesen, ihre Teilung nicht genauer verfolgt. De Lamaters diesbezügliche Angaben für *Anabaena* bedürfen noch der Stützung (De Lamater, S. 361). Autonom sich teilende „Nucleosomen", die Kernen entsprechen sollen, hat schon Hollande (1933—1935) beschrieben; die methodisch unzureichend gestützten Angaben erscheinen durch Guilliermond und Gavaudan widerlegt (obwohl gerade derartige Strukturen zu erwarten wären). Erst recht erscheinen neuere Angaben (Hollande und Hollande 1944) unbewiesen, denen zufolge die Nucleosomen auf spiremartigen Fäden aufgereiht sein sollen. Die ältere cytologische Literatur hat größtenteils nur mehr historischen Wert (vgl. Geitler 1936), abgesehen von dem wesentlichen und mehrfach gesicherten Befund, daß Desoxyribose-Nucleinsäure im Centroplasma vorkommt (Poljansky und Petruschewsky). Die Aufklärung weiterer Einzelheiten bleibt der Zukunft überlassen. Sicher ist schon jetzt, daß sich Chromosomen und Mitosen bei Cyanophyceen nicht werden auffinden lassen. Zu erwarten ist aber der Nachweis permanenter feulgenpositiver Einheiten.

Im Fall der Bakterien erscheint eine Aufklärung des Zellbaus infolge der meist geringen Zellgröße noch schwieriger. Die Frage, wieweit die Bakterien überhaupt eine systematisch homogene Gruppe darstellen — was sich vielleicht einmal als nicht ganz bedeutungslos erweisen wird — kann hier nicht einmal diskutiert werden. Aufschlußreiche Untersuchungen über den Bau des Cytoplasmaäquivalents stehen noch aus. Für die Assimilationspigmente führenden Purpurbakterien ist es aber erwiesen, daß keine Plastiden ausgebildet sind (farbstofftragende submikroskopische Partikel in Extrakten aus zertrümmerten Zellen von *Rhodospirillum* als „Chromatophoren" zu bezeichnen, wie Pardee, Schachman und Stanier tun, ist ungerechtfertigt). Ebenso fehlen anscheinend morphologisch distinkte Mitochondrien und ein Vakuom[1]. Als Kernäquivalente

---

[1] Allerdings identifizieren Mudd, Winterscheid, de Lamater und Henderson die „Granula" der Mycobakterien mit Mitochondrien.

treten je Zelle in Ein- oder Mehrzahl vorhandene, sich regelmäßig teilende feulgen-positive Körper auf, die Nucleoide (PIEKARSKI). Sie sind aber wohl ihrerseits zusammengesetzt. Nach DE LAMATER (vgl. auch DE LAMATER und MUDD) sollen in bestimmten Fällen drei chromosomenartige Körper vorhanden sein, die mitose-ähnlich mittels einer Spindel, an deren Polen auch Centrosomen liegen, auf-geteilt werden und die auch einen mitotischen Formwechsel durchmachen (Ver-kürzung, Kondensation in den mittleren Teilungsstadien). Ein solcher geregelter Verteilungsmechanismus sich identisch reproduzierender Strukturen ist — wie im Fall der Cyanophyceen — zu erwarten, — er bildet ja die Voraussetzung jeder erbgleichen Teilung; daß aber die bisher beschriebenen Strukturen wirklich Mitosen und die verteilten Körper Chromosomen sensu stricto sind, ist nicht bewiesen (vgl. auch BISSET, dessen Gegengründe allerdings nicht zwingend sind). Die Realisierbarkeit von Mitosen innerhalb so winziger Dimensionen ist auch aus Gründen des Heranreichens an die Molekülgröße wenig wahrscheinlich. Kern, Chromosom und Spindel sind so gut definierte Begriffe, daß eine viel weitergehende Strukturanalyse als die vorliegende gefordert werden muß, wenn diese Begriffe auf die Bakterienzelle übertragen werden sollen. Und wenn an-geführt wird, daß die kleinsten Kerne bei Pilzen zu ihren Dimensionen an die Kleinheit der Bakterien,,kerne" heranreichen, so ist doch festzuhalten, daß die Kerne der Pilze jedenfalls eine Membran und einen Nucleolus besitzen und sich ganz wie andere Kerne verhalten; unbeschadet ihrer eben einmal extrem geringen Größe. Daß andererseits bei Bakterien Chromosomen*äquivalente* vorhanden sind, legen auch die Befunde der Bakteriengenetik nahe, — falls man sie nicht auf das Gebiet der ,,Plasmagene" abschieben will (vgl. im übrigen die Zusammen-fassung PIEKARSKIs).

Die Bakteriencytologie befindet sich zur Zeit in lebhafter Entwicklung[1]. Daß Kernäquivalente vorhanden sind, steht fest. Wie große Ähnlichkeiten mit Zellkernen — der Begriff Kern im wohl definierten Sinn gebraucht — sich weiter-hin ergeben werden, läßt sich nicht voraussagen. Daß die Übereinstimmung so weit geht, daß nicht ein deutlicher Abstand gegenüber der Kern-Cytoplasma-Organisation anderer Organismen bestehen bliebe, ist unwahrscheinlich; jedenfalls ermöglichen die derzeitigen Befunde nicht, die bestehende Kluft zu überbrücken.

Über die Spirochäten liegen noch weit weniger verwertbare Angaben vor. Sicher ist aber auch für sie, daß in ,,Nucleoiden" lokalisiert Desoxyribose-Nuclein-säure vorkommt (SCHLOSSBERGER, JAKOB und PIEKARSKI).

# III. Zelle und Energide.

Unabhängig von der besonderen Organisation ihres Protoplasten handelt es sich auch im Fall der ,,Kernlosen" um *Zellen*, d. h. um morphologische und physiologische Einheiten, die in diesem Sinn durchaus vergleichbar mit den Zellen anderer Organismen sind. Diese Zellen sind auch fähig, höher organisierte Verbände zu bilden, so bei den Blaualgen Thalli aus mehrreihigen, verzweigten Fäden mit Scheitelzellwachstum, bei den Bakterien immerhin polarisierte Fäden; derartige Bildungen entsprechen durchaus den bei anderen Algen oder Pilzen verwirklichten. Wenn Bakterienzellen zwei oder vier Nucleoide besitzen oder das Centroplasma mancher Blaualgen vielleicht typisch aus einer größeren Zahl von feulgenpositiven Einheiten aufgebaut ist, so würde es sich allerdings in gewissem Sinn um vielwertige Zellen handeln; bei den polynucleoiden Bakterien scheint in die Spore tatsächlich nur ein einziges Nucleoid einzugehen. Bei den Blaualgen gibt es Arten *(Dermocarpa)*, die lange Zeit ohne Zellteilung wachsen, bis die Teilungen dann in einem Zug unter progressiver Zellverkleinerung

---

[1] Vgl. die ausführliche zusammenfassende Darstellung SCHUSSNIGS.

ablaufen, was also ganz der Vielfachteilung in einem vielkernigen Sporangium etwa einer Protococcale entspräche; manche Dauerzellen *(Anabaena, Gloeotrichia)* vergrößern sich unter Vermehrung der feulgenpositiven Strukturen exzessiv und liefern dann bei der Keimung unter Teilung ohne Wachstum entsprechend kleinere, d. h. wieder normalgroße vegetative Zellen. Es ist möglich, daß hier Vorgänge zugrunde liegen, analog denen, die sich im Rahmen der Kern-Cytoplasma-Organisation morphologisch viel klarer ausdrücken.

Gerade im Bereich der Kern-Cytoplasma-Organisation scheinen dem Zellbegriff unter Umständen gewisse Schwierigkeiten zu erwachsen. Es gibt Protisten, deren Vegetationskörper aus einheitlichem Plasma mit vielen Kernen

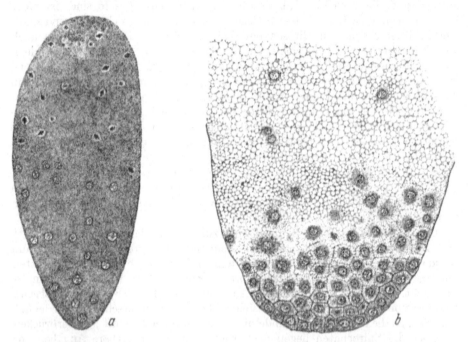

Abb. 1a u. b. Junger Embryo von *Stangeria paradoxa* (Cycadine). a Plasmodiales Stadium mit gleichmäßig verteilten Kernen bzw. Teilungsfiguren. b Alter, basaler Abschnitt mit beginnender Aufteilung in Zellen. Etwa 42fach. (Nach CHAMBERLAIN.)

besteht (Plasmodien der Myxomyceten, der Rhizochrysidalen u. a.), und es gibt Algen von siphonalem Bau; es können sogar Embryonen höherer Pflanzen, z. B. von Cycadinen, in dieser Weise aufgebaut sein, d. h. keine Kammerung im Sinne der Ausbildung morphologisch abgegrenzter Zellen besitzen (Abb. 1). Alle diese Bildungen, die übrigens in tierischen Geweben eine noch größere Rolle spielen, lassen sich durch den Begriff der SACHSschen Energide, wie ihn MAX HARTMANN gefaßt hat, unschwer verstehen: es handelt sich um Protoplasten ohne morphologisch markierte Grenzen, in denen aber zweifellos die gleichen physiologischen Beziehungen zwischen Kern und Plasma, die sich äußerlich in der Kern-Plasma-Relation ausdrücken, bestehen wie in abgegrenzten Zellen. Jeder Kern „beherrscht" ein entsprechendes Plasmaareal (wenn auch nicht immer ein und dasselbe). Alle diese Bildungen entstehen aus Zellen und können sich wieder in Zellen unterteilen, indem die Energiden sich abgrenzen; dies geschieht z. B. auch im weiblichen Gametophyten der Angiospermen. Es handelt sich nicht um mehrkernige Zellen, sondern um latent vielzellige, eben um *polyenergide* Organismen bzw. Entwicklungsstadien.

Ein einfaches, aber aufschlußreiches Beispiel bietet die Volvocale *Pyramidomonas montana*, die bei Kultur auf Agar monströse Formen mit mehreren Kernen, aber auch mehreren Geißelapparaten, Chromatophoren, Augenflecken usw. hervorbringt (Abb. 2). Es ist völlig klar, daß es sich nicht um mehrkernige Individuen, sondern um eine Summation solcher handelt, die durch Hemmung der Cytokinese (Plasmateilung) bei weiterlaufender Mitose und Teilung der plasmatischen Organellen entstanden sind. Natürliche solche Bildungen sind die Doppelzellen der Distomatinen, die mit der Längsachse vereinigte ganze Flagellaten darstellen. Eigenartige natürliche Mehrfachbildungen grundsätzlich gleicher Art finden sich bei dem Dinoflagellaten *Polykrikos*, dessen „Zellen" aus 2, 4 oder 8 „Zoïden" bestehen, d. h. aus Einheiten, die je einer Zelle samt Furchen, Geißelapparat usw. entsprechen (allerdings merkwürdigerweise nur die halbe

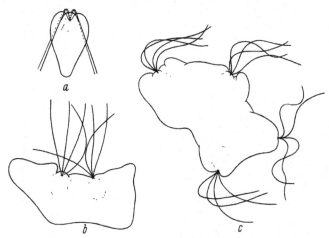

Abb. 2a—c. *Pyramidomonas montana* (Volvocale). a Freischwimmendes, normales Individuum; am Vorderende des birnförmigen Kerns das Basalkorn der Geißeln, daneben contractile Vacuolen. b, c Monströse Individuen, die 24 Std bzw. 10 Tage auf 4 %igem Nährsalzagar kultiviert wurden. (Nach GEITLER.)

Zahl von Kernen besitzen; vgl. z. B. DOFLEIN, GRASSÉ). Wie durch Hemmung der Cytokinese höherwertige Bildungen entstehen, lehrt weiterhin der Vergleich verschiedener Gattungen amöboider Chrysomonaden, bei denen alle Übergänge von Einzelindividuen, d. h. mononergiden Zellen, über primitive und differenziertere Filarplasmodien bis zu typischen Fusionsplasmodien ohne jede Zellabgrenzung vorkommen. Daß diese Plasmodien amöboid entwickelt sind, ist dabei unwesentlich. Grundsätzlich die gleiche fusionsplasmodiale, polyenergide Organisation zeigen junge Embryonen von Gymnospermen sowie Endosperme und weibliche Gametophyten von Angiospermen (und tierische Embryonen, wie die der Insekten mit superfizieller Furchung): es kommt — zunächst — zur Aufhebung der gewöhnlichen Korrelation zwischen Mitose und Cytokinese; die Protoplastentrennung wird erst später nachgeholt. Alle diese Bildungen sind homolog mit *viel*zelligen Gebilden: es handelt sich nur um einen anderen Modus der Durchführung der Vielzelligkeit. Dies ergibt sich anschaulich z. B. für siphonale Algen auch daraus, daß zu jedem Kern nicht nur eine bestimmte Menge Plasma, sondern auch eine bestimmte Menge Plastiden, Chondriosomen usw. „gehört" (Abb. 3). Die Beziehungen zwischen Kern und Plasma sind offensichtlich keine anderen wie in einem monoenergiden Protoplasten; — mit dem Unterschied allerdings, daß nicht immer dasselbe Plasma mit dem gleichen Kern agieren muß und daß ein Plasmabereich unter der Wirkung mehrerer Kerne stehen kann. Die Voraussetzung hierfür ist dadurch gegeben, daß das

Plasma über größere Bereiche hin gleichartig ist, d. h. die gleiche Beschaffenheit wie in untereinander nicht differenzierten abgegrenzten Zellen besitzt (treten Differenzierungen auf, so werden Wände eingeschaltet — Sporangien und Gametangien von *Vaucheria, Saprolegnia*). Die Energide ist somit

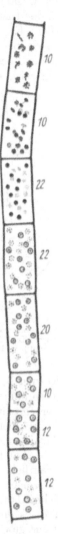

Abb. 3. Ausschnitt aus dem plasmatischen Wandbelag von *Vaucheria* innerhalb des Bereichs einer Teilungswelle: Kerne vorwiegend in Anaphase bis Telophase (dazu 4 Metaphasen und 1 späte Prophase); außer den Kernen auch Chromatophoren und Öltropfen dargestellt. — Flemm.-Benda, etwa 950fach. (Nach Geitler 1934.)

Abb. 4. Teilungswelle über Querwände hinweggreifend bei *Cladophora alpina:* von oben nach unten: Kerne synchron in Metaphase, darunter in Anaphase, früher und später Telophase; unten Querwandbildung. — Alk.-Eisess., Essigcarm., etwa 160fach. (Nach Geitler 1934.)

keine bloße gedankliche Konstruktion, sondern besitzt nachweisbare Realität (vgl. im übrigen ausführlicher Hartmann 1953, Geitler 1934, S. 13ff.).

In diesem Zusammenhang ist es von Interesse, daß es in cellulären wie polyenergiden Verbänden die gleiche Art des Ablaufs von Teilungswellen gibt. Bekannt ist das Verhalten in Endospermen der Angiospermen: das Verteilungsbild aufeinanderfolgender Stadien der Mitosen läßt ein entsprechendes Gefälle teilungsauslösender Stoffe erkennen; dabei kann sich auch die gleiche Teilungsfigur, je nachdem sie parallel oder senkrecht zum Gefälle steht, unterschiedlich verhalten (Abb. 5). Analoges gilt für siphonale Algen (Abb. 3). Grundsätzlich die gleiche Wanderung von Teilungsstoffen ereignet sich aber offenbar auch in den aus einkernigen Zellen aufgebauten spermatogenen Fäden der Characeen über *Querwände hinweg*, allerdings viel langsamer und unter synchroner Teilung von „Blöcken" hintereinander liegender Zellen, die Abkömmlinge einer Mutterzelle sind (GEITLER 1948/49). Über Querwände — die hier nicht Zellgrenzen

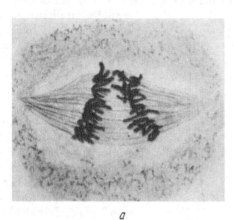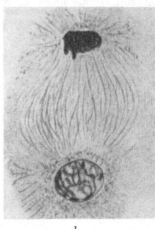

*a*     *b*

Abb. 5 a u. b. Ana- und Telophase aus dem Endosperm von *Iris pseudacorus*, die Spindeln senkrecht bzw. parallel zur Richtung des Teilungsgefälles eingestellt: Wirkung verschiedener Plasmabereiche auf die Spindel und Schnelligkeit der Rekonstruktion der Tochterkerne. — Stark vergr., Pikroform.-Essigs. (Nach JUNGERS.)

markieren — hinweggehende Teilungswellen treten auch bei Siphonocladalen auf (Abb. 4).

Nach allem bedarf es bei der Besprechung der Teile, die den Protoplasten aufbauen, keiner besonderen Unterscheidung zwischen echtzelligen und plasmodial-polyenergiden Organismen oder Entwicklungsstadien. Das gleiche gilt für *syncytiale* Verbände, d. h. polyenergide Bildungen, die durch Fusionen (nicht durch Teilungen) von Zellen bzw. Protoplasten entstehen; sie spielen bei Pflanzen allerdings nur eine untergeordnete Rolle, wenn man von Gefäßen, Milchsaftröhren u. dgl. absieht (anders bei Tieren!). Auf jeden Fall sollte zwischen Plasmodien und Syncytien unterschieden werden, da es sich um entwicklungsgeschichtlich verschiedene Dinge handelt (BELAR 1928, GEITLER 1934), obwohl beide zusammen cönocytische Organisation besitzen[1].

---

[1] Merkwürdige, meist nicht beachtete Sonderfälle der cellulären Ausbildung finden sich bei Flagellaten (PASCHER). Grundsätzlich besitzt eine Flagellatenzelle 1 Kern, 1 Geißelapparat, 1 Chromatophorenapparat, 1 Vacuolensystem. Es gibt aber Chrysomonaden mit zwar 1 Kern, aber im übrigen verdoppelten Organellen, also mit 2 Geißelapparaten usw. *(Didymochrysis)* und solche, die 1 Kern und 1 Geißelapparat besitzen, aber Chromatophor, contractile Vacuolen und „Mundleiste" verdoppelt haben *(Amphichrysis);* sie sind unvollkommene Doppelzellen (vollkommene Doppel- bzw. Vielfachzellen sind bei den oben erwähnten Distomatinen und bei *Polykrikos* verwirklicht). Für die interessanteste dieser Flagellatenformen, PASCHERS *Didymochrysis*, ist allerdings nicht klar bewiesen, daß sie nur *einen* Kern besitzt und daß es sich nicht vielleicht um eine bloß vorübergehende Teilungshemmung handelt.

# IV. Die Zelle als Ganzes.

## 1. Größe.

Die kleinsten Zellen treten bei Bakterien und Cyanophyceen auf (Mikrokokken 0,15 $\mu$, manche Blaualgen 0,5 $\mu$ im Durchmesser). Dies ist wohl kein Zufall: gerade ihnen fehlt die Differenzierung in Kern und Cytoplasma, die offenbar eine bedeutendere Größe voraussetzt. Die Werte für Zellen mit Kern-Cytoplasma-Differenzierung liegen im allgemeinen um ein bis zwei Größenordnungen höher, — sofern ihr Volumen ausschließlich oder vorwiegend durch Plasma bestimmt wird. Viel größer werden Zellen, die Reservestoffe speichern (Eizellen von Gymnospermen, Speichergewebe) oder — der gewöhnliche Fall erwachsener Pflanzenzellen — Zellsaftvacuolen bilden; ihr Volumen wird durch sekundäre Faktoren mitbestimmt. Im ersten Fall ist die Vergrößerung mit Plasmavermehrung unter Eiweißsynthese verbunden, im anderen, wenn er rein ausgeprägt ist, erfolgt kein echtes Wachstum. Auch besonders voluminöse Zellen mancher Protisten verdanken ihre Größe zum Teil der Zellsaftbildung.

Primär spielt bei der Realisierung einer bestimmten Zellgröße die Kern-Plasma-Relation eine wichtige Rolle. Was ihr zugrunde liegt, ist noch ungeklärt. Vielfach sieht es so aus, als ob das Volumen des Kerns das Plasmavolumen bestim-

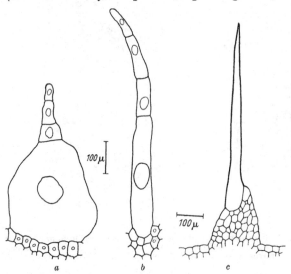

Abb. 6a—c. Wachstum von Trichomen unter Endopolyploidie. a, b *Bryonia dioica*: a Klebstoffhaar der Anthere mit 256ploider Basalzelle. b Filamenthaar mit 128ploider Basalzelle (die nächste Zelle 16ploid, Epidermis diploid). c *Mercurialis annua*, hochpolyploides einzelliges Trichom der Fruchtknotenwand. (Nach TSCHERMAK-WOESS und HASITSCHKA 1954.)

men würde. Das Kernvolumen selbst wird aber außer durch intranucleäre — chromosomale und extrachromosomale — Faktoren, die teils echte Wachstumsvorgänge darstellen, teils auf Hydratation beruhen (s. Abschn. V 4), auch durch das Plasma bestimmt (s. Abschn. V 2). Abgesehen von Stoffspeicherung und Zellsaftbildung geht die Vergrößerung im Zuge der Ausdifferenzierung pflanzlicher Zellen am häufigsten mit endomitotischer Polyploidisierung einher, die einen echten Wachstumsvorgang darstellt (Abb. 6; TSCHERMAK-WOESS und HASITSCHKA, zusammenfassend GEITLER 1953); sie wird offenbar vom Plasma gesteuert. Um über den Stand der „Zellwandanatomie" hinauszugelangen, ist neben plasmatischer im besonderen auch karyologische Anatomie zu betreiben. Im ganzen gesehen ist die Zellvergrößerung ein sehr komplexer Vorgang.

Nach unten zu ist der Zellgröße — wie der Größe der Organellen — durch die Annäherung an die Molekülgröße und die Größe der Molekülverbände eine Grenze gesetzt. Die Beachtung dieser Zusammenhänge macht manche anatomische Eigentümlichkeiten verständlich. Allgemein sind kleine Zellen nicht geometrische Verkleinerungen großer, sondern unkompliziertere Bildungen. Dies zeigt z. B. der Vergleich groß- und kleinzelliger Desmidiaceen und im besonderen ihres Chromatophors: bei kleinzelligen geht die Zahl der Chromatophorenlappen zurück — bei wenig sinkender Größe des einzelnen Lappens (Abb. 7) —, ebenso sinkt die Zahl der Pyrenoide; komplizierte Zell- und Chromatophorengestalten sind nur oberhalb einer gewissen Größe realisierbar. Bei den Diatomeen läßt sich die Auswirkung des Größenfaktors nicht nur beim Vergleich verschiedener Arten, sondern innerhalb des gleichen Klons erkennen: er drückt sich in der Vereinfachung des Zellumrisses, der Gestalt des Chromatophors, aber auch des Kerns aus, wobei wohl vielfach die Zunahme der Oberflächenspannungskräfte mitspielt; auch die Aufhängung des Kerns an Plasmafäden oder -lamellen erscheint an eine gewisse Größe gebunden (Abb. 8; eingehend bei GEITLER 1932). Gerade bei den Diatomeen wird die Bedeutung des Größenfaktors besonders sinnfällig dadurch, daß sich die Zellen infolge ihres starren Teilungsschemas, das Verkleinerung nach sich zieht, zu Tode teilen, wenn sie eine bestimmte Minimalgröße erreicht haben (GEITLER 1932). Was sich hier natürlich abspielt, läßt sich bei *Eudorina* experimentell durch Dauerbelichtung erzielen, die die Teilung gegenüber dem Wachstum so weit fördert, daß nicht mehr lebensfähige Zwergzellen entstehen (HARTMANN 1921).

## 2. Gestalt.

Die einfachste Gestalt freier Zellen ist die Kugel, d. i. die Form, die ein Flüssigkeitstropfen infolge seiner Oberflächenspannung annimmt, wenn er nicht unter der Wirkung weiterer Kräfte steht. Membranlose Zellen kugeln sich tatsächlich ab, wenn sonstige Kräfte dies nicht verhindern. Solche Kräfte stellen sich aber z. B. ein, wenn lokale Veränderungen der Oberflächenspannung erfolgen und Strö-

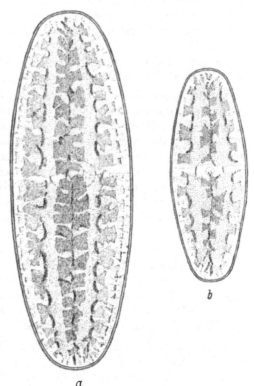

Abb. 7 a u. b. *Netrium digitus* (Desmidiacee), groß- und kleinzellige Varietät: Lappung der Chromatophorenleisten in Abhängigkeit von der Zellgröße. (Nach GEITLER 1934.)

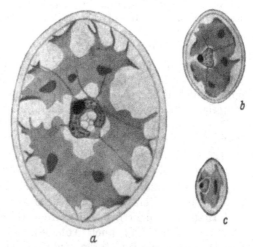

Abb. 8 a—c. Gestalt des Kerns und des Chromatophors in Abhängigkeit von der Zellgröße bei Diatomeen: verschieden große Zellen von *Cocconeis placentula* var. *lineata* (a, b) und var. *pseudolineata* (c). — Subl.-Alk., 1000fach. (Nach GEITLER.)

mungen hervorrufen, wie im Fall amöboid beweglicher Zellen. Anders geartete, starre Formveränderungen entstehen unter Mitwirkung von Skeletbildungen,

z. B. bei Flagellaten durch einen Achsenstab, der den an sich weichen Zell-
körper in die Länge spannt. Es genügt aber auch eine entsprechende submikroskopische Struktur des Plasmas selbst, vor allem wohl seiner Oberfläche,
um spezifische Formen hervorzubringen; so im Fall der nackten, langzipfelig-
tetraedrischen Schwärmer von *Hydrurus* und *Phaeodermatium* oder im Fall der
komplizierten Reliefbildungen der Gymnodiniaceen.

Zu den Skeletbildungen gehören auch die Zellmembranen, durch deren be-
sondere Ausgestaltung bestimmte Zellformen fixiert werden können. Im einfachsten Fall nehmen aber auch behäutete Zellen Kugelform an: der Turgor-
druck spannt die Membran allseitig gleich, sofern sie gleichmäßig elastisch ist.
Solche Turgorformen zeigen viele Cyanophyceen; stehen die Zellen in festem
Verband, so ergeben sich Zwangsformen zwischen Turgordruck und gegenseitiger
Behinderung; isoliert nehmen die Zellen Kugelform an. Wird die Membran im
Alter verdickt und unelastisch, so behält sie ihre Form bei, auch wenn die Zelle
abstirbt; sie stellt dann das Abbild eines bestimmten Lebensstadiums der
Zelle dar.

Ähnliche Verhältnisse herrschen in Parenchymen, die sich mit einem Seifen-
schaum vergleichen lassen. Bei künstlicher oder autonomer Lockerung des Ver-
bands kann Abkugelung erfolgen. In anderen Parenchymen tritt aber infolge
spezifischer Wachstumsvorgänge Wandbildung in bevorzugten Richtungen ein,
wodurch sich charakteristische, nach dem Bild der physikalischen Analogie unmittelbar nicht deutbare Abweichungen ergeben.

Allgemein sind auch Skeletbildungen Erzeugnisse des Protoplasten und als
solche der Ausdruck einer bestimmten, im submikroskopischen Bau des lebenden
Plasmas begründeten Organisation. In dieser ist auch die Polarität verankert,
die vielfach die Zellgestalt wesentlich mitbestimmt (vgl. Abschn. V 2). Angesichts
der gesamten bestehenden Formenmannigfaltigkeit läßt sich wohl feststellen, daß
bei ihrem Zustandekommen Kräfte, die in einfachen physikalischen Systemen
wirken, beteiligt sind, daß die Zellgestalt aber nicht allein ihr Produkt ist; — was
nicht bedeutet, daß sich Formbildungsprozesse außerphysikalisch oder akausal
abspielen. Im Falle reiner Turgor- oder Zwangsformen *fehlt* eine entsprechende
übergeordnete Organisation[1].

Daß im übrigen Zellgestalten bei gegebener Organisation durch Außenbedingungen in ihrer natürlichen Entwicklung beeinflußt werden können und entsprechend reagieren, ist selbstverständlich (Pilzhyphen in Abhängigkeit von der
Unterlage; Sklereiden, Spikularzellen u. a. mit „anarchischem" Wachstum in Parenchymen, usw.); erst recht gilt dies für pathologische Fälle (vgl. die Zusammenfassung bei Küster).

Die kompliziertesten Zellgestalten treten — im Verein mit hoher intracellulärer Differenzierung und bedeutender Größe — dort auf, wo die Zelle den ganzen
Organismus darstellt; — also nicht bei den in anderer Hinsicht am weitesten
differenzierten Vielzellern. Die Komplikation der Zellen gewisser Dinoflagellaten
und Diatomeen übertrifft bei weitem die etwa bei Phaeophyceen oder Kormophyten erreichte (vgl. auch Abb. 38). Die Gestaltung kann bis zur Imitation
von Differenzierungen führen, die sich bei Vielzellern *inter*cellulär abspielen (so
besitzen manche Dinoflagellaten Augenflecke mit Pigmentbecher und Linse sowie

---

[1] Nur sehr selten läßt sich eine *spezifische* Gestaltung rein mechanisch restlos verstehen.
So besitzen die Schwebestacheln von *Biddulphia mobiliensis* eine charakteristische Knickung
an bestimmter Stelle, die dadurch verursacht wird, daß die Stacheln junger Zellen während
ihrer Bildung innerhalb der Mutterzelle an bestimmten, durch die Raumverhältnisse gegebenen Stellen der Mutterzellwand anstoßen und sich hierbei abbiegen.

Nesselkapseln und contractile Befestigungsorganellen). Komplizierte Zellformen sind aber keineswegs allgemein funktionell deutbar; vielfach, z. B. bei den Desmidiaceen, bestehen solche Beziehungen offensichtlich überhaupt nicht.

# V. Die Bestandteile
## des in Kern und Cytoplasma differenzierten Protoplasten.
### 1. Allgemeines.

Wenn man von Bestandteilen der Zelle spricht, so ist es klar, daß es sich um eine gedankliche, aus praktischen Gründen vorgenommene Unterscheidung von Teilen handelt, die in Wirklichkeit eine physiologische Einheit bilden. Dies gilt bereits für die Unterscheidung von Kern und Plasma: beide sind für sich allein auf die Dauer nicht lebensfähig. Es ist andererseits begreiflich, daß nur die Analyse durch Behandlung der unterscheidbaren einzelnen Bestandteile die Grundlage für ein tieferes Verständnis des Ganzen bieten kann. Wie viele Bestandteile unterscheidbar sind, hängt bis zu einem gewissen Grad — und abgesehen von subjektiven Momenten — vom jeweiligen Stand der Methodik ab. Durch die Benützung der Elektronenoptik haben sich neue Gebiete eröffnet, deren Erforschung allerdings so sehr am Anfang steht, daß eine einigermaßen abgerundete und befriedigende Darstellung noch nicht möglich ist.

### 2. Cytoplasma.

Im allgemeinen unterscheidet man zwischen dem „undifferenzierten" oder „Grundplasma" (Hyaloplasma) und den in ihm eingelagerten Differenzierungen wie Plastiden, Mitochondrien u. a. oder ± ergastischen Einschlüssen. Die Unterscheidung war so lange verhältnismäßig leicht durchführbar, als nur die im Lichtmikroskop sichtbare Struktur erforschbar war und das Grundplasma als homogen-gelige Substanz erschien. Schwierigkeiten bestanden und bestehen allerdings auch in diesen Dimensionen insofern, als es fraglich ist, was man von kleinsten Einschlüssen im Plasma als noch zu diesem gehörig ansehen will (vgl. BELAR 1928, S. 61, auch KÜSTER). Durch die Einbeziehung des submikroskopischen Bereichs ist, wie schon früher zu vermuten war, erwiesen, daß dem „Grundplasma" eine hochdifferenzierte Architektonik zukommt (FREY-WYSSLING 1953, BRETSCHNEIDER). Wieweit sie artspezifisch, zellspezifisch und leistungsspezifisch ist, läßt sich vorläufig nicht bestimmen[1].

Mit dieser Struktur offenbar in engem Zusammenhang steht die Ausbildung von lichtoptisch gut bis eben noch erkennbaren Gebilden, den sog. „Plasmapartikeln", die in Zentrifugaten aus Homogenisaten gewonnen werden und die im histologischen Präparat als Mitochondrien und Mikrosomen, vielleicht auch als namenlose Elemente erscheinen. Diese Partikel sensu stricto sind offenbar selbstreproduktiv und bilden wohl „Populationen", die von Zelle zu Zelle weitergegeben werden, wobei vermutlich mit der Verschiedenheit des Leistungstyps der Zelle spezifische Veränderungen auftreten. Ihre Erforschung steht in den ersten Anfängen — dies gilt namentlich für das botanische Gebiet —, so daß sich noch keine endgültigen Lösungen ergeben[2].

So ungewiß unsere Vorstellungen über die submikroskopische Differenzierung des sog. „undifferenzierten" Plasmas im einzelnen sind, so sicher läßt sich behaupten, daß gerade sie es ist, die allen Differenzierungsvorgängen und den verschiedenartigen Leistungen der Zelle wie auch letzten Endes ihrer spezifischen

---

[1] HAGENAU und BERNHARD möchten auf Grund ihrer Beobachtungen an Blutzellen allgemein zell- und gewebespezifische Grundstrukturen annehmen.

[2] Die Ansätze wie die Problematik findet man klar herausgearbeitet in der übersichtlichen Darstellung LEHMANNS; vgl. auch FREY-WYSSLING 1953.

Gestalt zugrunde liegt. Auf *sie* gehen die lichtoptisch sichtbaren intra- und inter-cellulären Differenzierungen zurück. Dies zeigt mittelbar, aber deshalb nicht weniger gewiß, jede typische inäquale, differentielle Zellteilung, deren Wesen darin besteht, daß eine im Cytoplasma einer Mutterzelle entstandene, charakte-ristische Inhomogenität auf zwei Tochterzellen verteilt wird, die sich nun ver-schiedenartig weiterentwickeln. Am eingehendsten ist in dieser Hinsicht die tierische Eifurchung untersucht, bei der Keimbezirke verteilt werden. Was

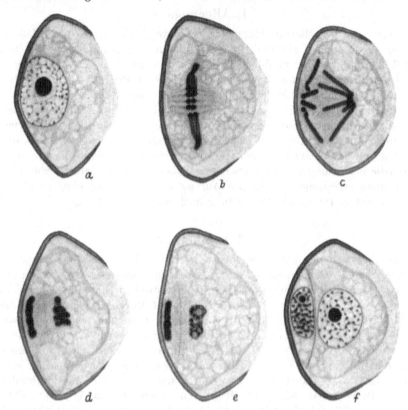

Abb. 9a—f. 1. Pollenmitose von *Gasteria cheilophylla:* wandständige Lage des Kerns (a) und daher der Spindel und Metaphaseplatte (b), Asymmetrie der Anaphase (c) und verschiedene Rekonstruktion der Tochterkerne (d—f). — Ausstrich, Flemm.-Benda, etwa 1150fach. (Nach Geitler 1934.)

dabei an verschiedenen Einschlüssen des Plasmas (Pigmente, Keimbahnkörper u. a.) oder Zuständen (z. B. verschiedenes Redoxpotential) sichtbar wird, sind die Auswirkungen einer verschiedenen Beschaffenheit des Grundplasmas, deren Erforschung die Aufgabe der nächsten Zukunft ist. Diese Verschiedenheit liegt allgemein auch der *Polarität* der Zelle zugrunde (A. Lang, Bünning 1952, 1953). Chemische Gradienten u. dgl., die dabei in Erscheinung treten, stellen, so wichtig sie sind, selbst wieder Auswirkungen der submikroskopischen plasmati-schen Architektonik dar, unbeschadet des Umstands, daß umgekehrt die jeweilige Plasmastruktur auch das Ergebnis ablaufender Prozesse ist (Bünning 1953, S. 181ff.)[1].

Auf botanischem Gebiet liefert ein eindrucksvolles Beispiel für inäquale Teilung einer polarisierten Zelle die 1. Teilung im Pollenkorn der Angiospermen (Abb. 9; Sax und Edmonds, Geitler 1935, Hagerup, Pinto-Lopes, La Cour

---

[1] Vgl. zur Problematik Danielli und Brown.

1949, SUITA, BRYAN). Infolge einer bestimmten groben Architektonik des Protoplasten nimmt der Kern eine bestimmte wandnahe Stellung ein, die Spindel wird senkrecht auf die Wandfläche orientiert und die Teilung verläuft daher inäqual. Ihr Ergebnis ist eine wandnahe, kleine Zelle, deren Plasma fast zur Gänze aus der umgewandelten Teilungsspindel hervorgeht und die arm an sichtbaren Einschlüssen sowie an Ribosenucleinsäure ist, und eine größere Zelle mit reichlichen Einschlüssen und ribosenucleinsäurereichem Plasma. Die beiden — genetisch identischen — Tochterkerne werden bereits in der Telophase verschiedenartig rekonstruiert und verhalten sich weiterhin verschieden, indem der der plasmaarmen wandständigen Tochterzelle eine lebhaftere Chromatinsynthese erfährt, während der andere in dieser Hinsicht eine mehr regressive Entwicklung durchmacht (vgl. auch Abb. 23); die kleine Zelle teilt sich später in die beiden Spermazellen, die andere übernimmt unter Vermehrung der im Plasma lokalisierten Ribosenucleinsäure die Stoffspeicherung im Pollenkorn und geht später zugrunde. Die Unterschiede der Tochterzellen sind quantitativer und qualitativer Natur, für das weitere Verhalten der primären Tochterkerne ist entscheidend, in *welches* Plasma sie gelangen (wird die Spindeleinstellung gestört, so unterbleibt die Differenzierung; SAX und EDMONDS, GEITLER 1935). *Voraus* geht eine intraplasmatische Differenzierung.

Derartige inäquale Teilungen sind weit verbreitet. Sie spielen sich bei der Bildung von Wurzelhaaren, Idioblasten u. a. ab (SINNOTT und BLOCH 1946, zusammenfassend A. LANG, BÜNNING 1953). Sie laufen auch bei der Bildung der Spaltöffnungsinitialen ab (zuletzt BÜNNING und BIEGERT). In diesem Fall spielt eine bedeutende Rolle eine polarisierte Plasmawanderung, die aber wohl noch nicht das wesentliche ausmacht, sondern die nur das unmittelbar sichtbare „Mittel" ist, die inäquale Teilung durchzuführen. Ob mit der Plasmaverlagerung auch eine Plasma*vermehrung* im Sinne einer Synthese neuen Plasmas erfolgt, wie BÜNNING annimmt, bedarf wohl noch des Beweises.

Intraplasmatische Differenzierung ist auch die Ursache für den inäqualen Verlauf der Cytokinese während der Meiose mancher pennaten Diatomeen, deren Ergebnis ein funktionstüchtiger und ein abortierender Gamet ist. Die grobmechanischen Hilfsmittel für die Durchführung des inäqualen Verlaufs der Teilung lassen sich an Plasma- und Kernverlagerungen, an der Schichtung von Einschlußkörpern u. dgl. erkennen; zugrunde liegt aber eine, zwar sicher erschließbare, jedoch unsichtbare Verschiedenheit im Grundplasma zwischen der Seite des Protoplasten, die der jüngeren Membranhälfte (der Hypotheka), und der Seite, die der älteren Membranhälfte (Epitheka) zugekehrt ist; die Förderung bzw. Hemmung ist artspezifisch konstant an die eine oder andere Seite gebunden (GEITLER 1951a, 1952). Eine bemerkenswerte Ausprägung der Zellpolarität findet sich bei *Oedogonium*, wo sich neben äußerlichen und plasmatischen morphologischen Merkmalen auch Gradienten nachweisen lassen; *jede* Zellteilung verläuft hier inäqual, die Anaphasefigur ist entsprechend unsymmetrisch, ganz ähnlich wie im Pollenkorn der Angiospermen (Abb. 10; TSCHERMAK). Die Tochterzellen können aber schließlich dennoch gleichwertig werden insofern, als sie weiterhin teilungs- und fortpflanzungsfähig sind; hierin besteht eine gewisse Übereinstimmung mit den Teilungsgefällen in Endospermen.

Die Plasmaarchitektonik bestimmt auch die *Richtung der Zellteilung,* — gleichgültig, ob sie äqual oder inäqual abläuft; der Kern bleibt dabei gänzlich unbeteiligt. Obwohl jeder Kern eine Polarität besitzt, die sich aus der Lage der Chromosomen in der vorhergegangenen Telophase ergibt und die in der Prophase auffallend als Polfeld in Erscheinung tritt (Abb. 11), stellt sich die Spindel, deren Lage z. B. bei den Kormophyten die Lage der Scheidewand bestimmt, unabhängig von der Kernpolarität „richtig" in bezug auf die Mutterzelle bzw. die Tochterzellen ein. Dies wird deutlich erkennbar, wenn in aufeinanderfolgenden Teilungen ein Richtungswechsel stattfindet, wie z. B. in Pollenmutterzellen: die Spindel der 2. Teilung wächst dann senkrecht zur Polaritätsachse des Kerns heran, die Chromosomen werden wie „durchgekämmt" umorientiert (Abb. 12[1]). Das gleiche erkennt man bei gleichbleibender Teilungsrichtung, z. B. in den Staubblatthaaren von *Tradescantia*, wenn der Kern durch Plasmaströmung eine Drehung erfahren hat. — In zellsaftreichen Kormophytenzellen wird der zunächst beliebig wandständige Kern zu Beginn der Prophase in Plasmasträngen dorthin verfrachtet, wo die zukünftige Scheidewand entstehen „soll"; hier baut das Plasma das Phragmosom

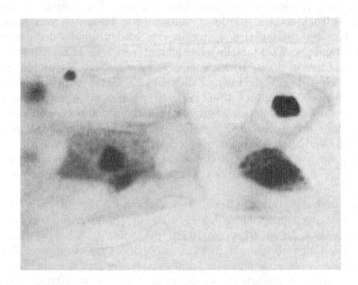

Abb. 10. Verschiedene Rekonstruktion der Tochterkerne von *Oedogonium* sp. (verschiedene Größe, Dichte der Struktur, Nucleolengröße); im Bild links der apikale Zellpol. — Flemm.-Benda, etwa 1270fach. (Nach Tschermak.)

Abb. 11 a—c.  Polfeld von der mittleren bis späten Prophase im Kern der 1. Pollenmitose von *Allium nutans.* — Ausstrich, Flemm.-Benda, etwa 1500fach. (Nach Geitler 1934.)

auf, dann erst entsteht die Spindel (Sinnott und Bloch 1941). Der Kern verhält sich ganz passiv, das Plasma allein bestimmt den Ort und die Orientierung.

Die Spindel kann aber auch überhaupt ohne Beziehung zur künftigen Scheidewand orientiert sein; so, wenn diese nicht in einem Phragmoplast entsteht[2].

[1] Für die Spindelstellung maßgebend erscheint weitgehend die äußere Gestalt der Zelle oder ihr zugrunde liegende plasmatische Faktoren (Einstellung der Spindellängsachse in der Längsrichtung der Zelle); an den Spindelstellungen in Pollenmutterzellen, vor allem in isodiametrischen, erkennt man, daß außerdem noch eine bestimmte innere Plasmaarchitektonik mitwirken muß, die sich *nicht* einfach in der Zellform ausdrückt. Bei manchen Araceen nehmen auch grobmechanische Gegebenheiten Einfluß auf die typisch abweichende Spindelstellung (Geitler und Vogl; Vogl).

[2] Über Scheidewandbildung und Cytokinese im allgemeinen vgl. Mühldorf (1951). — Im Fall pathologischer tierischer Eifurchungen kann die gewöhnliche Korrelation zwischen Mitose und Plasmateilung so weit aufgehoben werden, daß Zellteilungen überhaupt ohne Kern ablaufen (Fankhauser).

Bei *Microspora Loefgreni* liegt die intranucleär gebildete Spindel senkrecht zur Längsachse des Fadens und somit parallel zur Scheidewand (Abb. 13; vgl. KOSTRUN und die dort gegebene Statistik). Diese Stellung erscheint mechanisch verursacht durch die querellipsoidische Gestalt des Kerns, die ihrerseits durch

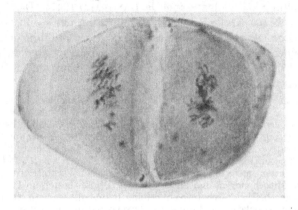

Abb. 12. Pollenmutterzelle während der II. Metakinese von *Anthericum* sp.: Umorientierung der Chromosomen während der Spindelbildung um 90° gegenüber der 1. Teilung (die rechte Platte wird sich senkrecht auf die Bildebene, die linke in der Bildebene einstellen). — Essigcarm. (Nach GEITLER 1935.)

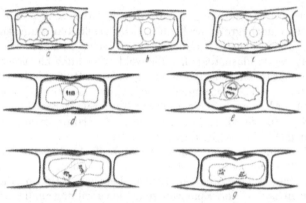

Abb. 13a—c. Spindelstellung bei der vegetativen Teilung von *Microspora Loefgreni* (Chlorophycee). a, b Zellen vor der Teilung mit Ruhekern in zwei um 90° gedrehten Ansichten (Kern wandständig). c Zelle vor der Teilung, Kern zentral. d—g Metaphase bis Telophase. — a—c Nach dem Leben, Original; d—g Alk.-Eisess., Essigcarm., nach KOSTRUN.

seine Einspannung in eine querlaufende Plasmamelle erzwungen wird. Die später in Richtung der Fadenlängsachse verschobene Lage der Tochterkerne kommt rein mechanisch durch die anaphasische Spindelstreckung zustande (KOSTRUN)[1].

## 3. Die Plasten.

### a) Einleitung; Mikrosomen.

Mit BELAR bezeichnet man als Plasten die differenzierten „cytoplasmatischen Gebilde, die nur aus ihresgleichen durch Zweiteilung entstehen". Dabei ist stillschweigend vorausgesetzt, daß es sich um Gebilde lichtmikroskopischer Dimensionen handelt. Es sind dies zunächst die *Plastiden*, die *Centrosomen*, die *Basalkörner* der Geißeln und, wie heutzutage mit Sicherheit behauptet werden kann,

[1] Daß dieser Fall aber nicht verallgemeinert werden darf, zeigt das Verhalten der linsenförmigen, in querverlaufende Plasmadiaphragmen eingespannten Kerne von *Spirogyra*Arten: die extranucleär entstehende Spindel bildet sich senkrecht zur Plasmalamelle und künftigen Scheidewand, also ebenso wie in Fällen von Phragmoplastenbildung (Abb. 37a).

die *Mitochondrien* (Chondriosomen). Der Golgi-*Apparat*, der zoologischerseits oft als permanente Struktur betrachtet wird, spielt im Pflanzenreich keine Rolle; Versuche, ihn mit dem Zellsaftvacuolensystem als „Vakuom" (Guilliermond, P. A. Dangeard) zu homologisieren, dürften fehlgeschlagen sein (vgl. die ausführliche Darstellung bei Küster; auch P. Dangeard 1947). Das Vakuom selbst ist offensichtlich *keine* permanente Struktur im Sinn eines selbständigen, autoreproduktiven Organells; es kann auf dasselbe in dieser einführenden Darstellung, unbeschadet seiner Wichtigkeit, nicht näher eingegangen werden und es muß, wie auch hinsichtlich Tonoplast, Haptogenmembran usw. auf die Einzeldarstellungen verwiesen werden[1].

Außer allgemein oder weit verbreiteten Plasten gibt es auch solche mit beschränkter Verbreitung. Hierher gehören die sog. „Blepharoplasten" der Trypanosomen, die sich bei der Zellteilung mitteilen und mit der Bildung des Basalkorns der Geißel in Beziehung stehen. Daß sie ihren Ursprung aus dem Zellkern nehmen, wie Schaudinn meinte, trifft nicht zu; doch sollen sie positive Nuclealreaktion zeigen; jedenfalls gibt es auch blepharoplastlose Stämme (vgl. im übrigen Doflein, S. 39). — Ferner sind Plasten die „Plättchen", die offenbar bei allen Diatomeen im Cytoplasma und meist in Kernnähe auftreten; sie sind fast so fixierungslabil wie Mitochondrien und lassen sich deshalb am besten im Leben beobachten. Sie werden nicht nur bei jeder vegetativen Teilung mitgeteilt, sondern werden auch durch die Gameten bei der Kopulation und durch die Auxosporen weitergegeben. Sie sind in Zahl und Aussehen weitgehend stabil und vom physiologischen Zustand der Zelle unabhängig. Chemismus und Funktion sind noch unbekannt; eine Proteingrundlage ist offenbar vorhanden (Gschöpf).

Den Plasten wären neuerdings zuzuzählen die *Mikrosomen*. Allerdings stehen diesem Vorgehen noch große praktische Schwierigkeiten entgegen, denn unter dem Namen „Mikrosomen" werden offenbar sehr verschiedene Gebilde verstanden: teils gewisse Plasmapartikel, die wohl tatsächlich kleinste autoreproduktive Einheiten darstellen, teils aber auch ergastische Bestandteile verschiedenster Art, die nur die eine Gemeinsamkeit besitzen, daß sie sehr klein sind (vgl. z. B. P. Dangeard). Aber auch für die erstgenannten ist der strikte Nachweis noch nicht geführt, daß sie allein durch identische Reduplikation entstehen (*angenommen* hat dies schon P. A. Dangeard 1919).

Jenen Mikrosomen, die wesentliche Bestandteile des Plasmas bilden und wohl universelle Verbreitung besitzen, ist eine relativ komplizierte Organisation zuzuschreiben, und sie besitzen offenbar wichtige Lebensfunktionen. Perner (1952a) wies an ihnen in der *Allium*-Epidermiszelle eine Proteingrundlage nach, machte das Vorhandensein von Ribosenucleinsäure wahrscheinlich und konnte an ihrer Oberfläche die Cytochromoxydase lokalisieren, die zum System der zelleigenen Atmungsfermente gehört. Diese Mikrosomen, die allerdings ungewöhnlich groß sind (2—4 $\mu$ im Durchmesser[2]), erscheinen also keineswegs als unorganisierte Lipoidtröpfchen, wie für Mikrosomen sonst meist angenommen wurde. Der Nachweis gelingt freilich nur mit sehr exakter Methodik und an Objekten, die eine eingehende morphologische Analyse des lebenden Protoplasten zulassen[3]. Die Verwendung von Fraktionen aus Homogenisaten allein birgt zahlreiche Fehlermöglichkeiten, zumindest bei Pflanzen (Perner 1952a, b, Perner und Pfefferkorn). Bei zoologischen Objekten wurden aber gerade mit dieser Methodik weitreichende Ergebnisse gewonnen (Lehmann, K. Lang). Sie deuten ebenfalls auf hohe Differenziertheit, Autoreproduktion und wichtige Funktion

---

[1] Relativ permanente Strukturen sind vielleicht die contractilen Vacuolen mancher Protisten.

[2] Bei Perner und Pfefferkorn, S. 102, wird ihre Größe aber mit 0,5—0,8$\mu$ angegeben (?).

[3] Wie kompliziert die Verhältnisse z. B. bei Vitalfärbung liegen, zeigen erneut die eingehenden Untersuchungen Drawerts mit Hilfe von Fluorochromierung: je nach der Sauerstoffspannung ist mit dem gleichen Farbstoff elektive Färbung der Mikrosomen oder der Mitochondrien möglich.

der Mikrosomen als Fermentträger hin. Als Mikrosomen werden hier allerdings meist *sub*mikroskopische Partikel bezeichnet und das Cytochromsystem erscheint dabei an die gröbere sog. Mitochondrienfraktion gebunden (K. LANG; auch MARQUARDT; daselbst weitere Literatur).

Nach allem scheint es gewiß, daß die Mikrosomen — oder wohl besser ein Teil derselben — nicht als „tote", unwesentliche Einschlüsse betrachtet werden dürfen; KÜSTERS Darstellung (1951) erscheint bereits überholt. Weitere gesicherte Einblicke werden sich vor allem an Objekten gewinnen lassen, die eine scharfe morphologische Unterscheidung von Mikrosomen und Mitochondrien in der lebenden Zelle ermöglichen. Abgesehen von Zellen höherer Pflanzen, wie die der *Allium*-Epidermis, erscheinen hierfür gewisse Diatomeen besonders geeignet (Abb. 14); ihre zellphysiologische und cytochemische Untersuchung wurde noch nicht in Angriff genommen.

## b) Mitochondrien.

Die Mitochondrien oder Chondriosomen sind stäbchen- bis fadenförmige, ovale oder fast kugelige Gebilde von ungefähr Bakteriengröße; sie lassen sich daher im Lichtmikroskop ohne weiteres erkennen (Abb. 15). Infolge ihrer hohen Fixierungslabilität, die wohl auf Lipoidreichtum beruht, werden sie nur durch bestimmte Fixierungsmittel, vor allem osmiumsäurehaltige Gemische, lebensgetreu erhalten, können aber — unter gewissen Kautelen — vital mit Janusgrün elektiv gefärbt werden (zuletzt DRAWERT), und sind im übrigen an geeigneten Objekten dank ihrer stärkeren Lichtbrechung gegenüber dem Grundplasma im Leben auch ungefärbt gut sichtbar[1].

Die Mitochondrien besitzen zähflüssige bis gelige Beschaffenheit, sind flexibel und oft deutlich formveränderlich. An günstigen Objekten, wie manchen Diatomeen mit großen, langen Mitochondrien, läßt sich erkennen, daß die

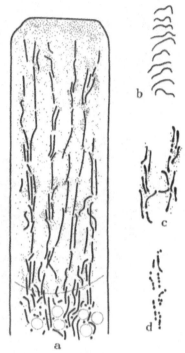

Abb. 14 a—d. *Pinnularia nobilis.* a Flächenbild der Mitochondrien-führenden peripheren, relativ unbewegten Plasmaschicht und der — in Wirklichkeit tiefer liegenden — Schichten dünnflüssigen Plasmas mit Mikrosomen. b Ein Mitochondrium, das zum Teil in das schnell bewegte Plasma geraten ist und schnelle, schlängelnde (passive) Bewegungen ausführt. c Durch Druck, d durch Lagern unter dem Deckglas geschädigte Mitochondrien. — Nach dem Leben. (Nach GEITLER 1937 b.)

Formveränderungen passiver Natur sind, d. h. durch Plasmaströmung hervorgerufen werden; auch dann, wenn sie lebhaft spirochätenartig verlaufen (Abb. 14 b; GEITLER 1937 b). Auch manche Chlorophyceen besitzen leicht zu unersuchende Mitochondrien (CHADEFAUD 1936). Solche Einzeller lassen eine

---

[1] Hierzu ist *kein* Phasenkontrast, sondern nur die richtige Handhabung des Mikroskops nötig. Es scheint nicht überflüssig darauf hinzuweisen, daß Vitalbeobachtungen bis in alle Einzelheiten auch mit gewöhnlicher Optik durchführbar sind und daß keineswegs erst die Phasenkontrastmikroskopie die Lebendbeobachtung feinerer Plasmastrukturen und des Aufbaus von Ruhekernen, Chromosomen, Spindel usw. ermöglicht hat, wie oft behauptet wird, geschweige, daß sie grundsätzlich Neues gebracht hat. Die Photographien der verschiedensten vitalen Feinstrukturen z. B. in den Werken BELARS, widerlegen diese Meinung zur Genüge (auch Abb. 25—27). Die Verwendung des Phasenkontrasts ist vielfach bequemer, — im übrigen handelt es sich um eine Modeerscheinung.

ungestörte Vitalbeobachtung sicherer zu als selbst z.B. die Staubblatthaare von *Tradescantia*.

Die Vitalbeobachtung erfordert gerade im Fall der Mitochondrien besondere Sorgfalt, da sie als erste geschädigt werden (vgl. neuerdings Perner und Pfefferkorn, Steffen). Viele Beschreibungen gehören offenbar in die Pathologie der Mitochondrien (unregelmäßige Spindelformen — Abb. 14c, Zerfall der Fäden in Kugeln — Abb. 14d u. a. m.). An *Pinnularia* läßt sich verfolgen, wie schon

mäßiger Druck tropfigen Zerfall hervorruft, — der übrigens reversibel sein kann, was wohl bedeutet, daß die Eiweißgrundlage intakt geblieben ist, und wie bei Sauerstoffmangel Gestaltsveränderungen eintreten (Geitler 1937b, Perner und Pfefferkorn). Vermutlich sind in gefärbten Präparaten auftretende derartige Formen fixierte vitale Artefakte.

Die Mitochondrien sind im Pflanzen- und Tierreich allgemein verbreitet und gehören zu den notwendigen Bestandteilen des Cytoplasmas. Daß sie in einzelnen Fällen nicht aufgefunden wurden, kann an fehlerhafter Methodik oder besonders geringer Größe liegen. Im Gegensatz zu den Mikrosomen, die histologisch einen Sammelbegriff darstellen, handelt es sich bei den Mitochondrien im allgemeinen um morphologisch wohl definierte, seit langem bekannte und oft untersuchte Gebilde sui generis[1].

Die Mitochondrien teilen sich zweifellos, werden von Zelle zu Zelle weitergegeben und besitzen somit Kontinuität, — obwohl gerade hierüber noch gründliche Untersuchungen fehlen und der schwer zu führende Beweis, daß ausnahmslos Mitochondrien nur aus Mitochondrien hervorgehen,

Abb.15a u. b. a Mitochondrien in Darmzellen des Hühnerembryos. b Mitochondrien und Plastidenanlagen — diese von Stärkekörnern aufgetrieben — in einer Zelle der Luftwurzel von *Chlorophytum Sternbergianum*. — a Etwa 2500fach, b etwa 1500fach. (Nach Meves aus Belar.)

noch nicht erbracht ist. Die bestehende oder zum Teil angenommene Kontinuität bedeutet aber nicht, daß ihre Anzahl immer gleichbleibt. Vielmehr schwankt die Zahl mit dem wechselnden physiologischen Zustand der Zelle; die Mitochondrien können daher nicht etwa als Träger von Plasmagenen nach dem Muster der zahlenkonstanten Chromosomen als Genträger in Anspruch genommen werden (doch wäre eine analoge Funktion auch ohne Zahlenkonstanz vorstellbar). Sie sind auch in der Zelle im allgemeinen regellos verteilt und werden bei der Cytokinese ohne einen Mechanismus, der auf genau zahlengleiche Verteilung abzielt, auf die Tochterzellen verteilt. Dennoch erscheint es keines-

---

[1] Über ihre angebliche Verwechselbarkeit mit Plastidenanlagen vgl. Abschn. c; die Diskussion hierüber bei Newcomer erscheint völlig überholt. Irreführend ist auch die Behandlung der Mitochondrien im Kapitel „Plastiden" bei Küster, ebenso die Vermengung beider Organellen bei P. Dangeard.

wegs abwegig, mit H. und R. LETTRÉ eine charakteristische Kern-Plasma-Mitochondrienrelation anzunehmen[1].

Die Verwendung von Homogenisaten bei elektronenoptischen Untersuchungen (Literatur bei NEWCOMER) hat allerdings insofern eine neue Situation geschaffen, als die Verbindung zwischen dem histologischen Bild und den „Plasmapartikeln" der Fraktion meist nicht hergestellt wurde. So wird oft einfach von „Grana mit Mitochondrienfunktion" gesprochen. An die so bezeichneten Gebilde erscheinen aber wichtige Enzyme, so der Zellatmung, gebunden — mit der Zellatmung wurden die Mitochondrien schon frühzeitig in Verbindung gebracht (JOYET-LAVERGNE) — und diese Gebilde zeigen elektronenoptisch einen komplizierten Aufbau, so im besonderen eine Membran, die aus kurzfaserigen und aus globulären Eiweißmolekülen aufgebaut ist; das übrige Eiweiß erscheint amorph (MÜHLETHALER, MÜLLER und ZOLLINGER). Sicher ist, daß die Mitochondrien der Fraktionen eine Proteingrundlage und hohen Lipoidgehalt besitzen (daher ihre Labilität) und außerdem Ribosenucleinsäure enthalten. Da im allgemeinen Identität zwischen den gröberen „Partikeln" der Fraktionen und den Mitochondrien des histologischen Bildes bestehen dürfte — das Gegenteil wäre in Anbetracht der zahlreichen Untersuchungen zu unwahrscheinlich, als daß es sich immer wieder ereignen sollte —, lassen sich die Befunde wohl auf die Mitochondrien im morphologischen Sinn anwenden. Und obwohl die wesentlichen Ergebnisse an tierischen Zellen gewonnen wurden, braucht auf Grund der morphologischen, entwicklungsgeschichtlichen und histochemischen Übereinstimmungen, die für die Mitochondrien der Tiere und Pflanzen seit langem erkannt wurden, kein Unterschied zwischen beiden gemacht zu werden. Eine Art von „Konfrontation" zwischen den Befunden der histologischen Technik und dem durch Nachweis der Atmungsfermente eruierten „Grana" nahmen übrigens BAUTZ und MARQUARDT an der intakten Hefezelle vor: es ergab sich dabei gute Übereinstimmung, d. h. die histologisch als Mitochondrien angesehenen Gebilde erwiesen sich als die Träger der Atmungsfermente. Allerdings sind gerade die Mitochondrien der Hefen morphologisch wenig charakteristisch. Und die Befunde stehen in einem gewissen Gegensatz zu den Angaben PERNERS (1952a), der bei *Allium* die Fermentträger als Mikrosomen ansieht und an den Mitochondrien keine entsprechende Reaktion fand (vgl. S. 138). Solche Widersprüche bestehen vielfach auch sonst: so will BRACHET ausdrücklich im Gegensatz zu CLAUDE als Träger der Enzyme und der Ribosenucleinsäure die Mikrosomen und nicht die Mitochondrien betrachten (vgl. NEWCOMER, S. 73).

Zusammenfassend zeigt sich, daß den Mitochondrien eine relativ komplizierte Feinstruktur zuzuschreiben ist und daß sie offenbar Träger wichtiger Zellfunktionen sind. Ihre Abgrenzung gegenüber „Mikrosomen" in Zentrifugaten ist allerdings *praktisch* nicht immer durchführbar und gegen zu weitgehende Schlußfolgerungen lassen sich grundsätzliche methodische Bedenken erheben (PERNER und PFEFFERKORN)[2]. Die Mitochondrienforschung befindet sich eben erst am Beginn einer lebhaften Entwicklung; eine sichere Unterbauung ihrer Ergebnisse ist wohl in nicht zu ferner Zukunft zu erwarten.

### c) Plastiden; Pyrenoide; Augenflecke.

Alle photosynthetisch autotrophen Pflanzen mit Kern-Cytoplasma-Differenzierung besitzen autonome, plasmatische Zellorganellen, die primär als Träger der Assimilationsfarbstoffe dienen, bei phylogenetisch weiter entwickelten Pflanzen aber auch in einer chlorophyllfreien, allein durch Karotinoide gefärbten oder auch in einer farblosen Modifikation auftreten. Alle zusammen bezeichnet man als *Plastiden* (Einzahl: die Plastide); die Assimilationsfarbstoffe führenden heißen *Chromatophoren* (Sonderfall die rein grünen Chloroplasten), die carotingelben oder -roten *Chromoplasten* — sie kommen schon bei Chloro- und Rhodophyceen vor, wie auch das Chromatoplasma der Blaualgen eine analoge

---

[1] Angaben über Bildung von Mitochondrien de novo aus dem Plasma (P. DANGEARD 1941, vgl. 1947, S. 60) halten einer Kritik nicht stand; denn daß die Mitochondrien experimentell reversibel unsichtbar zu machen sind und dann scheinbar aus dem Nichts auftauchen, ist richtig, daß aber ihre Proteinträgerstruktur dabei *nicht* erhalten geblieben ist, bleibt unbewiesen.

[2] Vgl. auch BOGEN, H. J.: Fortschr. Bot. **15**, 222ff. (1954).

Ausbildung nehmen kann; die farblosen nennt man *Leukoplasten* (auch diese besitzen aber im allgemeinen die *Fähigkeit* Chlorophyll zu bilden).

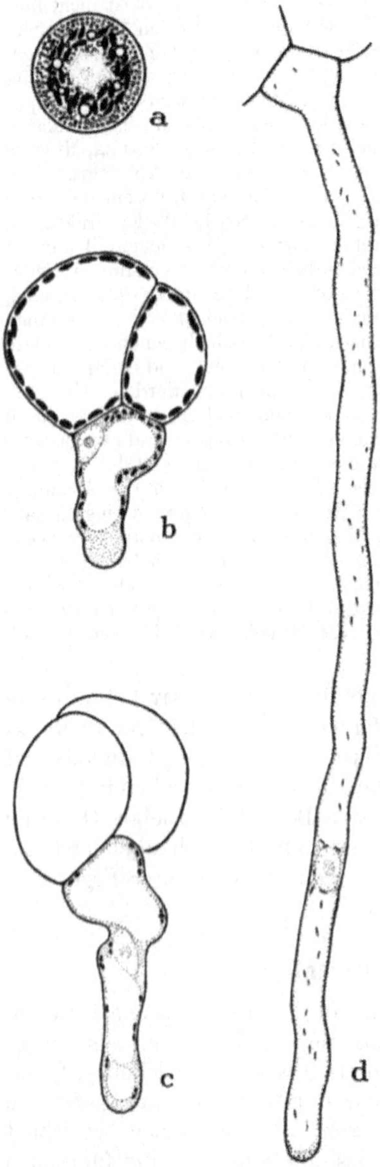

Abb. 16a—d. Regressive Entwicklung zu mitochondrienartigen Körpern der Chloroplasten, die bei der Keimung der Spore von *Equisetum maximum* in das Rhizoid gelangt sind. a Spore. b—d Aufeinanderfolgende Stadien des Rhizoidwachstums. — Nach dem Leben, etwa 510fach. (Nach Geitler 1934.)

Die Plastiden entstehen ausschließlich durch Zweiteilung aus ihresgleichen, — abgesehen von pathologischen Fällen des Zerfalls in mehrere Stücke oder pathologischen Fusionen (vgl. Küster). Die Plastiden sind autoreproduktive Organellen par excellence, ihre absolute Kontinuität ist eine völlig gesicherte Tatsache. Seit den Beweisen, die Schmitz (1882) und Schimper (1883, 1885) erbracht haben, hat sich die Richtigkeit der These immer wieder bewahrheitet und ihre allgemeine Gültigkeit hat sich von den Protisten bis hinauf zu den Angiospermen überzeugend erwiesen (Geitler 1934, Hartmann 1953, Weier und Stocking). Wenn dennoch bis in die neueste Zeit von einzelnen behauptet wird, daß Plastiden aus Mitochondrien hervorgingen (Literatur bei P. Dangeard und Küster, die diese überholte Kontroverse zu ernst nehmen), so beruht dies auf einer unkritischen Deutung und auf der losgelösten Betrachtung des Befundes, daß bei den Spermatophyten die in den Meristemen als Leukoplasten ausgebildeten embryonalen Chloroplasten morphologisch und mikrochemisch schwer von den Mitochondrien zu unterscheiden sind[1]. Da es sich um eine ausschließlich methodische Frage handelt, ist es verfehlt, von zweierlei Sorten von Mitochondrien — solchen, die zu Plastiden werden, und solchen, die bleiben, was sie sind — zu sprechen (Guilliermond, Mangenot und Plantefol). Nirgends sonst im Pflanzenreich gibt es eine derartige Verwechslungsmöglichkeit, Plastiden und Mitochondrien sind in allen Entwicklungsstadien klar getrennt, — übrigens auch bei *Anthoceros* (vgl. meine ausführlichere Darstellung 1934 sowie Hartmann 1953[2]). Auch Fälle regressiver Entwicklung von Plastiden zu mitochondrienartigen Gebilden (Abb. 16) bieten keine Gegenbeweise.

Die Plastiden besitzen bei einfach organisierten Protisten nur einen unbedeutenden Formwechsel; er besteht in Größenveränderungen des Chromatophors, die zwischen

---

[1] Mit sorgfältiger Methodik gelingt es übrigens auch hier (Rezende-Pinto).

[2] P. Dangeard hält an der Trennung von Mitochondrien und Plastiden zwar fest, führt aber als Beweggrund die — unhaltbare — Annahme an (S. 124), daß die Mitochondrien im Unterschied zu den Plastiden de novo aus dem Plasma entstünden (vgl. S. 141, Fußnote [1]).

maximaler Entfaltung bei optimalen Lebensbedingungen und entsprechender
Verkleinerung unter Ausbleichung in Dauerzellen schwanken. Von hier bis
zu dem reichen Formwechsel anatomisch hochdifferenzierter Algen und den
Verhältnissen bei den Blütenpflanzen bestehen alle Übergänge. Besonders
bei den Rhodophyceen zeigen die Chromatophoren in verschieden großen und
verschiedenartig funktionierenden Zellen des gleichen Thallus bedeutende und
charakteristische Unterschiede in Größe und Gestalt (vgl. z. B. das bekannte,
in fast alle Lehrbücher eingegangene Bild OLTMANNS' von *Ceramium*); in Scheitel-
zellen sind sie embryonal, d. h.
klein, einfach geformt und
± farblos (Abb. 18); in anderen
Fällen bestehen auffallende Unterschiede zwischen der Ausbildung in Assimilationszellen
und assimilatorisch inaktiven
Fortpflanzungszellen (Abb. 17).
Extreme Rückbildungen erfolgen naturgemäß allgemein in

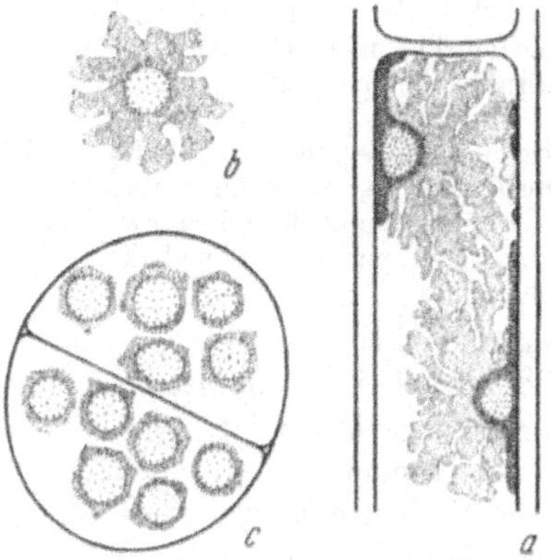

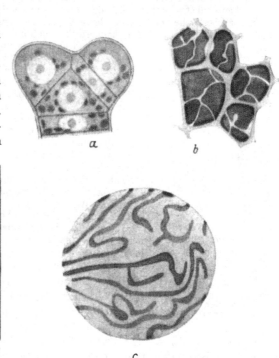

Abb. 17.                                          Abb. 18.

Abb. 17a—c. Verschiedene Ausbildung der Plastiden im Thallus der Rhodophycee *Rhodochorton floridulum.*
a Maximale Entfaltung in einer lebhaft assimilierenden Zelle (kombiniertes Flächen- und Querschnittsbild).
b In einer älteren Zelle. c In einem zweigeteilten Tetrasporangium; die Größe des zentralen Pyrenoids bleibt
unverändert. — Nach dem Leben, alle gleich stark vergr. (Nach GEITLER 1934.)

Abb. 18a—c. Entwicklung der Plastiden im Thallus der Rhodophycee *Ceramium rubrum.* a Vegetationspunkt,
die Plastiden sind kleine, fast farblose Scheiben. b Assimilationszellen der Rinde. c Jüngere Zentralfadenzelle. —
Nach dem Leben, alle gleich stark vergr. (Nach GEITLER 1934.)

**Spermien.** — Zu den entwicklungsgeschichtlich begründeten Veränderungen
kommen solche, die unmittelbare physiologische Reaktionen darstellen. Bis zu
einem gewissen Grad kann die Konfiguration des umgebenden Plasmas rein
mechanisch die Formbildung beeinflussen. Durch Plasmaströmungen können
vorübergehende und wechselnde Deformationen hervorgerufen werden (so an den
Chloroplasten der Blütenpflanzen), wobei die physikalisch-plastische Beschaffenheit der Plastiden unmittelbar sichtbar wird.

Die autonome Gestaltung hängt innerhalb gewisser Grenzen auch von der
absoluten Größe ab. Wie bei anderen Organellen und an ganzen Zellen erweist
sich in Fällen vergleichbarer Baupläne die Differenzierung an eine bestimmte
durchschnittliche Größe gebunden (Abb. 7, 8).

Die Chromatophoren werden im allgemeinen, abgesehen von ihrem Feinbau, als optisch homogen organisiert angesehen. Dies erweist sich aber als

unzutreffend, wenn man bedenkt, daß nur bestimmte Stellen *Pyrenoide* zu bilden vermögen; es müssen hierfür doch wohl bestimmte intraplastidale strukturelle Unterschiede maßgebend sein. Die Pyrenoide selbst sind, obwohl in der Hauptsache aus „toter" Eiweißsubstanz bestehend, die oft den Eindruck von Kristalloiden macht, doch wohl nicht so einfach organisiert, wie es früher schien.

Dies zeigen die verschiedenen artspezifischen Typen (radiäre, polarisierte Pyrenoide; einfache, zweiteilige, mehrfach zusammengesetzte usw.; Geitler 1926, Chadefaud 1936, 1941[1]). Diese Typen finden sich konvergent bei Organismen mit Stärkeassimilation und daher Stärke-umhüllten Pyrenoiden, wie bei solchen ohne Stärkehülle. Einen bemerkenswerten Typus bieten die Cryptomonaden: die Stärke wird bei ihnen nicht im Chromatophor, sondern an seiner Oberfläche kondensiert; dennoch sind die Pyrenoide Stärke-umhüllt, weil sie selbst nicht im Chromatophor liegen, sondern ihm bloß anliegen oder in ihm eine so exzentrische Lage einnehmen, daß sie nur von einem unsichtbar dünnen Häutchen desselben überdeckt sind (Abb. 19). Ähnlich verhalten sich manche Dinoflagellaten und Bangiaceen *(Asterocytis)*, bei denen die Stärke ebenfalls extraplastidal entsteht und trotz zentraler Lage des Pyrenoids *in* einem sternförmigen Chromatophor dieser so tiefe Einbuchtungen besitzt, daß die Stärke am Grund der Buchten pyrenoidnahe entsteht und das Aussehen eines Stärke-umhüllten Pyrenoids erreicht wird (Geitler 1924, 1926); bei manchen Dinophyceen liegen die Stärke-umhüllten Pyrenoide aber wohl nicht unmittelbar in den Chromatophoren, wenn auch in fixer Lagebeziehung zum Chromatophorenapparat (Geitler 1943a). In allen Fällen tritt die Rolle der Pyrenoide bei der Stärkebildung deutlich in Erscheinung. — Merkwürdige Lage- und Bauverhältnisse finden sich bei Diatomeen (Tschermak-Woess 1953).

Die Pyrenoide sind im übrigen teilungsfähig — der aktive Teil ist dabei offenbar die Plastidensubstanz —; sie können aber auch neu aus dieser gebildet werden. Bei ein und derselben Art kann beides, je nach den Außenbedingungen und der Schnelligkeit der Zellteilung, vorkommen (Geitler 1926).

Abb. 19. *Chroomonas caudata* (Cryptomonade), ganze Zelle mit Chromatophor und ihm innen anliegendem Pyrenoid, und Pyrenoid in Seiten- und Flächenansicht. — Nach dem Leben. (Nach Geitler.)

Differenzierungen innerhalb der Plastiden bestehen auch hinsichtlich der Carotinbildung. Bei bestimmten sternförmigen Algenchromatophoren ist sie in gesteigertem Maße auf den zentralen Teil, der das Pyrenoid enthält, beschränkt (Pascher und Petrova, Geitler 1944); an diesen Stellen müssen andere physiologische Bedingungen als an der Peripherie herrschen und möglicherweise liegt eine verschiedene Struktur zugrunde. Noch auffallender ist die konzentrierte Karotinbildung, die streng lokalisiert an den Enden lang-bandförmiger Chromatophoren in langgestreckten Zellen von *Spirotaenia* und ähnlichen Desmidiaceen *(Closteriospira*, unveröffentlicht) auftritt (Geitler 1943b).

Auch der *Augenfleck* (Stigma) der Flagellaten, Zoosporen, Gameten und Spermien wäre hier zu erwähnen, soweit er als Bestandteil des Chromatophors erscheint und mit ihm wächst und sich teilt (z. B. bei der Zweiteilung von *Pyramidomonas*); in anderen Fällen tritt er als ziemlich selbständiges, vermutlich mit dem Chromatophor homologes Organell auf (heterotrophe Dinoflagellaten, Euglenaceen); in wieder anderen wird er eigenartigerweise bei der Zell- und Chromatophorenteilung neu gebildet, während der Augenfleck der Mutterzelle zugrunde geht (Abb. 20; Geitler 1934, Küster).

Die Plastiden sind als plasmatische Gebilde gegenüber Schädigungen aller Art weitgehend empfindlich. Ihre Pathologie bildet daher ein umfangreiches Kapitel (vgl. besonders Küster). Aus ihrem natürlichen plasmatischen Milieu gerissen desorganisieren sie schnell und charakteristisch. Ein merkwürdiger Sonderfall ist aber bei Rotatorien gegeben, die als ausschließliche Nahrung den Inhalt lebender Dinoflagellaten, Chrysomonaden und eventuell auch grüner Flagellaten aussaugen und sich dabei deren Chromatophoren einverleiben: diese

---

[1] Ob die von Chadefaud als „Kinetosom" beschriebene, centrosomenartige Teilstruktur typisch für Pyrenoide ist, bleibt noch zu untersuchen.

Chromatophoren werden in der Magenwand über lange Zeiträume gespeichert und zeigen dabei keine der sonst so auffälligen im Mikroskop erkennbaren Degenerationserscheinungen, so bei *Ascomorpha* (MACK) und *Anapus* (unveröffentlicht); sie sehen vielmehr, sowohl was Struktur wie Färbung anlangt, ganz normal aus — abgesehen davon, daß vermutlich Fusionen vorkommen; sie produzieren im Licht, solange das Tier lebt, Sauerstoff; Assimilation konnte aber nicht nachgewiesen werden (MACK; vgl. außer der hier zitierten Literatur auch BEAUCHAMP, dessen botanische Angaben allerdings wohl noch der Nachprüfung bedürfen).

Daß die Plastiden als autoreproduktive Gebilde hochorganisiert sind und einen ihrer Funktion entsprechenden Feinbau besitzen, war seit langem zu vermuten und hat sich durch mikroskopische, polarisationsoptische und elektronen-optische Untersuchungen auch bewahrheitet (vgl. die Zusammenfassung WEIERS und STOCKINGs). Doch herrscht keine Einigkeit der Meinungen hinsichtlich der allgemeinen Bedeutung der festgestellten Strukturen. Wie die kritische Darstellung von HEITZ und MALY zeigt, fehlen noch Untersuchungen sämtlicher Entwicklungsstadien der Plastiden, vor allem der jüngsten, so daß sich nicht sagen läßt, welche Feinstruktur allgemein charakteristisch ist. Nach HEITZ und MALY sind von „allgemeinster Verbreitung und in jedem Entwicklungsstadium vorhanden" nur das Stroma — d. h. die lichtoptisch homogen erscheinende Grundsubstanz — und eine semipermeable Membran. Im *erwachsenen Chloroplasten* ist dazu ein System von Lamellen — die nach HEITZ und MALY auch Bänder sein könnten — feststellbar, die abwechselnd Protein- und Lipoidcharakter besitzen, und in

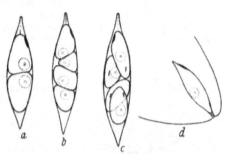

Abb. 20a—d. Verhalten des Augenflecks bei der Teilung von *Chlorogonium euchlorum* (Volvocale). a Nach dem ersten, b nach dem zweiten Teilungsschritt, c später, d die vorderste Tochterzelle mit altem und neugebildetem Augenfleck; der Augenfleck der Mutterzelle bleibt während der Teilungen erhalten, die neuen Augenflecke bilden sich in den Tochterzellen (c) unabhängig von ihm. — Etwa 500fach.
(Nach GEITLER 1934.)

den Lamellen oder Lamellenpaketen treten farbstofführende Grana auf, die aus senkrecht zur Flächenausdehnung der Plastide orientierten geldrollenartigen „Stößen" von Scheiben bestehen (HEITZ 1936, STRUGGER 1951, FREY-WYSSLING und MÜHLETHALER 1949). Im jungen Chloroplasten scheint die Chlorophyllbildung aber nicht an die Grana gebunden zu sein; vielfach fehlen Grana, wenigstens von lichtoptischen Dimensionen, überhaupt. Die Annahme STRUGGERs, daß die Grana Kontinuität besäßen, ist somit noch nicht sicher bewiesen, obwohl neueste Beobachtungen (PERNER 1954, STRUGGER 1953) wieder für sie sprechen[1]. — Der lamelläre Feinbau läßt sich vielfach schon im Mikroskop an desorganisierten Chromatophoren von Flagellaten und Algen in vergröberter Form erkennen.

### d) Centrosom (Cytozentrum); Basalkörner der Geißeln; Blepharoplasten.

Als Centrosomen bezeichnet man mehr oder weniger distinkte plasmatische Gebilde, die meist in Kernnähe liegen und in deutlicher Beziehung zur Bildung der Spindel stehen: während oder auch schon vor der Mitose teilt sich das Muttercentrosom, die Tochtercentrosomen wandern auseinander, zwischen ihnen wächst die Spindel aus, sie markieren von Anfang an die Spindelpole. Centrosomen finden sich bei den meisten Protisten und Tieren in verschiedener Ausbildung (ein einheitlicher Gebrauch der Ausdrücke Centrosom, Centriol, Centrosphäre wurde noch nicht erreicht; ohne Berücksichtigung der zoologischen Mannigfaltigkeit ist eine brauchbare Übersicht nicht möglich; vgl. BELAR 1926, DOFLEIN, SCHRADER). Im Pflanzenreich besitzen Centrosomen abgesehen von autotrophen Flagellaten viele Algen und Pilze — unter den Diatomeen nur manche; sie bzw. ihre Homologa treten auch bei der Spermienbildung von Moosen, Farnen und Cycadinen und wohl auch bei *Ginkgo*, also im

---

[1] Weitere Literaturangaben bei HEITZ und MALY.

Zusammenhang mit *begeißelten* Stadien auf (vgl. weiter unten), fehlen aber in anderen Zellen und sonst bei Kormophyten überhaupt (ausführlicher Geitler 1934). Eine Permanenz der Centrosomen als *distinkte* morphologische Bildungen besteht demnach in vielen Fällen *nicht*. Aus verschiedenen Erwägungen heraus läßt sich aber annehmen, daß ein *Cytozentrum* in jeder Zelle vorhanden ist (Belar 1928, Hartmann 1953); denn die Bipolarität der Teilungsspindel tritt ja überall in Erscheinung (abgesehen von Pathologien und phylogenetisch abgeleiteten Grenzfällen); das Teilungszentrum braucht aber nicht immer und überall als morphologisch distinktes Centrosom ausgebildet zu sein. Für diese Überlegungen sprechen auch die Übergänge, die sich beim gleichen Objekt zwischen auffallend distinkter und mehr diffuser Ausbildung bis zu völliger Unsichtbarkeit finden (Geitler 1934, ausführlich Hartmann 1953). In diesem Sinn lassen sich auch die Polkappen an den Spindelpolen bei den Angiospermen mit Centrosomen homologisieren, und es ist bemerkenswert, daß sich bei *Tradescantia* auch während der Kernruhe entsprechende vitale Strukturen feststellen lassen (Wada 1950—1952, Hartmann 1953). Im übrigen ist die Vermutung, daß distinkte Centrosomen nur an Spindeln mit zugespitzten Polen auftreten und daß die Stumpfpoligkeit, z. B. der meisten Angiospermenspindeln, mit dem Fehlen von Centrosomen zusammenhinge, trügerisch; *Monocystis* (Gregarine) besitzt ein sehr auffallendes Centrosom, aber stumpfpolige Spindeln; und bei Angiospermen treten auch spitzpolige Spindeln, aber niemals Centrosomen auf (Abb. 21).

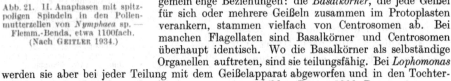

Abb. 21. II. Anaphasen mit spitzpoligen Spindeln in den Pollenmutterzellen von *Nymphaea* sp. — Flemm.-Benda, etwa 1100fach. (Nach Geitler 1934.)

Die Centrosomen können sich im übrigen vom Kern unabhängig machen. In pathologischen Fällen (kernlose tierische Eier) vermögen sich die Centrosomen wiederholt zu teilen und — chromosomenlose — Spindeln zu liefern; bei Amphibien kann die Entwicklung ohne funktionierende Chromosomen bis zur Blastula fortschreiten (Fankhauser).

Zwischen Centrosomen und Geißelapparat bestehen allgemein enge Beziehungen: die *Basalkörner*, die jede Geißel für sich oder mehrere Geißeln zusammen im Protoplasten verankern, stammen vielfach von Centrosomen ab. Bei manchen Flagellaten sind Basalkörner und Centrosomen überhaupt identisch. Wo die Basalkörner als selbständige Organellen auftreten, sind sie teilungsfähig. Bei *Lophomonas* werden sie aber bei jeder Teilung mit dem Geißelapparat abgeworfen und in den Tochterzellen — unter Mitwirkung des Centrosoms — neu gebildet (Belar 1926, Doflein).

Mit den Basalkörnern und weiterhin den Centrosomen sind offenbar homolog die *Blepharoplasten*, d. h. die plasmatischen Bildner und Träger der Begeißelung vom polyciliaten Typus, also der Zoosporen von *Oedogonium* und der Kormophytenspermien. Sie entstehen aus den Centrosomen oder unter Beteiligung derselben (vgl. zusammenfassend Schnarf, auch P. Dangeard, Küster).

Physiologische Untersuchungen über Centrosomen fehlen noch, ebenso wie elektronenoptische Beobachtungen über ihren Feinbau. Die funktionelle Bedeutung bei der Spindelbildung wie die Beziehungen zur Geißelbildung bleiben im einzelnen noch ungeklärt.

# 4. Zellkern.

## a) Allgemeines.

Im Unterschied zu den Plasten entstehen Zellkerne nicht nur durch Teilung, sondern auch durch Fusion aus ihresgleichen. Normalerweise geschieht dies bei jedem Sexualakt; außerdem gibt es mehr oder weniger pathologische somatische Fusionen (vgl. Küster, S. 266ff.).

Die Kernteilung (Mitose, Karyokinese) besteht im wesentlichen aus zweierlei Vorgängen: der autonomen Teilung oder Längsspaltung der Chromosomen, welcher identische Reproduktion der Chromonemen zugrunde liegt, und der Ausbildung eines Verteilungsapparats, der Spindel, die die geregelte Aufteilung der — unter charakteristischem Formwechsel manövrierfähig gemachten — Tochterchromosomen (Chromatiden, Spalthälften) auf zwei Tochterkerne bewirkt. Die beiden gewöhnlich gekoppelten Abläufe: Chromosomenreproduktion und Spindelbildung, lassen sich experimentell auseinanderlegen, so durch Colchicin, das die Spindelbildung hemmt, aber die Chromosomenteilung unalteriert läßt; sie sind ferner in pathologischen Fällen getrennt (Restitutionskernbildung im weitesten Sinn) und im normalen Entwicklungsgang findet Chromosomenteilung ohne Spindelbildung bei der Endomitose statt (GEITLER 1941, 1953); in allen Fällen ist das Ergebnis ein polyploider Kern. Während der normalen Meiose erfolgt einmalige Chromosomenteilung, aber zweimalige Spindelbildung. Bei Tieren — Hinterdarm von Dipterenlarven — laufen während der normalen Entwicklung nach wiederholten Endomitosen Kernteilungen mit Spindelbildung ohne Chromosomenteilung ab. Spindeln können im übrigen auch ohne Kern, so in kernlosen tierischen Eiern, auftreten und sogar mit Cytokinesen einhergehen (FANKHAUSER).

Von der Mitose, die streng geregelt abläuft und chromosomal identische oder — im Fall der normalen Meiose — nur im Allelenbestand verschiedene Tochterkerne ergibt, ist streng zu unterscheiden die Kernfragmentation, die einen in jeder Hinsicht regellosen Zerfall des Kerns ohne Chromosomenteilung und Spindelbildung in zwei oder mehrere, zufällig gleich- oder auch ungleichgroße Teile bewirkt (bekanntes Beispiel: die „Kerne", richtig Kerntrümmer in den ausgewachsenen Internodialzellen der Characeen; HEITZ 1932, GEITLER 1941 — da es sich um Kernfragmente handelt, sind diese Zellen *nicht* polyenergid). Als Kernfragmentationen sind auch die in alten Geweben auftretenden sog. Amitosen zu betrachten, sofern nicht überhaupt verkannte defekte Anaphasen vorliegen. Wirkliche Amitosen gibt es nur bei den Ciliaten und gewissen Protisten, und zwar in endopolyploiden Kernen; zugrunde liegt eine $\pm$ geordnete Genomsegregation, die im einzelnen noch der Aufklärung bedarf (K. GRELL 1950, 1953). Außerdem gibt es in gewissen ausdifferenzierten tierischen Geweben „Amitosen", deren Wesen noch unbekannt ist.

Die Ausdrücke „direkte" und „indirekte" Kernteilung sind demnach endgültig zu streichen. Es gibt nur *die* Kernteilung, d. i. die Mitose, — die übrigens so direkt wie nur möglich verläuft: denn alle chromosomalen Formwechselprozesse samt der Spindelbildung erscheinen als die denkbar ökonomischsten Behelfe, um die Herstellung zweier genetisch identischer Tochterkerne aus einem Mutterzellkern von einem Bau, wie er einem Kern eben grundsätzlich eigentümlich ist, zu gewährleisten.

### b) Kernstruktur.

Der Zellkern ist nach einfachster und allgemeingültiger Definition eine gegen das Cytoplasma abgegrenzte Ansammlung von Chromosomen, und zwar im allgemeinen von Chromosomen in maximal entfaltetem und daher physiologisch optimal aktivem Zustand. Entsprechend den ganz verschiedenen Leistungen erscheinen die Chromosomen im Ruhekern, der im physiologischen Sinn ein Arbeitskern ist, anders gebaut als im mitotischen Kern, wo sie maximal „kontrahiert" oder „kondensiert" sind, d. h. die Chromonemen maximal spiralisiert haben und deshalb den mechanischen Anforderungen der Mitose optimal entsprechen. Kommt der Spiralisierungsformwechsel pathologischerweise in Unordnung, so ergibt sich ein nicht entwirrbarer Chromosomenknäuel, mit dem die Spindelmechanik nichts mehr anzufangen weiß (Abb. 22.)

Es ist nicht möglich, den Chromosomenbau aus dem Bau des Ruhekerns zu verstehen; wohl aber ergibt sich umgekehrt aus den Baueigentümlichkeiten der Chromosomen und ihrem Verhalten in der Telophase das Verständnis für den Aufbau des Ruhekerns. Auf diesem Wege wurde die grundlegende Einsicht gewonnen (HEITZ), daß der Kern eine ganz bestimmte artspezifische — eventuell gewebespezifisch modifizierte — Architektonik besitzt und *nicht* aus undefinierbaren „Körnchen", „Schollen" u. dgl. besteht; Ausdrücke wie „Reticulum",

„Karyotingerüst", „Lininfäden" usw. sind nichtssagend und überholt. Erst recht sind Aussagen wie: „der Kern" färbt sich vital, unzureichend; es muß vielmehr festgestellt werden, ob sich die Euchromonemen, das Heterochromatin, die Grundsubstanz, die Chromomeren oder die Nucleolarsubstanz färben.

Es ist an dieser Stelle nicht möglich, eine auch nur einigermaßen erschöpfende Übersicht über das Gesamtgebiet der Karyologie zu geben[1]. Einleitend läßt sich etwa folgendes hervorheben.

Der Kern ist nach außen durch die Kernmembran abgegrenzt. Sie ist hochkompliziert gebaut und besitzt im Unterschied zu anderen Grenzschichten die merkwürdige Eigentümlichkeit, daß sie geformte Körper, z. B. Nucleolen, durchtreten lassen kann (SCHRADER, S. 119)[2].

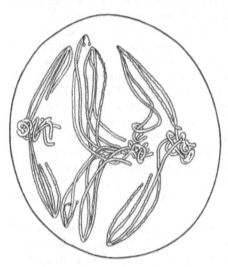

Abb. 22. Abnorme Pollenmutterzelle von *Paris quadrifolia* in I. Anaphase, die Chromatiden ohne Großspiralen, daher nicht trennbar. — Essigcarm. (Nach GEITLER.)

Die Kernwand umschließt die in Kürze nicht leicht definierbare Kerngrundsubstanz (Karyolymphe); in ihr liegen die Chromosomen bzw. ihre Derivate und dazu die Nucleolarsubstanz, die meist zu Nucleolen geformt auftritt. Die Nucleolen stehen mit der Eiweißsynthese im Cytoplasma in enger Beziehung (CASPERSSON), sie treten daher z. B. nicht in Spermakernen auf und ihre Größe schwankt mit der Leistung der Zelle.

Für die Struktur des Kerns ist bestimmend, in welchem Ausmaß die Chromosomen in der Telophase durch Hydratation, Abbau der Hüllsubstanz (Matrix) und der Desoxyribosenucleinsäure, Entspiralisierung der Chromonemen u. a. m. verändert werden. Vom Erhaltenbleiben der mitotischen Ausbildung in total oder partiell heterochromatischen Chromosomen bis zum völligen Unsichtbar-

werden der Chromosomen gibt es alle Übergänge. Daß die langgestreckten Grundbauelemente der Chromosomen, die Chromonemen, mit ihrer longitudinalen Chromomerendifferenzierung in jedem Ruhekern, also auch dann, wenn sie unsichtbar sind, erhalten bleiben, ist nicht zweifelhaft. Die Beweisführung beruht zwar auf Analogieschlüssen, doch sind sie zwingend. Es ist an sich schon schwer vorstellbar, daß bei dem einen Organismus die Chromosomen persistieren — hierfür gibt es unmittelbare Beweise —, beim anderen aber nicht; es verhalten sich aber auch Schwesterkerne (Abb. 23) und sogar im gleichen Kern verschiedene Chromosomen, ja Abschnitte des gleichen Chromosoms verschieden (so in extremen Chromozentrenkernen mit „Prochromosomen", deren Euchromatin in der Grundsubstanz optisch völlig „untergeht", — in der Prophase allerdings unmißverständlich wieder sichtbar wird; Abb. 24)[3]. Die chromosomale Strukturlosigkeit vieler Kerne im mikroskopischen Bild ist nur ein — im einzelnen erklärbarer — Grenzfall. Auch aus der Tatsache, daß vital strukturlos erscheinende Kerne erst nach Fixierung die erwarteten Strukturen zeigen, ergibt sich kein

---

[1] Es sei hier auf die ausführliche Darstellung TISCHLERs verwiesen.

[2] Daß sich nicht alle, oft recht flüchtige derartige Angaben verifizieren lassen, ist selbstverständlich.

[3] Noch bei BELAR (1928) finden sich die seit BOVERI erbrachten zahlreichen morphogenetischen Beweise besonders angeführt; ihre Besprechung erscheint hier überflüssig, zumal sie sich, seit HEITZ eine neue tragfähige Grundlage geschaffen hat, viel leichter führen ließen.

Gegenargument; die Frage der Artefaktbildung spielt, seitdem signifikante Artefakte sicher definierbar sind, keine Rolle mehr.

Das vital homogene Aussehen hat zweierlei Ursachen: entweder sind die chromosomalen Strukturen gleich stark lichtbrechend wie die Umgebung; dann läßt sich die reale Struktur durch kontrollierte Fixierung sichtbar machen; oder es ist tatsächlich keine solche Struktur vorhanden, d. h. die Chromosomen sind extrem abgebaut; dann ergibt auch die Fixierung keine Struktur oder liefert

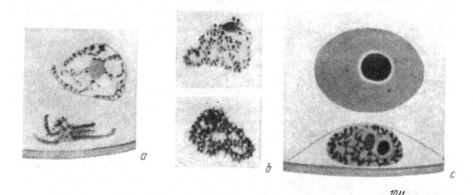

Abb. 23a—c. Verschiedene Rekonstruktion und Weiterentwicklung der Tochterkerne der 1. Pollenmitose von *Uvularia grandiflora*. a Frühe Telophase (leicht schräge Ansicht). b Mittlere Telophase. c Nach der Teilung (generative Zelle und vegetativer Kern). — Ausstriche, Flemm.-Benda. (Nach GEITLER 1935.)

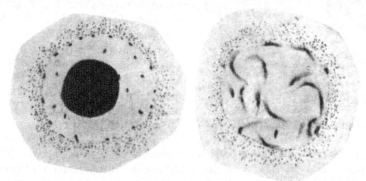

Abb. 24. Chromozentrenkern von *Citrullus vulgaris* und mittlere Prophase desselben: die Chromozentren erweisen sich als proximale Abschnitte der sich ausdifferenzierenden Chromosomen, deren distale, euchromatische Abschnitte schwach gefärbt erscheinen. — Flemm.-Benda. (Nach DOUTRELIGNE.)

nur uncharakteristische Eiweißgerinnsel (elektronenoptische Untersuchungen haben in dieser Hinsicht aus erklärlichen methodischen Gründen bisher versagt). Hier hat die Analogieschlußbildung einzusetzen. Die zahlreichen Lebendbeobachtungen gerade auf botanischem Gebiet haben eine Fülle von — auch ohne Phasenkontrast — vital sichtbaren und photographierbaren chromosomalen Strukturen erbracht (Kerne von Angiospermen — BELAR; vgl. besonders 1928, Abb. 3b, c; Pteridophyten — WADA 1940/41; HEITZ 1942, Chlorophyceen, Flagellaten, Diatomeen u. a.; Abb. 25—27)[1].

[1] Es trifft übrigens nicht zu, wenn für tierische Kerne behauptet wird (z. B. LEHMANN), sie besäßen im Leben keine *chromatische* Struktur. Dies gilt im allgemeinen nur für Eikerne. Deutlich und übereinstimmend mit dem Fixierungsbild strukturiert sind z. B. die lebenden, ungeschädigten Kerne von Lymphocyten (BELAR 1928, Abb. 3d, e), die verschiedener Gewebe von Crustaceen, die Muskelkerne von *Tubifex*, Kerne von Insektenlarven, Rotatorien, — ganz zu schweigen von den Riesenkernen der Dipterenlarven, vom Makronucleus der Ciliaten von den Kernen der Dinoflagellaten (Abb. 29d), Euglenen (Abb. 25c, 29e) usw.

Die chromosomalen Kernstrukturen lassen sich im Anschluß an Heitz auf wenige Typen zurückführen. 1. Die Chromosomen sind euchromatisch und werden bis auf ihre Chromonemen abgebaut; diese bleiben mit Matrix $\pm$ beladen und sind entsprechend färbbar. Ein solcher *Chromonemenkern* sieht vital und nach richtiger Fixierung gleichmäßig „körnig" aus (Abb. 25, 29 c, f—h); die scheinbaren Körnchen sind optische Querschnitte der Chromonemen (wieweit zusätzlich ein Chromomerenbau sichtbar werden kann, steht noch dahin); bei schlechter Fixierung entstehen netzige Strukturen (daher der Name „reticulärer" Kern). Durch Ammoniakbehandlung, die Entspiralisierung der Chromonemen bewirkt, lassen sich Meta- und Anaphasen direkt in Ruhekerne überführen, die sich von natürlichen in nichts als im Fehlen des Nucleolus unterscheiden (Kuwada und Nakamura). 2. Bestimmte Chromosomenabschnitte, manchmal auch ganze Chromosomen, bleiben im wesentlichen in mitotischer (kondensierter) Ausbildung erhalten: sie bestehen aus Heterochromatin, das im Ruhekern in Form von Chromozentren auftritt (Abb. 24, 30, 32; *Chromozentrenkerne*); oder die Chromozentren fließen zu Sammelchromozentren zusammen (Abb. 31). Auffallend viel Heterochromatin findet sich bei manchen *Fritillaria*-Arten (La Cour 1951), extrem viel bei *Navicula radiosa*, bei der fast kein Euchromatin ausgebildet ist und das Heterochromatin im Ruhekern ein einziges riesiges Sammelchromozentrum bildet (Abb. 28; Geitler 1951 b). Das Euchromatin kann in Chromozentrenkernen völlig unsichtbar werden (Abb. 24); oder es bleibt

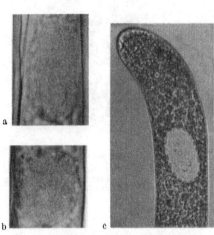

Abb. 25a—c. Chromonemakerne im Leben. a, b aus einem Staubblatthaar von *Tradescantia virginica* (Spitzenzelle mit 1 Nucleolus, intercalare Zelle mit 2 Nucleolen); c von *Euglena* sp. (die drei Kügelchen sind keine Nucleolen, sondern dem Kern anliegende Plasmaeinschlüsse). — a, b 900fach, c 400fach. (Nach Belar 1929 und 1926.)

chromonematisch sichtbar erhalten, dann entsteht ein Mischtypus aus 1. und 2. Fall (Abb. 29g). 3. Das Euchromatin wird völlig abgebaut und es ist kein Heterochromatin vorhanden; der Kern erscheint, abgesehen von Nucleolen, völlig leer. Es sind also nur die matrixlosen Chromonemen übriggeblieben (höhere Pilze, Rhodophyceen, Eizellen der Angiospermen u.a.; Abb. 28a, 29a, b). Solche Kerne zeigen manchmal wenigstens diffuse Nuclealfärbung, lassen also den Nachweis der Desoxyribosenucleinsäure, die für die mitotischen Chromosomen bezeichnend ist, noch zu; manchmal aber auch das nicht (für tierische Eizellen wurde aber der Nachweis geführt, daß es sich dann nur um extreme Grade von Verdünnung handelt; Alfert). 

Das Verständnis der Ursachen des konstitutionell verschiedenen Abbaus der Matrix bei verschiedenen Organismen steht noch aus; ebenso der Erscheinung, daß das Verhältnis von Gesamtvolumen der Chromosomen und Kernvolumen artspezifisch sehr verschieden sein kann; auch die Problematik des Heterochromatins im morphologischen, physiologischen und genetischen Sinn ist noch ungelöst (Vendrely). Schließlich steht auch die Analyse der gewebespezifischen Verschiedenheiten der Kernstrukturen und ihre Beziehung zur Leistung der Zelle noch in den ersten Anfängen (vgl. aber z. B. Grafl); gerade in dieser Hinsicht dürfte eine richtig betriebene karyologische Anatomie tiefere Einblicke vermitteln können; — aus dem Kernbau läßt sich vielfach schon morphologisch weit mehr erschließen, als gewöhnlich angenommen wird.

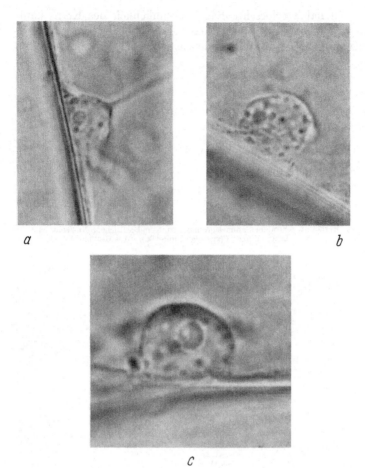

Abb. 26a—c. Chromozentrenkerne im Leben aus Trichomen junger Laubblätter von *Cucurbita pepo*; außer den Chromozentren ist je 1 Nucleolus sichtbar. — a, b 1200fach, c 1800fach. — Original.

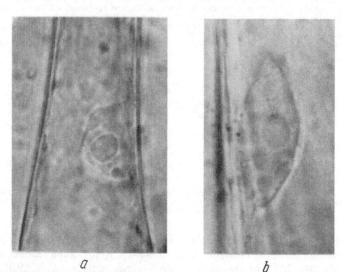

Abb. 27a u. b. Chromozentrenkerne aus Trichomen junger Laubblätter von *Sinapis alba* im Leben; Chromozentren in a im Profil, in b im Flächenbild sichtbar, dazu je 1 Nucleolus. — a 1200fach, b 1800fach; Original.

Eine nicht näher erforschte Baueigentümlichkeit des Kerns kann sich auch in der Lage des Nucleolus ausdrücken. Manchmal liegt er beliebig — in bezug auf den Kern, nicht auf das ihn kondensierende SAT-Chromosom. In bestimmten

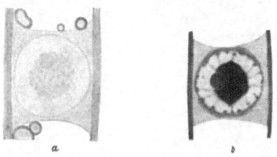

Abb. 28a u. b. Kern mit zentralem Sammelchromozentrum und seitlichem Nucleolus von *Navicula radiosa*, a im Leben, b nach Fixierung (Subl.-Alk.). (Aus Geitler 1934.)

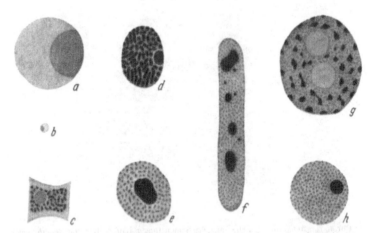

Abb. 29a—h. Bau verschiedener haploider und diploider Ruhekerne. a, b aus einer fast reifen Ascospore und einer Paraphyse von *Ascobolus immersus*: exzentrischer Nucleolus, homogene Grundstruktur. c Chromonemakern von *Diatoma vulgare*. d Kern mit fast metaphasisch erhaltenen Chromosomen und exzentrischem Nucleolus von *Peridinium* sp. e ähnlich, aber feiner strukturierter Kern von *Euglena sanguinea*. f Chromonemakern aus einer langgestreckten Zelle des Nucellus von *Ornithogalum Boucheanum* mit 4 Nucleolen und kleinen Chromozentren. g Chromonemakern mit Chromozentren aus der Antherenwand von *Lilium regale*. h Chromonemakern aus einem Perigonblatt von *Paris quadrifolia* ohne jedes Heterochromatin. — a, b nach dem Leben, etwa 1900fach, die anderen 1600fach, c, g Essigcarm., daher Nucleolen ungefärbt; d, e Subl.-Alk., f, h Flemm.-Benda. (Nach Geitler 1934, h Original.)

Abb. 30a—d. Chromozentrenkern von *Lactuca denticulata*. a Ruhekern. b Prophase, Eu- und Heterochromatin deutlich unterschieden. c Anaphase. d Telophase, Sichtbarwerden der Chromozentren an den proximalen Chromosomenabschnitten. — Alk.-Eisess., Essigcarm. (Nach Heitz 1928.)

Fällen liegt er aber konstant peripher, so bei Asco- und Basidiomyceten und bei Dinoflagellaten; erstere besitzen aber „leere", wenn auch nicht eiweißarme Kerne (Abb. 29a, b), letztere äußerst grob strukturierte (Abb. 29d); das Verhalten ist also unabhängig vom Erhaltungszustand der Chromosomen. Bei

*Cocconeis* (Abb. 8) liegt der Nucleolus immer an der konvexen Seite des Kerns. Im Fall von *Navicula radiosa* (Abb. 28) ist die subperiphere Lage aus der Verdrängung durch das zentrale Sammelchromozentrum mechanisch verständlich.

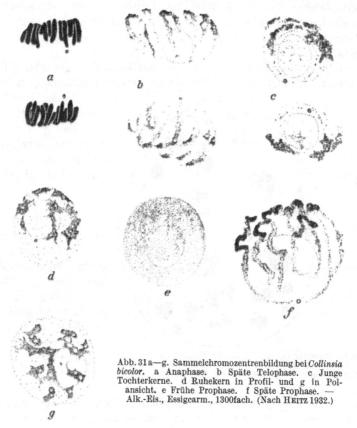

Abb. 31a—g. Sammelchromozentrenbildung bei *Collinsia bicolor*. a Anaphase. b Späte Telophase. c Junge Tochterkerne. d Ruhekern in Profil- und g in Polansicht. e Frühe Prophase. f Späte Prophase. — Alk.-Eis., Essigcarm., 1300fach. (Nach HEITZ 1932.)

## c) Größe, Gestalt, Lage in der Zelle.

Das Kernvolumen kann bei Einzellern wie bei Vielzellern individuell stark schwanken. Es wird, abgesehen von konstitutionellen Eigentümlichkeiten, von den Chromosomen und von extrachromosomalen Faktoren im Kern bestimmt und hängt weiterhin von cytoplasmatischen Einflüssen ab. Die *Chromosomen* wirken bestimmend durch ihre Zahl und durch ihr Eigenvolumen. Ihre *Zahl* vervielfacht sich durch Kernfusionen und endomitotisch. Die endomitotische Kernvergrößerung kann sehr beträchtlich sein; so sind 64ploide Kerne in bestimmten ausdifferenzierten Organen keine Seltenheit; 128ploid werden sie z. B. in den Korollhaaren von *Cucurbita pepo* (Abb. 32), in anderen Trichomen 256ploid, in Haustorien wohl noch höher endopolyploid (TSCHERMAK-WOESS und HASITSCHKA 1953, 1954; zusammenfassend GEITLER 1953). Dementsprechend nimmt auch das Zellvolumen zu (Abb. 6; sekundär steigt es noch zusätzlich durch Zellsaftvermehrung). Das Chromosomen*volumen* läßt sich — schon methodisch — schwerer fassen. Veränderungen können auf Vervielfachung der Chromonemen (Polymerie) beruhen, die aber noch nicht bewiesen ist, oder durch Zu- und Abnahme der Matrixsubstanzen hervorgerufen werden (vgl. die ausführliche Erörterung bei GEITLER 1953). In beiden Fällen handelt es sich um echte Wachstumsvorgänge. Die *extrachromosomalen* Veränderungen bestehen in Zunahme der Proteine im Kern (CASPERSSON 1947), — die auch rhythmisch erfolgen kann, so daß das Volumen auf das zwei-, vier-, achtfache usw. anwächst (SCHRADER und Leuchtenberger); auch hierbei handelt es sich um echtes Wachstum. Oder der Kern vergrößert sich durch Wasseraufnahme, wächst also nicht; davon können auch die Chromonemen und Chromozentren ergriffen werden (vgl. Geitler 1953). Demnach ist die Analyse des Kernwachstums recht schwierig. Vollends ist ein klarer Einblick in die physiologischen Beziehungen, die sich in der Kern-Plasmarelation ausdrücken, zunächst kaum zu gewinnen (vgl. Bünning, Hartmann 1953).

Je nach seinem Hydrationszustand ist der Kern flüssiger oder geliger. Daß er verformbar ist, zeigt sich unmittelbar an den Zwangsformen, die er bei Zug und Druck durch Plasmalamellen, im plasmatischen Wandbelag usw. annimmt. Autonome, artspezifische Gestalten, die von der Kugelform stark abweichen, finden sich bei manchen Protisten; doch tritt mit sinkender Zellgröße und offenbar infolge steigender Oberflächenspannung die Eigengestalt zugunsten der Kugelform zurück (Abb. 8).

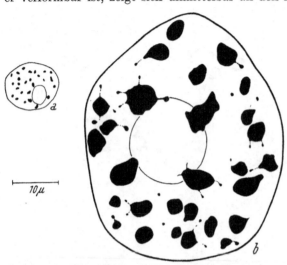

Abb. 32a u. b. Diploider (a) und 128ploider Kern (b) aus einem jungen und alten Trichom der Korolle von *Cucurbita pepo*. — Alk.-Eis., Essigcarm. (Nach Tschermak-Woess und Hasitschka 1953).

Die Lage des Kerns in der Zelle ist unbestimmt — so meist in den Geweben der höheren Algen und Kormophyten — oder ist artspezifisch im Zusammenhang mit einem bestimmten Bauplan des Protoplasten fixiert (Flagellaten, Diatomeen, Conjugaten; seine bestimmte Lage in speziell differenzierten Zellen, z. B. Trichomen höherer Pflanzen, ist damit nicht unmittelbar vergleichbar). Daß er sich autonom zu Stellen gesteigerter Leistung (Wundrand, Membranbildung) hinbewegt, ist unbewiesen. Offensichtlich wird er vom Plasma, das sich eben dorthin bewegt und sich hier ansammelt, verfrachtet (Küster, S. 185ff.). So erklärt sich auch das regelmäßige Aufsuchen der Querwandanlagen nach der Zellteilung, das sich nicht nur bei Pflanzen mit Phragmoplastenbildung, sondern auch bei Conjugaten und Diatomeen mit zentripetaler Wandbildung findet. Übrigens muß sich der Kern gar nicht an Zellorten mit gesteigerter Leistung befinden (typisch basale Lage in manchen Wurzelhaaren trotz deren Spitzenwachstum). Ob sich Kerne überhaupt aktiv zu bewegen vermögen, ist ungewiß. Jedenfalls fehlen Bewegungsorganellen (Angaben über „amöboide" Bewegungen beruhen auf Mißdeutungen fixierter Kerne). Auffallend sind allerdings die Kernwanderungen bei der Diploidisation der Hyphen höherer Pilze: hier verlagern sich die Kerne des einen Geschlechts in den Hyphen des anderen von den jüngeren zu den älteren Abschnitten entgegen der Plasmaströmung. Auch Gametenkerne wandern in Zygoten aufeinander zu. Als Bewegungsursache kämen wohl nur einseitige Änderungen der Oberflächenspannung in Betracht (Gäumann).

### d) Mitose und Chromosomen.

Allen Mitosen, gleichgültig ob es sich um solche von Protisten, Pflanzen oder Tieren handelt, ist gemeinsam, daß längsgespaltene, entsprechend manövrierfähige *Chromosomen* und *Spindeln* auftreten (vgl. S. 146). Angaben über Fehlen von Spindeln bei gewissen Protisten und in Pollenschläuchen haben sich als irrig erwiesen. Der mitotische Formwechsel der Chromosomen ist grundsätzlich überall der gleiche. Doch kann die Synchronisation von Chromosomenverkürzung

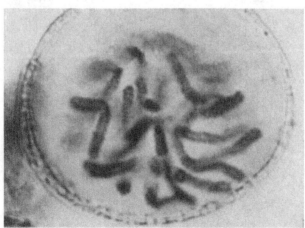

Abb. 33. Spiralbau außermeiotischer (somatischer) Chromosomen; Anaphase der 1. Pollenmitose von *Paris quadrifolia*, oben etwa 2800fach, unten etwa 2000fach. — NH₃-Alk., Alk.-Eis., Essigcarm.; oben nach GEITLER 1943c, unten Original.

und Spindelentwicklung manchmal auch anders als gewöhnlich verlaufen: so besitzen bei Coccidien (Sporozoen) und bei den in Radiolarien parasitischen Dinoflagellaten die Chromosomen während der Metaphase noch die Länge und das spiremartige Aussehen, das sie sonst in der Prophase zeigen (vgl. BELAR 1926)[1].

---

[1] Von dem aberranten Formwechsel gewisser tierischer Protisten (Hypermastiginen), bei denen die Chromosomen im *Ruhekern* in spätprophasischer Ausbildung persistieren, sei hier abgesehen (vgl. dazu K. GRELL 1953b).

Die Chromosomen sind grundsätzlich eindimensional gebaut, d. h. sie besitzen eine bestimmte longitudinale Differenzierung, die der linearen Anordnung der Gene entspricht. In den mittleren Mitosestadien ist das zugrunde liegende Bauelement, das Chromonema, spiralisiert (Abb. 33); die Chromosomen sind dann

entsprechend verkürzt und können auch „punktförmig“ aussehen (über den Spiralisierungsformwechsel vgl. Manton). Die artspezifische Größenschwankung ist oft sehr beträchtlich (Abb. 34); auch intraindividuelle Größenunterschiede kommen vor (vgl. Geitler 1953).

Stärker als durch die Chromosomen wird der Habitus der Mitose durch die verschiedene Ausbildung des Spindelapparats bestimmt. Die Spindel kann intranucleär entstehen, und es kann die ganze Mitose unter Erhaltenbleiben der Kernmembran ablaufen (manche Protisten und Metazoen). Im allgemeinen wird die Kernwand aufgelöst, so bei Kormophytenmitosen, und wenn auch wohl keine

Abb. 34a u. b.    Extreme Größenunterschiede metaphasischer Chromosomen. a Äquatorialplatte von *Spirodela polyrrhiza*. b Kleinstes und größtes Chromosom von *Haemanthus Kalbreyeri*. (Nach Blackburn.)

Vermischung des Kerninhalts mit dem Cytoplasma erfolgt (Becker 1936, 1938, Wada 1935, 1940/41), so handelt es sich doch nicht um eine intranucleäre Mitose sensu stricto[1]. Sicher sind viele Spindeln rein cytoplasmatischen Ursprungs,

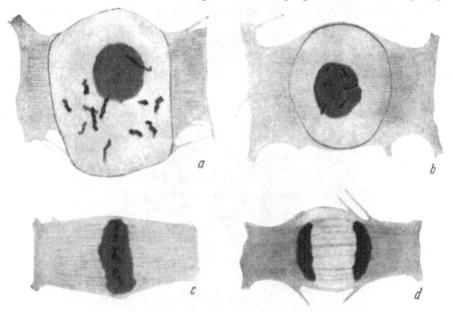

Abb. 35a—d. Späte Prophase bis mittlere Anaphase von *Spirogyra* „X“: Einwandern der Chromosomen in die Nucleolarsubstanz (a, b) und ihr Indistinktwerden von der Metaphase (c) an bis zur Anaphase (d). — Alk.-Eis., Essigcarm., etwa 900fach. (Nach Geitler.)

ja die extranucleäre Spindel kann — bei Hypermastiginen — dauernd durch die Kernwand getrennt von den Chromosomen bleiben, — so unwahrscheinlich dies klingt (vgl. Schrader, S. 36, Hartmann 1953, S. 285, Grell 1953b). Bei *Spirogyra* entsteht die Spindel extranucleär und wächst intranucleär weiter (Abb. 35, 37a). Bei Cocciden (Schildläusen) bildet sich intranucleär für jedes

[1] Nach Wada (1950—1952) soll aber die Kernwand im gewissen Sinn, z. B. auch bei *Tradescantia*, erhalten bleiben.

Chromosom getrennt eine Spindel in beliebiger Richtung; die Einzelspindeln werden erst nachträglich zentriert. Im übrigen kann die Spindel einfach sein oder — unter den Pflanzen bei den Diatomeen — aus Zentralspindel und Mantelfasern zusammengesetzt sein. Es gibt breite, stumpfpolige und schmale, spitzpolige Spindeln (Abb. 21, 35—37), und die einen wie die anderen können sich mit oder ohne Centrosom entwickeln.

Die Nucleolen werden normalerweise aufgelöst oder sie können in ± stattlichen Resten erhalten bleiben und auf die Tochterkerne verteilt werden (typisch für Cladophoraceen, Eugleninen und Dinoflagellaten). Bei *Spirogyra* kann sich ein Teil der Nucleolarsubstanz — die hier allerdings gewisse Besonderheiten zeigt — auf die Chromosomen niederschlagen und sie mehr (Abb. 35) oder weniger (Abb. 37) maskieren (GEITLER 1935b, GODWARD).

Wieweit die Spindelmechanik in allen Fällen gleich oder ähnlich ist, läßt sich zur Zeit nicht entscheiden. Das Problem der Kräfte, die zunächst die Chromatidenpaare in den Äquator bringen und dann die Chromatiden zu den verschiedenen Polen befördern — wobei sich deskriptiv eine große Zahl von Gesetzmäßigkeiten erkennen läßt (zusammenfassend ÖSTERGREN 1949b) — ist

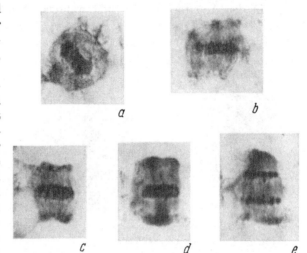

Abb. 36a u. b. Späte Prophase (a), die Chromosomen noch außerhalb des Nucleolus, Metaphase (b) und frühe Anaphase von *Spirogyra* sp.: der Rest der Nucleolarsubstanz verbackt mit den Chromosomen. — Subl.-Alk., Photo, etwa 1000fach. (Nach GEITLER.)

ebenso wie die Frage, in welcher Weise dabei die Spindel und ihre Bauelemente oder auch autonome Chromosomenbewegungen beteiligt sind, noch ungelöst (vgl. die umfassende, besonders kritische und aufschlußreiche Darstellung SCHRADERs; sie ergibt, daß mit einer bestimmten Kraft nicht alle Erscheinungen erklärt werden können; verschiedene Kräfte müssen ineinander greifen; — an botanischen Objekten allein läßt sich übrigens kein Einblick in die Fülle von Tatsachen gewinnen, die alle erklärt werden müssen). Es steht nicht einmal fest, ob sich die Spindel allgemein als ± modifiziertes Taktoid auffassen läßt, obwohl diese Annahme viel erklärt (ÖSTERGREN). Sicher ist jedenfalls, daß die Spindel eine längsorientierte Feinstruktur irgendwelcher Art besitzt und grundsätzlich bipolar organisiert ist (woran auch abnorme mehrpolige Spindeln nichts ändern)[1]. Der längsfaserige Bau ergibt sich — abgesehen davon, daß er sich aus der Chromosomenbewegung ablesen läßt und abgesehen von vielen anderen Indizien — zwingend aus der Doppelbrechung der Spindel und aus den signifikanten Fixierungsartefakten, die, wenn auch vergröbert, längsfaserigen Bau ergeben (SCHMIDT; INOUÉ 1951, 1953, — hier wird mittels eines besonders adaptierten Polarisationsmikroskops nicht nur die Doppelbrechung als solche, sondern die Faserstruktur selbst an der lebenden Spindel nachgewiesen, wobei sich wesentliche Übereinstimmung mit

---

[1] Auf die Unterscheidung von die Pole verbindenden continuous fibers und die zwischen Chromosomen und Polen in den Halbspindeln ausgespannten chromosomal fibers kann hier gar nicht eingegangen werden.

gut fixierten Spindeln ergibt). Die bekannten Mikrodissektionsversuche von
Chambers, die einen Längsfaserbau auszuschließen schienen, sind durch gegen-
teilige Befunde Carlsons widerlegt (vgl. auch Schrader). Daß in einer so
organisierten Spindel auch *transversale* Verschiebungen der Chromosomen erfolgen
können (Östergren 1949a), deutet auf den Taktoidcharakter oder jedenfalls
auf eine dynamische Struktur hin.

Die Voraussetzung des geregelten Verhaltens der Chromosomen in der Spindel
ist aber auch die, daß sie mit der Spindel in bestimmte Interaktion treten. Dafür
ist gesorgt dadurch, daß die Chromosomen eine besonders differenzierte Ansatz-
stelle *(Centromer, Kinetochor)* besitzen, mittels welcher in der Prophase An-
heftung an die Spindel erfolgt; hier werden die Chromosomen bzw. Chromatiden in den Äquator gezogen und während der Anaphase polwärts bewegt.

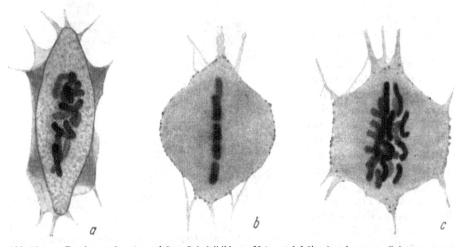

Abb. 37a—c. Prophase mit extranucleärer Spindelbildung, Meta- und frühe Anaphase von *Spirogyra crassa*;
in c, leicht schräg gesehen. „Parallelverschiebung der Chromatiden." — Subl.-Alk. (Nach Geitler).

Abweichungen von der typischen Synchronisation der Bewegungsabläufe kommen,
abgesehen von Störungen in der Meiose, selten vor. Doch trennen sich bei Diatomeen die Tochtercentromeren schon in der Meta- statt in der Anaphase (Geitler 1951b). Das Centromer, das selbst ein zusammengesetztes, polarisiertes
Organell ist (Lima de Faria 1949b, Tjio und Levan, Östergren), besitzt konstante Lage im Chromosom und unterteilt dieses in zwei Arme von bestimmter
Länge, bestimmter Chromomerenzahl, eventuell bestimmten heterochromatischen
Abschnitten und sekundären Einziehungen, welche teils funktionslos sind, teils
im Dienste der Nucleolenkondensation stehen (SAT-Chromosomen — Heitz
1931a; solche sind nicht nur bei Kormophyten und Tieren bekannt, sondern
kommen z. B. auch bei Cladophoraceen — Geitler 1936 — und bei *Spirogyra*
— Godward — vor).

Der Besitz eines lokalisierten Centromers entspricht dem Normalbau der Chromosomen.
Unerwarteterweise besitzen aber die Hemipteren, einige andere Insekten, Skorpione und
*Luzula* (Malheiros, Castro und Camara) sowie vermutlich *Spirogyra* (Abb. 37, Geitler
1930, Godward 1953) und Plasmodiophorales (Horne, Webb, vgl. auch Olive) Chromosomen, die über die ganze Länge — genauer entlang — einer Flanke sich an die Spindel
anheften („diffuses Centromer"). Eine noch ungelöste Schwierigkeit für die Vorstellung
besteht dabei darin, daß das Chromonema ja spiralisiert ist, also das an einer Flanke des
*Chromosoms* entwickelte diffuse Centromer, das doch wohl mit dem Chromonema in Zusammenhang stehen muß, in bezug auf dieses Lücken aufweisen und somit aus Teilstücken zusammengesetzt sein muß (Östergren 1949a). Vielleicht entspringt es dennoch lückenlos rundum,
wird aber gewissermaßen einseitig „durchgekämmt", so daß eine Insertion an der Flanke

nur vorgetäuscht wird. Das unvermittelte Auftreten solcher Chromosomen bei *Luzula* inmitten der Liliifloren mit Chromosomen vom Normalbau zeigt, daß es sich kaum um einen Bauunterschied grundsätzlicher Natur handeln kann; wie die Homologien liegen und was phylogenetisch primär ist, läßt sich allerdings noch nicht entscheiden (vgl. LIMA DE FARIA 1949b). Entwicklungsbiologisch betrachtet hat der Bauunterschied aber wichtige Folgen: wenn Chromosomenfragmentationen auftreten, was spontan geschehen kann und die Grundlage phylogenetischer chromosomaler Umbauten abgibt, ist ein „gewöhnliches" Chromosom ohne Spindelansatz verloren, da es an der nächsten Mitose nicht mehr teilnehmen kann; bei diffusem Spindelansatz kann aber jedes Bruchstück persistieren und, wie röntgeninduzierte Fragmentationen bei *Luzula* ergeben haben, persistieren solche Bruchstücke auch tatsächlich (CASTRO und NORONHA-WAGNER)[1].

In den grundsätzlichen Merkmalen aller Mitosen — autonome Chromosomenspaltung und Spindelbildung — unterscheiden sich auch die mit diffus-centromerischen Chromosomen nicht von allen anderen. In allgemeiner Hinsicht besteht tatsächlich Übereinstimmung zwischen sämtlichen Organismen mit Kern-Cytoplasma-Differenzierung. Ältere Versuche, „primitivere" Typen von Mitosen („Promitosen") zu unterscheiden und Mitosestammbäume zu entwerfen, sind überholt (BELAR 1926; kurz zusammengefaßt GEITLER 1934; mit Beschränkung auf die Protozoen GRELL 1953b). Es gibt nur *eine* Mitose, — mit vielen Varianten, die aber im phylogenetischen Sinn nicht interpretierbar sind; was sich zeigt, ist allein eine größere Mannigfaltigkeit bei den Protisten; bei den höheren Pflanzen und Tieren hat sich ein ziemlich invariabler Typus stabilisiert[2].

Der Zellkern ist das komplizierteste Organell der Zelle. Was dies bedeutet, wird so recht sinnfällig, wenn er eine Ausbildung von der Art der Riesenspeicheldrüsenkerne der Dipterenlarven annimmt. Grundsätzlich ist aber jeder Kern — abgesehen von der zusätzlichen Endopolyploidie und somatischen Chromosomenpaarung im Falle der erwachsenen Speicheldrüsenkerne (zur allgemeinen Orientierung vgl. HARTMANN 1953) — gleichartig gebaut. Für botanische Untersuchungen muß im allgemeinen das Pachytän genügen, in dem bei weitgehender Entspiralisierung der Chromonemen der Chromomerenbau maximal deutlich in Erscheinung tritt. Auch im „homogenen" Ruhekern liegt eine entsprechende Struktur zugrunde. Sie wird nicht nur wegen ihrer zu geringen Größe lichtoptisch nicht sichtbar, sondern bleibt auch deshalb unerkennbar, weil die natürliche vergröbernde Zusatzstruktur fehlt, die z. B. im Fall des Pachytäns durch Ausbildung einer lokalisierten feulgenpositiven Hülle um die Chromomeren die submikroskopische Struktur deutlich macht, und weil unter anderen physiologischen Bedingungen andere in der gleichen Grundstruktur liegende Ausbildungsmöglichkeiten realisiert sind, die sie indistinkt machen. Daher bewirkt die beste, d. h. lebensgetreueste Fixierung nicht, daß elektronenoptisch anderes als im Lichtmikroskop beobachtet werden kann, und es versagen daher auch Ultramikrotomschnitte; denn die Feinbauelemente sind, schon im Leben, miteinander „verklebt" oder „verbacken" und notwendigerweise indistinkt. Hier fehlen analoge Verfahren, wie sie z. B. zur lichtmikroskopischen Sichtbarmachung der Spiralchromonemen in Chromosomen durch richtig dosierte vitale Vorbehandlung mit destilliertem Wasser oder $NH_3$-Dämpfen u. a. ausgearbeitet wurden; auch *diese* Struktur kann das Elektronenmikroskop an keinem noch so lebensgetreu fixierten Chromosom sichtbar machen. Aus den gleichen Gründen zeigen die elektronenoptischen Bilder auch von Pachytänchromosomen bestenfalls gleich viel, aber keinesfalls mehr Einzelheiten als lichtmikroskopische. Es dürfte nur eine Frage der Zeit sein, bis brauchbare Macerationsverfahren gefunden werden,

---

[1] Analoge Beobachtungen liegen für Tiere vor.

[2] Über die augenblickliche Unsicherheit in der Deutung der Teilung der Nucleoide der Bakterien vgl. S. 125.

um elektronenoptisch die Analyse des Feinbaus von Kern, Chromosomen und Spindel ernstlich in Angriff zu nehmen. Anfänge auf zoologischem Gebiet liegen schon vor (CLAUDE und POTTER, POLLISTER und MIRSKY, MIRSKY und RIS).

# VI. Die Zellmembran.

Im Unterschied zur tierischen Organisation ist für die pflanzliche Zelle die Ausbildung einer festen, im ausgewachsenen Zustand toten Hülle, der Membran oder Zellwand, charakteristisch. Sie liegt dem Protoplasten an und steht mit ihm in engem Kontakt. Bei vielen Protisten kommen dagegen $\pm$ abstehende Gehäuse vor, die aus Membransubstanz aufgebaut sind, aber ihrer Entstehung nach andersartige Bildungen sind; gehäusebildende Protisten können außerdem auch eine Membran im eigentlichen Sinn ausbilden, so im Cystenzustand (z. B. *Trachelomonas*).

Stammesgeschichtlich zeigt es sich, daß die Zelle mit „toter" Wand auf membranlose Flagellaten mit plasmatischen Oberflächengrenzschichten zurück-geht. Zwischen primitiv und komplizierter gebauten nackten Oberflächen und ausgesprochenen Membranen gibt es alle Übergänge. Die Ausbildung einer Cellulose-, Pektin- oder Chitinmembran wird zunächst im geißellosen Ruhe-zustand erreicht, der im allgemeinen eine Dauerzelle darstellt; doch gibt es als Sonderentwicklungen auch festbehäutete Flagellaten (manche Volvocalen, Dino-flagellaten; für die Ableitung der typischen Algen-, Pilz- und Kormophytenzelle von den Flagellaten vgl. GEITLER 1934, S. 56ff.).

Im allgemeinen ist die Membranbildung an die Anwesenheit des Zellkerns gebunden. Doch können Membranen unter Umständen auch von künstlich kernlos gemachten Proto-plasten gebildet werden und, bei *Acetabularia* sowie in geringerem Ausmaß auch bei *Micrasterias*, an spezifischen Gestaltungen teilnehmen (zusammenfassend KÜSTER, S. 769, KALLIO). Kernlose Pollenkleinzellen können artspezifische Exinestrukturen hervorbringen (DRAHOWZAL). Vermutlich wirken vom Kern gelieferte Stoffe nach.

Die pflanzlichen Membranen sind im ausgewachsenen Zustand wohl im all-gemeinen als unbelebt zu bezeichnen. Doch ist nicht zu vergessen, daß die An-wendung des Begriffs auf einen Teil einer lebenden Zelle immer etwas willkürlich bleibt. Außerdem *entsteht* jede Membran aus Plasma; sie wird im Lauf ihrer Entwicklung allmählich plasma-unähnlicher, ist aber doch bis zu einem gewissen Grad noch lebendig, solange sie wächst, und stellt kaum ein bloßes Abscheidungs-produkt dar (vgl. die ausführliche Diskussion bei KÜSTER, S. 539ff.). Merk-würdiges Membranwachstum, das sich durch tote Flüssigkeitsräume getrennt vom Protoplasten abspielt, gibt es allerdings bei der Sporenbildung von *Sela-ginella* und *Isoëtes* (FITTING; vgl. auch KÜSTER, S. 540) sowie bei der Pollen-kornentwicklung von Oenotheraceen (zuletzt GEITLER 1937a); hier muß das vom Plasma gelieferte Baumaterial durch Diffusion an Ort und Stelle gelangen; nicht unwesentlich dürfte dabei sein, daß sich die Entwicklung nicht an einer gegenüber der Außenwelt freien Oberfläche, sondern im Innern eines Sporangiums abspielt.

Eine eigenartige Zwischenstellung zwischen lebenden Plasmahäuten und toten Membran-bildungen nehmen anscheinend die Querwände (nicht die Membranscheiden) der hormogo-nalen Cyanophyceen ein; sie besitzen wohl mehr den Charakter einer Pellikula als einer festen Membran. Dies läßt sich unter anderem daraus schließen, daß ihnen Plasmodesmen fehlen: Plasmodesmen durchbrechen sonst tote Membranen und stellen so die Kontinuität der Proto-plasten her; bei den Cyanophyceen sind sie sozusagen überflüssig (MÜHLDORF 1938).

Die frühesten Schritte der Membranbildung, d.h. der Übergang von der rein plasmatischen Wandanlage — sei es in einem Phragmoplasten, in einer Ringleiste oder als oberflächliche Bildung — bis zur festen Membran sind uns noch weit-gehend unbekannt. Dagegen haben physikalische und elektronenoptische Untersuchungen tiefere Einblicke in das spätere Wachstum ergeben und auch ein konkretes Verständnis des — lange vermuteten — submikroskopischen micellaren Feinbaus ermöglicht (z. B. FREY-WYSSLING, MÜHLETHALER und WYCKOFF). Die Micelle erscheinen dabei nicht als selbständige, isolierte Gebilde, sondern

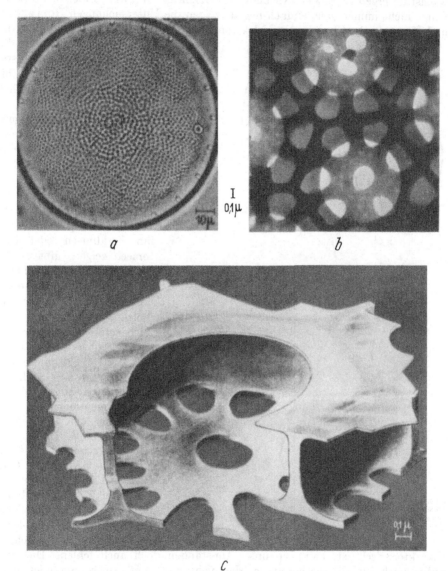

Abb. 38a—c. Membranbau einer Diatomee *(Actinocyclus Ehrenbergii)*. a Ganze Zelle in Schalenansicht, lichtoptisch 500fach. b Ausschnitt daraus, elektronenoptisch 24000fach. c Rekonstruktionsbild aus b, 36000fach.
(Nach HELMCKE und KRIEGER).

röntgenologisch als ungestörte Gitterbereiche, die untereinander durch Fadenmoleküle verbunden sind. Die submikroskopische Textur der Primärwand, die aus locker verwobenen Mikrofibrillen besteht, legt den Schluß nahe, daß sie während ihres Wachstums von Plasma durchtränkt ist (vgl. zusammenfassend KÜSTER, S. 539ff., 548ff., BÜNNING 1953, S. 111ff.). Auch die Art des Anschlusses neugebildeter Scheidewände an alte Wände spricht dafür (FREY-WYSSLING und MÜHLETHALER 1951). Im Sinn wenigstens partieller Lebendigkeit läßt sich auch das bekannte Haftenbleiben des Protoplasten an der Wand bei Plasmolyse verstehen. Im übrigen hat das Elektronenmikroskop auch die grundsätzliche Richtigkeit der Unterscheidung von Intussuszeptionswachstum — hierbei werden in gelockerte flächenhafte Bereiche neue Fibrillen eingeflochten — und Appositionswachstum ergeben (Frey-Wyssling und Stecher, Stecher, Preston; daß die Dinge nicht immer ganz einfach liegen, zeigen die Untersuchungen Roeløfsens und Houwinks). Ebenso haben elektronenoptische Untersuchungen die mit der mechanischen Leistung der Membran in Beziehung stehenden Feinstrukturen bis in Einzelheiten verständlich gemacht.

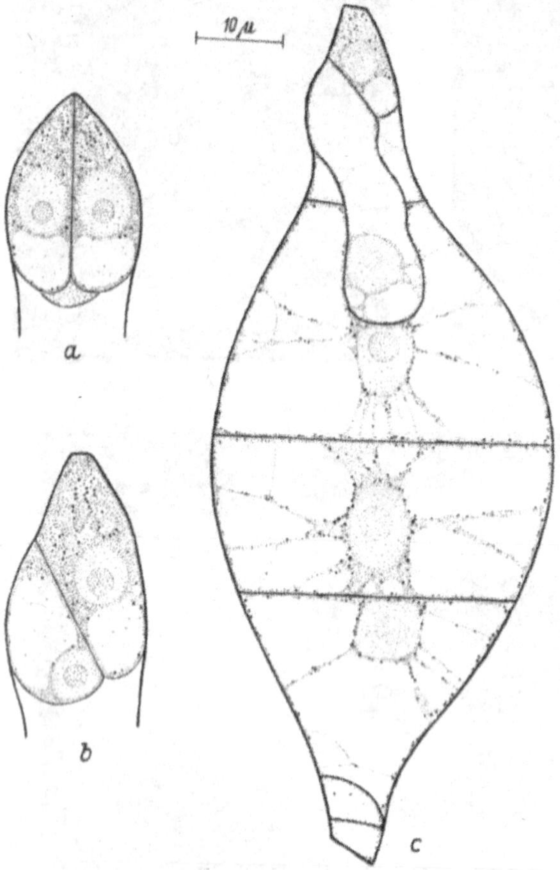

Wenn die Zelle den ganzen Organismus darstellt, kann die Membran, analog wie die Zellform und andere Organellen, eine außerordentlich hohe Differenzierung erfahren. Einen Membranbau von der Komplikation, wie ihn Peridinieen oder Diatomeen zeigen, gibt es bei Vielzellern nicht. Dabei handelt es sich zunächst nicht um Eigentümlichkeiten im molekular-micellaren Bereich, sondern um vergleichsweise gröbere Strukturen, die bereits lichtoptisch auffallen, wenn sich auch ihr Feinbau in submikroskopische Dimensionen verliert. Die Höhe der Differenzierung, die z. B. Diatomeenschalen erreichen, veranschaulicht Abb. 38 (vgl. Helmcke und Krieger).

Abb. 39a—c. Embryosack von *Monotropa hypopithys*, nach dem Leben. a, b Eiapparat in zwei Ansichten (in a die Eizelle durch die beiden Synergiden verdeckt). c Nach der Befruchtung, der einzellige Embryo in das Endosperm einwachsend. (Nach Geitler 1934.)

Obwohl die Membranbildung für die pflanzliche Organisation in ihrer typischen Ausprägung durchaus charakteristisch ist, kann sie unter Umständen doch auch fehlen. Dies gilt selbst noch für die Anthophyten, die nicht nur membranlose Spermien oder Spermienzellen hervorbringen, sondern bei denen auch plasmodiale (vgl. S. 126 ff.), also überhaupt nicht gekammerte polyenergide Embryonen (Abb. 1) und Endosperme, wenn auch nur als vorübergehende Stadien auftreten; die grundsätzliche Einheit der pflanzlichen und tierischen Organisation tritt hierin deutlich in Erscheinung. In dieser Hinsicht auffallende Verhältnisse herrschen auch im weiblichen Gametophyten der Angiospermen. Die definitiven Zellen des Embryosacks sind zunächst nur durch Kerne in gemeinsamem Plasma repräsentiert; erst nachträglich erfolgt die Abgrenzung zu Zellen. Eikern und oberer Polkern einerseits, die beiden Synergidenkerne andererseits sind vermutlich

Schwesterkerne. Die Membran zwischen den letzteren entsteht in einem Phragmoplasten und wächst dann um den abzugrenzenden Protoplasten bogenförmig
herum; die Membran der Eizelle scheint de novo zu entstehen (es liegen noch
wenige exakte Untersuchungen vor; vgl. SCHNARF). Diese Membranen sind im
übrigen zwar fest und z. B. durch Fixierung vom Protoplasten abhebbar, dürften
aber keine Pektin- oder Cellulosereaktion geben (allerdings liegen auch gegen-
teilige Angaben vor). Jedenfalls handelt es sich bei den Zellen des Eiapparates
um Bildungen, die aus einem polyenergiden Gametophyten hervorgehen und
die auch im erwachsenen Zustand hinsichtlich der Membranbeschaffenheit aus
dem Rahmen der Kormophytenorganisation herausfallen; sie sehen, schon rein
äußerlich betrachtet, merkwürdig genug aus (Abb. 39). Nach der Befruchtung
besitzt die Eizelle und besitzen selbstverständlich die Zellen des Embryos eine
typische Membran im botanischen Sinn.

Es ist andererseits bezeichnend, daß im Bereich der pflanzlichen Organisation
Bildungen wie siphonale Algen und Pilze möglich sind. In diesen Fällen setzt
sich die Membranbildung auch gegen die polyenergide Organisation durch:
nach außen ist die ungekammerte Pflanze von einer Membran umgeben, die
sich, abgesehen davon, daß sie nicht monoënergide Einzelprotoplasten umhüllt,
in nichts von einer anderen Algen- oder Pilzmembran unterscheidet.

Unbeschadet der Wichtigkeit der Zellwandbildung für die Pflanzen ist doch
auch bei ihnen eine Zelle nicht definiert durch „eine Menge Plasma, das von
einer Wand umgeben ist". Entgegen einer öfters vertretenen Meinung sind
polyenergide Algen und Pilze nicht „vielkernige Zellen" (z. B. KÜSTER); auf
Grund einer solchen Auffassung begäbe man sich der allgemeinsten Einsicht in
die Organisation der Lebewesen.

# Beschluß.

Die Grundlage jedes physiologischen Verständnisses bildet die Morphologie,
denn die physiologischen Abläufe „hängen" ja nicht „in der Luft". Die Anatomie
bietet die feste Unterlage zunächst im beschreibenden Sinn, weiterhin im vergleichenden und schließlich auch im physiologischen, indem sie physiologische
Probleme oder wenigstens ihre Ansätze so weit einbezieht, daß sie heuristisch
wertvoll wird. Eine strenge Grenze zwischen Anatomie, Entwicklungsgeschichte
und Entwicklungsphysiologie läßt sich auf einem höheren Stand der Forschung
gar nicht ziehen, wenn auch im Hinblick auf Arbeitsökonomie eine Trennung
möglich oder erwünscht sein kann. Ein tatsächlicher Widerspruch zwischen
kausaler und akausaler Betrachtung besteht nicht; nur aus Gründen der methodischen Zweckmäßigkeit und der Grenzen des Könnens und der Kenntnisse
muß die eine oder andere Betrachtungsweise betont werden. Viele rein anatomische Beobachtungen verlieren durch Außerachtlassung jedweder physiologischen Problemstellungen an Wert, wie andererseits physiologische, im vorliegenden Fall zellphysiologische Angaben zu wenig auswertbar sind, wenn sie der
exakten morphologischen Grundlage entbehren. Während aber die anatomische
Forschung in der Regel sich das Bewußtsein bewahrt hat, daß sie wohl die Grundlage darstellt, nicht aber das Ganze biologischer Erkenntnis ausmacht, trifft
das Umgekehrte weniger oft zu. So kann es kommen, daß aus einer gewissen
Geringschätzung der Fortschritte auf dem Gebiet der Anatomie physiologische
Daten auf anatomische Kenntnisse bezogen werden, die seit langem überholt
sind. Wie produktiv im allgemeinsten Sinn die Zellanatomie werden kann,
zeigt das Beispiel der Cytogenetik; wesentliche neue Ergebnisse hat sie auch auf
dem Gebiet der Strukturforschung des Zellkerns gebracht, das gleiche gilt für

Plastidenbau, Zellwandstruktur u. a. m. Als Zukunftshoffnung zeichnet sich ihre Bedeutung ab im Fall der Problematik der Zelldifferenzierung (vgl. dazu Bünning), der Mitosemechanik (Schrader), der Plasmapartikel und der Plasmastruktur überhaupt (Lehmann, Frey-Wyssling) und in vielen anderen Fällen. Mit Recht schreibt A. Lang über bestimmte neue anatomische Befunde (S. 344): „Obgleich es sich vorwiegend um ‚deskriptive' Untersuchungen handelt, kann ohne Übertreibung gesagt werden, daß sie an Bedeutung für die entwicklungsphysiologische Problematik keiner der experimentellen Arbeiten derselben Zeit nachstehen." Es ist zu erwarten, daß sich solche Feststellungen mehren werden; die Zellanatomie, richtig betrieben, ist nicht tot; sie befindet sich in lebhaftem Aufschwung.

# Literatur.

Alfert, M.: A cytochemical study of oogenesis and cleavage in the mouse. J. Cellul. Comp. Physiol. 36, 381—409 (1950).

Bautz, E., u. H. Marquardt: Die Grana mit Mitochondrienfunktion in Hefezellen. Naturwiss. 40, 531 (1953). — Beauchamp, P. de: Contribution a l'étude du genre Ascomorpha et des processus digestifs chez les Rotifères. Bull. Soc. zool. France 57, 428—449 (1932). — Becker, W. A.: Vitale Cytoplasma- und Kernfärbungen. Protoplasma (Berl.) 26, 439—487 (1936). — Recent investigations in vivo on the division of the plant cell. Bot. Rev. 4, 446 bis 472 (1938). — Belar, K.: Der Formwechsel der Protistenkerne. Jena 1926. — Die cytologischen Grundlagen der Vererbung. In Handbuch der Vererbungswissenschaft, Bd. I. Berlin 1928. — Beiträge zur Kausalanalyse der Mitose. III. Untersuchungen an Staubfadenhaarzellen und Blattmeristemzellen von Tradescantia virginica. Z. Zellforsch. 10, 73—134 (1929). — Bisset, K. A.: The interpretation of appearances in the cytological staining of bacteria. Exper. Cell Res. 3, 681—688 (1952). — Brachet, J.: Biochemical and physiological interrelations between nucleus and cytoplasm during early development. Growth Symp. 11, 309—324 (1947). — Bretschneider, L. H.: The fine structure of protoplasm. Survey Biol. Progr. 2, 223—257 (1952). — Bringmann, G.: Vergleichende licht- und elektronenoptische Untersuchungen an Oszillatorien. Planta (Berl.) 38, 541—563 (1950). — Über Beziehungen der Kernäquivalente der Schizophyten zu den Mitochondrien höher organisierter Zellen. Planta (Berl.) 40, 398—406 (1952a). — Die Organisation der Kernäquivalente der Spaltpflanzen unter Berücksichtigung elektronenmikroskopischer Befunde. Zbl. Bakter. II 107, 40—70 (1952b). — Bryan, H. D.: DNA-protein relations during microsporogenesis of Tradescantia. Chromosoma 4, 369—392 (1951). — Bünning, E.: Morphogenesis in plants. Survey Biol. Progr. 2 (1952). — Entwicklungs- und Bewegungsphysiologie der Pflanze. Berlin-Göttingen-Heidelberg 1953. — Bünning, E., u. F. Biegert: Die Bildung der Spaltöffnungsinitialen bei Allium cepa. Z. Bot. 41, 17—39 (1953).

Carlsson, J. G.: Microdissection studies of the dividing neuroblast of the grasshopper, Chortophaga viridifasciata (de Geer). Chromosoma 5, 199—220 (1952). — Caspersson, T.: The relations between nucleic acid and protein synthesis. Symposia Soc. Exper. Biol. 1, 127—151 (1947). — Cell growth and cell function. London 1950. — Castro, D. de, e N. Noronha-Wagner: Nota sobre a perpetuação de fragmentos cromosómicos em Luzula purpurea. Agronomia lusitana 14, 95—99 (1952). — Chadefaud, M.: Le cytoplasma des Algues vertes et des Algues brunes; ses éléments figurés et ses inclusions. Rev. algol. 8, 5—264 (1936). — Les pyrénoïdes des algues et l'existence chez les végétaux d'un appareil cinétique intraplastidal. Ann. des Sci. natur. Bot., Sér. XI 2, 1—13 (1941). — Claude, A., and T. S. Potter: Isolation of chromatin threads from the resting nucleus of leukemic cells. J. of Exper. Med. 77, 345—354 (1943).

Dangeard, P.: Cytologie végétale et cytologie générale. Paris 1947. — Dangeard, P. A.: Sur la distinction du chondriome des auteurs en vacuome, plastidome et sphérome. C. Acad. Sci. Paris 169, 1005 (1919). — Danielli, J. F., and R. Brown: Structural aspects of cell physiology. Symposia Soc. Exper. Biol. 6 (1952). — De Lamater, E. D.: A consideration of the newer methods for the demonstration of nuclear structure in bacteria and other micro-organisms. Mikroskopie (Wien) 7, 358—362 (1952). — De Lamater, E. D., and S. Mudd: The occurence of mitosis in the vegetative phase of Bacillus megatherium. Exper. Cell. Res. 2, 499—512 (1951). — Doflein, F., u. E. Reichenow: Lehrbuch der Protozoenkunde, 6. Aufl. Jena 1949—1953. — Doutreligne, Jenny: Chromosomes et nucléoles dans les noyaux du type euchromocentrique. Cellule 42, 29—100 (1933). — Drahowzal, Grete: Beiträge zur Morphologie und Entwicklungsgeschichte der Pollenkörner. Österr. bot. Z. 85, 241—269 (1936). — Drawert, H.: Vitale Fluorochromierung der Mikrosomen mit Janusgrün, Nilblausulfat und Berberidinsulfat. Ber. dtsch. bot. Ges. 66, 134—150 (1953).

FANKHAUSER, G.: Nucleo-cytoplasmic relations in amphibian development. Internat. Rev. Cytol. 1, 165—193 (1952). — FITTING, H.: Bau und Entwicklungsgeschichte der Makrosporen von Isoëtes und Selaginella und ihre Bedeutung für die Kenntnis des Wachstums pflanzlicher Zellmembranen. Bot. Ztg 58, 107—164 (1900). — FREY-WYSSLING, A.: Submicroscopic morphology of protoplasm, 3. Aufl. Amsterdam u. New York 1953. — FREY-WYSSLING, A., u. K. MÜHLETHALER: Über den Feinbau der Chlorophyllkörner. Vjschr. naturforsch. Ges. Zürich 94, 179—183 (1949). — Zellteilung im Elektronenmikroskop. Mikroskopie (Wien) 6, 28—31 (1951). — FREY-WYSSLING, A., K. MÜHLETHALER u. R. W. WYCKOFF: Mikrofibrillenbau der pflanzlichen Zellwände. Experientia (Basel) 4, 475—476 (1948). — FREY-WYSSLING, A., u. H. STECHER: Das Flächenwachstum der pflanzlichen Zellwände. Experientia (Basel) 7, 420—421 (1951).

GÄUMANN, E.: Über die Geschwindigkeit der Kernwanderung bei Pilzen. Ber. dtsch. bot. Ges. 59, 283—287 (1941). — GAVAUDAN, P., et N. GAVAUDAN: Quelques remarques sur la cytologie des Oscillariées. Bull. Soc. bot. France 80, 706—712 (1933). — GEITLER, L.: Über einige wenig bekannte Süßwasserorganismen mit roten oder blaugrünen Chromatophoren; zugleich ein Beitrag zur Kenntnis pflanzlicher Chromatophoren. Rev. algol. 1, 357 bis 375 (1924). — Zur Morphologie und Entwicklungsgeschichte der Pyrenoide. Arch. Protistenkde 56, 128—144 (1926). — Der Formwechsel der pennaten Diatomeen. Jena 1932. [Auch Arch. Protistenkde 78, 1—226 (1932).] — Grundriß der Cytologie. Berlin 1934. — Beobachtungen über die erste Teilung im Pollenkorn der Angiospermen. Planta (Berl.) 24, 361—386 (1935a). — Neue Untersuchungen über die Mitose von Spirogyra. Arch. Protistenkde 85, 10—19 (1935b). — Schizophyceen. In K. LINSBAUERS Handbuch der Pflanzenanatomie, Bd. VI. Berlin 1936. — Zur Morphologie der Pollenkörner von Clarkia elegans. Planta (Berl.) 27, 426—431 (1937a). — Chromatophor, Chondriosomen, Plasmabewegung und Kernbau von Pinnularia nobilis und einigen anderen Diatomeen nach Lebendbeobachtungen. Protoplasma (Berl.) 27, 534—543 (1937b). — Das Wachstum des Zellkerns in tierischen und pflanzlichen Geweben. Erg. Biol. 18, 1—54 (1941). — Koloniebildung und Beeinflussung der Unterlage bei zwei Dinococcalen. Beih. bot. Zbl., Abt. A 62, 160 bis 174 (1943a). — Lokalisierte Karotinbildung in langgestreckten Algenzellen. Österr. bot. Z. 92, 212—214 (1943b). — Über eine postmeiotische Teilungsanomalie und den Spiralbau der Chromosomen von Paris quadrifolia. Chromosoma 2, 519—530 (1943c). — Furchungsteilung, simultane Mehrfachteilung, Lokomotion, Plasmoptyse und Ökologie der Bangiacee Porphyridium cruentum. Flora (Jena) 137, 300—333 (1944). — Über die Teilungsrhythmen in den spermatogenen Fäden der Characeen. Österr. bot. Z. 95, 147—162 (1948/49). — Zelldifferenzierung bei der Gametenbildung und Ablauf der Kopulation von Eunotia (Diatomee). Biol. Zbl. 70, 385—398 (1951a). — Der Bau des Zellkerns von Navicula radiosa und verwandten Arten und die präanaphasische Trennung von Tochtercentromeren. Österr. bot. Z. 98, 206—214 (1951b). — Über differentielle Teilung und einen im Zellbau begründeten kopulationsbegrenzenden Faktor bei der Diatomee Cocconeis. Z. Naturforsch. 7b, 411 bis 414 (1952). — Endomitose und endomitotische Polyploidisierung. Protoplasmatologia VI C. Wien 1953. — GEITLER, L., u. ELISABETH VOGL: Über die Ursachen der rechtwinkeligen Durchschneidung von Scheidewänden. Naturwiss. 44, 377 (1944). — GODWARD, M. B. E.: On the nucleolus and nucleolar-organizing chromosomes of Spirogyra. Ann. of Bot. 14, 39—53 (1950). — GEITLERS nucleolar substance in Spirogyra. Ann. of Bot. 17, 403—416 (1953). — GRAFL, INA: Cytologische Untersuchungen an Sauromatum guttatum. Österr. bot. Z. 89, 81—118 (1940). — GRASSÉ, P. P.: Traité de Zoologie, I, 1. Paris 1952 — GRELL, K. G.: Der Kerndualismus der Ciliaten und Suktorien. Naturwiss. 37, 347—356 (1950). — Cytologische Untersuchungen an Aulacantha scolymantha. Arch. Protistenkde 98, 157—160 (1952). — Die Chromosomen von Aulacantha scolymantha HAECKEL. Arch. Protistenkde 99, 1—54 (1953a). — Der Stand unserer Kenntnisse über den Bau der Protistenkerne. Verh. dtsch. zool. Ges. 1953b. — GSCHÖPF, O.: Das Problem der Doppelplättchen und ihnen homologer Gebilde bei den Diatomeen. Österr. bot. Z. 99, 1—36 (1952). — GUILLIERMOND, A.: La structure des Cyanophycées. C. r. Acad. Sci. Paris 197, 182 bis 184 (1933). — GUILLIERMOND, A., G. MANGENOT et L. PLANTEFOL: Traité de cytologie végétale. Paris 1933.

HAGENAU, F., and W. BERNHARD: Le problème de l'ultrastructure du cytoplasme et des artéfacts de fixation. Exper. Cell. Res. 3, 629—648 (1952). — HAGERUP, O.: A peculiar asymmetrical mitosis in the microspores of Orchis. Hereditas (Lund) 24, 94—96 (1938). — HARTMANN, M.: Dauernde agame Zucht von Eudorina elegans. Arch. Protistenkde 43, 223—286 (1921). — Allgemeine Biologie, 4. Aufl. Stuttgart 1953. — HEITZ, E.: Heterochromatin, Chromozentren, Chromomeren. Ber. dtsch. bot. Ges. 47, 274 bis 284 (1929). — Die Ursache der gesetzmäßigen Zahl, Lage, Form und Größe pflanzlicher Nukleolen. Planta (Berl.) 12, 775—844 (1931a). — Nukleolen und Chromosomen in der Gattung Vicia. Planta (Berl.) 15, 495—505 (1931b). — Die Herkunft der Chromozentren. Dritter Beitrag zur Kenntnis der Beziehung zwischen Kernstruktur und qualitativer Verschiedenheit der Chromosomen in ihrer Längsrichtung. Planta (Berl.) 18, 571—636 (1932). — Chromosomenstruktur und Gene. Z. Abstammgslehre **70**, 402—447 (1935). — Untersuchungen über den Bau der Plastiden. I. Die gerichteten Chlorophyllscheiben der Chloroplasten. Planta (Berl.) **26**, 134—163 (1936). — Lebendbeobachtung der Zellteilung bei Anthoceros und Hymenophyllum. Ber. dtsch. bot. Ges. **60**, 28—36 (1942). — Heitz, E., u. R. Maly: Zur Frage der Herkunft der Grana. Z. Naturforsch. 8b, 243—249 (1953). — Helmcke, J.-G., u. W. Krieger: Diatomeenschalen im elektronenmikroskopischen Bild. I. Transmare-Photo G.m.b.H., Berlin 1953. — Hollande, A. Ch.: Remarques au sujet de la structure cytologique de quelques Cyanophycées. Archives de Zool. **75**, 145—184 (1933—35). — Hollande, A. Ch., et G. Hollande: Observations sur la structure cytologique de quelques Cyanophycées. Bull. Soc. bot. France **91**, 61—63 (1944). — Horne, A. S.: Nuclear division in the Plasmodiophorales. Ann. of Bot. **44**, 199—230 (1930).

Inoué, S.: A method for measuring small retardations of structures in living cells. Exper. Cell Res. **2**, 513—517 (1951). — Polarization optical studies on the mitotic spindle. I. The demonstration of spindle fibers in living cells. Chromosoma **5**, 487—500 (1953).

Joyet-Lavergne, Ph.: La recherche des zones d'oxydation dans la cellule végétale. C. r. Soc. Biol. Paris **110**, 918—920 (1932a). — Sur le pouvoir oxydant du chondriome dans la cellule vivante. C. r. Soc. Biol. Paris **110**, 552—553 (1932b).

Kallio, P.: The significance of nuclear quantity in the genus Micrasterias. Ann. bot. Soc. zool.-bot. fenn. „Vanamo" **24**, 1—122 (1951). — Kostrun, Gertrud: Entwicklung der Keimlinge und Polaritätsverhalten bei Chlorophyceen. Österr. bot. Z. **93**, 172—221 (1944). — Küster, E.: Die Pflanzenzelle, 2. Aufl. Jena 1951. — Kuwada, Y., u. T. Nakamura: Behaviour of chromonemata in mitosis. II. Artificial unravelling of coiled chromonemata. Cytologia **5**, 244—247 (1934).

La Cour, L. F.: Nuclear differentiation in the pollen grain. Heredity (Lond.) **3**, 319—337 (1949). — Heterochromatin and the organisation of nucleoli in plants. Heredity (Lond.) **5**, 37—50 (1951). — Lang, A.: Entwicklungsphysiologie. Fortschr. Bot. **12**, 340—441 (1949).— Lang, K.: Lokalisation der Fermente und Stoffwechselprozesse in den einzelnen Zellbestandteilen und deren Trennung. 2. Kolloquium Dtsch. Ges. Physiol. Chemie, 24—42. Berlin-Göttingen-Heidelberg 1952. — Lehmann, F. E.: Mikroskopische und submikroskopische Bauelemente. 2. Kolloquium Dtsch. Ges. Physiol. Chemie, 1—18. Berlin-Göttingen-Heidelberg 1952. — Lettré, H., u. Renate Lettré: Kern-Plasma-Mitochondrien-Relation als Zellcharakteristikum. Naturwiss. **40**, 203 (1953). — Lima de Faria, A.: The structure of the centromere of the chromosomes of rye. Hereditas (Lund) **35**, 77—85 (1949a). — Genetics, origin and evolution of kinetochores. Hereditas (Lund) **35**, 422—444 (1949b).

Mack, B.: Über ein eigentümliches Verhalten isolierter Plastiden in der Magenwand des Rotators Ascomorpha ecaudis Perty. Österr. bot. Z. **99**, 156—160 (1952). — Malheiros, Nydia, D. de Castro e A. Camara: Cromosomas sem centrómero localizado. O caso de Luzula purpurea Link. Agronomia lusitana **9**, 51—71 (1947). — Manton, Irene: The spiral structure of chromosomes. Biol. Rev. **25**, 486—508 (1950). — Marquardt, H.: Neue Ergebnisse auf dem Gebiet der Plasmavererbung. Umschau **52**, 546—549 (1952a). — Die Natur der Erbträger im Zytoplasma. Ber. dtsch. bot. Ges. **65**, 198—217 (1952b). — Mirsky, A. E., u. H. Ris: Isolated chromosomes. J. Gen. Physiol. **31**, 1—6 (1947). — Mudd, S., L. C. Winterscheid, E. D. de Lamater and H. J. Henderson: Evidence suggesting that the granules of mycobacteria are mitochondria. J. Bacter. **62**, 459—475 (1951). — Mühldorf, A.: Einige Betrachtungen zur Membranmorphologie der Blaualgen. Ber. dtsch. bot. Ges. **56**, 316—335 (1938). — Die Zellteilung als Plasmateilung. Wien 1951. — Mühlethaler, K., A. F. Müller u. H. N. Zollinger: Zur Morphologie der Mitochondrien. Experientia (Basel) **6**, 16—17 (1950).

Newcomer, E. H.: Mitochondria in plants. II. Bot. Review **17**, 53—89 (1951).

Östergren, G.: Luzula and the mechanism of chromosome movements. Hereditas (Lund) **35**, 445—468 (1949a). — A survey of factors working at mitosis. Hereditas (Lund) **35**, 525—528 (1949b). — Olive, L. S.: The structure and behavior of fungus nuclei. Bot. Review **19**, 439—586 (1953).

Pardee, A. B., H. K. Schachman and R. Y. Stanier: Chromatophores in Rhodospirillum rubrum. Nature (Lond.) **169**, 282—283 (1952). — Pascher, A.: Beiträge zur allgemeinen Zellehre. I. Doppelzellige Flagellaten und Parallelentwicklungen zwischen Flagellaten und Algenschwärmern. Arch. Protistenkde **68**, 268—304 (1929). — Pascher, A., u. J. Petrova: Über Porenapparate bei einer neuen Bangiale (Chroothece mobilis). Arch. Protistenkde **74**, 490—522 (1931). — Perner, E. S.: Zellphysiologische und zytologische Untersuchungen über den Nachweis und die Lokalisation der Cytochrom-Oxydase in Allium-Epidermiszellen. Biol. Zbl. **71**, 43—69 (1952a). — Über die Veränderungen der Struktur und des Chemismus der Zelleinschlüsse bei der Homogenisation lebender Gebilde. Ber. dtsch. bot. Ges. **65**, 235—238 (1952b). — Zum mikroskopischen Nachweis des „primären Granums" in den Leukoplasten. Ber. dtsch. bot. Ges. **67**, 26—32 (1954).—Perner, E. S., u. G. Pfefferkorn: Pflanzliche Chondriosomen im Licht- und Elektronenmikroskop unter Berücksichtigung ihrer morphologischen Veränderungen bei der Isolierung. Flora (Jena) **140**, 98—129 (1953). — PIEKARSKI, G.: Die Zellkernäquivalente der Bakterien. 2. Kolloquium Dtsch. Ges. Physiol. Chemie, 83—99. Berlin-Göttingen-Heidelberg 1952. — PINTO-LOPES, J.: On the differentiation of the nuclei in pollen grains. Portugal. Acta Biol., Ser. A **2**, 237—247 (1948). — POLJANSKY, G., u. G. PETRU-SCHEWSKY: Zur Frage über die Struktur der Cyanophyceen-Zelle. Arch. Protistenkde **67**, 11—45 (1929). — POLLISTER, A. W., and A. E. MIRSKY: The isolation of chromosomes from resting nuclei. Genetics **28**, 86 (1943). — PRESTON, R. D., and others: An Electron microscope study of cellulose in the wall of Valonia ventricosa. Nature (Lond.) **162**, 665 (1948).

REZENDE-PINTO, M. C. DE: Über die Genese und die Struktur der Chloroplasten bei den höheren Pflanzen. Ergebnisse und Probleme. Protoplasma (Berl.) **41**, 336—342 (1952). — ROELØFSEN, P. A., u. A. L. HOUWINK: Architecture and growth of the primary cell wall in some plant hairs and the Phycomyces sporangiophore. Acta bot. neerl. **2**, 218—225 (1953).

SAX, K., and H. W. EDMONDS: Development of the male gametophyte in Tradescantia reflexa. Bot. Gaz. **95**, 156—163 (1933). — SCHIMPER, A. F. W.: Über die Entwicklung der Chlorophyllkörner und Farbkörper. Bot. Ztg **41**, 105, 121, 137, 153 (1883). — Untersuchungen über die Chlorophyllkörner und die ihnen homologen Gebilde. Jb. wiss. Bot. **16**, 1—247 (1885). — SCHLOSSBERGER, H., A. JAKOB u. G. PIEKARSKI: Zur Systematik der Spirochäten. Naturwiss. **37**, 186—187 (1950). — SCHMIDT, W. G.: Doppelbrechung der Kernspindel und Zugfasertheorie der Chromosomenbewegung. Chromosoma **1**, 253—264 (1939). — SCHMITZ, F.: Die Chromatophoren der Algen. Verh. naturhist. Ver. preuß. Rheinl. **40** (1882). — SCHNARF, K.: Vergleichende Cytologie des Geschlechtsapparates der Kormophyten. Monograph. vergl. Zytologie **1** (1941). — SCHRADER, F.: Mitosis. The movement of chromosomes in cell division, 2. Aufl. New York 1953. — SCHRADER, F., and CECILIE LEUCHTENBERGER: A cytochemical analysis of the functional interrelations of various cell structures in Arvelius albopunctatus (DE GEER). Exper. Cell Res. **1**, 421—452 (1950). — SCHUSSNIG, B.: Handbuch der Protophytenkunde, Bd. 1. Jena 1953. — SCHWANITZ, F.: Bakterien haben einen Zellkern! Umschau **52**, 492—494 (1952). — SINNOTT, E. W., and R. BLOCH: Division in vacuolate plant cells. Amer. J. Bot. **28**, 225—232 (1941). — Comparative differentiation in the air roots of Monstera deliciosa. Amer. J. Bot. **33**, 587 bis 590 (1946). — STECHER, H.: Über das Flächenwachstum der pflanzlichen Zellwände. Mikroskopie (Wien) **7**, 30—36 (1952). — STEFFEN, K.: Zytologische Untersuchungen an Pollenkorn und -schlauch. 1. Phasenkontrastoptische Lebenduntersuchungen an Pollenschläuchen von Galanthus nivalis. Flora (Jena) **140**, 140—174 (1953). — STRUGGER, S.: Die Strukturordnung im Chloroplasten. Ber. dtsch. bot. Ges. **64**, 69—83 (1951). — Über die Struktur der Proplastiden. Ber. dtsch. bot. Ges. **66**, 439—453 (1953). — SUITA, N.: Studies on the male gametophyte in Angiosperms. II. Differentiation and behaviour of the vegetative and generative elements in the pollen grains of Crinum. Cytologia (Fujii Jub.-Bd.) **2**, 920—933 (1937).

TISCHLER, G.: Allgemeine Pflanzenkaryologie. In K. LINSBAUERS Handbuch der Pflanzenanatomie, 2. Aufl., Bd. II. Berlin 1934—1951. — TJIO, J. H., and A. LEVAN: The use of oxiquinoline in chromosome analysis. An. Estacion exper. Aula Dei **2**, 21—64 (1950). — TROLL, W.: Das Virusproblem in ontologischer Sicht. Abh. Gesamtgeb. wiss. Bot., herausgeg. von A. FREY-WYSSLING, A. SEYBOLD u. W. TROLL. Wiesbaden 1951. — TSCHERMAK, ELISABETH: Vergleichende und experimentelle cytologische Untersuchungen an der Gattung Oedogonium. Chromosoma **2**, 493—518 (1943). — TSCHERMAK-WOESS, ELISABETH: Über auffallende Strukturen in den Pyrenoiden einiger Naviculoideen. Österr. bot. Z. **100**, 160—178 (1953). — TSCHERMAK-WOESS, ELISABETH, u. GERTRUDE HASITSCHKA: Veränderungen der Kernstruktur während der Endomitose, rhythmisches Kernwachstum und verschiedenes Heterochromatin bei Angiospermen. Chromosoma **5**, 574—614 (1953). — Über die endomitotische Polyploidisierung im Zuge der Differenzierung von Trichomen und Trichozyten bei Angiospermen. Österr. bot. Z. **101**, 79—117 (1954).

VENDRELY, L.: The Heterochromatin problem in Cyto-genetics as related to other branches of investigation. Bot. Review **15**, 507—582 (1949).

VOGL, ELISABETH: Untersuchungen über die Teilungsrichtungen in Pollenmutterzellen und bei der Blaualge Chroococcus. Österr. bot. Z. **94**, 2—29 (1947).

WADA, B.: Mikrurgische Untersuchungen lebender Zellen in der Teilung. II. Cytologia **6**, 381—406 (1935). — Über die Spindelfigur bei der somatischen Mitose der Prothallienzellen von Osmunda japonica Thunb. Cytologia **11**, 353—368 (1940/41). — The mechanism of mitosis based on studies of the submicroscopic structure and the living state of the Tradescantia cell. Cytologia **16**, 1—26 (1950/52). — WEBB, C. B.: The cytology and life history of Sorosphaera veronicae. Ann. of Bot. **49**, 41—52 (1935). — WEIDEL, W.: Neue Erkenntnisse der modernen Virusforschung. Umschau **53**, 257—259 (1953). — WEIER, T. E., and C. R. STOCKING: The chloroplast: structure, inheritance and enzymology. II. Bot. Review **18**, 14—75 (1952).

# Das Wasser,
## seine physikalischen und chemischen Eigenschaften unter besonderer Berücksichtigung seiner physiologischen Bedeutung.

Von

## L. v. Erichsen.

Mit 12 Textabbildungen.

Ἄριστον μὲν ὕδωρ.
PINDAR.

Wählt man das Wasser als Objekt einer zusammenfassenden Betrachtung, so könnte es zunächst scheinen, als ließe sich diese Frage in wenigen Sätzen erschöpfend behandeln. Daß das aber nicht der Fall ist, daß der mit dem Wasser zusammenhängende Fragenkomplex vielmehr auch heute noch der endgültigen Lösung harrt, hat besonders die Entwicklung der letzten Dezennien auf das deutlichste gezeigt. So kann es nicht verwundern, wenn gerade im Zusammenhang mit dem Wasserproblem im weitesten Sinne die Namen der bekanntesten Forscher auf den Gebieten der Chemie, Physik, Physikalischen Chemie und der Biologie, wie RAMSAY, RÖNTGEN, NERNST, TAMMAN, BRAGG, EUCKEN u. a. immer wieder auftauchen.

Da es sich um ein komplexes Problem handelt, das hier unter besonderer Berücksichtigung des physiologischen Standpunktes zusammenfassend behandelt werden soll, seien zunächst die verschiedenen Gesichtspunkte herausgestellt, unter denen es betrachtet werden muß. Zugleich müssen einige notwendige Begriffsbestimmungen vorausgeschickt werden, um Mehrdeutigkeiten zu vermeiden, denen man bei der Darstellung von einschlägigen Teilproblemen nicht selten begegnet; weitere Definitionen werden, soweit erforderlich, bei der Besprechung besonderer Teilfragen ergänzend eingefügt.

Unter dem Ausdruck Wasser soll hier das reine flüssige Wasser verstanden werden, das nur aus Wasserstoff und Sauerstoff im Atomverhältnis 2:1 besteht, also völlig frei von fremden Elementen ist und der isotopischen Zusammensetzung nach zu 99,76 Gew.-% (SOUCI 1952) der chemischen Bruttoformel $(H_2O)_n$ entspricht, wogegen der Rest von 0,24 Gew.-% aus analogen Verbindungen der natürlichen Isotopen des Wasserstoffs und des Sauerstoffs (s. u.) besteht.

Der Begriff Eis soll die feste Form der vorstehend als Wasser definierten flüssigen Substanz bezeichnen, in welche diese bei Abkühlung mit oder ohne Druckerhöhung übergeht, wobei besondere Modifikationen (s. u.) als solche gekennzeichnet werden.

Je nachdem, zu welcher Frage das Verhalten des Wassers bzw. Eises in Bezug gesetzt werden soll, ist es zweckmäßiger, dafür den Charakter eines stofflichen Kontinuums zu wählen, etwa für hydrodynamische Fragen, oder die atomistische, molekulardimensionale Betrachtungsweise, die wiederum sich auf rein physikalische, kristallographische Erwägungen oder auf solche chemischer und valenzmäßiger Art stützen kann. Es liegt in der Natur der Sache, daß dazwischen Übergänge aller Art bestehen.

# Wasser und Eis als Stoffe im Bezug zur Umwelt.

Das Wasser wie das Eis stehen als Bestandteile der Erdoberfläche in einer Wechselwirkung mit den anderen Substanzen, welche insbesondere für die organisierte Substanz der lebenden Materie von integrierender Bedeutung ist. Diese Wirkung ist in erster Linie neben einer universellen Lösungsmittelfunktion die einer weitgehenden Temperaturkonstanz der Biosphäre und als notwendige Folge die weitgehende Beibehaltung des flüssigen Aggregatzustandes in derselben. Es ist hier nicht der Ort, auf die Bedeutung des letzteren für das biologische Geschehen näher einzugehen. Im einzelnen beruht diese Stabilisierung auf den nachstehend geschilderten Anomalien des Wassers bzw. Eises und des Überganges zwischen beiden Substanzformen.

Die zunächst trivialste Ursache liegt in der überaus großen Menge, in der das Wasser auf der Erdoberfläche vorkommt, die es zu rund 65% als Meerwasser bedeckt. Es ist damit die auf der zugänglichen Erdoberfläche in der größten Menge vorhandene chemische Verbindung. Die Erde scheint in dieser Hinsicht gegenüber den anderen Planeten eine Sonderstellung einzunehmen, die eine Entwicklung höher organisierten Lebens erst ermöglicht hat. Durch diesen großen mengenmäßigen Anteil spielt das Wasser seine dominierende Rolle im Wärmehaushalt der Biosphäre, also der Hydrosphäre selbst, der Oberfläche der Lithosphäre und des größten Teiles der Atmosphäre.

Für die Erhaltung des flüssigen Zustandes der Hauptmenge der Wasservorräte der Erde sind einige Eigenschaften des Wassers wesentlich. Die Volumenzunahme beim Übergang des Wassers in Eis am Gefrierpunkt, die es nur mit einigen wenigen Stoffen, wie etwa Wismut, gemeinsam hat, bewirkt, daß das Eis mit der Dichte 0,917 nicht im gefrierenden Wasser von der Dichte 1,000 untersinkt, sondern im Gegenteil eine Isolierschicht gegen weitere Abkühlung darstellt, da es als fester Körper die Wärme nur noch durch Leitung, aber nicht durch Konvektion weitergibt. Ein Zufrieren der natürlichen Gewässer in ihrer gesamten Menge ist dadurch nicht möglich.

Diese Ausdehnung beim Gefrieren ist auch insofern von Einfluß auf das biologische Geschehen, als damit unter entsprechenden Voraussetzungen eine Sprengwirkung verbunden ist. Die Umwandlung des festen Gesteinmaterials in feine bodenbildende Partikel geht in der Hauptsache zu Lasten derselben, allerdings auch zuweilen eine sprengende Zerstörung der wasserhaltigen, lebenden Zellsubstanz.

Die hohe spezifische Wärme, die wesentlich höher liegt als diejenige aller anderen Stoffe, befähigt das Wasser zur Aufnahme großer Wärmemengen ohne wesentliche Temperatursteigerung. Umgekehrt vermag es große Wärmemengen abzugeben, ohne dabei schnell abzukühlen. So wird auch die Temperaturregulierung im lebenden Organismus durch dessen hohen Wassergehalt weitgehend erleichtert.

Das Dichtemaximum des Wassers bei $4^0$ wirkt in der gleichen Richtung wie die Ausdehnung beim Gefrieren. Bei Erreichen dieser Temperatur, die im Meerwasser je nach dem Salzgehalt niedriger liegt, sinkt das Wasser von der abgekühlten Oberfläche ab und verhindert so eine Abkühlung der gesamten Wassermasse bis auf den Gefrierpunkt. Eine solche weitere Abkühlung wäre auch in diesem Falle vorwiegend nur durch Wärmeleitung möglich, da das kältere Wasser infolge seiner geringeren Dichte an der Oberfläche verbleibt.

Weiterhin liegen die Werte für die Erstarrungswärme mit 1,437 kcal/mol = 79,4 cal/g und für die Verdampfungswärme mit 11,109 kcal/mol = 539 cal/g gegenüber anderen Substanzen ungewöhnlich hoch. Es bedarf eines umfangreichen und lang dauernden Wärmeentzuges, um ein wasserhaltiges Milieu vollständig zum Gefrieren zu bringen und unter $0^0$ abzukühlen. Das gleiche gilt für eine Temperaturerhöhung, die zu einer stark wärmezehrenden Verdampfung bzw. Verdunstung führt.

Faßt man diese für das Wasser charakteristischen Eigenschaften zusammen, so resultiert daraus die eingangs angeführte thermisch stabilisierende Wirkung der Wassermassen der Erdoberfläche. Das Ausmaß dieses Effektes zeigt sich am deutlichsten am Gegensatz zum ariden Wüstenklima, das gegenüber dem sehr konstanten Seeklima tägliche Oberflächentemperaturunterschiede bis zur Größenordnung von fast $80^0$ aufweisen kann.

Auf ein weiteres nicht zu vernachlässigendes Moment weist Souci (1952) hin. Unter Annahme einer mittleren Tiefe der Weltmeere von 3800 m ergibt sich dafür ein mittlerer Bodendruck von 380 at, der infolge der außergewöhnlich hohen Kompressibilität des Wassers bei Temperaturen dicht oberhalb des Eispunktes zu einer Verdichtung des Wassers in größeren Tiefen führt, die den Meeresspiegel um rund 36 m tiefer liegen läßt, als es beim Vorliegen einer nicht oder nur sehr schwach kompressiblen normalen Flüssigkeit der Fall sein würde. Im letzteren Falle würde ein großer Teil der kontinentalen Landmassen vom Meere bedeckt sein, woraus sich zweifellos eine ganz andere Entwicklung der Flora und Fauna ergeben hätte.

Schließlich sei noch das überaus große Lösungsvermögen des Wassers erwähnt, das ebenfalls von Bedeutung für seine Wechselwirkung mit anderen Stoffen ist.

Die eben kurz angeführten Anomalien der Wassersubstanz sind durch deren ungewöhnlichen molekularen Aufbau bedingt, worauf weiter unten einzugehen sein wird. Ebenso werden dort auch weitere Abweichungen gegenüber dem Verhalten einer normalen Flüssigkeit zu erwähnen sein.

## Die Anomalien des Wassers im Vergleich mit normalen Substanzen.

Die physikalischen Eigenschaften von Wasser und Eis erscheinen noch anomaler, wenn man sie mit denen anderer Flüssigkeiten vergleicht, deren Molekularformel der des Wassers analog ist, oder mit solchen, die eine ähnliche Elektronenstruktur wie das Wassermolekül aufweisen. Das Wasser zeigt dann ein von demjenigen völlig abweichendes Verhalten, das es aufweisen müßte, wenn es lediglich ein einfaches Kondensat des praktisch monomolekularen Wasserdampfes wäre, so daß schon hieraus der Schluß gezogen werden muß, daß die einfache Formel $H_2O$ für das flüssige Wasser nicht zutreffen kann, was gleicherweise für das Eis gilt. Es ist vielmehr daraus zu folgern, daß hier statt dessen ein komplexes und stark veränderliches System vorliegt.

Tabelle 1. *Schmelz- und Siedepunkte verschiedener flüchtiger Nichtmetallverbindungen im Vergleich zum Wasser* (nach Vlès 1938).

| Verbindung | Formel | Schmelz- punkt $^0$C | Siedepunkt $^0$C |
|---|---|---|---|
| Wasser . . . . . . . . | $H_2O$ | 0 | $+100$ |
| Schwefelwasserstoff . . | $H_2S$ | $-83$ | $-61$ |
| Selenwasserstoff . . . . | $H_2Se$ | $-64$ | $-42$ |
| Chlordioxyd . . . . . | $ClO_2$ | $-76$ | $+9$ |
| Schwefeldioxyd . . . . | $SO_2$ | $-73$ | $-10$ |
| Stickoxydul . . . . . | $N_2O$ | $-102$ | $-88$ |
| Stickstoffdioxyd . . . | $NO_2$ | $-11$ | $+26$ |

Schon seit dem letzten Dezennium des vorigen Jahrhunderts vergleicht man die physikalischen Kennzahlen des reinen Wassers mit denjenigen von analog aufgebauten anorganischen flüchtigen Verbindungen, so mit Schwefel- und Selenwasserstoff, Stickoxyden usw., wobei sich die aus der Tabelle 1 ersichtlichen grundlegenden Unterschiede ergeben (Vlès 1938).

Alle diese Stoffe, die den Molekularaufbau $X_2Y$ gemeinsam haben, zeigen gegenüber dem Wasser wesentlich niedrigere Schmelz- und Siedepunkte und einen engeren Temperaturbereich des flüssigen Zustandes, so daß die Eigenschaften des Wassers selbst in der Tat als anomal bezeichnet werden müssen.

MECKE (1948 II) gibt eine übersichtliche Gegenüberstellung zwischen den sonstigen physikalischen Daten des Wassers und den dafür auf Grund der einfachen Formel $H_2O$ zu erwartenden Werten, die in der Tabelle 2 zu finden sind.

Tabelle 2. *Zusammenstellung der physikalischen Daten des Wassers und der auf Grund der chemischen Formel zu erwartenden Werte* (nach MECKE 1948 II).

| Eigenschaft | vorhanden | statt |
|---|---|---|
| Kritische Temperatur, $^0$C . . . . . . | 374 | $\sim$50 |
| Siedepunkt, $^0$C. . . . . . . . . . . | 100 | $\sim-100$ |
| Gefrierpunkt, $^0$C . . . . . . . . . . | 0 | $\sim-120$ |
| Verdampfungswärme, kcal/mol . . . . | 9,7 | $\sim$4 |
| Schmelzwärme, kcal/mol . . . . . . . | 1,4 | $\sim$0,5 |
| Spezifische Wärme, kcal/grad·mol . . $\Big\{$ | 18 | $\sim$9 |
|  | Minimum b. 35$^0$ | monoton mit der Temperatur ansteigend |
| Verdampfungsentropie, cal/grad·mol . | 26 | $\sim$19 |
| Dichte, g/cm$^3$ . . . . . . . . . $\Big\{$ | 1 | $\sim$0,5 |
|  | Maximum b. 4$^0$ | monoton mit der Temperatur fallend |
| Molvolumen, cm$^3$/mol . . . . . . . . | 18 | $\sim$40 |
| Volumenänderung beim Gefrieren, cm$^3$/mol . . . . . . . . . . . . . | Vergrößerung um 1,62 | Verkleinerung um 1,5 |
| Viscosität, c-Poise. . . . . . . . . . | 1,7 | $\sim$0,2 |
| Oberflächenspannung, dyn/cm . . . . | 75 | $\sim$7 |

Ähnliche Abweichungen von dem nach der Molekülformel zu erwartenden Verhalten zeigt eine ganze Reihe anderer Verbindungen, allerdings in weit weniger ausgeprägtem Maße. Dazu gehören z. B. die Alkohole und sonstigen hydroxylhaltigen organischen Verbindungen, Hydroxylamin, Cyanwasserstoff usw. Diese Ähnlichkeit ist, um es hier vorwegzunehmen, keine rein zufällige, sondern beruht auf Gemeinsamkeiten in der Wechselwirkung zwischen den Molekülen.

Es liegt auf der Hand, daß dem Wasser und den in ihren Eigenschaften sich ähnlich verhaltenden Stoffen nicht der einfache molekulare Aufbau zugeschrieben werden kann, wie er normalen Flüssigkeiten zu eigen ist.

Unter normalen Flüssigkeiten sind solche zu verstehen, die wie etwa Brom, Tetrachlorkohlenstoff oder Quecksilber zweifellos dicht gepackte Zusammenlagerungen von Molekülen darstellen, die ähnlich wie Schrotkugeln in einem Gefäß gelagert sind, sofern nicht die Wärmebewegung eine solche Lagerung dehnt und stört, wenn die Temperatur ansteigt (FORBES 1941).

Die ersten in dieser Richtung zielenden Gedankengänge liegen zeitlich bereits sehr lange zurück. TAMMANN (1926) verweist darauf, daß PARROT schon 1827 das Dichtemaximum des Wassers bei 4$^0$ auf eine Änderung der molekularen Zusammensetzung zurückgeführt hat. Nach solchen zunächst sehr allgemein gehaltenen Erwägungen griff WHITING (1884) das Problem im Rahmen der mathematischen Behandlung einer Kohäsionstheorie der Flüssigkeiten neuerlich auf, wobei es sich zeigte, daß das Wasser seinen für die meisten Flüssigkeiten gültigen Formeln nicht gehorchte, daß dafür die Zusatzannahme von Umordnungsvorgängen bei Temperaturänderungen erforderlich wurde. Aus wahrscheinlichkeitstheoretischen Überlegungen leitete WHITING ab, daß beim Gefrierpunkt das Wasser zum Teil Eigenschaften des festen Zustandes zeigt, während gleichzeitig dem Eis gewisse Merkmale des flüssigen Wassers zuzuschreiben sind. Unabhängig von WHITING untersuchte RÖNTGEN (1892) das Verhalten des Wassers näher und kam zu dem Schluß, daß die verschiedenen Anomalien des Wassers, so vor allem die Abnahme der Kompressibilität mit steigender Temperatur (normale Flüssigkeiten werden mit steigender Temperatur wegen der Aufweitung der Molekülabstände kompressibler), die Zunahme des Ausdehnungskoeffizienten mit dem Druck sowie die Abnahme der Viscosität mit steigendem Druck, durch die Annahme zu erklären sind, „daß das flüssige Wasser aus einem Aggregat von Arten verschieden constituirter Molecüle besteht. Die Molecüle erster Art, welche wir auch Eismolecüle nennen wollen, da wir ihnen gewisse Eigenschaften des Eises beilegen werden, gehen durch Wärmezufuhr in Molecüle zweiter Art über." Zwischen beiden Molekülarten wurde ein temperaturabhängiges, reversibles Gleichgewicht angenommen.

Diese Überlegungen von Whiting und von Röntgen sind für die weitere Entwicklung bis heute richtungweisend gewesen, sie konnten aber einen vorwiegend nur qualitativen Charakter tragen, solange die wirkliche Anordnung der Moleküle im Eis und im Wasser nur aus gewissermaßen makroskopischen Eigenschaften indirekt abgeleitet werden konnte. Infolgedessen schien etwa 1910 ein gewisser Abschluß der Auffassungen erreicht zu sein, den im Rahmen eines von der Faraday Society veranstalteten Symposiums über die Struktur des Wassers (Symposium 1910) der Chairman der Tagung in den Worten zusammenfaßte: "I would think of this discussion, one will soon find in the textbooks that while ice is trihydrol, and steam monohydrol, liquid water is mostly dihydrol with some trihydrol in it near the freezing point and a little monohydrol near the boiling point." Unter Hydrol wurde hierbei das einzelne Wassermolekül der Formel $(H_2O)_1$ verstanden, unter Dihydrol und Trihydrol mehr oder weniger fest verbundene, räumlich nicht näher definierte Aggregate aus Monohydrol. Alle drei Formen konnten sich in Abhängigkeit von der Temperatur ineinander umwandeln.

Mit der zuletzt ausgedrückten, sicherlich sehr vereinfachenden Annahme ist die bis dahin eigentlich befremdliche Tatsache, daß es neben nur einem gasförmigen Aggregatzustand deren zwei für die kondensierte Form der Materie gibt, verständlicher geworden. Der flüssige Zustand des Wassers und anderer Stoffe nimmt dadurch eine Art von Zwischenstellung zwischen dem ungeordneten Aufbau des Gases und dem im Grenzfall ideal geordneten Aufbau des kristallisierten Festkörpers ein. Die Flüssigkeit weist, am ausgeprägtesten im Falle des Wassers, kleinste geordnete Bezirke im Sinne der vorstehend postulierten höheren Hydrole auf, die aber im Gegensatz zum Kristall nicht fest zueinander orientiert sind. Der Übergang Eis ⇌ Wasser ⇌ Dampf entspricht demnach der Folge Fernordnung ⇌ Nahordnung ⇌ Unordnung. Die beiden Grenzzustände der idealen Ordnung bzw. der idealen Unordnung haben infolgedessen ihre weitgehende Aufklärung bereits erfahren, während das beim Übergangszustand noch keineswegs der Fall ist (Schäfer 1948). Insbesondere aber fehlt bis heute noch ein befriedigendes Verständnis für die Existenz eines scharfen Schmelz- und Siedepunktes, deren Vorhandensein durch die Annahme der eis- bzw. dampfartigen Anteile im flüssigen Wasser noch erschwert wird. Noch undurchsichtiger werden die Verhältnisse bei der Betrachtung der Wechselwirkung solcher nahgeordneter Zustände des Wassers mit hochpolymeren Stoffen der lebenden Materie (Wolf, Frahm und Harms 1937), worauf weiter unten einzugehen sein wird.

Die wesentlichen Fortschritte, die sich bezüglich des heute angenommenen Bildes von der Struktur des Wassers und des Eises ergeben haben, beruhen auf der Anwendung von Untersuchungsmethoden, die einen direkten oder zumindest Indizieneinblick in die molekularen bzw. atomaren Bereiche selbst gestattet haben. Dazu gehören in erster Linie die Anwendung der Röntgenstrahlen- und Elektronenbeugung einerseits und der Ultrarot- und Ramanspektralanalyse andererseits.

## Aufbau des Wassermoleküls.

Als grundlegend im Zusammenhang mit den hier betrachteten Fragen sind die Untersuchungen von Mecke (1932 und 1933) über das Ultrarotabsorptionsspektrum des Wasserdampfes zu betrachten. Aus seinen Messungen konnte er eindeutig ableiten, daß das Wassermolekül keinen gestreckten, sondern einen winkeligen Aufbau besitzt, daß also die Wasserstoffatome sich nicht an den gegenüberliegenden Enden eines Durchmessers des kugelförmig gedachten Sauerstoffatoms befinden, sondern an den Enden zweier Kugelradien, die miteinander einen Winkel von ungefähr 105⁰ bilden. Da diese Ergebnisse an Wasserdampf gewonnen sind, der nach Dampfdichtemessungen monomer ist, beziehen sie sich wirklich auf das einzelne Wasser- oder Monohydrolmolekül.

Auch bezüglich der sonstigen Details des einzelnen Wassermoleküls sind weitere Einzelheiten geklärt worden. So wurde aus Messungen der Röntgenstrahlenbeugung der scheinbare Durchmesser des Moleküls zu 2,72—2,77 Å (BERNAL und FOWLER 1933, MAGAT 1936, FOX und MARTIN 1940) abgeleitet, sehr ähnliche Werte wurden aus Untersuchungen über den Radius des Kristallwassers ermittelt (BERNAL und FOWLER 1933), ebenso durch Extrapolation der Dichte auf 0⁰ K (MAGAT 1936). Innerhalb des Moleküls selbst ergab sich aus den Messungen von MECKE ein Abstand zwischen den Mittelpunkten des Sauerstoff- und jeden Wasserstoffatoms zu 0,970 Å und ein Abstand der Mitten beider Wasserstoffatome von 1,539 Å (MAGAT 1936), woraus der erwähnte Valenzwinkel von rund 105⁰ resultiert.

Da die beiden Wasserstoffatome des Moleküls dessen Durchmesser gegenüber dem des Sauerstoffatoms allein nur in sehr geringem Maße vergrößern, kann das Wassermolekül mit guter Annäherung als ein fast kugelförmiger Dipol betrachtet werden. Für das Dipolmoment des Monohydrols haben SCHUPP und MECKE (1948) in Übereinstimmung mit GROVES und SUGDEN (1937) einen Wert von $1,83 \pm 0,04$ D gefunden, der höher liegt als der frühere Wert von 1,70 von HEDESTRAND (1929). Für die Dielektrizitätskonstante des Wassers ergibt sich ein Wert von rund 81, die wegen ihrer abnormen Höhe ebenfalls zu den Anomalien des Wassers zu rechnen ist, ebenso wie ihre starke Temperaturabhängigkeit (EUCKEN 1944, S. 515/516). Diese Eigenschaft eines Dipols ist für die Kennzeichnung des Wassermoleküls geeignet, wenn es sich um seine Lagebeziehungen zu einem entfernteren Objekt, also auch zu einem etwa homogenen äußeren elektrischen Felde handelt. Im Falle einer Nahwirkung ist es dagegen als Quadrupol aufzufassen (BERNAL und FOWLER 1933, KATZOFF 1934, MAGAT 1936). Das soll besagen, daß auf der Kugeloberfläche des Sauerstoffatoms in größtem gegenseitigem Abstand und ebensoweit von den beiden positiven Protonen entfernt, sich die beiden negativen Ladungen des Moleküls konzentrieren. Eine solche gegenseitige Lage bedeutet aber, daß diese vier Konzentrationspunkte der negativen bzw. positiven Ladung auf der Oberfläche des Wassermoleküls die Ecken eines Tetraeders abgrenzen, woraus sich für die gegenseitige Orientierung der Moleküle ganz besondere Verhältnisse ergeben (FORBES 1941). Zu diesem Aspekt der tetraedrischen Ladungsverteilung im Wassermolekül ist als weiteres, die Verknüpfung von Wassermolekülen ermöglichendes Moment, die sog. *Wasserstoffbrücke* hinzugekommen. Da diese gerade für das Wasser, sowohl bezüglich der Wechselwirkung mit anderen Molekülarten, als auch der Wassermoleküle untereinander, von außerordentlicher Bedeutung ist, soll anschließend näher darauf eingegangen werden.

## Die Wasserstoffbrücke als lockere Bindungsart.

Die Möglichkeit, daß durch Vermittlung eines Wasserstoffatoms analog zu den Elektronen der kovalenten Bindung eine Verknüpfung zweier in der Elektronenkonfiguration geeignet aufgebauter Moleküle erfolgen kann, ist erstmalig wohl von LATIMER und RODEBUSH (1920) dargelegt worden. Unter Benutzung der LEWISschen Betrachtungsweise kann danach das Wassermolekül H:Ö:H an den beiden einsamen Elektronenpaaren Wasserstoffatome anlagern oder auch seinerseits Wasserstoffatome abgeben. So übt ein Elektronenpaar auf ein H-Atom eines benachbarten Wassermoleküls eine Kraft aus, die zur Verknüpfung dieser beiden Wassermoleküle über die sog. Wasserstoffbindung (hydrogen bond) ausreicht:

$$\overset{\cdot\cdot}{\text{H}:\overset{\cdot\cdot}{\text{O}}:\text{H}} \quad + \quad \overset{\text{H}}{\underset{\overset{\cdot\cdot}{\text{H}}}{:\overset{\cdot\cdot}{\text{O}}:}} \quad \rightarrow \quad \overset{\text{H}}{\underset{\overset{\cdot\cdot}{\text{H}}}{\text{H}:\overset{\cdot\cdot}{\text{O}}:\text{H}:\overset{\cdot\cdot}{\text{O}}:}}$$

In analoger Weise kann an das noch freie Elektronenpaar des rechten Sauerstoffatoms noch ein weiteres Wassermolekül über eine Wasserstoffbrücke angelagert werden. Damit wird der Sauerstoff koordinativ vierwertig, in Bestätigung der Auffassung von Thomsen (1885), der aus der Hydratationswärme von Salzen eine dimere Struktur des Wassers (Dihydrol) forderte und diese auch auf Grund anderer Werte einer höheren Wertigkeit des Sauerstoffs zuschreiben zu müssen glaubte.

Da diese Bindung im Vergleich mit anderen Bindungsarten relativ schwach ist (weak bond) und zu Verwechslungen mit anderen festen Bindungsarten des Wasserstoffatoms führen kann, hat später Huggins (1937) die Bezeichnung Wasserstoffbrücke dafür vorgeschlagen. Die Abb. 1 veranschaulicht die Modifikation, die das Modell des Wassermoleküls dadurch erfahren muß.

Während bei der Außerachtlassung der Möglichkeit zur Wasserstoffbrückenbildung das Wassermolekül als in sich abgeschlossenes, nur zu besonders ausgeprägten

Abb. 1.   Einfaches winkliges Wassermolekül (links) und dasselbe mit Kennzeichnung der Ansatzstellen für Wasserstoffbrücken (rechts).

elektrostatischen Wechselwirkungen geeignetes Gebilde erscheint (links), zeigt es nunmehr einen ausgesprochen bindungsfreudigen Charakter (rechts).

Das Zustandekommen der Wasserstoffbrücke geht in der Weise vor sich, daß beim Zusammenstoß zweier Wassermoleküle diese nicht als solche bestehen zu bleiben brauchen, sondern bei geeigneter Stoßlage ein Proton auszutauschen vermögen.   Geeignete Stoßlagen werden dabei im allgemeinen solche sein, bei denen ein Wasserstoffatom des einen Moleküls ungefähr mit einer der Stellen negativer Ladung des getroffenen Moleküls in engste Nachbarschaft gerät. Das die Wasserstoffbrücke bildende Wasserstoffatom bzw. Proton gehört dann nicht starr zu dem einen oder anderen Molekül, sondern in quantenmechanischer Wechselwirkung zu beiden (Hückel 1948, S. 445). Sutherland (1940) bevorzugt gegenüber dieser quantenmechanischen Betrachtungsweise die klassische elektrostatische Auffassung, da sich der Abstand beider O-Moleküle bei der H-Brückenbildung nicht merklich ändert. Dieser Vorgang kann sich auch gleichzeitig zwischen mehreren Molekülen abspielen, so daß sich hieraus die Möglichkeit der Bildung von kettenförmigen oder andersartigen räumlich angeordneten Assoziaten ergibt.

Die Bindungsenergie der Wasserstoffbrücke ist nicht sonderlich hoch und beträgt nur 3 kcal/Mol (Cross, Burnham und Leighton 1937) bis 4,5 kcal/Mol (Robertson 1940, Forbes 1941).

Zu der Vorstellung über den leichten Protonenaustausch auf Grund der Bildung und des Aufbrechens von Wasserstoffbrücken ist man experimentell durch die Beobachtung über den Austausch von Wasserstoff und dem isotopen Deuterium (s. u.) geführt worden. Dieser vollzieht sich so außerordentlich schnell und leicht, daß seine Geschwindigkeit nicht meßbar ist. Dabei hat sich gezeigt, daß dieser Austausch ein allgemeines Kennzeichen der Wechselwirkung zwischen hydroxylhaltigen Molekülen ist, also auch zwischen Wasser und Alkoholmolekülen, aber auch mit Ammoniak und Aminen.

$$ROH + D_2O \rightleftharpoons ROD + DOH \quad \text{bzw.} \quad 2\,ROH + D_2O \rightleftharpoons ROD + H_2O.$$

Dahingegen tauschen die keine Wasserstoffbrücken bildenden Mercaptane ebenso wie der Schwefelwasserstoff keinen Wasserstoff aus (HÜCKEL 1948).

Die Schnelligkeit dieses Austausches im Falle des Wassers zeigt, daß der Wasserstoffbrücke im Mittel eine nur sehr begrenzte Existenzdauer zugeschrieben werden kann (BOSSCHIETER und ERRERA 1937). Sie ist aber immerhin groß im Vergleich zu der mittleren Stoßzahl zwischen den einzelnen Molekülen. Es ist daher anzunehmen, daß die Verknüpfung durch Wasserstoffbrücken anteilmäßig durchaus neben die rein elektrostatische Dipol- bzw. Quadrupolanziehung zu stellen ist. Es besteht allerdings noch keine völlige Übereinstimmung über die Festigkeit der Bindung durch H-Brücken. So will UBBELOHDE (1940) zwischen einer festen und einer normalen lockeren Form derselben unterschieden wissen.

Diese Möglichkeit ist nicht von der Hand zu weisen, da die H-Brücke ja einen Übergang zur kovalenten Bindung darstellt (ULICH 1941, S. 315—316), wie auch PAULING (1939, S. 71) das Vorhandensein einer gemischten Bindungsart interpretiert hat, wonach gerade bei den Wassermolekülen die Bindung der Protonen eine gewisse Tendenz zur Ionenbindung aufweist.

Da die Wasserstoffatome bei der Röntgenstrahlbeugung keine erkennbaren Interferenzen ergeben, ist man bei der Festlegung der Abstände derselben von den Sauerstoffatomen innerhalb der Brücke auf indirekte Rückschlüsse aus optischen Messungen angewiesen. Aus diesen wird abgeleitet, daß sich das Wasserstoffatom nicht in der Mitte zwischen den beiden das Molekülzentrum repräsentierenden Sauerstoffen befindet, sondern jeweils näher bei dem einen derselben, und zwar in einem Abstand von etwa 0,99—1,0 Å (ROBERTSON 1940, FOX und MARTIN 1940 und FORBES 1941), woraus sich ein Abstand von 1,76 Å zum anderen O-Atom ergibt.

In die Ebene projiziert, würde sich dann ein Abstandsschema etwa der folgenden Art ergeben:

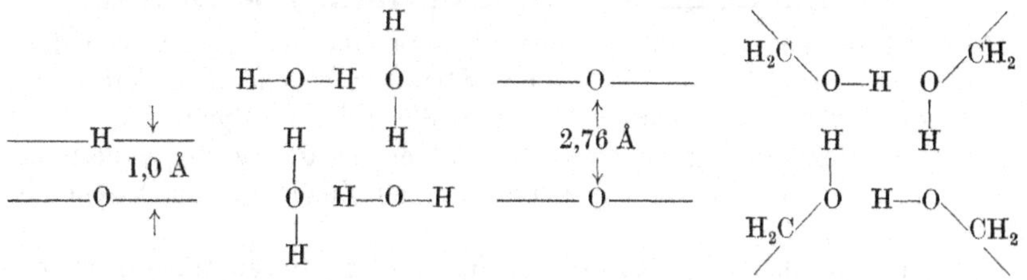

Man setzt im allgemeinen stillschweigend voraus, daß das Proton einer Wasserstoffbrücke auf der geraden Verbindungslinie zwischen den beiden dadurch verknüpften Sauerstoffatomen liegt. Diese Annahme braucht nicht unbedingt zuzutreffen, da schon weiter oben die relativ geringe Energie der Wasserstoffbrückenbindung erwähnt wurde.

UBBELOHDE (1940) läßt daher eine Verzerrung dieses Winkels durch geringe Energiebeträge zu und schreibt einer solchen, in gewissen Grenzen leicht deformierbaren Richtkraft mit schwacher Bindungsenergie eine nicht zu unterschätzende Bedeutung für den Aufbau der recht leicht umwandelbaren Eiweiß-strukturen im wäßrigen Milieu zu. Auf Grund von Röntgenbeugungsmessungen plädiert Katzoff (1934) ebenfalls dafür, eine Abwinkelung der normalerweise gestreckten Wasserstoffbrücke bis auf 150⁰ statt 180⁰ zuzulassen.

Zahlreiche Einzelheiten über die Ausbildung von Wasserstoffbrücken zwischen den verschiedensten organischen und anorganischen Verbindungen mit einer großen Zahl von Beispielen findet man bei Huggins (1937) und Lassettre (1937). Ihre Anführung würde den hier gegebenen Rahmen bei weitem überschreiten; auf einzelne in physiologischer Hinsicht besonders interessierende Details wird an geeigneter Stelle zurückzukommen sein.

Ehe anschließend auf die Verknüpfung der einzelnen Wassermoleküle zu höheren Einheiten eingegangen wird, seien die dafür maßgebenden Charakteristika des Einzelmoleküls nochmals kurz zusammengefaßt:

Das Wassermolekül hat unter allen Verbindungen mit gleicher Elektronenzahl bzw. mit gleichem Aufbauprinzip $X_2Y$ den bei weitem kleinsten Moleküldurchmesser, wobei das Gesamtmolekül als praktisch kugelförmig zu betrachten ist. Die Lage der Wasserstoffatome ist relativ zum Sauerstoff unsymmetrisch, wodurch das Molekül den Charakter eines Quadrupols erhält, dessen je zwei positive und negative Ladungen auf der Kugeloberfläche so angeordnet sind, daß sie miteinander den räumlichen Tetraederwinkel von rund 109⁰ bilden. Dieser Quadrupolcharakter hat eine erhebliche gegenseitige elektrostatische Anziehung zur Folge, wobei vier Moleküle in geeigneter Lage angezogen werden können. Die Anwesenheit von gerade zwei Wasserstoffatomen, also der Hälfte der tetraedrischen Koordinationsmöglichkeiten des Sauerstoffs, bietet die Möglichkeit, zwei weitere Moleküle mit Hilfe dieser Protonen über je eine Wasserstoffbrücke zu binden und selbst auf die gleiche Weise an zwei andere Moleküle gebunden zu werden. Von allen analogen chemischen Verbindungen hat es die höchste van der Waalssche Konstante a entsprechend seiner größten zwischenmolekularen Anziehung, die wiederum auf dem niedrigsten Wert für b, entsprechend dem kleinsten Abstand zwischen den Molekülmitten, beruht.

Das Wassermolekül ist also für sich genommen denkbar einfach gebaut. Gerade diese Einfachheit in Verbindung mit der lockeren tetraedrischen Verknüpfungstendenz bietet für die Entstehung der mannigfachsten Strukturen nahezu beliebige Möglichkeiten (Errera 1937 I und II), im Gegensatz zu komplizierten Molekülen, die nur wenige, sehr selektive Zusammenlagerungsmöglichkeiten zulassen, so wie das Sandkorn im Beton sich den verschiedensten Formgebungen fügt, was beim umfangreichen Quader nicht der Fall ist.

## Das Eis, seine Entstehung und Struktur in Abhängigkeit von Temperatur und Druck.

Da der feste Aggregatzustand des Wassers, also das Eis, wegen der darin herrschenden Fernordnung (s. o.) einer Strukturuntersuchung auf optischem, kristallographischem und insbesondere dem Wege der Röntgen- und Elektronenbeugung besonders gut zugänglich ist, liegen hierüber die meisten experimentellen Untersuchungen vor, aus denen sich in großen Zügen das nachstehende Bild ergibt.

Wenn das Eis sich sehr spontan und in seiner gesamten Masse bei sehr tiefer Temperatur bildet, so fehlt ihm die Zeit, eine Kristallgitterstruktur auszubilden, und es resultiert das sog. amorphe, glasige Eis (vitreous ice). Es entsteht z. B. dann, wenn das Wasser in sehr dünner Schicht durch flüssige Luft zum Erstarren gebracht wird. Darauf beruht zum Teil die relative Unempfindlichkeit organisierter wasserhaltiger, lebender Substanz gegenüber einer so schnellen Abkühlung, welche die Ausbildung von gröberen, strukturzerstörenden Eiskristallen nicht zuläßt. Eine analoge amorphe Form des Eises beschreibt Staronka (1939); sie bildet sich dann, wenn in einem durch flüssige Luft gekühlten Glasrohr unter geringem Druck Wasserdampf kondensiert wird. Das resultierende amorphe Eis erleidet beim Erwärmen auf —130⁰ eine ausgeprägte exotherme Umwandlung in eine trübe, kristallisierte Eismodifikation.

Dem stehen die Ergebnisse von KÖNIG (1942 und 1943) entgegen, der unter etwa analogen Bedingungen eine Eismodifikation erhielt, die sich bei der Untersuchung mittels Elektronenbeugung als kubische Kristallform erwies, deren Gitteraufbau mit der des *Cristobalits* identisch war.  Die Abb. 2 zeigt, daß sich eine solche Struktur ohne weiteres mit der tetraedrischen Anordnung von Wasserstoffbrücken als Bindeglied zwischen den einzelnen Sauerstoffatomen des Gitters vereinbaren läßt.

Aus den genaueren Untersuchungen von KÖNIG erklären sich vielleicht auch die Beobachtungen von STARONKA, da oberhalb von —140⁰ eine starke Sammelkristallisation unter Vergrößerung der Eiskristalle eintritt, welche die Trübung plausibel macht.

Oberhalb von —80⁰ tritt eine andere Kristallform des Eises auf, die einen mit dem des *Tridymits* identischen Gitteraufbau zeigt, so daß man diese Kristallform, die dem Eis unter normalen Verhältnissen bis zum Schmelzpunkt zu eigen

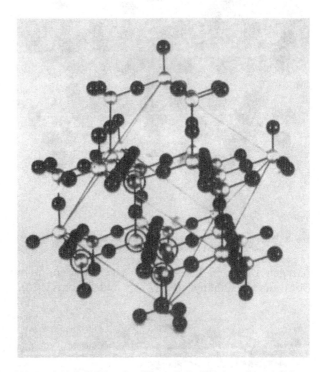

Abb. 2.  Cristobalitgitter des Eises.  Helle Kugeln = Sauerstoffatome; dunkle Kugeln = Wasserstoffbrücken.

ist, als Tridymiteis bezeichnet hat.  Den Gitteraufbau desselben, der in Übereinstimmung durch zahlreiche Bearbeiter untersucht und aufgeklärt worden ist (CHADWELL 1927, BERNAL und FOWLER 1933, KÖNIG 1942 und 1943), gibt schematisch die Abb. 3 wieder.  Aus der Anordnung der obersten Schicht der Wasserstoffatome ist zu ersehen, daß dieser normalen Kristallform des Eises eine hexagonale Struktur zu eigen ist.

Es handelt sich nun dabei, wie leicht zu ersehen, keineswegs um eine besonders dichte Anordnung der Wassermoleküle, sondern im Gegenteil um eine sehr sperrige Struktur mit großen Hohlräumen.  Zugleich ist aus der Abb. 3 analog zu der Abb. 2 eine bevorzugte Schichtenbildung mit besonders fester Verknüpfung durch ein dichtes Netz von Wasserstoffbrücken zu erkennen, deren Zahl zwischen je zwei Schichtebenen wesentlich geringer ist.  Unter besonderen Verhältnissen können dank derartigen Hohlraumbildungen auch die sog. festen Gashydrate sich bilden, in denen Gasmoleküle ringsum von einem recht stabilen Wassermolekülgerüst umgeben sind (v. STACKELBERG 1954).

Was läßt sich nun aus dieser heute wohl feststehenden Struktur des normalen Eises an Erklärungen für sein Verhalten gewinnen?  Das lockere Tridymitskelet bedingt das große Molvolumen bzw. die geringere Dichte gegenüber

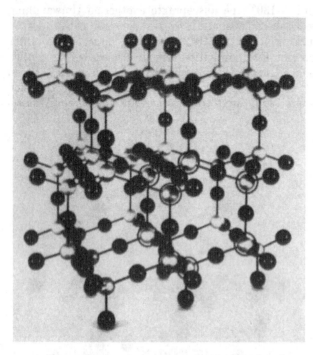

Abb. 3. Tridymitgitter des gewöhnlichen Eises. Helle Kugeln = Sauerstoffatome; dunkle Kugeln = Wasserstoffbrücken.

dem flüssigen Wasser. Die hexagonale Struktur spiegelt sich in der Kristallform natürlichen Eises wider, so etwa der Schneeflocken; auf gefrierenden Wasseroberflächen bilden sich häufig große tafelige Einkristalle mit Orientierung der Basisfläche zum Wasserspiegel aus, die beim Anschlagen eine sechsstrahlige Sprungstruktur zeigen (Schmidt und Baier 1935, S. 217/218).

Da die Wasserstoffbrücke mit ihrer geringen Bindungsenergie (s. o.) sich leicht bildet und wieder zerfällt, können kleine Kristalleinheiten schon bei enger Berührung durch relativ geringen Druck in kürzester Zeit verwachsen, womit auch die Regelation des Eises zusammenhängt (Schmidt und Baier 1935). So erklärt sich zwanglos die Verfestigung von Schnee zu Gletschereis und den Inlandeismassen Grönlands und der Antarktis. Parallel dazu geht die leichte Translationsmöglichkeit parallel zur Basisfläche (Bragg 1938), die eine Verschieblichkeit der Eissubstanz in sich zuläßt.

Auf diese Weise ist die Entstehung gewaltiger kompakter Eismassen gewissermaßen als äolisches Sedimentgestein mit homogenem Bindemittel möglich. Solches Sekundäreis besteht aus Kristalliten von Tridymitstruktur, die aber vorwiegend regellos zueinander orientiert sind; im Gegensatz zum primären Eis, das analog zu einem plutonischen Gestein aus der Flüssigkeit erstarrt ist, zeigt es daher im allgemeinen einen deutlichen Tyndalleffekt, als Folge des pseudoheterogenen Aufbaues, so daß es von vielen Autoren auch als kolloidales Eis bezeichnet wird.

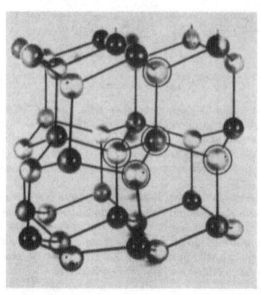

Abb. 4. Modell des Wurtzitgitters, nach dem die Sauerstoffatome im normalen Tridymiteis angeordnet sind. (Alle Kugeln stellen Sauerstoffatome dar.)

Das Vorhandensein von Schichtebenen bevorzugter Festigkeit, welche die intrakristalline Verschieblichkeit des Eises und damit dessen Plastizität ermöglichen, zeigt sich besonders deutlich, wenn man die Sauerstoffatome im Eis als Repräsentanten der einzelnen Wassermoleküle für sich betrachtet. Sie befinden sich entsprechend der koordinativen tetraedrischen Vierwertigkeit des Sauerstoffs auf den Gitterpunkten eines Wurtzitgitters (Abb. 4), in denen jedes Sauerstoffatom tetraedrisch von vier weiteren umgeben ist. Zur Verdeutlichung dieses raumzentrierten Tetraeders sind fünf Atomkugeln der Abb. 4 durch schwarze Kreise gekennzeichnet.

Über die gegenseitigen Abstände der benachbarten Wassermoleküle im Eis besteht keine völlige Übereinstimmung (CHADWELL 1927), jedoch hat die über-

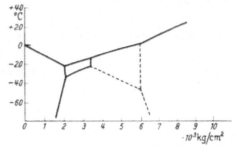

wiegende Mehrzahl der Messungen auf röntgenographischem Wege dafür den Wert des Wassermolekülradius von 2,76 Å ergeben (BERNAL und FOWLER 1933, KÖNIG 1942 und 1943 u. a.).

Neben der Beeinflussung durch die Temperatur unterliegt die Kristallstruktur des Eises auch einer solchen durch den äußeren Druck.

Da zu solchen eingreifenden Änderungen jedoch Drucke von Hunderten und Tausenden von Atmosphären erforderlich sind, die physiologisch höchstens experimentelle Bedeu-

Abb. 5. Zustandsdiagramm der bei hohen Drucken beständigen Eisarten. (Nach HÜCKEL 1948.)

tung haben können, sollen hier der Kürze halber nur die Existenzbereiche der vor allem von TAMMANN (1900 und 1923) experimentell untersuchten Eisarten in der Abb. 5 wiedergegeben werden. Diese Eisarten haben Schmelzpunkte und Dichten, die zum Teil über denjenigen des flüssigen Wassers liegen.

## Das Schmelzen des Eises.

Das Schmelzen des Eises vollzieht sich, wie eingangs erwähnt, unter anomalen Erscheinungen. Diese sind dadurch gekennzeichnet, daß einmal der calorische Betrag der Schmelzwärme mit 1400 cal/Mol (vgl. Tabelle 2) recht hoch ist, und daß zum anderen die Schmelze ein geringeres Volumen als der Kristall aufweist, also eine Kontraktion eintritt.

Aus dem ersten Faktum ist zu entnehmen, daß beim Schmelzvorgang bei weitem nicht alle Wasserstoffbrücken gelöst werden können, da hierfür der drei- bis vierfache Wärmebetrag erforderlich wäre (s. o.). Die zweite Erscheinung erklärt sich zwangsläufig aus der Weiträumigkeit des Tridymitgitters des Eises, das beim Schmelzen zusammenbricht und dichteren Lagerungsmöglichkeiten der Wassermoleküle Platz macht (ROBERTSON 1940 u. a.).

Abb. 6. Moleküllagerung bei dichtester Kugelpackung.

Berücksichtigt man den Molekülradius des Wassers von 1,37 Å, so müßte das Schmelzwasser eine Dichte von 1,84 statt 1 aufweisen, wenn die Moleküle sich beim Zerfall des Eisgitters in einer dichtesten Kugelpackung (Abb. 6) wie eine normale Flüssigkeit zusammenlagern würden (BERNAL und FOWLER 1933). Daß diese Voraussetzung nicht zutreffen kann, ergibt sich einmal aus der Differenz der vorstehend angeführten Energiebeträge, zum anderen aus Strukturuntersuchungen am Schmelzwasser (BERNAL und FOWLER 1933, VLÈS 1938, ROBERTSON 1940, DORSEY 1940, ULICH 1941, HÜCKEL 1948).

Auf ein Fortbestehen von eisähnlichen Gebilden im Wasser deutet auch das Vorhandensein eines Tyndalleffektes im Schmelzwasser hin (VLÈS 1938, DORSEY 1940, ULICH 1941). Zumindest in großer Temperaturnähe des Schmelzpunktes ist der Effekt auch dadurch erklärbar, daß durch statistische Temperaturfluktuationen innerhalb des Schmelzwassers örtlich Mikrounterkühlungsbereiche von sehr kurzer Lebensdauer auftreten, in denen sich das Tridymitgitter zurückbildet, so daß die Flüssigkeit optisch mikroheterogen bzw. kolloidal und damit lichtstreuend wird. Darauf beruhende ähnliche Effekte in Zweistoffgemischen sind von SMOLUCHOWSKI, EINSTEIN und von ZIMM quantitativ behandelt worden (ZIMM 1950).

## Die Struktur und das Verhalten des flüssigen Wassers.

Aus Vorstehendem geht hervor, daß beim Schmelzen des Eises das Tridymitgerüst als Ganzes zwar einem Zerfall unterliegt, daß jedoch kleinere Einheiten desselben offensichtlich auch im flüssigen Wasser oberhalb des Schmelzpunktes erhalten bleiben. Es ist daher vielfach üblich, aus diesem Grunde von „eisartigem" Wasser oder „zusammengebrochener Eisstruktur" bei tiefen Temperaturen dicht oberhalb 0° C zu sprechen.

Über die Größe dieser auch nach dem Schmelzen verbleibenden, dem Eis-

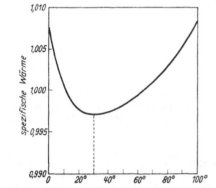

Abb. 7.   Temperaturabhängigkeit der spezifischen Wärme des Wassers. (Nach VLÈS 1938.)

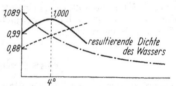

Abb. 8. Erklärung der Maximaldichte des Wassers (—) bei 4° durch Überlagerung der Dichten von Dihydrol (· — · — ·) und Trihydrol (———). (Nach VLÈS 1938.)

aufbau entsprechenden Aggregate soll vorerst nichts ausgesagt werden. Ihr Vorhandensein erklärt aber zwanglos die hohe spezifische Wärme des Wassers oberhalb des Schmelzpunktes. Wie die Abb. 7 zeigt, liegt sie durchweg weit höher als diejenige von normalen Flüssigkeiten, was als gewissermaßen in höhere Temperaturbereiche verschleppte Schmelzwärme des Eises aufzufassen ist (ULICH 1941, S. 53). Der Zerfall der primär beim Schmelzen entstandenen Aggregate zu kleineren Bruchstücken geht in einer begrenzten Temperaturspanne vor sich, woraus sich das Minimum der spezifischen Wärme bei rund 30° ergibt. Bereits unterhalb dieser Temperatur muß aber ein zunehmender Weiterzerfall der sekundären Bruchstücke eintreten, so daß die spezifische Wärme wieder ansteigt. Diese Form der Kurve macht daher das Auftreten von mindestens drei Arten verschiedener Aggregate von Wassermolekülen im Temperaturbereich des physiologischen Geschehens erforderlich.

VLÈS interpretiert auch das Dichtemaximum des Wassers bei 4° aus der Überlagerung der Dichten zweier ihr Mischungsverhältnis mit der Temperatur ändernden Aggregate von Wassermolekülen, die er als Dihydrol und Trihydrol ansieht (Abb. 8).

Wenn man von der weitgehend übereinstimmenden Meinung absieht, daß im Wasser in der Nähe von 0° gewisse Strukturelemente des Eises enthalten sein müssen, die sich auch leicht wieder zum Eiskristall zusammenlagern können, so herrscht bezüglich der Zähligkeit der bei verschiedenen Temperaturen vorherrschenden Wasseraggregate bei den zahlreichen Untersuchern dieser Frage der größte Widerstreit der Auffassungen und damit noch heute ein fast völliger Mangel an Klarheit.

Dieser Zustand ist dadurch bedingt, daß infolge des Fehlens einigermaßen starrer größerer Strukturen die Untersuchung durch Röntgenstrahlen- oder Elektronenbeugung zu keinen eindeutigen Meßergebnissen gelangen läßt (HÜCKEL 1948, S. 447). Unter dem Begriff Zähligkeit soll hier die Anzahl von Wassermolekülen $(H_2O)_1$ verstanden werden, die das jeweils betrachtete Aggregat aufbauen. Unter Aggregat bzw. Assoziat wiederum ist eine reversible (FRUHWIRTH 1937) Zusammenlagerung mehrerer Moleküle zu einem Mehrfachkomplex zu verstehen (MALSCH 1937, MECKE 1948 I).

Aus der Vielzahl der Meinungen werden nachstehend einige wenige herausgegriffen, die auf eingehenden experimentellen oder mathematischen Grundlagen beruhen.

Die ursprüngliche, auf dem Symposium (*Symposium* 1910) der Faraday Society allgemein vertretene Auffassung vom Mono-, Di- und Trihydrol ist in ihrer Einfachheit wohl als überholt zu betrachten. TAMMANN (1926) vertritt die Existenz eines im Gleichgewicht mit Hexahydrol stehenden Trihydrols bei $0°$ auf Grund der ermittelten Volumenisothermen. BERNAL und FOWLER (1933) nehmen als wesentlichen Bestandteil des Wassers eine Gruppierung an, die sich

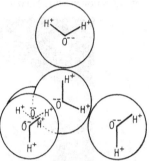

Abb. 9. Tetraedrischer Aufbau des fünfzähligen Wasseraggregates, (Nach BERNAL und FOWLER, 1933.)

in der Weise aufbaut, daß um ein zentrales Wassermolekül vier weitere tetraedrisch gelagert sind. Diese Gruppierung ist ja bereits im Eisgitter enthalten (vgl. Abb. 4). Bei niedriger Temperatur ergeben diese Tetraeder daher eine Art von verwackeltem und deformiertem Tridymitgitter, das bei höherer Temperatur in ein ähnlich mangelhaftes Quarzgitter übergeht. Es wird also der gesamten Wassermasse eine molekular grobkörnige (WOLF, FRAHM und HARMS 1937) oder quasi kristalline Struktur zugeschrieben (DORSEY 1940, ULICH 1941).

Dieser Wassertetraeder ist in sich durch Wasserstoffbrücken verfestigt (Abb. 9). Infolgedessen würden sich nach dieser Theorie, die sich zahlreiche Autoren zu eigen gemacht haben (MAGAT 1936, ULICH 1936, FINBACK und VIERVOLL 1943, VLÈS 1938), Wasseraggregate als ganze Vielfache eines Pentahydrols konstruieren lassen (FOX und MARTIN 1940). Durch Ramanmessungen z. B. von CROSS, BURNHAM und LEIGHTON (1937) hat sich die Anwesenheit gewisser Mengen tetraedrisch koordinierter Aggregate bei $0°$ bestätigen lassen, jedoch weisen die Messungen auch auf das Vorhandensein geringerer Zähligkeiten hin (GANZ 1936 und 1937, MECKE 1948 I und II).

Einen eingehenden Überblick über zahlreiche weitere Aggregationshypothesen bringen MOESVELD und HARDON (1931) sowie DORSEY (1940). Ihre Zitierung würde aus dem hier vorliegenden Rahmen herausfallen.

In betonten Gegensatz zu der Vorstellung vom tetraedrischen Pentahydrol hat sich neuerdings EUCKEN (1946, 1947, 1948 I, 1948 II, 1949) gestellt, indem er als wesentliche Aggregate im flüssigen Wasser in der Reihenfolge steigender Temperatur Achter-, Vierer-, Zweiergruppen und schließlich einzelne Wassermoleküle annimmt, deren Existenz er auf Grund einer Art von Fourieranalyse der Anomalien des Wassers ableitet. Die Achteraggregate sollen eine Struktur besitzen, wie sie die Abb. 10 links zeigt. Sie läßt sich ebenfalls als Baustein des Tridymiteises auffassen, wenn man sie mit den durch Kreise gekennzeichneten Sauerstoffatomen in der Abb. 3 vergleicht. Als Viereraggregate kommen dann

die Hälften der Achtergruppen in Betracht (Abb. 10 rechts). Durch weiteren
Zerfall gehen diese wiederum in Doppel- bzw. Einzelmoleküle über. Nähere
Details über diese weitgehend ausgebaute Theorie sind den Originalarbeiten zu
entnehmen.

Wenn das Wasser sich aus definierten wenigen Arten von mehrzähligen
Aggregaten aufbauen würde, wäre schließlich die Möglichkeit nicht außer acht
zu lassen, daß bezüglich der Einstellung der temperaturmäßigen Gleichgewichte
zwischen denselben gewisse Relaxationen zu bemerken wären. Darüber liegen
verschiedene Untersuchungen mit recht widersprechenden Resultaten vor.

Genaue Messungen rein physikalischer Daten, und zwar des Dampfdruckes,
der magnetischen Suszeptibilität, der Infrarotabsorption, des Brechungsindex

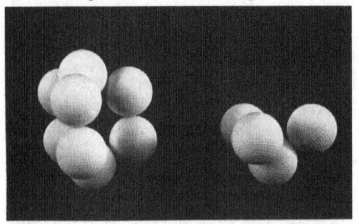

Abb. 10.  Acht- und vierzählige Wasseraggregate.  (Nach EUCKEN.)

und der Dichte haben gezeigt, daß solche Verzögerungen der Eigenschaftsänderung gegenüber der Temperatur unmeßbar klein sein müssen (DORSEY 1940).

Dem stehen biologische Versuche von BARNES und BARNES (1932) gegenüber,
welche die Alge *Spirogyra* in zweimal täglich erneuerten Wasserproben züchteten,
die einerseits frisch aus Eis erschmolzen, andererseits frisch aus Wasserdampf
kondensiert wurden. Im ersteren Falle wurde normales Wachstum verzeichnet,
während im Kondensat ceteris paribus ein Zusammenklumpen und Schrumpfen
der Protoplasten beobachtet wurde. Die gleichen Ergebnisse wurden mit *Euglena*
und *Amoeba* gefunden. Die beiden Autoren nahmen an, daß das Eisschmelz-
wasser vorwiegend aus Trihydrol besteht, das abgekühlte Kondensat dagegen
überwiegend aus Dihydrol, deren Gleichgewicht sich nur sehr langsam einstellt.
Das Trihydrol wurde dabei als für den Zellturgor bestimmend angesehen. Leider
haben sich die Versuche durch andere Forscher nicht reproduzieren lassen (DOR-
SEY 1940).

Bei der Gegensätzlichkeit der Auffassungen ist stets vor Augen zu halten, daß diese auf
der Schwierigkeit beruht, eindeutige zahlenmäßige Experimentalwerte für das offensichtlich
sehr komplexe Gefüge des flüssigen Wassers zu erhalten. Von der rein mathematischen
Seite her lassen sich die Meßdaten für die Anomalien durch die Annahme von immer mehr
differenzierten Molekülaggregaten mit passend gewählten Eigenschaften den rechnerisch
ermittelten Werten beliebig näher bringen, jedoch fehlt ihnen die Schlüssigkeit des Beweises
(HONIGMANN 1932).

Zusammenfassend ist bezüglich der Struktur des flüssigen Wassers zu sagen,
daß hierfür im Gegensatz zum Eis noch kein einheitliches, exaktes Bild existiert.
Bei der Sonderstellung des flüssigen Aggregatzustandes ist ein solches wahrscheinlich auch gar nicht zu erwarten. Das Wasser ist eher als ein dreidimensio-
nales Netzwerk von tetraedrischer, elastischer Verzweigung und im wahrsten

Sinne des Wortes fließenden Strukturen aufzufassen (STEWART 1939). Es sind an Stelle von Gleichgewichten bestimmter definierter Aggregate wahrscheinlich Wanderungen, Bildung und Zerfall der verschiedensten Mikrostrukturen anzunehmen, die sich mit dem tetraedrischen Koordinationsprinzip des Wassermoleküls vereinbaren lassen, die unter Wechsel der Individuen und des Ortes, unter Knüpfung und Aufbrechen von Wasserstoffbrücken, VAN DER WAALSschen und Dipolattraktionen, nach- und nebeneinander entstehen und vergehen. Bestimmend für diese Vorgänge ist dabei wahrscheinlich das Wechselspiel zwischen den ordnenden Bindungskräften einerseits und den destruktiven, in Größe und Richtung statistisch bestimmten thermischen Bewegungsenergien der Wassermoleküle. Die Ausbildung größerer Ordnungsbereiche muß infolgedessen bei niedriger Temperatur wahrscheinlicher sein als bei starker thermischer Molekularbewegung infolge höherer Temperatur.

Diese Vorstellungen finden eine gewisse Stütze in Modellversuchen von KAST und STUART, über die VOLKMANN (1948) berichtet. Es wurde dabei unter anderem mit runden, magnetischen Dipolscheibchen gearbeitet, die auf einer flachen Unterlage mit einer angemessenen Frequenz gerüttelt wurden. Die interessanten Versuche zeigten, gewissermaßen als zweidimensionale Projektion des dreidimensionalen Geschehens im flüssigen Wasser, ein Bild, das in großen Zügen dem eben zusammengefaßten entsprach. Bei sinngemäßer Umrechnung der beobachteten Stabilitäten der vorübergehend sich bildenden Aggregate ist den analogen Gebilden von molekularen Dimensionen im flüssigen Wasser eine mittlere Lebensdauer von rund $10^{-9}$ sec zuzuschreiben, womit sie im Mittel mehrere tausend thermische Stöße überdauern würden.

## Durch Bindungskräfte bedingte stoffliche Eigenschaften des flüssigen Wassers.

Dank den auch im flüssigen Wasser auftretenden, die Einzelmoleküle untereinander verknüpfenden Bindungskräften der Dipolanziehung und der Wasserstoffbrückenbildung setzt das Wasser, als Stoff gesehen, den seine Gestaltung zu verändern strebenden Einflüssen einen größeren Widerstand entgegen, als ihn eine normale Flüssigkeit aus Molekülen gleicher Größe aufweisen würde.

Die Höhe der Oberflächenspannung (vgl. Tabelle 2), die rund dreimal so hoch liegt wie die von Tetrachlorkohlenstoff mit 26,77 ($CCl_4$ besitzt wie das Wasser Tetraederstruktur und gleiche Elektronenzahl, bildet aber keine Wasserstoffbrücken), erklärt sich auf diese Weise zwanglos (RAMSAY 1894, REHBINDER 1926).

Das gleiche gilt für die Viscosität, die selbst im Vergleich zu den Alkoholen, die ja ebenfalls zur Bildung von Wasserstoffbrücken befähigt sind (HUGGINS 1937, LASSETTRE 1937, ASTBURY 1940, FORBES 1941, SUHRMANN und WALTER 1951), sehr hoch liegt (GIRARD und ABADIE 1936, WOLF, FRAHM und HARMS 1937).

Da zwischen der intermolekularen Bindungsfestigkeit und der Temperatur offenbar kein linearer Zusammenhang besteht, zeigen ebenso wie die Viscosität (TAMMANN 1937) auch die Kompressibilität und der Brechungsindex keine lineare Temperaturabhängigkeit (MAGAT 1936 und 1937), wobei sich besonders im Bereich von etwa 30—60⁰ eine stärkere Krümmung bemerkbar macht (TAMMANN 1937). Die scharfen Unstetigkeiten in dieser Temperaturzone, die VLÈS (1938) aus den Meßwerten von MAGAT herausliest, scheinen einer mehr subjektiven Auffassung zu entsprechen.

Im ultraroten Spektrum macht sich hier ein allmählich stärkeres Nachlassen der intermolekularen Verknüpfungen am Rückgang der dafür charakteristischen Absorptionsbanden bemerkbar (GANZ 1936). Die gegebenenfalls für physiologische Fragen wichtigen Viscositätszahlen des reinen Wassers sind von DORSEY

(1940, S. 183) kritisch überprüft und zusammengestellt worden. Sie sind auszugsweise in der Tabelle 3 wiedergegeben.

Tabelle 3. *Viscosität $\eta$ des Wassers in cP zwischen 0 und 100°* (Nach Dorsey 1940.)

| Temperatur, °C | 0 | 5 | 10 | 15 | 20 | 25 | 30 |
|---|---|---|---|---|---|---|---|
| $\eta$, cP | 1,794 | 1,519 | 1,310 | 1,145 | 1,009 | 0,894 | 0,800 |
| Temperatur, °C | 35 | 40 | 45 | 50 | 55 | 60 | 65 |
| $\eta$, cP | 0,720 | 0,653 | 0,607 | 0,549 | 0,507 | 0,470 | 0,437 |
| Temperatur, °C | 70 | 75 | 80 | 85 | 90 | 95 | 100 |
| $\eta$, cP | 0,407 | 0,381 | 0,357 | 0,336 | 0,317 | 0,299 | 0,284 |

Im vorstehenden wurde das Wassermolekül $(H_2O)_1$ durchweg als kleinste Einheit und Baustein größerer Gefüge in Form von Eis und Wasser betrachtet. Der folgende Abschnitt soll sich mit der wechselseitigen Beeinflussung von Wasser und darin enthaltenen Fremdbestandteilen befassen. Als Fremdbestandteile sind hier solche zu verstehen, deren stöchiometrische Zusammensetzung nicht der Formel $(H_2O)_n$ genügt.

## Die Dissoziation des Wassers.

Für verschiedene, auch praktisch bedeutsame Erscheinungen, insbesondere für das Verhalten schwacher Säuren und Basen in wäßriger Lösung, fällt die nach der Gleichung

$$H_2O \rightleftharpoons H^+ + OH^-$$

erfolgende Eigendissoziation des Wassers in gleiche Mengen entgegengesetzt geladener Ionen ins Gewicht. Die genauere Prüfung dieses Dissoziationsvorganges des Wassermoleküls hat gezeigt, daß sich das von einem solchen abgespaltene $H^+$-Ion wahrscheinlich sofort an ein zweites Molekül $H_2O$ anlagert, so daß es richtiger zu sein scheint, den Ionisationsvorgang durch die Gleichung

$$2 H_2O \rightleftharpoons H_3O^+ + OH^-$$

auszudrücken (Eucken 1944, S. 216).

Zu dem letzteren Gleichgewichtsschema kommen auch Wolf, Frahm und Harms (1937), indem sie das Dihydrol als das bei normalen Temperaturen stabilste Aggregat auffassen. Die Bildung von Ionen mit koordinativ dreiwertigem Sauerstoff kann danach unmittelbar von den Doppelmolekülen her geschehen, wobei in der zweiten Gleichung $(H_2O)_2$ an Stelle von $2 H_2O$ zu setzen ist. Es herrscht auf jeden Fall gegenwärtig Übereinstimmung darüber, daß das H-Ion nicht als freies Proton im Wasser vorliegt, sondern an ein Wassermolekül gebunden ist.

Die Ionisationskonstante des Wassers ist von zahlreichen Forschern nach den verschiedensten Methoden bestimmt worden (Bjerrum 1923, Ölander 1929, Harned und Geary 1937, Harned und Cook 1937, Ebert 1949 u. a.). Die von Harned und Robinson (1940) sorgfältig gesichteten und überprüften Werte dafür sind zusammen mit analogen Messungen von Bjerrum und Unmack (1929) in der Tabelle 4 zusammengefaßt.

Der $p_H$-Wert für neutrales Wasser kann durch Halbierung der in der Tabelle 4 enthaltenen Zahlen für ($-\log K$) erhalten werden. Ein abermals anomales Verhalten zeigt das Wasser, wenn man die Beweglichkeit seiner Ionen in einem äußeren elektrischen Felde mißt. Die Wasserstoff- und Hydroxylionen bewegen sich dann um Größenordnungen schneller, als sie es nach theoretischen Überlegungen, die auf dem Ionenradius und der Viscosität des Wassers beruhen, tun dürften. Diese Erscheinung ist erklärlich, wenn man eine Art von GROTTHUSSchem Wanderungsmechanismus annimmt (ULICH 1933, S. 181 ff.). Man geht dabei von der Überlegung aus, daß H$^+$ und OH$^-$ ja lösungsmitteleigene Ionen sind. Ein in Richtung der Kathode wanderndes H-Ion wird dabei mit großer Wahrscheinlichkeit auf das Sauerstoffatom von Wassermolekülen treffen, die eine dem Felde entsprechende bevorzugte Orientierung besitzen. Es kann sich dabei an den Sauerstoff anlagern, der dafür in entgegengesetzter Richtung ein anderes Proton abgibt usw. usf. Durch die so ersparten Teilwege kommt die scheinbar hohe Wanderungsgeschwindigkeit zustande. BERNAL und FOWLER (1933) haben diese Frage auch von der quantenmechanischen Seite her behandelt und sind gleichfalls zu quantitativ befriedigenden Ergebnissen gekommen.

Tabelle 4.
*Dissoziationskonstante des Wassers bei verschiedenen Temperaturen.*

| Temperatur °C | — log K | |
|---|---|---|
| | nach HARNED und ROBINSON (1940) | nach BJERRUM und UNMACK (1929) |
| 0 | 14,9435 | 14,926 |
| 5 | 14,7338 | — |
| 10 | 14,5346 | — |
| 15 | 14,3463 | — |
| 18 | — | 14,222 |
| 20 | 14,1669 | — |
| 25 | 13,9965 | 13,980 |
| 30 | 13,8330 | — |
| 35 | 13,6801 | — |
| 37 | — | 13,590 |
| 40 | 13,5348 | — |
| 45 | 13,3960 | — |
| 50 | 13,2617 | — |
| 55 | 13,1369 | — |
| 60 | 13,0171 | — |

# Lösungen von Fremdionen in Wasser.

Einen ganz erheblichen Eingriff in das Gefüge des Wassers bedeutet das Einbringen von Fremdionen, also von elektrolytisch dissoziierenden Verbindungen. So ist es seit langem bekannt, daß durch den Zusatz solcher Stoffe in den meisten Fällen das Volumen der Lösung nicht entsprechend größer wird, sondern bis zu gewissen Konzentrationen sogar abnimmt. Ebenso hat die eigentümliche Tatsache seit jeher Aufmerksamkeit erregt, daß in Elektrolytlösungen die Viscosität des Wassers nicht, wie zu erwarten, erhöht, sondern verringert wird. Die Erscheinung wird meist als negative Viscosität bezeichnet, exakter ist es aber wohl, von einer negativen spezifischen Viscosität $\eta_{sp} = \dfrac{\eta_L - \eta_W}{\eta_W}$ zu sprechen (ULICH 1936, EUCKEN 1947 und 1948 I). Ursprünglich wurde der erstere Effekt der Kontraktion beim Auflösen von Elektrolyten ausschließlich durch eine Elektrostriktion der Wassermoleküle erklärt, wonach im elektrischen Felde der Ionen eine Anziehung und dichtere Zusammenpressung der Wassermoleküle erfolgt (DRUDE und NERNST 1894). Auch nach neueren Untersuchungen ist wenigstens ein Teil dieser Erscheinungen durch die Elektrostriktion zu erklären (GEFFCKEN 1931, MOESVELD und HARDON 1931).

Die hierbei auftretenden Kräfte sind ganz enorm, wenn man berücksichtigt, daß im extremen Falle des Kristallwassers das scheinbare Molvolumen des Wassers bis zu 25% niedriger liegen kann als im ionenfreien flüssigen Wasser (EUCKEN 1948 I).

Weitgehende Übereinstimmung herrscht darüber, daß beide Erscheinungen der Volumen- und Viscositätsverringerung in der Hauptsache auf in den Ordnungszustand der Wassermoleküle eingreifende Änderungen durch die Ionen zurückzuführen sind. Es liegt nahe, daß hierfür quantitative Angaben noch unsicherer fundiert sind als solche über den Charakter von Ordnungszuständen im reinen Wasser.

Die Ionen umgeben sich entsprechend ihrem Radius und ihrer Ladung (GANZ 1936 und 1937 I und II) mit einer gewissen Zahl von Wassermolekülen (SUHRMANN und BREYER 1933).

Für die Zahl der so festgehaltenen Wassermoleküle gilt allerdings kein festes stöchiometrisches Verhältnis (SCHREINER 1934), so daß man je nach der benutzten Meßmethode dafür zu verschiedenen Werten kommt (DAHR 1914, FRICKE 1922, EUCKEN 1948 I). Je nach der Bedeckungsdichte der Ionen mit Wassermolekülen in Abhängigkeit von der Konzentration und der Temperatur sind die Ionenhydrate noch verformbar, während sie bei vollkommener Bedeckung als starre Kugeln aufzufassen sind. Die Hydratationszahlen lassen sich aus der Hydratationsenergie und -entropie ermitteln (KORTÜM 1941), die dafür von ULICH (1936) und KUJUMZELIS (1938) mitgeteilten Werte zeigt die Tabelle 5.

Es fällt besonders bei den zweiwertigen Kationen die hohe Hydratationszahl auf. Da daraus ein großes starres Volumen des Ionenhydrats resultiert, zeigen wässerige Lösungen von Erdalkaliionen eine erhöhte Viscosität, also positive spezifische Viscosität.

Tabelle 5. *Hydratationszahlen verschiedener Anionen und Kationen.* (Nach ULICH 1936 und KUJUMZELIS 1938.)

| Anionen | | Kationen | |
|---|---|---|---|
| Ion | Hydratations-zahlen | Ion | Hydratations-zahlen |
| Cl⁻ | 3 | Li⁺ | 6 |
| Br⁻ | 2 | Na⁺ | 4 |
| J⁻ | 0,5 | K⁺ | 2,5 |
| NO⁻³ | 2 | Ca⁺⁺ | 8—10 |
| | | Sr⁺⁺ | 8 |
| | | Ba⁺⁺ | 6—8 |

Nach EUCKEN (1948 I) braucht sich die Hydratation bzw. Solvatation nicht auf eine einschichtige Hülle zu beschränken. So soll das Magnesiumion $Mg^{++}$ außer der ersten, sehr fest gebundenen Hydrathülle eine weitere besitzen, die nur locker festgehalten wird und daher schon bei mäßiger Erwärmung der thermischen Bewegungsenergie zum Opfer fällt. Ähnliches gilt für das Li-Ion. Es steht der Annahme nichts entgegen, daß die Solvatation solcher Ionen sogar noch weiter geht und das Hydratwasser sukzessive in das praktisch freie Wasser übergeht. Auch das würde zur Viscositätserhöhung beitragen.

Neben die Solvatation der Ionen zu Ionenhydraten tritt als zweiter, die Viscositäts- und Volumenänderungen unter dem Einfluß gelöster Elektrolytionen verursachender Faktor deren destruktive Wirkung auf die Nahordnung des flüssigen Wassers (SUHRMANN und BREYER 1933).

Die Vertreter von Theorien, die sich auf das Vorhandensein bestimmter Molekülaggregate im Wasser stützen, schreiben dem elektrischen Felde der Ionen nicht nur eine Nahwirkung auf die Hydrathülle, sondern auch eine Fernwirkung zu, von EUCKEN als Lösungsmitteleinfluß bezeichnet, die wesentlich über deren Bereich in das reine Wassermilieu hineinreicht (STEWART 1937, EUCKEN 1948 I). Dadurch werden z. B. von den Achteraggregaten nach EUCKEN größere Mengen zerstört und zerfallen in die kleineren Aggregate (vgl. Abb. 10). Da letztere weniger sperrig sind als die Ausgangsgebilde, können sie sich dichter lagern, und verursachen so die beobachtete Kontraktion, während aus dem gleichen Grunde die Viscosität geringer wird, soweit diese Wirkung nicht durch die umfangreichen Ionenhydrate überkompensiert wird. Analoges gilt für die zahlreichen anderen Aggregationstheorien und Hypothesen des Wassers, so daß es sich erübrigt, darauf an dieser Stelle im Detail einzugehen. Ihnen ist gemeinsam, daß eine Verminderung der Sperrigkeit der Molekülanordnungen, deren Vorhandensein neben Kompressibilitätsmessungen (CHADWELL 1927) auch Ultraschalluntersuchungen (CLAEYS, ERRERA und SACK 1937) bewiesen haben, durch die elektrostatische Fernwirkung der Ionen eintritt.

# Wäßrige Lösungen von Nichtelektrolyten.

Eine Strukturbeeinflussung, wie sie soeben für wäßrige Elektrolytlösungen beschrieben wurde, ist selbstverständlich auch in Lösungen von Nichtelektrolyten zu verzeichnen. Einerseits ist hier zwar der volumenverringernde Einfluß der Elektrostriktion im allgemeinen verhältnismäßig schwach oder nicht nachweisbar. Andererseits sind aber außer den Elektrolyten nur solche Substanzen in Wasser überhaupt in merklichen Mengen echt oder kolloidal löslich, die selbst über Atomgruppierungen verfügen, die zur Bildung von Wasserstoffbrücken befähigt sind. Dazu sind in erster Linie nachstehende Konfigurationen oder prosthetischen Gruppen zu rechnen (HUGGINS 1937, LASSETTRE 1937, ASTBURY 1940):

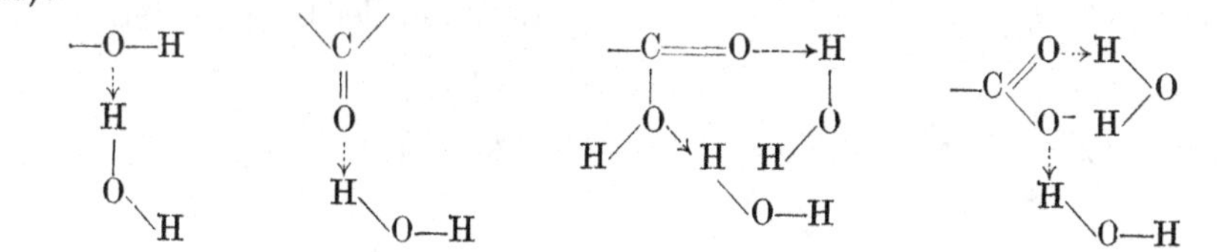

Hydroxylgruppe   Carbonylgruppe     Carboxylgruppe     Ionisierte Carboxylgruppe

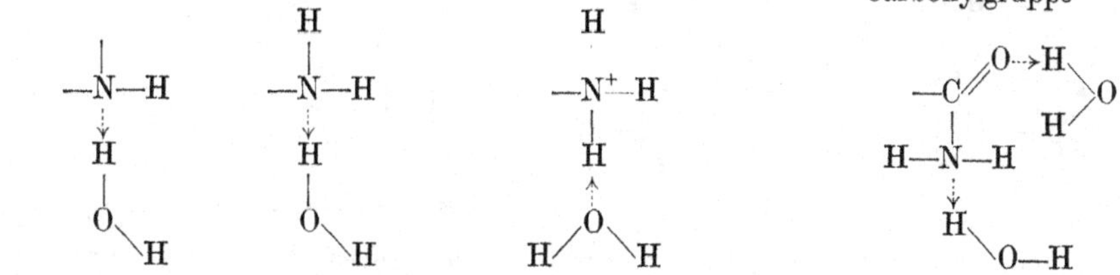

Iminogruppe   Aminogruppe   Ionisierte Aminogruppe   Amidgruppe

Man erkennt darin die grundlegenden reaktionsfähigen Bestandteile derjenigen Substanzen, die in der lebenden Zelle neben dem Wasser selbst die integrierende Rolle spielen.

Es ist das kein Zufall, nachdem das Leben erst einmal auf der Basis des wäßrigen Milieus begonnen hatte. Lebensvorgänge konnten sich dann zwangsläufig nur noch unter Verwendung von vollkommen oder intermediär in Wasser löslichen Stoffen abspielen.

Die Stoffe, die im Wasser gelöst werden, können flüssig (z. B. Alkohole) oder fest (z. B. Stärke, Gelatine) sein. Bildet eine Flüssigkeit mit Wasser eine Lösung, so tritt beim Vermischen eine negative Wärmetönung auf, wenn dadurch in erster Linie die Entstehung von Eigenassoziaten jeder der beiden Komponenten begünstigt wird. Treten dagegen Mischassoziate auf, so zeigt sich eine positive Wärmetönung, meist unter Kontraktion (MECKE 1948 II). Die Abb. 11 gibt ein schematisches Bild einer solchen Mischassoziation, aus dem gleichzeitig die Verzweigung der Alkoholketten am Wassermolekül zu erkennen ist.

Solche Gemische zeigen unter anderem ein Minimum der Viscosität bei einem Molverhältnis von 1 Alkohol zu 4 Wasser (HATSCHEK 1929, S. 129—155). Auch für die Grenzlöslichkeit von Wasser in den höheren Alkoholen spielt dieses Molverhältnis eine wesentliche Rolle (v. ERICHSEN 1952).

Schon bei so relativ einfach aufgebauten Systemen sind die Bindungs- und Strukturverhältnisse kaum noch zu übersehen, und man muß sich mit schematischen Vorstellungen über dieselben begnügen. Für die hochmolekularen Naturstoffe, wie Stärke oder gar Eiweiß, gilt das in erhöhtem Maße.

Immerhin konnten BAWN, HIRST und YOUNG (1940) zeigen, daß kleinere Stärkebausteine etwa von der Größenordnung der Dextrine *nicht* durch Verknüpfung mittels Wasserstoffbrücken zum eigentlichen Stärkemolekül verbunden werden. Eine recht feste Bindung von Wassermolekülen an den freien Hydroxylgruppen der Kohlenhydrate ist dagegen wohl anzunehmen. Für das Eiweiß ist das Vorhandensein von Wasser sogar von integrierender Bedeutung. Es zeigt seine eigentlichen Eigenschaften, seine wirkliche Beschaffenheit und Funktionen erst in Gegenwart von Wasser. UBBELOHDE (1940), ASTBURY (1940) u. a. unterscheiden dabei locker und fester gebundenes Wasser. ASTBURY schreibt das erstere der multipolaren Beschaffenheit des Eiweißmoleküls zu, während der fester gebundene Anteil wahrscheinlich über Wasserstoffbrücken mit Carboxyl- und anderen geeigneten funktionellen Gruppen verknüpft ist.

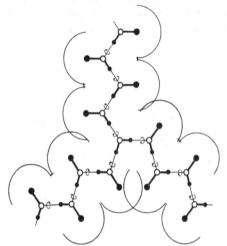

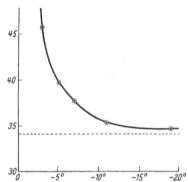

Abb. 11. Verzweigung von Alkoholketten am Wassermolekül des Mischassoziates. ● = CH₃; ○ = O; ● = H. (Nach KAST und PRIETZSCHK 1941.)

Abb. 12. Anteil des beim Abkühlen von Gelatinegelen nicht auskristallisierenden Eises. (Nach LUYET und GEHENIO 1940.)

Innerhalb der Eiweißmoleküle selbst sollen sich ebenfalls zahlreiche Wasserstoffbrücken zwischen solchen Gruppen befinden.

Auch BERNAL (1940) macht solche Unterschiede zwischen aussalzbarem und fester gebundenem Wasser, das nur durch erhöhte Temperatur zu entfernen ist. Seine Untersuchungen lassen vermuten, daß z. B. das Eiweißmolekül des Tabakmosaikvirus in Lösung eine durch Wasserstoffbrücken fest haftende Wasserhülle von etwa 10 Å besitzt. Aus Dampfdruck- und Kompressibilitätsmessungen über feinstem, feuchtem Holzmehl der Sitkafichte (Picea falcata) leitet BARKAS (1940) auch für die komplexe Holzsubstanz eine dicht gepackte Schicht von durch Wasserstoffbrücken gebundenem Wasser ab.

Daß die Festigkeit und Zahl der Eiweiß und Wasser verbindenden Wasserstoffbrücken ganz erheblich sein muß, geht außerdem aus Untersuchungen von LUYET und GEHENIO (1940) hervor. Ihre Befunde zeigt die Abb. 12, aus der zu ersehen ist, daß rund 35% des Wassergehaltes von Gelatinegelen beim Abkühlen nicht auskristallisieren. LUYET (1938) konnte selbst Gelatinelösungen mit 90% Wassergehalt ohne Kristallbildung bis zum glasigen Erstarren abkühlen, während man reines Wasser bisher höchstens bis —40° unterkühlen konnte (DORSEY 1940), ehe es spontan erstarrt.

Schließlich ist in Ergänzung der Frage der Wasserstoffbrücken zwischen Wasser und organischen Verbindungen noch eine Form dieser Bindung zu erwähnen, welche im Gegensatz zu den bisherigen Beispielen die Wasserlöslichkeit nicht erleichtert, sondern herabsetzt, meist sogar in sehr starkem Maße. Sie wird als Scherenbindung (Chelatbildung) bezeichnet und beruht auf der intramolekularen Bildung einer Wasserstoffbrücke. Eine solche wird dann leicht entstehen, wenn etwa eine Hydroxylgruppe und eine geeignete Sauerstoffunktion sich in

o-Stellung an einem Benzolring befinden, so etwa bei der o-Methoxybenzoe-
säure (I) oder beim Methylsalicylat (II und III):

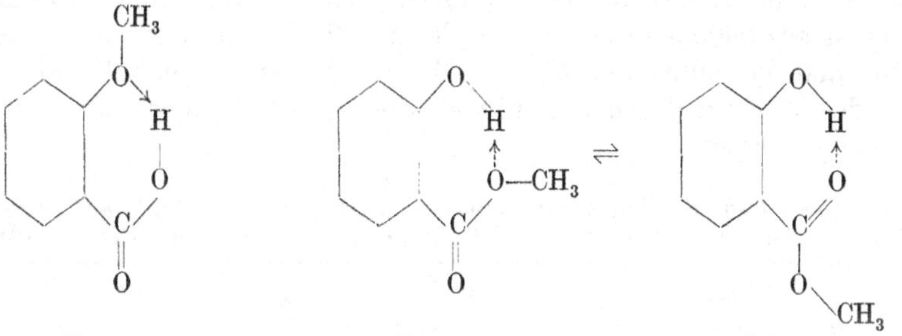

I II III

Das Methylsalicylat ist sogar imstande, zwei Arten von Scherenverbindungen
(II und III) zu bilden.

Das Vorliegen solcher Chelate läßt sich bei unerwartet geringer Wasserlöslichkeit vermuten und durch Messung des Ultrarot-Absorptionsspektrums (Fox und MARTIN 1940)
oder der magnetischen Suszeptibilität (ANGUS und HILL 1940) nachweisen. Mit der geringen
Wasserlöslichkeit geht ein ausgeprägter lipophiler Charakter der Scherenverbindungen parallel.
Es erscheint somit nicht ausgeschlossen, daß gerade die sehr hochmolekularen Eiweißmoleküle
an verschiedenen Stellen des Moleküls teils hydrophile Zentren besitzen, die also zur Bildung
von äußeren Wasserstoffbrücken befähigt sind, zum anderen Teil aber lipophile, fett- und
lipoidlösende Abschnitte, welche diese Eigenschaft der inneren Wasserstoffbrücke, der Chelatbildung verdanken.

## Normales und schweres Wasser.

Alle vorstehend kurz dargestellten Eigenheiten des festen und flüssigen
Wassers galten, wie eingangs betont, für natürliches Wasser, dessen Zusammensetzung überwiegend der Formel $H_2O$ entspricht. H bezeichnet darin Wasserstoffatome von der Massenzahl 1, O Sauerstoffatome mit der Massenzahl 16.
Im natürlichen Wasser sind aber außerdem noch Wasserstoffatome der Masse 2
(schwerer Wasserstoff, Deuterium, Symbol D), ebenso die Isotopen des Sauerstoffs mit den Massenzahlen $^{17}O$ und $^{18}O$ enthalten (SEELWOOD 1941, MATTAUCH
und FLAMMERSFELD 1949, SOUCI 1952).

Daraus ergeben sich folgende neun verschiedene Arten von Wassermolekülen, aus denen
sich das bisher betrachtete Wasser zusammensetzt:

| | | |
|---|---|---|
| $H^{16}O\ H$ | $H^{17}O\ H$ | $H^{18}O\ H$ |
| $H^{16}O\ D$ | $H^{17}O\ D$ | $H^{18}O\ D$ |
| $D^{16}O\ D$ | $D^{17}O\ D$ | $D^{18}O\ D$ |

Seit einiger Zeit ist auch das radioaktive Wasserstoffisotop Tritium (Symbol T) mit der
Massenzahl 3 präparativ zugänglich, so daß sich neuerdings an Schauplätzen von Atom-
bombenexplosionen Wassermoleküle mit T-Atomen an Stelle von H vorfinden können.
Damit verdoppelt sich die Zahl der möglichen Arten von Wassermolekülen abermals, so
daß das Wasser sich maximal aus 18 verschiedenen Molekülarten aufbauen kann.

Da für experimentelle Zwecke in erster Linie die D- und T-haltigen Wassermoleküle in Betracht kommen, sind in der untenstehenden Tabelle 6 die wesent-
lichsten Eigenschaften derselben wiedergegeben.

Das $D_2O$ ist zu rund $^1/_{5000}$ im natürlichen Wasser enthalten, selbstverständlich
auf Grund des leichten H-Austausches zwischen den Wassermolekülen (s. o.)
nach dem Massenwirkungsgesetz in der Hauptsache als HDO. Seine Konzentration wechselt mit der Herkunft des Wassers (SOUCI 1952) und kann durch
Messung der Oberflächenspannung oder auf optischem Wege (BROIDA 1954)
schnell und genauer ermittelt werden.

Die Tabelle 6 zeigt, daß zwischen schwerem und normalem Wasser nicht zu vernachlässigende Unterschiede bestehen. Da auch das Molvolumen des ersteren um etwa 0,3% höher liegt als das von natürlichem Wasser (ROBINSON und BELL 1937), so ist anzunehmen, daß die Größe und die Lebensdauer der weiter oben (Struktur und Verhalten des flüssigen Wassers) besprochenen Mikroordnungsbereiche des Wassers durch die störend abweichenden Eigenschaften des $D_2O$ beeinträchtigt werden.

Tabelle 6. *Unterschiede der Daten von normalem* ($H_2O$), *schwerem* ($D_2O$) *und überschwerem Wasser* ($T_2O$). (Nach ROBINSON und BELL 1937, SEELWOOD 1941, EUCKEN 1949.)

| Wasserart | $H_2O$ | $D_2O$ | $T_2O$ |
|---|---|---|---|
| Schmelzpunkt, °C . . . . | 0 | +3,8 | +10 |
| Siedepunkt, °C . . . . . | +100 | +101,4 | +102,8 |
| Spezifisches Gewicht. . . | 1,0000 | 1,1078 | 1,33 |
| Subl.-Wärme des Eises bei 0°, kcal/mol . . . . | 12,170 | 12,631 | — |

Um festzustellen, in welchem Maße der Gehalt an schwerem Wasser sich auf das Wachstum von lebenden Zellen auswirkt, haben BARNES und LARSON (1933) ähnliche Versuche mit $D_2O$-haltigem Wasser ausgeführt, wie sie von BARNES und BARNES (1932) für Schmelz- und Kondenswasser beschrieben wurden. Stark angereichertes Schwerwasser erwies sich für die Alge *Spirogyra* als schnell tödlich, während sie sich in nur wenig angereichertem Wasser (D = 1,000061) besonders gut entwickelt haben soll. Da hier wie bei den Kontrollversuchen mit nur je einem Zellfaden gearbeitet wurde, kommt den Versuchen wohl nur qualitative Bedeutung zu. Interessanter sind Versuche an Fermentreaktionen im $D_2O$-Milieu, da sie spezifischer sind als die Summe der Lebenserscheinungen von Zellen. Dabei fanden BARNES und LARSON (1933), daß die Pankreasamylase in schwerem Wasser viel langsamer verzuckert als in normalem Wasser, so daß das Erythrodextrinstadium ceteris paribus erst nach 8 statt nach 6 min erreicht wurde. Auch in Fermentationsversuchen mit Zymase in $D_2O$ wurde eine um rund 10% geringere $CO_2$-Entwicklung beobachtet.

Den Versuchen kommt der Wert zu, nachgewiesen zu haben, daß zwischen normalem und schwerem Wasser physiologisch beachtliche Unterschiede bestehen. Ob dagegen in Schwerwasser sich überhaupt lebende Substanz normal verhalten kann, geht aus ihnen nicht hervor. Dafür wäre es erforderlich, die Objekte nicht aus normalem in schweres Wasser überzuführen, sondern von vornherein in letzterem über mehrere Generationen zu züchten, damit zelleigener Wasserstoff und der des Wassermilieus identisch werden. Auf solche Lebewesen wird normales Wasser vermutlich ebenso schädlich wirken wie bei den umgekehrten Verhältnissen der Versuche von BARNES und LARSON.

An dieser Stelle sei schließlich noch erwähnt, daß auch das Tritium in außerordentlich geringen Mengen ($T:H \sim 10^{-14}$) in der Natur vorkommt; es wird hier unter dem Einfluß der kosmischen Strahlung aus dem $^{14}N$ der Atmosphäre gebildet und gelangt so in den Wasserkreislauf. Es kann, ähnlich wie $^{14}C$, zur radiometrischen Altersbestimmung biologischen Materials herangezogen werden (LIBBY 1946, FIREMAN 1953, GROSSE and KIRSHENBAUM 1954).

## Schlußbemerkungen.

Es herrscht weitgehende Übereinstimmung darüber, daß in Lösungen von Elektrolyten und Nichtelektrolyten eine erhebliche Strukturänderung des Wassers erfolgt. Sie beruht einerseits auf der Bedeckung der Fremdmoleküle oder Ionen mit Wassermolekülen auf Grund elektrostatischer Anziehung oder Wasserstoffbrückenbildung. Solches, zwar vorwiegend an keine festen Strukturen, sondern an gelöste Teilchen gebundenes Wasser, wäre wohl im weiteren Sinne als

„bound water" zu betrachten. Zum anderen aber wird vor allem durch Ionenfelder auch die Fernorientierung im Wasser gestört, wobei sich wesentliche Eigenschaften, wie Viscosität usw. ändern können.

Es ist daher der Auffassung von VLÈS (1938) die Zustimmung nicht vorzuenthalten, wenn er feststellt, daß von vielen Bearbeitern die Physikochemie der Lösungen auch heute noch so behandelt wird, als wenn das Wasser ein unveränderliches Lösungsmittel wäre; er stellt fest, daß die auf VAN T'HOFF zurückgehende Methode, das Verhalten verdünnter wäßriger Lösungen mit dem gasförmigen Zustand zu vergleichen, nicht mit den experimentellen Befunden in Einklang zu bringen ist und als überholt betrachtet werden kann. In dieser Hinsicht sind in den letzten Jahren erhebliche Fortschritte zu verzeichnen gewesen.

Bei der weiteren experimentellen Bearbeitung des Wasserproblems wäre eines noch wünschenswert: Alle bisherigen für Wasser gewonnenen Angaben beziehen sich, soweit ersichtlich, auf solches der natürlichen Zusammensetzung, das ja isotopisch und damit eigenschaftsmäßig heterogen ist. Da aber gerade in kristallinen oder irgendwie kristallähnlichen Substanzen schon kleinste Störstellen, die von Bausteinen mit abweichenden Eigenschaften besetzt sind, bedeutende Fernwirkungen entfalten können, werden Messungen an reinstem $H_2^{16}O$ sicherlich in dem einen oder anderen Falle ein übersichtlicheres Bild ergeben, als es bis heute für den Sammelbegriff Wasser vorliegt.

## Literatur.

ANGUS, W. R., and W. K. HILL: Diamagnetism and hydrogen bond. Trans. Faraday Soc. **36**, 923—927 (1940). — ASTBURY, W. T.: The hydrogen bond in protein structures. Trans. Faraday Soc. **36**, 871—880 (1940).

BARKAS, W. W.: Wood-water relationships. V. The hydrostatic compressibility of the wood-water aggregate. Trans. Faraday Soc. **36**, 824—834 (1940). — BARNES, H. T., and T. C. BARNES: Biological effect of associated water molecules. Nature (Lond.) **129**, 691 (1932). — BARNES, T. C., and E. J. LARSON: Further experiments on the physiological effect of heavy water and of ice water. J. Amer. Chem. Soc. **55**, 5059—5060 (1933). — BAWN, C. E. H., E. L. HIRST and G. T. YOUNG: The nature of the bonds in starch. Trans. Faraday Soc. **36**, 880—885 (1940). — BERNAL, J. D.: Diskussionsbemerkungen in der General Discussion „The Hydrogen Bond". Trans. Faraday Soc. **36**, 886 (1940). — BERNAL, J. D., and R. H. FOWLER: A theory of water and ionic solution, with particular reference to hydrogen and hydroxyl ions. J. Chem. Physics **1**, 515—548 (1933). — BJERRUM, N.: Dissoziationskonstanten von mehrbasischen Säuren und ihre Anwendung zur Berechnung molekularer Dimensionen. Z. physik. Chem. **106**, 219—242 (1923). — BJERRUM, N., u. A. UNMACK: Elektrometrische Messungen mit Wasserstoffelektroden in Mischungen von Säuren und Basen mit Salzen. Math. Fysiske Meddelelser **9**, 1—206 (1929). — BOSSCHIETER, G., et J. ERRERA: Spectre d'absorption infrarouge de l'eau liquide, solide et en solution. C. r. **204**, 1719—1721 (1937). — BRAGG, Sir W.: Ice. Bericht des Roy. Inst. of Gr. Brit. weekly even. Meet., 18. März 1938, S. 1—37. — BROIDA, H. P.: Schnelle und genaue Analyse von $D_2O$—$H_2O$-Gemischen mittels optischer Spektroskopie. Angew. Chem. **66**, 146 (1954).

CHADWELL, H. M.: The molecular structure of water. Chem. Rev. **4**, 375—398 (1927). — CLAEYS, J., J. ERRERA and H. SACK: Absorption of ultrasonic waves in liquids. Trans. Faraday Soc. **33**, 136—141 (1937). — CROSS, P. C., J. BURNHAM and P. A. LEIGHTON: The Raman spectrum and the structure of water. J. Amer. Chem. Soc. **59**, 1134—1147 (1937).

DAHR, N.: Verbindung des gelösten Körpers und des Lösungsmittels in der Lösung. Z. Elektrochem. **20**, 57—81 (1914). — DORSEY, N. E.: Properties of ordinary water-substance. New York: Reinhold Publishing Corporation 1940. — DRUDE, P., u. W. NERNST: Über Elektrostriktion durch freie Ionen. Z. physik. Chem. **15**, 79—85 (1894).

EBERT, L.: Bemerkung zur zweiten Dissoziationskonstante des Wassers. Mh. Chem. **80**, 788—789 (1949). — ERICHSEN, L. v.: Die kritischen Lösungstemperaturen in der homologen Reihe der primären normalen Alkohole. Brennst.-Chem. **33**, 166—172 (1952). — ERRERA, J.: Structure of liquids studied in the infra-red. Trans. Faraday Soc. **33**, 120—129 (1937). (I). — La structure de l'eau dans l'infra-rouge. J. Chim. physique **34**, 617—626 (1937). (II). — EUCKEN, A.: Grundriß der Physikalischen Chemie. 6. Aufl. Leipzig: Akademische Verlags-Gesellschaft 1944. — Zur Kenntnis der Konstitution des Wassers. Nachr.

Akad. Wiss. Göttingen, Math.-physik. Kl., Math.-physik.-chem. Abt. **1946**, 38—48. — Der Einfluß gelöster Stoffe auf die Konstitution des Wassers. Nachr. Akad. Wiss. Göttingen, Math.-physik. Kl., Math.-physik.-chem. Abt. **1947**, 33—36. — Ionenhydrate in wäßriger Lösung. Z. Elektrochem. **51**, 6—24 (1948). (I). — Assoziation in Flüssigkeiten. Z. Elektrochem. **52**, 255—269 (1948). (II). — Unterschiede zwischen den thermisch-kalorischen Eigenschaften des schweren und leichten Wassers. Nachr. Akad. Wiss. Göttingen, Math.-physik. Kl., Math.-physik.-chem. Abt., **1949**, 1—11.

FINBACK, C., u. H. VIERVOLL: The structure of liquids. II. The structure of liquid water. Tidsskr. Kjemi, Bergvesen og Metallurgi **3**, 36—40 (1943). — FIREMAN, E. L.: Measurements of the (n, ³H) cross section in nitrogen and its relationship to the tritium production in the atmosphere. Physic. Rev. **91**, 922—926 (1953). — FORBES, G. S.: Water: Some interpretations more or less recent. J. Chem. Educat. **18**, 18—24 (1941). — FOX, J. J., and A. E. MARTIN: Investigations of infrared spectra (2,5—7,5 μ). Absorption of water. Proc. Roy. Soc. Lond. **174**, 234—262 (1940). — FRICKE, R.: Über Molekel- und Ionenhydrate. Z. Elektrochem. **28**, 161—181 (1922). — FRUHWIRTH, O.: Versuche zur Deutung der Assoziation des Wassers auf Grund dielektrischer Polarisationsmessungen. Mh. Chem. **70**, 157—167 (1937).

GANZ, E.: Über das Absorptionsspektrum von Wasser, wäßrigen Lösungen und Alkoholen zwischen 0,70—0,95 μ). Ann. Physik [5] **26**, 331—348 (1936). — Über das Absorptionsspektrum von wässerigen Lösungen zwischen 0,70—0,90 μ. Z. physik. Chem. [B] **35**, 1—10 (1937). (I). — Über das Absorptionsspektrum von flüssigem Wasser zwischen 2,5 μ und 6,5 μ. Ann. Physik [5] **28**, 445—457 (1937). (II). — GEFFCKEN, W.: Über die scheinbaren Molvolumina gelöster Elektrolyte. Z. physik. Chem. [A] **155**, 1—28 (1931). — GIRARD, P., et P. ABADIE: Interactions moléculaires et structure des liquides. C. r. **202**, 398—400 (1936). — GROSSE, A. V., and KIRSHENBAUM, A. D.: The natural tritium content of atmospheric hydrogen. Physic. Rev. **93**, 250—251 (1954). — GROVES, L. G., and S. SUGDEN: The dipole moments of vapours. J. Chem. Soc. (Lond.) **1935**, 971—974.

HARNED, H. S., and M. A. COOK: The ionic activity coefficient product and ionization of water in univalent halide solutions. A numerical summary. J. Amer. Chem. Soc. **59**, 2304—2305 (1937). — HARNED, H. S., and C. G. GEARY: The ionic activity coefficient product and ionization of water in barium chloride solution from 0 to 50⁰. J. Amer. Chem. Soc. **59**, 2032—2035 (1937). — HARNED, H. S., and R. A. ROBINSON: A note on the temperature variation of the ionization constants of weak electrolytes. Trans. Faraday Soc. **36**, 973—978 (1940). — HATSCHEK, E.: Die Viscosität der Flüssigkeiten. Dresden und Leipzig: Theodor Steinkopff 1929. — HEDESTRAND, G.: Die Berechnung der Molekularpolarisation gelöster Stoffe bei unendlicher Verdünnung. Z. physik. Chem. [B] **2**, 428—444 (1929). — HONIGMANN, E. J. M.: Die Möglichkeit der Bildung komplexer Moleküle. Naturwiss. **20**, 635—639 (1932). — HÜCKEL, W.: Anorganische Strukturchemie. Stuttgart: Ferdinand Enke 1938. — HUGGINS, M. L.: Hydrogen bridges in organic compounds. J. of Org. Chem. **1**, 407—456 (1937).

KAST, W., u. A. PRIETZSCHK: Die molekulare Struktur des unterkühlten und des glasigen Äthylalkohols. Z. Elektrochem. **47**, 112—116 (1941). — KATZOFF, S.: X-ray studies of the molecular arrangements of liquids. J. Chem. Physics **2**, 841—852 (1934). — KÖNIG, H.: Elektroneninterferenzen an Eis. Nachr. Akad. Wiss. Göttingen, Math.-physik. Kl., Math.-physik.-chem. Abt., **1942**, 1—6. — Eine kubische Eismodifikation. Z. Kristallogr. [A] **105**, 279—286 (1943). — KORTÜM, G.: Elektrolytlösungen. Leipzig: Akademische Verlagsgesellschaft 1941. — KUJUMZELIS, T. G.: Über die Änderung der Struktur des Wassers durch Ionen. Z. Physik **110**, 742—759 (1938).

LASSETTRE, E. N.: The hydrogen bond and association. Chem. Rev. **20**, 259—303 (1937). LATIMER, W. M., and W. H. RODEBUSH: Polarity and ionization from the standpoint of the Lewis theory of valence. J. Amer. Chem. Soc. **42**, 1419—1433 (1920). — LIBBY, W. F.: Atmospheric helium three and radiocarbon from cosmic radiation. Physic. Rev. **69**, 671—672 (1946). — LUYET, B. J.: The existence of a vitreous state in gelatine gels. Physic. Rev. [2] **53**, 323 (1938). — LUYET, B. J., and P. M. GEHENIO: Life and death at low temperatures. Normandy, Miss.: Biodynamica 1940.

MAGAT, M.: Recherches sur le spectre Raman et la constitution de l'eau liquide. Ann. Physique [11] **6**, 108—193 (1936). — Raman spectra and the constitution of liquids. Trans. Faraday Soc. **33**, 114—120 (1937). — MALSCH, J.: Die Struktur der Dipolflüssigkeiten. Ann. Physik [5] **29**, 48—60 (1937). — MATTAUCH, J., u. A. FLAMMERSFIELD: Isotopenbericht. Tübingen: Verlag der Z. Naturforsch. 1949. — MECKE, R.: Intensitätswechsel im Rotationsschwingungsspektrum des Wasserdampfes. Naturwiss. **20**, 657 (1932). — Das Rotationsschwingungsspektrum des Wasserdampfes. Z. Physik **81**, 313—331 (1933). — Zur Thermodynamik der Wasserstoffbrückenbildung. Z. Elektrochem. **52**, 107—110 (1948). (I). — Kräfte in Flüssigkeiten. Z. Elektrochem. **52**, 269—282 (1948). (II). — MOESVELD, A. L. TH.,

u. H. J. Hardon: Zur Kenntnis der Elektrostriktion. Z. physik. Chem. [A] 155, 238—256 (1931).

Ölander, A.: Studien über Brombernsteinsäure I. Einleitende Untersuchungen über das Ionenprodukt des Wassers und einige andere Dissoziationskonstanten. Z. physik. Chem. [A] 144, 49—72 (1929).

Pauling, L.: The nature of the chemical bond, 1. Aufl. Ithaca, New York: Cornell University Press 1939.

Ramsay, W.: Die Komplexität und Dissoziation von Flüssigkeitsmolekeln. Z. physik. Chem. 15, 106—116 (1894). — Rehbinder, P.: Wasser als oberflächenaktiver Stoff. Oberflächenaktivität und Adsorptionskräfte. II. Z. physik. Chem. 121, 103—126 (1926). — Robertson, J. M.: The formation of intermolecular hydrogen bonds. Trans. Faraday Soc. 36, 913—921 (1940). — Robinson, R. A., and R. P. Bell: The thermal molal volume of water and deuterium oxide in dioxan solution. Trans. Faraday Soc. 33, 650—652 (1937). — Röntgen, W. C.: Über die Konstitution des flüssigen Wassers. Ann. Physik 45, 91—97 (1892).

Schäfer, K.: Die Struktur der Flüssigkeiten. Z. Elektrochem. 52, 245—254 (1948). — Schmidt, W., u. E. Baier: Lehrbuch der Mineralogie. Berlin: Gebr. Bornträger 1935. — Schreiner, E.: Zur Hydratation einwertiger Ionen. Z. anorg. Chem. 135, 333—369 (1924). — Schupp, R. L., u. R. Mecke: Dielektrische Präzisionsmessungen an Lösungen assoziierender Stoffe. Z. Elektrochem. 52, 54—60 (1948). — Seelwood, P. W.: Heavy water. J. Chem. Educat. 18, 515—520 (1941). — Souci, S. W.: Die Chemie und Physik des reinen Wassers. Ber. Internat. Balneol. Kongr. d. ISMH 1952, S. 16—24. — Stackelberg, M. Frh. v.: Feste Gashydrate. Z. Elektrochem. 58, 25—39 u. 40—45 (1954). — Staronka, L.: Otrzymywanie bezpostaciowej odmiany wody przez kondensację pary wodnej w niskiej temperaturze. Ann. Soc. Chim. Polon. 19, 201—212 (1939). — Stewart, G. W.: Effect of ionic forces shown by the liquid structure of alkali halides and their aqueous solutions. Trans. Faraday Soc. 33, 238—247 (1937). — The variation in the structure of water in ionic solutions. J. Chem. Physics 7, 869—877 (1939). — Suhrmann, R., u. F. Breyer: Untersuchungen im ultraroten Absorptionsspektrum über die Änderung des Lösungsmittels durch die gelöste Substanz. Z. physik. Chem. [B] 20, 17—53 (1933). — Suhrmann, R., u. R. Walter: Über die Beeinflussung des Lösungsvermögens von Methanol für n-Hexan und verwandte Kohlenwasserstoffe durch Veränderung seines Assoziationszustandes. Abh. Braunschw. Wiss. Ges. 3, 135—152 (1951). — Sutherland, G. B. B. M.: The investigation of hydrogen bonds by means of infra-red absorption spectra. Trans. Faraday Soc. 36, 889—897 (1940). — Symposium 1910: Symposium on the constitution of water. Trans. Faraday Soc. 6, 71—123 (1910).

Tammann, G.: Über die Grenzen des festen Zustandes. Ann. Physik [4] 2, 1—31 (1900). — Aggregatzustände (Die Änderung der Materie in Abhängigkeit von Druck und Temperatur), 2. Aufl. Leipzig: L. Voß 1923. — Zur Kenntnis der molekularen Zusammensetzung des Wassers. Z. anorg. Chem. 158, 1—16 (1926). — Die abnormen Abhängigkeiten der Eigenschaften des Wassers von der Temperatur und dem Druck. Z. anorg. Chem. 235, 49—61 (1937). — Thomsen, J.: Über das Molekulargewicht des flüssigen Wassers. Ber. dtsch. chem. Ges. 18, 1088 (1885).

Ubbelohde, A. R.: Diskussionsbemerkung in der General Discussion „The Hydrogen Bond". Trans. Faraday Soc. 36, 886 (1940). — Ulich, H.: Elektrische Leitfähigkeit von Flüssigkeiten und Lösungen; Hand- und Jahrbuch der Chemischen Physik (Herausgeg. A. Eucken und K. L. Wolf), Bd. 6/II. Leipzig: Akademische Verlagsgesellschaft 1933. — Neue Anschauungen über die Besonderheiten des Wassers und der wässerigen Lösungen. Z. angew. Chem. 49, 279—288 (1936). — Kurzes Lehrbuch der Physikalischen Chemie, 3. Aufl. Dresden u. Leipzig: Theodor Steinkopff 1941.

Vlès, F.: Les données actuelles sur la constitution et les propriétés physico-chimiques de l'eau. Arch. Physique Biol. 15, 33—85 (1938—1942). — Volkmann, H.: Flüssigkeitsstrukturen. Physikalische Chemie; Bd. 30 der Reihe Naturforschung und Medizin in Deutschland 1939—1946 (Fiat Review of German Science); S. 151—182. Wiesbaden: Dieterichsche Verlagsbuchhandlung 1948.

Whiting, T.: A theory of cohesion. Diss. Harvard University, Cambridge, Mass. 1884. — Wolf, K. L., H. Frahm u. H. Harms: Über den Ordnungszustand der Moleküle in Flüssigkeiten. Z. physik. Chem. [B] 36, 237—287 (1937).

Zimm, B. H.: Opalescence of a two-component liquid system near the critical mixing point. J. Physic. Coll. Chem. 54, 1306—1317 (1950).

# Water content and water turnover in plant cells.

By

## Paul J. Kramer.

With 5 figures.

## Introduction.

Water is quantitatively the most abundant constituent of actively growing, metabolizing cells, and reduction of the water content of plant cells below normal results in cessation of growth and interference with most physiological processes. Further reduction in water content below a certain critical point which varies with the species and the condition of the cells, results in death from dehydration. The importance of water can be better appreciated if we list its various roles in plants. First of all water is an essential constituent of living protoplasm, often constituting 80 to 90% of its total weight in active cells. Water is also a reagent in various chemical reactions, particularly in hydrolytic reactions and in photosynthesis, and it is the solvent in which all movement of materials into and out of cells and from cell to cell occurs. Most cell walls are permeated with water, hence it forms a continuous system extending throughout the plant which probably serves as a pathway for considerable lateral translocation. In these ways it affects the rate of all metabolic processes. Water also plays an essential role in maintaining the form and structure of herbaceous plant tissues through maintenance of cell turgidity.

The water content of the various cells and tissues is in a continual state of change, varying from hour to hour, day to day, and season to season. To understand fully how and why such changes in water content occur requires consideration of the chemical and physical characteristics of various parts of the cell, of the forces which hold water in cells, and of the principles which govern water movement between cells and their environment. It is quite impossible to review all of the important literature in this article, but the important theories and viewpoints are presented. Some of these topics will be discussed in detail in other sections of the Handbook and the reader is also referred to CRAFTS, CURRIER and STOCKING (1949) for a more detailed discussion of some of them.

## The water content of plant cells.

The wide range of water contents characteristic of various types of plant tissue are illustrated by the data in Table 1. Regions composed chiefly of rapidly enlarging cells such as young roots and the inner leaves of lettuce often have water contents in excess of 90% of their fresh weight and herbaceous stems often contain only slightly less water. Mature leaves frequently contain less water than herbaceous stems and the data of WILSON, BOGGESS and KRAMER (1953) show that the root systems of sunflower have a lower water content than the leaves and stems. This fact, and also the diurnal fluctuations in water content characteristic of an herbaceous plant, are shown in Fig. 1. The water content of tree trunks is lower than that of leaves, but varies considerably with species, season, and position in the tree (BUSGEN 1911, GIBBS 1935, RABER 1937). HUCKEN-PAHLER (1936) reported the water content of phloem of *Pinus echinata* to be about 66% of the fresh weight, that of the outer two annual rings to vary from 50% at 1 ft. to nearly 75% at 25 ft. above ground level, while the center of the stem contained 52% of water at 1 ft. and 57% at 19 ft., the highest level at which it could be sampled.

Seeds of some species are relatively high in water, but many seeds normally dry down to less than 10% moisture without loss of viability, and will survive much longer if kept dry than if allowed to fluctuate in moisture content (CROCKER and BARTON 1953). Many mosses, liverworts, ferns, and lichens contain less than 10% of water in an air-dry condition and can survive for months in this dehydrated condition. One of the more interesting and puzzling physiological phenomena is the fact that many seeds can be air dried and some can even be dried over concentrated sulphuric acid for weeks without injury, but if they are allowed to germinate the seedlings are killed by even slight dehydration. According to ILJIN (1953) resistance to dehydration is related to cell size, shape, ratio of volume to surface, osmotic pressure, and rate of dehydration.

## Forces holding water in cells.

Water is held in cells principally by osmotic and imbibitional forces. Osmotic forces exist because the addition of solutes to water decreases the concentration of water molecules and the solute molecules or ions also exert attractive forces on them. This causes a decrease in activity or free energy which results in a decrease in diffusion pressure of the water. Osmotic forces are effective only where a solution is separated from pure solvent or from a solution of different concentration by membranes permeable to water, but not to the solute. Imbibition occurs because of the attraction for water of various hydrophilic colloidal materials such as cellulose and proteins, and common examples are seen in water uptake by seeds, cell walls, wood, and gelatin. The water absorbed by these materials is oriented over the surface (adsorbed) and held in the microcapillaries, largely by surface forces which are strong enough to decrease the activity of the water molecules. Relatively dry colloidal materials may hold small amounts of water with forces of hundreds or even thousands of atmospheres, and its diffusion pressure is decreased correspondingly. An important characteristic of imbibition is that it results in swelling of the imbibing material, by forcing apart the micelles, fibrils and other submicroscopic units of which it is composed. Some water occurs as water of hydration, combined with molecules or ions of the solute, and it differs from water held by surface forces chiefly in the fact that it combines with the hydrated material in definite proportions, as five molecules of water per molecule of copper sulfate and six molecules per molecule of sucrose. Water held firmly by surface forces or as water of hydration is sometimes termed "bound water", because it is held so firmly that it cannot act as a solvent and cannot be frozen at ordinary temperatures. The nature and physiological significance of bound water will be discussed in the next chapter of this volume.

The important fact with respect to cell water relations is that the activity of water molecules is reduced in these several ways. This reduction in activity can be expressed as a reduction in specific free energy (BROYER 1947) or a reduction in diffusion pressure, but among plant physiologists it is more often expressed as an increase in suction force, suction tension, or diffusion pressure deficit (MEYER 1945, MEYER and ANDERSON 1952).

Because the term diffusion pressure deficit (commonly abbreviated as DPD) will be used frequently in this paper it will be defined at this time. Water under any particular conditions of temperature and pressure has a certain diffusion pressure or fugacity. Treatments such as the addition of solutes, the action of surface forces, decrease in temperature, decrease in turgor and wall pressure, and the imposition of tension all lower the diffusion pressure as compared to that of water in the original condition. It is difficult to measure the actual diffusion pressure of water, but the amount by which the diffusion pressure of water in a solution is reduced is termed its diffusion pressure deficit or DPD. The DPD of a solution is equal to its osmotic pressure, but in plant cells the DPD of the cell sap is greatly modified by the wall pressure and may decrease from a value equal to the osmotic pressure of the cell sap to zero as the turgor and wall pressure increase. The DPD of a cell is equal to the osmotic pressure of a solution in which it neither gains nor loses water. Methods of measuring it are discussed later in this paper.

Besides these purely passive physical forces which cause the osmotic intake and retention of water in plant cells some physiologists think that water is absorbed and held by "active" forces resulting from the expenditure of energy by the protoplasm itself. This active or nonosmotic absorption of water is thought to resemble the accumulation of ions by cells, at least to the extent that it occurs against a diffusion gradient and depends on energy provided by aerobic respiration. Much additional work needs to be done on the occurrence and nature of this type of water absorption and it will be discussed later in this paper, and in more detail in connection with water accumulation in Volume 2 of this series.

## Where water is held in cells.

At equilibrium the water in a plant cell is distributed among the several structures of the cell according to their relative diffusion pressure deficits. Some is held in the walls by imbibitional forces, some is held in the protoplasm (cytoplasm, nucleus, and plastids) by similar forces, but in most cells the larger part occurs in the vacuoles where it is held by osmotic forces. Obviously the water in these various parts of the cell is in dynamic equilibrium, and change in the diffusion pressure deficit of water in one part of the cell as when starch is hydrolyzed to sugar, or vice versa, results in change in the water content of other parts of the cell.

**Water in the cell wall.** The amount of water held in cell walls depends on their thickness, structure, and chemical composition. The middle lamella is composed principally or entirely of hydrophilic, colloidal, pectic compounds and the primary wall is also colloidal in nature. The cellulose which composes the framework of the primary wall is less hydrophilic than the pectic compounds often associated with it. STAMM (1944) states that only about 8% of water is held by surface bonds on the micelles of pure cellulose. In cellulose walls much water is held in the spaces between the micelles and fibrils. These spaces vary in size from less than 10 to 100 m$\mu$ in width and are interconnected, forming extensive channels in pure cellulose walls such as the secondary walls of cotton fibers. During maturation pectic compounds, lignin, suberin, cutin, and other substances are often deposited in the intermicellar spaces. Cell walls often show considerable changes in thickness with variation in water content. CRAFTS (1931) reported a 50% reduction in volume during drying of the walls of phloem cells from potato stolons.

As secondary thickenings of the walls occur they lose most of their elasticity and if cutin or suberin is deposited on or in the cellulose framework of the walls their permeability to water and their ability to imbibe it is greatly decreased. The mobility of water in cell walls is of interest in connection with water movement through plants. The water held by hydrogen bonding on the cellulose chains is relatively immobile, as is that in the smallest microcapillary spaces between fibrils, but that in the larger capillary spaces between the cellulose fibrils probably moves quite freely. Considerable water movement through the cell walls must occur in the wood, across the cambium, and in the mesophyll tissues of the leaves at the ends of the veins (STRUGGER 1938, ZIEGENSPECK 1945). Movement in cell walls may also be important in the passage of water from the epidermal cells to the xylem elements in roots. Because of the appreciable resistance of protoplasmic membranes to water movement it is possible that some of the water which enters roots by passive absorption moves along the cell walls to the xylem because this probably constitutes the part of least resistance. The possibility of movement of water and solutes along the walls of cells in roots has been discussed recently by HYLMO (1953).

**Water in the protoplasm.** In growing tissue water often constitutes 90% or more of the protoplasm, hence even in many fully vacuolated cells considerable water occurs in the protoplasm. In meristematic tissue and other cells with small vacuolar volume most of the water may be associated with the protoplasm. The amount of water present and the manner in which it occurs has important effects on such properties as fluidity, plasticity, and elasticity of protoplasm. There has been much speculation concerning the structure of protoplasm and detailed discussions may be found in GUILLIERMOND (1941), SEIFRIZ (1942), in a recent review by VIRGIN (1953), FREY-WYSSLING (1953), and in later sections of this volume. Several investigators have pointed out that cytoplasm shows both the fluidity characteristic of liquids and the elasticity characteristic of solids, *i.e.* it behaves both as a solid and a liquid. This seems consistent with the concept that it consists of two continuous phases, a liquid phase consisting of water, and a solid phase which consists largely of proteins and lipoidal material. According to FREY-WYSSLING (1953) the solid phase consists of a network of very thin strands of polypeptide chains with side chains, connected to each other by hydrogen bridges, ionic attraction, and van der Waals forces (SPONSLER and BATH 1942). The meshes of this network or molecular framework are occupied by water, oil droplets, and other insoluble materials, and various solutes. The high water content of cytoplasm results from the width of the spaces between the polypeptide framework. While some of the water is bound to the protein chains and side chains the spaces between the chains in fully hydrated protoplasm are so large that much of it is held so loosely that it can move fairly freely. SPONSLER, BATH and ELLIS (1940) concluded from x-ray and infra-red absorption data with gelatin that water molecules are held on protein chains by hydrogen bonding to oxygen and nitrogen atoms, both in the backbone of the chain and to lesser extent on the side chains. Their studies on gelatin indicate that there is a considerable increase in spacing and increased freedom of movement of water molecules as the water content increases from 30 to 35%, but little change occurs from 35 to 90% of water. Presence of a large amount of loosely bound water contributes to the fluidity and relatively low viscosity characteristic of most active protoplasm. Reduction in water content leaves only the more firmly bound portion of the water, causing decrease in fluidity and increase in viscosity and resulting finally in a gel condition such as is found in the protoplasm of dormant seeds. FREY-WYSSLING (1953) emphasizes the fact that all of the various components of cytoplasm must be oriented in an orderly fashion at all times. Maintenance of this arrangement or structure is essential to life and is dependent on energy supplied by respiration. Cytoplasm can never become a true sol, because if all the bonds are broken and it becomes liquified it no longer has the properties of living matter. Under certain conditions the cytoplasm may be regarded as a homogeneous system of proteins, lipids, and solutions, but equilibrium between these phases is precarious and often upset, resulting in what is sometimes termed "demixing" (Entmischung) or the concentration of one phase in almost pure form, as water in the vacuoles. Some details of this concept of protoplasmic structure have been criticized (LEPESCHKIN 1950), but no equally acceptable alternative has been proposed.

The nucleus and plastids ordinarily hold a small part of the total water content of the cell. It seems safe to assume that they hold water in the same ways in which it is held by the cytoplasm, but in much smaller amount. As they are surrounded by differentially permeable membranes they constitute tiny osmotic units within the protoplasm. Evidence has recently been presented showing that mitochondria also are surrounded by morphologically distinct membranes which may be differentially permeable (FARRANT, ROBERTSON and WILKINS 1952, PALADE 1953).

**Water in the vacuole.** The vacuoles of a plant cell vary in number, shape and size. In meristematic tissues they occur as numerous, tiny, rod-shaped, or thread-like structures scattered throughout the cytoplasm. As cells mature the vacuoles increase in size and coalesce to form a large central vacuole. In most mature plant cells they occupy a considerable part of the cell volume and contain most of the water. The vacuolar sap contains numerous organic compounds and is the principal site of accumulation of inorganic salts. Some of these substances are absorbed from the environment, others may have been synthesized by the protoplasm, while substances such as tannins and calcium oxalate crystals are probably merely byproducts of cell metabolism. GUILLIER-MOND (1941) and FREY-WYSSLING (1953) emphasize the role of vacuoles as accumulation regions into which are excreted various products of metabolism and even foreign substances such as dyes. Some of these materials, especially the inorganic salts and sugars occur in true solution, but others such as proteins, tannins, mucilages and dextrins occur in the colloidal condition. As a result the viscosity of the vacuolar sap is normally about twice that of water (FREY-WYSSLING 1953) and sometimes it even forms a gel (GUILLIERMOND 1941).

The physical and chemical characteristics of the vacuole will be discussed in more detail in later sections of this volume.

**Vacuolar versus protoplasmic sap.** Some attempts have been made to determine how much of the water in a cell occurs in the protoplasm and how much in the vacuole. CHIBNALL (1923) exposed spinach leaves to ether, then expressed the sap, which he believed came almost entirely from the vacuoles. The water remaining in the press cake was regarded as being held by the wall and the protoplasm. This seems questionable because injury or death of the protoplasm almost certainly is accompanied by the loss of considerable water from its structure. MASON and PHILLIS (1939) pressed out all of the sap obtainable at about 14,000 lbs. per sq.in. from carefully stacked cotton leaves, then froze the residue and expressed the remaining sap. The sap obtained before freezing was regarded as vacuolar sap, that obtained from the frozen residue was believed to come from the disintegrating protoplasm. The concentration of potassium and chlorine in the vacuolar sap was only about one-fifth as great as in the socalled protoplasmic sap. MASON and PHILLIS (1939) estimated that only 30% of the total water in these cells occurred as vacuolar sap. There is serious doubt if this method gives the separation of sap claimed for it. Application of pressure almost certainly displaces some of the loosely bound water occupying the larger spaces between the cytoplasmic fibrils and this inevitably becomes mixed with the vacuolar sap which is filtering through. It also appears probable that some of the solutes are filtered out of the vacuolar sap as it passes through the finer structure of the cytoplasm because it is very unlikely that the concentration of chlorine and potassium is actually five times greater in the cytoplasm than in the vacuole.

A number of attempts have been made to measure the volume of the vacuole in living cells, but with divergent results. It would be inferred from the data of MASON and PHILLIS that the vacuoles occupy less than 30% of the total volume of the parenchyma cells of cotton leaves. On the other hand SCARTH and LEVITT (1937) estimated that the vacuoles occupy 42 to 58% of the total volume of the cortical parenchyma cells of apple twigs, and CRAFTS, CURRIER and STOCKING (1949) state that in cells of storage parenchyma the vacuole often occupies 70 to 90% of the total cell volume. It seems probable that in most mature, active plant tissue the volume of water in the vacuoles amounts to over half of the total cell volume. In meristematic tissues vacuoles are small and most of the water in such tissues must occur in the protoplasm where it is held by imbibitional forces. BUTLER (1953a) estimated that in wheat roots at least 15% of the total root volume is occupied by water which occurs outside of the protoplasts, probably largely in the cell walls. In another study it was estimated that 70 to 75% by weight of the water in young wheat roots occurs in the vacuoles, 5% in the cytoplasm and 20 to 25% in the walls (BUTLER 1953b). The water in the walls and possibly that in the cytoplasm forms a continuous system.

## Water balance within the cell.

It obviously is impossible for the wall, protoplasm, and vacuole to attain equality with each other in respect to water content, but they do tend to approach equality with respect to the osmotic and imbibitional forces with which water is held, i.e. they tend to approach equilibrium with respect to the free energy or diffusion pressure deficit of the water in their various parts. Any change in diffusion pressure of one part of the cell will be followed by readjustment in water content of other parts of the cell until a new diffusion pressure deficit equilibrium is attained. For example, as water begins to evaporate from the wall of the mesophyll cells of a transpiring leaf a diffusion pressure deficit develops in the walls, causing water to move from the adjacent protoplasm into the wall. This in turn produces a diffusion pressure deficit in the protoplasm which causes movement of water from the vacuole into the protoplasm. The diffusion pressure deficit developed in this cell will cause an influx of water from neighboring cells or xylem elements. If a cell possessing a diffusion pressure deficit is supplied with water, absorption will occur until the diffusion pressure deficits of all its parts are completely satisfied.

If starch or some other insoluble compound is converted into a soluble compound the diffusion pressure deficit of the vacuole will be increased and water will move from the protoplasm and wall into the vacuole until a new equilibrium of diffusion pressure deficit is attained. Conversion of solute into an insoluble compound, as sugar to starch and nitrate into protein, its removal by translocation, or its utilization, as in respiration, will result in decrease in diffusion pressure deficit and loss of water. Lundegårdh (1950) has discussed the importance of such changes in wheat roots.

Osterhout (1945) reported that the chloroplasts of *Nitella* and *Spirogyra* sometimes lose water to other parts of the cell and show noticeable decrease in volume. Following immersion of Nitella cells in 0.5 M. NaCl solution water moved out of the vacuole into the cytoplasm, causing shrinkage of the former and swelling of the latter. Contraction of the vacuoles, accompanied by swelling of the protoplasm, but with little or no change in total cell volume has been reported by others. Bogen (1951) found that a dilute solution of LiCl causes vacuolar contraction in cells of onion epidermis, but if the solution is made hypertonic with sucrose cap plasmolysis occurs. Küster (1950) stated that so much water sometimes moves out of the vacuoles into the cytoplasm that the vacuoles solidify and break up. Crafts, Currier and Stocking (1949) discuss these shifts in water content and suggest that at least some of them are caused by changes in the imbibitional forces of the protoplasm, but it appears likely that osmotic forces are involved in the vacuolar contraction observed in Nitella cells by Osterhout. These and other disturbances of the usual osmotic relationships indicate how easily the sensitive internal water balance of the various parts of a cell can be disturbed.

Changes in rate and nature of metabolic processes and in kind of metabolic products, as well as changes in the environment of a cell, affect the diffusion pressure deficit of the various parts differently. Changes in $p_H$ and in kinds of ions present in the environment may also cause internal changes in water distribution by affecting the imbibitional capacity of the protoplasm differently from that of the wall, or by affecting the permeability of cell membranes. A condition of static equilibrium therefore is seldom attained and never long maintained, and water probably moves back and forth among the various parts of the cell more or less constantly as a result of the tendency to maintain an equilibrium in the diffusion pressure deficits of the various parts of the cell.

## Changes in total water content of cells and tissues.

Quantitative measurement of changes in water content of individual cells in situ usually is impossible but the magnitude of such changes can be inferred from measurement of the changes in water content of samples of tissue large enough to be weighed easily. Water content varies with the age of the tissue, whether active or dormant, woody or herbaceous, and in any given tissue it varies from day to night, from season to season, and with the water supply. Some data on water content were presented in an earlier section of this paper (Table 1 and Fig. 1), but some additional data will now be presented.

**Age of tissue.** The water content of meristematic cells appears to be lower than that of cells which are enlarging. This is shown in Table 2, based on a study of the growing regions of roots of barley *(Hordeum vulgave)* and loblolly pine *(Pinus taeda)*. Water content is low in the meristematic region because the volume of the vacuoles is very small, but during cell enlargement the volume of the vacuoles increases more rapidly than the volume of dry matter, hence the water content increases. As the cells differentiate and mature the walls thicken, the proportion of dry matter increases and the water content decreases somewhat. Finally, because of changes in solute content and probably because of changes in the water-holding capacity of the protoplasm accompanying senescence, cells in the older tissues begin to lose water to younger tissues, and often dry out and die. Young cotton bolls can absorb water even from a wilted

Table 1. *Water content of various plant tissues expressed as percentage of fresh weight.*

|  | Plant parts | Water content | Authority |
|---|---|---|---|
| Roots | Barley, apical portion | 93.0 | KRAMER and WIEBE 1952 |
|  | *Pinus taeda*, apical portion | 90.2 | HODGSON 1953 |
|  | *Pinus taeda*, mycorrhizal roots | 74.8 | HODGSON 1953 |
|  | Carrot, edible portion | 88.2 | CHATFIELD and ADAMS 1940 |
|  | Sunflower, av. of entire root system | 71.0 | WILSON *et al.* 1953 |
| Stems | Asparagus stem tips | 88.3 | DAUGHTERS and GLENN 1946 |
|  | Sunflower, 7 wks. old av. of entire stems | 87.5 | WILSON *et al.* 1953 |
|  | *Pinus banksiana* | 48–61 | RABER 1937 |
|  | *Pinus echinata*, phloem | 66.0 | HUCKENPAHLER 1936 |
|  | *Pinus echinata*, wood | 50–60 | HUCKENPAHLER 1936 |
|  | *Pinus taeda*, twigs | 55–57 | McDERMOTT 1941 |
| Leaves | Lettuce, inner leaves | 94.8 | CHATFIELD and ADAMS 1940 |
|  | Sunflower, av. of all leaves on 7 wks. old plant | 81.0 | WILSON *et al.* 1953 |
|  | Cabbage, mature | 86.0 | MILLER 1938 |
|  | Corn, mature | 77.0 | MILLER 1938 |
| Fruits | Tomato | 94.1 | CHATFIELD and ADAMS 1940 |
|  | Watermelon | 92.1 | CHATFIELD and ADAMS 1940 |
|  | Strawberry | 89.1 | DAUGHTERS and GLENN 1946 |
|  | Apple | 84.0 | DAUGHTERS and GLENN 1946 |
| Seeds | Sweet corn, edible | 84.8 | DAUGHTERS and GLENN 1946 |
|  | Field corn, dry | 11.0 | CHATFIELD and ADAMS 1940 |
|  | Barley, hullless | 10.2 | CHATFIELD and ADAMS 1940 |
|  | Peanut, raw | 5.1 | CHATFIELD and ADAMS 1940 |

plant, probably because of the high imbibitional forces of the sligthly vacuolated young cells resulting in development of very high diffusion pressure deficits (ANDERSON and KERR 1943). Apparently young rapidly growing stem tips can also develop higher diffusion pressure deficits than older parts of the plant. WILSON (1948), for example, found that tomato stem tips continued to elongate even though the stems as a whole shrank because of an internal water deficit. The stem tips of uprooted plants sometimes continue to grow a few days at the expense of water in older parts of the plants.

**Growing versus dormant tissue.** Decrease in water content of the cells almost always occurs as tissues cease to grow and become dormant. Many investigators have reported that the moisture content of twigs and buds decreases in the autumn as growth ceases and they become dormant. An example of this is the work of STARK (1936) on apple stems shown in Fig. 3 of the section on bound water, where other examples are cited. Equally characteristic is the decrease in moisture content of maturing seeds. An example of this is shown in Fig. 2, taken from the work of SHIRK (1942) on rye and wheat seeds.

Table 2. *Water content of various regions of a growing root. Data for barley from* KRAMER and WIEBE *(1952), that for Pinus taeda is unpublished data of the same authors.*

| Distance from root tip in mm | Water content as percentage of fresh weight | |
|---|---|---|
|  | Barley | *Pinus taeda* |
| 0–2 | 89.4 | 82.4 |
| 2–4 | 93.4 | 89.9 |
| 4–6 | 94.1 | 91.3 |
| 6–8 | 94.0 | 92.2 |
| 8–10 | 94.2 | 91.6 |
| 18–20 | 93.5 | 90.2 |
| 48–50 | 93.2 | 87.0 |

**Diurnal variations in water content.** Marked diurnal variations in water content occur in the cells of transpiring plants, the water content decreasing during the day and increasing at night. These variations have been studied by many workers and much of the data have been summarized by MAXIMOV (1929) and KRAMER (1949). Typical curves for an herbaceous plant growing in well watered soil are shown in Fig. 1 from WILSON, BOGGESS and KRAMER (1953). The sharp decreases in water content observed during the day are obviously the result of heavy water loss by transpiration which causes withdrawal of water from the stem and roots more rapidly than it can be absorbed. As a result a water deficit is produced in the plant which usually is satisfied during the night when absorption generally exceeds water loss (KRAMER 1937). In many instances cells of leaves are able to absorb water from fruits, resulting in midday reduction in rate of growth or even shrinkage of fruits when transpiration is rapid (see KRAMER 1949). Apparently cells of leaves can develop higher diffusion

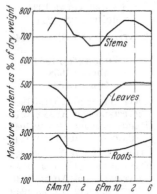

Fig. 1. Diurnal fluctuations in water content of various parts of field grown sunflower plants. Water content is expressed as percentage of dry weight. From WILSON, BOGGESS and KRAMER (1953).

pressure deficits than cells of fruits. Mature cotton bolls also lose water to the leaves during midday periods of rapid transpiration, but young cotton bolls do not (ANDERSON and KERR 1943).

SHREVE (1916) reported diurnal variations in water holding capacity of cactus tissue, with a maximum during the day and a minimum at night. The diurnal variations in water content of guard cells are well known, but not completely explained. PHILLIS and MASON (1945) reported marked diurnal variations in water content of discs of cotton leaf tissue floated on water, but WEATHERLEY (1947) attributed these to the fact that water loss exceeded absorption at midday, because they were eliminated when the discs were floated on water in a humid atmosphere.

Autonomic changes in water content. In addition to the diurnal variations in water content already discussed, other changes have been observed which are more or less rhythmic and not directly related to environmental conditions. For example the stomates of plants placed in darkness often continue to open and close for several days because of changes in turgor of the guard cells (SAYRE 1926). Periodicity in transpiration has also been observed under constant environmental conditions (BIALE 1941, MONTERMOSO and DAVIS 1942) and detopped root systems show diurnal periodicity in amount of exudation (GROSSENBACHER 1939, HAGAN 1949, HEIMANN 1950, ENGEL and FRIEDERICHSEN 1952). ENDERLE (1951) observed a morning maximum and nightly minimum in growth and turgor of carrot in tissue culture which persisted after 4 months in darkness ROSENE (1941) observed rhythmic variations in water uptake by onion roots.

The protoplasmic streaming of slime molds (KAMIYA 1950) and of root hairs (GOLDACRE 1952) is said to show a definite rhythm. STÅLFELT (1947) and VIRGIN (1951) observed that the viscosity of protoplasm in Elodea cells is higher at night than during the day, and also shows fluctuations of a few hours duration which persist for about 3 days in darkness. Various turgor movements of plants also show rhythmic periodicity (BÜNNING 1953). These autonomic rhythmic variations in water content and other cell properties probably are caused directly or indirectly by variation in amount and nature of metabolic processes which in turn affect the structure, permeability, and hydration of the protoplasm.

# Water content and physiological processes.

The water content of cells must be relatively high for normal growth and metabolism. Cell enlargement occurs only with an adequate supply of water and the fact that reduction in growth accompanies an internal water deficit is well known. LOOMIS (1934) concluded that growth of corn is controlled primarily by the supply of water to the growing stem tip and this was confirmed by THUT and LOOMIS (1944). Decrease in water content also results eventually in decrease in or cessation of processes such as photosynthesis and respiration, but the rates of these processes do not always decrease in proportion to the decrease in water content (ILJIN 1923). A small decrease in water content below the maximum sometimes is accompanied by an increase in photosynthesis, but

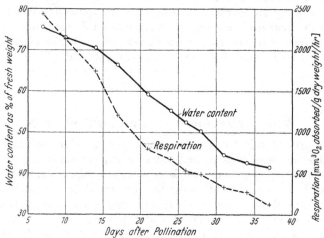

Fig. 2. Relation between water content and rate of respiration of maturing rye seed. Wheat seed show the same relationship. From SHIRK (1942).

further decrease invariably results in reduction and finally in cessation of photosynthesis. PARKER (1952) reported that as Austrian pine needles were allowed to dry rapidly the rate of respiration first decreased, then showed a temporary increase followed by decrease to zero and similar behavior has been observed during dehydration of plant tissue by other investigators (MONTFORT and HAHN 1950). The decrease in respiration accompanying decrease in water content during maturation of seeds is shown in Fig. 2. The respiration of dry seeds is very low and increases very slowly with increase in water content up to a critical point (about 15% in corn, oats, and wheat) where further increase in water content results in very rapid increase in rate of respiration. This is shown in Fig. 2 of the section on bound water where the causes of this relationship are discussed. APPLEMAN and BROWN (1946) found that an increase in moisture content of only 5% of the fresh weight of corn seeds resulted in a 5-fold increase in aerobic respiration and a 10-fold increase in anaerobic respiration. The practical application of the relationship between moisture content and respiration of seeds is very important in connection with storage of grain and other seeds because high moisture content results in high respiration, causing "heating" and spoilage.

Decrease in water content of plant cells and tissues usually produces important changes in their chemical and physical properties. In general hydrolytic processes are increased and synthetic processes are hindered by water deficit, resulting for example in hydrolysis of starch and proteins, but SPOEHR (1919) found that dehydration of certain succulents caused accumulation of polysaccharides and CLEMENTS (1937) found that hemicellulose accumulated in soybeans subjected to drought. LEVITT and SCARTH (1936), NORTHEN (1943), STOCKER (1948) and SIMONIS (1952) are among those who have reported that dehydration produces changes in permeability and viscosity of protoplasm. NORTHEN and STOCKER believed that dehydration causes disassociation of the protoplasm which results in the physical changes observed and also activates enzymes, causing increased respiration and hydrolytic activity. SIMONIS (1952) found that although drought reduced leaf area it often resulted in increased water content of the leaves, possibly because changes in protoplasmic structure enabled it to bind more water. Changes in chemical composition might themselves affect physiological processes. For example, the conversion of starch to sugar might be accompanied by an increase in respiration in spite of a decrease in water content and the increase in respiration might cause a decrease in apparent photosynthesis although little or no change in rate had actually occurred. Further discussion of the relation between water content and physiological activity can be found in CRAFTS, CURRIER and STOCKING (1949), KRAMER (1949), LEVITT (1951b), HUBER (1951) and RICHARDS and WADLEIGH (1952).

## Water movement into and out of cells.

The data presented in the section on changes in water content indicate that large amounts of water pass in and out of a typical plant cell during its life. In part this is a slow movement associated with cell enlargement and the changes associated with maturation and senescence, but rapid movement of considerable volumes of water also occur daily in many tissues of a plant during periods of heavy transpiration. The volume of water passing through a rapidly transpiring plant in a day may exceed the total water volume of the plant. Under such conditions the water content of many of the cells of the roots and leaves turns over several times per day. Another group of cells which undergo considerable changes in water content each day are the guard cells of the stomates. The speed with which water can enter or leave a cell is frequently demonstrated in the laboratory during experiments involving plasmolysis or deplasmolysis where changes in volume of 20% or more often occur within an hour. ROSENE (1950a) estimated that a root hair often absorbs its own volume of water in 3 or 4 minutes.

Cell membranes. Water entering or leaving the vacuole must pass through the wall and the layer of cytoplasm which lines it. As stated previously the walls of most plant cells are relatively permeable to water and solutes, except where deposition of cutin or suberin has occurred. As mentioned earlier, cellulose walls contain numerous spaces between the micelles and fibrils. These water filled spaces obviously permit relatively free passage of water and solutes. SKENE (1943) estimated that cell walls are 50 to 80,000 times as permeable to sucrose as the protoplasm and LEVITT, SCARTH and GIBBS (1936) found that removal of the protoplasts from the walls caused little change in permeability to water.

The layer of cytoplasm lying between the wall and vacuole usually is regarded as a differentially permeable membrane, hindering or preventing the passage of solutes and offering considerable resistance to the passage of water. The layer of cytoplasm is not homogenous, but possesses definite membranes at both the outer and inner surfaces. That next to the wall is called the plasmalemma or ectoplast, that next to the vacuole the tonoplast, while the body of the cytoplasm is sometimes termed the mesoplasm. These membranes, particularly the tonoplast, appear to be high in lipoids with a framework of protein molecules oriented with their hydrocarbon chains in the lipoid layer and their polar groups in the water layer in such a manner that they produce a sieve effect. Because these membranes are composed of both hydrophilic and hydrophobic materials the arrangement of which is suggestive of both a sieve and a mosaic structure, it is often stated that the membranes show characteristics of both the sieve and the solution type of permeability. RUHLAND has long claimed, however, that differences in permeability can be explained by a relatively simple ultrafilter or sieve theory (RUHLAND and HEILMAN 1951). The reader is referred to BROOKS and BROOKS (1941), DAVSON and DANIELLI (1952), FREY-WYSSLING (1953), and the recent review by VIRGIN (1953) for more detailed discussion of the structure of these membranes and of the various theories of permeability.

It is often supposed that most of the resistance to water movement occurs in the plasmalemma and tonoplast, but this may not be true. Some years ago HUBER and HÖFLER (1930) and DE HAAN (1933) claimed that resistance to water movement is a property of the entire layer of cytoplasm and not merely of the surface membranes, and HÖFLER (1950) and SEEMAN (1950a, 1953) have presented further evidence in support of this view. HÖFLER (1950) pointed out that permeability to water is affected by the thickness of the layer of cytoplasm and removal of everything outside of the tonoplast increases the rate of water movement into and out of the vacuole, but rupture of the plasmalemma does not produce any sudden increase in permeability. SEEMAN found permeability to water affected by salts, $p_H$, and temperature in a manner indicating that the entire mass of protoplasm is involved rather than merely a surface layer. HOPE and ROBERTSON (1953) doubt if the plasmalemma has a lower permeability than the mass of cytoplasm and regard the tonoplast as the principal physiological barrier to solutes. BUTLER (1953a) also is doubtful if the outer surface of the cytoplasm is less permeable to solutes than the interior. On the other hand there is some evidence that the entrance of sugar into plant cells is controlled by enzymes located in the outer surface of the cytoplasm (BROWN 1952, STREET and LOWE 1950). It has also been suggested that enzymes or other protein molecules act as "carriers" in nonosmotic water uptake (GOLDACRE 1952, ROSENBERG and WILBRANDT 1952). This would not require any special impermeability of the plasmalemma, but would assume a much greater amount of metabolic activity than has previously been attributed to cell membranes. Evidently more research is needed before we can decide where the principal physiological barrier to water and salts is located in the layer of cytoplasm lying between the wall and the vacuole.

## Factors affecting permeability to water.

According to older views cell membranes were relatively passive barriers and permeability was explainable in terms of molecular size, solubility, and electrical charge, but this relatively simple concept now seems inadequate. As just mentioned, it is now believed that various enzymatic activities occur in the surface of the cytoplasm and ROSENBERG and WILBRANDT (1952) link permeability to enzyme activity and the synthesis of carrier proteins, while GOLDACRE (1952) suggests that transport across membranes depends on protein molecules which undergo changes in adsorptive capacity by folding and unfolding. Permeability to water, like permeability to solutes, is considerably affected by changes in metabolic activity as well as by various chemical and physical factors of the environment. This is to be expected if permeability is controlled by the cytoplasm with its complex framework of hydrated protein molecules and its even more complex surface membranes. The structure is maintained by the expenditure of energy released in respiration and is affected by the electrical charge of ions in the environment and by the extent of its hydration.

**Permeability measurements.** Numerous measurements of the permeability of cells to solutes have been made, but relatively few measurements of their permeability to water are available. Permeability to water usually is measured by observing the rate of change in volume during plasmolysis or deplasmolysis (Höfler 1917, Huber and Höfler 1930, Levitt, Scarth and Gibbs 1936). Huber and Höfler (1930) found the permeability of cells of Maianthemum to vary about 50-fold, but the permeability to water was about 120 to 140 times the permeability to urea and 10,000 times that for sucrose. Huber (1933) found the epidermal cells of Vallisneria to be 30 or 40 times more permeable than the mesophyll cell of the same leaves and Url also (1952) found the epidermal cells

Table 3. *Permeability of various plant cells to water. The data of* Palva *and* Levitt, Scarth *and* Gibbs *are for a difference in osmotic pressure of one atm. The data of* Rosene *represent observed rates of water intake from micropotometers.*

| Material | Permeability in $\mu^3/\mu^2$/min. | Investigator |
|---|---|---|
| Cells of onion bulb scale . . . . | 0.3 | Levitt, Scarth and Gibbs 1936 |
| Alga, *Tallypellopsis stelligera* . . | 1.08 | Palva 1939 |
| Onion roots . . . . . . . . | 0.84 | Rosene 1941 |
| Root hairs of 10 spp. . . . . . | 0.3–4.4 | Rosene and Walthall 1949 |
| Old radish root hairs . . . . . | 0.47–1.94 | Rosene 1950a |
| Young radish root hairs . . . . | 4.46–1.16 | Rosene 1950a |

of the plants he studied more permeable than those beneath them. Some permeability measurements are summarized in Table 3, showing the wide range of permeability encountered in plant tissue. During measurement of water intake by individual epidermal cells and root hairs Rosene (1941, 1950a) and Rosene and Walthall (1949) observed differences in permeability of root hairs of different ages and lengths, and differences on a single hair, the tip usually being more permeable. Fischer (1950) found that permeability to water increased with age in 9 species of plants which he studied.

**Hydration.** Increased hydration of the protoplasm might be expected to increase permeability because it usually increases the size of the water-filled spaces between the structural units and decreases the firmness of water binding in the spaces, but it also has been claimed that the accompanying swelling of the hydrophilic components of cell membranes might decrease permeability by blocking the pores (Bogen 1940, 1941; Myers 1951). De Haan (1933) reported that increase in water content of onion epidermis was accompanied by increase in permeability and Homès (1943) found this also to be true of dahlia roots. Aykin (1946) found that permeability to water of carrot tissue, discs cut from Nuphar leaves, and animal membranes was decreased by dehydration. Levitt, Scarth and Gibbs (1936) found that permeability of protoplasts from onion cells increased during deplasmolysis, and attributed the change to increased pore size caused by stretching of the membranes under the influence of turgor pressure. Probably the mechanical stretching of the membranes and their increased hydration both tend to increase permeability.

**Solutes.** The effects of solutes on permeability of cell membranes to water and solutes has been extensively studied. The reader is referred to Brooks and Brooks (1941), Höber (1945) and Davson and Danielli (1952) for more extensive discussions than space permits in this paper.

It is generally agreed that calcium decreases permeability to water and sodium and potassium increase it (see Heilbrunn 1952, for example); Baptiste (1935) reported that soaking discs of carrot and potato tissue in hypotonic solutions of KCl, NH⁴Cl, and NaCl increased permeability, and soaking in $MgCl_2$ and $CaCl_2$ decreased it. De Haan (1933) found concentration an important factor, low concentrations of $Ca(NO_3)_2$ and $Co(NH_3)_6Cl_3$ decreasing and high concentrations increasing permeability, while all concentrations of $NaNO_3$ tested increased permeability. Bogen (1940) found that although 0.01 M solutions of several salts increased permeability of epidermal cells to urea and glycerine 0.1 M solutions decreased permeability. Prell (1953) found that 0.02 M solutions of Li, K, and Co nitrate usually increased permeability to several solutes. Guttenberg and Beythien (1951) found that dilute $CaCl_2$ decreased permeability while dilute KCl increased it, but a mixture of the two salts had no effect. Seeman (1953) emphasized the imortance of the time factor in permeability studies because he found that calcium, magnesium and strontium, caused an early increase in permeability, followed by a decrease. Urea and glucose always decreased permeability to water probably because of dehydration, while electrolytes affect it through the presence of variously hydrated ions which probably also alter the electrical charge on the cell membranes. It is to be expected that the structure and hence the permeability of membranes with a protein framework would be affected by the $p_H$ and this has been found true by various investigators (L. and M. Brauner 1943, Seeman 1950a, Guttenberg and Meinl 1952).

**Gases.** Little work has been done on the effects of oxygen concentration on permeability, but it has been demonstrated that reduction in aerobic respiration reduces accumulation of water and it probably also reduces permeability. Work of Hoagland and Broyer (1942) indicates that permeability to water of the roots of tomato is greatly reduced in an oxygen-free atmosphere, and Rosene (1950b) and Rosene and Bartlett (1950) found that absence of oxygen decreased the intake of water by roots and root hairs. Kramer (1940), Hoagland and Broyer (1942), and Chang and Loomis (1945) all reported that a high concentration of carbon dioxide drastically reduces the permeability of roots to water, presumably because of decreased permeability of the protoplasmic membranes. This can scarcely be attributed entirely to $p_H$ effects, because the $p_H$ of the system would not be reduced sufficiently to greatly affect cell permeability.

**Temperature.** Almost without exception increase in temperature within the noninjurious range results in increased permeability to water and solutes. The effects vary considerably with the species of plant and with the temperature range studied, and probably with the method of study used. Stiles (1924), Heilbrunn (1952) and others have stated that the $Q_{10}$ for permeability to water is about 2 to 3, but studies by plasmometric methods indicate that it probably is lower than the $Q_{10}$ for permeability to solutes. In several species studied by Seeman (1950b) the $Q_{10}$ for water ranged from 1.3 to 1.6 but it is doubtless much higher than this in some species. Levitt and Scarth (1936) reported that cells of cold-hardened plants are more permeable to water than those of nonhardened plants. Furthermore, several workers have noted that sudden cooling of the root system results in more severe wilting and interference with water absorption than if the root systems are allowed to cool slowly to the same temperature (Kramer 1942, Barney 1950, Böhning and Lusanandana

1952). Perhaps rapid cooling causes an increase in viscosity and decrease in permeability of the root cells to water which partly disappears after a few days at low temperature. If the roots are cooled slowly the protoplasm of some species may become adjusted to the low temperature without passing through a period of drastically reduced permeability. This situation deserves further investigation.

The effects of temperature, particularly of low temperature, on permeability probably vary considerably among various species. It has been reported by several investigators that absorption of water by plants from cold habitats or cold climates is reduced less by cooling the soil than absorption by plants from warm habitats (Döring 1935, Brown 1939, Kozlowski 1943). Water absorption by warm season crops such as cotton and watermelons was reduced much more by cooling the soil than was water absorption by a cool season crop such as collard *(Brassica oleracea acephala)*, probably because of a greater decrease in permeability of the root cells to water when cool (Kramer 1942). The effects of temperature on permeability to water are therefore of considerable ecological importance (Whitfield 1932, Clements and Martin 1934, Michaelis 1934). They are also occasionally important with respect to cultivated plants. Schroeder (1939), for example, found that greenhouse cucumbers were injured by cold soil and watering with cold water.

**Radiation effects.** In general exposure to radiation of various wavelengths appears to increase the permeability of cells to water and solutes (Heilbrunn 1952). Weber (1929) found that cells of sun leaves of *Ranunculus ficaria* had a lower permeability to water and higher viscosity than cells of shade leaves, but this may not be a direct effect of light. Brauner and Brauner (1940) reported that visible light increased the water permeability of carrot tissue and Lepeschkin (1930, 1948) and others have reported that exposure to visible light increases permeability to solutes. Virgin (1951) made an extensive study of the effect of light on viscosity and reported that it is decreased in the same manner that permeability is increased by illumination. This is further evidence of the dependence of permeability on protoplasmic structure. According to Biebl (1947) and Virgin (1952) the effects of light on viscosity and permeability can be sharply localized in a limited area by illuminating a narrow band of cells in a piece of tissue. Treatment with nonlethal doses of ultraviolet radiation increases permeability, but the effects disappear within 24 hours (Toth 1949). Heilbrunn and Daugherty (1933) found that irradiation with ultraviolet caused loss of calcium from the cortical protoplasm of amoeba and suggested that loss of calcium from the surface layer might be responsible for the increase in permeability following irradiation of tissue. It is difficult to distinguish cause and effect in this instance, but it seems that loss of calcium must have occurred as the result of injury to the protoplasm which in itself might have caused changes in permeability.

**Toxic substances.** Injury to cells of a reversible nature might either increase or decrease permeability to water, depending on how the protoplasmic structure is affected, but if injury progresses so far as to result in death a marked increase in permeability to both water and solutes occurs. This supports the view expressed earlier that the permeability of protoplasmic membranes depends on characteristics which are maintained only by the continual expenditure of energy provided by respiration. Any factor which seriously affects respiration might be expected to affect permeability. Anesthetics and respiration inhibitors are examples of substances which in low concentrations might decrease permeability, eut in concentrations high enough to cause permanent injury will increase it.

Unfortunately few data are available concerning the effects of these substances on passive permeability. GUTTENBERG and BEYTHIEN (1951) found that substances which inhibit growth usually also decrease permeability to water. The effect of coumarin and of two concentrations of indoleacetic acid on permeability are shown in Fig. 3. Although it will be shown in a later section of this paper that respiration inhibitors reduce or prevent water uptake by root systems and pieces of plant tissue it is uncertain how much of this can be attributed to changes in permeability of the cell membrane.

According to GÄUMANN and JAAG (1947), the fungus *Fusarium lycopersici* produces a toxic substance which causes an irreversible increase in permeability of the cells of infected tomato plants. As a result so much water is lost that the plants wilt, not from lack of water, but from loss of differential permeability by the cell membranes. YARWOOD (1947) found a similar increase in permeability of bean leaves infected with powdery mildew and rust.

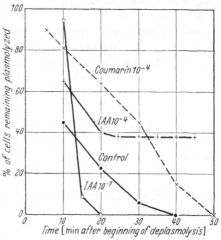

Fig. 3. Effects of high and low concentration of indoleacetic acid (IAA) and of the growth inhibitor coumarin on permeability to water as measured by the rate of deplasmolysis of epidermal cells of *Rhoeo discolor*. Concentrations are in grams per cubic centimeter. From GUTTENBERG and BEYTHIEN (1951).

## Osmotic movement of water.

For over a century botanists generally have based their treatment of water movement in plants on the osmotic theory first proposed by DUTROCHET in 1837. According to this theory plant cells are regarded as osmometers in which the protoplasm acts as a differentially permeable membrane between the vacuolar sap and the external solution or adjacent cells, and water movement is a purely physical process. The only important modification of this concept was the gradual realization that water movement really occurs along gradients of diffusion pressure or decreasing free energy of water, rather than along gradients of osmotic pressure. Some of the early progress toward this new concept is found in papers by URSPRUNG and BLUM (1916), THODAY (1918) and HÖFLER (1920). The variations in terminology used in discussions of osmotic phenomena by various authors are confusing. As WALTER (1952) wrote recently, „Fast jeder Autor benutzt eine andere Bezeichnungsweise und mitunter widerspricht er sich selbst". All of them were searching for a satisfactory term to express the condition of the water in the cells with respect to water in their environment. WALTER (1931) proposed the term "Hydratur" to refer to the condition of the water in cells and tissues in terms of its vapor pressure. Other terms suggested include suction force, suction tension, suction potential, turgor deficit, diffusion pressure deficit, osmotic potential difference, and net influx specific free energy. The last mentioned term was proposed by BROYER (1947) for a thermodynamic treatment of water relations. This writer will use the diffusion pressure terminology of MEYER (1938, 1945) because the advantages of wide use and general familiarity seem to outweigh the objections offered by WALTER (1952) and by LEVITT (1951a).

The cause of water movement in an osmotic system can be regarded as the difference in free energy or diffusion pressure of the water in different parts of the system, because water tends to diffuse from regions of higher to regions of lower free energy or diffusion pressure. The diffusion pressure of water is increased by increase in temperature or pressure and decreased by the addition of solutes. It is impractical to measure the diffusion pressure itself, but it usually is relatively easy to measure the amount by which the diffusion pressure of the water in a cell or other system differs from that of pure water at the same temperature and atmospheric pressure. This difference is known as the diffusion pressure deficit, often abbreviated as DPD. The decrease in diffusion pressure of water in a solution caused by the addition of solutes is equal to the osmotic pressure (OP) and can be calculated from the vapor pressure or freezing point depression, either of which can be measured accurately. In plant cells an added factor is the resistance of cell walls to extension, resulting in the development of wall pressure (WP) which is numerically equal to the turgor pressure (TP), but exerted in the opposite direction.

The relations among these various osmotic quantities of a cell in the range from full turgor to incipient plasmolysis can be represented as follows:

$$\text{Cell DPD} = \text{OP} - \text{TP (TP} = \text{WP)}.$$

Where the surrounding tissues exert pressure on the cells under consideration the diffusion pressure of the water is increased correspondingly and the equation becomes:

$$\text{DPD} = \text{OP} - \text{TP-Pressure of surrounding cells}.$$

Tension or negative pressure often develops in transpiring plants and this reduces the diffusion pressure of the cell sap. The equation to represent this is as follows:

$$\text{DPD} = \text{OP} - (- \text{TP}).$$

Although this terminology is widely used various modifications are proposed from time to time. Among those who have recently discussed modifications in terminology or concepts are ALGEUS (1951). BROYER (1950, 1951), BURSTROM (1948), HAINES (1951), HYGEN and KJENNERUD (1952), LEVITT (1951a, 1953), THODAY (1952) and WALTER (1952). BROYER (1947, 1950, 1951) has proposed a mathematical treatment of cell water relations in terms of specific free energy which emphasizes the fact that water and solutes tend to move along gradients of decreasing free energy. The rigorous mathematical treatment made possible by this approach would be more useful if the experimental data on which it is used were more accurate. For most physiological studies it is just as satisfactory, and to most botanists much more intelligible, to express free energy gradients in terms of diffusion pressure gradients expressed in atmospheres.

**The water balance.** According to the osmotic theory of cell water relations cells gain water if their diffusion pressure deficit is higher than that of their environment and lose water if it is lower than that of their environment. The relations of cell volume, OP, TP, and DPD are shown graphically in Fig. 4 taken from HÖFLER (1920). In a cell at incipient plasmolysis the DPD is equal to the OP because no turgor pressure exists. If such a cell is placed in water inward diffusion occurs, increasing the volume of water in the vacuole and resulting in the development of turgor pressure and decrease in OP because of dilution of the vacuolar sap. As mentioned previously, application of pressure causes an increase in DP of water, hence the DPD of the cell gradually decreases as the TP increases and becomes zero when the TP just balances the OP. At this point a dynamic equilibrium exists in which no net movement of water occurs in either direction, although the water concentrations and osmotic pressures are very different inside and outside of the cells.

It is the DPD (suction force or suction potential of some authors) rather than the OP which determines whether a cell or tissue will gain or lose water from its environment, hence this term deserves special comment. As already shown, it can vary from zero to a value equal to the OP at incipient plasmolysis and if the cell is placed under tension the DPD can exceed the OP because $DPD = OP - (-TP)$. It is assumed that in most plant cells the osmotic and imbibitional forces tend to be in equilibrium and measurement of the OP of the cell sap will give a satisfactory measure of the DPD. Some exceptions may occur, however, where the volume of imbibing material is large in relation to the volume of vacuolar sap. KERR and ANDERSON (1944) found that as cotton seeds matured their DPD began to exceed their OP, probably because their imbibitional forces became dominant over their osmotic forces and the OP of the expressed sap no longer determined their DPD.

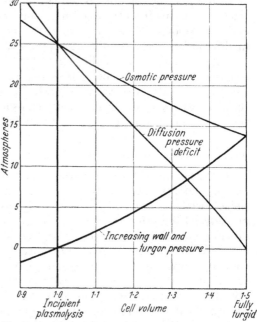

Fig. 4. Graphic presentation of changes in osmotic quantities of a cell accompanying changes in cell volume. Osmotic pressure decreases with increase in volume because of dilation of the cell sap. Diffusion pressure deficit decreases with increasing turgidity and reaches zero when the turgor pressure becomes equal to the osmotic pressure. Modified from MEYER and ANDERSON (1952) after HÖFLER (1920).

The DPD of large masses of plant tissue can be measured either by observing changes in volume or changes in weight of tissue after immersion in sucrose or mannitol solutions of various concentrations. The OP of the solution in which no change in weight or volume occurs is assumed to be equal to the DPD of the tissue because a cell or group of cells neither gains nor loses water in a solution having the same DPD. URSPRUNG (1923) developed a method suitable for thin tissue such as epidermal layers, flower petals, and thin leaves. Strips are cut, their length determined, and they are immersed in solutions of various concentrations for a few hours, then remeasured to determine in which solution minimum change in length occurs. The DPD of bulky tissue such as potato tuber or beet roots can be measured by the method described by MEYER and WALLACE (1941). Discs or small cylinders of tissue are weighed, immersed in a series of solutions of sucrose or mannitol for a few hours, then reweighed to determine in which solution the minimum change in weight occurs. Both methods give only approximations of the actual DPD and are open to some criticism (ERNEST 1934), but are the best means available for determining the force with which a cell can absorb water.

## Active or nonosmotic absorption of water.

In recent years some doubt has arisen concerning the adequacy of simple osmotic diffusion to explain some of the results obtained in studies of cell water relations and it has been suggested that cells probably are capable of accumulating water against a diffusion pressure gradient by an "active" or nonosmotic process dependent on metabolic energy, much as they accumulate solutes. This doubt arose because it was found occasionally that the osmotic pressure of the solution required to prevent water absorption was greater than the osmotic pressure of the cell sap. This seemed to indicate that the expression DPD = OP − TP did not hold, presumably because some factor was bringing about water absorption in addition to the OP of the cell, and this additional factor was termed active or nonosmotic absorption.

Bogen (1953) and Bogen and Prell (1953) have attempted to analyze the importance of this factor by studying the deplasmolysis of individual cells. The degree of plasmolysis should depend on the ratio of the osmotic pressure of the cell sap to the osmotic pressure of the external solution, but they consistently found deviations from the expected values which they attributed to metaosmotic and nonosmotic forces. In summary they suggested that water absorption might consist of (1) osmotic absorption, (2) metaosmotic absorption caused by chemical and physical binding of water, especially in the vacuole, (3) electro-osmosis, and (4) nonosmotic forces related directly to cell metabolism. If it is agreed that water moves along gradients of decreasing free energy it scarcely seems necessary to distinguish between osmotic and metaosmotic forces because they both operate by decreasing the free energy of the water in the cell.

For convenience in presentation, the evidence for the occurrence of water intake by nonosmotic processes will be discussed under several subheadings.

**Discrepancies between cryoscopic and plasmolytic measurements.** Bennet-Clark, Greenwood, and Barker (1936) found the osmotic pressures of certain tissues when measured plasmolytically were much higher than the osmotic pressures of the sap expressed from the same kind of tissue. They regarded this difference as evidence that the protoplasm secretes water into the vacuole instead of merely acting as a passive differentially permeable membrane. Similar discrepancies were observed by a number of other investigators including Mason and Phillis (1939), Roberts and Styles (1939), Bennet-Clark and Bexon (1940), Lyon (1942) and Currier (1944), but it is thought that many of them are the result of errors inherent in the methods used. Errors in the plasmolytic method include adhesion of the cytoplasm to the wall, penetration of the plasmolyzing solute, and difficulty in measuring cell volume. Even greater errors probably occur when sap is expressed under pressure for cryoscopic determinations because of filtration of solutes and release of dilute solution from the cytoplasm and cell walls.

Levitt (1947) believed that most of the claims of the occurrence of nonosmotic absorption mentioned above resulted from errors of measurement and incorrect interpretation of the data. He calculated the energy required to maintain a pressure by a nonosmotic mechanism and concluded that a pressure gradient of more than one or two atmospheres would require an impossibly high expenditure of energy. Levitt's calculations were criticized by Bennet-Clark (1948) and Myers (1951) because they thought the value for permeability used by him was too high, and by Spanner (1952) because he used too large a cell area and calculated the energy required improperly. Levitt (1953) seems to have answered these specific objections, though the writer can scarcely accept his claim that "there is not a shred of evidence" to support the occurrence of nonosmotic water uptake.

**Respiration and water intake.** If nonosmotic water absorption occurs against a diffusion pressure gradient the energy must be supplied by respiration. There is adequate evidence that water intake is somehow related to respiration because it is reduced by treatments which reduce respiration. Steward, Stout and Preston (1940) concluded that aerobic respiration, protein synthesis, water absorption and mineral accumulation are closely related processes. Aerobic conditions were found to be essential for water uptake in potato tissue by REIN-DERS (1938, 1942) and BRAUNER, BRAUNER and HASMAN (1940), for oat coleoptiles by KELLY (1947) and for root hairs by ROSENE (1950b) and by ROSENE and BARTLETT (1950).

Respiration inhibitors have been found to reduce or prevent water absorption by tomato root systems (VAN OVERBEEK 1942), onion roots (ROSENE 1947), oat coleoptiles (KELLY 1947), potato tuber (HACKETT and THIMANN 1952) and epidermal cells of Rhoeo and dandelion *(Taraxacum otticinale)* (BOGEN 1953). On the other hand LEVITT (1948) found that KCN did not inhibit auxin-induced water uptake by potato and BOGEN found a type of water uptake by epidermal cells of dandelion and Rhoeo which was not inhibited by KCN, but was inhibited by azide and arsenite, and was stimulated by glucose-1-phosphate.

BONNER, BANDURSKI and MILLERD (1953) proposed that water intake and respiration are linked by transfer of energy through adenosinetriphosphate, perhaps in a mechanism in which auxin molecules function as a part of the transport system. BOGEN (1953) found the special type of water uptake which he studied was affected by various chemicals in the same manner as protoplasmic streaming and suggested that it was not linked directly to respiration, but was related to cell metabolism in an indirect manner. LEVITT (1947, 1953) claims that it would be impossible for a cell to release enough energy to maintain a nonosmotic gradient of more than one or two atmospheres, but the values on which these calculations are based are scarcely reliable enough to give very precise results.

There seems to be no doubt that some relationship exists between water uptake and respiration, but the nature of this relation still remains uncertain. It may be a direct expenditure of energy for nonosmotic accumulation as proposed by BONNER, BANDURSKI and MILLERD (1953) or it may be related indirectly through effects on permeability of the protoplasmic membranes and extensibility of the cell walls. The effects of deficient oxygen and excess carbon dioxide on permeability were mentioned earlier. BRAUNER, BRAUNER and HASMAN (1940) observed that water permeability of unaerated potato tuber was only 63% of aerated tuber after 20 hours. BRAUNER and BRAUNER (1943) also observed that the tensility and plasticity of cell walls of potato tissue are considerably greater under aerobic than under anaerobic conditions. They attributed at least part of the increased water intake under aerobic conditions to the decrease in wall pressure produced by increased extensibility of the walls. The role of wall extensibility will be discussed further in connection with auxin effects.

It seems that permeability of protoplasmic membranes must inevitably be affected by treatments which change the rate of respiration because their structure depends on a continuous supply of energy. This makes it difficult to determine whether the reduction in water uptake caused by a respiration inhibitor or the increase caused by a stimulator results from change in rate of nonosmotic water accumulation or from change in permeability of cell membranes and extensibility of cell walls. BOGEN (1953) pointed out that an increase in permeability of cell membranes to water will not affect the total amount of water absorbed, but only increase the rate at which it is absorbed. Only a gain or loss of solute which causes a change in concentration of the vacuolar sap will cause a change in the water content at equilibrium.

LUNDEGÅRDH relates water uptake to respiration in another manner. Uptake of salt by cells is linked to anion respiration and the uptake of salts causes osmotic absorption of water. Metabolic activity also can cause loss of water by decrease in concentration of solutes, as when nitrate is converted into protein and sugar is used in respiration (LUNDEGÅRDH 1950).

**Effects of auxin on water intake.** It has been shown repeatedly that auxin increases the uptake of water by plant tissue and our attention will be given to the various explanations offered for this. REINDERS (1938, 1942) observed that indoleacetic acid increased water uptake by potato tissue under aerobic conditions and concluded from the decrease in dry weight that respiration was also increased. REINDERS thought that the increased water uptake was caused by increased solute concentration in the cell sap, resulting from hydrolysis of starch to sugar. COMMONER and MAZIA (1942) and COMMONER, FOGEL and MULLER (1943) thought that auxin increased water uptake because it increased salt accumulation in the cells. These explanations are incorrect because VAN OVERBEEK (1944), LEVITT (1947), HACKETT (1952) and BRAUNER and HASMAN (1952) all found that auxin causes a decrease in osmotic pressure of the cell sap at the same time that it causes an increase in water content.

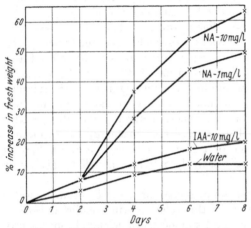

Fig. 5. Effects of indoleacetic acid (IAA) and naphthalene-acetic acid (NA) on water uptake by discs of potato tissue. Fresh weight increases are expressed as percentages of weights at beginning of experiment. From HACKETT and THIMANN (1952).

Ts'o and STEINBAUER (1953) found that indoleacetic acid greatly increases the water uptake of segments of pea stem, but maleic hydrazide reduces the auxin effect about 50%. Addition of fumaric acid greatly reduces maleic hydrazide inhibition, probably because it circumvents the inhibiting action of maleic hydrazide on respiration. HASMAN (1953) found that di-n-amylacetic acid increases aerobic water uptake by potato tissue, possibly by activating auxin, but 6-aminoundecane inhibits it, apparently by destroying differential permeability of the cell membranes. HACKETT (1952) found that α-naphthalene-acetic acid causes much greater water uptake by potato tissue than indoleacetic acid (see Fig. 5), but BONNER, BANDURSKI and MILLERD (1953) found that with good aeration indoleacetic acid increases the water uptake of slices of dormant tuber of Jerusalem artichoke (*Helianthus tuberosus*). BOGEN (1953) observed what he regarded as another type of nonosmotic water uptake by epidermal cells of dandelion and Rhoeo because it was not affected by indoleacetic acid.

Although there is general agreement that auxin causes increased water uptake there is no such agreement concerning the manner in which this is brought about. Having discarded the idea that it causes an increase in concentration of cell solutes, the other possible effects are (1) an increase in permeability of cell membranes, (2) increased extensibility of the cell walls, and (3) direct stimulation of active or nonosmotic water absorption. Any or all of these effects might be related to the increase in respiration which usually accompanies increased auxin supply.

BRAUNER and HASMAN (1949) found that indoleacetic acid increases the permeability of potato tissue to water within 6 hours and GUTTENBERG and

BEYTHIEN (1951) and GUTTENBERG and MEINL (1952) found that permeability is almost doubled by it at $10^{-7}$ gm. per liter (see Fig. 4). BRAUNER and BRAUNER (1943) found that indoleacetic acid increases the extensibility of cell walls and BRAUNER and HASMAN (1949) concluded that the auxin-induced increase in water intake during the first few hours is caused by increased permeability; after 24 to 48 hours the increased extensibility of the cell walls becomes the most important factor. All investigators emphasize the necessity of good aeration for auxin to affect water uptake. BRAUNER and HASMAN (1952) concluded that auxin increases water intake principally by increasing respiration which in turn causes modification of the cell wall characteristics. They were unable to decide whether or not nonosmotic water uptake occurred in their experiments, but if so it must have been very small in quantity. LEVITT (1953) also attributes the increase in water uptake caused by auxin to increased extensibility of the cell walls. This permits continued osmotic uptake of water because wall pressure never becomes equal to the osmotic pressure so long as the walls are extending plastically.

In contrast THIMANN (1951) suggested that auxin brings about "the direct metabolic pumping of water" into cells, but HACKETT and THIMANN (1953) concluded that water uptake is not directly linked to respiration because the two processes are not affected to the same extent by auxin or respiration inhibitors. BONNER, BANDURSKI and MILLERD (1953) proposed that auxin is directly involved in the use of respiratory energy in absorption of water, the addition of auxin increasing the rate of respiration only if water is absorbed. This situation suggests the existence of a "water respiration" analogous to the "salt respiration" of LUNDEGÅRDH. Both these investigators conducted long-time experiments lasting for several days, during which growth or permanent cell enlargement probably occured. This certainly is true of the potato tissue used by THIMANN and HACKETT, where, as they state, they studied water uptake associated with cell enlargement. BURSTROM (1953) has pointed out the difficulty of identifying a nonosmotic factor in water uptake by enlarging cells because the wall pressure is also changing, and he is unable to find any evidence of nonosmotic water uptake in the elongating portion of roots. The data of BONNER et al. (1953) perhaps are more convincing, though the tissues used may have made some growth. BURSTROM and FREY-WYSSLING (1950) agree that auxin increases cell enlargement by stimulating partial dissolution of local areas of growing cell walls and deposition of new wall material which permits cell extension. Perhaps auxin functions by activating enzymes which soften the walls by digestive action while increased metabolic activity and protoplasmic streaming speeds up deposition of new wall material. At any rate there is strong evidence indicating that the effect of auxin is closely linked to respiration. The data thus far presented seems, however, to be subject to more than one interpretation.

It seems that there is not enough information available to explain the relation of auxin to nonosmotic uptake of water, but it appears very probable that its chief role in osmotic uptake of water is to increase the extensibility of the cell walls, thereby keeping the wall pressure so low that water absorption continues. When the walls lose their ability to expand the wall pressure increases rapidly and soon equals the osmotic pressure, causing the DPD to fall to zero and water absorption to cease. Auxin also definitely increases permeability to water as measured by rate of deplasmolysis and this must be important in short-time experiments.

AUDUS (1949) questioned whether auxin acts separately in the various phenomena in which it is supposed to be involved or through a single master system and concluded that the latter is more probable. It likewise seems more probable that auxin acts on permeability, cell wall extensions and related processes indirectly through its effect on metabolism rather than by direct effects on these processes.

**Electroosmosis.** Another mechanism sometimes proposed to explain at least part of the water uptake of cells and tissues is electroosmosis. It has been demonstrated that water can move across membranes and even through pieces of twig (Stern 1919) under the influence of an applied electric current, and differences in electrical potential have been measured frequently across tissues and organs, and even between the outside and inside of large cells. Blinks (1940), for example, found potential differences of 70 to 80 millivolts between the vacuole and the external surface of Halicystis cells, the outer surface being positive to the interior. He found that the potential was modified by salts, oxygen, temperature, and respiration inhibitors. The rate of water movement across a membrane depends on the potential across the membrane and the direction is toward the negative side, or inward in plant cells. Hope and Robertson (1953) recently reviewed the work on the bioelectric properties of plant cell membranes, but did not add to our information concerning water movement.

Among those who have proposed electroosmotic theories of water uptake are Bennet-Clark and Bexon (1946), Brauner and Hasman (1946, 1947), Studener (1947), Arens (1949) and Bauer (1952). Brauner and Hasman and Studener suggested that electroosmosis is responsible for some water movement because they observed that a higher concentration of sugar than of salt is required to prevent water uptake by plant tissue. For example, Brauner and Hasman (1947) found that the apparent DPD of beet tissue was about 1.15 atmospheres higher in sucrose than in $CaCl_2$. This difference was regarded as a measure of an electroosmotic component of water uptake eliminated by addition of an electrolyte, but it seems that the difference might just as well be attributed to the effects of ions on the hydration and permeability of the protoplasm. Brauner and Hasman (1946) attributed only about 10% of water intake of potato tissue to electroosmosis. Lundegårdh (1944) observed potential differences of about 100 millivolts between the surface and the interior of wheat roots, but believed that it had little effect on water movement. Blinks and Airth (1951) found that applied electrical potentials of up to 1500 millivolts produced no appreciable water movement in Nitella cells. The available evidence certainly indicates that electroosmosis plays little or no part in the uptake of water by plant cells.

**Relative importance of osmotic and nonosmotic absorption of water.** The preceding discussion has shown that the absorption of water by plant cells is not as simple a process as it was once supposed to be. There seems to be little doubt that osmotic movement along DPD gradients can account for the major changes in water content which occur in mature plant tissue. Such changes are large, often amounting to 15 or 20% of the fresh weight of the tissue, they can occur quite rapidly (for example the recovery of wilted leaves), and they are reversible. On the other hand it has been clearly demonstrated that water uptake by plant tissue is reduced or prevented by oxygen deficiency and respiration inhibitors, and increased by auxin. Furthermore cells and tissues sometimes appear to absorb water against a DPD gradient, that is they absorb water from solutions with an OP higher than the apparent OP of their own cell sap. These observations are generally presented as evidence for nonosmotic absorption of water, but they are not conclusive proof that respiratory energy is being used directly in bringing about the intake of water. Changes in metabolic activity almost certainly affect osmotic water movement indirectly because they affect the permeability of cell membranes and the extensibility of cell walls, and a considerable part of the supposed nonosmotic water absorption might be explained by these indirect effects.

The most convincing examples of nonosmotic water absorption are those involving the addition of auxin to potato and Jerusalem artichoke tissue. At least one or two days are required for the effect of auxin to become very important and growth (permanent cell enlargement) almost certainly occurs. Water uptake by growing tissue is modified by the existence of a low wall pressure, as pointed out by BURSTROM (1953) and it therefore occurs under quite different conditions from water uptake in mature tissue where plastic wall extension no longer occurs. Thus although cell enlargement is accompanied by water intake it is not dependent on it so long as plasmolysis does not occur, at least not until plastic cell wall extension ceases. It might almost be said that growing cells absorb water because they are growing, rather than that they grow because they absorb water. Because of this situation it probably is impossible to demonstrate nonosmotic absorption in growing tissue.

Although the author does not believe that nonosmotic absorption of water is an important factor in plant cell water relations it is possible that a limited amount may occur. The mechanisms suggested thus far are highly hypothetical. Possibly conditions sometimes exist which result in small amounts of electroosmotic water movement. OSTERHOUT (1947) suggested that as a result of local metabolic activity solutes might accumulate at one end of a cell, causing intake at that end and loss at the opposite end where the osmotic pressure is lower. Possibly some sort of secretion occurs into the vacuole. FRANCK and MAYER (1947) proposed a scheme in which chemical energy is used to transport materials across a membrane against a diffusion gradient. This involves cross-transfer of reactants and products of a chemical reaction which occurs at one surface and is reversed at the other surface. GOLDACRE (1952) proposed that protein molecules might carry adsorbed water and solutes across cell membranes, unfolding at the outer surface and adsorbing them, then folding at the inner surface and "shedding" the adsorbed water and solutes into the vacuole. BONNER, BANDURSKI and MILLERD (1953) suggested that auxin itself might act as the carrier for water. Streaming of the protoplasm ought to speed up such transport. More attention probably should be given to the research on water transfer in animals where the kidney presents particularly interesting problems. ROBINSON (1953) in a recent review regards nonosmotic movement of water as of much greater importance in animals than in plants.

It is obvious that much more research will be necessary before any final conclusions can be reached concerning the mechanism of nonosmotic water absorption or its importance in the water economy of plants. Meantime we should maintain an open mind and an impartial attitude toward the sometimes contradictory evidence which is being presented to us.

## Literature.

ALGEUS, S.: Views on turgor pressure and wall pressure. Physiol. Plantarum (Copenh.) 4, 535–541 (1951). — ANDERSON, D. B., and T. KERR: A note on the growth behavior of cotton bolls. Plant Physiol. 18, 261–269 (1943). — APPLEMAN, C. O., and R. G. BROWN: Relation of anaerobic to aerobic respiration in some storage organs with special reference to the Pasteur effect in higher plants. Amer. J. Bot. 33, 170–181 (1946). — ARENS, K.: The "active membrane"—an hypothesis to explain the transfer of solutes in plants, as depending upon respiration. Rev. canad. de Biol. 8, 157–172 (1949). — AUDUS, L. J.: The mechanism of auxin action. Biol. Rev. **24**, 51–93 (1949). — AYKIN, S.: The relations between water permeability and suction potential in living and non-living osmotic systems. Rev. Fac. Sci. Univ. Istanbul **11**, 271–295 (1946).

BAPTISTE, E. C. D.: The effect of some cations on the permeability of cells to water. Ann. of Bot. **49**, 345–366 (1935). — BARNEY, C. W.: Effects of soil temperature and light intensity on root growth of loblolly pine seedlings. Plant Physiol. **26**, 146–163 (1951). — BAUER, L.: Über den Wasserhaushalt der Submersen. I. Zur Frage der Saugkräfte der Submersen. Protoplasma (Wien) **41**, 178–188 (1952). — BENNET-CLARK, T. A.: Non-osmotic water movement in plant cells. Disc. Faraday Soc. **3**, 134–139 (1948). — BENNET-CLARK, T. A., A. D. GREENWOOD and J. W. BARKER: Water relations and osmotic pressures of plant cells. New Phytologist **35**, 277–291 (1936). — BENNET-CLARK, T. A., and D. BEXON: Water relations of plant cells. New Phytologist **39**, 337–361 (1940). — BIALE, J. B.: Periodicity in transpiration of lemon cuttings under constant environmental conditions. Proc. Amer. Soc. Hort. Sci. **38**, 70–74 (1941). — BIEBL, R.: Strahlenwirkung auf das pflanzliche Protoplasma. Mikroskopie (Wien) **2**, 328–335 (1947). — BLINKS, L. R.: The relations of bioelectric phenomena to ionic permeability and to metabolism in large plant cells. Cold Spring Harbor Symp. Quant. Biol. **8**, 204–215 (1940). — BLINKS, L. R., and R. L. AIRTH: The role of electroosmosis in living cells. Science (Lancaster, Pa.) **113**, 474–475 (1951). — BOGEN, H. J.: I. Ionenwirkung auf die Permeabilität von *Rhoeo discolor*. Z. Bot. **36**, 65—106 (1940). — II. Ionenwirkung auf die Permeabilität von *Gentiana cruciata*. Planta (Berl.) **32**, 150–175 (1941). Über Kappenplasmolyse und Vakuolenkontraktion. I. Die Wirkung von LiCl und Neutralrot und ihre Abhängigkeit von der Konzentration und dem osmotischen Wert in der Außenlösung. Planta (Berl.) **39**, 1–35 (1951). — Beiträge zur Physiologie der nichtosmotischen Wasseraufnahme. Planta (Berl.) **42**, 140–155 (1953). — BOGEN, H. J., u. H. PRELL: Messung nichtosmotischer Wasseraufnahme an plasmolysierten Protoplasten. Planta (Berl.) **41**, 459–479 (1953). — BÖHNING, R. H., and B. LUSANANDANA: A comparative study of gradual and abrupt changes in root temperature on water absorption. Plant Physiol. **27**, 475–488 (1952). — BONNER, J., R. S. BANDURSKI and A. MILLERD: Linkage of respiration to auxin-induced water uptake. Physiol. Plantarum (Copenh.) **6**, 511–522 (1953). — BRAUNER, L., and M. BRAUNER: Further studies of the influence of light upon the water intake and output of living plant cells. New Phytologist **39**, 104–128 (1940). — The relations between water-intake and oxybiosis in living plant-tissues. II. The tensility of the cell wall. Rev. Fac. Sci. Univ. Istanbul **8**, 30–75 (1943). — BRAUNER, L., M. BRAUNER and M. HASMAN: The relation between water-intake and oxybiosis in living plant-tissues. Rev. Fac. Sci. Univ. Istanbul **5**, 266–309 (1940). — BRAUNER, L., u. M. HASMAN: Untersuchungen über die anomale Komponente des osmotischen Potentials lebender Pflanzenzellen. Rev. Fac. Sci. Univ. Istanbul **11**, 1–37 (1946). — Weitere Untersuchungen über die anomale Komponente des osmotischen Potentials lebender Pflanzenzellen. Rev. Fac. Sci. Univ. Istanbul **12**, 210–254 (1947). — Über den Mechanismus der Heteroauxinwirkung auf die Wasseraufnahme von pflanzlichem Speichergewebe. Bull. Fac. Med. Istanbul **12**, 57–71 (1949). — Weitere Untersuchungen über den Wirkungsmechanismus des Heteroauxins bei der Wasseraufnahme von Pflanzenparenchymen. Protoplasma (Wien) **41**, 302–326 (1952). — BROOKS, S. C., and M. M. BROOKS: The permeability of living cells. Protoplasma-Monogr. **19** (1941). — BROWN, E. M.: Some effects of temperature on the growth and chemical composition of certain pasture grasses. Missouri Agric. Exper. Sta. Res. Bull. **1939**, 299. — BROWN, R.: Protoplast surface enzymes and absorption of sugar. Int. Rev. Cytol. **1**, 107–118 (1952). — BROYER, T. C.: The movement of materials into plants. I. Osmosis and the movement of water into plants. Bot. Rev. **13**, 1–58 (1947). — On the theoretical interpretation of turgor pressure. Plant Physiol. **25**, 135–139 (1950). — An outline of energetics in relation to the movement of materials through a two-phased system. Plant Physiol. **26**, 598–610 (1951). — BÜNNING, E.: Lehrbuch der Pflanzenphysiologie, Vols. 2 and 3. Berlin-Göttingen-Heidelberg: Springer 1953. — BÜSGEN, M.: Studien über den Wassergehalt einiger Baumstämme. Z. Forst- u. Jagdwiss. **43**, 137–154 (1911). — BURSTROM, H.: A theoretical interpretation of the turgor pressure. Physiol. Plantarum (Copenh.) **1**, 57–64 (1948). — Studies on growth and metabolism of roots. IX. Cell elongation and water absorption. Physiol. Plantarum (Copenh.) **6**, 262–276 (1953). — BUTLER, G. W.: Ion uptake by young wheat plants. II. The "apparent free space" of wheat roots. Physiol. Plantarum (Copenh.) **6**, 617–635 (1953a). — Ion uptake by young wheat plants. I. Time course of the absorption of potassium and chloride ions. Physiol. Plantarum (Copenh.) **6**, 594–616 (1953b).

CHANG, H. T., and W. E. LOOMIS: Effect of carbon dioxide on absorption of water and nutrients by roots. Plant Physiol. **20**, 221–232 (1945). — CHATFIELD, C., and G. ADAMS: Proximate composition of American food materials. U. S. Dept. Agric. Circ. **549**, 1–191 (1940). — CHIBNALL, A. C.: A new method for the separate extraction of vacuole and protoplasmic material from the leaf cells. J. of Biol. Chem. **55**, 333–342 (1923). — CLEMENTS, H. F.: Studies in drought resistance of the soy bean. State Coll. Washington Res. Studies

5, 1–16 (1937). — CLEMENTS, F. E., and E. V. MARTIN: Effect of soil temperature on transpiration in *Helianthus annuus*. Plant. Physiol. 9, 619–630 (1934). — COMMONER, R., and D. MAZIA: The mechanism of auxin action. Plant Physiol. 17, 682–685 (1942). — COMMONER, B., S. FOGEL and W. H. MULLER: The mechanism of auxin action. The effect of auxin on water absorption by potato tuber tissue. Amer. J. Bot. 30, 23–28 (1943). — CRAFTS, A. S.: Movements of organic materials in plants. Plant Physiol. 6, 1–42 (1931). — CRAFTS, A. S., H. B. CURRIER and C. R. STOCKING: Water in the physiology of the plant. Waltham, Mass.: Chronica Botanica Co. 1949. — CROCKER, W. A., and L. V. BARTON: Physiology of seeds. Waltham, Mass.: Chronica Botanica 1953. — CURRIER, H. B.: Water relations of root cells of *Beta vulgaris*. Amer. J. Bot. 31, 378–387 (1944).

DAUGHTERS, M. R., and D. S. GLENN: The role of water in freezing foods. Refrig. Engng. 52, 137–140 (1946). — DAVSON, H., and J. F. DANIELLI: The permeability of natural membranes. Cambridge Eng. Univ. Press 1952. — DÖRING, B.: Die Temperaturabhängigkeit der Wasseraufnahme und ihre ökologische Bedeutung. Z. Bot. 28, 305–383 (1935). — DUTROCHET, H. J.: Memoires pour servir a l'histoire anatomique et physiologique des végétaux et des animaux. Paris: J. B. Bailliere 1837.

ENDERLE, W.: Tagesperiodische Wachstums- und Turgor-Schwankungen an Gewebekulturen. Planta (Berl.) 39, 570–588 (1951). — ENGEL, H., and I. FRIEDERICHSEN: Weitere Untersuchungen über periodische Guttation etiolierter Haferkeimlinge. Planta (Berl.) 40, 529–549 (1952). — ERNEST, E. C. M.: The effect of intercellular pressure on the suction pressure of cells. Ann. Bot. 48, 915–918 (1934).

FARRANT, J. L., R. N. ROBERTSON and M. J. WILKINS: The mitochondrial membrane. Nature (Lond.) 171, 401 (1953). — FISCHER, H.: Über protoplasmatische Veränderung beim Altern von Pflanzenzellen. Protoplasma (Berl.) 39, 661–676 (1950). — FRANCK, J., and J. E. MAYER: An osmotic diffusion pump. Arch. of Biochem. 14, 297–313 (1947). — FREY-WYSSLING, A.: Physiology of cell wall growth. Annual Rev. Plant. Physiol. 1, 169–182 (1950). — Submicroscopic morphology of protoplasm. Houston: Elsevier Publ. Co. 1953.

GÄUMANN, E., u. O. JAAG: Die physiologischen Grundlagen des parasitogenen Welkens. II. Ber. schweiz. bot. Ges. 57, 132–147 (1947). — GIBBS, R. D.: Studies of wood. II. The water content of certain Canadian trees, and changes in the water-gas system during seasoning and flotation. Canad. J. Res. 12, 727–760 (1935). — GOLDACRE, R. J.: The folding and unfolding of protein molecules as a basis of osmotic work. Int. Rev. Cytol. 1, 135–164 (1952). — GROSSENBACHER, K. A.: Autonomic cycle of rate of exudation of plants. Amer. J. Bot. 26, 107–109 (1939). — GUILLIERMOND, A.: The cytoplasm of the plant cell. Waltham, Mass.: Chronica Botanica Co. 1941. — GUTTENBERG, H. v., u. A. BEYTHIEN: Über den Einfluß von Wirkstoffen auf die Wasserpermeabilität des Protoplasmas. Planta (Berl.) 40, 36–69 (1951). — GUTTENBERG, H. v., u. G. MEINL: Über den Einfluß von Wirkstoffen auf die Wasserpermeabilität des Protoplasmas. II. Über den Einfluß des $p_H$-Wertes und der Temperatur auf die durch Heteroauxin bedingten Veränderungen der Wasserpermeabilität. Planta (Berl.) 40, 431–442 (1952).

HAAN, I. DE: Protoplasmaquellung und Wasserpermeabilität. Rec. Trav. bot. néerl. 30, 234–335 (1933). — HACKETT, D. P.: The osmotic change during auxin-induced water uptake by potato tissue. Plant Physiol. 27, 279–284 (1952). — HACKETT, D. P., and K. V. THIMANN: The action of inhibitors on water uptake by potato tissue. Plant Physiol. 25, 648–652 (1950). — The nature of the auxin-induced water uptake by potato tissue. Amer. J. Bot. 39, 553–560 (1952). — The nature of the auxin-induced water uptake by potato tissue. II. The relation between respiration and water absorption. Amer. J. Bot. 40, 183–188 (1953). — HAGAN, R. M.: Autonomic diurnal cycles in the water relations of nonexuding detopped root systems. Plant Physiol. 24, 441–454 (1949). — HAINES, F. M.: The dynamics of cell expansion by turgor. Ann. Bot. N. S. 15, 219–250 (1951). — HASMAN, M.: Investigations of the water exchange of potato tissue under the effect of 6-aminoundecane and di-n-amylacetic acid. Physiol. Plantarum (Copenh.) 6, 187–198 (1953). — HEILBRUNN, L. V.: An outline of general physiology. Philadelphia: W. B. Saunders Company 1952. — HEILBRUNN, L. V., and K. DAUGHERTY: The action of ultraviolet rays on amoeba protoplasm. Protoplasma (Wien) 18, 596–619 (1933). — HEINMANN, M.: Abhängigkeit des Blutungsverlaufs von Beleuchtung und Blattzahl. Planta (Berl.) 40, 377–390 (1952). — HOAGLAND, D. R., and T. C. BROYER: Accumulation of salt and permeability in plant cells. J. Gen. Physiol. 25, 865–980 (1942). — HODGSON, R. H.: A study of the physiology of mycorrhizal roots on *Pinus taeda*. M. A. Thesis, Duke Univ., 1953. — HÖBER, R.: Physical chemistry of cells and tissues. Philadelphia: P. Blakiston Son & Co. 1945. — HÖFLER, K.: Die plasmolytisch-volumetrische Methode und Anwendbarkeit zur Messung des osmotischen Wertes lebender Pflanzenzellen. Ber. dtsch. bot. Ges. 35, 706–726 (1917). — Ein Schema für die osmotische Leistung der Pflanzenzelle. Ber. dtsch. bot. Ges. 38, 288–298 (1920). — New facts on water permeability. Protoplasma (Wien) 39, 677–683 (1950). — HOMÈS, M. V.: Perméabilité a l'eau et turgescence de la cellule végétale. Bull. Soc. roy. bot. belg. 75, 70–79

(1943). — HOPE, A. B., and R. N. ROBERTSON: Bioelectric experiments and the properties of plant protoplasm. Austral. J. Sci. 15, 197–203 (1953). — HUBER, B.: Beiträge zur Kenntnis der Wasserpermeabilität des Protoplasmas. Ber. dtsch. bot. Ges. 51, 53–64 (1933). — Wasserumsatz und Stoffbewegungen. Fortschr. Bot. 12, 227–249 (1951). — HUBER, B., and K. HÖFLER: Die Wasserpermeabilität des Protoplasmas. Jb. wiss. Bot. 73, 351–511 (1930). — HUCKENPAHLER, B. J.: Amount and distribution of moisture in a living short-leaf pine. J. Forestry 34, 399–401 (1936). — HYGEN, G., and J. KJENNERUD: Osmotic relations during cell expansion. Physiol. Plantarum (Copenh.) 5, 171–182 (1952). — HYLMO, B.: Transpiration and ion absorption. Physiol. Plantarum (Copenh.) 6, 333–405 (1953).

ILJIN, W. S.: Der Einfluß des Wassermangels auf die Kohlenstoffassimilation durch die Pflanzen. Flora (Jena) 116, 360–378 (1923). — Causes of death of plants as a consequence of loss of water: conservation of life in desiccated tissues. Bull. Torrey Bot. Club 80, 166–177 (1953).

KAMIYA, N.: The protoplasmic flow in myxomycete plasmodium as revealed by a volumetric analysis. Protoplasma (Wien) 39, 344–357 (1950). — KELLY, S.: The relations between respiration and water uptake in the oat coleoptile. Amer. J. Bot. 34, 521–526 (1947). — KERR, T., and D. B. ANDERSON: Osmotic quantities in growing cotton bolls. Plant Physiol. 19, 338–349 (1944). — KOZLOWSKI, T. T.: Transpiration rates of some forest tree species during the dormant season. Plant Physiol. 18, 252–260 (1943). — KRAMER, P. J.: The relation between rate of transpiration and rate of absorption of water in plants. Amer. J. Bot. 24, 10–15 (1937). — Causes of decreased absorption of water by plants in poorly aerated media. Amer. J. Bot. 27, 216–220 (1940). — Species differences with respect to water absorption at low soil temperatures. Amer. J. Bot. 29, 828–832 (1942). — Plant and soil water relationships. New York: McGraw-Hill Book Co. 1949. — KRAMER, P. J., and H. H. WIEBE: Longitudinal gradients of $P^{32}$ absorption in roots. Plant Physiol. 27, 661–674 (1952). — KÜSTER, E.: Über Vakuolenkontraktion in gegerbten Zellen. Protoplasma (Berl.) 39, 14–22 (1950).

LEPESCHKIN, W. W.: Light and the permeability of protoplasm. Amer. J. Bot. 17, 953–970 (1930). — Influence of temperature and light upon the exosmosis and accumulation of salts in leaves. Amer. J. Bot. 35, 254–259 (1948). — Über die Struktur und den molekularen Bau der lebenden Materie. Protoplasma (Wien) 39, 222–243 (1950). — LEVITT, J.: The thermodynamics of active (non-osmotic) water absorption. Plant Physiol. 22, 514–525 (1947). — The role of active water absorption in auxin-induced water uptake by aerated potato discs. Plant Physiol. 23, 505–515 (1948). — Toward a clearer concept of osmotic quantities in plant cells. Science (Lancaster, Pa.) 113, 228–231 (1951a). — Frost, drought, and heat resistance. Annual Rev. Plant Physiol. 2, 245–268 (1951b). — Further remarks on the thermodynamics of active (non-osmotic) water absorption. Physiol. Plant. 6, 240–252 (1953). — LEVITT, J., and G. W. SCARTH: Frost-hardening studies with living cells. II. Permeability in relation to frost resistance and the seasonal cycle. Canad. J. Res., Sect. C 14, 285–305 (1936). — LEVITT, J., G. W. SCARTH and R. D. GIBBS: Water permeability of isolated protoplasts in relation to volume change. Protoplasma (Wien) 26, 237–248 (1936). — LOOMIS, W. E.: Daily growth of maize. Amer. J. Bot. 30, 594–601 (1934). — LUNDEGÅRDH, H.: Bleeding and sap movement. Ark. Bot. Ser. A 31 (2), 1–56 (1944). — The translocation of salts and water through wheat roots. Physiol. Plantarum (Copenh.) 3, 103–151 (1950). — LYON, C. J.: A non-osmotic force in the water relations of potato tubers during storage. Plant Physiol. 17, 250–266 (1942).

MASON, T. G., and E. PHILLIS: Experiments on the extraction of sap from the vacuole of the leaf of the cotton plant and their bearing on the osmotic theory of water absorption by the cell. Ann. Bot., N. S. 3, 531–544 (1939). — MAXIMOV, N. A.: The plant in relation to water. English trans. by YAPP. London: Allen a. Unwin 1929. — MCDERMOTT, J. J.: The effect of the method of cutting on the moisture content of samples from tree branches. Amer. J. Bot. 28, 506–508 (1941). — MEYER, B. S.: The water relations of plant cells. Bot. Review 4, 531–547 (1938). — A critical evaluation of the terminology of diffusion phenomena. Plant Physiol. 20, 142–164 (1945). — MEYER, B. S., and D. B. ANDERSON: Plant Physiology. New York: D. van Nostrand 1952. — MEYER, B. S., and A. M. WALLACE: A comparison of two methods of determing the diffusion pressure deficit of potato tuber tissue. Amer. J. Bot. 28, 838–843 (1941). — MICHAELIS, P.: Ökologische Studien an der alpinen Baumgrenze. IV. Zur Kenntnis des winterlichen Wasserhaushaltes. Jb. wiss. Bot. 80, 169–247 (1934). — MILLER, E. C.: Plant physiology. New York: McGraw-Hill Book Co. 1938. — MONTERMOSO, J. C., and A. R. DAVIS: Preliminary investigation of the rhythmic fluctuations in transpiration under constant environmental conditions. Plant Physiol. 17, 473–480 (1942). — MONTFORT, C., u. H. HAHN: Atmung und Assimilation als dynamisches Kennzeichen abgestufter Trockenresistenz bei Farnen und höheren Pflanzen. Planta (Berl.) 38, 503–515 (1950). — MYERS, G. M. P.: The water permeability of unplasmolyzed tissues. J. of Exper. Bot. 2, 129–144 (1951).

NORTHEN, H. T.: Relationship of dissociation of cellular proteins by incipient drought to physiological processes. Bot. Gaz. **104**, 480–485 (1943).

OSTERHOUT, W. J. V.: Water relations in the cell. J. Gen. Physiol. **29**, 73–78 (1945). — Some aspects of secretion. I. Secretion of water. J. Gen. Physiol. **30**, 439–447 (1947). — OVERBEEK, J. VAN: Water uptake by excised root systems of the tomato due to non-osmotic forces. Amer. J. Bot. **29**, 677–683 (1942). — Auxin, water uptake and osmotic pressure in potato tissue. Amer. J. Bot. **31**, 265–269 (1944).

PALADE, G. E.: An electron microscope study of the mitochondrial structure. J. Histochem. a. Cytochem. **1**, 188–211 (1953). — PALVA, P.: Die Wasserpermeabilität der Zellen von *Tolypellopsis Stelligera*. Protoplasma (Berl.) **32**, 265–271 (1939). — PARKER, J.: Desiccation in conifer leaves: anatomical changes and determination of the lethal level. Bot. Gaz. **114**, 189–198 (1952). — PHILLIS, E., and T. G. MASON: Studies in foliar hydration in the cotton plant. VI. A gel theory of cell water relations. Ann. of Bot. **9**, 297–334 (1945). — PRELL, H.: Untersuchungen über die Aufnahme von Anelektrolyten in Zellen di- und polyploider Pflanzen. Planta (Berl.) **41**, 480—508 (1952).

RABER, O.: Water utilization by trees, with special reference to the economic forest species of the north temperate zone. U. S. Dept. Agric. Misc. Publ. **1937**, 257. — REINDERS, D. E.: The process of water intake by discs of potato tuber tissue. Proc. Kon. Akad. Wetensch. Amsterdam **41**, 820–831 (1938). — Intake of water by parenchymatic tissue. Rec. Trav. bot. néerl. **39**, 1–140 (1942). — RICHARDS, I. A., and C. H. WADLEIGH: Soil water and plant growth. Agronomy, Bd. 2, S. 73–251, New York: Academic Press. 1952. — ROBERTS, O., and S. A. STYLES: An apparent connection between the presence of colloids and the osmotic pressure of conifer leaves. Sci. Proc. Roy. Soc. Dublin **22**, 119–125 (1939). — ROBINSON, J. R.: The active transport of water in living systems. Biol. Rev. Cambridge Philos. Soc. **28**, 158–192 (1953). — ROSENBERG, TH., and W. WILBRANDT: Enzymatic processes in cell membrane penetration. Int. Rev. Cytol. **1**, 65–92 (1952). — ROSENE, H. F.: Comparison of rates of water intake in contiguous regions of intact and isolated roots. Plant Physiol. **16**, 19–38 (1941). — Reversible azide inhibition of oxygen consumption and water transfer in root tissue. J. Cellul. a. Comp. Physiol. **30**, 15–30 (1947). — Ageing and the influx of water into radish root-hair cells. J. Gen. Physiol. **34**, 65–73 (1950a). — The effect of anoxia on water exchange and oxygen consumption of onion root tissues. J. Cellul. a. Comp. Physiol. **35**, 179–193 (1950b). — ROSENE, H. F., and L. E. BARTLETT: Effect of anoxia on water influx of individual radish rood hair cells. J. Cellul. a. Comp. Physiol. **36**, 83–96 (1950). — ROSENE, H. F., and A. M. J. WALTHALL: Velocities of water absorption by individual root hairs of different species. Bot. Gaz. **111**, 11–21 (1949). — RUHLAND, W. u. U. HEILMANN: Über die Permeabilität von *Beggiatoa mirabilis* für Anelektrolyte bei Narkose mit den homologen Alkoholen $C_1 - C_9$ als Beitrag zur Ultrafiltertheorie. Planta (Berl.) **39**, 91–120 (1951).

SAYRE, J. D.: Physiology of the stomata of *Rumex patientia*. Ohio J. Sci. **26**, 233–267 (1926). — SCARTH, G. W., and J. LEVITT: The frost-hardening mechanism of plant cells. Plant Physiol. **12**, 51–78 (1937). — SCHROEDER, R. A.: The effect of root temperature upon the absorption of water by the cucumber. Missouri Agric. Exper. Sta. Res. Bull. **1939**, 309. — SEEMAN, F.: Zur $c_H$-Abhängigkeit der Wasserpermeabilität des Protoplasmas. Protoplasma (Wien) **39**, 147–175 (1950a). — Der Einfluß der Wärme und UV-Bestrahlung auf die Wasserpermeabilität des Protoplasmas. Protoplasma (Wien) **39**, 535–566 (1950b). — Der Einfluß von Neutralsalzen und Nichtleitern auf die Wasserpermeabilität des Protoplasmas. Protoplasma (Wien) **42**, 109–132 (1953). — SEIFRIZ, W.: The structure of protoplasm. Ames Iowa State College Press 1942. — SHIRK, H. G.: Freezable water content and the oxygen respiration in wheat and rye grain at different stages of ripening. Amer. J. Bot. **29**, 105–109 (1942). — SHREVE, E. B.: An analysis of the causes of variations in the transpiring power of cacti. Physiol. Res. **2**, 73–127 (1916). — SIMONIS, W.: Untersuchungen zum Dürreeffekt. I. Morphologische Struktur, Wassergehalt, Atmung und Photosynthese feucht und trocken gezogener Pflanzen. Planta (Berl.) **40**, 313–332 (1952). — SKENE, M.: The permeability of the cellulose cell wall. Ann. of Bot. **7**, 261–273 (1943). — SPANNER, D. C.: The suction potential of cells and some related topics. Ann. of Bot. **16**, 379–407 (1952). SPOEHR, H. A.: The carbohydrate economy of cacti. Carnegie Instn. Publ. **1919**, 287. — SPONSLER, O. L., and J. D. BATH: Molecular structure in protoplasm. In: The structure of protoplasm. Ames Iowa State College Press 1942. — SPONSLER, O. L., J. D. BATH and J. W. ELLIS: Water bound to gelatin as shown by molecular structure studies. J. Physic. Chem. **44**, 996–1006 (1940). — STÅLFELT, M. G.: The influence of light upon the viscosity of protoplasm. Ark. Bot. (Stockh.) Ser. A **33** (4), 1–17 (1946). — STAMM, A. J.: Surface properties of cellulosic materials. In L. E. WISE, Wood chemistry, 449–550. New York: Rheinhold 1944. — STARK, A. L.: Unfrozen water in apple shoots as related to their winter hardiness. Plant Physiol. **11**, 689–711 (1936). — STERN, K.: Über elektroosmotische Erscheinungen und ihre Bedeutung für pflanzen-physiologische Erscheinungen. Z. Bot. **11**,

561–604 (1919). — Steward, F. C., P. R. Stout and C. Preston: The balance sheet of metabolites for potato discs showing the effect of salts and dissolved oxygen on metabolism at 23⁰ C. Plant Physiol. 15, 409–447 (1940). — Stiles, W.: Permeability. London: Wheldon a. Wesley 1924. — Stocker, O.: Beiträge zu einer Theorie der Dürreresistenz. Planta (Berl.) 35, 445–465 (1948). — Street, H. E., and J. S. Lowe: The carbohydrate nutrition of tomato roots. II. The mechanism of sucrose absorption by excised roots. Ann. of Bot. 14, 307–329 (1950). — Strugger, S.: Die lumineszenzmikroskopische Analyse des Transpirationsstromes in Parenchymen. I. Die Methode und die ersten Beobachtungen. Flora 133, 56–68 (1938). — Studener, O.: Über die elektroosmotische Komponente des Turgors und über chemische und Konzentrationspotentiale pflanzlicher Zellen. Planta (Berl.) 35, 427–444 (1947).

Thimann, K. V.: Studies on the physiology of cell enlargement. Growth, Symposium 10, 5–22 (1951). — Thoday, D.: On turgescence and the absorption of water by the cells of plants. New Phytologist 17, 108–113 (1918). — Turgor pressure and wall pressure. Ann. of Bot. 16, 129–131 (1952). — Thut, H. F., and W. E. Loomis: Relation of light to growth of plants. Plant Physiol. 19, 117–130 (1944). — Toth, A.: Quantitative Untersuchungen über die Wirkung der UV-Bestrahlung auf die Plasmapermeabilität. Österr. bot. Z. 96, 161–195 (1949). — Ts'o, P., and G. P. Steinbauer: Effect of maleic hydrazide on auxin-induced water uptake by pea stem segments. Science (Lancaster, Pa.) 118, 193–194 (1953).

Url, W.: Permeabilitätsverteilung in den Zellen des Stengels von Taraxacum officinale und anderer krautiger Pflanzen. Protoplasma (Wien) 40, 475–501 (1951). — Ursprung, A.: Zur Kenntnis der Saugkraft. VII. Eine neue vereinfachte Methode zur Messung der Saugkraft. Ber. dtsch. bot. Ges. 41, 338–343 (1923). — Ursprung, A., and G. Blum: Zur Methode der Saugkraftmessung. Ber. dtsch. bot. Ges. 34, 525–539 (1916).

Virgin, H. I.: The effect of light on the protoplasmic viscosity. Physiol. Plantarum (Copenh.) 4, 255–357 (1951). — An action spectrum for the light induced changes in the viscosity of plant protoplasm. Physiol. Plantarum (Copenh.) 5, 575–582 (1952). — Physical properties of protoplasm. Annual Rev. Plant Physiol. 4, 363–382 (1953).

Walter, H.: Die Hydratur der Pflanze und ihre physiologisch-ökologische Bedeutung. Jena: Gustav Fischer 1931. — Kritisches zur Darstellung der osmotischen Zustandsgrößen in den verschiedenen Lehrbüchern der Botanik. Planta (Berl.) 40, 550–554 (1952). — Weatherley, P. E.: Note on the diurnal fluctuations in water content of floating leaf disks. New Phytologist 46, 276–278 (1947). — Weber, F.: Plasmolysezeit und Lichtwirkung. Protoplasma (Wien) 7, 256–258 (1929). — Whitfield, C. J.: Ecological aspects of transpiration. II. Pikes Peak and Santa Barbara regions: edaphic and climatic aspects. Bot. Gaz. 94, 183–196 (1932). — Wilson, C. C.: Diurnal fluctuations in growth in length of tomato stem. Plant. Physiol 23, 156–157 (1948). — Wilson, C. C., W. R. Boggess and P. J. Kramer: Diurnal fluctuations in the moisture content of some herbaceous plants. Amer. J. Bot. 40, 97–100 (1953).

Yarwood, C. E.: Water loss from fungus cultures. Amer. J. Bot. 34, 514–520 (1947).

Ziegenspeck, H.: Fluoroskopische Versuche an Blättern über Leitung, Transpiration und Abscheidung von Wasser. Biol. generalis (Wien) 18, 254–326 (1945).

# Bound water.

By

## Paul J. Kramer.

With 4 Figures.

## Introduction.

The concept of "bound water" is based on observations by numerous workers that a variable fraction of the total water in many organic and inorganic colloidal systems does not behave as it would if it existed in the same condition as water in a container, i.e. as "free water". This so-called bound water does not act as a solvent, it remains unfrozen at temperatures many degrees below zero centigrade, it is not easily removed by pressure, and it appears to be unavailable for certain physiological processes. Although it seems to have been observed first in inorganic systems its possible importance in the life processes of living organisms was soon appreciated and numerous studies have been made. Space does not permit discussion of all of these studies but enough papers will be cited to indicate various views which have existed as well as some opinions concerning the physiological significance of bound water.

FOOTE and SAXTON (1916, 1917) while studying gels of silica, alumina, ferric hydroxide, and a mixture of lamp-black and water by the dilatometer method found that a portion of the water in each of these systems remained unfrozen. This they termed "combined water". BOUYOUCOS (1917) used the dilatometer method to measure the fraction of soil water which remained unfrozen at —40⁰ C. He termed this fraction "unfree water" and regarded it as unavailable for plant growth. According to GORTNER (1938) the first suggestion that water might exist in a special condition in living organisms was made by BALCAR, SANSUM, and WOODYATT (1919). They proposed as a theoretical explanation of fever that there might be a bound — free water equilibrium in the body tissues, disturbance of which would produce pathological symptoms such as fever, but they knew of no technique by which such an equilibrium could be studied. NEWTON (1922) concluded from pressure-dehydration studies that winter wheat plants exposed to low temperatures bound water on colloidal material in their tissues and devised a technique to measure the amount of bound water. This was described by NEWTON and GORTNER (1922). RUBNER (1922) independently concluded that protoplasmic colloids in animal tissue bind water and developed a calorimetric method for measuring it.

During the next 20 years a large amount of research was done on methods of measuring bound water and on its physiological importance in relation to cold and drought resistance. Most of this work was done in North America and Russia as the bound water concept was never very popular with plant physiologists in western Europe. Unfortunately most of the Russian work is available only as abstracts, but their results seem to be similar to those obtained elsewhere and failure to refer directly to it will not distort our conclusions. It is impossible to cite all of the published papers, but enough representative ones will be discussed to present various methods and viewpoints.

# Definitions of bound water.

It is difficult to define or describe bound water with any great precision or in a few words. As WEISMANN (1938) wrote, „Was mit dem Ausdruck gebundenes Wasser gemeint ist, läßt sich nicht mit wenig Worten sagen." In practice bound water must be defined in terms of the method used to measure it. Over a dozen techniques have been described to measure bound water and these will be discussed later. The methods most commonly used in plant physiological research depend on the assumptions that bound water does not function as a solvent and that it does not freeze at a low temperature, usually arbitrarily set at $-20^\circ$ C. These are somewhat arbitrary definitions laid down for convenience in comparing the amounts of bound water in various systems. Obviously the amount of bound water in any system may vary somewhat with the method used to measure it. For this and other reasons the concept has been severely criticized. WALTER (1931) complained that no sharp distinction is possible between free and bound water, and GREENBERG and GREENBERG (1933) and HILL (1930) found so little bound water in the material they studied that they concluded that it could be of little physiological importance. GORTNER (1932) has a good review of the earlier work and the contradictary views on this subject. In contrast to these views OKUNTSOV and TARASOVA (1952) claim their studies indicate that little or no truly free water exists in leaves of barley, wheat, or begonia.

Perhaps the nature of bound water can be understood better after consideration of the forces responsible for its existence. Such consideration makes it obvious that no sharp distinction between bound and free water is possible, and that the distinction which is made is based on the magnitude of the forces rather than on any difference in the nature of the forces with which molecules are held. Even in the liquid form water molecules are associated together in a more or less regular structure by means of hydrogen bonds. A hydrogen bond is an hydrogen atom strongly associated with two electronegative atoms, holding them together. In water this bonding occurs between the oxygen atoms with the result that all of the molecules in a mass of water may be regarded as weakly associated together. If water molecules come in contact with polar surfaces which have a greater electrical attraction for them than they have for each other they are adsorbed on these surfaces. Such water molecules are held more firmly and packed more closely than they are in liquid water. This adsorbed water is equivalent to much of what is often termed imbibed water; and the more firmly held part of it constitutes "bound" water. GORTNER (1938, p. 304) states that, "We may conclude therefore that the forces which bind water on the surface of the lyophilic colloids are of the same nature as the forces which immobilize water molecules in the ice crystal lattice. However, there is evidence that these forces on a surface or at an interface may be of greater magnitude than the forces of association of water molecule for water molecule or the forces which tend to arrange water molecules in the ordinary ice crystal lattice." Therefore at least part of the molecules of the bound water film may be expected to have an activity which is less than the activity of the $H_2O$ molecule in the ordinary ice lattice. There is no doubt that the forces holding water on the surfaces of hydrophilic colloids are very great because at least part of this water cannot be removed by the forces of ice crystal formation at $-20^\circ$ C. or lower, or in some instances by heating to temperatures of several hundred degrees. Proteins are particularly effective in binding water because they bind it in two places, by the hydrophilic end groups of the side chains and by the oxygen and nitrogen atoms of the peptide linkages (BULL, 1948, p. 234; also LLOYD 1933). SPONSLER, BATH, and ELLIS (1940) were able to calculate with considerable accuracy the amount of water which could be bound by a protein from its amino acid content.

Another example of bound water is that associated with hydrated molecules and ions. Sucrose, for example, is supposed to have six molecules of water associated or bound to each sucrose molecule (SCATCHARD, 1921) and other molecules and ions are hydrated to greater or lesser degrees. CHANDLER (1941) believed that too much emphasis is placed on water binding by colloids and too little on crystalloids because thermodynamic studies show relatively little orientation of the molecules in the dilute colloidal systems which he studied. WEISMANN (1938) also felt that the methods used do not distinguish between the water bound by colloids and by solution and that too much emphasis is placed on colloids.

Table 1. *A comparison of three methods of measuring the bound water in a gum arabic sol and in corn leaf sap* (SAYRE 1932).

| Determination Gum arabic | Percentage of total water which is bound | | |
|---|---|---|---|
| | Cryoscopic Method | Dilatometric Method | Calorimetric Method |
| 1 | 10.84 | 13.11 | 10.59 |
| 2 | 15.59 | 9.17 | 11.97 |
| 3 | 12.23 | 11.80 | 14.12 |
| 4 | 13.00 | 11.32 | 11.07 |
| 5 | 11.10 | 13.68 | 13.87 |
| 6 | 10.92 | 13.59 | 15.80 |
| Mean | 12.28 | 12.11 | 12.90 |
| Corn leaf sap | 5.8 | 12.7 | 14.8 |

Another way of defining bound water is in terms of lowered "activity". Activity refers to the thermodynamically effective concentration of a substance in contrast to its molar concentration. The presence of solutes or the adsorption of water molecules on surfaces lowers its activity, as shown by a decrease in vapor pressure. BRIGGS (1931) measured the vapor pressure of water associated with various hydrophilic colloids as a function of water content and found that it decreased with decreased water content because smaller amounts of water were held more firmly.

It is hoped that this discussion has made it clear that no sharp distinction between bound and free water is possible. The amount of bound water

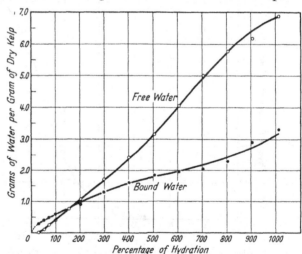

Fig. 1. Effect of change in total water content (percentage of hydration) on relative amounts of bound and free water present in pieces of kelp stipe (CHRYSLER 1934).

present depends on the method used to measure it (Table 1); the total amount of water present in the system (Fig. 1); and the environment (Table 2). At low water contents the total amount of bound water present may decrease though the percentage of total water which is bound usually increases (CHRYSLER 1934). CRAFTS, CURRIER, and STOCKING (1949) suggest that at low water contents water is held very firmly in monomolecular films by hydrogen bonds, at intermediate contents the layers are thicker and less firmly bonded, and as saturation is approached the additional increments of water are held very loosely by capillary condensation.

Table 2. *Relation of bound water content to season and environment.*

| Investigator | Species | Season or Treatment | Bound Water as % of total $H_2O$ | Bound $H_2O$ as % of dry wt. |
|---|---|---|---|---|
| Meyer 1932 | Pinus rigida | Aug. | 27.0 | 43.2 |
| | | Jan. | 28.2 | 37.9 |
| Schopmeyer 1939 | P. taeda | Moist soil | 22.10 | 38.65 |
| | | Dry soil | 30.8 | 36.9 |
| Schopmeyer 1939 | P. echinata | Moist soil | 18.16 | 34.15 |
| | | Dry soil | 25.00 | 37.16 |
| Greathouse 1934 | Varieties of clover | | | |
| | Ohio (hardy) | Jan. | 16.5 | 50.7 |
| | French (nonhardy) | Jan. | 12.7 | 44.0 |
| | Ohio | March | 14.18 | 55.1 |
| | French | March | 5.42 | 23.68 |
| | Ohio | April | 10.32 | 55.79 |
| | French | April | 9.14 | 49.57 |

Bound water might also be defined in terms of physiological processes in the sense that it is or is not free to take part in certain processes. For example, the rate of respiration of wheat seeds increases with increase in moisture content,

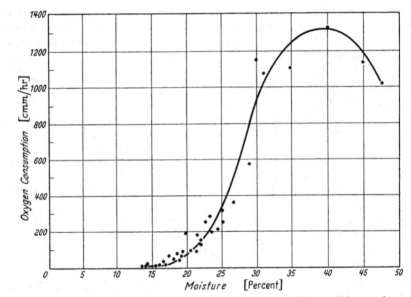

Fig. 2. Relation between moisture content of oats and rate of respiration. Note rapid increase in respiration with increase in moisture content above 16% (Bakke and Noecker 1933).

but not in a regular fashion (Bailey and Gurjar 1918). Increase in moisture content had little effect on respiration up to 14.5%, but a further increase of only 1% resulted in a 200% increase in respiration. Bakke and Noecker (1933) found a similar situation in oats, as shown in Fig. 2. This situation will be discussed further in a later section.

These observations suggest that most of the water below a certain critical moisture content is held too firmly to take part in physiological processes and only after an excess of free water is present can respiration and other processes begin to increase in rate. Somewhat similar relationships have been observed for the germination of fungus spores and the growth of bacteria. Such behavior probably is characteristic of most structures and organisms which can be dehydrated without injury. Only after the imbibitional forces of the cell colloids are satisfied does sufficient water become available as a solvent and a reagent to permit an increase in rate of physiological processes.

## The measurement of bound water.

Soon after the concept of bound water was introduced various workers developed a number of methods for measuring the amount of bound water in various types of materials. GORTNER (1938) listed more than a dozen methods which have been used to measure bound water. Some of these merely demonstrate the presence of bound water and could not be used for quantitative measurements, others can be used only on certain kinds of tissue, and still others are difficult to perform. After briefly discussing each of these methods the three more commonly used ones will be described in detail.

1. The Cryoscopic Method. This is based on the assumption that bound water does not act as a solvent when solute is added to it. The water content of a sample of sap is determined and enough sucrose is added to produce a molar solution. The depression of the freezing point of the sap is determined and if it is higher than it should be this means that part of the water is unavailable to act as a solvent. This method of measuring bound water was worked out by NEWTON and GORTNER (1922) and will be discussed in detail later.

2. The Dilatometric Method. When water freezes the volume of ice is greater than the previous volume of water. If a system containing a known volume of water is frozen and the volume increase is less than expected, then a part of the water must have remained unfrozen. This method has been used by numerous workers, including FOOTE and SAXTON (1916) on inorganic systems, BOUYOUCOS (1917) on soil, and JONES and GORTNER (1932) on organic gels. It will be discussed in more detail later.

3. The Calorimetric Method. The calorimetric method was used by MÜLLER-THURGAU (1880, 1886) to study the freezing point of plant tissues and was applied to the measurement of bound water by RUBNER (1922) and THOENES (1925). It is also called the heat of fusion method because it depends on the fact that when 1 gram of water as ice melts, 79.75 calories of heat are absorbed from the environment. If tissue of known water content is cooled well below freezing (usually to — 20° C.) and then thawed out in a calorimeter, the heat absorbed can be calculated and compared with that which would have been absorbed if all of the water present in the tissue had been frozen. The procedure and the formula for calculating the amount of bound water by this technique will be given later.

4. The Direct Pressure Method. Differences in the amount of water expressed from various materials under a given pressure have been considered to indicate differences in the amount of water bound by the tissues. NEWTON (1922) found that he could express nearly twice as much sap per gram of leaf tissue from the leaves of the wheat variety containing the least bound water as from the leaves of the variety containing the most bound water as measured cryoscopically. MEYER (1928) found that the amount of water expressed from a gram of pine needles decreased as the needles became cold hardened and he regarded the amount of water pressed out of the unfrozen tissue as free water, while that forced out after freezing was regarded as bound water. MEYER (1932) later concluded that these differences were of little significance in relation to cold resistance. LLOYD and MORAN (1934) found that a large amount of water could be expressed from gelatin gels at pressures up to 8000 lbs. per square inch, (563 kg.p.sq.cm.) but between 8000 and 38,000 lbs. (563 to 2665 kg. p.sq.cm.) very little water was expressed, although appreciable amounts were present.

5. Refractometric Method. The amount of bound water in various colloidal systems has been measured by the use of the refractometer (DUMANSKII 1933). According to GORTNER this method is accurate and exact and it has been used by several workers in Russia (SIMINOVA 1939).

6. Polarimetric Method. This method is of limited usefulness because it can be used only if the solute is optically active and the colloid is optically inactive or can be removed

before making the reading. It was used by Koets (1931) to study the selective adsorption of water out of sucrose solutions of various concentrations by silica gels. Water adsorption was independent of concentration over the range of sucrose solutions which he studied.

7. The X-ray Method. Attempts have been made to determine the presence of bound water by use of X-rays because the presence of shells of oriented water molecules ought to give X-ray patterns similar to those produced by ice. Barnes and Hampton (1935) concluded from study of X-ray diffraction patterns that it is possible to decide when all the water is bound in a gelatin gel by the presence or the absence of certain characteristic ice bands. This method seems to have rather limited applicability because it merely indicates the presence of water associated in a certain lattice structure, but does not indicate the amount present.

8. Infra-red Absorption. Many years ago Coblentz (1911) stated that the infra-red transmission curve of a gelatin film did not resemble that for water, but indicated that the water was present as "water of constitution". Later studies of 35% gelatin sols showed the appearance of infra-red absorption bands characteristic of hydrogen bonds (Sponsler, Bath, and Ellis 1940). This method is probably useful in studies of the manner in which water is bound, but is not likely to be applicable to measurement of bound water in plant materials.

9. Heat of wetting. It is well known that when colloids imbibe water heat is released because water molecules lose part of their kinetic energy when they are adsorbed on interfaces and this often is demonstrated in teaching laboratories by measuring the temperature rise when dry starch is wetted in a vacuum bottle. Gortner (1938, p. 299) has summarized data showing that at low water contents 300 to 400 calories are released per gram of water adsorbed by silica gel and gelatin. When liquid water is converted to ice only 80 calories per gram are released, hence it may be assumed that the water molecules are bound somewhat more firmly by these gels than in ice. Dumanskii and Voitsekhovskii (1948) found this method to give satisfactory results with starch.

10. Specific Heat Method. Hampton and Mennie (1932, 1934) observed that part of the water in a colloidal gel has a specific heat which is abnormally low and they regarded this portion as "bound" water. By measuring the specific heats of various components of the system it is possible to calculate the amount of bound water.

11. Drying Method. Data obtained by drying materials at various temperatures shows that the measurable amounts of moisture are held so firmly by colloidal gels that they are retained even at high temperatures. Data obtained by Nelson and Hulett (1920) (also see Gortner 1938, pp. 293—294) show that wheat flour contains 1% and protein (edestin) 1.9% more water on a dry weight basis when heated to about 200° C. than is removed by drying in a vacuum oven at 100° C. It has been suggested that presence of this bound water results in too high values for oxygen and hydrogen in the analysis of proteins. Todd and Levitt (1951) measured the bound water content of fungus mycelium as the difference between the water content attained at equilibrium in a desiccator and the content when oven dried.

12. Osmotic Pressure Method. Levitt and Scarth (1936) calculated the bound water content of cells from plasmolytic determinations of the osmotic pressure of plant cells. They found considerable amounts in hardened tissue of catalpa and liriodendron, less in unhardened tissue, but none at all in hardened or unhardened cabbage. They believed the bound water occurs in the vacuolar sap rather than in the protoplasm.

13. Dielectric Constant Method. Marinesco (1931) concluded that appreciable amounts of the water in various sols and gels have a dielectric constant much lower than that of free water. This presumably is because the water molecules adsorbed on colloidal surfaces are held so firmly that they behave as a rigid system.

14. Vapor-pressure Method. The amount of lowering of the vapor pressure of free water by addition of a known amount of a nonelectrolyte is known. If addition of a known amount of sucrose results in an abnormally large decrease in vapor pressure this indicates that a certain amount of the water is bound and cannot act as a solvent. Hill (1930), using this method found very small amounts of bound water in blood, casein, and egg white, and concluded that bound water is of negligible significance. Briggs (1931, 1932) using another method of computing bound water from vapor pressure measurements found somewhat higher values, but lower than those obtained by other methods. Gortner (1938, p. 301) claims the vapor pressure method yields lower values than other methods because it assumes that an equilibrium is attained between the gel and the dispersion medium (water) when in fact gels never are strictly in equilibrium with their dispersion medium.

Weismann (1938) complained that these various methods really cannot differentiate between water bound by colloids and by solutes. He regards the calorimetric and dilatometric methods and the method of BRIGGS (1931) as measuring the same water, that which resists dehydration, and the results therefore ought to be comparable. The results obtained by several other methods including the cryoscopic method of NEWTON and GORTNER (1922) depend on experimental conditions and the method of calculation, hence the results obtained are not comparable with those obtained by the first mentioned methods. LEVITT (1951) claimed that if bound water has any role in hardiness it is the water bound in the protoplasm, not that in the cell or the plant as a whole, hence only that in the protoplasm should be measured. Unfortunately the methods commonly used either measure bound water in the plant sap (cryoscopic method) or in the entire plant.

Of these various methods only three have been widely used on biological materials, the cryoscopic, the dilatometric, and the calorimetric methods. The techniques of each of these methods will now be described in detail, for the most part following the methods of SAYRE (1932) who made a careful study of the three methods.

The Cryoscopic Method. This method was developed by NEWTON and GORTNER (1922) and is based on the assumption that bound water does not dissolve sucrose. The method requires a refractometer, apparatus for determining the freezing point of sap, and a press and press cylinder for expressing plant sap.

An important prerequisite for the successful utilization of plant sap in physiological studies is assurance that representative samples of sap are being obtained. For chemical analyses and measurements of such properties as conductivity and osmotic pressure the most satisfactory method is that which yields the largest amount of sap per unit of tissue (BROYER and HOAGLAND 1940). This usually requires prefreezing of the tissue, application of high pressures, and use of relatively small samples. MEYER (1928) found that although freezing in an ice-salt mixture and with solid $CO_2$ give similar yields of sap from pitch pine needles in the summer, freezing with solid $CO_2$ gives a much higher yield in the winter.

NEWTON and MARTIN (1930) regard prefreezing as undesirable for bound water measurements, a view later concurred in by MEYER (1932). They recommend storing the tissue in containers packed in buckets of crushed ice until the juice is expressed. The tissue is ground as finely as possible in a meat grinder, and the sap expressed immediately in a hydraulic press. This operation is best performed in a refrigerated room but if this is impossible the tissue, apparatus, and sap should be kept near freezing during the operation. NEWTON and MARTIN apply pressure very slowly to keep the sap flowing gradually until the pressure gauge on the press begins to show pressure, when the operation is stopped. The writer believes that it is preferable to apply a fixed pressure of perhaps 2000 lb. per sq. in. (140 kg. p. sq.cm.) for a given period of time such as 5 minutes. All who have studied the problem report an increase in yield of sap with increase of pressure up to at least several thousand pounds per square inch. While 2000 lb. per sq. in. may suffice for most herbaceous material, a pressure of 10,000 lb. (700 kg. p. sq.cm.) or more is sometimes necessary to obtain representative samples of sap from highly lignified leaves. BROYER and HOAGLAND (1940) found that as the sample size was decreased from 125 to 50 grams proportionately more of the total water was expressed. This occurs because in large samples less sap is expressed from the center of the tissue mass than from the outside. Care should be taken to arrange the tissue in the press cylinder so pressure will be applied uniformly over the entire mass. It is desirable to conduct a preliminary study in order to decide on the best method of obtaining sap before starting measurements on the sap. Perhaps the most important precaution is to use exactly the same technique, time, and pressure in obtaining all samples which are to be compared. Having obtained a sample of plant sap the freezing point depression is obtained in the usual manner (see WALTER 1931, or CRAFTS, CURRIER, and STOCKING 1949, pp. 90—91, for a summary of methods). Determine the refractive index of the sap with a good refractometer. Assuming that the total solids in the sap have the same refractive index as sucrose the percentage of solids can be obtained very rapidly from a table of refractometric readings and sugar contents found in many handbooks or it can be obtained directly from a refractometer equipped with a sugar scale (GORTNER and HOFFMAN 1922). Later experience has shown that this is not true of sap from wheat leaves (NEWTON and MARTIN 1930) or from corn leaves, the refractometer reading being too high for these saps, but it is true for sap from corn stems (Sayre 1932). It is therefore necessary to determine the actual dry matter content by oven drying samples of the various saps and correcting the refractometer readings accordingly. After determining the depression of the freezing point and the content of solid matter, weigh out very accurately enough sap to contain just 10 gm. of water and add 3.4218 gms. of sucrose (one hundredth of a mole). After the sucrose is dissolved (0.5 hr. will suffice) the freezing point depression of the sap plus sucrose is determined in exactly the same manner as that of the original sample of sap. If all the water were available as a solvent this should result in a one molar solution having a depression of the freezing point of 2.085° C., but if part of it is bound the freezing point depression will be greater than expected. These determinations should be replicated two or three times.

The percentage of bound water in the sap is calculated by the formula of Newton and Gortner (1922). In working form this is as follows:

$$\frac{D_1 - (D + 2.085)}{D_1 - D} \times 89.2 = \text{Bound water}$$

$D$ = freezing point depression of the sap

$D_1$ = freezing point depression of sap plus sucrose

2.085 = depression of freezing point of a weight molar sucrose solution

89.2 = percentage of free water in a weight molar solution of sucrose, assuming that the sucrose exists as a hexahydrate.

Grollman (1931) criticized this formula because it fails to take into account any dissolved salts in the original sap. When a correction is made for salts the bound water values are decreased and sometimes even become negative values.

Sayre (1932) preferred to eliminate accurate weighing of the sap and sucrose. After determining the refractive index and freezing point depression he measured out 10 ml. of sap and added about 3 gm. of sucrose, then determined the refractive index and freezing point depression of this mixture. The actual amount of sugar added was determined from the refractive index. While this increases the amount of calculating it eliminates two careful weighings. The bound water content is calculated by the same formula, but the value 2.085 must be adjusted to correspond to the actual amount of sugar added and 89.2 must be adjusted similarly. Sayre found this method preferable where numerous determinations were being made because graphs could be made from which variable values could be read.

*Advantages and Disadvantages of the Cryoscopic Method.* The cryoscopic method is particularly applicable to liquid or semiliquid material and of course cannot be used with nonliquid systems. Among its disadvantages are the fact that it measures only the amount of water binding colloids in expressed sap, but not that in the cell walls which in some tissues bind large amounts. Furthermore, it measures only the water which is not free to dissolve sucrose and it takes no account of the fact that freezing and addition of sucrose probably alter the free-bound water ratio. Although, measurements made by this method have shown good correlation with drought resistance in various species of grasses and varieties of wheat (Newton and Martin 1930), and with cold resistance in wheat varieties it is not recommended for measurement of the total bound water in plant tissue (Levitt 1951).

The Dilatometer Method. This method is based on the fact that when water freezes it expands about 9% in volume. Therefore if a sample of material of known water content is immersed in a liquid with a low freezing point which is immiscible with water and inclosed in a system in which small changes in volume can be measured, the expansion accompanying freezing can be measured and the amount of water which froze can be calculated. From this the amount of unfrozen or bound water can be calculated.

The essential parts of a dilatometer are a bulb to hold the material and a capillary tube in which the volume changes can be observed. Jones and Gortner (1932) used one with a bulb about 1.5 × 3 cm. with the capillary tube coming off from the middle perpendicular to its length. One end of the bulb was sealed, the other end closed with a carefully fitted glass stopper. SAYRE (1932) used BABCOCK cream test bottles which are convenient for liquid samples, but difficult to load and unload with pieces of solid tissue. JONES and GORTNER used toluene and SAYRE petroleum ether as the immiscible liquid. Each dilatometer must be calibrated for the amount of water and temperature range used. The reader is referred to the original papers for further details. A freezing bath or chamber with variable temperature control is required to cool the dilatometer.

The dilatometer method is simple and accurate and can be used for either solid or liquid systems. The chief difficulty encountered is in removing air from the system. This is very difficult with plant tissue and even though all air is removed bubbles often are formed during the freezing process by air coming out of solution as the temperature decreases. One of its chief advantages is the fact that the amount of frozen and unfrozen water can be observed at various temperatures.

*The Calorimetric Method.* This method was first used by MÜLLER-THURGAU (1880) in studying the freezing of plant tissue and was modified by RUBNER (1922) and THOENES (1925) for measurement of bound water. It has been used in essentially the form described here by ROBINSON (1928), ST. JOHN (1931), SAYRE (1932), MEYER (1932), CHRYSLER (1934), and GREATHOUSE and STUART (1934). This method also is based on the assumption that bound water does not freeze at some arbitrarily chosen temperature, usually — 20⁰ C. The tissue is frozen and then thawed in a calorimeter and during thawing the amount of unfrozen or bound water is calculated from the difference between the observed and calculated rise in temperature.

The equipment necessary to determine bound water by the calorimeter method consists of a freezing bath or cabinet in which the temperature can be maintained at — 20 to — 25⁰ C., a calorimeter which can be made from a dewar flask or pint vacuum bottle, some freezing tubes, and low temperature thermometers. ROBINSON (1928), GREATHOUSE (1935) and SCHOPMEYER (1939) froze their material in tinfoil cups inclosed in glass vials and transferred the cups and inclosed material to the calorimeter. Tubes of thin brass 10 to 15 cm. in length and 2.5 cm. in diameter, closed at both ends with rubber stoppers, can be used for freezing large samples of plant material. SAYRE (1932) recommends that one end of the tube should have a larger diameter than the other to facilitate dislodging the sample but MEYER (1932) and the writer have found cylindrical tubes satisfactory.

To make a series of determinations carefully weighed samples of plant tissue of convenient size, usually 20 or 25 grams, are placed in the freezing tubes or other containers. The tubes are immersed in a freezing bath at the desired temperature (usually — 20⁰ C.) for at least 3 or 4 hours or overnight to insure that the entire sample is in equilibrium with the bath. The tubes are then removed from the bath, one at a time, and the frozen sample in each transferred into the calorimeter. If the tube is warmed slightly by holding it in the hands, the sample can usually be driven out with a wooden plunger. A pint vacuum bottle containing 250 ml. of water which has come to equilibrium with the walls of the bottle makes an excellent calorimeter. It should be equipped with a calorimeter thermometer reading in 0.02⁰ over a range of about 15 to 35⁰ C. As soon as the sample is immersed in the water the calorimeter is stoppered and allowed to stand for a short time (3 min. is adequate), then the contents are stirred by shaking the calorimeter and the temperature recorded when an equilibrium is attained.

For calculation of the bound water the specific heat of the dry plant tissue and a correction factor for the calorimeter are necessary. The specific heat of the dry matter can be determined by substituting dried material for the fresh samples and benzene for water. Benzene is preferable to water for this purpose because its lower specific heat results in a greater change in temperature in the calorimeter than would occur with water, and because dry plant material will remain immersed in the benzene. The specific heat of dry plant tissue is in the neighborhood of 0.3. The correction factor for the calorimeter is determined by substituting a piece of ice of about the same volume and known weight for the sample. The procedure used is otherwise the same as with samples of plant tissue. It is easy to calculate what the equilibrium temperature ought to be for a piece of ice of known weight and any deviation from this represents heat lost by warming the walls of the calorimeter and the thermometer rather than being used in warming the sample.

The amount of free water — Wf (water which formed ice) — in the system is calculated as follows:

$$Wf = \frac{FNSw(T-Te) - [WdSd(Te-Ts) + WwSw(Te-Ts)]}{Q - [(Sb-Si)\ (Tm-Ts)]}$$

F = calorimeter correction factor
N = Number of grams of water in calorimeter
Sw = Specific heat of water over the temperature range T–Te
T = Original temperature of the water in the calorimeter
Te = Equilibrium temperature of water in the calorimeter
Wd = Weight of dry matter in the sample
Wb = Weight of bound (unfrozen) water in the sample
Wf = Weight of ice (free water) in the sample
Ww = Weight of all water in the sample, determined by oven drying representative samples
Sd = Average specific heat of dry matter in temperature range of experiment
Si = Average specific heat of ice in temperature range
Sb = Average specific heat of bound water in temperature range
Ts = Temperature of sample = temperature of freezing bath
Tm = Melting point of ice = freezing point of water
Te = Equilibrium temperature
Q = Heat of fusion of ice at Tm = 79.75
Bound water, Wb = Ww–Wf

Ts and Tm are below zero and should be used in the equation with negative signs.

The arguments on which this equation is based can be found in papers by Meyer (1932), Sayre (1932) and others, and will not be repeated here. Gortner (1938, p. 286) gives a simplified equation in which it is assumed that the specific heat of ice is 0.5 and that of water below 0° C. is 1.0. In fact the specific heat of water increases below 0° C. and that of ice decreases as it is cooled and for very precise work these facts should be taken into account.

This method has been found very satisfactory by those who have used it. One of its advantages is that determinations can be made on uninjured tissue. The only objection is that if the tissue contains much solute the bound-free water equilibrium may shift as the temperature is lowered and additional water may crystallize out. If all comparisons are made on tissue cooled to the same temperature this will not cause any serious errors.

*A Comparison of Various Methods of Determining Bound Water.*—Sayre (1932) has made a careful comparison of these methods, using an 18.6% sol of gum arabic. The results are summarized in table 1. It appears that the three methods gave essentially the same results on the gum arabic. Sayre also attempted to compare the three methods on sap expressed from corn leaves. The calorimeter method gave 14.8%, the dilatometer method 12.7%, and the cryoscopic method only 5.8%. The writer would expect the cryoscopic method to give lower results on plant tissue because it measures only the water bound by hydrophilic colloids and solutes in the sap while the other methods include that bound in the cell walls and protoplasm.

Dumanskii and Voitsekhovskii (1948) compared the cryoscopic, heat of fusion, specific gravity, drying, and heat of wetting methods on potato starch. They obtained bound water contents ranging from 31 to 38% by these methods, but 52—55% by the dilatometer method. The discrepant values obtained with the dilatometer method were attributed to errors caused by air bubbles.

## Amounts of bound water in plant tissue.

The amounts of bound water found in plant tissues vary widely in the species, the condition under which the plants are grown, and the methods used to measure it. Table 3 from Sayre (1932) shows the bound water content of a number of

Table 3. *Total, free, and bound water content of various materials determined by the calorimetric method. From* SAYRE *(1932).*

| Material | Date | Total water % of fresh wt. | Bound water % of total | Bound water as % of dry wt. | Freezing temperature ° C. | Treatment of tissue |
|---|---|---|---|---|---|---|
| Buckeye twigs | Feb. | 56.0 | 53.2 | 67.8 | —23 | Shavings |
| Buckeye twigs | Feb. | 54.0 | 54.2 | 64.2 | —23 | Whole pieces |
| Maple twigs | March | 44.0 | 53.6 | 42.2 | —24 | Shavings |
| Pine needles | March | 56.0 | 27.3 | 34.8 | —23 | Ground |
| Pine needles | March | 56.0 | 26.4 | 33.7 | —23 | Minced |
| Pine needles | April | 56.8 | 28.2 | 37.0 | —22 | Minced |
| Pine needles | April | 56.8 | 35.0 | 46.1 | —12 | Minced |
| Corn leaf blade | Aug. | 72.2 | 21.6 | 56.1 | —25 | Minced |
| Corn leaf blade | Aug. | 72.2 | 21.3 | 55.4 | —25 | Ground |
| Corn leaf sheath | Aug. | 78.3 | 13.3 | 47.9 | —25 | Ground |
| Corn stem, upper | Aug. | 78.3 | 11.2 | 31.0 | —25 | Ground |
| Corn stem, lower | Aug. | 80.9 | 7.9 | 33.5 | —25 | Ground |

different materials determined by the calorimetric method. These data suggest that twigs and pine needles contain a higher percentage of bound water than leaves and stems of herbaceous plants. The amount of bound water was the same in ground and minced tissue, suggesting that mechanical injury of the tissue has little effect on results obtained with this method. DAUGHTERS and GLENN (1946) measured the bound water content of a number of fruits and vegetables calorimetrically and some of their results are summarized in table 4. As expected the values are low, except for sweet corn which contains a large amount of water-binding colloids.

Table 4. *Bound water content of fresh fruit and vegetables when harvested for preservation by freezing. From* DAUGHTERS *and* GLENN *(1946).*

| Plant part | Total Water as % of fresh wt. | Bound Water as % of fresh wt. | Plant part | Total Water as % of fresh wt. | Bound Water as % of fresh wt. |
|---|---|---|---|---|---|
| Asparagus tips | 88.34 | 3.48 | Pea seeds | 80.40 | 4.97 |
| Sweet corn kernels | 84.78 | 33.40 | Raspberry fruits | 82.02 | 5.58 |
| Green beans | 90.63 | 1.03 | Strawberry fruits | 89.14 | 4.72 |

The water content of the plant definitely affects the amount of bound water in its tissues. Several workers have reported that as the total water content decreases the proportion which is bound increases. CHRYSLER (1939) found that as the total content of kelp stipe was increased the amount of bound water present also increased, but not proportionally, hence the percentage of water which was bound decreased with increasing moisture content although the absolute amount of bound water increased (see fig. 1). STARK (1936) on the other hand found the bound water content of apple twigs to increase in the winter although the total water content decreased (see fig. 3).

The apparent bound water content of plant tissue often varies with the method used to measure it, as shown in table 1 from data of SAYRE (1932). The temperature of freezing also affects the apparent bound water content as shown by data for pine needles from SAYRE and for red clover roots from GREATHOUSE (1935), shown in table 5.

Table 5. *Effect of temperature at which tissue is frozen on the bound water content as measured calorimetrically.*

| Investigator | Material | Total water as % of fresh wt. | Freezing Temperature ° C. | Bound water as % of total H₂O | Bound water as % of dry wt. |
|---|---|---|---|---|---|
| SAYRE 1932 | Pine needles | 56.8 | —12 | 35.0 | 46.0 |
|  |  |  | —22 | 28.2 | 37.0 |
| GREATHOUSE 1935 | Unhardened clover roots | 83.5 | —10 | 20.53 | 103.90 |
|  |  |  | —20 | 9.18 | 46.45 |
| GREATHOUSE 1935 | Hardened clover roots | 82.3 | —15 | 15.43 | 78.06 |
|  |  |  | —25 | 13.65 | 69.09 |
| GREATHOUSE 1935 | Unhardened clover roots | 81.8 | —10 | 14.47 | 72.16 |
|  |  |  | —20 | 7.36 | 38.66 |
|  |  |  | —30 | 1.99 | 11.76 |
|  |  |  | —40 | 1.95 | 11.54 |
|  |  |  | —50 | 1.43 | 10.59 |

# Physiological significance of bound water.

As soon as the concept of bound water was introduced to biologists it attracted wide attention as a possible means of explaining various physiological phenomena. Bound water seemed particularly promising to ecologists, physiologists and agronomists as a basis for cold and drought resistance of plants. Both types of injury involve dehydration of the protoplasm and it was thought that if a part of the water is bound firmly by the hydrophilic colloids of plants this ought to protect them from excessive dehydration by ice formation. The interest in bound water approached the proportions of a "fad", the output of papers being quite large in the late twenties and early thirties, then decreasing during the next decade and now a paper on bound water in plants is rare except in Russia. The reasons for the decrease in interest will become apparent after reading the following sections.

For convenience the work on the physiological significance of bound water will be discussed in several separate sections.

*Cold Resistance.* — For many years there has been a lively interest in the biochemical basis of the large differences in cold resistance found among various species and varieties of wild and cultivated plants. As much of the injury from cold was attributed to dehydration during freezing, interest began to center on factors which might increase resistance to dehydration during freezing. ROSA (1921) found that the amount of water per 100 g. of tissue frozen at —5 to —6° C. was greater in unhardened than in hardened vegetables and the increase in amount of unfreezable water was associated with an increase in the amount of hydrophilic colloids, especially soluble pentosans. This study was made by the dilatometer method and appears to be the first one in which increased cold resistance was correlated directly with the amount of bound water. The same year STRAUSBAUGH (1921) suggested that increase in cold resistance of plums was correlated with an increase in imbibitional forces with which water was held, but he made no measurements of bound water. In the following year NEWTON (1922) reported that differences in winter hardiness of wheat varieties were well correlated with the amount of sap that was retained by the leaves when subjected to a fixed pressure, the leaves of the hardiest varieties retaining the most water. MEYER (1928) found that there was relatively little change in water content of leaves of pitch pine (*Pinus rigida* Mill.) from summer to winter, but much less of the total sap could be expressed at a given pressure from the cold-resistant winter leaves. The increased water retaining power of the cold-resistant leaves was attributed to an increase in colloidal gels capable of binding water. DUNN and BAKKE (1926) attempted to measure the amount of hydrophilic colloids in hardened and unhardened tissue by measuring the amount of dye adsorbed by finely ground tissue. They found a good correlation between the amount of dye adsorbed by ground apple twigs and cold resistance of the trees from which the twigs were collected. Later DUNN (1937) concluded that this test was of doubtful value because no reliable correlation was found between cold resistance and dye adsorption by individual leaves or plants. ROBINSON (1927, 1928) adapted the calorimetric method to measurement of bound water in insects and found an increase in bound water content of certain insects as they were cold-hardened by exposure to low temperature.

Many other studies of the relation between bound water content and cold resistance have been made with varying and contradictory results. LEVITT (1941, 1951) has cited most of the work, and only a few papers will be mentioned here. MANEY (1931) found no correlation between the amount of water frozen and variety differences in cold resistance of apple. STARK (1936) found

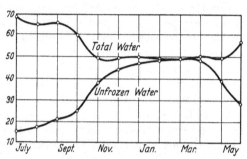

Fig. 3. Percentages of total and unfrozen or bound water in apple twigs at various seasons of the year (STARK 1936).

the bound water content of apple twigs to increase during the winter and total water content to decrease (see fig. 3), but he found no significant differences in bound water content between resistant and non-resistant varieties. STEINMETZ (1926) and WEIMER (1929) found no correlation between bound water content and relative cold resistance of alfalfa, but GRANDFIELD (1943) found that cold-resistant alfalfa plants contained more bound water than non-resistant ones. GREATHOUSE and STUART (1934) reported that more bound water was found in the roots of cold-resistant varieties of red clover than in the non-cold-resistant varieties, but VAN DOREN (1937) could find no correlation between bound water content of wheat leaves and hardiness.

It had been hoped that differences in bound water content might serve as a useful means of selecting cold-resistant strains, but these contradictory results soon shattered that hope. MARTIN (1927) stated that although many cold resistant varieties of wheat contained more bound water than non-resistant varieties, the most practical method of testing cold resistance was to expose the plants to low temperatures. This view is generally held today. MEYER (1932) measured the bound water content of pitch pine needles by the calorimetric method and found more free and bound water present per gram of dry weight in the summer than in the winter. He concluded that there was no evidence of any relation between bound water content and cold resistance in pitch pine. LEVITT (1941) after reviewing the work on bound water in relation to cold resistance admitted that many weaknesses and inconsistences exist, but believed that the concept should not be discarded entirely.

It seems that during cold hardening a number of detectible changes in plant tissue commonly but not invariably occur. Among those which have been observed are decrease in total water content, increase in bound water, increase in osmotic pressure and conductivity of expressed sap, increase in permeability, increase in true viscosity and decrease in structural viscosity (Kessler and Ruhland 1938, Scarth 1944), and changes in chemical constituents of the cells. Scarth suggested that increase in hydrophilic colloids might be the most important factor in cold hardening and Siminovitch and Briggs (1953) found a close correlation between changes in water soluble protein nitrogen and change in cold resistance in the living bark of black locust (Robinia pseudoacacia). If increase in these constituents really increases cold resistance it probably is because they increase the ability of the protoplasm to endure freezing without injury rather than by increasing resistance to dehydration.

It is difficult to separate cause and effect, but it seems likely that most of the changes in the more easily measured properties of plant tissue are largely the result of fundamental physicochemical changes produced by low temperature rather than the causes of resistance to injury from low temperature. The increase in osmotic pressure is partly caused by increase in soluble carbohydrates and partly by decrease in water content, and decrease in total water content seems to be responsible for much of the increase in conductivity and bound water which sometimes is observed. In some plant tissues much of the bound water occurs in the cell walls where it can scarcely protect the protoplasm and the small amounts which occur in the plant sap can scarcely be effective protection against dehydration of the protoplasm. It seems probable that the changes in bound water content usually result from changes in total water content and chemical composition and are not themselves the cause of increased cold resistance.

*Drought Resistance.* — As soon as bound water was reported to be correlated with cold resistance in some instances, studies were started to find if it was also correlated with drought resistance. Newton and Martin (1930) made an extensive study of this relationship in various grasses and wheats. As shown in table 6, they found the bound water content of the expressed sap to be higher

Table 6. *Relation between usual habitat, drought resistance and bound water content of grasses. Bound water was determined cryoscopically on expressed sap. From* Newton *and* Martin *(1930), Table XXVII.*

| Species | Drought Resistance | Usual Habitat | Bound Water as % |
|---|---|---|---|
| *Bouteloua gracilis,* Blue grama | High | Dry plains | 22.7 |
| *Stipa comata,* Western spear grass | High | Dry plains | 12.4 |
| *Agropyron cristatum,* Crested wheat grass | High | Dry plains | 11.7 |
| *A. tenerum,* Western rye grass | High | Dry plains | 10.3 |
| *Bromus inermis,* Awnless brome grass | High | Dry plains | 10.3 |
| *Agropyron smithii,* Western wheat grass | High | Dry plains | 7.7 |
| *Poa pratensis,* Kentucky bluegrass | Low | Well watered areas | 5.3 |
| *Phleum pratense,* Timothy | Low | Well watered areas | 4.5 |
| *Calamagrostis canadensis,* Blue joint grass | Low | Well watered areas | 3.5 |
| *Panicularia grandis,* Tall manna grass | Low | Wet sites | 3.3 |
| *Beckmannia erucaeformis,* Slough grass | Low | Sloughs | 2.3 |

in grasses from dry habitats than in those from moist habitats. The bound water content of the drought resistant varieties of wheat was also higher than that of less drought resistant varieties. They concluded that bound water content was a more reliable indicator of drought resistance than osmotic pressure. Whitman (1941) also studied drought resistance of prairie grasses in four different grassland types. He likewise found that as the habitat became drier the percentage of total water which was bound increased consistently, but this was caused by the decrease in total water content rather than by increase in water binding capacity. The absolute amount of bound water per gram of dry matter was not well correlated with dryness of the habitat and the differences in absolute amounts of bound water found among species were not related to differences in drought resistance.

KORSTIAN (1933) found some evidence of a correlation between drought resistance and bound water content in leaves of several tree species. SCHOP-MEYER (1939) measured changes in total water content, bound water, osmotic pressure and solute concentration of loblolly pine (*Pinus taeda* L.) and shortleaf pine (*Pinus echinata* Mill.) seedlings growing in soil near field capacity and near permanent wilting. No significant differences in amount of bound water per unit of dry weight were found between the species or between moist and dry soil (see Table 2). The percentage of total water which was bound was higher in the afternoon and in dry soil, but this was caused by decrease in total water content.

WELCH (1938) studied changes in bound water content of drying bryophyllum leaves. As the total moisture content decreased the percentage of total water which was bound increased, but there was no increase in absolute amount of bound water. WELCH concluded that the extraordinary ability of bryophyllum leaves to retain water could not be explained in terms of bound water. MIGAHID (1945) on the other hand, reported that the bound water content was higher in xerophytes than in mesophytes, and the bound water content of individual plants of a species was higher when the plants were grown in a dry habitat than when grown in a moist habitat.

Some of the objections which apply to bound water as an important factor in cold resistance also apply to it as a cause of drought resistance. The quantity of water bound in cell walls is likely to be higher in xerophytes than in meso-phytes, but this can scarcely protect the protoplasm against dehydration. As MAXIMOV (1929) stated some years ago, true drought resistance of the type found in many lichens, mosses, and ferns, and seeds and in a few flowering plants depends on the ability of the protoplasm to endure dehydration without injury rather than on prevention of dehydration. The slow rate of dehydration of bryophyllum leaves, cactus stems, and some other succulent tissues results from the presence of thick layers of cutin and few stomates which greatly reduce water loss. If these protective coverings are removed the water storage tissue dries out very rapidly and is killed. In some instances the colloidal constituents of these plants have a high water retaining capacity, but most of the water is held so loosely that it evaporates rapidly if exposed to the air. Only after the moisture content has fallen below that necessary for growth, or even for the maintenance of life, does the force with which the remaining water is held become great enough to seriously curtail water loss.

*Bound Water in Seeds.*—Although air-dry seeds usually contain 10% or more of water on an oven-dry basis they are inactive physiologically, as indicated by their low rate of respiration. As shown in fig. 2, when the moisture content increases above a certain critical value the rate of respiration suddenly begins to rise very rapidly. It has been suggested that this is because most of the water in an air-dry seed is so firmly bound by colloids that it cannot act as a reagent in hydrolytic or other processes. SHIRK and APPLEMAN (1940) reported that 93% of the water in wheat seeds soaked one-fourth hour was unfreezable at —25° C., i.e., was bound, and after soaking 5 hours the unfreezable water had decreased only to 50% of the total water. As the percentage of free water increased from 7 to 50% the rate of respiration increased from 0.5 to 89.4 cu. mm. of $O_2$ per gm. dry wt. per hour (see fig. 4). SHIRK (1942) reported that during maturation of wheat and rye seeds a very large decrease in free water content occurred and only a slight decrease in bound water, hence in a mature seed most of the water was bound. The rate of respiration during maturation decreased at almost the same rate as the decrease in amount of freezable water. These and other similar observations seem to indicate rather definitely that bound water can be regarded as largely unavailable for physiological processes.

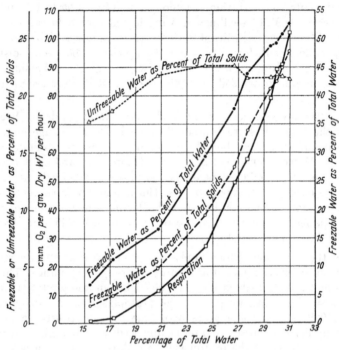

Fig. 4. Relation between freezable (free) water content and rate of respiration of Minhardi wheat (SHIRK and APPLEMAN 1940).

This unavailability of the water held by dry seeds extends to the fungi which often grow on them if they are too moist. Apparently some of these fungi can utilize water which is unavailable for the seed itself as they often attack corn which is too dry to sprout. Nevertheless, if corn is dried to about 12% of moisture so little water is available that the fungi which commonly cause spoilage cannot grow (SEMENIUK and GILMAN 1944).

The situation existing in seeds supports the view of the role of bound water proposed in the discussion of its role in drought and cold resistance. If bound water is unavailable for use in physiological processes of seeds it probably is equally unavailable for physiological processes in other plant structures. The amount of bound water present depends largely on the amount of hydrophilic colloids in the tissue and variations in the bound water content produced by varying intensities of cold and drought are caused by changes in total water content and chemical composition, rather than as a direct response to drought or low temperature.

It is well known that dry seeds are usually injured much less by freezing than those high in moisture content. MCROSTIE (1939), and LIPMAN (1936) were able to cool several kinds of seeds to 1.35° to 4.2° K. for 44 hours without injury after first drying them for two weeks in a partial vacuum over sulfuric acid. Perhaps dry seeds are uninjured at low temperatures because the remaining water is bound so firmly that it cannot be crystallized. Of course the low metabolic rate accompanying reduction in moisture content may also increase resistance to injury from freezing.

*Miscellaneous Observations.*—BOUTHILLIER and GOSSELIN (1937) found an appreciable amount of bound water (0.57 g. per g. dry wt.) in the brown alga, fucus. The amount varied inversely with the concentration of sea water, increasing when it was diluted and decreasing when it was made twice the normal concentration. FRIEDMAN and HENRY (1938) reported more bound water in spores than in vegetative cells of bacteria. They thought the decreased amount of free water in spores might account for their greater heat resistance, but it seems possible that it results from a change in protoplasmic structure which also gives increased resistance to heat and cold. LUYET and GEHENIO (1938) reported that moss plants containing less than 30% of water could be frozen in liquid air and thawed slowly without unjury, but at higher moisture contents slow thawing resulted in considerable injury while rapid thawing was accompanied by no injury regardless of the moisture content of the material. Apparently at moisture contents below 30% the water in these mosses was so firmly bound that it did not form crystals even at the temperature of liquid air. OUCHINNIKOV (1951) suggested that the shell of loosely bound water surrounding an enzyme hinders its activity by preventing close contact with the substrate.

LIPMAN (1936) showed that bacterial spores would survive temperatures near absolute zero. Later he found that several species of fungi and bacteria survived cooling to the temperature of liquid air while actively growing (LIPMAN 1937). He attributed this to the fact that water was so firmly bound by their protoplasm that it could not be crystallized, hence no serious disruption of structure occurred.

Measurements have been made of bound water in various animal tissues, dairy products, leather, peat, and a variety of other organic and inorganic materials. Most of this work has no direct bearing on the physiological significance of bound water and therefore will not be discussed.

## Summary.

The concept of "bound water" is based on observation that a variable fraction of the total water in some plant tissues does not act as a solvent and remains unfrozen at temperatures of $-20°$ C. or lower. Much of this water is held on the surfaces of hydrophilic colloids, but some is associated with hydrated ions and molecules. Bound water can be measured by several methods, but those used most frequently are the calorimetric, dilatometric, and cryoscopic methods. The calorimetric method seems most satisfactory for plant tissue.

The amount of bound water found in plant tissue varies with the species used, the environment in which the plants are grown, and the method used to measure it. The bound water content seems to be higher in woody than in herbaceous plants, higher in the winter than in the summer, and higher in plants from dry habitats than in those from moist habitats.

Attempts to explain cold and drought resistance of plants as the result of a high bound water content have not been satisfactory. It seems probable that changes in bound water content usually result from changes in total water content and in chemical composition and are not themselves the cause of cold or drought resistance. Much of the bound water in plants occurs in cell walls where it can scarcely affect the protoplasm, and water held so firmly that it cannot function as a solvent can scarcely take part in physiological processes. Bound water may be of some importance in seeds, spores, and other air-dry plant structures, but it probably is of little significance in growing plants.

## Literature.

BAILEY, C. H., and A. M. GURJAR: Respiration of stored wheat. J. Agricult. Res. 12, 685—713 (1918). — BAKKE, A. L., and N. L. NOECKER: The relation of moisture to respiration and heating in stored oats. Iowa Agricult. Exper. Stat. Res. Bull. 1933, 165. — BALCAR, J. O., W. D. SANSUM and R. T. WOODYATT: Fever and the water reserve of the body. Arch. Int. Med. 24, 116—128 (1919). — BOUTHILLIER, L. P., and G. GOSSELIN: Contribution to the study of bound water in marine algae in vivo. Naturaliste Canad. 64, 65—80 (1937). — BOUYOUCOS, G. J.: Measurement of the inactive or unfree moisture in the soil by the dilatometer method. J. Agricult. Res. 8, 195—217 (1917). — BRIGGS, D. R.: Water relationships in colloids. I. Vapor pressure measurements in elastic gels. J. Physic. Chem. 35, 2914—2929 (1931). — Water relationships in colloids. II. "Bound" water in colloids. J. Physic. Chem. 36, 367—386 (1932). — BROYER, T. C., and D. R. HOAGLAND: Methods of sap expression from plant tissues with special reference to studies on salt accumulation by excised barley roots. Amer. J. Bot. 27, 501—511 (1940). — BULL, H. B.: Physical Biochemistry. New York: John Wiley & Sons 1948.

CHANDLER, R. C.: Nature of bound water in colloidal systems. Plant Physiol. 16, 273—291 (1941). — CHRYSLER, H. L.: Amounts of bound and free water in an organic colloid at different degrees of hydration. Plant Physiol. 9, 143—155 (1934). — COBLENTZ, W. F.: The role of water in minerals. J. Franklin Inst. 172, 309—355 (1911). — CRAFTS, A. S., H. B. CURRIER and C. R. STOCKING: Water in the physiology of plants. Waltham, Mass.: Chronica Botanica Co. 1949.

DAUGHTERS, M. R., and D. S. GLENN: The role of water in freezing foods. Ref. Engineering 52, 137—140 (1946). — DUMANSKII, A.: Die Bestimmung der Menge des gebundenen Wassers in dispersen Systemen. Kolloid-Z. 65, 178—184 (1933). — DUMANSKII, A. V., i R. V. VOITSEKHOVSKII: Methods of determination of the hydrophilism of disperse systems. Kolloid-Z. 10, 413—422 (1948). See Chem. Abstr. 43, 7781 (1949). — DUNN, S.: Value of the dye-adsorption test for predetermining the degree of hardiness. Plant Physiol. 12, 869—874 (1937). — DUNN, S., and A. L. BAKKE: Adsorption as a means of determining relative hardiness in the apple. Plant Physiol. 1, 165—178 (1926).

FOOTE, H. W., and B. SAXTON: The effect of freezing on certain inorganic hydrogels. J. Amer. Chem. Soc. 38, 588—609 (1916). — FRIEDMAN, C. A., and B. S. HENRY: The bound water content of vegetative and spore forms of bacteria. J. Bacter. 35, 11 (1938).

GORTNER, R. A.: The role of water in the structure and properties of protoplasm. Annual Rev. Biochem. 1, 21—44 (1932). — Outlines of biochemistry. Second Edition. New York: John Wiley & Sons 1938. — GORTNER, R. A., and W. F. HOFFMAN: Determination of moisture content of expressed plant tissue fluids. Bot. Gaz. 74, 308—313 (1922). — GRANDFIELD, C. O.: Food reserves and their translocation to the crown buds as related to cold and drought resistance of alfalfa. J. Agricult. Res. 67, 33—47 (1943). — GREATHOUSE, G. A.: Unfreezable and freezable water equilibrium in plant tissues as influenced by sub-zero temperatures. Plant Physiol. 10, 781—788 (1935). — GREATHOUSE, G. A., and N. W. STUART: A study of the physical and chemical properties of red clover roots in the cold-hardened and unhardened condition. Maryland Agricult. Exper. Stat. Bull. 1934, 370. — GREENBERG, D. M., and M. M. GREENBERG: Ultrafiltration II. "Bound" water (hydration) of biological colloids. J. Gen. Physiol. 16, 559—569 (1933). — GROLLMAN, A. J.: The vapor pressure of aqueous solutions with special reference to the problem of the state of water in biological fluids. J. Gen. Physiol. 14, 661—683 (1931).

HAMPTON, W. F., and J. H. MENNIE: Heat capacity measurements on gelatin gels. Canad. J. Res. 7, 187—197 (1932); 10, 452—462 (1934). — HILL, A. V.: The state of water in muscle and blood and the osmotic behavior of muscle. Proc. Roy. Soc. Lond., Ser. B 106, 477—505 (1930).

JONES, I. D., and R. A. GORTNER: Free and bound water in elastic and non-elastic gels. J. Physic. Chem. 36, 387—436 (1932).

KESSLER, W., and W. RUHLAND: Weitere Untersuchungen über die inneren Ursachen der Kälteresistenz. Planta (Berl.) **28**, 159—204 (1938). — KOETS, P.: Water adsorption on silica gel. Proc. Acad. Sci. Amsterdam **34**, 420—426 (1931). — KORSTIAN, C. F.: Physico-chemical properties of leaves and leaf sap as indices of the water relations of forest trees. Ext. Compt. Rend. Congrès de Nancy, 1932, de l'Union internationale des Instituts de Recherches forestières, S. 312—325 (1933).

LEVITT, J.: Frost killing and hardiness of plants. Minneapolis: Burgess Publ. Co. 1941. — Frost, drought, and heat resistance. Annual Rev. Plant Physiol. **2**, 245—268 (1951). — LEVITT, J., and G. W. SCARTH: Frost-hardening studies with living cells. Canad. J. Res., Sect. C **14**, 267—305 (1936). — LIPMAN, C. B.: Normal viability of seeds and bacterial spores after exposure to temperatures near the absolute zero. Plant Physiol. **11**, 201—205 (1936). — Tolerance of liquid-air temperatures by spore-free and very young cultures of fungi and bacteria growing on an agar media. Bull. Torrey Bot. Club **64**, 537—546 (1937). — LLOYD, D. J.: The movements of water in living organisms. Biol. Rev. **8**, 463 (1933). — LLOYD, D. J., and T. MORAN: Pressure and the water relations of proteins. I. Isoelectric gelatin gels. Proc. Roy. Soc. Lond., Ser. A **147**, 382—395 (1934). — LUYET, B. J., and P. M. GEHENIO: The survival of moss vitrified in liquid air and its relation to water content. Bio-dynamica **2**, No 42, 1—7 (1938).

MANEY, T. J.: Correlation of bound water in apple wood with hardiness. Ann. Rep. Iowa Exper. Stat. 1931, 97—98. — MARINESCO, N.: Polarisation dielectrique et structure des colloides. J. Chim. Physic. **28**, 51—91 (1931). — MARTIN, J. H.: Comparative studies of winter hardiness in wheat. J. Agricult. Res. **35**, 493—535 (1927). — MAXIMOV, N. A.: Internal factors of frost and drought resistance in plants. Protoplasma (Berl.) **7**, 259—291 (1929). — McROSTIE, G. P.: The thermal death point of corn from low temperatures. Sci. Agricult. **19**, 687—699 (1939). — MEYER, B. S.: Seasonal variations in the physical and chemical properties of the leaves of the pitch pine with especial reference to cold resistance. Amer. J. Bot. **15**, 449—472 (1928). — Further studies on cold resistance in evergreens, with special reference to the possible role of bound water. Bot. Gaz. **94**, 297—321 (1932). — MIGAHID, A. M.: Binding of water in xerophytes and its relation to osmotic pressure. Bull. Fac. Sci. Fouad I Univ. **25**, 83—92 (1945). See Biol. Abstr. **20**, 3815 (1946). — MÜLLER-THURGAU, H.: Über das Gefrieren und Erfrieren der Pflanzen. Landwirtsch. Jb. **9**, 133—189 (1880); **15**, 453—610 (1886).

NELSON, O. A., and G. A. HULETT: The moisture content of cereals. J. Industr. Engng. Chem. **12**, 40—45 (1920). — NEWTON, R. A.: A comparative study of winter wheat varieties with especial reference to winter killing. J. Agricult. Sci. **12**, 1—19 (1922). — NEWTON, R., and R. A. GORTNER: A method for estimating hydrophilic colloid content of expressed plant tissue fluids. Bot. Gaz. **74**, 442—446 (1922). — NEWTON, R., and W. M. MARTIN: Physico-chemical studies of the nature of drought resistance in crop plants. Canad. J. Res. **3**, 336—427 (1930).

OKUNTSOV, M. M., and E. N. TARASOVA: State of water in plants. Dokl. Akad. Nauk. SSSR. **83**, 315—317 (1952). See Chem. Abstr. **46**, 8722 (1952). —. OUCHINNIKOV, N.N.: Effect of loosely-bound water on the activity of catalase. Biochimija **16**, 205—208 (1951).

ROBINSON, W.: Water binding capacity of colloids, a definite factor in winter hardiness of insects. J. Econ. Entomol. **20**, 80—88 (1927). — Relation of hydrophilic colloids to winter hardiness of insects. Colloid Symposium Monogr. **5**, 199—218 (1928). — ROSA, T. J.: Investigations on the hardening process in vegetable plants. Missouri Agricult. Exper. Stat. Bull. 1921, 48. — RUBNER, M.: Über die Wasserbindung in Kolloiden mit besonderer Berücksichtigung des quergestreiften Muskels. Abh. preuß. Akad. Wiss., Physik.-math. Kl. 1922, 3—70.

SAYRE, J. D.: Methods of determining bound water in plant tissue. J. Agricult. Res. **44**, 669—688 (1932). — SCATCHARD, G.: The hydration of sucrose in water solution as calculated from vapor-pressure measurements. J. Amer. Chem. Soc. **43**, 2406—2418 (1921). — SCARTH, G. W.: Cell physiological studies of frost resistance: a review. New Phytologist **43**, 1—12 (1944). — SCHOPMEYER, C. S.: Transpiration and physico-chemical properties of leaves as related to drought resistance in loblolly pine and shortleaf pine. Plant Physiol. **14**, 447—462 (1939). — SEMENIUK, G., and J. C. GILMAN: Relation of molds to the deterioration of corn in storage. A review. Proc. Iowa Acad. Sci. **51**, 265—280 (1944). — SHIRK, H. G.: Freezable water content and the oxygen respiration of wheat and rye grain at different stages of ripening. Amer. J. Bot. **29**, 105—109 (1942). — SHIRK, H. G., and C. O. APPLEMAN: Oxygen respiration in wheat grain in relation to freezable water. Amer. J. Bot. **27**, 613—619 (1940). — SIMINOVITCH, D., and D. R. BRIGGS: Studies on the chemistry of the living bark of the black locust tree in relation to frost hardiness. IV. Effects of ringing on translocation, protein synthesis and the development of hardiness. Plant Physiol. **28**, 177—200 (1953). — SIMONOVA, E. F.: Change of hydrophilic properties of roots and leaves of sugar beets. Kolloid-Ž. **5**, 749—754 (1939). — SPONSLER, O. L., J. D. BATH and J. W. ELLIS: Water bound to gelatin as shown by molecular structure studies. J. Physic. Chem. **44**, 996—1006 (1940). — Stark, A. L.: Unfrozen water in apple shoots as related to their winter hardiness. Plant Physiol. **11**, 689—711 (1936). — Steinmetz, F. H.: Winter hardiness in alfalfa varieties. Minnesota Agricult. Exper. Stat. Techn. Bull. **1926**, 38. — St. John, J. L.: The temperature at which unbound water is completely frozen in a biocolloid. J. Amer. Chem. Soc. **53**, 4014—4019 (1931). — Strausbaugh, P. D.: Dormancy and hardiness in the plum. Bot. Gaz. **71**, 337—357 (1921).

Thoenes, F.: Untersuchungen zur Frage der Wasserbindung in Kolloiden und tierischem Gewebe. Biochem. Z. **157**, 174—186 (1925). — Todd, G. W., and J. Levitt: Bound water in Aspergillus niger. Plant Physiol. **26**, 331—336 (1951).

Van Doren, C. A.: Bound water and electrical conductivity as measures of cold resistance in winter wheat. J. Amer. Soc. Agronom. **29**, 392—402 (1937).

Walter, H.: Die Hydratur der Pflanze und ihre physiologisch-ökologische Bedeutung. Jena: Gustav Fischer 1931. — Weimer, J. L.: Some factors involved in the winter-killing of alfalfa. J. Agricult. Res. **39**, 263—283 (1929). — Weismann, O.: Eine theoretische und experimentelle Kritik der "bound water-theorie". Protoplasma (Berl.) **31**, 27—68 (1938). — Welch, W. B.: Water relations in bryophyllum calycinum subjected to severe drying. Plant Physiol. **13**, 469—487 (1938). — Whitman, W. C.: Seasonal changes in bound water content of some prairie grasses. Bot. Gaz. **103**, 38—63 (1941).

# The chemistry of plant cytoplasm[1].

By

## Sam G. Wildman[2] and Morris Cohen[3].

With 5 figures.

## I. The localization of cytoplasmic constituents.

### A. The heterogeneity of submicroscopic cytoplasm.

When living plant protoplasm is observed with a microscope, the protoplasmic inclusions—nuclei, plastids, mitochondria, ergastic droplets, and other smaller particles ranging down to the limits of light microscopy—often can be seen in active cyclosis. The sluggish manner in which the inclusions change their position as the protoplasm courses around the cell suggests that they are being carried along by a somewhat viscous matrix, which appears visibly different from the vacuole. The hyaline character of the matrix also suggests that it is homogeneous. However, this impression of homogeneity disappears when the matrix is examined with the ultramicroscope, for some particles measuring about 0.1 micron and others barely detectable by this instrument, that is, about 50 to 100 Å in diameter, have been discerned (GAIDUKOV 1906). Aside from the nucleus, the non-vacuolar portion of the protoplast is the cytoplasm of the cytologist.

In keeping with the plan of this volume, which is to treat separately the nucleus, plastids, mitochondria, and ergastic products, we propose here to restrict our discussion of the chemistry of the cytoplasm to what is known in regard to the chemical constitution of the *submicroscopic* cytoplasm. We do this with the full realization that the treatment is artificial, because knowledge of this portion of the cytologist's cytoplasm has little meaning for an understanding of the life of the cell until it can be integrated with the function, metabolism, and composition of the visible inclusions. Indeed, if it is considered that during cyclosis compounds may be produced by the coming together of reactants and enzymes in "solution" in the cytoplasm and that many of the reactions which are identifiable *in vivo* and *in vitro* probably occur at interfaces between microscopic and submicroscopic phases, it becomes obvious that the visibly clear, homogeneous portion of the cytoplasm is, in reality, heterogeneous. The degree of heterogeneity would very much depend upon the location, movement, development, and metabolic condition of the inclusions.

In the final analysis, the greatest difficulty facing anyone having the temerity to describe the composition of cytoplasm as a prelude to understanding its function, lies in the invisibility of the components of hyaline cytoplasm. With a microscope we can see that the nucleus and other living inclusions occur only in the protoplasm and not in the vacuole or cell wall. The cytologist thus provides

[1] This work was facilitated by the financial support we have received from the United States Atomic Energy Commission and Cancer Research Funds of the University of California. We are greatly indebted to Drs. ALBERT SIEGEL, HOWARD BOROUGHS, and IRVING RAPPAPORT for their helpful suggestions during the preparation of this manuscript.

[2] Department of Botany, University of California, Los Angeles.

[3] Fellow of the American Cancer Society, 1953–1954, as recommended by the Committee on Growth of the National Research Council.

rather rigid definitions for the biochemist to judge the character of some of the materials he may isolate from the dispersed contents of ruptured cells.

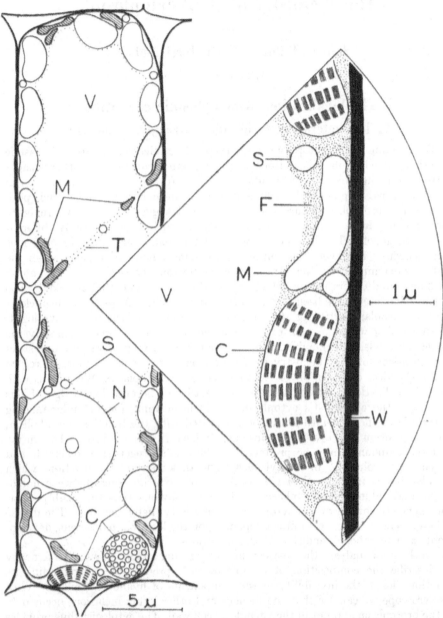

Fig. 1. Diagrammatic section through a generalized leaf palisade parenchyma cell, approximately to scale, except for cell and organelle membranes. N, nucleus with central nucleolus; C, chloroplasts, some showing arrangement of their internal lamellate grana in side and face view (lamellae traversing chloroplast and confluent with grana omitted for clarity—see HEITZ 1936, STRUGGER 1951, COHEN and BOWLER 1953); M, rodlike mitochondria, shaded for emphasis (internal structure not shown—see PALADE 1953, SJÖSTRAND 1953); S, spherical sphaerosomes (PERNER 1953); F, nucleoprotein particles (Fraction I), the even distribution shown here being somewhat arbitrary; T, cytoplasmic trabecula; V, vacuole; W, cell wall; ergastic inclusions omitted.

We only need to mention here the revolution which has occurred in our understanding of respiration, as exemplified by the localization of certain oxidative processes in the mitochondria, to appreciate the fruitfulness of a combined cytological-biochemical approach to the metabolism of living matter. The invisibility of submicroscopic cytoplasmic components, in contrast, has prevented light microscopic cytology from providing the biochemist with any substantial catalogue of materials that might be encountered in the hyaline cytoplasm.

With the increasing use and understanding of the capabilities of the electron microscope and the improvements in ultra-thin sectioning (SELBY 1953), we cannot escape the feeling that the problem of invisibility is slowly, but surely, being overcome. Certainly, a tremendous potential for resolving the structural elements of cytoplasm resides in the continued and expanded use of this instrument. [Of equal importance to cytochemistry will be the further development of specific electron stains (LAMB *et al.* 1953).] It is on account of this philosophy that we intend to restrict further our discussion of cytoplasm, as far as possible, to those instances where biochemical findings can be correlated to some degree with the direct observation of cytoplasm. By adopting this approach, we hope

Table 1. *Approximate relative volumes and numbers per cell of intracellular organelles and submicroscopic particulates.*

|  | Relative volume | Number per cell |
|---|---|---|
| Fraction I particulates | 1 | 100,000,000 |
| Sphaerosomes . . . . | 150,000 | 300 |
| Mitochondria . . . . | 1,000,000 | 700 |
| Chloroplasts . . . . . | 15,000,000 | 100 |
| Nuclei . . . . . . . | 200,000,000 | 1 |

to relieve somewhat the tedium which would otherwise result from having to couch in new words what already has been said better by others with regard to the chemical analysis of cytoplasm. Most of all, we would hope to leave the reader with the impression that though there are many obstacles in the road ahead, cytochemical signs have in fact been erected which point towards the possibility of a far more comprehensive understanding of the functional organization of the cytoplasm.

Since much of our discussion of the composition of submicroscopic cytoplasm will center on the palisade and spongy parenchyma cells of leaves, a composite diagrammatic section through a typical palisade cell is presented in Fig. 1 to serve as orientation for this discussion. The diagram is based, in part, on recent information acquired by observation of thin sections of tobacco leaf tissue with the electron microscope (COHEN unpublished). An effort has been made to present the protoplasmic inclusions in their proper relative sizes. The comparatively vast expanse of the central vacuole limits the protoplasm mostly to a narrow peripheral region of the cell, which in turn is enclosed by the cell wall. The submicroscopic cytoplasm, as we have defined it, is depicted by the stippled area of the protoplasm. However, the dots have been drawn approximately to such a scale as to represent a specific, particulate, nucleoprotein, which, on the evidence to be provided in a later section of this chapter, we believe occurs in the cytoplasm of the tobacco leaf. This nucleoprotein has been named Fraction I protein (WILDMAN and BONNER 1947, SINGER *et al.* 1952). In Table 1 are given the very approximate numbers and relative volumes of various cellular particulates of leaf cells.

## B. Criteria for the *in situ* localization of substances.

GLICK (1949, 1953), BRADFIELD (1950), GOMORI (1952), DOUNCE (1952), DEMPSEY and WISLOCKI (1946), SCHNEIDER and HOGEBOOM (1951), HOLTER (1952), and DANIELLI (1946, 1953), among others, have led a movement to establish certain cytochemical criteria which should be satisfied before the localization *in situ* of a substance in protoplasm can be said to have been achieved. Briefly, for a particular intracellular chemical substance, the fixation and treatment prior to the application of cytochemical reagents should prevent the diffusion and partial or selective degradation of that substance. The subsequent addition of a specific cytochemical reagent or, for enzymes, a specific substrate, should result in the deposition of a microscopically visible, non-diffusible product at all the physiologically normal, *in vivo* sites of that constituent. Needless to say there is very little published work which can unequivocally meet such extreme standards. Also, with the present paucity of suitable analytical methods, there is a limit to the amount of information which can be gleaned from the observation of *in vivo* protoplasm. Consequently, cytochemistry is forced to rely, to some extent, on observations made on materials after they have been prepared from homogenized tissues. However, the results and interpretations of this approach deserve the closest scrutiny, for, as many workers have pointed out, it is important to evaluate the relative contributions of the various tissues usually dispersed in the homogenate. The degree of homogeneity of a particular fraction, based on its physical and chemical characters, and its relation to some moiety of the *in situ* protoplasm, must also be taken into account.

Diffusion artifacts occurring in *in situ* studies can under certain circumstances be minimized, but the effects of diffusion may be accelerated during homogenization owing to mixing and to the destruction of semipermeability of inclusion membranes (A. L. Dounce, personal communication to Danielli 1953). The inclusions themselves may in fact be destroyed. Subsequent washing during the fractionation procedure may be to no avail in reducing surface contamination of one fraction by another, and the integrity of a fraction may have suffered considerable damage because of the mixture of vacuolar and cytoplasmic environments. Reactants or enzymes and substrates which *in vivo* may be normally separated by the hierarchy of protoplasmic organization[1] are also thrown together. Nevertheless, this approach has been particularly effective in the setting up of *in vitro* enzymatic model systems which may be indicative of *in vivo* functional capacity.

In the following consideration of the cytochemical literature of submicroscopic cytoplasm the reader should then weigh the results and their implications in the light of the probable shortcomings of the methods applied.

## C. The intracellular detection of chemical elements in the cytoplasm.

It is probably a safe assumption to say that any element absorbed by a root hair from the soil solution and which subsequently appears in the vacuole or in still another cell, must have at least a transitory existence in the submicroscopic cytoplasm. Likewise, in the laying down of the middle lamella and subsequent layers of cellulose, or in the reproduction of intracellular organelles and their growth, the constituents of these structures must have arrived by way of the cytoplasm, either by diffusion and cyclosis or by temporary adsorption on, or incorporation into, the submicroscopic particulate moiety.

Attempts to study the *in situ* distribution of the chemical elements of cytoplasm have met with little success. Mineral elements as a group can be localized by microincineration (Uber 1940), and emission electron microscopy of the

---

[1] How important this organization is to the physiology of the cell may be illustrated by the mode of snapdragon pollen germination following stratification of intracellular protoplasm by centrifugation (Beams 1943).

resultant spodogram can identify the *in situ* calcium and magnesium without distinguishing between them (SCOTT 1943). Soft X-ray ($\lambda = 8$–$12$ Å) absorption by thin tissue sections permits the determination of dry weights of small intracellular volumes (ENGSTRÖM 1950, 1952, 1953) having cross sectional diameters as little as $1\,\mu$ (ENGSTRÖM 1949). The relative densities of the components of onion root tip cells (ENGSTRÖM and LINDSTRÖM 1950) and developing pollen mother cells (ENGSTRÖM 1952) have been visualized in X-ray micrographs. Measurement of the degree of absorption at element-specific absorption edges, or discontinuities, of the X-ray absorption spectrum of a cytological site leads to the quantification of the carbon, nitrogen, oxygen, phosphorus, and sulfur corresponding to those absorption edges (ENGSTRÖM 1953). Neither this method nor emission electron microscopy have yet been applied to plants.

The application of radioautography (FITZGERALD *et al.* 1953) is also limited as regards cytological localization of elements in that only a few elements give the short tracks in the photographic emulsion necessary for sufficient resolution: $H^3$ less than $1\,\mu$ (FITZGERALD *et al.* 1951), $S^{35}$ 1–1.5 $\mu$ (HOWARD and PELC 1951 b), and $P^{32}$ about $2\,\mu$ (HOWARD and PELC 1951 a). The use of $C^{14}$ gives even less resolution (ANDRESEN *et al.* 1952, FITZGERALD and ENGSTRÖM 1952), but has been histochemically applied to *Hydrangea* leaves (DUGGER and MORELAND 1953). It has been suggested that $H^3$, stably bound to carbon atoms, would not only improve resolution, but might be more easily introduced into the compound than would $C^{14}$ (FITZGERALD *et al.* 1951). The future application of this type of tritium compound may provide cell physiology with one of its most important tools. There are few cytochemical methods for the *in situ* detection of elements which in solution occur as ions. These methods are usually complicated by diffusion phenomena, despite the use of freeze-dry methods (GLICK 1948, GOMORI 1952).

The cytochemical detection of potassium, devised by MACALLUM (1905), has been criticized by him and others (LLOYD 1925, DOWDING 1925, POPPEN *et al.* 1953) because of insufficient knowledge about the diffusion of potassium, the reagent, and the effects of washing. Potassium is not detectable in nuclei or chloroplasts, but is found in the cytoplasm and vacuoles where it often appears to be concentrated at membrane surfaces.

Zinc has been detected in animal cytoplasm and cytoplasmic granules by means of dithizone, which forms colored complexes with Zn, Mn, Fe, Co, Cu, Ni, Pd, Ag, Cd, In, Sn, Pt, Au, Hg, Tl, Pb, and Bi (MAGER *et al.* 1953). This method awaits application to plant cytochemistry. Likewise awaiting such application is the formation of colored insoluble 8-hydroxyquinoline metal derivatives, which also exhibit fluorescent colors (GLICK 1949).

Knowledge obtained by analysis of tissue which has been frozen, thawed, and pressed can, at best, only localize elements relative to the vacuole and nonvacuolar protoplasm. This method would be subject to the variation imposed by possible base exchanging properties of the protoplasm through which the vacuolar contents are expressed (OLSEN 1948 b). Perhaps a more reliable separation of vacuole and protoplasm is achieved in the special cases of *Nitella* and certain unicellular algae where the vacuole is very large and the contents may be obtained by dissection.

Certain elements can be localized by implication through detection of compounds of which they are integral constituents, although the possibility of adsorptive contamination exists. Phosphorus, for example, is an integral part of nucleic acid; carbon, hydrogen, oxygen, sulfur, and nitrogen are the principal elements found in proteins and other substances of protoplasm. Certain metals

Table 2. *Elements which have been directly localized in plant cytoplasm or non-vacuolar protoplasm.*

| Element | Plant | Method | Reference |
|---|---|---|---|
| Hydrogen (H³) | *Torula utilis* | Autoradiography | FITZGERALD *et al.* 1951 |
| | *Scenedesmus* sp. | Autoradiography | CHAPMAN-ANDRESEN 1953 |
| Nitrogen (N) | *Nitella* (cell wall and protoplasm) | Dissection | HOAGLAND and DAVIS 1923 |
| | Beech leaves (protoplasm) | Freeze-thaw-pressure | OLSEN 1948a |
| Sodium (Na²⁴) | *Nitella* (protoplasm) | Dissection | BROOKS 1951 |
| | *Valonia macrophysa* (protoplasm) | Dissection | BROOKS 1953 |
| Magnesium (Mg) | *Nitella* (cell wall and protoplasm) | Dissection | HOAGLAND and DAVIS 1923 |
| | Beech leaves (protoplasm) | Freeze-thaw-pressure | OLSEN 1948a |
| Aluminum (Al) | *Nitella* (cell wall and protoplasm) | Dissection | HOAGLAND and DAVIS 1923 |
| | Corn, cabbage, and redtop roots | Cytochemistry | McLEAN and GILBERT 1927 |
| Silicon (SiO₂) | *Nitella* (cell wall and protoplasm) | Dissection | HOAGLAND and DAVIS 1923 |
| Phosphorus (PO₄) | *Nitella* (cell wall and protoplasm) | Dissection | HOAGLAND and DAVIS 1923 |
| | Beech leaves (protoplasm) | Freeze-thaw-pressure | OLSEN 1948a |
| Phosphorus (P³²) | *Nitella* (protoplasm) | Dissection | BROOKS 1951 |
| | *Vicia faba* root tip | Autoradiography | HOWARD and PELC 1951a |
| Sulfur (SO₄) | *Nitella* (cell wall and protoplasm) | Dissection | HOAGLAND and DAVIS 1923 |
| Sulfur (S³⁵) | Bean root | Autoradiography | HOWARD and PELC 1951b |
| Potassium (K) | *Lilium, Tulipa, Equisetum, Spirogyra,* fungus parasite of *Spirogyra* | Cytochemistry | MACALLUM 1905 |
| | Wheat grain and white spruce leaves at various stages | Cytochemistry | DOWDING 1925 |
| | Potato (various tissues) | Cytochemistry | PENSTON 1931 |
| | Beech leaves (protoplasm) | Freeze-thaw-pressure | OLSEN 1948a |
| Potassium (K⁴²) | *Nitella* (protoplasm) | Dissection | BROOKS 1951 |
| Calcium (Ca) | *Nitella* (cell wall and protoplasm) | Dissection | HOAGLAND and DAVIS 1923 |
| Zinc (Zn) | Orange leaves | Spodogram treated with sodium nitroprusside | REED and DUFRÉNOY 1935 |

are localized by the detection, either *in situ* or in homogenate fractions, of proteins to which they are stably bound (*e.g.*, cobaltoprotein—BALLENTINE and STE-PHENS 1951) or of enzymatic activities in which they take part as prosthetic groups. The elements of water are, of course, no problem, although the locali-zation in the cytoplasm is by inference rather than by *in situ* demonstration. A summary of some of the elements which have been directly localized in plant cytoplasm or in non-vacuolar protoplasm is given in Table 2.

## D. The localization of low molecular weight organic compounds in cytoplasm.

Since protoplasm is the seat of a myriad of chemical reactions involving syntheses, degradations, and rearrangements, we should expect to find localized in the cytoplasm an enormous array of low molecular weight compounds. Such a list, if compiled, would probably have to embrace every known natural product found in plants. Even though such substances as the colored anthocyanins may appear to be confined to the vacuole, it is still difficult to conceive of these materials not having been in contact with the cytoplasm at some time during their manufacture and metabolism. Since the second part of this discussion is concerned with the vast array of compounds that might be encountered in cytoplasm, we shall confine our discussion at this point to the specific localiza-tion of substances in submicroscopic cytoplasm. If a local concentration gradient were found around a particular structural element, it would then be reasonable to implicate that element in the metabolism of the compound being studied. While much effort has been expended on this approach, the specific localization of a low molecular weight compound in the cytoplasm remains an extremely difficult proposition. A good illustrative example of the difficulties which might be encountered with almost any low molecular weight compound is the case of ascorbic acid, or vitamin C.

The cytochemical detection of ascorbic acid is dependent upon the ability of this substance to reduce silver ion to metallic silver. The deposition of this silver has been found to occur almost anywhere in the cell, depending upon a great many variables. These include the following (WEIER 1938, NAGAI 1950, 1952a, 1952b, METZNER 1952, CHAYEN 1953): the plant species, its environment, the $p_H$ of the test reagent, and the condition of the tissue at the beginning of the test (whether or not it is living; if living, whether or not under conditions inhibitory to chlorophyll formation; if dead, the way in which it was killed; $p_H$ of the cell sap; intracellular ascorbic acid concentration; technique—whole mounts, sections, mode of fixation, if any, vacuum infiltration, or freed cell contents). Whether or not ascorbic acid is the cytochemical reductant responsible has been studied chromatographically by NAGAI (1951), who found that with most plants only the vitamin was involved, but that in broad bean and in *Vicia sativa* 3,4-dioxyphenylalanine (Dopa) also took part in the reaction.

Chloroplasts *in situ* are capable of precipitating silver under a wide range of conditions. Yet, when they are freed by dissection (WEIER 1938) or fractionated from homogenates (JONES and HAMNER 1953), ascorbic acid does not seem to be appreciably associated with them.

NAGAI (1950), studying the intracellular distribution of silver precipitation in test media varying in $p_H$ from 2 through 9, found that in lower epidermal cells of *Vicia faba* the optimum $p_H$ for chloroplasts was from 4–6 (range of 3–9), whereas the optimum for the vacuole and cytoplasm was 8–9 (range of 4–9). There were slight variations in the intensity of the reaction and $p_H$ range depending upon the plant, tissue, and stage of development. In *Rumex acetosa*, perhaps because of a higher vacuolar $p_H$, the optima for chloroplasts and cytoplasm were $p_H$ 7 and 9, respectively. NAGAI suggests that colloidal silver may migrate to and be deposited on chloroplasts because of opposite surface charges. He attributes this precipitation, generally called the "Molisch reaction", to the unbuffered test media employed.

Somewhat similar are the results of METZNER (1952), who segregated the data according to the $p_H$ of the infiltrated buffer and to the site of silver precipitation, that is, nucleus, cytoplasm, and chloroplast grana, stroma, and surface. Precipitation in the chloroplast stroma and on nuclei is greatest at $p_H$ 5-6, which suggested to METZNER that something other than ascorbic acid, perhaps an enzyme, may be involved. The maximum and minimum precipitation occur at the red maximum and the minimum of the chlorophyll absorption spectrum only if $AgNO_3$-treatment begins under the monochromator. METZNER does not choose between microsomes or mitochondria as cytoplasmic sites of ascorbic acid, pointing out that although mitochondria do not precipitate silver, ascorbic acid might still be associated with them without being detectable because of the glutathione likewise present. NAGAI (1952a, 1952b) has shown that ascorbic acid can be present without giving a positive test.

While NAGAI (1950) also indicated that mitochondria never precipitate silver, CHAYEN (1953) found that if bean and onion root tips were fixed in such a way to preserve the microscopically visible, fine, intracellular structure, the silver precipitation occurred only on "... cytoplasmic granules, probably mitochondria". Any other type of fixation caused deposition on mitotic chromosomes, at the periphery of interphase nuclei, and in the cytoplasm. CHAYEN suggested that this was probably the result of diffusion of ascorbic acid from the sites of the destroyed mitochondria.

The intracellular localization of ascorbic acid involves the early death of the cell and the reduction of silver ion by a reducing agent which probably is diffusible. This may account for precipitation which very often occurs at or close to protoplasmic interfaces. It would, therefore, seem difficult in the light of the preceding statements to say whether or not ascorbic acid, in life, occurs in the submicroscopic cytoplasm. The distribution may, indeed, be solely a matter of local $p_H$ conditions. One should certainly not overlook the possible ubiquitous distribution of the vitamin in the cell. Its likely role in respiration (WAYGOOD 1950) is possibly indicative that ascorbic acid and, incidentally, other highly soluble and diffusible metabolic products may exist in the living cell cytoplasm as microscopic atmospheres emanating from their specific centers of synthesis, and that, with the currents and eddies of cyclosis, the outer portions of these atmospheres are dispersed to other parts of the cell, where they either remain free or are bound (and, perhaps, their activities masked), or are further transformed. This may explain the apparent "difficulties" encountered in the *in situ* cytochemical detection of ascorbic acid.

The cytochemical localization of other low molecular weight substances is even more ambiguous than that of ascorbic acid. Some of them are fluorescent (GOODWIN 1953, BEST 1948). Attempts to locate thiamine and riboflavin by their fluorescence have only resulted in showing them to be concentrated in the outer layers of the cytoplasm or in the cell walls (SOMERS *et al.* 1945a, 1945b), the latter site being, *a priori*, an unlikely habitat for such metabolically active substances. Such a distribution might be the result of both their water solubility and the drastic action of the alkaline ferricyanide pretreatment necessary for their fluorescence.

Methods, which have been subjected to the same criticism, have also been proposed for the detection of free α-amino acids (CHU *et al.* 1953), sugars (OKAMOTO *et al.*, as mentioned in GOMORI 1952), *etc.* The detection of polysaccharides, consisting of 8–12 glucose units each, has been a basis for the localization of phosphorylase activity *in situ* (DYAR 1950).

While we may register despair that such methods are probably still incapable of providing the precise location of low molecular weight substances in protoplasm, the prodigious effort which has gone into this approach has paid off handsomely in regard to our knowledge concerning the distribution and concentration of substances at the tissue level of organization in plants and animals.

## E. Cytoplasmic proteins: *in situ* cytochemistry.

Attempts to locate cytoplasmic protein *in situ* have, in general, met with greater success than efforts directed towards low molecular weight compounds. One reason is that the cytoplasm is rich in proteins. The fact that proteins are such large molecules with greatly reduced capacity for diffusion more often allows them to be fixed in position by such methods as freeze-drying or chemical fixatives. Even though fixation may appear to be successful to the extent that protoplasmic inclusions are satisfactorily preserved, the cytoplasmic proteins may have been greatly altered in physical and chemical properties. Chemical fixation, for example, may cause cytoplasm to undergo a great loss in mass (BRATTGÅRD and HYDÉN 1952). Proteins may aggregate to microscopically visible dimensions, enzymatic activity may be destroyed, and chemical groupings demonstrable in unfixed cytoplasmic proteins may not remain intact (DANIELLI 1953). While freeze-drying is thought to be one of the mildest methods of fixation, the fact that many of the reagents used in cytochemistry must be dissolved in water may undo the advantage of this type of fixation (GOMORI 1952).

For fine rendition of structure both in light and electron microscopy, $OsO_4$ seems to be best and, at least for electron microscopy, the most widely used fixative. (See STRANGEWAYS and CANTI 1927 and BAKER 1950 for effect of $OsO_4$ in dark field and light microscopy, respectively, and PALADE 1952 for effect of buffered $OsO_4$ in electron microscopy.) Yet, according to DANIELLI's (1953) tabulation of fixatives and their effects on various chemical groups, $OsO_4$ not only destroys the integrity of all groups for which its effect is known, but also renders enzymes inactive. Neutral formaldehyde, often similarly used as a fixative, presents a more easily stained protoplasm for light microscopy (BAKER 1950), and nearly as fine fixation for electron microscopy (ROZSA and WYCKOFF 1951). It is one of the most innocuous fixatives from the standpoint of cytochemistry according to DANIELLI's tabulation.

Localization of proteins in cytoplasm is usually either of two kinds: non-specific and indirectly specific[1]. Proteins, without reference to kind, may be localized by the use of certain reagents which react with specific constituent groups of the protein molecule. Examples of such protein localization in the cytoplasm of plants are given in Table 3. Proteins may also be revealed through the localization of nucleic acids with which the proteins may be more or less closely associated. This is accomplished by staining (TURCHINI 1949, FLAX and HIMES 1952, KORSON 1951, KURNICK 1952, CHAYEN and NORRIS 1953, and SWIFT 1953) and by means of UV-microspectrophotometry (DANIELLI 1947, COMMONER 1949, MELLORS 1950, CASPERSSON 1950, and SWIFT 1953).

[1] An exception is the hemoglobin protein found in the root nodules of certain legumes (SMITH 1949).

Table 3. *Methods of specific localization of in situ protein constituent groups.*
*The few applications to plants are indicated under comments.*

| Method and selected references | Groups detected | Comments |
|---|---|---|
| Sakaguchi (GLICK 1949, GOMORI 1952) | Arginine | Applied to wheat grain and very young seedlings (GLICK and FISCHER 1946b); drastic (alkaline) |
| Romieu (GLICK 1949, GOMORI 1952) | Tryptophane | Drastic (acid) |
| Millon (GLICK 1949, GOMORI 1952) | Tyrosine | Drastic (acid) |
| After freeze-drying or chemical fixation, use of appropriate non-chromogenic blocking reagents followed by the following classes of chromogenic reagents (DANIELLI 1950, 1953): | | A very flexible technique promising wide application; diffusion artifacts can only occur if proteins themselves diffuse, and tests for this are easily performed; modifications permit color intensification; adaptable to electron microscopy (LAMB et al. 1953) |
| Diazonium hydroxides | Histidine, tryptophane, tyrosine | |
| Aromatic nitro compounds | Tyrosine, SH, S-S, $NH_2$, COOH, CHOH | |
| Aromatic aldehydes | SH, $NH_2$ | |
| 1-(4-chloromercuriphenyl-azo)-naphthol-2 (BENNETT 1951) | SH | Applied by ROBERTS (1952) to *Phaseolus vulgaris* roots and embryos |
| | S-S | Same method but subsequent to blocking of SH followed by reduction of S-S to SH (BARRNETT and SELIGMAN 1953) |
| Blue azo dye method (WEISS and SELIGMAN 1953) | $NH_2$ | Involves a $p_H$ of 8.5 and aqueous acetone |
| Ninhydrin (GLICK 1949, GOMORI 1952) | α-amino | Noncorrosive but heating necessary |
| Mono or ditetrazolium salts or resazurin | Reducing sites | Applied to *Phaseolus vulgaris* roots, *Acer* stems, and *Zea mays* embryos (ROBERTS 1952) |

The overlapping absorption maxima in the 2600 Å region of both ascorbic acid and purine and pyrimidine rings of nucleic acids may complicate the results (CHAYEN 1953). Both methods have been subjected to confirmation by means of ribonuclease and desoxyribonuclease (KORSON 1951, JACOBSON and WEBB 1952, BRACHET 1953; but see criticisms of the techniques by DANIELLI 1947, POLLISTER et al. 1951, and CHAYEN and NORRIS 1953). Table 4 lists examples of the detection *in situ* of nucleic acid in plant cytoplasm.

With the refinement of methods of fractionation of plant homogenates, it may be possible to obtain protein fractions in a sufficiently pure condition so that when used as antigens they will elicit highly specific antibodies. These antibodies, coupled to fluorescein isocyanate, may then be applied to tissue sections in order to "stain" specifically the original protein antigen *in situ*. This type of experiment has already been done in animal cytochemistry (COONS and KAPLAN 1950, COONS et al. 1951).

Table 4. *Localization of pentose nucleic acid in situ in submicroscopic plant cytoplasm.*

| Plant | Method | Annotated references |
|---|---|---|
| Rye embryo | Non-nuclear protoplasm obtained by centrifugation in non-aqueous solvent mixtures of different specific gravities | BEHRENS 1938—localization in non-nuclear protoplasm |
| | UV-microspectrophotometry | CASPERSSON and SCHULTZ 1939—"with an additional absorption indicating a high concentration of proteins" (due mainly to tyrosine and tryptophane) |
| Onion bulb epidermis | UV-microspectrophotometry; this paper discusses the bases for and the techniques of quantitative cytophotometry (see also CASPERSSON 1950) and cytochemistry of nucleic acids | RASCH and SWIFT in SWIFT 1953 —low cytoplasmic absorption with 260–270 m$\mu$ plateau; discussion and citation of literature on presence of cytoplasmic DNA in plant microsporocytes and in animals |
| *Allium* root tip | UV-microspectrophotometry | CASPERSSON and SCHULTZ 1939— maximum at 2600 Å in meristematic cytoplasm, minimum at 2600 Å in older non-dividing cells (as in living *Tradescantia* pollen mother cells—CASPERSSON 1950); absorption at about 2800 Å in both cases (due mainly to tyrosine and tryptophane) |
| Corn microsporocyte | Metachromatic staining for RNA and DNA by azure B (= azure A phthalate); specificity tested with ribonuclease and desoxyribonuclease (FLAX and HIMES 1950); involves Carnoy fixation (with attendant loss in mass, according to BRATTGÅRD and HYDÉN 1952) and staining at $p_H$ 4 | FLAX and HIMES 1952—a critical microspectrophotometric investigation of the method with respect to variation in dye concentration, $p_H$, and temperature, and to protoplasmic structure |
| Spinach root tip periblem | UV-microspectrophotometry after 45% acetic acid maceration of tissues for several days | CASPERSSON and SCHULTZ 1940 |
| Bean root tip | Ribonuclease pretreatment followed by Unna's methyl green-pyronin stain | CATCHESIDE and HOLMES 1947— cytoplasm fails to stain pink, as it does without ribonuclease treatment (see also DALTON and STRIEBICH 1951, for special application of method to electron microscopy of animal tissues) |
| Yeast | UV-microspectrophotometry | CASPERSSON 1950—high in living actively growing cells but low in old cultures |

Degradation of certain substances by means of solvents or enzymes has been applied to the localization of lipids, pentose nucleoprotein, and protein in nerve cells by X-ray microradiography (BRATTGÅRD and HYDÉN 1952).

Enzyme localization is, of course, of extreme importance and significance. There are, however, no specific chemical groupings which would serve to distinguish one enzyme from another. Consequently, one approach is to locate

Table 5. *Enzymes which have been ascribed to submicroscopic plant cytoplasm*[1,2].

| Enzyme (and metal constituent or prosthetic group, as given by Sumner and Somers 1953) | Plant (and method) | Notes | Reference |
|---|---|---|---|
| Acid phosphatase(s)[3] | Pea root tip (*in situ*, by method of Bayliss *et al.* 1948) | Mostly close to nucleus | Dyar 1950 |
| | Wheat grain and epicotyl and roots of germinated grain (*in situ* with various substrates) | Intensity varies with tissue and substrate | Glick and Fischer 1946a |
| | Spinach (*in situ* by Gomori method) | Also "at nuclei" and druses (also, by electron microscopy of homogenate, at surface of a few plastids) | Macdowall 1953 |
| | *Allium cepa* root tips and *Pisum sativum* rootlets and cotyledons (*in situ* by Gomori modification) | Positive test in cytoplasm with high (1.0%) concentration of Na-$\beta$-glycerophosphate, suggested to be a diffusion artifact from the nucleus | Tandler 1953 |
| | Tobacco leaves (fractional centrifugation) | Withstood 144,000 g. for 3 hours | Boroughs 1953 (see also Boroughs 1954); Eggman 1953 |
| Aldolase | Tomato and sugar beet leaves (fractional centrifugation) | Withstood 18,000 r.p.m. for 15 minutes | Tewfik and Stumpf 1949 |
| Alkaline phosphatase(s)[3] ($+Mg^{++}$ or $Co^{++}$ or $Zn^{++}$) | Unspecified root tip smears (Takamatsu-Gomori method) | Reaction less intense in mitotic cytoplasm than in meiotic | Bhattacharjee and Sharma 1951 |
| | Onion root tip (*in situ* by diazo dye or Takamatsu-Gomori method) | Also in nuclei; in contrast to animal nuclear enzyme, that of plants rapidly destroyed by usual alcohol fixation | Danielli 1953 |
| | Onion and corn root (*in situ* with alcohol fixation and various substrates other than glycerophosphate) | In nuclei and nucleoli in 18 to 24 hours, in cytoplasm in 48 hours | Ross and Ely 1951 |
| | *Iris germanica* leaf (*in situ* by alcohol fixation and method of Menten *et al.* 1944) | "In the chloroplasts or in the surrounding cytoplasm at the interface" | Yin 1945 |
| Amylase | Pea seedling (water homogenate—$p_H$ 6.8) | Withstood 60,000 g. for 30 minutes | Stafford 1951 |
| Ascorbic oxidase (copper enzyme) | Pea seedling (water homogenate—$p_H$ 6.8) | Withstood 60,000 g. for 30 minutes | Stafford 1951 |

Table 5. (Continued.)

| Enzyme (and metal constituent or prosthetic group, as given by SUMNER and SOMERS 1953) | Plant (and method) | Notes | Reference |
|---|---|---|---|
| Carbonic anhydrase (zinc enzyme) | Several higher plants (homogenate in phosphate buffer-glucose medium) | Withstood unspecified centrifugation which sedimented chloroplasts; no activity in whole or broken chloroplasts | BRADFIELD 1947 |
| | Several higher plants (homogenate) | Withstood 15,000 g. for 15 minutes | WAYGOOD and CLENDENNING 1951 |
| Catalase (prosthetic group: iron-containing hematin) | Spinach leaves (centrifugal and chemical fractionation) | In non-Fraction I soluble protein | WILDMAN and BONNER 1947 |
| Cytochrome oxidase (prosthetic group: iron-containing hematin) | Peanut cotyledons (centrifugal fractionation) | Associated with 20 mμ particles (withstood 13,000 g. for 15 minutes) | NEWCOMB and STUMPF 1952 |
| Dehydrogenases—isocitric, malic, glutamic, alcohol | Spinach leaves (centrifugal and chemical fractionation) | In non-Fraction I soluble protein | WILDMAN and BONNER 1947 |
| Fatty acid oxidase | Peanut cotyledons (centrifugal fractionation) | Associated with 20 mμ particles (withstood 13,000 g. for 15 minutes) | NEWCOMB and STUMPF 1952 |
| Peroxidase (prosthetic group: iron-containing hematin) | Spinach leaves (centrifugal and chemical fractionation) | In non-Fraction I soluble protein | WILDMAN and BONNER 1947 |
| Phosphorylase | Pea root tip (*in situ*; glucose-1-PO$_4$ substrate, followed by polysaccharide staining) | Without reference to organelles other than nuclei; however, in meristem, intracellular distribution in cytoplasm different from that of acid phosphatase (see above) | DYAR 1950 |
| | Pea seedling (water homogenate—p$_H$ 6.8) | Withstood 60,000 g. for 30 minutes | STAFFORD 1951 |
| | Germinated seeds of soybean and castor bean, and broad bean embryos (*in situ*) | Only in plastids | YIN and SUN 1949 |
| | Potato leaf (centrifugation of homogenate) | Non-plastid cytoplasm | WEIER and STOCKING 1952 |
| Succinic dehydrogenase | Peanut cotyledons (centrifugal fractionation) | Associated with 20 mμ particles (withstood 13,000 g. for 15 minutes) | NEWCOMB and STUMPF 1952 |

*Table 5.* (Continued.)

| Enzyme (and metal constituent or prosthetic group, as given by Sumner and Somers 1953) | Plant (and method) | Notes | Reference |
|---|---|---|---|
| Tyrosinase (copper enzyme) | *Nicotiana tabacum* (homogenate in unbuffered sucrose solution) | Withstood 19,000 g. for 20 minutes (supernatant "devoid of microscopically visible particulates") | du Buy *et al.* 1950 |
| | Many dicots (0.1 M $K_2HPO_4$ homogenate) | Withstood 16,000 g. for 30 minutes | Webster 1952 |
| | Spinach leaves (centrifugal and chemical fractionation) | In non-Fraction I soluble protein | Wildman and Bonner 1947 |

[1] Lang (1952) has tabulated the literature on enzymes associated with animal tissue homogenates according to the proportions of activity found in the centrifugally fractionated nuclei, mitochondria, microsomes, and supernatant.

[2] Where enzymatic activity is reported to be associated with microsomal or soluble fractions of homogenates, it is important to evaluate the results from a cytological standpoint. For example, Webster (1953) has indicated that in the *in vitro* enzymatic synthesis of glutamine by bean seedling homogenate fractions, 61% of the activity is associated with the mitochondria if the preparation of the homogenate is in 0.45 M sucrose —0.05 M potassium phosphate buffer at $p_H$ 7.3, which preserves mitochondria in an apparently intact condition (see Millerd and Bonner 1953); but, when the homogenate is prepared in 0.1 M buffer at $p_H$ 7.1, 95% of the activity lies in the soluble fraction. On the other hand, the possibility should not be overlooked that adsorption of an enzyme onto an intracellular organelle may occur because of insufficient buffering capacity in the homogenate preparation medium, as apparently is the case with wheat catalase (Hagen and Jones 1952).

[3] With animal tissues, in contrast to the results of Danielli (1953), which are based, in part, on the detection of the organic moiety of the phosphate substrate, the likelihood of diffusion artifact staining of nuclei in *in situ* tests for alkaline phosphatase is suggested by the work of Martin and Jacoby (1949; superimposed sections), Novikoff (1951), and Goetsch and Reynolds (1952; localization without removal of paraffin embedding material to prevent lateral diffusion—see also Ruyter and Neumann 1949 and, for acid phosphatase, Goetsch and Raynolds 1951). The same problem for acid phosphatase would seem to be resolved by the parallel study of rat liver tissue *in situ* (activity mostly in nuclei) and of fractionated homogenates (activity mostly in mitochondria and microsomes), as well as mixtures of the two, where it is pointed out that fixation used for *in situ* detection will, in homogenates, destroy the visible organization of mitochondria, perhaps doing likewise to microsomes, and thus make the release of diffusible enzyme possible (Palade 1951).

enzymes by the specific precipitation of the products of their activity in a microscopically visible form. The difficulties attending this technique have already been mentioned. In fact, most of our information on the locale of enzymes in protoplasm has come from homogenate experiments in which an attempt has been made to cytologically identify the protoplasmic fraction bearing the enzymatic activity. Many plant enzymes appear as "soluble" proteins and, hence, are presumed to be located in the invisible cytoplasm. A tabulation of these enzymes is found in Table 5.

## F. Cytoplasmic proteins: fractionation of homogenates.

In attempting to describe the protein composition of the hyaline cytoplasmic moiety, we are forced to rely to a large extent on information derived from the homogenate type of experiment (see Chibnall 1939 for work previous to 1939).

However, the possibility of a combined cytological-biochemical approach to the protein composition of cytoplasm is now much nearer at hand with the advent of thin-sectioning techniques coupled with the tremendous magnifications afforded by the electron microscope. Indeed, a start has already been made in this direction and on this account we will restrict our description of cytoplasm composition to work on the tobacco leaf, for this is the only instance that we know of in plants where a conscious attempt has been made to establish cytological criteria which can be correlated to some degree with the findings derived from the fractionation of homogenates.

The cytology of the submicroscopic cytoplasm. It is regrettable that the electron microscopy of plant tissues has, in general, lagged behind that of animal tissues, especially as exemplified by the magnificent work of SJÖSTRAND (1953), where *in situ* intracellular particulates have been defined down to the level of 45 Å size. The cause of this lag may, perhaps, be traced to the three dimensional fixation difficulties presented by the very large plant cell vacuole and its usually acid contents. As was indicated earlier, the ultramicroscope seems to have shown a gap in the size of intracellular particulates in plant cells between the lower limits of light microscopy and close to the lower limits of ultramicroscopy, that is, about 50–100 Å (GAIDUKOV 1906, not confirmed by GUILLIERMOND 1941). The results of the electron microscopy of tobacco leaf tissue sections must be considered preliminary and mainly suggestive, but such work also appears to indicate a gap in organized structures between the mitochondrial level (0.5 micron) and cytoplasmic structures of about 0.02 micron (COHEN unpublished)[1]. Consequently, if materials isolated from homogenates are subjected to electron microscopy and are found to greatly exceed this lower figure, they should not be considered as typical of *in situ* cytoplasm. Conversely, failure to find fairly large quantities of materials at the 0.02 micron size level might indicate that homogenization had affected the cytoplasm in an unfavorable way.

Problems associated with the preparation of tobacco leaf cytoplasmic proteins by homogenization. Before we consider the protein composition of tobacco leaf cytoplasm, it is desirable to consider first some of the difficulties associated with the homogenization technique for preparing plant proteins. In the living cell, the vacuole occupies a large portion of the total volume of the cell. In many plant tissues, tobacco leaves being a particularly striking example, the contents of the vacuole are quite acid, the degree of acidity being a function of the physiological age-status of the tissue. On the other hand, the cytoplasmic $p_H$ is generally thought to be close to neutrality (MEYER and ANDERSON 1952)[2]. While the vacuolar reservoir of acidity is effectively prevented by the tonoplast from making contact with the cytoplasm, one of the consequences of homogenization is to dilute enormously all of the protoplasmic constituents into an acid environment. While the effects of dilution are not assessable, the effect of acid on the cytoplasm is to cause immediately the rapid and irreversible aggregation of the cytoplasmic proteins (SINGER *et al.* 1952, EGGMAN 1953). Consequently, precautions must be taken to insure the immediate neutralization of the vacuolar

---

[1] This is in contrast to the situation in animal cells (PORTER *et al.* 1945, and PALADE 1952).

[2] SMALL (1946) points out that there is very little evidence for this textbook generality, but it would appear that the single experiment on which the generality is based was performed with the least damage to the cytoplasm of all the work on cytoplasmic $p_H$ published until that time. By the microinjection of dyes into root hairs of the water plant, *Limnobium spongia*, the $p_H$ of the cytoplasm in cyclosis was found to be 6.9 ± 0.2 (CHAMBERS and KERR 1932). The surrounding medium, however, was pond water at a $p_H$ of 7.6 to 8.0.

acid at the time the protoplasm is released from the ruptured cells, and this entails the use of strong buffers (Eggman 1953).

A second problem to consider is the unfortunate fact that all materials contained in the protoplasm are brought into immediate contact as the result of the violent mixing forces engendered by homogenization. The plant cell wall presents a rather formidable barrier to the release of protoplasm which necessitates high shearing forces for its rupture (Crook 1946, Wildman and Bonner 1947). Since there seems to be no other convenient method for the extraction of leaf protoplasm other than homogenization, the possibility must always be borne in mind that violent agitation may degrade fragile particulate structures such as chloroplasts to the extent that some of their materials may be reduced to a size invisible by light microscopy and thus be included unwittingly with the cytoplasm. The following method for the preparation of cytoplasmic proteins minimizes much of the effects of dilution and acidity. Evidence will be presented to support the idea that these proteins are not derived from the degradation of the larger structures of tobacco leaf protoplasm.

**The preparation of tobacco leaf cytoplasmic proteins.** The fractionation of tobacco leaves by the scheme illustrated in Fig. 2 provides large yields of cytoplasmic proteins (Wildman and Bonner 1947, Wildman et al. 1949, Singer et al. 1952, Eggman et al. 1953). The method leads to quite reproducible results, both in terms of yield and composition of the cytoplasmic proteins. The colloid mill is an efficient means to insure that most of the cells of the leaf are ruptured during homogenization. Eggman (1953) has found that the protoplasm can be maintained in a neutral condition only when strong buffers are employed as the extraction medium. According to his experiments, when two parts by weight of leaves are homogenized in the presence of 1 part by volume of buffer, a $p_H$ 7 buffer with a molarity of 0.5 is required to prevent the $p_H$ of the extracted protoplasm from becoming more acid than $p_H$ 6.8. Experiments in this laboratory (unpublished) have shown that when 1 part tobacco leaves are homogenized in the presence of 4 parts of 0.05 M phosphate buffer, a buffer commonly used in the homogenization of plant tissues, the $p_H$ of the extracted protoplasm may drop to $p_H$ 6.2–6.3, and the yield of cytoplasmic proteins be less than 30% of that obtained when the protoplasm is kept close to neutrality.

After removal of unbroken cells and cell walls, the cell-free protoplasm is further fractionated by spinning at high speed in a preparative ultracentrifuge. The supernatant solution is a clear, sparkling, non-opalescent material rich in protein. In fact, about 50% of the total protein nitrogen of the tobacco leaf is to be found in this clear solution. Using the proportions of leaves to buffer described above, solutions of cytoplasmic proteins containing 1% or more by weight are obtained.

Because of their soluble character, the composition of the cytoplasmic proteins may be ascertained by conventional methods of protein analysis such as electrophoresis and analytical ultracentrifugation, and it is largely from this type of analysis that our present information on the protein composition of cytoplasm is derived (Frampton and Takahashi 1946, Wildman and Bonner 1947, Wildman et al. 1949, Singer et al. 1952, Eggman et al. 1953, Commoner et al. 1952, Pirie 1950). The scanning patterns which occupy the bottom of Fig. 2 are representative of the behavior of the cytoplasmic proteins of tobacco leaves during electrophoresis and analytical ultracentrifugation when homogenates are maintained near neutrality. However, comparison of the two methods of analysis shows at once that the analytical ultracentrifuge is in this instance capable of far greater resolution of the cytoplasmic protein mixture than is electrophoresis. Consequently, the following description of the protein composition of cytoplasm will be based on results obtained with the ultracentrifuge.

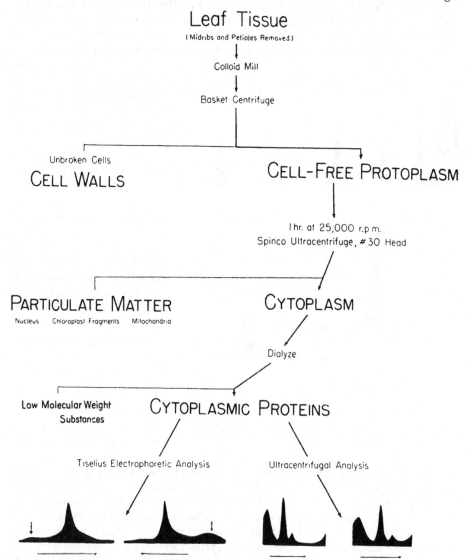

## Leaf Tissue
(Midribs and Petioles Removed)

Colloid Mill

Basket Centrifuge

Unbroken Cells
## Cell Walls

## Cell-Free Protoplasm

1 hr. at 25,000 r.p.m.
Spinco Ultracentrifuge, #30 Head

## Particulate Matter
Nucleus    Chloroplast Fragments    Mitochondria

## Cytoplasm

Dialyze

Low Molecular Weight
Substances
## Cytoplasmic Proteins

Tiselius Electrophoretic Analysis

Ultracentrifugal Analysis

Fig. 2. Procedure for the fractionation of cytoplasmic proteins from leaf homogenates; electrophoretic starting boundaries indicated by vertical arrows, migrational directions by horizontal arrows. [After WILDMAN and JAGENDORF (1952)]. See ALBERTY (1953) and NICHOLS and BAILEY (1949) for discussions of electrophoresis and analytical ultracentrifugation, respectively.

   In contrast to electrophoresis, where resolution of a protein mixture depends upon the different electrical charges carried by each protein in a mixture, resolution by centrifugation depends upon the mass of each protein in a mixture. Thus, when tobacco cytoplasm is subjected to the intense gravitational fields achieved by modern centrifuges, the schlieren optical scanning pattern reveals much of interest in regard to the composition of cytoplasm. Three components are resolved, two of which comprise about 90% of the entire mixture. The middle

component, constituting about 35% of the total protein mixture, is characterized
by a remarkably symmetrical peak which is indicative of a protein possessing
a high degree of homogeneity with respect to molecular weight. SINGER *et al.*
(1952) have called this characteristic protein of tobacco cytoplasm, Fraction I
protein. In the cytoplasmic protein mixture, Fraction I protein has a sedimentation constant of 19 Svedberg units (SINGER *et al.* 1952, COMMONER *et al.* 1952);
other calculations suggest a molecular weight of about 375,000 (EGGMAN *et al.*
1953). Fraction I protein is highly susceptible to denaturation and aggregation

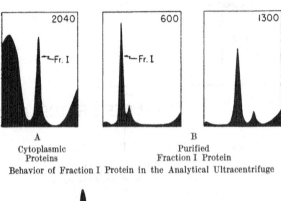

A
Cytoplasmic
Proteins

B
Purified
Fraction I Protein

Behavior of Fraction I Protein in the Analytical Ultracentrifuge

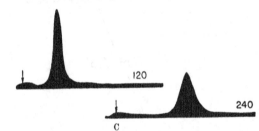

C
Electrophoretical Behavior of Purified Fraction I Protein

Fig. 3 A–C. Analytical ultracentrifuge patterns of whole cytoplasm (A)
and of Fraction I protein fractionated from it (B); migration is from
left to right for indicated period in seconds after rotor had attained a
speed of 50,220 r.p.m. (C), electrophoretic descending boundaries of
ultracentrifugally fractionated Fraction I after 120 and 240 minutes.
[After EGGMAN *et al.* (1953).]

by acid (SINGER *et al.* 1952)
and homogenates which have
been allowed to depart from
neutrality invariably show
reduced amounts of Frac-
tion I protein.

The minor component,
which moves in the gravitational field ahead of Fraction I protein, has a sedimentation constant of about
24 S. Evidence has been
presented to show that this
small amount of protein is
apparently an aggregate derived from Fraction I protein
(EGGMAN *et al.* 1953). The
trailing component making
up the third peak in the ultracentrifuge scanning pattern
has the least mass and a
mean sedimentation constant
of about 4 S. The peak lacks
symmetry which indicates
this fraction of cytoplasm to
be a heterogeneous mixture
of proteins.

**The isolation of Fraction I protein.** Because Fraction I protein is the protein
of largest mass and constitutes such a large part of the total proteins contained
in cytoplasm, much effort has been devoted towards its isolation and physical,
chemical, and biological characterization. EGGMAN (1953, EGGMAN *et al.* 1953)
has developed a centrifugal method whereby Fraction I protein can be largely
freed of the other proteins contained in cytoplasm. After extraction according
to the methods described, the cytoplasm whose composition is depicted in Fig. 3a,
is spun at approximately 150,000 g. for 3 hours in a preparative ultracentrifuge.
A pellet of protein is deposited, which, when redissolved in neutral buffer, resedi-
mented again and redissolved, and then subjected to analytical ultracentrifugation behaves according to the scanning patterns shown in Fig. 3b. Less than
10% of the 4 S trailing components is apparent. Instead, almost all of the area
of the scanning pattern is occupied by Fraction I protein. Furthermore, purification has not changed the sedimentation behavior of Fraction I protein, except
to promote the formation of the 24 S component. As shown in Fig. 3c, purified
Fraction I protein migrates in the electrophoresis apparatus as a material possessing a high degree of homogeneity with respect to electrical charge. The yield

of Fraction I protein of about 90% purity, as judged by analysis in the ultracentrifuge, often amounts to as much as 5 mg. per gram fresh weight of tobacco leaves.

**Fraction I protein as a possible cytological entity.** We may inquire at this point as to whether purified Fraction I protein meets the cytological criteria established for the size of particles we might expect to encounter in intact cytoplasm. Purified Fraction I protein, when examined with the electron microscope, is found to consist mainly of nearly spherical particles having diameters of about 0.01–0.02 microns, as shown in Fig. 4. Interestingly enough, before direct visualization, the size of Fraction I particles had been predicted to be in this range of magnitude as the result of extensive analysis of Fraction I protein in the analytical ultracentrifuge (EGGMAN 1953). Thus, two independent methods of analysis have converged to support the idea that Fraction I protein is in a size range small enough not to impinge upon the gap in particle sizes apparent in thin-section electron microscopy[1].

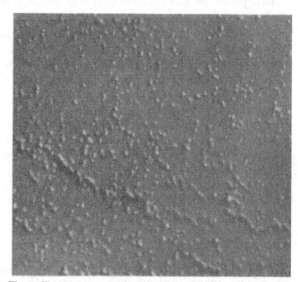

Fig. 4.   Electron photomicrograph of Fraction I after ultracentrifugal fractionation; magnification 22,000 ×.

**The chemical properties of Fraction I protein.** Purified Fraction I protein contains nucleic acid of the ribose type (EGGMAN 1953, EGGMAN *et al.* 1953). However, the amount of nucleic acid associated with the protein is variable with different preparations. When the organic phosphorus content of different preparations of Fraction I protein has been determined, the amount of phosphorus has varied from 0.15 to 1.5%. EGGMAN *et al.* (1953), suggest that the variation in nucleic acid (estimated from phosphorus determinations) is a reflection of the physiological age and nutritional status of the plant leaves at the time of harvest. Analysis of the constituents of the nucleic acid obtained from purified Fraction I protein reveals ratios of guanine to adenine of 1.4, cytidylic acid to uridylic acid of 1.1, and total purines to total pyrimidines of 1.3. EGGMAN *et al.* (1953) also provide suggestive evidence that all of the ribonucleic acid found in the soluble form after homogenization of tobacco and spinach leaf tissue is associated with Fraction I protein.

**Evidence that Fraction I protein is not derived from larger structures.** We have already alluded to the danger of homogenization degrading the visible structures of protoplasm such that some of their materials are reduced to the size of

---

[1] It is doubtful that the 50–100 Å particles detected by GAIDUKOV (1906) could represent Fraction I protein because the ultramicroscope is incapable of resolving the approximately $10^8$ Fraction I particles which calculations show are crowded into the cytoplasm of an average leaf cell. In addition, the refraction of this protein would probably be insufficiently different from that of the surrounding medium to be detectable by this means [see discussion of ultramicroscopy by FREY-WYSSLING (1953)].

cytoplasmic particles. However, isotope experiments performed by JAGENDORF (1954), appear to rule out the possibility that Fraction I protein could be derived from large particles. Tobacco leaves were allowed to incorporate $N^{15}H_4Cl$. Upon homogenization, the protoplasm was fractionated into chloroplasts, an enriched mitochondrial fraction, and cytoplasmic proteins. Fraction I protein was isolated from the cytoplasm. The mitochondrial fraction was found to incorporate $N^{15}$ into proteins at a considerably faster rate than any other protein fraction of the cell. Practically no incorporation occurred in the proteins of chloroplasts. The cytoplasmic proteins incorporated the label significantly, but at a much slower rate than the mitochondrial fraction, and Fraction I protein was found to incorporate the label at an even slower rate than the other proteins of cytoplasm.

That Fraction I protein could be derived from mitochondria seems very unlikely in view of the great difference in the labeling of the two materials. The remote possibility that weakly labeled Fraction I protein could be derived from essentially unlabeled chloroplasts appears to be eliminated from experiments with tobacco mosaic virus (JAGENDORF 1954). Infection with the virus does not affect the rate nor the degree of labeling of any protoplasmic fraction except Fraction I protein. Chloroplasts still fail to incorporate appreciably the label, and the mitochondria proteins retain a high but unaltered rate of isotope incorporation. Likewise, the smaller, non-Fraction I proteins, as an unfractionated group, are not markedly affected by the presence of virus. However, the rate at which Fraction I protein incorporates $N^{15}$ is greatly accelerated by the presence of virus, the increase in rate often being twice the rate of Fraction I protein obtained from uninfected leaves. Evidently, solubilization of chloroplasts can be eliminated as a source of Fraction I protein, and we feel that the isotope evidence provides further assurance that Fraction I protein is an integral component of the cytoplasm.

**Proteins other than Fraction I protein in cytoplasm.** As revealed by ultracentrifugal analysis, about 50% of the total cytoplasmic proteins consists of a heterogeneous mixture having a mean sedimentation constant of about 4 S. Unfortunately, our information on these proteins virtually ends with the last statement. As yet, the electron microscope has failed to resolve these proteins. That they contain different kinds of enzymatic activities has been demonstrated (BONNER and WILDMANN 1946, WILDMAN and BONNER 1947) but attempts to quantitatively resolve the protein mixture have not yet met with any measure of success. Phosphatase activity, once considered to be a property of Fraction I protein (WILDMAN and BONNER 1947), has been shown by BOROUGHS (1953) and EGGMAN (1953) to belong more properly among the smaller proteins of tobacco cytoplasm. In fact, to date Fraction I protein must be considered as being non-enzymatic, although it is difficult to believe that future experiments will fail to uncover a specific catalytic function for this major protein component of cytoplasm.

**The composition of the soluble proteins obtained from various leaves.** SINGER *et al.* (1952), have examined the soluble protein fraction obtained after the homogenization of several different kinds of leaves. Their results are shown in Fig. 5. No cytological criteria are available as with the tobacco leaf. Nevertheless, the cytoplasmic proteins could be prepared by the method described above for tobacco leaves, and the centrifuge results show a striking similarity in protein composition, suggesting a close correspondence in the basic protein composition of different cytoplasms. Each of the different soluble protein preparations gives evidence of containing a major fraction of protein whose sedimentation behavior, ranging from 16–19 S, is similar to that of tobacco leaf Fraction I protein. Depending upon the leaf, from 23–50% of the total protein mixture consists of this kind of protein.

**The soluble proteins of plant tissues other than leaves.** Because of the absence of cytological and suitable physico-chemical criteria, it is very difficult to assess the composition of the cytoplasm of plant tissues different from leaves. Soluble protein extracts have been prepared from such diverse sources as potato tubers (FRAMPTON and TAKAHASHI 1946, LEVITT 1952), Jerusalem artichoke callus

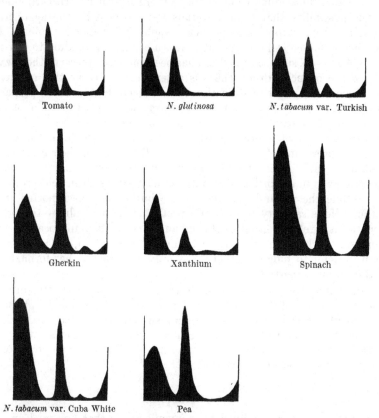

Fig. 5. Analytical ultracentrifuge patterns (migration from left to right) of leaf cytoplasmic proteins of various plants. [After SINGER et al. (1952).]

and crown gall (G. CH. CAMUS, personal communication; see also CAMUS, as reported by GAUTHERET 1953); *Avena* coleoptiles, etiolated pea stems and leaves, and tobacco leaf petioles (CAMPBELL 1951). While much has been learned about the behavior of these proteins, as for example their behavior during electrophoresis, the protein extracts have not been subjected to enough analysis to permit further speculation in regard to their precise location in protoplasm.

Seed proteins have for years occupied the close attention of many investigators, and a vast literature has been accumulated (see, for example, JENSEN 1952, SNELLMAN and DANIELSSON 1953, DJANG et al. 1952, and review by BONNER 1950). In spite of much information on their chemical composition and physical behavior, the lack of cytological evidence to denote whether they belong in the cytoplasm has precluded their inclusion in this discussion. It is interesting, however, that NEWCOMB and STUMPF (1952) have described a soluble protein from germinating peanut cotyledons which catalyzes the oxidation of fatty acids. Under the electron microscope, the protein is found to consist of spherical particles possessing average diameters of 0.02 micron.

"**Microsomes**" **isolated from homogenates of plant tissues.** We view with some alarm the increasing number of reports of "microsomal" fractions which have appeared in the recent literature on plant proteins and enzymology (WEBSTER 1953, STAFFORD 1951, LEVITT 1952). Other than the fact that the "microsomes" were brought down at centrifugal forces which are considered to sediment cyto-logically demonstrable animal microsomes, there is still no critical evidence to dismiss the possibility that such materials were formed by the aggregation of much smaller sized cytoplasmic particles, such as Fraction I protein. In the fractionation of animal cell homogenates, it is a common procedure to disrupt the cells in 0.8 M sucrose, since this medium has been shown to preserve the enzymatic integrity of the mitochondria. There is a tendency to carry this method over to plant tissues with slight modifications, and indeed, highly active preparations of mitochondria are often obtained. However, when fractionation of smaller particles is attempted from the same homogenate after removing the mitochondria, we feel that a word of caution is appropriate before the term "microsome" be assigned indiscriminately to such fractions. By virtue of their large vacuoles, plant cells are greatly different from animal cells. More often than not, disruption of plant cells, even in the presence of buffers, may still result in acid protoplasmic extracts. While the $p_H$ and molarity of the buffer may be reported, the real test of the buffer's capacity to maintain protoplasmic neutrality—the $p_H$ of the homogenate—is rarely given. [See HAGEN and JONES (1952) for locale of catalase in protoplasm of wheat leaves as a function of homogenate $p_H$.] On this account it is often difficult to judge whether a "microsome" fraction has formed by acid aggregation of small particles.

We would imagine that no single fractionation scheme is yet suitable for isolation of all of the different classes of particles to be found in plant proto-plasm. Instead, we feel that more attention should be devoted to the develop-ment of methods which will meet the peculiar requirements for isolation of each class of particles and which at the same time can be critically evaluated from both a cytological and cytochemical standpoint.

# G. Conclusions.

The hyaline cytoplasm, defined for convenience as that non-nuclear portion of protoplasm separated from the vacuole by the tonoplast and invisible under the light microscope, is an extremely complex mixture of substances ranging in size from ions to macromolecular nucleoproteins. A previous lack of cytological criteria to establish what kinds of materials might be encountered *in situ* in undisturbed cytoplasm is gradually being overcome. The newer techniques of electron microscopy are providing the plant physiologist with an important tool in his attempt to ascribe function to structure by way of the chemical con-stitution and biological activity of submicroscopic entities, the presence of which he has suspected in his work with tissue homogenates. Of equal importance, is the possibility of discovery of unsuspected structures heretofore unresolved by centrifugal fractionation of homogenates. While only the barest beginning has been made in a combined cytological-biochemical approach to an under-standing of the composition and function of cytoplasm, there is every reason to believe that the way has been paved for rapid, future progress in this difficult area of cytochemistry.

# Literature.

ALBERTY, R. A.: Electrochemical properties of the proteins and amino acids, in H. NEU-RATH and K. BAILEY, eds., The proteins. Chemistry, biological activity, and methods, Vol. 1, pt. A. New York: Academic Press 1953. — ANDRESEN, N., C. CHAPMAN-ANDRESEN and R. HOLTER: Autoradiographic studies on the amoeba *Chaos chaos* with $^{14}$C. C. r. Labor. Carlsberg, Sér. chim. **28**, 189–220 (1952).

BAKER, J. R.: Cytological technique. London: Methuen 1950. — BALLENTINE, R., and D. G. STEPHENS: The biosynthesis of stable cobalto-proteins by plants. J. Cellul. a. Comp. Physiol. **37**, 369–387 (1951). — BARRNETT, R. J., and A. M. SELIGMAN: Investigations of the histochemical localization of disulfides. (Abstract.) J. Histochem. a. Cytochem. **1**, 392–394 (1953). — BAYLISS, M., D. GLICK and R. A. SIEM: Demonstration of phosphatases and lipase in bacteria and true fungi by staining methods and the effect of penicillin on phosphatase activity. J. Bacter. **55**, 307–316 (1948). — BEAMS, H. W.: Ultracentrifugal studies on cytoplasmic components and inclusions. Biol. Symposia **10**, 71–90 (1943). — BEHRENS, M.: Über die Lokalisation der Hefenucleinsäure in pflanzlichen Zellen. Hoppe-Seylers Z. **253**, 185–192 (1938). — BENNETT, H. S.: The demonstration of thiol groups in certain tissues by means of a new colored sulfhydryl reagent. Anat. Rec. **110**, 231–247 (1951). — BEST, R. J.: Studies on a fluorescent substance present in plants. Part 3: The distribution of scopoletin in tobacco plants and some hypotheses on its part in metabolism. Austral. J. Exper. Biol. a. Med. Sci. **26**, 223–230 (1948). — BHATTACHARJEE, D., and A. K. SHARMA: Phosphatase in mitotic and meiotic cycle of plant chromosomes. Sci. a. Cult. **17**, 268–269 (1951). Original not seen. (Abstract.) Biol. Abstr. **26**, 2112 (1952). — BONNER, J.: Plant proteins. Fortschr. chem. organ. Naturstoffe **6**, 290–310 (1950). — BONNER, J., and S. G. WILDMAN: Enzymatic mechanisms in the respiration of spinach leaves. Arch. of Biochem. **10**, 497–518 (1946). — BOROUGHS, H.: Studies on the acid phosphatases of green leaves. Ph. D. thesis, California Institute of Technology, Pasadena 1953. — Studies on the acid phosphatases of green leaves. Arch. of Biochem. a. Biophysics **1954**. — BRACHET, J.: The use of basic dyes and ribonuclease for the cytochemical detection of ribonucleic acid. Quart. J. Microsc. Sci. **94**, 1–10 (1953). — BRADFIELD, J. R. G.: Plant carbonic anhydrase. Nature (Lond.) **159**, 467–468 (1947). — The localization of enzymes in cells. Biol. Rev. **25**, 113–157 (1950). — BRATTGÅRD, S.-O., and H. HYDÉN: Mass, lipids, pentose nucleoproteins and proteins determined in nerve cells by X-ray microradiography. Acta radiol. (Stockh.) Suppl. **94**, 48 pp. (1952). — BROOKS, S. C.: Penetration of radioactive isotopes, $P^{32}$, $Na^{24}$ and $K^{42}$ into *Nitella*. J. Cellul. a. Comp. Physiol. **38**, 83–93 (1951). — The penetration of radioactive sodium into *Valonia* and *Halicystis*. Protoplasma (Wien) **42**, 63–68 (1953).

CAMPBELL, J. M.: Electrophoretic studies of leaf proteins. Ph. D. thesis, University of California, Los Angeles 1951. — CASPERSSON, T., and J. SCHULTZ: Pentose nucleotides in the cytoplasm of growing tissues. Nature (Lond.) **143**, 602–603 (1939). — Ribonucleic acids in both nucleus and cytoplasm, and the function of the nucleolus. Proc. Nat. Acad. Sci. U.S.A. **26**, 507–515 (1940). — CASPERSSON, T. O.: Cell growth and cell function. A cytochemical study. New York: Norton 1950. — CATCHESIDE, D. G., and B. HOLMES: The action of enzymes on chromosomes. Symposia Soc. Exper. Biol. **1**, 225–231 (1947). — CHAMBERS, R., and T. KERR: Intracellular hydrion concentration studies. VIII. Cytoplasm and vacuole of Limnobium root-hair cells. J. Cellul. a. Comp. Physiol. **2**, 105–119 (1932). — CHAPMAN-ANDRESEN, C.: Autoradiographs of algae and ciliates exposed to tritiated water. Exper. Cell Res. **4**, 239–242 (1953). — CHAYEN, J.: Ascorbic acid and its intracellular localization with special reference to plants. Intern. Rev. Cytol. **2**, 78–131 (1953). — CHAYEN, J., and K. P. NORRIS: Cytoplasmic localization of nucleic acids in plant cells. Nature (Lond.) **171**, 472–473 (1953). — CHIBNALL, A. C.: Protein metabolism in the plant. New Haven: Yale Univ. Press 1939. — CHU, C. H. U., M. H. FOGELSON and C. A. SWINYARD: A histochemical test for alpha amino acids. (Abstract.) J. Histochem. a. Cytochem. **1**, 391–392 (1953). — COHEN, M., and E. BOWLER: Lamellar structure of the tobacco chloroplast. Protoplasma (Wien) **42**, 414–416 (1953). — COMMONER, B.: On the interpretation of the absorption of ultraviolet light by cellular nucleic acids. Science (Lancaster, Pa.) **110**, 31–40 (1949). — COMMONER, B., P. NEWMARK and S. D. RODENBERG: An electrophoretic analysis of tobacco mosaic virus biosynthesis. Arch. of Biochem. a. Biophysics **37**, 15–36 (1952). — COONS, A. H., and M. H. KAPLAN: Localization of antigen in tissue cells. II. Improvements in a method for the detection of antigen by means of fluorescent antibody. J. of Exper. Med. **91**, 1–13 (1950). — COONS, A. H., E. H. LEDUC and M. H. KAPLAN: Localization of antigen in tissue cells. VI. The fate of injected foreign proteins in the mouse. J. of Exper. Med. **93**, 173–188 (1951). — CROOK, E. M.: The extraction of nitrogenous materials from green leaves. Biochemic. J. **40**, 197–209 (1946).

DALTON, A. J., and M. J. STRIEBICH: Electron microscopic studies of cytoplasmic components of some of the cells of the liver, pancreas, stomach, and kidney following treatment with ribonuclease. (Abstract.) J. Nat. Canc. Inst. 12, 244–245 (1951). — DANIELLI, J. F.: Establishment of cytochemical techniques. Nature (Lond.) 157, 755–757 (1946). — A study of techniques for the cytochemical demonstration of nucleic acids and some components of proteins. Symposia Soc. Exper. Biol. 1, 101–113 (1947). — Studies on the cytochemistry of proteins. Cold Spring Harbor Symp. Quant. Biol. 14, 32–39 (1950). — Cytochemistry. A critical approach. New York: John Wiley 1953. — DEMPSEY, E. W., and G. B. WISLOCKI: Histochemical contributions to physiology. Physiologic. Rev. 26, 1–27 (1946). — DJANG, S. S. T., C. D. BALL and H. A. LILLEVIK: The isolation, fractionation, and electrophoretic characterization of the globulins of mung bean *(Phaseolus aureus)*. Arch. of Biochem. a. Biophysics 40, 165–174 (1952). — DOUNCE, A. L.: The interpretation of chemical analyses and enzyme determinations on isolated cell components. J. Cellul. a. Comp. Physiol. 39, Suppl. 2, 43–74 (1952). — DOWDING, E. S.: The regional and seasonal distribution of potassium in plant tissues. Ann. of Bot. 39, 459–474 (1925). — DU BUY, H. G., M. W. WOODS and M. D. LACKEY: Enzymatic activities of isolated normal and mutant mitochondria and plastids of higher plants. Science (Lancaster, Pa.) 111, 572–574 (1950). — DUGGER jr., W. M., and D. E. MORELAND: An autoradiographic technique for detailed studies with plant tissue. Plant Physiol. 28, 143–145 (1953). — DYAR, M. T.: Some observations on starch synthesis in pea root tips. Amer. J. Bot. 37, 786–792 (1950).

EGGMAN, L.: The cytoplasmic proteins of green leaves. Ph.D. thesis, California Institute of Technology, Pasadena 1953. — EGGMAN, L., S. J. SINGER and S. G. WILDMAN: The proteins of green leaves. V. A cytoplasmic nucleoprotein from spinach and tobacco leaves. J. of Biol. Chem. 205, 969–983 (1953). — ENGSTRÖM, A.: Microradiography. Acta radiol. (Stockh.) 31, 503–521 (1949). — Use of soft X-rays in the assay of biological materials. Progr. Biophysics a. Biophysical Chem. 1, 164–196 (1950). — X-ray absorption methods in histochemistry. Labor. Invest. 1, 278–285 (1952). — X-ray methods in histochemistry. Physiologic. Rev. 33, 190–201 (1953). — ENGSTRÖM, A., and B. LINDSTRÖM: A method for the determination of the mass of extremely small biological objects. Biochim. et Biophysica Acta 4, 351–373 (1950).

FITZGERALD, P. J., M. L. EIDINOFF, J. E. KNOLL and E. B. SIMMEL: Tritium in radioautography. Science (Lancaster, Pa.) 114, 494–498 (1951). — FITZGERALD, P. J., and A. ENGSTRÖM: The use of ultraviolet-microscopy, Roentgen-ray-absorption, and radioautographic techniques in the study of neoplastic disease. A discussion of these cytochemical techniques. Cancer (N. Y.) 5, 643–677 (1952). — FITZGERALD, P. J., E. SIMMEL, J. WEINSTEIN and C. MARTIN: Radioautography: theory, technic, and applications. Labor. Invest. 2, 181–222 (1953). — FLAX, M. H., and M. H. HIMES: A differential stain for ribonucleic and desoxyribonucleic acids. (Abstract.) Anat. Rec. 108, 529 (1950). — Microspectrophotometric analysis of metachromatic staining of nucleic acids. Physiologic. Zool. 25, 297–311 (1952). — FRAMPTON, V. L., and W. N. TAKAHASHI: Electrophoretic studies with the plant viruses. Phytopathology 36, 129–141 (1946). — FREY-WYSSLING, A.: Submicroscopic morphology of protoplasm. Amsterdam: Elsevier 1953.

GAIDUKOV, N.: Über die ultramikroskopischen Eigenschaften der Protoplasten. Ber. dtsch. bot. Ges. 24, 192–194 (1906). — GAUTHERET, R.-J.: La culture des tissues végétaux. 3. Applications à la pathologie. Nature (Paris) 1953, No 3214, 53–57. — GLICK, D.: Techniques of histo- and cytochemistry. New York: Interscience 1949. — A critical survey of current approaches in quantitative histo- and cytochemistry. Internat. Rev. Cytol. 2, 447–474 (1953). — GLICK, D., and E. E. FISCHER: Studies in histochemistry. XVII. Localization of phosphatases in the wheat grain and in the epicotyl and roots of the germinated grain. Arch. of Biochem. 11, 65–79 (1946a). — Studies in histochemistry. XVIII. Localization of arginine in the wheat grain and in the epicotyl and roots of the germinated grain. Arch. of Biochem. 11, 81–87 (1946b). — GOETSCH, J. B., and P. M. REYNOLDS: Obtaining uniform results in the histochemical technic for acid phosphatase. Stain Technol. 26, 145–151 (1951).— GOETSCH, J. B., P. M. REYNOLDS and H. BUNTING: A modification of the Gomori method avoiding the artifact staining of the nucleus. (Abstract.) J. Nat. Canc. Inst. 13, 255–256 (1952). — GOMORI, G.: Microscopic histochemistry. Chicago: Univ. of Chicago Press 1952. — GOODWIN, R. H.: Fluorescent substances in plants. Annual Rev. Plant Physiol. 4, 283–304 (1953). — GUILLIERMOND, A.: The cytoplasm of the plant cell. Waltham, Mass.: Chronica Botanica 1941.

HAGEN, C. E., and V. V. JONES: Factors in the determination of the intracellular localization of enzymes. Bot. Gaz. 114, 130–134 (1952). — HEITZ, E.: Gerichtete Chlorophyllscheiben als strukturelle Assimilationseinheiten der Chloroplasten. Ber. dtsch. bot. Ges. 54, 362–368 (1936). — HOAGLAND, D. R., and A. R. DAVIS: The composition of the cell sap of the plant in relation to the absorption of ions. J. Gen. Physiol. 5, 629–646 (1923). — HOLTER, H.: Localization of enzymes in cytoplasm. Adv. Enzymol. 13, 1–20 (1952). — HOWARD, A., and S. R. PELC: Nuclear incorporation of P$^{32}$ as demonstrated by autoradiographs. Exper. Cell Res. **2**, 178–187 (1951a). — Synthesis of nucleoprotein in bean root cells. Nature (Lond.) **167**, 599–600 (1951b).

Jacobson, W., and M. Webb: The two types of nucleoproteins during mitosis. Exper. Cell Res. **3**, 163–183 (1952). — Jagendorf, A. T.: Personal communication, 1954. — Jensen, R.: On isolation and amino acid composition of "β-globulin" extracted from the seeds of barley *(Hordeum vulgare)*. Acta chem. scand. (København) **6**, 771–781 (1952). — Jones, R. W., and K. C. Hamner: The intracellular distribution of ascorbic acid in turnip leaves. Plant Physiol. **28**, 314–316 (1953).

Korson, R.: A differential stain for nucleic acids. Stain Technol. **26**, 265–270 (1951). — Kurnick, N. B.: Histological staining with methyl-green pyronin. Stain Technol. **27**, 233–242 (1952).

Lamb, W. G. P., J. Stuart-Webb, L. G. Bell, R. Bovey and J. F. Danielli: Specific stains for electron microscopy. Exper. Cell Res. **4**, 159–163 (1953). — Lang, K.: Lokalisation der Fermente und Stoffwechselprozesse in den einzelnen Zellbestandteilen und deren Trennung. Colloq. dtsch. Ges. physiol. Chem. **2** (1951, Mikroskopische und chemische Organisation der Zelle), 24–47 (1952). — Levitt, J.: Two methods of fractionating potato tuber proteins and some preliminary results with dormant and active tubers. Physiol. Plantarum **5**, 470–484 (1952). — Lloyd, F. E.: The cobalt sodium hexanitrite reaction for potassium in plant cells. Flora (Jena) **118/119**, 367–385 (1925).

Macallum, A. B.: On the distribution of potassium in animal and vegetable cells. J. of Physiol. **32**, 95–128 (1905). — Macdowall, F. D. H.: Absence of acid phosphatase from chloroplasts of spinach and iris. Plant Physiol. **28**, 317–318 (1953). — Mager, M., W. F. McNary jr. and F. Lionetti: The histochemical detection of zinc. J. Histochem. a. Cytochem. **1**, 493–504 (1953). — Martin, B. F., and F. Jacoby: Diffusion phenomenon complicating the histochemical reaction for alkaline phosphatase. J. of Anat. **83**, 351–363 (1949). — McLean, F. T., and B. E. Gilbert: The relative aluminum tolerance of crop plants. Soil Sci. **24**, 163–175 (1927). — Mellors, R. C., R. E. Berger and H. G. Streim: Ultraviolet microscopy and microspectroscopy of resting and dividing cells: studies with a reflecting microscope. Science (Lancaster, Pa.) **111**, 627–632 (1950). — Menten, M. L., J. Junge and M. H. Green: A coupling histochemical azo dye test for alkaline phosphatase in the kidney. J. of Biol. Chem. **153**, 471–477 (1944). — Metzner, H.: Die Reduktion wäßriger Silbernitratlösungen durch Chloroplasten und andere Zellbestandteile. Protoplasma (Wien) **41**, 129–167 (1952). — Meyer, B. S., and D. B. Anderson: Plant physiology, 2nd ed. New York: Van Nostrand 1952. — Millerd, A., and J. Bonner: The biology of plant mitochondria. J. Histochem. a. Cytochem. **1**, 254–264 (1953).

Nagai, S.: Experimental studies on the reduction of silver nitrate by plant cell. I. Dynamic process in reduction and precipitation. J. Osaka City Univ. Inst. Polytech., Ser. D, Biol. **1**, 33–44 (1950). — Experimental studies on the reduction of silver nitrate by plant cell. II. Nature and responsibility of substances which cause the reduction. J. Osaka City Univ. Inst. Polytech., Ser. D, Biol. **2**, 1–8 (1951). — Nagai, S., and E. Ogata: Experimental studies on the reduction of silver nitrate by plant cell. III. Further evidences on the rôle of ascorbic acid in the Molisch reaction. J. Osaka City Univ. Inst. Polytech., Ser. D, Biol. **3**, 37–45 (1952a). [Original not seen. Abstracted by O. Härtel, Protoplasma (Wien) **42**, 491–492 (1953).] — Experimental studies on the reduction of silver nitrate by plant cell. IV. The reaction in etiolated seedlings. J. Osaka City Univ. Inst. Polytech., Ser. D,,Biol. **3**, 46–55 (1952b). [Original not seen. Abstracted by O. Härtel, Protoplasma (Wien) **42**, 491–492 (1953).] — Newcomb, E. H., and P. K. Stumpf: Fatty acid synthesis and oxidation in peanut cotyledons, in W. D. McElroy and B. Glass, eds., Phosphorus metabolism, Vol. 2. Baltimore: Johns Hopkins Press 1952. — Nichols, J. B., and E. D. Bailey: Determinations with the ultracentrifuge, in A. Weissberger, ed., Physical methods of organic chemistry, 2nd ed., Vol. 1, pt. 1. New York: Interscience 1949. — Novikoff, A. B.: The validity of histochemical phosphatase methods on the intracellular level. Science (Lancaster, Pa.) **113**, 320–325 (1951).

Olsen, C.: The mineral, nitrogen and sugar content of beech leaves and beech leaf sap at various times. C. r. Labor. Carlsberg, Sér. chim. **26**, 197–230 (1948a). — Adsorptively bound potassium in beech leaf cells. C. r. Labor. Carlsberg, Sér. chim. **26**, 361–367 (1948b).

Palade, G. E.: Intracellular localization of acid phosphatase. A comparative study of biochemical and histochemical methods. J. of Exper. Med. **94**, 535–548 (1951). — A study of fixation for electron microscopy. J. of Exper. Med. **95**, 285–298 (1952). — An electron microscope study of the mitochondrial structure. J. Histochem. a. Cytochem. **1**, 188–211 (1953). — Penston, N. L.: Studies of the physiological importance of the mineral elements in plants. III. A study by microchemical methods of the distribution of potassium in the potato plant. Ann. of Bot. **45**, 673–692 (1931). — Perner, E. S.: Die Sphärosomen (Mikrosomen) pflanzlicher Zellen. Sammelreferat unter Berücksichtigung eigener Untersuchungen.

Protoplasma (Wien) 42, 457–481 (1953). — Pirie, N. W.: The isolation from normal tobacco leaves of nucleoprotein with some similarity to plant viruses. Biochemic. J. 47, 614–625 (1950). — Pollister, A. W., J. Post, J. G. Benton and R. Breakstone: Resistance of ribonucleic acid to basic staining and ribonuclease in human liver. (Abstract.) J. Nat. Canc. Inst. 12, 242–243 (1951). — Poppen, K. J., D. M. Green and H. T. Wrenn: The histochemical localization of potassium and glycogen. J. Histochem. a. Cytochem. 1, 160–173 (1953). — Porter, K. R., A. Claude and E. F. Fullam: A study of tissue culture cells by electron microscopy. J. of Exper. Med. 81, 233–246 (1945).

Reed, H. S., and J. Dufrénoy: The effects of zinc and iron salts on the cell structure of mottled orange leaves. Hilgardia 9, 111–141 (1935). — Roberts, L. W.: A study of the tetrazolium reaction in plant tissues. Abstract of Ph. D. thesis (Univ. of Missouri, 1952) in Diss. Abstr. 12, 363 (1952). — Ross, M. H., and J. O. Ely: Alkaline phosphatases in fixed plant cells. Exper. Cell Res. 2, 339–348 (1951). — Rozsa, G., and R. W. G. Wyckoff: The electron microscopy of onion root tip cells. Exper. Cell Res. 2, 630–641 (1951). — Ruyter, J. H. C., and H. Neumann: A critical examination of the histochemical demonstration of the alkaline phosphomonoesterase. Biochim. et Biophysica Acta 3, 125–135 (1949).

Schneider, W. C., and G. H. Hogeboom: Cytochemical studies of mammalian tissues: the isolation of cell components by differential centrifugation: a review. Cancer Res. 11, 1–22 (1951). — Scott, G. H.: Mineral distribution in the cytoplasm. Biol. Symposia 10, 277–289 (1943). — Selby, C. C.: Microscopy. II. Electron microscopy: a review. Cancer Res. 13, 753–775 (1953). — Singer, S. J., L. Eggman, J. M. Campbell and S. G. Wildman: The proteins of green leaves. IV. A high molecular weight protein comprising a large part of the cytoplasmic proteins. J. of Biol. Chem. 197, 233–239 (1952). — Sjöstrand, F. S., and J. Rhodin: The ultrastructure of the proximal convoluted tubules of the mouse kidney as revealed by high resolution electron microscopy. Exper. Cell Res. 4, 426–456 (1953). — Small, J.: $p_H$ and plants. New York: Van Nostrand 1946. — Smith, J. D.: The concentration and distribution of haemoglobin in the root nodules of leguminous plants. Biochemic. J. 44, 585–591 (1949). — Snellman, O., and C. E. Danielsson: An experimental study of the biosynthesis of the reserve globulins in pea seeds. Exper. Cell Res. 5, 436–442 (1953). — Somers, G. F., and M. H. Coolidge: Location of thiamin and riboflavin in wheat grains. Science (Lancaster, Pa.) 101, 98–99 (1945a). — Somers, G. F., M. H. Coolidge and K. C. Hamner: The distribution of thiamine and riboflavin in wheat grains. Cereal Chem. 22, 333–340 (1945b). — Stafford, H. A.: Intracellular localization of enzymes in pea seedlings. Physiol. Plantarum 4, 696–741 (1951). — Strangeways, T. S. P., and R. G. Canti: The living cell in vitro as shown by dark-ground illumination and the changes induced in such cells by fixing reagents. Quart. J. Microsc. Sci. 71, 1–14 (1927). — Strugger, S.: Die Strukturordnung im Chloroplasten. Ber. dtsch. bot. Ges. 64, 69–83 (1951). — Swift, H.: Quantitative aspects of nuclear nucleoproteins. Internat. Rev. Cytol. 2, 1–76 (1953).

Tandler, C. J.: The use of cobalt in the histochemical technic for acid phosphatase. J. Histochem. a. Cytochem. 1, 151–153 (1953). — Tewfik, S., and P. K. Stumpf: Carbohydrate metabolism in higher plants. II. The distribution of aldolase in plants. Amer. J. Bot. 36, 567–571 (1949). — Turchini, J.: La détection cytochimique des constituants hydrocarbonés provenant de l'hydrolyse des acides nucléiques cellulaires. Exper. Cell Res. Suppl. 1, 105–110 (1949).

Uber, F. M.: Microincineration and ash analysis. Bot. Review 6, 204–226 (1940).

Waygood, E. R.: Physiological and biochemical studies in plant metabolism. II. Respiratory enzymes in wheat. Canad. J. Res., Sect. C, Bot. Sci. 28, 7–62 (1950). — Waygood, E. R., and K. A. Clendenning: Intracellular localization and distribution of carbonic anhydrase in plants. Science (Lancaster, Pa.) 113, 177–179 (1951). — Webster, G. C.: The occurrence of a cytochrome oxidase in the tissues of higher plants. Amer. J. Bot. 39, 739–745 (1952). — Enzymatic synthesis of glutamine in higher plants. Plant Physiol. 28, 724–727 (1953). — Weier, T. E.: Factors affecting the reduction of silver nitrate by chloroplasts. Amer. J. Bot. 25, 501–507 (1938). — Weier, T. E., and C. R. Stocking: The chloroplast: structure, inheritance, and enzymology. II. Bot. Review 18, 14–75 (1952). — Weiss, L. P., and A. M. Seligman: Histochemical demonstration of protein-bound amino groups. (Abstract.) J. Histochem. a. Cytochem. 1, 386–387 (1953). — Wildman, S. G., and J. Bonner: The proteins of green leaves. I. Isolation, enzymatic properties and auxin content of spinach cytoplasmic proteins. Arch. of Biochem. 14, 381–413 (1947). — Wildman, S. G., C. C. Cheo and J. Bonner: The proteins of green leaves. III. Evidence of the formation of tobacco mosaic virus protein at the expense of a main protein component in tobacco leaf cytoplasm. J. of Biol. Chem. 180, 985–1001 (1949). — Wildman, S., and A. Jagendorf: Leaf proteins. Annual Rev. Plant Physiol. 3, 131–148 (1952).

Yin, H. C.: A histochemical study of the distribution of phosphatase in plant tissues. New Phytologist 44, 191–195 (1945). — Yin, H. C., and C. N. Sun: Localization of phosphorylase and of starch formation in seeds. Plant Physiol. 24, 103–110 (1949).

# II. The chemical composition of constituents found in plant cells.

Since protoplasm is the seat of countless chemical reactions involving syntheses, degradations, and rearrangements, we should expect that a great proportion of the enormous array of participating compounds would, at one time or another, even if only for an instant, be present in the submicroscopic cytoplasm. Even with substances which appear to be confined within the vacuole, colored anthocyanins, for example, and within the chloroplast membrane, the chlorophyll and carotenoid pigments, for example, it is difficult to conceive of wholly non-cytoplasmic metabolic pathways for their synthesis.

## A. The elements.

Almost forty of the elements have been detected in plants, although only 15 of these, C, H, O, P, K, N, S, Ca, Fe, Mg, B, Mn, Zn, Mo, and Cu, appear to be essential to the growth and reproduction of plants. Of these essential elements, oxygen, nitrogen, hydrogen, and carbon (as $CO_2$) are, in part, present as dissolved gases. As ions, the essential elements occur either singly, for example $K^+$, $Mg^{++}$, $Mn^{++}$, etc., or in combination with other elements, for example, $PO_4^{\equiv}$, $SO_4^{=}$, etc. Nitrogen can occur in various oxidation levels, as $NO_3^-$, $NO_2^-$, or $NH_4^+$. Carbon, hydrogen, oxygen, nitrogen, and sulfur, in covalent linkage, are constituents of the organic compounds, to a very few of which iron and magnesium are coordinated.

Ions may also be complexed with other ions or with naturally occurring chelating substances in solution. Certain metal ions (Ca, Mg, Mn, Ba, Fe, Cu, Co, Ni, Zn, Cd, K, Al, Cr, V, rare earths) may activate enzyme-substrate systems through more or less stable coordination with the enzyme or its prosthetic group, and, at the same time, some of them, because they may exist in more than one valence state, may take part in intermediate electron transfer in coupled series of enzymatic oxidation-reduction reactions.

Hydrogen, as a proton, plays an especially important role in cell metabolism. As an expression of the buffering systems in the cell, it acts as a regulator of enzymatic activity, perhaps because of the general effect of hydrogen ion concentration on amphoteric substances, of which enzymes are a special case. Its positive charge is responsible for certain inter- and intramolecular attractions, the ubiquitous "hydrogen bonds", which, though relatively weak, apparently contribute (together with the VAN DER WAALS' forces of similar magnitude) to the maintenance of organization in protoplasmic structures and aggregates, from the convolutions of the polypeptide chains of proteins and the oriented association of cellulose polysaccharide chains into cellulose microfibrils down to the level of associated water molecules. The electron transfer which occurs in metabolic oxidations and reductions is often simultaneously accompanied by hydrogen transfer.

Water in the cell is not merely a passive medium in which chemical substances are dispersed and chemical reactions may conveniently occur, but is a frequent direct participant in anabolism and catabolism, and, because of its polar structure and its properties of adsorption and ionization, is a determinant in the mobilities and physical and chemical activities of the soluble and colloidal substances of the cell.

The quadrivalence of carbon and the ability of this element to share pairs of electrons with other carbon atoms, as well as with nitrogen, oxygen, and sulfur atoms, have not only provided the great chemosynthetic potential of the living cell, but also the wherewithal for the prolific activities of the organochemical laboratories during the past century. Indeed, thousands of compounds have been isolated from plants, and their physical and chemical characterizations have filled many published volumes. It will be our objective here, however, to show the types of ramification into chemostructural series of greater and greater complexity which occur in the manufacture of cytoplasm, rather than merely to compile a list of phytochemicals.

## B. The structural diversity of phytochemicals.

Table 6 has been arranged to show how the kinds of carbon skeletons which are encountered in the compounds which have been isolated from plants, range from the simplest condition of carbon atoms being joined to each other in linear combination—perhaps, in this sense, the least complicated phytochemical might be the ethylene gas, $CH_2:CH_2$, found in ripening fruit—to the increased complexity afforded not only by union with oxygen, nitrogen, and sulfur in the skeleton, but also by branching and cyclization of the skeletons.

Further diversification of the skeletons may occur in several ways; namely, by (1) the incorporation of substituent groups, (2) increase in chain length, (3) variation in degree and position of unsaturation, (4) intermolecular linkages, and (5) cyclization. Table 7 is illustrative of the various classes of compounds, which, depending on their end and side groups, are the alcohols, aldehydes, amines, etc., each class being basic to vast arrays of derivatives. Thus, for example, the fatty acids (Table 8) are derivatives involving merely the lengthening of the linear carbon chain through the addition of carbon and hydrogen, the alteration of degree and position of unsaturation, and the addition of carbonyl oxygen or hydroxyl groups at various positions on the chain.

Intermolecular linkage through sulfur, oxygen, or phosphoric acid, as depicted in Table 9, offers still another means of elaboration. Indeed, the glycoside, peptide, and phosphoric acid linkages, because they permit not only relatively enormous chain lengths through repetitive linkage, but also innumerable orders in which any of a number of possible substituents may appear in the chain, are responsible for the great number of different kinds of polysaccharides, proteins, and nucleic acids, respectively. The repetitive peptide linkage of 20 different amino acids, for example, could lead, theoretically, to the formation of over $2.4 \times 10^{18}$ different molecules, and each molecule would still weigh but a small fraction of the weight of most known protein molecules.

Elaboration by intermolecular linkage is further illustrated by the fact that the fatty acids of Table 8 most commonly occur in ester linkage with glycerol, a trihydric alcohol; and these esters constitute the oils and fats, which are most prevalent in fruits and seeds.

The triglycerides are triesters with saturated (6 or more carbon atoms) and unsaturated (10 or more carbon atoms) fatty acids containing even numbers of carbon atoms. Since most fats and oils are mixtures of triglycerides to which any three of the fatty acids present in a particular fat or oil may be esterified on the basis of maximum heterogeneity, it is probable that a fat or oil will not only contain a simple triglyceride, as for example, tripalmitin,

$$H_2C \cdot O \cdot CO \cdot (CH_2)_{14} \cdot CH_3$$
$$|$$
$$HC \cdot O \cdot CO \cdot (CH_2)_{14} \cdot CH_3,$$ but also several different kinds of mixed
$$|$$
$$H_2C \cdot O \cdot CO \cdot (CH_2)_{14} \cdot CH_3$$

Table 6. *Some of the kinds of carbon skeletons encountered in the structure of protoplasmic constituents.*

---

I. Carbon only
  A. Straight chain

n-heptane $CH_3 \cdot CH_2 \cdot CH_2 \cdot CH_2 \cdot CH_2 \cdot CH_2 \cdot CH_3$

  B. Cyclic

catechol

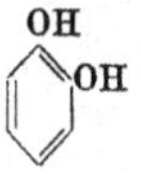

  C. Isoprenoid (branching)
    1. Non-cyclic

angelic acid $CH_3 - CH = \overset{\overset{CH_3}{|}}{C} - COOH$

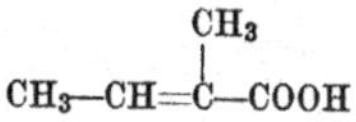

    2. Cyclic

cadinene

---

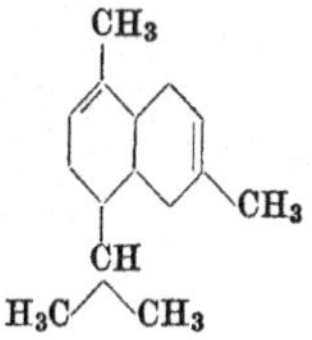

II. Carbon + oxygen
  A. Straight chain

ethyl acetate $CH_3 \cdot CH_2 \cdot O \cdot CO \cdot CH_3$

  B. Cyclic

peonin

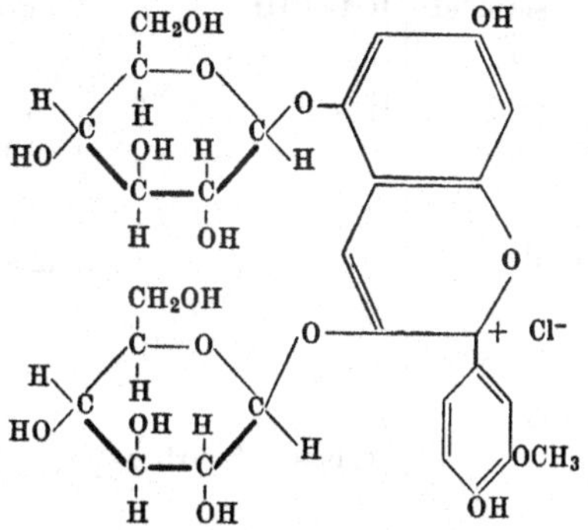

---

III. Carbon + nitrogen (+ oxygen)
  A. Straight chain

arginine $\overset{H_2N}{\underset{HN}{>}} C \cdot NH \cdot CH_2 \cdot CH_2 \cdot CH_2 \cdot CH(NH_2) \cdot COOH$

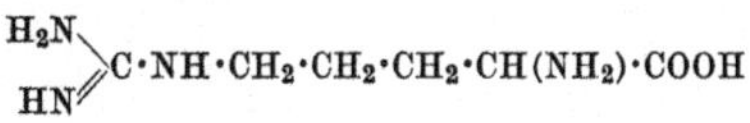

  B. Cyclic

nicotine

---

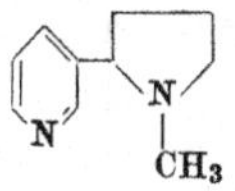

IV. Carbon + sulfur (+ nitrogen)
  A. Straight chain

allyl sulfide $CH_2 : CH \cdot CH_2 \cdot S \cdot CH_2 \cdot CH : CH_2$

  B. Cyclic

biotin

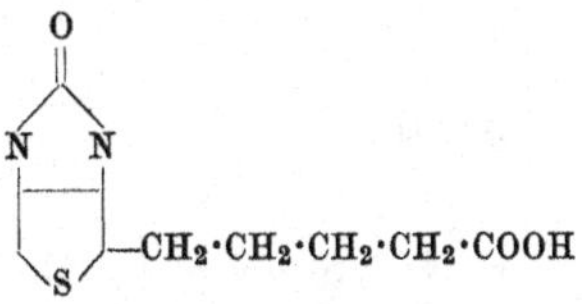

Table 7. *The variation in phytochemical structure introduced by end and side group elaboration.*

| Type of elaboration | General formula | Examples | |
|---|---|---|---|
| alcohol | $R \cdot OH$ | isoamyl alcohol | $(CH_3)_2 \cdot CH \cdot CH_2 \cdot CH_2OH$ |
| phenol derivative | 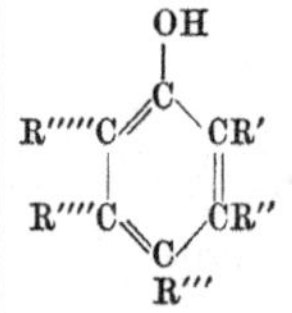 | catechol | 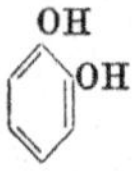 |
| aldehyde | $R \cdot CHO$ | acetaldehyde | $CH_3 \cdot CHO$ |
| ketone | $R \cdot CO \cdot R'$ | carvone | 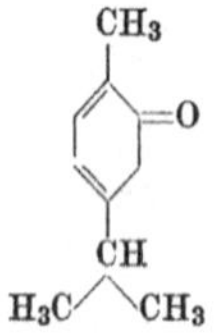 |
| monobasic acid | $R \cdot COOH$ | formic acid | $HCOOH$ |
| dibasic acid | $HOOC \cdot R \cdot COOH$ | traumatic acid (wound hormone) | $HOOC \cdot CH:CH \cdot CH_2 \cdot CH_2 \cdot CH_2 \cdot CH_2 \cdot$ $CH_2 \cdot CH_2 \cdot CH_2 \cdot CH_2 \cdot COOH$ |
| keto acid | | phenylpyruvic acid | 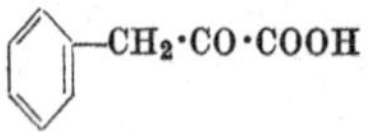 $-CH_2 \cdot CO \cdot COOH$ |
| fatty acid | $R \cdot COOH$ [R has $\geq 5$ carbons] | linderic acid | $CH_3 \cdot CH_2 \cdot CH_2 \cdot CH_2 \cdot CH_2 \cdot CH_2 \cdot CH_2 \cdot$ $CH:CH \cdot CH_2 \cdot CH_2 \cdot COOH$ |
| amino acid | $R \cdot CH(NH_2) \cdot COOH$ | histidine | 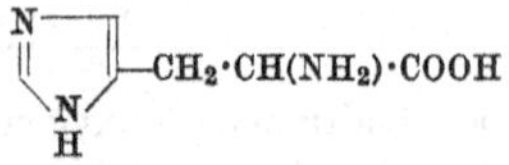 $-CH_2 \cdot CH(NH_2) \cdot COOH$ |
| amide | $R \cdot CO \cdot N\!\!<\!\!^{R'}_{R''}$ | nicotinamide | 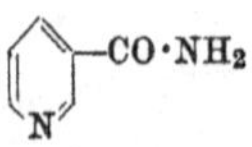 $-CO \cdot NH_2$ |
| amine | $R \cdot N\!\!<\!\!^{R'}_{R''}$ | putrescine | $H_2N \cdot CH_2 \cdot CH_2 \cdot CH_2 \cdot CH_2 \cdot NH_2$ |
| nitrile | $R \cdot C:N$ | 3-indolylacetonitrile | 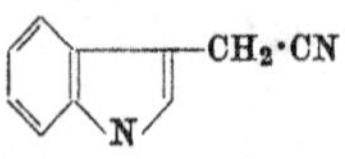 $-CH_2 \cdot CN$ |
| sulfhydril | $R \cdot SH$ | cysteïne | $HS \cdot CH_2 \cdot CH(NH_2) \cdot COOH$ |
| isothiocyanate | $R \cdot N:C:S$ | allyl isothiocyanate | $CH_2:CH \cdot CH_2 \cdot NCS$ |

Table 8. *Fatty acids as examples of elaboration of a basic carbon skeleton by increase in chain length and by variation in degree of saturation and side group substituents.*

| Class | No. of carbon atoms | Fatty acid | Positions of double or triple bonds | Positions of keto or hydroxy groups |
|---|---|---|---|---|
| | | **Saturated** | | |
| $C_n H_{2n+1} COOH$ | 6 | caproic | | |
| | 8 | caprylic | | |
| | 10 | capric | | |
| | 12 | lauric | | |
| | 14 | myristic | | |
| | 16 | palmitic | | |
| | 18 | stearic | | |
| | 20 | arachidic | | |
| | 22 | behenic | | |
| | 24 | lignoceric | | |
| | 26 | cerotic | | |
| | | **Unsaturated** | | |
| $C_n H_{2n-1} COOH$ | 10 | obtusilic | 4:5 | |
| | 10 | caproleic | 9:10 | |
| | 12 | linderic | 4:5 | |
| | 14 | tsuzuic | 4:5 | |
| | 16 | palmitoleic | 9:10 | |
| | 18 | petroselinic | 6:7 | |
| | 18 | oleic | 9:10 | |
| | 22 | erucic | 13:14 | |
| | 26 | ximenic | 17:18 | |
| | 30 | lumequeic | 21:22 | |
| $C_n H_{2n-3} COOH$ | 18 | linoleic | 9:10, 12:13 | |
| $C_n H_{2n-5} COOH$ | 18 | elaeostearic | 9:10, 11:12, 13:14 | |
| | 18 | linolenic | 9:10, 12:13, 15:16 | |
| $C_n H_{2n-7} COOH$ | 18 | parinaric | 9:10, 11:12, 13:14, 15:16 | |
| $C_n H_{2n-3} COOH$ | 18 | tariric | 6:7 | |
| $C_n H_{2n-9} COOH$ | 18 | isamic | 9:10, 11:12, 17:18 | |
| | | **Hydroxy** | | |
| $C_n H_{2n} O_4$ | 14 | ipurolic | | 3, 11 dihydroxy |
| | 18 | dihydroxystearic | | 9, 10 dihydroxy |
| $C_n H_{2n-2} O_3$ | 18 | ricinoleic | 9:10 | 12 hydroxy |
| | | **Keto** | | |
| $C_n H_{2n-2} O_3$ | 18 | lactarinic | | 6 keto |
| $C_n H_{2n-8} O_3$ | 18 | licanic | 9:10, 11:12, 13:14 | 4 keto |
| | | **Cyclic** | | |
|  $\bigcirc$ $-(CH_2)_n COOH$ | | | | |
| saturated | 6 | aleprolic | | |
| | 10 | aleprestic | | |
| | 12 | aleprylic | | |
| | 14 | alepric | | |
| | 16 | hydnocarpic | | |
| | 18 | chaulmoogric | | |
| unsaturated | 18 | gorlic | 6:7 | |

Table 9. *Examples of diversity in structure arising from the linkage of heterogeneous groups.*

| Type of elaboration | General formula | Examples |
|---|---|---|
| sulfide linkage | $R \cdot S \cdot R$ | allyl sulfide  $CH_2:CH \cdot CH_2 \cdot S \cdot CH_2 \cdot CH:CH_2$ |
| disulfide linkage | $R \cdot S \cdot S \cdot R$ | cystine  $HOOC \cdot CH(NH_2) \cdot CH_2 \cdot S \cdot S \cdot CH_2 \cdot CH(NH_2) \cdot COOH$ |
| ester linkage | $R \cdot O \cdot CO \cdot R$ | ethyl anthranilate  $CH_3 \cdot CH_2 \cdot O \cdot CO$ 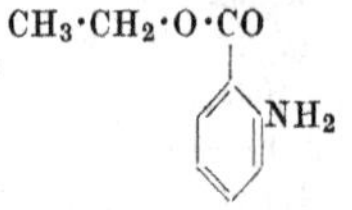 |
| ether linkage | $R \cdot O \cdot R$ | colchicine  (probable) 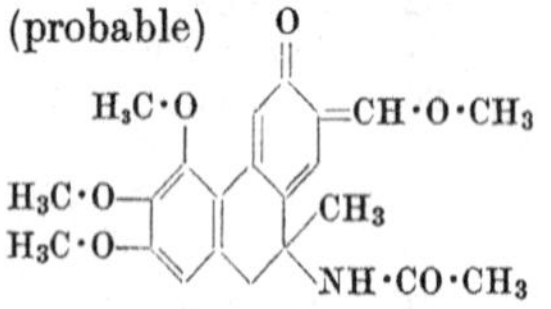 |
| glycosidic linkage | R–O-sugar  or    R–O-sugar residue | sucrose 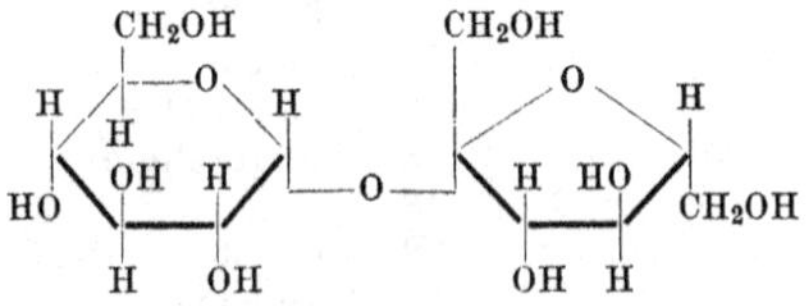 |
| peptide linkage | $R \cdot CH \cdot CO \cdot NH \cdot CH \cdot R'''$<br>  $\quad$ R'$\qquad$ R''<br>(may be repetitive, as in polypeptides and proteins) R' and R'' are amino acid residues | glutathione  $HOOC \cdot CH(NH_2) \cdot CH_2 \cdot CH_2 \cdot CO \cdot NH \cdot CH \cdot CO \cdot NH \cdot CH_2 \cdot COOH$<br>$\qquad\qquad\qquad\qquad\qquad\qquad\qquad$ CH$_2 \cdot$SH |
| phosphoric acid linkage | R<br> \|<br>sugar residue–R'<br> \|<br>(phosphoric acid)$_n$<br> \|<br>sugar residue–R''<br> \|<br>R'''<br>(as in prosthetic nucleotide groups and as repetitive linkage in nucleic acids) R' and R'' are pyrimidine or purine bases | portion of ribonucleic acid chain 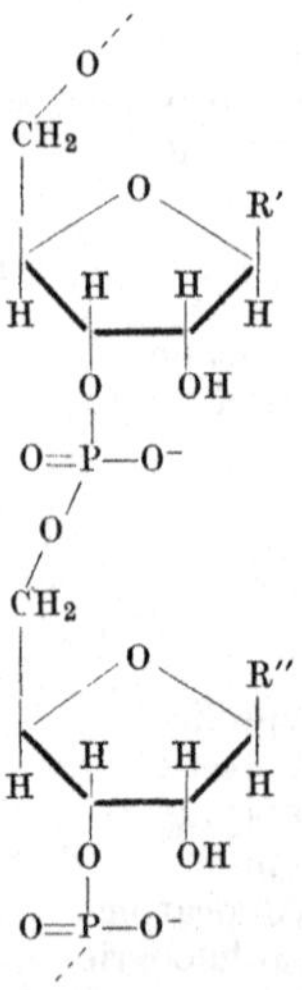 |

triglycerides, which may be symbolically represented by the following:

$$H_2C \cdot O \cdot R_1 \qquad H_2C \cdot O \cdot R_1 \qquad H_2C \cdot O \cdot R_2$$
$$HC \cdot O \cdot R_2 \qquad HC \cdot O \cdot R_4 \qquad HC \cdot O \cdot R_2 \quad \text{etc.,}$$
$$H_2C \cdot O \cdot R_3 \qquad H_2C \cdot O \cdot R_1 \qquad H_2C \cdot O \cdot R_n$$

where $R_1$, $R_2$, $R_3$, $R_4 \ldots R_n$ are the $n$ different fatty acids present in the particular oil or fat. The amounts of each kind of glyceride will, of course, depend on the molar ratios of the fatty acids esterified.

The further elaboration of fats through the phosphoric acid esterification of diglycerides (mainly of stearic, palmitic, oleic, linoleic, and linolenic acids) leads to a class of compounds known as the phospholipids. These are of two general types; namely, (1) the Ca, Mg, and K salts of phosphatidic acid and (2) the esters of phosphatidic acid with choline (lecithin series) and ethanolamine (cephalin series).

$$H_2C \cdot O \cdot R_1 \qquad\qquad H_2C \cdot O \cdot R_1 \qquad\qquad H_2C \cdot O \cdot R_1$$
$$HC \cdot O \cdot R_2 \qquad\qquad HC \cdot O \cdot R_2 \qquad\qquad HC \cdot O \cdot R_2$$
$$\underset{\overset{\|}{O}}{} \qquad\qquad \underset{\overset{\|}{O}}{} \qquad\qquad \underset{\overset{\|}{O}}{}$$
$$H_2C \cdot O \cdot \underset{OH}{P} \cdot OH \quad\quad H_2C \cdot O \cdot \underset{OH}{P} \cdot O \cdot CH_2 \cdot CH_2 \cdot \overset{+\ominus}{N}(CH_3)_3 \quad\quad H_2C \cdot O \cdot \underset{OH}{P} \cdot O \cdot CH_2 \cdot CH_2 \cdot NH_2$$

| phosphatidic acid | α-lecithin | α-cephalin |
|---|---|---|

Lipositols are phospholipids containing, for example with soybean lipositol, inositol (a cyclic, 6-carbon, polyhydric alcohol with a role similar to that of glycerol in the fats and oils), galactose, phosphoric acid, ethanolamine tartrate, and fatty acids in a molar ratio of $1:1:1:1:2$.

Although some straight chain hydrocarbons occur in plants, for example, ethylene gas ($CH_2 : CH_2$), $n$-heptane ($C_7H_{16}$) in essential oils, and higher members of the paraffin series, such as $n$-hentriacontane ($C_{31}H_{64}$), in cytoplasmic waxes, it is the branching of the isoprene unit, $\overset{H_2C}{\underset{H_3C}{>}}C - \overset{H}{C} = CH_2$, which, in the form of open, as well as cyclized, directly linked chains, with varying degrees of saturation, affords the maximum possibility for structural diversity in the hydrocarbons and hydrocarbon skeletons which are so ubiquitous in the plant kingdom. Thus, the isoprenoid unit is the building block of the terpenes ($C_{10}H_{16}$) and sesquiterpenes ($C_{15}H_{24}$) and their alcohol, aldehyde, ketone, acid, and other derivatives, all of which are volatile constituents of essential oils[1]. Some of the phytochemicals displaying the isoprenoid configuration are shown in Table 10, which has been arranged to illustrate variation in oxidation level in a given basic structure (in a vertical direction), as for example, ocimene, geraniol, geranial, and geranic acid, and also (in a horizontal direction) variation in degree of saturation and cyclization.

---

[1] In addition to these compounds and the isoprenoid isoamyl alcohol (Table 7), there are a great number of other important non-isoprenoid constituents of essential oils, among which are the following: $n$-heptane (Table 6); formaldehyde, acetaldehyde, and acetone (Table 20); pelargonic aldehyde, $CH_3 \cdot (CH_2)_7 \cdot CHO$; pelargonic acid, $CH_3 \cdot (CH_2)_7 \cdot COOH$; allyl sulfide (Table 6); allyl isothiocyanate (Table 7); benzaldehyde, salicylaldehyde, and salicylic acid (Table 11); various esters, such as ethyl acetate. $CH_3 \cdot CH_2 \cdot O \cdot CO \cdot CH_3$, and benzyl acetate, methyl salicylate, and ethyl anthranilate (Table 11); certain methyl ether phenol

derivatives, such as anisaldehyde, anethole, and eugenol (Table 11); and indole,

Table 10. *The isoprenoid unit as a*

## Hydrocarbons

ocimene

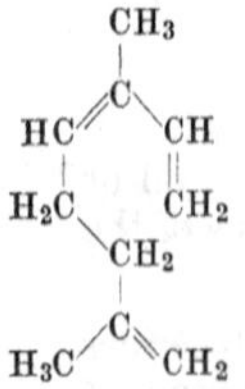

CH₃
|
C
HC   CH
H₂C   CH₂
CH₂
H₃C   CH₂

Δ³-menthene

CH₃
|
CH
H₂C   CH₂
H₂C   CH
C
CH
H₃C   CH₃

p-cymene

CH₃
|
C
HC   CH
HC   CH
C
CH
H₃C   CH₃

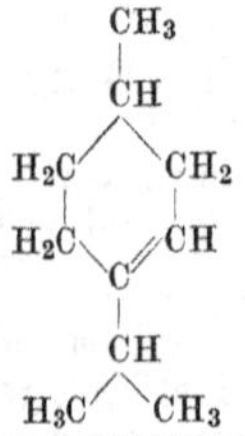

α-thujene

CH₃
|
C
HC   CH
H₂C   CH₂
C
CH
H₃C   CH₃

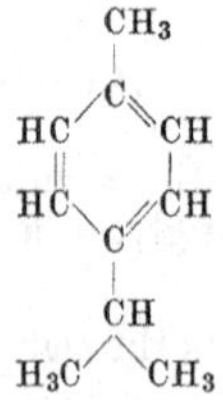

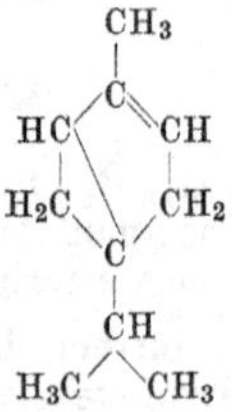

## Alcohols

geraniol

CH₃
|
C
H₂C   CH
H₂C   CH₂OH
CH₂
H₃C   CH₂

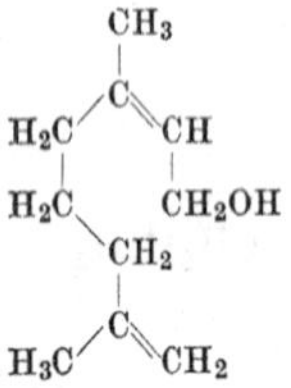

menthol

CH₃
|
CH
H₂C   CH₂
H₂C   CHOH
CH
CH
H₃C   CH₃

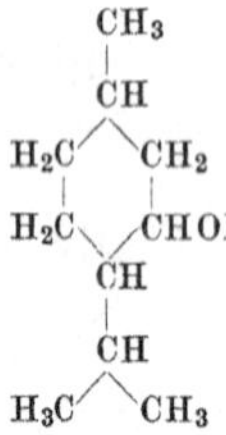

thymol
(a phenol derivative)

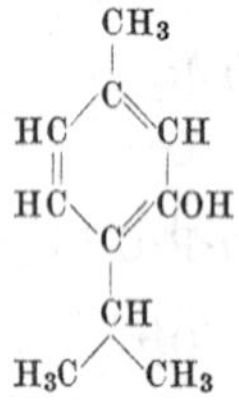

CH₃
|
C
HC   CH
HC   COH
C
CH
H₃C   CH₃

thujyl alcohol

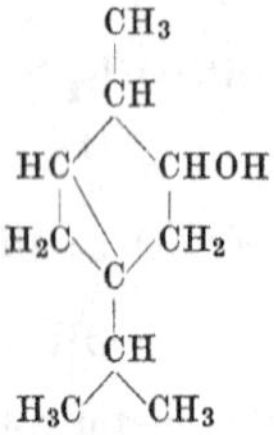

CH₃
|
CH
HC   CHOH
H₂C   CH₂
C
CH
H₃C   CH₃

## Aldehydes

geranial

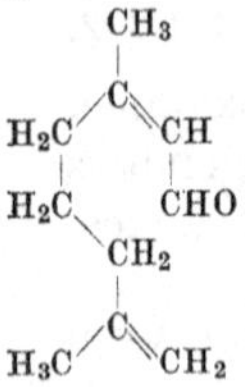

CH₃
|
C
H₂C   CH
H₂C   CHO
CH₂
H₃C   CH₂

cuminaldehyde

CHO
|
C
HC   CH
HC   CH
C
CH
H₃C   CH₃

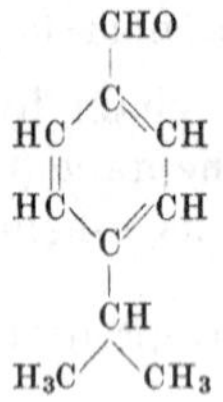

## Ketones and acids

geranic acid

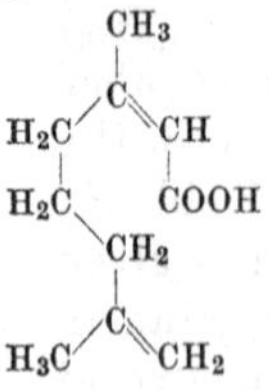

CH₃
|
C
H₂C   CH
H₂C   COOH
CH₂
H₃C   CH₂

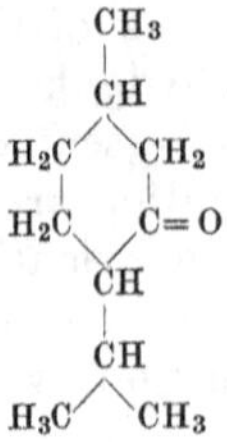

menthone

CH₃
|
CH
H₂C   CH₂
H₂C   C=O
CH
CH
H₃C   CH₃

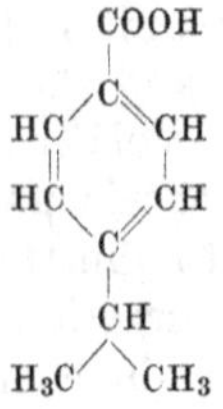

cumic acid

COOH
|
C
HC   CH
HC   CH
C
CH
H₃C   CH₃

thujone

CH₃
|
CH
HC   C=O
H₂C   CH₂
C
CH
H₃C   CH₃

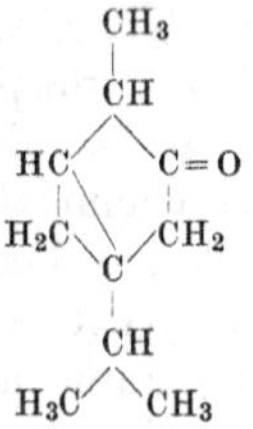

## Miscellaneous higher terpenes

farnesol

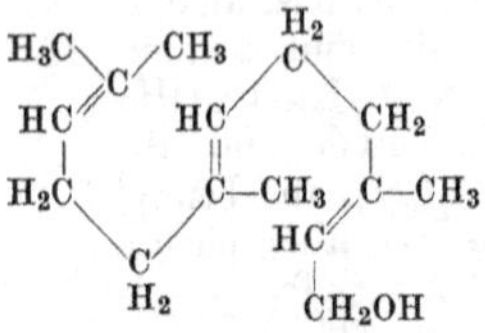

squalene

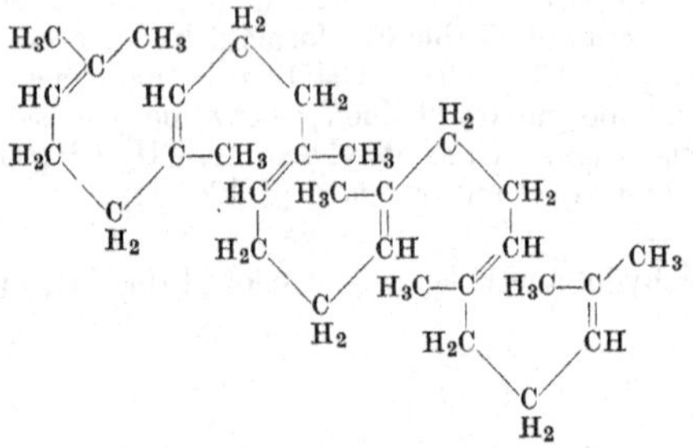

phytol

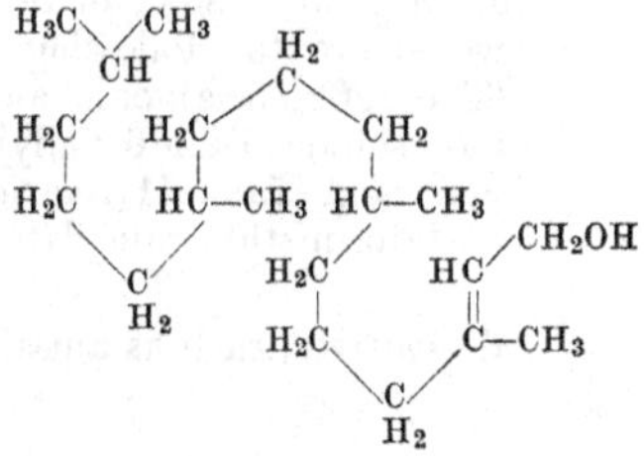

*building block of structural elaboration.*

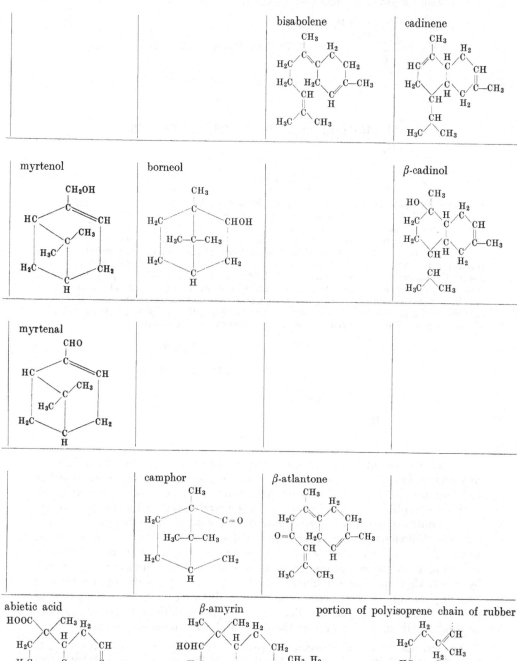

Abietic acid is a non-volatile diterpene constituent of pine resin. The diterpenoid phytol is an ester substituent of chlorophyll and also provides the open chain hydrocarbon portion of the vitamins E ($\alpha$-tocopherol, a seed oil) and $K_1$, a foliar oil.

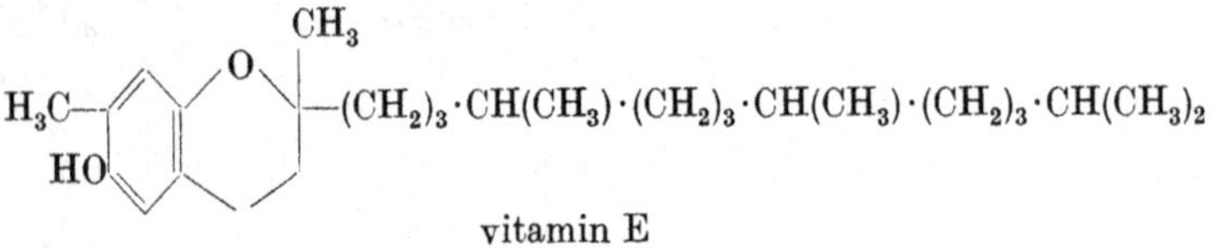

$$CH_3$$
$$H_3C- \quad O \quad -(CH_2)_3 \cdot CH(CH_3) \cdot (CH_2)_3 \cdot CH(CH_3) \cdot (CH_2)_3 \cdot CH(CH_3)_2$$
$$HO$$

vitamin E

$$O$$
$$-CH_2 \cdot CH : C(CH_3) \cdot (CH_2)_3 \cdot CH(CH_3) \cdot (CH_2)_3 \cdot CH(CH_3) \cdot (CH_2)_3 \cdot CH(CH_3)_2$$
$$-CH_3$$
$$O$$

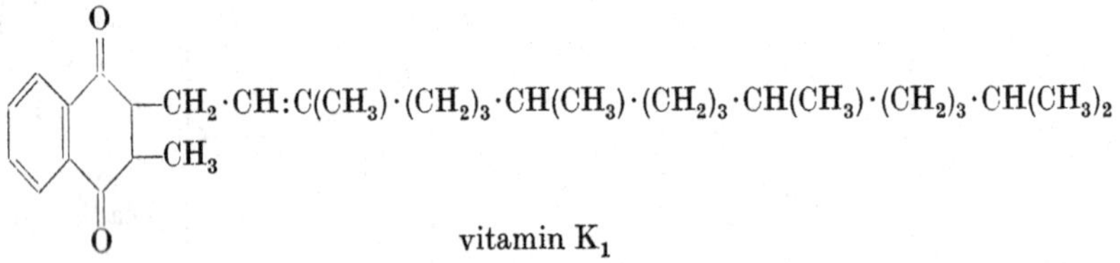

vitamin $K_1$

A triterpene, squalene, which may be considered equivalent to a symmetrical linkage of two farnesol skeletons, is found in certain plant oils, while another triterpene, $\beta$-amyrin, occurs in ester linkage in the latex of some plants. The carotenes (tetraterpenes) and xanthophylls (alcohol, methoxy, fatty acid ester, keto, and ketohydroxy tetraterpene derivatives) are the carotenoid pigments of plastids. The elasticity of rubber and the plasticity of gutta percha reside in the *cis* and *trans* configurations, respectively, of their straight polyisoprenoid chains, the former consisting of some 500 to 5000 unit chains, the latter, of perhaps 100 unit chains. The isoprene configuration also appears as a significant structural feature of the sterols, which are alcohols having a complex structure and occurring free in the plant or as esters and glycosides, and which are obtained along with the lipids in ether extractions of plant tissue. The most common of the phytosterols is $\gamma$-sitosterol.

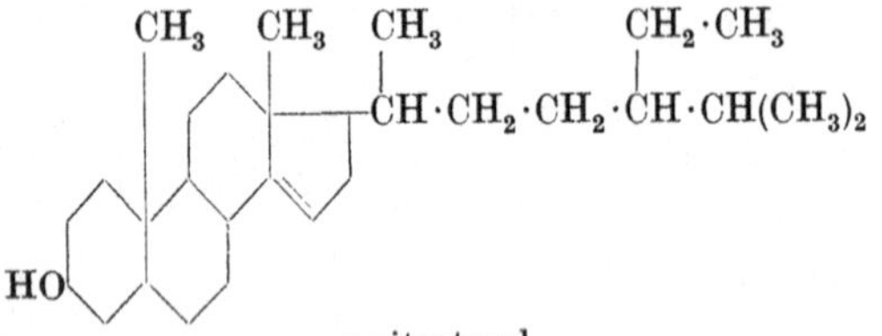

$$CH_3 \quad CH_3 \quad CH_3 \qquad CH_2 \cdot CH_3$$
$$-CH \cdot CH_2 \cdot CH_2 \cdot CH \cdot CH \cdot CH(CH_3)_2$$
$$HO$$

$\gamma$-sitosterol

Because substitution of various groups at different positions on the benzene ring and fusion of the benzene ring with other cyclic structures are possible, and because many of such substituent groups and fused cyclic structures are, in turn, susceptible to variation in the ways already described in Tables 7 and 9, the aromatic compounds possess tremendous potential for elaboration in many different directions. Some of these are arranged in Table 11 in such a manner as to exhibit various types of substitution on the ring (in a vertical direction) and various end or side groups in those substituents (in a horizontal direction), by which they may be classified. Of course, the very nature of the benzene ring implies that several reactive groups of different kinds may therefore be present. Thus, hydroxyanthranilic acid is a phenol derivative possessing both an amine and an acid group in the *ortho* and *meta* positions, respectively.

Styrene, the isoprenoid aromatic compounds, and many of the aromatic alcohols, aldehydes, acids, esters, and ethers are constituents of the essential oil-resin-balsam group of compounds. Certain of the alcohols, aldehydes, and acids occur as glycosides. Because they are highly susceptible to oxidation, the dihydric phenols, catechol and hydroquinone, as well as *ortho* dihydroxy derivatives, such as gallic acid, chlorogenic acid, and 3,4-dihydroxyphenylalanine (dopa), may have the ability to act as hydrogen carriers in terminal enzymatic oxidations in plants, catechol + polyphenoloxidase and hydroquinone + laccase being two known aerobic oxidative enzyme systems possibly playing such a role. Certain of the phenol derivatives (catechol, phloroglucinol, and resorcinol) and aromatic acids

## Table 11. The elaboration of the benzene ring by substitution.

| Hydrocarbons | Phenol derivatives | Alcohols | Aldehydes | Acids | Esters | Ethers | Amines | Amino acids |
|---|---|---|---|---|---|---|---|---|
| styrene $CH:CH_2$ | | benzyl alcohol (an ester constituent) $CH_2OH$  | benzaldehyde (a glycoside constituent) $CHO$  | benzoic acid (an ester constituent) $COOH$  | benzyl acetate $CH_2 \cdot O \cdot CO \cdot CH_3$  | | | |
| | | cinnamyl alcohol (an ester constituent) $CH:CH \cdot CH_2OH$ | cinnamic aldehyde $CH:CH \cdot CHO$ | cinnamic acid (an ester constituent) $CH:CH \cdot COOH$ | | | | phenylalanine $CH_2 \cdot CH(NH_2) \cdot COOH$ |
| | | | | phenylpyruvic acid $CH_2 \cdot CO \cdot COOH$ | | | | |
| | 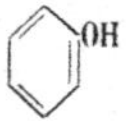 catechol $OH$ $OH$ | saligenin (a glycoside constituent) $CH_2OH$ 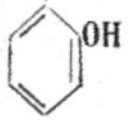 $OH$ | 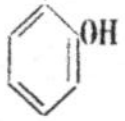 salicylaldehyde $CHO$ $OH$ | salicylic acid $COOH$ 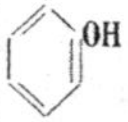 $OH$ | methyl salicylate $CH_3 \cdot O \cdot CO$ 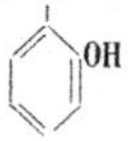 $OH$ | | | |
| | | | | anthranilic acid $COOH$ 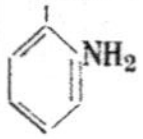 $NH_2$ | ethyl anthranilate $CH_3 \cdot CH_2 \cdot O \cdot CO$ 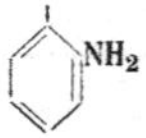 $NH_2$ | | | kynurenin $CO \cdot CH_2 \cdot CH(NH_2) \cdot COOH$ 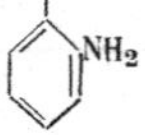 $NH_2$ |

*Table 11.* (Continued.)

| Hydro-carbons | Phenol derivatives | Alcohols | Aldehydes | Acids | Esters | Ethers | Amines | Amino acids |
|---|---|---|---|---|---|---|---|---|
| | quinone | | | hydroxyanthranilic acid | | anisaldehyde  | p-aminobenzoic acid 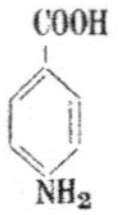 | |
| | hydro-quinone | | | p-hydroxybenzoic acid (a glycoside constituent) | | anethole | hordenine | tyrosine |
| | resorcinol 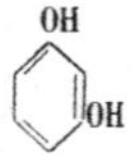 | | | p-coumaric acid | | vanillin | | |
| | | | | protocatechuic acid | | veratric acid | | |

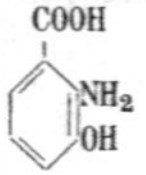

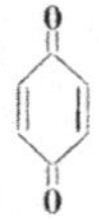

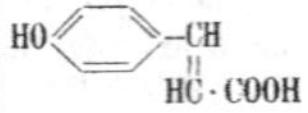

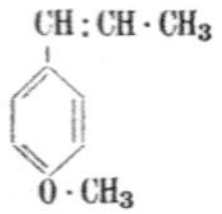

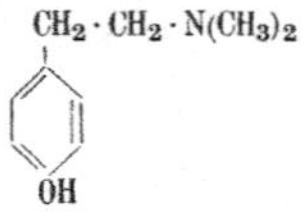

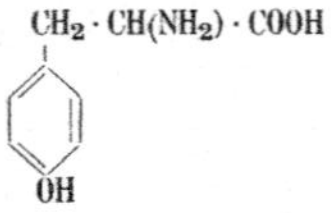

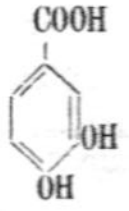

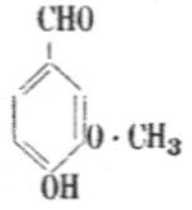

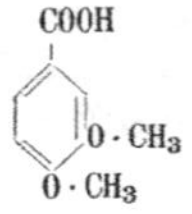

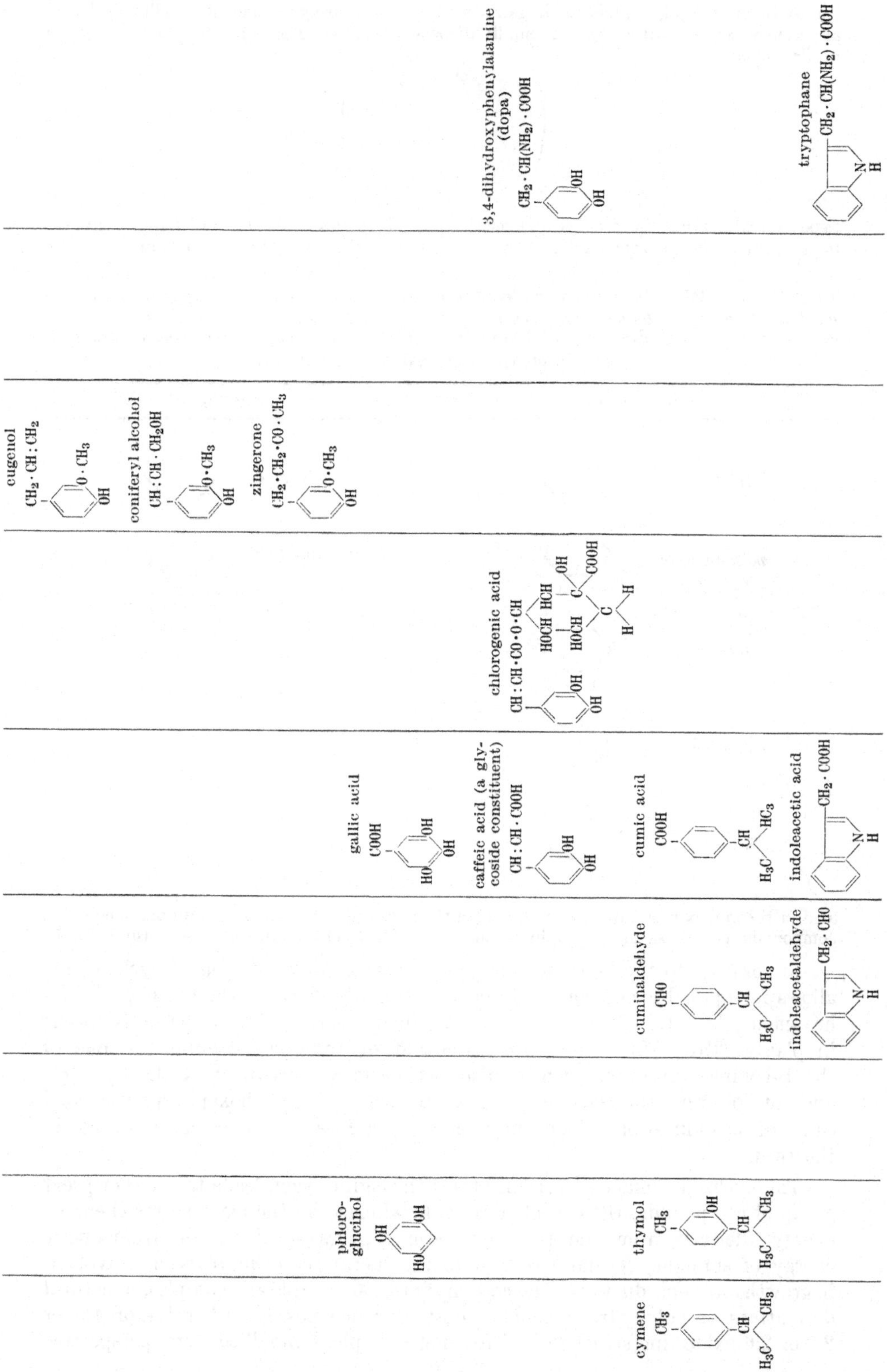

3,4-dihydroxyphenylalanine (dopa)
$CH_2 \cdot CH(NH_2) \cdot COOH$

tryptophane
$CH_2 \cdot CH(NH_2) \cdot COOH$

eugenol
$CH_2 \cdot CH : CH_2$
$O \cdot CH_3$
$OH$

coniferyl alcohol
$CH : CH \cdot CH_2OH$
$O \cdot CH_3$
$OH$

zingerone
$CH_2 \cdot CH_2 \cdot CO \cdot CH_3$
$O \cdot CH_3$
$OH$

chlorogenic acid
$CH : CH \cdot CO \cdot O \cdot CH$

gallic acid
$COOH$

caffeic acid (a gly-coside constituent)
$CH : CH \cdot COOH$

cumic acid
$COOH$

indoleacetic acid
$CH_2 \cdot COOH$

cuminaldehyde
$CHO$

indoleacetaldehyde
$CH_2 \cdot CHO$

phloro-glucinol

thymol
$CH_3$

cymene
$CH_3$

(protocatechuic acid, caffeic acid, gallic acid, and a pentaglucoside of digallic acid) are apparently components of the colloidally dispersed tannins. The polymerization of eugenol or flavanone

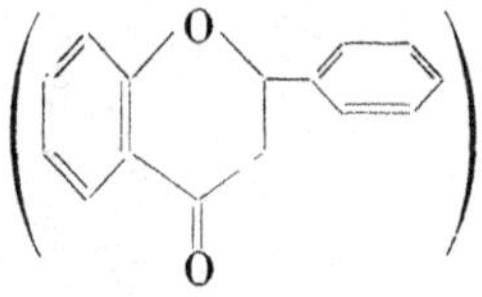

type derivatives has been suggested as the possible pathway of cell wall lignin synthesis in the plant. The benzene nucleus is also present in certain amino acids, in *p*-aminobenzoic

Table 12. *The naturally occurring cyclohexanehexols and derivatives in the mnemonic arrangement based on d-quercitol as a 2-desoxy-D-gluco-trans 2,6,1 compound, as suggested by* PIGMAN *and* GOEPP *(1948); short vertical lines represent hydroxyl groups, open squares represent carboxyl groups, and hydrogens not shown.*

| Cyclohexanehexols | | Derivatives | |
|---|---|---|---|
| *d*-quercitol | 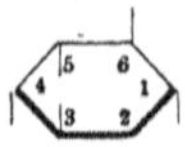 | *l*-quinic acid | 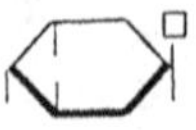 |
| *meso*-inositol | 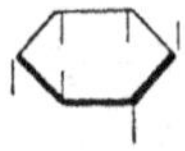 | shikimic acid | 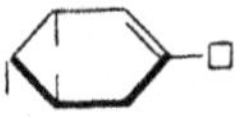 |
| *d*-inositol | 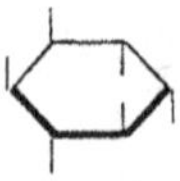 | | |
| *l*-inositol | 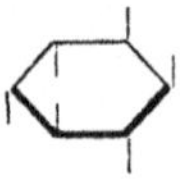 | | |
| scyllitol | 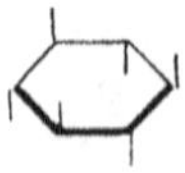 | | |

acid (a B-complex vitamin), and, as a part of the indole nucleus, in a great number of complex compounds as well as in tryptophane and the indolylaceto compounds (see also Table 7).

In contrast to the benzene ring, the saturated 6-carbon ring of cyclohexane allows for optical isomerism in its polyhydroxy derivatives, the inositols (cyclohexanehexols). Of all such possible isomers, only a few apparently occur in plants (Table 12). The most important of these is *meso*-inositol (one of the B-complex vitamins), which also occurs as a constituent of the lipositols and of phytin, the calcium-magnesium salt of the hexaphosphoric acid ester of *meso*-inositol, which may serve as a means of phosphate storage in the seed.

The cyclized linkage of carbon atoms through oxygen leads to another great group of compounds, the carbohydrates, in which form the plant stores chemical energy (derived from the photosynthetic conversion of the electromagnetic energy of sunlight) for later release to the energy-requiring reactions involved in growth and reproduction. The carbohydrates are simple and complex structural derivatives of polyhydric alcohols, most of which possess either 1,4 or 1,5 or 2,5 or 2,6 oxide ring structures. Thus, for example, a modified, semi-perspective

Haworth representation of D-glucose is

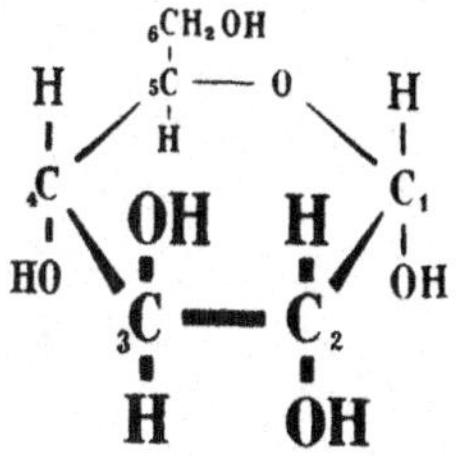

or, in abbreviated form,

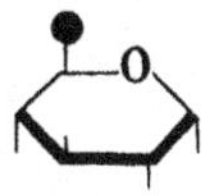

As with the inositols, optical isomerism permits a great diversity among the monosaccharides, which for the naturally occuring 3-carbon to 7-carbon compounds and their alcohol, anhydride, and acid derivatives is depicted in Table 13. L-ascorbic acid, vitamin C, is structurally related to L-sorbose and occurs in the plant together with its keto and oxidized forms.

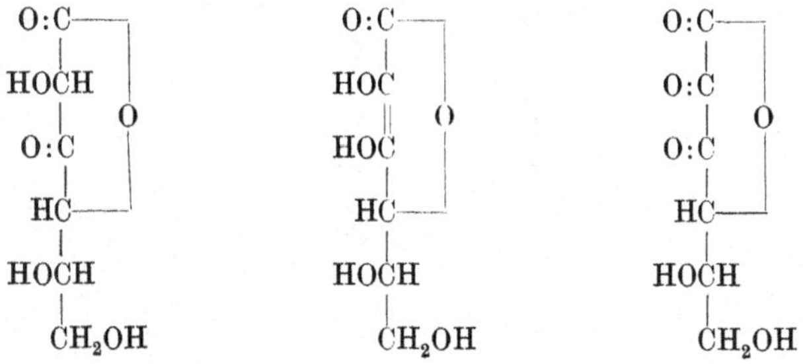

keto form of ascorbic acid      L-ascorbic acid      dehydroascorbic acid

The oligosaccharides, representing two or more sugar residues linked together through an oxygen bridge, are represented in Table 14. Sucrose, of course, is the most prominent carbohydrate to be found in plant tissues, sometimes accounting for as much as 20% of the total weight of a sugar beet. Along with the polysaccharides, which are presented in Table 15, the oligosaccharides arise presumably by virtue of the property of monosaccharides to undergo enzymatic phosphorylation and then, in the presence of suitable phosphorylases, transglycosidation. This property permits great variety in the combinations, substitutions, and rearrangements involved in the polymerization, to different degrees, of mono- and oligosaccharides to polysaccharides. Cellulose, the primary constituent of the plant cell wall, and starch, an osmotically inactive, storage carbohydrate accumulated through photosynthesis, are two polysaccharides whose occurrence is ubiquitous throughout most of the plant kingdom. Glycogen, the principal polymer of glucose in most animal tissues, is also present in certain plants.

Sugars also occur in glycosidic linkage with non-sugar compounds to form the plant glycosides, examples of classes of which appear in Table 16.

By incorporation of nitrogen into the carbon skeleton, the plant produces a wide variety of compounds, many of which, such as the proteins, coenzymes, nucleic acids, etc., are known to be of great physiological importance in plant metabolism.

A special class of heterogeneous, nitrogen-containing substances, whose function in plant metabolism is, however, still unknown, are the alkaloids, which are represented by the compounds arranged in Table 17. Although showing

Table 13. *Optical isomerism in the monosaccharides; oxygen in rings so indicated; short vertical lines with solid circles are —OH groups; other lines with solid circles are —CH₂OH groups, with solid squares, —CH₃ groups, and with open squares, —COOH groups; hydrogens not shown. "*" indicates no, or rare, occurrence free in nature.*

| | Monosaccharides | Oligosaccharide and polysaccharide forms of sugar | Natural substances of which sugar is a hydrolytic constituent | Sugar alcohols and anhydrides; uronic acids |
|---|---|---|---|---|
| Trioses | dihydroxyacetone phosphate $CH_2[O\cdot PO(OH)_2]\cdot CO\cdot CH_2OH$ | | | glycerol* |
| | 3-phosphoglyceraldehyde $CHO\cdot CHOH\cdot CH_2[O\cdot PO(OH)_2]$ | | | |
| | 1,3-diphosphoglyceraldehyde $CHOH\cdot CHOH\cdot CH_2[O\cdot PO(OH)_2]$ $\dot{O}\cdot PO(OH)_2$ | | | |
| Tetroses | | | | i-erythritol 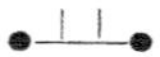 |
| Pentoses | L-arabinose* | 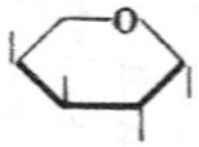 araban | glycosides, gums, pectic materials, hemicelluloses, bacterial polysaccharides | |
| | D-arabinose* | 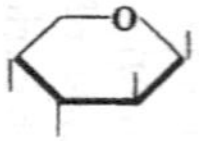 | glycosides in the genus *Aloe* | D-arabitol  |
| | D-ribose* |  | vitamins, coenzymes, nucleic acids (RNA) | ribitol 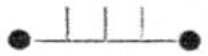 |
| | 2-desoxy-D-ribose* | 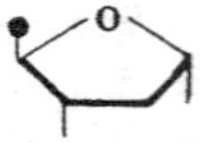 | nucleic acids (DNA) | |
| | D-xylose* | 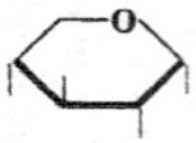 xylan | glycosides, gums, hemicelluloses | |
| Methyl pentoses | 6-desoxy-D-glucose* | 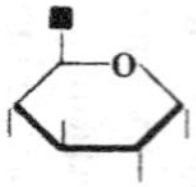 | quinovin (a glycoside) | |
| | L-fucose* |  | sea weed cell wall constituent, gums | |
| | D-fucose* 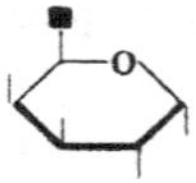 | | | |

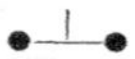

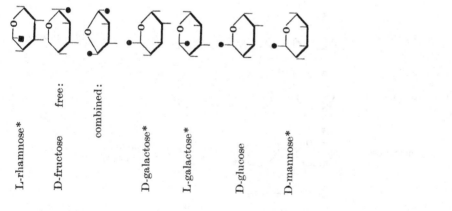

| | | | |
|---|---|---|---|
| **Hexoses** | L-rhamnose* | | |
| | D-fructose free: | gentianose, sucrose; inulin | |
| | combined: | | |
| | D-galactose* | melibiose, raffinose, stachyose; galactan | agar, gums, and mucilages |
| | L-galactose* | | agar, gums, and mucilages |
| | D-glucose | sucrose, maltose, etc.; cellulose, starch, lichenin | |
| | D-mannose* | mannan | orchid tuber mucilage, white spruce hemicellulose |
| | D-galacturonic acid* (in gum arabic, saponins, glycosides, oligosaccharides) | | |
| | galactitol | | |
| | D-glucuronic acid* (in pectins, pectic acid, mucilages) | | |
| | 1,5-D-mannitan (styracitol) | | |
| | D-mannuronic acid* (in alginic acid) | | |
| | D-mannitol (in plant exudates) | | |
| | 1,5-D-sorbitan (polygalitol) | | |
| | D-sorbitol | | |
| | D-perseitol | | |
| | D-volemitol | | |
| **Heptoses** | D-mannoheptulose | | |
| | Sedoheptulose | | |

Table 14. *Monosaccharides in glycosidic linkage* (symbols same as for table 13).

| | Oligosaccharides | | Polysaccharide and higher oligosaccharide forms of sugar |
|---|---|---|---|
| Disaccharides | cellobiose* | 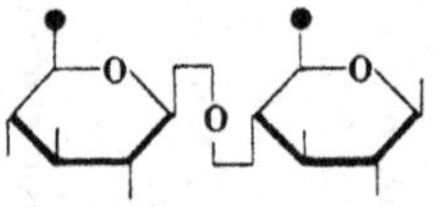 | cellulose |
| | chitobiose* | 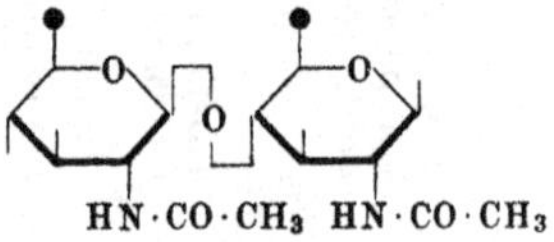 HN·CO·CH₃  HN·CO·CH₃ | chitin |
| | gentiobiose* | 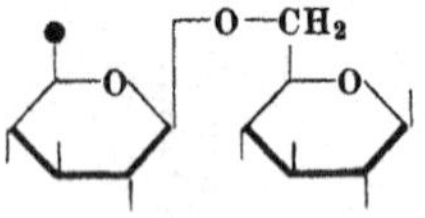 O—CH₂ | gentianose |
| | maltose* | 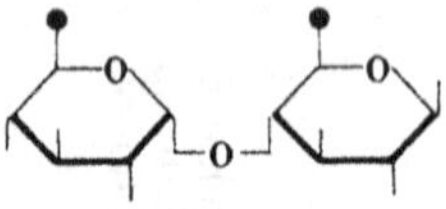 | starch |
| | melibiose | 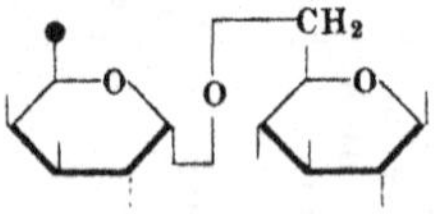 CH₂ (free in plant exudations) | raffinose |
| | sucrose | 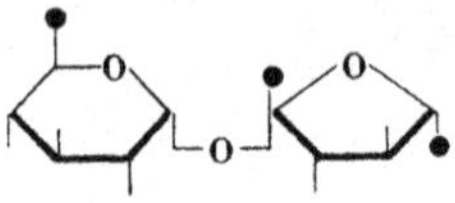 | melezitose ? gentianose, raffinose |
| | trehalose | 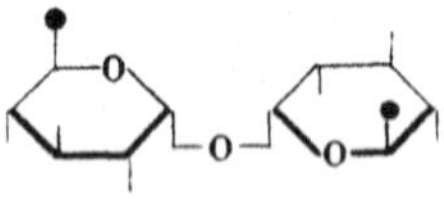 | |
| | turanose* | 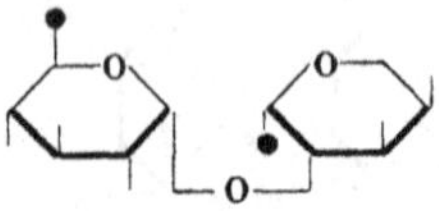 | melezitose |
| Trisaccharides | gentianose | 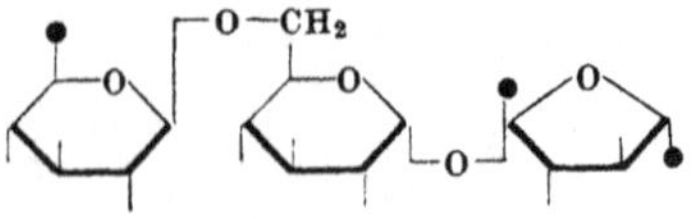 O—CH₂ | |
| | melezitose | 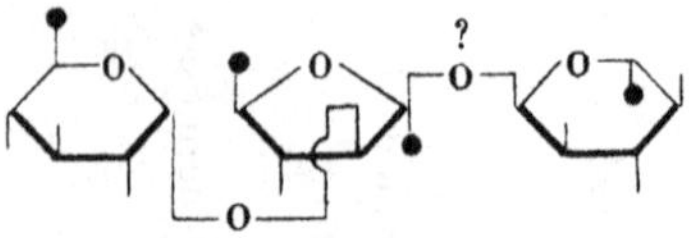 (in plant exudations) | |
| | raffinose | 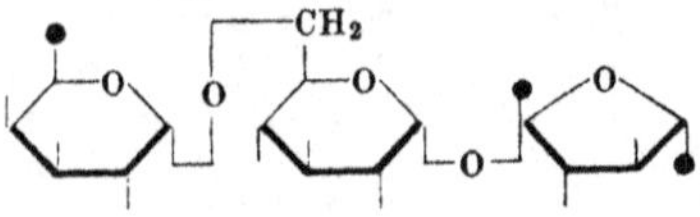 CH₂ | |
| Tetrasaccharide | stachyose | 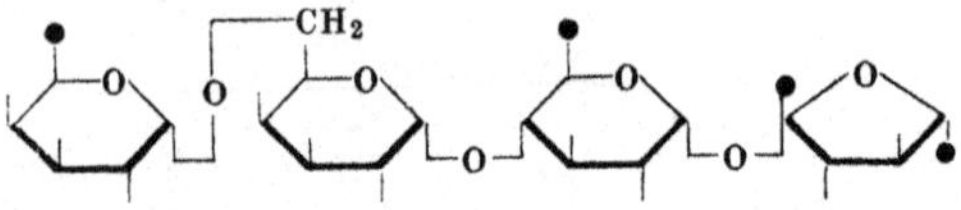 CH₂ | |

Table 15. *Diversity in structure resulting from polymerisation, as illustrated by the plant polysaccharides.*

| Basic monosaccharide unit | Polysaccharide | Straight or main chain glycosidic linkages | If present, glycosidic linkages at branching points | Comments |
|---|---|---|---|---|
| L-arabofuranose | araban | α-1,5 | α-2,1 or α-3,1 | in peanut or apple, associated with pectin; branching points at alternate residues |
| D-xylopyranose | xylan | β-1,4 | 1,3 | in esparto grass there is one terminal L-arabofuranose per 18 xylose units |
| D-fructofuranose | inulin | 1,2 | | a possible terminal glucose unit in a sucrose type linkage may be present, with another glucose unit attached to the 1 or 3 carbon atom of an undetermined fructose unit |
| | irisan | 1,2 | 2,4 | completely branched |
| | asparagosan graminan sinistran | 1,2 | ? | branched |
| | tritican | 1,2 | 2,6 | highly branched with possible terminal glucose unit |
| | phlean | 2,6 | | apparently a closed loop |
| D-galactopyranose | galactan | β-1,4 | | pectic galactan of seeds of *Lupinus albus* |
| | agar | β-1,3 | | calcium salt; 3,6 anhydro rings may occur; one terminal L-galactopyranose-6-sulfate in 1,4 linkage |
| D-galactopyranose-4-sulfate | carrageenan | side chains are α-1,3 | ? | side chains attached to acid-resistant residue branched by way of carbon atoms 1, 3, and 6 and high in L-galactopyranose |
| D-galactopyranose-6-sulfate | galactan present in *Iridophycus* | 1,3 | | |
| D-galacturono-pyranose | pectic acid | α-1,4 | | |
| | pectin | α-1,4 | | with varying amounts of methyl ester groups |
| D-glucopyranose | cellulose | β-1,4 | | repeating cellobiose units |
| | starch: amylose | α-1,4 | | repeating maltose units |

*Table 15.* (Continued.)

| Basic monosaccharide unit | Polysaccharide | Straight or main chain glycosidic linkage | If present, glycosidic linkages at branching points | Comments |
|---|---|---|---|---|
| | starch: amylopectin | α-1,4 | α-1,6 | tentatively, a central, β-malt amylase-resistant, branched dextrin, consisting of chains of repeating maltose units, 8–9 glucose units per chain, with branches of repeating maltose units, 15–18 glucose units per branch, which are hydrolyzable by β-malt amylase to maltose residues only |
| | glycogen | α-1,4 | α-1,6 | similar to amylopectin, except that chains of branched central dextrin only 4 glucose units each and hydrolyzable terminal branch chains contain 5–7 glucose units each |
| | lichenan | mostly β-1,4 some β-1,3 | | |
| | yeast dextran | β-1,3 | | about 29 glucose units per chain |
| | laminaran | β-1,3 | | |
| | barley glucan | β-1,6 | | 7 to 11 glucose units per chain (gentiobiose type linkage) |
| D-mannopyranose | mannan | β-1,4 | | |
| D-mannurono-pyranose | alginic acid | β-1,4 | | |

Table 16. *Representative plant glycosides* (symbols same as for table 13).

| Class | Examples |
|---|---|
| phenols | arbutin 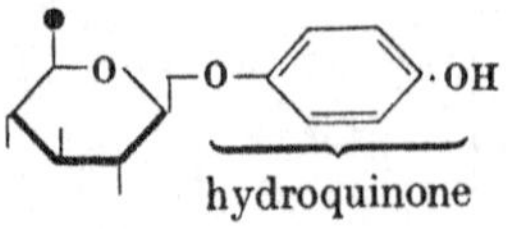 |
| alcohols | coniferin 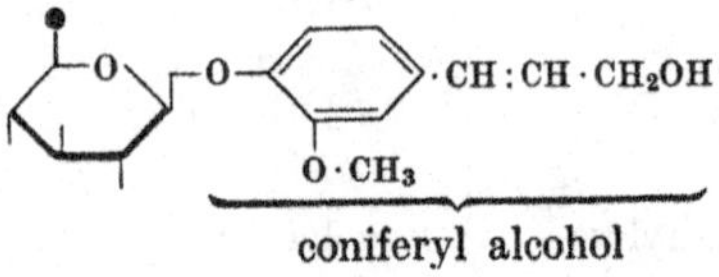 |

hydroquinone

coniferyl alcohol

*Table 16.* (Continued.)

| Class | Examples |
|---|---|
| aldehydes | **amygdalin** 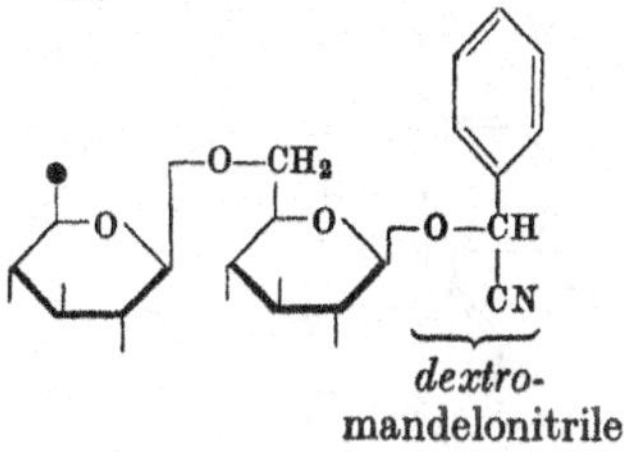  **linamarin** 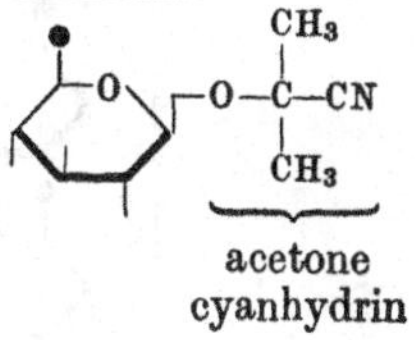 |
| | *dextro-*mandelonitrile    acetone cyanhydrin |
| acids | **gaultherin** 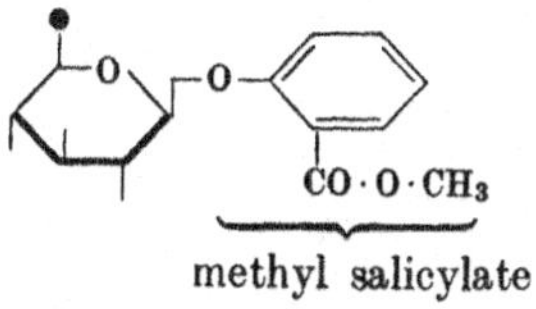 |
| | methyl salicylate |
| oxycumarin derivatives | **aesculin** 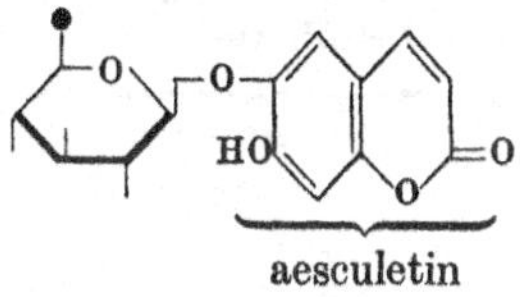 |
| | aesculetin |
| oxyanthraquinone derivatives | **ruberythric acid** 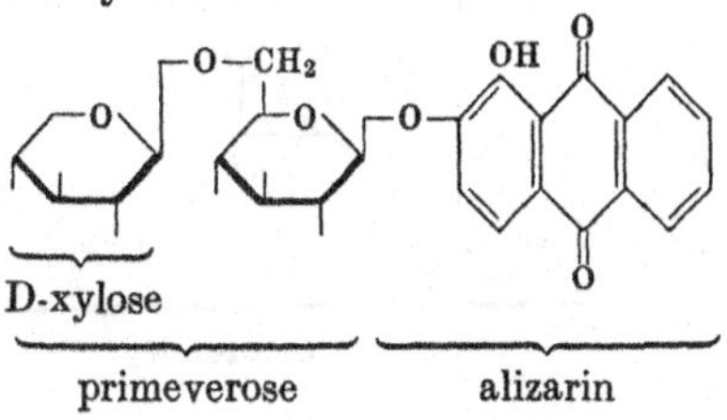 |
| | D-xylose |
| | primeverose    alizarin |
| oxyflavone derivatives | **rutin** 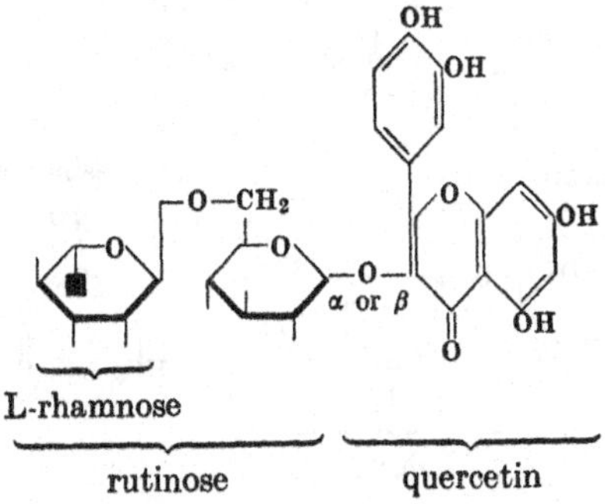 |
| | L-rhamnose |
| | rutinose    quercetin |

Table 16. (Continued.)

| Class | Examples |
|---|---|
| anthocyanins | peonin |

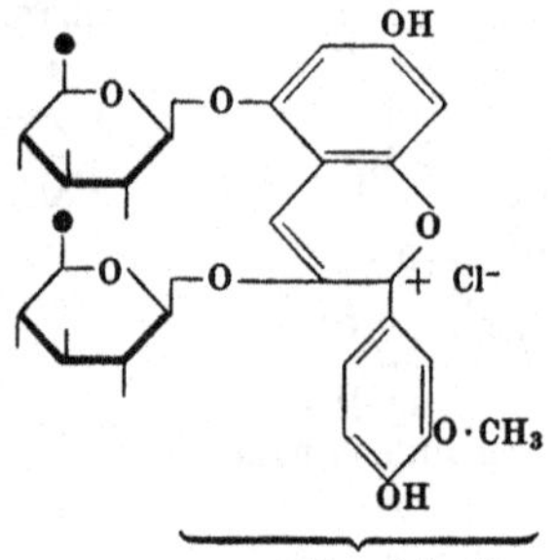

peonidin chloride

cardiac glycosides     thevetin

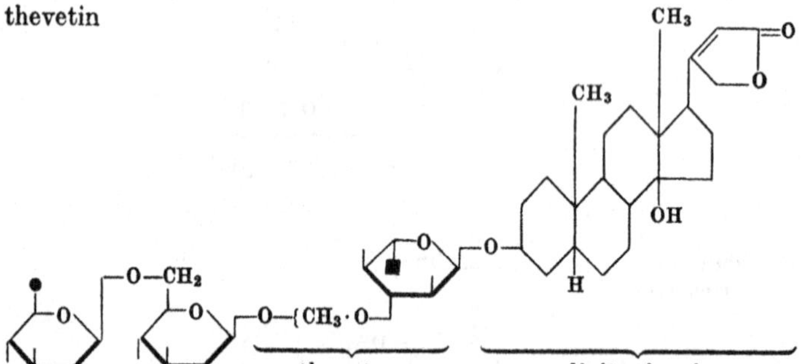

thevetose          digitoxigenin

digitalis saponins     trillin

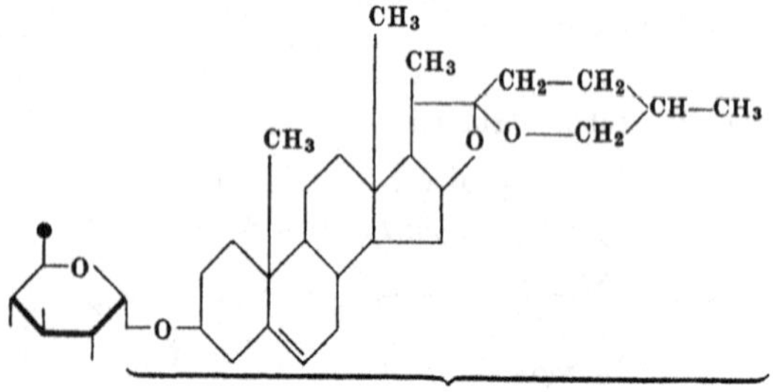

diosgenin

mustard oils     sinigrin (a thioglycoside); furanose configuration unconfirmed since 1914; β-linkage probable

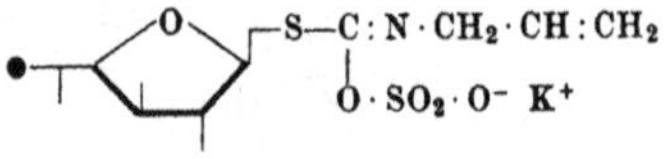

others     indican                    adenosine (an N-glycoside)

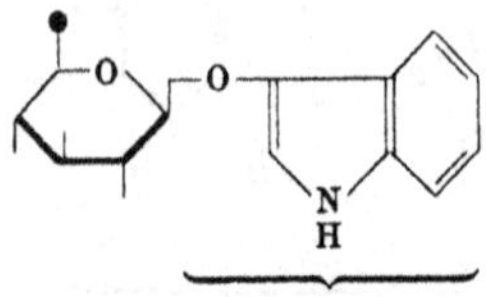

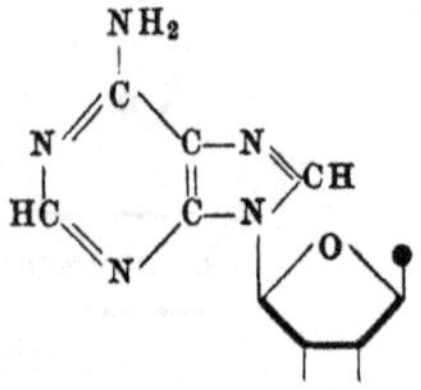

indoxyl

Table 17. *Nitrogen heterocyclics: plant alkaloids.*

| Heterocyclic nuclei on which class of alkaloid based | Examples | Structurally similar compounds |
|---|---|---|
| pyridine  | trigonelline (a betaïne) 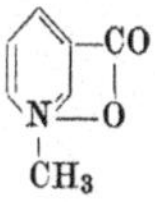 <br><br>pyridine + pyrrolidine nuclei:<br>nicotine 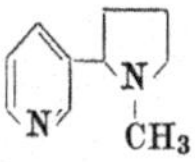      nornicotine 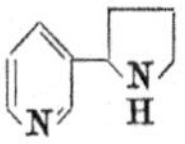 <br><br>pyridine + piperidine nuclei:<br>anabasine 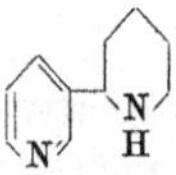 | nicotinic acid 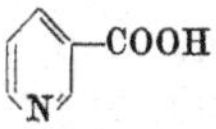    nicotinamide 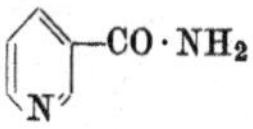 <br><br>pyridoxal 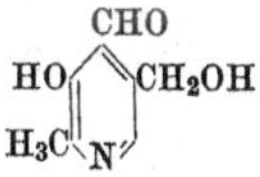     pyridoxamine 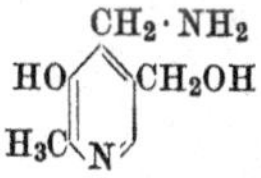 <br><br>pyridoxine 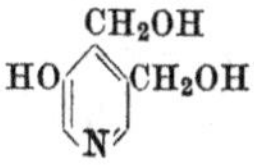 |
| piperidine  | coniine 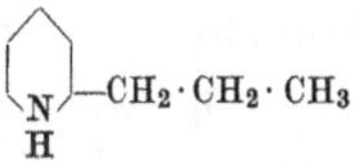       pelletierine 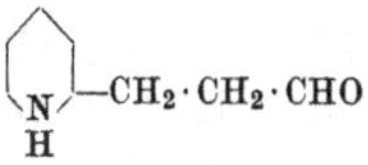 <br><br>piperine 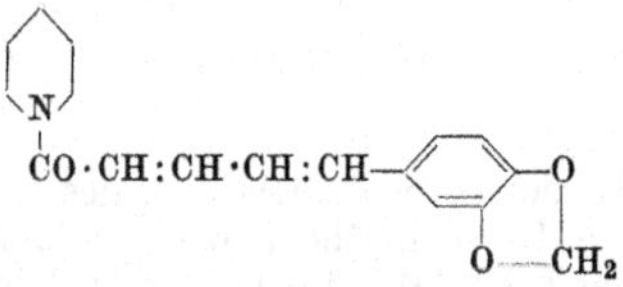 | |
| pyrrolidine 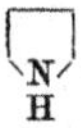 | pyrrolidine 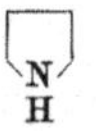   stachydrine (a betaïne) 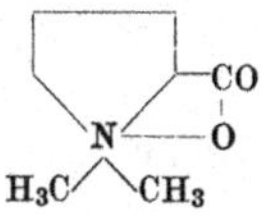 <br><br>cuskhygrine 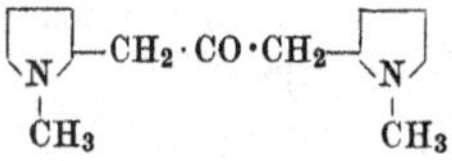 | pyrrole nuclei, as present in heme (see Table 19)  |
| quinoline  | quinine 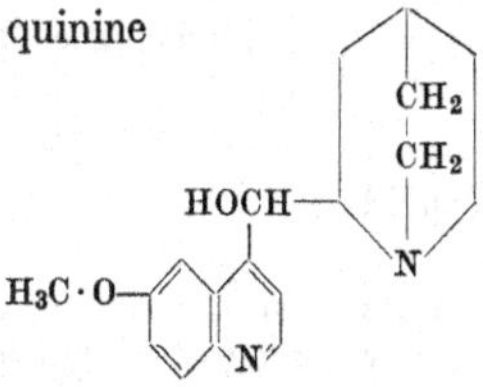 | |

*Table 17.* (Continued.)

| Heterocyclic nuclei on which class of alkaloid based | Examples | Structurally similar compounds |
|---|---|---|
| isoquinoline  | papaverine 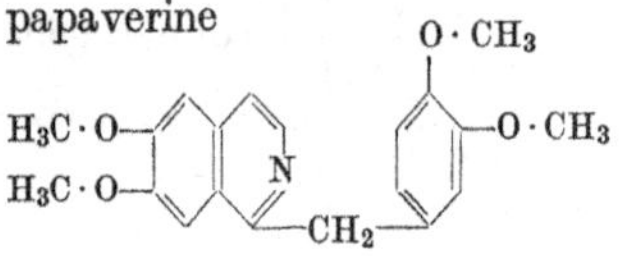 | |
| tropane 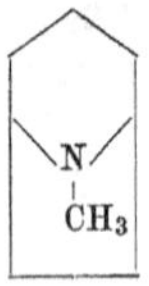 | cocaine 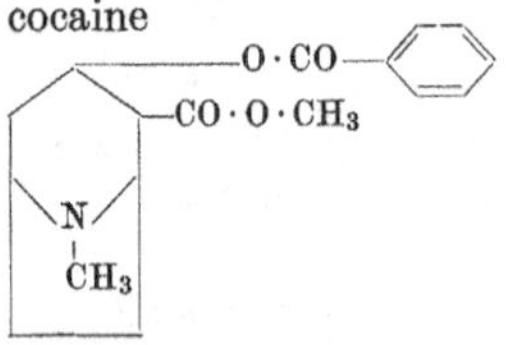 | |
| other | caffeine<br>theophylline } see Table 18<br>theobromine | |

great diversity in structure, they are present in only a relatively few families of angiosperms, mostly dicots. Two members of the B complex group of vitamins, nicotinic acid (niacin) and pyridoxine (vitamin $B_6$), show structural features in common with the alkaloid trigonelline, which is also a betaïne; nicotinamide, pyridoxal, and pyridoxamine are coenzyme substituents.

Another group of compounds characterized by heterocyclic carbon and nitrogen atoms are the pyrimidine and purine bases (Table 18), which are so important as constituents of nucleic acids, relatively high molecular weight compounds which are intimately associated with, if not identical with, the hereditary material of life and, as conjugated to proteins, a characteristic feature of plant viruses.

The nucleic acids contain pyrimidine and purine bases, ribose or desoxyribose (Table 13), and phosphoric acid in a molecular ratio of 1:1:1 in the phosphoric acid-linked chain of nucleosides (base + sugar, see "adenosine" in Table 16). A complete nucleotide is exemplified by yeast adenylic acid.

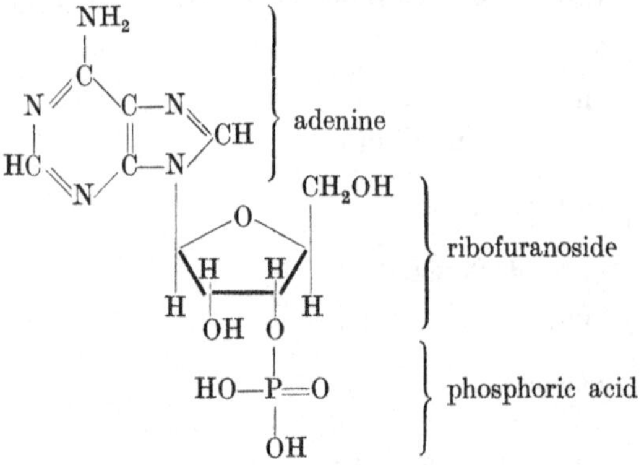

yeast adenylic acid

A portion of a chain of ribonucleotides, where $R_1$, $R_2$, etc., represent any of the nucleic acid bases shown in Table 18, is depicted in Table 9. The two general categories of nucleic acids

Table 18. *Nitrogen heterocyclics: pyrimidine and purine bases.*

| Basic nucleus | Bases in nucleic acids and prosthetic groups | |
|---|---|---|
| pyrimidine 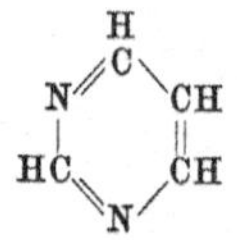 | cytosine (in nucleic acids) 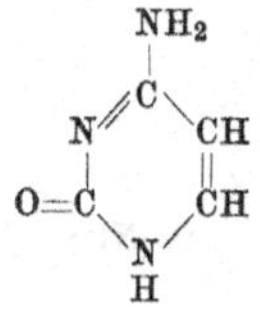 | thymine (in desoxyribonucleic acids) 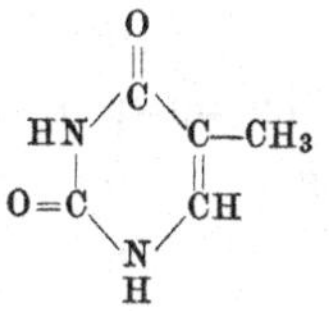 |
| | uracil (in ribonucleic acids) 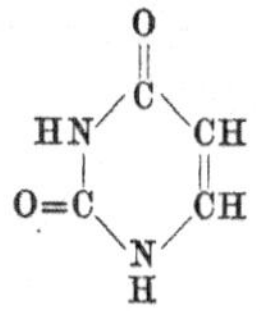 | thiamine chloride (vitamin $B_1$, a constituent of cocarboxylase) 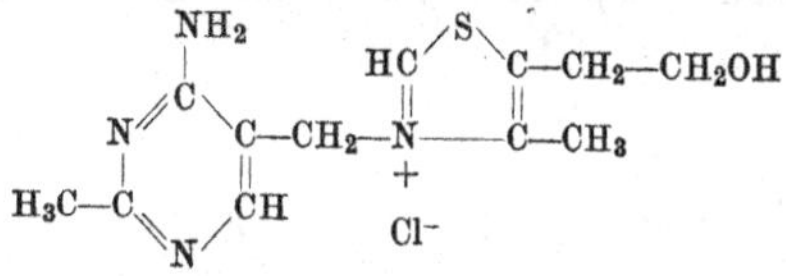 |
| purine (condensed pyrimidine and iminazole rings) 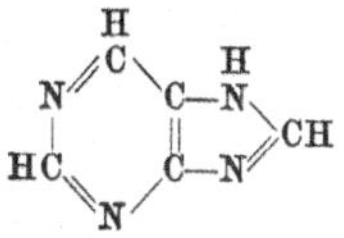 | adenine (free and also in nucleic acids, nucleosides, and flavin-adenine and phosphopyridine nucleotide coenzymes) 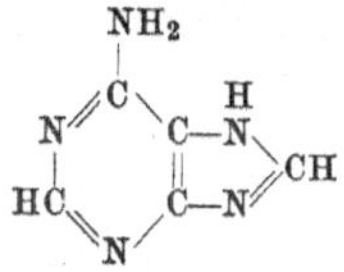 | guanine (free and also in nucleic acids and nucleosides) 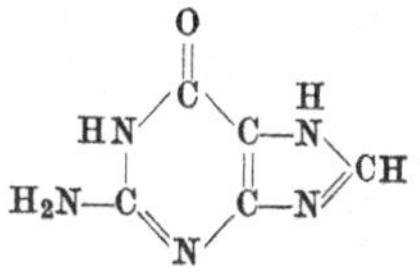 |

Other bases

| xanthine (free) 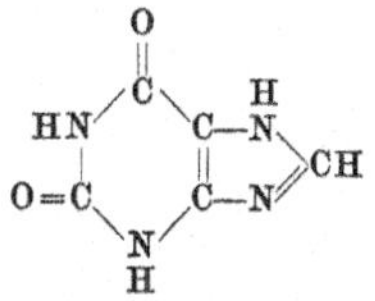 | hypoxanthine (free and in nucleosides) 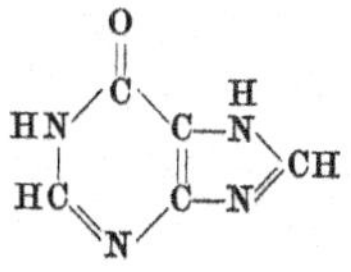 | uric acid (free) 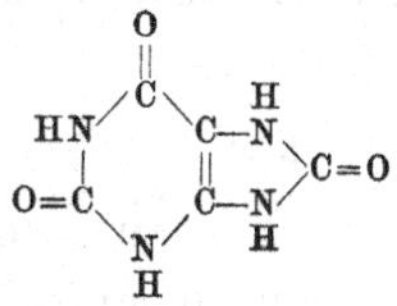 |
|---|---|---|

Alkaloid bases:

| caffeine 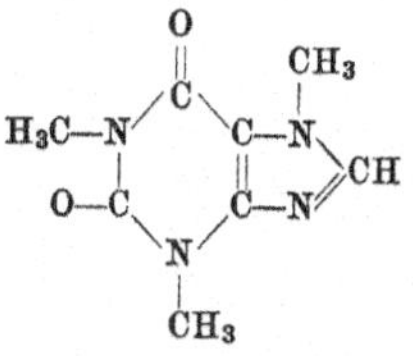 | theobromine 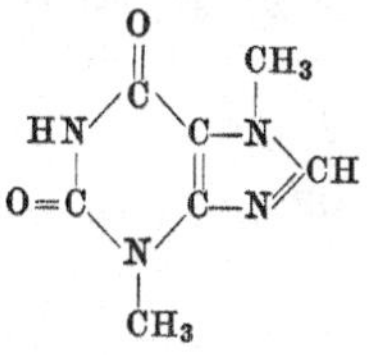 | theophylline 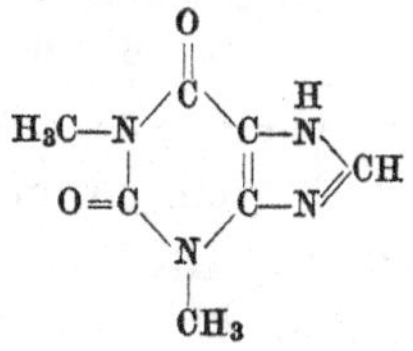 |
|---|---|---|

Table 19. *Nitrogen heterocyclics as linked constituents of prosthetic groups and coenzymes.*

| Prosthetic group or coenzyme | Examples of enzymes and auxilliary compounds with which found |
|---|---|
| adenine flavin dinucleotide 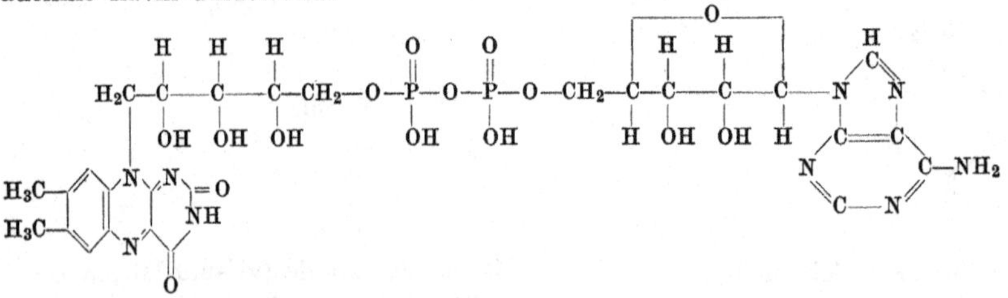 | glucose aerodehydrogenase |
| diphosphopyridine nucleotide (DPN, coenzyme I) 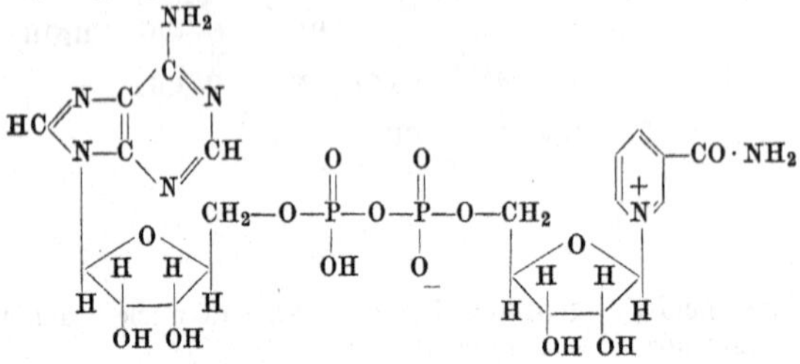 | dehydrogenases, malic for example |
| triphosphopyridine nucleotide (TPN, coenzyme II) 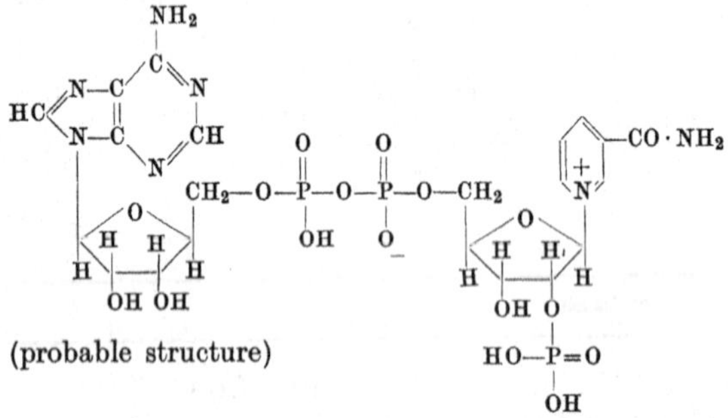 (probable structure) | dehydrogenases, isocitric for example |
| adenosine triphosphate (mung bean) 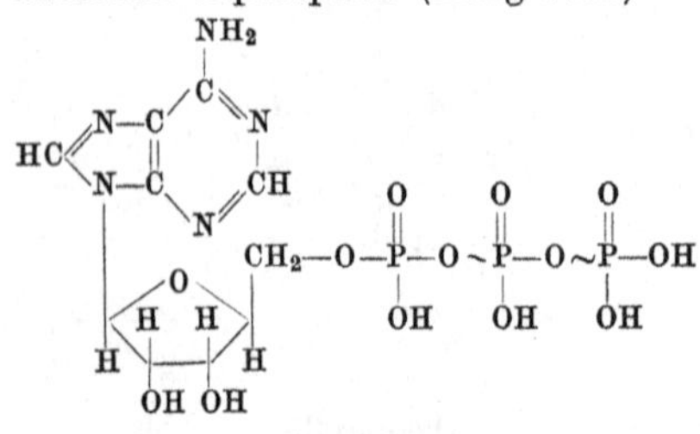 | phosphokinases (enzymes catalyzing the transfer of high energy phosphates) |
| flavin mononucleotide (riboflavin phosphate ester) 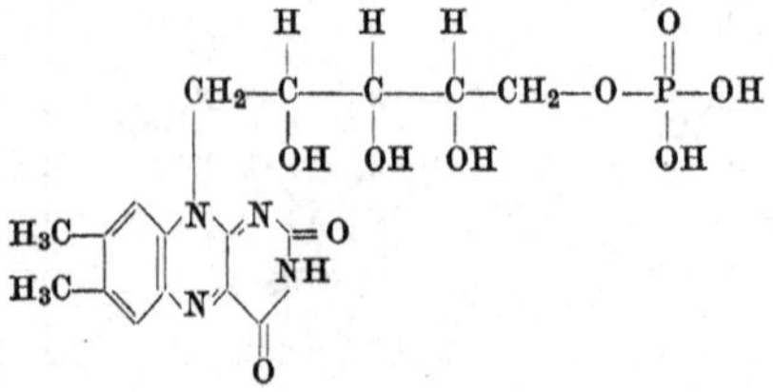 | cytochrome reductase |

*Table 19.* (Continued.)

| Prosthetic group or coenzyme | Examples of enzymes and auxilliary compounds with which found |
|---|---|
| pyridoxal phosphate 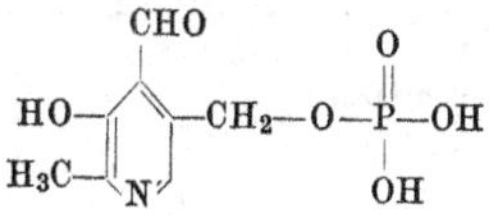 | transaminases, amino acid decarboxylase |
| cocarboxylase (thiamine pyrophosphate) 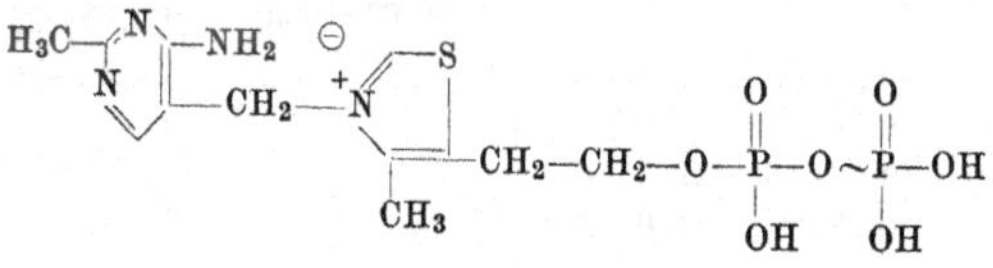 | carboxylase |
| heme 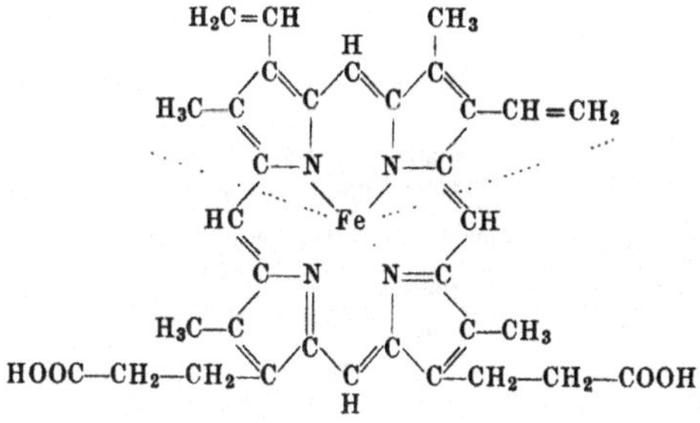 | catalase, peroxidase, cytochrome c |

in plants are the desoxyribonucleic acid (DNA) of the nucleus and the ribonucleic acid (RNA) of cytoplasm and nucleus. The kinds of DNA or RNA probably depend on the order in which the various nucleotides appear in the nucleic acid chain. The ratios of the bases in the polynucleotide are apparently not equivalent, as, for example, a molecular ratio of approximately $1.63:1.16:1.14:1$ of guanine, adenine, cytidylic acid, and uridylic acid, respectively, in Fraction I nucleoprotein.

Thiamine chloride (vitamin $B_1$) is a pyrimidine united to a substituted thiazole ring, and, as the diphosphate, is the coenzyme, cocarboxylase (Table 19), of carboxylase. The purine adenine is an integral part of the structure of several coenzymes (Table 19) which function in the transfer of electrons during the oxidative reactions of respiration. Adenosine-5-monophosphate is a remarkable compound which functions as a means of conserving the energy derived from the glycolysis and oxidation of carbohydrate during respiration, the energy being stored in the form of high energy phosphate bonds, as shown for adenosine triphosphate in Table 19. Free purines can also be isolated from plants, as, for example, caffeine from the coffee bean and theophylline from tea leaves.

Other important phytochemicals containing heterocyclic carbon and nitrogen nuclei are riboflavin (see "flavin mononucleotide" in Table 19) or vitamin $B_2$, vitamin $B_{12}$ (a trivalent cobalt complex), folic acid (a B complex vitamin),

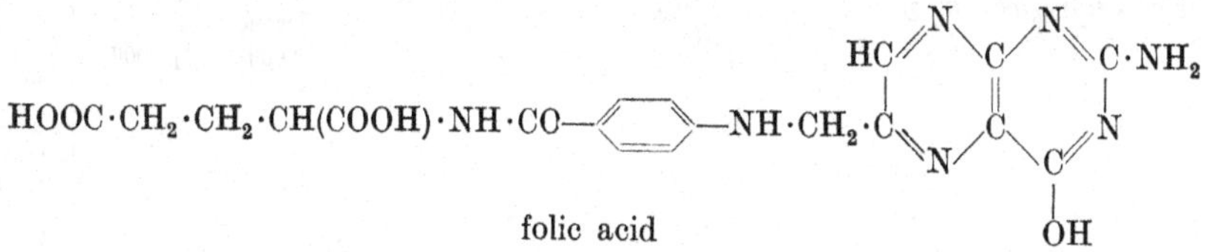

folic acid

Table 20. *Basic structural relationships and end and side group*

| Alcohols | Ketones and aldehydes | Monobasic acids | Aldehydic and hydroxy acids | Polybasic acids | Keto acids |
|---|---|---|---|---|---|
| methyl (in esters and ethers)<br>$CH_3OH$ | formaldehyde<br>$HCHO$ | formic<br>$HCOOH$ | | | |
| ethyl<br>$CH_3 \cdot CH_2OH$ | acetaldehyde<br>$CH_3 \cdot CHO$ | acetic<br>$CH_3 \cdot COOH$ | glycolic<br>$CH_2OH \cdot COOH$<br><br>glyoxylic<br>$CHO \cdot COOH$ | oxalic<br>$COOH \cdot COOH$ | |
| glycerol (in esters)<br>$CH_2OH \cdot CHOH \cdot CH_2OH$ | acetone<br>$CH_3 \cdot CO \cdot CH_3$<br>dihydroxyacetone phosphate<br>$CH_2[O \cdot PO(OH)_2] \cdot CO \cdot CH_2OH$<br>3-phosphoglyceraldehyde<br>$CHO \cdot CHOH \cdot CH_2[O \cdot PO(OH)_2]$<br>1,3-diphosphoglyceraldehyde<br>$\overset{|}{C}HOH \cdot CHOH \cdot CH_2[O \cdot PO(OH)_2]$<br>$O \cdot PO(OH)_2$<br>acrolein<br>$CH_2 : CH \cdot CHO$ | propionic<br>$CH_3 \cdot CH_2 \cdot COOH$ | 2-phosphoglyceric<br>$COOH \cdot CH[O \cdot PO(OH)_2] \cdot CH_2OH$<br>3-phosphoglyceric<br>$COOH \cdot CHOH \cdot CH_2[O \cdot PO(OH)_2]$<br>1,3-diphosphoglyceric<br>$\overset{|}{C}O \cdot CHOH \cdot CH_2[O \cdot PO(OH)_2]$<br>$O \cdot PO(OH)_2$ | malonic<br>$COOH \cdot CH_2 \cdot COOH$ | pyruvic<br>$CH_3 \cdot CO \cdot COOH$<br>phosphoenolpyruvic<br>$COOH \cdot$<br>$C[O \sim PO(OH)_2] : CH_2$ |
| | | isobutyric<br>$(CH_3)_2CH \cdot COOH$ | L-malic<br>$COOH$<br>$HO\overset{|}{C}H$<br>$H\overset{|}{C}H$<br>$\overset{|}{C}OOH$    L-tartaric<br>$COOH$<br>$HO\overset{|}{C}H$<br>$H\overset{|}{C}OH$<br>$\overset{|}{C}OOH$ | succinic<br>$COOH \cdot CH_2 \cdot CH_2 \cdot COOH$<br>oximinosuccinic<br>$\qquad NOH$<br>$\qquad \|$<br>$COOH \cdot CH_2 \cdot C \cdot COOH$<br>fumaric<br>$HC \cdot COOH$<br>$\|$<br>$HOOC \cdot CH$ | oxaloacetic<br>$COOH \cdot CH_2 \cdot CO \cdot COOH$ |
| isoamyl<br>$(CH_3)_2CH \cdot CH_2 \cdot CH_2OH$ | | isovaleric<br>$(CH_3)_2CH \cdot CH_2 \cdot COOH$<br>β-methylcrotonic<br>$(CH_3)_2C : CH \cdot COOH$ | | glutaric<br>$COOH \cdot (CH_2)_3 \cdot COOH$ | α-ketoisovaleric<br>$(CH_3)_2CH \cdot CO \cdot COOH$<br>α-ketoglutaric<br>$COOH \cdot CH_2 \cdot CH_2 \cdot CO \cdot$<br>$COOH$ |
| | αβ-hexylenealdehyde<br>$CH_3 \cdot CH_2 \cdot CH_2 \cdot CH : CH \cdot CHO$ | caproic<br>$CH_3 \cdot CH_2 \cdot CH_2 \cdot CH_2 \cdot$<br>$CH_2 \cdot COOH$ | citric<br>$COOH$<br>$H\overset{|}{C}H$<br>$HO\overset{|}{C} \cdot COOH$<br>$H\overset{|}{C}H$<br>$\overset{|}{C}OOH$   isocitric<br>$COOH$<br>$H\overset{|}{C}OH$<br>$H\overset{|}{C} \cdot COOH$<br>$H\overset{|}{C}H$<br>$\overset{|}{C}OOH$ | cisaconitic<br>$COOH$<br>$\overset{|}{C}H$<br>$\|$<br>$\overset{|}{C} \cdot COOH$<br>$H\overset{|}{C}H$<br>$\overset{|}{C}OOH$<br>adipic<br>$COOH \cdot (CH_2)_4 \cdot COOH$ | oxalosuccinic<br>$COOH$<br>$\overset{|}{C}=O$<br>$H\overset{|}{C} \cdot COOH$<br>$H\overset{|}{C}H$<br>$\overset{|}{C}OOH$<br>α-ketoisocaproic<br>$(CH_3)_2 \cdot CH \cdot CH_2 \cdot CO \cdot$<br>$COOH$<br>α-keto-β-methyl-<br>n-valeric<br>$CH_3 \cdot CH_2 \cdot CH(CH_3) \cdot$<br>$CO \cdot COOH$ |

[1] Amino acids marked with an asterisk (*) are constituents of plant proteins. D-amino acids occur

*diversity of some phytochemicals of unusual physiological interest.*

| L-amino acids[1] | | Betaïnes | Amines |
|---|---|---|---|
| | | | monomethylamine $CH_3 \cdot NH_2$ |
| glycine* $CH_2(NH_2) \cdot COOH$ | | betaïne $\begin{array}{c} CH_2\text{---}CO \\ \mid \quad\quad O \\ N\text{---}O \\ H_3C \diagup \; \diagdown CH_3 \\ \mid \\ CH_3 \end{array}$ | dimethylamine $(CH_3)_2NH$ |
| alanine* $CH_3 \cdot CH(NH_2) \cdot COOH$ <br> cysteïne $HS \cdot CH_2 \cdot CH(NH_2) \cdot COOH$ | serine* $CH_2OH \cdot CH(NH_2) \cdot COOH$ <br> cystine* $S \cdot CH_2 \cdot CH(NH_2) \cdot COOH$ <br> $S \cdot CH_2 \cdot CH(NH_2) \cdot COOH$ | | trimethylamine $(CH_3)_3N$ |
| aspartic* $COOH \cdot CH_2 \cdot CH(NH_2) \cdot COOH$ <br> homoserine $CH_2OH \cdot CH_2 \cdot CH(NH_2) \cdot COOH$ <br> homocystine $S \cdot CH_2 \cdot CH_2 \cdot CH(NH_2) \cdot COOH$ <br> $S \cdot CH_2 \cdot CH_2 \cdot CH(NH_2) \cdot COOH$ <br> threonine* $CH_3 \cdot CHOH \cdot CH(NH_2) \cdot COOH$ | asparagine $H_2N \cdot CO \cdot CH_2 \cdot CH(NH_2) \cdot COOH$ <br> djenkolic $S \cdot CH_2 \cdot CH_2 \cdot CH(NH_2) \cdot COOH$ <br> $CH_2$ <br> $S \cdot CH_2 \cdot CH_2 \cdot CH(NH_2) \cdot COOH$ <br> homocysteïne $HS \cdot CH_2 \cdot CH_2 \cdot CH(NH_2) \cdot COOH$ <br> α-aminobutyric $CH_3 \cdot CH_2 \cdot CH(NH_2) \cdot COOH$ | | putrescine $H_2N \cdot CH_2 \cdot CH_2 \cdot CH_2 \cdot CH_2 \cdot NH_2$ <br> isobutylamine $(CH_3)_2CH \cdot CH_2 \cdot NH_2$ |
| valine* $(CH_3)_2CH \cdot CH(NH_2) \cdot COOH$ <br> glutamic* $COOH \cdot CH_2 \cdot CH_2 \cdot CH(NH_2) \cdot COOH$ <br> methionine* $CH_3 \cdot S \cdot CH_2 \cdot CH_2 \cdot CH(NH_2) \cdot COOH$ <br> canavanine $H_2N \cdot C \cdot NH \cdot O \cdot CH_2 \cdot CH_2 \cdot CH(NH_2) \cdot COOH$ <br> $\parallel$ <br> $NH$ <br> citrulline $HN \cdot CH_2 \cdot CH_2 \cdot CH(NH_2) \cdot COOH$ <br> $H_2N \cdot C : O$ | ornithine $H_2N \cdot (CH_2)_3 \cdot CH(NH_2) \cdot COOH$ <br> glutamine $H_2N \cdot CO \cdot CH_2 \cdot CH_2 \cdot CH(NH_2) \cdot COOH$ <br> proline* $\begin{array}{c} CH_2\text{---}CH_2 \\ \mid \quad\quad \mid \\ CH_2 \; CH \cdot COOH \\ \diagdown N \diagup \\ \mid \\ H \end{array}$ <br> hydroxyproline* $\begin{array}{c} HOCH\text{---}CH_2 \\ \mid \quad\quad \mid \\ CH_2 \; CH \cdot COOH \\ \diagdown N \diagup \\ \mid \\ H \end{array}$ | stachydrine $\begin{array}{c} CH_2\text{---}CH_2 \\ \mid \quad\quad \mid \\ CH_2 \; CH\text{---}CO \\ \diagdown N \diagup \quad O \\ H_3C \; \diagdown CH_3 \end{array}$ <br> betonicine and turicine $\begin{array}{c} HOCH\text{---}CH_2 \\ \mid \quad\quad \mid \\ CH_2 \; CH\text{---}CO \\ \diagdown N \diagup \quad O \\ H_3C \; \diagdown CH_3 \end{array}$ | isoamylamine $(CH_3)_2CH \cdot CH_2 \cdot CH_2 \cdot NH_2$ |
| arginine* $HN \cdot CH_2 \cdot CH_2 \cdot CH_2 \cdot CH(NH_2) \cdot COOH$ <br> $H_2N \cdot C : NH$ <br> α-aminoadipic $COOH \cdot (CH_2)_3 \cdot CH(NH_2) \cdot COOH$ <br> leucine* $(CH_3)_2CH \cdot CH_2 \cdot CH(NH_2) \cdot COOH$ <br> isoleucine* $CH_3 \cdot CH_2 \cdot CH(CH_3) \cdot CH(NH_2) \cdot COOH$ | histidine* $\begin{array}{c} N\text{---} \\ \diagdown\text{---}CH_2 \cdot CH(NH_2) \cdot COOH \\ N \\ \mid \\ H \end{array}$ <br> lysine* $H_2N \cdot (CH_2)_4 \cdot CH(NH_2) \cdot COOH$ | trimethylhistidine $\begin{array}{c} N\text{---} \\ \diagdown\text{---}CH_2\text{---}CH\text{---}CO \\ N \quad\quad H_3C\text{---}N\text{---}O \\ \mid \quad\quad\quad H_3C \diagup \diagdown CH_3 \\ H \end{array}$ <br> ergothioneine $\begin{array}{c} HS\text{---}N\text{---} \\ \diagdown\text{---}CH_2\text{---}CH\text{---}CO \\ N \quad\quad H_3C\text{---}N\text{---}O \\ \mid \quad\quad\quad H_3C \diagup \diagdown CH_3 \\ H \end{array}$ | |

infrequently as components of products of microbial metabolism.

*Table 20.*

| Alkohols | Ketones and aldehydes | Monobasic acids | Aldehydic and hydroxy acids | Polybasic acids | Keto acids |
|---|---|---|---|---|---|
|  |  |  |  |  | phenylpyruvic 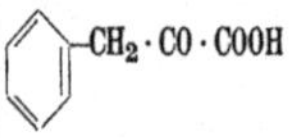<br>⬡—$CH_2 \cdot CO \cdot COOH$ |
|  | indoleacetaldehyde<br>—$CH_2 \cdot CHO$ 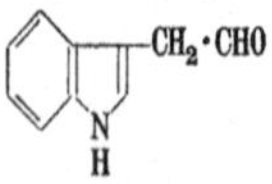 | indoleacetic<br>—$CH_2 \cdot COOH$ 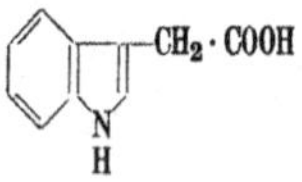 |  |  |  |

[1] Amino acids marked with an asterisk (*) are constituents of plant proteins.

biotin (a B complex vitamin—see Table 6), and indoleacetic acid, a plant auxin, and its derivatives (Tables 7 and 20).

Thirty-four different α-amino acids, nineteen of them being constituents of plant proteins, have been isolated from plants. Because of the great interest attached to the synthesis and metabolism of the amino acids, their arrangement in Table 20 is such as to indicate structural similarities to other, mostly aliphatic, compounds of metabolic importance.

The salient features of Table 20 may be noted briefly as follows: Alcohols ordinarily occur as esters and only rarely in a free condition in higher plants; *e.g.*, ethyl alcohol under conditions of oxygen stress, isoamyl alcohol in oil of peppermint, and the higher aliphatic alcohols in some waxes. (Also present in waxes are the higher aliphatic hydrocarbons, ketones, acids, and esters.) Some of the aldehydes and ketones and some hydroxy, polybasic, and keto acids are known intermediates in carbohydrate anabolism and catabolism. The amino acids and their amide derivatives, asparagine and glutamine, are, aside from their importance in protein metabolism, of additional significance to other activities of the cell, as, for example, the role of tryptophane in the synthesis of indoleacetic acid and the participation of amino acids and certain α-keto acids in reversible enzymatic transaminations. An amino acid with the amino group in the β-position to the carboxyl group, β-alanine, is a part of pantothenic acid (a B complex vitamin), $CH_2OH \cdot C(CH_3)_2 \cdot CHOH \cdot CO \cdot NH \cdot CH_2 \cdot CH_2 \cdot COOH$. The betaïnes shown in the table are, structurally, methylated amino acid derivatives in which the α-amino nitrogen is pentavalent; they are most prevalent in young and actively growing tissues, are thought to be by-products of protein metabolism, and may also possibly participate in enzymatic transmethylations. Choline, $(CH_3)_3^+ N \cdot CH_2 \cdot CH_2OH$, like the betaïnes, a possible methyl donor, occurs both free in the plant and as a constituent of the phospholipid lecithin. The other substituted amines are probably products of protein decomposition in the plant, mono-, di-, and trimethylamine being gases at ordinary temperatures. Urea, $NH_2 \cdot CO \cdot NH_2$, and guanidine, $HN:C(NH_2)_2$, may also be end products of protein catabolism.

Although the amino acids occur free in the plant in appreciable amounts, varying with the tissue and its physiological age and nitrogen supply, a great part of amino nitrogen is present in the form of proteins. The proteins are non-dialyzable molecules of high molecular weight (colloidal dimensions), which provide much of the structure of cytoplasm as well as a participating control of metabolism and heredity. The great variety of proteins in any given cell corresponds both to the number and to the order in which the amino acids appear in the protein chain molecule. The activities of the proteins depend not only on their amino acid constitution, but also on their surface and intramolecular

(Continued.)

| L-amino acids [1] | | Betaïnes | Amines |
|---|---|---|---|
| phenylalanine*<br>⬡—CH₂·CH(NH₂)·COOH | 3,4-dihydroxyphenylalanine (dopa)<br>⬡—CH₂·CH(NH₂)·COOH<br>HO  OH | | |
| tyrosine*<br>⬡—CH₂·CH(NH₂)·COOH<br>HO | N-methyltyrosine<br>⬡—CH₂·CH·COOH<br>HO  H₃C·NH | | hordenine<br>⬡—CH₂·CH₂·N(CH₃)₂<br>HO |
| tryptophane*<br>⬡—CH₂·CH(NH₂)·COOH<br>N<br>H | kynurenin<br>⬡—CO·CH₂·CH(NH₂)·COOH<br>⬡—NH₂ | hypaphorine<br>⬡—CH₂–CH–CO<br>N  H₃C–N—O<br>H  H₃C  CH₃ | |

D-amino acids occur infrequently as components of products of microbial metabolism.

charges, which reflect the amphoteric nature of amino acids, and on the configuration of the protein, as determined by (1) the stereochemistry of amino acid residues linked together by peptide bonds, —CO·NH—, and (2) the manner

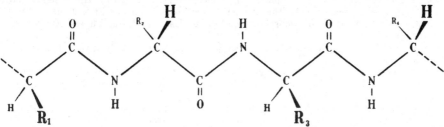

Polypeptide chain (semi-perspective), where $R_1$, $R_2$, $R_3$, etc., are each the portion of an amino acid residue attached to its α-carbon atom.

in which the minor and major helices or folds of the polypeptide chain, together with the associations between neighboring protein molecules, are stabilized by hydrogen bonding or by —S—S— linkages or, possibly, by salt, ester, or ether linkages between folds.

As short and long term reserves in, for example, leaves and seeds, respectively, proteins are the ready means for hydrolytic mobilization of amino acids at leaf senescence or seed germination for translocation to points of protein resynthesis.

The arbitrary classification of extracted plant proteins into albumins, globulins, prolamins (gliadins), glutelins, histones, and protamines is based on their solubilities in water, dilute acid, dilute alkali, dilute neutral salt solutions, and 70–80% alcohol. Whether or not such chemical separations have any physiological significance has yet to be demonstrated. As with any means of extraction, the question arises as to whether the method employed has not altered the substance extracted.

A special group of proteins, the enzymes, may consist entirely of protein; urease, which catalyzes the decomposition of urea, is an example. Other enzymes are proteins (apoenzymes) conjugated with other simple or complex molecules (prosthetic groups), which are required for the activity of the catalyst. Catalase is an example in which four heme molecules are conjugated to a protein to form an enzyme which catalyzes the decomposition of $H_2O_2$. Still other enzymes are incapable of acting as catalysts in the absence of an appropriate coenzyme (see Table 19).

Hemoglobin, a protein conjugated of heme (Table 19) and globin, occurs in the protoplasm of cells of certain leguminous root nodules containing certain strains of *Rhizobium* spp. Lipoproteins that occur in plant cytoplasm outside of the mitochondria and chloroplasts have not been demonstrated. A plant protoplasmic mucoprotein, a great part of which is in the chondriome, and which would appear to be a protein-polysaccharide conjugate, has been detected cytochemically; in contrast to animal mucoproteins, it contains no aminohexoses. A protein compound which, in containing up to 5% hexosamine and only about 11.5% nitrogen, more closely resembles some of the animal mucoproteins, has, however, been isolated from *Aspergillus*.

## Literature.

BALDWIN, E.: Dynamic aspects of biochemistry, 2nd ed. Cambridge: Cambridge University Press 1952. — BARRON, E. S. G.: Thiol groups of biological importance. Adv. Enzymol. 11, 201–266 (1951). — BONNER, J.: Plant biochemistry. New York: Academic Press 1950.

COLE, W. H., ed.: Some conjugated proteins. New Brunswick (New Jersey): Rutgers University Press 1953.

DAVIDSON, J. N.: The biochemistry of the nucleic acids, 2nd ed. London: Methuen 1953. — DEUEL jr., H. J.: The lipids, vol. I: chemistry. New York: Interscience 1951.

FOSTER, J. W.: Chemical activities of fungi. New York: Academic Press 1949. — FREY-WYSSLING, A.: Submicroscopic morphology of protoplasm, 2nd English ed. Amsterdam: Elsevier 1953

GORTNER, R. A.: Outlines of biochemistry, 3rd ed., edited by R. A. GORTNER jr., and W. A. GORTNER. New York: Wiley 1949.

HAUROWITZ, F.: Chemistry and biology of proteins. New York: Academic Press 1950. — HEILBRON, I., and H. M. BUNBURY: Dictionary of organic compounds, 4 vols. London: Eyre & Spottiswoode 1953.

JAMES, W. O.: Reaction paths in plant respiration. Endeavour 13, 155–162 (1954).

KARRER, P.: Organic chemistry, 4th English ed. Amsterdam: Elsevier 1950.

MANSKE, R. H. F., and H. L. HOLMES, eds.: The alkaloids. Chemistry and physiology. New York: Academic Press 1950 (vol. I), 1952 (vol. II), 1953 (vol. III). — McELROY, W. D., and B. GLASS, eds.: Phosphorus metabolism, a symposium on the role of phosphorus in the metabolism of plants and animals. Baltimore: Johns Hopkins Press 1951 (vol. 1), 1952 (vol. 2). — McILROY, R. J.: The plant glycosides. London: Arnold 1951.

NEURATH, H., and K. BAILEY, eds.: The proteins, vol. I, part A. New York: Academic Press 1953.

PIGMAN, W. W., and R. M. GOEPP jr.: Chemistry of the carbohydrates. New York: Academic Press 1948.

SIMONSEN, J. L.: The terpenes, 2nd ed., edited by J. L. SIMONSEN and L. N. OWEN. Cambridge: Cambridge University Press 1947 (vol. I), 1949 (vol. II), 1952 (vol. III). — STEELE, C. C.: An introduction to plant biochemistry, 2nd ed. London: Bell 1949. — SUMNER, J. B., and K. MYRBÄCK: The enzymes. Chemistry and mechanism of action. New York: Academic Press 1950 (vol. I, pt. 1), 1951 (vol. I, pt. 2, and vol. II, pt. 1), 1952 (vol. II, pt. 2). — SUMNER, J. B., and G. F. SOMERS: Chemistry and methods of enzymes. New York: Academic Press 1953.

THOMAS, M.: Plant physiology. London: Churchill 1947. — TONZIG, S.: I muco-proteidi e la vita della cellula vegetale. Saggio di una cito-fisiologia dell'acqua. Padova: Libreria Universitaria, Randi 1941. [Original not seen. Abstract in Biol. Abstr. 22, 2562–2563 (1948).]

WHISTLER, R. L., and C. L. SMART: Polysaccharide chemistry. New York: Academic Press 1953.

# Microscopic and submicroscopic structure of cytoplasm.

By

## William Seifriz.

With 42 figures.

## 1. Form.

The shapes of cells and of organisms is an expression of simple physical forces and of a complex inner organization. One or the other of these two influences may dominate. Thus, an ovum is spherical for the same reason that a water drop-let is spherical, surface tension forces determine the shape. But the form of a non-spherical cell is determined primarily by inner organizational forces, which may be similar to those of a crystal or far more intricate. The irregular form of a plasmodium, the invaginated disk of a red blood cell, and the bilateral symmetry of man, are determined by an inner structural pattern. A complex external form is the first evidence we have that there is an inner directive force, an organization, in living matter.

The term "organization", so frequently applied to protoplasm, denotes more than structure. It implies orderliness in activity as well as in pattern. It need not preclude the contention of D'ARCY THOMPSON (1942) that "no organic forms exist save such as are in conformity with physical and mathematical laws". In making this statement, THOMPSON realized that an interpretation in mathematical terms does not constitute a full explanation.

Whether or not protoplasm is wholly amenable to physical forces and mathe-matical laws cannot be said with finality, but it is well to assume that it is, until the last question, What is life?, is reached.

A simple and convincing example of D'ARCY THOMPSON's hypothesis is the shapes of cells in mass. Plant cells frequently assume the shape which plastic spheres assume when subjected to uniform pressure. Much work has been done on the problem of cell shape by LEWIS (1923, 1933, 1937) and others (MATZKE, 1927). It led to a lively discussion on the form assumed by plastic spheres under pressure. The conditions which cells in mass must meet are: 1. stackability, 2. hexagonal cross-section, 3. minimum surface for maximum volume, 4. stable equilibrium.

The cube, the rhombic, and the pentagonal dodecahedron, and the tetra-kaidecahedron all satisfy the first condition, that of stackability. They pile without voids. Dodecahedra and tetrakaidecahedra meet the second requirement by having an hexagonal cross-section. The tetrakaidecahedron best fulfills the third condition, that of space economy. The figure which most nearly approaches the sphere, is a cube truncated by an octahedron, having eight hexagonal and six quadrilateral facets with two kinds of dihedral angles (Fig. 1 A). There was considerable argument as to the stackability of the tetrakaidecahedron, first dis-cussed by Lord KELVIN. The controversy ended in Lord KELVIN's favor as MATZKE (1927) and others have shown. However, when the simplest possible experiment to ascertain the geometrical form of spherical bodies in mass was done by Buffon —

who packed peas tightly in a jar, soaked them in water and waited until the interstices were filled by swelling — the peas were found to be rhombic dodecahedra, twelve sided bodies. This is what one would expect because stacked spheres fit together so that a central one rests upon three with six surrounding it at its equatorial plane and three resting on top, *i.e.*, the original one is surrounded by twelve.

After years of patient observation and reconstruction, LEWIS (1937) found pith cells to be 14-hedral. Thus are cells, like bubbles, liquid drops, and all semifluid bodies which fill space in aggregates, of one and the same form. The orthic tetrakaidecahedron is the shape toward which cells in masses tend (Fig. 1 B).

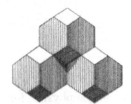

Fig. 1A.  Orthic tetrakaidecahedra.

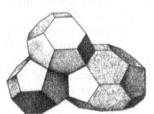

Fig. 1 B.  Irregular polyhedra — 14 hedron above, 13 hedron left, 15 hedron right — showing stackability of irregular forms
(F. T. LEWIS.)

The concept that cell shape is an expression of external and internal factors is not accepted by all, yet it is none the less true that there are external and internal forces which influence cell development and therefore together determine form. NEEDHAM (1936) and WADDINGTON (1940, 1951) accept this concept. The latter expresses the interrelationship when he says: neither the old systematic morphology, nor the mathematisation of morphology in D'ARCY THOMPSON's work, nor yet the biochemical studies of purified substances and enzyme systems, are adequate, the unification of morphology and biochemistry is needed.

Each investigator must reach his own conclusion as to where the emphasis is to be laid, whether on the mathematical-physical interpretation of THOMPSON (1942), or the "meaningful" interpretation of biological form as set forth by "artistically inclined scientists", who see in form "an aesthetic design" which is "not in harmony with the evergrowing and so often unrelated details of cellular physiology and biochemistry" BLOCH (1941, 1952) and WEYL (1952) hold the latter view, which was the view Goethe held. TROLL (1926) saw meaning, in the sense of function if not of purpose, in the simplest shapes of bodies.

The forces from within which determine growth and form may be the same physical ones with which we are familiar in the non-living world, though at times they appear to take other paths. Thus, the five-fold symmetry of biological forms, is never found in crystals. This does not mean that the form is incapable of mathematical expression. Many forms common in living matter are not found in crystals, such as the reniform, or kidney-shape, of the violet leaf, yet they can be mathematically expressed. The reniform shape of the violet leaf is a sine curve, with the formula:

$$r = \sin \Theta / 2 .$$

The form which an *Amoeba* assumes defies mathematical expression. One would be inclined therefore to say that it is wholly determined from within, yet any one of a number of experiments will prove that external factors play a part.

One of the most interesting problems in growth and form, which is also a problem in protoplasmic structure, is the spiral tendency so common in plants and animals.

## 2. Spirality.

Spiral form and movement are of common, perhaps universal, occurrence in life. Trees show a spiral twist (SEIFRIZ 1933a, 1933b, PRESTON 1948). Vines twine usually to the right but sometimes to the left, often refusing to twine in the opposite direction. The arrangement of leaves on the stem of *Costus*, of seeds in a sunflower head, and of spines on a cactus, is a spiral one, so also is the arrangement of the microscopic knobs on the spores of *Laccaria* which LOCQUIN (1945) has shown to be dextrorotary. The cellulose in a bast fiber wall is spirally oriented (Fig. 2).

The spiral tendency in plants has been rather extensively studied in connection with phyllitaxy where the spirality may be both clockwise and counter-clockwise (ALLARD 1942, 1946). Animals as well as plants show spiral tendencies: the shells of mollusks, the umbilical cord in mammals, and the horns of animals are spirally formed.

Not only are the form and arrangement of plant and animal parts spirally oriented, but growth (CASTLE 1936, 1942, PRESTON 1948) and movement are also spiral. Men walk, when lost in the forest, in "ever decreasing circles". They walk and drive in a spiral when blind folded (SCHAEFFER 1927, 1931).

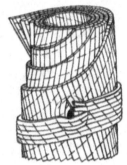

There are, as before, two possible interpretations of the spiral tendency in organisms, environment and protoplasm. That the environment, in the form of wind, is responsible for the spiral form of trees has often been assumed, but without justification. Wind cannot twist old wood nor roots. The spiral twist in plants is established when the cellulose is first deposited in the tissue.

A second environmental factor which has been held responsible for the spiral tendencies in growth and movement is the rotation of the earth. Its influence,

Fig. 2. Spiral wrapping in a bast fiber (G. SCARTH).

known as the Coriolis effect, takes several forms. The rotation of the earth exerts a pull on air and water. This pull is greatest at the equator, where it sends whirlpools to the right in the north, and to the left south of the equator. Whether the pull of the earth determines the dextro- or laevo-rotation of plants is as yet an unproved speculation. I have long opposed it, though there is some evidence to support it. It is said that *Dioscorea batatas* twines to the left and *D. alata* to the right in the Rio de Janeiro Botanic Garden, and from my own experience I know that *D. batatas* twines to the right and *D. alata* to the left in Philadelphia greenhouses. But one must be careful in the identification of species in this genus, for it is known to have closely related species which differ in their twining, some being left-handed and some right-handed. Several twiners are known to turn to the left in both Philadelphia and Rio de Janeiro, among them are *Ipomoea*, *Delichos* and *Thunbergia*.

It is of interest to the student of protoplasm to realize that different species within a genus may differ in the direction of their twining, and that whereas most plants turn to the left, some few, notably in the genus *Dioscorea*, turn to the right, and some are indifferent.

The spiral tendency in organisms is a basic quality of the protoplasm (SEIFRIZ 1933 b). That this is true is supported by the torsion which protoplasm undergoes while growing (KAMIYA and SEIFRIZ 1953); and also in the general occurrence of spirality in chromosomes (Fig. 3), a now widely accepted fact (KAUFMANN 1926, IWATA 1940, MATSUURA 1938, GEITLER 1938).

Dextro- and laevo-symmetry in organisms may be attributed to certain basic molecules in protoplasm, possibly to the twist in the carbon atom or to the now accepted spiral form of protein molecules (Fig. 2).

That spiral growth is a heritable mutation (DE VRIES, page 561, 1911) attributable to a single gene (PLAGGE 1938, BOYCOTT *et al.* 1930, STURTEVANT 1923) is now accepted by some observers. Its occurrence in the snail *Limnaca* has been used as an illustration of maternal inheritance. If *D* is the dominant gene for dextro-spirality and *d* the recessive allelomorphic gene for laevospirality, then if a *DD* female is crossed with a *dd* male, all $F_1$ progeny have shells with a righthanded twist, and if a *dd* female is crossed with a *DD* male, all $F_1$ progeny have shells with a left-handed twist. If, now, crosses are made among the respective $F_1$ offspring, the $F_2$ progeny will have the genotype of the mother (Da); all are

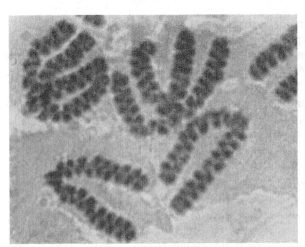

Fig. 3. Spiral chromosomes. (MATSUURA, heretofore unpublished.)

righthanded spirals. Mendelian segregation is not observed until the third generation when the *dd* allelomorph present in the females of the $F_2$ generation is revealed in the $F_3$ generation as a left-handed spiral. We are thus able to attribute three qualities to the spiral habit: it is located in a single gene, it shows Mendelian segregation, and it is an example of maternal inheritance.

## 3. Cellular plan.

Cellular organization is frequently cited as a necessary property of life. It is inconceivable that a cell can carry on the multiplicity of reactions which go on within it without a specific organization. We are forced, however, to confess that very little of this organization is visible.

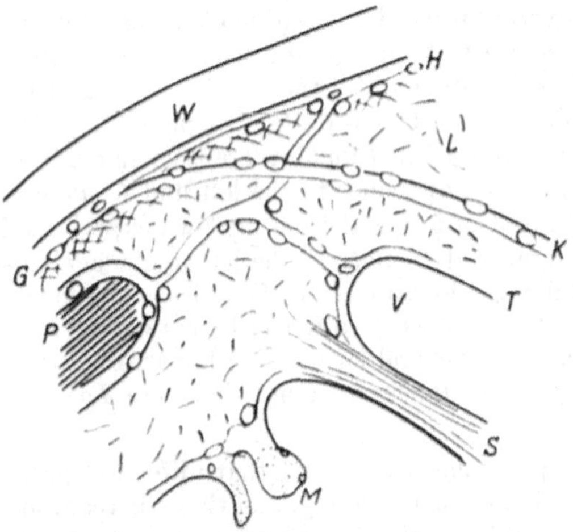

Fig. 4. The plan of a cell: *W* cell wall; *H* ectoplast or protoplasmic surface film; *G* plasmagel layer or cortical endoplasm; *L* plasmasol or liquid endoplasm; *K* kinoplasm; *T* tonoplast; *V* vacuole; *S* transvacuolar strand; *M* myelin processes; *P* plastid (G. SCARTH).

In the main, mature cells of the higher plants possess a large central vacuole. Surrounding it, pressed against the cell wall, is a layer of cytoplasm. It consists of an inner granular plasm, an outer layer of clear hyaloplasm, and a surface membrane. In the plasmodium of a slime mould the hyaloplasmic layer may be very broad (LEWIS 1942). Delicate strands of protoplasm often traverse the vacuole (Fig. 4). The position of the nucleus is sometimes permanent, as is true of the centrally located nucleus of *Spirogyra*.

There is a further indication of cellular organization in the polarity which is evident in cells.

SCARTH (1942) has given a description of gross cellular organization which is best described by a figure (Fig. 4).

# 4. The fine structure of protoplasm.

Turning now to the structure of protoplasm, that which is microscopically visible, we are immediately confronted by two facts: first, that much of what we see is of little significance—the fat droplets, for example, are important as stored food but play no active role in the mechanisms of life. The second somewhat discouraging fact is that when the particle which we see is obviously of great vital importance, as is the nucleus, it looks very much the same in all cells, though we know that certain basic differences must exist. Nucleus, fat droplets, vacuoles, alveoli, plastids, mitochondria, microsomes, and other granules, dispersed in a moderately viscous, hyaline, ground substance, is the usual picture of protoplasm wherever found.

Fig. 5. The protoplasmic emulsion flowing out of an egg of *Fucus*.

## The protoplasmic emulsion.

Viewed under the low power of the microscope, protoplasm appears to be a suspension of granules. These granules were called *microsomes* and were by ALTMANN (1890) presumed to represent the ultimate living units of protoplasm. He compared them to bacteria. The cell thus became a colony of minute, living granules.

The idea that the smaller protoplasmic inclusions are bacteria appears again and again in biology. Mitochondria were once so regarded and were therefore assumed to have a certain autonomy of their own apart from the cell. The rapid motion of the rod-shaped mitochondria in fibroblasts suggests that these inclusions get about quite independently. One could say the same for the fat droplets in plant cells but here autonomous motility is out of the question. Mitochondria may be living entities in the same sense as are chromosomes but they are not bacteria nor do they possess autonomy of motion; they are propelled by protoplasmic forces (SEIFRIZ 1953).

When protoplasm is viewed under high magnification it is seen to contain many globules; some are of fatty substances and some are small vacuoles filled with liquid (Fig. 5). Even the smallest of the particles, the "granules", were found by SPEK (1924) to be liquid, for they fuse on coming into contact with each other. In short, the visible structure of protoplasm is an emulsion. That this emulsion is fundamental and to be regarded as the basic structure of protoplasm has been advocated many times and in several forms. The most renowned of these speculations was BÜTSCHLI's alveolar hypothesis.

## The alveolar structure.

BÜTSCHLI (1894) observed a symmetrical honeycomb or checkerboard pattern in protozoa (Figs. 6 and 8). He regarded this structure as fundamental. It seemed to be made up of tightly compressed globules which, owing to pressure,

become angular, their geometrical shape being that of dodeca- or tetrakaidecahedra. How alveolar protoplasm results from an emulsion under pressure is
indicated in Fig. 7, where A, schematically represents the usual protoplasmic
emulsion; B, alveolar spheres; C or D alveolar protoplasm; and E another possible

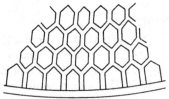

Fig. 6. Alveolar protoplasm in the ciliate
*Euplotes.*

orientation which, in F, readily leads to a false
picture of chromatin granules on a linen thread.

The criticism that has been directed against
the alveolar hypothesis cannot be based on the
contention that the structure is an artifact.
This criticism arose from work by HARDY and
H. FISCHER (1899), who showed that an alveolar
structure can be produced in gelatin by treatment
with certain reagents. It was, therefore, main-
tained that the structure is artificially produced in protoplasm. This is not
true, as Fig. 8 reveals. BÜTSCHLI's error did not lie in the observation of artifacts but merely in claiming a universal occurrence and fundamental value
for the alveolar structure
in living material.

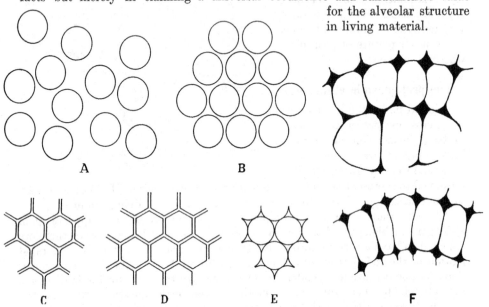

Fig. 7 A—E. Five possible arrangements of the globules of an emulsion.    Fig. 7 F. A false picture of chromatin
granules on a linen thread
(CHAMBERLAIN).

The alveoli of protoplasm are probably minute vacuoles, for vacuoles are very
abundant in protoplasm, more so than is usually realized. The alveolar structure is
but one form of an emulsion (SEIFRIZ 1930).

## The emulsion hypothesis.

It was once remarked (WILSON 1925) that the visible picture of protoplasm is
a coarse representation of that which lies beyond; therefore, to describe what one
sees rather than speculate on what one could not see is good science. Those
young biologists who first dared, as the physicists and chemists had long since
done, to look beyond the microscope, were called severely to account by their
elders, who continued to regard protoplasm as an emulsion. One must, in all
fairness, grant that there are some sound reasons for doing this.

CLOWES (1916) found a striking analogy between the behavior of emulsions and the permeability of the plasma membrane. He developed an ingenious hypothesis of cell permeability based on the behavior of emulsions. Loeb had found that sodium chloride increases and calcium chloride decreases the permeability of cells. There was, therefore, an "antagonism" between the two. When present in proper proportions there is no toxic effect. This proportion is that which exists in such physiologically balanced solutions as sea water and blood. CLOWES found that sodium (the hydroxide) causes the formation of emulsions of the oil-in-water type, whereas calcium (the chloride) brings about reversal to the water-in-oil type. Sodium and calcium, in the proportion of 100 molecules of the former to 1 or 2 of the latter, which is approximately the proportion in which these elements occur in sea water and in blood, balance one another. The mixture has no effect on emulsions. CLOWES viewed the plasma membrane as a fine emulsion. When oil is dispersed in water, the membrane is more permeable to water-soluble substances, i.e. salts, when water is dispersed in oil, the membrane is less permeable, except to fat solvents. The former condition is presumably produced by sodium, and the latter by calcium. Normally, said CLOWES, the

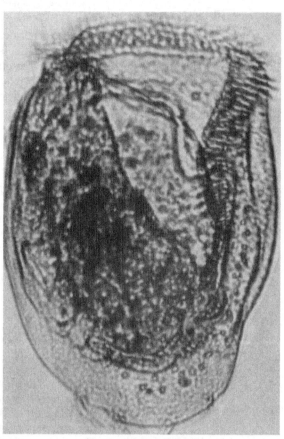

Fig. 8. The ciliate *Euplotes.*

protoplasmic emulsion is in a state of equilibrium near the reversal point, for it is bathed in a balanced solution. It is, therefore, readily thrown one way or the other by a change in concentration of the salts in the surrounding medium.

The best support of CLOWE's hypothesis is the fact that most oil emulsions are at the reversal point when sodium and calcium are present in the proportion in which they occur in physiologically balanced solutions. The hypothesis met with adverse criticism on the grounds that there is no evidence that the protoplasmic membrane is a fine emulsion, and that it is the hydroxyl ($OH^-$) anion rather than the sodium cation which reverses and holds emulsions in the oil-in-water state (SEIFRIZ 1923). In reply to these criticisms, DIXON and CLARK said that an electrical stimulus affects emulsions in the same way as it does protoplasm. An electric current will cause an emulsion, originally almost impermeable to ions and water-soluble substances, to become fairly permeable, which is the same effect that electric stimuli have on living tissues; viz., they increase permeability. DIXON and CLARK, therefore, concluded that an hypothesis of the structure of the plasma membrane which explains two such apparently unconnected and remarkable phenomena as the antagonistic action of certain ions on permeability (the hypothesis of CLOWES) and the permeability changes produced by electric stimuli (the work of DIXON and CLARK) deserves serious consideration. This is true, yet it may simply mean that two rather diverse types of systems (an emulsion and a living jelly) show similar responses to certain environmental changes.

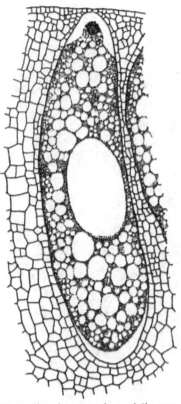

Fig. 9. Vacuolate protoplasm of the egg cell of *Ceratozamia* (C. J. CHAMBERLAIN).

We are forced to discard the emulsion hypothesis of permeability control in spite of two substantial facts in its support, because of the following reasons. There is no direct evidence whatever of phase reversal in the visible protoplasmic emulsion. It is very unlikely that protoplasm could exist as a living substance if fat were the continuous phase—metabolic reactions take place in aqueous media. As the stability of an emulsion increases with decrease in size of the dispersed particles, owing to a great increase in the surface tension of the stabilizing membrane, the ultra-microscopic protoplasmic emulsion (if it exists) would be extremely difficult to reverse.

The biologists were at first supported in their emulsion hypothesis of protoplasmic structure by the chemists. WO. OSTWALD (1919) was of the opinion that gelatin, and jellies in general, were fine emulsions, from which notion came the term "emulsoids". His evidence was the fact that the viscosity of both emulsions and protein solutions increase with increase in concentration of the dispersed phase. The "emulsoid" concept of gel structure never was accepted by able chemists. HATSCHEK (1917) finally dispelled all doubt in proving that jellies are not "emulsoid", i.e., emulsion-like, in structure. He mathematically analyzed the theory of gels as systems of two liquid phases, and concluded that "the theory that gels consist of two liquids must be pronounced untenable". ELLIS and DONNAN, previously pointed out that an oil emulsion is a model suspension colloid.

In denying the notion that protoplasm in its finer, fundamental, structure, is an emulsion, I in no way deny the obvious fact that in its coarser, visible structure protoplasm is an emulsion. There is no question as to the reality of the visible protoplasmic emulsion; it may assume any one of a number of forms (Fig. 7), from the symmetrical arrangement of uniform alveoli postulated by BÜTSCHLI (Fig. 6 and 8), to the scattered distribution of vacuoles of unequal size such as occur in many egg cells (Fig. 9). A very striking example of the protoplasmic emulsion is to be seen in Spirogyra (SEIFRIZ 1931), where the emulsion is an exceedingly fine one, the two phases of it being almost indistinguishable owing to insufficient optical differentiation. This difficulty is overcome by using dark-field illumination.

Biologists have long held that there is a continuous framework of some sort which is the structural background of protoplasm. Life in a dispersion (solution) of isolated units, no matter how complex the mixture, is inconceivable. This theoretical concept received support from observations on fixed and stained material where there could be seen a structure variously described but in each instance consisting of a meshwork or entanglement of fibers forming a three-dimensional net or sponge. The idea of continuity in protoplasmic structure is thoroughly sound and is supported by ample evidence, though much, yet by no means all, of the cytological support given to it is faulty.

## The fibrillar hypothesis.

The fibrillar hypothesis, advanced by FLEMMING and others, ascribes to proto-plasm the structure of an entanglement of fibrils. FLEMMING elevated these

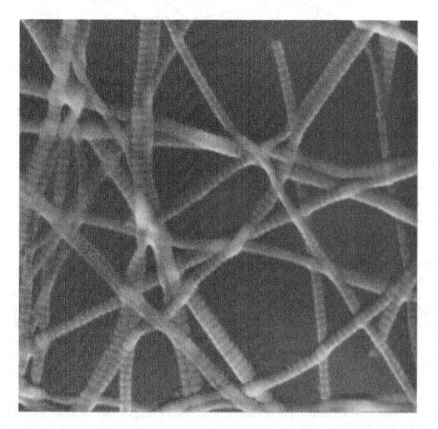

Fig. 10. Electron photograph of collagen (cowhide) × 40,000: chrome shadowed (R. BORASKY).

*fibrillae*, as did ALTMANN his granules, above the lowly station of mere structural units and viewed them as the seat of the energies on which life depends. The drawings of FLEMMING of connective tissue, of HEIDENHAIN of muscle and spinal ganglion cells, and of preparations by STRASBURGER depict a fibrillar structure. Such a structure is characteristic of, and visible, in certain living tissues. Electron microscope photographs show the construction of sinew to be that of an aggre-gation of molecular fibers (Fig. 10).

### The reticular hypothesis.

Linear structural units may be oriented so as to form an entanglement such as exists in a brush heap, or they may be arranged in a more orderly manner in the fashion of a three-dimensional net. Earlier controversies centered around these two possibilities.

The reticular, fibrillar, net and spongelike structures seen in fixed protoplasm are true fibrous coagula but may involve an emulsion caught in the coagulum. For example, chromatin granules on a linin thread, a structure that has played a prominent role in modern cytological and genetical theories of nuclear behavior, is readily reproduced by a distorted emulsion of large globules with a minimum of dispersion medium. The granules would then be the interstices where several globules approach each other, and the linin thread would be the connecting

Fig. 11. A brush heap of fibrous molecules.

strands of the dispersion medium. This is shown by comparing a typical drawing of chromatin material (Fig. 7 F) showing chromatin granules on a linin thread, with one or more of the possible configurations of an emulsion.

So far we have dealt with fibers in fixed protoplasm. Though these may be artifacts, yet coagulated material may still tell something of the true and normal structure. Thus, when blood clots the coagulum is seen to be fibrous. The same is true of coagulated latex and milk. The fibrous structure in the clot indicates the probability of a similar structure in the unclotted material. Tendon, muscle, and nerve are visibly fibrous; the fibers may be dissected out. There is every probability that the visible fibers are formed of finer fibrillae. The likelihood of protoplasm being fibrous in its molecular structure is indicated by the properties which are common to it and other material known to be fibrous. Among such properties are elasticity, tensile strength, and non-Newtonian viscosity. A specific example will illustrate my supposition.

Of two soap solutions, one of low concentration and low viscosity, and one of high concentration and high viscosity, the former was elastic and the latter not; the former held a small metal particle in suspension, whereas the latter could not support the same particle. It would seem, therefore, that the elastic yet thin soap solution possessed a structure that would account for its elastic qualities and for its ability to support a metal particle, and that the thicker yet inelastic soap lacked such a structure. This supposition was supported by microscopic examination. The elastic soap solution contained long, slender crystals, whereas the other soap resembled chalk dust. We have in the behavior and structure of these two soaps the basis of all generally accepted hypotheses of the structure of jellies. Elastic colloidal systems are built of linear crystalline units. Their intermeshing gives elasticity and rigidity to liquids which yet flow freely. This is structurally possible if we regard the framework of fibers as not fixed but labile, capable of readjustment, comparable to a loosely put together brush heap (Fig. 11).

Investigations on the structure of cellulose give an insight into modern interpretations of the mechanisms underlying the behavior of colloidal jellies, including protoplasm.

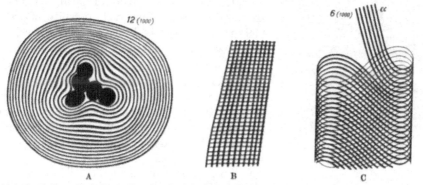

Fig. 12 A–C.  A. Layering in the epidermis cell of the fruit shell of *Ocimum*.  B. Cross striations in a cell wall. C. Six banded spiral in a *Dipteracanthus* mucilage tube (CARL v. NAEGELI).

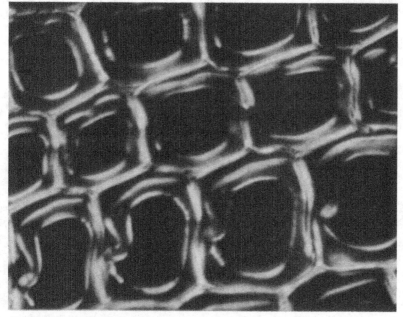

Fig. 13.  Birefringence of cell walls seen through crossed nicol prisms. The anisotropic layers are brilliantly illuminated. The isotropic layers are dark; the dark layers may, however, be composed of anisotropic material viewed perpendicular to the axis of the fibers (R. D. PRESTON).

## Natural cellulose.

Historically, the polypeptide chain takes priority over the linear cellulose molecule, but it was the latter which led directly to modern concepts of the molecular fiber. Early botanists anticipated the wood fiber by reasoning along the lines followed by WILLIAM BRAGG, who, wishing to convince his audience of Nature's wisdom in using fibrous molecules for construction, said, tall trees are able to withstand heavy wind pressures, giving but not breaking, because wood is built of fibers. The most obvious evidence of this fact is the economic

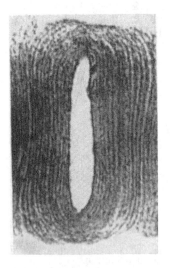

use to which certain plants are put; *Linum*, *Musa*, *Agave*, and numerous other genera yield long and strong fibers from which rope and fabric of high tensile strength are made. Botanists have long accepted the fibrous nature of natural cellulose as an uncontroversial fact, yet, in spite of the evidence, some few, as PRESTON (1948) says, have remained adamant (BAILEY and KERR 1935).

We must first recognize that cellulose is not a chemical entity but a family of substances. The nearest approach to a pure cellulose in nature is cotton fiber, which is 90 per cent "alpha" cellulose. Among the less highly polymerized celluloses is the intermediate product, hemi-cellulose, present in young plant tissues.

Fig. 14 A. Lamellae in a cotton fiber, swollen in cuprammonium hydroxide: the layered structure is of the secondary wall in a 25-day fiber: × 950 (C. HOCK).

Cellulose is laid down in the cell wall by protoplasm. The structure resulting NAEGELI (1864) first fully described. He referred to a striated structure of lamellae (A. Fig. 12), the crossing of the striae at an angle of 45⁰ (B. Fig. 12), and the spiral twisting of cellulose threads (C. Fig. 12). All these observations have been substantiated by later work involving different methods of research. Maceration and swelling reveal the lamellated structure of natural cellulose (Fig. 14 A).

Fig. 14 B. Segmented striae in a plant cell wall, in dark-field.

Fig. 14 C. Segmented striae in fossil coal, in dark-field (R. THIESSEN).

The concentric layers in cotton fibers indicate the time required to deposit the cellulose, for there are as many layers as there are days which the fiber requires to reach maturity.

That the secondary wall of the cell is lamellated is vividly substantiated by polarized light. The lamellae are seen to be of optically different material, isotropic layers alternating with anisotropic ones (Fig. 13).

The lamellation observed between nicol prisms is a relatively coarse one. Even NAEGELI, with simpler optical equipment, saw a finer lamellation (Fig. 12)

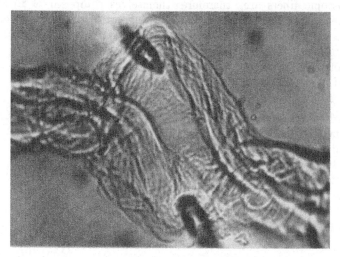

Fig. 15. Fibrillae in macerated wood, separated by micro-needles (C. HOCK).

Fig. 16. Electron photograph of cellulose fibers in *Valonia* (R. D. PRESTON).

which is well brought out by dark-field observations, both in new wood and in coal (Fig. 14 B and C). There is also an indication of segmentation, *i.e.*, of micellae, in the dark-field picture of the cell wall and coal (Fig. 14 B and C).

Maceration (RITTER 1928) and microdissection (SEIFRIZ and HOCK 1936, and HOCK 1942) reveal a fine fibrillar structure of wood (Fig. 15). The electron microscope shows this structure to persist down to molecular dimensions (Fig. 16). The microscopic fibers are, therefore, themselves composed of fibrils just as they, in their turn, form larger fibers. The smallest of the fibers is the cellulose molecule, a chain of anhydrous glucose rings joined by an oxygen bridge, each glucose ring being the mirrored image of its neighbour, *i.e.*, it is rotated through 180° (Fig. 17 C).

The molecular pattern of cellulose was arrived at by a series of postulates which began with SPONSLER. In the first spatial model of cellulose (SPONSLER and DORE 1926) the glucose residues were supposed to be linked alternately by 1,1-glucosidic and 4,4-ether-like bonds (Fig. 17 A). HAWORTH (1929) then modified this model (Fig. 17 B), MEYER and MARK (1928), by taking into account chemical

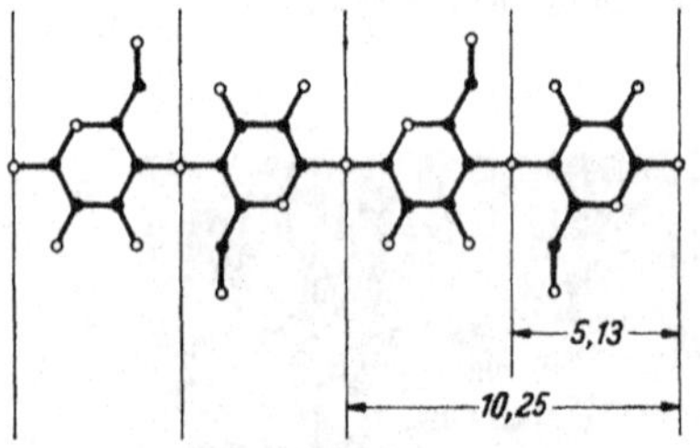

Fig. 17 A. Three dimensional model of the cellulose chain (O. L. SPONSLER and DORE).

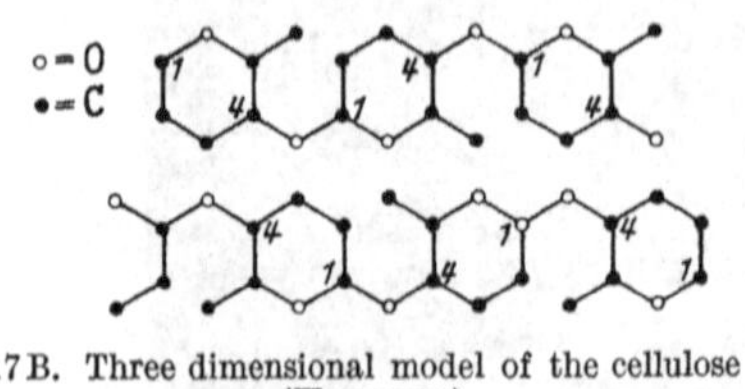

Fig. 17 B. Three dimensional model of the cellulose chain (HAWORTH).

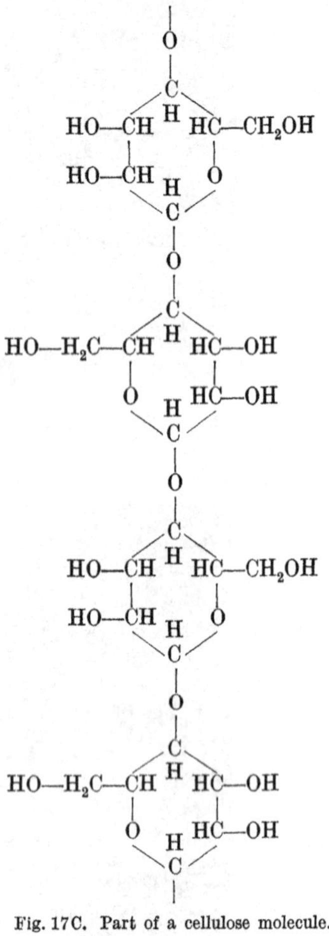

Fig. 17 C. Part of a cellulose molecule.

as well as X-ray data, proposed a monoclinic unit cell (Fig. 18) with the dimensions: $a = 8.35$ Å; $b = 10.3$ Å; $c = 7.9$ Å; $\beta = 84°$. The foregoing structure of the cellulose molecule has been challenged by numerous authors for various reasons.

The cellulose chain varies between 500 and a 1000 or more Å in length, which means at least 100 glucose units in each chain. KRAEMER (1938), on the basis of molecular weight determinations, established a possible maximum length of the chain at 1.7 $\mu$ which would mean over 2000 glucose units. A molecule, such as this, of varying size and weight, was unknown to classical chemistry.

A property of natural cellulose, widely recognized, is its *micellar pattern*. NAEGELI's *micellae* (Fig. 19 A), or molecular aggregates, when postulated by him for colloidal jellies, were subjected to severe criticism, but they have now become well established as a structural feature of cellulose. If by micel is merely meant a form of molecular aggregation, then the micellar structure may be universal, for homogeneity is rare in any system, heterogeneity is the rule, certainly in natural products. In surface films, where one would expect to find a uniform distribution of molecules, JOLY (1950) finds aggregation, leaving the film full of holes. The surface aggregates are micels of a kind.

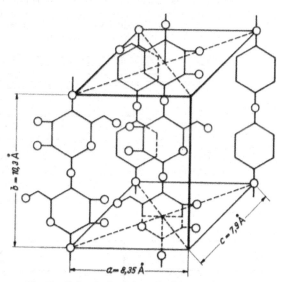

Fig. 18. Unit cell of cellulose (K. MEYER and H. MARK).

In cellulose, the micels were at first presumed to be distributed like bricks in a wall (Fig. 19 B). NAEGELI's picture of the fine structure of the cell wall indicated that bricks in a wall is the probable orientation. There arose, however, much difficulty in regard to this orientation. There were no intermicellar bonds. Within the micel there are several types of bonds (a and b, Fig. 19 B) one of which, within the chain, is a primary valence bond. The question, what holds the micels together, was answered by assuming the presence

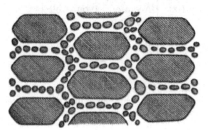

Fig. 19 A. NAEGELI'S concept of micels in the cell wall.

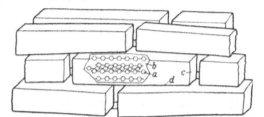

Fig. 19 B. A possible distribution of micels in natural cellulose: a, b, c, and d indicate intermolecular forces.

of a cementing material ("Kittsubstanz"). The tensile strength of cellulose would then depend as much on the intermicellar cement as on the intramicellar cellulose. The unlikelihood of such a situation soon became apparent (SEIFRIZ 1934). The optical observations of FREY-WYSSLING (1940) added further evidence for the reality of micels. The problem then became one of interpretation, and the postulation of a structure which will satisfy all conditions: the parallel alignment of long molecular threads, their aggregation into fascicles such as to produce optical heterogeneity, the presence of some form of intermicellar bond which will account for high tensile strength yet not obliterate the segmented appearance, and the maintenance of sufficient crystallinity to account for the double refraction of cellulose.

An interesting and early postulate was that of a "brush-heap", a structure which met certain conditions very well, notably the elastic properties of jellies. But in a brush-heap there

Fig. 20. Micels in a brush-heap.

is no evidence of micellae or aggregation. This was allowed for by the plan pictured in Fig. 20. A random distribution of micellae may occur in organic substances of low tensile strength such as regenerated cellulose or cellophane. For natural cellulose and silk fiber which are of high tensile strength, an arrangement as in A Fig. 21 would be ideal. The long molecular chains would be arranged in an ordered fashion parallel with the fiber axis, with pronounced overlapping. This arrangement provides the maximum possible strength in the direction of the fiber axis, as the mutual cohesive force of the long chains is thus fully utilized. To rupture the fiber it is necessary to cause the chains to slip past one another against this cohesion. But there would be no clear-cut termination of a micel, or any micels at all for that matter. This difficulty was settled by the plan shown in Fig. 21 B.

Fig. 21 A and B. A Overlapping cellulose fibers. B Overlapping fibers forming micels.

Cellulose in the form of flax, displays an excellent orientation of micels parallel to the fiber axis, and has a strength comparable to the best steel. Tensile strength values agree well with X-ray analyses of symmetry, as the following table from MARK (1943) shows.

| | kg. per sq. mm. | | kg. per sq. mm. | | kg. per sq. mm. |
|---|---|---|---|---|---|
| Best steel . . . . . | 170 | Cotton . . . . . . | 28 | Acetate, rayon ordi- | |
| Copper wire . . . . | 40 | Silk . . . . . . . | 35 | nary . . . . . | 18—20 |
| Cast iron . . . . . | 20 | Flax over . . . . . | 100 | Acetate, rayon well | |
| Aluminium . . . . | 10 | Viscose, ordinary . | 25 | oriented . . . . . | 60 |
| Lead . . . . . . . | 3 | Viscose, well oriented | 80 | Rubber, ordinary . | 15—20 |
| Wood, in the direc- | | | | Rubber, well oriented | 60 |
| tion of the fibers. | 8—15 | | | | |

The problem is greatly helped through consideration of the crystalline properties of cellulose. That such properties are present is indicated by the birefringence of cellulose (Fig. 13) and the excellent spectrograms which are obtained (Fig. 22 A). Particularly striking are the distinctions, such as PRESTON (1948) has found between the cellulose walls of closely related species (Fig. 22 B).

Crystals possess a structural unit known as the *elementary* or *unit cell*. The space occupied by the unit cell is the smallest parallelopiped within a crystal that still has the properties of the material as a whole (Fig. 18). From the volume of this parallelopiped, the density of the material, and AVOGADRO's number, it is possible to calculate the number of molecules in the elementary or unit cell of a crystal. In the case of sodium chloride, it is four. When this method came to be applied to organic material (SPONSLER 1926) it was found that the number of anhydrous glucose groups ($C_6H_{10}O_5$) associated with the unit cell was very small, namely four. It was, therefore, concluded by HESS (1928) that the molecular weight of cellulose must be low, as originally assumed by others. But organic chemists found it very hard to reconcile the known properties of cellulose with low molecular weight. Later studies all indicate a high molecular weight. FREUDENBERG (1932) estimated, on the basis of demolition kinetics, that there are fifty glucose residues to a chain and therefore the molecular weight is 8,100. STAUDINGER (1932) from viscosity determinations judged the molecular weight of purified cotton to be 190,000 with 35,000 for rayon. HAWORTH and MACHEMER (1932) estimated a mean molecular weight of 30,000. Determinations by STAMM (1930), by the cenrifuge method of SVEDBERG, indicate a value of 40,000 which is a likely average. Certainly the cellulose molecule must be a large one, but as identical X-ray diffraction patterns are obtained from materials which unquestionably have fairly low molecular weights, the unit cell of cellulose must be small. The contribution of SPONSLER (1926) consisted in demonstrating that there is no inconsistency between long chain molecules and a small unit cell.

According to MARK the unit cell for native cellulose has the dimensions 8.35 A.U., 10.3 A.U., 7.9 A.U., whereas SPONSLER gave 10.7 × 10.3 × 12.2. These seemingly different values really amount to the same thing. The identity period along the fiber is the same in both, namely, 10.3, and this agrees with the dimensions of the cellobiose molecule. The elementary cell of cellulose consists, then, of four anhydrous glucose groups (two anhydrous cellobiose molecules) each pair of which is one link in a cellulose chain. A model of a unit cell would preferably contain 10 $C_6H_{10}O_5$ groups, *i.e.*, two glucose rings at each corner and two in the center of the cube; but as each unit cell, when entirely surrounded by others, shares each vertical edge with three other unit cells, then the number of $C_6H_{10}O_5$ groups which can be allocated to each individual unit cell is four, *i.e.*, two on one edge, and the two on the center axis which are not shared.

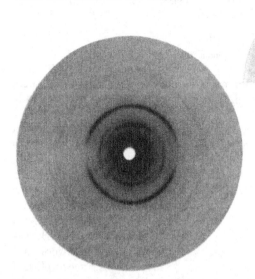

Fig. 22 A. X-ray diagram of the wall of *Valonia*, beam parallel to the direction of the more abundant set of cellulose chains. CuKα radiation, slit 0,5 mm, diameter specimen film distance 3 cm.

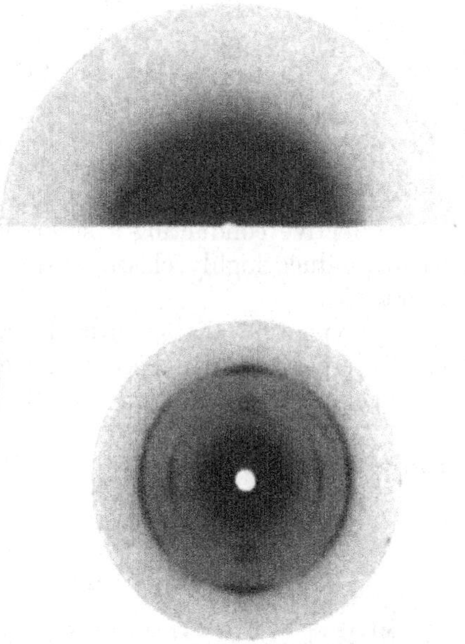

Fig. 22 B. X-ray patterns of *Spongiomorpha* (above) and *Cladophora* (below): note striking difference in crystallinity (R. D. PRESTON).

The story of cellulose applies, in a more or less modified form, to many organic compounds. Thus, the chitins, no matter what their source, give essentially the same X-ray photographs as do the celluloses, which indicates that chitin consists of analogous chain molecules, based on the acetyl-glucosamine residue (Fig. 23).

The dimensions of the unit cell of acetyl-glucosamine are, according to MEYER (1942):

$9.40 \times 19.25 \times 10.46$ Å (fiber-axis).

This means that the length of the residue in the direction of the fiber-axis is 5.23 Å as compared with 5.14 for the $\beta$-glucose unit in cellulose, which is further evidence of the close similarity in size and form of structural units among organic substances.

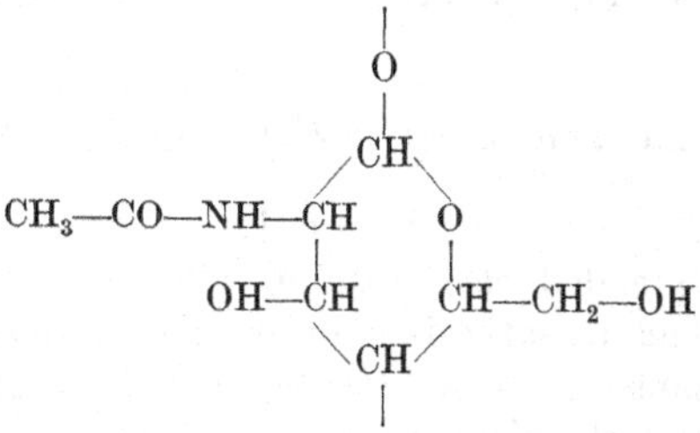

Glycogen has, like starch, yielded only amorphous photographs.

$$CH_3\!-\!CO\!-\!NH\!-\!\underset{\displaystyle |}{CH}\quad O$$

Fig. 23. The acetyl-glucosamine residue (ring) of the chitin chain molecule.

Knowledge of the structure of cellulose, chitin, and rubber lead to a better understanding of polymers, but protoplasm is basically a protein complex; it is only through the proteins that we can arrive at an understanding of the structure of tissues, of muscle, nerve, sinew, and protoplasm.

## Proteins.

There is no more familiar term in chemistry than *polypeptide chain* (Fig. 24). In proportion as the polypeptide chain grew did protein chemistry grow. So well established has the concept of chain molecules become in protein and polymer chemistry, that it would be difficult to think of either without the linear molecule. The molecular chain has, however, had its opponents with much to support their view. The protein fiber has held its ground so well because so much can be done with it. One can twist fibers into threads and form a rope or tendon; one can weave them into a living fabric, wrap them into long bundles and obtain

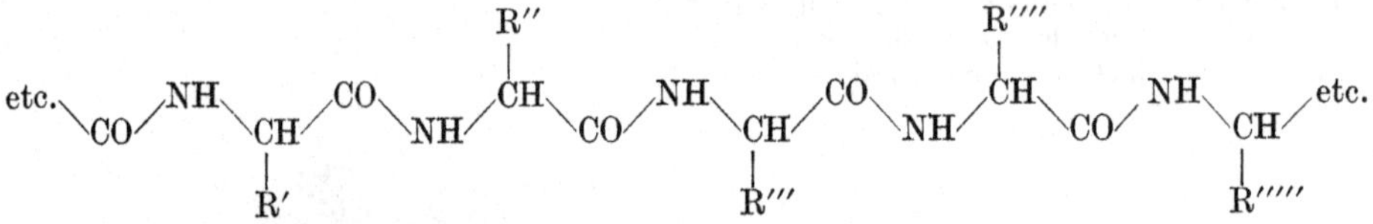

Fig. 24. The polypeptide chain.

thereby effective conducting systems for nerve impulses; or fold and coil them so as to produce highly elastic material. None of these systems can be made with granules.

The polypeptide chain, with the amino acid radical as a link and lateral bonds for cross ties, can become an exceedingly complex structure, when the simple backbone of a protein (Fig. 25) becomes a molecule of insulin (Fig. 26).

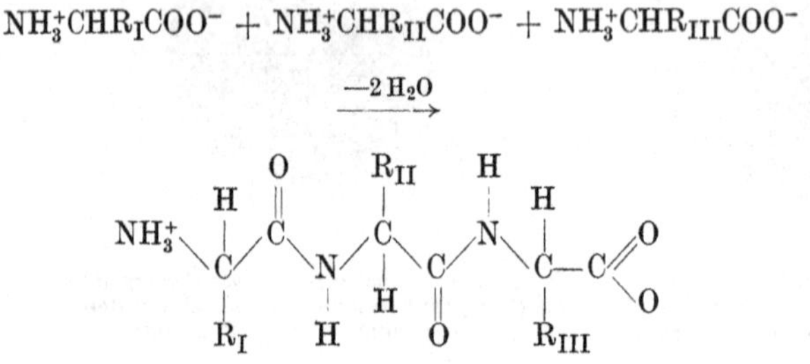

$$NH_3^+CHR_ICOO^- + NH_3^+CHR_{II}COO^- + NH_3^+CHR_{III}COO^-$$

$$\xrightarrow{-2H_2O}$$

Fig. 25. A peptide chain, the backbone of a protein.

Though the concept of a polypeptide chain dates from the classical researches of EMIL FISCHER (1906), and though excellent work was done by his followers, not much was accomplished until SPONSLER (1928), MEYER and MARK (1928), and others (ASTBURY 1934) applied cellulose concepts to proteins. A polypeptide chain is made up of several hundred α-amino acid residues. In the case of silk fibroin, these are mostly the two simplest α-amino acids, glycine $NH_2$—CH—COOH

with H below,

and alanine $NH_2$—CH—COOH. When the chains are stereochemically fully

with $CH_3$ below,

extended each residue will occupy a length along the fibre-axis of 3.5 Å.U. This hypothesis is in excellent agreement not only with the pattern and dimensions of the X-ray photograph of silk fibroin, but also with its general physical and chemical properties. MEYER and MARK (1928) suggested that the residues of glycine and alanine follow each other alternately in the polypeptide chains of silk.

That the polypeptide chain is not usually a fully extended one, nor even always a partially extended one, is now generally agreed (Fig. 27); thus, it may be a tight coil, but that it is a sphere (SVEDBERG 1926—33) or a hollow sphere (WRINCH 1937), and remains so is not accepted as a primary state.

The controversy between the linear and the spherical protein molecule finds a partial solution in the coiled chain, which originated in the helical molecule. A variety of these have been suggested varying chiefly in the nature of the intramolecular bonds. Such a bond could be the now familiar hydrogen bond (ASTBURY 1940, HUGGINS 1945). A hydrogen-bonded helical configuration of the polypeptide chain would probably involve resonance between the carbon-oxygen and carbon-nitrogen positions. Of five possible arrangements of the amino acid

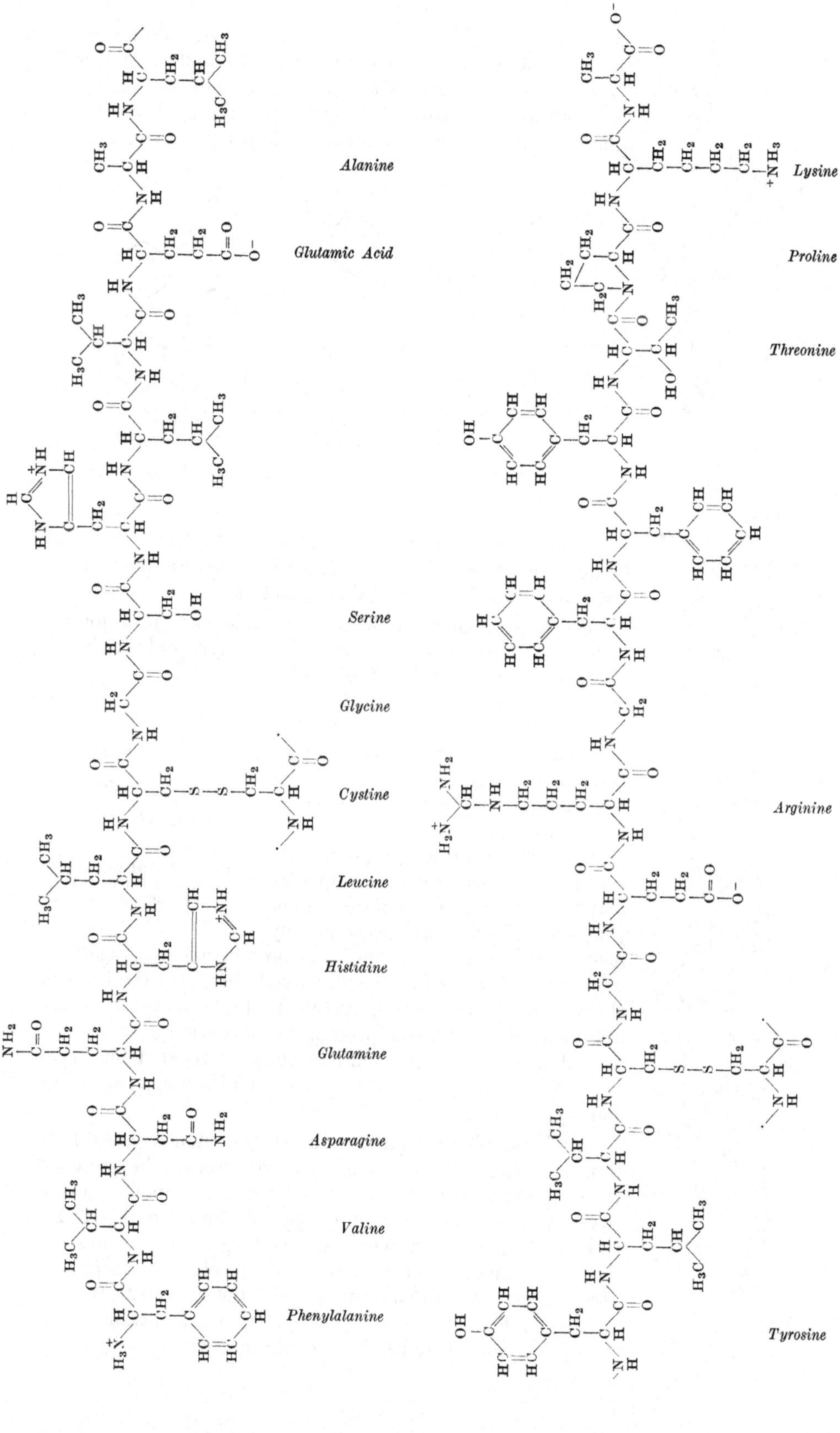

Fig. 26. A peptide chain from the protein molecule of insulin (K. U. LINDERSTRØM-LANG).

Fig. 29. The helical pattern of a protein molecule (L. PAULING and R. B. COREY.)

residues, two are assumed to satisfy basic assumptions, namely a helix with 3.7 residues per turn, and a helix with 5.1 residues per turn. The 3.7 residue helix is thought to represent the structure of α-keratin, α-myosin, hemoglobin, other globular proteins, and synthetic polypeptides; and the

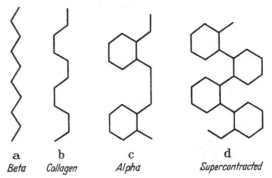

| a | b | c | d |
|---|---|---|---|
| Beta | Collagen | Alpha | Supercontracted |

Fig. 27. Suggested schema for the 4 principal states of fibrous proteins (W. T. ASTBURY).

5.1 residue helix represents the super-contracted form of keratin and myosin (PAULING *et al.* 1951). Several possibilities are represented in Figs. 27 d, 28, and 29.

The super-contracted helical molecule may thus form a globular body though yet retaining the polypeptide chain.

The protein chain has fulfilled an important function and there is no need to discard it, even though we may have to admit that it is not always present and that it may become spherical or itself be built of spherical units. In any case, a denial of the protein chain is no longer possible in the face of electron micro-photographs which show the molecular fiber (Fig. 10). What X-ray spectrography made possible in earlier studies of protein structure, electron microscopy is doing for present day studies. On the basis of the earlier X-ray studies, ASTBURY (1934) stated that the structure of all fibrous proteins must be analogous—they are all based on long chain-molecules aggregated into submicroscopic, imperfectly crystalline bundles lying parallel or spirally inclined.

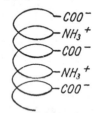

Fig. 28. The helical molecule of a protein ion at the isoelectric point (K. MEYER).

An X-ray photograph of ordinary unstretched wool differs from the same wool when stretched; on releasing, the stretched wool rapidly recovers its original length and simultaneously the first X-ray photograph re-appears (compare Fig. 22 B). Hair, nail, spine, horn, whalebone, etc., give rise to substantially the same X-ray diffraction photograph, which is that of keratin. ASTBURY (1943) continues with the statement, that we are dealing here with a reversible intramolecular transformation of the fiber substance, *keratin*; this protein normally exists as an isomer (α-keratin) which can be changed into a longer modification (β-keratin) simply by pulling on it. This is the mechanical basis of the remarkable long-range elasticity of mammalian hairs: the mechanism is that of a molecular spring, "a conclusion that is surely of the deepest significance in the study, not only of hairs, but of the movements of all sorts of living things". Possibly the primary contribution of ASTBURY is the manner in which he assembled the polypeptide chains into a grid. "We must also conclude from the analysis of numerous X-ray photographs in the light of a mass of physico-chemical data that the main poly-peptide chains, which run lengthways along the fibre, are linked side-to-side by interactions and combinations between their respective side-chains (the R-groups in the general formula given above), so as to build up, as the fundamental keratin unit, a kind of polypeptide 'grid'" (Fig. 30).

ASTBURY correlates these conclusions with the problems of physiologists. "All X-ray studies of protein fibre structure are only by way of appren-ticeship to a much more serious task, that of elu-cidating the molecular mechanism of muscular activity. Muscle, to be sure, has been photographed frequently enough by X-rays—the pioneer efforts are again due to HERZOG and his colleagues—but the immediate problem now is to discover the structure of the muscle protein *myosin*, which has been shown by BOEHM and WEBER (1932) to be mainly responsible for the diffraction pattern given by muscle itself, whether living or dead. The first thing one notices about the X-ray photo-graph of muscle is its striking resemblance, not only in general arrangement of spots and arcs but also in actual dimensions, to that of un-

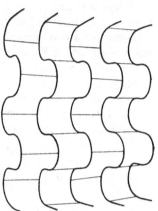

Fig. 30. Skeletal model illustrating a polypeptide "grid" of keratin or myosin in the *a* configuration. (W. T. ASTBURY).

stretched mammalian hairs, which suggests at once that the main-chains of the myosin molecule are in a folded or α-configuration." ASTBURY (1937) con-cludes these „adventures of the X-ray analyst" by pointing out that nature has used the general plan which he describes for making many proteins. He says: "The principal intra-molecular periodicities of keratin, both longitudinal and lateral, closely resemble those found in the crystal structure of unaltered pepsin. This agreement is not at all so fantastic as it seems at first sight, for SVEDBERG'S work (1934) points strongly to a uniformity of size among the fundamental protein units.—This earlier conclusion of ASTBURY'S did not remain as harmon-ious as he anticipated, for SVEDBERG'S units are globular, not fibrous, in form. SVEDBERG (1933) assumed a spherical form for all protein molecules because such a shape best satisfied his results in ultracentrifugation. One must, therefore, tentatively accept the possibility that sedimentation constants and molecular weights suggest a spherical shape for certain protein molecules. But it is not the spherical molecules alone for which SVEDBERG'S work is known. He postulated a scheme of simple multiples for the molecular weights of the proteins. Though he termed the scheme "rather fantastic" and "hesitated somewhat" to advance it, his previous experiments on ultracentrifugal determinations of native proteins supported it. One gets the impression, he says, that protein molecules are built up by successive aggregations of definite units.

"The number of iron atoms per molecule has been calculated on the assump-tion that the iron content is equal for all the proteins containing hemin and that the molecular weights are simple multiples of 34,500. The remarkable agreement between the observed and calculated values makes this hypothesis highly probable." The steps between the molecular weights become larger and larger as the weight increases. Especially the gap from 2,500,000 to 5,000,000 is very striking. No protein molecule has so far been found in this large interval. Svedberg had previously suggested (1926, 1933) that the basic unit of protein structure is 17,600, or about half of 34,500, and that corpuscular proteins in solution fall into groups that are multiples of 17,600. Egg albumin, insulin, pepsin, zein, Bence Jones protein, etc., all have molecular weights of roughly twice this number; haemoglobin and serum albumin have four times; edestin, excelsin, and other seed globulins sixteen times, and so on, up to weights of millions. Furthermore, molecules of higher weight can be split into units of submultiple weight.

There was an historical background to Svedberg's reasoning, for just as the benzene ring had played its role in aromatic chemistry so could some analogous intramolecular conformation play its role in protein chemistry. It was this molecular group which Svedberg hypothecated.

Following Svedberg, Wrinch (1937a) gave support to the globular proteins in another way. By an ingenious mathematical analysis she constructed hollow spheres using a socalled "cyclol bond" involving peptide residues. According to the theory, peptide chains are capable of uniting to form two- and three-dimensional nets, three peptide residues forming a ring as in II, Fig. 31.

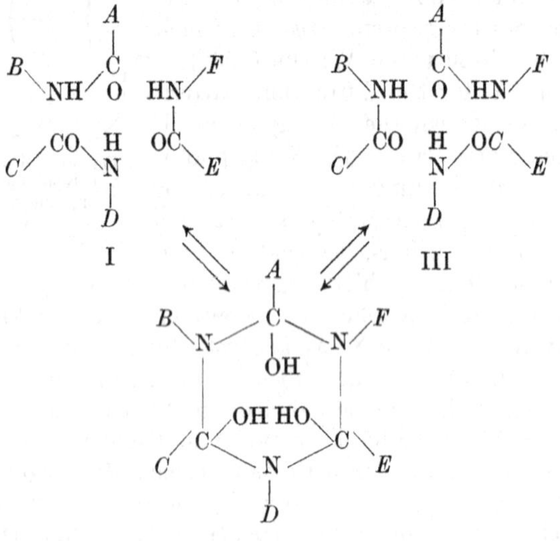

Fig. 31. Peptide residues capable of forming 2 and 3-dimensional nets (D. Wrinch).

$A$, $C$, and $E$ represent residues of the type $R \cdot CO \cdot NH \cdot CHR$—; $B$, $D$, and $F$ are residues of the type $R \cdot NH \cdot CHR \cdot CO$—. A hollow sphere results (Fig. 32).

Numerous doubts arose as to the possibility and the usefulness of the cyclol theory. Mathematically it could not be attacked. Only by chemical analyses and structural considerations could a basis for criticism be found. The first doubt was based on structural difficulties. Mechanical properties, such as elasticity and tensile strength, would be difficult of interpretation. To attempt to satisfy these physical properties with a spherical molecule is rather like asking a weaver of cloth to make his fabric of sand instead of threads. Furthermore, a spherical protein molecule in solution will mean Newtonian behavior, yet if there is one firmly established property of many protein solutions, it is their anomalous flow.

Wrinch replied to some of these criticisms by pointing out that cyclol fabrics have patches of "stickiness" in virtue of particular arrangements of R-groups containing such groupings as $-COO^-$, $-NH_3^+$, $-CO-NH_2$, $-OH$, $C=O$, $=NH$, and $\equiv N$. "The presence of many non-protein molecules such as carbohydrates, phosphatides, etc. in cytoplasm, which presumably fulfill important functions in virtue of their positions in the organized structure, suggests that the native protein surfaces must have a very large capacity to anchor foreign molecules. In this connection, it is worth attention that a completely cyclized lactim residue in the form $=N-CHR-C(OH)=$ has as many sites as the linear form $-NH-CHR-CO-$."

The WRINCH (1937b) theory of the hollow sphere is not without merit, but the SVEDBERG globular protein molecule leaves more leeway for harmony between the linear and the spherical hypotheses. If the spherical molecule can be unwrapped a fiber will result. The final blow to the cyclol bond came from the chemists who stated that compounds such as II in Fig. 31 are unknown, and a consideration of the thermal data shows that it is improbable that they should occur in considerable quantity in equilibrium with free peptides. As observation of the optical activity showed no such

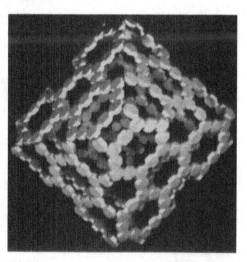

Fig. 32. A hollow spherical protein molecule (D. WRINCH).

reaction, either in neutral aqueous solution or in solutions of the chlorhydrates or of alkali salts, cyclol formation does not occur.

The answer to the problem of the relation between fibrous and globular proteins was indicated by earlier speculations on coiled molecular chains. Research on the rubber molecule was a material factor in this direction. And now we have good chemical evidence on the conversion of globular to oriented fibrous proteins (NUTTING, et al. 1944), which indicates that both fibrous and globular proteins are constructed to a common plan, or at least that some common factor is involved in their method of synthesis. BERGMANN and NIEMANN (1937, 1935—36) state that the number of kinds of amino-acid residues in a protein molecule and the number of each of the various residues are expressible in mathematical form. On this basis, the minimum molecular weights of chicken egg albumin, cattle haemoglobin, cattle fibrin, and silk fibroin are found to correspond to 288, $2 \times 288$, $2 \times 288$, and $9 \times 288$ residues respectively, which is equivalent to saying that chemical analysis places not only two globular proteins but also two fibrous proteins in one and the same scheme of multiple molecular weights, namely, the SVEDBERG scheme, founded on ultracentrifugal studies of the globular proteins.

BANGA and SZENT-GYÖRGYI (1940) at one time arrived at a very satisfying conclusion in saying that when nature wishes to build or to give tensile strength, she uses the fiber, as in the muscle protein myosin, but when she wants a nutritional protein she uses one with a spherical molecule such as albumin. But it seems that this view must be modified, for spherical molecules may be converted into linear ones, and the latter may become spherical by coiling.

NUTTING *et al.* (1944), working with artificial protein fibers, state that globular proteins, such as casein and those from soy beans, can be unfolded into extended polypeptide chains, a fact which "has been demonstrated with a dozen globular proteins".

## Micels.

A number of the problems involved in the study of cellulose reappear in the proteins; one of these is the question of micels. They were originally postulated by NAEGELI for gelatin and protoplasm. FREUNDLICH (1932) accepted them as widely present throughout the organic world. He thought, as I have said, that heterogeneity, not homogeneity, is the law of nature. Molecular aggregation is to be expected. LOEB (1922) was of the contrary opinion. He admitted the presence of micels in gelatin only during setting when scattered molecular aggregates are formed, but as the gelatin sets the micels are eliminated, with the result that in the solid jelly there is a homogeneous distribution of molecules. This could be true and may represent the condition in a clear soap jelly, but only rarely are soaps clear jellies. Soap curds are obviously micellar. One ordinarily does not associate micels with the crystalline state but RILEY and OSTER (1951) describe them in crystalline ribonucleic acid. Here the micels are small having but seven molecules each, therefore comparable in size to the unit cell of crystals.

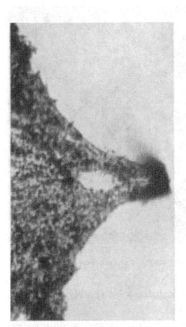

Fig. 33. The elasticity of protoplasm.

## Elasticity.

One cannot work long on protoplasm without realizing that the living substance is at times highly elastic (SEIFRIZ 1926), and possesses other related properties such as torsion (KAMIYA 1954). One may demonstrate the elasticity of protoplasm by tearing it with micro-needles (Fig. 33), or by watching a severed strand snap back. Very striking is the recoil of a long and much stretched pseudopod on a fibroblast in culture when torn loose from the substratum. The biologist must explain the elasticity of protoplasm just as the physicist or polymer chemist must explain the elasticity of gelatin (FREUNDLICH and SEIFRIZ 1923), wool, or rubber. This problem has been a most interesting one, especially in the case of rubber where the stretch is sometimes 800 per cent.

The simplest possible interpretation of a structural framework which will account for elasticity in protoplasm and polymers in general is a brush-heap of fibers (Fig. 11). MEYER (1942) states that all biological systems with high and reversible elasticity are a net of fibrous molecules, the plan of which he suggests in Fig. 34. What MEYER here does is put the old cytological reticulum theory on a molecular basis. He also, as did TEORELL (1935), interprets the selective anion permeability of cells, *e.g.* erythrocytes, in terms of a net of basic protein chains.

There is rather general agreement among the polymer chemists that elasticity in organic systems indirectly involves cross linkage. The function of the cross links is to strengthen the material and suppress plastic flow. "A textile fibre having long-range elastic characteristics consists of flexible chains reinforced with a proper balance of cross links." It is well established that in wool the disulfide groups, of the amino acid cystine, form cross links between molecular chains. The same situation exists in rubber; cross linkage changes raw rubber into highly elastic vulcanized rubber.

Cross links, notably absent in cellulose where there does not seem to be any place to fasten a lateral tie, are certainly present in abundance in proteins, otherwise our understanding of the protein chain would be a false one. Cross linkage has been made use of to explain elasticity (FREY-WYSSLING 1940, SEIFRIZ 1920), continuity in structure (SEIFRIZ 1936), and the non-Newtonian behavior of protoplasm (SEIFRIZ 1952). In the face of so much evidence it is surprising to read: "Cross linkage is never found in tissue with an active metabolism . . ." (PRYOR 1952). The statement is too sweeping. One must differentiate between covalent cross linkages such as disulfide bonds, and the much weaker but still very important cross linkages due to hydrogen bonding (Fig. 35). The presence of a large number of covalent cross linkages is certainly unfavorable to rapid movement within

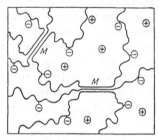

Fig. 34. The submicroscopic structure of protoplasm; *M* micel; ⊖ bound anions; ⊕ free actions (K. MEYER).

tissues. Structures, like keratin, which have a great many such linkages are relatively inert. However, myosin contains cystine disulfide linkages, although far fewer than keratin. Probably most contractile tissues such as muscle, contain cross linkages due to hydrogen bonding. It is true that if myosin, in the alpha form, is to be described by an alpha helix (PAULING and

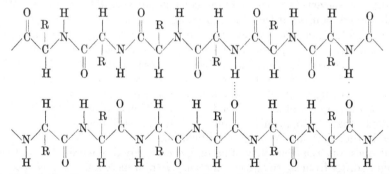

Fig. 35. Hydrogen bonding. A hydrogen atom is shared by amino acids in each of the parallel polypeptide chains. The figure illustrates the general plan of cross linkage in the parallel alignment of protein fibers.

COREY 1951) then the hydrogen bonds from the peptide linkages are all within the chain and not cross linkages between chains. However, apart from the peptide chain itself, there can be, and almost certainly is, hydrogen bonding between some of the charged and polar groups of the amino acid side chain groups attached to different peptide chains. Although little has been rigorously proved in regard to the extent or nature of the cross linkages in such structures as muscle, there must be some cross linking to maintain the general stability of the structure, but they would not be numerous enough or strong enough to interfere with the capacity of the contractile elements in the tissue to coil and uncoil rapidly.

## Coacervates.

While fibers held the attention of the polymer chemists, the colloid chemists were busy with systems called *coacervates*, in which fibers played no part, and yet these systems are capable of activities which heretofore were attributed only to fibrous systems.

The distinguishing characteristic of the colloidal state of matter is its oneness in the kind and magnitude of the charge of its suspended particles; the distinguishing characteristic of a coacervate is its dual nature. Among lyophobic colloidal suspensions, the particles are either all negative, as in gold, or all positive, as in lead. To mix the two means to precipitate them. But in a coacervate the particles are of opposite charge. When coming into contact they neutralize each other, partially at least if not fully, but no precipitate is formed. Such behavior indicates that the suspended particles are hydrated organic matter. The $p_H$ value is critical. A good coacervate may be made with gelatin and gum arabic at a $p_H$ below 4.8. Gum arabic is negative at all $p_H$ values whereas gelatin is positive below its isoelectric point, which is at $p_H$ 4.8. Electrophoretic experiments show that coacervate particles may be positive or negative, but for optimal coacervation the droplets should be uncharged. BUNGENBERG DE JONG (1929, 1935, 1936a, 1936b) coined the word "coacervate" to define a system in which there is a delicate balance between solvation and dehydration, between solvation and electric charge, between repulsion and attraction, and the co-existence of apparently contradictory properties such as the insolubility in water of a fluid system with an aqueous medium. It is these properties which make coacervates similar to protoplasm.

BUNGENBERG DE JONG (1936) points to a number of analogies between coacervates and protoplasm, such as: the coacervate is a colloid-rich liquid which is not miscible with the equilibrium liquid: vacuoles appear in the coacervate in response to a rise in temperature or when an electric field is applied; oil drops are taken up by coacervates just as living cells engulf such drops. BULL (1943), on the other hand, points out that there are some obvious differences between coacervates and protoplasm, thus, coacervates are true liquids in that the rate of flow is Newtonian, i.e., it is proportional to the applied force, which is not true of protoplasm. Coacervates also show no preferential permeability. This latter criticism is merely one way of saying that coacervates are not as versatile as protoplasm. They do, however, resemble protoplasm in some respects, and in a general way it may be said that protoplasm is a coacervate; it may also be said, as has often been done, that protoplasm is a colloid. The two statements are true but misleading in that they are an over-simplification.

One questions the feasibility of naming each newly discovered "colloid" which is found to differ from those already known. We now have not only "coacervates", but "complex", "compound", "autocomplex", "bicomplex", and "tricomplex" coacervates. DERVICHIAN (1949) warns against the misuse of these terms.

When positive and negative colloidal particles of unlike chemical nature, such as a protein and nucleic acid, attract each other by electrical forces and join to form larger phase units, fusion being prevented by hydration layers on each particle, then a *complex* coacervate results; physically this is a soft fluid gel. If one kind of colloidal particle in such a system, for instance the positive ones, are replaced by suitable cations, we have an autocomplex coacervate. But why these distinctions when the simplest coacervate, the standard gelatin-gum arabic system is already bicomplex?

### Rubber.

A discussion of rubber might seem out of place in a chapter on the structure of protoplasm, yet the two substances have much in common, more than do cellulose and protoplasm, and it was work on cellulose which laid the foundation for the stereochemistry of the proteins and of protoplasm. Furthermore, rubber, like cellulose, is a secretion product of protoplasm.

The identity of the cellulose fiber is no longer questioned. The reality of the protein, and protoplasmic, fiber is at times in doubt, but at other times it is real beyond question. The rubber molecular fiber is also not well established but it is accepted as a possible and plausible structural unit, possessing great length, extraordinary recoil, great adjustability, crystalline when oriented, and "fluid" when relaxed. The qualities of rubber are in so many ways the qualities of protoplasm that a knowledge of one must inevitably help in an understanding of the other.

Part of the rubber problem is the nature of bonding. One needs in rubber, as in protoplasm, a bond sufficiently fluid to permit ready shifting yet give a fair amount of rigidity. The bonds must be capable of changing from a loose to a firm grip. Several types of bonds will satisfy these requirements: already mentioned are resonating bonds, and the hydrogen bond (Fig. 35). Particularly adapted to rubber is the "fluid joint" (Fig. 36) (TRELOAR 1943, 1947, HUGGINS 1946). The bond in a folded polypeptide chain has many drawbacks, among them the fact

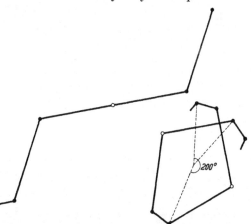

Fig. 36. A possible structure of polyoxyethylene (polyethylene oxide), $(—CH_2—CH_2—O—)_n$. A projection of a portion of the helical chain is shown at the lower right. Above and to the left is the arrangement obti. ned by unrolling the helix on to a plane; this shows the different pitch of the C—C and C—O—C portions of the chain (M. L. HUGGINS).

that the angle of the bond cannot be changed; therefore folding, as is so readily accomplished in pleated material, is out of the question.

The fluid bond first arose in rubber chemistry because it was necessary to find a mechanism which will permit an extensibility of 900 percent. The problem of extensibility involves all elastic polymers; the answer may not however be the same in each elastic system.

Among substances composed of long chain molecules, such as polyethylene, in which X-ray diffraction data reveal crystalline arrangements, some (HUGGINS 1946) have a planar zigzag arrangement of the atoms in the chain and others do not. In the former, when a chain is extended as far as possible, the expected bond distances ($\sim 1.53$ Å for C—C) and bond angles ($\sim 110^0$ for $<$ C—C—C) are maintained. In contrast, the chain bonds in the molecule of polyoxyethylene, or polyethylene oxide $(—CH_2—CH_2—O—)_n$, are not all alike. On stretching, the molecules open, after the manner of a spiral (Fig. 36).

One generally accepted view of the chain-like nature of the rubber molecule is the Weber model:

$$—CH_2—C(CH_3) = CH—CH_2—CH_2—C(CH_3) = CH—.$$

GUTH (1943) pointed out that the restoring force in stretched rubber, like the pressure exerted by a gas, is associated with varying entropy of the material, rather than with changing internal energy. He postulated single long flexible molecules as representing the structural elements of all rubber-like materials. The long polyisoprene chain in natural rubber is a typical flexible molecule which can be brought into a great many configurations of essentially equal internal energy by twisting it at the single carbon bonds. Such a molecule, abserved under thermal agitation, is likely to be found in a highly twisted or coiled state. In an ideal flexible molecule all configurations would have exactly the same energy. Any model of a rubber molecule must permit formulating a quantitative theory of the elastic and thermo-elastic properties of the bulk mate-rial; for instance, it must give an adequate idea of the form of the stress-strain curve.

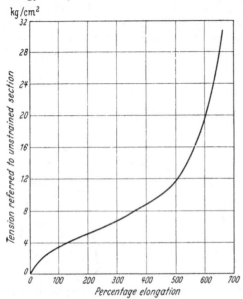

Fig. 37. Force extension curve for vulcanized rubber (L. R. G. TRELOAR).

Fig. 37 is a typical stress-strain curve for rubber, with the char-acteristic double curvature. The last upturn of the curve is, eventually, accentuated by crystallization, but it begins well before the onset of crys-tallization, and must be accounted for. The curve is different from the simple linear form which one would expect from a single molecule. To understand the curve, one must have a theory of the structure of an assemblage of molecules as well as of the individual molecule. This is also the problem which confronts the protoplasmologist, he must have a suitable protein molecule and he must assemble them so as to explain the physical, particularly the rheological (SEIFRIZ 1952) properties of protoplasm. A possible interpretation for rubber is given by TRELOAR (1947) in Fig. 38.

JAMES and GUTH (1944) point out that a network of molecular chains makes up a large part of bulk rubber, but it is not satisfactory as a model, for a three-

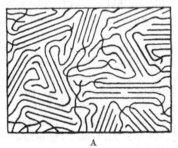
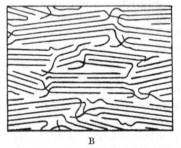

Fig. 38 A and B. Molecular structure of rubber: parallel bundles are crystallites; *A*. unstretched, *B*. stretched.

dimensional network of idealized flexible chains, stretching through space between points, is in thermal agitation. The lateral forces exerted inside rubber will be essentially the same as those inside a liquid of rubber-like molecules, except at some relatively few firm junctions in the network. The molecules will, in the main, jostle and move past each other, exerting on each other forces similar to those in a liquid, in effect, a hydrostatic pressure.

In such a model of rubber there is a coexistence of fluidity and rigidity, which is a characteristic of all non-Newtonian colloidal systems including protoplasm. When the mathematical expression of the theory is plotted with an experimental stress-strain curve, the two are found to coincide. The same is true of the theoretical and experimental curves of entropy.

The specific behavior of rubber molecules and the fact that every polymer of high molecular weight has the possibility of being rubber-like, has led chemists and biologists to compare muscle and protoplasm to rubber. One would not designate the state of aggregation of muscle as liquid, but it is certainly not solid. Physiologists have made the observation that microscopic worms can perforate muscle fibrillae and slip through without leaving any hole or other damage subsequently visible. The viscosity of muscle determined by the damping of a tuning fork, corresponds with that of a very viscous liquid.

The distinction between relaxed and contracted muscle is much like that between relaxed and contracted rubber. In both, the relaxed form is "amorphous" in that the main valency chains are not fully extended, and are connected with neighboring chains by liquid bonds which shift easily and determine consistency. On contraction, crystallization takes place, a fact confirmed by X-ray studies for both rubber and muscle.

The distinction between relaxed and normally contracted muscle MEYER (1942a) attempts to elucidate in terms of rubber chemistry. He says: "In artificially stretched muscle there is a reorientation of structural units, from an orientated and improbable into a more probable original state." Active contraction is however of quite another nature. "The contracted muscle behaves as a normal solid body: the 'liquid' linkages are converted into 'solid' linkages, which can only be ascribed to a chemical change." Just what is meant by a chemical change is not said.

That contribution of rubber chemistry to the rheology of protoplasm which is of most interest to us is the fluid bond. It not only permits stretching rubber 900% but it converts protoplasm from the static mechanical framework postulated by a biophysicist or stereochemist, into a dynamic mechanism.

The fluid bond is well described by TRELOAR (1947, 1949) and GUTH (1943). Apparently it was HALLER who first had the idea that thermal vibrations would lead to distortion of bond lengths and angles, as a result of which a chain-like molecule would not be expected to be perfectly stiff and straight. *As bond lengths and angles must be at least approximately maintained*, it seemed likely that the thermal energy would lead to larger vibrations and rotations in a plane normal to the main valence chain than in the direction of the chain. *i.e.*, that *rotations about bonds* (Fig. 36) should *take place much more freely than deformations of bonds or valence angles*.

## Protoplasm.

The extent to which the foregoing concepts can be applied to protoplasm is the problem which confronts us, it is one of the great problems in science. Biologists have long been aware that protoplasm is in a state of delicate balance which, when upset, results in death. Something of this delicate balance is found in coacervates. The analogy is in many ways close. Colloidal aggregates in protoplasm are both acidic and basic and may, therefore, be of both positive and negative charge. We can, however, with equal satisfaction, turn to other systems. Thus, the mechanism of reactions among high polymers is a complicated affair. It is not the weight or chain length of the molecule which complicates matters, but branching, cross linking, and transfer. These qualities are lacking in coacervates.

If any one of the elementary steps in a poly-reaction, such as polymerization, is considered, it will be found to be a relatively simple process. Thus, when —OH

and —COOH, or —NH₂ and —COOH join, water is split off and an ester or amide
link is formed (FLORY 1944). The activation energy involved is between 10,000
and 15,000 cal. per mol. The collision probabilities are of the usual order of
magnitude, $10^9$ to $10^{11}$. The functional end groups react independently and the
efficiency of collisions between them is amenable to statistical treatment. There
is, therefore, as FLORY (1944) and MARK (1943) make clear, nothing extraordinary
in any one stage of a poly-reaction to which we can attribute the extraordinary
reactions known to occur in living matter. When the total picture is viewed
something of the complexity of the situation is seen. *Polymorphism* is one feature
of this complexity.

Diamond and graphite are examples of polymorphism; another is the four
different stable lattice arrangements of the normal paraffins which differ only

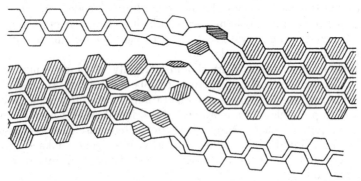

Fig. 39. Cellulose chains which, when in parallel alignment, form micels or crystallites (above left and below right).
The same chain may leave a micel and pass through a disordered, amorphous, or intermicellar area (center) before
entering another micel (H. MARK).

in the angles between chains, and the crystallographic directions; α- and β-keratin
is still another example. Two such modifications are known to occur in guttapercha, and three, possibly four, in cellulose. Still another form of polymorphism
is that existing in the relaxed and the stretched forms of rubber. In the unstretched state a diffuse X-ray diffraction ring is obtained; upon stretching a typical
fiber, or crystalline, pattern is obtained. By measuring the identity periods, it is
possible to distinguish a variety of elastomers, *e.g.*, rubber and polyisobutylene.
This would not be easy by chemical analysis. MARK (1943) calls attention to the
use to which X-ray patterns can be put in the qualitative analysis of cellulose.
He illustrates this in Fig. 39, in which are pictured crystalline and amorphous
areas with one and the same cellulose chain passing from a crystalline through an
amorphous to a second crystalline region. There are thus degrees of crystallization and degrees of randomness in cellulose, rubber, and other high polymers
some of which cannot be brought into a fully crystalline state.

These are only a few examples of polymorphism of which the number is
infinite. J. B. LEATHES (1926) gives a striking analysis of the complexity of
protein structure. He takes "a very simple case": suppose the protein is a chain
of only fifty links; if all the links are different, the number of possible permutations is $\lfloor 50$, a collossal value. But if we reduce the number, and if one link recurs
ten times, 4 recur four times, and 10 recur twice, then the number of possible
permutations will be, $\dfrac{\lfloor 50}{\lfloor 10} \times (\lfloor 4)^4 \times (\lfloor 2)^{10}$. In such a protein of 50 links, of
which only 19 are different, the number of possible arrangements of its parts will
be $10^{48}$. Light takes 300,000 years to travel the length of the Milky Way. This
distance, expressed in Angstrom units, of which 100,000,000 equal a centimeter,

will be less than $10^{32}$. It is thus clear how great the variations in disposition of the parts of a protein molecule may be and how far we are from being able to map out such a structure.

The search of the biophysicist for a nonliving system which possesses some of the properties of the living substance is our reason for this discussion on the properties of colloidal matter, coacervates, polymers, and elastomers. In the past, many of us have laid great emphasis on fibers as the structural units of elastic organic systems, of cellulose, proteins, and protoplasm. It is now difficult to readjust one's thinking and to find in non-fibrous material a mechanical basis for elasticity and such vital processes as muscular contraction, nerve conductance, and proto-plasmic streaming. The older ideas that a brush heap is elastic and a sand pile not, that a fabric or a rope has tensile strength and a starch paste not, that a nerve fiber will conduct an impulse with greater ease and rapidity than will a non-fibrous system, are all sound concepts, but some hypotheses based on a mechanism of fibers may not be true (SCHMIDT 1928, FREY-WYSSLING 1949, 1953, GOLDACRE 1950, LOEWY 1949). That protoplasmic streaming involves the contraction and expansion of folded polypeptide chains now seems unlikely (SEIFRIZ 1953).

Viewing an activity of protoplasm as one in which fibers are not involved, does not mean repudiating all evidence of the existence and use of fibers in living systems. No one, who is fully informed, questions the fibrous structure of wood and of muscle and nerve. Possibly certain fibers are but temporary, being linear aggregates of spherical particles. This SZENT-GYÖRGYI (1947) believes to be true of muscle, yet the fibers are real and function as such. A segmented structure may actually be seen in collagen fibers (Fig. 10).

Having long emphasized the fibrous and elastic nature of living matter, I have been occasionally disturbed by the complete absence of elasticity or any indication of a fibrous structure in protoplasm. A needle may traverse the protoplasm without great hindrance; moving not as through a liquid but as through a soft paste. I assumed the condition to be pathological, but the protoplasm later returned to normal activity exhibiting the customary elasticity. I am now of the opinion that protoplasm, when plastic, i.e., devoid of elasticity, is quite normal, but the state is neither an active nor usual one; active streaming, so characteristic of healthy slime mold protoplasm, is reduced to a minimum and in places wholly absent. Such a state is probably assumed by slime mold protoplasm shortly before sclerotizing or before spore formation. In short, protoplasm can be paste-like, devoid of elastic qualities, and free of fibers.

Fiber formation is undoubtedly a transient process in protoplasm, being of greater or lesser permanency depending on the tissue, relatively permanent in muscle, nerve, and tendon, relatively temporary in the plasmodium of a slime mold.

It is one of the enigmas of living matter how protoplasm can flow, churn, and mix so thoroughly, apparently with utter disregard to any structural plan, and yet retain an organization which is not only specific for life, but for each species, if not for each cell.

The question now arises, is there any material in protoplasm which is peculiar to it? Most substances isolated are available in pure form in the chemical laboratory. To be sure, most of them are prepared from tissues, yet, gradually more and more of them are capable of synthesis from the chemical elements.

The search for some protein constituent of protoplasm which can be singled out as the basic protein of life, is an old one, but it continues. HANSTEIN distinguished between the active living protoplasm and the passive, lifeless "metaplasm". SACHS called the former "energid" and the latter "energid products".

The term "metaplastic" (or "paraplastic") still persists for metabolic products which are generally recognized as lifeless. Other biologists speculating on this distinction have wandered far into the philosophical field by postulating special vital bodies which give to protoplasm the properties of life. BUFFON and VERWORN conceived of gigantic molecules termed "biogens" which were supposed to be the life-giving elements of protoplasm, the rest of the material being presumably nonliving. SPENCER postulated "physiological units"; and ALTMANN, "bioblasts". In this category, though referring more particularly to hereditary units, belong the "gemmules" of DARWIN, the "pangens" of DE VRIES, the "plastidules" of HAECKEL, the "biophores" of WEISMANN, and the "genes" of modern geneticists. NAEGELI's "idioplasm" theory although speculative like the others, has the advantage of resting on a chemical foundation which is in harmony with certain known facts. He attributed hereditary traits to specific molecular aggregations. This led directly to his micellar theory of protoplasmic structure. LLOYD and SCARTH (1927) brought back into use STRASBURGER's old terms of "kinoplasm" or active plasm, and "trophoplasm" or nutritive plasm. BENSLEY (1942, 1943) has isolated extractions from liver, one of which he called *plasmosin* and presumed it to be a protein of vital significance to the cell. LOEWY (1952) has isolated a myosin-like substance from slime molds which he regarded as a "morphoplastic" system, *i.e.*, a system capable of structural changes under the influence of adenosin triphosphate and related nucleotides.

All such attempts are worthy ones, because they add to our knowledge of the chemical constitution of protoplasm, but we cannot yet assume that any one protein complex is of greater importance than another (BENSLEY 1942, 1943).

## Protoplasmic inclusions.
### Particles.

The particles in protoplasm may be studied morphologically (GUILLIERMOND 1941), chemically, or as parts of mechanisms. Here I wish to speak of cell particles as indicators of mechanisms.

All particles in protoplasm, move and appear sometimes to possess autonomous motility. This is as true of fat droplets as of mitochondria. For example, the color granules in *Stentor igneus* make sudden comet-like or saltatory journeys and often glide forward as if moving in channels which guide them. They travel as individuals, at an estimated rate of $2^1/_2$ m$\mu$ per second. The entire interior of *Stentor* is at times active with movements, the particles resembling a crowd in which the individuals are going in diverse directions, stopping and starting, and all this activity goes on without evident movement of the protoplasm.

The trajectories of the particles in *Parafelliculina amphora* are both straight and sharply curved. Particles $^1/_4$ to 3 m$\mu$ in diameter move 10 to 100 times this distance in single trips, at the rate of 2 to 3 m$\mu$ per second.

TOSESUKI describes the locomotory movement of the nucleolus in *Salvinia natans*. The nucleolus may revolve 90 degrees on either axis. Occasionally, two nucleoli move in different paths at different rates.

MAX SCHULTZE called attention to the flowing, gliding, and crawling of granules within the plasm of filose rhizopods, some going in while others are coming out, and this may occur in the smallest of threads. ANDREWS (unpublished information) has described similar movement of particles in *Formanifera*. When granules meet they may go around one another or, after a pause, one may follow the other, or both resume their original directions. In wide pseudopods granules may swarm about like people in a crowd, at times halting and trembling, but at length moving to the end of the pseudopod. Often the granules are still, and then, on resuming movement, reverse their direction.

A significant observation is the movement of particles in staminate hairs *independent of protoplasmic flow*. Whereas commonly all granules move at about the same speed, stop for a while, and then continue in the same, or in the opposite direction, granules may pass without check or they may meet head on, stop, and then the one follow the other (Fig. 40).

HEIDENHAIN's observations are particularly apropos here. He (1911) says that for many minutes together the cytoplasm may remain quiet while granules pass through it. Many granules may move *in rows* in one direction, or go in opposite directions. (Note the many rows of particles in the broad band of cytoplasm in Fig. 40).

DEGNER observes similar movements in the pseudopodia of shrimps. The granules move outward and inward, in parallel rows, in the same direction or in opposite directions; some grains may leave a row and move to another row to travel in the opposite direction while the grains left behind continue on as before. Particles in adjacent rows going in the same direction, do not all go at the same rate. DEGNER concludes: "currents of plasm cannot be so diverse and complex as to explain the movements seen."

After making similar observations relative to the movement of particles of whatever nature in plant cells in culture, I (SEIFRIZ 1953) advanced the hypothesis that the propagation of forces *identical to those which traverse nerve fibers* are respon-

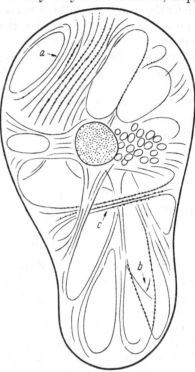

Fig. 40. A plant cell, showing granules moving along fine protoplasmic threads. Note particularly the alignment of particles in what appears to be a continuous sheet of protoplasm.

sible for the movement of protoplasmic particles. The particles do not move under their own power nor are they necessarily carried by the protoplasmic stream.

## Chloroplasts.

Among the larger protoplasmic inclusions are the *chloroplasts*, the structure of which is fairly well settled, at least as to the grana which appear to be distributed in oriented lamellae, consisting of alternating protein and lipid disks (THOMAS, et al., 1952, 1953, STEIMANN 1952, LEYON 1953).

The existence of a membrane around chloroplasts continues to be a controversial problem. STRAUSS (1953) discusses the subject and concludes that there is a membrane with lipids on the outer and protein on the inner side.

## Mitochondria.

The function of mitochondria has long been a subject for discussion, with little conclusive evidence until the biochemists introduced their methods of study. The work done is excellent, but possibly we have a repetition of the old cytological enthusiasm for the vital significance of mitochondria. Much is now again attributed to them and to the microsomes. LINDBERG and ERNSTER (1952) are of the opinion that mitochondria contain all of the iron-porphyrin respiratory enzymes of the cell, and the microsomes are the bearers of the cytoplasmic nucleic acid. This means that if microsomes are the precursors of mitochondria, as was once thought to be true, then these granules change their function as they age.

The mitochondria thus attain an extraordinary role in life. They contain the entire mechanism for the liberation of energy from the substrates, and the means to fix the energy thus set free in energy rich bonds. "The energy from the primary substances containing such bonds is transferred reversibly by the mitochondrial system to ATP" (HYDÉN 1954). By concentrating the respiratory and energy-transferring enzymes and coenzymes in mitochondria, cells achieve a maximum efficiency.

This would appear to be sufficient responsibility for mitochondria, but additional functions are ascribed to them. They are said to possess secretory activity, to take part in lipid metabolism, and (the sarcosomes) serve as a source of ATP (SACKTOR 1953).

### Gene.

The particle, if it is such, which represents the ultimate among particles is the gene. The reality of the gene is still not a settled subject. STERN (1952) regards it as a biological concept without chemical analogy, whereas WATSON and CRICK give the gene a detailed pattern based on the structure of DNA.

## Protoplasmic threads.

There can be no doubt of the existence of fibers in protoplasm, and threads in cells. As the protoplasm in cellular threads is often actively streaming, we are confronted with a difficult task, namely, to interpret motion in a thread which possesses tensile strength. This can be done in several ways; we can ascribe motion to the inner plasm and attribute tensile strength to the outer cortical layer. Resort to coacervates in another answer (BUNGENBERG DE JONG 1932); these remarkable systems possess no fibers yet hold together, and do not mix with water even though their dispersion medium is an aqueous solution.

The structure of living threads, and of protoplasm in general, is the subject of another chapter. Here we are interested only in the presence, kind, and function of protoplasmic strands in cells, or from cells.

Strands of protoplasm are of numerous kinds. There are those consisting of whole cells, such as the fibers of muscle, nerve, and tendon. Then there are protoplasmic strands within cells, often of striking appearance as they stretch across a vacuole, with the protoplasm, chloroplasts, mitochondria, and fat particles all actively moving (Fig. 40). Of great botanical interest are the intercellular threads or bridges which play a significant role in life. And, finally, there are surface processes ranging from myeline threads of lipoid material to the protoplasmic strands found in abundance on the surface of cells, on fibroblasts in culture and on plasmolyzed protoplasts.

Two types of living cellular fibers which have been the subject of much controversy are the spindle fibers and the intercellular connections known as *plasmodesmata*. The reality of spindle fibers is not our present problem.

Protoplasmic bridges between cells have been accepted and emphatically denied. At present, their acceptance is rather general. Intercellular protoplasmic connections were always a justifiable assumption in order to account for coordinated activity among the cells in a body. It has been assumed that the body fluids act as a *liaison* between plant cells. Animal cells in tissues are in contact, *i.e.* protoplasm touches protoplasm, consequently, cellular bridges are unnecessary.

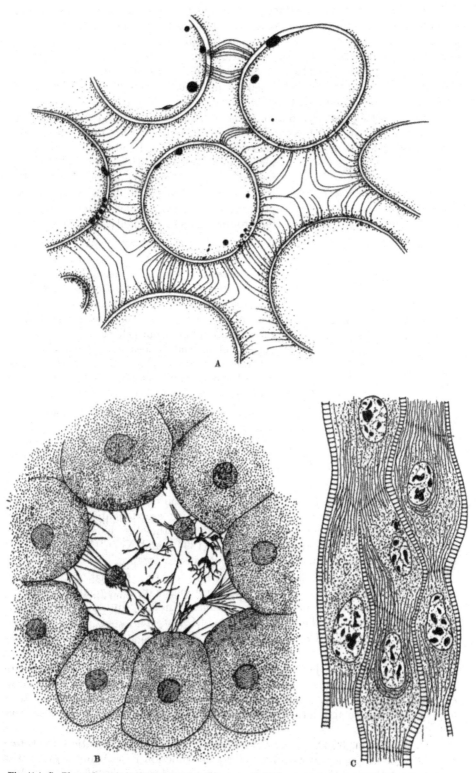

Fig. 41 A–C. Plasmodesmata between cells of *A. Diospyros*, *B.* blastomeres of *Asterias*, *C.* human epidermis
(A—C. J. CHAMBERLAIN, B—E. A. ANDREWS, C—M. IDE 1889).

This could all be true but the fact still remains that there *are* protoplasmic connections between cells, in both plant and animal tissues (Fig. 41 A–C). STRASBURGER (1901), and most early botanists (GARDINER 1884), never doubted the existence of plasmodesmata. Discussion of them centered on such problems as whether or not the connecting thread was continuous or whether the ends of the two threads from the two separate protoplasts merely touch each other (STRASBURGER 1901, KRENITZ 1902).

When cells are plasmolyzed an innumerable number of threads are sometimes formed (Fig. 42). It can be assumed, and was so assumed, that these threads are the intercellular protoplasmic bridges which are stretched and not broken during plasmolysis. This may at times be true, but the formation of these threads may arise in another way, namely, through adhesion of the protoplasm to the cell walls, a more likely happening, for protoplasm is often highly glutinous.

An excellent and full treatment of the subject of cellular bridges is given by MEYER (1920). He refutes one assumption, namely that animal tissues lack plasmodesmata because they have no need of it, the surface of one cell touching that of the next. He illustrates protoplasmic threads extending from cell to cell in a great variety of animal tissues, *e.g.* human skin and cat intestinal muscle. These figures are from the works of many authors (IDE 1889, HOLMGREN 1905, HEIDENHAIN 1911).

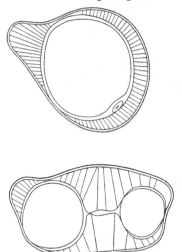

Fig. 42. Protoplasmic threads in plasmolyzed cells (R. CHODAT).

A further interesting question in connection with plasmodesmata is the possibility of cells in different kinds of tissues establishing connections. Sperm fuse with eggs of their own species only. Pollen will sometimes fertilize flowers of closely related species. Rare cases exist of two plant genera, but always in the same family, crossing, *e. g.* radish and cabbage, teosinte and maize. Frequently, however, the opposite extreme occurs. Identical cells will often not fuse, as is true of two amoebae which are daughters of the same parent by fission. Two myxomycete plasmodia, on the other hand, fuse if of the same species. DE BARY and later LISTER (1894), and NOLL (1897) found it impossible to get myxamoebae of different species to fuse.

As for higher plants, when the shoot of one species is grafted on to another species, there must be some mingling of the protoplasm of the two, and this might well include intercellular protoplasmic bridges. In fact, BUDER (1911) found plasmodesmata between the epidermal cells of *Cytisus purpureus* and the underlying parenchyma cells of *Laburnum vulgare*. MEYER reports the same for other grafts.

The cellular bridges just considered are true plasmodesmata but there are numerous other forms of intercellular connections which are somewhat more specific. Such are the protoplasmic strands which pass through the sieve plates of phloem tubes. In another category are the intercellular connections in a colony of *Volvox*.

All such cellular bridges serve a function, as lines of communication, to transport food, or as pseudo-nerve fibers. The last function is probably that of the protoplasmic bridges between the blastomeres of *Asterias blastulae* (Fig. 41 B).

## Myelin Processes.

Myelin threads are renowned among protoplasmic filaments, though a function cannot well be attached to them. They are of lipoid or fat and can be readily made artificially. As the surface of protoplasm is presumably covered with oil, myelin processes are likely to protrude at any time. The introduction of sperm will cause processes to arise at the surface of the egg where they may possibly serve in guiding sperm, though this is pure speculation.

KÜSTER (1926) describes delicate cytoplasmic processes protruding into the vacuole which GICKLHORN (1931) regards as of myelin material. These filaments stain intensely with the lipoid soluble chrysoidin and rhodamin B (STRUGGER 1935).

## Organization.

The word "organization" has been frequently used by biologists to describe the orderliness of protoplasmic activity. Cell functions are synchronized. There is a harmony in nature which must have a structural basis. Any one who has observed the remarkable phenomenon of mitosis cannot help but be impressed with the regularity of the process and the orderliness of the distribution of the chromosomes. Many substances and processes play an important role in life without entering into the structural organization of the cell. Such relatively superficial components of the cell are the food particles, myelin threads, mitochondria, and chloroplasts. Lying beyond them is an organization of structural units which is best comprehended when visualized in terms of a continuity of structural parts. Protoplasm holds together.

## Literature.

ALLARD, H. A.: Some aspects of the phyllotaxy of tobacco. J. Agricult. Res. **64**, 49 (1942). — Clockwise and counterwise spirality in tobacco. J. Agricult. Res. **73**, 237 (1946). — ALTMANN, R.: Die Elementarorganismen und ihre Beziehungen zu den Zellen, 1890; see also, Die Granulatheorie und ihre Kritik. Arch. f. Anat. **1893**. — ASTBURY, W. T.: Fundamentals of fiber structure. London 1933. — X-ray studies of protein structure. Cold Spring Harbor Symp. Quant. Biol. **2**, 15 (1934). — Protein structure from the viewpoint of X-ray analysis. C. r. Lab. Carlsberg, Ser. chim. **22**, 45 (1937). — ASTBURY, W. T., and S. DICKINSON: X-ray studies of the molecular structure of myosin. Proc. Roy. Soc. Lond. Ser. B **129**, 307 (1940).

BAILEY, I. W., and T. KERR: The visible structure of the secondary wall. J. Arnold Arboretum **16**, 273 (1935). — BANGA, I., and A. SZENT-GYÖRGYI: Structure-Proteins. Science (Lancaster, Pa.) **92**, 514 (1940). — BENSLEY, R. R.: Chemical structure of cytoplasm. Science (Lancaster, Pa.) **96**, 389 (1942). — BENLSEY, R. R., et. al.: Frontiers in Cytochemistry. Biol. Sym., Vol. 10, ed. N. L. Hoer. Lancaster, Pa. 1943. — BERGMANN, M.: Harvey Lect. **31**, 56 (1935/36). — BERGMANN, M., and C. NIEMANN: Newer biological aspects of protein chemistry. Science (Lancaster, Pa.) **86**, 187 (1937); see also J. of Biol. Chem. **115**, 77 (1936); **118**, 301 (1937). — BLOCH, R.: Wound healing in higher plants. Bot. Rev. **7**, 110 (1941). — Goethe, idealistic morphology and science. Amer. Sci. **40**, 313 (1952). — BOEHM, G., and H. H. WEBER: Das Röntgendiagramm von gedehnten Myosinfäden. Kolloid-Z. **61**, 269 (1932). — BOYCOTT, A. E., C. DIVER, S. L. GARSTANG and F. M. TURNER: The inheritance of sinistrality in Limnaea peregra. Philosophic. Trans. Roy. Soc. Lond., Ser. B **219**, 51 (1930). — BUDER: Studien Laburnum. Z. Abstammungslehre **5**, 209 (1911). — BÜTSCHLI, O.: Investigations on microscopic foams and on protoplasm. London 1894. — BULL, H. B.: Physical biochemistry. New York **1943**. — BUNGENBERG DE JONG, H. G.: Actualités scientifiques et industrielles, Exposé de Biologie, La Coacervation, Paris 1936. — BUNGENBERG DE JONG, H. G., and J. BONNER: Phosphatide auto-complex coacervates. Protoplasma **24**, 198 (1935). — BUNGENBERG DE JONG, G. H., and W. A. L. DEKKER: Über Koazervation. Kolloid-Beih. **43**, 143 (1936).

CASTLE, E. S.: The influence of certain external factors on spiral growth. J. Cellul. a. Comp. Physiol. **7**, 445 (1936). — Spiral growth in phycomyces. Amer. J. Bot. **29**, 664 (1942). — CLOWES, G. H.: Protoplasmic equilibrium. J. Phys. Chem. **20**, 407 (1916). — Antagonistic electrolyte effects. Science (Lancaster, Pa.) **43**, 750 (1916).

DERVICHIAN. D. G.: Colloidal systems. Research (Lond.) 2, 211 (1949). — DE VRIES, H.: The Mutationstheorie. Leipzig 1901.

FISCHER, E.: Untersuchungen über Aminosäuren, Polypeptide und Proteine. Ber. dtsch. chem. Ges. 39, 530 (1906). — Synthese von Depsiden und Gerbstoffen. Ber. dtsch. chem. Ges. 52, 809 (1919). — FISCHER, H.: Fixierung, Färbung und Bau des Protoplasmas. Jena 1899. — FLORY, P. J.: Network structure and the elastic properties of rubber. Chem. Rev. 35, 51 (1944). — FREUDENBERG, K.: The relation of cellulose to lignin in wood. J. Chem. Ed. 9, 1171 (1932). — FREUNDLICH, H.: Kapillarchemie. Leipzig 1932. — FREUNDLICH, H., and W. SEIFRIZ: Über die Elastizität von Solen und Gelen. Z. phys. Chem. 104, 233 (1923). — FREY-WYSSLING, A.: The submicroscopic structure of the cytoplasm. J. Roy. Microsc. Soc. 60, 128 (1940). — Physiochemical behavior of cytoplasm. Research (Lond.) 2, 300 (1949). — Submicroscopic morphology of protoplasm. Amsterdam and New York 1953.

GARDINER.: On the continuity of the protoplasm through the wall of vegetable cells. Arb. bot. Inst. Würzburg 3, 1 (1884). — GEITLER, L.: Chromosomenbau. Protoplasma Monogr. 14 (1938). — GOLDACRE, R. J., and I. J. LORCH: Folding and unfolding of protein molecules in relation to cytoplasmic streaming. Nature (Lond.) 166, 497 (1950). — GUILLIERMOND, A.: The cytoplasm of the cell. Waltham, Mass. 1941. — GUTH, E.: The problem of the elasticity of rubber. Pub. No 21, Amer. Assoc. Adv. Sci. 103 (1942); 21, 103 (1943).

HATSCHEK, E.: Analysis of the theory of gels as systems of two liquid phases. Trans. Faraday Soc. 12, 17 (1917). — HAWORTH, W. N.: The constitution of sugar. New York 1929. — HAWORTH, W. N., and H. MACHEMER: Molecular structure of cellulose. J. Chem. Soc. 1932 II, 2270. — HEIDENHAIN, M.: Plasma und Zelle. Jena 1911. — HESS, K.: Die Chemie der Zellulose. Leipzig 1928. — HOCK, C.: Microscopic structure of the cell wall. The structure of protoplasm, ed. W. SEIFRIZ. Ames, Iowa 1942. — HÖBER, R.: Physical Chemistry of Cells and Tissues. Philadelphia 1945. — HOLMGREN, E.: Zur Kenntnis der zylindrischen Epithelzellen. Arch. mikrosk. Anat. 65, 280 (1905). — HUGGINS, M. L.: Comparison of the structures of stretched linear polymers. J. Chem. Phys. 13, 37 (1945). — A new approach to the theory of rubber-like elasticity. J. Polymer Res. 1, 1 (1946). — HYDÉN, H.: Minerva Med. 1, 3 (1952). Ann. Rev. Physiol. 16, 17 (1954).

IDE, M.: Nouvelles observations sur les cellules epithéliales. Cellule 5, 321 (1889). — IWATA, J.: Studies on chromosome structure. Jap. J. of Bot. 10, 365 (1940).

JAMES, H. M., and E. GUTH: Theory of the elastic properties of rubber. Phys. Rev. 59, 111 (1941). — JENNINGS, H. S.: The behavior of lower organisms. Carnegie Publ. 16, 129 (1904).

KAMIYA, N., and S. ABE: Bioelectric phenomena in the myxomycete plasmodium. J. Coll. Sci. 5, 149 (1950). — KAMIYA, N., and W. SEIFRIZ: Torsion in protoplasm. J. Exper. Cell Res. 1954. — KAUFMANN, B. P.: Chromosome structure. Amer. J. Bot. 13, 59, 355 (1926). — KIENITZ, G.: Neue Studien über Plasmodesmen. Ber. dtsch. bot. Ges. 20, 93 (1902). — KRAEMER, E. O., and W. D. LANSING: The molecular weights of cellulose and cellulose derivatives. J. Phys. Chem. 39, 153 (1935); see also Industr. Engin. Chem. 30, 1200 (1938).

LANGMUIR, I.: The constitution of solids and liquids. J. Amer. Chem. Soc. 39, 1848 (1917). LEATHES, J. B.: Function and design. Science (Lancaster, Pa.) 64, 387 (1926); see also Lancet 1925, 803, 853, 957, 1019. — LEWIS, F. T.: The typical shape of polyhedral cells in vegetable parenchyma, etc. Proc. Amer. Acad. Arts a. Sci. 58, 537 (1923). — Mathematically precise features of epithelial mosaics. Anat. Rec. 55, 323 (1933). — The shape of compressed spheres. Science (Lancaster, Pa.) 86, 609 (1937). — LEYON, H.: Exper. Cell Res. 4, 371 (1953). — LILLIE, R. S.: Transmission of activation in passive metals. Science (Lancaster, Pa.) 48, 51 (1918). — The recovery of transmissivity in passive iron wires. J. Gen. Physiol. 3, 107 (1920). LISTER, A.: A monograph of the mycetozoa. London 1894. — LLOYD, F. E., and G. W. SCARTH: The structural organization of plant protoplasm in the light of micrurgy. Protoplasma 2, 189 (1927). — LOCQUIN, M.: Le developpment des spirales gauches á la surface des spores de Laccari. Soc. Linné de Lyon 14, 41 (1945). — LOEB, J.: Proteins and the theory of colloidal behavior. New York 1922. — LOEWY, A.: A theory of protoplasmic streaming. Proc. Amer. Phil. Soc. 93, 326 (1949). — An actomyosin-like substance from the plasmodium of a myxomycete. J. Cellul. a. Comp. Physiol. 40, 127 (1952).

MARK, H., et al.: The chemistry of large molecules, Eds. R. E. BURK and O. GRUMMITT. New York 1943. — MATSUURA, H.: Chromosome studies on trillium. Cytologia 9, 243 (1938). — MATZKE, E. B.: Analysis of the orthic tetrakaidecahedron. Bull. Torrey Bot. Club 54, 341 (1927). — MEYER, A.: Analyse der Zelle. Jena 1920. — MEYER, K. H.: Natural and synthetic high polymers. New York 1942 a. — MEYER, K. H., and H. MARK: Über den Bau des kristallisierten Anteils der Cellulose. Ber. dtsch. chem. Ges. 61, 593 (1928). — MEYER, K. H., and A. J. A. VAN DER WYK: Molecular processes during deformation of rubber-like elastic bodies. J. Polymer Res. 1, 49 (1946). — MIRSKY, A. E., et al.: Nature (Lond.) 169, 128 (1952).

NEEDHAM, J.: Order and life. Cambridge 1936. — NOLL, F.: Niederrhein. Ges. Natur. Heilk., Bonn, Sitzgsber. 1897. — NUTTING, G. C., F. R. SENTI, and M. J. COPLEY: Conversion of globular to fibrous proteins. Science (Lancaster, Pa.) **99**, 328 (1944). OSTWALD, W.: Elektrische Eigenschaften halbdurchlässiger Scheidewände. Z. phys. Chem. **6**, 71 (1890). — OSTWALD, WO.: Handbook of colloid chemistry. London 1919. PAULING, L., R. B. COREY and H. BRANSON: The structure of proteins. Proc. Nat. Acad. Sci. U.S.A. **37**, 205 (1951). — PLAGGE, E.: Gen-bedingte Prädeterminationen bei Tieren. Naturwiss. **26**, 4 (1938). — PRESTON, R. D.: The organization of the cell wall. Philosophic. Trans. Roy. Soc. Lond., Ser. B **224**, 131 (1934). — Spiral growth and spiral structure. Biochim. et Biophysica Acta **2**, 155 (1948). — PRYOR, M. G. M.: Deformation and flow in biological systems, ed. FREY-WYSSLING. New York and Amsterdam 1952.

RILEY, D. P., and G. OSTER: An X-ray diffraction investigation of aqueous systems of desoxyribonucleic acid. Biochim. et Biophysica Acta **7**, 526 (1951). — RITTER, G.: Composition and structure of the cell wall. Ind. Eng. Chem. **20**, 941 (1928). — The morphology of cellulose fibers. Paper Trade J. **1935**.

SACKTOR, B.: The mitochondria of the house fly. J. Gen. Physiol. **36**, 371 (1953). — SCARTH, G.: Structural differentiation of cytoplasm. The structure of protoplasm, ed. W. SEIFRIZ. Ames, Iowa 1942. — SCHAEFFER, A. A.: Spiral movement in Amoebae. Anat. Rec. **34**, 115 (1927). — On molecular organization in amoeban protoplasm. Science (Lancaster, Pa.) **74**, 47 (1931). — SCHMIDT, W. J.: Die Ergebnisse der NAEGELIschen Micellarlehre. Naturwiss. **16**, 900 (1928). — SEIFRIZ, W.: Elasticity as an indicator of protoplasmic structure. Amer. Naturalist **60**, 124 (1920). — Phase reversal in emulsions and protoplasm. Amer. J. Physiol. **66**, 124 (1923). — The alveolar structure of protoplasm. Protoplasma **9**, 177 (1930). — The structure of protoplasm. Science (Lancaster, Pa.) **73**, 648 (1931). — Twisted trees and the spiral habit. Science (Lancaster, Pa.) **77**, 50 (1933a); see also, More about the spiral habit. Science (Lancaster, Pa.) **78**, 361 (1933b). — The origin, composition, and structure of cellulose in the living plant. Protoplasma **21**, 129 (1934). — Protoplasm. New York 1936. — Deformation and flow in biological systems, ed. FREY-WYSSLING. New York and Amsterdam 1952. — Mechanism of protoplasmic movement. Nature (Lond.) **171**, 1136 (1953). — SEIFRIZ, W., and C. HOCK: The structure of paper pulp fibers, Paper Trade J., Tech. Assoc. May 7, 1936; Feb. 1, 1940. — SPEK, J.: Zum Problem der Plasmastrukturen. Z. Zellenlehre **1**, 278 (1924). — SPONSLER, O. L.: The molecular structure of the cell wall. Amer. J. Bot. **15**, 525 (1928). — SPONSLER, O. L., and W. H. DORE: The crystal structure of glucose. J. Amer. Chem. Soc. **53**, 1639 (1931); see also Colloid Sympos. Monogr. **4**, 174 (1926). — STAMM, A. J.: The state of dispersion of cellulose in cuprammonium solvent. J. Amer. Chem. Soc. **52**, 3047 (1930). — STAUDINGER, H.: Die hochmolekularen organischen Verbindungen — Kautschuk und Cellulose. Berlin 1932. — The formation of high polymers. Trans. Faraday Soc. **32**, 97 (1936). — STEIMAN, E.: Exper. Cell Res. **3**, 367 (1952). — STERN, H., et all.: J. Gen. Physiol. **35**, 559 (1952). — STRASBURGER, E.: Über Plasmaverbindungen pflanzlicher Zellen. Jb. wiss. Bot. **36**, 493 (1901). — STRAUS, W.: Bot. Rev. **19**, 147 (1953). — STURTEVANT, A.: Inheritance of cioling in Limnaea. Science (Lancaster, Pa.) **58**, 269 (1923). — SVEDBERG, T., and coworkers: Molecular weight determinations. J. Amer. Chem. Soc. **55**, 2834 (1933). — SZENT-GYÖRGYI, A.: Chemistry of muscular contraction. New York 1947.

TEORELL, T.: A quantitative theory of membrane permeability. Proc. Soc. Exper. Biol. a. Med. **33**, 282 (1935). — THOMAS, J. B., et al.: Biochim. et Biophys. Acta **8**, 90 (1952); **10**, 230 (1953). — THOMPSON, D'ARCY: On growth and form. Cambridge 1942. — TREOLAR, L. R. G.: Physics of rubber elasticity, Oxford 1949; see also Trans. Faraday Soc. **39**, 241 (1943); **43**, 277 (1947). — TROLL, W.: Goethe, Morphologische Schriften. Jena 1926. —

WADDINGTON, C. H.: Organizers and genes. Cambridge 1940. — The character of biological form, in aspects of form, a symposium, ed L. L. WHYTE. New York 1951. — WEYL, H.: Philosophy of mathematics and natural science. Princeton 1952. — WILSON, E. B.: The cell in development and inheritance. New York 1925. — WRINCH, D. M.: Conversion of globular to fibrous proteins. Nature (Lond.) **137**, 411 (1936). — On the pattern of proteins. Proc. Roy. Soc. Lond., Ser. A **160**, 59 (1937a). — The cyclol hypothesis and the „globular" proteins. Proc. Roy. Soc. Lond. Ser. A, No 907, **161**, 505 (1937b).

# The physical chemistry of cytoplasm.

By

## William Seifriz.

With 27 figures.

The title of this chapter is a very comprehensive one. It covers much of modern physiology, biochemistry, biophysics, and what BULL calls physical biochemistry. The book by H. BULL (1943) is one of the most satisfactory among recent publications devoted to the physical chemistry of living systems. There are also such special texts as those by HÖBER (1945), GLASSER (1944), GORTNER (1949) and my own (SEIFRIZ 1936).

## 1. Viscosity.

Every activity of the cell is in some way associated with the viscosity of protoplasm. The streaming of protoplasm, the flow of sap, the contractility of muscle, mitosis, growth, metabolic activities, all are influenced by the viscosity of protoplasm. Biologists have been fully aware of the significance of viscosity. The emphasis laid on it during the past half century is indicated by the number of familiar names associated with it: CHAMBERS (1924), CONKLIN (1917), FRY (1934), HEILBRONN (1922), HEILBRUNN (1926b), LEWIS (1942), MAST (1931), NĚMEC (1901), NORTHEN (1938, 1949), PFEIFFER (1936), SCARTH (1924a), WEBER (1922), and SEIFRIZ (1929).

"Everything flows", wrote HERACLEITUS, and LUCRETIUS observed that some liquids flow more rapidly than others. The slower movement of olive oil, he thought, might be due to "elements more large or else more crooked and intertangled". LUCRETIUS thus became the first polymer chemist. How near right he was is shown by recent work which indicates that in a solution containing very long, threadlike molecules, normal movement is impeded, and high viscosity results. One can calculate the molecular weight of substances, such as rubber, presuming a very simple and direct relation to exist between viscosity and molecular size.

When liquids flow it is particles which flow, atoms, molecules, ions, and molecular aggregates, and in so doing they meet other particles. Interference, or internal friction, results. This property of fluids, which is a *resistance to flow*, is *viscosity*.

The study of the laws of flow has become a science in its own right; it is called *rheology*, and includes a variety of phenomena.

Sir ISAAC NEWTON first gave mathematical expression to the viscosity of a liquid in a simple law of *fluidity* expressed in terms of the pressure necessary to produce a specific *rate* of flow. In his Principia (1665) he states that:

$$v = \varphi F d$$

in which $v$ is the velocity of flow, $\varphi$ the fluidity, and $F$ the shearing stress on one of two planes separated by the distance $d$. NEWTON did not bother to prove his theory experimentally. It remained for POISEUILLE to do this.

Poiseuille expressed viscosity in terms of the *volume* of flow of a liquid through a capillary within a given time. The experimental equation became:

$$v = \frac{p\,r^4}{l}\,K,$$

where $v$ is the volume of liquid discharged in unit time, $p$ the pressure, $r$ the radius and $l$ the length of the capillary, and $K$ a constant which is characteristic of each liquid. Temperature remains the same, being specified for each experiment. Poiseuille did not deduce coefficients of viscosity. This was done later and led to the value $\pi/8$ for the constant $K$. As it is more convenient to measure time than to keep within unit time, the formula became:

$$v = \frac{\pi\,p\,r^4\,t}{8\,l\,\eta}\,.$$

The absolute viscosity is here expressed in terms of a *coefficient* $\eta$.

## Definitions.

There is no need here for a complete list of the terms and their definitions which have come into use and received official sanction by rheologists. The reader is referred to Vol. 1, p. 510 of the Journal of Rheology for 1930, or to Bingham's treatise (1930) where definitions for all terms are given. It is, however, necessary to warn the reader that certain of these definitions may be ignored because they are misleading, and others will have to be ignored by the biologist because they are difficult enough to determine in known solutions. To work with comparable accuracy when determining the viscosity of protoplasm is impossible. Resistance to flow, or viscosity, is expressed in c.g.s. units called *poises*, which are dyne-secs. per cm.². If one imagines two planes of unit area and unit distance apart, moving one over the other at unit velocity (Fig. 1), then the tangential force, or *shearing stress*, necessary to accomplish this is one poise; in this case:

$$\text{force} = \frac{\eta\,A\,u}{d}$$

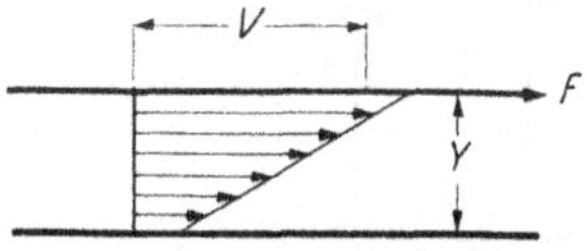

Fig. 1. Diagram illustrating viscosity by the shear of one plane against another.

in which, $A$ is the area of the imaginary planes separated by the distance $d$; $u$ is the velocity difference in cm. per sec., and $\eta$ the *coefficient of viscosity*.

Coefficients of viscosity expressed in poises are usually decimal values, thus 0.01 for water. A whole number is more convenient, so the *centipoise* was devised, and water becomes 1 (at 20.20° C.), and glycerine goes from 8.3 poises to 830 centipoises.

Some absolute viscosity values expressed in poises are:

| | | | | | |
|---|---|---|---|---|---|
| Water | 0.018 | at 0° C. | Olive oil | 0.84 | at 20° C. |
| Water | 0.01 | at 20° C. | Glycerin | 8.30 | at 20° C. |
| Benzene | 0.0065 | at 20° C. | Glucose | 27,000.00 | at 67° C. |

Viscosity may be expressed as *internal friction*, due to molecules gliding by each other; but it may be defined in a number of other ways, for example, in terms of the energy necessary to maintain a constant velocity gradient.

Some liquids, like most solids, do not yield immediately to an applied tangential force, a definite minimum *shearing stress* being necessary to start flow (O–E for curve C, Fig. 2 A) or, if flow does start immediately (B, Fig. 2 A) then a certain pressure is required before a constant viscosity value is attained. The shearing stress, in dynes per sq. cm., required to produce the continuous flow of a liquid or deformation of a solid is the *yield value*.

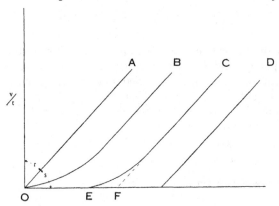

Fig. 2 A. Viscosity graphs: ordinates, $v/t$, rate of flow; abscissae, shearing stress: graph $A$, pure, Newtonian, flow — $r$ is viscosity and $s$ fluidity: graph $B$, quasi-viscous, non-Newtonian, flow: graph $C$, similar to $B$ but with greater shearing stress; $O$–$E$, the stress necessary to start flow, $O$–$F$ the total stress: graph $D$, a theoretical possibility never experimentally realized.

A number of distinctions are made by the rheologist which seem useless, such as the attempt to differentiate between viscosity and consistency. But there are distinctions which must be recognized, even though the biologist can not hope to cope with them. Such a one is that between *intrinsic* and *specific* viscosity.

STAUDINGER (1932) made the first modification of classical ideas on viscosity. Following the knowledge that most mixtures do not obey NEWTON'S law, he derived and defined the *specific viscosity* of a solution, establishing the relationship:

$$\eta_{sp} = \frac{\eta - \eta_0}{\eta_0}$$

$\eta_0$ being the viscosity of the solvent. The specific viscosity takes account of the contribution of the solute to the viscosity of the solution, the viscosity of the solvent being taken as unity. The justification for the foregoing assumptions is seen in the fact that the specific viscosity of equal concentrations of a given polymer in many solvents is very nearly the same, and independent of the solvent viscosity, $\eta_0$.

We look for understanding through simplification yet with each further insight into the mechanism responsible for the resistance to flow we find another factor. Thus, the specific viscosity is dependent on the concentration, $c$; hence the *reduced viscosity*, $\eta_{sp}/c$, is introduced, thus including the contribution of the solute to the viscosity of the solution per unit concentration. This quality could be expected to be independent of the concentration, and therefore a true molecular constant, which is true in many cases such as solutions of compact protein molecules and low molecular weight hydrogen polymers. EWART (1948) states that for coiling polymer molecules of high molecular weight the reduced viscosity

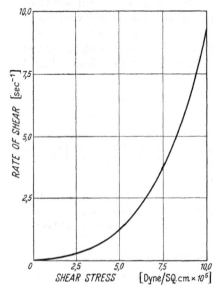

Fig. 2 B. Experimental flow curve of rubber at 140 C, showing the dependence of rate of flow on shear stress (L. R. G. TRELOAR).

varies considerably with concentration, due to interactions between molecules. It approaches a constant value in very dilute solutions where these interactions vanish.

And now we come to another expression; the foregoing limiting value which the reduced viscosity approaches in dilute solutions is known as the *intrinsic viscosity*, with the symbol $\eta$.

This brings us back to where we started, with a coefficient of viscosity $\eta$, and this is what one would expect where, as in the solution laws, we state an ideal situation, which is approached in dilute solutions.

Unfortunately, the end is not yet in sight, for we may express the reduced viscosity with $\eta_{sp}/c$, and still have left an *inherent viscosity* to ponder over.

In this connection one may ask is Newtonian flow a criterion of ideality in a solution. This may not be true. Non-Newtonian flow can be the result of either purely hydrodynamic effects or of forces between the various particles constituting the solution. One can imagine, and perhaps realize too, solutions which are non-Newtonian solely because of the contribution of hydrodynamic effects. An example would be a fairly dilute tobacco mosaic virus solution. Because of the fact that the rods would tend to orient in high velocity gradients, the apparent viscosity coefficients should be lower than for low velocity gradients where the particles are presumably randomly oriented. This should happen even at such great dilution that virus-virus interaction is impossible. A suspension of completely inert fine glass fibers should behave the same way.

The concept of ideality is borrowed from thermodynamics. In an ideal mixture the activity coefficients of all constituents would be unity. Frankly, were such a concept carried over to the problem of viscous flow, one should be able to predict the coefficient of viscosity of a liquid from its thermodynamic properties. An ideal mixture is one in which the viscosity is some relatively simple function of composition and such a mixture apparently does not exist.

Indeed, we may question whether or not the straight line graph of a Newtonian fluid remains straight if extended indefinitely. It is quite conceivable that it will not, and thus force us to the conclusion that no liquid is Newtonian.

The ideal liquid must be interpreted in quite a different way from the ideal or perfect gas. The latter obeys the law: $PV = RT$. But the ideal liquid is one in which the internal forces at any internal section are always normal to the section even during motion. The forces are purely pressure forces, and as there can be no tangential forces, there is no friction. Such a liquid system does not exist, for in addition to pressure forces, tangential or shearing stresses always come into play thus giving rise to fluid friction (DOUGHERTY 1937). This entire field of variable viscosity is well handled by BULL (1943).

We shall do well to return to NEWTON's simple equation for fluidity, the unit for which is now called the *rhe*, being the reciprocal of the poise. This unit has been brought up to date by the following equation:

$$n_{sp} = \frac{n - n_0}{n_0}.$$

I have told of the foregoing distinctions solely in order that the reader may have some idea of the complexity of a situation which at first appears to be a rather simple one. The biologist and medical worker can be content with the coefficient of viscosity $\eta$ and the knowledge that it will, at low stresses, be a variable in most organic solutions.

From the foregoing it is obvious that many organic solutions deviate from NEWTON's law of fluidity. The first indication of this came in such solutions as gelatin and led to the expression *non-Newtonian behavior*.

## Non-Newtonian behavior.

If a solution of gelatin is compared to one of glycerine of about the same viscosity, it will be found that the glycerine obeys NEWTON's law, the viscosity graph is a straight line (A, Fig. 2), whereas the gelatin solution does not obey NEWTON's law, the graph is a curve (B or C, Fig. 2). The gelatin solution is *non-Newtonian*. Most organic fluids are non-Newtonian in behaviour; thus, rubber shows anomalous viscosity (Fig. 2 B).

HERACLEITUS was the first rheologist, NEWTON the first to give mathematical expression to fluidity, and BINGHAM (1930) the father of the science of Rheology.

I have always felt that rheologists banded together to gain the necessary strength to oppose the classicists whose laws were founded on the ideal behavior of ideal solutions at infinite dilution. It was the very "lawlessness" of a solution of gelatin which appealed to the rheologists, as it did to the colloid chemists.

Unfortunately, in founding a nomenclature of rheology, BINGHAM introduced the old and familiar word *plasticity* to characterize a new situation, the significant feature of which is the very antithesis of that which distinguished plasticity in the original sense. Plasticity is that property of solids which permits them to be molded under pressure, or more exactly, deformed under a shearing stress, and yet hold their own shape when the shearing stress is removed. Clay thus becomes the perfect plastic substance because it is easily shaped in the hands of the potter, and once given shape retains it. In both the BINGHAM and the popular sense, plastic substances possess *yield value* and *mobility*. Yield value is the shearing stress required to start deformation, and mobility is flow after the yield value has been exceeded, being proportional to the rate of deformation.

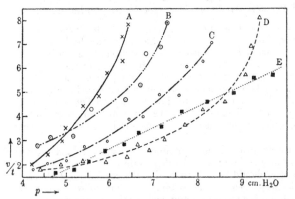

Fig. 3. Curves of protoplasmic deformation flow rate, $v/t$, plotted against the shearing force, cm.³ H₂O: *A, C, D*, protoplasmic drops of *Physalis* pericarps; *B*, denuded protoplasts of Orchis maculata after treatment in 1. KNO₃; *E*, straight line graph of glycerin. (From H. H. PFEIFFER.)

Newtonian liquids possess no yield value, *i.e.* no force or shearing stress which must be applied to start flow, but non-Newtonian liquids, and plastic solids require an initial pressure to start flow, *i.e.* both possess a yield value. But the plastic solid, by definition, shows no return, whereas the non-Newtonian fluid is elastic, and therefore does not hold its new form. We may therefore say that gelatin is plastic because it possesses a yield value, but as it cannot be molded and yet show no tendency to return, it cannot be plastic in the original meaning of the word.

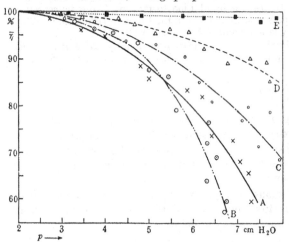

Fig. 4. Absolute viscosity values expressed as percentages of the initial values plotted against shearing force; *A, B, C* protoplasmic drops of opened cells of *Chara fragilis* after treatment in salts; *D, E*, model of droplets of sodium stearate and glycerin. (From H. H. PFEIFFER.)

At the inaugural meeting of the Society of Rheology (SEIFRIZ 1929) I ventured the suggestion that all non-Newtonian solutions of high molecular weight polymers would prove to be elastic. Two soaps illustrated this situation well. Both were sodium stearate yet when one was mixed in water at high concentration it was Newtonian, showed no elasticity, no rigidity, and was amorphous, *i.e.* granular in microscopic appearance; whereas the other soap sample was non-Newtonian at low concentration, highly elastic, rigid though fluid, and crystalline in appearance, *i.e.* showed long fibrous crystals when microscopically examined. I know of no exceptions to the rule that non-Newtonian liquids are elastic, HARDY's hydrocarbon oils notwithstanding.

That protoplasm should prove to be non-Newtonian was expected. Its elasticity, double refraction, and high protein content all pointed in this direction; but proof did not come until PFEIFFER (1937) was able to "persuade" protoplasm to move through a capillary under pressure. He also determined the modulus of the shearing stress necessary to deform a protoplast (1940). Both methods showed protoplasm to be non-Newtonian (Figs. 3 and 4).

As viscosity is proportional to shearing stress and to rate of flow, one may determine either and from it derive the other: the volume of flow is cm.$^3$/sec. and the shearing stress is the pressure in dyne/cm.$^2$.

The two sets of graphs (Figs. 3 and 4) by PFEIFFER are sufficient to reveal the anomalous behaviour of protoplasm; like most protein solutions it is non-Newtonian. Note that in each chart the graph for glycerine is a straight line. The other curves in Fig. 3, which are the mean rates of shear, $DQ/R^3$ in sec$^{-1}$, as ordinates, plotted against the shearing stress exerted on the protoplasm, $PRp/2L$ in dyne/cm.$^2$, as abscissae, are not straight lines.

PFEIFFER further proved the non-Newtonian nature of protoplasm by measuring the differences in viscosity before and after forced flow. The amplitude in Brownian motion of the protoplasmic particles was the measure of viscosity. The curves in Fig. 4 were the result.

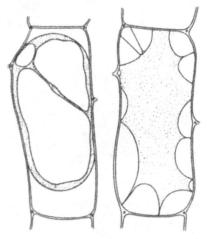

Fig. 5. Plasmolyzed cells of convex and concave contours.

## Methods and values.

Viscosity measurements of non-living fluids are easily made by measuring the rate of fall of lead shot or the rise of bubbles; or by timing the flow of liquid through a capillary and comparing the rate with water or another standard. If the dimensions of the capillary and the pressure applied are known, the viscosity may be calculated in terms of poises. The capillary method first made it possible for POISEUILLE to give mathematical expression to NEWTON's law of fluidity. Viscosity measurements of protoplasm are far more difficult.

### Osmosis.

The osmotic method for determining the consistency of cell contents was used by EWART (1905). He compared the viscosity of the cell sap with isotonic solutions, and found values between $\eta = 0.01$ and $0.02$ at $20^0$ C. In cells rich in sugars the viscosity of the cell sap rose to $\eta = 0.06$.

### Plasmolysis.

The configuration of protoplasts after plasmolysis was said to be an indicator of protoplasmic consistency. The normal cell, when put in hypertonic sugar solutions, plasmolyzes either with a smooth and convex surface and no adhesion to the cellulose wall, or with concave depressions and adhesion to the cell wall at many points (Fig. 5).

Considerable error was involved in the foregoing deductions. The convex surface of a plasmolyzed protoplast indicates not low viscosity but low adhesive qualities, and the concave protoplast with many fine attached threads indicates not high viscosity but a high degree of adhesiveness (SCARTH 1923). The presence and duration of protoplasmic threads may be used to indicate glutinosity of the protoplasmic membrane.

Glutinosity or tackiness, elasticity, surface tension, and tensile strength are not correlated with viscosity. There is no reason why they should be.

## Gravity.

Plant cells sometimes contain freely suspended starch grains called *statoliths* which normally lie at the bottom of a cell. They were, by HABERLANDT (1903), assumed to serve as gravitational sense organs. If a cell is turned through 180° the starch grains slowly fall. By exposing a living cell to view, repeatedly reversing the plant, and noting the time of fall of the statoliths, an indication of the consistency of the protoplasm can be obtained. The living cell becomes a gravity viscometer in which viscosity is determined by the time of fall of a sphere through a known length of the fluid. The viscosity value is calculated from STOKES' law:

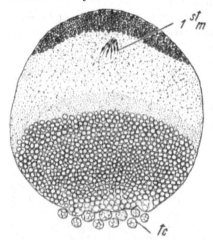

Fig. 6. A stratified ovum of the mollusk *Cumingia:* the germinal vesicle with nucleolus and chromosomes is at the centripetal pole (1st $m$ = first meiotic spindle); the test cells, *tc*, (follicular) are at the centrifugal pole. (From E. G. CONKLIN 1917.)

$$V = \frac{2r^2(D-d)g}{9\eta}$$

where $V$ is the velocity of fall, $r$ the radius of the particle, $D$ the specific gravity of the particle, $d$ the specific gravity of the medium, $g$ the gravity constant, and $\eta$ the viscosity of the medium.

NĚMEC (1901) found the time of fall, and therefore the protoplasmic consistency, to increase with decreasing temperature. He thought dehydration to be the responsible factor. This need not, however, have been the case.

HEILBRONN (1912) compared the rate of fall of statoliths in *Phaseolus* and *Vicia* cells with that of freed starch grains in water; the average velocity in the cell was 4.37 $\mu$/min., or 0.073 $\mu$/sec. Statoliths liberated in water showed an average velocity of 43 seconds through 74 $\mu$, or a velocity in $\mu$/sec. of 1.71. The viscosity of the protoplasm becomes therefore $\frac{1.71}{0.072}$, or 23.7 times that of water. This value is the usual minimum of the protoplasm of an active cell; it is about the viscosity of concentrated sulphuric acid or a thin oil.

## Centrifugal force.

The centrifugal method of determining protoplasmic consistency is simply the gravity method in which centrifugal force is substituted for gravity. If the dispersed particles in protoplasm are of two kinds and of different weights, they may be stratified by centrifuging (Fig. 6).

When the eggs of fish, or of the mollusk *Cumingia*, are centrifuged the visible granular substances of the protoplasm are displaced; pigment and yolk are carried to the outer pole, and oil drops to the inner pole, with a band of clear protoplasm lying between. If the eggs are more thoroughly centrifuged four zones are formed. Some of the earliest work of this sort was done by MORGAN, CONKLIN (1917), NĚMEC (1926), and later by HEILBRUNN (1920). HEILBRUNN (1926) found that a force of 311 times gravity will stratify eggs in 90 seconds, and a force of 4,968 times gravity will stratify in 5 seconds. He obtained a value of 0.043 poises for the viscosity of *Cumingia* egg protoplasm. HARVEY was able to pull eggs apart with a force of 6,595 times gravity. This means that great injury can be caused by centrifuging, which NĚMEC (1926) demonstrated in the dividing cell; the spindle disappears and the chromosomes are reduced to a spherical shape. The difficulties involved in injury and pathological changes are ones to which all methods of determining protoplasmic consistency are subject.

A further limitation of the centrifuge method is the impossibility of determining local differences in consistency between minute regions of the protoplast. No single viscosity value can be given for a cell. Especially true is this of unicellular organisms such as a protozoan the internal anatomy of which may be quite complex. A similar situation exists in the intricate structure of the mitotic figure of a dividing cell.

There are numerous other difficulties in determining protoplasmic viscosity by the centrifuge method. Particles moving through finely granular protoplasm will give one value; those moving through hyaline protoplasm will give a different value. COSTELLO (1934) realized this and raised the absolute viscosity of the protoplasm to 0.118 poises. He further pointed out that a viscosity value of any solution containing granules obtained by the centrifuged method will depend upon the equation used, several of which are:

$$\eta_s = \eta \frac{1 + 0.5f}{(1-f)^4} \qquad \text{(EINSTEIN)} \quad \eta = 0.016$$

$$\eta_s = \eta \, (1 + 4.5f) \qquad \text{(HATSCHEK)} \quad \eta = 0.044$$

$$\eta_s = \eta \frac{1}{1 - \sqrt[3]{f}} \qquad \text{(HATSCHEK)} \quad \eta = 0.033$$

$$\eta_s = \eta \frac{1}{(1 - f - 3/2 \, f^{5/3})^{5/2}} \qquad \text{(SMOLUCHOWSKI)} \quad \eta = 0.008$$

$$\eta_s = \eta \frac{1}{\left(1 - \dfrac{f}{f_\infty}\right)} \qquad \text{(BINGHAM and DURHAM)} \quad \eta = 0.059$$

The foregoing factors explain why such contradictory results on protoplasmic consistency have been obtained. In contrast to the low values of some workers (twice the consistency of water for sea-urchin eggs), CONKLIN (1917) demonstrated a high viscosity in dividing *Crepidula* eggs. One need only recall that the coefficient of viscosity of sugar, heated sufficiently to flow at 109° C., is 28,000, which is equivalent to a comparative viscosity of 2,800,000 times that of water. HEILBRUNN, in recognizing the fact that the consensus of opinion is in favor of a moderately high viscosity of protoplasm, regards the prevalent conception of protoplasm as a highly viscous material as due in part to an "illusion" based on the failure to appreciate the influence of the "stiff" protoplasmic membrane which holds cells "rigid".

## Microdissection.

Microdissection (CHAMBERS 1924, PETERFI 1923, and DE FONBRUNNE 1932), as a method for determining protoplasmic consistency yields, at best, a scale of viscosity values which are only approximate. The criteria used to determine values are the distance from a moving needle at which protoplasmic particles are disturbed—the greater the distance the more viscous the medium—the retention of furrows made by a needle—the longer a furrow persists the greater the viscosity—the rapidity at which a deformed protoplast regains its original shape—viscosity is inversely proportional to the rapidity—and the ease with which extended protoplasmic threads are reincorporated into the protoplasmic mass. It is not a difficult task to closely estimate at least ten degrees of consistency, e.g. between water and firm gelatin (SEIFRIZ 1920).

The chief advantages of the microdissection method lies in its delicacy. The minutest region of the protoplast can be explored and local differences in viscosity estimated. The technique permits an approximation of values of the viscosity of the cytoplasm of various regions in a dividing cell, of astral rays, polar areas, cortex, nuclei, chromosomes (CHAMBERS 1924), and the protoplasmic membrane.

The consistency of nuclei varies greatly. It may be fluid, as in epithelial cells, or a soft but tenacious, dough-like jelly, as in amphibian *erythrocytes* (SEIFRIZ 1926).

The protoplasmic membrane, both cytoplasmic and nuclear, may be of high consistency, elastic, of high tensile strength (SEIFRIZ 1920), or it may become quite fluid as at the tip of an advancing pseudopodium in an *Amoeba* or myxomycete.

### Streaming.

There is no more obvious way to tell the viscosity of a liquid than to measure the rate at which it flows at constant temperature and pressure, but the method when applied to protoplasm becomes a pitfall. Stimulation to streaming in *Elodea* cells after treatment in water with a slight trace of copper, or in a solution of strontium and barium salts, in saponin and alcohol (SEIFRIZ 1936), may mean a lowering of viscosity, but these agents may stimulate, *i.e.* activate the motive force without any change in viscosity.

A correlation between rate of streaming and degree of viscosity holds only if all other factors, particularly the motive force, remain constant, and this is often not true. In the rhythmic reversal in flow of the protoplasm in certain slime molds, as in any shuttle movement, the protoplasm must, when reversing its direction of flow, come to a standstill. The speed, therefore, changes constantly, from zero to a maximum. This variation in rate is due to change in pressure, to applied motive force, and not to viscosity. There is no evidence that any change in viscosity need occur during the change in rate of protoplasmic flow.

### Brownian movement.

BAAS-BECKING (1929) made painstaking measurements of Brownian movement by noting the distance traversed every 15 seconds by granules in the outer protoplasmic layer of *Spirogyra* cells. By means of a detailed mathematical analysis, the applicability of EINSTEIN's viscosity equation to individual trajectories was tested, and thus the viscosity of the protoplasm estimated. BAAS-BECKING concluded that the layer in question is heterogeneous, with a consistency which varies between that of water to several hundred times that of water.

### Electromagnetism.

HEILBRONN (1922) developed a technique for the measurement of protoplasmic consistency by means of the electromagnetic attraction of a metal particle embedded in protoplasm. The force, in amperes, necessary to set the particle in motion is an indicator of the viscosity of the protoplasm. The method consisted in allowing living plasmodia to ingest small iron particles $(30\,\mu)$, then attracting the particle by an electromagnet.

FREUNDLICH and SEIFRIZ (1923) independently developed a similar technique for determining the elasticity of protoplasm and of non-living colloidal systems. The method as originally applied consisted in placing a minute nickel particle in colloidal solutions with the aid of a micromanipulator which permits handling particles as small as $7\,\mu$. The particle remains in suspension in very dilute solutions. It is then attracted by an electromagnet and the rate at which the particle moves determined. The method has been applied to living protoplasm (SEIFRIZ 1936), and both viscosity and elasticity determined. The endoplasm of the *Echinarachnius* egg was found to have the consistency of glycerine, a value of $\eta$ of about 8.00.

If the substance is elastic, the maximum distance which the particle can travel and yet return to its original position, is an indicator of the elasticity of the medium.

### Electrophoresis.

Ascertaining protoplasmic consistency by timing the rate of electrophoretic flow of protoplasmic particles is one which presents great difficulties. If the likelihood of the current causing a change in consistency is ignored or allowed for, the viscosity of the protoplasm can be calculated on the basis of the formula: $V = \dfrac{\zeta H \cdot D}{4\pi\,\eta}$, in which $V$ is the velocity, $\zeta$ the potential difference at the boundary of the particle and the medium, $D$ the dielectric constant of the medium, $H$ the fall in potential, and $\eta$ the coefficient of viscosity.—The formula is often expressed as $P.D. = \dfrac{4\eta\,V\eta}{K\,X}$, in which P.D. is the potential difference between particle and medium, $\eta$ the coefficient of viscosity, $V$ the velocity of particle in cm. per sec., $K$ the dielectric constant, and $X$ the potential gradient, *i.e.*, the drop in potential in E.S.U. per cm.

The electrophoretic wandering of protoplasmic particles has been observed; both cytoplasmic and nuclear particles migrate to the positive pole. TAYLOR (1925) and others, have demonstrated the electrophoretic migration of particles in the protoplasm of the slime mold *Stemonitis*. A direct current of 0.02 amperes causes migration of the small microscopic particles toward the anode. A current of $4 \times 10^{-6}$ amperes brings about electrophoresis of the ultramicroscopic particles; some move toward the anode, some to the cathode, and some not at all.

### Changes in protoplasmic consistency.

#### Normal changes.

No true understanding of protoplasmic consistency is possible without a full realization of the wide range through which it may pass. Change in metabolic activity is likely to be accompanied by change in consistency. In general, low viscosity is coincident with high physical and chemical activity, and high viscosity characteristic of a reduced physiological activity. Protoplasm when dormant over a long period of time may become quite hard and dry and as brittle as thin glass; this condition is realized by the sclerotium of a slime mold.

The protoplasm of the bread mold, *Rhizopus* may flow out of a ruptured filament with ease when streaming is actively underway, but when quiet for some time, the protoplasm may be of high consistency, and then forced out of a broken hypha, just as one squeezes oil paint from an artist's tube.

**Mitosis.** Pronounced rhythmic changes in protoplasmic consistency are coincident with mitosis. The pioneer worker on this subject was NĚMEC (1899). He found an increase in consistency starting during prophase and persisting through meta- and anaphyse. At the completion of mitosis, the viscosity reverts to the original fluid state. NĚMEC was able to discern regional differences in the viscosity of the protoplast in mid-mitosis. He emphasizes the rigidity of the mitotic figure as a whole, characterizing it as a unified system which is not easily distored, and from which it is impossible to tear out any of its parts by moderate centrifuging. Firm jelly-like qualities and elasticity characterize

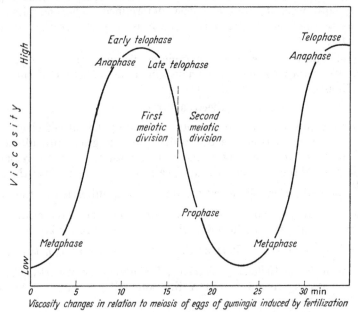

Viscosity changes in relation to meiosis of eggs of gumingia induced by fertilization

Fig. 7. Viscosity changes during meiosis. (Modified from H. B. FRY.)

protoplasm, in the opinion of NĚMEC. Both are prerequisites to an interpretation of the mechanism of cell division, and both have an important bearing on the problem of protoplasmic structure.

Much work has been done on protoplasmic viscosity changes during mitosis (NĚMEC 1899, CHAMBERS 1924, HEILBRUNN 1915, SEIFRIZ 1920, FRY and PARKS 1934). There is general agreement among a number of workers that there is an increase in the viscosity of protoplasm during the early stages of mitosis. The work is well summarized by FRY (1934) in one drawing (Fig. 7).

There is little use in discussing the significance of viscosity fluctuations during mitosis until later students of this problem agree as to the actual values. We now know that there are viscosity changes which coincide with mitotic changes, that there are as many cycles of changes in viscosity as there are mitotic cycles. The most logical deduction is that viscosity is always low at the beginning of a mitotic cycle when the structures concerned are being organized, and that viscosity is high at the end of a mitotic cycle and remains so until the succeeding one makes its appearance (Fig. 7).

But we must also recognize that there are regional differences in consistency. Astral rays, polar areas, and spindle region are of low viscosity, the ray substance undergoing centripetal flow. I (1920) was able to puncture a sea-urchin *(Tripneustes)* egg when in midmitosis, and with microneedles force, by pressure, the liquid substance from the large clear spindle region, out of the egg so as to form a globule on the surface; and then, by pressure on the globule, force the substance back into the spindle region of the still intact mitotic figure.

**Reproduction.** Reproduction involves changes in viscosity. With spore production in the slime mold there is a uniform increase in consistency in the plasmodium. LLOYD (1926) demonstrated an increase in viscosity in the protoplasts of the green alga *Spirogyra* at the time of contact of the two gametes preceding copulation. He found also a viscosity gradient in the male gamete, the consistency is higher at the posterior end and low in the region of fusion. These observations have an important bearing on the mechanism of conjugation.

The foregoing normal control of protoplasmic viscosity indicates a remarkable capacity of the living substance to change its consistency without a change in environment. The capacity of living matter to maintain the *status quo* is given the name *homeostasis*. But it is just as remarkable for protoplasm to change from one state to another without evident cause. A reason one can surmise, but not the cause. The mechanism is probably associated with enzymatic activity.

That this last statement is true is indicated by some work of STÅLFELT (1949). He observed the effects of water-soluble extracts from leaves, fruits, tubers, etc., on the viscosity of the protoplasm of *Elodea* leaves. STÅLFELT concludes that protoplasm possesses an equilibrium system which maintains a relatively uniform viscosity. The system consists of two components, one of which increases and one of which decreases protoplasmic consistency. These substances have a high physiological activity and presumably occur in very minute quantities. Normal variations in viscosity which center about a mean value are maintained by the regulatory mechanism.—If I may hazard a guess, the two regulatory substances postulated by STÅLFELT may be adenosin triphosphate and adenylic acid.

## Induced changes.

Studies on the effects of environmental factors on the viscosity of protoplasm are so numerous that only the briefest review is possible here.

**Temperature.** The viscosity of protoplasm should change with temperature, in the same way as do proteins, *i.e.* a decrease with moderate rise in temperature, but this is not always true. BASS-BECKING (1929) finds an increase in viscosity with rise in temperature, followed by a rapid decline (Fig. 8). It is quite likely that protoplasm has a temperature-viscosity control mechanism which differs from that of simple non-living systems.

Studies on the anesthetic affects of temperature (SEIFRIZ 1950) give a clear picture of protoplasmic viscosity behavior (Fig. 9).

All proteins ultimately clot with temperature rise. Protoplasmic streaming slows down as temperature rises, and stops between 30 and 40° C. At 42° death occurs, due to coagulation, never to liquefaction.

The temperature coefficients of protoplasm, at 10 degree intervals, have been found to vary within the range of the coefficients ($Q_{10}$) of egg albumin, and of chemical and physiological reactions in general, the larger coefficients being for the lower and the smaller for the higher ranges of temperature. This is true of the viscosity-temperature coefficients of blood, plasma, and serum.

Some work of my own, however, indicates that the $Q_{10}$ of protoplasm is not an ordinary one, but with exceedingly high values.

**Hydration.** The taking in of water by the cell results in viscosity changes, but the expected decrease with hydration does not always occur. Electrolytes in the imbibed aqueous medium may raise the swelling capacity of dispersed colloidal particles, and an increase in viscosity with hydration results.

**Electrolytes.** The influence of salts on protoplasmic consistency has been extensively studied. It is an accepted rule in biology that monovalent metals decrease and divalent metals increase the viscosity of protoplasm. This is, in general, true, but the rule has many exceptions. The literature on salt antagonism (LOEB 1922) will acquaint the reader with the complexity and confusion of the problem. Calcium ordinarily gelates protoplasm. HEILBRUNN (1930) and others have written extensively on calcium as the element which is responsible for much that goes on in life, particularly in relation to precipitation and gelation phenomena. They assume that calcium clotting stimulates. With this I disagree, clotting can never stimulate, activate, as in fertilization. My own view is a contrary one, not so much in regard to calcium, which does increase protoplasmic consistency, but in regard to coagulation as a form of stimulation.

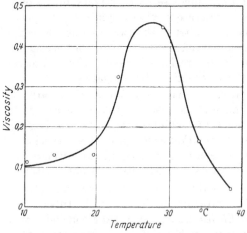

Fig. 8. Graph showing the effect of temperature on the viscosity of protoplasm. (From BAAS-BECKING.)

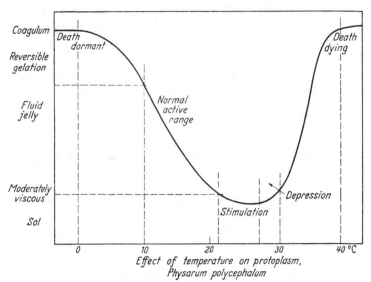

Fig. 9. Effect of temperature on protoplasm.

In all my work (1950) coagulation is a depressing, a quieting reaction, as one would expect it to be. To clot blood or albumin is to inactivate it.

The uncertainty and controversy over salts effects are apparent everywhere in the literature (LOEB 1922, CLOWES 1916, SEIFRIZ 1936).

Acids generally throw the normal state of protoplasm to the more viscous, and alkalies to the more fluid side.

**Anesthetic agents.** Here, again, there is much confusion. Anesthetic agents are said to reduce protoplasmic viscosity when in low, non-lethal concentrations. Higher concentrations of ethyl-alcohol, ether, and chloroform cause an increase in viscosity.

My own results (1949) are simply put: all anesthetic agents, of whatever nature, gelate protoplasm, whereas stimulants solate protoplasm. Some scientists do not like general rules; the exceptions, they say are too many. But general rules guide and the foregoing is a general rule. Exceptions occur, not so much among different anesthetic agents as at different times with different specimens. Such is the unpredictability of protoplasm.

**Electricity.** As early as 1864 KÜHNE (1864) used electrical stimulation as a procedure for the study of reversible viscosity changes in protoplasm. Later BAYLISS (1918) and TAYLOR (1925) studied the effect of mild electric shock on the Brownian movement of the ultramicroscopic particles in protoplasm. On application of the current there is a cessation of movement which is resumed when the electric stimulation is stopped. Gelatin has occurred, but is reversible. These experiments explain anesthesia by electric current (SEIFRIZ 1950).

**Radiation.** Much work has been done in biology and medicine on the effect of radiation on tissues (DUGGAR 1936, BERSA 1926). Mild dosages of Roentgen rays cause no change. Increased dosage will cause drastic damage. Work on all forms of radiation, X-rays, radium, and the modern forms of radiation from the cyclotron can as yet be expressed only in vague generalities particularly the work on protoplasm.

**Mechanical disturbances.** Mechanical irritation may cause pronounced changes in the consistency of protoplasm which vary considerably. Pressure may cause collapse of the mitotic figure of the dividing echinoderm egg due to solation of the protoplasm (SEIFRIZ 1924b). More striking still is the explosion of protozoa and Fucus eggs when suddenly punctured with a needle. The behavior is closely analogous to the thixotropic collapse of a gel (FREUNDLICH 1942).

Supersonic waves are a special form of purely physical disturbance which has a pronounced effect on protoplasm. Much has been attempted but very little successful work done.

## 2. Gel properties of protoplasm.
### Elasticity.

When work on the viscosity of protoplasm was abundant (circa 1920–1930) it became evident that the elastic properties of protoplasm are a far better indicator of the structure which lies beyond the visible (SEIFRIZ 1926).

The best way to prove that protoplasm is elastic is to stretch it (Fig. 10). It returns, slowly to be sure, to its original form when released.

Cell parts as well as the cytoplasm are elastic. The dominance of protein material makes this inevitable. The freed nucleus of the red blood cell of the amphibian *Cryptobranchus* possesses extraordinary elasticity. It may be stretched to 20 times its original length and yet return.

It is possible to find in the literature statements such as, "fluid protoplasm is inelastic, living protoplasm must be looked upon as a colloidal sol and not as a colloidal gel". Just what would this author say of the hard, brittle, dry sclerotium of a slime mould? It is true that protoplasm can at times be of low viscosity, and it can at rare times be wholly devoid of elasticity; it then becomes

a starch-like paste of high viscosity. The usual state of protoplasm when metabolically active is that of an elastic and moderately viscous fluid jelly. As v. MOHL
said a century ago, protoplasm is a "viscid mass"; and DUJARDIN preceded
him with an even better characterization of protoplasm as a "living jelly, glutinous, and insoluble in water".

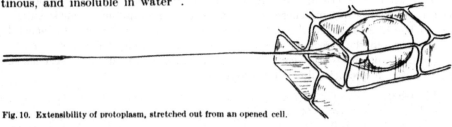

Fig. 10. Extensibility of protoplasm, stretched out from an opened cell.

## Torsion.

Viscosity gives little insight into the structure of protoplasm whereas certain
other qualities, such as elasticity and birefringence, are indicators of what lies
beyond. Another of these indicators is torsion.

Frequently, a casual observation reveals what is only later discovered and
proved. The serpentine path of growth of a thread of slime mould protoplasm
shown in Fig. 11 is evidence
of torsion in protoplasm.

That protoplasm has a
spiral twist in it I suggested

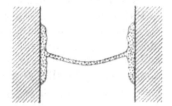

Fig. 11.                                Fig. 12.

Fig. 11. Serpentine growth of a thread of slime mold due to protoplasmic torque.

Fig. 12. A protoplasmic thread suspended in the air between two masses of protoplasm creeping on agar walls.
(From KAMIYA and SEIFRIZ 1954.)

when seeking the cause of spiral growth and arrangement in plants and animals.
That such an interpretation of spirality in organisms is justified has been demonstrated by KAMIYA (1954). He suspended a protoplasmic thread between two agar
blocks (Fig. 12) and allowed it to grow. For a while after the preparation was made
the protoplasmic thread hung down in the form of a catenary, i.e. in the form
presented by a chain suspended between two terminals, showing that the
rigidity of the protoplasm is overcome by its own weight. On careful observation
the protoplasmic thread can be seen to tighten and relax rhythmically with the
same period as that of the protoplasmic flow. The thread can also be seen to
pulsate through change in its diameter.

When the plasmodium is left under the foregoing conditions for a longer period
of time, say 2 or 3 hours, it is noticed that the hanging thread gradually takes
a form deviating from a catenary line. This form resembles that of a contorted
rope. Fig. 13 shows the stages through which protoplasm passes when twisting
and establishing a state of torsion. The thread first hangs loosely, then a loop
is formed one is beginning in the upper left corner of Fig. 13 and completed in

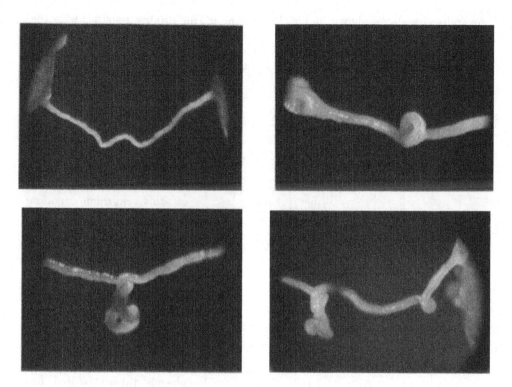

Fig. 13. Four views of the same protoplasmic strand undergoing spiral twisting due to torsion while growing (N. KAMIYA).

the upper right corner. The twisting continues and may reverse itself. In the lower right corner (Fig. 13) there are two right-handed and one left-handed loop in the same thread.

In order to make a quantitative analysis of the spiralling nature of protoplasm, KAMIYA arranged the ingenious set-up pictured and described in Fig. 14. If a cylindrical scale is placed around a moist chamber, and a protoplasmic thread with a mirror is placed in the center, and if a beam of light is projected vertically from below, the mirror will reflect the beam on to the

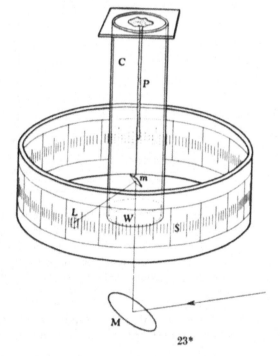

Fig. 14. General setup for measuring rotation of the lower end of a hanging protoplasmic thread. From the free end of the thread a tiny mirror is hung by means of a delicate rod. The mirror hangs at an angle of 45° with the horizontal. The thickness of the protoplasmic thread and the size of the mirror are drawn somewhat greater than the actual proportion. *S:* cylindrical scale; *P:* protoplasmic thread; *C:* moist chamber; *W:* water at the bottom of the moist chamber to prevent the material from drying; *m:* hanging, inclined mirror; *L:* light spot projected by the beam reflected at *m* onto the cylindrical scale; *M:* mirror to reflect the beam in a vertical direction. (From KAMIYA and SEIFRIZ 1954.)

cylindrical scale, as shown in Fig. 14. As the mirror rotates, together with the free end of the thread, the reflected beam also rotates. By reading the position of the bright spot projected on the scale at short intervals, it is not difficult to trace rather exactly the changes in the orientation angle of the reflected beam.

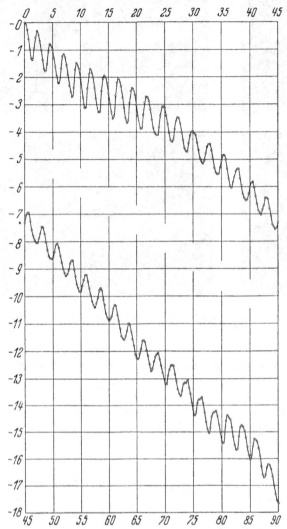

Fig. 15. Time course of rotation of the free end of the protoplasmic thread. The undulating curve declines wave by wave showing the dominance in clockwise rotation over counter-clockwise rotation. (From KAMIYA and SEIFRIZ 1954.)

Frequently a clockwise rotation reverses to the counter-clockwise rotation without coming all the way back to the initial state of orientation from which the preceding counter-clockwise rotation started. Repetition of this rhythmic motion naturally results in an increasingly greater twist of the protoplasmic thread in the counter-clockwise direction.

The dominant direction of twist is more frequently counter-clockwise but not always so. Fig. 15 represents a striking example of the clockwise rotation in which the free end of the protoplasmic thread twisted for 17.7 revolutions in the clockwise direction during 90 minutes.

Perhaps the most unexpected feature of this work is the discovery that the streaming of the protoplasm in the thread is not the cause, nor result, nor necessarily correlated with the torque, even though the two are frequently synchronized.

In Fig. 16 are represented the torsion and change in length of the thread, and simultaneously the time durations of the upward and downward streaming. It is interesting to see that there are 4 reversals in the streaming direction during one period of twisting motion, i.e. two durations of upward flow and two alternate durations of downward flow. From this fact one might believe that the period of twisting motion must be twice as long as that of the protoplasmic streaming. Further experiments, however, showed that there seems to be no causal and directly coupled relation between the flow of the endoplasm and the torsion of the thread.

To prove this more strikingly, we conducted some experiments using the technique of KAMIYA's (1942) double-chamber. In Fig. 17 is shown somewhat schematically the double-chamber consisting of compartments $A$ and $B$. In $A$ there is a protoplasmic blob ($a$), and in $B$ a hanging strand ($b$) with a small mirror ($m$) attached to it. As is seen in the figure, the strand of protoplasm ($b$) is connected with the protoplasmic mass ($a$) through the air-tight septum between the two compartments. If we apply compression or suction to the compart-

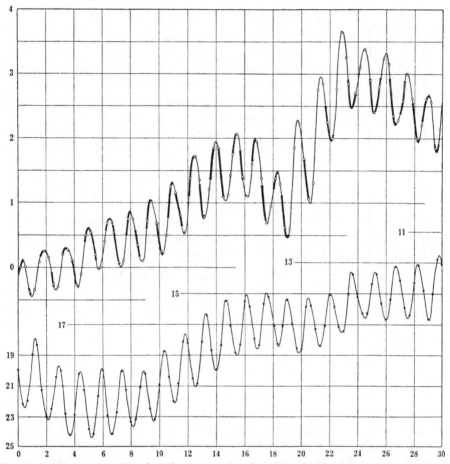

Fig. 16. Graphical representation of rotation (open circles) of the free end of the thread, of elongation of the thread (solid circles) and of the streaming duration in the thread. Heavy line on the rotation curve shows the duration in which the streaming took place upwardly in the thread while thin lines show the opposite case. Temp.: 26° C. (From KAMIYA and SEIFRIZ 1954.)

ment $A$ while compartment $B$ is kept open to the atmosphere, the endoplasmic flow along the thread in the septum is strikingly affected. Thus, by changing the air pressure in $A$ we can control the speed and direction of the endoplasmic flow inside the hanging strand very accurately. By this means it is possible to accelerate, stop or reverse the flow, and yet we find that such modifications of the flow exert no direct effect whatsoever on the twisting motion of the protoplasmic thread.

The remarkable and fundamental conclusion to the foregoing experiments, and to other work by KAMIYA (1942—1950), is the fact that there are 3 kinds of rhythms, all synchronized during the usual normal activity of the plasmodium, but which have apparently nothing to do with each other, but are dependent upon and controlled by a fourth rhythm. These three rhythms are a 50 second rhythm in direction of flow (KAMIYA 1950), a parallel rhythm in electrical potential (KAMIYA and ABE 1950), and a rhythm in torque. The nature of the fourth controlling rhythm is wholly unknown. When we know this fourth rhythm we shall probably be as close to a vital force as a biophysicist is likely to get. Personally, I believe the responsible force will be found to be of the same general nature as that responsible for nerve activity (SEIFRIZ 1953).

## Thixotropy.

SZEGVARY and SCHALEK (FREUNDLICH 1942) discovered that many gels collapse when mechanically disturbed. The phenomenon was called *thixotropy*. The collapse is followed immediately by resetting.

That protoplasm should prove to be thixotropic was anticipated. If an echinoderm egg in mid-mitosis is disturbed by a needle, the mitotic figure suddenly collapses, and may, if the shock is not too great, rebuild itself.

The study of thixotropy led to other properties of gels such as *dilatancy*, and a phenomenon which FREUNDLICH called *rheopexy*. The last name refers to the great reduction in the time required for the sol-gel transformation of a thixotropic system to take place when it is gently rolled or tapped.

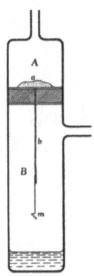

Fig. 17.   Double-chamber for controlling the protoplasmic streaming into or out of the hanging thread by regulating the air pressure in compartment. *A* Compartment *B* opens to the atmosphere. *a:* protoplasmic mass. *b:* hanging thread of protoplasm. *m:* tiny, hanging mirror. (From KAMIYA and SEIFRIZ 1954.)

## Swelling and shrinking.

One could say that the physiology of protoplasm is essentially the mechanics of swelling and shrinking (SEIFRIZ 1946). Osmosis, imbibition, permeability control, secretion, excretion, oedema, all are functions of the swelling and shrinking of protoplasm. Much is involved but we shall have to be content with the foregoing reference.

## Contractility.

Contractility is a property which medical men like to ascribe to muscle alone. To this I disagree; protoplasm in its simplest form is contractile. Were it not, it is unlikely that muscle would be. Myosin, the contractile protein in muscle, is regarded as a muscle protein and not likely to be found in plant protoplasm or a myxomycete. Contractile proteins are in all protoplasm. As contractility is more highly developed in muscle, it is best considered there.

The extent to which muscle is or can be elastic, is controversial. The anatomists base their view of a non-elastic muscle on the simple fact that muscle cannot be stretched. The medical chemists base their belief in an elastic muscle on the known fact that muscle contains elastic material. Perhaps both views are correct because a muscle fiber is a bundle of fibrils of two kinds, inelastic collagen and elastic myosin. Muscle must "give" somewhat, otherwise the strain to which it is often put would cause it to snap or tear loose. To these comments must now be added those of the chemist who says that myosin, the muscle protein, is highly elastic. GUTH (1947) states that rubber may be considered a simple prototype of muscle. The two have much in common; thus, resting muscle, like rubber, develops heat when stretched rapidly.

The effect of ATP on myosin fibers is identical to its influence on muscle fibers. Contraction takes place in ATP with a salt concentration of 0.05 molar.

If this salt concentration is increased to 0.25 molar and the ATP removed, elongation takes place, which may be stopped by lowering the salt concentration. If, now, ATP is reapplied, contraction occurs again.

The contractile property of actomyosin with ATP was originally shown by SZENT-GYÖRGYI (1941). The fibers used by him, however, lacked structural continuity. The slightest tension imposed on them caused elongation, rather than a contraction. "The individual molecules may have been contracting, but there was no intermolecular continuity" (HAYASHI 1953). Surface-spread actomyosin, compressed to form a fiber, possesses continuity, proved by its ability to lift a load.

Such fibers immersed in a solution of ATP contract due to kinking and knotting of the fibrils. The fibers show strong birefringence, positive with respect to the long axis. They also show strong ATP-ase activity.

No speculation is necessary to carry these observations over to cellular behavior. Muscle contractility is a highly unique and extraordinary quality, but contractility did not arrive late in the evolution of living matter, it was present in primitive protoplasm. The early descriptions of protoplasm emphasize its contractility; HAECKEL recognized it, so also KÜHNE (1864), v. MOHL, and DUJARDIN.

A striking example of contractility in protoplasm is the rhythmic manifestation of it in plasmodia (KAMIYA 1942), vacuoles, and cilia, which in higher organisms, becomes the familiar rhythmic pulsations of the heart, diaphragm, intestines, and uterus.

Protoplasmic and muscular contractions may involve the tightening of coiled or folded molecular chains, and they may be correlated with phosphorylation and entropy changes. What is true of the one is likely to be true of the other.

Research on the chemistry of muscle has been extensively carried out by FLETCHER, SZENT-GYÖRGYI (1953), HOPKINS, HILL (1934) and MEYERHOF.

If one leaves a comparatively light load on a muscle it will slowly fall quite a distance. A muscle of 15 cm. in the living state will, if isolated and left in sea water under a small load, relax slowly to 25 cm. Muscle may be a special case, just as is any form of protoplasm, but HILL (1951) finds that the thermoelastic behavior of living muscle is quite obedient to physical laws; it, like rubber, is a perfectly reversible system at physiological temperatures. There are those of the same opinion who remind us, however, that muscle is an "independent" heat machine, i.e., not a passive piece of material.

## 3. Surfaces.

Chemistry may be divided into point chemistry, line chemistry, and surface chemistry; and the greatest of these is surface chemistry. It is at surfaces that substances come into contact and where reactions occur. It is not chemical reaction which the physicist has in mind when he speaks of surface activity, rather is it energy revealed in some such form as tension.

Though I run the risk of minimizing the role of surface tension in protoplasmic behavior, I should like to point out that biologists, in making use of surface forces to explain nearly every conceivable form of cell behavior, have often indulged in speculations for which there is no justification. That surface forces are involved in cellular processes such as amoeboid motion, protoplasmic streaming, the migration of chromosomes, cellular fission, muscular activity, and nerve conduction, is a safe deduction, but that changes in surface tension are the cause of these processes is very doubtful. In short, surface forces play a tremendous role in the life of the cell, but surface tension is not the *sine qua non* of all cell activities.

Some misunderstanding has arisen through failure to distinguish between surface energy and surface tension. The former is the broader concept. It is as real as the surface tension which is a manifestation of it. The true reason why a chapter on surface energy so quickly degenerates into a discussion of surface tension is the difficulty with which the former is expressed and the ease with which the latter is measured and interpreted (W. K. LEWIS, *et al.* 1942).

Every surface, interface, or phase boundary is a source of potential energy which is available for work. This *energy* is expressed in *ergs* per *square* cm.; the *tension* is expressed in *dynes* per *linear* cm.

## Surface tension films.

Because of the crowding of the molecules at the surface of a liquid, due to unequal intermolecular attraction, a layer of tightly packed molecules is formed. This is the renowned surface tension film.

In a heterogeneous system substances will be concentrated in the surface in accordance with GIBBS' principle. This principle is a mathematical expression of surface condensation or selective adsorption (see Adsorption).

Early experimental determinations of the amount of solute adsorbed in a surface yielded results of the order of magnitude predicted, but recently McBAIN and DAVIES (1950) have made more accurate estimates. They have measured the amount of solute adsorbed at the surfaces of aqueous solutions of para-toluidine, amyl alcohol, and camphor, and find that it is about twice that predicted by the GIBBS' equation. The layer becomes saturated at quite low concentrations.

An important factor in the formation of monomolecular surface films is spreading. Obviously, if an oil will not spread on water, no film can be formed. Why some liquids spread and others not, is a problem of long standing. It is a problem similar to wetting. LANGMUIR (1917) assumed that spreading is determined by molecular polarity (Fig. 18 A). HARKINS (1937) denied this and stated that if the surface energy is decreased by the substance applied, the latter will spread and a film will be formed, but if there is an increase in free energy, the substance will not spread. A drop of hydrocarbon oil will spread over a large water surface because in the process of spreading there is a decrease in free energy. The spreading will continue until the oil film is reduced to a single molecule in thickness. Decrease in free energy, therefore, and not molecular polarity determines spreading. Polarity plays its part in determining molecular orientation.

## Monolayers.

Dipolar molecules orient at interfaces (LANGMUIR 1917). A soap, at an oil-water interface, will align its molecules with the metal end in the water phase (Fig. 18 B). In a similar manner are protein molecules oriented at surfaces, though the orientation usually distorts the molecule or causes a breakdown into smaller kinetic units (JOLY 1948). JOLY (1939) has shown that if protein is spread under a substantial surface pressure, the film obtained will have different properties from films spread against lower pressures. The latter are characteristically insoluble, inert, with minimum thickness and low surface viscosity. Films at extremely low pressures behave as two-dimensional gases with characteristic Newtonian viscosity. With greater pressure and constraint, they pass into a liquid phase with increasing viscosity which becomes non-Newtonian. Under still higher pressures, the molecules become tightly packed to form a solid film. If the pressure is further increased, the film collapses, becomes folded and insoluble. The folding of an insoluble film is known as the DEVAUX effect.

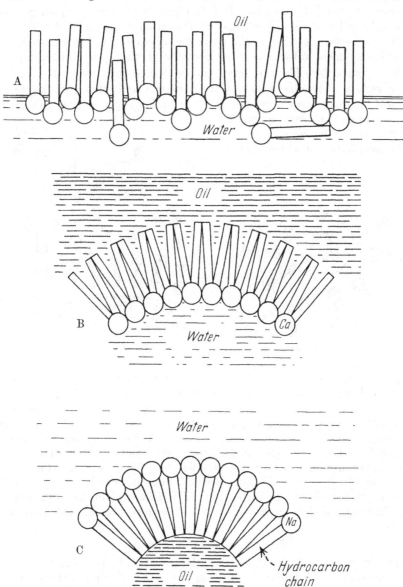

Fig. 18 A–C.   A, the orientation of polar molecules at an oil-water interface: B, a sodium soap stabilizing an oil-in-water emulsion; C, a calcium soap stabilizing a water-in-oil emulsion.

The orientation of molecules to form films, whether of oil, soap, or protein, must occur at every interface in the cell. This is not an assumption, because the more substantial films can be isolated. Where two or more parallel streams of protoplasm touch each other while flowing, yet do not intermix, an oriented monolayer must separate them.

## Multilayers.

When chemists speak of monolayers they imply that the surface molecules are packed into a film which is but one molecule thick. Available chemical or adsorption bonds are made use of; there is no means or place for additional molecules to attach themselves. But biological and colloidal films go beyond the orderly and well defined monolayer. On occasion the biologist will be interested in monolayers, as when lipoids cover the surface of a cell, but cell membranes are ordinarily more substantial than monolayers, they are multilayered.

Proof that surface films are often thicker than a single molecule is to be had from a variety of familiar forces. One of these is hydration. No one questions the great depth of the hydration layer around a sugar molecule, or the layer of water molecules surrounding a protein particle. Where a hydrated salt ion may have but a single layer

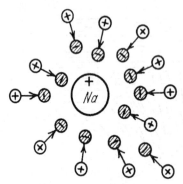

Fig. 19 A.   A hydrated ion.

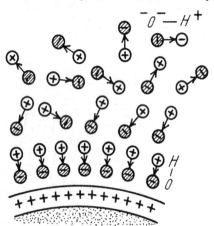

Fig. 19 B.   A hydrated protein particle.

of water molecules (Fig. 19 A) a protein particle may have a hydrated layer several molecules thick, the water molecules pile up (Fig. 19 B).

The molecules in multilayers may be haphazardly arranged, or they may be a composite of monolayers, as in the film of a soap bubble. This film is an interesting example of an ordered multimolecular layer. If the soap is sodium oleate, the metal ends of the molecules will face each other and the fatty ends face each other. Between each row of metal ends will be a layer of water (Fig. 20). The double row of oily ends constitutes a well lubricated interface the two surfaces of which glide readily upon each other. This fact explains the kaleidoscope of colors in soap bubbles. Such multilayers determine the selective permeability of cells.

## The protoplasmic membrane.

The structural pattern of the cell surface membrane has been the subject of much specualtion. OVERTON (1895) said it was an oil film. Others broadened this concept to include hydrophilic substances, i.e. proteins as well as fats, arranged in a mosaic pattern.

My own picture of the cell membrane is that of a sponge permeated by a variety of substances coming from within. I was led to this model by some work on latex. The surface of a latex glubule is probably not pure protein, as postulated by some rubber chemists, but part protein and part hydrocarbon, the same oil which fills the core of the globule and which forms the bulk of crude crepe rubber. A porous protein covering permeated by oil from within is a likely pattern of the latex globule surface and of cell membranes.

Booij (1949) postulates an ingenious model of the cell surface. He follows Bungenberg de Jong (1932) who thought the protoplasmic membrane to be a

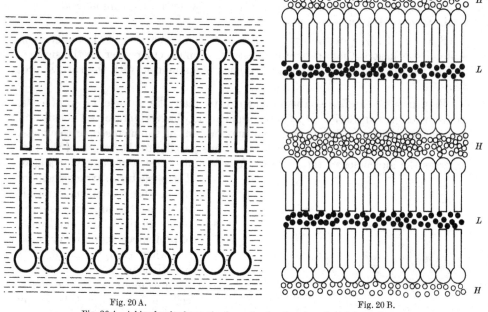

Fig. 20 A.                                   Fig. 20 B.

Fig. 20 A. A bimolecular layer of polar molecules, forming a lamellar micel.
Fig. 20 B. Multilamellar system of lipoids with interlayers of hydrophilic (H) and hydrophobic material (L).
(From W. J. Schmidt.)

tri-complex of protein, phosphatide, and a cation. Booij's hypothesis is best presented in the form of his own diagram (Fig. 21). Of it, he quite frankly says: a theory of permeability must be regarded as an attempt to synthesize the many, and often conflicting data. The protoplasmic membrane shows three general characteristics: 1. it is a "statistical" sieve; 2. it is a solvent for substances poorly soluble in water; 3. it is an ion-exchange layer.

We must conclude that the possibilities of variations in the cell surface are enormous. The membrane may be a uni-, bi-, pauci- or multimolecular layer. The orientation of molecules will follow certain rules, such as those pictured in Fig. 20, but numerous factors will interfere with a simple plan. The number of double bonds is a determining factor. Booij adds the lecithin phosphatidic

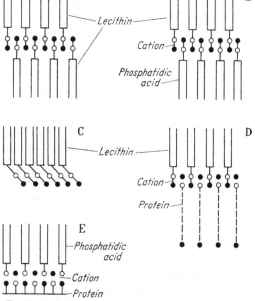

Fig. 21. A selection of possible membrane structures. More layers may be present and the apolar chains may point to the medium as well as to the cell's interior.
(From H. L. Booij.)

acid ratio which will vary from species to species, and the amount of apolar substances within the membrane. The kind of proteins will also vary, even from cell to cell. In spite of these "bewildering possibilities" we can, I think, continue to make our models, real or hypothetical, of the cell surface layer with some satisfaction and confidence. After all, we have experimental facts. SOLLNER (1955) has made cell models which simultaneously accumulate anions and cations against concentration gradients.

SCARTH (1942) is of the opinion that a free oil film on cells is unlikely, in spite of conclusions that the cell lipoids form a bimolecular layer on the surface. Rather do the cell lipoids enter into colloidal combination with the surface proteins.

Birefringence studies enabled SCHMIDT to plan the surface layer of cells. He says (SCHMIDT 1937): "the outer segment of the vertebrate cell consists of transverse alternating layers of lipoid and protein of submicroscopic thickness. The lipoid molecules form multimolecular films parallel to the length of the rod, whereas the molecules of the protein "foils" (Folien) are transversely arranged.

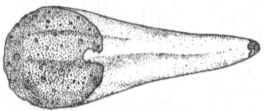

Fig. 22. The nuclear membrane lifted off from a coagulated amoeba nucleus.

This structural arrangement of lipoid and protein molecules is found in all regions of cytoplasm where the linear macromolecules of protein are regularly oriented and where the quantity of lipoid is sufficient.

A composite nature of the cell surface is postulated by SJÖSTRAND, even for mitochondria where he found two protein membranes, each 45 Å thick separated by 70 Å of lipid material. SJÖSTRAND's evidence of double membranes in cells rests on optical polarization analyses, as does the membrane model developed by MITCHISON (1952).

The nuclear membrane has received much attention (Fig. 22), so also the membrane of the cell vacuole, the tonoplast. Neither of these two protoplasmic membranes has been questioned as much as has the surface film on the cell.

The tonoplast vacuole membrane is an inner cytoplasmic surface functioning differently from the outer surface but having many properties in common with the latter one. One of these is the formation of surface processes which protrude into the large vacuole (KÜSTER 1931).

Speculations on the thickness of the plasma membrane are numerous. It is probable that the outer layer is a bi- or at least a pauci-molecular oil film. The evidence from the electron-microscope supports a two-layered plasma membrane.

My own opinion, based on microdissection studies, is that the protoplasmic surface may be so fluid, as at the tip of a pseudopod when flowing forward, that no membrane can be distinguished. But when quiet, the cell surface may become a tough, elastic, semi-rigid layer, sufficiently thick to be isolated as a delicate film.

As for the chemical composition of the cell membrane, little can be said other than the usual insistence on fat and protein. However, JONAS (1954) in a study of phosphate adsorption by erythrocytes, and the uptake of radio-active orthophosphate by erythrocytes concludes that adsorption proceeds according to the FREUNDLICH adsorption isotherm, that the exact positions in the surface of an erythrocyte can be postulated, i.e. where lipoid, calcium, protein molecules, and adsorbed phosphate ions are situated, and where the carbon, oxygen, hydrogen, and phosphorus atoms are located in the stromatin chain, all expressed in Ångström units. Such precision goes beyond that of the biologist. We can only hope the evidence justifies it.

## Permeability.

The plasma membrane and selective cell permeability are indissociably linked. To speak of one is to imply the other. The control of cellular permeability is one of the most variable of biological phenomena, and the most confusing to biologists. All that can be done with it is to state the experimental facts and then postulate a possible mechanism which will explain the particular facts under consideration, as established in work on a particular cell. The subject has been extensively and ably reviewed by numerous biologists (BROOKS 1941, COLLANDER 1923, HÖBER 1945, JACOBS 1924).

We start with a few simple facts: fat solvents such as chloroform enter cells very rapidly; water soluble salts enter very slowly, seldom attaining full equilibrium. But fat solvents are not essential to life, indeed they are harmful, whereas salts are very essential to all forms of life.

OVERTON (1900) was justified, on the basis of his experiments, in saying that all substances which are fat soluble enter cells rapidly, whereas non-fat soluble substances do not enter rapidly. But OVERTON knew that among all essential nutrient substances none is fat soluble, whereas salts which are the life of the plant, are not fat soluble. He met this criticism by saying that the fats he had in mind were lipoids and these are partially water soluble. RUHLAND (1908, 1912, 1913) clarified the situation by pointing out that the lipoid and ultrafilter theories are possible explanations of some permeability phenomena but neither one of them meets all situations. Thus, a number of dyes, such as methylene green and thionin, enter cells readily but are not fat soluble; and similarly, the fat-soluble dyes, cyanosin and Bengal rose, will not enter the living cell at all.

The classical hypotheses of cell permeability fall into three categories: mechanical, e.g. the cell membrane is a sieve; chemical, i.e. substances enter by chemical combination or solubility; and electrical, i.e. substances enter because of an electric charge. CZAPEK'S surface tension hypothesis falls into the first category. The OVERTON lipoid solubility hypothesis is a chemical explanation. The electrical explanation takes numerous forms.

OSTWALD was among the first to claim that a positively charged membrane will let negative ions pass through but not positive ones. This idea is strongly opposed by BEUTNER (1950).

Another aspect of the permeability problem is that discussed by JACOBS (1924). OVERTON (1900) had previously noticed that non-polar molecules enter cells rapidly, e.g. hydrocarbons such as benzene. If, however, there are polar groups (e.g. OH, COOH, $NH_2$) the rate of entrance is much reduced. Ethyl alcohol contains no polar group and enters rapidly. Tetrahydric alcohol, however, enters cells very slowly.

The best way to view the permeability problem is to recognize a few general hypotheses and realize that there are many exceptions. There is no need, however, to recognize the exceptions as a rule. Thus, because CO enters the cell as a molecule it is folly to say that all substances enter the cell as molecules. There are no molecules but only ions of the essential elements in the soil. GRAY (1931) expresses the whole matter clearly and tersely when he says, though there is evidence that the cell is not equally permeable to anions and cations, yet one must recognize the fact that both ions enter.

Biological hypotheses of permeability control are rather simple postulates; each well explains one experimental fact. But the protoplasmic surface is a dynamic system, not a mere sieve. PETERS (1929) has said, much goes on in a cell surface, there is structure in the chemical sense, and activity in the physical sense. PETERS (1930) makes a statement which is certain to meet medical opposition. He says: "The surfaces presented by the more highly organized molecules and particularly the proteins, I regard as a kind of central nervous system." I think he is quite right, though I (SEIFRIZ 1953) should say the same for all the surfaces in a cell, of which there are many.

## 4. Adsorption.

Few physical-chemical phenomena have played a greater role in biological thinking than adsorption. The subject is so vast a one that the reader must consult the original literature for an adequate comprehension of the subject.

FREUNDLICH (1932) was the older authority on adsorption and its biological application; today, perhaps it is EMMETT (1945) in the field of chemistry, and CASSIDY (1954) in the field of physics.

A discussion of adsorption must start with the GIBBS principle. GIBBS expressed his principle mathematically but it can as well be stated in words: a dissolved substance is positively adsorbed if it lowers surface tension and negatively adsorbed if it raises it. This means that a substance in solution will be more concentrated in an interface if it lowers the tension there and less if it raises the tension.

The biological applications of adsorption are so many that again I must ask the reader to refer to the literature [SEIFRIZ (1936), FREUNDLICH (1932), EMMETT (1945), and McBAIN (1950)]. There is universal agreement that, in general, surface adsorption is non-stoichiometric. The ratio between the amount of material adsorbed and the total weight of the adsorbent will naturally depend upon the ratio of surface to volume in the particles of adsorbent. However, in certain types of chemical adsorption the question may well be raised as to whether there is some type of stoichiometry between the chemically adsorbed molecules and the solid adsorbent. For example, if carbon monoxide is chemically adsorbed on iron catalysts at low temperatures, there seems to be considerable evidence that each iron atom on the surface will be capable of holding one chemically bound carbon monoxide molecule. Similarly, one can well imagine that in biological systems, there might be certain active centers that would hold molecules of adsorbate in a 1:1 ratio, provided the adsorbate is sufficiently tightly held by the surface. Such a behavior would perhaps be considered as stoichiometric if limited to a consideration of the active parts of the surface. In general, however, one usually thinks of adsorption as being non-stoichiometric.

It is customary to divide adsorption into physical adsorption and chemical adsorption. Physical adsorption is a non-specific variety that takes place through VAN DER WAALS' forces. This type of adsorption will occur between any gaseous or liquid adsorbate and any solid to a certain extent. It is well illustrated by the adsorption of nitrogen on various solids at and below room temperature. As a matter of fact, the adsorption of nitrogen at $-195^0$ C. has now been developed into a method for measuring the surface area of all finely divided materials, from paint pigments to dried bacteria.

Chemical adsorption is defined as that variety of adsorption in which the adsorbate molecules are held to the surface of the solid by chemical rather than physical forces. This type of adsorption is operative in the holding of hydrogen on various finely divided metals at and above room temperature. It can also be illustrated by the adsorption of carbon monoxide on metals such as iron, nickel, and cobalt. The heat of chemical adsorption is ordinarily in the range 10.50 kilocalories per mole, whereas the heat of physical adsorption is usually no more than 50% greater than the heat of liquefaction of the adsorbate.

Usually it is possible to distinguish between physical adsorption and chemical adsorption for a particular system. For example, nitrogen is held exclusively by physical adsorption on iron catalysts below $100^0$ C.; and chemical adsorption above this temperature. Such chemical adsorption is believed to be the initial step in the production of synthetic ammonia over iron catalysts at about $500^0$ C. in modern industrial processes.

Sometimes both chemical adsorption and physical adsorption occur in the same temperature range. It then becomes very difficult to ascertain exactly the amount of physical adsorption compared to the amount of chemical adsorption that may be taking place. Undoubtedly this situation arises in biology. It is certainly possible for water to be held by physical adsorption on most organic materials. However, it is probably also true that water is held on to certain groups by chemical rather than by physical adsorption. There is evidence, for example, that water vapor is adsorbed more strongly by cellulose than corresponds to purely physical adsorption. Adsorption of water vapor on certain inorganic gels, such, for example, as the silica gel, is combined so intimately with the surface of the gel, that it may have a binding in the range of 50—75 kilocalories per mole of water. On the other hand, a considerable amount of the adsorption of water vapor on silica gels can be shown to be of a physical variety that is very easily removed by evacuation at room temperature.

## 5. Optical properties.

The optical properties of protoplasm are more striking and reveal more of the structure of living matter than is generally realized. When, in the days of abundant research on the viscosity of protoplasm, it became evident that viscosity tells little of protoplasmic structure, search was made for certain optical properties in living matter which would give some evidence of structural organization. Double refraction, particularly flow birefringence, were the properties sought. Success was immediate in certain kinds of tissue. Nerve, muscle, and tendon are strikingly doubly refractive. There was, however, some success in less specialized forms of protoplasm. SCHMIDT (1937) found an encysted *Amoeba* to show double refraction. Further evidence then came from PFEIFFER (1949, 1951), INOUÉ (1951), SWANN (1952), MONROY (1947), and others.

The possible crystalline nature of protoplasm has long been a debatable subject. Those who viewed protoplasm as an emulsion (WILSON 1940, BÜTSCHLI 1894) and ascribed to it low viscosity, Newtonian behavior, and water miscibility, have denied that protoplasm is crystalline, whereas those who viewed protoplasm as a soft jelly, usually liquid enough to flow, elastic, non-Newtonian in behavior, imbibing water with avidity but not soluble in it, and possessing a specific continuity in structure, accepted the idea of crystalline protoplasm, particularly flow birefringence. Though they did not have optical properties in mind, the pioneer workers on protoplasm, v. MOHL and DUJARDIN, were clear in their description of protoplasm.

The double refraction of certain crystals, notably felspar, is a familiar optical phenomenon the study of which is undertaken with the aid of *Nicol prisms* which, when mounted, constitute a *polariscope*.

Some crystals, like felspar, split rays of light incident to their surface into two separate refracted rays, for which reason they are said to be *doubly refractive*. Their physical properties differ in different directions. Their symmetry is not pure; they are therefore, said to be *anisotropic*. Perfectly symmetrical crystals are optically *isotropic*, because their properties are the same in all directions. Such crystals are relatively few, being mostly of the cubic class as is sodium chloride.

When viewing isotropic material between crossed Nicol prisms, rotation of the upper prism will completely blackout all light, whereas when anisotropic material is so viewed it will exhibit striking double refraction (Fig. 23).

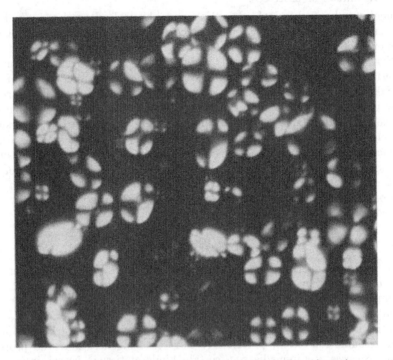

Fig. 23. Double refraction of wheat starch (500 ×). (Author unknown.)

Certain materials of biologic interest, such as rubber, possess a particularly unique property, namely, when stretched, or in any manner deformed, the amorphous rubber becomes crystalline; the irregular arrangement of their atoms assumes a symmetry, and double refraction results. There are, therefore, three fundamental structural and optical states of matter: amorphous without symmetry of any kind, isotropic with perfect (cubic) symmetry, and anisotropic with partial symmetry.

To establish the mere existence of double-refraction in living matter is usually sufficient, but the crystallographer will want precise quantitative measurement. TRELOAR (1949) describes the method well.

Double refraction in crystals is due to a specific and orderly arrangement of atoms. In colloidal matter, where submicroscopic molecular aggregates, even microscopic particles, determine the optical properties of the system, the presence of double refraction is assumed to be proof of rod-shaped particles or micels. Such particles are likely to be polar.

Just as rubber assumes crystalline properties when deformed, so also do colloidal suspensions become doubly refractive when stirred or poured. This property is known as *flow birefringence*. The polar micels orient in two directions. Orientation in three directions is typical of the solid state only.

The dyes benzopurpurin and vanadium pentoxide show flow birefringence, but only after aging; optical properties then appear indicating that the particles are now rod-like. The particles may attain microscopic dimensions, being often more than $1\,\mu$ long; their lateral dimensions remain amicronic, *i.e.* ultramicroscopic. Double refraction is noticeable in a vanadium pentoxide sol even when extremely dilute, only $^1/_{30}$ milligram per liter. A variety of optical effects characterize such solutions, iridescent color and the Faraday-Tyndall effect being among them.

The occurrence of light effects is the simplest means of determining whether or not a sol contains spherical or non-spherical particles. When a sol is stirred and eddy currents set up, the partial orientation of a non-spherical particles results in a stronger or weaker Tyndall light depending upon the position of the particles with respect to the illuminating ray and the observer.

If the particles in a doubly refractive sol are rod-shaped and polar, they will display electrical properties, and the sol will possess a measurable *dielectric constant*. The particles will orient in an electric field, and the sol will become doubly refractive. The dielectric constant, like double refraction, increases with age in a vanadium pentoxide sol. The value of the dielectric constant of a freshly prepared sol is equal to that of water, about 80; in an older sol the value may be as high as 400.

The physical basis of flow birefringence is not as simple as the above explanation implies. There are two explanations of the orientation of linear particles, the mechanical one of Maxwell and the electrical one of Kerr. The former is based on the hydrodynamics of the motion of a particle in a continuous liquid medium, operating with a Maxwellian representation of forces of compression and expansion acting under a $45^0$ angle to the flow gradient, and tending to orient the particle with its long axis parallel to the expanding force.

The Kerr effect explains why some substances become optically anisotropic when they are brought into an electrostatic field, and hence show birefringence. The direction of the electric field corresponds in this case to the optical axis of the birefringent crystals.

Numerous other hypotheses have been presented including one by O'Konski and Zimm (1950). They, studying electrical orientation and relaxation effects in aqueous colloids, particularly tobacco mosaic virus, find that there are orienting mechanisms other than those due to permanent or induced dipoles. The authors are of the opinion that the results of Lauffer (1946), who studied the birefringence produced in aqueous tobacco mosaic virus solutions by 60-cycle sinusoidal fields, and of others who observed in bentonite aquasols birefringence which varied in magnitude and sign with the frequency of the applied sinusoidal voltage, are to be explained by the distortion of the ionic atmosphere surrounding the particles, and the interactions between the particles and their ionic atmospheres.

Orientation-birefringence has been established in many cases including strain-birefringence in gels under shear (Guth 1944). However, flow birefringence is not always a simple matter of orienting rods; there are other factors including interference, one form of which is Brownian motion. Where anisometric particles are large, they, even at low velocity gradients, orient with the long axis parallel to the stream lines, which is the position of minimum dissipation of

energy and therefore of minimum viscosity and maximum birefringence. Where
the anisometric particles are small, and realizable shear rates are insufficient,
the disorientation caused by Brownian motion will overcome the tendency to
orient.

EINSTEIN equated the Brownian disorientation of spherical particles, and
BURGERS did the same for rod-shaped particles. In the case of spherical particles
we have:

$$\frac{A^2}{t} = \frac{RT}{4\pi N r^3 \eta}$$

where $A$ is one-third of the square of the mean angle of rotation in time $t$, $r$ is
the radius of the spheres, and $\eta$ the viscosity of the liquid dispersion medium;
the other terms have their usual significance.

Brownian disorientation may be assessed as a rotatory diffusion constant, $D$,
which is given by:

$$D = \frac{RT}{8\pi \eta r^3} ;$$

while for rod-shaped particles we have:

$$D = \frac{3KT}{8\pi \eta} \left( \frac{\log l/r - 0.8}{l/2^3} \right).$$

Ultimately, a state of equilibrium is reached between hydrodynamic orientation and Brownian disorientation. Orientation will increase with particle size
and with rate of shear. Disorientation will decrease with particle size and with
viscosity of the dispersion medium but will increase with rise of temperature,
because of increased heat motion of the particle and reduction of viscosity of
the medium in which they rotate.

One final comment should be made before turning to the biological application of double refraction. The expression "polarized light" in reference to
refracted light indicates a relationship between an optical property such as refractive index and an electrical property such as polarization. This relationship is the physical basis of double refraction.

The biologist strives for quantitative results and precision with as much
zeal, if not success, as the physicist, but any one at all familiar with living material will know that the foregoing treatment of birefringence cannot be observed
in full by the biologist. He should, however, be aware of the complexity of the
problem. It is often enough to establish the reality of a physical quality to
justify a point of view. At the beginning of this chapter I said there were two
views of protoplasm, one that of a low viscosity, optically inactive, solution;
the other a high viscosity optically active colloidal, i.e. high polymer, dispersion.
It is only necessary to prove the existence of double refraction to substantiate
the latter view. Outstanding in this respect is the work of SCHMIDT (1937) and
PFEIFFER (1940).

That certain living tissues are birefringent is established beyond all doubt.
Connective and contractile tissue such as striped muscle, tendon, and nerve
are strongly anisotropic, so also the products of living tissues such as cellulose,
hair, wool, and silk.

The work of MONROY and MONTALENTI (1947); HUGHES and SWANN (1952);
and INOUÉ and DAN (1951) leave little doubt as to the reality of double refraction in living material when observed with a well adjusted polarization microscope. Work by these and other scientists has also proved that not only is protoplasm under favorable conditions birefringent, but also the nucleus, chromosomes,
spindle fibers, chloroplasts, mitochondria, and centrosome. YASAKI and SINOTÔ

(1952) have shown the nucleus and chromosomes to be doubly refractive (Fig. 24). PFEIFFER (1951, 1953) made photoelectric recordings of the anisotropy of *Drosophila* chromosomes, and found that stretching of the chromosomes influenced the interband spaces which become positively birefringent, leaving the bands negatively anisotropic. PFEIFFER's (1951) refractometric determinations of cells in mitosis show minimum values at the beginning of metaphase and the end of anaphase. SCARTH (1927) showed that double refraction in chloroplasts disappeared on solation and returned on regelation.

Double refraction was certain to prove instrumental in settling the controversy over the reality of the spindle fibers. INOUÉ and DAN (1951) obtained a

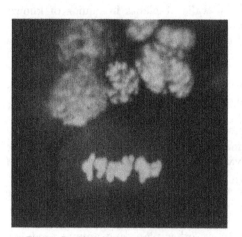 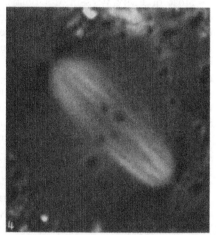

Fig. 24. Double refraction of chromosomes (Y. YASAKI and Y. SINOTÔ).　　Fig. 25. Double refraction of spindle fibers (S. INOUÉ and K. DAN).

photograph of spindle fibers (Fig. 25). They reject older theories of the structure of the aster and conclude that positive longitudinal birefringence clearly indicates that the rays are well defined entities probably protein in nature. This conclusion disposes of the false contention that spindle fibers are artifacts.

There is a pitfall in double refraction studies to which I should like to call attention. Certain of the pictures offered by biologists as proof of double refraction in living material show the edge or the surface of the cell or nucleus to be brightly illuminated. The pictures closely resemble typical cases of simple diffraction. Double refraction may be present but the photograph is no proof of it. The double refraction of points, lines, edges, and surfaces is rarely convincing proof of birefringence in the material. The difficulty can be partially straightened out by immersing the material in a medium of the same refractive index. On the other hand, it is at surfaces that molecules become readily and prominently oriented, and therefore it is here that one is more likely to find double refraction. JOLY (1949) has devoted much attention to the properties of surfaces, including birefringence during flow with reference to the effects of heat, acidity, and electrolytes.

If, for a moment, we reverse our view of this subject, and consider the effect of polarized light on organisms we find little evidence and much discussion. The ancients unknowingly indulged in the belief of an effect by polarized light on man; it is implied in such terms as moon-madness and lunacy. Today it is generally admitted that polarized light does have an effect on life but its usual natural source is too feeble to exercise any notable influence. CASTLE (1942) finds this true in the phototropic effect of polarized light on the growth of the fungus *Phycomyces*. Far more striking is the response to differences in intensity between two beams from two sources of light opposed at 180°. The response is due to differences in the reflection losses at the cell surface and consequently in the relative intensities of certain rays of light within the cell.

## Double refraction in non-living protoplasmic products.

Much work has been done on non-living materials which are the secretion products of protoplasm. Preeminent among these is starch, probably the first organic material in which double refraction was demonstrated.

Most of FREY-WYSSLING's polarization studies have been on the structure of cellulose. He (1953) immersed the cell walls of plants in liquids of known index of refraction. When a crystal is immersed in a liquid the index of refraction of which is equal to that of one of the three indices of the crystal, then the boundary between liquid and crystal, viewed in the light of Nicol prisms, disappears provided the plane of polarized light is in line with the direction of that particular crystal index which is identical with the index of the oil. A series of mixtures of amyl-benzyl alcohol and cinnamon clove oils of increasing refractive indices were used. In this way, FREY-WYSSLING showed that cell walls are built of submicroscopic rod-shaped particles which he identifies with the NÄGELI micels. The long axis of the micels corresponds to the direction of the greatest refractive index; the latter value, therefore, gives the orientation of the micels in the wall.

That the products of protoplasm are strongly birefringent is indirect evidence that protoplasm is the same. Not only starch, cellulose, rubber (when stretched), chitin, hair, and wool are doubly refractive, but so also are most organic substances of high molecular weight, and many other plant and animal products. Rod-shaped particles, and flow birefringence have been described for quinine, lithium urate, barium malonate, and fibrin. EDSALL (1930) finds that myosin solutions are anisotropic when flowing and, therefore, presumably possess rod-shaped micels.

The most interesting and satisfying feature of the work on birefringence is the fact that it correlates with other properties. If there is birefringence we can expect to find elasticity, non-Newtonian behavior and crystalline rod-shaped particles or *crystallites*, and do, as in certain soap solutions. Because protoplasm is elastic it is certain to be non-Newtonian, and to show flow birefringence.

The cause of birefringence in protoplasm is usually ascribed to the orientation of crystalline particles, though it has been associated with lipoid substances in erythrocytes (SCHMIDT 1937). The foregoing emphasis on the correlation of optical and physical properties leads to other phenomena related to double refringence. One of these is *rheodichroism*.

*Rheodichroism* owes its origin to the different powers of light absorption of oriented particles when viewed in parallel direction perpendicular to the axis of alignment. The apparatus and the mathematics involved are given by ZOCHER (1925) and PFEIFFER (1940, 1953). PFEIFFER has demonstrated that at constant viscosity, swelling rate, temperature, and flow speed, the intensity of rheodichroism is expressed by curves of the same general type as double refraction, *i.e.* at first linear, then flattening out as the rate of flow speed increases.

Dyes, *e.g.* rhodanine B and neutral red, are necessary to reveal rheodichroism in protoplasm. The dye stains the lipid fraction of the oriented protoplasmic particles. The particles responsible are fluorescent.

## Polarity.

The foregoing section on optical properties gives emphasis to the role played by polar particles in determining flow birefringence, non-Newtonian behavior, and such properties as elasticity and rigidity. Polarity has also long played a part in biological thinking whether it is the polarity of an egg, a man, or a willow twig.

In protoplasm, polarity reveals its presence in many ways. If protoplasm is subjected to a mild electric current the visible particles frequently line up in compact parallel rows covering a large area.

The chemist is interested in the polarity of molecules and, naturally, the question arises, is molecular polarity the basis of cellular and organismal polarity? The answer cannot be categorically given, but it may well be true that molecular polarity and molecular orientation determine the polarity of the cell and the organism as a whole. Certainly the orientation of molecules and particles will be as much involved in life processes as it is in non-living material.

Polar molecules are of wide distribution. The water molecule is dipolar. Proteins, with $NH_3^+$ and $COO^-$ radi les, yield dipolar ions. Such electric polarity will cause particles to orient in an electric field and if they are linear, birefringence will result.

With the aid of an electric field it is possible to determine the degree of polarity, and the center of gravity of the two unlike charges on a linear molecule; the position is the point of electrical balance. It is further possible to determine the abundance and position of the ionized groups, for example, the sulfhydrol group. The particles responsible for rheodichroism have been called *leptones* by PFEIFFER (1940, 1953). He finds the term convenient because it refers not to those numerous particles, globules, plastids, and chondriosomes in protoplasm but to that group of ultramicroscopic particles which are responsible for flow birefringence. LAUFFER (1946) says that the hydrodynamic view of rheodichroism, like birefringence, and non-Newtonian flow, is that these phenomena are the result of particle orientation without linkage between micels.

## Color.

Color is a difficult study. When it is stated that there is no blue pigment in any blue feather, a realization of the role of colloidal structure in determining color becomes evident. There is also no blue pigment in blue eyes, blue snow, ice, thin milk, sea and sky.

The blue, purple, and red color of samples of colloidal gold, all prepared in the same way, is an interesting but as yet unsolved problem. In making colloidal gold by electrical dispersion one may control amperage, heat, $p_H$, salt concentration, and specific ionic effect, and still the color varies. It appears that the blue color is a structural one, *i.e.* colloidal, whereas the red color is determined by the nature of the environment of the particles. Recent work by TURKEVICH *et al.* (1954) still leaves the matter in doubt.

ZIEGENSPECK and PFEIFFER (1954) have made a study of color in polarized light in plant tissues with striking effects.

X-rays and ultra-violet light enter into our discussion primarily as a method rather than as a property of protoplasm. Using X-rays LANGELAAN (1937) made a microscopic analysis of multiperiod gratings or grids in muscle, nerve, and connective tissues. The grid structure in tissue persists from the microscopic to the Ångstrom dimensional order.

The expression "grid" is indissociably associated with ASTBURY (1940) whose extensive studies of the crystalline properties of tissues through the use of X-ray photography led him to a grid structure of muscle and wool.

Ultraviolet photography has been much used in the past and excellent pictures of mitoses obtained (LUCAS 1930), but without significant results.

## Fluoromicroscopy.

The fluorochrome method was apparently first developed by HAITINGER (1938) and later applied to biological material.

Cells are treated with *fluorochromes, e.g.*, berberine sulfate or auramine o, and observed with a luminescence microscope. Bacteria thus treated are clearly observed as bright golden rod shaped bodies against the dark background. STRUGGER (1931, 1949) has been particularly active in the use of fluorochromes such as acridine orange. Forms of coriphosphin are fluorochromes which make nucleus and cytoplasm luminescent. YASAKI has for some twenty years studied luminescent phenomena in organisms. The fresh water shrimp, *Xiphocaridina compressa*, is rendered luminescent by bacteria living in its body. Using a fluoromicroscope, YASAKI and SINOTÔ (1952) labelled chromosomes and nuclei with aluminum-morine. They find that in root tip tissue, these structures shine as brilliant greenish yellow bodies, whereas the cytosome is faintly luminous. In animal tissue *(Locusta migratoria)* and salivary gland chromosomes, meiotic chromosomes and nuclei in the first and second *spermatocyte* divisions, groups of proteins are exposed on the surface or lie within the molecular complex. Through such experiments as these, initiated by studies on birefringence, it will be possible to learn more of the inner configuration of protein molecules. We are only at the beginning of such studies, and only beginning to realize how great is the biological application. The discrimination which living systems show in their choice of stereoiosomeric forms, which was PASTEUR's brilliant discovery, is a consequence of enzymic activity. This is organic catalysis coupled with optical activity.

## 6. Salts.

A discussion of salts is probably more suited for a chapter on the chemistry of protoplasm but salts play a prominent physical role in life. Of this botanists are fully aware, as in osmotic balance, but precisely because salts are so often presented only as nutrients or as agents in maintaining osmotic balance, do I wish to add this note.

Salts are in large measure responsible for electrical balance, and for homeostatic stability in general, in all living systems. I can best present my note by quoting OVERTON.

OVERTON (1902), in describing his experiments on the indispensability of sodium, says: Hitherto I believed, in common with most physiologists, that sodium chloride served solely to maintain the osmotic pressure of tissue fluids, and that nature prefers this substance only because it is cheap to come by and comparatively harmless. This view has now become untenable, and there is no doubt that sodium ions possess a very specific function in muscular contraction and in the conduction of excitation through muscle, and perhaps also through nerve.

"If we now ask exactly what part sodium might play during the conduction of excitation, there appear to be only two alternative possibilities. Either sodium affects the surface properties of the muscle fibers, that is, it acts upon a superficial layer of the sarcoplasm without actually penetrating the fibers, or there must be an exchange between cations which are present inside the muscle fibers (most probably potassium ions) and sodium ions in the surrounding medium."—

OVERTON concludes: "On the whole I am inclined to favour the second suggestion ...".—What OVERTON has said of muscle and nerve (KATZ 1952) is equally true of plant protoplasm.

## 7. Electrical forces.

That electrical forces play a prominent role in life has long been contended, and generally accepted. Errors in experiment and interpretation have been made, GALVANI's being one of the first, but today these forces can be measured.

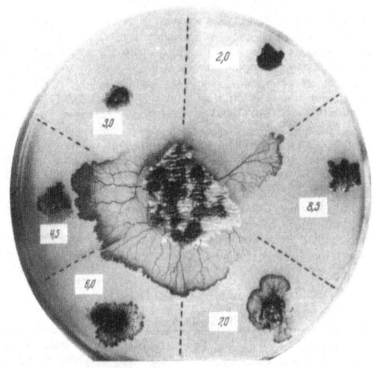

Fig. 26. The chemotaxis of a slime mould, being a growth orientation to $p_H$.

To be sure, they may be electrical only because we measure them with electrical instruments. Perhaps we should simply say that there exists a difference in energy level between any two parts on a living body. However, that this difference is one in electric potential is certainly a justified conclusion in such organisms as the electric eel (ALBE-FESSARD 1951). It is also justified where differences in ionic concentrations exist, and this means throughout the living world. Oxidation-reduction equilibria involve a transfer of electrons and these yield an electric potential which can be measured. Electrical forces are, therefore, present in living matter and must be recognized. Just how they arise and what their function is cannot always be stated. Also, the biologist often uses descriptive terms such as "wound potentials", which are meaningless to the physicist.

*Magnetic forces* apparently have no effect on living matter, nor do they arise in living matter.

*Static electricity* was said by Sir OLIVER LODGE to accelerate the growth of plants, but American workers could not corroborate his results.

*Galvanotaxis*, which is the autonomous movement of organisms toward the pole of an electric circuit, is believed to be a guiding force in directing leucocytes, amoebae, myxomycetes, and ciliated protozoa to the cathod or anode (VERWORN 1896, COEHN and BARRATT 1905, MAST 1931, WATANABE, KODATI and KINOSHITA 1938, and ANDERSON 1951). I cannot speak for all cases, but so far as myxomycetes are involved, I find all "galvanotaxis" to be *chemotaxis* with $p_H$ as the determining factor. How striking the orientation of a slime mold is to $p_H$ is shown by ANDERSON (1951). Note how in Fig. 26 the plasmodium exhibits maximum growth on the acid side of neutrality, and how one branch creeps along the dividing line between $p_H$ 2 and 8.5, *i.e.* along a line of about $p_H$ 5.

## Phase-boundary potentials.

Phase-boundary potentials and *"membrane" potentials* are a controversial subject. The latter have been long supported with much research by MICHAELIS (1925), and denied as having anything to do with the membrane as such by BEUTNER (1944, 1950). BEUTNER holds that membranes serve only to establish and maintain phase boundaries, and it is at phase boundaries where electric potentials are set up. The MICHAELIS-BEUTNER controversy has been best answered by SOLLNER (1942, 1953).

*Electrophoresis* (cataphoresis) is a much studied phenomenon, perhaps even more so in biology than in colloid chemistry where it was discovered. It, and the associated phenomena of electro-osmosis, stream potential, and the Dorn effect, were to a large extent in the hands of FREUNDLICH (1932) during the first years of this century.

Electro-osmosis may well occur in living systems. It is the flow of a liquid, usually water, through a capillary or porous wall. If potentials exist across porous membranes, water will flow toward the electronegative side. It was once thought that protoplasmic streaming is an example of electro-osmosis. It is quite possible that the one-way flow of water through the walls of the intestines is an electro-osmotic phenomenon.

## Electrophoresis.

The migration of particles suspended in a liquid medium and subjected to an electric field was first observed by REUSS in 1807, and later by FARADAY, DuBOIS-REYMOND, and QUINCKE. Before the study of particle migration had progressed very far, the similar movement of a solution through capillaries under the influence of an electric field was noted. The two phenomena were first given mathematical interpretation by QUINCKE and experimental precision by HELMHOLTZ. FREUNDLICH (1925) established the relationship between the *zeta* or electrokinetic potential of particles and the thermodynamic, or NERNST, potential of electrodes.

Following QUINCKE's suggestion that charges are adsorbed, HELMHOLTZ assumed that a *double layer* of ions is held at the interface between the solid, whether capillary wall or particle, and the liquid which touches it. He expressed the potential of this double layer in terms of the volume of liquid flowing through a capillary in a given time. The formula became:

$$v = \frac{\zeta \, r^2 \, E}{4 \, \eta \, l}$$

where $v$ is the velocity or volume of liquid flowing per second through a capillary of radius $r$ and length $l$, under the influence of the external e.m.f., $E$; $\eta$ is the coefficient of viscosity of the medium and $\zeta$ the potential at the surface of the capillary wall.

The dielectric constant, $D$, was not taken into account by HELMHOLTZ, if indeed it was understood at this time, and was later added by LAMB. For $E/l$ a potential gradient $H$ was substituted. The formula now is:

$$v = \frac{\zeta r^2 H D}{4\eta}.$$

It expresses the rate in terms of volume per unit time of liquid moving through a capillary under the influence of an electric field.

As the rate of movement of particles in an electric field is controlled by essentially the same factors involved in the movement of liquid through a capillary, the electro-osmotic formula may be revised for particles:

$$u = \frac{\zeta H D}{4\eta}$$

where $u$ is the electrophoretic velocity in centimeters per second.

A very interesting development occurred in ideas on the nature of the distribution of the charges (ions) surrounding colloidal particles in which FREUNDLICH (1926), SMOLUCHOWSKI, GOUY and DEBEYE took part. But it is chiefly to GOUY (1925) that the newer concept of a *cloud* or *field* of ions, rather than a compact double layer, surrounds colloidal particles.

The role of electrophoresis in physiology is not well known, yet some facts stand out as undeniable. One is the fact that all living cells migrate in an electric field; and *all*, with no known normal exceptions, move to the anode, *i.e.* the cell is negatively charged.

Many correlations between physiological activities and zeta potentials on cells have been made. Bacteria are said to show a difference in potential, and therefore mobility, in the rough and smooth forms.

The discovery that certain smooth intestinal bacteria have no electric mobility at any $p_H$ values, whereas the corresponding rough variants possess a high mobility which is dependent on $p_H$, is an unexpected discovery.

I once made the suggestion that the then four blood types would probably be distinguished by their surface charge, established by electrophoretic migration. LANDSTEINER objected, on the grounds that the coagulation tests by which blood types are established are not a simple matter of surface charge but of protein specificity (ABRAMSON and MOYER 1942). This is undoubtedly true, but it is also true that bacterial agglutination, which is a kind of coagulation, is to a large extent determined by surface forces. Bacteria agglutinate at their *critical potential* which NORTHRUP (1921, 1922) established at 11 mv. The situation may be merely a matter of two concomitant properties running parallel to each other.

One of the most ingenious and fundamental relationships between the zeta potential and a biological property ever made is that of MOYER. He (1934a) studied the electric mobilities of the latex particles of *Euphorbias*. *Hevea brasiliensis*, which is the commercial source of rubber, is not available in the greenhouses of temperate regions, but an excellent substitute is to be found among those plants known as the spurges which are species of *Euphorbia*. When a stem of *Euphorbia* is cut, it exudes latex. Latex is a fine suspension of hydrocarbon (rubber) in an aqueous medium of salts, carbohydrates, and proteins. MOYER first established migration curves, plotting rate of mobility against acidity (Fig. 27). The latex when collected is of a fairly constant $p_H$. Those species known to be taxonomically related yielded cataphoretic migration curves of identical form which crossed the line of no migration at almost precisely the same $p_H$ value, *i.e.*, they had the same isoelectric points; whereas those species taxonomically not closely related yielded curves of different form, with other isoelectric points. An occasional species which did not fall into any one group, proved to be an isolated form of questionable relationship on taxonomic grounds.

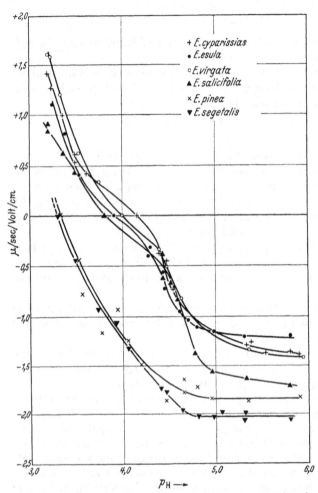

Fig. 27. Electrophoretic migration curves of the latex particles of *Euphorbia:* note that four species form one closely related group, and two species form another group. (From L. MOYER.)

Further study brought other interesting facts to light. The geographic distribution of the species was such as to agree with the protein relationship, and chromosome number showed similar conformity (MOYER 1934b).

Some data on the electrophoresis of droplets of protoplasm, here mentioned for the first time, indicate an isoelectric point at $p_H$ 4.1, and a typical mobility curve,

A summary of work on electrophoresis has been published by ABRAMSON *et al.* (1939, 1942).

*Nerve impulses* would not appear to be a botanical study, yet in reality they are. Applied to animal material they constitute a large chapter in medical science (HILL 1934). Applied to plants they are a still new and difficult problem.

The "reflex arc" (MOLISCH 1929), and action potentials (BURR 1934) have been measured in the sensitive plant Mimosa. There is no reason why an action potential should not be a response to a stimulus in an elementary form of protoplasm such as a thread of slime mold (KAMIYA and ABE 1950).

*The plant viewed as a nervous system* is not an acceptable view to the medical profession, but if a "reflex arc" may be measured in Mimosa, an action potential in *Nitella*, and a rhythmic change in potential in myxomycetes in harmony with a rhythm in flow, it is difficult to escape the view that the plant is a nervous system; after all, it is a living system. The attention which PETERS (1930) has given this subject is particularly satisfactory.

*Rhythmic potential changes* occur in the slime mold *Physarum*. They parallel the rhythm in flow. KAMIYA's work on this problem has been quite extraordinary (1942, 1950).

*Electroencephalography* represents the ultimate in electrophysiology (BERGER 1933, 1936). As it applies to the brain, it can not have an immediate application to plant life, but the potential changes which occur in mammals are likely to find, in time, their counterpart in plants.

*Electric fields* applied to protoplasmic behavior is a highly speculative subject, but if the field postulated by the physical chemist is accepted, and no more done to it than to permit a wave of depolarization to traverse a thread the surface of which is the field in question, then we have a theory of several electrical phenomena which are characteristic of living matter; *i.e.*, nerve conduction, muscular activity, and protoplasmic streaming (SEIFRIZ 1953a and b). That we deal with "nerve" conduction in an elementary thread of protoplasm is beautifully illustrated by the behavior of moving particles in a cell. These granules, suspended in protoplasmic threads which are less in diameter than the particles they carry, exhibit all the peculiarities in movement which is to be seen in muscle fibers, *i.e.* the protoplasmic thread is elementary muscle and nerve fiber.

Of protoplasm it may be said, if there is motion there will be an electrical potential, and this justifies PETERS (1930) remark that the cell is a nervous system.

## Literature.

ABRAMSON, H. A., and L. S. MOYER: The electrical charge of red blood cells. J. Gen. Physiol. **19**, 601 (1936). — ABRAMSON, H. A., *et al.*: Electrophoresis, a symposium. Ann. N. Y. Acad. Sci. **39**, 105 (1939). — ABRAMSON, H. A., L. S. MOYER and M. H. GORIN: Electrophoresis of proteins. New York 1942. — ALBE-FESSARD, D.: Données sur les caractères de la commande centrale de la décharge chez la torpille et chez la raie. Arch. Sci. physiol. **5**, 197 (1951). — ANDERSON, J. D.: Galvanotaxis of slime mold. J. Gen. Physiol. **35**, 1 (1951). — ASTBURY, W. T.: Fundamentals of fibre structure. London 1923. — X-ray studies of molecular structure of myosin. Proc. Roy. Soc. Lond. Ser. B **129**, 307 (1940). — Proteins. Chem. a. Industr. Eng. **1941**, 491.

BAAS-BECKING, L. G., *et al.*: Proc. Kon. Nederl. Akad. Wetensch. B **25**, 1 (1929). — BAYLISS, W. M.: General physiology. London 1918. — BĚLEHRÁDEK, J.: Protoplasmic viscosity as determined by a temperature coefficient. Nature (Lond.) **118**, 478 (1926). — BERGER, H.: Über das Elektrencephalogramm des Menschen. Arch. f. Psychiatr. **99**, 555 (1933); **104**, 678 (1936). — BERNSTEIN, M. H., and D. MAZIA: The desoxyribonucleoprotein of sea urchin sperm. Biochim. et Biophysica Acta **10**, 600 (1953). — BEUTNER, R.: Bioelectricity, Medical physics, ed. O. GLASSER, p. 35, Chicago 1944. — The alleged membrane potential produced by diffusion in nerve and muscle fibers. Nature (Lond.) **166**, 197 (1950). — BINGHAM, E. C.: Fluidity and plasticity. New York 1922. — Some fundamental definitions of rheology. Rheology **1**, 507 (1930). — BLUM, H. H.: Time's arrow and evolution. Princeton 1951. — BOOIJ, H. L.: Faraday Soc. Disc. **143** (1949). — BROOKS, S. C., and M. BROOKS: The permeability of living cells. Berlin 1941. — BÜTSCHLI, O.: Investigations on microscopic foams and in protoplasm. London 1894. — BULL, H. B.: Physical biochemistry. Chicago 1943. — BUNGENBERG DE JONG, H. G.: Die Koazervation und ihre Bedeutung für die Biologie. Protoplasma **15**, 110 (1932). — BURR, H. S.: An electrometric study of Mimosa. Yale J. Biol. a. Med. **15**, 823 (1943).

CASSIDY, H. G.: Adsorption for organic chemists. New Haven, Conn. 1954. — CASTLE, E. S.: Spiral growth in Phycomycetes. Amer. J. Bot. **29**, 664 (1942). — CHAMBERS, R.: The physical structure of protoplasm. V. General cytology, ed. COWDRY. Chicago 1924. — CLOWES, G.: Protoplasmic equilibrium. J. Phys. Chem. **20**, 407 (1916). — COATES, C. W., R. T. COX and L. P. GRANATH: The electric discharge of the electric eel, Electrophorus electricus. Zoologica, N. Y. Zool. Garden **22**, 1 (1937). — COEHN, A., and W. BARRATT: Über Galvanotaxis vom Standpunkte der Physikalischen Chemie. Z. allg. Physiol. **5**, 1 (1905). — COLLANDER, R.: The permeability of plant protoplasts to non-electrolytes. Trans. Faraday Soc. **33**, 985 (1937). — CONKLIN, E. G.: Effect of centrifugal force on the structure and development of the eggs of Crepidula. J. of Exper. Zool. **22**, 311 (1917). — COSTELLO, D. P.: The hyaline zone of the centrifugal eggs of Cuminga. Biol. Bull. **66**, 257 (1934).

DANIELLI, J. F.: Cytochemistry. New York 1953. — DAVIES, H. G.: The ultra-violet absorption of chick fibroblasts. Exper. Cell Res. **3**, 453 (1952). — DENUES, A. R.: Electron microscopy of residual chromosomes. Exper. Cell Res. **4**, 333 (1953). — Science (Lancaster,

Pa.) **113**, 203 (1951). — DIANNELIDIS, T., and K. UMRATH: Über das elektrische Potential bei den Myxomyceten. Protoplasma **42**, 312 (1953). — DOUGHERTY, R. L.: Hydraulics. New York 1937.

EMMETT, P. H.: Gas adsorption. Industr. a. Engin. Chem. **37**, 639 (1945). — EWART, A. J.: On the physics and physiology of protoplasmic streaming in plants. Oxford 1903. — EWART, R. H.: Adv. Colloid Sci. (Rubber) **2**, 197 (1948).

FREUNDLICH, H.: Capillarchemie. Leipzig 1932. — Some mechanical properties of sols and gels. The structure of protoplasm, p. 85, ed. W. SEIFRIZ. Ames, Iowa 1942. — FREUNDLICH, H., and G. ETTISCH: Das elektrokinetische und das thermodynamische Potential. Z. phys. Chem. **116**, 26 (1925). — FREUNDLICH, H., and W. SEIFRIZ: Über die Elastizität von Solen und Gelen. Z. phys. Chem. **104**, 233 (1923). — FREUNDLICH, H., et al.: Strömungsdoppelbrechung von Farbstofflösungen. Z. phys. Chem. **115**, 119 (1923). — FREY-WYSSLING, A.: Submicroscopic morphology of protoplasm and its derivatives. New York and Amsterdam 1953. — FRY, H. J., and M. E. PARKS: Mitotic changes and viscosity. Protoplasma **21**, 473 (1934).

GECKLER, P. R., and R. F. KIMBALL: Effect of X-rays on micronuclear number in Paramecium. Science (Lancaster, Pa.) **117**, 80 (1953). — GICKLHORN, J.: Intracellulare Myelinfiguren. Protoplasma **15**, 90 (1931). — GLASSER, O.: Medical physics. Chicago 1944. — GORTNER, R. A.: Outlines of biochemistry. New York 1949. — GOUY, A.: Sur la constitution de la charge électrique à la surface d'un électrolyte. J. de Phys., Ser. 4, **457** (1925). — GRAY, J.: Experimental cytology. Cambridge 1931. — GUTH, E.: Surface chemistry, ed. F. R. MOULTON, p. 103. Washington 1943. — The elasticity of rubber and of rubber-like materials. Amer. Assoc. Adv. Sci. **21**, 103 (1944).

HAITINGER, M.: Fluorescenzmikroskopie und ihre Anwendung in Histologie und Chemie. Leipzig 1938. — HARKINS, W. D.: Surface chemistry symposium. Amer. Assoc. Adv. Sci. **7**, 19 (1937). — HAYASHI, T.: Surface-spread protein fibers as a basis for cell structure and cell movement. Amer. Naturalist **87**, 209 (1953). — HEIDENHAIN, M.: Plasma und Zelle. Jena 1911. — HEILBRONN, A.: Eine neue Methode zur Bestimmung der Viskosität lebender Protoplasten. Jb. wiss. Bot. **61**, 284 (1922). — HEILBRUNN, L. V.: J. of Exper. Zool. **43**, 313; **44**, 255 (1926). — The Coll. Chem. of protoplasm: Protoplasma Monogr. I. Berlin 1928. — The surface precipitation reaction of living cells. Proc. Amer. Philos. Soc. **69**, 295 (1930). — HEILWEIL, I. J., et al.: Electron microscopy of chromosomes. Science (Lancaster, Pa.) **116**, 12 (1952). — HILL, A. V.: The mechanics of the contractile elements of muscle, p. 490, in Deformation and flow in biological systems, ed. A. FREY-WYSSLING. Amsterdam 1952. — HILL, A. V., et al.: Physical and chemical changes in nerve activity. Suppl. to Science (Lancaster, Pa.) **79**, (1934). — HÖBER, R., et al.: The physical chemistry of cells and tissues. Philadelphia 1945. — HUGHES, A.: The mitotic cycle. London 1952. — HYDEN, H.: Physical properties of protoplasm. Ann. Rev. Physiol. **16**, 17 (1954).

INOUÉ, S., and K. DAN: Birefringence of the dividing cell. J. of Morph. **89**, 423 (1951).

JACOBS, M.: Permeability of the cell to diffusing substances. General cytology. Chicago 1924. — JOLY, M.: J. Colloid. Sci. **5**, 49 (1950). — JOLY, M., and E. BARBU: Etude par la birefringence d'écoulement de la dénaturation thermique de la sérumalbumine. Bull. Soc. Chim. biol. Paris **31**, 1642 (1949). — JONAS, H.: Observations on the mechanism of phosphate uptake by rabbit erythrocytes. Biochim. et Biophysica Acta **13**, 241 (1954).

KAMIYA, N.: Physical aspects of protoplasmic streaming, p. 199, in The structure of protoplasm, ed. W. SEIFRIZ. Ames, Iowa 1942. — The rate of protoplasmic flow. Cytologia **15**, 183 (1950). — KAMIYA, N., and W. SEIFRIZ: Torsion in a protoplasmic thread. Exper. Cell Res. **6**, 1 (1954). — KAMIYA, N., and S. ABE: Bioelectric phenomena in the myxomycete plasmodium. J. Colloid. Sci. **5**, 149 (1950). — KATZ, B. J.: The properties of the nerve membrane. Symposia Soc. Exper. Biol. **6**, 16 (1952). — KENNEDY, J. L., et al.: A new electroencephalogram associated with thinking. Science (Lancaster, Pa.) **108**, 527 (1948). — KÜSTER, E.: Beiträge zur Kenntnis der Plasmolyse. Protoplasma **1**, 73 (1931). — KUHNE, W.: Untersuchungen über das Protoplasma und die Contractilität. Leipzig 1864. — KURNICK, W. B., and I. H. HERSKOWITZ: Polyteny in Drosophila nuclei based on desoxyribonucleic acid. J. Cellul. a. Comp. Physiol. **39**, 281 (1952).

LANGELAAN, J. W.: Über mikroskopische Gitteranalyse. Extrait des Arch. néerl. Physiol. **22**, (1937). — LANGMUIR, I.: Properties of liquids. J. Amer. Chem. Soc. **39**, 1848 (1917). — LAUFFER, M.: Currents in biochemical research. New York 1946. — LAWRENCE, A. S. C., et al.: Anomalous viscosity and flow birefringence. J. Gen. Physiol. **27**, 233 (1944). — LEWIS, W. K., L. SQUIRES and G. BROUGHTON: Industrial chemistry of colloidal and amorphous materials. New York 1942. — LEWIS, W. R.: The relation of the viscosity changes of protoplasm to amoeboid locomotion and cell division. The structure of protoplasm, ed. W. SEIFRIZ, p. 163. Ames, Iowa 1942. — LINDBERG, O., and L. ERNSTER: On the mechanism of phosphorylative transfer in mitochondria. Exper. Cell Res. **3**, 209 (1952). — LLOYD, F.: Maturation and conjugation in Spirogyra. Trans. Roy. Canad. Inst., Toronto 5, 151 (1926). — LOEB, J.: Proteins and the theory of colloidal behavior. New York 1922. — LUCAS, F. F.: The architecture of living cells. Proc. Nat. Acad. Sci. U.S.A. 16, 599 (1930).

MAST, S. O.: The nature of the action of electricity in producing response and injury in Amoeba proteus (Leidy) and the effect of electricity on the viscosity of protoplasm. Z. vergl. Physiol. 15, 309 (1931). — MAST, S. O., and F. M. ROOT: Observations on feeding of Amoeba on rotifers, nematodes and ciliates, and their bearing on the surface-tension theory. J. of Exper. Zool. 21, 33 (1916). — MAZIA, D., and K. DAN: The isolation and biochemical characterization of the mitotic apparatus. Proc. Nat. Acad. Sci. U.S.A. 38, 826 (1952). — McBAIN, J. W.: Colloid Sci. Boston 1950. — MEEUSE, A.: Plasmodesmata. Bot. Rev. 7, 249 (1941). — MICHAELIS, L.: Contribution to the theory of permeability of membranes for electrolytes. J. Gen. Physiol. 8, 33 (1925). — MITCHISON, J. M.: The structure of the cell membrane. Symposia Soc. Exper. Biol. 6, 105 (1952). — A symposium on radiobiology. New York 1952. — MOLISCH, H.: The reflex arc in Mimosa. Nature (Lond.) 123, 562 (1929). — MONROY, A., and G. MONTALENTI: Variations of the submicroscopic structure of the cortical layer of fertilized and parthenogenetic sea urchin eggs. Biol. Bull. 92, 151 (1947). — MOYER, L. S.: Species relationships in Euphorbia as shown by the electrophoresis of latex. Amer. J. Bot. 21, 293 (1934a). — Electrophoresis of latex and chromosome number in poinsettias. Bot. Gaz. 95, 675 (1934b).

NEMEČ, B.: Über die Wahrnehmung des Schwerkraftreizes bei den Pflanzen. Jb. wiss. Bot. 36, 80 (1901). — NORRIS, C. H.: Elasticity studies on the myxomycete. J. Cellul. a. Comp. Physiol. 14, 117 (1939). — NORTHEN, H. T.: Protoplasmic structure in Spirogyra. Bot. Gaz. 100, 238, 619 (1938/39). — Effects of methylene blue on the structural viscosity of protoplasm. Trans. Amer. Microsc. Soc. 59, (1940). — NORTHROP, J. H., et al.: The stability of bacterial suspensions. J. Gen. Physiol. 4, 639, 655 (1921); 5, 127 (1922).

O'KONSKI, C., and B. H. ZIMM: Method for studying orientation in colloids. Science (Lancaster, Pa.) 111, 113 (1950). — OVERTON, E.: Studien über die Aufnahme der Anilinfarben durch die lebende Zelle. Jb. wiss. Bot. 34, 669 (1900).

PÉTERFI, T.: Das mikrurgische Verfahren. In Handbuch der mikrobiologischen Technik. Berlin 1923. — PETERS, R. A.: Surface structure in the integration of cell activity. Trans. Faraday Soc. 26, 797 (1930). — PFEIFFER, H. H.: Experimental cytology. Leiden 1940. — Experimental researches on the non-Newtonian nature of protoplasm. Cytologia, Fujii Jubilee 1937, 701. — Photoelektrische Anisotropie-Messungen an Riesenchromosomen von Drosophila. Cellule 54, 4 (1951). — Polarisationsoptische Untersuchungen am Spindelapparat mitotischer Zellen. Cytologia 16, 194 (1951). — Neue Versuche zur Leptonik intermitotischer Zellkerne. Exper. Cell Res. 2, 279 (1951). — Recent advances in rheodichroism of protoplasm. Proc. Internat. Congress Rheology, Oxford 1953. — POMERAT, C. M.: Rotating nuclei in tissue cultures of adult human nasal mucosa. Exper. Cell Res. 5, 191 (1953).

RIS, H.: The chemistry of chromosomes. Sym. on cytology. Michigan State College 1951. — RIS, H., and A. E. MIRSKY: Isolated chromosomes. Exper. Cell Res. 2, 263 (1951). — RUHLAND, W.: Beiträge zur Kenntnis der Permeabilität der Plasmahaut. Jb. wiss. Bot. 46, 1 (1908). — Studien über die Aufnahme von Kolloiden durch die pflanzliche Plasmahaut. Jb. wiss. Bot. 51, 376 (1912). — Zur Kritik der Lipoid- und der Ultrafiltertheorie der Plasmahaut. Biochem. Z. 54, 59 (1913).

SCARTH, G. W.: Adhesion of protoplasm to cell wall. Trans. Roy. Soc. Canada 17, 137 (1923). — Cations and the viscosity of protoplasm. Quart. J. Exper. Physiol. 14, 115 (1924). — Colloidal changes and protoplasmic contraction. Quart. J. Exper. Physiol. 14, 99 (1924). — The structural organization of plant protoplasm. Protoplasm 2, 189 (1942). — SCHMIDT, W. J.: Doppelbrechung von Chromosomen und Kernspindeln und ihre Bedeutung für das kausale Verständnis der Mitose. Arch. exper. Zellforsch. 19, 352 (1936). — Die Doppelbrechung von Karyoplasma, Cytoplasma und Metaplasma. Protoplasma Monographien, Nr 11. Berlin 1937. — Polarisationsoptische Analyse der Verknüpfung von Protein und Lipoidmolekeln, erläutert am Außenglied der Sehzellen der Wirbeltiere. Pubbl. Staz. Zool. Napoli 23, 159 (1940). — SEIFRIZ, W.: Protoplasm. New York 1936. — Viscosity values of protoplasm as determined by microdissection. Bot. Gaz. 70, 360 (1920). — The viscosity of protoplasm: Molecular physics in relation to biology. Bull. Nat. Res. Council 69, 229 (1929). — Elasticity as an indicator of protoplasmic structure. Amer. Naturalist 60, 124 (1936). — Swelling and shrinking of protoplasm. Trans. Faraday Soc. B 42, 259 (1946). — The effects of various anesthetic agents on protoplasm. Anesthesiology 11, 24 (1940). — The rheological properties of protoplasm, in deformation and flow in biological systems, ed. FREY-WYSSLING. Amsterdam 1952. — Mechanism of protoplasmic movement. Nature (Lond.) 171, 1136 (1953). — Le Mécanisme des mouvements protoplasmiques. Rev. Cytol. et Biol. végét. 14, 31 (1953). — SEIFRIZ, W., and H. L. POLLACK: Eine Theorie über den Mechanismus der Erregung. Protoplasma 39, 55 (1949). — STÅLFELT, M. G.: The lability of the protoplasmic viscosity. Physiol. Plantarum 2, 341 (1949). — STAUDINGER, H.: Die hochmolekularen Organischen Verbindungen Kautschuk und Cellulose. Berlin 1932. — STERN, H., et al.: Some enzymes of isolated nuclei. J. Gen. Physiol. **35**, 559 (1953). — STRUGGER, S.: Über Plasmolyse mit Kaliumrhodanid. Ber. dtsch. bot. Ges. **49**, 453 (1931). — Vitalfärbung pflanzlicher Zellen. Protoplasma **24**, 108 (1935). — Praktikum der Zell- und Gewebephysiologie der Pflanze. Heidelberg 1949. — Über die Struktur der Proplastiden. Ber. dtsch. bot. Ges. **66**, 439 (1953). — SWANN, M. M.: The nucleus in fertilization. Symposia Soc. Exper. Biol. **6**, 89 (1952). — The mitotic cycle, ed. A. HUGHES. London 1952. — SZENT-GYORGYI, A.: Nature of life. New York 1948. — Chemical physiology of contraction in muscle. New York 1953.

TAYLOR, C. V.: Cataphoresis of ultramicroscopic particles in protoplasm. Proc. Soc. Exper. Biol. a. Med. **22**, 533 (1925). — TRELOAR, L. R. G.: The physics of rubber elasticity. Oxford 1949. — TURKEVICH, J., G. GARTON and P. C. STEVENSON: The color of colloidal gold. J. Coll. Sci. Suppl. **1**, 26 (1954).

VERWORN, M.: Untersuchungen über die polare Erregung der lebendigen Substanz durch den konstanten Strom, III. Mitteilung. Arch. ges. Physiol. **62**, 415 (1896). — VINCENT, W. S.: The isolation and chemical properties of nucleoli of starfish oocytes. Proc. Nat. Acad. Sci. U.S.A. **38**, 139 (1952). — VLÉS, F.: Propriétés optiques des muscles. Paris 1911.

WATANABE, A., M. KODATI and S. KINOSHITA: Über die negative Galvanotaxis der Myxomyceten-Plasmodien. Bot. Mag. (Tokyo) **52**, 441 (1938). — WEBER, F.: Die Viskosität des Protoplasmas. Naturwiss. **21**, 113 (1922). — Plasmolyseform und Protoplasmaviskosität. Österr. bot. Z. **73**, 261 (1924). — WILSON, E. B.: The cell in development and inheritance. New York 1940. — WINKLE, Q. VAN, et al.: Electron microscopy of isolated chromosomes. Science (Lancaster, Pa.) **115**, 711 (1952).

YASAKI, Y., and Y. SINOTÔ: Studies on nuclei and chromosomes by fluoromicroscopy. Cytologia **17**, 336 (1952).

ZIEGENSPECK, H., u. H. H. PFEIFFER: Botanische Mikro-Farbfotografie im polarisierten Licht. In H. STÖCKLER, Die Leica in Beruf und Wissenschaft, 4. Aufl., S. 245—252. Frankfurt a. M.: Umschau-Verlag 1954. — ZIRKLE, R. E., and W. BLOOM: Irradiation of parts of individual cells. Science (Lancaster, Pa.) **117**, 487 (1953). — ZOCHER, H.: Freiwillige Strukturbildung in Solen. Z. anorg. u. allg. Chem. **147**, 91 (1925).

# Pathology.

By

## William Seifriz.

With 26 figures.

The pathological state of an organism, a tissue, or cell may be determined by seeking visible morphological changes, or by making chemical analyses, such as a determination of the sugar or protein in blood. The latter method is usually the more refined but often it is impossible, as in the diagnosis of cancer which is still a matter of histological examination. The biochemist has made many valuable contributions to the physiology and pathology of tissues but the pathology of protoplasm is still primarily a matter of observation, not analysis.

The changes which protoplasm may undergo as the result of a pathological condition are many and varied. Some of them are of such common occurrence that they cannot be correlated with any given toxic agent or injury factor; others are specific for a particular abnormal environment or poison.

## Gross morphology.

The first readily observable pathological change which a cell may undergo is one in form. Thus, an amoeba constantly changes shape as a normal procedure, but an alteration in the composition of the surrounding medium may cause variation from the normal.

Plant cells are usually enclosed in a heavy wall of cellulose. The cell cannot, therefore change its shape readily, but the protoplast within may do so, particularly at the time of plasmolysis.

## Plasmolysis.

The variability in form which a plasmolyzed protoplast may assume is familiar to every botanist. But there is no reason to believe that each of these forms indicates a pathological condition. Some certainly do, others are merely an expression of a normal variation in the physical state of the protoplasm at the time of plasmolysis.

The shape of a plasmolyzed protoplast has been used as an indication of the viscosity of protoplasm. Differences in the form of the plasmolyzed protoplast have long been recognized as pathological (KÜSTER 1929, 1937). But one must be cautious in judging, for the variety of forms which the normal protoplast can assume on plasmolysis is great. Among the plasmolyzed cells shown in Fig. 1, *A*, *B*, *C*, and *D* are from normal cells; *E* is probably and *F* certainly an abnormal cell; *D* may indicate abnormality, but more likely is it the result of a high concentration of salt or sugar causing rapid plasmolysis.

That form indicates degree of viscosity has been suggested (WEBER 1924, 1925) and contested (SCARTH 1924, SEIFRIZ 1929). The objection is justified; the form of a plasmolyzed protoplast does not indicate viscosity. No matter what the viscosity, if the protoplasm is not tacky, not glutinous, and will not adhere to the cell wall, the viscosity may be ever so high yet the plasmolyzed protoplast will round up as in Fig. 1-*A*; conversely, if the protoplasm is sticky, the viscosity may be very low yet the form in Fig. 1-*B* will result.

As a young investigator I correlated viscosity and surface tension, maintaining that increased viscosity meant increased surface tension. It was CZA-PEK (1911) who told me that the correlation was false, that viscosity and surface tension are independent variables; this is equally true of tackiness, or stickiness, and viscosity; they are independent variables. It is the adhesive quality of

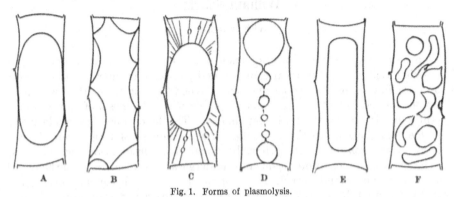

Fig. 1. Forms of plasmolysis.

protoplasm, not viscosity, which determines the form of a plasmolyzed protoplast. KÜSTER (1935) was of this opinion.

CHOLODNY (1924) and PRÁT (1922) have studied the effects of various plasmolyzing solutions on the shape of protoplasts.

Further observations by KÜSTER (1929) which are worthy of note, are his descriptions of one-sided plasmolysis, and "pearl" formation. If a cell is mechanically injured on one side, *e.g.*, by an electric shock, it will, on subsequent

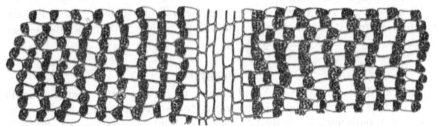

Fig. 2. One sided plasmolysis in the leaf of *Catharinea*. (From BRILLIANT.)

plasmolysis, shrink from the cell wall only on that side; traumatic shock conditions the protoplasm.

The formation of "pearls" or droplets frequently occurs in plasmolyzed cells (Fig. 1-*D*). It, too, may or may not be due to a pathological condition.

One-sided plasmolysis is of not uncommon occurrence, and it is not always due to injury. It is the expression of internal protoplasmic conditions arising from the normal metabolic activity of cells (Fig. 2).

The cells surrounding a necrotic area plasmolyze to one side (Fig. 3).

KÜSTER (1929) tells of another unusual habit which is apparently a species characteristic. The plasmolyzed hair cells of *Gynura* always assume an hourglass form, with the protoplasm free of the lateral walls and adherent to the upper and lower transverse walls. This fact brings up an interesting question: can the form of plasmolyzed cells characterize a species? It may, but surely only to a limited degree.

That plasmolysis itself sets up a pathological state is shown by the work of HÖFLER (1928). He demonstrates the change which the protoplast undergoes during plasmolysis. Adhesion decreases; as a result, the protoplasmic miniscus changes.

Fig. 3. Plasmolysis of cells surrounding a necrotic area in a leaf of *Elodea*. (From WEBER.)

*Cap plasmolysis* ("Kappenplasmolyse") has become a familiar term through HÖFLER's (1928) interest in it. When a salt slowly enters a cell, after the solution has plasmolyzed the cell, the effect of the salt on the protoplasm causes swelling.

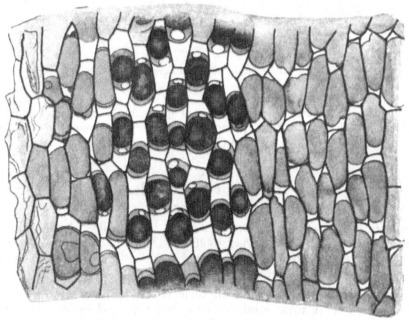

Fig. 4. Cap plasmolysis. (From STRUGGER, after HÖFLER 1934.)

The protoplasm and vacuole are losing water by exosmosis, while the protoplasm is swelling by taking in water by imbibition. The result is shrinkage of the vacuole and swelling of the surrounding protoplasm which produces an appearance quite different from the usual plasmolysis; the protoplasm rests like two *caps* on the vacuole, the tonoplast sharply delimited. That adjoining cells may behave differently in their plasmolysis is strikingly shown by HÖFLER.

STRUGGER (1949) describes a modified form of cap plasmolysis. Potassium rhodanid has a pronounced swelling effect on the plasma colloids. If *Allium* epidermal cells are plasmolyzed in a molar solution of KCNS, shrinking is rapid, and a typical cap plasmolysis results. Entrance of the salt into the protoplasm now follows and swelling results, but only of the surrounding cytoplasm while the vacuole remains plasmolyzed. The cell appears to be undergoing deplasmolysis, when in reality, the terminal cytoplasmic caps are swelling by imbibition. Within half an hour the cell cavity is again full with a plasmolyzed vacuole in the center.

The foregoing discussion of the pathology of plasmolysis is a very brief one in comparison to the voluminous literature on the subject. One can go further by reading KÜSTER (1929, 1935) HÖFLER (1928, 1932) whose investigations on *Kappenplasmolyse* are extensive; and COLLANDER (1920) whose studies on plasmolysis are part of his work in permeability.

### Amoeboid movement.

Amoeboid movement is rarely an indication of pathological protoplasm but occasionally it must be so, when protoplasm, never known to move about, assumes amoeboid activity. Such was the case with an echinoderm egg.

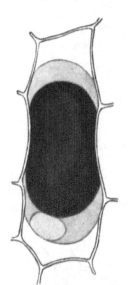

Fig. 5. Cap plasmolysis adjoining normal plasmolysis. (After HÖFLER 1928.)

Fig. 6. Amoeboid movement of an echinoderm egg.

The ovum was stimulated by the presence of sperm, but being unripe could not be fertilized. The sperm fluid (presumably an exudate) brought on amoeboid activity which lasted three minutes (Fig. 6).

# Surface phenomena.
## Permeability changes.

In spite of a relatively wide natural range, permeability changes are still excellent indicators of a pathological condition. The rate of entrance of alcohol into a cell, determined by the time required to kill, will indicate what changes in permeability a previous reagent has brought about.

However, those who interpret all toxic effects in terms of permeability err in two ways. A toxic substance may injure the cell surface, or it may do little surface damage and produce severe injury within the cell. To lay emphasis on one of these and forget the other is fautly reasoning. It is, of course, true that the first contact between poison and protoplasm is at the surface of the cell, but it is also true that two substances may enter at the same rate and only one of them prove to be toxic. A comparable illustration is that of the monosaccharides which were once thought to enter the cell at different rates, but are now known to enter at the same rate, one, glucose, being consumed by the cell, and the other, mannose, left unused.

Here, metabolism, not permeability, determines choice. However, permeability changes do occur and are often a primary factor in the pathological state of a cell. Davson and Danielli (1943) believe this to be true of drug action.

A thorough treatise on pathological permeability changes would fill a volume, and we should in the end have little that is definite. There are a few general laws which appear to hold. It is generally agreed that monovalent cations increase permeability by solating protoplasm, and divalent cations decrease permeability by gelating protoplasm, but there are numerous exceptions (Brooks 1944, Collander 1921, Jacobs 1924). Usually, surface active substances, such as alcohol, chloroform, and saponin, increase permeability rapidly.

## Visual surface changes.

Observable alterations in the cell surface are among the most obvious criteria of a pathological condition. The protrusion of papillae, the

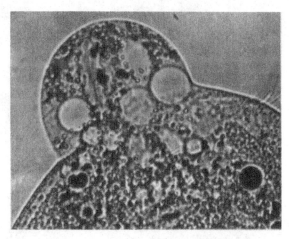

Fig. 7. Blister on the protozoan *Euplotes*.

formation of filose processes, the development of surface blisters; the disruption of the cell surface such as will cause the loss of protoplasm, are all pathological surface changes readily recognized, but to distinguish between these abnormal pseudopods, papillae, blebs, blisters, and syneretic sweat globules is difficult, though, they do differ in form, contents, and lethal extent.

The most obvious surface change is the formation of a large globular protrusion. These are of several types, some are single, huge blisters (Fig. 7), others are pseudopods formed where these do not normally exist (Fig. 8). The cause may be an external agent reducing surface tension at one point, or the responsible factor may be wholly internal.

Fig. 8. Blebs formed on an amphibien (*Cryptobranchus*) erythrocyte brought on by puncture with a needle.

Surface globules containing cytoplasm are often pinched off and set free (Fig. 9). Colloidal silver will produce such blisters. If the surface globule is clear, free of cytoplasm, the exudate which fills it is protoplasmic fluid or cell sap (Fig. 10). Such an exudation is the result of syneresis, or sweating by contraction of the protoplasmic gel. Syneretic vesicles are formed on protoplasm when it is bathed in epinephrine hydrochloride (Lepow 1938). Frequently, both granular blisters and hyaline vesicles are formed, with many intermediate stages.

The variety of pathological shapes which protoplasm can assume are very great. One of these has come to be known in our laboratory as "cauliflower", formed on the end of a pseudopod (Fig. 11). It is a characteristic injury when a slime mold is bathed in its own slime, or exotoxin (Seifriz 1944). The same treatment will produce large sized warts when a thread of protoplasm is bathed in this exotoxin (Fig. 12). To a slime mold, as to most organisms, its own excretions are poisonous.

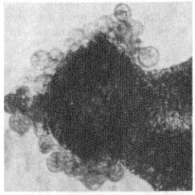

Fig. 9. Globules pinched off and set free from the surface of the plasmodium of a myxomycete.

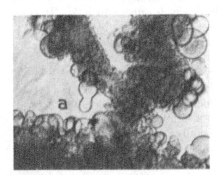

Fig. 10. Hyaline vesicles (a) produced by syneresis.

Fig. 11. Bumpy abnormality formed at the tip of a pseudopodium.

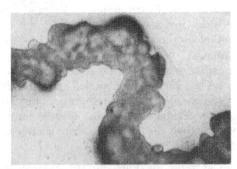

Fig. 12. Warts formed on a thread of myxomycete protoplasm bathed in the toxic excretion of the plasmodium.

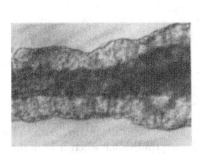

Fig. 13.

Fig. 14.

Figs. 13 and 14. Oedematic swelling of the cortical layer of protoplasm.

## Cortical changes.

The syneresis of protoplasm may occur over the entire surface of a plasmodium, forming a broad band or border of uniform width (Fig. 13). As the excreted fluid contains protein, it soon coagulates and is then discarded. If the toxic effect has not been too great, an inner membrane is formed leaving the remainder of the protoplasm alive.

A broad peripheral band of abnormal protoplasm may also result from excessive absorption of water. Such swelling is comparable to oedema (Fig. 14). The cortical protoplasm loses control of its powers of imbibition.

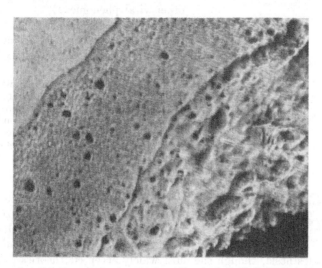

Fig. 15. Three zones formed in the cortical region of a plasmodium treated with U.V. light: outer zone is a plastic paste; intermediate zone is of dying protoplasm; inner region (black) is living protoplasm.

Frequently, cortical swelling is zoned. The broad outer band is of hyaloplasm and soon coagulates to form a soft inelastic paste. The second layer is of granular cytoplasm which, on coagulating, retains some elasticity. The third inner layer is living protoplasm (Fig. 15).

Fig. 16. Papillae formed on the surface of an *echinoderm* egg in the presence of sperm.

A typical alcohol effect is the formation of a great number of surface papillae. The papillae bear a close resemblance to myelin threads, and may, like these, be lipoidal in nature.

Papillae arise normally on the surface of ripe echinoderm eggs; their function is unknown, though they may serve to guide the sperm. Similar papillae of varied and fantastic shapes arise on the surface of unripe ova when in the presence of sperm (Fig. 16). These and the normal papillae may arise as a result of the catalytic action of sperm fluid.

# Internal changes.

## Viscosity.

Change in viscosity is a common diagnostic quality of a pathological condition, but it can be misleading. Viscosity changes have been used to explain more physical phenomena in cells than any other change which protoplasm can undergo: these include, mitosis, protoplasmic streaming, salt effects, and many other vital phenomena.

The range in the viscosity of protoplasm is great, from that of a thin, actively flowing, liquid, to the infinite viscosity of a sclerotium (SEIFRIZ 1929).

LLOYD (1926) refers to an interesting and normal difference in the viscosity of the protoplasm of the male and female protoplasts of Spirogyra, the former adheres to the wall of the copulation bridge and the latter does not. LLOYD makes an error here; again it is the stickiness of the two gametes which differs, not necessarily the viscosity.

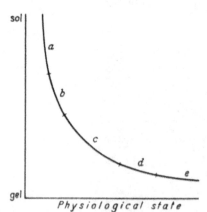

Fig. 17. Comparison of the physiological state of protoplasm with its colloidal state. Increased liquefaction results with stimulation; increased gelation results with depression: *a* death, *b* stimulation, *c* normal, *d* depression, *e* death.

## Gelatinization.

CLAUDE BERNARD said that anesthesia occurred because the anesthetized protoplasm coagulated. Had he said the protoplasm gelatinizes he would have had an excellent theory of anesthesia, though the most widely accepted theories today are those involving lipoid solubility (OVERTON) and adsorption (WARBURG).

However, whatever our theory, it is nonetheless true that anesthetized protoplasm does gelatinize, i.e., gelate reversibly (SEIFRIZ 1941). This being true, it was assumed that stimulants have the opposite effect, which they do (SEIFRIZ 1951). We may, therefore, draw a theoretical curve (Fig. 17).

In general, all depressants gelatinize protoplasm if death does not occur; otherwise irreversible gelation, i.e. coagulation, follows. In contrast, all stimulants solate protoplasm. If stimulation is excessive, liquefaction and complete disintegration of the protoplasm follow. The living substance loses its continuity in structure and goes into solution in the surrounding aqueous medium. Caffeine will cause the liquefaction of protoplasm.

Fig. 18.            Fig. 19.

Fig. 18. Granular coagulated protoplasm at *b*: living protoplasm at *a*.

Fig. 19. Fibrous coagulated protoplasm.

The xanthine derivatives, caffeine, theobromine, and theophylline, are all stimulants, and all cause a remarkable liquefaction of the protoplasm of slime molds. The opiate, heroin, which is also a stimulant, not a depressant as ordinarily thought, has the same solating effect on protoplasm as do the purines. Sodium and potassium salts (the nitrates and nitrites more than the chlorides) lower the viscosity of protoplasm greatly, and both ions are known to produce stimulation. Benzedrine, noted for its highly stimulating effect on man, liquefies the protoplasm of slime molds.

## Coagulation.

At death, protoplasm may solate and completely disintegrate, or gelate. If liquefaction does not go too far, recovery is possible. If the gelation is gelatinization rather than coagulation, the protoplasm will recover (Fig. 17).

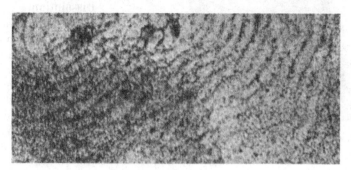

Fig. 20 A.

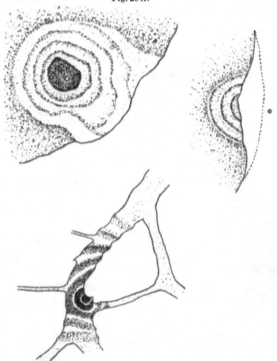

Fig. 20 B.

Fig. 20 A and B. Coagulated protoplasm at death becomes stratified (URAGUCHI).

The protoplasmic coagulum at death may be finely granular, coarsely granular (Fig. 18), fibrous (Fig. 19), or stratified (Fig. 20), the last is a LIESEGANG phenomenon. The toxic agent is in large measure responsible for the type of coagulum. URAGUCHI (1941) has pictured some excellent banding in protoplasm.

## Fragmentation.

A common form of pathological change is fragmentation. The protoplast breaks up into innumerable globules (Fig. 21). Pathological fragmentation may be caused by colloidal silver. Death usually follows but not necessarily so.

There is a form of fragmentation of a myxomycete plasmodium which is normal. Just before fruiting the entire plasmodium breaks up into "cells". The habit may tie in with the behavior of the *Acrasiales*.

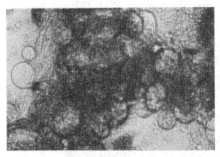

Fig. 21. Fragmented protoplasm due to colloidal silver.

## Dissolution.

The dissolution of protoplasm involves a complete collapse in structure which permits the protoplasm to go into solution in the surrounding medium. It is a familiar occurrence among protozoa where it often takes the form of a sudden and complete disintegration of the cell. The cell literally explodes. What really happens here is a collapse due to an elementary form of nervous shock (if one may use such a term for protoplasm where no elaborate nervous system exists). An egg of *Fucus* when suddenly punctured with a needle may explode.

## Wound barriers.

Psychologists use the term *homeostasis* to indicate the capacity of an organism to maintain a *steady state*. Protoplasm exhibits such a capacity, one example of which is its ability to stop the spread of a wound. It establishes a *wound barrier* whenever this is at all possible (Fig. 22).

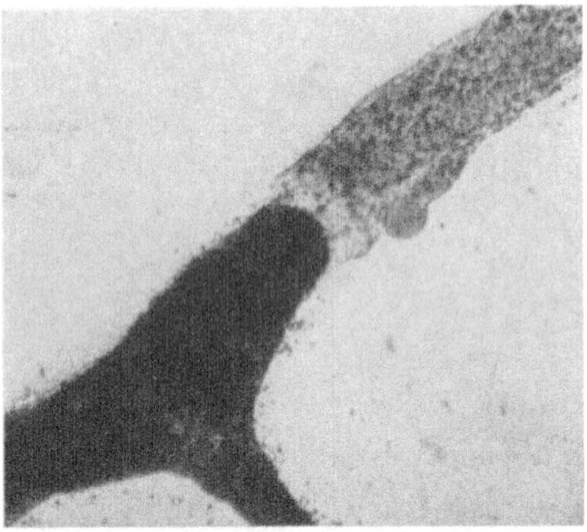

Barriers which save life are numerous in the living world, one of the most renowned being the blood-brain barrier. This latter mechanism has to do with the distribution of substances between blood and the central nervous system. Of it FRIEDE-MANN (1942) says, the existance of a barrier between blood and brain was first suggested by the observation that certain substances when injected into the vascular system fail to produce any change in the C-N-S, though they have pronounced effects after intracerebral injection. The problem is one in capillary permeability. The impermeability of the cerebral capillaries to pathogenic agents is important for many questions involving the C-N-S.

Fig. 22. A wound barrier formed between the dead protoplasm above and the living protoplasm below: the scar membrane prevents further spreading of the hypodermically injected sodium pentothal.

## Islands of injury.

The expression, *islands of injury*, has been introduced here to indicate a pathological condition which points to a very fundamental fact, namely, the apparently homogeneous plasmodium of a slime mold is not homogeneous at all. The protoplasm in various parts of the plasmodium differs in many ways, in salt concentration, in rhythmic movement (KAMIYA 1950) and in its sensitivity to poisons. As a result, islands of injury (Fig. 23) are often scattered throughout the plasmodium when a toxic substance is added. Thus does a continuous protoplasmic mass, even though in a state of flow everywhere, show the same differentiation in physiological state as does a tissue wherein groups of cells are more or less sensitive to a toxic agent than are an adjoining group of cells (Fig. 3).

Among other structural changes in protoplasm which may indicate a pathological condition but which must be judged with caution, is pronounced vacuolization. This is a typical pathological change, but excessive vacuolization is so often seen in the normal cell that it is difficult to say with certainty when vacuolization is abnormal.

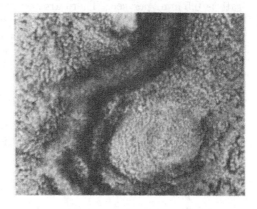

### Cell inclusions.

Cell inclusions will obviously show a great variety of pathological changes. An abnormal chloroplast, especially when large as is the chromatophore of *Spirogyra* (Fig. 24), is frequently the first indication that the cell is not normal.

Fig. 23. An island of injured protoplasm: the protoplasm in the island is dead; that in the large channel is living.

Abnormalities in mitosis, particularly chromosome aberrations, are so numerous that no listing of them will here be attempted. Also, an abnormal mitotic figure such as a tripolar egg (Fig. 25) may be due to environmental, *i.e.* pathological factors, or to gene disturbances. One must, throughout a study of the pathology of the cell, be careful to distinguish between morphology and genetics, and remember that neither dominates, rather is there a balance between the two. There is in every cell a self perpetuating balance or normalcy, which is upset by pathological conditions (Fig. 26). A review by SEIFRIZ (1939), and the extensive work by KÜSTER (1929, 1935) deal with many pathological conditions.

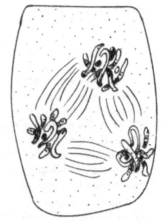

### Protoplasmic streaming.

Still another uncertain criterion of pathological protoplasm is a change in rate or type of streaming.

Fig. 24. A degenerating *Spirogyra* chromatophore.

Fig. 25. Tripolar division in an eggcell.

Cyclosis can take many forms in cells and vary so much in rate that to say a change is an indication of abnormality is not always or usually true. In addition to streaming, there are several forms of cellular movement: amoeboid, euglenoid, ciliary, and gliding. There is also the movement of cell parts, of chromosomes and mitochondria.

Protoplasmic flow may be a general agitation, a churning motion, a circulation or rotation, a tidal or back and forth movement, and a fountain or sleeve type of streaming. Only when the normal path reverts to another can it be regarded as pathological. The usual type of flow in an *Elodea* cell is a circulatory one, following the long axis of the cell, commonly referred to as cyclosis. Under abnormal conditions the protoplasm may rotate laterally, turning like a barrel.

The greater error is made when a change in rate of flow is assumed to be pathological, involving a change in viscosity. The rate of flow in a slime mold plasmodium may vary from zero to 1.25 mm. per sec. as a maximum; the average rate is 0.5 mm. per sec. There are two major factors involved in protoplasmic streaming, viscosity and pressure. The latter usually determines the rate of flow. As protoplasmic streaming may vary greatly without change in viscosity, rate of flow *per se* cannot be regarded as an indication of a pathological state.

Fig. 26. Abnormal chromosomes distribution due to a pathological condition.

## Fusion.

A phenomenon which throws some light on the physiology of pathology is that of the fusion between two protoplasts. Two amoebae may crawl over each other and never show any tendency to fuse, whereas egg and sperm of the same species when in healthy condition will always fuse. The same is true of the plasmodia of myxomycetes; under the proper conditions they invariably fuse, if of the same species. Strong vigorous streaming is established at the point of fusion. Occasionally, fusion does not occur. In this case an amusing situation may arise; two plasmodia will apparoach each other and when at a distance of 1 or 2 mm., both progress no further but maintain this distance for quite some time; then, they may fuse at one point or they may retreat.

The problem, why two protoplasts do or do not fuse is often attributed to surface tension, but this is, I believe, rarely the explanation, and it certainly cannot explain why two plasmodia maintain a uniform distance from each other. Slime molds excrete a mucous from which fact they derive their common name. When the coating of slime forms an appreciable layer on the surface, an other approaching plasmodium will not cross it. Where the surfaces of two rapidly advancing pseudopods are clean, fusion is immediate. Slime molds in culture never cross their own slimy tracks if, in their search for food, they can avoid them.

There is another case of intermixing which involves fusion of the protoplasm of two seperate species though the fusion is of a restricted nature. Burgell (1905) and Küster (1935) describe the fusion of the protoplasm of two Phycomycetes, *Chaetocladium* as a parasite on *Mucor* as host. The fusion is through window or sieve-like openings. Other fusions similar to the above have been described by a number of investigators, *e.g.*, by Jones and Drechsler (1920).

## Pathological chemistry.

One of the most significant yet difficult problems in biology and medicine is the question why one substance is toxic and another not, especially when the two are closely related. One may evade the problem by saying that one substance is toxic because protoplasm, or some protein in protoplasm, combines with that substance but not with the non-toxic one. This is true, but the question "why?" still remains unanswered.

A complete and wholly satisfactory physical and chemical interpretation of the pathological effect of poisons is not to be had, but an approach to it is possible. The absence of a theory and the difficulties of a research problem have not yet deterred scientists from either speculation or experimentation; indeed, all well established facts have been arrived at only after passing through the uncertainty of preliminary exploration. This is the stage in which we now are in regard to toxicity. However, though there is much uncertainty, there are some facts which can be nicely interpreted, and there are a few general rules which hold fairly well.

## Inorganic salts.

Common experience is sufficient to reveal the fact that among the elements of the periodic table the lighter ones are less toxic than the heavier ones; in short, toxicity increases as one descends from the top to the bottom of the table. There is also an increase in toxicity, but less consistent, as one goes from left to right in the table. These truths must have a physical-chemical background.

The properties of metallic ions which are involved in toxicity are the following:

| | |
|---|---|
| 1. Position in the periodic table | 7. Solubility |
| 2. Atomic weight | 8. Adsorption |
| 3. Diameter | 9. Surface tension |
| 4. Valence | 10. Electronic configuration |
| 5. Mobility | 11. Capacity to form complexes |
| 6. Hydration | 12. Affinity to protoplasm |

1. The position in the periodic table is obviously not in itself a physical-chemical property, but is determined by one or more of these properties. It is a convenient means of referring to the aggregate of properties which determine an elements position in the periodic table.

2. Atomic weight is also a property which in itself cannot be responsible for the toxic effect of an element but it is an indication of other properties which may have a bearing on toxicity.

3. Diameter will be a factor in the ease with which an ion can enter a cell. It is further an indication of the number of electronic shells.

4. Valence is a property of primary importance in determining toxicity for it determines ionic charge. The precipitating and coagulating powers of an element are in direct proportion to valence. If the toxicity of an element is due to its coagulating power, then valence plays its role in toxicity. Combining power may, however, be more significant than precipitating power in determining toxicity.

5. Ionic mobility undoubtedly plays a role in toxicity; the more active, more rapid ion, will enter a cell more readily, which is true of hydrogen.

6. Hydration is a factor of such significance in toxicity that I shall consider it in more detail later.

7. Solubility is a rather obvious factor, but functions only in a relative sense; if solubility is very low a substance cannot be highly toxic.

8. Adsorption on to the cell surface and on to protoplasmic particles is a factor of great significance, again deserving separate consideration.

9. A change in surface tension may cause a pathological condition but one must be cautious in applying this concept too freely.

10. Electronic configuration involves atomic complexity. How far one can assume that atomic intricacy is in itself a factor in determining toxicity is a

question but it is generally true that with increase in electronic orbits, and in the number of electrons in inner orbits, there is an increase in toxicity.

11. The capacity of an element to form complexes is an indication of its tendency to form coordinate-covalent salts with cell proteins and therefore of its power as a toxic agent.

12. Protoplasmic affinity for a metal may be merely a question of habit, *i.e.* the protoplasm has long been adapted to assimilating and using a certain metal in metabolic processes. If plants have needed and used potassium but not sodium from time immemorial, it is probable that sodium will prove to be more toxic than potassium. This is certainly true of lithium.

That the toxicity of metals increases from top to bottom and from left to right in the periodic table, is shown to be true in the following table of relative toxicities:

Group Ia: Na, K < Rb < Cs
Group Ib: Cu < Ag < Au
Group IIa: Ca < Sr < Ba
Group IIb: Zn < Cd < Hg

In general, the order of toxicity of ions follows some physical or chemical property. Thus, the three metals Rb < Sr < Y are toxic in the same order as their atomic volume and radius:

Vol. Radius ($10^{-8}$ cm.)

$Rb^{+1}$ — 56.2 — 1.49
$Sr^{+2}$ — 34.5 — 1.27
$Y^{+3}$ — 19.5 — 1.06

When a metal falls out of a toxicity order, as lithium occasionally does, one must search for some other explanation. Sometimes this explanation is to be found in the chemical affinity of the metal for a protein, or again a biological characteristic such as habit, or in a misinterpretation of a colligative or other property. Thus, atomic radii, determined by X-ray measurements of crystal lattices, increase in the order:

$Li^+$ (0.78 Å), $Na^+$ (0.98 Å), $K^+$ (1.33 Å), $Rb^+$ (1.49 Å), $Cs^+$ (1.65 Å)

This is, in the main, the order of toxicity for some organisms, *e.g.* myxomycetes, but not for other organisms, *e.g.* echinoderm eggs (RUNNSTRÖM 1951).

Another caution: the relation between an actual ionic radius and an apparent and effective one is determined by the radius of the ion plus the hydration sheath. Because of its greater tendency for undergoing hydration, the effective size of the lithium ion is greater than that of the sodium ion. Important, too, is the question whether the hydration sheath on a metal ion shields the proto-plasm from the chemical action of the metal.

Solubility is a factor in toxicity which may upset a relation established on some other basis. The following table of relative solubilities indicates this:

| | Cl | $NO_3$ | $SO_4$ |
|---|---|---|---|
| Li — | 19.0 | 12.0 | 3.12 |
| Na — | 6.2 | 10.3 | 1.37 |
| K — | 4.7 | 3.1 | 0.63 |
| Rb — | 7.5 | 3.6 | 1.80 |
| Cs — | 3.9 | 1.2 | 4.95 |

One can with profit consider a number of colligative and other properties of ions and show the bearing of each on toxicity. Thus, the intensity of the electric field surrounding a metallic cation is certain to be a factor.

It is given by the expression:

$$E = \frac{1}{4\pi\varepsilon_0} \frac{Ze}{r^2}.$$

$E$ = electric field at surface of ion; $Z$ = valence of metal; $e$ = charge (coulombs) of the electron; $r$ = ionic radius

The above generalizations hold. Periodic toxicity depends on effective ionic radii, volumes, mobilities, charge, and valence, and on chemical combination with the protoplasmic proteins. Where metals, such as Au, Hg, and Tl are in contradiction to the general rule, we must search for other causes, and sometimes they can be found. Thus, the order based on atomic weight and radius is upset by Cu which is more toxic than Rb, and Ag which is more toxic than Cs; but Cu, Ag, and Au stand apart in column I of the periodic table and are always grouped together by the chemist. The orbit just within the outer valence orbit has 18 electrons in all three of these metals, but only eight electrons in the five other monovalent ions.

Well known examples of anomalous behavior in the vertical groups are $H^+$ in group I, and the metals Be and Mg in group IIa. The toxicities of Be, Mg, and Ca are in the order $Be^{+2} > Mg^{+2} > Ca^{+2}$. The anomaly in the case of $H^+$ is obvious, as already stated, and that of the three bivalent ions is a question of a chemical property not yet understood.

Such reasoning is the only possible approach to the toxicity problem. It is complicated and uncertain, but hardly more so than the complexities of relatively simple salt solutions, one of which is *complex ion formation*.

Those ions which tend to form complexes are metals like Zn, which do not have, as in the case of Cu, Ag, and Au already cited, the stable octet configuration of 2s and 6p electrons in the electron shell immediately below the valence electrons, but contain in addition 10d electrons.

The alkali metal ions, $Li^+$, $Na^+$, $K^+$, $Rb^+$, and $Cs^+$, in group Ia of the periodic table, and the alkaline earth metal ions, $Be^{+2}$, $Mg^{+2}$, $Ca^{+2}$, $Sr^{+2}$, and $Ba^{+2}$, in group IIa, all have stable octets and therefore do not form complex ions or other coordination compounds. The true heavy metals, on the other hand, such as $Cu^{+2}$, $Ag^+$, $Au^{+1\,\&\,+3}$, in group Ib, and $Zn^{+2}$, $Cd^{+2}$, and $Hg^{+2}$, in group IIb, form complex salts with certain proteins and other organic compounds as well as complex ions with certain inorganic donors, in order to complete their stable octet configurations.

As I have pointed out, the metals in the Ib complex forming series, Cu, Ag, Au are far more toxic than the non-coordinating alkali metals Li, Na, K, Rb, and Ca: and, in group II, the members of the IIb complex forming series, Zn, Cd, Hg, are much more poisonous than the IIa metals, Be, Mg, Ca, Sr, and Ba.

From the above comparison, it can be induced that the heavy metal salts of the proteins, in which coordinate-covalent bonding occurs in addition to ordinary salt bonding, are more stable, form more rapidly and therefore kill the protoplasm more rapidly than the purely salt-like alkali and alkaline earth compounds. In other words, the heavy metals are more effective in precipitating protoplasm.

If the metals of salts, chlorides or nitrates, are arranged in the order of their toxic effects on protoplasm the series becomes:

$$Li^+ < Na^+ < K^+ < Rb^+ = Cs^+$$
$$Ca^{++} < Mg^{++} < Sr^{++} < Ba^{++}$$
$$La^{+++} < Pb^{++} < Au^{+++} < Ag^+ < Th^{++++}$$

It is wiser to group the ions into three lyotropic series and to stagger them, for such an arrangement better illustrates the facts, *e.g.*, the bivalent Ca is less toxic than monovalent Rb and Cs, and trivalent La is considerably less toxic than divalent Sr and Ba.

## Organic substances.

Toxic gases and organic substances present a somewhat different problem from that of salts.

Having found protoplasm to react in an identical manner when in the presence of carbon dioxide and nitrous oxide, I sought a record of their isosteric properties and found the following table:

*The isosteric properties of carbon dioxide and nitrous oxide.*   (From I. LANGMUIR.)

|  | $CO_2$ | $N_2O$ |
|---|---|---|
| Number of exterior electrons . . . . . . . . . . . | 22 | 22 |
| Molecular weight . . . . . . . . . . . . . . . | 44 | 44.02 |
| Viscosity, at $20^0$ C. and 1 atm . . . . . . . . | $148 \times 10^{-6}$ | $148 \times 10^{-6}$ |
| Critical pressure (atm) . . . . . . . . . . . . | 77 | 75 |
| Critical temperature . . . . . . . . . . . . . | $31.9^0$ | $35.4^0$ |
| Heat conductivity, at $100^0$ C. . . . . . . . . . | 0.0506 | 0.0506 |
| Density of liquid, at $-20^0$ C. . . . . . . . . | 1.031 | 0.996 |
| Density of liquid, at $10^0$ C. . . . . . . . . . . | 0.858 | 0.856 |
| Refractive index of liquid, $D$ line $16^0$ C. . . . . . | 1.190 | 1.193 |
| Dielectric constant of liquid, at $0^0$ C. . . . . . . | 1.582 | 1.598 |
| Magnetic susceptibility of gas, at $16^0$ C. and 40 atm . . | $0.12 \times 10^{-6}$ | $0.12 \times 10^{-6}$ |
| Solubility in $H_2O$, at $0^0$ C. . . . . . . . . . | 1.780 | 1.305 |
| Solubility in alcohol, at $15^0$ C. . . . . . . . . . | 3.13 | 3.25 |

Gases are in a class by themselves. To interpret the pathological effects of organic compounds we must look in another direction, namely, molecular pattern. The meaning of this is best shown by two experiments. The order of toxicity of the barbiturates is that shown in the following plan, which is also the order of complexity of molecular pattern, from the most toxic, sodium pentothal, to the least toxic, barbital.

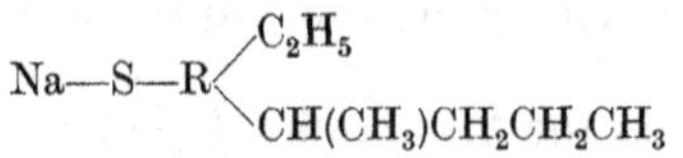

$$Na\!-\!S\!-\!R \diagup^{C_2H_5}_{CH(CH_3)CH_2CH_2CH_3}$$

Sodium pentothal

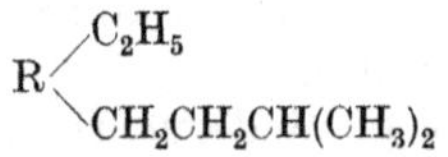

$$R \diagup^{C_2H_5}_{CH_2CH_2CH(CH_3)_2}$$

Amytal

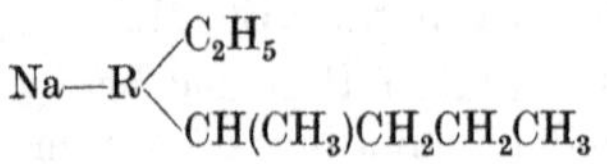

$$Na\!-\!R \diagup^{C_2H_5}_{CH(CH_3)CH_2CH_2CH_3}$$

Nembutal

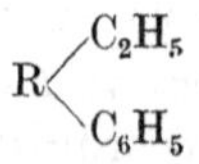

$$R \diagup^{C_2H_5}_{C_6H_5}$$

Phenobarbital

$$R \diagup^{C_2H_5}_{C_2H_5}$$

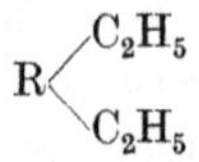

Barbital

With the loss of the elements sulphur and sodium, and decrease in length of the carbon chain, *i.e.*, with a lessening in the polarity of the molecule, toxicity diminishes.

A similar situation is found in the purine group. Xanthine is non-toxic. Theophylline and theobromine are rather toxic and caffeine very toxic. If we study the molecular pattern of these four substances we see that xanthine is free of $CH_3$ radicles, theophylline and theobromine have two each but in different positions, and caffeine has three. Theophylline and theobromine are equally toxic on slime mold protoplasm, but in different ways (SEIFRIZ 1951).

```
    NH—C=O                          CH₃N----C=O
     |    |                           |      |
  O=C    C—NH                      O=C    C—NH
     |    ‖  \                        |      ‖  \
     |    ‖   >CH                     |      ‖   >CH
    NH—C   N                        CH₃N----C   N
     Xanthine                        Theophylline

    NH—C=O                          CH₃N----C=O
     |    |                           |      |
  O=C    C—NCH₃                    O=C    C—NCH₃
     |    ‖  \                        |      ‖  \
     |    ‖   >CH                     |      ‖   >CH
  CH₃N—C   N                        CH₃N----C   N
     Theobromine                      Caffeine
```

The study of the relationship between chemical structure and pharmacological action has led to the synthesis of many therapeutic agents (GOODMAN and GILMAN, p. 7, 1940). Sometimes this relationship is very broad. Certain basic structural features are associated with narcotic activity, and the activity is lost if the slightest change is made in the configuration of the molecule. In some instances specificity of structure is of such fundamental importance that optical isomers differ in their pharmacological action. On the other hand, variations in structure may sometimes be made without loss of pharmacological properties.

It is at least clear that molecular pattern is a deciding factor in toxicity.

Pathological effects of a variety of forms are described by CURRIER (1949). Among the responses of plant cells to herbicides, including, for example, 2,4-dichlorophenoxyacetic acid, CURRIER lists the following: cap and other forms of irregular clumping of chloroplasts, rupture of protoplasts, cytolysis, swollen nuclei, granular cytoplasm, vacuolization, and stimulation to streaming. These are all familiar pathological effects. The important point in such an investigation is the specificity of the reaction, i.e. to what extent each pathological change is due to a specific toxic agent. CURRIER believes all effects are due to the hydration caused by the herbicides used.

GOLDACRE (1952) has shown that general anesthetic agents which are fat-soluble cause a reversible increase in the area of the surface of an amoeba, resulting in a lifting of the membrane. Response to touch can then only occur when the membrane is pushed into the granular cytoplasm. This, GOLDACRE believes, suggests that the membrane acts as an enzyme influencing a non-diffusible substrate in the granular cytoplasm.

The work of MARSLAND bears on this problem. He showed that long-chain paraffins become anesthetic agents when attached as a cap to the membrane of the cell, but are inactive when injected into the cytoplasm.

I should like to conclude, with added emphasis, that many socalled specific effects are not specific at all. The "lithium effect" is obtained with a dozen different substances of great variety. Any one substance, such as auxin, may do quite different things depending on the site of influence, e.g., whether root or stem. Colchicine, the pathological effect of which was discovered by HAVAS (1937a, 1937b), will produce bulbous hypertrophy of plant roots which resemble the tumors induced by B. tumefaciens; yet, AMOROSO (1935) and others have shown that cholchicine inhibits the growth of malignant tumors in animals. Thus does cholchicine function in opposite ways in plants and animals.

## Literature.

BRILLIANT, B.: Les formes de la plasmolyse. C. r. Acad. Sci. URSS 1927. — BROOKS, S. C., and M. M. BROOKS: The permeability of living cells. Protoplasma monograph, Nr. 19. Ann Arbor, Michigan 1944. — BURGELL, H.: Untersuchungen über Phycomyces. Flora (Jena) 108, 353 (1905).

CHOLODNY, N.: Über Protoplasmaveränderungen bei Plasmolyse. Biochem. Z. 147 (1924). — CLARK, A. J.: The mode of action of brugs on cells. London 1933. — The action of narcotics on enzymes and cells. Trans. Faraday Soc. 33, 1057 (1937). — Discussion on pharmacological action. Proc. Roy. Soc. Lond. Ser. B 121, 580 (1937). — COLLANDER, R.: Elektrolytische Vorgänge bei der Plasmolyse. Pflügers Arch. 185, 224 (1920). — Permeabilität pflanzlicher Protoplasten. Jb. wiss. Bot. 60, 354 (1921). — CURRIER, H. B.: Responses of plant cells to herbicides. Plant Physiol. 24, 601 (1949). — CZAPEK, F.: Bestimmung der Oberflächenspannung der Plasmahaut. Jena 1911.

DAVSON, H., and J. F. DANIELLI: The permeability of natural membranes. Cambridge 1943.

FRIEDEMANN, U.: Blood-brain barrier. Physiologic. Rev. 22, 125 (1942).

GOLDACRE, R. J.: The action of anesthetics on amoeba and the mechanism of the response to touch. Symposia Soc. Exper. Biol. 6, 128 (1952). — GOODMAN, L., and A. GILMAN: The pharmacological basis of therapeutics. New York 1940.

HÖFLER, K.: Über Kappenplasmolyse. Ber. dtsch. bot. Ges. 45, 73 (1928). — Das Permeabilitätsproblem. Ber. dtsch. bot. Ges. 49, 79 (1931). — Plasmolyseform. Protoplasma 16, 189 (1932).

JACOBS, M. M.: Permeability of the cell. Cowdry general cytology, p. 99. Chicago 1924. — JONES, F. R., and C. DRECHSLER: Crown-wart of alfalfa. J. Agricult. Res. 20, 295 (1920).

KAMIYA, N.: The rate of protoplasmic flow. Cytologia 15, 194 (1950). — KÜSTER, E.: Pathologie der Pflanzenzelle. Protoplasma Monogr., Nr. 13. Jena 1929. — Die Pflanzenzelle. Berlin 1935.

LEPOW, S.: Some reactions of protoplasm to alkaloids. Protoplasma 31, 161 (1938). — LLOYD, F. E.: Maturation and conjugation in Spirogyra. Trans. Roy. Canad. Inst., Toronto 5, 151 (1926).

PRÁT, S.: Plasmolyse und Permeabilität. Biochem. Z. 128 (1922).

RUNNSTRÖM, J., and T. GUSTAFSON: Developmental physiology. Ann. Rev. Physiol. 13, 57 (1951); see also S. HÖRSTADIUS, Pubbl Staz. zool. Napoli, Suppl. 21, 131 (1949).

SCARTH, G. W.: Colloidal changes and protoplasmic contraction. Quart. J. Exper. Physiol. 14, 99 (1924). — SEIFRIZ, W.: The viscosity of protoplasm: Molecular physics in relation to biology. Bull. Nat. Res. Council 69, 229 (1929). — Pathological changes in protoplasm. Protoplasma 32, 538 (1939). — A theory of anesthesia based on protoplasmic behavior. Anesthesiology 2, 300 (1941); see also: The effect of various anesthetic agents on protoplasm. Anesthesiology 11, 24 (1950). — Exotoxins from slime molds. Science (Lancaster, Pa.) 100, 74 (1944). — The microchemistry of toxicity. Mikrochem. Mikrochim. Acta 36, 1114 (1951). — SEIFRIZ, W., and M. URAGUCHI: The toxic effects of heavy metals on protoplasm. Amer. J. Bot. 28, 191 (1941). — STRUGGER, S.: Praktikum der Zell- und Gewebephysiologie der Pflanze. Heidelberg 1949.

URAGUCHI, M.: Rhythmic banding in protoplasm. Cytologia 11, 332 (1941).

WEBER, F.: Plasmolyseform und Plasmaviskosität. Österr. bot. Z. 73, 261 (1924). — Beurteilung der Plasmaviskosität nach der Plasmolyseform. Z. wiss. Mikrosk. 42, 146 (1925).

# Einschlüsse[1].

## Von
## Kurt Steffen.

Mit 1 Textabbildung.

Die in diesem Abschnitt beschriebenen Einschlüsse sind ihrer chemischen Zusammensetzung, ihrer Genese, Funktion und ökologischen Bedeutung nach uneinheitlich. Sie befinden sich zum mindesten zeitweilig im Cytoplasma, doch ist ihre Entstehung im Cytoplasma nicht in allen Fällen gesichert. Die hier gegebene Übersicht erhebt keinen Anspruch auf Vollständigkeit.

## 1. Feste Proteine.

Eiweiß in fester Form kommt im Pflanzenreich hauptsächlich in drei verschiedenen Modifikationen vor: als Aleuronkörner, Eiweißkristalloide und Eiweißspindeln. Daneben werden Eiweißkörper unbestimmter Herkunft und Zusammensetzung beschrieben wie z. B. die Stachelkugeln bei *Nitella* und *Opuntia* (HÄRTEL 1951, WEBER, KENDA und THALER 1952b). Nach HÄRTEL sollen die Stachelkugeln bei *Nitella* Proteinkörper, nach PRÁT (1948) Lipoproteidkomplexe sein. HÄRTEL nimmt an, daß die Stachelkugeln plastidogenen Ursprungs und Produkte eines exzessiven Stoffwechsels seien.

Nach FREY-WYSSLING (1953) bauen sich Aleuronkörner aus globulären Molekülen auf, während die Eiweißspindeln aus dem fibrillären, zusammengelagerten Struktureiweiß entstanden sind.

### a) Aleuronkörner.

Die in vielen Früchten und Samen als Reservestoffe vorkommenden Aleuronkörner treten in zwei Erscheinungsformen auf: einschlußfrei oder mit Einschlüssen. Ihre Größe variiert von 3—5 μ bei Getreide bis zu 20 μ bei *Ricinus*. Im kompliziertesten Fall bestehen die Aleuronkörner aus einer Grundsubstanz, in die Eiweißkristalloide und Globoide in wechselnder Zahl eingelagert sind. Die Globoide bestehen aus Phytin, einem Ca-Mg-Salz der Inosithexaphosphorsäure, das als Phosphatreserve dient. Es ist schwer zu mobilisieren und dadurch einer anorganischen Phosphatspeicherung überlegen (zur quantitativen Bestimmung des Phytins vgl. HELLOT und MACHEBOEUF 1947). Die Aleuronkörner können durch Anthocyane gefärbt sein (z.B. beim Mais nach CHAZE 1933, 1934), die bei der Samenreifung aus Oxyflavonen entstehen (vgl. auch SPIESS 1904, 1905). Eine wabige Struktur der Grundsubstanz wird bei Untersuchung in nicht wasserfreiem Medium vorgetäuscht (MUSCHIK 1953). Da die Aleuronkörner in wässerigem Medium verquellen, ist die Untersuchung nur im wasserfreien Einschlußmittel (z.B. Glycerin, spezifisches Gewicht 1,26) und mit nicht wasserhaltigen Reagenzien möglich.

Wegen der technischen Schwierigkeiten ist über die Entstehung der Aleuronkörner wenig bekannt. Ihre Bildung erfolgt weder in den Zellen eines Gewebes gleichzeitig noch synchron in der Einzelzelle.

---

[1] Das Manuskript wurde im Mai 1954 abgeschlossen.

Diskutiert wurde ihre Entstehung aus dem Cytoplasma, aus Vakuolen und aus Plastiden (zuletzt Wieler 1943, Quilichini 1948, Rezende-Pinto 1952, Muschik 1953). Die Auffassung von Dangeard (1921a, b u. c), wonach die zentrale Vakuole sich in kleinere Vakuolen zerteilt, in denen dann die Aleuronkörner entstehen, hat sich als nicht richtig erwiesen. Das Sichtbarwerden der entstehenden Aleuronkörner kann nicht allein durch einen Entquellungsvorgang erklärt werden, da Aleuron bereits in Samen mit noch hohem Wassergehalt auftritt. Unwahrscheinlich geworden ist auch die Auffassung von Wieler (1943), wonach die Aleuronkörner als Niederschlag entstehen. Das wabige Aussehen der Grundsubstanz sollte durch die Bildung von Mikrosphäriten bedingt sein. Gerade diese wabige Struktur ist aber jetzt durch die Arbeit von Muschik als Artefakt erwiesen.

Von Muschik selbst wird für einschlußfreie Aleuronkörner folgende, bisher unbewiesene Theorie vertreten: bei der Sojabohne werden in den Maschenräumen eines fibrillären oder molekularen Eiweiß-Lipoidnetzwerks einwandernde Substanzen (z.B. Aminosäuren) an bestimmten Attraktionsbereichen zu Reserveeiweißen zusammengefügt. Dabei sollen sich im Zuge der Entmischung die lipophilen Komponenten in einer Grenzschicht um das entstehende Aleuronkorn sammeln. Durch das Heranwachsen der Aleuronkörner entsteht nach Herauslösen dieser das charakteristische Plasmanetz, das der Anlaß zur Theorie der vakuolären Entstehung der Aleuronkörner war.

Frey-Wyssling (1953) versucht die von Pfeffer (1872), Wakker (1888) und Lüdtke (1889) gemachte Beobachtung, wonach zuerst die Einschlüsse gebildet werden, wie folgt zu erklären: Die Aleuronkörner sind flüssige Vakuolen, in denen die Bestandteile bei aktiver Dehydratation nach ihrer Löslichkeit ausfallen. Zunächst die Globoide, dann die Kristalloide und schließlich als Füllmasse zwischen beiden: die Grundsubstanz. — Oxalatdrusen in Aleuronkörnern sind von Lüdtke für *Vitis*, *Amygdalus* und *Acer* und von Guilliermond (1941) für *Oenanthe Phellandrium* beschrieben worden.

### b) Eiweißkristalloide.

Die Eiweißkristalloide unterscheiden sich von echten Kristallen durch ihre Imbibitions- und Quellfähigkeit. Sie sind in zahlreichen Erscheinungsformen für niedere und höhere Pflanzen beschrieben worden (ausführliche Literatur bei Küster 1951). Am leichtesten zugänglich sind die würfelförmigen Kristalloide in den peripheren Schichten der Kartoffelknolle (Eicke und Köhler 1943). Besonders große Eiweißkristalloide wurden von Paetow (1931) in den Pollenkörnern der Dilleniacee *Wormia fruticosa* gefunden. Am häufigsten treten Kristalloide in den Aleuronkörnern auf. Auch in Kernen werden sie beobachtet (zuletzt von Ryzkov 1951, Kenda, Thaler und Weber 1951, Miličić 1954).

### c) Eiweißspindeln.

Küster (1951) möchte die Eiweißspindeln durch ihre Flexibilität und ihren faserigen Bau von den Kristalloiden unterscheiden. Jedoch ist diese durch lichtmikroskopische Untersuchung gestützte Abgrenzung nicht mehr streng durchführbar. So wurden bereits von Weber (1953a) für *Pereskiopsis* Eiweißpolyeder beschrieben und für Viruseinschlußkörper gehalten. Neuerdings ist es nun Steere und Williams (1953, dort technische Angaben und weitere Literatur) gelungen, die Kristallkörper aus den Haaren von mit Tabakmosaikvirus infizierten Pflanzen zu isolieren und im Elektronenmikroskop zu untersuchen. Dabei zeigte sich, daß die Kristalloide, die den von Weber (1953a) beschriebenen gleichen, aus zahllosen 300 mµ langen, wahrscheinlich parallelisierten Virus-partikelchen und einer unbekannten, verdampfbaren Substanz bestehen.

Die von MOLISCH (1885) bei *Epiphyllum truncatum* entdeckten Eiweiß-spindeln sind inzwischen bei einer großen Anzahl von Kakteen beschrieben worden (Literatur bei WEBER 1951b, WEBER, KENDA und THALER 1953). Ihre Form ist sehr variabel, es werden z.B. dünne, flexible Fäden, Spindeln, Ringe, zu Achterfiguren gedrehte Ringe und zopfähnlich verflochtene Strukturen ab-gebildet (WEBER und KENDA 1952a). Ihr Bau scheint jedoch im allgemeinen faserig zu sein („Eiweißfibrillenbündel" nach KÜSTER 1948, 1951, vgl. auch PFEIFFER 1941). Die Parallelorientierung der Einzelkristalle, die zur Aggregat-bildung führt, wird mit der Anordnung in einem Taktoid verglichen (WEBER, KENDA und THALER 1952a). Die Eiweißkristalle zeigen Doppelbrechung (positiv zur Längsachse) und geben mit dem MILLON-schen Reagens Eiweißreaktion. Sie dürften im all-gemeinen in kugel-, scheiben-, spindel- oder ring-förmigen, cytoplasmatischen X-Körpern entstehen (WEBER, KENDA und THALER 1953) und diese über-dauern. Bei *Pereskiopsis pititache* (WEBER 1953a) wurde in den X-Körpern Vakuolenbildung und in diesen als Ausnahme das Auftreten von Eiweiß-polyedern beobachtet, sonst dürften in ihnen Eiweiß-stäbchen entstehen, die zum Teil aus den X-Körpern herausragen können (WEBER, KENDA und THALER 1952a). Innerhalb des X-Körpers kann es zuweilen zu einer zonierten Aggregatbildung von Eiweiß-stäbchen kommen (Abb. 1, WEBER und KENDA 1952b).

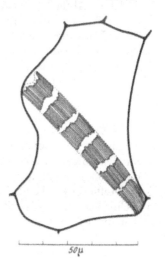

Abb. 1. *Opuntia subulata*. Blatt-epidermiszelle. Die gebogene Virus-Eiweißspindel zeigt Zonenstruktur. Nach WEBER und KENDA 1952b.

Die beschriebenen Proteinkristalle sind zwar bei den Kakteen nicht selten, kommen jedoch nicht bei allen Individuen vor. Sie werden für Viruseinschluß-körper gehalten (WEBER, KENDA und THALER 1952a und 1953 und MILIČIĆ 1954). Dafür sprechen: 1. Pfropf- und Impfversuche (MIKOSCH 1908, ROSENZOPF 1951a u. b und WEBER 1954). 2. Ähnlichkeit in Form und Verteilung mit bekannten Viruseinschlußkörpern. 3. Vergesellschaftung der kristalloiden Einschlußkörper-chen mit protoplasmatischen Einschlüssen, die den X-Bodies viruskranker Pflan-zen gleichen, und 4. das Persistieren der Kristalloide beim Vergilben, was gegen ihren Reservestoffcharakter spricht (WEBER 1953b). Das Auftreten von Eiweiß-kristalloiden, X-Körpern, Stomata-Anomalien (WEBER 1951a, KENDA und WEBER 1953) und Kaktorubin (WEBER und KENDA 1952a) kann als Symptom einer Viruserkrankung genommen werden. Ob die bei *Impatiens* (KÜSTER 1948) häufig, jedoch nicht immer, vorkommenden Eiweißspindeln durch Virusinfektion ent-stehen, ist noch unentschieden (WEBER 1953c). Cytoplasmaballen, die als X-Körper zu deuten wären, fehlen den Balsamineen. Die bei *Drosera*-Arten (BRAT, KENDA und WEBER 1951) stets vorkommenden Rhabdoide und die Eiweißspindeln von *Valerianella* (WEBER 1940, ROSENZOPF 1949) möchte WEBER (1953c) für arteigene, nicht durch Virusinfektion bedingte Eiweiß-kristalloide halten.

In durch Virusinfektion bedingten Wundtumoren werden kugelige, arginin-haltige Proteineinschlußkörper (Sphaerula) beobachtet (LITTAU und BACK 1953). Diese dürfen aber nicht ohne weiteres mit den beschriebenen Eiweiß-körpern identifiziert werden, da sie nicht doppelbrechend sind.

## 2. Polysaccharide.

Als im Plasma auftretende Reservestoffe mit beschränktem Vorkommen wären zu nennen: die Paramylonkörner der Phytoflagellaten, die sog. Florideenstärke und das Glykogen niederer Organismen.

### a) Paramylon.

Paramylon kommt bei *Euglena* in Zylinderform mit ellipsoidischem Querschnitt und bei *Phacus* in Scheibenform vor. Bei den stab- und prismenförmigen Paramylonkörnern sind die Mizellen parallel der Längsachse, bei scheibenförmigen parallel der Abflachung tangential angeordnet (KAMPTNER 1952). Die Paramylonkörner können durch Apposition wachsen (CHADEFAUD 1937) und vergrößern sich im Laufe des Tages durch die Assimilationstätigkeit (SCHILLER 1952). Bei *Euglena acus* beobachtete SCHILLER Querteilung der stabförmigen Paramylonkörner. Paramylon ist nicht verkleisterbar und mit Jod nicht anfärbbar (KAMPTNER 1952).

### b) Florideenstärke.

Die Florideenstärke wird wahrscheinlich im Plasma gebildet, sie ist stark doppelbrechend [KÜSTERs widersprechende Angabe (1951, S. 424) ist irrig] und färbt sich unter Verquellung mit Jod weinrot-violett (KYLIN 1937, 1943a). Nach COLIN (1934) und BARRY und Mitarbeiter (1949) nimmt die Florideenstärke eine Mittelstellung zwischen Stärke und Glykogen ein. Es läßt sich nicht mit Sicherheit ausschließen, daß die Stärkekörner an den Chromatophoren gebildet und sekundär verlagert werden (CZURDA 1928).

### c) Glykogen.

Glykogen kommt bei Bakterien, Cyanophyceen und Pilzen als Reservestoff vor. Die bisher untersuchten Glykogene sind zum Teil untereinander und von tierischen Glykogenen im Molgewicht, im Verzweigungsgrad und in der Zahl der die Äste aufbauenden Glukosemoleküle unterschieden (vgl. z.B. STEPANENKO und AFANASIEVA 1947, STAUDINGER 1948).

Nach ILLINGWORTH, LARNER und CORI (1952) schwankt der Endgruppengehalt und damit der Verzweigungsgrad innerhalb einer Species erheblich und ist (z. B. beim süßen Mais) vom Reifezustand des Gewebes abhängig.

Nach MEYER und FULD (1949) soll das Glykogen einer bestimmten Maisvarietät („Golden Bantam") von tierischem Glykogen nicht unterschieden sein, nach NORTHCOTE (1953) ist das Hefeglykogen dem der Rattenleber sehr ähnlich. In gelöster Form kommt das Polysaccharid oft bei Algen vor (vgl. z.B. HENRY 1949, FREDRICK 1951). KYLIN (1943b) betont, daß sich das Glykogen der Cyanophyceen von tierischem Glykogen unterscheidet und der Florideenstärke nahesteht. In den Cyanophycinkörnchen der Cyanophyceen liegt es wahrscheinlich als Glykoproteid vor (DRAWERT 1949/50, dort weitere Literatur, und VON ZASTROW 1953). Aus der Anfärbbarkeit der Cyanophycinkörnchen mit Trypaflavin möchte v. ZASTROW keine Schlüsse über den eventuellen NS-Gehalt dieser Körnchen ziehen. (Zur färberischen Differenzierung der Cyanophyceenzelle vgl. BAUMGÄRTEL 1920, GEITLER 1925; histochemische Nachweismethoden des Glykogens bei ARZAC und FLORES 1949, LISON 1949, MARTIN 1949, PACAUD 1949, PRITCHORD 1949, ERÄNKÖ 1950; spektrophotometrischer Nachweis mit Hilfe der Infrarotabsorption bei LEVINE, STEVENSON und BORDNER 1953).

# 3. Volutin (Polymetaphosphat).

Die von GRIMME (1902) und MEYER (1904) nach dem Vorkommen bei *Spirillum volutans* als Volutin beschriebenen Granula sind mit den NEISSERschen Körperchen, den BABESschen Körperchen und den metachromatischen Granula gleichzusetzen. Zur Identifizierung des Volutins werden in der Literatur angegeben: die von MEYER verwendete Methylenblau-verdünnte Schwefelsäuremethode, die NEISSERsche Diphtheriefärbung, die Toluidinblaufärbung und die Löslichkeit in heißem Wasser und verdünnten organischen Säuren. In Pikrinsäure und Fettsolventien ist Volutin unlöslich (weitere Untersuchungsmethoden bei NAGEL 1948 und KRÜGER-THIEMER 1954). Cytochemisch wurde Volutin entweder für Ribosenucleinsäure oder Metaphosphat gehalten, was zu großer Verwirrung geführt hat (vgl. NAGEL 1948). Heute faßt man Volutin meist als Metaphosphat auf (WIAME 1946).

Volutin kommt bei den meisten Mikroorganismen vor, bei einigen Bakterien, bei Algen, der Hefe und den meisten anderen Pilzen und vielleicht bei Protozoen. Den höheren Pflanzen fehlt es. Die Angaben über die Lokalisation des Volutins sind widerspruchsvoll: bei Bakterien (zuletzt KRÜGER-THIEMER 1954, KRÜGER-THIEMER und LEMBKE 1954) und Diatomeen soll es im Plasma, bei Pilzen und *Phytomastigoda* in den Vakuolen vorkommen.

Die Volutingranula der Mykobakterien sind kugelförmige, ellipsoidische, ziemlich starre Gebilde variabler Größe, die von den Mitochondrien bei reichlichem Nährstoff- und Phosphatangebot als energiereiche „Polymetaphosphate" gebildet und als Energie- und Phosphatreserven (vgl. auch HOFFMANN-OSTENHOF und WEIGERT für Hefe 1952) abgelagert werden. Daraus geht schon hervor, daß das Volutin kein integrierender Bestandteil der Mykobakterienzelle ist. Bei Phosphor- und Energiemangel werden die Granula enzymatisch abgebaut. „Die Eigenschaften der Metachromasie, der Basophilie, der starken Elektronenstreuung und der Verdampfbarkeit bei starker Elektronenbestrahlung beruhen auf hohem Gehalt der Granula an Polymetaphosphaten" (KRÜGER-THIEMER 1953). — Da Mitochondrien und Granula isotope Gebilde sind, und beide außerdem bei Bakterien im Lichtmikroskop gerade noch sichtbar sind, ist die Zuordnung der Bezeichnungen schwierig, und Verwechslungen sind verständlich. Die charakteristischen Unterschiede zwischen Bakterienmitochondrien und Granula sind in Tabelle 1 zusammengestellt. Bei der Hefe färben sich bei niederem $p_H$-Wert Metaphosphatgranula an (LINDEGREN 1949, SCHMIDT 1951, EBEL 1952), die nach HARTMAN und LIU (1954) meistens vorübergehend mit Mitochondrienäquivalenten assoziiert sind, zuweilen aber auch in der zentralen Vakuole vorkommen sollen. Der Volutingehalt in jungen teilungsfähigen Zellen ist größer als in alten (LINDEGREN 1951, HARTMAN und LIU 1954). Auch bei der Hefe hat die enge Assoziation der Metaphosphatgranula mit den Chondriosomen dazu geführt, letztere für das Volutin zu halten (Literatur bei NAGEL 1948, LIETZ 1951). LIETZ (1951) sieht das Volutin als extranukleäre Nukleolen an.

Bei den Cyanophyceen sollen nach DRAWERT (1949/50) und v. ZASTROW (1953) synonym sein: die Volutinkörner MEYERs, die Epiblasten von BAUMGÄRTEL (1920), die Metachromatinkörner und die basophilen Grana von BRINGMANN (1950, 1952). Letzterer konnte nachweisen, daß die basophilen Grana Metaphosphat- und RNS-haltig sind. Er lehnt aber den Ausdruck Volutin als unnötig ab, da diese Körper gleichzeitig DNS enthalten und somit zum Karyoidsystem zu rechnen sind (vgl. auch HERBST 1954). v. ZASTROW kommt dieser Auffassung sehr nahe, wenn sie feststellt, daß die Metachromatinkörner Partikel der Zentralsubstanz sind, die entweder schon in der Zelle vorhanden waren oder durch Denaturierung (Dughi 1946) entstanden sind. Es müßte geprüft werden, ob etwa auch bei den Cyanophyceen die Metaphosphatgranula nur im Bereich der Zentralsubstanz entstehen und sekundär verlagert werden.

Bei *Acetabularia* (Stich 1953) soll die Metaphosphatbildung in Abhängigkeit von der Photosynthese erfolgen.

Tabelle 1. *Darstellungsmethoden und Eigenschaften von Mitochondrien und Granula bei Mykobakterien und anderen Mikroorganismen nach* Krüger-Thiemer *1954.*

| | Mitochondrien | | Granula | |
|---|---|---|---|---|
| | nativ | fixiert | nativ | fixiert |
| Hellfeldbild (ungefärbt) | stark lichtbrechende Körper | | | |
| Phasenkontrastbild | dunkle Körper | | | |
| Ultraviolettbild (260 m$\mu$) | starke Absorption | | Absorption unsicher (fehlt bei reinen Metaphosphatgranula) | |
| Elektronenmikroskopisches Bild | schwache bis mittlere Elektronenstreuung, unscharf begrenzt | | starke Elektronenstreuung, scharf begrenzt | |
| Wirkung schwacher Elektronenbestrahlung (5—10 $\mu$Å) | langsame, formbeständige Verkohlung | | keine wesentliche Veränderung | |
| Wirkung starker Elektronenbestrahlung (über 30 $\mu$Å) | schnelle, formbeständige Verkohlung | | Schmelzen und Verdampfen mit Rückstand (Ringe oder wabige Strukturen) | |
| Dichte nach Vakuumtrocknung und formbeständiger Verkohlung | $0,8 \pm 0,2$ g/cm³ (wie übriges Cytoplasma) | | $1,3 \pm 0,2$ g/cm³ | |
| Vorkommen, Form und Konsistenz | regelmäßig in lebenden Zellen, rundlich, gelartig | | nährbodenabhängig, ziemlich starre Kugeln oder Ellipsoide, manchmal eingedellt | |
| Membran | nicht bewiesen | | unwahrscheinlich | |
| Bestandteile | Enzymeiweiß, Nucleinsäuren, Phospholipoide | | Polymetaphosphate Nucleinsäuren (?) | |
| Löslichkeit in heißem Wasser und schwachen organischen Säuren | | | gut löslich | |
| Toluidinblau | — | | — | metachrom. |
| Gealtertes Methylenblau | — | | — | rötlich |
| Neisser-Färbung | — | | — | blauschwarz |
| Schwermetallfärbungen | — | — | — | + |
| Nadi-Reaktion | + | ± | — | — |
| Janusgrün-B-Reduktion | + | — | — | — |
| Tetrazoliumsalzreduktion | + | ± | — | — |
| Kaliumtelluritreduktion | + | — | — | — |
| Fast-Green-Anilin (Harmans Mitochondrienfarbstoff) | — | + | — | — |
| Säure-Hämatin (Phospholipoidnachweis nach Baker) | — | + | | |
| Sudanschwarz-3-Zitrat | — | + | | |

# 4. Lipoid- und Fetttropfen.

Lipoid- und Fetttropfen kommen im Pflanzenreich entweder als Reserve-stoffe oder als Entmischungsprodukte vor. Mit den gebräuchlichen Fettfär-bungen lassen sich die Fette und Lipoide nicht getrennt erfassen. Feinere histo-chemische Untersuchungsmethoden sind schwierig und nicht immer spezifisch (vgl. ROMEIS 1948, CAIN 1950, LENNERT und WEITZEL 1952, MEYER-BRUNOT 1952). Ansätze zu einer fluoreszenz-optischen Analyse sind gemacht (BOERNER 1952, dort weitere Literatur).

Überschreiten die im Plasma gebundenen Lipoidmoleküle eine bestimmte Kon-zentration, so ballen sie sich analog den wasserhaltigen Vakuolen zu nunmehr sicht-baren Lipoidtropfen zusammen. Die Lipoidtropfen besitzen keine Struktur. — Für pflanzliche Fette bezeichnend ist der Gehalt an ungesättigten Fettsäuren. Da die Glyceride solcher Säuren flüssig sind (fette Öle), so sind auch die mikro-skopisch sichtbaren Fetttropfen flüssig bis halbflüssig und im physikochemischen Sinne homogen. Sie sind meist optisch isotrop; Anisotropie wurde bei Lipoid-sphärokristallen (z. B. Cholesterinestern PFEIFFER 1951) beobachtet. Fett in kristallinischer Form wurde z. B. im Endosperm von *Elaeis* als kurze Kristall-nadeln (MOLISCH 1913) und in Form von Sphäriten bei *Bryum* (LORCH 1931) gefunden.

Fett- und Lipoidtropfen treten fast immer im Plasma auf, im Zellkern wurde niemals mikroskopisch nachweisbares Fett gefunden. Bei den Diatomeen *Anomoeoneis bohemica* (HÖFLER 1940 und 1943) und *Nitzschia putrida* (BARG 1943) wurde Fett in der Vakuole beobachtet, jedoch nicht entschieden, ob es auch dort entstanden ist. Gefärbte Öltropfen (sog. Chromolipoide) treten bei niederen (z. B. Farbsaum bei *Pilobolus*) und höheren Pflanzen häufig auf. So färbt z. B. Carotin die kleinen Öltropfen in den Blumenkronblättern der gelben *Ranunculus*-Arten (KÖSTLIN 1924).

Die Fette sind hochwertige, energiereiche Reservestoffe. Sie sind osmotisch unwirksam und leicht oxydierbar. In 80% aller Phanerogamensamen kommt Fett als Hauptreservestoff vor (PAECH 1950, dort weitere Literatur). Im Nähr-gewebe ist es allerdings meist in optisch nicht auflösbarer Emulsion, seltener in größeren Tropfen vorhanden. Von manchen Bäumen (sog. „Fettbäumen", z. B. *Tilia, Juglans, Quercus*) wird Fett während des Winters in den Markstrahlen gespeichert.

Fett- und Öltropfen können bei abnormen Stoffwechselverhältnissen in großer Menge entstehen (fettige Degeneration), oder sie treten besonders bei prämortalen Zuständen infolge Entmischung des Lipoid-Proteinkomplexes (Lipophanerosis) des Plasmas, der Plastiden oder der Chondriosomen auf. Sie unterscheiden sich von den Reservelipoiden dadurch, daß sie nicht wieder mobili-siert werden können. Zu dieser Kategorie scheinen auch die Ölkörper der Leber-moose zu gehören, denn auch sie werden nicht wieder aufgelöst und daher als Exkrete betrachtet. Die bei höheren Pflanzen (z. B bei *Iris*) beobachteten Elaio-plasten werden zum Teil für Lipoidballungen (GUILLIERMOND 1941), zum Teil für Plastiden mit Sonderfunktion gehalten (FAULL 1935, vgl. auch REZENDE-PINTO 1952).

Daß auch ätherische Öle im Plasma und nicht in einer resinogenen Schicht, wie TSCHIRCH meinte, gebildet werden, ist für die Magnoliaceen, Lauraceen, Piperaceen, Calycanthaceen und Aristolochiaceen gesichert (LEHMANN 1926, LEEMANN 1927 und 1928). Meistens findet die Bildung in Spezialzellen, den Exkret-behältern statt. Jedoch fließen die gebildeten Tropfen nicht einfach zu größeren Öltropfen zusammen, sondern sammeln sich bei intracellulärer Ölspeicherung in einer Öltasche, die blasenartig in das Zellinnere vorspringt und durch das eindringende Öl allmählich ausgeweitet wird (Lehmann 1926, Leemann 1928). Über die chemische Zusammensetzung der Öltaschensubstanz ist nichts Gesichertes bekannt (über den Mechanismus des Durchtritts durch die Membran vgl. Sperlich 1939 und Paech 1950). Nach Paech (1952) werden im Cytoplasma von *Asarum* neben den Tropfen des ätherischen Öles weniger lipophile Tröpfchen gefunden, von denen Paech glaubt, daß sie in ätherisches Öl umgewandelt werden. Die Bildung des ätherischen Öles ist in den letzten Phasen auch postmortal möglich (Esdorn 1951).

Auch das später zwischen Kutikula und Zellmembran abgeschiedene Öl äußerer Drüsenhaare und das in die Harzgänge der Koniferen austretende Harz (Hannig 1922, 1930, Franck 1923) wird zunächst in Tropfenform im Cytoplasma gebildet (Wenzl 1935a u. b, Doetsch 1937, Sperlich 1939, Trapp 1949).

# 5. Calciumoxalat.

Calciumoxalat kommt im Pflanzenreich als Mono- und Dihydrat vor (Literaturübersicht bei Netolitzky 1929, Pobeguin 1943, Küster 1951 und v. Philipsborn 1952a u. b). Ältere Angaben über das Vorkommen von Trihydrat haben sich durch Debye-Scherrer-Röntgen-Diagramme als unrichtig herausgestellt. Die Unsicherheit und die Verwechslung mit dem Trihydrat war dadurch bedingt, daß das Dihydrat zusätzlich etwa 0,5 Mol Wasser als zeolithisches Wasser aufnehmen kann (v. Philipsborn 1952a). Das Monohydrat (= Whewellit der mineralogischen und medizinischen Literatur) ist mit einer maximalen Doppelbrechung von 0,160 stärker doppelbrechend als das Dihydrat (Weddelit) mit einer maximalen Doppelbrechung von 0,021 (die widersprechenden Angaben von Küster 1951, S. 440 sind unrichtig). Das Monohydrat kristallisiert monoklin, das Dihydrat bildet tetragonale Kristalle. Die Unterscheidung ist durch das Röntgendiagramm möglich.

In der botanischen Literatur wird das Calciumoxalat in folgenden Erscheinungsformen beschrieben (Frey 1929):

I. Einzelkristalle.
  A. Große, gut sichtbare Einzelkristalle.
     a) Isodiametrisch (Dihydrat: Oktaeder; Monohydrat: rhomboederähnliche Kombinationen aus Prisma und Pinakoid).
     b) Prismatisch (Mono- und Dihydrat, mit Tendenz zur Zwillingsbildung: Styloide).
  B. Einzelkristalle an der Grenze der Sichtbarkeit (Kristallsand). Der Kristallsand von *Atropa belladonna* besteht nach dem Röntgendiagramm aus Monohydrat (Pseudotetraeder) (v. Philipsborn 1952a).

II. Aggregate.
  A. Lockere Aggregation von parallelisierten Monohydratnadeln (Raphiden).
  B. Selten vorkommende Sphärokristalle (Monohydrat).
  C. Kristallsterne (Drusen). Großflächige mit noch erkennbarem Umriß der tetragonalen Dipyramide (Dihydrat), stärker gezackte mit kleineren Flächen (Monohydrat).

Das Dihydrat ist metastabil, es geht außerhalb der Zelle stets in Monohydrat über. Umwandlung von Di- in Monohydrat wurde für die lebende untere Blattepidermis von *Rhoeo* beschrieben (Frey 1925, Frey-Wyssling 1938). Dihydrat bildet sich in Lösungen mit niedrigem osmotischem Druck bei Übersättigung, in der Pflanze besonders in wasserspeicherndem Gewebe (Frey 1929, Frey-Wyssling 1935). In Idioblasten wird nur Monohydrat gebildet. Es ist nicht sicher, ob das Calciumoxalat im Plasma entsteht und sekundär in die Vakuolen hineingelangt, wie z. B. Netolitzky (1929) annimmt. Einschluß von organischen Bestandteilen in die Kristalle ist möglich.

Nach FREY (1929) kann man von einer aktiven Exkretion nur bei der lokalisierten Bildung des Calciumoxalats in Idioblasten sprechen. Bei fehlender Lokalisation oder an Gewebsgrenzen fällt das Calciumoxalat als Reaktionsprodukt zweier Ionen spontan und unabhängig vom lebenden Protoplasten aus. Die von außen passiv zugeführten Ca-Ionen fangen die gebildete Oxalsäure weg. Die Oxalsäure regeneriert bis zu einem Gleichgewicht, das aber immer wieder durch neue Ca-Ionen gestört wird. Da die überschüssigen Ca-Ionen schädlich sind, kann man große Mengen ausgeschiedenen Calciumoxalats als Exkretion in erweiterter Fassung des Begriffes bezeichnen (PAECH 1950). Es muß allerdings betont werden, daß bei Calciummangel (beim Austreiben des Laubes oder bei der Samenkeimung) das Löslichkeitsprodukt unterschritten wird und ein Teil des Calciums wieder in Lösung geht. PAECH (1950) nennt deswegen das Calciumoxalat ein Reservestoffexkret.

Das Vorkommen bestimmter Kristallformen ist für manche Familien ein konstantes Merkmal (NETOLITZKY 1929), jedoch wurde auch der Übergang von einer Erscheinungsform in die andere beobachtet (z.B. bei Solanaceen KREUSCH 1933).

Zur quantitativen Bestimmung des Calciumoxalatgehaltes vergleiche BAKER (1952) und MASON (1952).

Über Vorkommen, Verteilung und Unterscheidung der selten vorkommenden Calciumsalze: Ca-Sulfat, -Carbonat, -Malat, -Tartrat und -Citrat vergleiche NETOLITZKY (1929), FREY (1929), MOLISCH (1923) und KÜSTER (1951).

# 6. Kieselkörper.

Kieselkörper, die wahrscheinlich aus reiner Kieselsäure bestehen (NETOLITZKY 1929) können anscheinend in lebenden und toten Zellbestandteilen, in und außerhalb der Zelle entstehen. Die Inhaltskörper treten gegenüber den Wandinkrustationen an Häufigkeit zurück, finden sich jedoch bei vielen Monokotylenfamilien wie Gramineen, Scitamineen, Palmen und Orchideen, selten bei Dikotylen (besonders bei Chrysobalaneen; ausführliche Liste bei NETOLITZKY 1929). Die Inhaltskörper können entweder in allen Zellen eines Gewebes (z.B. Chrysobalaneen) oder in Spezialzellen, z.B. den Kristallzellreihen auftreten. In diesem Fall enthalten die die Leitbündel begleitenden Zellreihen statt Oxalat Kieselsäure. Form und Größe der Zelleinschlüsse sind verschieden. Die Kieselkörper können klar oder durch eingeschlossene organische Substanz trübe erscheinen. Sie sind die Ablagerungsform eines mit dem Transpirationsstrom eingeschleppten Stoffes.

## Literatur.

ARZAC, J. P., and L. G. FLORES: The histochemical demonstration of glycogen by silver complexes. Stain Technol. 24, 25—31 (1949). — BAKER, C. J. L.: The determination of oxalates in fresh plant material. Analyst (Lond.) 77, 340—344 (1952). — BARG, T.: Über den Fettgehalt der Diatomeen. Ber. dtsch. bot. Ges. 61, 13—27 (1943). — BARRY, V. C., T. G. HALSALL, E. L. HIRST and J. K. V. JONES: The polysaccharides of the *Florideae*. Floridean starch. J. Chem. Soc. (Lond.) 1949, 1468—1470. — BAUMGÄRTEL, O.: Das Problem der Cyanophyceenzelle. Arch. Protistenkde 41, 50—148 (1920). — BRAT, L., G. KENDA u. FR. WEBER: Rhabdoide fehlen den Schließzellen von *Drosera*. Protoplasma (Wien) 40, 633—635 (1951). — BRINGMANN, G.: Vergleichende licht- und elektronenmikroskopische Untersuchungen an Oscillatorien. Planta (Berl.) 38, 541 bis 563 (1950). — Über Beziehungen der Kernäquivalente von Schizophyten zu den Mitochondrien höher organisierter Zellen. Planta (Berl.) 40, 398—406 (1952). — CAIN, A. J.: The histochemistry of lipoides in animals. Biol. Rev. Cambridge Philos. Soc. 25, 73—112 (1950). — CHADEFAUD, M.: Recherches sur l'anatomie comparée des Euglé-niens. Botaniste (Paris) 28, 86—185 (1937). — CHAZE, J.: Sur la présence de pigments anthocyaniques ou de composés oxyflavoniques dans les grains d'aleurone de certaines graminées. C. r. Acad. Sci. Paris **196**, 952—955 (1933). — Sur le mode de formation des grains d'aleurone dans les graminées et sur la production dans ceux-ci de composés oxyflavoniques et anthocyaniques. C. r. Acad. Sci. Paris **198**, 840—842 (1934). — Colin, H.: Sur l'amidon des floridées. C. r. Acad. Sci. Paris **199**, 968—970 (1934). — Czurda, V.: Morphologie und Physiologie des Algenstärkekornes. Beih. bot. Zbl., 1. Abt. **45**, 97—270 (1928).

Dangeard, P.: Sur la formation des grains d'aleurone dans l'albumen du ricin. C. r. Acad. Paris **173**, 857—859 (1921a). — Sur la formation des grains d'aleurone dans l'albumen du ricin pendant la germination. C. r. Acad. Sci. Paris **173**, 1401—1403 (1921b). — L'évolution des grains d'aleurone en vacuoles ordinaires et la formation des tannins. C. r. Acad. Sci. Paris **172**, 995—997 (1921c). — Doetsch, R.: Beitrag zur Kenntnis der Bildung von ätherischem Öl. Diss. Zürich 1937. — Drawert, H.: Zellmorphologische und zellphysiologische Studien an Cyanophyceen. I. Mitteilung. Literaturübersicht und Versuche mit Oscillatoria Borneti Zukal. Planta (Berl.) **27**, 161—209 (1949/50). — Dughi, R.: Contribution à l'étude des inclusions cellulaires des Cyanophycées. Rev. gén. Bot. **53**, 412, 461, 510 (1946).

Ebel, J. P.: Recherches sur les polyphosphates contenus dans diverses cellules vivantes. IV. Localisation cytologique et rôle physiologique des polyphosphates dans la cellule vivante. Bull. Soc. Chim. biol. Paris **34**, 498—504 (1952). — Eicker, R., u. E. Köhler: Beobachtungen an den Eiweißkristallen der Kartoffelsorte „Juli". Protoplasma (Wien) **38**, 64—70 (1943). — Eränkö, O.: Demonstration of glycogen and lipides in the cytoplasm of human neutrophilic leucocytes. Nature (Lond.) **165**, 116—117 (1950). — Esdorn, I.: Untersuchungen über den Einfluß verschiedener Faktoren auf den ätherischen Ölgehalt an absterbenden Pflanzen. Phytopath. Z. **17**, 433—443 (1951).

Faull, A. F.: Elaioplasts in *Iris:* a morphological study. Arnold Arboretum J. **16**, 225—267 (1935). — Franck, A.: Über die Harzbildung in Holz und Rinde der Koniferen. Bot. Archiv **3**, 173—184 (1923). — Fredrick, J. F.: Preliminary studies on the synthesis of polysaccharides in the algae. Physiol. Plantarum **4**, 621—626 (1951). — Frey, A.: Calciumoxalat-Monohydrat und Trihydrat in der Pflanze. Diss. Zürich 1925. — Calciumoxalat-Monohydrat und Trihydrat. In Linsbauers Handbuch der Pflanzenanatomie, Bd. III/1a Berlin: Gebrüder Bornträger 1929. — Frey-Wyssling, A.: Die Stoffausscheidung der höheren Pflanzen. Berlin 1935. — Submikroskopische Morphologie des Protoplasmas und seiner Derivate. Protoplasma-Monogr. **15** (1938). — Submicroscopic morphology of protoplasm. Amsterdam, Houston, London u. New York: Elsevier Press 1953.

Geitler, L.: Synoptische Darstellung der Cyanophyceen in morphologischer und systematischer Hinsicht. Beih. bot. Zbl., 2. Abt. **41**, 163—294 (1925). — Grimme, A.: Die wichtigsten Methoden der Bacterienfärbung in ihrer Wirkung auf die Membran, den Protoplasten und die Einschlüsse der Bacterienzelle. Diss. Marburg 1902. — Guilliermond, A.: The cytoplasm of the plant cell. Waltham, Mass: Chronica Botanica Co. 1941.

Härtel, O.: Die Stachelkugeln von *Nitella*. Protoplasma (Wien) **40**, 526—540 (1951). — Hannig, E.: Untersuchungen über die Harzbildung in Koniferennadeln. Z. Bot. **14**, 385 bis 421 (1922). — Über den Mechanismus der Sekretausscheidung bei den Drüsenhaaren von *Pelargonium*. Z. Bot. **23**, 1004—1014 (1930). — Hartman, P. E., and Ch. Liu: Comparative cytology of wild *Saccharomyces* and a respirationally deficient mutant. J. Bacter. **67**, 77 bis 85 (1954). — Hellot, R., et M. Macheboeuf: Les proteides de la graine d'arachide *(Arachis hypogea)*. VI. Mémoire. Identification des impuretés phosphorées. Bull. Soc. Chim. biol. Paris **29**, 817—822 (1947). — Henry, M. H.: Contribution à la recherche des glucides solubles et des lipedes chez les floridées. Rev. gén. Bot. **56**, 352—363 (1949). — Herbst, Fr.: Über die Kernäquivalente von Aphanothece caldariorum P. Richt und Pseudoanabaena catenata Lauterb. Ber. dtsch. bot. Ges. **67**, 183—187 (1954). — Höfler, K.: Aus der Protoplasmatik der Diatomeen. Ber. dtsch. bot. Ges. **58**, 97—120 (1940). — Über Fettspeicherung und Zuckerpermeabilität einiger Diatomeen und über Diagonalsymmetrie in Diatomeenprotoplasten. Protoplasma (Wien) **38**, 71—104 (1943). — Hoffmann-Ostenhof, O., u. W. Weigert: Über die mögliche Funktion des polymeren Metaphosphats als Speicher energiereichen Phosphats in der Hefe. Naturwiss. **39**, 303—304 (1952).

Illingworth, B., J. Larner and G. T. Cori: Structure of glycogens and amylopectins. I. Enzymatic determination of chain length. J. of Biol. Chem. **199**, 631—640 (1952).

Kamptner, E.: Eine polarisationsoptische Untersuchung an Paramylonkörnern von *Euglena* und *Phacus*. Österr. bot. Z. **99**, 556—588 (1952). — Kenda, G., I. Thaler u. Fr. Weber: Kernkristalloide in Stomatazellen. Protoplasma (Wien) **40**, 624—632 (1951). — Kenda, G., u. Fr. Weber: Stomata-Anomalie von *Opuntia*-Virusträgern. Österr. bot. Z. **100** (1953). — Köstlin, H.: Zur physiologischen Anatomie gelber *Ranunculus*-Blüten. Bot. Archiv **7**, 325—346 (1924). — Kreusch, W.: Über Entwicklungsgeschichte und Vorkommen des Calciumoxalats in Solanaceen. Beih. bot. Zbl., 1. Abt. **50**, 410—431 (1933). — Küster, E.: Über die Eiweißspindeln von *Impatiens*. Biol. Zbl. **67**, 27—31 (1948). — Die Pflanzenzelle,

2. Aufl. Jena: Gustav Fischer 1951. — KRÜGER-THIEMER, E: Untersuchungen an den Granula von Mykobakterien. Diss. Kiel 1954. — KRÜGER-THIEMER, E., u. A. LEMBKE: Zur Definition der Mykobakteriengranula. Naturwiss. **41**, 146—147 (1954). — KYLIN, H.: Anatomie der Rhodophyceen. In LINSBAUERS Handbuch der Pflanzenanatomie, Bd. II, Abt. 6. Berlin: Gebrüder Bornträger 1937. — Zur Biochemie der Rhodophyceen. Kgl. Fysiogr. Sällsk. Lund Förh. **13**, 51—63 (1943a). — Zur Biochemie der Cyanophyceen. Kgl. Fysiogr. Sällsk. Lund Förh. **13**, 64—77 (1943b).

LEEMANN, A.: Contribution, a l'étude de l'*Asarum europaeum* L. avec une étude particulière sur le développement des cellules sécrétrices. Bull. Soc. bot. Genève, Ser. 2, **19**, 92—173 (1927). — Das Problem der Sekretzellen. Planta (Berl.) **6**, 216—233 (1928). — LEHMANN, C.: Studien über den Bau und die Entwicklungsgeschichte von Ölzellen. Planta (Berl.) **1**, 343—373 (1926). — LENNERT, K., u. G. WEITZEL: Zur Spezifität der histologischen Fettfärbungsmethoden. Z. wiss. Mikrosk. **61**, 20—29 (1952). — LEVINE, H., J. R. STEVENSON and R. H. BORDNER: Identification of glycogen in whole bacterial cells by infrared spectrophotometry. Science (Lancaster, Pa.) **118**, 141—142 (1953). — LIETZ, K.: Beitrag zur Hefecytologie. Arch. Mikrobiol **16**, 275—302 (1951). — LINDEGREN, C. C.: The yeast cell, its genetics and cytology. St. Louis: Educational Publishers 1949. — The relation of metaphosphate formation to cell division in yeast. Exper. Cell Res. **2**, 275—278 (1951). — LISON, L.: Sur la réaction de BAUER appliquée à la recherche histochimique du glycogène. C. r. Soc. Biol. Paris **143**, 117—118 (1949). — LITTAU, V. C., and L. M. BLACK: Spherical inclusions in plant tumors caused by a virus. Amer. J. Bot. **39**, 87—95 (1952). — LORCH, W.: Anatomie der Laubmoose. In LINSBAUERS Handbuch der Pflanzenanatomie, Bd. VII/1. Berlin: Gebrüder Bornträger 1931. — LÜDTKE, FR.: Beiträge zur Kenntnis des Aleuronkorns. Diss. Erlangen-Berlin 1889.

MARTIN, B. F.: A method for demonstrating the presence of alkaline phosphatase and glycogen in the same section. Stain Technol. **24**, 215—216 (1949). — MASON, A. C.: The determination of small amounts of calcium in plant material. Analyst (Lond.) **77**, 529—533 (1952). — MEYER, A.: Orientierende Untersuchungen über Verbreitung, Morphologie und Chemie des Volutins. Bot. Z. **62**, 113—152 (1904). — MEYER, K. U., u. M. FULD: Recherches sur l'amidon 44. Le glycogène de Zea Mais variété „Golden Bantam". Helvet. chim. Acta **32**, 757—761 (1949). — MEYER-BRUNOT, H. G.: Zur histochemischen Unterscheidung von gespaltenem und ungespaltenem Fett. Z. wiss. Mikrosk. **60**, 476—480 (1952). — MIKOSCH, C.: Über den Einfluß des Reises auf die Unterlage. In LINSBAUER, Wiesner-Festschrift Wien 1908. — MILIČIĆ, D.: Viruskörper und Zellteilungsanamolien in *Opuntia brasiliensis*. Protoplasma (Wien) **43**, 228—236 (1954). — MOLISCH, H.: Über merkwürdig geformte Proteinkörper in den Zweigen von *Epiphyllum*. Ber. dtsch. bot. Ges. **3**, 195—202 (1885). — Mikrochemie der Pflanze, 3. Aufl. Jena: Gustav Fischer 1923. — MUSCHIK, M.: Untersuchungen zum Problem der Aleuronkornbildung. Protoplasma (Wien) **42**, 43—57 (1953).

NAGEL, L.: Volutin. Bot. Review **14**, 174—184 (1948). — NETOLITZKY, FR.: Die Kieselkörper. Die Kalksalze als Zellinhaltskörper. In LINSBAUERS Handbuch der Pflanzenanatomie, Bd. III/1a. Berlin: Gebrüder Bernträger 1929. — NORTHCOTE, D. H.: The molecular structure and shape of yeast glycogen. Biochem. J. **53**, 348—352 (1953).

PACAUD, A.: Remarques techniques relatives à la recherche histochimique du glycogène. Bull. Histol. appl. **26**, 153—156 (1949). — PAECH, K.: Biochemie und Physiologie der sekundären Pflanzenstoffe. Berlin: Springer 1950. — Die Differenzierung der Ölzellen und die Bedeutung des ätherischen Öles bei Asarum europaeum. Z. Bot. **40**, 53—66 (1952). — PAETOW, W.: Embryologische Untersuchungen an Taccaceen, Meliaceen und Dilleniaceen. Planta (Berl.) **14**, 441—470 (1931). — PFEFFER, W.: Untersuchungen über die Proteinkörner und die Bedeutung des Asparagins beim Keimen der Samen. Jb. wiss. Bot. **8**, 429 bis 574 (1872). — PFEIFFER, H. H.: Experimentelle Beiträge zur submikroskopischen Feinbaukunde (Leptonik) undifferenzierten Cytoplasmas. Ber. dtsch. bot. Ges. **59**, 288—295 (1941). — Untersuchungen an Lipoidtropfen in Explantaten in vitro. Protoplasma (Wien) **40**, 48—53 (1951). — PHILIPSBORN, H. V.: Über Calciumoxalat in Pflanzenzellen. Protoplasma (Wien) **41**, 415—424 (1952a). — Die Entwicklung der mikroskopischen Methoden in der Mineralogie und deren Bedeutung für die allgemeine Mikroskopie und für die Technik. In Handbuch der Mikroskopie in der Technik, Bd. 4. Frankfurt 1952b. — POBEGUIN, T.: Les oxalates de calcium chez quelques angiospermes. Ann. Sci. natur. Bot., Sér. 11, **4**, 1—95 (1943).— PRÁT, S.: The cell-inclusions in *Nitella*. Věstník Král. Čseske Spoleonosti Nauk. Tř. Mat.-Přirod. **1947**, 1—16 (1948). — PRITCHORD, J. J.: A new histochemical method for glycogen. J. of Anat. **83**, 30—31 (1949).

QUILICHINI, R.: Die Cytoplasmabestandteile einiger Leguminosensamen und ihre Entwicklung während der Keimung. C. r. Acad. Sci. Paris **226**, 690—692 (1948).

REZENDE-PINTO, M. C. DE: Über die Genese und die Struktur der Chloroplasten bei den höheren Pflanzen. Ergebnisse und Probleme. Protoplasma (Wien) **41**, 336—342 (1952). — ROMEIS, B.: Mikroskopische Technik, 15. Aufl. München: Leibniz 1948. — ROSENZOPF, E.:

Beiträge zur Kenntnis der Eiweißspindeln. Diss. Univ. Graz 1949. — Sind Eiweißspindeln Virus-Einschlußkörper? Phyton (Horn, N.-Ö.) 3, 95—101 (1951a). — Hemmung der Eiweißspindel-Bildung durch UV-Bestrahlung. Phyton (Horn, N.-Ö.) 3, 102—103 (1951). — Ryzkov, V. L.: Einige morphologische und chemische Eigentümlichkeiten des Zellkernes bei *Rhinanthus minor* Ehrh. und anderen Scrophulariaceen. Dokl. Akad. Nauk SSSR., N. S. 78, 363—365 (1951).

Schiller, J.: Über die Vermehrung des Paramylons und über Alterserscheinungen bei Eugleninen. Österr. Bot. Z. 99, 413—420 (1952). — Schmidt, G.: The biochemistry of inorganic pyrophosphates and metaphosphates. In W. D. McElroy u. B. Glass, Phosphorous metabolism, Bd. I, S. 443—470. Baltimore, Maryland: Johns Hopkins Press 1951. — Sperlich, A.: Das trophische Parenchym. B. Exkretionsgewebe. In Linsbauers Handbuch der Pflanzenanatomie, Bd. 1/IV. Berlin: Gebrüder Bornträger 1939. — Spiess, K. v.: Über die Farbstoffe des Aleurons. Österr. bot. Z. 54, 440—446 (1904). — Die Aleuronkörner von *Acer* und *Negundo*. Österr. bot. Z. 55, 24—25 (1905). — Staudinger, H.: Über natürliche Glykogene. Makromolekulare Chem. 2, 88—108 (1948). — Steere, R. L., u. R. C. Williams: Identification of crystallin inclusion bodies extracted intact from plant cells infected with tobacco mosaic virus. Amer. J. Bot. 40, 81—84 (1953). — Stepanenko, B. N., u. E. M. Afanasieva: On the iodine reaction of glycogenes of various origin. Biochimija 12, 111—112 (1947). — Stich, H.: Der Nachweis von Metaphosphaten in normalen, verdunkelten und Trypaflavin-behandelten Acetabularien. Z. Naturforsch. 8b, 36—44 (1953).

Trapp, I.: Neue Untersuchungen über Bau und Tätigkeit der pflanzlichen Drüsenhaare. Diss. Gießen 1949. — Tschirch, H.: Die Harze und die Harzbehälter, 2. Aufl. Leipzig 1934.

Wakker, J. H.: Studien über die Inhaltskörper der Pflanzenzelle. Jb. wiss. Bot. 19, 423—496 (1888). — Weber, F.: Eiweißspindeln von *Valerianella*. Protoplasma (Wien) 34, 148—152 (1940). — Viruskörper fehlen den Stomazellen. Protoplasma (Wien) 40, 635 bis 638 (1951a). — Trypanoplasten-Viruskörper von *Rhipsalis*. Phyton (Horn, N.-Ö.) 3, 273 bis 275 (1951b). — Eiweißpolyeder in *Pereskiopsis*-Virusträgern. Protoplasma (Wien) 42, 283 bis 286 (1953a). — Eiweiß-Spindeln (Viruskörper) in vergilbenden *Pereskia*-Blättern. Österr. bot. Z. 100, 319—321 (1953b). — Sind alle Pflanzen mit Cytoplasma-Eiweißspindeln Virusträger? Phyton (Horn, N.-Ö.) 5, 189—193 (1953c). — Kakteen-Virus-Übertragung durch Pfropfung. Protoplasma (Wien) 43, 382—384 (1954). — Weber, F., u. G. Kenda: Cactaceen-Virus-Eiweißspindeln. Protoplasma (Wien) 41, 111—120 (1952a). — Die Viruskörper von *Opuntia subulata*. Protoplasma (Wien) 41, 378—381 (1952b). — Weber, F., u. G. Kenda u. I. Thaler: Viruskörper in Kakteenzellen. Protoplasma (Wien) 41, 277—286 (1952a). — „Stachelkugeln" in *Opuntia*. Phyton (Horn, N.-Ö.) 4, 98—100 (1952b). — Eiweißspindeln und cytoplasmatische Einschlußkörper in Pereskiopsis. Protoplasma (Wien) 42, 239—245 (1953). — Wenzel, H.: Osmotische und Permeabilitätserscheinungen an Labiaten-Drüsenhaaren. Protoplasma (Wien) 23, 187—202 (1935a). — Untersuchungen über die Exkretbildung in den Drüsenhaaren der Labiaten. Jb. Bot. 81, 807—828 (1935b). — Wieler, A.: Der feinere Bau der Aleuronkörner und ihre Entstehung. Protoplasma (Wien) 38, 21—63 (1943). — Wiame, J. M.: Basophilie et metabolisme du phosphore chez la levure. Bull. Soc. Chim. biol. Paris 28, 1—5 (1946).

Zastrow, E. M. v.: Über die Organisation der Cyanophyceenzelle. Arch. Mikrobiol. 19, 174—205 (1953).

# Chemistry of the nucleus.

By

## J. A. Serra.

The usual distinctions between plants and animals vanish when dealing with the chemical and physical properties, or become restricted to those parts of the cells, such as the plastids, which are almost exclusive of one kind of cells. This is especially true of the nucleus and for this reason it would not be practical to separate data obtained on plants from those secured in animal tissues. On the other hand, certain types of data are more easily obtained in animal cells, with the consequence that results on plant cells must be completed by those reported for animal tissues. The unity of cell life allows that for many chemical and physical problems the data on plants and animals may be discussed together. Therefore, in this chapter and the following two ("Submicroscopic structure of the nucleus" and "Physical chemistry of the nucleus") we will treat of plant and animal nuclei indistinctly, trying to point out the differences between both kinds of cells only in so far as they become of importance for the subject under discussion.

## A. Nucleus-cytoplasm relations[1].

It would almost be unnecessary to discuss at the chemical level the differences between nucleus and cytoplasm and a concept of the cell nucleus if this was always distinct as a morphologically easily observable organ of the cell. In Bacteria, certain Fungi and the greater Viruses, the morphological distinction of the nucleus is difficult or its organization is so different from that usually seen in higher plants that a more general concept is necessary in order to include all the cases. It is admitted by many that in the apparent exceptions chromosomes are produced by the nuclear bodies at one period or another of cell life. The concept that the nucleus is that part of the cell which in cell division gives rise to chromosomes has become generally accepted. However, it is yet under discussion if true chromosomes, with a structure similar to that found in higher plants are present in Bacteria, in Yeasts, and certainly they do not exist as such in the Viruses.

To solve these difficulties a chemical concept has been more or less explicitly admitted by the generality of those dealing with these problems. According to such a concept the nucleus would be that part of the cell where thymonucleic acid or thymonucleoproteids are present. This, however, seems to be not too wise a definition as it is not sufficiently proven that thymonucleic acid cannot occur in cytoplasmic formations and also that all differentiations concerning typical nuclear gene localization are thymonucleic in composition. Some recent work, confirming previous observations, tends to show that indeed thymonucleic acid may exist outside the nucleus as a general case and that cytologists should be careful in not interpreting results about a positive Feulgen reaction in the cytoplasm as being due to faulty technique or to the presence of aldehyde con-

---

[1] For a brief introduction on the microscopical morphology of the nucleus, with the nomenclature here used, see the following chapter "Submicroscopic structure of the nucleus".

taining compounds other than desoxyribonucleic acid. That the presence of "diffuse" and ordinarily "unfixable" thymonucleic acid not revealed by classic procedures is more than a theoretical possibility has been demonstrated in the cytoplasm of microsporocytes of plant cells (SPARROW and HAMMOND 1947), probably also in cytoplasm granules of root tip meristems (CHAYEN and NORRIS 1953), and in other cells (former citations in TISCHLER 1934 and WILSON 1937). There is also strong evidence that in animal cells the cytoplasm of the amphibian oocyte contains the greater amount of thymonucleotides, or of polymerized desoxyribonucleic acid existing in the cell (HOFF-JÖRGENSEN and ZEUTHER 1952, SZE 1953). It is possible, but it actually seems doubtful, that extra-nuclear thymonucleotides are ultimately of nuclear origin, the opposite view being equally plausible in many cases; rather it seems likely that the synthesis of these nucleotides is effected by nuclear-cytoplasmic interactions. Probably it will be found in future work that desoxyribonucleotides, mostly in an "unfixable" form, are generally present also in the cytoplasm. Therefore, no cytochemical concept of the nucleus based on the presence of thymonucleic acid alone is possible and to date no other cell components are known to exist exclusively in the nucleus. Such a qualitative difference in composition would at most be possible only at a much more refined level of chemical distinction, when the composition not of groups of compounds but of individual proteins and proteids could be studied. Gross chemical differences, at the qualitative level, between nucleus and cytoplasm seem unlikely since both parts make the integrated biological unit of work which is the cell, exchanging their components in metabolism.

The only general concept of the nucleus at present tenable not only for higher plants and animals as also for Bacteria and the higher Viruses is the organizational one, at the biological level. The nucleus or the nucleoids are morphologically recognizable cell differentiations related to the long term life of the cell, especially to cellular syntheses and particulate or gene heredity typically following mendelian ratios and which, at least in some stages, usually contain thymonucleotides or thymonucleoproteids. That the nucleus is not absolutely necessary for growth has been shown by experiments on the relatively great though unicellular umbrella-like green alga *Acetabularia mediterranea*, in which the half of the cell deprived of the only nucleus, which lies in the rhizoid, continues to grow for some time and possesses a certain capacity for regeneration (HÄMMERLING 1934). Newer observations on the same material (BRACHET 1952a, b, c, BRACHET and CHANTRENNE 1951, 1952) demonstrate that $CO_2$ is photosynthetically incorporated into aminoacids during about two weeks after enucleation, at the same rate in both the nucleated and the non-nucleated halves. About 1 month after enucleation $CO_2$ incorporation has fallen only to a rate 70% of that of the nucleated part and this is true not only for the proteins as a whole but also for those of the chloroplasts and of the centrifugable cytoplasm granules or microsomes (in the protozoon *Amoeba* the microsomes' proteins disappear before those of the ground cytoplasm after enucleation). Oxygen consumption is also equal for both halves of *Acetabularia* (CHANTRENNE and BRACHET 1952).

These results are in accord with similar observations on animal cells, amphibian eggs and *Amoeba proteus* (BRACHET 1952c, for a summary). Non-nucleated fragments of amoeba show almost no change in $O_2$ uptake but their contents of ribonucleotides rapidly decline while the nucleated part of the cell keeps its relative amount of ribonucleic acid. $^{32}P$ uptake is also very different in nucleated and non-nucleated halves; 10 days after enucleation the $^{32}P$ uptake in non-nucleated fragments is only about 5% that of the nucleated parts. These findings have been interpreted (BRACHET 1952c) as evidence for a direct intervention of the nucleus in the coupling between oxidations and phosphorylations in the cell. However, in *Acetabularia* the difference in $^{32}$P uptake between nucleated and non-nucleated parts depends on the age of the alga, old individuals showing a smaller difference, the incorporation being only 1.45–1.85 times greater in the nucleated half while young ones show ratios 5.0–6.5 times greater (BRACHET 1952a). The adenosine-triphosphatase content of nucleated and non-nucleated parts of amoeba keeps almost the same, while acid phosphatase rapidly declines in the non-nucleated half. This demonstrates that the role of the nucleus is an indirect one and that interactions between nucleus and cytoplasm are also necessary for this function.

Several other components of the cell, enzymes, basic and non-basic proteins have been found not to differ markedly in the first days after enucleation and the more characteristic difference occurs in ribonucleic acid. It is necessary, however, to be cautious in generalizing about this difference and as yet it seems unjustified to conclude that the nucleus is the site of ribonucleotides' synthesis. As in the case of phosphorylations, the right conclusion should be that the nucleus in some species or some cells is indispensable for the maintenance of a typical concentration of ribonucleic acid in the cell, this probably being linked to phosphorylations and cozymase synthesis (BRACHET 1952c). It seems likely that the nucleus affects these functions rather by altering the cell-permeability through the cessation of the synthesis of certain protoplasm or enzyme constituents, possibly phosphoprotein or phospho-lipido-protein in nature. As yet no single function other than particulate mendelian heredity has been found to be exclusive of the nucleus (see also HAUROWITZ and CRAMPTON 1952) and also no group of compounds has been localized exclusively in the nucleus although thymonucleic acid is the chemical that most nearly fulfils this demand. Chemical differences known today between nucleus and cytoplasm reside, on one side, in the relative concentrations of the compounds present and on the other hand surely in the fine composition of some proteins, proteids, nucleic acids or other groups of compounds which the actual state of cytochemistry is yet unable to reveal.

Nucleus and cytoplasm work, at the chemical level, as an integrated biological unit. During active cell life interactions are always taking place between nucleus and cytoplasm, some of which may be of the type schematically represented by: $A$ (outside of the cell, nutrient) $\rightarrow \|$ cell membrane $A + BC$ (in cytoplasm) $\rightarrow ABC \rightarrow \|$ nuclear membrane $ABC + D$ (in the nucleus) $\rightarrow AC$ in the nucleus $+ BD$ which $\rightarrow \|$ passes to the cytoplasm. A part of these interactions may be concerned with apoenzyme production or with furnishing basic patterns, specific parts of polypeptide chains or of nucleic acid chains. These interactions, however, remain to be studied and generalizations, as demonstrated by the above case of phosphorus metabolism, may be attempted only after observing many different species of cells.

# B. Methods and their limitations.

The state of some chapters of science strongly depends on the methods used. This is particularly valid for the chemistry of the cell nucleus, the knowledge of which not only is incomplete but also reflects the inaccuracy of the methods employed, sometimes without due regard to criticism. Two kinds of methods are used to study nuclear composition, *viz.* the analytical ones proper of chemistry and which may be applied only after the nucleus is isolated or its components extracted from the cell, and the chemocytological, usually called cytochemical or histochemical, which deal with intact cells or tissue sections. Both kinds of methods have their limitations and errors. The analytical methods are quantitative but necessitate of previous isolation procedures which may alter the original composition of the nucleus, while the histochemical methods generally are only qualitative or, despite many attempts to overcome this limitation, at most only grossly quantitative. A brief mention will be made of the chief methods used and their errors.

## a) Isolation and extraction procedures.

Isolation procedures (summary Dounce 1952) were inaugurated by Smith (1856) who isolated nuclei from skin and tumours with acetic acid. This work, however, was overlooked until the researches of Brunton (1870) and principally of Miescher (1871) to whom we are indebted for the foundations of modern methods. Miescher digested pus cells with hydrochloric acid and pepsin, which left the nuclei apparently intact, and from these nuclei and spermatozoid heads prepared a strongly acidic compound which he named "nuclein". Similar isolation and extraction methods were afterwards used in the work of Kossel and collaborators (summary Jones 1920, Kossel 1928) who have contributed the basic knowledge on nucleic acids.

Isolation of nuclei may be either direct, each nucleus isolated from the cytoplasm directly under the microscope, for instance with a micromanipulator; or indirect or mass isolation, in bulk from the tissues. Owing to the time necessary for individual isolation and the inherent difficulties the direct isolation procedure is not suitable for chemical analysis by the great majority of the actual analytical methods, although it may become the procedure of choice in future when more sensitive analytical methods are devised. By coupling extraction methods with microchromatography and special photometric methods it may be possible to work with individual nuclei (for a microextraction method applied to whole cells see for instance Edström 1953). Mass isolation has been more frequently applied to cells of animal origin, as plant cells with a cellulose envelope are more difficult to work out. The usual steps in mass isolation of nuclei are: 1. Differential maceration of cytoplasm and nucleus, the former being disintegrated, either mechanically or chemically by dissolution or enzyme attack, while the nucleus ideally should remain intact. 2. Separation of the nuclei from the macerated cytoplasm, by filtration and/or differential centrifugation and sedimentation. Filtration usually is done with large mesh or cheese cloths and employed in the former phases of the procedure to separate out the greater cell debris and entire cells. Centrifugation is performed by suspending the macerated cells in liquids of suitable specific gravity so as to sediment only the nuclei or only the cytoplasmic components. Generally in practice the separation is far from complete. Nuclei usually are lighter and sediment at lower specific gravities than cytoplasmic debris. The appropriated spec. grav. of the liquid must be empirically found and generally lies in the range 1.35–1.45, most frequently 1.41–1.42. Fine adjustment of the spec. grav. of the liquid and of the centrifugal force, usually between 1,000–6,000 g, may bring about a reasonable separation.

The weak side of mass isolation resides in the maceration and the liquids for suspension of cell fragments. Operation at $0°$ C. in a cold room, generally employed, minimizes autolysis and chemical action upon the cell. Even with the less aggressive solvents, however, decomposition of sensitive compounds, adsorption of cytoplasmic components by the nucleus, loss of nuclear components and reaction between these and the cytoplasm cannot be entirely avoided and are so much to be feared as generally the extent to which they were effective is not known. During isolation some cytoplasmic constituents may diffuse into the nucleus and be there adsorbed, and conversely nuclear components may pass to the macerated mass or the suspension liquids. It is necessary that in future work the extent of these alterations be thoroughly investigated before attempting any generalizations. Without entirely subscribing to the rather pessimistic view that mass isolation methods are at present almost impossible to evaluate or interpret (Danielli 1953), it must be emphasized that isolation and extraction procedures only give indications about the qualitative and quantitative aspects of "some" nuclear components, not necessarily about the state of these components within the living nucleus, let alone of their functions in normal cell life. At present the latter are merely hypothetical extrapolations which, sometimes owing perhaps to they being repeatedly presented with insistence, are supposed to have become acceptable theories.

Two kinds of solvents are now in use during isolation procedures: aqueous and nonaqueous. Aqueous solvents give rise to great losses of nuclear substances which may amount to more than half the weight of the nucleus as compared with nuclei isolated in non-aqueous solvents (Daly et al. 1952, Dounce 1952, Danielli 1953). The latter also give rise to severe losses of lipidic and possibly lipo-proteic components. Even physiological saline, sucrose and other supposedly "indifferent" media like gum arabic (DOUNCE 1952) should extract some nuclear components. In general it may be said that aqueous media should be preferred when the fat components are being studied and organic solvents shall be employed for the other components. Enzymes may be affected by both and direct isolation or at most mass isolation in "indifferent" media (sucrose, gum arabic also dissolve enzymes from the nucleus) should be attempted.

In the following, three examples of isolation procedures in non-aqueous and aqueous solvents, one for plant the others for animal tissues are briefly described (more particulars: GLICK 1949, BEHRENS 1932, 1938, STONEBURG 1939, DOUNCE and LAN 1943, DOUNCE 1943a, 1950, HOERR 1943, LASAROW 1943, ALLFREY et al. 1952, WILBUR and ANDERSON 1951, VENDRELY 1952, STERN and MIRSKY 1953). Procedure of STERN and MIRSKY (1952) based on that of BEHRENS: 6 g. of wheat germ as obtained from milling firms, after thorough extraction with petrol ether, are suspended in 300 ml. of fresh petrol ether and ground in a ball mill for 48 hours. Suspend the ground tissues in 500 ml. of a mixture of cyclohexane-carbon tetrachloride (benzene inactivates some enzymes) adjusted to a specific gravity of 1.395 and centrifugate. Supernatant is fraction A, sediment is fraction B + C. Separate and pour the supernatant in $^1/_3$ its volume of petrol ether. By centrifugation a sediment of mainly cytoplasmic debris and a few nuclei is obtained (control at the microscope and stain by acetorcein-fast green, which colours cytoplasm green, nuclei red). Fractions B + C suspended in cyclohexane-carbon tetrachloride mixture adjusted to 1.447 spec. grav. are centrifuged at about 2,000 g. for 60 minutes. Supernatants collected, suspended in about $^1/_3$ their volume of petrol ether, centrifuged and the operation repeated. Combined supernatants are fraction B, which consists mainly of nuclei and starch granules, besides cytoplasmic fragments and whole cells. Sediments are fraction C, of starch granules, nuclei, cytoplasmic fragments and debris. Yield: 13.3% fraction A, 35% B, 51.7% C. Purification of fraction B: Fraction suspended in mixture (cycloh.-carbon tetr.) adjusted to spec. grav. between 1.416–1.420 and centrifuged at about 6,000 g. Cytoplasmic fragments sediment at spec. grav. higher than 1,420. Yield about 5% of initial weight. This purified fraction B has about 85% purity as tested by the distribution of $\beta$-amylase activity.

Procedure of VENDRELY (1952): Animal tissues cooled to —25⁰ C. as soon as possible after remotion. Mince into fine sections with a razor blade and suspend in about 10 weights of M/3 citric acid at a temperature of + 2⁰ C. Shake in a mechanical shaker. Filter on tissue-paper. Centrifugate in a cooled centrifuge 5 minutes at 3,500 rev. per minute (RPM) and collect the sediment by suspending in M/18 citric acid. Centrifugate at 2,500 RPM. Repeat the operation at least 3 times, each time with lower speed centrifugation, and control at the microscope the elimination of cytoplasmic debris. A final centrifugation for 5 minutes at 1,000–1,200 RPM gives an homogenous suspension of pure nuclei.

Procedure of SCHNEIDER and PETTERMANN (1950; from ALLFREY 1954): All operations at 2⁰ C and solutions previously cooled to this temp. 50 g. fresh calf thymus finely minced with scissors. Gently homogenize in a Waring blendor for 4 min. at 35 RPM with 50 ml 0.5 M sucrose plus 400 ml. 0.25 M sucrose- 0.0018 M $CaCl_2$ solution (the latter designated as sucr-$CaCl_2$ sol.). The homogenete is filtered through a double layer of gauze (Johnson & Johnson type I) and then through a single layer of flannelette. Centrifugate the filtrate at 700 g. (2,000 RPM) for 10 min. Discard supernate. Sediment resuspended in 90 ml. sucr-$CaCl_2$ sol. and allowed to settle for 10 min. in a 100 ml. cylinder. Decant supernate carefully through double thickness of gauze and discard the clumps of nuclei, fiber, and cells at the bottom of the cylinder. Centrifugate the filtrate at 700 g. for 10 min., which sediments the nuclei. Wash these sediments with about 50 ml. sucr-$CaCl_2$ sol. Finally suspend in another 25 ml. of this sol. Yield about 625 mg. nuclei, only with negligible contamination by whole cells and debris.

Of the numerous extraction procedures only one will be mentioned (MIRSKY and POLLISTER 1946). 50 g. of untreated wheat germ thoroughly extracted with petrol ether, dried at room temperature and washed with 0.14 M NaCl solution. Extract with 1 M NaCl and vigorously stir. Clarify by centrifugation of extracts at 10,000 RPM. Pour into 10 volumes of distilled water which precipitates a fibrous mass containing nucleoprotein that previously was supposed to be a homogenous protein (called chromosin) but now is recognized to be a mixture containing chiefly thymonucleohistone.

## b) Histochemical methods.

As these methods are used also for the cytoplasm and have been sufficiently summarized and discussed in special monographs (LISON 1936, RIES 1938, GLICK 1949, MOOG 1952, DANIELLI 1953), only a brief mention of them will be made here. The methods involve a final observation of the cell at the microscope and so they should better be named chemo-cytological, reserving the designation of histo- and cytochemistry for the chemistry of the cell components as studied by analytical methods. In chemocytological procedures the chemical entities are finally characterized by optical properties which may be: 1. colour proper of the groups of compounds, either in the visible as in pigments (carotenes, melanins, hemo-globins) or by fluorescence in the ultraviolet; 2. absorption spectrum in the visible, the ultra-violet or soft X-rays; 3. formation of a characteristic precipitate or coloured compound; 4. extraction by solvents and/or enzymes; 5. microradioautography; 6. emission spectro-photometry and emission electron micrography; 7. microincineration. All or the majority of these methods have been applied to the qualitative demonstration of characteristic groups of compounds or of elements and their quantitative determination, but the degree to which they attain a quantitative standard is variable. Most quantitative at present is mass deter-mination by X-ray microradiography, either of the elements (Engström 1946, 1947, 1950, Engström and Lindström 1947) or, coupled with extraction, of certain components (Bratt-gård and Hydén 1952), which allows determinations of constituents expressed in terms of dry cell weight with an error better than 5% (Glick, Engström and Malmström 1951). When coupled with extraction the method is subject to the errors inherent to using solvents or enzymes in the complicated biological systems of the cell. Dry mass, water content and thickness may also be determined by interference microscopy (Barer 1953).

Photometric determinations in the ultraviolet (Caspersson 1936, 1940, 1950, Hydén 1943, Thorell 1947) and the visible (Pollister and Ris 1947, Pollister and Moses 1949, Ris and Mirsky 1949, Lison and Pasteels 1950, Sandritter 1952, Moses 1952) have been widely employed in the last decade or so to determine nucleic acids and proteins in the cell nucleus, giving rise to a special chapter of histochemistry called "cytophotometry". The degree of accuracy of the measures depends on many variables and generally is poor. Sources of error have been discussed chiefly by Caspersson (1936), Thorell (1947) and Moses (1952; see also addendum to Allfrey et al. 1952). Much of the work has been done by employing the Feulgen reaction or colorations with basic stains and pyronin-methyl green. As is well known, usually nuclei do not stain uniformly by these procedures, cases of inter-phase nuclei being uniformly coloured by the Feulgen reaction (Ris and Mirsky 1949) possibly deriving from the retaining of formalin of the fixative by the nucleus, since proteins only very slowly lose the combined aldehyde on washing (see also below). The discontinuity in staining gives rise to such enormous errors when whole nuclei are studied by cytophoto-metry that the procedure necessitates of several empirical corrections and does not fulfil the standards of a quantitative method. Many works employing the combined method of photometry and staining are vitiated by these errors and the conclusions are without a suf-ficient basis. To employ cytophotometry in these cases, besides other conditions concerning the fulfilment of the Beer-Lambert law by the reaction or the staining employed, it is necessary either to restrict the measured surface to the minimum possible in relation to the wave length employed or to proceed to integration of many small measures in different points of the nucleus (Lison and Pasteels 1950). To measure a nucleus with discrete stained bodies as if it was uniformly coloured is to run the risk of working in vain. It is to be hoped that cytophotometric work will be thoroughly revised with more accurate methods.

Another question concerns the specificity of the Feulgen reaction and the localization of the formed coloured product in relation to the original site of the thymonucleic acid in the nucleus. Recent controversies are not yet completely settled but the general consensus is that the reaction is well localized in what concerns the chromosomes, although it may be that thymonucleotides are present also in the nucleoplasm and are lost during ordinary fixation and/or the Feulgen hydrolysis (summary and references Di Stefano 1948, Milo-vidov 1949, ch. 12, Danielli 1953). On the other hand, thymonucleotides of the cytoplasm generally are lost during the hydrolysis for a Feulgen reaction and may diffuse to the nucleus, where they contribute to a false coloration of supposed chromatin granules (Chayen and Norris 1953). The use of other stains and especially methyl green-pyronin mixtures has also been under active discussion (summaries Brachet 1953, Singer 1952). The conclusion about methyl green is that it stains in part by chemical affinity to acids (nucleic) and in part according to the degree of polymerization and the physical state of the thymonucleic acid and the proteins to which this is linked. According to the pretreatments and the phy-siological state of the cell one or another of these causes may predominate.

Enzymes have also been in wide use for the obtention of data on nuclear chemistry by selectively degrading and eliminating some compounds, chiefly nucleic acids but also proteins (summaries Kaufmann 1950, Sanders 1952, Brachet 1953, Danielli 1953). Except the case of ribonuclease and in part also thymonuclease, generally enzymes are not as yet suffi-ciently selective or active as would be necessary to serve as fine dissecting instruments of analysis. An important condition concerns the purity of the enzyme as tested in test tube and another always present is the possibility that in such biological systems as the nucleus enzymes may have a specificity and action different from those found in test tube.

These uncertainties have led to strong criticism of the use of enzymes in cytochemistry (DANIELLI 1953, ch. 1). Their use must be accurately controlled by other means. Selective extraction may in certain cases advantageously substitute enzyme treatments, as purified enzymes are costly. The extractability depends not only on the compounds present in the nucleus as also on the fixation. Sodium chloride in molar concentration (MIRSKY and POL-LISTER 1946) has been used to extract nucleoproteins and hydrochloric, perchloric and tri-chloroacetic acids for the remotion of nucleic acids (summ. and refs. ATKINSON 1952). The general conclusion is that these means are not specific when mucopolysaccharides are present.

## C. Chemical composition of the nucleus.

In this chapter only recent data will be considered more thoroughly. For older data and the historical foundations of nuclear chemistry the works of KIESEL (1930) and MILOVIDOV (1949) may be consulted. First the nucleus will be considered as a whole — see Table 1 — and then particular properties of the different parts will be described. When mass isolation is used generally no distinction may be made between nucleoplasm, chromatic threads, nuclear membrane and nucleoli, although chromatic threads and the nucleoli can be separately isolated. The difficulties in assessing the degree of purity of these nuclear parts will be even greater than for the nucleus as a whole. Results of chemical analysis and histochemistry will be considered together.

### a) The nucleus as a whole (Table 1).
### Inorganic constituents.

Some of the inorganic constituents are believed to be combined in the nucleus with proteins or with nucleic acids. However, as they are readily diffusible and without any doubt exchanged during physiological changes in the cell between nucleus and cytoplasm, it is to be expected that the contents of the nucleus in these constituents be highly variable according: 1. to the species or strain, 2. ambiental conditions, especially food or composition of the water for aquatic species, 3. the physiology of the cell, 4. the state of the nucleus, either in resting stage or in mitosis. Therefore, different observers may obtain widely divergent results in what concerns inorganic nuclear constituents and this indeed is what a perusal of the literature reveals. Work trying to relate the physiology of the cell to the inorganic constituents of the nuclei remains almost entirely to be done. Another difficulty in locating inorganic constituents is that in many cases they may be in a masked state not revealed by histochemical tests and necessitating previous microincineration.

Of the alkaline metals nothing special is known concerning their distribution between nucleus and cytoplasm. It has been stated by some authors that the nucleoplasm is poor in ash and therefore also in alkaline metal salts; the data, however, should be revised having in mind the above considerations about relations with cell physiology. Alkaline earth metals, especially calcium and magnesium have been found in several types of nuclei and the chromosomes are particularly rich in these metals during mitosis (references in HORNING 1952). However, these data are not always generalizable; in *Drosophila* the nuclei seem to be poor in calcium (references POULSEN and BOWEN 1952) and amphibian oocyte nuclei are devoid of ionic calcium since the addition of this causes a disruption of the nucleoplasm (CALLAN 1952). It seems that the general conclusion about the contents of the nucleus in calcium and magnesium is that it is highly variable according to the case under study and the four conditions referred to above.

Table 1. *Composition of the nucleus as a whole.*

DNA = desoxynucleic acid; RNA = ribonucleic acid; T Prot = total proteins; B Prot = basic proteins; NB Prot = non-basic proteins; T Lip = total lipids; P Lip = phospholipids. + indicates presence in unknown or variable amounts; — indicates absence; $\pm$ indicates dubious presence or absence. Absolute amounts are expressed in $\mu\mu g$/per diploid nucleus. $1\,\mu\mu g = 10^{-12}$ g. When not otherwise stated, % refers to the dry nucleus weight with its lipids.

| Constituents | | | | |
|---|---|---|---|---|
| Inorganic | Highly variable: Na, K, Ca, Mg, Fe, Cu<br>Must exist as coenzymes: Ca, Mg, Fe, Mn, Zn | | | |
| Nucleic acids | DNA: 5–10%<br>or up to 25–30%<br>6–60 $\mu\mu g$ | | RNA: Generally $^1/_{10}$–$^1/_3$ of the proportion of DNA, in plants up to 90%<br>0.3–1.5 $\mu\mu g$ | |
| Proteins | T Prot: 50–80%<br>24–100 $\mu\mu g$ | B Prot: 20–40% of the T Prot (exceptionally up to 90%—protamines)<br>10–40 $\mu\mu g$ | NB Prot: 60–80% of the T Prot (exceptionally less)<br>10–60 $\mu\mu g$ | |
| Lipids | T Lip: 8–12 or up to 40%<br>5–80 $\mu\mu g$ | | P Lip: 90% of the T Lip<br>4–70 $\mu\mu g$ | |
| Enzymes | Phosphatases: alkaline — or $\pm$; acid — or $\pm$; adenosinetriphosphatase and other nucleotidases — or $\pm$; acid DNAase +; nucleosidases + | | Glycolytic in general +; cytochromoxidase +; catalase — or $\pm$; esterases and proteases in general +; $\beta$-amylase + or —; arginase + or $\pm$; uricase — | |
| Other general components | Glycolytic metabolites; probably general: ascorbic acid [0.05—0.5% ( ?)], glutathione [0.1–1.0% ( ?)]. | | | |
| Special components | Several, peculiar to the species tissues or to the kind of cell. | | | |

By autoradiographic technique copper has not been found in greater concentration in the nucleus than in the cytoplasm, while in some kinds of cells the nucleus shows iron in greater concentration than the cytoplasm (Poulsen and Bowen 1952). Other data point out that the nucleolus can accumulate iron (Horning 1952). However, probably these facts have validity only for some species and under some conditions. A series of cations should be present in the nucleus if it is admitted that the enzymes recorded for the nucleus work there as in test tube and particularly in what concerns activators; these are, according to Poulson and Bowen (1952): Zn, Co, Fe, or Cu required by aldolase; Mn, Ni or Co required by arginase; Mg, Mn, or Zn required by enolase; Mn, Mg, Fe, or Co for alkaline phosphatase; Ca for acid phosphatase; Ca or lanthanides for succinic dehydrogenase. The most probable of these activators are Zn, Fe, Mn, Mg, and Ca. The general conclusion is that the state of our knowledge of the nuclear inorganic constituents is far from satisfactory and that much more work is needed.

## Nucleic acids.

It is generally admitted that there are in the nucleus of plant and animal cells two types of nucleic acid, desoxynucleic (DNA) or thymonucleic, and ribonucleic (RNA) or zymonucleic. However, no demonstration has been given that other nucleic acids do not exist in the nucleus or that the two chief types of nucleic acids, *viz.* the desoxyribo- and the ribonucleic cannot be in part substituted in certain cells or conditions by other kinds of nucleic acids.

The view held until recently was that these nucleic acids have a molecule formed of 4 nucleotides, each constituted by a purine or a pyrimidine base, a pentose sugar and a phosphoric acid residue (refs. LEVENE and BASS 1931, BREDERECK 1938, DAVIDSON 1950, CHARGAFF *et al.* 1953). The known bases for DNA were adenine and guanine (purines), cytosine and thymine (pyrimidines). To these latter 5-methylcytosine and hydroxymethylcytosine have been added more recently (WYATT 1951, 1952) as pyrimidine bases of thymonucleic acid. The bases for ribonucleic acid are adenine and guanine as purines, and cytosine and uracil as pyrimidines. It has been reported that some bases may be lacking in DNA and RNA from certain sources (refs. BENDICH 1952). The sugars are desoxyribose and ribose respectively in thymo- and zymonucleic acids. As relatively few sources have been used to obtain quantities of nucleic acids suitable to chemical analysis, it is possible that in future other bases remain to be discovered. On the other hand, it is not known if sugars other than desoxyribose are present in the nucleotides of desoxyribonucleic acids from different species.

Quantitative analysis, electrometric titration and determination of the molecular weight had led to the assumption (refs. in JONES 1920, LEVENE and BASS 1931, GULLAND 1947, WYATT 1952) that DNA and RNA are alike in their constitution, their molecules being tetranucleotides. This hypothesis formed the basis of the until relatively recently accepted tetranucleotide theory, which has been questioned since 1945 (GULLAND *et al.*) and now is recognized as being only a first approximation to the complicated, and as yet largely unknown, structure of nucleic acids. The analysis of nucleic acids has met with great difficulties not only in the separation of the bases, which only recently has benefitted from paper chromatography technique, as also because the hydrolysis is not easy. Purine bases are easily split from DNA and RNA with dilute acids while pyrimidines require stronger hydrolysis leading in many cases to degradation. Even taking into account these difficulties, the analyses lead to the conclusion that the four bases do not exist in equimolecular proportions in DNA from bovine tissues which shows (mean values from 6 series of analyses, reported from six authorships by WYATT 1952): adenine 1.14, guanine 0.90, thymine 1.12, cytosine 0.82, methylcytosine (from only 3 series of analyses) 0.0497. RNA also shows its bases in proportions different from the equimolecular ratio postulated by the tetranucleotide hypothesis. It should be noted that methylcytosine exists in wheat germ DNA in greater proportion than in the analysed animal tissues (BENDICH 1952).

Moreover, a series of analyses of calf thymus DNA obtained from 21 individuals gave very similar compositions, favouring the view that DNA has a constant mean composition within each species but may vary from species to species (CHARGAFF *et al.* 1953 and refs.). The mean constitution found for calf thymus DNA was (in per cent, with a 0.2 standard error): adenine 29.0, guanine 21.2, cytosine (+ methylcytosine) 21.2, thymine 28.5. The molecular ratios purines (adenine + guanine)/pyrimidines (thymine + cytosine), as well as adenine/thymine and guanine/cytosine are all equal to unity, within the experimental error. This is interesting in view of the now held as most probable molecular structure for DNA, which postulates a double helical chain with bondings adenine-thymine and guanine-cytosine (WATSON and CRICK 1953—see the following chapter, p. 453). Fractionation of several individual samples from calf thymus DNA (CHARGAFF *et al.* 1953) with mild procedures, chiefly by sodium chloride extraction, led to the conclusion that the constancy of the mean composition probably is no more than the result of the sum of "a whole spectrum" of individually different DNAs with slightly different compositions. It is to be noted that in all fractions the ratios purines/pyrimidines are almost equal to unity, as required by the structural hypothesis, which gives strength to the conclusion of CHARGAFF *et al.* (1953).

This poses the question of the relations between composition and molecular structure of DNA as revealed by its X-ray diagram (of sodium thymonucleate). From earlier data (ASTBURY 1938, ASTBURY and BELL 1940; refs. SERRA 1942) it has been inferred that P atoms were located along a backbone of sugar-P-sugar at 3.4 Å distances, which are close to the period 3.38 Å found in distended polypeptide chains, with which the polynucleotide chains could combine. Now it has been recognized that there are several other spacings (OSTER 1952, WATSON and CRICK 1953, FRANKLIN and GOSLING 1953, WILKINS *et al.* 1953, ARNDT and RILEY 1953). The now held as the most probable view is that the polymerized Na-thymonucleate is formed of a double helical chain, each coiled around the same axis and both right-handed helices, with the atoms alternating. The phosphate groups are to the outside, while the bases remain in the inside and the two helices are held together by hydrogen bonds between neighboring purines and pyrimidines (adenine with thymine and guanine with cytosine). The spacings are influenced by the water content. In the more hydrated form the vertical component of inter-base distance is 3.4 Å and the pitch of the helix is 34 Å.

On the other hand, it has been found that the old structural formula for RNA as a single backbone of phosphoric acid-sugar-phosphoric acid-sugar does not account for the fact that electrometric titration between $p_H$ 2 and 12 of the deaminated derivative (see GULLAND 1947) proved that for each four P atoms one secondary and three primary dissociations existed, while according to the old structural hypothesis only primary dissociations should be found (except for the terminal P of a chain, which should have also a secondary one). It should be noted, however, that despite electrophoretic and other criteria of purity, no definitive proof has been adduced that the structure of nucleic acids is uniform and the finding of secondary dissociation groups could correspond to differences in structure between several fractions of a preparation not otherwise separable. The coupling of these data with the results of the X-ray studies remains to be done.

The divergences from the tetranucleotide hypothesis and the differences in composition found in preparations from several sources have been acclaimed as a demonstration of the specificity of nucleic acids, which by being different according to the species could constitute a basis of genetic specificity. However, in future it is possible that the composition of nucleic acids will be found to result somewhat different according to the physiology of the cells and therefore according to the tissue and perhaps the diet. This point is related to the question of the constancy of nucleic acid (DNA) per chromosome set (BOIVIN et al. 1948, VENDRELY and VENDRELY 1949, and detailed discussions in VENDRELY 1952, MIRSKY and RIS 1951 and POLLISTER 1952). According to a view which has been widely in vogue for some time, DNA should be constant per haploid set of chromosomes for each species and several analyses of the mean value of DNA in several tissues tended to demonstrate the point and to be in favour of equating the genes of the chromosomes to DNA; the constancy of DNA while the content in proteins of the nucleus is variable should derive from the constancy of the nuclear genes. This idea, however, represents an oversimplification of the role of the genes, since it is chiefly the quality, not the quantity, of the genetic material that interests to gene specificity and the chromosomes could very well increase their matter proportionally to cell growth without dividing. It is not known in what measure the methods employed (aqueous isolation) may concur to this result and, on the other hand, conflicting data have been presented (LISON and PASTEELS 1951) in the sea urchin embryo, in which the nuclei of different blastomeres may have different amounts of DNA. Although these data have been obtained by cytophotometric methods and therefore are open to criticism, the idea of DNA constancy is yielding to the impact of more data (see discussion to the paper of POLLISTER 1952). Recently the constancy of DNA has been linked with that of arginine (VENDRELY and VENDRELY 1953).

In plants it has been found that DNA does increase during the interphase and just previous to meiosis and mitosis in *Lilium* microsporocyte development (OGUR et al. 1951) and the common observation of cytologists that in mitosis there is a great increase in chromaticity must have general validity, demonstrating an increase in DNA. The general conclusion about this question of the constancy of DNA is that the mean amount per nucleus tends to be the same in different tissues in nuclei not engaged in mitosis or meiosis and that during cell division or development the amount may change to many times the usual mean interphase amount, of the order of a few $10^{-12}$ g. per diploid nucleus ($6–60 \times 10^{-12}$ g./nucleus. This mean content represents something peculiar to chromosome structure and may or may not be related to the number of genes. In terms of nuclear weight, DNA may represent anything from about 5% (wheat germ— STERN and MIRSKY 1952) to about 10% (DOUNCE 1952) or at most 25–30% (ALLFREY et al. 1953), which in the order of magnitude agree with the value of 10% estimated by CASPERSSON (1936).

To continue this brief description of facts and concepts relative to the content of the nucleus as a whole in nucleic acids, the question of the presence of ribonucleic acid shall be mentioned. The old concept of KOSSEL that ribonucleic acid (RNA) was characteristic of plants and DNA of animals has been found to be wrong and the analyses demonstrate the presence of ribose in the nucleus. In animal nuclei, amounts of RNA of the order of $0.3–1.5 \times 10^{-12}$ g. have been demonstrated (VENDRELY 1952, SCHNEIDER 1947) corresponding to about from $1/_{10}$ to $1/_3$ of the amount of DNA. The RNA seems to be more intimately associated with the non-histone proteins of the chromosomes than DNA, which is extracted with histone by 1 M NaCl (MIRSKY and RIS 1947). In Protozoa *(Paramaecium)* the amount of RNA should be about the same, or greater, than that of DNA (MOSES 1950). The mean amount of RNA per nucleus should be more variable than that of DNA (DI STEFANO *et al.* 1952) and particularly in Lepidoptera "chromatin elimination" cytologically observable in metaphase should be the result of extrusion of RNA and protein from the chromosomes and the same phenomenon in a less visible form may be common during normal anaphase and give rise to interchromosomic fibrillae (RIS and KLEINFELD 1952, JACOBSON and WEBB 1951).

In plant nuclei RNA has been found for instance in wheat germ in amounts almost as great as those of DNA (4.6% DNA and 4.1% RNA of the weight of each nucleus—STERN and MIRSKY 1952). The nucleus occupies in these cells about $1/_2$ of the cell volume and has also about $1/_2$ of the cell RNA. This relatively high proportion of RNA probably is common in plant nuclei. In *Lilium* microsporocytes and microspores it has been found that RNA is present in each cell in an amount 2 to 9 times that of DNA (OGUR *et al.* 1951) and probably in cells with the smaller amounts RNA is distributed with almost equal concentration between nucleus and cytoplasm, as is the case in wheat germ. On the other hand, an increase in RNA paralleling that of DNA has been found to occur in mitosis and meiosis (OGUR *et al.*).

The distribution of RNA in the nucleus may be somewhat different from that of DNA. As no Feulgen reaction is given by the nucleolus, which exhibits a great absorption at 260 m$\mu$ characteristic of purines and pyrimidines, the conclusion has been drawn that the nucleoli possess RNA in relatively high concentration. Besides in the chromatin of the chromosomes, RNA probably is also present in the nucleoplasm, as will be referred to below.

The functions of RNA in the nucleus probably are manifold, one of them being similar to that of cytoplasmic RNA, serving to neutralize the basic parts of unfolded polypeptide chains in process of being synthesized or folded.

To end this account on nucleic acids the state of these compounds in the cell nucleus will be briefly considered. The view was generally held that nucleic acids were chiefly combined with proteins of the basic type in salt-like combinations of their phosphoric acid rests with basic amino groups of lateral polypeptidic chains, especially from arginine and in a lower scale from histidine and lysine. Analytical results of comparing the DNA and arginine contents of isolated nuclei (VENDRELY and VENDRELY 1953) give support to this opinion, as in non-mitotic nuclei there is a proportionality between DNA and arginine. However, from in vitro research (DOUNCE 1952) it is known that DNA may combine with non-basic proteins by its phosphate groups, easily forming insoluble complexes with ordinary proteins at or below the isoelectric $p_H$ point. A lowering of the $p_H$ of isolated nuclei from 7 to 6 causes considerable dehydration and condensation of "chromatin", which should indicate that the DNA is not mostly, or at least entirely, complexed with histone since this, being highly cationic at $p_H$ 6–7 and nucleic acid strongly anionic, should not be affected by such a small $p_H$ lowering. Another fact which points out in the same direction is that to extract the nucleohistone from nuclei with NaCl some degree of autolysis is necessary, while the complex afterwards easily dissolves in molar saline (Dounce 1952). However, this could be due to the histone being also linked in the intact nucleus to other compounds, probably lipids, forming nucleo-histone-lipid complexes. The conclusion is that the true state of nucleic acids in living nuclei is yet unknown although it is probable that they generally are present there under the form of nucleo-protein complexes either with histone, with non-histone proteins, or forming more complicated ternary or quaternary complexes of nucleic acid-histone-other proteins-lipids. In what measure DNA or RNA may also exist as free compounds remains to be seen, although it seems that by their strongly acidic properties they should not be physiologically tolerable as such and should at least form potassium or sodium salts. The tendency to regard DNA alone as entirely forming the chromosomes and ultimately the genes, as yet has no factual basis in histochemistry and probably will not resist the test of time.

## Proteins.

Proteins exist in the nucleus in many cases probably under the form of complexes with nucleic acids, forming nucleoproteins. Some workers seem to admit implicitly that all nuclear proteins are nucleoproteins, which at best is an unbased generalization and one indeed improbable. In fact the protein content of nuclei isolated in non-aqueous solvents is much greater than the content in nucleic acids of the same nuclei and therefore only doing it in improbably low proportions the nucleic acids could combine with all proteins. The conclusion emerges that nuclear proteins may exist under the form of nucleoproteins combined with DNA and RNA or under other forms, as simple proteins or of complexes with lipoids (lipoproteins and especially phosphatidoproteins) or with phosphorus, etc. Nucleo-lipido-protein complexes may also be possible.

The amount of protein of nuclei may be appraised only by employing non-aqueous solvents as isolation media, since aqueous solvents extract a part of the nuclear proteins (citric acid extracts from 18 to 55% of protein) and cyto-photometric methods, especially those utilizing Millon's reaction, are inaccurate (Allfrey et al. 1952, addendum). In animal cells isolated in non-aqueous media about 60–80% of the nucleus should be of protein (values of Allfrey et al. 1952 corrected for the probable amount of RNA). The mean protein amount in different tissues of an individual is much more variable than the corresponding amount of DNA. It should be noted, however, that in these analyses the protein has been obtained by difference after the DNA and RNA have been considered and that the nuclei were lipid extracted. The absolute amount of protein should be of the order of $24–100 \times 10^{-12}$ g., that is approximately 4–10 times the amount of DNA. By the same differential procedure, lipid free wheat germ nuclei should have about 90% protein. Other nuclei (calf thymus gland) lipid free yielded about 54% isolated protein (Mayer and Gullick 1942).

The general conclusion about the content of the whole nucleus in total proteins is that these form about 60–90% of the dry weight of the nucleus without lipids, or 50–80% when the latter are also considered. This statement, however, has lesser meaning than that on the nucleic acids, which form only two kinds of compounds (with possible minor variants) while proteins may vary almost "ad infinitum". It was believed that the proteins of the nucleus were chiefly of the basic type, rich in diaminoacids, and especially arginine but it is now recognized that several other kinds of proteins and even non-individualized polypeptide compounds may also be present. The basic protein types are protamines and histones. The former are richer in arginine + lysine and sometimes also in histidine, the three basic aminoacids. Arginine + lysine form about 70–90 % of the protamines and 20–50 % of the histones. Some basic proteins are particularly rich in lysine, while in the majority it is arginine that predominates. The line of separation of protamines and histones is more or less arbitrary and transitions are possible, for instance in Molluscs (HULTIN and HERNE 1948). Two species of the same genus may show greater differences in the diaminoacids content than two other species of different genera. The presence of certain aminoacids, and particularly those with sulphur (cystine, cysteine, methionine), the diacid ones (glutamic and aspartic) and the aromatic and indolic (phenylalanine, tyrosine, tryptophane) has not been established in protamines, which seem devoid of these aminoacids and to possess only less than ten aminoacids besides the dibasic ones (KOSSEL 1928, KIESEL 1930, LLOYD and SHORE 1938, HULTIN and HERNE 1948, GREENBERG et al. 1951). Histones, on the contrary, are more complex in their composition and may possess almost all the aminoacids. It is sometimes said that they are devoid of the aromatic and indolic aminoacids but this is not true (KIESEL 1930, MIRSKY and POLLISTER 1946, DALY et al. 1951) although they may be present in small amounts. Histones are complicated proteins with the majority of the existing natural aminoacids present in their molecule and may have an infinity of possible configurations of their chains. Protamines, on the other hand, may represent a special case of simplification of proteins which are carried by the sperm as a kind of reserve of dibasic aminoacids (SERRA 1942).

It is interesting to note that, despite the fact that nucleic acids may complex with proteins of several kinds, the content of arginine of the nucleus is in many cases parallel to the DNA content (VENDRELY and VENDRELY 1953). In erythrocyte nuclei the ratio DNA/arginine is of the order of 4.3(bull)–5.4(perch), with the majority of values in the range $5.1 \pm 0.2$. The absolute amounts of DNA per nucleus vary from 1.7 to $6.4 \times 10^{-12}$ g. and those of arginine from 0.34 to $1.49 \times 10^{-12}$ g. but nevertheless the proportion keeps more or less constant. In spermatozoids the ratio is different, but this may be accounted for by a transformation to basic proteins with more lysine and histidine (VENDRELY and VENDRELY 1953). It is not known in which measure the method of nuclear isolation (citric acid) may be responsible for an apparent constancy in composition or if this represents a real parallelism between the overall basic aminoacid content of the nucleus and its DNA content, which could derive from the bonding of phosphate groups chiefly to arginine.

Fractionation of the nuclear proteins has given, besides histones or protamines, other proteins of a non-basic character, also designated in cytochemical work (CASPERSSON 1936, 1940) as "higher proteins" or tryptophane-containing proteins, although these designations are inaccurate since histones may also have tryptophane (SERRA 1947a, MIRSKY and POLLISTER 1946, see also LLOYD and SHORE 1938) and show a complicated aminoacid composition. Ultraviolet spectrophotometry data may be analyzed to show the difference between both kinds of protein (CASPERSSON 1940, HYDÉN 1943), basic and non-basic, and the combination of results of MILLON's reaction and the arginine reaction (SERRA 1944, 1946) also gives an idea of the relative concentrations of both classes of proteins. Isolated thymus nuclei dissolved in NaOH (3 %) may be fractionated by isoelectric precipitation with acetic or mineral acids (HCl, $H_2SO_4$) which leave the basic proteins into solution, from which they may be precipitated by 70 % alcohol in presence of ammonium chloride (MAYER and GULICK 1942).

By these methods the nuclei were fractionated in about 34 % histone, or about 36 % of the nuclear nitrogen, while to nucleic acid belonged about 32 % of the nuclear nitrogen. About 28 % of the dry nucleus weight corresponded to nonhistone proteins in two fractions, one of about 19% of the nuclear weight precipitating at $p_H$ 5.8–6.15, the other extracted by 5% NaCl, both fractions containing sulphur (MAYER and GULICK 1942). The properties of these fractions indicate that they are in part a protein mixture containing sulphur linked to carbon by a double bond. In part they may be also mixed with decomposition products of histones. Their double bond S probably derives from SH groups of the nucleoplasm (SERRA 1947a) which were decomposed by the treatment with NaOH, that is known to attack SH groups (see SERRA and ALBUQUERQUE for refs. on this latter point, in other compounds). It is to be noted that for these analyses the nuclei were isolated in organic solvents and therefore lipid free.

From other attempts to fractionate nuclear proteins, preparations have been obtained to which the authors gave special designations. STEDMAN and STEDMAN (1943, 1947) isolated from nuclei an acidic protein which they believed to form the essential part of the chromosomes and therefore named chromosomin. The relative amount of this protein material has been given differently in different papers though it seems that it could be of the order of 10–20% on a dry weight basis. In view of the method of extraction and from what is known on the heterogeneity of the nucleus it must be concluded, however, that chromosomin is either a protein mixture or a part of some nuclear proteins split by the extraction procedure (SERRA 1943). Another case of fractionation (MIRSKY and POLLISTER 1946, MIRSKY and RIS 1947, 1951) has given rise to a new designation, chromosin, which was desoxyriboprotein without detectable contamination by RNA. This nucleoprotein should be formed of DNA + histone + non-histone protein and was isolated from plant, bacteria, and animal nuclei by extraction with molar solutions of NaCl. The histone and DNA are largely dissociated when the chromosin is dissolved in M NaCl while in 0.02 M NaCl they unite into a complex with partial depolymerization of nucleic acid. This designation of chromosin has been discontinued in recent work of the authors as it was recognized to be merely a not well defined mixture. It has given place to the recognition separately (same authors) of DNA, RNA, histone and a non-histone or "residual protein" which was left after the other components were separated from "chromosomes" isolated in 1 M NaCl. Apparently the former "chromosin" was almost equal to the isolated "chromosomes". As the isolation was carried on in aqueous media, considerable losses should occur. Fractionation of the "chromosomes" isolated from lymphocytes with 1 M NaCl gave 90–92% of the mass of the "chromosomes" as desoxyribose-nucleohistone and the remainder 8–10%, under the form of coiled threads, is the residual protein containing 12–14% of RNA and 2–3% DNA. In the most recent work it seems that the authors have also more or less abandoned the ideas about this fractionation in favour of supposing that DNA, and not the residual protein, is the genetic material properly and the proteins would be non-essential constituents of the chromosomes. This is an interesting example of how theoretical conceptions may influence the march of analytical work.

Another example of fractionation (ENGBRING and LASKOWSKI 1953) is that of chicken erythrocyte nuclei from which an acid extracted histone fraction with several histones, a neutrally extracted mixture of proteins, and an alkali extracted electrophoretically homogenous protein were obtained. This latter protein contained lipid and carbohydrate and probably the apparent homogeneity is rather the result of the attack by alkali, its chemical properties being not in favour of ultimate homogeneity. This work gave results which are closer to what is to be expected from biological considerations, namely a great complication of the nuclear proteins which probably are also different according to the tissue and the physiological state of the cell. It should be noted, on the other hand, that these results have been secured by an aqueous method of nuclei isolation and all the limitations of these methods apply here.

According to some data (references BRACHET 1944) proteins or peptidic compounds containing SH groups are also present in the nucleus and an important role has been ascribed to sulfhydril compounds during mitosis and cellular syntheses. This subject, however, necessitates of a thorough revision based on data obtained with new methods. By a specific colour test for sulfhydril the nuclei of several tissues have been found either not to have SH groups or to have them in smaller concentrations than the cytoplasm (BARRNETT and SELIGMAN 1952). This test reveals chiefly SH groups bound to proteins; for glutathione-SH see below, under the heading "Other compounds".

As shown by these citations, the knowledge on fractionation of nuclear proteins is yet in a very fluid and incomplete state. Great complications should be anticipated from what is known about nuclear structure, as pointed out by SERRA (1949). Each structure may have its proteins, namely those of the nucleoplasm, those of the chromosomes, of the nucleolus, and possibly also of the nuclear membrane, which may be designated respectively as nucleoplasmins, matricins and genonemoproteins, enchylemins, nucleolins, etc. In chromosomes the part which deposits during division and constitutes the matrix of cytologists should also be different from the so-called chromonemata, each part with their own proteins.

Attempts to characterize nuclear proteins by enzyme (protease) attack have not yet given final results, probably owing to the great complexity of the mixture of proteins within the nucleus (refs. KAUFMANN 1953).

## Lipids.

Lipids are largely unknown as nuclear components. Generally an extraction with organic solvents is performed, either concomitantly with isolation or after this procedure. It seems as if to many researchers it was granted that lipids are nuclear components without importance, the emphasis being placed on nucleoproteins and more recently upon DNA alone. Several attempts to demonstrate lipids in the nucleus by histochemical tests generally have given uneven results, of uncertain interpretation (former references GULICK 1941). Analytical procedures also gave equivocal results (see KIESEL 1930). More recently, however, lipids have been demonstrated in the nucleus by histochemical tests, namely in the nucleolus (ALBUQUERQUE and SERRA 1951) and chemical analyses have shown a high content of lipids in the nucleus, from about 8% to 40% of the dry weight, parallel to that of the tissues in general (STONEBURG 1939). In rat liver nuclei isolated by the citric acid procedure 10.5–11.0% of lipids have been found (DOUNCE 1942). About 4 to 8% of the phosphorus of chicken erythrocyte nuclei were found to be lipid P according to SCHMIDT and TANNHAUSER (ENGBRING and LASKOWSKI 1953). By histochemical tests, in the majority of the nuclei only the nucleolus and some special zones of the chromosomes associated with nucleolar formation have histochemically demonstrable lipids and namely phospholipids (ALBUQUERQUE and SERRA 1951), but the presence of lipids demonstrable by general lipid stains has also been found very probable in the nucleoplasm of oocyte nuclei adjacent to chromosomes (SERRA 1947). Fractionation of the lipids of nuclei (WILLIAMS et al. 1945) resulted that in rat liver about 90% of them are phospholipids and especially lecithin and cephalin (respectively 61.7% and 24.0% of the total lipids). STONEBURG (1939) has also found a relatively great amount of cholesterin.

Such a great amount of lipids cannot be the result of cytoplasmic contamination, adsorption and/or diffusion during isolation. On the other hand, the analyses should have general validity since several tissues (muscles, tumours, liver, etc.)

have been analysed and always a relatively great concentration of lipids has been found. Therefore, lipids and especially phospholipids must be normal constituents of the nucleus, not only of the nucleolus, which cannot contain the whole mass of phosphatids present in the nucleus, as also of the nucleoplasm, although here they have not yet been histochemically demonstrated with certainty. Nuclear lipids shall be chiefly combined with protein forming lipoproteins, and probably also with ribonucleic acid, in complexes not very different from those found in mitochondria. However, only in the nucleolus the phosphatids are in such a relative amount or in a kind of combination capable of being demonstrated by known histochemical tests and probably only in this nuclear constituent are there lipo-nucleo-proteins more similar to those of mitochondria. The other lipids of the nucleus must be in such a state that do not give lipid histochemical reactions, forming stable lipoid complexes probably of the type of alternating layers surrounded by a water barrier, as have been postulated for cytoplasm proteins (Chargaff 1949). It has been postulated that nuclear lipids may play a role in individualizing chromosomes and in separating out phases (Serra 1947a).

## Enzymes.

Among nuclear constituents the enzymes have modernly attracted much attention, a fact which in part results from current conceptions about how the genes work. It is generally admitted that genes work by means of primary gene products and for many authors these would be enzymatic in nature. However, it is easily seen that enzymes with their great molecules cannot be directly synthesized as replicas of the genes and that probably only compounds with smaller molecules, perhaps highly specific not very long polypeptide chains may be synthesized in contact with the genes of the chromosomes. It is possible that enzymes may afterwards be synthesized in the nucleoplasm, and chiefly in the cytoplasm, from these basic chains, the cytoplasm being very probably the chief center of protein synthesis and where coenzymes of exogenous origin first penetrate from outside the cell. Therefore, enzymes only with much difficulty may be supposed to appear as primary gene products. On the other hand, it is also plausible that primary gene products act rather by shifting the balance of a complicated system of enzymes and active compounds, than by producing a certain amount of one or several secondary gene products which realize the phenotype (Serra 1949a).

The fact that a certain enzyme is found or not in isolated nuclei is not decisive about the *in situ*, and with greater reason *in vivo*, activity of that enzyme in the nucleus. We have already discussed above the possibilities of adsorption, diffusion and inactivation of nuclear compounds during the procedure of isolation. These artifacts are, of course, relatively much more important when dealing with compounds like the enzymes, which generally exist in smaller amounts and have a great chemical activity, than with compounds existing in greater concentrations. Histochemical tests can aid in elucidating some doubts but the methods must be evaluated very critically before valid conclusions can be drawn. On the other hand, besides the mere presence, for the admission that a given enzyme is active as such in the nucleus it is necessary to demonstrate: 1. that the enzyme is truly individualized in the nucleus and does not represent an artificial scission product which happens to have enzymatic activity, a subject which bears relations to apoenzyme specificity; and 2. that the demonstrated enzyme is really active in nature and is not inactivated in the living nucleus by inhibitors. The converse problem, the presence of enzymes in the living nucleus which cannot be demonstrated by the crude methods at hand is, of course, yet much more pertinent.

These difficulties are well illustrated by the case of alkaline phosphatase, which has given rise to an enormous literature (summaries NOVIKOFF 1952, DANIELLI 1953).

Generally the Gomori-Takamatsu technique is used to demonstrate histochemically this enzyme (an alternative technique is that of the azo dyes—see DANIELLI 1953). After a suitable fixation by acetone or 80% alcohol and embedding in paraffin through benzene, keeping the temperature below $60^0$ C., or alternatively by freezing-drying, the sections are incubated with a medium buffered at $p_H$ 9.4 containing a phosphate (generally glycerophosphate) plus a calcium salt to precipitate the phosphate liberated according to the reaction $ROPO_3H_2 \underset{\longleftarrow}{\overset{\pm\ H_2O}{\longrightarrow}} ROH + HOPO_3H_2$. The calcium phosphate is transformed to cobalt phosphate by treatment with a cobalt salt, and afterwards to cobalt sulphide with an ammonium sulphide solution. Cobalt sulphide, which is brown or black, is easily visible at the microscope. The difficulties in interpreting the findings result from diffusion of the enzyme proper, of the liberated phosphate, or of calcium phosphate and these difficulties may act in different degrees according to the tissue under study. The last two causes of error may be minimized by refinements of technique, while the first may escape control, as the enzyme may diffuse into the nucleus and be absorbed there when the cell dyes.

Nuclei isolated in non-aqueous media contain a smaller concentration of enzyme than those isolated in aqueous media or histochemically studied. In histochemical tests generally the nucleus, especially the chromosomes and in these the chromatic bands (salivary gland chromosomes) give a heavier sulphide precipitate than the cytoplasm, while in nuclei isolated in organic solvents, which are supposed not to attack the enzyme and not to let it diffuse back or forth from or to the nucleus, the activity in alkaline phosphatase of the nucleus is only a fraction of that of the cytoplasm (between 1 and 4% of the activity of the tissue as a whole, according to ALLFREY et al. 1952 b). These results contrast with those obtained in nuclei isolated by the citric acid technique, which may present concentrations of alkaline phosphatase greater than those of the cytoplasm. If these results are taken at their face value, they should mean that the histochemical tests do not reveal the precise location of alkaline phosphatase, although they may give an indication about the tissue as a whole. The liberated phosphate produced by enzymic activity in the cytoplasm and the calcium-phosphate in formation would be adsorbed by the nuclei, and especially the chromosomes, so as to give there an intense reaction. On the other hand, it may also be presumed that the organic solvents used during isolation could differentially inactivate alkaline phosphatase, that of the nucleus being more inactivated than that of the cytoplasm. This possibility seems rather remote and is disproved by controls of the whole tissue (ALLFREY et al. 1952 a). The general conclusion is that in the tissues as yet investigated by the isolation method, alkaline phosphatase seems either almost absent or to exist in the nuclei in lower concentrations than in the tissue as a whole. Despite positive histochemical tests (SHARMA et al. 1953) these conclusions should also be valid for plant tissues. Histochemical results should be thoroughly revised with improved methods, for instance an incubation mixture with 40% acetone (FREDRICSSON 1952, see also HERMAN and DEANE 1953) which is reported to give no coloration to nuclei grace to it strongly reducing the solubility of calcium phosphate while enzyme activity is only slightly affected, or with azo dye technique (DANIELLI 1953). It must be emphasized that all modern studies about nuclear enzymes point out that the concentrations and even the presence of the enzymes are very variable according to species, tissue and cell particularities, and variable also according to the state and the physiology of each cell. Therefore, the conclusions are only valid for the kind of cells under study. The same should in fact largely apply to other nuclear constituents, the pretense constancy of DNA amount being only a first approximation result.

Another point of general interest when the results of enzyme studies after isolation procedures are compared with those of histochemical tests is the possibility of enzymes existing in an inactive form which is activated by some step of the employed method. For instance, it has been observed that the use of a sonic oscillator enhances the activity of aldolase and catalase in isolated nuclei from 14 to 280% of the original (DOUNCE 1952).

The case of acid phosphatase is similar to that of alkaline phosphatase, as the same factors may cause false adsorptions and inactivation by the fixatives (RABINOVITCH, JUNQUEIRA and FAJER 1949, other refs. ABOLINŠ 1952).

In what concerns other phosphatases and especially nucleotidases the results have been variously interpreted (cf. ALLFREY et al. 1952b, DOUNCE 1952, NOVI-KOFF 1952, and also WACHSTEIN et al. 1952). According to the data obtained from nuclei isolated in organic solvents, adenosinetriphosphatase, 3 and 5-adenylic acid phosphatase and DNA-phosphatase (neutral or alkaline) either do not exist in nuclei or are present there in lower concentration than in the cytoplasm. An interesting case is that of acid desoxyribonuclease, which acts at $p_H$ 5.2 and, contrary to ordinary or neutral nuclease which would be special to the pancreas, having a digestive function, the DNAase with near $p_H$ 5.2 optimum exists in all tissues so far examined and occurs in greater concentration in the cytoplasm than in the nucleus; this latter generally has a concentration of only 20–80% that of the cytoplasm (ALLFREY and MIRSKY 1952). On the other hand, the enzymes capable of metabolizing nucleosides may exist in the nucleus in higher concentrations, and especially nucleoside phosphorylase, adenosine deaminase and guanase. As was said in Section A, the nucleus could be involved in controlling oxidative phosphorylation by synthesizing di- and tri-phosphopyridine nucleotides.

To end this discussion about the existence and concentration of enzymes which could be involved in nucleic acid metabolism we would like to point out that many similar discussions generally are vitiated by the implicit postulate that an enzyme, to be active in a certain part of the cell, must be found there in higher concentrations than elsewhere in the same tissue or cell. However, it may happen that either 1. the enzyme diffuses forth and back from the cytoplasm to the nucleus and concentrates in this latter for only small intervals when it is being employed, or 2. that the presence of inhibitors or enzyme inactivants plays the decisive role in controlling the activity of the enzyme almost independently of its concentration.

Esterases exist in all nuclei in smaller or greater concentrations and probably proteases are also present there, as the nucleus easily undergoes autolysis (DOUNCE 1952). $\beta$-amylase activity was found to be smaller in nuclei than in the cytoplasm of wheat germ (ALLFREY et al. 1952a). Arginase usually is also present but according to the type of the cell its concentration is variable and does not bear a simple proportionality to the content of the tissue as a whole in this enzyme (ALLFREY et al. 1952, BLASCHKO et al. 1952). Uricase is not present in the studied nuclei in detectable concentration. The same may be said about catalase, which generally has a lower concentration in nuclei or may be absent. Succinic dehydrogenase, and therefore the Krebs' cycle as a whole if this is always so uniform as usually is believed, are lacking in nuclei, but some parts of the cycle may be present since malic dehydrogenase has been found there (DOUNCE 1952). Cytochrome C is doubtful and cytochrome oxidase seems to be present while flavoprotein is absent. Several glycolytic enzymes are present and probably glycolysis is the chief energy yielding mechanism of nuclei (DOUNCE 1952). However, the majority of these data necessitate of being corroborated by improved methods.

Other components of the nucleus which may turn to have enzymatic activity are the ribonucleotides, although probably these are more active in the cytoplasm (BINKLEY 1952). Particularly the enzyme cysteinylglycinase of pig kidney, which hydrolyses the dipeptide cysteinylglycine (formed in enzymatic glutathione disintegration) would be a pentosenucleic acid; however, it is difficult to warrant the complete absence of protein contamination. Probably in future it will be found that several enzymatic actions in the cell are due to non-protein compounds or to compounds adsorbed on non-specific protein.

No general and valid conclusions about enzymes of the nucleus and their role in cell life can be drawn at present, as the state of the investigations in this difficult domain of histochemistry and biochemistry is far from satisfactory. It seems that the chief energy yielding system for the nucleus is glycolysis and that phosphorylations linked with nucleotide coenzymes formation are controlled by the nucleus. The chief role of the nucleus, however, must be in relation to the formation of relatively small but highly specific compounds, probably polypeptidic in nature and possibly also of some specific nucleic acid chains, by means of which gene specificity is exerted; however, this latter role is as yet entirely hypothetical. It is also necessary to remind that generalizations from a study of only a few types of nuclei are not legitimate in this domain, since the enzymic content of the nucleus may be highly variable according to the tissues and the physiological state of the cell.

## Other compounds.

Besides nucleic acids, proteins, lipids and enzymes, the nucleus also has several other kinds of compounds, chiefly as metabolites or physiological agents but also as particular components of each tissue or of each species. Structurally, proteins, nucleic acids and lipids are the principal nuclear components. However, other compounds may be of importance in nuclear physiology. Metabolites of the glycolytic cycle must be present in all nuclei since glycolysis, as said above, seems to be the chief energy-yielding process of nuclear respiration, that appears to be entirely, or at least predominantly, of the anaerobic type. On the other hand, newer data seem to support the idea that ascorbic acid and glutathione, which are of importance in cytoplasm physiology, also exist in the nucleus in concentrations similar to those found in the cytoplasm (in the studied cases relative amounts are in the nucleus about 75–90% of those of the cytoplasm) and indeed probably they should be of general occurrence in the nucleus. To these compounds applies what has been said above about diffusion to and from the nucleus, when dealing with the inorganic constituents. From the histochemical point of view doubt has been cast on the occurrence of these compounds in the nucleus (CHAYEN 1953, for the case of ascorbic acid) but negative results could be due to washing out of the substance during preparation of the tissues or the histochemical reaction itself. Isolation by suitable procedures, f.i. in organic media, should be conclusive in this respect, but this claim must yet be more substantiated before the results are accepted as definitive. With some reserve in mind the following may be said about the existence and role of these diffusible compounds in the nucleus.

Ascorbic acid has been shown to occur in nuclei of calf thymus and liver cells, isolated in organic media (STERN and TIMONEN 1954). It also occurs in relatively great concentrations in plant meristems, where it is preserved by usual fixatives and probably plays a role in mitosis (CHAYEN 1952, 1953). It has been demonstrated in lily anthers that ascorbic acid increases during mitosis

(Stern and Timonen 1954). In this case the amount of ascorbic acid passes from about $750 \mu M/100$ g. of lyophilized anthers to about $1300 \mu M$ during microspore mitosis. Its role in nuclear physiology, however, remains problematic, although it may be presumed that it functions as hydrogen acceptor and donor. Only by studying more complex systems than those generally used, of one-to-one or step-by-step relations, for in vitro enzymic reactions may this point and others of nuclear physiology be clarified. Hormones, steroids and other active substances may in the cell be involved in metabolism at the same time that enzymes, substrates and mediators.

With basis on work performed about a decade ot two ago, glutathione and other SH-compounds have been assumed to be present in significative concentrations in the nucleus and an important role in nuclear physiology has been presumed for these compounds (refs. Brachet 1944, chap. V). The results of newer methods have somewhat restricted these conclusions, as it was demonstrated that the histochemical reactions employed, at least for SH groups bound to proteins, were not well localized and that, instead of showing a greater concentration in these groups than the cytoplasm, the nucleus does not possess SH-bound groups or only in a lower concentration (Barrnett and Seligman 1952). Now by isolation in organic media a relatively high content of glutathione has been found in calf thymus and liver nuclei (Stern and Timonen 1954) and the problem once more returns to discussion. In these tissues the content of glutathione of the nucleus is about the same as in the cytoplasm (of the order of $0.01-0.02 \mu M$ glut./1 g. of tissue). It seems — if it can be generalized — that the nucleus has a significant concentration of glutathione although its SH fixed to proteins may be in smaller concentration than in the cytoplasm; that is, dipeptide-SH should be relatively more abundant in the nucleus than polypeptide-SH. The role of this glutathione may be a formative one, in the synthesis of proteins, especially their cystine and cysteine residues, or it may act as donor-acceptor of hydrogen and in the last role to glutathione applies what has been said above with respect to nuclear ascorbic acid.

Obviously, this list of glycolytic compounds, ascorbic acid and glutathione does not exhaust the, as yet poorly studied, subject of the general nuclear components in addition to the "classic" nucleic acids, proteins, lipids and enzymes. However, no further data are available on this point. Almost the same happens with the peculiar components, which are different according to the species or the tissue. In the case of chicken erythrocytes it has been shown (Allfrey et al. 1952b, Stern et al. 1951) that hemoglobin exists in the nucleus (expressed as heme-Fe the nucleus has a concentration of hemoglobin about $1/3$ that of the cell in general). That this was not due to contamination has been demonstrated by mixing with other nuclei, which do not incorporate hemoglobin. Muscle nuclei (beef heart) do not have myoglobin, present in the rest of the cell. In plants it is to be presumed that general cell components may also occur in the nucleus. In meristems of bean roots a fluorescence, excited by 3650 Å light, has been demonstrated, besides in the cytoplasm, also in the nucleoli, probably corresponding to pterins, which have absorption curves in the ultraviolet similar to those of nucleic acids (Chayen 1952). Catechol, which is present in the meristems in concentrations like those of total nucleic acid $(2.1 \times 10^{-5}$ g. per root meristem against $2.5 \times 10^{-5}$ g. for total nucleic acid — Chayen 1952) possibly may also occur within the nucleus. From these examples it follows that the picture of nuclear composition may be much more complicated than the simple descriptions in terms of the three or four chief classes of components might at first sight suggest.

# Some old "chemical concepts" about the nucleus.

Some old designations about real or pretense chemical entities of the nucleus are yet found in cytological literature; so a brief review of their current status may be of some service.

*Nuclein* is the old name for a step in the fractionation of nucleoproteins, according to MIESCHER, KOSSEL and others. It would be: Nucleoprotein = Nuclein + Histone or Protamine. And Nuclein = Nucleic acid + Protein. In reality nucleoproteins may be formed not only from basic as also from neutral or even acidic proteins with nucleic acids. Modernly it has been adduced that DNA could be combined in the nucleus in part not as a histone salt, or at least that many of the phosphate groups of DNA should be present as potassium salts and could combine with other proteins with which they form an insoluble complex at $p_H$ below the isoelectric point of the protein. When the $p_H$ is lowered to 6.0 a dehydration and condensation of a complex soluble in NaCl and similar to chromatin should result. When the $p_H$ is lowered to 4.0 the complex becomes insoluble in sodium chloride at this $p_H$ (DOUNCE 1952). It seems possible that in future it will be found that the matter of the chromosomes which increases during mitosis is more than a complex of basic protein and DNA and that it has also non-basic protein besides RNA, thus demonstrating the soundness of the old concept of nuclein. Today, however, the attentions are turned to simpler matters, to DNA alone or histones alone, etc.

*Chromatin.* According to FLEMMING's original (1880) definition, this should be the substance in the nucleus which tincts by treatment with stains. It should exist in the chromosomes only when the nucleus divides while in resting stage it would be present in the membrane, the reticulum, the nucleolus and also in the intermediary substance. This wide definition has been narrowed down by subsequent writers and the concept evolved with the advance of knowledge on nuclear composition (refs. MILOVIDOV 1949). Of course, no single substance could be implied in the original definition and the staining with any dye is not a suitable criterium. After the discovery of the Feulgen nuclear reaction this has been selected to define chromatin and the modern concept is that it is a substance or group of substances which in part form the chromosomes and are composed in part of thymonucleic acid, so that chromatin always gives a positive Feulgen reaction.

In this way chromatin acquired a semi-morphological and semi-chemical meaning and the use of this concept is justified by the fact that present knowledge on the chemical composition of chromosomes and the parts of resting stage nucleus which give a positive Feulgen reaction is incomplete. When this knowledge will enable a complete description of that matter which accumulates in the chromosomes during mitosis and in part disintegrates as a fixable structure in anaphase and telophase, the need for the use of the term chromatin will have disappeared. It is admitted that chromatin is chiefly a complex or group of complexes formed of DNA plus histone, having also non-basic proteins and RNA in smaller concentrations.

The old designations of *basichromatin* for the Flemming chromatin colourable by basic stains and of *oxychromatin* for the other part of the chromatin which tincts by acidic stains are no more in use. Oxychromatin should be, to some authors, a synonym of linin.

*Karyotin* is a term employed for the reticulum which forms in the nucleus after fixation and should be an artifact (KLEIN 1930). According to other authors (MILOVIDOV 1949) it would be the chromatin of the resting stage nucleus. It seems not to be a necessary designation, as chromatin, completed by reticulum, or fixation reticulum, and chromomeres or chromocentres, are sufficient terms in a morphological sense and chemically no definite sense can be given to karyotin, as different from chromatin. The reticulum should rather correspond to a structure of fibrils of the chromosomes plus some more dense parts of the nucleoplasm.

*Plastin* is an old designation for the residual material of the nucleus or the cell after nuclein extraction and of course has no meaning besides that of a mixture resulting from certain preparative procedures.

*Linin* would correspond to *achromatin*, another substance existing in the nucleus and its structures which does not give the reactions of chromatin. Surely no single substance has such properties and the name must not be used when it is aimed at any precision.

*Chromosomin* is not strictly an old concept, as it has been employed only after 1940 (STEDMAN and STEDMAN 1943). It would be an acidic protein of the chromosomes. New studies are necessary to prove its real existence and, in positive case, to isolate it and determine its composition (see also p. 426).

*Chromosin* is also a modern designation (MIRSKY and POLLISTER 1946) for a desoxyribose nucleoprotein complex which should form the chromosomes. The designation has been discontinued since then by its authors (see above "Proteins", p. 426).

## b) Metabolism of nuclear constituents.

Since this subject lies somewhat outside the scope of this chapter, only a very brief reference to it will be made (more literature in BENDICH 1952, DALY *et al.* 1952, SMELLIE *et al.* 1953, STICH and HÄMMERLING 1953). In animals the precursors of nuclear proteins and lipids should be the same as those of cell proteins and lipids in general, with the obliged amino acids for the former, more or less different according to the species. In what concerns the nucleic acids, the pentoses probably may arise from several sources, ribose phosphate from phosphogluconic acid. Combinations of 2-carbon and 3-carbon sugars, one of which is phosphorylated, should be the major source of the pentoses in general. Hexoses seem not to be the major source of pentoses. As with other nucleic acid components, however, it must be born in mind that only very few species have been investigated in this respect and therefore generalizations are somewhat premature.

The at a time much debated question of the interconversion of RNA into DNA and vice-versa has now lost some of its interest since it has been found that both may derive their components from relatively simple compounds ever present in the actively metabolizing cell. It has been found in recent work that possibly DNA may appear from RNA, while the reverse would be more difficult. However, the results may have other interpretations. It seems that one of the sugars may be converted into another by intermediary of the degradative pathway. The phosphorus of nucleic acids may be derived from inorganic phosphate or from several sources of organic phosphorus, which contribute to the common pool of cell phosphates. $^{32}P$ is rapidly incorporated into RNA and more slowly into DNA when the cells are not dividing. The several C and N atoms of the purine and pyrimidine bases may be derived from different sources. The constituent atoms of these bases are in dynamic equilibrium with a common pool of $CO_2$, formate, $NH_3$, glycine and serine of the body in the investigated species (mammals and birds). The bases themselves, when administered to experimental animals have been found to be incorporated with very different efficiencies. Adenine is efficiently incorporated while guanine is used only after much dilution. The pyrimidines are not significantly utilized by the rat. Orotic acid (uracil-4-carboxylic acid, pyrimidine found in milk) is an efficient precursor of pyrimidines in *Neurospora* and the rat.

In what respects the turnover rate of DNA and RNA, the results found with $^{14}C$ and $^{15}N$, at the same time that with $^{32}P$, indicate a marked difference between non-mitotic and mitotic tissues. In tissues without a significant rate of mitoses RNA turnover is relatively rapid, while DNA turnover is slow, the former about

2–5 times greater according to the component of the acids under consideration. In regenerating or mitotically active tissues DNA is more actively metabolized, almost attaining for some precursors ($^{15}$N glycine into purines) the same rate as RNA. The lower, although significant, turnover of DNA in non-mitotic tissues probably is not related to the DNA being the gene substance, as some authors seem to admit more or less implicitly in their discussions, since genes must be active in the non-mitotic nucleus (the so-called metabolic nucleus or nucleus of synthesis) and such an activity should imply some metabolism. From the cytological point of view the regions which in the resting stage retain their DNA are called heterochromatic and at least in some species it is proven that heterochromatic regions are more or less devoid of gene activity, or have an activity of only a special polygenic character. Therefore, a slow turnover of DNA could be linked with genetic inertness and probably this will be the conclusion to be reached in future work. It seems that there are two fractions of DNA, one with a rapid and the other with a slow turnover, only the second being linked with genetic inertness, the former pertaining to euchromatic regions but in amount much smaller, in the non-mitotic nucleus, than the second fraction. DNA should have especially a structural role, as contrasted with a metabolic role for RNA.

It has also been stated that RNA of the nucleus has a greater turnover than the RNA of cytoplasm (SMELLIE *et al.* 1953) and the nucleolus partakes into this rapid turnover. The proteins of the nucleus have an incorporation rate of $^{15}$N-glycine, $^{14}$C-formate and $^{35}$S-methionine that is comparable to the rate in cytoplasmic proteins. Of several protein fractions of the nuclei, histone shows the slowest incorporation of $^{15}$N, while $^{14}$C was incorporated at the same rate in histone as in the other fractions. On the other hand, the rate of incorporation of $^{15}$N-glycine in the nuclear proteins is greater in actively metabolizing tissues (DALY *et al.* 1952) paralleling a similarly greater incorporation into cytoplasmic proteins in such tissues. As the fractionation of nuclear proteins has not yet been pursued so far as to permit closer location of particular fractions into each structure, it is not known if the proteins of chromonemata, etc. show also a great rate of synthesis or only the nucleoplasm and nucleolar proteins. Another important point to be noted is that nuclear proteins and RNA on one side, and proteins and DNA on the other, are synthesized at different rates, as can be judged from the rate of incorporation of $^{14}$C and $^{15}$N (SMELLIE *et al.* 1953). This finding does not favour the idea that RNA presides in some manner to protein synthesis.

### c) The separate nuclear constituents (Table 2).

After a general survey of the nuclear composition in which the nucleus has been analyzed as a whole or the several fractions were obtained with basis on chemical properties rather than on morphological distinctions (although in some cases chemical fractionating was intended to follow these distinctions), we shall now briefly summarize the data on each nuclear component (see also Table 2).

### Nuclear membrane.

That this is a real membrane and not merely a boundary between nucleus and cytoplasm has been known from old microdissection experiments. Data on the structure and composition of the nuclear membrane have been gathered chiefly from animal cells (refs. for oocyte nuclei CALLAN 1950, 1952) but the membranes of plant nuclei (SCOTT 1950) concord in structure with these and therefore the results seem to have general validity. However, the treatments may be

Table 2. *Composition of the separate nuclear parts.*

NA = total or undiscriminated nucleic acids; DNA = desoxynucleic acid; RNA = ribonucleic acid; Prot = proteins in general; B Prot = basic proteins; NB Prot = non-basic proteins; Lip = lipids in general; P Lip = phospholipoids; Enz = enzymes; + indicates presence; — indicates absence; ± indicates variable presence or absence. %s are referred to the dry weight of each nuclear part.

| | Inorganic | NA | DNA | RNA | Prot | B Prot | NB Prot | Lip | P Lip | Enz |
|---|---|---|---|---|---|---|---|---|---|---|
| Nuclear membrane | ++ highly variable Na, K, Ca, Mg, Fe, Cu, Zn | | | | Inner layer: + (elastin like) outer layer: + (with lipids) | | | Outer layer: + (with proteins) | | |
| Nucleoplasm | | + | + at times | + or ++ | ++ | + or ++ | ++ | + 5-10% or up to 40% | + 4-9% or up to 35% | ++ |
| Chromosomes kalymma | ± Ca, ± Mg, at times several of the nucleoplasm | ++ 30-60% | ++ 30-50% (exceptionally less) | + 3-20% (exceptionally more) | ++ 30-60% | ++ 30-50% | + 10-20% (exceptionally up to 40%) | (?) | (?) | ± |
| chromonemata | Ca (?) | + (?) | + (?) | + (?) | ++ | + (?) | ++ | — | — | |
| Nucleolus | + (?) | ++ 5-10% | — (at times +?) | ++ 5-10% | ++ 60-80% (?) | ++ 30-50% (?) | ++ 20-30% (?) | ++ 10-30% (?) | ++ 10-30% (?) | ± |
| Spindle | (?) | ++ 5-10% (?) | — | + 5-10% (?) | ++ 60-80% (?) | + 5-10% (?) | ++ 50-70% (?) | + (?) 10-20% (?) | + (?) 10-20% (?) | (?) |

in part responsible for the greater or lesser evidence of the nuclear membrane and its thickness (CHAYEN 1952). Negative results with electron micrography (ROZSA and WYCKOFF 1950) probably are due to faulty fixation.

From the chemical point of view, the nuclear membrane should be formed of two protein layers, an outer one more porous and an innermost more elastic. In oocytes the outer layer has also lipids, staining with Sudan black and giving a positive alkaline phosphatase reaction (CALLAN 1952), which probably has only the meaning of an adsorption — see above "Enzymes". From the fact that alcohol and chloroform and phosphate ions at 0.2 M concentration over a wide $p_H$ range disrupt this outer layer, it follows that the lipids and proteins should be disposed in a kind of mosaic or meshwork which forms the pores. The inner layer is protein in nature and by its insolubility in distilled water, dilute and concentrated saline solution and acids it is neither nucleoprotein nor histone. Pepsin and trypsin rapidly disintegrate this inner layer. It may be stained by orcein and fuchsin-resorcin, which are stains of elastin; this is the basis for supposing that the protein of the inner layer is elastin-like. The inner layer is the chief one in giving strength to the nuclear membrane and probably the outer layer is in part a cytoplasmic production, establishing the transition between cytoplasm and nucleus.

It should also be noted that both layers of the nuclear membrane may be revealed in electron micrographs of cells treated with buffered osmic acid (OBER-LING et al. 1953). These photographs show an inner layer more complete than the outer one. The absorption of osmium by both layers may be interpreted as favouring the existence of lipids not only in the outer as also in the inner layer, at least in some cells.

## Nucleoplasm.

This designation is to be preferred to that of nuclear sap because in some cells, and especially old ones, the nucleus may have a gel like consistency. Even in apparently fluid nuclei, as those of animal oocytes, the nucleoplasm has a structural phase (CALLAN 1952). The chemical composition of the nucleoplasm must be inferred chiefly from indirect data on the nucleus as a whole by discounting the compounds belonging to the other nuclear components. This kind of differential result is strongly influenced by the views on the relations between chromosomes and the bulk of the nucleus, according to it is admitted that: 1. the chromosomes almost fill the entire nucleus, or 2. the chromosomes are reduced in the non-mitotic nucleus to chromonemata and some matrix or chromatin while nucleoplasm is a more fluid or more gelified phase between them. The second view will be adopted here for the normal type of nucleus; only exceptionally, as for instance in the salivary glands of Sciara, may the chromosomes contain almost all the nucleoplasm.

The nucleoplasm has been considered to be mainly a sol or gel of proteins. This probably is the composition of only the simplest cases, as found in animal oocytes, which should have two colloidal phases, one of them fluid and the other structural. This latter may be dissected out and torn into fragments (CALLAN 1952). The fluid phase of oocyte nucleoplasm cannot contain much ionic calcium and magnesium since a very small (0.0015 M and lower) concentration of these ions causes coagulation and disruption of the nucleoplasm (CALLAN 1952). However, these data cannot be generalized without further inquiry as it is to be expected that the composition of the nucleoplasm will reflect the physiological changes in the cell and therefore should be different according to the tissues and their metabolism. It is to be presumed that the nucleoplasm must have at least also some lipids, since the relatively high concentration of the lipids found by direct analysis in many nuclei (see above, "The nucleus as a whole") cannot be entirely in the nucleoli and the chromatic granules. Probably this is true also of oocyte's nucleoplasm. The composition of the nucleoplasm lipids should be similar to that of the whole nucleus. It is likely that nucleoplasm lipids form some kind of pellicle around the chromosomes and are important for the separation of phases occurring within the nucleus. Glutathione should also be a general component of the nucleoplasm since it, or similar SH-compounds exist in the nucleus in concentrations equal or only somewhat lower as in the cell in general (Stern and Timonen 1954).

The proteins of the nucleoplasm have not yet been studied separately by chemical analyses. They must be a complex mixture of nucleoplasmins of variable composition according to the interactions with the cytoplasm on one side and the chromonemata and nucleoli on the other. Their study could give some interesting indications on the nature of these interactions.

In what concerns other possible nucleoplasmic components, the presence of nucleic acids is presumed to be very probable in most cases. In the published photographs of living and well fixed plant nuclei (for instance Chayen 1952) a general absorption at 2650 Å is notorious in the homogenous regions of the nucleus from where chromatic threads or blocks are absent. It may be that this absorption is also due to ascorbic acid and (in bean roots at least, where these compounds are known to occur in the nucleolus) possibly to pterin compounds (Chayen 1952). In animal nuclei of living fibroblasts also a general absorption at 2650 Å exists which in part should derive from compounds other than nucleic acids (Walker and Yates 1952) while, on the other hand, amphibian oocyte nuclei seem not to have demonstrable amounts of nucleotides in their nucleoplasm (Callan 1952). It follows as a general conclusion that nucleotides may or may not be present in the nucleoplasm according to the state of the cell and that other substances absorbing at 2650 Å may also be present there, as for instance ascorbic acid.

That nucleic acid must, at least in some cases, exist in the nucleoplasm is inferred from the fact that nucleic acids, especially RNA but also very probably DNA are synthesized in the cytoplasm, or by interaction between this and the nucleus (see Section A). This poses the question of the existence of DNA, possibly in an unfixable form, in the nucleoplasm. Some cases have been reported of nuclei showing a generalized Feulgen reaction (Ris and Mirsky 1952) as if they were an almost homogenous vesicle of DNA. This appearance is very different from the usual image after alcohol-acetic or chromic-osmic fixatives and has been obtained after fixation with formol-containing fluids. It remains to be solved if the formol or other fixative ingredients have not been sufficiently washed out (formol combines with proteins, especially their basic groups, from which it is difficult to remove) and have contributed to a generalized reaction; or if DNA is in fact more or less uniformly distributed in at least some types of nuclei where it is fixable only by certain fixatives and especially formol containing ones. The former alternative seems to be more probable but nevertheless indirect evidence points out that in some states (especially in mitotic prophase and ana- and telophase)[1] the nucleoplasm should also have thymonucleotides.

---

[1] Of course, in anaphase it is the periplasm or mixoplasm (protoplasm immediately surrounding the chromosomes) and not properly the nucleoplasm, that only exists as such when a nuclear membrane incloses a nucleus, which has the nucleic acids then commencing to leave the chromosomes.

## Chromosomes and genes.

Chromosomes properly, or complete chromosomes, are those of the mitotic and meiotic nucleus or of some particular kinds of nuclei such as the salivary glands of Dipteran larvae, which are believed to be permanently in a special type of mitosis (endomitosis with fusion of homologous). In the non-mitotic nucleus the chromosomes are represented only by some of their components. According to the view here adopted the chromosomes reduce down to the chromonemata plus some chromatin "points" and blocks, the greater ones being of heterochromatin, during the typical non-mitotic phase (for more details see the following chapter, p. 446). The chromosomes of the middle mitotic and meiotic phases, when they are in a state of contraction and intensely take basic stains, are formed of the chromonemata which came from the resting stage, plus some materials which deposit on, and probably in part also within the chromonemata (between the fibrillae) to give the complete or typical chromosome. In salivary chromosomes these materials are restricted to chromatic bands and the chromocentres, when these are present.

One of the views opposite to this is that chromosomes essentially remain as such in the non-mitotic nucleus. This would require that the only changes in composition should consist in a doubling of the chromosome matter during each mitosis, presumably in the middle phases, which is contrary to the majority of the observations, especially the morphological ones. Generally those which sustain this view start from the postulate that the chromosomes are formed of genes and these are composed of DNA alone. Now it is evident from the morphological observations (*in vivo*, phase contrast) that the chromosomes change of aspect in telophase and this cannot be only due to hydration. The reverse phenomenon, of greater chromaticity etc. is observed in prophase and here again this cannot result only from dehydration and condensation of chromosomes without any addition of matter except water. This is obvious in the case of salivary chromosomes, which it would be absurd to pretend to derive from mitotic chromosomes simply by hydration or gain in fluid phase. Without any doubt, there is gain and loss of chromosome matter respectively in prophase and telophase. This matter is called matrix or kalymma and the skeleton or permanent structures are the chromonemata (SERRA 1947a, RATTENBURY and SERRA 1952). Without this very elementary and classic distinction or any other equivalent it is not possible to treat of the chemical composition of chromosomes in cytologically intelligible terms.

The composition of chromonemata from the resting nucleus, without any matrix, is unknown. However, to judge from the salivary chromosomes, they may be simply proteic. The salivary chromosomes are formed of parts (chromomeric discs) with DNA, basic proteins and non-basic proteins (quantitatively probably about $1/10$ of the former and the same of basic proteins, the rest of non-basic proteins) and of the inter-chromomeres, without demonstrable amounts of DNA and basic proteins, but with non-basic proteins (SERRA 1947a).

If the chromatic material (DNA plus histone and possibly other proteins and RNA) of the bands may be equaled to the matrix of ordinary chromosomes, it follows that chromonemata properly may be devoid of DNA or RNA and the genes are not necessarily nucleoproteic in nature (discussion in SERRA 1949a, p. 496–504). The opinions of cytologists in this respect are yet far from unanimous, and more observations are necessary without sacrifice to theoretical *a priori* pseudo-postulates. Chromonemata in some species do not give the Feulgen reaction (some snail oocytes—SERRA *et al.* 1944, 1947a). However, they may have DNA in smaller concentration than the sensitivity of reaction, or the DNA may be in a masked or unfixable state, as happens to lipoids, or they may have RNA. Only future work may decide about these possibilities. As yet no conclusive proof has been adduced that genes are nucleoproteins, let alone that they are simply DNA or RNA and the evidence is rather in favour of they being proteins in their simplest composition and nucleoproteins when working. The evidence obtained from studies on viruses is too indirect to be of great service in this connection and it is being recognized that virus multiplication is rather a function of the parasitized cell, though catalyzed by the virus.

As the fibrous structure of the chromonemata is not a compact one, it follows that nucleoplasmic components may always be present within the chromonema; this is evident in certain cases when the chromosomes occupy almost the whole nucleus (salivary glands of *Sciara*). The nucleoplasmic components within the chromonemata may be more or less similar to the nucleoplasmic general composition; their proteins may be called enchylemins (Serra 1949).

The other part of the chromosomes, the kalymma or matrix, is composed of chromatin, which, as now inferred, is formed chiefly of DNA plus histones, and also of non-histone proteins and RNA. Quantitatively, the matrix makes the bulk of the chromosomes, the chromonemata forming perhaps 10% or less of the metaphase chromosome. Therefore, the analyses of "chromosomes" as yet made, reflect rather the composition of the matrix than that of the complete chromosome. From these analyses it may be concluded that the matrix contains (referred to the dry weight) about 30–50% DNA, 3–20% RNA, 30–50% histones and 10–20% or up to 40% non-histone proteins. Lipids should also be present in the matrix either as proper constituents or as contaminants from the nucleoplasm. However, their state or their amount, or both, do not allow an unequivocal demonstration of them in chromosomes with existing histochemical methods.

It is not known if there are any enzymes proper of the chromosomes or if only adsorbed enzymes may be present there. Another interesting question concerns the differences between euchromatic and heterochromatic regions of the chromosomes. It has been claimed that heterochromatin differs from euchromatin in having "lower" (*i.e.* histone) proteins and perhaps RNA (Caspersson 1941) but this could apply only to the chromocentre of *Drosophila* salivary glands, which is a special case. Probably the differences between eu- and heterochromatic regions reside chiefly in the chromonemata corresponding regions, heterochromatic zones retaining their matrix in non-mitotic stages while in euchromatic ones it is disintegrated (Serra 1942, 1947a). It seems likely that this is correlated with relative genic inertness.

## Nucleolus.

By virtue of their not giving the Feulgen reaction it is admitted that generally nucleoli do not have DNA. Instead, as they strongly absorb at 2560 Å it has been concluded that nucleoli contain RNA. Basic and non-basic proteins have also been characterized in the nucleolus (refs. Serra 1947a). According to cytophotometric determinations, which in the case of the nucleolus, a relatively more or less homogenous body, may be much more accurate than in the nucleus as a whole, the RNA amount of the nucleolus in *Zea mays* should be about 10% of the amount of total protein (Pollister, cit. Carlson *et al.* 1951). By ultraviolet photometry it has been concluded that the nucleolus may show different ratios RNA/proteins, perhaps according to genetical constitution (Caspersson *et al.* 1940, Schultz *et al.* 1940).

With respect to other constituents, it has been shown both from indirect evidence and histochemical tests that the nucleolus always contains a relatively high amount of phospholipids (ALBUQUERQUE and SERRA 1951) and probably other lipids are also present (SERRA 1947a). The lipids generally are not homogenously distributed throughout the entire nucleolus and may form, besides an outer membrane of pellicle, also rings and inclusions. Many nucleoli are vacuolated or possess spherular or crystalloidal inclusions, the vacuoles having a lesser density than the rest of the nucleolus. By treatment with buffered osmic acid the nucleolus shows at the electron microscope several inclusions under the form of a net or a twisted and coiled filament, sometimes also as a cortex enclosing a non-osmiophilic vacuole (BERNHARD et al. 1952, OBERLING et al. 1953). Possibly these images correspond to the nucleoloneme previously demonstrated by silver impregnation or coloration by a ferric alumen-potassium ferricyanide-pyrogallic acid method (ESTABLE 1930, ESTABLE and SOTELO 1951). However, the nucleoloneme should not be of lipids as it resists fixation in ether containing fixatives; it could happen that a protein complementary structure resists to this fixation and gives the nucleoloneme images. Probably these lipidic or osmiophilic parts of the nucleolus represent aspects of a separation of phases process taking place when the nucleolus is engaged in syntheses and which gives rise to the microscopically observable figures of phospholipine. Probably also this is in relation to vacuolation and budding as seen for instance during oocyte growth (SERRA and LOPES 1941). It is not known what part the nucleolar zones of the chromosomes take in giving rise to such figures.

Of other constituents, it has been reported that oocyte nucleoli give negative reactions for glutathione, carbohydrates, iron, peroxydase and phenolase (GERSCH 1940, CALLAN 1952). Alkaline and acid phosphatase reactions have been stated to be positive (CALLAN 1952) or negative (RABINOVITCH 1949); the above discussion about these enzymes (see "Enzymes") applies here. Pterins or similar compounds have been found to occur in nucleoli of bean root meristems (CHAYEN 1952). Probably these are special components without general occurrence.

## Achromatic spindle.

Although typically it is not a component of the nucleus, the achromatic spindle is closely associated with chromosome partition to the two daughter cells in meiosis and mitosis; so we will discuss its constitution, if only very briefly. In some cases the spindle may be of intranuclear origin also. It is always an apparently fibrous structure which develops between two small sphaerulae (centrosomes, in animals and lower plants) or two greater or smaller polar surfaces (attraction plates, plants). It has been supposed that the spindle is formed by mixing of the nucleoplasm with the cytoplasm, proteins with basic isoelectric point of the nucleoplasm combining with cytoplasmic proteins with acidic isoelectric point (RYBAC 1950). Also it has been admitted that a transformation from globular to fibrous proteins takes place, probably of an enzymatic nature, and that the fibrils are more or less stabilized by phospholipids, which would condition the adhesion of the centromere of the chromosomes to the spindle fibres (SERRA 1949b).

In a renewed version of the hypothesis of a globular to fibrous transformation it is admitted that the making of S—S (disulfide) bridges plays an important role in the formation of the achromatic spindle (MAZIA 1953). It was previously assumed that SH groups would be operative in spindle formation (see DUSTIN 1949). The isolation of the spindle plus the chromosomes in dividing sea urchin eggs may be obtained by dissolving the cytoplasm with a digitonin solution, or by the use of stronger detergents if the cells are previously treated with hydrogen peroxyde. It has been stated that the isolated spindles are mainly composed of protein, having also some RNA and other components. Quantitatively, the mitotic apparatus (spindle and asters plus chromosomes) makes about 2% of the cell weight (MAZIA 1953). Besides protein some ribonucleic acid has been found in it (MAZIA 1953), but this may belong to trapped cytoplasm or nucleoplasm. Probably lipids will be found also there in accordance with indirect inferences (SERRA 1949b). Thioglycolic acid has proved capable of dissolving isolated mitotic spindles and from this and the fact that peroxyde has the opposite effect the probable conclusion is that a partial transformation from —SH to —S—S— groups takes place in spindle formation as it is admitted, in a much greater scale, in the transformation of ordinary cellular proteins into keratin, for instance in the hair roots. The —SH to —S—S— transformation should reverse when the spindle disintegrates. Both transformations could be due either to specific enzymes or to more general alterations in the redox potencial of the cell. The latter alternative seems more probable, with the folding and unfolding of the spindle proteins as a particular case of the more general folding and unfolding of protein molecules which may be the basis of other cell processes like secretion and osmotic work (GOLDACRE 1952). Some —S—S— bridging will assist in stabilizing the fibrils but this cannot be the primary cause of fibril formation. From old microdissection experiments it is known that the living spindle may easily liquefy if disturbed with a needle (CHAMBERS 1924).

### d) Changes during mitosis.

Chemical changes during mitosis may underlie the morphological phenomena of the prophasic appearing of chromosomes, the individualization and separation of two chromatids in each chromosome, formation of the spindle, and the telophasic reconstitution of the non-mitotic nucleus with nuclear membrane and nucleoli formation and disintegration of a greater or lesser part of the chromatin. In meiosis, besides these, there is also chromosome pairing. Some of these phenomena are rather associated with changes at the colloidal level, of a physicochemical nature, than at the chemical level, and will be discussed in the respective chapter. In some cases also the fundamental data have already been discussed above when dealing with membrane, chromosomes, nucleoli and spindle. It should also be said that our knowledge of the mitotic chemical changes is yet less certain than that of the static chemical properties of the several nuclear components.

Two opposite views have been held in what concerns the variation of nucleic acid amount during mitosis. The majority of cytomorphologists and many cytochemists believe that DNA increases during prophase and decreases during telophase, it being synthesized during the passage from interphase to prophase and partially degraded in telophase; others suppose that the DNA is suddenly doubled in metaphase when two sets of chromosomes separate. Representatives of the first and the second current have used similar methods with opposite results. Such a discrepancy may be explained by two circumstances: differences in the relatively great errors inherent to the cytophotometric method used, and differences in material. For instance, PASTEELS and LISON (1950) found that the synthesis of DNA occurs during telophase while FAUTREZ and FAUTREZ (1953) say that the greater increase is at the beginning of interphase and RIS (1947) affirms that total nucleic acid content is constant during mitosis except for a doubling at metaphase, while DNA increases in prophase. In meiosis, according to the same author (RIS 1947) there would be an increase in total nucleic acid and in DNA from pre-leptotene to pachytene. All the remarks about the errors of cytophotometry apply here and only after future work with more critical methods may a conclusion be reached on this point of DNA and RNA variation during mitosis and meiosis. Until then it seems preferable to rely upon general cytological observations. It must be born in mind that a part of DNA and RNA in some mitotic and meiotic stages certainly is found in an unfixable form and that when ultraviolet photometry is used substances other than nucleic acids may absorb in the band of 2600 Å.

By the use of reliable chemical methods (OGUR et al. 1951) it has been found that in *Lilium longiflorum* the DNA per cell (and per nucleus if it is supposed to be all in the nucleus, which may not be true) increases during interphase from meiosis to post-meiotic mitosis and much more rapidly just before mitosis or at mitosis (the data do not allow a decision on this particular). The rate of increase at mitosis is about 30 times as great as that during interphase. The DNA mean content (in $\mu\mu$g per mother cell and per pollen grain—$1\mu\mu$g $= 10^{-12}$g.) is in late meiotic microsporocyte 253, in earliest microspore 53, in microspore just before pollen mitosis 103, in microspore at earliest pollen mitosis 250, and in pollen at anthesis 373. It should be noted that from meiosis to the earliest microspore a diploid nucleus has given rise to 4 haploid ones. The change in RNA content per cell or per pollen grain (in $\mu\mu$g) demonstrated by the same data (OGUR et al. 1951) is yet greater than DNA change. The mean amounts are: late meiotic microsporocyte 602, earliest microspore 116, microspore just before pollen mitosis 190, microspore in early pollen mitosis 592, pollen at anthesis 1704. The ratio RNA/DNA is in the 5 stages respectively 2.4, 2.2, 1.8, 2.4 and 4.6. There is no way of separating cytoplasmic from nuclear RNA. The increase in RNA is even greater than that of DNA, the data show that both acids can be synthesized *de novo* from extra- or intra-cellular sources, without an obligatory conversion of RNA into DNA and vice-versa. Data obtained by the cytophotometric method in *Tradescantia paludosa* essentially agree with these conclusions (BRYAN 1951).

By a biochemical method SPARROW et al. (1952b) found somewhat different values in *Trillium erectum* L. The values in $10^{-12}$ g. ($\mu\mu$g) per pollen mother cell are respectively for DNA and RNA: mid pachytene $220 \pm 33$ and $190 \pm 77$; late pachytene $290 \pm 34$ and $200 \pm 38$; diplotene $330 \pm 31$ and $190 \pm 30$; first division diakinesis-anaphase I $320 \pm 77$ and $200 \pm 74$; second division late anaphase II-early interphase $380 \pm 95$ and $280 \pm 48$. The variation in nucleic acid content is less obvious than in the determinations of OGUR et al. in a different species. It must be noted, however, that the determinations in *Trillium* were not made late during pollen maturation and refer to a less extense period of the plant life. Nevertheless, increases in RNA and chiefly in DNA are apparent although statistically not yet demonstrated owing to the small number of determinations.

Another question concerns the evolution of proteins during mitosis and meiosis. By cytophotometry (RIS 1947) it has been adduced that the ratios nucleic acid/total protein as well as nucleic acid/histone and non-histone proteins remain constant throughout the whole mitotic cycle, from interphase to telophase. The ratio histone/non-histone proteins would be 0.46–0.48. This should be reinvestigated by direct analytical methods. By histochemical tests it has been demonstrated (SERRA 1947a) that basic proteins are deposited on the chromonemata during prophase. These proteins are a part of the chromatin. The arginine reaction seems to indicate that a part of the chromatin basic proteins comes from the nucleoplasm to the chromosomes but quantitative data are lacking. It seems that in mitosis the cell is able to increase its nuclear content in basic proteins by conversion of other amino acids into dibasic ones. According to cytophotometric data (Bryan 1951) the proteins would increase during microsporogenesis of *Tradescantia* parallelly to DNA but at a much faster rate. By ultraviolet spectrometry it has been suggested that a loss of proteins from the nucleus should occur in prophase and a synthesis in telophase (Caspersson 1941) but these conclusions were reached chiefly from indirect data on the chromosomes and necessitate of being confirmed.

By staining methods and ribonuclease use it has been concluded that ribonucleoproteins appear to be shed from the chromosomes during anaphase in the periplasm between the two moving groups of chromosomes (Jacobson and Webb 1951) and the same has been shown to occur in the case of butterflies' oocytes during meiosis, in which the sheding of ribonucleic acid occurs already at metaphase (Ris and Kleinfeld 1952). Probably such alterations in the composition of chromosomes may occur, although rarely, also in prophase, giving rise to interchromosome fibrils and to stickiness. This latter, however, may also derive from other causes (depolymerization of DNA, solvatation of the matrix nucleoproteins).

The alterations in lipids during mitosis are not known. The nucleolus reappears in telophase according to two chief types (Rattenbury and Serra 1952) and then phosphatids form a part of the pre-nucleolar substance (Albuquerque and Serra 1951). It is concluded that separation of phases must occur at the spots were nucleolar matter is appearing, by which phosphatids are separated from lipo-proteins, forming complexes more rich in lipoid, probably phosphatid-proteins-RNA. It seems also that lipids must play an important role in the separating out of chromosomes from the nucleoplasm.

## Literature.

See at the end of the chapter: "Physical chemistry of the nucleus" p. 499.

# Fine structure of the nucleus.

By

## J. A. Serra.

With 11 figures.

As a necessary preliminary to a discussion of fine structure, a brief review of essential data on the microscopic structure of the nucleus will be attempted. Since no unanimity has yet been reached on certain points of this subject some ambiguity, and especially among those readers more devoted to the chemical and physical aspects of Biology, may result if each author does not state precisely what he is referring to.

Owing to their showing different degrees of organization and in part also to having different orders of dimensions, it is convenient to treat separately on the one hand the nucleus of higher plants, which includes the Cormophyta and some Algae and, on the other hand, that of Fungi, Yeasts and Bacteria. The more complex Viruses probably are comparable in this respect to Bacteria and as a discussion of their nuclei would not be based on good microscopical data no attempt will be made to deal with them separately.

## A. Microscopic morphology of the nucleus.

**Higher plants.** The true structure of the nucleus, that is of the living nucleus, is now again being intensively studied, with the aid of phase contrast microscopy. The chief results as yet obtained demonstrate an important point, namely that the structures revealed by using the best cytological fixatives and staining methods more or less closely correspond to the living. A certain coarsening and some distortion of structures are inevitable in these methods but their degree may now be evaluated with some accuracy. It is necessary to have in mind that a part of the cell components may be lost during fixation and the subsequent handling of the tissues and this may influence the results of staining procedures.

First a distinction must be made between the dividing and the non-dividing or quiescent nucleus. The dividing nucleus is simpler and typically is represented only by the chromosomes. In exceptional cases the nucleoli may show some degree of persistence. As the nuclear membrane disintegrates during mitosis and meiosis (except in endomitosis and related phenomena) no true nucleus, as an organizational differentiation of the cell separated from the cytoplasm by a boundary, exists in the middle stages of nuclear division. This fact, together with the genetic interpretation of mendelian heredity, has led to the conclusion that the chromosomes are the essential parts of the nucleus and that they preserve their individuality during the non-dividing stage. The chromosomes must remain organized along their long axis during the stage of their optical invisibility. Only this assumption reasonably explains the fact that chromosomes present the same fine pattern of organization along their axis during numerous successive cell divisions, as can be seen with more detail in salivary glands of Diptera. However, it does not follow from this that in the non-dividing nucleus other nuclear components are not necessary for nuclear functions; the contrary must be true since other nuclear components are differentiated with a constancy that demonstrates their importance in cell life.

Chromosomes are generally believed to have always a similar structure whatever their degree of development. In some groups it is possible to observe great differences in size of the chromosomes among the cells of one individual, for instance between root meristems and oospheres or, in a more typical case, between cerebral ganglia and malpighian tubules and salivary glands of Dipteran larvae. The majority of cytologists accept as a postulate that the greater chromosomes display the same structures as the mitotic ones, more visibly only because they are unravelled; or in other words, it is postulated that true growth of the chromosomes, in the sense of the introduction of new matter, either takes place only in thickness and not in length, or if longitudinal true growth also occurs it faithfully preserves the existing structure. Contrary to this opinion it has been proposed (CASPERSSON 1941) that the parts between chromomeres or chromatic bands of salivary chromosomes and the corresponding structures of meiotic and mitotic chromosomes are inserted during their growth or in prophase and that subsequently they should disintegrate at the beginning of each mitotic cycle, so that metaphase chromosomes are reduced to the essential constituents, which would be the chromomeres. This view has no morphological basis and no facts except some spectrophotometric data which may be interpreted in other ways, have been shown to demonstrate it. So we will assume the former point of view, which permits one to draw conclusions about the structure of ordinary mitotic chromosomes from the facts known for the greater, more easily observable chromosomes of cells of greater size.

In many meiotic prophase, and in oosphere chromosomes (even more distinctly in salivary chromosomes) there may be seen alternating regions of greater and lesser chromaticity, which can correspond respectively to a greater and a lesser diameter of the chromosome. However, the greater diameter of more chromatic parts has no general validity and chiefly results from differential spiralization, the chromosomes being more tightly twisted at the chromomeres, which therefore give the visual impression of beads. In a view which now appears naïve these beads have been equated to genes (discussion and figures in SERRA 1947a, 1949a). If ordinary chromosomes have all of the structures which are seen in the greater chromosomes then it follows that in the direction of the long axis of the chromosome there are alternating achromatic and chromatic regions or bands, these latter showing in meiotic prophase a tighter helical coiling correlated with a precocious deposition of matrical nucleoproteins at these points. Salivary chromosomes have about the same diameter in chromatic and achromatic bands.

Another important question concerns the existence of permanent and temporary components of the chromosomes [see also chapter "Chemistry of the nucleus, b) Chromosomes and genes"]. The two alternative views, which are corollaries of the postulate of the cytological counter-part of genetical continuity are 1. that metaphasic chromosomes result from resting stage ones by condensation of the same matter in a smaller space, without addition of new components, and 2. that during prophase chromatin is deposited on the chromosomes, which show also a shortening. The latter is the generally accepted view and it presupposes that the shortening of the chromosomes at the same time as chromatin is deposited, is effected not by disintegration of some parts of the chromosomes but by helical coiling at microscopic and submicroscopic levels, as well as by condensation (greater density of the chromosome as a whole by loss of water and possibly other non-essential components accompanied by other changes at the colloidal and molecular levels, for example gelification and polymerization—SERRA 1947b). In telophase complementary changes occur, disintegration of the chromatin component of the chromosomes takes place. By extension of this interpretation it has been assumed that in some cases chromosomes may be deprived of chromatic matter without losing their full genetic properties. These essential parts of the chromosomes may be called chromonemata or genonemata, although the first designation is somewhat improper since they are not chromatic. The transitory materials which are deposited and disintegrate more or less completely according to the stage of nuclear division may be called chromatin, kalymma or matrix, terms which have been employed with several meanings by cytologists.

In this concept the chromosomes have a variable composition according to the stage of cell division but the chromonemata have a constant mean composition (barring statistical fluctuations and more important differences which may constitute the basis of some mutations—see fig. p. 463). Chromonemata grow between divisions by lateral apposition; changes in length occur only by insertion of additional segments from the same chromonema (duplication) or by translocation of segments from other chromonemata. It is conceivable that in some cases (salivary chromosomes destined to perish in pupation) intercalary growth could also take place by lengthwise replication of existing structures, but this is an hypothesis which remains to be more closely examined. On the other hand, the amount of chromatin is variable (see preceding chapter for definition and composition) according to the nucleus and the stage of mitosis and meiosis, being at a maximum in the middle phases and at a minimum in leptotene of meiosis and non-dividing nuclei. As discussed in the previous chapter ("Chromosomes and genes") the chromonemata may be simply protein in composition or may have also some ribonucleic acid (RNA); if so they have no chromatin, which is characterized by the presence of DNA (desoxyribonucleic acid). Reasons have been presented (SERRA 1949a) favouring the view that genes (or, better, genetically active regions of the chromonemata) may be simply protein or protein-RNA and so the chromonemata in their simplest state would not contain DNA.

Some morphological characteristics of chromosomes, such as primary constrictions with the centromere, secondary constrictions, and nucleolar zones may be understood on the basis of chromonema differentiations. In general, the chromosome has only one region by which it can firmly become bound to the spindle; this region may be called the centromere, lying in the primary constriction. This differentiation probably results from the production by the chromonemata of compounds more akin to those of the spindle. In some plant species, however, the spindle attachment is an almost general property of the chromosomes (MALHEIROS et al. 1947). Secondary non-nucleolar constrictions may result from inverted duplications or perhaps from the inactivation of formerly functional nucleolar zones there located, the second alternative being less probable. Secondary nucleolar constrictions or, better, nucleolar zones have a special metabolic activity involving the early synthesis of phosphatids (ALBUQUERQUE and SERRA 1951, RATTENBURY and SERRA 1952) and are another interesting chromonema differentiation. All these differentiations are to be understood in terms of the accumulation in a restricted chromonematic zone of properties formerly scattered throughout the chromonema.

We now turn to the non-dividing or so called resting stage nucleus. How the chromosomes of the dividing nucleus are related to the structures of the non-dividing nucleus has been object of discussion and two extreme views have been held with more or less justification—see Fig. 1, p. 448. According to Fig. 1A the nucleus, excluding the nucleolus, is formed by the chromosomes alone, which have swollen to fill the whole nucleus, the heterochromatic parts remaining in a more or less condensed state. The nuclear membrane itself could be conceived, in a logical extension of hypotesis A, as being no more than the outer pellicle of the externally placed chromosomes. In B the chromosomes are represented in the non-dividing nucleus only by very thin chromonematic threads (or chromatic, according to some authors—POLLISTER 1952) and the rest is filled by "nuclear sap". In this view a membrane as an individualized structure should be present. A third view, which embodies some features of these two opposite views and corresponds more closely to the known facts, is represented in Fig. 1C. In this view the chromonemata become swollen but do not fill the entire nucleus and heterogeneity within the nucleus is due not only to the existence of heterochromatin but also to the presence of points where chromonemata touch one another and nucleo-proteins are localized.

It has been observed that non-dividing nuclei may present two extreme types (SERRA 1947b): a thread or reticulate type and an "homogenous" or, at most, finely dotted type, the latter having in many cases round or elongated masses which correspond to chromocentres and prochromosomes. Intermediate types are common. These structures may be observed in vivo (MARTENS 1927, DELAY

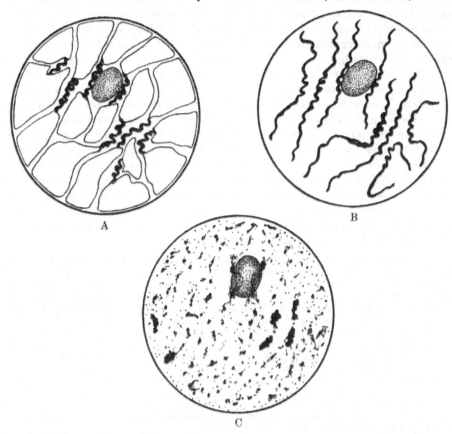

Fig. 1 A–C. Schemes of the non-dividing nucleus. See text. In C a nucleus of the almost homogenous type, with 4 heterochromatic non-nucleolar regions and 2 heterochromatic regions adjacent to the nucleolar zones. The points, blocks and some threads are of chromatin. A and B redrawn from POLLISTER, 1952, Exp. Cell Res. Suppl. 2, 59; C original.

1949 and refs.). Several authors have considered various other types (DELAY 1949 for a summary) but the essential structures always derive from the absence or presence and relative abundance of two kinds of elements: threads, which tend to form a reticulum, and points or blocks, which when greater visibly correspond to a part of the chromosome and may be called prochromosomes. The blocks are seen to give the chromatin reactions and may be called chromo-centres (HEITZ 1932, 1934) whether they originate from only one or a few, or from several chromomeres. In both types of nuclei there may be chromatic granules of dimensions smaller than the chromocentres, disposed in thread type nuclei more or less in relation to the knots of the reticulum or adhering to them after fixation. It is not known whether chromatic granules can exist free from the reticulum or the chromonemata. Probably the majority of chromatic granules are no more than chromomeres or parts of chromosomes which have kept their chromatin since the previous telophase. In many cases the chromomeres of adjacent chromosomes fuse into larger bodies, as also do the chromocentres. In scheme C of Fig. 1 it is supposed that chromatic granules may have two origins, (1) as chromomeres (single or aggregate), and (2) independently of the chromomeres as small chromatic bodies which by fixation adhere to the reticulum. The latter case will be more common at the beginning of mitosis and meiosis.

The reticulum itself will be the result of two processes: coalescence of fibres from the chromonemata, and organization of the nucleoplasm in vesicles or in reticulate replicas of the net formed by the chromonemata (see also in the following chapter: "Physico-chemical and colloidal properties"). This refers to *in vivo* structure. During fixation the reticulate structure may become coarser, due to protein precipitation over the network already present as a vital structure. In scheme C of Fig. 1, it is supposed that the chromonemata become strongly hydrated in regions where chromatic blocks do not maintain a marked gel state. This hydration brings about a slight dispersion of the chromonemata which possibly become divided into smaller fibres of only a limited number of polypeptide fibrils. Dispersion and hydration of the chromonemata proteins cause microscopic invisibility. On the other hand, in extreme thread type nuclei, a relatively great amount of chromatin remains on the chromonemata, which appear almost as chromosomes in many points.

This scheme is based on the great majority of cytological observations. Another very different interpretation, which supposes that the chromatinic compound DNA is uniformly distributed in the resting nucleus, has already been mentioned (see preceding chapter "Separate nuclear constituents. Nucleoplasm"). Special cases are those of greatly enlarged nuclei of animal oocytes, which may have their counter-part in plant oospheres, and of salivary chromosomes. All these represent nuclei engaged in active syntheses while chromosomes are visible. In oocytes, the diplotene chromosomes gradually lose their chromaticity and stretch to many times their ordinary length. In some cases the Feulgen reaction disappears (*e.g.* the snail *Cepaea nemuralis*—SERRA *et al.* 1945), while in others it remains. Only the negative cases are crucial in allowing the conclusion that in some cases chromonemata either do not have DNA or have it in amounts below the sensitivity of the test. This simple fact lends support to the conclusion that chromonemata and genes have less DNA than the Feulgen test can reveal. It follows that in some species at least, chromosomes may exist without DNA or with it in an amount per chromomere or visible grain below the sensitivity of the Feulgen test (amount per granule and not concentration in the nucleus, since we are dealing with bodies in a semi-solid system). It seems that this simple logic has not been apprehended by some (for instance ALFERT 1950). In great oocytes of snails the Feulgen reaction was not given by the whole mass of chromosomes packed to one pole of the nucleus by centrifugation (SERRA *et al.* 1945, 1947a) while in other species, for instance batracians (BRACHET 1940), or sharks (DODSON 1948), some DNA remains even in fully stretched chromosomes, which may mean that the disintegration of the non-essential parts of the chromosomes has not been pursued to its limit. The final decision about the important question of chromonemata existing without DNA depends upon the sensitivity of the Feulgen reaction. These questions are now becoming more acute as it is revealed that ultra-violet absorbing substances other than nucleic acids may be present in the nucleus (see preceding chapter, p. 432).

Between the chromonemata and the fibrils of each chromonema there is a dispersing phase, in some cases very fluid in others more compact and gel-like, which is the nucleoplasm, and which may be called nuclear sap only when very aqueous. This protoplasm of the nucleus certainly plays an important rôle and the genetic properties of the chromonemata must be realised through interaction with the nucleoplasm.

Another nuclear organite is the nucleolus or nucleoli, whose maximum number per nucleus in higher plants and animals is the same as that of nucleolar zones in the chromosomes (HEITZ 1931, RESENDE 1937, 1938). It has been admitted that a "nucleolar organizer" under the form of chromatic bodies is present at the chromosome zones associated with nucleolus formation but this view is untenable, at least in its orthodox form (RESENDE 1939, 1940, RESENDE *et al.* 1944, RATTENBURY and SERRA 1952), since in many cases no such bodies are present in the nucleolar chromosomes or any other chromosome of the nucleus. A special property of the nucleolar zones has been shown to be the precocious attachment of the phospholipids (ALBUQUERQUE and SERRA 1951).

A clear zone around the nucleolus, or nucleolar corus (CHAYEN 1952) is visible in meristem and staminal hair cells and another clear zone may also exist in the middle of the nucleus. These zones correspond to nucleoplasm with a lower amount of substances absorbing in the ultraviolet, in the band of 2650 Å.

**Cyanophyceae, Fungi and Bacteria.** It has long been known that Fungi, some Algae, and Bacteria may exibit types of internal cell structure apparently very different from those of cells from higher plants and animals. The evidence now at hand shows that during division these cells form more or less typical chromosomes with longitudinal differentiations like those of higher plants. The known genetical behaviour of some Fungi and Algae (*Neurospora*, Yeasts, *Chlamydomonas*, etc.) demonstrates that genes located in the chromosomes along their length, as in maize, *Drosophila*, etc. exist in these forms, although some rather difficult problems of genetical segregation remain to be solved (LINDEGREN 1953 and refs.). In Bacteria, too, chromosomes have been portrayed which are more or less similar to those of higher plants. The difficulties in observing such minute and confusing structures, however, are well apparent when considering the different interpretations given to similar figures in Yeasts (*cf.* for instance PINTO-LOPES 1948, LINDEGREN 1952, WINGE 1951, SUBRA-MANIAN 1950). Two trends are generally apparent in studies of the nuclear structure of these difficult materials: one attempting to find only structures homologous to those of higher plant cells, the other supposing that new structures or at least the same structures but combined in new forms exist in the lower plants. No unanimity of opinion has yet been reached and sometimes the same appearances are interpreted in opposite ways (see for instance the discussion by BISSET of the paper of DE LAMATER 1951).

For Cyanophyta, figures have been published which resemble the bacterial nucleus with aspects of division of globous corpuscles or of a filament, but no regular distribution mechanism as in mitosis is apparent (NEUGNOT 1950). If there are one or more fibrous structures with the genetic materials, it is not clear how they are equationally passed to the two daughter cells. On the other hand, the evidence now available (DRAWERT 1948, BRINGMANN 1950, 1952, v. ZASTROW 1953 and refs.) points to Cyanophyceae having a pigmented chromoplasm and a central body without pigment but which has both desoxyribose and ribose nucleic acids (BRINGMANN 1952, v. ZASTROW 1953). The central body also gives histochemical tests for phosphates and lipids (BRINGMANN). Probably both are in the form of phosphatids, as found in higher plant cells in the nucleoli (ALBUQUERQUE and SERRA 1951). These observations may signify that the central body is equivalent to a diffuse nucleus with its nucleolus (another homology, that of mitochondria for the phosphates and lipids, has been proposed by BRING-MANN). Cells capable of division should have their central body as a united formation well delimited from the chromoplasm, while in ageing cells the central body becomes separated into several scattered particles and finally, in old vacuolated cells, the basophily may completely disappear (v. ZASTROW). During this process the material of the central body lies at the periphery of the incipient vacuoles and the appearance may resemble that of fungal nuclei. Filamentous bodies and particles resulting from this involution of the central body could give rise to the appearance of chromosomatic bodies and false mitotic figures. Glycogen is evenly distributed in the central body and chromoplasm.

By analogy to what happens in Bacteria, Fungi and Algae which have been genetically studied, it should be pointed out that mendelian heredity also occurs in Cyanophyceae, and in such a case chromosomic structures should also exist. The observations revealing a homogeneous distribution of nucleic acids do not preclude the existence of chromosomic structures formed of protein, embedded in a common matrix, and a common nucleolus. That is, protein fibrous structures should lie embedded in substance formed of chromatin and nucleolar material together, which becomes scattered when the cell ages and finally may become exhausted. This, as yet merely a hypothesis, could be tested in future work by employing protein colour tests and histochemical reactions.

In Fungi, the best evidence at hand points out that chromosomes and a spindle similar to that of animal cells, with two centrosomes, differentiate during mitosis and meiosis (review and refs. OLIVE 1953). For Yeasts yet another interpretation is advocated, according to which the nucleus is formed of a large central vacuole containing the chromosomes associated in pairs, while a special centrosome and an elongated block of centrochromatin occupy one of the poles of this vacuolated nucleus. This is very unlike the nucleus of other Fungi and the majority of the investigators are of the opinion that the vacuole is a true cell vacuole, the "chromosomes" merely inclusions of the vacuole, the centrochromatin is the block of chromatin (heterochromatin) of the nucleus, while the "centrosome" plus the block of chromatin should be the true nucleus (refs. OLIVE 1953). It is possible that nucleolar persistency, frequent in lower plants, and the almost permanent association of nucleolar chromosomes and heterochromatin with the nucleolus obscure the figures, which are already difficult to interpret owing to the small size of the nucleus.

Resting nuclei of Fungi may be interpreted in terms similar to those of higher plants, with two chief types, thread and disperse, and both in an expanded or a contracted form (PINTO-LOPES 1948) depending on the hydration. Extremely expanded or hydrated nuclei have a great vacuole with the other structures standing at its periphery, while extremely contracted nuclei show only a small chromatic sphere which results from the fusion of the nucleolus and chromatin. The latter is similar to the structure of the sperm nucleus, except perhaps for the nucleolus.

There are two extreme views on the resting nucleus of Bacteria. According to the one it may be composed of several disperse nucleoids, while according to the other it may be in the form of a discrete chromosome always in course of division, except in spores. On the other hand, some accept that generally the bacterian cell is plurinucleated while others suppose that as a rule it is uninucleated and the cells disposed in linear succession. The more generally held opinion is that Bacteria have nuclear structures homologous to those of Fungi (ROBINOW 1942, PIEKARSKI 1950, DE LAMATER et al. 1952). However, the opposite view, that so called bacterian chromosomes are chromatic bodies which do not divide longitudinally and pretense spindles are in fact cell walls, continues to be supported (BISSET 1950, 1951, 1952). The only conclusion as yet possible is that Bacteria may be uni- or plurinucleate, that the nucleus is resolved into filamentous, helicoidal chromatic bodies at mitosis, and that typical spindles may or may not be formed. It seems also that part of the genetic materials stay at the surface of the cell, where the action of added specific DNA (for instance in *Pneumococcus*) seems to take place.

## B. Fine structure of the nucleus.

The fine or submicroscopic structure of the nucleus is in part deduced directly from electron microscope pictures and polarization microscope data, but is chiefly inferred from indirect data on the molecular and colloidal structures of compounds which form the basis of living matter[1]. Electron microscope studies (refs. of methods Bretschneider 1952a) have proved rather deceiving in some respects for the interpretation of the finest structure of protoplasm, chiefly because the range of sizes from the molecular to that of the smaller viruses in as yet practically (though theoretically not completely) inaccessible to this kind of microscopy. In this domain and in the molecular one, inferences from X-ray data and results from other kinds of physical methods are the only source of basic structural knowledge of the cell. In the present chapter, no attempt will be made to review fully the basic knowledge of molecular architecture and colloidal chemistry since this has been ably done in recent publications (Frey-Wyssling 1938, 1953; for earlier work Mark and Meyer 1938, Aschoff et al. 1938). For this reason only some more recent interpretations which may be of relevance for our subject will be briefly mentioned here.

### a) Structural components.

Structural compounds of protoplasm, that is, compounds which by themselves or in combination constitute the architectural basis of living matter are chiefly the lipids, proteins and nucleic acids. Steroids and carbohydrates may also play qualitatively an important structural role but quantitatively they occupy in this respect a lower level. The molecular architecture of lipids, carbohydrates and steroids has been known for some decades with a great degree of accuracy. This is not true for the proteins and nucleic acids, whose molecular structures have been under discussion ever since X-ray data were first obtained on them, and which are once again being interpreted in radically different manners. As is well known, proteins may show distended or almost distended $\beta$-type polypeptide chains (silk fibroin, naturally distended keratin of feathers and artificially stretched hairs) while other structural types correspond to more or less folded $\alpha$-chains[2]. The folding has been variously interpreted, from the hypothesis of Astbury and Bell (1937, 1941), passing through the hypothesis of the cyclols (Wrinch 1937, 1940), to that of globular foldings (Pauling 1940). Now, with former X-rays data and the consideration of atomic models as a basis, a helical structure has been proposed (Pauling and Corey 1953a and refs.) which is the extension to the molecular level of the valence angles proper of C—N—C chains—Fig. 2. According to the helical hypothesis the $\alpha$-helix with about 3.6 aminoacid residues per helix turn should constitute the basis of the structure of $\alpha$-keratin, hemoglobin and other globular proteins. This helix axis also follows a helical path with a pitch of 66 Å which, as a consequence, determines that three polypeptide chains twist around one another to form a rope, and six chains twist around a seventh in the same way. Hair and other keratin formations will be composed of 7-strand protein cables parallelly disposed. The space between the ropes will be occupied by single chains with their helical coiling. Hydrogen bonds N—H ... 0 at distances of $2.80 \pm 0.12$ Å will maintain the chains in place as was already postulated for former less ordered types of folding (Pauling 1940).

Helical structures have recently been proposed also for the nucleic acids (Watson and Crick 1953a, b, Franklin and Gosling 1953, Wilkins et al. 1953, Arndt and Riley 1953, Pauling and Corey 1953—see also in the previous chapter, p. 424 and 428). The hypothesis which now is supposed to fit the X-ray and analytical data better is that of a double helical structure with

---

[1] The methods will not be discussed here as they were considered in previous chapters.
[2] See also "Proteins" in the chapter by W. Seifriz: Microscopic structure of cytoplasm.

two helices twisted around the same axis and each formed by the phosphate-sugar part of mononucleotides as a kind of ribbon—Fig. 3. The helices are internally united by hydrogen bonds formed between their nitrogenous bases. The two chains are both right-handed helices but their atoms alternate so that in each chain there is a nucleotide at 3.4 Å intervals, with phosphate and sugar groups to the outside of the structure. The whole structure has a repeat period or turn of 34 Å and the distance of the P atom to the fibre axis is 10 Å. This should be the model of Na thymonucleate form B, the most hydrated and less ordered, obtained at 75% relative humidity and retaining about 40% water, while the A form is more crystalline (FRANKLIN and GOSLING 1953 b). Form A should also have the same double helical fibre but with 11

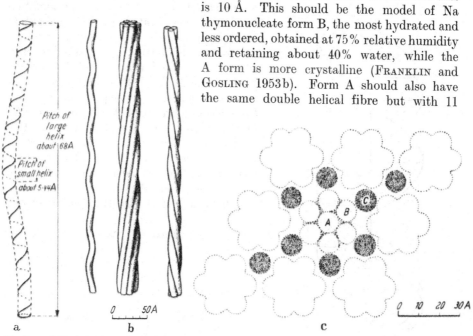

a                          b                                                  c

Fig. 2a–c. Helical structure of α-proteins. a An elementary helix with pitch of about 3.6 amino acid residues and 5.44 Å, forming a compound helix with a pitch of 12.5 times the smaller one or 68 Å. b A simple helix of about 10 Å diameter (space occupied by the main polypeptide chain and the side chains) and one rope of 3 and another of 7 strands. The three-strand rope has a lead of 200 Å, the seven-strand cable one of 400 Å. These are proposed for the structure of feather rachis keratin in the proportion of 2 to 1 respectively of three-strand and seven-strand ropes. c Cross-section through cables of 7 strands as those of Fig. 2 b and interstitial helices of the type of simple helix, forming a compact body. In nature the space is more efficiently filled than suggested by the drawing.
According to PAULING and COREY, Nature (Lond.) **171**, 59 (1953).

nucleotides per turn with a helix pitch of 28 Å instead of 34 Å for form B, a distance P-fibre axis of 9 Å and a vertical inter-base distance of 2.55 Å instead of the 3.4 Å characteristic of the B form (FRANKLIN and GOSLING 1953 b).

The model of WATSON and CRICK fits the analytical data. It postulates that the only possible pairing of the bases to form H-bonds is always that of a purine with a pyrimidine base. The planes of the bases are perpendicular to the fibre axis but a variety of positions is possible for each base or each nucleotide along the fibre. There would be no space for two purines or two pyrimidines to be in front of one another. This pairing requires that the amount of purines and of pyrimidines be equal, within the experimental error, in analyses of pure DNA and this has been verified in practice (CHARGAFF *et al.* 1953). If only the favoured forms among the possible tautomeric forms are physiologically produced then adenine should pair only with thymine and guanine only with cytosine or the methyl cytosines (5-methyl and 5-hydroxymethyl). In fact it has been found that both ratios adenine/thymine and guanine/cytosine (+ methyl cytosines) are almost equal to 1 (CHARGAFF *et al.* 1953, WYATT 1952). Difficulties are encountered when the reduplication of the polynucleotide chains is envisaged. In order that a new chain may be linked to a single polynucleotide ribbon it is necessary that the two helices of each elementary fibre become free, which is difficult since both are postulated to be necessarily plectonemic and therefore cannot be separated unless they untwist (WATSON and CRICK 1953 b). The difficulty would not exist if polynucleotide chains are reproduced by direct replication through binding of mononucleotides at the proper sites of single pre-existent free polynucleotide chains which could always exist in a statistical aggregate of DNA fibres. Or it may happen that reproduction does not take place directly over preexisting chains and is instead mediated by enzymes of the cytoplasm or nucleoplasm. These difficulties derive from the idea that DNA is the ultimate genetic material upon which gene specificity is based and would disappear if this hypothesis were abandoned.

The present state of knowledge on the structure of DNA seems not yet too satisfactory and possibly in the near future it will be recognized that the twin helical structure is only a first approximation to a more complicated natural aggregation involving the existence of several elementary polynucleotide fibrils side by side in each fibre, as postulated in the case of the α-folded proteins, and that DNA reduplication is mediated by proteins. The postulated helical-fibril basis, however, both for proteins and DNA, seems to be a real advance in this domain of molecular architecture, important for the understanding of the submicroscopic structure of living matter. It seems more than a mere coincidence that chromosomes also have a helical structure, either of one

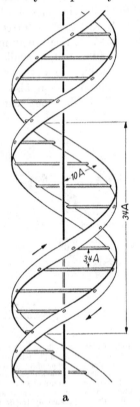

Fig. 3a and b. Helical model of DNA structure, according to Watson and Crick, Cold Spring Symp. Quant. Biol. 18, 123 (1953). a The twin helices with ribbons formed of P-deoxyribose, and rodsof opposite nitrogen bases. b Structure of a "rod" of a purine and a pyrimidine with the bonding by hydrogen bonds. The C atoms at lower left and right belong to the sugar.

or of two helices (so called minor and major spirals); the same basic structural principle appears to be involved at the sub-microscopic and at the microscopic levels and the causes operating at both levels may also be similar.

More recently (Dekker and Sachman 1954) a model for the structure of the native DNA macromolecule has been proposed which, at least in part, avoids some of these difficulties. This model is an extension of that of Watson and Crick. The absence of evidence for triesterified phosphate groups makes unlikely any hypothesis of a branched structure. On the other hand, titration studies indicate that 3,000–5,000 end groups exist per DNA macromolecule and this fact is better accounted for by supposing that the two polynucleotide chains which twist around a common axis to form the elementary fibre of Fig. 3a do not run continuously throughout a whole macromolecule but are interrupted at intervals, at about each fiftieth nucleotide. Interruptions in one strand should be staggered in relation to those of the other, so that no extreme weakness of fiber strength would result. In this manner, minimum basic units of about 50 nucleotides, with average molecular weight of 15,000, should exist in the macromolecule. The unravelling of the two strands can be obtained by heating, for instance, or by mild heating in presence of urea, which are methods for breaking hydrogen bonds. Partial repolymerisation can be secured after hydrogen bond splitting, while hydrolysis leads to breakage of the polynucleotide chains, after which repolymerisation is difficult. Although this model represents an improvement compared with previous ones, difficulties are still apparent when biological reduplication of DNA is envisaged. As said above, it could perhaps be supposed that reduplication begins by free terminal single strands, but it seems also probable that it is mediated by proteins or through enzymes which break hydrogen bonding of pieces of the two strand fibres.

At a super-molecular level, some efforts have been made to find directly by electron microscopy the basic elements postulated by the diverse theories of the structure of protoplasm (review BRETSCHNEIDER 1952b). As this is a subject to be considered especially when dealing with the structure of ground cytoplasm, we will not discuss it here in any detail (see chapter: W. SEIFRIZ, Microscopic structure of cytoplasm).

A fluid colloidal system theory reminiscent of the physico-chemical theories of two or three decades ago has been put forward recently (LEPESCHKIN 1950, refs. BRETSCHNEIDER 1952b), while the other theories admit a structural basic frame. The chief hypotheses are: 1. the attachment point theory (FREY-WYSSLING) in which the basic structures are polypeptide chains with chemical bonding at the attachment points, 2. a chromidial theory in which fibrillar protein filaments of 500 Å and nucleic acid granules of 2,000–3,000 Å alternate in rows (MONNÉ), 3. a biosome theory of globular elements disposed in branching chains and with nodules (LEHMANN), 4. a lepton theory of fibrillar protein elements with a mean thickness of 150 Å with nodules disposed in a kind of string of pearls (BRETSCHNEIDER), and 5. a fibrillar-foliar theory supposing that the basic structure may be either fibrillar as postulated in the attachment-point hypothesis or, more frequently, foliar or lamellar, forming repeating strata of protein lamellae alternating with other "stabilizing" compounds, lipoids, nucleic acids, etc. (SERRA 1943b). All of these hypotheses will be strongly affected by the developments on the molecular structure of proteins and nucleic acids. If a helical fibre structure is solidly demonstrated both for proteins and nucleic acids, the basic structural elements of protoplasm would be the polypeptide and polynucleotide "ropes" composed of a limited number (about 5–20) of elementary fibrils. Electron pictures recently obtained after osmium fixation (BERNHARD et al. 1952, OBERLING et al. 1953) suggest that lipoids play an important role in the structure of the basic protoplasmic elements, possibly in a manner similar to that formerly postulated (SERRA 1943b). The foliar structure is particularly clear in the ergastoplasm around the nucleus.

It seems that the nucleus might have the same basic structure in its nucleoplasm, although in many cases more fluid, as the cytoplasm, the same basic structural elements being found in both.

## b) Nuclear membrane.

Despite some data pointing to the nuclear membrane being only a limiting boundary between cytoplasm and nucleus (refs. LEPESCHKIN 1938), it seems safe to conclude that these represent extreme cases of a very thin membrane and that a nuclear membrane as an autonomous structure really exists in the typical nucleus. It is possible that the distinctness of the membrane may be influenced by a separation of phases which more or less hydrates or dehydrates its layers; in root meristems, the membrane is more distinct in prophase than in telophase or interphase (CHAYEN 1952). In electron microscope pictures two layers, joined at many contact points with nodules (Fig. 4; see also Figs. 5, 6, 11), are generally recognizable in the nuclear membrane (see pictures by BERNHARD et al. 1952). The outer layer, which has been found to contain lipoids, shows in sections many discontinuities which, in some cases, correspond to round pores, (oocytes—CALLAN 1952 and refs.) found to be of 400 Å diameter

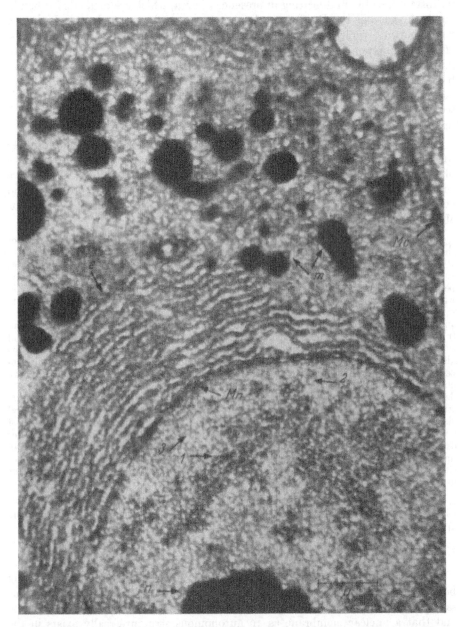

Fig. 4. Thin section (2,000 Å) of a liver cell of frog fixed in buffered osmic acid and included in esterwax. Electron micrograph, magn. 18,000, orig. magn. 5,300. Courtesy of Prof. OBERLING and Dr. F. HAGUENAU. *Mc* cell membrane; *m* mitochondria; *e* "ergostoplasmic" structures (secretion Anlagen) in layers. In the nucleus a central nucleolus *n*, a double nuclear membrane *Mn*, and nucleoplasm with *1* a gross structural phase corresponding to microscopically visible chromatic threads, *2* a finer chromosomic structure in continuity with the gross structure (see also Fig. 8) and *3* a meshwork formed by the fine nucleoplasmic structural phase together with precipitated parts of the fluid nucleoplasmic phase. Artefacts due to osmium precipitation are possible. *cf.* figs 5 and 6.

and with centers distant about 1000 Å from one another. In other cases 70 Å pores have been found (BRETSCHNEIDER 1952a). The inner layer is more continuous and consists of protein (elastin like); in electron micrographs of oocytes no heterogeneity has been found in its structure (CALLAN 1952). The thickness of the inner membrane in oocytes has been found to be 150 Å and that of the outer one 300 Å. In other cells, however, both layers may be of about the same thickness. These observations agree with old views which held that the nuclear membrane is a double layer (SCHACHT 1852, HOFMEISTER 1867—see LEPESCHKIN 1938, p. 27).

The nuclear membrane as so defined, with two layers, is not very different from other cell membranes, which usually have a porous or meshwork texture with pores of about $500 \pm 200$ Å. The case of the inner nuclear membrane layer being homogenous, however, seems atypical in that an homogenous membrane could hinder the metabolic changes with the cytoplasm and nothing of that kind is encountered in tracer studies, although it has been found that egg albumin and bovine plasma albumin do not enter the nuclear membrane of oocytes (CALLAN 1952). Former data obtained with the polarization microscope indicated that the boundary of the nucleus would have an optically negative spherite texture caused by protein chains running in a tangential direction (SCHMIDT 1939, refs. FREY-WYSSLING 1948) or a double layer of lipo-protein and protein (MONNÉ 1942). It is this latter interpretation which has afterwards been found to hold true by electron photomicrography.

From other nuclear properties, namely the great elasticity and the possibility of a marked increase in volume it follows that the inner layer is not rigid. It may be about $20 \pm 10$ protein layers thick, and it is not probable that a truly compact structure is formed by so many layers. A resolution of this structure can be expected in future work, probably a fibrillar structure with finely interwoven elastin polypeptide chains. In some electron micrographs, (BERNHARD et al. 1952, OBERLING 1953) it seems that the inner membrane layer may show discontinuities compensated by the outer layer (Figs. 4–11). This could account for membrane permeability to large molecules, or the mechanism suggested in CALLAN (1952) could also come into action: passing from straight polypeptide and polynucleotide chains through fine pores below the present practical resolution of the electron microscope.

## c) Nucleoplasm.

Until recently it was generally supposed that the nucleoplasm is structureless, a more or less characteristic sol (FREY-WYSSLING 1948). By centrifugation, a fluid which could be designated as karyolymph can be separated from a structural framework which would correspond to the reticulum. The nucleoplasm of oocytes appears to be formed of two colloidal phases (CALLAN 1952). A certain organization is given by a structural phase with some rigidity, which can be manipulated with needles and lacerated into smaller fragments which maintain their form for a while. Another, fluid, phase is dispersed between the meshes of the structural phase.

Electron micrographs (BERNHARD et al. 1952, OBERLING et al. 1953) of osmium fixed cells reveal a network in the nucleoplasm more or less similar to that of cytoplasm but generally finer in texture (Figs. 4, 5, 11). The absorption of osmium by these structures suggests that lipoids are present, possibly in combination with proteins, and so stabilizing the network, which contains nodules as described above for the structure of protoplasm in general. As shown by the relative freedom with which nucleoli and chromosomes may be moved away through the nucleoplasm, the meshwork formed by the proteins, lipids and probably also nucleic acids easily breaks down and forms again. Nucleoplasm therefore probably has at least three or four phases as shown in Figs. 4, 5 and 11 (see also p. 476 of the following chapter).

Normal non-mitotic nuclei probably have part of their nucleoplasm within the fibrillar frame of the chromonemata, which are shown in some points of electron pictures by denser spots with meshes of thicker interwoven fibres. In some cases double filaments with spiralization are visible in these spots

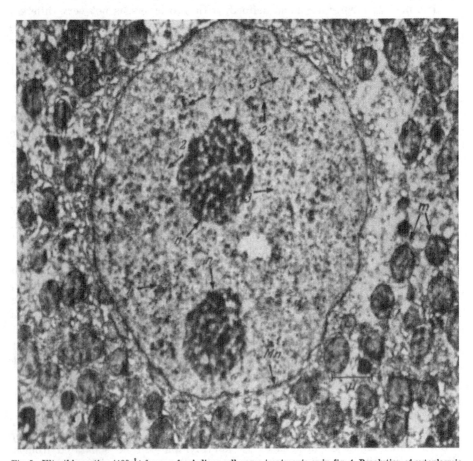

Fig. 5. Ultrathin section (400 Å) from a frog's liver cell, same treatments as in fig. 4. Resolution of cytoplasmic and nuclear structures by the finer section, especially the nucleoli *n* in which a kind of filament forming a meshwork is seen. Other designations as in Fig. 4. Magn. 13,000. (Unpublished electron photograph, courtesy of Dr. Haguenau and Prof. Oberling.)

(Yasurumi *et al.* 1953). The filaments break down to 80—160 m$\mu$ granules. The closer structural relations of the three components which occupy the bulk of the nucleus, fluid nucleoplasm, structural nucleoplasm and chromonemata, remain to be clarified in future work.

### d) Chromosomes and genes.

It is obvious from their microscopic aspect that the chromosomes have a more or less marked fibrous structure. Inferences based on genetical work about the location of the genes along the long axis of the chromosomes also point to a linear structure. Chromosomes without fixation or after certain types of fixation are highly extensible.

In the absence of calcium, the chromosomes from oocytes may be stretched to twice their original length without losing their elasticity and rupture is obtained only after they are stretched to more than 8.1 times their original length (DURYEE 1937, 1941). Salivary chromosomes are not so markedly extensible, showing perfect elasticity only when distension does not pass 1.1–2.0 times their original length (more particulars SERRA 1949a, p. 506). However, they may be stretched to 3–4 times their length or more without rupture (GEITLER 1938, PFEIFFER 1940, 1941, BUCK 1942). It is believed that salivary and oocyte chromosomes

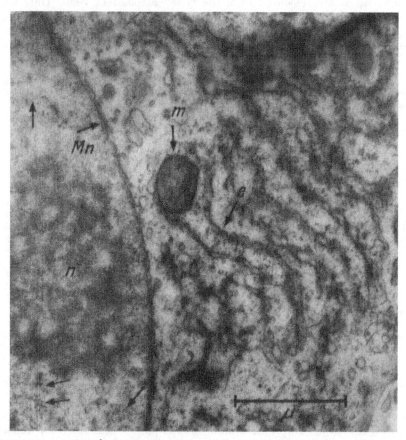

Fig. 6. Ultrathin section (400 Å), frog liver, treatments as in fig. 4. Resolution of "ergastoplasmic" structures *e* which now appear as channels or as double layers with other matter inside. Numerous granules are seen, which may be osmiated secretion or structural elements of the cytoplasm. A foliar structure is seen in the "ergasto- plasm" and the external layers of the mitochondria *m*. In the nucleus a thinner nucleolus *n* without contrast, a double nuclear membrane *Mn* and helical threads in the nucleoplasm (arrows) probably corresponding to the finest helices of chromonemata (diameter about 50–60 Å). Similar helices are seen in the cytoplasm. *cf.* fig. 8. Magn. 30,000. (Unpublished photograph of Dr. ROUILLE, by courtesy of Dr. HAGUENAU.)

develop from mitotic ones by natural distension; mitotic chromosomes should therefore be highly extensible, and this is verified in practice in chromonema regions where there is not too much chromatin. In salivary chromosomes it is observed that the inter-chro- momeres are much more extensible than the chromomeres; in fact, almost the whole exten- sibility is due to the non-chromatic interbands. This signifies either that the chromatin of the chromatic bands hinders stretching, or that the lack of extensibility of the chromatic bands is due to the fact that the molecular structure of the chromosome at these points is already in a naturally distended state (see Fig. 7, p. 463).

These simple physical data demonstrate the fibrous nature of the chromosomes and that their skeleton materials are not in a state of natural distension but rather that some degree of folding at the sub-microscopic level must exist, since oocyte and salivary chromosomes show only a very loose helical coiling whose stretching could not allow for the increase in length. Among the structural compounds of the nucleus only nucleic acids, especially DNA, and proteins may be considered as a basis for the fibrous structure of chromosomes. This problem has been attacked by chemical means and, at the sub-microscopic level, by polarization microscopy. There is no doubt that DNA, proteins, and RNA exist in the chromosomes; differences of opinion concern only the question which of them is, or are, essential from the genetical point of view. Only recently the view that DNA is the carrier of genetic specificity has been held with strong conviction, and because of this the partisans of this view seem to have lost interest in the protein part of the chromosomes (see the preceding chapter "Chromosomes and genes").

A brief examination of the evidence in favour of the belief that DNA is istelf the genetic material reveals that the unique facts demonstrating this are derived indirectly from bacteriophage studies (refs. and summaries in the several articles of Cold Spring Harbor Symp. Quant. Biol. 18, 1953). It has been demonstrated that some phages infect bacteria by injecting into the bacterium only part of their contents, 80% of which, by weight, is composed of DNA. The other 20% could be accounted for as loss during the injecting by the virus of its contents. The part which remains outside the bacterium has the appearance of an empty membrane with an injecting tube attached to the host. It is possible by osmotic shock to force the phage to discard its content similarly to when it meets a host, and the amount of proteins in the discarded product, wich differ in composition from the membrane proteins by the tests so far performed, may be only as much as 5% of the total product, that is $1/_{20}$ of the DNA by weight (HERSHEY 1953). Because of these difficulties in characterizing a protein fraction in the injectable product it has been concluded by some (not by HERSHEY, for instance—see Discussion of his article) that DNA alone is the genetic carrier of phage specificity which induces self multiplication within the bacterium, in what is called a "vegetative phase". On the other hand, it has been concluded that the N and P of phage parents amounts to only 0.1% of that of the total nucleoprotein of the progeny phages (KOZLOFF 1952), which means that the material of the new phages is almost entirely of bacterian origin. Manifestly the crucial point of the discussion lies in the presence or absence of an "impurity" in the injected product, which may or may not be identical with the protein of the membrane which remains outside. The injected part may be composed of a kind of minimum reproductive phase together with much DNA which is needed as a structural component for the production of more protein of the membranes and the reproductive phases. Only further experimental work can clarify this important question.

Plant viruses generally do not contain DNA but contain RNA instead. (It is now believed that RNA is always present in these viruses, but this may turn out in future to be a simplified view, because in earlier work the existence of viruses consisting simply of protein or lipo-protein has also been reported.) By indirect evidence and probably also by analogy with the phages, it has been argued that in these cases the RNA could be within the protein envelope (MARKHAM 1953). As it is known that RNA can amount to only a few per cent of the virus weight it follows that a very inefficient mechanism of virus reproduction, similar to that of the phages with their possible "impurity" accompanying the injected DNA, exists in plant viruses if RNA is the genetic material. For the moment these are perplexing questions which demand further experimentation and better analyses.

Another point in favour of the belief that DNA and/or RNA are the carriers of genetic specificity is the transformation of bacterial types from non-infective to infective and from one colonial form to another by the junction of specific DNA to the medium (refs. EPHRUSSI-TAYLOR 1951, HOTCHKISS 1951, 1952). However, the same facts can be interpreted in another way, the DNA serving as specific substrate for genic action which can only be synthesized by the deficient bacterial cell once it is supplied from the exterior (refs. SERRA 1949a). Possibly the supplied DNA will be employed at the periphery of the bacterium and does not penetrate into the cell to be incorporated in the respective nucleus, nucleoids or chromosomes.

Summing up, it may be said that several facts point to DNA or RNA being carriers of genetic specificity. The same facts, however, do not exclude other alternatives: a) that accompanying proteins, possibly in very small amounts, are the specific compounds and DNA or RNA only structural subordinate compounds or, when in greater amounts, matrix components which include the significant genetic compounds; b) that the genetic material is a nucleoprotein formed of a small amount of nucleic acid combined with a small part of protein, the rest, if present, as in phages, being a kind of reserve to serve in reproduction; c) that DNA is an essential component of the cytoplasm or the membrane, but not of the gene substance of the chromosomes or nucleus. As yet no decision can be taken in favour of one of these possibilities but the facts seem to point to at least a part of genetic specificity being linked to nucleic acids acting in the cytoplasm. The nucleic acids might have specific configurations impressed on them when they were synthesized in contact with the chromosome proteins and might carry genetic specificity from the nucleus to the cytoplasm.

The chromosomes without any doubt are always formed of protein and the question is whether DNA is or is not always present there (see preceding chapter, p. 439, "Chromosomes and genes"). To interpret the seemingly contradictory facts, it has been postulated that the genes of the chromosomes (chromogenes) may be simply protein when their composition is at its simplest, or linked to DNA and RNA, under the form of nucleoprotein, when functioning or multiplying (SERRA 1949a). Nucleic acids should carry with them genetic specificity after they leave the chromosome proteins. This would also explain the case of ghosts of bacteriophages, that is membranes and tubes without their DNA (plus the "impurities", if any), being able to lyse the bacteria and perhaps more efficiently than the whole phage, though they have lost the power of multiplying.

When salivary or oocyte chromosomes are examined no doubt can subsist that the basis of the fibrous nature of chromosomes are the proteins which run in fibrils along the chromosome. After treatment with nucleases, chromosomes remain apparently almost intact, their bands and interbands having the original appearance, although the bands have lost all their chromaticity (SERRA 1943a). Polarization microscopy (refs. SCHMIDT 1937) has revealed that chromosomes of plant cells in good conditions are optically positive with respect to their length. Non-mitotic nuclei and certain anaphase chromosomes show negative anisotropy. Artificially stretched fibres of DNA (sodium salt) or of squashes of nuclei show negative anisotropy, while protein fibres are positively anisotropic along their length (SCHMIDT 1937, FREY-WISSLING 1948). According to the parts played by the two components, proteins and nucleic acids, and their more or less marked arrangement in fibres, the chromosome may show positive or negative anisotropy. This is further complicated by the helical coiling of almost all chromosomes, which may cause positive anisotropy. These factors may concur to give a dubious result, as shown by the case of the chromatic salivary bands, in which negative anisotropy was found in visible light and interpreted (SCHMIDT 1937, 1941) as showing that DNA fibrils run parallel to protein fibrils in the bands, while dichroism measurements in the ultraviolet (CASPERSSON 1940b) demonstrated that the orientation of polynucleotide chains parallel to the chromosome long axis could only be very slight. The previously found negative anisotropy of sperm heads (SCHMIDT 1937, 1941) also does not apply directly to the case of chromosomes and must be interpreted as being due to the stretching of this structure, with an excess of DNA, during spermatogenesis. In this phase it is seen that a previously spheroid nucleus generally acquires an elongated form.

These facts may be interpreted by admitting that in chromatic bands and chromomeres in general there is a chromatin kalymma of nucleoproteins (with RNA also but chiefly with DNA and histone) in the form of a more or less compact gel of elongated fibrils. It is an open question whether any DNA or RNA are present also inside the chromonema proper, that is in the part of the chromosome which would remain after all kalymma is removed, and, of course, this is a question of basic importance for the nature of the chromogenes, which may be different from plasmagenes and from bacteriophages and viruses. As said above, the chromomeric kalymma may hinder the distension of the chromomeres and so explain their lack of marked extensibility; however, another interpretation is also possible, namely that the protein chains are already more or less distended at these points; this would facilitate the visualization of the doubling of the genetically specific structures found in the chromomeres, which may be both polypeptide and polynucleotide chains or only the former—see Fig. 7.

Besides possibly playing a role in the supply of energy to the genes, the functions, mainly structural, of the kalymmatic or external nucleoproteins at the chromomeres and in the whole mitotic chromosomes are probably twofold: they function in reduplication, by individualising bundles of chromonema fibrils into visible chromosomes; and they protect the genetically active parts against chemical attack by unphysiological compounds and mechanical injury during cell movements and especially in karyokynesis, in which chromosome parts without kalymma would tend to break. This latter property would derive from the chromonema fibrils being of necessity not uninterrupted from end to end of the chromosome, since the energy to break the chromosome would then be much higher than is physiologically probable in crossing-over and is provided in certain types of radiation-induced breaks (discussion Serra 1949a, p.510). The chromonema is composed of sub-units which bear genetic specificity and may or may not be linked by non-genic parts, the interchromomeres. It must be noted that it is not necessary that each chromonema sub-unit be a gene; on the contrary, it seems that generally each gene may correspond to more than one genetically active zone (Serra 1942, 1949a, p.424).

According to this interpretation the genes would not be corpuscles or macromolecules. The thinnest individualized chromosomes as yet photographed in the electron microscope (from oocytes—Tomlin and Callan 1951, see Callan 1952) had a diameter of about 200 Å, which demonstrates that visible chromosomes are much hydrated. In ordinary mitotic chromosomes, a part of them may also be occupied by nucleoplasm imprisoned between the fibrils. The importance of calcium for the stretching of chromosomes, for their distinctness and their physical state in general also points to the hydration being greater when chromosomes are less visible or less well individualized. Salivary chromosomes, observed under the electron microscope (Pease and Baker 1949, Schultz et al. 1949), show a greater density at the bands, where spheroidal particles from 500–1500 Å have been photographed. These bodies must be shrunk particles of nucleoproteins and are not the genes, which, if they had this size, would occur in a number far greater than that of from 2,000–10,000 which is ordinarily postulated. Ordinary mitotic chromosomes under the electron microscope show predominantly their coating of chromatin and only at the stages when this coat is at a minimum or naturally absent may the basic chromosome structures be observed. Pachytene maize chromosomes (Buchholz 1947) show a continuous chromonema thread of fibrils in which are embedded or to which are attached much more dense spheroidal bodies of 850–6600 Å diameter. These bodies very probably are the chromatin particles of the chromonematic kalymma. To judge from the 200 Å cross section reported for oocyte chromosomes, the number of elementary chains (polypeptidic and perhaps also polynucleotidic) disposed at 10 Å distance will be about $4 \times 10^2$, which seems a relatively low number compared to the $10^4$ expected for a chromonema just visible under the ordinary

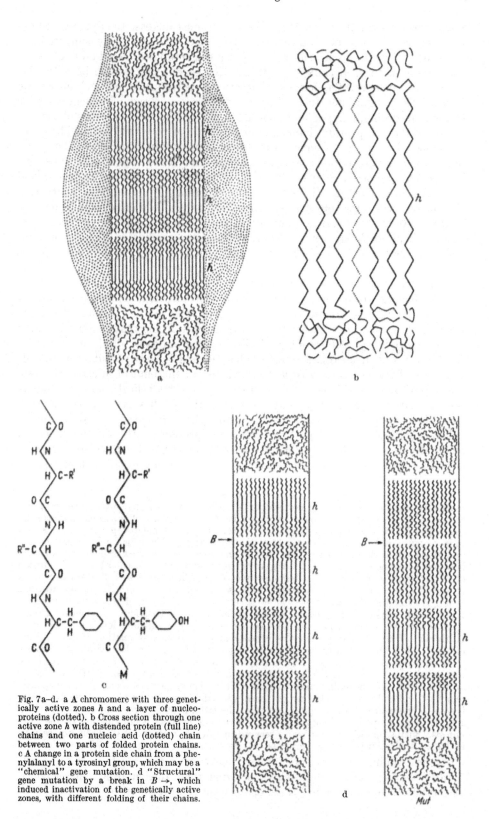

Fig. 7a–d. a A chromomere with three genetically active zones $h$ and a layer of nucleoproteins (dotted). b Cross section through one active zone $h$ with distended protein (full line) chains and one nucleic acid (dotted) chain between two parts of folded protein chains. c A change in a protein side chain from a phenylalanyl to a tyrosinyl group, which may be a "chemical" gene mutation. d "Structural" gene mutation by a break in $B \rightarrow$, which induced inactivation of the genetically active zones, with different folding of their chains.

microscope (SERRA 1949a, p. 511). Hydration must increase the apparent thickness of the chromosome to a greater extent than was previously supposed.

Another hypothesis of gene structure has been proposed, based on DNA being the genetic material (STERN 1947, 1952). The gene is supposed to be a "modulation" of nucleoprotein, formed of interlaced polypeptide and polynucleotide chains. The modulation, in analogy to modulation of audio frequency signals or sound recording, is supposed to consist in different relative positions of the side groups of the polypeptide chains and of the bases of polynucleotides. The genes are supposed to be stereoisomeres with the same analytical composition. The polynucleotide and polypeptide chains are supposed to coil in visible or submicroscopic helices. This hypothesis may easily be brought into accord with the views now held on DNA helical structure. However, no experimental or observed facts, apart from the analogy with gross physical systems, have as yet been discovered to demonstrate it. It is easily seen, on the other hand, that not only polynucleotide as also polypeptide chains would become specifically "modulated". Both proteins and DNA or RNA could act as carriers of genetic specificity and this explains at the same time the case of the bacteriophages injecting their DNA and the fact that the fundamental structure of chromosomes is of proteins.

The folding of the chromonema fibrils (the kalymma must be essentially a gel formed of non-orientated or poorly orientated fibrils) is inferred in part from facts about their extensibility and in part from what is known about the microscopically visible coiling. Helical coiling exists in all chromosomes so far studied and in meiosis a double helical coiling is clearly visible in suitable materials. In some cases a very minute helix at the limit of microscopic resolution is also discernible in mitotic chromosomes. This helical coiling represents the adaptation of a fibrillar body to occupy the small space compatible with ordered distribution to the spindle poles of daughter chromosomes. It is remarkable but not unexpected that nucleic acids and proteins, both essential biological materials, are also coiled as helical fibrils at the molecular level, which represents the same form of adaptation although it derives from atomic bonding possibilities.

The understanding of why biological fibrillar materials should have helical coilings at molecular, sub-microscopic and microscopic levels does not say anything, however, about the actual causes of these coilings. Leaving aside the molecular coiling, which derives from the properties of the constituent atoms themselves, the sub-microscopic and microscopic (minor and major) coilings probably have the same origin. The helical microscopic coiling may derive from the presence of two half chromosomes both formed of elements coiled in helices of the same torsion (both dextrogyre or sinistrogyre) at a molecular level. The torsion of fibrils coiled in helices at a sub-microscopic level obliges them to coil again in a helix at an higher level until the microscopic level is attained. In its turn the sub-microscopic coiling is caused in the same way by the molecular helical coiling of the polypeptide (and perhaps also polynucleotide) chains of the chromonemata. Therefore, the helical chromosome should be formed of at least 4 helical elements at two different levels and the cycle of division of the microscopic helices is retarded one division in relation to the division of the 2 sub-microscopic helices. In meiosis again another pair of helices appears from a retarded cycle of division of II meiosis. All these helices must be plectonemic since the chromosome divides in the helical state and the helices must uncoil if the two halves are to separate. The uncoiling process begins with the distension of chromosomes in anaphase-telophase and ends in the following prophase when the old-spiral-prophase (RESENDE 1946) is seen to give place to a new-spiral-prophase; the latter should correspond to the new division, or more properly individualization of daughter helices, in anticipation of the next mitosis. Microscopically it has been observed in suitable material that chromosomes may anticipate their division for the next one or two mitoses.

Helical coiling at a microscopic level should therefore result from the existence of two similarly coiled helices deriving their torsion from helices coiled at sub-microscopic levels and these again from the torsion at a molecular level.

This process of coiling may be assisted by a deposition of long chain molecules of polypeptides and polynucleotides on the chromonemata. At the beginning of mitosis and meiosis these chains may perhaps be deposited in a more or less ordered form along the length of the chromonema; afterwards the deposition becomes less ordered or altogether disordered so that the kalymma shows no obvious orientation of its particles. The chromosomes then continue to shorten down to the limits found in mitosis and especially in meiosis, by a gelifying process which involves both the kalymma and the chromonemata (SERRA 1947b).

To what extent these theoretical deductions fit the facts may be seen from published electron micrographs (HOWANITZ 1953) showing the "spiral" structure of chromosomes. It is to be noted that the preparation of these chromosomes partially dissolved the chromatin and left chiefly the chromosome skeleton. In some pictures, (for instance HOWANITZ' Fig. 9) several helical coilings of successively lower orders may be seen (Fig. 8). At least 4 or 5 orders of helices are visible. According to these data, the diameter of the strands in red blood cell nuclei range from 40–50 Å to about 50 times this number and the helix of each order has about 2 times the diameter of the preceding order. If proteins were the base of this structure the minor helices should have a pitch of about 66–68 Å and multiples of this should be found in successive orders, that is about 134, 268, 536, 1072, 2144, 4288 Å, the latter two already in the limit of microscopic visibility. Hydration and inclusion of other materials (DNA, RNA, nucleoplasm components) could increase the diameter of these helices to double or more in each order.

*Heterochromatin*, cytologically defined as the parts of the chromosomes which in telophase and interphase retain their coat of kalymma in a more condensed state than the remainder of the chromosome (euchromatin), has sometimes been stated to have a structure different from that of euchromatin. As suggested in the preceding chapter (see "Chromosomes and genes"), it seems that the differences derive from the genetic materials themselves existing in the chromonemata of heterochromatic regions and not from their having two kinds of chromatin. Only in *Drosophila* and perhaps some other species of Diptera does the heterochromatin of the salivary chromocentre show a loose, less compact texture than euchromatic parts. In *Chironomus*, the heterochromatin of the salivaries has a compact structure similar to that of euchromatin and in mitotic chromosomes heterochromatic regions can only with difficulty be distinguished from euchromatic ones during the middle mitotic phases. A somewhat greater degree of compactness, which may be revealed by gradual physical coloration (or differentiation) methods, discernible even at metaphase, is probably the result of a more marked gelification as a consequence of the gelifying process acting at the heterochromatic regions during a longer period. The case of the *Drosophila* salivary chromocentres is explained by the presence of compounds which do not become well fixed in acetic acid and it has been suggested (SERRA 1949a) that this is caused by the inclusion of a relatively great amount of nucleoplasm within the chromocentre. It is possible that in other cases the heterochromatic regions also bear relations to the nucleoplasm, different from those of euchromatic zones.

The essential characteristics of heterochromatic zones probably derive from their relative genetic inertness, which al.ows nucleoproteins to remain attached to the chromonemata after telophase in greater amounts than in euchromatic regions. (It is obvious that heterochromatic zones have a different nucleoprotein metabolism.) If this hypothesis is valid, no structural differences at a supra-molecular and sub-microscopic level should exist between eu- and heterochromatic zones or such differences should concern the mutations themselves which caused the genetic inertness and which may bring about fine changes of structure not easily apparent. Position effects, especially those of the centromere and the nucleolar zones may also cause heterochromatization linked with relative genetic inertness (SERRA 1949a and refs.).

With respect to the *structure of the centromere* and the primary constrictions, in well analysed cases a pair of chromatic granules and adjoining fibrils may be

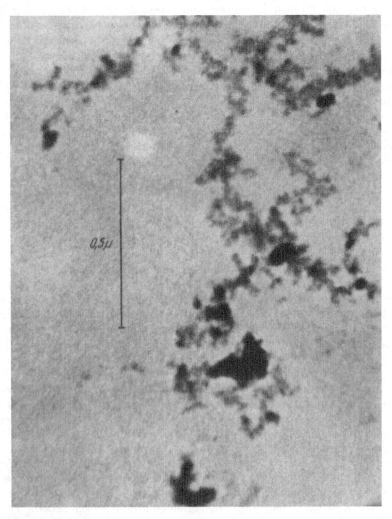

Fig. 8. Helical structure of chicken erythrocytes' residual chromosomes. Seven orders of helices are seen, the lower one of diameter about 50 Å. Magnification about 88,000. From HOWANITZ, Wasman J. 11, 1 (1953).

seen in each chromonema in this region (LIMA DE FARIA 1949, 1950). These granules, or the centromere as a whole, have been equated (DARLINGTON) to kinetogenes, which would govern the attachment of the chromosome to the spindle and divide only at metaphase. Generally the centromere is Feulgen positive both in the chromomeres and the interchromomeres at pachytene (LIMA DE FARIA 1949b) but this may be different at meta- and anaphase, when it is attached to the spindle fibrils. At pachytene no structural qualitative differences are discernible between the centromere and other chromosome regions. The differences are, therefore, either submicroscopic and have not yet been cleared up or, more probably, they reside in the functioning of these regions, which at certain stages present a composition of the respective kalymma different from adjacent regions, as has been demonstrated for nucleolar zones (ALBUQUERQUE and SERRA 1951).

In what concerns the structure of *secondary constrictions*, and especially of the nucleolar zones, there has been some controversy about the existence of chromo-

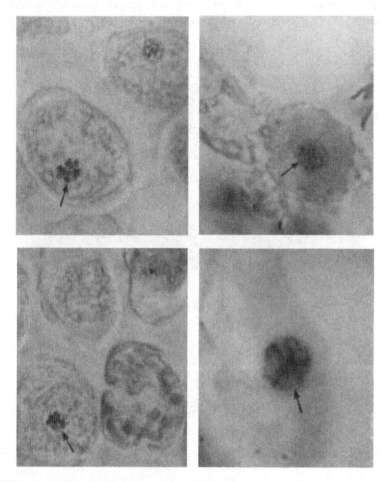

Fig. 9. Cells from the ovary of *Nothoscordum* (Liliaceae) showing a filamentous body (nucleoloneme) inside the nucleolus (arrow) by iron-pyrogallic staining. (Courtesy of Prof. ESTABLE, unpublished photographs.)

meres and of DNA at these chromosome regions. Of the simple, non-nucleolar secondary constrictions it is only known that they represent points where kalymma chromatin is not linked to the chromonemata, except when the chromosomes are at their greater contraction. Possibly they derive from reverse repeats or from nucleolar zones which during the evolutionary history of the species have lost their power to condensate a nucleolus. Nucleolar zones have a variable chromaticity, which has given rise to the designation of olistherozones (RESENDE 1946) and their distension is also variable. Their composition is different from anaphase onwards and it is the relative lack of kalymma and the presence of phosphatids (ALBUQUERQUE and SERRA 1951) which give rise to the peculiar aspects of the zone. In larger nuclei (salivaries in animals) small nucleoli may be seen inside

30*

the chromosomes or laterally attached to them. The difference between nucleolar zones and the rest of the chromosome lies in the special capacity of the zones for binding phosphatids and perhaps other compounds, which is a property of the chromonema at this region and not necessarily correlated with a discernible difference in fine structure.

### e) Nucleolus.

In many cases the nucleolus appears as a homogenous body. Nucleoli with vacuolar or other inclusions, however, are not infrequent. In some cases several

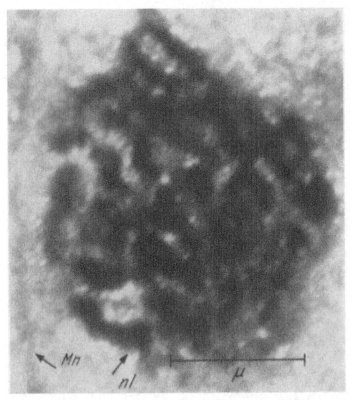

Fig. 10. A 2,000 Å section of a cell from rat liver, treatments as in fig. 4. Magn. 36,000. Nucleolus with a filamentous body *nl.* See resolution of this structure in fig. 11. (Courtesy Prof. OBERLING and Dr. HAGUENAU.)

concentric zones may also be observed in nucleoli (RATTENBURY, to be publ.), particularly with special staining methods (lipids, or nucleolar aceto-carmin of RATTENBURY). The inclusions may also be crystalloidal. In oocytes it is easy to observe the vacuolation *in vitro*, before fixation, and to follow the emission of nucleolar buds, which afterwards dissolve in the nucleoplasm (SERRA and LOPES 1945). The passage of figured nucleoli to the cytoplasm of similar cells in rapid growth has also been repeatedly reported. A pellicle at the surface of the nucleolus may also be visible. All these zones, vacuoles and the pellicle may become apparent with staining for phospholipids (ALBUQUERQUE and SERRA 1951). In living *Tradescantia* staminal hairs photographed at 2650 Å with an exposure not damaging the cells, the nucleoli have been found to have an outer more absorbing zone and a poorly absorbing core (CHAYEN 1952). Similar zones can also be seen *in vivo* in other plant cells.

Within the nucleolus, structures of the nucleolar zone may be observed, with chromomeres and interchromomeres, especially in large nucleoli of salivary glands of Dipteran larvae. However, with differential staining, for instance pyronin-methyl green, these structures may also become apparent in several plant nucleoli. Other interesting intranucleolar structures have been reported, visible either in ordinary microscopy (ESTABLE and SOTELO 1951) or with the

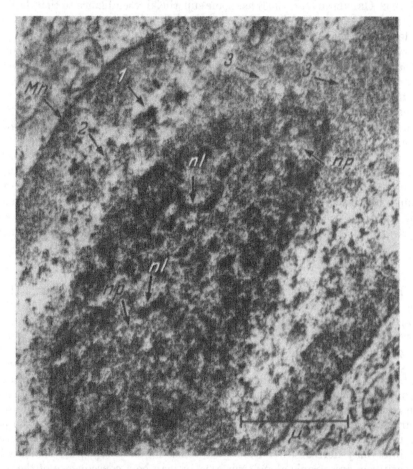

Fig. 11. Thinner section (400 Å) showing resolution of the nucleolar structures. The nucleolar filament *nl* (comp. with fig. 10) is now seen to be formed of helical structures and granules, the rest of the nucleolus is of nucleoloplasm *np* and has a finer structure of filaments and granules. Possibility of osmic artefacts not excluded. Other designations of the nucleoplasm (*1, 2, 3*) as in fig. 4. No nucleolar membrane is seen. Magn. 30,000, orig. magn. 10,300. (Unpublished photographs, courtesy of Dr. HAGUENAU and Prof. OBERLING.)

electron microscope (BERNHARD *et al.* 1952, OBERLING *et al.* 1953)—Figs. 5, 9, 10, 11. The former authors have designated the structures they describe as the nucleoloneme, with the form of a filamentous body with nodules, curls and twists. It is said to be a permanent component of all nucleoli in vegetal and animal cells and never formed de novo but appearing by transverse division of preexisting nucleolonemes. During cell division the nucleoloneme is said to fragment and apply itself against chromosomes, with which it passes to the two daughter cells. These structures applied against chromosomes may be composed of nucleolar and prenucleolar matter, which forms layers in certain types of nuclei (RATTENBURY and SERRA 1952), and the question of the continuity of the nucleoloneme remains to be clarified. It is also necessary to investigate what connections, if any, exist between the nucleoloneme and the nucleolar zones sometimes existing within the nucleoli. The nucleoloneme may be stained with iron alum-potassium ferrocyanide-pyrogallol but may also be observed *in vivo*, according to ESTABLE and SOTELO (1951).

The structure seen in electron pictures after osmium fixation (BERNHARD *et al.* 1952, OBERLING *et al.* 1953) has some superficial resemblance to that described as the nucleoloneme but it occupies a greater nucleolar volume and may take the form of a central vacuole with a denser layer at the periphery, as is also observed in phospholipid preparations. Probably these structures represent the resolution by electron microscopy of the parts seen by ordinary microscopy after phospholipid staining (ALBUQUERQUE and SERRA 1951), since they become heavily impregnated by osmium. These structures may represent a separation of phases and may be an index of nucleolar activity. Further resolution of them has now been obtained by electron photography of finer sections (down to 400 Å) of animal cells fixed in buffered osmium tetroxide (HAGUENAU, in coll. with OBERLING and ROUILLE, person. comm.). In these photographs—Figs. 5, 11— it is apparent that the nucleolus consists of a filamentous part and of nucleoloplasm. Both these parts show a finer structure of helicoidal filaments and of granules (see espec. Fig. 11), which may be respectively of proteins and/or nucleic acid, and of lipoprotein globular particles. The photographs do not reveal a pellicle, which surely exists to separate the nucleolus from the rest of the nucleus.

## C. Spindle structure.

The spindle of mitosis and meiosis has been the subject of many studies with the polarization microscope (refs. SCHMIDT 1937, FREY-WYSSLING 1948, 1953). These have proved that a fibrous structure exists *in vivo* and not only after fixation. Phase-contrast microscopy has also clearly shown a spindle in living cells (HUGHES and SWANN 1948). The spindle and its fibres as well as the aster fibrils are positively birefringent *in vivo*. Each microscopically visible fibril may be formed of smaller elongated particles linked between themselves (HUGHES and SWANN 1948). The spindle poles are positively birefringent spherites with respect to their radii. In living sea urchin eggs it has been shown that the birefringency is at a maximum at about $^1/_3$ from the distance between the poles and the equator of the spindle (SWANN 1952). The positive birefringence of the fibrils may derive either from their consisting of polypeptide chains which are positively anisotropic along their axis, or may be a consequence of the orientation of elongated micelles without intrinsic birefringence. As the chemical tests and indirect evidence about cytoplasmic composition indicate that the spindle is formed at least in part from proteins, it has been admitted that the spindle fibrils were positively birefringent because they have polypeptide chains more or less orientated along the fibril axis.

The fibrils may extend from pole to pole or only to the chromosomes and probably many of them end loose among the others. The spindle as a whole forms a kind of gel structure, with the fibres in part behaving like channels with a more fluid matter inside. The beginning of the anaphase movement is marked by a swelling of the spindle at its equatorial plane, perpendicularly to the pole axis (LESSLER, discussion to article of INOUÉ 1952). This probably corresponds to the transformation of fibrous to globular proteins which has been postulated by several authors (for instance SERRA 1949b, MAZIA 1953).

The globularization process may be operative in the anaphase movement of the chromosomes, which has also other components, such as a contraction of the spindle fibres. The division of the chromosomes, as is demonstrated by the existence of endomitosis, does not depend on the spindle although it can be more or less activated by it.

It is not certain whether the spindle is formed of protein fibrils alone, with a partial transformation of SH to S—S groups (MAZIA 1953) as in the keratinization process (see preceding chapter: "Chemistry of the nucleus, Achromatic spindle"), or whether phospholipids or perhaps ribonucleic acid lie in molecularly orientated patterns among the spindle fibres (SERRA 1949b). The latter hypothesis would explain the appearance of channels, with a more gelified cortex including more fluid matter inside. It would also explain the linking of the spindle fibres to the chromosomes, which would also have phospholipids at the centromeres (or in the whole chromosome kalymma, if there are multiple attachment points, as in *Luzula* and other plants and in Hemiptera, etc.). On the other hand, the influence of narcotics and high temperatures on spindle functioning would also be more easily explained (see next chapter, p. 488).

Colchicine treatment of the spindle *in vivo* (INOUÉ 1952) soon causes a decrease in birefringence of the asters and fibres and a shortening of the spindle as a whole. Shortening continues until finally the fibres almost disappear as if they were emptied of their contents. Colchicine would affect the linkages between the fibril micelles (INOUÉ 1952), which are broken. However, it is not known how this can agree with the hypothesis of S—S group formation playing an important rôle in spindle structure, since it is not known how colchicine may affect the SH to S—S transformation. It may be that the action of colchicine is more general, the maintaining of the spindle being not a passive process as is that of keratin fibre formation, but rather an active one depending on an energy supply to maintain the proteins unfolded, as postulated for other cell processes, such as secretion (GOLDACRE 1952, DANIELLI 1953). In that case colchicine would poison some enzymes or would combine with active groups involved in the process, and the same is true to a greater or lesser extent of other c-mitotic substances (see following chapter "B. Mitosis and meiosis").

## Literature.

See at the end of the chapter "Physical chemistry of the nucleus", p. 499.

# Physical chemistry of the nucleus.

By

## J. A. Serra.

With 2 figures.

Application of Physical Chemistry knowledge and especially of its Colloidal Science branch to biological problems has given rise to a great bulk of work particularly in the first third of the actual century, which culminated in monographies like those of HEILBRUNN (1928) and LEPESCHKIN (1938) for the cell in general and that of SCHAEDE (1929) for the plant nucleus. More modernly biological research based on Physical Chemistry has concerned chiefly the cytoplasm in what respects the physical state and cellular activities such as permeability, contractility and some elaborations and, on the other hand, cell physiological functions in general such as respiration, enzyme action and the like (see for instance the book by HÖBER *et al.* 1945). Meanwhile the physical chemistry of the nucleus has been somewhat neglected in favour of chemistry and fine structure, which mainly have attracted the attention of investigators in the last twenty years. The physical chemistry of the nucleus is in many points similar to that of the cytoplasm and as this is treated in another chapter of the Handbook[1] no attempt will be made to deal here with general physical chemistry of protoplasm and only the phenomena peculiar to the nucleus will be mentioned. A brief account of physical properties will also be given.

## A. Non-mitotic nucleus.

Several of the physical and physico-chemical properties may be studied only in the non-mitotic nucleus since during mitosis the nucleus generally loses its limiting boundary and reduces to the chromosomes, whose dimensions and state are not suitable for such studies. In mitosis we are confronted rather with dynamic problems than with the physical state, which concerns only the chromosomes. Of course, physical and physico-chemical knowledge on the nucleus has to be gained from studies *in vivo* and this adds to the difficulties of the employed methods (for a review of these methods: HEILBRUNN 1928, HÖBER *et al.* 1945).

### a) Physical properties[1].

The nucleus of plant cells generally has a *specific gravity* (spec. grav.) greater than the cytoplasm. In animal cells it may not be the same, particularly in cells rich in reserves, such as the oocytes; in these cells the nucleus comes in centrifugation to the centripetal pole, together with some ground cytoplasm, and only the fat inclusions are lighter than it. Of course, the place occupied by the nucleus in the centrifuge gradient after a prolonged centrifugation depends not only on its properties as also on the relative spec. grav. of the several cell constituents, whose properties may vary according to the cell physiology. For instance in oocytes the nucleus is particularly hydrated while the cytoplasm is rich in inclusions of greater density.

---

[1] See also the previous chapters; *cf.* W. SEIFRIZ, "The physical chemistry of cytoplasm" (also for definitions and methods).

The determination of spec. grav. in isolated nuclei may be bound to great errors owing to the loss of components during isolation (see chapter "Chemistry of the nucleus—Methods", p. 416). In mass isolation the nuclei of wheat germ sediment in cyclohexane-carbon tetrachloride mixtures of spec grav. 1.416–1.420 while cytoplasmic debris sediment at spec. grav. higher than 1.420 (STERN and MIRSKY 1952). These nuclei have no lipids, which are dissolved by the isolation fluid and probably are also dehydrated, which results in a greater spec. grav. On the other hand, nuclei isolated by the citric acid methods may lose up to 50–60% of their protein content, showing a lower spec. grav.

In *Spirogyra* it has been found that nuclei immediately after isolation had a spec. grav. of $1.094 \pm 0.007$, while cytoplasmic droplets showed a spec. grav. of $1.076 \pm 0.016$ (PFEIFFER 1935; MILOVIDOV 1949). Nuclei of animal cells generally show spec. grav. greater than this, of the order of 1.38–1.39. In plants in general it is to be expected that the spec. grav. of the nucleus will not be much greater than 1, in agreement with the greater hydration (vacuolation) of plant cells. Centrifugation demonstrates that within the nucleus the order of increasing spec. grav. is: nucleoplasm, reticulum, chromatin, nucleolus. In oocytes it has been calculated that the nucleoplasm has a spec. grav. (at 20° C.) of 1.04 while that of the nucleolus is 1.14 (HEILBRUNN 1928). These values must be near those typical of plant cells. The nucleolus is easily displaced by centrifugation and may pass through the nuclear membrane. It naturally occupies the lower part of the nucleus when this is less dense and permits the dislocation of its constituents.

The *viscosity* of the nucleus must not be very different from that of cytoplasm, which is given as ranging from 0.09 to 0.19 ($\eta$, absolute viscosity) in slime molds (HEILBRUNN, from HEILBRUNN 1928), 0.018 in oocytes and 0.24 in *Vicia* stem parenchyme. This compares with the viscosity of water, which is 0.01, and is much lower than that of olive oil (0.99) or glycerol (10.69). A 3% solution of egg albumin has a viscosity of 0.012. In general it may be said that the viscosity of protoplasm is from 1.2 to 30 times that of water. The nucleus may have very different viscosity values according to its state and generally it has a visible structure. Therefore, the viscosity of the nucleus refers to the more or less homogenous caryoplasmic regions. From other authors' observations it has been guessed (HEILBRUNN 1928, MILOVIDOV 1949) that the nucleus has viscosities of the order of 0.02 to 0.07, not much greater than 2 to 7 times that of water. These determinations have recently been more or less neglected while emphasis has been laid on the knowledge of structure and diversity rather than on fluid state and homogeneity.

It is probable that viscosity changes like those which are observed in cytoplasm also exist in the nucleus following the action of mechanical or other physical, or of chemical factors. For the cytoplasm it is known that pressure, mechanical stimulation, or shaking (LEPESCHKIN 1937, KAHL 1951) bring about a decrease in viscosity. Light also affects viscosity and may be the chief factor of a diurnal rythm which is found in the viscosity of plants (VIRGIN 1953). Temperature also has a marked effect on plasm viscosity, with two maxima, at low and high temperatures, the latter caused by coagulation. Between these limits viscosity has a minimum at about normal ambient temperatures and then another smaller maximum. These phenomena have been interpreted in favour of a primary structural effect on the points of attachment of protein chains and a secondary effect on plasm lipids (FREY-WYSSLING 1938, KAHL 1951). It seems probable that in reality a part of the effects of shaking are due to thixotropy phenomena and to a separation of lipoid and protein phases; the latter could also be operative in temperature effects.

Among the numerous chemicals whose effects on plasm viscosity have been tested colchicine and c-mitotic substances have a special interest (refs. STÅLFELT 1949). Colchicine is one of the substances which more markedly affect plasm viscosity but the opinions about the direction in which the alterations take place are diverse, either of slowly increasing the viscosity after several hours (STÅLFELT 1949) or of decreasing it (BEAMS and EVANS 1940, see STÅLFELT). From what is known on the action of colchicine in bringing about spindle destruction during mitosis it would seem more probable that a decrease in viscosity should be induced by colchicine treatment and the discrepancies probably result from the materials under test, non-mitotic and mitotic cells possibly reacting in different ways, and from the region of the cell where the test is performed. This point merits further investigation. It would also be desirable that viscosity and other physical determinations be made also in the nucleus, although this generally is more distinctly structural than the cytoplasm and therefore measurements should be made only in relatively homogenous nucleoplasm regions. The influence of several chemicals on the nucleus has been studied chiefly with regard to changes of structure. Changes of protoplasm viscosity during mitosis have been imputed to the pentose nucleic acid content of the cytoplasm (CARLSON 1946) but this is based only on the similarity between the curves of viscosity and RNA content.

Related with the viscosity is the question of *elasticity and plasticity* of the nucleus, which in part depend on the properties of the nuclear membrane. As stated in previous chapters (s. p. 435 and 455, "Nuclear membrane") the membrane is formed of two layers, the inner one of elastin-like composition and properties. At the same time, the content of the nucleus generally is relatively fluid and its structure sufficiently deformable to allow a great plasticity. This is not true of some differentiated cells, in which the nucleus is in a condensed or dehydrated condition. Only in more fluid nuclei, which indeed constitute the general case, can a great elasticity and plasticity be expected; these nuclei show an almost spherical form.

The boundary of the nucleus may be disorganized and nuclei induced to fuse by means of several narcotics, lipid solvents, chloral hydrate and also by physical actions such as alternatives of compression and decompression or high temperatures (refs. MILOVIDOV 1949). However, much of this domain remains to be clarified as no distinction has in many cases been attempted between fusion of nuclei properly, which is the fusion of normal non-mitotic nuclei, and restitution of mitotic daughter nuclei fusing at anaphase or telophase. It seems that it is the latter phenomenon which is chiefly responsible in these experiments for the constitution of viable polyploid nuclei, true fusion of non-mitotic nuclei ordinarily bringing about non-viable conditions.

*Hydration* and *dehydration* is studied in plant cells generally in connexion with plasmolysis phenomena. The latter, however, concern chiefly the cytoplasm and will not be considered here. In oocyte nuclei conflicting evidence has been published (cf. DURYEE 1940 and CALLAN 1952) concerning the validity of the Hofmeister series. In oocytes swelling or hydration would be successively more efficient in the order: chloride, sulphate, acetate, citrate, thiocyanate, iodide, sulphide, hydroxide (DURYEE) but this could not be confirmed (CALLAN). It is possible that this disparity in results depends upon the state of the studied cells. Some of the anions destroy the structure while others (bicarbonate and chloride) preserve it. On the other hand, for the cations it has been found that hydration of the nuclear sap in equimolecular solutions is minimum for K salts, in the order $Cs > Rb > K < Na < Li$ (CALLAN 1952). This is the same as for the cytoplasm and must be interpreted in a similar manner (FREY-WYSSLING 1938, 1953) as caused by differences in the radius of hydration of the two ions of the alkaline chlorides, hydration layers being almost equal in KCl, while in LiCl and NaCl the cations have greater layers and the contrary happens in RbCl and CsCl. The nucleus does not present any special problem in this respect.

An interesting property of some hydrated nuclei is their sensitivity towards calcium and strontium salts. In oocytes (CALLAN 1952) dilute solutions of these salts (0.05 to 0.003 M for amphibian oocytes) coagulate the nuclear sap, causing flocculation and irreversible precipitation. More concentrated solutions (about 2 to 4 times) induce an "explosive" disorganization of the nucleus. Magnesium and barium salts also have coagulation action, which is preceded by some degree of liquefaction, and the action is not explosive. Calcium ions have also a role in increasing the refraction of salivary chromosomes in relation to the nucleoplasm and so rendering them more apparent.

Nuclear *permeability* as distinct from the permeability of the cell in general has not been much studied. The permeability of the nuclear membrane seems to depend chiefly on the properties of the inner layer since it is not visibly influenced by the presence of phosphate ions. The outer layer is disrupted by this ion at 0.2 M concentration on a wide $p_H$ range and is also partially dissolved by alcohol followed of chloroform. Copper ions hinder the latter action (CALLAN 1952). Permeability has been demonstrated in isolated oocyte nuclei for water, sugars, dextrin and polymerized nucleic acids, but not for egg albumin, bovine plasm albumin, glycogen and gum acacia. Electrolytes may easily penetrate the nuclear membrane. The membrane should have, however, a preferential permeability for anions, but the opinions are not unanimous on this point (cf. DURYEE 1940 and CALLAN 1952). The properties of isolated nuclei may be different from those of *in situ* living nuclei, whose permeability may be linked with active metabolic interchange of substances with the cytoplasm (see also in the chapter "Chemistry of the nucleus", "Nucleus-cytoplasm relations" and "Metabolism of the nucleus" and the chapter "Fine structure of the nucleus. Nuclear membrane").

## b) Physico-chemical and colloidal properties.

The $p_H$ of plant nuclei (chiefly of the nucleoplasm) generally lies in the range 4.0 to 5.8, with the more frequent values between 4.2 and 5.0. Animal cells generally have nuclei with higher $p_Hs$, from 6.4 to 7.6 (refs. MILOVIDOV 1949, Tables VI and V). $p_H$ changes of about 0.5 may be observed in the chromosomes and their immediate neighbourhood during mitosis, very probably linked with the deposition of nucleic acid and basic proteins on the chromonemata and their disintegration in telophase. In relation to the cytoplasm, generally the nucleus is somewhat more alkaline.

The *isoelectric point* (IEP) of plant nuclei corresponds to a very acidic $p_H$, generally between $p_H$ 3.0 and 4.2 (refs. MILOVIDOV 1949, Table X). Individual differences may be caused by the different composition of mitotic and non-mitotic nuclei, according to what has been said in the chapter "Chemistry of the nucleus". In agreement with these findings, in cataphorese experiments the nucleus either remains in place or goes towards the anode since the IEP of the cytoplasm generally is more alkaline, between $p_H$ 4.0 and 6.5. This difference may derive from the higher content of the nucleus in nucleic acids but may be reversed in certain states. The nucleolus has also been observed to dislocate towards the anode. On the other hand, the chromosomes and the resting nucleus do not dislocate in strong electrostatic fields, when the cells are not subjected to cataphorese (BAJER 1950), which demonstrates that the nucleus and the chromosomes are not electrically charged in normal conditions and the charge in cataphoresis is a result of the electric current itself.

These properties are of importance in vital nuclear staining, which depends on the penetration of the coloured part of the stain in the nucleus and of both physical and physico-chemical affinities. The chromatin and chromosomes show greater affinity for basic stains but may also colour for purely physical reasons of density of structure, as a kind of ultra-filtration effect (greater molecules being retained by the meshwork of the structure). The nucleolus, as a dense component also stains in part by effect of its density of structure and may take acidic stains. Methylene blue and neutral red are the most usually employed vital stains, eosin, erythrosin, gentiana violet and Bismarck brown being also frequently used for vital staining.

With respect to the *colloidal properties*, it has been alternatively admitted that the nucleus may be taken as a kind of vesicle filled of fluid with the nucleoli and some other corpuscles floating inside, or that it has a gel-like structure (refs. LEPESCHKIN 1938, MILOVIDOV 1949). According to the hypothesis of fluidity the nucleus should have a coarsely disperse part of chromatin granules, nucleoli and other granules, and a colloidally disperse part in which the micelles cannot be seen even under the ultramicroscope owing to their hydrophilous character (LEPESCHKIN 1938). Modernly it is generally admitted that the nucleus, as also the cytoplasm, is always structural and that no simple colloidal system as yet artificially obtained can reproduce its structure.

To the more hydrated kind belong the oocyte nuclei and even in these, evidence of a structure has been adduced. If these nuclei are isolated in phosphate buffer (2.3% $KH_2PO_4$ and 1.5% $K_2HPO_4$ in suitable proportions to give the desired $p_H$) of $p_H$ 6.3 to 6.8 the membrane distends but the nucleoli and chromosomes remain as in life, owing to the fact that they are held in place by a structural component which does not distend nor coagulate. At lower $p_H$ (5.8) the membrane also distends while the sap coagulates. Above $p_H$ 6.9 the membrane distends and the nucleoli and chromosomes sink to the bottom, owing to disorganization of the structural component. The reality of this structural component has been demonstrated by microdissection with needles out of the nucleus, the fragments do not immediately losing their shape (CALLAN 1952). No structure, however, can be distinguished at the ultramicroscope in both the structural and the disperse phases and the evidence for a structure in one of them is only indirect.

From these observations (see also Figs. 4–6 and 11, p. 456–459 and 469 of preceding chapter) it may be concluded that three phases at least can be distinguished in the nucleoplasm of the more hydrated kind of living nuclei: a) a microscopically visible structural phase composed of chromosome remnants, possibly also of nucleic acids and proteins not immediately composing the chromosomes and the nucleoli; b) a structural colloidal phase, responsible for maintaining the visible structures in place, whose structure actually is invisible at the ultramicroscope, probably owing to its hydrophilic character; and c) a disperse colloidal phase which is responsible for immediate hydration and dehydration in suitable $p_H$ and salt media. In nuclei of ordinary non-mitotic type in which the chromosomes as such are not visible, while in oocytes they are, a part of the chromosomes is also hydrated so as to become invisible. This phase, though in principle different from the structural colloidal phase of the nucleoplasm, may be intimately mixed with the latter so as to become almost indistinguishable from it.

In less hydrated nuclei, and especially in the nucleus of old cells from plants, a more or less gelified structure may be found, but unless pycnosis or other irreversible pathological states result, hydration may bring about a structure similar to that of ordinary, swollen, nuclei. Probably the same phases exist in the gelified nuclei as in the hydrated ones, though this may be more difficult to demonstrate experimentally.

In electron photomicrographs obtained after osmium vapour fixation (OBERLING et al. 1953) a fine meshwork containing at certain points coarser structures is visible—see Figs. 4–6 and 11 of preceding chapter. Probably these represent respectively the structural colloidal phase, and the chromosomic visible and invisible parts. The disperse phase should fill in the meshwork of the structural phases; in fixation it more or less adheres to the fibrils, lamels, and granules of the structural phases and generally has particles below the practical resolution power of the electron microscope.

## c) Structural properties.

Some of the alterations induced by physical or chemical agents on the non-mitotic nucleus direct or indirectly influence the visible structural phase and therefore are observable at the microscope. These may be called structural changes *sensu lato*, including both those of the non-mitotic and of the mitotic nucleus. The usual distinction between "structural" changes, as referred only to those affecting the chromosomes, and the other cell changes, called "physiological", has no basis since both are physiological, and cytologically visible alterations by definition are structural. All cytologically visible changes, as distinct from, though causally connected with, colloidal or molecular and atomic changes, will be treated as structural. Many of the causative agents induce when in greater doses a state of pycnosis (POLITZER 1934), which is characterized by a deep uniform taking of stains, especially basic ones, and a marked degree of stickiness. The state of pycnosis is similar to those assumed by other parts of cells or entire cells in process of being digested or phagocyted, respectively within the cell itself or by phagocytes. When exhibited by the nucleus it may be supposed to represent an attack by the cytoplasm to an unhealthy nucleus. Fundamentally pycnosis represents a "Mischung" process, a junction of phases which previously existed separately and finally the whole nuclear mass is a mixture where thymonucleic (and probably also ribonucleic) acid predominates. Several causes and different ways may lead to the same final result of pycnosis, which therefore may appear following the action of diverse physical and chemical agents.

**Radiation effects.** The cytological action of radiations has been studied chiefly with respect to the chromosomes and mitosis. However, effects on the non-mitotic nucleus may also be well apparent, especially when high doses of radiation are employed. Vacuolation and swelling of the nucleus, alterations in its staining pattern and increase of the nucleolus are among the reported changes induced by high doses (SPARROW and RUBIN 1952). Neutrons provoke in resting nuclei several phenomena (RESENDE 1951) such as hypochromaticity by rarefaction of thymonucleic acid (chromatolyse), fragmentation of the nucleus (karyoclasis), hyperchromaticity by the Feulgen reaction, pinching out of small fragments from the nucleus and complete pulverization followed by the vanishing of the nucleus. These changes are observed immediately after irradiation with neutrons which lasted for 3 hours (cf. Est. III–IV of RESENDE 1951). Fragmentation often is preceded by chromatic rarefaction. Some of the aspects of resting stage nuclei show an explosive extrusion of nuclear substance, as if a small bit of the nuclear content was forced out of the nucleus in front of the neutron beam itself (RESENDE, oral comm.). In what measure the smear method may or may not have exaggerated these aspects, remains to be seen.

It seems that this variety of aspects cannot be understood on the basis of only one primary action of neutrons. They also cannot be the remote consequence of chromosome changes during mitosis, since fixation of roots (*Allium cepa, Aloë mitriformis*) was done immediately after irradiation and the time this has lasted was not sufficient for the occurrence of so many mitoses. Many hours or days after neutron action there were yet conspicuous changes of the chromosomes and of resting stage nuclei, such as hyper- and hypochromaticity, pycnosis, fragmentation and absence of nuclei. After this time, as immediately after irradiation, some hypochromatic nuclei show areas of greater chromaticity, as if chromocentres were produced, while these are not visible in the controls. Another characteristic is the irregularity of action, adjacent cells showing very different aspects, from normal to diverse pathological appearances.

At a time it was supposed that the action of radiations on the nucleus was exerted chiefly, or at least in a great measure, by depolymerization of the desoxyribonucleic acid (DNA), which is readily accomplished *in vitro* in solutions of Na-nucleate, and this could afterwards lead to a more profound degradation of DNA in the nucleus. However, irradiation also causes protein denaturation and associated changes in viscosity and solubility so that *a priori* several components of the nucleus may be thought to be affected. On nucleic acids X-rays doses of $10^6$ r fragment the polynucleotides with liberation of inorganic phosphate and a smaller quantity of purine bases, deamination, and ring fission of purine and pyrimidine bases (SCHOLES and WEISS 1952). These changes are similar to those caused by hydrogen peroxyde in the presence of ferrous salts, which is in favour of the conclusion that the primary action of X-rays on DNA aqueous solutions is the splitting of water molecules ($H_2O \rightsquigarrow H + OH$) which produces very active free radicals OH like those of a ferrous salt-$H_2O_2$ system ($Fe^{++} + H_2O_2 \rightarrow Fe^{+++} + OH^- + OH$) (refs. LEA 1947, SCHOLES and WEISS 1952, SPARROW et al. 1952a). The presence of $O_2$ increases the effects of X-rays or $\gamma$ radiation by producing $H_2O_2$ and hydroperoxyl radical $HO_2$ (reactions: $H + O_2 \rightarrow HO_2$ and $HO_2 + H \rightarrow H_2O_2$). The remotion of H radicals by these reactions tends to increase the action of radiation since then H and OH radicals do not react with one another to give again $H_2O$ ($H + OH \rightarrow H_2O$) and more free radicals OH result available for other reactions.

The energy imparted by the radiation to a molecule may be transferred from atom to atom or to neighbouring molecules. Within the molecule it may be passed on along it until a bond requiring less energy to be broken is attained and a reaction occurs. For instance, in aminoacids irradiation specifically causes deamination. In the case of large molecules the energy rapidly distributes to all the bonds and generally no single bond is broken if bonding energy is greater than the mean energy per bond supplied by the radiation. It has been calculated that to destroy a cell no more than $10^{-6}$ to $10^{-8}$ of the molecules present in the cell need to be ionized (SPARROW et al. 1952a). This fact may be accounted for by admitting the transfer of energy between adjacent molecules and the distribution within the molecules themselves. The free radicals produced and especially the oxidizing ones are able to cause exothermic reactions which propagate the action of the radiation in a reaction chain, so that one ionization may change many molecules. When the molecules are large there is an increased probability that they absorb the energy of a photon, either directly or by transfer from adjacent molecules, but at the same time greater energies are needed to break a high number of bonds, so that as a rule they are less sensitive to radiation. Great molecules and those which strongly absorb radiation act as protective agents against radiation (refs. SPARROW et al. 1952a).

There is now no doubt that in great part the effects of radiation on cells are of the indirect type, due to water splitting and to peroxyde formation, or secondary reactions derived from these. Until some years ago the theory generally admitted was that of the direct effect on a "target", or direct "hit", according to which the effect is caused by the absorption of energy by the molecule or cell part itself which is changed (cf. also p. 483). Both effects, direct and indirect are more or less efficient according to the system. For instance, in irradiation of enzymes in aqueous solutions generally the indirect or water effect is almost as effective as the direct one, exerted on the enzyme molecules (carboxypeptidase—DALE et al. 1949). On the other hand, in solutions of tobacco mosaic virus the efficiency of the direct effect is greater than that of the indirect one by a coefficient of 4000 to 1 (SPARROW and RUBIN 1952). When inactivation of viruses is studied the "target theory" sufficiently explains the events because the indirect effect is much smaller.

Cytological effects of radiations on the resting nucleus may be direct or indirect and their interpretation, in view of the known complex action of radiations on in vitro organic systems, is not so simple as it was believed about a decade ago. Several chemicals are known to produce on the chromosomes, and the nucleus in general, effects similar to those of radiations and a linear relation between dose and effect is also found for instance in the case of the mustards. The linear dose-effect relation was one of the chief arguments in favour of the target theory. As a direct action of the mustards and other radiomimetic chemicals is not possible, their effect being produced by chemical reaction, this argument loses much of its weight and a thorough revision of the interpretation of the biological actions of radiations has become a necessity.

It seems probable that these actions depend on diverse possible mechanisms for each type of final effect such as mutation, chromosome breakage, or hypochromaticity in the non-mitotic nucleus. For instance, the simple proportionality of mutations may derive from relations between the energy of the radiation and that of the bonds which ultimately are broken in the attained macromolecules or chains (see below). The target theory has been modified so as to include the new concept of indirect action being almost as frequent or perhaps more so than the direct effect. It is not necessary that the "hit" be within the sensitive molecule or group of atoms but rather it suffices that the energy be conveyed to sensitive spots in order that the event can take place.

In regard to the interpretation of the cytologically observable changes of the resting nucleus induced by radiation, probably more than one component is involved. Hypochromaticity may be caused by depolymerization and subsequent more or less intense degradation of the DNA, or of the protein part to which it is linked, or both. The depolymerized or degraded DNA is not fixable and therefore is lost in preparations made by the usual methods immediately after irradiation. It must be noted, however, that by the methyl green staining no evidence of depolymerization has been found in meiotic nuclei (SPARROW et al. 1952b) but this may be due to the methods used or to the fact that chromosomes and not resting stage nuclei were irradiated. Hyperchromaticity, on the contrary, as pycnosis in general, corresponds to the dedifferentiation of phases, as said above, with indiscriminate mixing of DNA to the other components. Besides these processes, active transfer of the damaged compounds, especially nucleoproteins through the now permeable nuclear membrane, to the cytoplasm where they may be digested, probably may also occur as in the case of chemicals.

Fragmentation or karyoclasis of the resting nucleus may be due to the conjugation of two processes, rupture of the nuclear membrane and a sudden increase in the osmotic pressure of the nucleoplasm with an outflow of the nuclear content to the cytoplasm. When the rupture of the membrane is localized in small areas only little extrusions of the nuclear content should result, as may be seen in cases of neutron radiation (RESENDE 1951). The membrane soon recovers its form owing to it being elastic. The rise in osmotic pressure may result not only from depolymerization and further degradation of DNA, RNA (ribonucleic acid) and possibly in part also of polypeptides, as also from a process of phase demixing (phase separation) of several compounds, chiefly of phosphatids and proteins from phosphatido-proteins. Phosphatids are known to cause swelling of the protoplasm. Although open to experimental analysis, the role of these latter has as yet no observational basis. The hypothesis of DNA degradation is based on the observed hypochromaticity, which has been stated (RESENDE 1951) to often precede nuclear fragmentation. The irregularity of the action may depend upon the physiological state of the cells and especially their capacity for almost instantaneous repair of neutron damage, and perhaps also on the primary reactions of neutrons being not homogenously distributed in the cells.

Another process by which fragmentation may occur is the protrusion of a mass, many times in the form of a mushroom (LEVAN et al. 1951), which afterwards may be pinched off out of the nucleus. In these cases the nuclear membrane has not been sufficiently damaged and in part preserved its elasticity.

**Effects of chemicals.** Effects of chemicals and drugs have been studied, like those of radiations, chiefly in the mitotic nucleus, in connection with genetic changes. This aspect of cytological research has suffered, if from anything, rather from superabundance of ad hoc observations than from scarcity of them. The same does not happen with the non-mitotic nucleus, in which duely controlled and cytologically reliable observations on this point are far from abundant. Trypaflavin is one of the most active compounds capable of inducing nuclear changes (refs. BAUCH 1947, 1948, RESENDE 1951). The changes are similar to those induced by neutrons and radiations in general, with hypo- and hyper-chromaticity, fragmentation and pinching out of small fragments which generally are ring shaped. Total pulverization of the nucleus is much less frequent than in the case of neutron radiation (RESENDE 1951). The interpretation of these aspects is the same as for the action of radiations, probably with several primary actions of a chemical nature on nucleic acids, proteins and proteids, with attack to the integrity of the nuclear membrane, and phase separation followed by liber-ation of compounds which cause nuclear swelling, such as phosphatids, and cations and anions previously combined with structural components. A unitary action on nucleic acids has been postulated (LETTRÉ 1950) on the basis of nucleic acid antagonizing the action of trypaflavin, but this now seems an oversimplifi-cation. Active transfer of damaged nucleoproteins from the nucleus to the cyto-plasm might also be possible.

Mustards are among the most active radiomimetic drugs, or compounds which imitate the effects of radiations. Changes caused in the resting nucleus probably are similar to those of trypaflavin, though the mustards, to judge from the genetic and chromosomic effects, must yet be much more efficient than trypaflavin. The alterations induced by the mustards in the resting nucleus are yet to be analysed. The chemical action of these compounds on the protoplasm may be manifold. The ethylenonion formed by the mus-tards (ethylenimonion in the nitrogen mustards, ethylensulfonion in the mustard gas $R—Z\,(CH_2CH_2 \cdot Cl) \to R—^+Z—CH_2CH_2 + Cl^-$, in which R may be an hydrocarbon radical and Z is an atom of S or N) is so strongly reactive with water and a series of chemical groups which are known to exist in living matter, such as groups amino, imino, sulfhydril, phenol, organic phosphates, etc., that no primary action has as yet been singled out from such a complex. Possibly reactions with water, similar to those above mentioned for the radiations play an important role in bringing about the cytological changes induced by the mustards. Other drugs with radiomimetic effects will be referred to below (p. 486).

There are a host of other chemicals capable of inducing mutations, among them formaldehyde, $H_2O_2$, sulfamides, potassium thyocianate, chloral hydrate, ammonia (refs. SERRA 1949, chs. XX, XXI). The effects of these chemicals generally are inconstant, do not giving a positive result with all species and in all circumstances. For instance, some act by oral administration, others by injection and others necessitate of directly acting upon the sex cells themselves or their precursors. It is to be presumed that many more would exert a mutative action if they came in contact with the gametes or their mother cells. The cyto-logical action of these compounds has not been investigated, however, although it is to be presumed that they are able to cause chromosome breakage and the corresponding phenomena of hypo- and hyperchromaticity, stickiness and frag-mentation in the non-mitotic nucleus. It has been stated that extracts of bulbs of *Allium cepa* (RESENDE 1951, KECK *et al.* 1951) and of *Allium sativum* (KECK *et al.* 1951) cause several anomalies in the structure of resting and mitotic nuclei of root tips of *Allium cepa* itself. The extracts bring about in the non-mitotic nucleus not only hypochromaticity, as also chromatin loss and its diffusion to the cytoplasm, where it is observed under the form of bigger or smaller droplets and masses preserved by the fixatives. By the histochemical arginine reaction it has been demonstrated that the droplets and masses which may be seen in the immediate vicinity of the nuclear membrane when diffusion or perhaps active transfer begins, contain also basic proteins of the histone type (KECK *et al.* 1951). Therefore, it is not only DNA as also histone, probably under the form of nucleohistones, which are dissolved away from the nucleus by the effect of *Allium* aqueous extracts. It seems probable that RNA also comes out with the other compounds. The active substances of the extracts, which *inter alia* contain desoxyribonuclease and other enzymes, are yet unknown. Diallylsulphide, however, causes the same phenomena, in aqueous solution, as onion extracts and it may be one of the active compounds.

Penicillin is another drug whose cytological effects have modernly been tested (LEVAN *et al.* 1951). In non-mitotic nuclei of *Allium* roots penicillin caused the habitual changes of hypochromaticity, irregular chromaticity, extrusion of a part of the nuclear content, often in the form of a mushroom, and also nuclear fragmentation, stickiness and pycnosis. The most pronounced of these changes are followed of lethality. Another substance which has been used rather extensively to unroll chromosome helices and whose actions on the non-mitotic nucleus are also marked is potassium cyanide. Cyanides give a strong alkaline reaction in aqueous solutions and their immediate action is chiefly due to alkalinity, although the poisoning of enzymes may also be a factor to be considered in this respect. They cause swelling of the non-mitotic nucleus, followed of chromatin dissolution and stickiness of what rests undissolved. Despiralization and stickiness of the chromosomes are the other conspicuous changes (LEVAN *et al.* 1952). It seems probable that many more substances would be found to act on the structure of the nucleus if they could be forced to enter the cells.

Less drastic actions of chemicals on the structure of the nucleus, namely on the heterochromatin content, have been studied in *Impatiens balsamina* and *Sinapis alba* (HÖVERMANN 1951). Root tips excised from sterile seeds were cultured in nutrient solutions with addition of 14 natural aminoacids, glucosamin, asparagin, leucylglycine and peptose. Both species show nuclei of the euchromocentric type, with numerous spots of heterochromatin or chromocentres. Changes of the size of the nucleus, as well as of the size and chromaticity (staining by gentian violet) of the chromocentres, were recorded. In general the size of the nucleus was the same or somewhat less than the controls and the nuclear membrane was also somewhat less apparent. The size of the chromocentres, however, and their chromaticity was increased by many of the aminoacids and by the amines and peptone. The greatest effects were shown by arginine in both species and leucine, cysteine and asparagine in only one or the other. Some of the compounds also increased the stainability of the nucleolus and many also favoured a higher rate of mitosis in the tissues.

These changes are particularly interesting when compared with the small diminution or the constancy of the nuclear volume, and do not represent a generalized nuclear growth. Former observations also pointed out that improved nutrition brought about an increase in the chromaticity of the nuclei in several species (refs. HÖVERMANN 1951) and especially in *Drosera* tentacles after administration of several foods, meat, bread, etc. Age and infection also increased chromaticity. This, however, must always be compared with nuclear volume if it has to have any meaning. Despite the lack of quantitatively more exact data which should compare a possible lowering in the nuclear volume with the increase in chromaticity, it seems safe to conclude that the latter is not the result of shrinking of the chromocentres with diminution of the volume of the nucleus, since chromocentres in general concomitantly do increase in size.

It seems that the general interpretation of these structural changes is that a better nutritional level and especially the presence of aminoacids causes increase in the chromatin content, probably DNA (Feulgen reaction was not applied) and histones, in the non-mitotic nucleus. More exact work, both in the measurement of chromaticity and size, as well as in the use of histochemical tests is needed, chiefly to check current ideas about the constancy of DNA in the nucleus.

It is necessary to remark that the effects of drugs may depend very much on the species and strain under test; for instance, colchicine induces fragmentation in resting nuclei of animal cells but not in plant cells (LEVAN 1951). Conclusions, therefore, are limited to the species in which the data were obtained.

**Heat and cold.** Cold treatments (see also below p. 484: "Chromosome derangements — Heat and cold") may cause hypochromaticity and irregular chromaticity
of the non-mitotic nucleus. The changes have not been studied in detail by
modern methods. Heat treatments generally lead to similar phenomena, frequently
followed of stickiness and pycnosis. Decrease in the volume of the nucleus may
also be observed below and above the normal range of temperature (RESENDE
*et al.* 1944) which is different according to the strain. Contracted nuclei probably
derive chiefly from those affected by extreme temperatures during mitosis.

# B. Mitosis and meiosis.

In this section changes induced in the chromosomes and in other cell parts
engaged in mitosis, by physical and chemical agents, as well as theories trying
to interpret in physico-chemical terms or in these plus "structural agents" the
phenomena of mitosis and meiosis will be dealt with. The changes caused in
the chromosomes, as yet analysed, are almost only the structural ones, morphologically apparent, since the dimensions of the chromosomes do not allow the
direct study of physical and colloidal alterations within them.

## a) Chromosome derangements.

**Effects of radiations.** Radiations act on the chromosomes by producing breaks
and other morphologically apparent changes (refs. LEA 1946, works by several
authors in Symp. Chromosome Breakage 1953). Owing to the breaks having
important genetic and evolutionary consequences, the other changes have received
much less attention. However, it seems probable that, at least in some cases,
the breaks are a consequence of the so-called physiological changes, immediately
not apparent but finally attaining the bonds which maintain together the units
of the chromosome. For instance, changes initiated in the kalymma might cause
the breakage of the chromonemata, either by mechanical implication of the
stickiness of daughter chromatids, which may cause so-called matrix-bridges
and corresponding breaks when greater adhesion prevails, or by the same chemical processes attacking the chromonemata. That this is more than a mere
possibility is demonstrated: 1. by the case of the "sticky" gene in maize, which
brings about not only stickiness of the chromosomes as also genetical unstability
due to chromosome breakage, 2. by the fact that the same agents may induce
both breaks and the other changes, and 3. by the observation that breaks may
be produced in the chromosomes by previously irradiating the non-mitotic
nucleus when changes of the "physiological" and "morphological" types are
brought about, some of the potential breaks being kept as such until mechanical
stress terminates the process.

The separation between so-called structural chromosome changes, generally
taken as synonymous to changes causing breaks and rearrangements, and the
"physiological" changes has in part derived from the acceptance of the target
theory. The observed breaks, as inferred from the number of fragments and
rearrangements, are only a fraction of those effectively produced, since many
undergo restitution of the primitive chromosome sequence with fusion of the
broken ends. Indirect methods, however, allow a calculation of the number
of primary breaks from the observed rearrangements (refs. LEA 1946, SERRA
1949c). The calculated number of these breaks per roentgen is about the same
in *Drosophila* and *Tradescantia* and it has been found that more densely ionizing
radiations or particles are more effective, the increasing order of efficiency

being X-rays, α particles, neutrons. The proportion of primary breaks which undergo restitution is also different according to the type of radiation. In *Tradescantia* pollen grain mitosis it has been computed that in mean about 17 ionizations on a 0.3 μ long densely ionizing end trail of a photon or on the path of a proton (LEA 1946) are necessary to produce a break.

Another question concerns the most sensitive states, in which a maximum of breaks are produced per unit of energy. According to the species and the tissues under experiment these states have been found variously to be localized in resting stage, in prophase (early or advanced) or even in metaphase (refs. SERRA 1949, SPARROW 1951). For not too great doses, irradiation during middle mitotic or meiotic stages generally does not bring about immediate breaks; these may appear, however, in the following division (see, however, SPARROW 1951 for a contrary result in *Trillium*). The conclusion is that in many cases the existence and the degree of condensation of a kalymma of chromatin over the chromonemata may protect them against the appearing of breaks but not against the breaks themselves, which may appear in the next division. The process must be analysed in its two phases, the breakage proper and the restitution and this has been duly done only in a few cases.

The number of primary breaks produced per roentgen is constant within certain limits, that is the calculated breaks are directly proportional to dose, while the number of rearrangements is theoretically proportional to dose$^2$ and in practice a proportionality to dose$^{3/2}$ is followed owing to many breaks restituting or causing lethality (refs. LEA 1946). The distribution of the dose in time may affect the restitution process when the chromosomes are not practically quiescent, as happens in sperm heads. Temperature and other conditions, addition of metal salts, presence of $O_2$, have a greater or lesser effect according to the material under study. The simple proportionality of the primary breaks (equation in case of a one-hit curve $x = a(1 - e^{-kd})$, in which $x$ is the number of primary events, $a$ and $k$ are constants, $d$ is dose and $e$ is the base of natural logarithms) has been assumed to be a demonstration of the target theory, which postulates that the effect is produced by one "hit" of the particle or photon within a certain volume or target. However, the same final result may be obtained if several, instead of only one hit are necessary to cause the effect, provided that the processes necessitate of a mean amount of energy approximately constant per primary event. In the case of chromosome breaks direct evidence points out, as said above, that it is not one but about 17 ionisations which are utilized in bringing about breakage. The hits or the points where the dissipation of energy begins may be relatively far (at microscopic distances of about 1 micron) from the points where the break is produced, transfer of energy being accomplished by mechanisms above referred to (see "Radiation effects in the non-mitotic nucleus", p. 477). In reality the only part of target theory which has stood the test of time is the concept of the event resulting from the dissipation of energy by a charged particle or photon while the explanations of the processes by which the final result is obtained have turned out to be much more complicated, but also more flexible and typical of complex biological systems, than it was previously supposed.

Besides breaks, radiations are capable of causing other alterations in the chromosomes, among them changes in spiralization, hypochromaticity, partial loss of chromatin substance (so called erosion — see LEVAN et al. 1949) and chromatic agglutination or stickiness. Irradiation of meiosis cells (pachytene and leptotene-zygotene) may interfere with major spiral formation and induce a shortening to half or less of the normal chromosome length at anaphase I. Major helices (spirals) apparently do not develop or develop only irregularly in irradiated *Trillium* cells (SPARROW et al. 1952 b).

The interpretation of these changes is difficult. No direct effect on DNA polymerization that could be revealed by the methyl-green test has been found in these experiments. Generally DNA depolymerization, easily brought about in test tube irradiation of DNA solutions, as well as depolymerization of histones and nucleoproteins, was supposed to be the most probable cause of chromosome cytological changes (as also in the case of the non-mitotic nucleus). With the actual evidence at hand other mechanisms have been invoked to account for spiralization changes, namely release of calcium ions, or of nucleotides (Sparrow et al. 1952b). In reality, the effects of radiation may be indirect, by influencing nucleic acid synthesis, or protein synthesis, or both, and so the deposition of kalymma nucleoproteins. A simple slowing down in the tempo of the prophase processes may suffice to induce greater gelification and shortening of the chromosomes. Alternatively, the lack of kalymma may cause a lower degree of helical coiling, the chromosomes do not shorten so strongly as when the normal amount of matrix is present. Cytological figures obtained after radiation experiments remain to be analysed more thoroughly with respect to these possibilities. As the quantity of DNA per nucleus (cytophotometry — see the reserves on this method, p. 418, chapter "Chemistry of the nucleus — Methods") seems to be the same in irradiated and normal cells, it is probable that it is the duration of prophase and the gelification process of the chromosomes which are chiefly affected. Radiation may also influence chromosome division and therefore the helicization (see p. 464 of the preceding chapter).

Hypochromaticity and partial loss of chromatin substance have the same meaning as discussed in the case of the resting stage nucleus (see p. 477). Neutrons easily induce these changes, as well as stickiness, that is chromatic agglutination (Resende 1951). The latter phenomenon has been interpreted as caused by depolymerization of DNA, but the new data are in favour of another explanation. It is possible that some cases of stickiness depend upon a partial de-gelification, or a peptisation of the kalymma chromatin, brought about by any process which affects DNA, histone or the interface between nucleoplasm and chromosomes. At this interface it may exist a kind of membrane, similar to the plasma membranes in not having an internal (from the side of the chromosome) limit but perfectly distinct from the nucleoplasm. Lipids should play a role in the differentiation of this interface as in other cell membranes (Serra 1947a). It seems logical to admit that the processes which lead to stickiness may begin either in the nucleoplasm or the chromosomes and that several causes may bring about the same end effect, of lack of distinction in the boundary nucleoplasm-chromosomes. Stickiness is the most frequent anomaly induced by non-physiological treatments of the cells and frequently appears spontaneously, without known interference with the normal ambient of the plant.

**Heat and cold.** Experiments on the action of the most varied physical and chemical agents upon the course of mitosis and the appearance of the chromosomes have been performed in quite great numbers. Generally such observations cannot be interpreted unequivocally in terms of causality and frequently the explanations tried by the authors are no more than mere assumptions based on current cytological opinion, which has fluctuated rather typically in turn of some preferred focal points: electrical charges, colloidal states, proteins, DNA. It is clear, however, that in so complicated biological systems very different causes may induce the same final cytological result and the converse may also be true, the same causes giving rise to different cytological aspects according to the state of the cell, or cell part, under study.

High and low temperatures cause hypochromaticity, changes in fine structure, loss of chromatin substance and, especially low temperatures, a strong shortening of the chromosomes. Low temperature treatment has become a standard method for the obtention of metaphase plates proper for the counting of chromosomes when these are numerous and/or relatively long. It would be tempting to interpret the shortening at a colloidal level, as dehydration or deswelling of the chromosome, and especially chromatin, gel. However, temperatures at which the shortening is observed, generally in the neighbourhood of $0^0$ C. for species of the temperate zone or about $-5$ to $-10^0$ C. for cold climate species, are somewhat higher than those expected to cause dehydration of the plasm with formation of ice crystalls. This would lead to an increase in the concentrations of salts in the cell plasm and to dehydration of the chromosomes. Formation of ice crystalls (refs. HEILBRUNN 1928, BELEHRÁDEK 1935) is believed to be correlated with the death of the cell, while shortening of the chromosomes do not, necessarily. Therefore dehydration of the chromosomes should begin before ice appearing and would be a consequence of vacuole formation in the cytoplasm previous to a similar vacuolation in the chromosomes, which is produced at temperatures lower than those allowing the prosecution of mitosis.

However, other explanations of the cold-provoked shortening of the chromosomes are also possible. It seems very probable that lowering of the temperature below the optimum will bring about a slowing down of the mitosis processes, especially of half-chromosomes separation and spindle functioning, without markedly affecting the helical coiling and gelification processes. This would lead to a shortening as observed in meiosis, where a more prolonged prophase is correlated with chromosomes much shorter (by a factor of 2 to 10) than in the corresponding mitoses. Only a timing of the course of mitosis in cold experiments, which is yet to be determined, could answer this question. Probably in pratice both causes, dehydration and delay, may act, according to the species and to the conditions of the experiment. The possible role of chromosome division on helicization should also be remembered in this connection (see also p. 464 of prec. chapter).

Generalized or non-localized hypochromaticities caused by abnormal temperatures have the same explanation as in the resting nucleus (see p. 479). Cases of hypochromaticity of certain zones in certain chromosomes of the set, however, have another meaning. These belong to the above referred fine structural changes and in some cases are accompanied by loss of chromatin substance, which confers to the affected chromosome regions a lesser diameter and visible changes in the helical coiling.

Because there is relative lack of chromatic matter and therefore of DNA in these regions, this phenomenon has also been designated as nucleic acid cold-starvation of the chromosomes (DARLINGTON and LA COUR 1940). This designation implies that DNA synthesis is slowed down by the cold treatment and the affected zones do not get their full share of DNA, in competition with the other chromosome parts. This interpretation has been challenged for instance in the case of *Trillium* species (RESENDE *et al.* 1944) where the hypochromatic (and hypochromatinic) regions should correspond to nucleolar zones which are revealed by the cold treatment, temperatures of $+6^0$ C. to about $-6^0$ C. being perfectly normal to these plants of cold climates. Between $-6^0$ C. and $-7^0$ C. the chromosomes shorten down to about half their normal length, which is maintained from $-6^0$ C. to $+25^0$ C. A similar decrease in chromosome length but without differentiation of hypochromatinic zones takes place at 25 to 26$^0$ C. From $+6^0$ to $+25^0$ C. the special zones, which have a morphology similar to nucleolar zones in other plants, are covered with a coat of chromatin that does not allow their recognition. This has been related (RESENDE *et al.* 1944) to the ecology of these cold plants and to the number of nucleoli, which in *Trillium sessile* may be of 10, as many as the chromosomes. Other species of temperate climates, such as *Aloe mitriformis* and *Vicia faba* showed the shortening at low and high temperatures immediately outside the limits of normal appearing of the chromosomes (0$^0$ to 34$^0$ C. in *Vicia*, $-3^0$ to 36$^0$ C. in *Aloe*) but the secondary constriction zones (called by RESENDE olistherozones or olisthero-chromatic zones, from olisthero = labil, as their chromaticity is variable even in normal conditions) were always visible at any temperature.

From these observations and others made by independent authors it must be concluded that, according to the species or strain, the low temperatures either 1. may bring about visibility of secondary constrictions, chiefly of nucleolar ones, in the case of cold climate plants, or 2. may cause irregular hypochromaticity, a lower level of deposition or binding of chromatin (DNA plus histone and possibly RNA and other kalymma components) on the affected regions of the chromonemata, as discussed in the case of radiation induced changes. It has not yet been proved that the hypochromatinic zones in alternative 2. are localized in determined regions of the chromosomes, while in alternative 1. they should be. If this was proved also for plants of alternative 2. then special segments, more sensitive to cold treatment, should exist in the chromosome set, either by DNA (plus other kalymma components ?) "starvation" or by one or a combination of several of other possible causes, such as: lower affinity of the affected chromonema zones to bind chromatin; different degrees of polymerization of protein kalymma components, followed by partial dispersion in the nucleoplasm; lack of a chromosome boundary with loss of chromatin; failure of correspondence between the integrative and disintegrative metabolism of DNA or other chromatin components at the special regions.

Below and above the limits of temperatures causing shortening of the chromosomes, mitoses begin to show pathological marks, which commence by stickiness, irregularity of the chromosome boundary and lead to alveolation, generalized stickiness and fusion of several chromosomes and pycnosis (refs. POLITZER 1934, BELEHRÁDEK 1935).

**Chemical agents.** Numerous experiments have been made on the influence of the most varied *ad hoc* chemicals upon chromosome structure and appearance. Generally it is not possible to give a precise interpretation to such experiments, which in many cases suffer from lack of controls, due regard to the effects of osmotic changes and of $p_H$, and separation between the momentary and the long-term effects, the latter largely exerted through metabolic changes. This is the more important as it is known that hypo- and hypertonicity alone (v. MÖLLENDORF 1938) may markedly influence the course of mitosis and the appearance of the chromosomes. On the other hand, many a case of a substance being not active in mitosis or being less active than others may be due to the cell being less permeable to it, and this according to the state of resting or division (HUGHES 1952b). In general it is to be expected that in isomolecular solutions the cations and anions will show the effects already described above (see "Physical properties, hydration", p. 474). Calcium and magnesium, and especially the former, cause a marked decreas ein the elasticity of chromosomes (CALLAN 1952). When the concentration of calcium is increased, irreversible coagulation may occur, as happens with the nucleoplasm.

Several classifications have been tried on the cytological action of diverse chemicals upon the chromosomes and mitosis in general, but without much result, as in many cases a same compound is able to induce various unrelated effects and in others the same final effect is obtained by compounds of very different nature. In what respects the chromosomes, several orders of effects may be distinguished at a first analysis: 1. shortening, generally linked with dehydration and plasmolysis phenomena; 2. swelling, followed of vacuolization and dissolution, first of the chromatin and then of the chromonemata also, provoked by hydration, and the latter effects by alkaline solutions; 3. stickiness of both chromosome halves, with the hindering of chromatid separation and production of so-called matrix anaphase bridges, in many cases non distinguishable in aspect and consequences from the "true" or chromonema bridges; 4. extension of the stickiness to cause fusion between different chromosomes and this may lead to generalized pycnosis, as described in old work on the pathology of mitosis (POLITZER 1934); 5. phenomena of chromosome fragmentation, called rhexis in earlier work on mitosis pathology.

As already said, pycnotic changes, or their commencement, which are stickiness and probably changes in chromaticity, may lead to rhexis or fragmentation and both are interrelated though they may be distinguished from a morphological point of view. On the other hand, it has been assumed that cross linking between protein fibres of the chromosomes (chromonemata) could result from the action of stickiness-inducing reagents such as the mustards and diepoxides, and so true or chromonema bridges should originate concomitantly with matrix or purely chromatinic bridges (GOLDACRE et al. 1949, Ross 1950—refs. LUDFORD 1953 and also papers in Symp. Chromosome Breakage 1953). The effects on other parts of the cells engaged in mitosis concern chiefly the spindle and will be treated in the following section.

The majority of the tested chemicals (refs. LUDFORD 1953) may have actions from types 1. to 4. on the chromosomes. The number of those which may also show action of type 5. is more limited. For instance, faulty separation of the chromosomes due to stickiness, lagging of chromosomes (possibly also defficient functioning of the centromere) and chromosome bridges may be induced by hypo- and hypertonic sea water, several salt solutions, high $CO_2$ tension, neutral red, trypaflavin, atcbrine, ethylene glycol, ether, several phenols and amines, etc. (lists and refs. p. 7–8 of LUDFORD 1953 and p. 238 of LEVAN 1951). To these, several purine derivatives may also be added (BIESELE et al. 1952). Ethylene glycol, lithium salts, and iron and chrom salts would be specific inducers of stickiness (LEVAN 1951 and refs.) but this necessitates of being confirmed by tests in several species. These drugs may act more sensitively in different stages of mitosis or of meiosis, but this is also highly variable according to the strains of plants which are being tested and their conditions of culturing.

Among the chemicals with fragmentation action or rhexis of the chromosomes the mustards are the prominent ones, followed of di-epoxides, dimethyl sulphate, benzoquinone, hydroquinone, pyrogallol, several cyclic hydrocarbons, aliphatic and aromatic arsenicals, caffeine, purine derivatives, etc. (lists LEVAN 1951, p. 238 and LUDFORD 1953, p. 8; see also BIESELE et al. 1952). These are "radiomimetic" compounds, whose action is similar in some effects to that of radiations and are considered to be also general mutagenics, producers of mutations.

Attempts to correlate chemical composition of the active compounds with their cytological action, both of the type of changes in chromaticity-stickiness-pycnosis and of the type of rhexis, have failed to give an unequivocal conclusion. The hypothesis that DNA, or proteins metabolism, or both, in the chromosomes were impaired has been in vogue at a time when these were the only two classes of compounds known to be present in the chromosomes. Blockage of the SH groups was also considered one of the chief mechanisms of mitosis inhibition in general (BRACHET 1944). Now that RNA is known to be also a chromatin component and that it is presumed with factual basis that lipids play a role in the differentiation of the chromosomic limiting surface, interference with these components may also be assumed. On the other hand, several systems of enzymes are also involved and these may also be supposed to be attacked by the active compounds, either in the chromosomes or in the nucleoplasm or cytoplasm adjacent to the chromosomes when there is no nuclear membrane (periplasm or mixoplasm). Here applies what has been said on the action of mustards and other drugs on the resting nucleus (see p. 480–481).

## b) Aberrations of the spindle and other mitotic derangements.

Next to changes in the chromosomes the most important derangements of mitotic cells concern the spindle apparatus. References have already been made to the composition, structure and functioning of the spindle (see p. 441 and

470–471 of the preceding chapters), as well as to some of its derangements induced by colchicine and the interpretation of the action of this chemical. Colchicine and some of its derivatives are yet the most powerful drugs with a spindle inhibiting action (refs. LETTRÉ 1950, 1951). It was supposed that this action should be related to its group Phenyl—C—C—N but this opinion is to be revised on account of the improvement in the knowledge on the molecular constitution of colchicine, which is now known to have a tropolone structure (LOUDON, COOK and LOUDON—refs. LUDFORD 1953) instead of a phenantrene one as was previously admitted. The active group in the molecule would be perhaps Phenyl—C—C—C—N, but this necessitates confirmation. Colchicine and other inhibitors of the spindle may act by one or both of two mechanisms: 1. by selectively competing with a cell compound for a common step of a physiologically important reaction, necessary for the integrity of the spindle; or 2. by attacking some of the structural compounds of the spindle, which are peptised. Possibility 2. seems more in accordance with the known facts but the other alternative must not be overlooked when studying different types of chemicals in this respect.

Colchicine has been seen to dissolve or peptize the formed spindle and not merely to hinder the formation of a new one (INOUÉ 1952), as if the fibrils were emptied of their content. A simple question of thixotropy of long micelles does not seem to be involved in this action and it rather appears that chemical bonds are being broken, possibly by combination of one of the materials of the fibres with colchicine. As pointed out in the previous chapter ("Spindle", p. 471) it is necessary to have in mind not only the proteins as also phosphatids and perhaps RNA when attempting to interpret the action of colchicine. Fundamental data on the combination of the alkaloid with possible components of the spindle are yet lacking. It has been postulated (ÖSTERGREN 1944, SERRA 1945) that colchicine and other substances of similar action act by their liposolubility as narcotics. They should associate with lipophilic protein side chains inducing a folding of the chains so that fibrous proteins become more or less corpuscular. This mechanism has not been proved in vitro and rests a mere assumption. It would not explain how several inorganic salts (lead nitrate, for instance) and simple hypotonicity induce also spindle disorganization. If it would be proven by future work that S—S bridges play an important role in spindle formation, the action of colchicine could turn out to be chiefly on these bridges but this as yet is only speculative.

Other substances with a marked colchicine action are narcotine, α-phenylmescaline, diethyldithiocarbamate, chloral hydrate, podophyllin, iodoacetamide, aminopterin, several quinones, organic mercurials, naphtalene derivatives, allyl mustard oil, cadmium salts, etc. (table in LUDFORD 1953; see also DUSTIN 1949).

There are substances which predominantly induce arrest at metaphase without markedly destroying the spindle. The separation of the half-chromosomes is more or less hindered although in the milder cases mitosis may ultimately be completed. These chemicals act chiefly upon the chromosomes, causing some degree of stickiness and possibly hinder the normal functioning of the centromere, especially its division. In greater doses they may also attack the spindle. The chromosomes acquire the typical colchicine-mitotic (abbreviated to c-mitotic) aspect, with both chromatids united by the centromere, strongly shortened, repelling one another so that each chromosome seems a kind of cross figure. Colchicinic substances also may somewhat hinder the centromere division. To the group of arresters at metaphase without much injury to the spindle belong ether, urethane, stilboestrol, cyanides and fluorides and nickel nitrate, among others (refs. LUDFORD 1953). Trypaflavin in small concentrations belongs also here. The final explanation of the action of these chemicals is similar to that of stickiness and colchicinic substances in general.

As a general remark on the grouping of the mitotically active substances according to their type of action, it must be said that one and the same substance may have several effects, for instance first upon chromosome stickiness, then upon the spindle, according to the concentration of the employed solutions. Specificity of action like that of colchicine is exceptional.

Besides chemicals, physical agents such as the radiations and heat and cold also interact with the spindle. In many cases the spindle is yet more sensitive than the chromosomes to radiations. The primary actions in this effect may be several, as discussed for the non-mitotic nucleus and chromosomes (p. 477, 482). Spindle formation and functioning depend on a balanced system of physiological processes any one of which may be damaged by the agent. High temperatures bring about destruction of the spindle (RIS 1949). This is more easily interpreted as an action upon the lipids, especially the phosphatids of the spindle, as for protoplasm in general, with its decrease of viscosity at high temperatures, before coagulation. Other hypotheses, however, especially of enzymic action, are also possible.

### c) Mitotic movements and theories of mitosis.

Mitosis generally is a coordinate system of movements and changes of the cell leading to division into two daughter cells, in which the cytoplasm and the nucleus are concomitantly involved. However, within normal conditions of cell life, the cytoplasm may fail to divide and the chromosomes fulfil their division without the nuclear membrane disintegrating. This is endomitosis. A similar phenomenon may be experimentally provoked when the spindle is hindered from functioning but in this case usually the nuclear membrane disintegrates. A restitution nucleus with twice the original number of chromosomes then results if the chromosomes were not prevented from dividing. From old experiments it is also known that oxygen lack, low temperatures and narcotics may have this effect (LOEB, HERTWIG and HERTWIG, WILSON—refs. LUDFORD 1953). Endomitosis is dealt with below (p. 495) in somewhat greater detail.

The converse phenomenon also may be observed, although only in abnormal conditions, namely division of the cytoplasm without presence of the nucleus or with this present but not dividing. This is the case of eggs of several animals, which continue to cleave until the blastula stage after the nucleus is removed or impeded from dividing by irradiation. This and the facts mentioned in the previous section demonstrate that nuclear and cytoplasmic divisions are largely independent from one another, although the cytoplasm and the nucleus may interact, as is shown by the slowing down of chromosome division when the spindle is affected, by the peculiar appearance of chromosomes in endomitosis, and, on the other hand, by the cases of formation of small "supernumerary" spindles when one or a few chromosomes, by accident, remain outside the main mitotic figure.

In analysing physico-chemical factors that may be involved in mitosis, attention should be given therefore to: 1. conditions of spindle formation; 2. anaphasic movements and relations of the chromosomes with the spindle; 3. cleavage of the cell into two daughter protoplasts; 4. integration of nuclear and especially chromosomic phenomena with correlative cytoplasmic changes.

**Conditions of spindle formation.** These include the appearing of the spindle and the polarization. About the former enough has already been said in the two previous chapters (see p. 441 and 470) in what respects the structural aspects. The spindle normally is bipolar, with one aster or one end plate in each pole. By experimental means it is possible to induce multipolarity, as for instance by changes from hyper- to hypotonicity and vice-versa. The cytoplasm of several cells, for instance eggs without centrosomes, has the capacity of forming asters. Birefringency rapidly increases from the center of the aster through the spindle and then declines to the equator (SWANN 1952). The aster or the end plates behave like centers of coordination of fibre formation and they may form *de novo* when conditions are proper. The means by which the cell normally controls aster formation so that there are only two poles probably is a structural one, by possessing centrosomes already with a power of centers of fibre formation much greater than anything else in the cell, or by substitution of the centrosomes by other cytoplasm inclusions (mitochondria aggregates, Golgi bodies, lipo-chondria ?) when a spindle is induced to form in cells without functional centrosomes, as seems to be typical of plant cells.

The energy necessary for spindle formation may be that of glycolysis, as for other mitotic processes. It is known that substances like Victoria blue, which impede aerobic cell respiration, do not hinder the course of cell division (LETTRÉ 1951). However, they may retard the beginning of subsequent mitoses. A part of the glycolytic energy may go to movements of the cytoplasm and of the cell surface which in some cells (for instance those in tissue culture) are very apparent during division, with formation of protuberances, small mushroom protrusions, etc., like ameboid movements (LETTRÉ 1951). If the hypothesis of the transformation of SH to S—S groups (MAZIA 1953, DUSTIN 1949) in spindle fibre formation is correct, then it follows that an higher redox potential should favour the appearing of aster fibrils, while low redox would have the opposite effect, since the conversion cysteinyl $\rightarrow$ cystinyl is induced by a high niveau of oxidations in the cell. The fact that normal cell division may progress under anaerobic conditions does not favour the hypothesis. It seems, therefore, that other mechanisms of spindle formation, with phosphatids playing a role, may be compatible with anaerobic conditions. In the actual state of knowledge on the energetics of cell division it is not possible to distinguish between the energy necessary for spindle formation and for the other processes involved in mitosis, or among several of the latter individually.

From the body of data now available, the more probable conclusion which at present seems legitimate is that spindle formation is not merely a physico-chemical process or processes which, by viscosity changes, form a polarized system of fibrils and channels. There is also at this level an active organization of heterogeneity, while viscosity and birefringency changes may be accompanying processes or indexes, not necessarily the immediate causes of spindle formation. The process by which cells form the fibrils, either of the spindle, or muscular, or keratinic probably is a general one serving also for several other cell activities, osmosis, secretion, etc. (GOLDACRE 1952, DANIELLI 1953), which could be linked with phosphorylation-dephosphorylation, glycolysis and other energy-yielding processes. The morphology as well as the physico-chemistry of such processes rest to be studied. In these processes the centrosomes or the end plates may function either as centers of diffusion of active substances or as points or regions where one or several substances from gradients existing in the cell concentrate so as to attain a level proper for fibre formation. The second alternative seems more probable, as it agrees with the observation of the appearing of new centrosomes and asters in certain cells. Changes in viscosity and hydration may be operative in bringing about the necessary concentration of one or several gradients at a point, for fibre formation. The substances implied in the gradients may be enzymatic and are not necessarily produced by the chromosomes, since spindles may appear when there are no chromosomes. In certain stages of cell life, fibres easily form and orient themselves to more solid bodies such as the chromosomes, in a kind of tactism. The whole cytoplasm partakes in these activities.

**Anaphasic movements.** The anaphasic movement of chromosomes (general refs. SCHRADER 1944) has been variously interpreted as due: 1. to repulsion of the chromosomes; 2. attraction between centrosomes and chromosomes; 3. elastic pull of the chromosomes by the fibres (or equivalent structures, formed of elongated micelles and the like) of the spindle; 4. by action of elongation of the spindle as a whole and of the "Stemmkörper", formed between the two daughter groups of chromosomes in anaphase. Generally mechanism 4. is supposed to act jointly with one of the other causes, especially with 3. Mechanisms 1. and 2. should be explained by long range forces of a physical nature, since there are no chemical bonding forces which may reach microscopic distances of the order of several micra. At most a sum of very many dipole moments or forces of a similar order of magnitude, as Van der Waals forces, more or less intermediary between the chemical bonds properly and physical forces of molecular cohesion, could act at such distances.

Among the physical forces, electric and magnetic ones may be assumed *a priori* as possible. From the gross analogy between the spindle fibrils and the lines of force in a magnetic field it has been suggested by ancient cytologists (FOL, STRASBURGER, refs. WILSON 1937, MILOVIDOV 1949) and several modern observers (refs. MILOVIDOV 1949) that the chromosomes behave in anaphase like small magnets. However, magnetic fields have no action on growth or the appearance of mitoses. Positive results may be explained in other ways, as temperature and other differences (MILOVIDOV 1949 and refs.). Another way of testing this hypothesis is by analysis of the anaphasic movement of the chromosomes towards the poles of the spindle, as referred to in the following, together with the hypothesis on electrical forces.

Electrical forces constitute the other most plausible hypothesis to explain mechanisms 1. and 2. This hypothesis has been repeatedly put forward in cytological literature (refs. SCHRADER 1944) but its bases are also poor. The be-

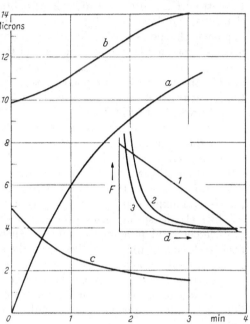

Fig. 1. Anaphasic separation of chromosomes. Average curves of chromosome separation (*a*), spindle half length (*b*), and length of one single set of chromosome fibres (*c*), according to HUGHES and SWANN [J. of Exper. Biol. 25, 45 (1948)]. *min*, time in minutes, ordinates in microns. In the insert the three types of curves expected: *1* from elastic pull, linear relationship between force $F$ and distance $d$; *2* and *3* respectively from relations of the type $F = k/d^2$ and $F = k/d^3$. These types must be compared with the length of the fibres (*c*).

haviour of chromosomes in an electrostatic field demonstrates that, at least in *Tradescantia* staminate hairs, the anaphasic chromosomes are not electrically charged and contrary data may have been the result of cataphoresis causing a charge in the chromosomes (KAMIYA 1937, BAJER 1950). On the other hand, the analysis of the anaphasic movement of each group of daughter-chromosomes, as well as of each pair of chromatids, reveals (HUGHES *et al.* 1948, BAJER *et al.* 1951) that the progression of the movement with time is not in accordance with what was to be expected from electrical repelling forces, either between two charged spheres ($F = k/d^2$, where $F$ is force, $k$ is a constant and $d$ distance), or between two charged plates ($F = k \{1 - [d/(d^2 + r^2)^{\frac{1}{2}}]\}$ with letters having the same meaning as before and $r$ the radius of the plates). The movement also does not follow the curve expected from the action of dipoles ($F = m/d^3$, in which $m$ is total charge and $d$ distance). The curves expected from electric charges or dipoles (plotting $F$ against $d$) are of types 2. and 3. of the insert in Fig. 1, while the movement of the chromosomes follows approximately a simple straight line (HUGHES and SWANN 1948, RIS 1949, BAJER 1950), with the velocity gradually decreasing as the chromosomes approach the spindle poles or, which is the same, at the measure the chromosomes become more and more separated.

This type of curve may be explained on the basis of an elastic force increasing from the poles to the equator of the spindle—type 1. of the insert in Fig. 1. These forces obeying Hooke's law are of the type $F = A - k \cdot d$, in which $F$ is force, as above, $A$ is the maximum force, in this case immediately after division of the centromeres at the very beginning of anaphase, $k$ a constant, and $d$ distance. In plotting $F$ against $d$ a straight line is obtained as observed in the chromosomes (Hughes and Swann 1948). It must be noted that when the chromosomes approach the poles of the spindle an effect of retardation is apparent, which may be explained by the non-linearity of the force at this level, or more probably by the increasing difficulties in motion as the chromosomes become packed against the centrosome or polar plates. The type of curve to be expected by this linear relation between $F$ and $d$ must be compared with the length of chromosome fibres—(c) in Fig. 1.

Another factor which, as clearly revealed in the cine photos, was contributing to the separation of the chromosomes, is the elongation of the spindle as a whole, that is separation of the spindle poles. From the published graphs (Hughes and Swann 1948) it may be guessed that about $1/4$ to $1/3$ of the anaphase separation is due to this cause and the rest to elastic pull—(b) in Fig. 1. This pull is best explained by admitting that the fibrillar structures of the spindle act like elastic springs whose force is directly proportional to stretching. The physico-chemical processes underlying such a behaviour may be several, from true stretching of fibrillar bodies when the spindle forms, followed of contraction to the primitive unstretched state, to the contraction of fibrillar structures which are in process of being folded to globular particles. It would also not be surprising if it turned out that a kind of channels in which viscous matter was flowing towards the poles of the spindle should be the responsibles for the elastic pull.

These data, obtained chiefly in chicken fibroblasts in tissue culture, may or may not be generalizable. One of the habitual failures of cytologists, rather than of cytology, is the full realization that biological processes, even the most fundamental ones at the cellular level, may be particular to a certain species or a certain type of cells. This is valid for mitotic processes in general, where several puzzling exceptions are found (Schrader 1944), such as preferential segregation, pulling of special chromosomes in meiosis "against" the spindle pole, etc. In the case of anaphase movements, however, other well analysed species have given results in agreement with those above related. In *Tradescantia* staminal hairs (Barber 1939, and especially Bajer 1950) the curves obtained were similar to those of chicken fibroblasts and in embryonic cells of Hemiptera and meiotic divisions of the grasshopper *Chortophaga* (Ris 1943, 1949) besides curves of the type of elastic pull, also evidence was found for the action of spindle elongation on the anaphase movement. The first steps in the separation, however, were due to the elastic pull only. It seems that the interpretation of elastic forces plus spindle elongation is generalizable to many species, although it is not excluded that other mechanisms may also be operative in certain cases.

Another question concerns the relations between the spindle and the chromosomes. This, however, at present is almost only in the morphological stage and therefore will not be treated here in any detail. The chromosomes may be related or attached to the fibrillar structures of the spindle by one or a few limited regions, the centromeres, in analysed cases with a chromomeric and interchromomeric structure (see preceding chapter, p. 466) or this property may be manifest in the whole chromosome, the "fibres" taking then the appearance of sheets. This type of generalized relation has been known since several years ago in Hemiptera (refs. SCHRADER 1944) and has recently been found also in plants (first in *Luzula*, afterwards in other Juncaceae, *Spirogyra*, *Scenedesmus* etc.). Probably it will be found in the majority of cases where chromosomes are small, while for long chromosomes the limited centromere will be the rule. As pointed out (SERRA 1942, 1945) the primitive case must be non-localized centromere, which shall be found particularly in Protista, the evolution in the sense of localized or limited relations (case of a centromere) being a specialization; afterwards in certain species a generalized type of relation could have been evolved again, as in Juncaceae (for other possible types of relation see scheme of LA COUR 1953).

It is possible that the relations between the centromere, or the chromosome as a whole, and the spindle root in a similarity of composition. The "pellicle" of the chromosomes, that is their boundary towards the nucleoplasm, may play the decisive role in this respect. It has been assumed that lipids, especially phospholipids, existing in the spindle and at the level of the centromere or the pellicle may be the material substrates of the relations (SERRA 1949 b). In histochemical tests the centromere region and the spindle show in fact evidence of having phospholipids (in meta- and anaphase—SERRA, unpublished).

Other compounds may also be involved in the attachment or this may be due to a type of relation in which the compounds are of a complementary or reactive nature. The pulling of the chromosomes in the anaphase movement must be due to the "shortening" of the fibrillar structures (by folding or by disintegration, etc. as said above) at the poles of the spindle rather than at the centromere, since it is observed that the attachment fibrils-chromosomes remains apparently intact in aspect when the chromosomes are moving and only later, at the beginning of telophase, do the fibrils disappear at the centromere level. This, however, may also be not generalizable. The *Luzula* type of general attachment may be due (SERRA 1945) to the matrix in these cases incorporating a certain proportion of compounds, possibly phosphatids, as those of the spindle, giving rise to a lipid rich pellicle. This as yet pure speculation is open to test by chemocytological methods.

**Theories of mitosis.** Several hypotheses have been advanced which endeavour to causally explain the onset of mitosis and its sequence. Some of the so-called theories of mitosis only try to explain the anaphasic movement of chromosomes; in reality, it seems that for their authors mitosis is no more than a separation of two half-chromosomes after metaphase, while the other phenomena, reduplication of the fundamental structures of the chromonemata, appearing of the chromosomes at the beginning of prophase, individualization of two daughter chromosomes, helical coiling, congression of the chromosomes in the equatorial plate, behaviour of the correlative alterations in the nuclear membrane, nucleolus and cytoplasm, simply are neglected. Any valid theory of mitosis must explain all these phenomena and not merely a single one.

The old concept of mitosis being a consequence of an unbalance between nucleus and cytoplasm, which when the nucleus/cytoplasm ratio had fallen short of a certain value was re-established by cell division, is now regarded as just only another way to say that each kind of cell has its own characteristic relative nuclear mass. The nucleus may grow by endomitosis without fusion, endomitosis with fusion (salivary glands of Diptera), inflation of the membrane with growth of nucleoli and stretching of the chromosomes (oocytes) and simply by increase of its volume with augmentation of the mass of nucleoplasm, nucleolus and possibly multiplication of the fundamental structures of the chromonemata. The several ways in which the nucleus answers the growth of the cytoplasm demonstrate that the behaviour of the nucleus is not the result of a change in a single variable, but rather that it results from an integration of stimuli, different according to the degree of differentiation and the state of the cell. The causality of these phenomena must be much more complex than it is found in the usual physico-chemical systems, in a manner typical of biological systems, which rarely fail to depend upon organization with its implications of space relations and time sequences. With these considerations in mind it becomes apparent that simple physico-chemical explanations should be tried only as first approximations to a valid causal analysis of mitosis.

Hydration-dehydration theories have time and again been advocated by cytologists to explain several of the mitotic phenomena. Hydration alterations are correlated with viscosity and streaming changes, which manifest in different parts of the cell during mitosis (refs. Schrader 1944). Old explanations of mitosis as a succession of gelation and peptization waves, with their correlative viscosity changes (Wassermann, see refs. Schrader 1944) have now only a historical value. The existence of preferential segregation in some plants and animals (a striking case is that of the fly *Sciara* in whose first spermatocytes the paternal chromosomes, without pairing, are pushed away from the single pole of the spindle to a bud which is cast off — Metz 1936) demonstrates that no single gradient may be responsible for chromosome movement alone. The inadequacy of the gelation-peptization hypothesis becomes even more striking when the complexities of the several other mitosis phenomena are considered.

Viscosity of the main mass of the cytoplasm decreases during prophase, while in the region of the spindle it increases, to reach a maximum in the metaphase spindle, which has a gel structure of high viscosity surrounded by fluid-like cytoplasm with low viscosity. During anaphase and telophase the cytoplasm becomes more viscous but the spindle slightly decreases in viscosity. During interphase the viscosity remains relatively high (data of several authors, refs. Swann 1952b). Other cyclic changes have also been found to be more or less clearly associated with mitosis, as variation in refractive index, in $p_H$, in temperature coefficient (Vles, Fauré-Fremiet, Ephrussi, refs. Swann 1952 and cits. Hughes 1952c). A slight change in $O_2$ uptake has also been observed in dividing eggs, respiration being at a maximum during meta- and anaphase (Zeuthen, refs. Swann 1952b). SH groups (reduced glutathione) also reach a maximum at meta- and anaphase (Brachet 1944). Another property with cyclic changes, which is at a maximum in metaphase, is cortical birefringence, while for cortical viscosity and permeability to water the contrary happens, both being at a minimum during meta- and anaphase (diagram in Swann 1952b). On differences of permeability has been founded the well known scheme of R. S. Lillie (see for instance Schrader 1944, p. 53) of electrical charges distributed in the mitotic cell in a way different from that of the resting cell. This scheme does not agree with the recent analyses of anaphasic movement. A pertinent question to be considered in all these changes is in what measure they are generalized or, as said above, correspond only to particular cases. It seems that some will be found to have a general value while others are special to each kind of cell.

On the other hand, all these variations clearly are no more than indices or accompanying changes of the mitotic transformations which are going on in the cell, and not causes of the mitosis phenomena. The complete mitosis depends upon a sequence of stimuli which may be counteracted at several points during the development of nuclear division. The appearing, shortening and visible "division" or cleavage of the chromosomes during prophase (or the cleavage in the previous ana- or telophase) has been supposed to take place in a large measure by the deposition of kalymma nucleoproteins on the chromonemata (Serra 1947b). However, other cell processes, particularly separation of phases — in which lipids in general, and particularly phospholipids, may play a role — may also be involved in these phenomena and the helical coiling probably will be explained by the helical structure of proteins and nucleic acids (see previous chapter, p. 452). The shortening will be a consequence of elementary helical coiling and division, and of a gelification process of the kalymma chromatin and then of the chromosome as a whole. Dehydration must play a role in the latter phenomenon. To these prophasic changes the, at the beginning gradual and afterwards rapid, disintegration of the nucleolus, as well as the very rapid disintegration of the nuclear membrane must be added as prophasic phenomena, probably depending on the correlated liberation of particular enzymes and on separation-of-phases.

All these phenomena, plus the congression in metaphase plate and the anaphasic movements, followed of the telophasic complex changes constitute a kind of epigenetic chain of events which may be stopped at any time by the lack of one or several of the factors—perhaps some chemical, others physical or organizational—that make possible the next step. Mitotic stimulants, for which the name "structural agents" has also been proposed, have been postulated to account for the onset of mitosis. These structural agents should be produced by the chromosomes (SWANN 1952b). However, no proof of the production of such agents and no hint on their nature are at present known. It seems probable indeed that some substances are released by the cell to induce the onset of prophase; such substances may be called mitostimulins. Recently the view that DNA could play the essential role in this as in a host of other cell phenomena has fallen in the favour of many cytologists but this is a trend that, being in the nature of an oversimplification, is likely to be substituted by more complex, and also more realistic, opinions. On the basis of $P^{32}$ uptake and fixation by acetic-alcohol it has been concluded that in bean root meristems about 8 hours pass between the synthesis of DNA and the onset of division, while the whole mitosis—interphase—mitosis cycle takes about 30 hours, of which 4 in division, 12 between the end of division and the commencement of $P^{32}$ uptake into DNA, 6 in DNA synthesis, and the remaining 8 hours in other steps preliminary to division (HOWARD and PELC 1953), perhaps in the building of other metabolites or structures necessary for division. Limitations of the experiments did not allow the determination of non fixable DNA, which is the form under which probably it exists first in the cytoplasm and afterwards in the nucleus but they do show that DNA synthesis is not immediately followed of mitosis and that DNA alone is not the immediate mitotic stimulant *par excellence*, at least in *Vicia faba*.

From this case it follows that the reality in what concerns possible mitotic stimulants is much more complex than the extreme uniters may be tempted to suppose. In animal cells steroids are among the compounds which form several of the hormones capable of inducing mitotic activity in some tissues. Hormones from the hypophysis may induce the onset of meiosis (MORICARD 1941). It is possible that in plant cells steroid components also play a role as mitotic stimulants. These stimulants, however, act only when the cell has attained a metabolic and growth state compatible with the beginning of division. This is equivalent to say that to enter into mitosis, of the endomitotic or of the spindle-mitotic types, a cell must: 1. have passed through a cycle of synthesis in which some essential components, at least the chromonema fundamental structures, DNA and basic proteins, and perhaps several other compounds have been anabolized in sufficient amounts; 2. possess a minimum amount of mitostimulins, either a) of endogenic or intracellular origin, or b) which came from the humors or the surrounding tissues. This by its relation to cancer is a subject of great importance but as yet very difficult to investigate.

**Endomitosis.** Owing to the importance it has acquired in modern morphological caryology we will refer here to this kind of chromosome division, although from the point of view of Physical Chemistry our knowledge on it is practically nil. Morphological phenomena of endomitosis have been the subject of many papers (modern reviews GEITLER 1948, 1953; WHITE 1952). Essentially it is observed that the chromosomes divide without concomitant nuclear membrane disintegration and therefore only one nucleus results, with the double, the quadruple, etc. of the original number of chromosomes, according to the endomitotic cycles the nucleus performed. This is polyploidy which, by its origin, is called endopolyploidy (GEITLER 1953 and refs.).

It is said that endopolyploidy and endomitosis may be operative in tissue differentiation but this seems to be true only in so far as differentiation is accompanyed or preceded by growth. Endomitosis is to be understood rather on the basis of nuclear growth than as serving for tissue differentiation. Nuclear growth, accompanying or not differentiation, may take several forms, according to the cells are or are not elaborating certain secretions or cytoplasmic formations and according to the state of the nucleus. So nuclear growth may take place by (see also p. 489 and 493): 1. Increase of the nucleoplasm, in part simply by hydration, and of the nucleoli, without multiplication of the chromosomes, which meanwhile are in meiotic prophase and stretch to a greater length, as in oocytes. 2. Multiplication of the fundamental structures of the chromonemata, accompanyed by a more or less parallel increase in nucleoplasm and nucleolar matter, typical of ordinary interphase nuclei engaged in a small growth cyclus or possibly in secretion and other elaborations without very marked nuclear growth, but found also in other nuclei. 3. Endomitotic processes, that is multiplication of the fundamental structures of the chromonemata with final individualization of two separated chromosomes; this may happen either with a marked charging of the structures with chromatin and a greater or lesser degree of helicization and then it is typical endomitosis; or may take place without such detectable changes in chromaticity and helicization and in this case it is atypical endomitosis, as a kind of gradation from mode 2, in which there is no endomitosis proper and only potential endopolyploidy. 4. Endomitosis followed of chromonemata multiplication, or the latter only (in many cases it is not possible to decide between these alternatives, which really is not of much relevance for an understanding of the phenomena) coupled with somatic pairing leading to fusion of homologous; this is typical of Dipteran salivary gland cells but is found also in other tissues of these larvae having big cells. True polyploidy results only from mode 3 but modes 2 and 4 give rise to potential polyploidy, which may become real if chromosomes afterwards differentiate from nuclei which underwent these modes of growth.

The common denominator to all these phenomena is nuclear growth, which is more or less synchronous to or even perhaps somewhat preceeding that of the cytoplasm. The differences apparently derive from

a) the state of meiosis or of "resting" of the nucleus,

b) the synthesis of chromatin and its relative amount, or else the binding of chromatin by the chromonemata,

c) the existence of somatic pairing.

The first alternative of a) conditions nuclear growth without chromonema multiplication, as in oocytes, while at the same time nucleoli and nucleoplasm show an active growth accompanying that of the cytoplasm and interacting with this.

Conditions b) cause when chromatin is synthesised in a suitable amount, or when it becomes bound to chromonemata in the right proportion, a typical endomitosis, while if the contrary happens only chromonema multiplication without typical endomitosis takes place. By isotopic studies of endomitosis it should be possible to distinguish between some of these alternatives and particularly if and when it is the synthesis or the binding of chromatin which is the operative factor.

Another alternative, listed under c), is operative in Diptera, causing fusion of the homologous with chromonema multiplication in this state. Although more data are necessary in this respect, it seems that in Dipteran salivaries initially there are one or two cycles of endomitosis and afterwards only chromonema multiplication without typical endomitotic phenomena.

Physico-chemical data which would allow a beginning of explanation of these differences in nuclear growth, according to the cells and their conditions, are yet entirely lacking. It could be presumed that the balance of nucleoproteins (or of a significant component of these, which may be a protein or one of the nucleic acids) between the cytoplasm and the nucleus is the chief factor operative in deciding what type of nuclear growth will be followed. If the cytoplasm was consuming a great amount of both proteins and nucleic acids, the chromonemata would not multiply, as in oocytes. If, on the other hand, nucleic acid was employed in the cytoplasm in greater proportion than proteins, chromonema multiplication could yet take place, while the appearing of typical endomitotic processes should depend on nucleic acids being available for the chromonemata. Dipteran somatic pairing would be of the nature of meiotic pairing, which is dealt with in the following section.

However, this hypothesis on the role of nucleoproteins has only a precarious basis and, as an oversimplification, seems to have only a small probability of being partially sufficient, if indeed it is necessary. Chromonema multiplication may be an autocatalytic phenomenon in which the appropriate compounds are produced by nucleus-cytoplasm interaction (see p. 415 of "Chem. of the nucleus"). Separation of daughter-chromatids from previously multiplied fundamental chromonema structures may be more complex, depending not only on nucleoproteins as also on mitostimulants, etc. The non-disintegration of the nuclear membrane may also depend on some enzymes not being activated, but nothing is known about it. For one of the chief differences between endomitotic growth and mitosis, see above (p. 489) "Conditions of spindle formation".

## d) Meiotic phenomena.

Meiosis essentially differs from ordinary mitosis in showing only a division of chromosomes with two anaphase separations. One of the separations follows the pairing of chromosomes, which constitute bivalents. It has been said that meiosis is a precocious mitosis begun before the chromosomes have divided, but in suitable materials divided chromosomes may be seen in leptotene. However, these chromosomes may look again as single in zygotene and early pachytene. There are in fact conditions in early I meiosis that cause a singleness of the chromosomes and this may hypothetically result from a relative lack of chromatin or of affinity between the chromonemata and chromatin, which only comes to individualize two chromatids in each chromosome during pachytene. What causes this lack is unknown but may be related to a relatively small but observable growth which takes place in all cells before the onset of meiosis. The influence of growth on the lack of chromatin in the chromosomes is clearly apparent in the case of the great growth period of female sexual cells.

*Pairing of the chromosomes* is one of the chief phenomena of meiosis. Its great specificity, which is shown not only by the pairing of homologous chromosomes as also of chromomeres when these are discernible, is paralleled by crystallization and other phenomena of atomic and molecular aggregation, at the physico-chemical level. However, the forces acting in crystal formation, as those of chemical bonding, are only short range ones which manifest at angström, not at micron, distances. To explain the attraction three kinds of hypotheses have been advanced: purely physical, physico-chemical and finally electrical, and structural.

A purely physical hypothesis is that of the pulsating chromomeres (FABERGÉ 1942) based on the Guyot-Bjerknes effect, of the pulsating and oscillating spheres in a fluid. A pulsating sphere periodically changes of volume with a change $V$ and with a frequency $f$ and the flow it causes tends to induce a pulsation in similar spheres at distances $a$ in the field of pulsating flow. Two pulsators will attract one another if they are pulsating with the same frequency and in the same phase and will repeal one another if, pulsating with the same frequency, they are in opposite phases (difference of phase of 180°). Two pulsators with different frequencies will in mean neither attract nor repeal one another. The force between two pulsators in the phase of attraction will be $F = \varrho \, \dfrac{E \, E'}{4 \pi a^2}$ in which $\varrho$ = density of the fluid, $E$ and $E'$ the energies of the two pulsators and $a$ distance between the centers of the spheres. $E = \sqrt{2\pi \cdot f \cdot V}$. As other values are constant, the formula may be written $F = K/a^2$; the force is inversely proportional to the square of distance and a long-range one. The resistance to the movement of each pulsator is, according to STOKES' law, $R = 6\pi \nu v r$ or $R = Cv$ as other values are constant, in which $\nu$ is viscosity of the fluid, $v$ velocity and $r$ the radius of the particle. The velocity $v$ will increase as the pulsators approach one another and $K = Ca^3/3t$, in which $t$ is the time spent in transposing the distance $a$. The total energy radiated by each pulsator is $P = cK/4\pi r^2$, in which $c$ is the velocity of propagation in water = $14.5 \times 10^4$ or $P = \pi \nu c \varrho a^3/2tr$. It is calculated that the frequencies $f$ possible for pulsators capable of existing in homologous chromosomes are of the order of $10^9$–$10^{10}$. On the other hand, supposing that the units of homology (pulsators) are at mean distances of $4\mu$ ($a = 2 \times 10^{-4}$ cm.) and that the time of pairing is 24 hours, with the viscosity of water, the total necessary energy will be $P = 0.007$ ergs for particles of radius $r = 25.10^{-7}$ and volume changes $V = 6.5 \times 10^{-17}$; $P = 7.10^{-4}$ ergs for $r = 25.10^{-6}$ and $V = 6.5 \times 10^{-14}$; and $P = 7.10^{-5}$ ergs for $r = 25.10^{-5}$ and $V = 6.5 \times 10^{-11}$. The latter two values are not impossible for cell processes and only limit the pulsators to a radius of about $0.25\,\mu$ or, taking into account the more or less cylindrical form of the chromosome, to a radius of $0.5\,\mu$. The pulsating of the units of pairing would correspond to the vibration usually found in chemical molecules.

These generally have higher frequencies but to the great complexity of the molecular aggregates which are the genes could correspond lower frequencies. The correspondence in phase should result from a resonance principle largely found in crystalls, etc. and homologous genes would easily enter into resonance while heterologous ones could only vibrate at characteristic different frequencies (FABERGÉ 1942). While this effect is a possible one it has not yet been put to a test in the timing of the pairing process and the assumptions on the vibration of the genes, as also some simplifications, have yet to be based on actual facts.

The hypothesis of dipole moments (FRIEDRICH-FREKSA 1940) is based on the assumption that nucleic acid is located only at certain points over a continuous protein structure of the chromosomes. This is true of salivary chromosomes but is dubious in meiosis. However, it may be assumed that in the latter case a greater concentration of nucleic acid at certain points (chromomeres) has the same effect, namely the appearing of dipole moments transversally to the long axis of the chromosomes. By some unknown means the dipoles must become parallel to one anothers and directed perpendicularly to the chromosome long axis. These *ad hoc* assumptions are not necessary if the deposition of chromatin with an excess of nucleic acid and — charges takes place on an almost reduplicated chromosome but with the two halves not yet separated and possessing + charges (SERRA 1947a).

The force of attraction between chromosomes would be $F = D^2/\varepsilon l r^3$ in which $D$ is the sum of all dipoles, each of a magnitude $d$, $\varepsilon$ the dielectric constant of the medium, $l$ chromosome length, $r$ the distance between chromosomes. The force $F$ will be greater in longer than in shorter chromosomes because $D^2$ increases more rapidly with length than $l$ itself. Each dipole is of the order of magnitude of $10^{-18}$ electrost.units.cm. If it is admitted that about $10^6$ dipoles exist in each chromosome with $\varepsilon$ of about 80, and $l$ of $20\,\mu = 2 \times 10^{-3}$ cm. and $r$ of $1\,\mu$, the force is $F = 10^{-24}/(80 \times 2.10^{-3} \times 10^{-12}) = 6 \times 10^{-12}$ dyn. The energy spent in an approximation of the two homologous from $1\,\mu$ to $0.1\,\mu$ would be about $10^{-14}$ erg. This force allows that a cylindrical body of $20\,\mu$ and $0.1\,\mu$ thick will move in a medium with a viscosity ranging from that of water to about 20 times this value with a velocity of about $1\,\mu$ in 60 minutes, but the velocity would be of about $1\,\mu$ in $1/_2$ minute if the force is of $5 \times 10^{-11}$ dyn. (Fig. 2), which is also possible for the great number of dipoles which may form in a chromosome.

Distances of $1\,\mu$ are supposed to be found in practice and have been derived from figures of the bouquet stage in grasshoppers (FRIEDRICH-FREKSA 1940). However, no such stage is found in all species and attraction may have to manifest itself at distances 10 to 20 times greater. Then forces of $10^{-15}$—$10^{-16}$ instead of $10^{-12}$ dyn should result and the movement would be very slow. The timing of the pairing process is not yet determined so it is not possible to compare the theoretical curves (Fig. 2) with the real movement. Another consideration, however, makes the hypothesis unlikely. The theory postulates a smaller, but nevertheless positive, attraction force between non-homologous chromosomes while in practice, for instance in salivaries, a repulsion is observed between the strictly non-homologous neighbouring chromomeres. This demonstrates that the specificity of pairing is more exacting than forces known in the physical phenomena with only mere quantitative variation. Incidentally it should be noted that the hypothesis of dipoles also tries to explain the shortening and the stretching of chromosomes by alternatively supposing that dipoles orient all to one side or to opposite sides of a chromosome. Coupled with the hypothesis of division (SERRA 1947a, Fig. 5) this would explain why after their division the chromosomes elongate and on the other hand why they shorten when double, but not yet fully divided into two separate halves—Fig. 2.

A structural hypothesis is that of flexible fibrils (SERRA 1947a) or perhaps of loops similar to those observed in oocyte chromosomes. From these chromosomes "hairs" and loops protrude which are not chromatic and differ in form, magnitude and/or sequence from one to another point of the chromosome. It suffices to admit that similar but not so easily observable lateral protrusions, flexible and which do not hinder too much the movements of the chromosomes, exist in other chromosomes in leptotene — to have a structural basis for pairing. This would result from the tendency of similar protrusions to fuse, while dissimilar ones in form and composition would impede the fusion. The fibrils and loops would extend to several micra besides the chromosomes, in all directions and therefore long-range forces would appear to be involved, which indeed were short range ones for the fusion of long and flexible structures. Lateral protrusions should be formed by interaction between the nucleoplasm and every chromosome locus and would be as specific as the genes or the genetically active regions in the chromosomes.

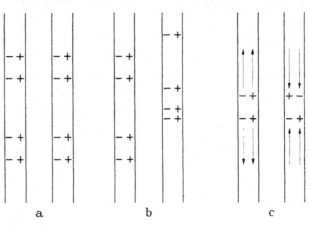

a        b        c

Although at present this is an hypothesis with a narrow factual basis it seems more likely than others of a purely physical character. Organization must yet act at this level of long range pairing and only at submicroscopic levels structure will cease to be effective. to be replaced by atomic and molecular forces. Only future research may be conclusive on this point. It must be remarked that so called somatic pairing, without chiasmata and taking place at distance, also exists and should be explained by the same theory as meiosis pairing.

Other processes typical of meiosis are crossing-over, quiasma forma-

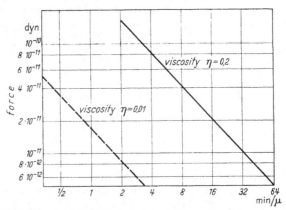

Fig. 2a–c. Above: Distribution of dipoles along chromosomes, (a) in an homologous pair, (b) in two heterologous chromosomes; in (c) elongation and shortening of the chromosomes according to the distribution of their dipoles. Below: Relation between force and velocity to be expected for a cylinder of $20\,\mu$ length and $0.1\,\mu$ radius moving $1\,\mu$ in a direction perpendicular to the axis. In abscisses the time necessary to move $1\,\mu$. The two curves are for two media, one supposed to be water, viscosity $\eta = 0.01$, and the other a fluid with 20 times this viscosity. [From FRIEDRICH-FREKSA: Naturwiss. 28, 376 (1940).]

tion, apparent repulsion of homologous from diplotene through diakinesis, and centromere functional singleness until the II division. Research on these phenomena as yet is only in a morphological state. Repulsion of the homologous may be explained by the negative charge of chromatin but this has no factual basis and remains for future studies.

## Literature[1].

ABOLINŠ, L.: A method for photoelectric comparison of the histochemical reaction of phosphomonoesterases. Acta endocrinol. (Copenh.) 9, 161–187 (1952). — ALBUQUERQUE, R. M., e J. A. SERRA: Nucleolar composition and the nucleolar zones of the chromosomes.

---

[1] Citations for the three chapters: "Chemistry of the nucleus", "Fine structure of the nucleus" and "Physical chemistry of the nucleus".

Portugal. Acta Biol. A **3**, 187–194 (1951). — ALFERT, M.: A cytochemical study of oogenesis and cleavage in the mouse. J. Cellul. a. Comp. Physiol. **36**, 381–406 (1950). — ALLFREY, V.: Amino acid incorperation by isolated thymus nuclei. I. The role of desoxyribonucleic acid in protein synthesis. Proc. Nat. Acad. Sc. U.S.A. **40**, 881–885 (1954). — ALLFREY, V., and A. E. MIRSKY: Some aspects of the desoxyribonuclease activities of animal tissues. J. Gen. Physiol. **36**, 227–241 (1952). — ALLFREY, V., H. STERN and A. E. MIRSKY: Some enzymes of isolated cell nuclei. Nature (Lond.) **169**, 128 (1952a). — ALLFREY, V., H. STERN, A. E. MIRSKY and H. SAETREN: The isolation of cell nuclei in non-aqueous media. J. Gen. Physiol. **35**, 529–554 (1952b). — ARNDT, U. W., and D. P. RILEY: Structure of air-dry sodium thymonucleate. Nature (Lond.) **172**, 803–804 (1953). — ASCHOFF, L., E. KÜSTER u. W. J. SCHMIDT: Protoplasma-Monogr. **17** (1938). — ASTBURY, W. T., and F. O. BELL: X-ray study of thymonucleic acid. Nature (Lond.) **141**, 747–748 (1938). — ATKINSON, W. B.: Differentiation of nucleic acids and acid mucopolysaccharides in histologic reactions by selective extraction with acids. Science (Lancaster, Pa.) **116**, 303–305 (1952).

BAJER, A.: Electrical forces in mitosis-I. Acta Soc. Bot. Polon. **20**, 709–737 (1950). — BAJER, A., and A. Z. HRYNKIEWICS: Notes on the anaphase mechanism and the energy of chromosome movement. Acta Soc. Bot. Polon. **21**, 5–15 (1951). — BARBER, H. N.: The rate of movement of chromosomes on the spindle. Chromosoma **1**, 33–50 (1939). — BARER, R.: Determination of dry mass, thickness, solid and water concentration in living cells. Nature (Lond.) **172**, 1097–1098 (1953). — BARRNETT, R. J., and A. M. SELIGMAN: Histochemical demonstration of protein-bound sulfhydril groups. Science (Lancaster, Pa.) **116**, 323–327 (1952). — BAUCH, R.: Trypaflavin als Typus der Chromosomengifte. Naturwiss. **34**, 346 to 347 (1947). — Irreversible Chromosomenschädigungen durch Trypaflavin. Planta (Berl.) **35**, 536–554 (1948). — BEHRENS, M.: Untersuchungen an isolierten Zell- und Gewebsbestandteilen. Abt. I. Isolierung von Zellkernen des Kalbherzmuskels. Z. physiol. Chem. **209**, 59–74 (1932). — Zell- und Gewebetrennung. In E. ABDERHALDENS Handbuch der biologischen Arbeitsmethoden. Abt. V, Teil 10, S. 1387–1389. Berlin: Urban & Schwarzenberg 1938. — BELEHRÁDEK, J.: Temperature and living matter. Protoplasma-Monogr. **8** (1935). — BENDICH, A.: Studies in the metabolism of the nucleic acids. Exper. Cell Res. Suppl. **2**, 181–197 (1952). — BERNHARD, W., F. HAGUENAU and CH. OBERLING: L'ultra-structure du nucléole de quelques cellules animales, révélée par le microscope électronique. Experientia (Basel) **8**, 58 (1952). — BIESELE, J. J., R. E. BERGER, M. CLARKE and L. WEISS: Effects of purines and other chemotherapeutic agents on nuclear structure and function. Exper. Cell Res. Suppl. **2**, 279–300 (1952). — BINKLEY, F.: Evidence for the polynucleotide nature of cysteinylglycinase. Exper. Cell Res. Suppl. **2**, 145–157 (1952). — BISSET, K. A.: The cytology and life-history of Bacteria. Edinburgh: Livingstone 1950. — Genetic mechanisms in bacteria and bacterial viruses. II. Cold Spring Harbor. Symp. Quant. Biol. **16**, 373–379 (1951). — Bacterial cytology. Internat. Rev. Cytol. **1**, 93–106 (1952). — BLASCHKO, H., W. JACOBSON and F. K. SANDERS: Enzyme systems of cells. In Cytology and Cell Physiology, edit. G. H. BOURNE, p. 322–372. Oxford: Clarendon Press 1952. — BOIVIN, A., R. VENDRELY and C. VENDRELY: La spécificité des acides nucléiques chez les êtres vivants. Coloque internat. Rockefeller—C.N.R.S. sur les unités biologiques douées de continuité génetique. Centre Nat. Rech. Sc., Paris, p. 67–78, 1949. — BRACHET, J.: La localization de l'acide thymonucléique pendent l'oogenèse et la maturation chez les Amphibiens. Archives de Biol. **51**, 151 to 165 (1940). — Embryologie chimique. Paris: Masson & Cie. 1944. — Le rôle du noyau cellulaire dans les oxydations et les phosphorylations. Biochem. et Biophysica Acta **9**, 221–222 (1952a). — Quelques effects des inhibiteurs des phosphorylations oxydatives sur les fragments nucléés et énucléés d'organismes unicellulaires. Experientia (Basel) **8**, 347 (1952b). — The role of the nucleus and the cytoplasm in synthesis and morphogenesis. Symposia Soc. Exper. Biol. **6**, 173–200 (1952c). — The use of basic dyes and ribonuclease for the cytochemical detection of ribonucleic acid. Quart. J. Microsc. Sci. **94**, 1–10 (1953). — BRACHET, J., and H. CHANTRENNE: Protein synthesis in nucleated and non-nucleated halves of *Acetabularia mediterranea* studied with carbon-14-dioxide. Nature (Lond.) **168**, 950 (1951). — Incorporation de C¹⁴O₂ dans les protéines des chloroplasts et des microsomes de fragments nucléés et anucléés d'"Acetabularia mediterranea". Arch. internat. Physiol. **60**, 547–549 (1952). — BRATTGÄRD, S.-O., and H. HYDÉN: Mass, lipids, pentose nucleoproteins and proteins determined in nerve cells by X-ray microradiography. Acta radiol. (Stockh.) Suppl. **94**, 48 p. (1952). — BREDERECK, H.: Nucleinsäure. Fortschr. Chem. org. Naturstoffe **1**, 23 (1938). — BRETSCHNEIDER, L. H.: Electron-microscopic investigation of tissue sections. Internat. Rev. Cytol. **1**, 305–322 (1952a). — The fine structure of protoplasm. Surv. of Biol. Progr. **2**, 223–256 (1952b). — BRINGMANN, G.: Vergleichende licht- und elektronenmikroskopische Untersuchungen an Oszillatorien. Planta (Berl.) **38**, 541–563 (1950). — Über Beziehungen der Kernäquivalente von Schizophyten zu den Mitochondrien höher organisierter Zellen. Planta (Berl.) **40**, 398–406 (1952). — BRUNTON, T. L.: J. Anat. a. Physiol., Ser. II **3**, 91 (1870). Zit. A. HUGHES 1952. — BRYAN, J. H. D.: DNA-protein relations during microsporogenesis of Tradescantia. Chromosoma 4, 369–392 (1951). — BUCHHOLZ, J.T.: Chromosome structure under the electron microscope. Science (Lancaster, Pa.) 105, 607–610 (1947). — BUCK, J. B.: Micromanipulation of salivary gland chromosomes. J. Hered. 33, 3–10 (1942). — BUCK, J. B., and R. D. BOCHE: Some properties of living chromosomes. Collecting Net. 13, 2 p. (1938).

CALLAN, H. G.: Some physical properties of the nuclear membrane. Proc. 6. Internat. Congr. Exper. Cit. 1947. Exper. Cell Res. Suppl. 1, 48 (1949). — A general account of experimental work on amphibian oocyte nuclei. Symposia Soc. Exper. Biol. 6, 243–255 (1952). — CARLSON, J. G.: Protoplasmic viscosity changes in different regions of the grasshopper neuroblast during mitosis. Biol. Bull. 90, 109–121 (1946). — CARLSON, J. G., and R. D. McMASTER: Nucleolar changes induced in the grasshopper neuroblast by different wavelengths of ultraviolet. Exper. Cell Res. 2, 434–444 (1950). — CASPERSSON, T.: Über den chemischen Aufbau der Strukturen des Zellkerns. Skand. Arch. Physiol. (Berl. u. Lpz.) Suppl. 73, 151 p. (1936). — Methods for the determination of the absorption spectra of the cell structures. J. Roy. Microsc. Soc. 60, 8–25 (1940a). — Die Eiweißverteilung in den Strukturen des Zellkerns. Chromosoma 1, 562–604 (1940b). — Nucleinsäurekette und Genvermehrung. Chromosoma 1, 605–619 (1940c). — Studien über den Eiweißumsatz der Zelle. Naturwiss. 29, 33–43 (1941). — Cell growth and cell function. New York: Norton (1950). — CASPERSSON, T., and J. SCHULTZ: Ribonucleic acids in both nucleus and cytoplasm, and the function of the nucleolus. Proc. Nat. Acad. Sci. U.S.A. 26, 507–515 (1940). — CHAMBERS, R.: The physical structure of protoplasm as determined by microdissection and injection. In E. V. COWDRY (Editor) General cytology. Chicago: Univ. Chicago Press 1921. — CHANTRENNE-VAN HALTEREN, M. B., et J. BRACHET: La respiration de fragments nucléés et énucléés d'Acetabularia mediterranea. Arch. internat. Physiol. 60, 187 (1952). — CHARGAFF, E.: Recent work on lipoproteins as cellular components. Exper. Cell Res. Suppl. 1, 24–31 (1949). — CHARGAFF, E., C. F. CRAMPTON and R. LIPSHITZ: Separation of calf thymus deoxyribonucleic acid into fractions of different composition. Nature (Lond.). 172, 289–292 (1953). — CHAYEN, J.: The structure of root meristem cells of Vicia faba. Symposia Soc. Exper. Biol. 6, 291–305 (1952). — Ascorbic acid and its intracellular localisation, with special reference to plants. Internat. Rev. Cytology 2, 77 (1953). — CHAYEN, J., and K. P. NORRIS: Cytoplasmic localization of nucleic acids in plant cells. Nature (Lond.) 171, 472–473 (1953). CLAUDE, A., and J. S. POTTER: Isolation of chromatin threads from the resting nucleus of leukemic cells. J. of Exper. Med. 77, 345–354 (1943).

DALE, W. M., L. H. GRAY and W. J. MEREDITH: The inactivation of an enzyme (carboxypeptidase) by X and gamma radiation. Philosophic. Trans. Roy. Soc. Lond., Ser. A 242, 33–62 (1949). — DALY, M., V. G. ALLFREY and A. E. MIRSKY: Uptake of glycine-N$^{15}$ by components of cell nuclei. J. Gen. Physiol. 36, 173–179 (1952). — DALY, M., A. E. MIRSKY and H. RIS: The amino acid composition and some properties of histones. J. Gen. Physiol. 34, 439–450 (1951). — DANIELLI, J. F.: Cytochemistry, a critical approach. New York: John Willey & Sons, London: Chapman & Hall 1953. — DARLINGTON, C. D., and L. LA COUR: Nucleic acid starvation of chromosomes in Trillium. J. Genet. 40, 185–213 (1940). — DAVIDSON, J. N.: The biochemistry of the nucleic acids. Methuen's monographs on biochemical subjects. London a. New York 1950. — DEKKER, C. A., and H. K. SCHACHMAN: On the molecular structure of deoxyribonucleic acid: an interrupted two-strand model. Proc. Nat. Acad. Sci. U.S.A. 40, 894–909 (1954). — DE LAMATER, E. D.: A new cytological basis for bacterial genetics. Cold Spring Harbor Symp. Quant. Biol. 16, 381–412 (1951). — DE LAMATER, E. D., M. E. HUNTER and S. MUDD: Current status of the bacterial nucleus, in Chemistry and physiology of the nucleus. Exper. Cell Res. Suppl. 2, 323–345 (1952). — DELAY, C.: Recherches sur la structure des noyaux quiescents chez les phanérogames. Thèse. Rev. Cytol. et Cytophys. végét. 9, 169–223; 10, 103–229 (1949). — DI STEFANO, H. S.: A cytochemical study of the Feulgen nucleal reaction. Chromosoma 3, 282–301. — Proc. Nat. Acad. Sci. U.S.A. 34, 75–80 (1948). — DI STEFANO, H. S., A. D. BASS, H. F. DIERMEIER and J. TEPPERMAN: Nucleic acid patterns in rat liver following hypophysectomy and growth hormone administration. Endocrinology 51, 386–393 (1952). — DODSON, E. O.: A morphological and biochemical study of lumpbrush chromosomes of vertebrates. Univ. California Publ. Zool. 53, 281–314 (1948). — DOUNCE, A. L.: Enzyme studies on isolated cell nuclei of rat liver. J. of Biol. Chem. 147, 685–698 (1943a). — Further studies on isolated cell nuclei of normal rat liver. J. of Biol. Chem. 151, 221–233 (1943b). — The enzymes. Vol. I. Edit. by J. B. SUMMER and K. MYRBÄCK. New York: Academic Press 1950. — The enzymes of isolated nuclei. The chemistry and physiology of the nucleus. Exper. Cell Res. Suppl. 2, 103–119 (1952). — DOUNCE, A. L., and T. H. LAN: Isolation and properties of chicken erythrocyte nuclei. Science (Lancaster, Pa.) 97, 584–585 (1943). — DRAWERT, H.: Zellmorphologische und zellphysiologische Studien an Cyanophyceen. I. Planta (Berl.) 37, 161–182 (1949). — DURYEE, W. R.: Isolation of nuclei and non-mitotic chromosome pairs from frog eggs. Arch. exper. Zellforsch. 19, 171–176 (1937). — The chromosomes of the amphibian nucleus. In: The Univ. of Pennsylvania Bicentennial Confer. on Cytology, Genetics and Evolution, Philadelphia, p. 129 to 141, 1941. — Dustin jr., P.: Mitotic poisoning at metaphase and —SH proteins. Proc. 6. Internat. Congr. Exper. Cytol., Exper. Cell Res. Suppl. 1, 153–155 (1949).

Edström, J. E.: Ribonucleic acid mass and concentration in individual nerve cells. A new method for quantitative determinations. Biochem. et Biophysica Acta 12 361–386 (1953). — Engbring, V. K., and M. Laskowski: Protein components of chicken erythrocyte nuclei. Biochem. et Biophysica Acta 11, 244–251 (1953). — Engström, A.: Quantitative micro- and cytochemical elementary analysis by roentgen absorption spectrography. Acta radiol. (Stockh.) Suppl. 63, 106 (1946). — Engström, A., u. B. Lindström: Histochemical analysis by X-rays of long wavelengths. Experientia (Basel) 3, 191 (1947). — A new method for determining the weight of cellular structures. Nature (Lond.) 163, 563 (1949). — A method for the determination of the mass of extremely small biological objects. Biochem. et Biophysica Acta 4, 351 (1950). — Ephrussi-Taylor, H.: Genetic aspects of transformation of Pneumococci. Cold Spring Harbor Symp. Quant. Biol. 16, 445–456 (1951). — Estable, C.: Estructura del nucléolo y algunas experiencias tendentes a demonstrar su significacion biologica. Congr. Médico del Centenario, Montevideo 9, 558–566 (1930). — Estable, C., and J. R. Sotelo: Una nueva estructura celular: el nucleolonema. Publ. Inst. Inv. Cienc. Biol. Montevideo 1, 105–126 (1951).

Fabergé, A. C.: Homologous chromosome pairing. The physical problem. J. Genet. 43, 121–144 (1942). — Fautrez, J., and N. Fautrez-Firlefyn: Desoxyribonucleic acid content of the cell nucleus and mitosis. Nature (Lond.) 172, 119–120 (1953). — Fautrez-Firlefyn, N., and J. Fautrez: Intranuclear proteins of the oocytes in Artemia salina L. Nature (Lond.) 172, 163–164 (1953). — Fell, H. B., and A. F. Hughes: Mitosis in the mouse: a study of living and fixed cells. Quart. J. Microsc. Sci. 90, 355–380 (1949). — Franklin, R. E., and R. G. Gosling: Molecular configuration in sodium thymonucleate. Nature (Lond.) 171, 740–741 (1953a). — Evidence for 2-chain helix in crystalline structure of sodium desoxyribonucleate. Nature (Lond.) 172, 156–157 (1953b). — Fredricsson, B.: A modification of the histochemical method for demonstration of alkaline phosphatase in which the non-specific reactions and the diffusion phenomena are reduced. Anat. Anz. 99, 97–106 (1952). — Frey-Wissling, A.: Submikroskopische Morphologie. Protoplasma-Monogr. 15 (1938). — Submicroscopic morphology of protoplasma and its derivatives, 2nd ed. New York: Elsevier Publ. Co. 1953. — Friedrich-Freksa, H.: Bei der Chromosomenkonjugation wirksame Kräfte usw. Naturwiss. 28, 376–379 (1940).

Geitler, L.: Chromosomenbau. Protoplasma-Monogr. 14 (1938). — Ergebnisse und Probleme der Endomitoseforschung. Österr. bot. Z. 95, 277–299 (1948). — Endomitose und endomitotische Polyploidisierung, in Protoplasmatologia, Handb. d. Protoplasmaforsch. Wien: Springer 1953. — Glick, D.: Techniques of histo- and cytochemistry. New York a. London: Interscience Publ. 1949. — Glick, D., A. Engström and B. G. Malmström: A critical evaluation of quantitative histo- and cytochemical microscopic techniques. Science (Lancaster, Pa.) 114, 253 (1951). — Goldacre, R. J.: The folding and unfolding of protein molecules as a basis of osmotic work. Internat. Rev. Cytol. 1, 135–164 (1952). — Gomori, G.: Pitfalls in histochemistry. Ann. New York Acad. Sci. 50, 968–981 (1950). — Gomori, G., and R. D. Chessick: Histochemical studies of the inhibition of esterases. J. Cellul. a. Comp. Physiol. 41, 51–62 (1953). — Greenberg (edit.) u. Mitarb.: Amino acids and proteins. Springfield, Ill.: Ch. C. Thomas 1951. — Gulland, J. M.: The structures of nucleic acids, in Nucleic acids. I. Symposia Soc. Exper. Biol. 1, 1–14 (1947).

Hämmerling, J.: Über formbildende Substanzen bei Acetabularia mediterranea. Roux' Arch. 131, 424–462 (1934). — Haurowitz, F., and C. F. Crampton: The role of the nucleus in protein synthesis, in The chemistry and physiology of the nucleus. Exper. Cell Res. Suppl. 2, 45–54 (1952). — Heilbrunn, L. V.: The colloid chemistry of protoplasm. Protoplasma-Monogr. 1 (1928). — Heitz, E.: Die Ursache der gesetzmäßigen Lage, Form und Größe pflanzlicher Nucleolen. Planta (Berl.) 12, 775 (1931). — Die Herkunft der Chromozentren. Planta (Berl.) 18, 571–639 (1932). — Über α- und β-Heterochromatin sowie Konstanz und Bau der Chromosomen bei Drosophila. Biol. Zbl. 54, 588–609 (1934). — Herman, E., and W. Deane: A comparison of the localization of alkaline glycerophosphatase, as demonstrated by the Gomori-Takamatsu method in frozen and in paraffin sections. J. Cellul. a. Comp. Physiol. 41, 201–223 (1953). — Hershey, A. D.: Functional differentiation within particles of bacteriophage $T_2$. Cold Spring Harbor Symp. Quant. Biol. 18, 135–139 (1953). — Höber, R., D. I. Hitchcock, J. B. Bateman, D. R. Goddard and W. O. Fenn: Physical chemistry of cells and tissues. Philadelphia: Blakiston Son & Co. 1945. — Hoerr, N. L.: Methods of isolation of morphological constituents of the liver cell. Biol, Symp. 10, 185–231 (1943). — Hoevermann, I.: Über Strukturveränderungen an Chromozentrenkernen unter dem Einfluß von Aminosäuren und ähnlichen Substanzen. Planta (Berl.) 39, 480–499 (1951). — Hoff-Jörgensen, E., and E. Zeuthen: Evidence of cytoplasmic desoxyribonucleosides in frog's eggs. Nature (Lond.) **169**, 245 (1952). — HORNING, E. S.: Microincineration and the inorganic constituents of the nucleus, in Cytology and cell physiology, p. 287–321. Oxford: Clarendon Press 1952. — HOTCHKISS, R. D.: Transfer of penicillin resistance in pneumococci by the desoxyribonucleate derived from resistant cultures. Cold Spring Harbor Symp. Quant. Biol. **16**, 457–461 (1951). — The biological nature of the bacterial transforming factors, in Chemistry and physiology of the nucleus. Exper. Cell Res., Suppl. **2**, 383–390 (1952). — HOWANITZ, W.: Chromosome structure. I. Analysis of spiral or nodule fragmentations. Wasman J. Biol. **11**, 1–22 (1953). — HOWARD, A., and S. R. PELC: Synthesis of desoxyribonucleic acid in normal and irradiated cells etc. Symp. Chromosome Breakage, Suppl. to Heredity, Vol. 6, p. 261–276. London: Oliver and Boyd 1953. — HUGHES, A. F.: Some historical features in cell biology. Internat. Rev. Cytol. **1**, 1–7 (1952a). — Inhibitors and mitotic physiology. Symposia Soc. Exper. Biol. **6**, 256–264 (1952b). — The mitotic cycle. London: Butterworth 1952c. — HUGHES, A. F., and M. M. SWANN: Anaphase movements in the living cell. A study with phase contrast and polarized light on chick tissue cultures. J. of Exper. Biol. **25**, 45–70 (1948). — HULTIN, T., and R. HERNE: Amino acid analysis of a basic protein fraction from sperm nuclei of some different Invertebrates. Ark. Kemi (Stockh.), Ser. A **26**, 1–8 (1948). — HYDÉN, H.: Protein metabolism in the nerve cell during growth and function. Acta physiol. scand. (Stockh.) **6**, Suppl. 17, 136 p. (1943).

INOUÉ, S.: The effect of colchicine on the microscopic and submicroscopic structure of the mitotic spindle, in Chemistry and physiology of the nucleus. Exper. Cell Res. Suppl. **2**, 305–314 (1952).

JACOBSON, W., and M. WEBB: The two types of nucleoproteins during mitosis. Exper. Cell Res. **3**, 163–183 (1951). — JONES, W.: Nucleic acids, their chemical properties and physiological conduct. Monographs on biochemistry. New York a. London: Longman Greens 1920.

KAHL, H.: Über den Einfluß von Schüttelbewegungen auf Struktur und Funktion des pflanzlichen Plasmas. Planta (Berl.) **39**, 346–376 (1951). — KAMIYA, N.: Untersuchungen über die Wirkung des elektrischen Stromes auf lebende Zellen. I. Cytologia, Fujii **2**, 1036 to 1042 (1937). — KAUFMANN, B. P.: An evaluation of the applicability in cytochemical studies of methods involving enzymatic hydrolysis of cellular materials. Portugal. Acta Biol. A Goldschmidt-Vol., 813–830 (1950). — Cytochemical studies of the action of trypsin. III. The course of deformation of salivary-gland chromosomes. Exper. Cell Res. **4**, 408–425 (1953). — KIESEL, A.: Chemie des Protoplasmas. Protoplasma-Monogr. **4** (1930). — KOSSEL, A.: The protamines and histones. Monographs on biochemistry. New York a. London: Longmans 1928. — Protamine und Histone. Einzeldarstellungen aus dem Gesamtgebiet der Biochemie, Bd. 2. Berlin: Franz Deuticke 1929. — KOZTOFF, L. M.: The fate of the infecting virus particle, in Chemistry and physiology of the nucleus. Exper. Cell Res. Suppl. **2**, 367–383 (1952).

LA COUR, L. F.: The Luzula system analysed by X-rays. Symp. Chromosome Breakage. Suppl. Heredity (Lond.) **6**, 77–81 (1953). — LASAROW, A.: The chemical structure of cytoplasm as investigated in Bensley's laboratory during the past ten years. Biol. Symp. **10**, 9–26 (1943). — LEA, D. E.: Actions of radiations on living cells. Cambridge: University Press 1946. — LEPESCHKIN, W. W.: Kolloidchemie des Protoplasmas, 2. Aufl. Dresden: Theodor Steinkopff 1938. — LETTRÉ, H.: Über Mitosegifte. Erg. Physiol. **46**, 379–396 (1950). — Zellstoffwechsel und Zellteilung. Naturwiss. **38**, 490–496 (1951). — LEVAN, A.: Chemically induced chromosome reactions in Allium cepa and Vicia faba. Cold Spring Harbor Symp. Quant. Biol. **16**, 233–243 (1951). — LEVAN, A., and T. LOFTY: Naphtalene acetic acid in the Allium test. Hereditas (Lund) **35**, 337–374 (1949). — LEVAN, A., and J. H. TJIO: Penicillin in the Allium test. Hereditas (Lund) **37**, 306–324 (1951). — LEVAN, A., and K. H. FRH. v. WANGENHEIM: Potassium cyanide in the Allium test. Hereditas (Lund) **38**, 298–313 (1952). — LEVENE, P. A., and L. W. BASS: Nucleic acids. New York: Chemical Catal. Co. 1931. — LIMA DE FARIA, A.: The structure of the centromere of the chromosomes of rye. Hereditas (Lund) **35**, 77–85 (1949a). — Genetics, origin and evolution of kinetochores. Hereditas (Lund) **35**, 424–444 (1949b). — The Feulgen test applied to centromeric chromosomes. Hereditas (Lund) **36**, 60–74 (1950). — LINDEGREN, C. C.: The structure of the yeast cell. Symposia Soc. Exper. Biol. **6**, 277–289 (1952). — Gene conversion in Saccharomyces. J. Genet. **51**, 625–637 (1953). — LISON, L.: Histochimie animale. Paris: Gauthier-Villars 1936. — LISON, L., et J. PASTEELS: Études histophotométriques sur la teneur en acide desoxyribonucléique des noyaux au cours du développement embryonnaire chez l'oursin Paracentrotus lividus. Archives de Biol. **62**, 2–44 (1951). — LLOYD, D. J., and A. SHORE: Chemistry of the proteins. London: Churchill 1938. — LUDFORD, R. J.: Chemically induced derangements of cell division. J. Roy. Microsc. Soc. **73**, 1–23 (1953).

MALHEIROS, N., D. DE CASTRO and A. CÂMARA: Cromosomas sem centrómero localizado. O caso da *Luzula purpurea* Link. Agronomia Lusitana **9**, 51–71 (1947).—MARKHAM, R.: Chemistry of some functional components of viruses. Cold Spring Harbor Symp. Quant. Biol. **18**, 141–148

(1953). — Martens, M. P.: La structure vitale du noyau. Observation vitale de la caryokinèse. C. r. Acad. Sci. Paris 184, 615, 728 (1927). — Mayer, D. T., and A. Gullick: The nature of the proteins of cellular nuclei. J. of Biol. Chem. 146, 433–440 (1942). — Mazia, D.: Cell division. Scientific American. 189, No 2, p. 53–63, Aug. 1953. — Metz, C. W.: Factors influencing chromosome movements in mitosis. Cytologia, Fujii 7, 219–231 (1936). — Miescher, F.: Über die chemische Zusammensetzung der Eiterzellen. In F. Hoppe-Seyler, Medizinisch-chemische Untersuchungen, Bd. 4, S. 441. Berlin 1871. Sowie in W. Jones, Nucleic acids (2nd ed.), p. 2. New York a. London: Longmans Green 1920. — Milovidov, P. F.: Physik und Chemie des Zellkernes. Protoplasma-Monogr. 20 (1949). — Mirsky, A. E., and A. W. Pollister: Chromosin, a desoxyribose nucleoprotein complex of the cell nucleus. J. Gen. Physiol. 30, 117–148 (1946). — Mirsky, A. E., and H. Ris: Isolated chromosomes. The chemical composition of isolated chromosomes. J. Gen. Physiol. 31, 1–6, 7–18 (1947). — Möllendorf, W. v.: Zur Kenntnis der Mitose. IV. Der Einfluß von Hypo- und Hypertonie auf den Ablauf der Mitose sowie auf den Wachstumsrhythmus von Gewebekulturen. Z. Zellforsch. 28, 512–546 (1938). — Molè-Bajer, J.: Influence of hydration and dehydration on mitosis. I. a. II. Acta Soc. Bot. Polon. 21, 73–94 (1951); 22, 33–44 (1953). — Monné, L.: Double refraction of the nuclear membrane. Ark. Zool., Ser. B 34, Nr 2 (1942). — Struktur- und Funktionszusammenhang des Cytoplasmas. Experientia (Basel) 2, 22 (1946). — On the induced formation of chromosome-like structures within the cytoplasm of mature sea urchin eggs. Ark. Zool., Ser. A 42, Nr 44 (1949). — Moog, F.: Histochemistry. Surv. of Biol. Progr. 2, 197–221 (1952). — Moricard, R.: Hormones and meiosis synthesis of thymon. acid in the eggs of mammals. Exper. Cell Res. Suppl. 1, 137–142 (1949). — Moses, M. J.: Nucleic acids and proteins of the nuclei of Paramecium. J. of Morph. 87, 493–536 (1950). — Quantitative optical techniques in the study of nuclear chemistry, in The chemistry and physiology of the nucleus. Exper. Cell Res. Suppl. 2, 75–94 (1952).

Neugnot, D.: Contribution à l'étude cytochimique des Cyanophycées par application des techniques de mise en evidence de l'appareil nucléaire chez les bactéries. C. r. Acad. Sci. Paris 230, 1311–1313 (1950).

Oberling, Ch., W. Bernhard, A. Gautier et F. Haguenau: Les structures basophiles du cytoplasme et leurs rapports avec le cancer. Presse méd. 61, 719–724 (1953). — Ogur, M., R. O. Erickson, G. U. Rosen, K. B. Sax and C. Holden: Nucleic acids in relation to cell division in Lilium longiflorum. Exper. Cell Res. 2, 73–89 (1951). — Olive, L. S.: The structure and behaviour of fungus nuclei. Bot. Review 19, 439–586 (1953). — Östergren, G.: Colchicine mitosis, chromosome contraction, narcosis and protein chain folding. Hereditas (Lund) 30, 429–467 (1944). — Oster, G.: Discussion to the paper of G. R. Wyatt 1952. Exper. Cell Res. Suppl. 2, 215–216 (1952).

Pauling, L., and R. B. Corey: Compound helical configurations of polypeptide chains: structure of proteins of the α-keratin type. Nature (Lond.) 171, 59–61 (1953a). — Structure of the nucleic acids. Nature (Lond.) 171, 346 (1953b). — Pease, D. C., and R. F. Baker: Preliminary investigation of chromosomes and genes with the electron microscope. Science (Lancaster, Pa.) 109, 8–10 (1949). — Pfeiffer, H. H.: Mikrurgische Versuche in polarisiertem Lichte zur Analyse des Feinbaues der Riesenchromosomen von Chironomus. Chromosoma 1, 526–530 (1940). — Mikrurgisch-polarisationsoptische Beiträge zur submikroskopischen Morphologie von Larven-Speicheldrüsenchromosomen von Chironomus. Chromosoma 2, 77–85 (1941). — Piekarski, G.: Haben Bakterien einen Zellkern? Zur Definition des Zellkerns. Naturwiss. 37, 201–205 (1950). — Politzer, G.: Pathologie der Mitose. Protoplasma-Monogr. 7 (1934). — Pollister, A. W.: Nucleoproteins of the nucleus, in The chemistry and physiology of the nucleus. Exper. Cell Res. Suppl. 2, 59–70 (1952). — Pollister, A. W., and C. Leuchtenberger: Nucleotide content of the nucleus. Nature (Lond.) 163, 360 (1949). — Pollister, A. W., and A. E. Mirsky: The nucleoprotamin of the trout sperm. J. Gen. Physiol. 30, 101–148 (1946). — Pollister, A. W., and M. J. Moses: A simplified apparatus for photometric analysis and photomicrography. J. Gen. Physiol. 32, 567–577 (1949). — Pollister, A. W., and H. Ris: Nucleoprotein determination in cytological preparations. Cold Spring Harbor Symp. Quant. Biol. 12, 147–154 (1947). — Poulson, D. F., and V. T. Bowen: Organization and function of the inorganic constituents of nuclei, in The chemistry and physiology of the nucleus. Exper. Cell Res. Suppl. 2, 161–179 (1952).

Rabinovitch, M.: Nucleolus and "nucleolus associated chromatin" acid phosphatase reaction. Nature (Lond.) 164, 878 (1949). — Rabinovitch, M., and L. C. Junqueira and A. Fajer: A chemical and histochemical study of the technic for acid phosphatase. Stain Technol. 24, 147–156 (1949). — Rattenbury, J. A., and J. A. Serra: Types of nucleolus reconstitution in telophase and the question of the "nucleolar organizer". Portugal. Acta Biol. A 3, 239–260 (1952). — Resende, F.: Über die Ubiquität der Sat-Chromosomen bei den Blütenpflanzen. Planta (Berl.) 26, 757 (1937). — Nucleoli and Sat-chromosomes. Bol. Soc. Broteriana, II. s. 13, 391–424 (1938). — Chromosome structure as observed in root tips. Nature (Lond.) 144, 481 (1939). — Über die Chromosomenstruktur in der Mitose der

Wurzelspitzen. II. Chromosoma **1**, 486–520 (1940). — Sur la constitution probable de l'olistherozone nucléolaire. Portugal. Acta Biol. A **1**, 265–269 (1946). — Karyokynesis. Portugal. Acta Biol. A **2**, 1–33 (1947). — Agentes modificadores do metabolismo celular. I. Bol. Soc. portug. Cie. nat. II. s. **3**, 181–211 (1951). — RESENDE, F., A. DE LEMOS PEREIRA and A. CABRAL: Sur la structure des chromosomes dans les méristèmes radiculaires. III. Portugal. Acta Biol. A **1**, 9–46 (1944). — RIES, E.: Grundriß der Histophysiologie. Berlin: Parey 1938. — RIS, H.: A quantitative study of anaphase movement in the aphid *Tamalia*. Biol. Bull. **85**, 164–178 (1943). — The anaphase movement of chromosomes in the spermatocytes of the grasshopper. Biol. Bull. **96**, 90–106 (1949). — RIS, H., and R. KLEINFELD: Cytochemical studies on the chromatin elimination in Solenobia (Lepidoptera). Chromosoma **5**, 363–371 (1952). — RIS, H., and A. E. MIRSKY: Quantitative cytochemical determination of desoxyribonucleic acid with the Feulgen nucleal reaction. J. Gen. Physiol. **32**, 125–146 (1949a). — The state of the chromosomes in the interphase nucleus. J. Gen. Physiol. **32**, 489–502 (1949b). — ROBINOW, C. F.: A study of the nuclear apparatus of Bacteria. Proc. Roy. Soc. Lond., Ser. B **130**, 299–324 (1942). — ROZSA, G., and R. W. G. WYCKOFF: The electron microscopy of dividing cells. Biochem. et Biophysica Acta **6**, 334–339 (1950). — RYBAK, B.: Recherches sur la constitution du fuseau achromatique. Bull. Soc. Chim. biol. Paris **32**, 703–718 (1950). — SANDERS, K.: Special methods, in Cytology and cell physiology, edit. by G. BOURNE, 2nd ed., p. 20–83. Oxford: Clarendon Press 1952. — SANDERSON, A. R.: Maturation and probable gynogenesis in the liver fluke, Fasciola hepatica L. Nature (Lond.) **172**, 110–112 (1953). — SANDRITTER, W.: Eine quantitative färberische histochemische Bestimmungsmethode der Nucleinsäure in Geweben. Z. wiss. Mikrosk. **61**, 30–37 (1952a). — Über den Nucleinsäurestoffwechsel in Plattenepithel und kleinzelligen Bronchialkarcinomen. Frankf. Z. Path. **63**, 387–422 (1952b). — Über den Nucleinsäuregehalt in verschiedenen Tumoren. Frankf. Z. Path. **63**, 423–446 (1952c). — SCHAEDE, R.: Die Kolloidchemie des pflanzlichen Zellkernes in der Ruhe und in der Teilung. Erg. Biol. **5**, 1–28 (1929). — SCHMIDT, G., and S. J. TANNHAUSER: J. of Biol. Chem. **161**, 83–92 (1945). — SCHMIDT, W. J.: Die Doppelbrechung von Karyoplasma, Zytoplasma und Metaplasma. Protoplasma-Monogr. **11** (1937). — Einiges über optische Anisotropie und Feinbau von Chromatin und Chromosomen. Chromosoma **2**, 86–110 (1941). — SCHNEIDER, W. C.: Nucleic acids in normal and neoplastic tissues. Cold Spring Harbor Symp. Quant. Biol. **12**, 169–178 (1947). — SCHOLES, G., and J. WEISS: Chemical action of X-rays on nucleic acids and related substances in aqueous systems. Exper. Cell Res. Suppl. **2**, 219–241 (1952). — SCHRADER, F.: Mitosis. Columbia Biol. Ser. 14. New York: Columbia Univ. Press 1944. — SCHULTZ, J., T. CASPERSSON and L. AQUILONIUS: The genetic control of nucleolar composition. Proc. Nat. Acad. Sci. U.S.A. **26**, 515–523 (1940). — SCHULTZ, J., R. MACDUFFEE and T. F. ANDERSON: Smear preparations for the electron microscopy of animal chromosomes. Science (Lancaster, Pa.) **110**, 5–7 (1949). — SCOTT, F. M.: Perforation of the surface membranes of nuclei and plastids by nucleodesmata and plasmodesmata. Bot. Gaz. **111**, 252–261 (1950). — SERRA, J. A.: Relations entre la chimie et la morphologie nucléaire. Bol. Soc. Broteriana, II. s. **16**, 83–135 (1942). — Sur la composition protéique des chromosomes et la réaction nucléale de Feulgen. Bol. Soc. Broteriana **17**, 203–211 (1943a). — La structure du protoplasme à l'échelle moléculaire. An. Fac. Farm. Univ. Porto **5**, 1–41 (1943b). — Improvements in the histochemical arginine reaction and the interpretation of this reaction. Portugal. Acta Biol. A **1**, 1–7 (1944). — Mitose e meiose. Acta I Reun. Biol., Lisboa, **1**, 47–96 (1945). — Histochemical tests for proteins and amino acids; characterization of basic proteins. Stain Technol. **21**, 5–18 (1946). — Composition of chromonemata and matrix and the role of nucleoproteins in mitosis and meicsis. Cold Spring Harbor Symp. Quant. Biol. **12**, 192–210 (1947a). — Contributions to a physiological interpretation of mitosis and meiosis. I. a. II. Portugal. Acta Biol. A **2**, 25–43, 45–90 (1947b). — A cytophysiological theory of the gene, gene mutation and position effect. Portugal. Acta Biol. A Goldschmidt-Vol., p. 401–562 (1949a). — The parallelism between the chemical and the morphological changes in the chromosomes during mitosis and meiosis. Exper. Cell Res. Suppl. **1**, 111–122 (1949b). — Moderna genética, geral e fisiológica. Imprensa de Coimbra 1949c. — SERRA, J. A., e R. M. ALBUQUERQUE: Natureza e causas da coloração canário em lãs brancas. II. $p_H$, alcalinidade e sugo. Publ. Junta. Nal. Prods. Pecuár. Lisboa, No 5 (in public). — SERRA, J. A., et A. QUEIROZ-LOPES: Données pour une cytophysiologie du nucléole. I. L'activité nucléolaire pendant la croissance de l'oocyte chez des Helicidae. Portugal. Acta Biol. A **1**, 51–91 (1945). — SHARMA, A. K., A. MOOKERJEA e C. GHOSH: Alkaline phosphatase technique in plant chromosomes. Portugal. Acta Biol. A **3**, 341–354 (1953). — SINGER, M.: Factors which control the staining of tissue sections with acid and basic dyes. Internat. Rev. Cytol. **1**, 211–255 (1952). — SMELLIE, R. M. S., W. M. MCINDOE u. J. N. DAVIDSON: The incorporation of $^{15}N$, $^{35}S$ and $^{14}C$ into nucleic acids and proteins of rat liver. Biochem. et Biophysica Acta **11**, 559–565 (1953). — SPARROW, A. H.: Radiation sensitivity of cells during mitotic and meiotic cycles. Ann. New York. Acad. Sci. **51**, 1508–1540 (1951). — SPARROW, A. H., and M. R. HAMMOND: Cytological evidence for the transfer of desoxyribose nucleic acid from nucleus to cytoplasm in certain plant cells. Amer. J. Bot. **34**, 439–445 (1947). — Sparrow, A. H., and B. A. Rubin: Effects of radiations on biological systems. Surv. Biol. Progr. **2**, 1–52 (1952a). — Sparrow, A. H., M. J. Moses and R. J. Dubow: Relationships between ionizing radiation, chromosome breakage and certain other nuclear disturbances. Exper. Cell Res. Suppl. **2**, 245–262 (1952b). — Stålfelt, M. G.: Effect of heteroauxin and colchicine on protoplasmic viscosity. Exper. Cell Res. Suppl. **1**, 63–78 (1949). — Stedman, E., and E. Stedman: Chromosomin, a protein constituent of chromosomes. Nature (Lond.) **152**, 267–269 (1943). — The chemical nature and functions of the components of cellular nuclei. Cold Spring Harbor Symp. Quant. Biol. **12**, 224–236 (1947). — Stern, H., V. Allfrey, A. E. Mirsky and H. Saetren: Some enzymes of isolated nuclei. J. Gen. Physiol. **35**, 559–578 (1951). — Stern, H., and A. E. Mirsky: The isolation of wheat germ nuclei and some aspects of their glycolytic metabolism. J. Gen. Physiol. **36**, 181–200 (1952). — Soluble enzymes of nuclei isolated in sucrose and non-aqueous media. J. Gen. Physiol. **37**, 177–187 (1953). — Stern, H., and S. Timonen: The position of the cell nucleus in pathways of hydrogen transfer: cytochrome C, flavoproteins, glutathione, and ascorbic acid. J. Gen. Physiol. **38**, 41–52 (1954). — Stern, K. G.: Nucleoproteins and gene structure. Yale J. Biol. a. Med. **19**, 939–949 (1947). — Problems in nuclear chemistry and biology, in Chemistry and physiology of the nucleus. Exper. Cell Res. Suppl. **2**, 1–16 (1952). — Stich, H., u. J. Hämmerling: Der Einbau von $^{32}$P in die Nucleolarsubstanz des Zellkernes von *Acetabularia mediterranea*. Z. Naturforsch. **8b**, 329–333 (1953). — Subramanian, M. K.: The problem of haploidy in yeasts. J. Ind. Inst. Sci. A **32**, 29–40 (1950). — Swann, M. M.: Structural agents in mitosis. Internat. Rev. Cytol. **1**, 195–210 (1952a). — The nucleus in fertilization, mitosis and cell division. Symposia Soc. Exper. Biol. **6**, 89–104 (1952b). — Symposium on chromosome breakage. Heredity (Lond.) Suppl. **6** (1953). — Sze, L. C.: Changes in the amount of desoxyribonucleic acid in the development of *Rana pipiens*. J. of Exper. Zool. **122**, 577–602 (1953).

Tischler, G.: Allgemeine Pflanzenkaryologie. In Handbuch der Pflanzenanatomie, 2. Aufl., Bd. II. Berlin: Gebrüder Bornträger 1934. — Thorell, B.: Studies on the formation of cellular substances during blood cell production. London: Henry Kimpton 1947. 120 p.

Vendrely, C.: L'acide désoxyribonucléique du noyau des cellules animales. Son rôle possible dans la biochimie de l'hérédité. Bull. biol. France et Belg. **86**, 1–18 (1952). — Vendrely, C., and R. Vendrely: Sur la teneur individuelle en acide désoxyribonucléique des gamètes d'oursins *Arbacia* et *Paracentrotus*. C. r. Soc. Biol. Paris **143**, 1386–1387 (1949). — Vendrely, R., and C. Vendrely: Arginine and deoxyribonucleic acid content of erythrocyte nuclei and sperms of some species of fish. Nature (Lond.) **172**, 30–31 (1953). — Virgin, H. J.: Physical properties of protoplasm. Annual Rev. Plant Physiol. **4**, 363–381 (1953).

Wachstein, M., and E. Meisel: Histochemical demonstration of 5-nucleotidase activity in cell nuclei. Science (Lancaster, Pa.) **115**, 652–653 (1952). — Walker, P. M. B., and H. B. Jates: Ultraviolet absorption of living cell nuclei during growth and division. Symposia Soc. Exper. Biol. **6**, 265–276 (1952). — Watson, J. D., and F. H. C. Crick: A structure for deoxyribonucleic acid. Nature (Lond.) **171**, 737–738 (1953a). — The structure of DNA. Cold Spring Harbor Symp. Quant. Biol. **18**, 123–131 (1953b). — White, M. J. D.: Nucleus, chromosomes and genes, in cytology and cell physiology. Oxford: Clarendon Press, 2. edit. 1952. — Wilbur, K. M., and N. G. Anderson: Studies on isolated cell components. I. Nuclear isolation by differential centrifugation. Exper. Cell Res. **2**, 47–57 (1951). — Wilkins, M. H. F., W. E. Seeds, A. R. Stokes and H. R. Wilson: Helical structure of crystalline deoxypentose nucleic acid. Nature (Lond.) **172**, 759–762 (1953). — Wilkins, M. H. F., A. R. Stokes and H. R. Wilson: Molecular structure of deoxypentose nucleic acids. Nature (Lond.) **171**, 738–740 (1953). — Wilson, E. B.: The cell in development and heredity, 3rd ed. New York: McMillan 1937. — Winge, O.: The relation between yeast cytology and genetics: a critique. C. r. Labor. Carlsberg, Ser. Physiol. **25**, 85–99 (1951). — Wyatt, G. R.: Recognition and estimation of 5-methylcytosine in nucleic acids. Biochemic. J. **48**, 581–583 (1951). — Specificity in the composition of nucleic acids, in The chemistry and physiology of the nucleus. Exper. Cell Res. Suppl. **2**, 201–215 (1952).

Yasuzumi, G., Z. Sugioka u. A. Tanaka: Electron microscope observations on the metabolic nucleus. Biochem. et Biophysica Acta **10**, 11–17 (1953).

Zastrow, E. M. v.: Über die Organisation der Cyanophyceenzelle. Arch. Mikrobiol. **19**, 174–205 (1953).

# Plastid structure, development and inheritance.

By

## S. Granick.

With 21 figures.

## I. Introduction.

The most generalized of the plastids is the chloroplast. Because of the importance of the chloroplast it will be the plastid type whose properties will be considered here in detail. In this review, two general aspects of the biology of the chloroplast will be considered, namely: the structure of the chloroplast and its relation to photosynthetic function; and the origin, growth, and inheritance of this cytoplasmic body. Since the editors plan to present a section on the chemical composition of the chloroplasts in another volume of the Handbook, considerations of the proteins, enzymes, pigments, starch, etc. of the chloroplast have been omitted except as they may be pertinent to the immediate discussion.

In the analysis of the structural details of the chloroplast, the physiologist is getting closer to one of the subjects that most intrigues him, that is, the relation between plastid structure, light absorption, and the conversion of light energy to chemical energy. Important advances in the study of plastid structure by electron microscopy have been made within recent years. The lamellae or disks which make up the grana provide evidence of a highly organized lipoprotein structure; however, the more intimate details of how chlorophyll and carotenoid molecules are oriented with respect to the lipoproteins, are still lacking. Nor is it yet known, what is the organization of enzymes involved in the first chemical stages that bring about the decomposition of water during photosynthesis.

Not only is the physiologist interested in how a unit of a cell functions, but he also desires to know how that functioning unit is formed, and what factors of heredity and environment control the production of that unit. The chloroplast is a uniquely suitable unit for inquiry along these lines. The chloroplast represents a complex organization of enzymes which serves not only to carry on photosynthesis, but also serves for self-duplication, and for the synthesis of at least some of the end steps in the biosynthetic chains of the chlorophylls, carotenoids, proteins, fats, and carbohydrates that are contained within this body. Development and growth of the plastid is controlled both by factors within the plastid, and by factors outside the plastid. The activity of the plastid is known to be indirectly affected or controlled by a large number of nuclear genes. Evidence is also accumulating that hereditary factors within the chloroplast, and also within the cytoplasm i.e., outside the chloroplast, take part in this control. In addition, processes of cellular differentiation, as yet poorly understood, affect the development of the plastid.

Depending on conditions outside the plastid, one or another of the end products within the plastid may be suppressed, modified, or even increased. Thus, mutation of some nuclear genes may suppress chlorophyll production leading to the formation of yellow chloroplasts. Or mutation of a nuclear gene may lead to the lack of carotenoids so the chloroplast is bluish green. Or because of processes of cellular differentiation, plastids, for example, in root cells or epidermal cells, may lack green and yellow pigments resulting in colorless bodies, the leucoplasts. The leucoplasts may store considerable quantities of starch and may then be designated as amyloplasts. If the pigments are lacking and protein formation is abnormally great even to the extent that protein crystals may form within the plastid, then the plastid is designated as an aleurone plastid. Or fat and steroid formation may be abnormally high and the plastid is called an elaioplast. If the chlorophyll and starch disappear from the plastid and carotene increases to the extent that it may crystallize as in the carrot root, then the plastid is designated as a chromoplast or yellow plastid. From these examples, it is apparent that various types of plastids, all so different in their final composition, may arise from a plastid which originally had the potentiality of forming a full complement of chloroplast substances. Factors external to the plastid may thus profoundly affect the expression of the plastid. A major task that lies before the physiologist is to dissect out these various influences on the plastid, perhaps through attempts to grow the plastid in tissue culture. Part of the answer to the problem of cellular differentiation may lie in the understanding of how the plastids, mitochondria, etc. become specialized.

The plastid is a large cytoplasmic body whose morphological features may be readily observed directly by examination of the intact cell, or the plastids may be readily isolated and their composition studied. In contrast to the smaller cytoplasmic granules such as the mitochondria, important physical and chemical changes in the plastids can be easily recognized, such as changes in size and shape, and the presence or absence of the chlorophylls, carotenoids, and starch. The studies on self-duplication of the plastids, the inheritable factors within them and the changes which occur in them during cellular differentiation are not only of importance in themselves, but may well serve to illuminate the properties of cytoplasmic granules which are at or below the limits of visibility of the light microscope. The various kinds of cytoplasmic granules represent complex organizations of enzymes that perform specialized functions. One may ask similar questions about them as about the chloroplast. Are they self-duplicating entities? What type of organization for self-reproduction do they possess? In what way do the nuclear genes influence the development and activities of these bodies?

Because the plastid is a self-duplicating body, the problems connected with its inheritance may well be similar to those connected with virus multiplication within the cytoplasm. For example, if plastids could infect cells or if hosts such as sucking insects could transmit plastids, the analogy between virus and plastid would be close. The analogy would be still closer if a plastid which had a chlorophyll defect and which could outgrow normal plastids was transmitted to a normal cell, resulting in a decrease in the photosynthetic ability of this cell.

In this review no attempt has been made to cite all the literature, nor to present an historical development of the subject. Rather, references have been selected with a view toward illustrating specific points of the discussion. Several reviews of the older literature with extensive references are those by WEIER (1938a), KÜSTER (1935), and SCHÜRHOFF (1924).

During the latter half of the 19th century, excellent studies were made on chloroplast structure and inheritance. It was then discovered that the chloroplasts contain grana and that plastids arise from pre-existing plastids, facts which have been confirmed and more firmly established by recent techniques. Particularly noteworthy were the contributions of MEYER (1883), SCHIMPER (1883), SCHMITZ (1884), and SACHS (1887).

# 1. Classification of plastids.

Plastids may be subdivided into two general groups: the *chromoplasts*, chroma-tophores or colored plastids, and the *leucoplasts* or colorless plastids. The *chromoplasts* may also be subdivided into two groups: those which are photosynthetically active and those which are not. This classification is useful because, merely by visual inspection, one may note the specialization of structure and function that may occur in plastids in different organs of a plant as well as in different cells of the same plant organ. By way of introduction, a brief survey of the various kinds of plastids is presented.

# 2. Photosynthetically active chromoplasts.

The photosynthetically active *chromoplasts* contain chlorophyll and carotenoid pigments. The most common plastid of this kind is the *chloroplast*. In this green plastid the chlorophylls generally comprise about two-thirds or more of the total pigments; xanthophylls represent about four-fifteenths and carotene about one-fifteenth. About 25–35% of the dry weight of the chloroplast is lipid; about 40–50% is protein and about 4–7% is chlorophyll.

Brown or yellow plastids, *phaeoplasts*, occur in the *Phaeophyta*, *Pyrrophyta*, and *Chrysophyta*. Here the brown carotenoids mask the chlorophyll color. According to STRAIN (1949), the yellow or brown color of the phaeoplasts appears to result from the physical state or condition of the pigments or from their geometrical arrangement in the plastids rather than from a preponderance of the carotenoids. For example, exposure of the plants to heat causes the plastids to turn green. Extraction of heated plants or of fresh plants with alcohol yields green solutions in which the chlorophylls predominate. The unequivocal evidence that in the brown algae fucoxanthin can absorb light and transfer the energy to chlorophyll *a* (DUTTON *et al.* 1943) suggests that this yellow pigment must be situated spatially quite close to the chlorophyll molecules.

Red-colored plastids or *rhodoplasts* often occur in the *Rhodophyceae*, especially the *Florideae*. The red color is due to the phycobilin protein, phycoerythrin (STRAIN 1949, LEMBERG and LEGGE 1949) which may be in sufficient abundance to mask the chlorophyll color. Phycocyanin, the blue phycobilin, may also occur to a smaller extent in some species of red algae and the color of plastids of such species may be the result of light absorption by chlorophyll, by phyco-erythrin, and by phycocyanin, as they may occur in varying proportions under differing environmental conditions (RABINOWITCH 1945). From studies of photo-synthesis, it is known that the light which is absorbed by the phycobilins is transmitted rather efficiently to chlorophyll *a* (HAXO and BLINKS 1950, FRENCH and YOUNG 1952). Within the plastids, the phycobilins fluoresce very little. However, if the cells are injured, the phycobilins readily diffuse out of the plastids and are then intensely fluorescent in visible light. In order to account for the absence of fluorescence in the uninjured rhodoplast, it is necessary to assume that the phycobilin molecules are arranged so that the light energy which is absorbed by them is transmitted to chlorophyll. If the spatial organization is disrupted, the light energy is not transmitted but is rather emitted in part as fluorescent energy by the tetrapyrrole prosthetic groups of the phycobilin.

In the blue-green algae, and in bacteria which photosynthesize in the infra-red, no organized plastids are observed. The available evidence suggests that grana are present in them which contain the carotenoids and chlorophylls. In the blue-green algae, phycoerythrin and phycocyanin may be present in the cells.

Whether these proteins are localized in the grana is unknown. The purple or red sulfur or non-sulfur bacteria *(Thiorhodaceae* and *Athiorhodaceae)* contain carotenoids which mask the bacteriochlorophyll color (RABINOWITCH 1945).

## 3. Chromoplasts devoid of photosynthetic activity.

These chromoplasts in general contain carotenoids but lack chlorophylls. They are red to yellow in color and are of highly variable shape. Carotenoids most often appear to be produced in plastids that have a photosynthetic function. The colors of flowers and fruits are often due to chromoplasts. Striking color changes occur as the green fruits and flowers mature and chlorophyll disappears. For example, the yellow chromoplasts of the petals of the buttercup *Ranunculus ficaria* first develop as chloroplasts.

The new colors that arise are often due to the formation of special carotenoids, not normally found in the green part of plants; such are the lycopene of tomato fruits, prolycopene in the berries of *Arum orientale,* and capsanthin of red peppers. Many of the xanthophylls of fruits and flowers occur with their hydroxyl groups esterified with fatty acids. In contrast, xanthophyll esters rarely occur in chloroplasts (STRAIN 1949). In autumn leaves of *Ginkgo biloba* or *Fagus silvatica* colored droplets containing carotenoids are present; these droplets may represent plastids which have degenerated so that only the fatty material containing the carotenoids remains (MÖBIUS 1937). Colored carotenoid droplets might also arise by solution in oil droplets of carotenoid molecules produced elsewhere, or they might represent vacuoles which have concentrated the more water-soluble carotenoids.

Carotenoids also occur in fungi and in bacteria. Are these carotenoids synthesized in specific cytoplasmic bodies that may represent vestiges of plastids or do the colored droplets arise independently of a plastid-like apparatus? In fungi, acidic carotenoids similar to astaxanthin are often present; the xanthophylls characteristic of higher plants are not found in them (GOODWIN 1952).

In *Neurospora crassa,* a complex mixture of carotenoids has been isolated by HAXO (1949), which includes lycopene, carotene, and spirilloxanthin as major components. GARTON et al. (1951) have observed that if growth of *Phycomyces Blakesleeanus* occurred in visible light, then the yield of carotenoids was doubled. In bacteria which contain carotenoids, the carotenoids are mostly xanthophylls. Bacteria lack the acidic carotenoids that are characteristic of fungal carotenoids.

As an example of a yellow chromoplast structure, one may consider the studies on the chromoplast of the carrot root. These chromoplasts develop from leucoplasts which contain starch. As the carotene increases in concentration, the starch disappears. The older chromoplasts appear as large flat plates in the form of parallelograms or as needles with a typically crystalline appearance; they are birefringent and dichroic (FREY-WYSSLING 1935). Analysis of the large chromoplasts by STRAUS (1954) indicates that carotene may make up 20–56% of the plastid on a dry weight basis. When the carotene content of the plastid is about 30%, the other lipids amount to 30–40%, about 15% may be protein and 0.5% may be ribose-nucleic acid. After extraction of the dried chromoplasts with ether STRAUS observed a residue consisting of very fine rectangular fibrils or rods which presumably represented the protein component and which appeared to have a lamellar structure. It would be of interest to investigate with the electron microscope the details of the lamellar structure and the regions of carotene localization in these bodies.

# 4. Leucoplasts.

The term leucoplast is applied to all mature colorless plastids. (The immature colorless plastids which occur in meristematic tissues are designated as proplastids and will be discussed in the section on inheritance of plastids.) Leucoplasts are of several kinds depending on the predominant material which they contain. When starch is predominant, as in storage organs such as the potato, the leucoplast is called an *amyloplast*. When oil is predominant, the plastid is called an *elaioplast*. When protein granules or crystals are predominant, the plastid may be designated as an *aleurone-plast*.

*Amyloplasts* represent mature plastids which are filled with starch, and are generally found in storage tubers, cotyledons, and endosperm. Starch is characteristic of the colored and colorless plastids of green algae and the higher plants; it is not formed in the cytoplasm apart from the plastids suggesting that the enzymes for starch deposition are localized primarily in these bodies. In algae such as *Spirogyra* and *Chlamydomonas*, starch is deposited primarily around the pyrenoid, although under conditions such as nitrogen starvation, other portions of the plastid may contain starch. On the assumption that starch is only formed in plastids, the proplastids of meristematic tissues may be distinguished from mitochondria if starch grains can be observed in the proplastids.

The deposition of carbohydrates other than starch may not necessarily be within the plastid. For example, in red algae, floridean starch is said to be produced at the surface of the plastid. In *Euglena*, paramylum grains are considered to be produced outside of the plastids but generally in the neighborhood of the pyrenoids. In many *Peridineae* starch is said to be produced inside, and in others, outside the plastid.

Two types of starch grains may be distinguished, the *assimilation* starch, and *reserve* starch. Assimilation starch is characteristic of actively photosynthesizing chloroplasts; the starch grains may be formed in large numbers per plastid but they remain small because they are continually dissolving and the sugars are being transported elsewhere.

Reserve starch is characteristic of storage organs and is most familiar for example as the large starch grains of the potato tuber. The starch grain of the potato is generally made up of a series of concentric layers successively deposited about a center or hilum. As starch continues to be deposited around the starch grain in concentric layers, the amyloplast membrane and accompanying stroma may become greatly distended. The amyloplast may rupture and remain only on one side of the starch grain. It is noteworthy that in the latter case, new starch will only be formed where the amyloplast membrane and stroma remain in contact with the starch grain.

According to FREY-WYSSLING (1953), in the starch grain the formation of a new concentric layer begins as a dense region of high refractive index, then it becomes less dense (*i.e.*, more hydrated) outwards until deposition of that particular layer ceases. When such a starch grain is examined in polarized light, it appears as a spherite cross. This fact indicates that the chains of starch molecules lie more or less parallel and extend from one concentric layer outward to the next concentric layer. Starch consists of two components. A major component is the highly branched $\alpha$-amylose which strains red with iodine. The minor component (20–30%) is $\beta$-amylose, a helical chain structure that stains blue with iodine. In a layered starch grain, $\alpha$-amylose molecules appear to be deposited in the inner denser and more refractive portion of a concentric layer whereas $\beta$-amylose helices accumulate in the outer, looser portions of the layer. For a detailed description of the enzymology and structure of starch, the reader may refer to the review by Hassid (1954). The concentric structure of the starch grain is due to a periodic layering connected presumably with a fluctuation in available substrates, in enzymes, in $p_H$, etc. At times this periodic layering may be correlated with alterations in night and day as observed by Meyer (1920).

*Elaioplasts* may be defined as plastids which develop a preponderance of oil. In most monocotyledons, a fatty oil appears in old chloroplasts which lose their chlorophyll. In epidermal cells of *Orchidaceae* and *Liliaceae*, the oily disorganized plastids fuse to form a droplet which has also been called an "elaioplast". Oil droplets are undoubtedly also formed independently of plastids. In many brown algae, the reserve storage product is oil instead of carbohydrate, and this oil appears to be formed in the pheoplasts (Mangenot 1923, Sharp 1934).

*Aleurone-plasts* or proteinoplasts may be defined as colorless plastids which contain much protein. For example, the leucoplasts of *Phajus grandifolius* may contain a parallel cluster of needle-like protein crystals. Proteinoplasts have been studied in epidermal cells of *Helleborus corsicus* (Härtel and Thaler 1953). Protein crystals and granules are produced in seeds of numerous plants especially in those seeds which also form large amounts of oil, as for example, in the seeds of *Ricinus* and the Brazil Nut. According to Mottier (1921), protein or aleurone formation involves the activity of permanent plastid primordia. These plastid primordia are considered to aggregate in large numbers in vacuoles where their combined products unite to form aleurone grains. The formation of proteins within the plastids is one of the markedly active properties of the plastid. This becomes evident when it is considered that more than half of the protein of a leaf parenchyma cell may reside in the chloroplasts. However, the formation of proteins is not exclusively a function of plastids. Protein crystals may be formed in the cytoplasm (independent of plastid action?) as for example in cells poor in starch in peripheral layers of potato, or even in nuclei as in *Lathraea squamaria* and in many *Scrophulariaceae* and *Oleaceae* (Meyer 1920). Likewise, protein granules (aleurone grains) of *Soja* cotyledons do not appear to be formed within plastids (Muschik 1953). Protein crystals in epidermal cells may also represent virus material or may be the result of virus action.

# II. General aspects of chloroplast structure.

## 1. Organization of cells in the mesophytic leaf.

Leaves are plant organs specialized for photosynthesis. In order to recall the location of the chloroplasts in the leaf, let us briefly examine the flat leaf of the dicotyledonous plant. In general, the leaf is an organ whose structure is specialized so that it provides the maximum area in proportion to mass for the absorption of sunlight. However, with respect to its surfaces, the leaf structure is a compromise between two opposing tendencies; namely, the tendency to absorb $CO_2$ by exposing the photosynthesizing cells to the atmosphere, and the tendency to protect the leaf cells from being desiccated by the atmosphere.

The mesophytic leaf (Fig. 1) consists of one to several somewhat compact layers of cells, the palisade parenchyma, near the upper surface of the leaf, and several layers of loosely arranged mesophyll cells, the spongy parenchyma, below them. These cell layers are enclosed above and below by a layer of epidermal cells which represent the outer covering of the leaf. All the parenchyma cells are arranged rather loosely so as to be in contact at some cellular surface with the air spaces of the leaf. The air spaces serve for the rapid exchange of gases, diffusion being very rapid in the gaseous state as compared with diffusion in the liquid state (GODDARD 1945). The spongy mesophyll cells, in contrast to the palisade cells, are often in fairly close association with vascular bundles and bundle ends (*i.e.* the leaf veins of phloem and xylem), through which substances are transported in the dissolved state to and from the leaves (HABERLANDT 1914). The development of the intercellular spaces and the compactness of cellular arrangement are in part related to the auxin content of the developing mesophyll

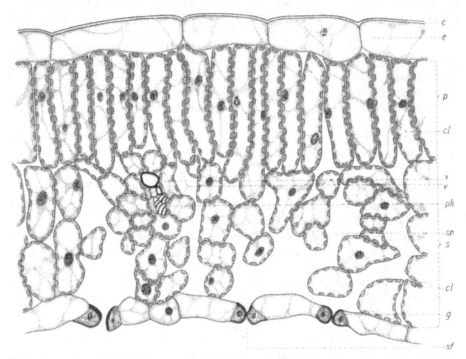

Fig. 1. Cross section of leaf of *Lonicera tatarica* illustrating cuticle, *c*; epidermis, *e*; palisade mesophyll, *p*, with chloroplasts, *cl*; vein, *v*, with xylem, *x*, and phloem, *ph*; spongy mesophyll, *s*, with air spaces, *sp*, and chloroplasts, *cl*; lower epidermis with stomata, *st*, and guard cells, *g*. By permission from *Botany, a Textbook for Colleges* J. B. HILL, L. D. OVERHOLTS and H. W. POPP. McGraw-Hill Co., 1950.

cells, and the effect of light in destroying the auxin (GOODWIN 1937, WENCK 1952). Thus, leaves developing in strong sunlight may have more compact mesophyll tissue with fewer air spaces probably because auxin is destroyed more rapidly under these conditions. On the other hand, leaves developing in the shade may have palisade cells which appear more like spongy mesophyll.

## 2. Chloroplasts of the mesophyll cells.

The mesophyll cells contain numerous chloroplasts which are green, and are shaped like plano-convex lenses. The chloroplasts are approximately $5\,\mu$ in diameter and $2$–$3\,\mu$ thick. They lie embedded in cytoplasm and appressed with their broad sides parallel to the cell walls. The flat lamellae or disks in the grana of the chloroplast are arranged so that they also face the adjacent cell wall (Fig. 8 A). In general, the chloroplasts tend to lie in those regions of the cell which are adjacent to intercellular air spaces (HABERLANDT 1914).

In 215 species of plants, MÖBIUS (1920) found that 75 per cent of the plastids had a long diameter between $4$–$6\,\mu$. In *Tropaeolum majus* the average dimensions of the chloroplast were 3.9 by $2.9\,\mu$ with an average volume of approximately $9.4\,\mu^3$ (MEYER 1918). In *Ricinus communis*, HABERLANDT (1914) estimated that the average palisade mesophyll cell contained about 36 chloroplasts and the spongy mesophyll cell contained about 20. Per square millimeter of leaf area, there were a total of 403,000 chloroplasts of which 92,000 or 18 percent were in the spongy mesophyll cells. In the palisade cells, the chloroplasts may often be so closely packed as to form an almost flat compact layer, lining all the cell walls of the cell. The cells of many plants appear to have the ability to regulate the number of plastids per cell. If the chloroplasts are few, division of the chloroplasts may take place. Under adverse physiological conditions, do all the chloroplasts degenerate equally or do a few survive at the expense of the others? This question has yet to be answered. In certain strains of maize, EYSTER (1929) noted that if cells have large plastids, there are fewer plastids per cell than if cells contain smaller plastids. In the immature leaf, the chloroplasts multiply by fission, but fission ceases at an early stage. In the tomato leaf the number of chloroplasts increases only by about 30% from the time the leaves are one third their maximum size to the fully expanded leaves, while on the other hand, protein content of the chloroplasts increases over ten fold during this time (GRANICK 1938b).

## 3. Effects of light on chloroplasts of mesophyll cells.

There are a number of effects of light on the development and composition of chloroplasts which may be briefly summarized here.

Visible light is essential for the formation of chlorophyll. As determined by studies of action spectra, the effective wave lengths are those at about 430 and 640 m$\mu$ in the leaf. The light energy absorbed at these wave lengths brings about the conversion of protochlorophyll to chlorophyll and changes the pale yellow tiny proplastids to pale green ones (FRANK 1946, SMITH *et al.* 1948b).

The most important action of visible light is, of course, to bring about photosynthesis. The light used in photosynthesis is that primarily absorbed by chlorophyll *a* at 445 and 680 m$\mu$ in the leaf. Once photosynthesis has started in the immature pale green plastids, the chloroplast rapidly enlarges and becomes green. At the same time, the chlorophylls, carotenoids, lipids and enzymes increase and become organized structurally to form the grana and lamellae or disks. In a leaf, a single chloroplast may absorb as much as 30–60% of the light incident on a leaf (SEYBOLD 1933).

Carotenoids appear to function in photosynthesis (WARBURG und KRUPPAHL 1954). Although the red light absorbed by chlorophyll is utilized for photosynthesis, it is utilized with maximum efficiency only if blue light, even of very low intensity, is simultaneously absorbed by the carotenoids (460–500 m$\mu$). Studies on carotenoid synthesis in fungi suggest that specific wave lengths of light also may be involved to some degree in carotenoid synthesis (GARTON *et al.* 1951). Whether carotenoid synthesis in chloroplasts is mediated to some degree by visible light remains to be determined.

Light may also be absorbed by accessory pigments of chloroplasts to be transmitted to chlorophyll *a* and be used in photosynthesis (p. 534).

High light intensity may cause a direct destruction of chlorophyll and this action may in part explain the higher chlorophyll content in leaves developing in the shade as compared to leaves developing in the sun. Under certain conditions carotenoids may also be destroyed by the action of intense light (GALSTON 1950).

Other effects of light on chloroplasts may result indirectly from the action of light on auxin (SKOOG 1951). Not only does the intensity of light affect the size and number of the palisade cells as compared to the mesophyll cells, but it also affects the size and pigment content of the chloroplasts. In general, leaves of the same plant which develop in the shade of other leaves are darker green and thinner and larger than leaves which develop in the sun. The chloroplasts of such "shade" leaves are larger and richer in chlorophyll and they contain a higher proportion of chlorophyll $b$ to chlorophyll $a$ than do the "sun" leaves. Thus, WILLSTÄTTER and STOLL (1913) found the "sun" leaves of *Sambucus nigra* to contain 0.80% chlorophylls $a+b$ per gm. dry weight, and shade leaves to contain 1.18%. Similarly, in *Platanus acerifolia*, the "sun" leaves contained 0.68% and the "shade" leaves 1.12% chlorophylls $a+b$. The effects of high light intensity on the chloroplasts are in part due to the effect of light in bringing about auxin destruction. Blue light (460–480 m$\mu$) appears to be most effective in destroying auxin. According to GALSTON (1950), the mechanism suggested for the destruction of auxin by light is that of a photo-oxidation of auxin resulting from the light absorbed by riboflavin and possibly also by $\beta$ carotene.

An effect of near infra-red light has recently been observed on leaf development which would affect chloroplast development. Illumination of etiolated bean plants with light of 650 m$\mu$ results in the expansion of the etiolated leaves, and presumably an increase in the protochlorophyll and carotene content of the leaves. Under these conditions it is probable that the pale yellow plastids enlarge as the etiolated leaves enlarge. Illumination with 735 m$\mu$ light can inhibit the promoting effect of the 650 m$\mu$ light (LIVERMAN *et al.* 1955).

Still another effect of light is to orient the plastids within the cells. In a number of plants there appears to be a well developed tendency for the chloroplasts to become arranged in the cells so that at a low intensity of light the broad surface of the plastid will face the light; and at high light intensity the edge of the plastid will face the light. Thus, at low light intensity, the chloroplast makes full use of the light for photosynthesis, and at high light intensity possibly avoids the destructive effects of excess light (SENN 1908, SCHANDERL und KAEMPFERT 1933, ZURZYCKI 1953). Blue light, possibly absorbed by carotenoids, but not red light, is effective in the phototaxis of the chloroplasts (VOERKEL 1933).

## 4. Epidermal cells.

The differentiation of the leaf markedly influences the development of the chloroplasts as is evident in the epidermal cells of the leaf. A layer of epidermal cells covers both the upper and lower surfaces of the thin leaf. These cells contain small plastids which are in general poorly developed and devoid of green and yellow pigments. Also, unlike the mesophyll cells, the epidermal cells are thinner and have a waxy cuticle on their outer surface, which reduces evaporation of water from the leaf. How this differentiation, which results in the formation of the epidermal cells, limits the development of the plastids in these cells, is unknown. Whether differentiation effects of this kind may at times play a role in bringing about certain kinds of leaf variegations is unknown.

## 5. Guard cells.

The guard cells and the chloroplasts within them represent a mechanism for regulating the size of the stomatic orifice through which direct gaseous exchange occurs between the intercellular mesophyll spaces and the outer air.

The guard cells are generally characterized by special thickenings of the walls that border the stomatic orifice (Haberlandt 1914). These anatomical features cause the stomata to open when the guard cells become turgid and to close when they lose their turgidity. The guard cells, in contrast to other epidermal cells, have functional green plastids which generally contain starch, even at a time when the parenchyma cells lack starch. The osmotic value of the other epidermal cells is relatively constant and always lower than that of the guard cells. The osmotic value of the guard cells, however, is not constant. It is lowest when the stomata are closed and highest (some 2–10 atmospheres more than in the epidermal cells) when the stomata are opened.

When the guard cells of a leaf are exposed to light in the morning, their $p_H$ increases, the osmotic pressure of their cell sap increases, their starch content decreases, and the stomata open (Miller 1938). In parenchyma cells, however, illumination brings about an increase in the starch content of the chloroplasts rather than a decrease. At night, the stomata of the majority of plants become closed, but not necessarily so completely as entirely to prevent gaseous exchange with the outside air. In the presence of factors which bring about rapid evaporation, such as strong sunlight or low humidity, only when the stomata are almost closed do they become effective in regulating water loss. If the supply of water reaches a certain minimum which causes the guard cells to lose turgidity in spite of their osmotic pressure, they may close regardless of other factors. The control of the opening and closing of the guard cells thus is a result of osmotic changes. The osmotic changes are probably primarily influenced by the sugar-starch equilibrium or, more precisely, by the glucose-1-phosphate-phosphorylase-starch system which is presumably localized in the chloroplasts. The precise mechanism by which light increases the osmotic pressure of the stomatal cells via this enzyme system is yet to be investigated.

## 6. Chloroplasts in algae.

Chloroplasts are well-defined cytoplasmic bodies which are in general green and contain the chlorophyll and carotenoid pigments. However, in the red and brown algae, the green chlorophyll color may not be evident because of the presence of other intensely absorbing pigments in the plastids. Chloroplasts are present in all cells that carry on photosynthesis. Recent studies have shown that the photosynthetic bacteria and the blue-green algae likewise possess organized bodies that contain the chlorophylls and carotenoids.

In the blue-green algae, chlorophyll appears to be diffusely distributed throughout the peripheral cytoplasm of the cell. Particulate colored material which contains the photosynthetic pigments may be separated from the photosynthetic bacterium *Rhodospirillum rubrum* (Pardee et al. 1952) and from blue-green algae like *Synechococcus cedorum* (Calvin and Lynch 1952). Such material appears to represent "grana" or grana-like structures (Geitler 1937).

Especially among the algae is there a great diversity of number and of form of chloroplasts (Fig. 2). In some species of algae there may be one or two chloroplasts in a cell (Fig. 2A, E); in other species there may be some ten or more chloroplasts per cell (Fig. 2C). The algal chloroplasts are usually placed parietally just beneath the cell wall, embedded in cytoplasm, as in the higher plants. The chloroplasts may be bell-shaped as in *Chlorella* or *Chlamydomonas*, or spiral-shaped as in *Spirogyra*, or form an irregular network, as in *Oedogonium* (Fig. 2A); or they may be suspended in the center of the cell as in the stellate-shaped chloroplasts of *Zygnema*. The shape of the chloroplast depends in part on the method of growth. In some chloroplasts, pseudopodial-like extensions occur, as in *Zygnema* or *Rhodochorton* to form a stellate shaped chloroplast; or the pseudopods may extend to form a plastid network, as in *Oedogonium*. In other types of chloroplasts holes may be formed by differential rates of growth of the expanding plastid. In still others, colorless regions of the plastid stroma may separate the green areas, as in *Chlorella* growing in the dark in the presence of glucose.

The effect of light intensity on the orientation of the chloroplasts is clearly observed in certain algae. For example, the flat, band-shaped chloroplast of *Mesocarpus* which is suspended lengthwise in the center of the cell tends to be

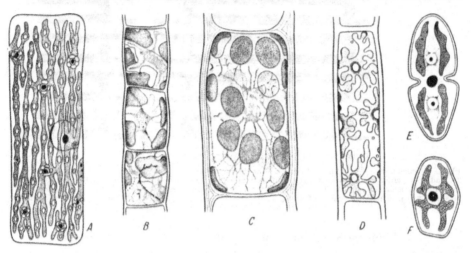

Fig. 2A–F. Various types of algal chloroplasts. A *Oedogonium* showing pyrenoids as dark dots surrounded by starch sheaths and starch grains (as white circles) lying within the stroma or protein matrix of the plastid. B *Leptonema fasciculatum.* C *Pilayella varia.* D *Rhodochorton floridulum.* E and F *Euastrum dubium*, front and side views. By permission from *Introduction to Cytology.* L. W. Sharp. McGraw-Hill Co., 1934.

perpendicular to the light at intermediate intensities; at high light intensities the chloroplast moves so as to be parallel to the light beam. In *Vaucheria*, which is coenocytic, the chloroplasts accumulate only in the illuminated region of the filaments but if the light intensity is too great, they move away from this region (SENN 1908).

## 7. Summary.

In the cells of fully expanded leaves of higher plants, several to many chloroplasts per cell may be present. The average size of a chloroplast is about $5\,\mu$ in diameter and $2$–$3\,\mu$ thick. The development, number per cell, and size of chloroplasts are influenced by external factors, such as light, which has varied effects. Internal factors such as differentiation also markedly influence plastid development.

In algae, there may be only one or two or several chloroplasts per cell, depending on the species. Various shapes of chloroplasts are encountered depending on the species.

All of the cells that carry on photosynthesis contain chloroplasts in which are concentrated the chlorophylls and carotenoids. Even in blue-green algae and in purple sulfur bacteria, in which no well-defined chloroplasts have been seen with the light microscope, recent studies have revealed tiny bodies in which the photosynthetic pigments are segregated.

## III. Details of chloroplast structure.

The general features of the chloroplast of the higher plants are the semi-permeable membrane which surrounds the chloroplast, the stroma in which are embedded the grana and the osmiophilic granules, the fine lamellae in the stroma which extend from one granum to another, and the discs of the grana (Fig. 3 and 9).

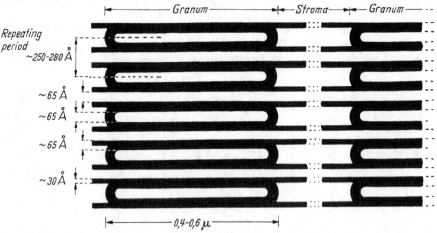

Fig. 3. Interpretation of structure of the *Aspidistra* chloroplast according to STEINMANN and SJÖSTRAND (1955). Two adjacent grana are shown connected with each other through a series of very fine lamellae, ~ 30 Å thick, extending through both grana. The pile of disks in a granum are represented by wide flattened tubes 4000 to 6000 Å in width. The disks consist of an upper and lower membrane, each being 35–40 Å thick and enclosing a space about 65 Å in height.

### 1. Chloroplast structure as revealed by the light microscope.

The literature on the structure of the chloroplast, as revealed by the light microscope, has been amply reviewed in recent years by WEIER (1938a), RABINO-WITCH (1945), WEIER and STOCKING (1952a), and FREY-WYSSLING (1953).

In intact cells of higher plants such as the leaf parenchyma cells of tomato and spinach, etc., the chloroplast appears to be saucer-shaped with its concavity facing the cell vacuole. The chloroplast may appear homogeneous or may be seen to contain fine granules, the grana. On injury of the cell, the chloroplast may take on a more or less coarsely granular appearance, or at times it may appear foamy and no distinct grana may be seen.

By tearing the cells apart in an isotonic or hypertonic sucrose solution (GRANICK 1938a) or better still in a polyethylene glycol solution (McCLENDON and BLINKS 1952), or in a high molecular weight substance like polyvinyl pyrrolidone, the chloroplasts float out into the solution, and may be seen to retain their smooth appearance for several minutes or longer, gradually taking on a granular appearance. If the chloroplasts are floated out into distilled water or isotonic saline, they may take on a distinctly granular appearance within several minutes; at the same time vesicles or blebs are observed to form on their surface. One or several blebs which develop seem to have their origin on the concave side of the chloroplast where a starch vacuole appears to be localized. Smaller, numerous blebs which form over the surface of the chloroplast may represent the swelling of units originally surrounded by membranes, disks that have come from the loosening of the grana structure, and myelin membranes and precipitation membranes that have formed on injury. A limiting osmotic membrane surrounds the chloroplast; this is inferred from the fact that chloroplasts, maintaining a coherent structure, can be removed from the cells without adhering cytoplasmic films, and that the chloroplasts show osmotic behavior (KNUDSON 1936, GRANICK 1938a).

The granules which appear in the chloroplasts were first reported in 1883 by MEYER (1883), who called them "grana". According to SCHIMPER (1883) all chloroplasts of pteridophytes and spermatophytes contain grana. These grana are especially clearly seen in orchids like *Acanthephippium* and in many *Crassulaceae*. In *Anthoceros* and in many algae which have large chloroplasts no grana are seen. The grana were later rediscovered by DOUTRELIGNE (1935) and by HEITZ (1936) (Fig. 4) in various higher plants but not in the chloroplasts of some algae like *Spirogyra* or *Mougeotia*. HEITZ observed that the grana were not spheres but rather had a disk shape. The grana are 0.5–2 $\mu$ in diameter, the size depending on the species. The grana of "sun" plants were noted to be smaller than those of "shade" plants. The grana are smaller in chloroplasts of the upper palisade cells and larger in chloroplasts of the lower spongy mesophyll cells. In *Elodea* the grana were found to be larger in August and smaller in January (BEAUVERIE 1938[1]).

Often no grana can be observed in the chloroplasts of living cells. WEIER (1938b) observed that both homogeneous and granular chloro-

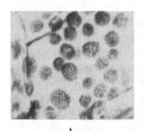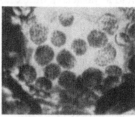

A                                       B

Fig. 4 A and B. A Grana in the chloroplasts of an intact cell of a water plant. HEITZ (1936). B Hand section through the leaf of a land plant examined in 6% sucrose solution. HEITZ (1936).

plasts occur in healthy plants. Granular chloroplasts were present in plants which were growing in the shade, and in young leaves. Homogeneous chloroplasts were found in older leaves and in plants growing in the sun. If the grana are to be considered as pre-formed units, as most evidence indicates, then the appearance of homogeneous chloroplasts might be due to a gradual change of refractive index at the phase boundaries between the grana and stroma in these chloroplasts. For example, starch grains in uninjured living plastids are probably not readily seen for the same reason; there may exist a gradual transition region, represented by a gradual change of refractive index between the dense starch grain and the plastid stroma instead of the sharp transition as seen in the injured plastid. On injury, homogeneous-appearing chloroplasts may become granular. In such cases it is probable that the injured chloroplasts absorb water, and swell; the grana may thus spread apart. The refractive index transition would thus become great over a short distance and the grana would then become visible.

In tobacco leaves WEIER (1938b) has observed homogeneous-appearing chloroplasts that remained non-granular even when the cells were killed with chloroform or ether. DE REZENDE-PINTO (1948) described chloroplasts of *Cotyledon umbilicus*, *Sedum album*, *Taraxacum officinale*, and *Aspidistra elatior* which contained a coiled thread structure with coarse granulations along the thread. He has suggested that the grana may be held together on a strand, the "plastonema", arranged in a helical fashion within the chloroplast. The "plastonema" is said to appear more distinct after the leaves have been in darkness for 2 days. According to this worker, the starch vacuole then becomes filled with sugar and when the chloroplast is mounted in distilled water it swells and the plastonema can be seen.

In the fern, *Isoetes*, the chloroplast of the mature sporophyte appears to contain grana that are connected to each other by threads (STEWART 1948).

[1] See also P. 544.

## 2. Chloroplast structure and composition as revealed by the polarizing microscope.

The chloroplasts of algae such as *Mougeotia, Mesocarpus, Spirogyra, Chlamy-domonas*, and the large chloroplast of *Anthoceros* consist of a single granum, with relatively coarse lamellae. Significant details of grana organization have become available from studies on single chloroplasts with the polarizing microscope, that in part confirm and even supplement details of electron microscopy. SCARTH (1924) first reported that the chloroplast was birefringent. Later, KÜSTER (1934) and MENKE (1934) observed that under crossed nicols these chloroplasts, when viewed on edge, were negatively birefringent with reference to the thickness of the plastid. For example, in *Closterium moniliferum* MENKE (1938a) found a negative uniaxial form-birefringence in the living cell. This suggested that layers of material were present in the chloroplast; the layers had a refractive index higher than the medium in which the layers were embedded; the layers or lamellae were parallel to the long axis of the plastid. At the death of the cell, the birefringence of the chloroplast increased markedly. The increase in birefringence might be explained if the embedding phase would contract away from the lamellae and thus cause the phase boundary to be more abrupt between the lamellae and the medium on either side of the lamellae. To obtain confirmatory evidence of lamellar organization (MENKE und KÜSTER 1938) gold crystals were developed within the chloroplasts of *Elodea* and spinach. The dichroism resulting from the deposited platelets of gold indicated a lamellar organization.

Further evidence of a lamellar structure was presented by MENKE and KOYDL (1939). The fixed *Anthoceros* chloroplast, when photographed on edge in U.V. light, appeared to consist of very thin lamellae. Electron microscope studies indicate that a single lamella is too thin to be distinguished in U.V. light. It is probable that what was observed was a partial unleafing of a series of groups of lamellae at a frayed edge of the plastid.

Suggestions about the chemical organization of the chloroplast may also be obtained from studies with polarized light. That layers of protein are present in the *Mougeotia* chloroplast is suggested by studies of FREY-WYSSLING and STEINMANN (1948). The rather thick plate-like chloroplast of *Mougeotia* is almost as wide and as long as the cell itself and may be considered to represent a single granum. The refractive index of the dense layer component of the chloroplast may be determined by embedding the fixed chloroplast in liquid embedding mixtures of acetone and methylene iodide of gradually increasing refractive indices (*i.e.* from 1.36 to 1.74). When the chloroplast was fixed in ZENKER's fluid (picric acid and $HgCl_2$), the birefringence of the chloroplast changed along a hyperbolic curve as the refractive index of the embedding medium was increased. This is the behavior to be expected of a layered structure in which the thickness of the layers is small compared with the wave length of the visible light used. At refractive index 1.58, the chloroplast became isotropic; this is the point where the layered component has the same refractive index as the embedding medium. Since acetone removed the lipids, the layered component was concluded to consist of protein.

Evidence that the lipids are also part of the layered component was suggested by fixing the *Mougeotia* chloroplast with $OsO_4$ (FREY-WYSSLING und STEIN-MANN 1948). With this fixative the lipids become partially insoluble in organic solvents. In addition to the variable birefringence due to the layers, a constant intrinsic birefringence was now found which is considered to be due to an oriented lipid component. MENKE (1938a) obtained evidence for a lipid component which was oriented so that the long chain lipid molecules were in a direction perpendicular to the layers. STRUGGER (1936) had found that Rhodamin B was a vital stain for chloroplasts. When the chloroplast of *Closterium moniliferum* was stained with Rhodamin B by MENKE (1938a), the stained chloroplasts were optically dichroitic, a result which was interpreted to indicate that the dye molecule, which has a planar structure, becomes oriented parallel to the lipid molecules and that the long axes of the lipid molecules lie perpendicular to the protein layers.

After extraction of the lipid components, MENKE found no negatively uniaxial form-birefringence in contrast to the findings of FREY-WYSSLING and STEINMANN. Chloroplasts in dried cells, or isolated chloroplasts that have dried, or chloroplasts in plasmolyzed cells, become positively uniaxial due to the overcompensation of the form birefringence by the intrinsic birefringence presumably of the lipid molecules; this fact also would appear to support the idea of orientation of the long axes of the lipid molecules in the direction of the thickness of the chloroplast.

From studies of optic dichroism of the chloroplast it should be possible to decide whether chlorophyll molecules have a specific orientation in the chloroplast. The color and intensity of absorption of a planar dye molecule like chlorophyll depend on its orientation with respect to the polarized light used (BRANCH and CALVIN 1941). The absorption of light by such a molecule is greatest when the electric vector of the light vibrates parallel to the major electronic oscillation in the plane of the molecule and not perpendicular to the plane. Crystals containing oriented dye molecules will appear differently colored depending on the direction of the polarized light entering the crystal. Such crystals are said to exhibit optic dichroism. MENKE (1938a) found that the dichroism of the chloroplasts, observed in white light, was very weak in very young pale green chloroplasts; the dichroism was not detectable in older chloroplasts. With red light of 681 m$\mu$ the dichroism sometimes appeared positive. However, FREY-WYSSLING and WUHRMANN (1947) could not detect optic dichroism. One interpretation of this result is that there may be either little or no specific parallel orientation of chlorophyll molecules; *i.e.*, that the chlorophyll molecules have a more or less random orientation in the chloroplast. Other interpretations, however, are also compatible with the absence of optic dichroism. For example, chlorophyll molecules may be oriented alternately, so that one lies parallel and the adjacent one lies perpendicular to the layer. Still another interpretation may be that chlorophyll molecules are oriented in small packets, but each packet would be oriented at random with respect to the others; such a distribution would also result in the absence of optic dichroism.

The evidence from polarized light studies may be summarized as follows: The chloroplast-granum is a layered structure composed of a dense component and a less dense or more fluid component. The dense component is made up of protein and of lipid molecules. The long chain lipid molecules are highly oriented (MENKE 1938a) or not highly oriented (FREY-WYSSLING und STEINMANN 1948); the orientation is such that the lengths of lipid molecules lie perpendicular to the layers. Whether the lipid molecules form a layer independent of the protein, or are part of the protein layer itself cannot be decided from these studies. No marked orientation of chlorophyll molecules can be detected. It seems reasonable to assume that the birefringent protein layers are synonymous with the disk membranes or lamellae as seen in the electron microscope.

### 3. Electron microscope studies[1].

The electron microscope has proved to be a powerful tool for investigating details of the structure of the chloroplast. Comparative studies of the action of fixatives together with examination of thin sections have already provided important data. Considerable advances are to be expected in this field in the next few years. Differential solvent extraction and enzyme digestion studies should provide detailed information on the localization of specific compounds

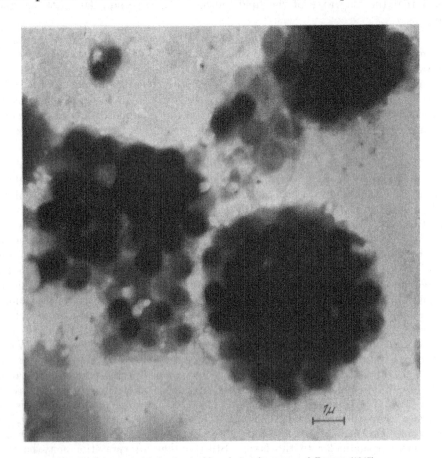

Fig. 5. Grana in isolated spinach chloroplasts. GRANICK and PORTER (1947).

in the grana. Studies of meristematic tissues of the higher plants should provide information on the origin and development of proplastids.

In the higher plants, electron microscope studies reveal the grana clearly, and confirm the idea that they are morphological units. For example, in an isolated spinach chloroplast (Fig. 5) 40 to 60 grana are present (GRANICK and PORTER 1947). The grana, when vacuum dried, appear as dense wafer-shaped bodies some 6000 Å in diameter and 800 Å thick, embedded in a protein-containing matrix, the stroma. In individual chloroplasts the grana appear to be rather uniform in size, but may vary somewhat from one chloroplast to another. In general, the density of the dried granum is quite high, but some grana are

---

[1] In the photographs the unit of length, $1\,\mu$, is represented by a bar. $1\,\mu = 10,000$ Å; $1\,m\mu = 10$ Å; $1$ cm. $= 10,000\,\mu$ or $10^8$ Å.

observed to have a lesser density than most. When the grana are extracted with methanol, a residue probably of protein remains which appears to make up about half of the original material. Similar results have been obtained with isolated tulip chloroplasts by ALGERA *et al.* (1947). These authors also report

the presence in their material, shadowed with metal, of a large number of rather uniform granules of about 25 m$\mu$ diameter both in the stroma and scattered all over the object film whose significance is not yet known.

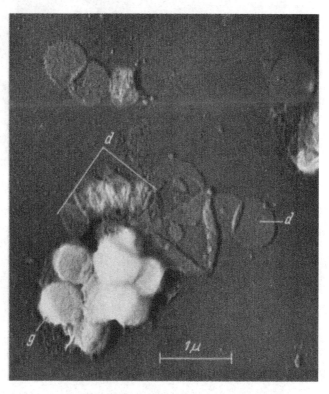

Fig. 6. Grana (*g*) and disks (*d*) of spinach chloroplasts, shadowed with gold. GRANICK and PORTER (1947).

DE DEKEN-GRENSON (1954) has studied the changes in the plastid on exposure of an etiolated leaf to the light. The leucoplasts are described as possessing a finely granular structure devoid of grana or of a primary granum. On exposure to light for a few days, the protein content of the plastid increases by 47 per cent at the same time that grana and the green color develop in the plastid.

In the preparations of spinach chloroplasts, thin membranes and blebs of various dimensions may be seen. KAUSCHE and RUSKA (1940) had first seen them in 1940, but at that time their significance could only be conjectured.

Later the large bladder-like membrane was identified by FREY-WYSSLING and MÜHLETHALER (1949) as the membrane which surrounds the chloroplast. Frequently, disks were observed which were of low density, but whose diameter was that of the dense grana themselves

Fig. 7. *Aspidistra* granum isolated in distilled water showing disks. STEINMANN (1952).

(Fig. 6); occasionally, the disks were seen to be spread out in a manner which suggested that a granum was made of a series of these disks.

The number of disks per cylindrical granum of *Aspidistra* (Fig. 7) may be estimated from STEINMANN's photograph (1952) to be about 30; the vacuum-dried disks were estimated to have a thickness of about 70 Å. LEYON (1953) has examined the isolated chloroplasts of *Beta saccharifica* and *Aspidistra* and found about 15–60 disks per granum. The diameter of the disk was 0.3–0.6 $\mu$.

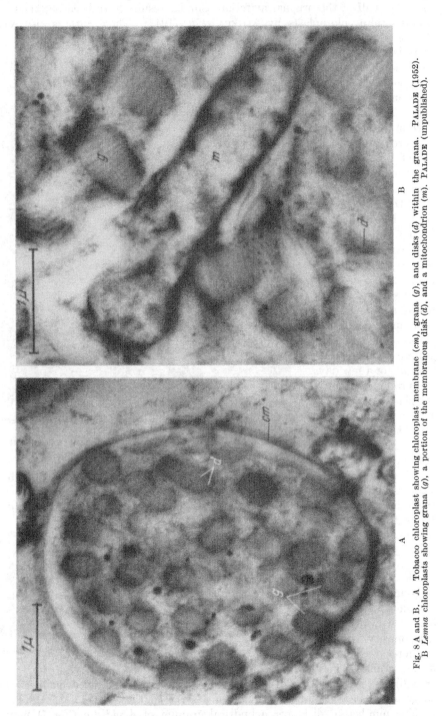

Fig. 8 A and B. A Tobacco chloroplast showing chloroplast membrane (*cm*), grana (*g*), and disks (*d*) within the grana. Palade (1952). B *Lemna* chloroplasts showing grana (*g*), a portion of the membranous disk (*d*), and a mitochondrion (*m*). Palade (unpublished).

Thin sectioning techniques, and buffered osmic acid fixation introduced by Palade (1952), have provided more detailed information of chloroplast structure. Flagellates are fixed in 1 per cent $OsO_4$ in a 0.028 M acetate-veronal buffer of $p_H$ 7.0–8.0 for 1–4 hours. Organisms have a tendency to swell during

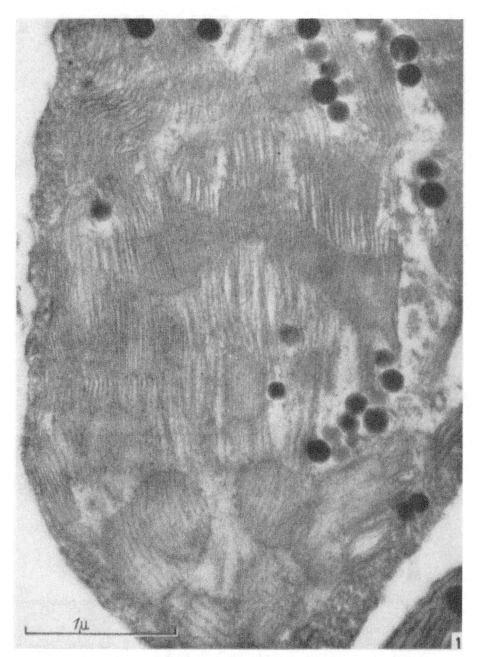

Fig. 9. *Aspidistra* chloroplasts with grana, disks, lamellae of stroma and osmiophile granules.
STEINMANN and SJÖSTRAND (1955).

fixation or post-fixation treatment. Swelling may be prevented partially by increasing the tonicity of the fixative with 0.15 M sucrose (WOLKEN and SCHWERTZ 1953). The fixed specimens are dehydrated through a series of ethanol concentrations in about a half hour. The material is then embedded in n-butyl methacrylate and polymerized by 2 per cent 2,4-dichlorobenzoyl peroxide. This treatment does not appreciably affect the cell dimensions. Sectioning may be

$1\mu$

Fig. 10 A. Chloroplast of barley. The grana are connected with each other by lamellae which extend through the stroma.

accomplished with glass knives on microtomes that are capable of cutting sections approximately $0.05\,\mu$ thick. The plastic remains in the section and is not removed. A modification of this technique by STEINMANN and SJÖSTRAND (1955) has given excellent results. Leaves of *Aspidistra elatior* were cut in 0.1 mm slices and fixed for 5 hours in 1% $OsO_4$ at $0^0$ C. The $OsO_4$ was dissolved either in the veronal buffer $p_H$ 7 or in 0.25 M sucrose. After fixation, the slices were fragmented in a blendor. The cell walls were removed by centrifugation. The supernate was a dispersion of chloroplasts which were dehydrated, embedded in methacrylate and sectioned.

Tobacco chloroplasts which have been fixed and sectioned (PALADE 1953) are seen to be made up of 40–80 grana per chloroplast (Fig. 8 A). The grana appear as cylinders rather than as flat wafers. The grana have not collapsed, because the embedding medium remains in the section. Each granum is $0.3$–$0.4\,\mu$ in diameter and appears to contain 10–15 disks. Lemna chloroplasts appear to contain somewhat taller grana (Fig. 8 B), approximately $0.7\,\mu$ in diameter, and each granum contains about 20 disks.

An interesting feature of the photograph of the tobacco chloroplast is the fact that the disks in the grana appear to lie parallel to the chloroplast membrane. The arrangement of the disks in *Aspidistra elatior* appears to be similar (LEYON 1953).

The studies of STEINMANN and SJÖSTRAND (1955) on the chloroplasts of *Aspidistra elatior* reveal that a granum in this type of chloroplast is represented by a long column made up of a stack of disks (Fig. 9). The width of the disks and therefore the width of the column is somewhat variable. Likewise, the height of the column is variable; a granum column may be made up of a stack of 20–30 disks or even more than 100 disks. Two adjacent grana may be connected through a series of fine lamellae which extend through both grana. The fine lamella is 30 Å thick, the dense lamella that makes up a

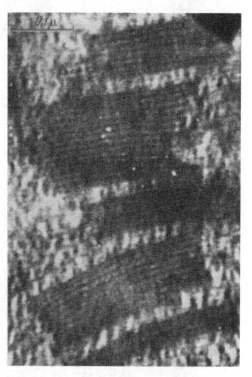

Fig. 10 B. Higher magnification of grana indicating that individual molecules of the disk are of about 30 Å diameter. VON WETTSTEIN (1955).

membrane of a disk is 35 Å thick. The space enclosed by the upper and lower membranes of a disk is about 65 Å. The arrangement of the fine lamellae, spaces and disks in the granum is given in Figure 3. The repeat period in a granum is about 250–270 Å.

VON WETTSTEIN (unpublished) has examined the chloroplast of barley *(Hordeum distichum)*. The lamellated grana (Fig. 10 A) are seen to be connected with each other by several lamellae which extend through the stroma. The inset of Figure 10 B indicates that the lamella is about 30 Å thick. Individual molecules of this dimension are apparent in the print.

A rather frequent component of the chloroplasts of leaves such as spinach, *Aspidistra*, etc. are osmiophilic granules which according to LEYON (personal communication) are present in all stages of development and sometimes in large quantities (LEYON 1954b). (See also Fig. 9.)

In many algae, in contrast to the higher plants, the chloroplast represents a single granum, *i.e.*, the chloroplast is made up of a pile of dense parallel membranes, the membranes being separated from each other by less dense material.

In a good electron microscope preparation a dense membrane is seen to be made up of two dense membranes or lamellae which are closely apposed with only a small space between them. In the egg cell of *Fucus* the dense lamella consists of four fine membranes or layers (Leyon and von Wettstein 1954) (Fig. 11 A).

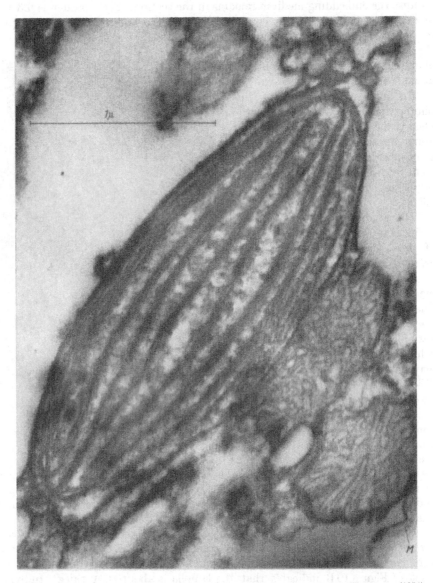

Fig. 11 A. Brown chloroplast in the egg cell of *Fucus vesiculosus*. Leyon and von Wettstein (1954).

The extended sheets or lamellae were first observed in the chloroplasts of *Spirogyra* and of *Mougeotia* by Steinmann (1952). The "lamellae" were reported to have a thickness of 70 Å, but since the embedding medium of polymer was removed from the sections it is uncertain what this dimension represents.

The photographs of the chloroplast sections of two algal flagellates, *Euglena* and *Poteriochromonas*, made by Wolken and Palade (1953), clearly show the sheets of dense parallel lamellae which extend from one end of the chloroplast to the other.

In *Poteriochromonas stipitata* (WOLKEN and SCHWERTZ 1953) which is an almost spherical unicellular flagellate Chrysomonad of $7-8\,\mu$ diameter there are present two lens-shaped chloroplasts which flank the nucleus. The average chloroplast is $3.26\,\mu$ long by $0.81\,\mu$ thick. The two chloroplasts are connected to each other by a membrane that lies closely apposed to the anterior part of the nucleus. The number of dense lamellae per chloroplast is about 10 (Fig. 11 B). The dense lamella is 390 Å thick and the interspace is 396 Å thick. On the hypothesis of a disk structure, the disk thickness including its upper and lower membrane would be approximately 650 Å thick (Fig. 15). The authors have calculated that a chloroplast of *Poteriochromonas* would contain about $1.10 \times 10^8$ chlorophyll molecules. The area of the 10 dense lamellae is $135 \times 10^{-8}$ cm², or if considered as two membranes, the area of the disk membranes would be $271 \times 10^{-8}$ cm². The area of the flat porphyrin "head" of the chlorophyll molecule is approximately 225 Å². If one assumes that the flat porphyrin heads were fitted together to form a continuous flat monomolecular film in the disk membranes, then the area occupied would be $248 \times 10^{-8}$ cm². Thus the chlorophyll molecules, if arranged as suggested, would just fill the membranes of the disks with a monofilm of chlorophyll. When *Poteriochromonas* is placed in the dark, the cells lose their green color and the plastids become shrunken and the lamellae disappear.

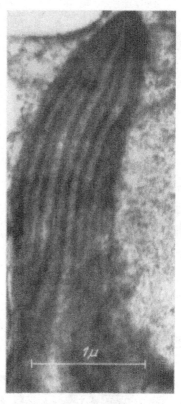

Fig. 11 B. Portion of the single-granum chloroplast of *Poteriochromonas stipitata*. The chloroplast is twisted so that the upper portion is cut perpendicular to the lamellae, and the lower portion is cut more or less parallel to the lamellae. The chloroplast abuts the nucleus at the right. WOLKEN and PALADE (1953).

The chloroplasts of the brown alga *Fucus vesiculosus* have been studied by LEYON and VON WETTSTEIN (1954). The spindle-shaped chloroplasts may be considered to represent a single granum containing parallel lamellae. The pinching-off of lamellae in the division of grana in *Fucus* plastids has been documented by the studies of VON WETTSTEIN (1954).

A chloroplast of an egg cell of *Fucus* measures $1.4-3.3\,\mu$ in length and contains 8 to 10 lamellae. The outermost lamella runs parallel to the plastid membrane and appears to extend completely around the plastid. There are two optically clear "vacuoles", one at either end of the plastid, which lie within the boundary of the outermost lamella. The lamella is about $370 \pm 80$ Å thick (Fig. 11A).

In the cells of the vegetative meristematic tip of *Fucus*, the plastid is 2.9 to $4.0\,\mu$ long and contains 8 to 10 extended lamellae, each about $280 \pm 70$ Å in thickness; each lamella is made up of two membranes. In old cortical cells the plastid is $4-8\,\mu$ long and contains 12 to 17 lamellae per plastid, the thickness per lamella being $280 \pm 90$ Å.

*Euglena gracilis* var. *bacillaris* is an elongated unicellular organism about $70 \times 20\,\mu$ (PRINGSHEIM 1948). It contains 8–12 chloroplasts, $1.23\,\mu$ in diameter $\times$ 6.5 $\mu$ in length; on the average, the chloroplast contains 21 lamellae (Fig. 12 A). The lamellar thickness is 242 Å and the interspace thickness is 374 Å. This chloroplast contains $1.02 \times 10^9$ chlorophyll molecules. Here also the area of the dense double membrane that constitutes a lamella would just approximate the area taken up by flat porphyrin units packed so as to form a flat continuous monomolecular film. When grown in the dark on anorganic substrate, *Euglena* loses its chlorophyll and the lamellae disappear. After exposure to some 4 hours of light, lamellae become recognizable in the plastids. The lamellae at this time were relatively few in number and of low density. With longer exposure to light the lamellae increased in number and density to a maximum by 72 hours. Mitochondria (Fig. 12 C) were not observed to be

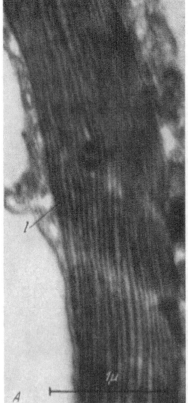

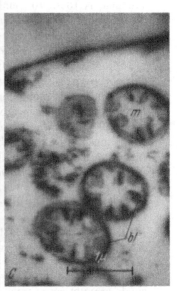

Fig. 12 A–C. *Euglena* chloroplast. A. Showing parallel lamellae (*l*) in a longitudinal section of a single-granum chloroplast. B. Showing organization of pyrenoid (*p*) with the lamellae traversing the pyrenoid. Wolken and Schwertz (1953). C. Cross section of mitochondria (*m*) showing pocket-like bladders (*bl*) projecting into the mitochondria. At the top is seen the outer edge of the *Euglena* membrane. Palade (unpublished).

converted to plastids. According to Pringsheim (1948) *Euglena* which has been grown in the dark on organic substrates contains plastids which are disk-shaped yellowish structures similar to chloroplasts, though smaller. Thus the ability to make small amounts of carotenoids must still be present in plastids developing in darkness, although a lamellar structure is no longer evident.

Leyon (1954a) reported that the chloroplasts of *Spirogyra*, *Mougeotia*, *Closterium*, and *Enteromorpha* contain dense extended layers of lamellae. The lamellae have a thickness of $80 \pm 20$ Å. In *Closterium lunula* the sheets of lamellae, although generally layered parallel to the length of the cell, may bend upon themselves as seen in a cross section of the pyrenoid (Fig. 13A, B).

The cup-shaped chloroplast of *Chlamydomonas reinhardi* has been examined by Sager and Palade (1954). As fixing fluid, an acetate-veronal buffer containing 1 per cent $OsO_4$ at $p_H$ 8–8.5 gave best results. The chloroplast has an undulating surface, a feature which explains the frequent presence in sections of several chloroplast fragments. The chloroplast is surrounded by a continuous membrane about 100 Å thick (Fig. 14 A). This membrane encloses some 40 single membranes, equivalent to 20 "lamellae"; it also encloses the carotenoid-containing droplets [*i.e.*, the eyespot (Fig. 14 B)], the pyrenoid surrounded by a sheath of starch plates (Fig. 14 A, D), and additional starch grains. The chloroplast is not uniformly green, but has areas that are less green. This is correlated with the fact that the lamellae are not continuous; narrow, irregular spaces, filled with a fine granular material, comparable to the stroma of higher plants, separate the stacks of lamellae. The stacks are less individualized than the grana of

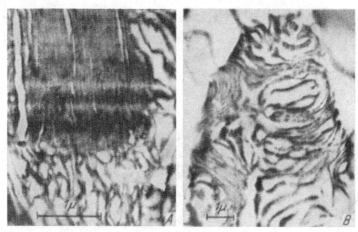

Fig. 13 A and B.  A. Section of part of a pyrenoid of *Closterium acerosum*. LEYON (1954 a).  B. Arrangement of the lamellae in a pyrenoid of *Closterium lunula*. LEYON (1954 a).

higher plants. The authors consider that the chloroplast of *Chlamydomonas* may represent a transitional form between the one-granum chloroplast of *Euglena* and the grana-containing chloroplasts of higher plants. The fine fixation has permitted further details to be detected. The coarse dense lamellae of other algae are here seen to be represented by two distinct dense membranes (Fig. 14 B and C). The dense membranes appear to be attached in pairs to points on the chloroplast membrane (Fig. 14 C). Each membrane is approximately 50 Å thick; the less dense space between them is about 240 Å. Sections through the chloroplast suggest that a disk is made up of a less dense region which is enclosed by the two dense membranes, so the total thickness of a disk would be 340 Å (Fig. 15).

A yellow mutant of *Chlamydomonas reinhardi* was also examined. This mutant, like the higher plants, produces chlorophyll only in the light. Wild type *Chlamydomonas* does not require light for chlorophyll formation and its disk structure is essentially the same when grown in light or dark. The yellow mutant of *Chlamydomonas*, however, requires light to form chlorophyll. In the light it turns green and develops the typical disk structure. However, if the yellow mutant is not exposed to light, the colorless plastid is still present, but the dense membranes of the disk are no longer to be seen. The pale yellow plastid is not lacking in density. Starch grains are present, vesicles and tubules are seen especially near the plastid membrane, near the eyespot and in the region of the pyrenoid. Do these vesicles and tubules perhaps represent rudiments

34*

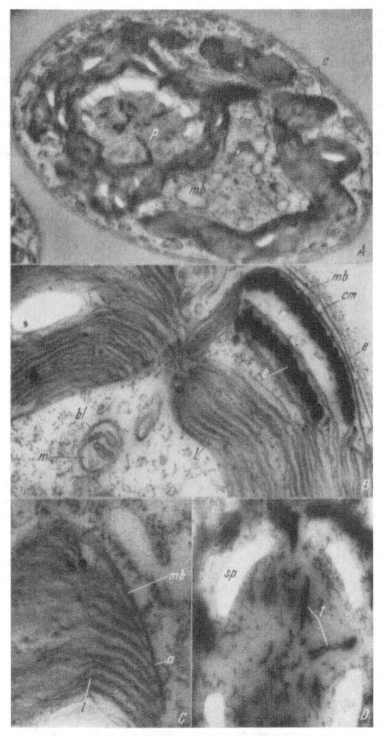

Fig. 14A–D. *Chlamydomonas rheinhardi* electron micrographs. SAGER and PALADE (1954). A. Chloroplast (*c*), although continuous, appears in sections due to its undulating surface. Chloroplast membrane (*mb*), and pyrenoid (*p*) are shown. B. Anterior end of *Chlamydomonas*, showing cytoplasmic membrane (*cm*), chloroplast membrane (*mb*), eye-spot region with two layers of carotenoid granules at *e* and *e*, mitochondrion in cross section (*m*), with pocket-like projection (*bl*), and lamellae (*l*) consisting of two membranes. C. Chloroplast membrane (*mb*) to which are attached the thin single membranes at (*a*). D. Pyrenoid in the lower half of the cell. The white areas (*sp*) are the regions from which starch has fallen out. Tubular elements (*t*), possibly extensions of the disks project between the starch plates (*sp*) into the central area.

of the disk structure seen in the green plastid[1]? It has been observed that if chlorophyll is present then the dense membranes are present; if chlorophyll is absent, the dense membranes are absent. This fact suggests the possibility that chlorophyll may be a component of the dense membrane.

From the blue green unicellular alga, *Synechococcus cedorum*, CALVIN and LYNCH (1952) obtained dense "grana" with diameters of 2200 Å. The cells were ground with alumina and centrifuged at high speed. Carotenoids and chlorophyll were associated with the grana; the phycocyanin was in solution.

*Rhodospirillum rubrum*, a photosynthetic bacterium, also contains granules in which are concentrated bacteriochlorophyll, carotenoids, proteins, and a small amount of pentose nucleic acid (PARDEE 1952). The "grana" had a rather uniform diameter of about 1100 Å. If the cells were transferred to the dark on organic substrates, they grew, but lost their pigments. No "grana" were found in such cells.

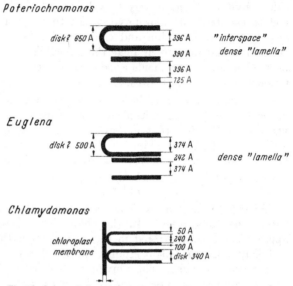

Fig. 15. Interpretation of the approximate dimensions of the lamellae, membranes, interspaces and disks of *Poteriochromonas*, *Euglena*, and *Chlamydomonas*.

The following hypothesis may perhaps explain the appearance of homogeneous chloroplasts, the "plastonema" and the granular chloroplasts. In young rapidly growing chloroplasts, the disks or lamellae may be as extensive as in the algal chloroplast of *Euglena* which represents a single granum (Fig. 16a). Then intermediate stages may occur in which the disks are more or less continuous, with

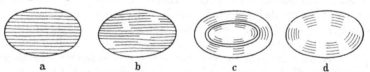

a          b          c          d

Fig. 16a–d. Transitions from the single-granum chloroplast (a) of algae to the partially discontinuous granum (b) as in some algae, to the incompletely separated grana (c) representing a plastonema thread, to the multiple grana (d) of the chloroplast of higher plants.

some areas of the chloroplasts being empty of them (Fig. 16b). A later stage may result if the disks are pinched-off incompletely so that disks of one granum still would be connected by several or more disks to an adjacent granum (Fig. 16c). This latter stage would result in the appearance of a "plastonema" thread, *i.e.*, of grana connected by threads. The last stage would be represented by a distinct pinching-off of the disks to form distinctly separated grana (Fig. 16d).

## 4. Chlorophyll localization in chloroplasts.

Since the time of ENGELMANN (1894) it has been known that there is no photosynthesis without chlorophyll. More recently it has been shown that in the HILL reaction the rate of release of $O_2$ by isolated chloroplasts from young

---

[1] See also p. 544.

barley seedlings is directly proportional to their content of chlorophyll (SMITH *et al.* 1947, 1948, 1952). The action spectrum in the HILL reaction also has been identified with that of chlorophyll (CHEN 1952).

An important step of photosynthesis is the absorption of light by chlorophyll. It has been shown, with various algal species that a portion of the light energy may be absorbed by other chloroplast pigments and then transmitted to chlorophyll *a*. Light energy may be absorbed by chlorophyll *b* of green algae and transmitted to chlorophyll *a*. In the brown algae, light energy may be absorbed by fucoxanthin and chlorophyll *c* for transmission to chlorophyll *a*. In the red and blue-green algae the phycobilins can act in this manner. In the purple bacteria, carotenoids appear to absorb and transfer light energy to bacteriochlorophyll (SMITH 1949, DUYSENS 1951, STRAIN 1951, RABINOWITCH 1951).

There also appears to be evidence that light absorbed by one chlorophyll *a* molecule can be transmitted to other chlorophyll *a* molecules. It has been variously estimated that energy may be transmitted through a hundred or even several thousand chlorophyll molecules to some enzyme molecule and still be effective in photosynthesis (THOMAS *et al.* 1953). In order to transmit light energy it would appear likely that the molecules of photosynthetic pigments should be closely packed and well oriented.

What is the present evidence in regard to the distribution of the chlorophylls within the chloroplast? In the chloroplast of the higher plants chlorophyll appears to be concentrated in the grana rather than in the stroma. JUNGERS and DOUTRELIGNE (1943) studied the amyloplasts of potato which turn green on exposure to light. The amyloplast covering a portion of the starch grain was observed to turn green, and at the edges of this plastid a single layer of grana was observed. They reported that here the individual granum can be seen to be green; the matrix appears colorless against the white background of the starch grain. WEIER and STOCKING (1952b) have noted that tobacco chloroplasts may swell slightly even when isolated in sugar solutions; the grana may then move apart and only the grana are seen to contain chlorophyll. To demonstrate that chlorophyll is in the grana, this pigment was used as a photosensitizer to bring about reduction of $Ag^+$ to $Ag^0$ (THOMAS *et al.* 1954). Presumably, silver would be deposited in the neighborhood of chlorophyll. The chloroplasts of higher plants were disintegrated weakly with a magnetostriction oscillator under nitrogen. The "grana" were then treated with $AgNO_3$ in the presence of light. It was observed with the electron microscope that silver was deposited on the proteinaceous parts of the grana and not in the stroma. In *Euglena* and *Poteriochromonas*, if the green chloroplast is considered to consist of one granum without stroma, then there is no doubt that chlorophyll is localized in the granum.

Where in the granum is chlorophyll localized? Indirect evidence suggests that chlorophyll might be localized in the dense lamellae (*i.e.*, disk membranes). For example, plastids which require light for the development of chlorophyll, as in *Euglena* or in a mutant *Chlamydomonas*, contain no lamellae and are not green. When the cells are placed in the light, chlorophyll develops in them and lamellae appear. Studies are needed with solvents and enzymes to obtain further evidence on chlorophyll localization.

The concentration of chlorophyll in chloroplasts is relatively high, 0.016 to 0.1 M (Table 1). In the coarse lamellae of *Euglena* and *Poteriochromonas* WOLKEN and SCHWERTZ (1953) have calculated that the flat porphyrin rings of chlorophyll, if laid side by side, would form monomolecular layers just equivalent to the surface area of the lamellae, if the lamellae are considered as a double membrane. In *Chlamydomonas* and presumably in higher plants the disk membranes are only 35–50 Å thick. The disk membranes would constitute perhaps only about 25 per cent of the volume of a granum. WOLKEN and SCHWERTZ (1953) have summarized the available data of chloroplast volume and chlorophyll concentration as presented in the table.

Table 1. *Volume and chlorophyll content of chloroplasts* (WOLKEN and SCHWERTZ, 1953).

| Organism | Volume of chloroplast ml. | Chlorophyll molecules per chloroplast | Concentration of chlorophyll molecules in moles per liter |
|---|---|---|---|
| Elodea densa . . . . . . . . . | $2.8 \times 10^{-11}$ | $1.7 \times 10^9$ | 0.1 |
| Mnium . . . . . . . . . . . . | $4.1 \times 10^{-11}$ | $1.6 \times 10^9$ | 0.065 |
| Euglena gracilis . . . . . . . . | $6.6 \times 10^{-11}$ | $1.02 \times 10^9$ | 0.025 |
| Poteriochromonas stipitata . . . | $1.1 \times 10^{-11}$ | $0.11 \times 10^9$ | 0.016 |

If a disk membrane is 50Å thick and 5000 Å in diameter, then some 90,000 chlorophyll molecules can be fitted into a monofilm of equivalent area, assuming juxtaposition of the porphyrin units. Considering the volume of the disk membrane, this would represent an approximately 0.15 molar concentration of chlorophyll. If one assumes values for the concentration of chlorophyll molecules in the chloroplasts of *Elodea* and *Mnium* as given in the Table 1 and further assumes that the grana occupy only one half the volume of the chloroplast, and that the membranes of the disks make up only 25 per cent of the volume of the grana, then the concentration of chlorophyll in the disk membranes of *Elodea* would be 0.8 molar and for *Mnium* 0.5 molar. Assuming that chlorophyll has a density of 1.0, this would mean that as much as half of the disk volume was made up of chlorophyll. Various other calculations result in similarly high values, on the hypothesis that chlorophyll is concentrated in the disk membranes of the grana [for further discussion see RABINOWITCH (1945)].

Chlorophyll in organic solvents has an absorption maximum in the red region of the spectrum at 6650 Å, but in the chloroplast the absorption maximum is at 6810 Å. Also chlorophyll fluorescence in the plastid is very low compared to the intense red fluorescence of chlorophyll in organic solvents. These facts suggest that chlorophyll is not in solution, (*i.e.*, not monomolecularly dispersed). Rather it seems probable that chlorophyll is an aggregated state, both because of the high concentration of chlorophyll in the chloroplast and because of the evidence of energy transmission from one chlorophyll molecule to another. If chlorophyll molecules are packed with the edges of the porphyrin planes adjacent to each other, or with the porphyrin planes facing each other, then the chloroplasts should exhibit a strong dichroism in polarized light. This is not found. See p. 521 for further discussion.

## 5. Pyrenoids.

The pyrenoids are dense, spherical bodies, usually surrounded by a sheath of starch grains, and embedded in the matrix of certain chloroplasts. The pyrenoids are present in a great majority of the green algae, the *Chlorophyceae*. They are also found in the chloroplasts of some diatom species, several *Xanthophyceae, Dinophyceae, Bangiales, Nemalionales,* and *Ectocarpales*; but they are absent from *Microspora*, from certain *Siphonales* and from the great majority of brown and red algae. The chloroplast of the primitive moss *Anthoceros* possesses a pyrenoid (McAllister 1914). Other bryophytes and the higher plants lack pyrenoids. Czurda (1928) has written an extensive review on pyrenoids and algal starch.

There may be one or more pyrenoids per chloroplast, depending on the species. For example, *Chlamydomonas* and *Zygnema* contain one pyrenoid per plastid; *Spirogyra* may contain several pyrenoids per plastid. Pyrenoids are said to be rich in protein since they generally stain more intensely with protein stains. Characteristic of many of the pyrenoids is the formation of a starch sheath surrounding the pyrenoid. This fact suggests that possibly the appropriate enzymes and conditions for starch synthesis are localized in this region of the pyrenoid. Although starch may form in other parts of the cell, starch is deposited in the region of the pyrenoid first and disappears from there last. For example in *Spirogyra*, starch is formed around the pyrenoids but minute grains of stroma starch may also be formed independently of the pyrenoids.

In some cases pyrenoids are said to be duplicated by fission (Fig. 17), although this may be only apparent. In *Spirogyra*, Szejnman (1933) has followed the

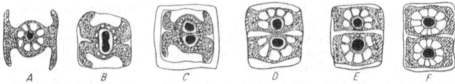

Fig. 17A–F. Division of plastid and pyrenoid (area in black) in *Hyalotheca mucosa* with its surrounding starch masses. By permission from *Introduction to Cytology*. L. W. Sharp. McGraw-Hill Co., 1934.

development of pyrenoids. In a hanging drop he was able to observe the same cell for a number of days. Growth of the ribbon chloroplast was found to attain 5 per cent of its length per hour. The growth was partly intercalary. The most intense growth occurred during the night which followed a cellular division. During the morning hours pyrenoids could be seen to form *de novo*. Sometimes, doubling of the length of the chloroplast doubled the number of pyrenoids. No evidence was obtained that the pyrenoids arose from pre-existing ones by division.

In *Euglena* (Wolken and Palade 1953) electron photomicrographs reveal the pyrenoid (Fig. 12B) as a dense central region through which extend dense lamellae. The spaces between the lamellae in this region may contain a higher concentration of protein which holds the lamellae more firmly together during fixation and sectioning. No carbohydrate sheath surrounds the pyrenoid. Frequently, the pyrenoid protrudes at the surface of the chloroplast and its protrusion is adjacent to a vacuole which may or may not be connected with the formation of the insoluble carbohydrate grains, paramylum, in this organism.

In *Chlamydomonas*, the pyrenoid is more typical, since a sheath of starch plates is deposited around the pyrenoid (Sager and Palade 1954). The pyrenoid is a relatively large, non-laminated body, 1.5–2 $\mu$ diameter, embedded in the posterior part of the chloroplast. Tubular elements, possibly extensions of the disks, project between the starch plates into the central area. These tubules are characterized by irregular thickenings along their walls (Fig. 14D). In the yellow mutant no starch plates are found, but in the posterior region of the cell bits of the tubular elements with irregular thickenings are present.

Leyon (1954a) has examined the pyrenoids of *Spirogyra*, *Closterium*, *Cladophora*, and *Enteromorpha* in the electron microscope. The lamellae of the chloroplast appear to continue their parallel path through the pyrenoid as seen in *Spirogyra*; or a number of groups of parallel lamellae may fold back on themselves as in *Closterium lunula* (Fig. 13 B). Starch appears to be deposited between the lamellae, adjacent to the pyrenoid. The pyrenoid of *Closterium acerosum* (Fig. 13 A) was dissected out of the plastid and was found to contain chlorophyll.

## 6. Eye-spot or stigma.

An eye-spot is an orange-red colored dot or streak region that is often present in the anterior portion of motile forms of many algae. A number of species may contain an eye-spot in a median or posterior portion of the cell, rather than in the anterior portion. An eye-spot is present in motile forms of green algae, either unicellular or colonial, and also in the zoospores and gametes of vegetatively non-motile forms. Eye-spots have been recorded in a few non-motile forms such as *Asterococcus*; these species are considered as relics of a motile ancestral condition. The eye-spots of the colonial *Volvox* are larger in the anterior cells of the colony and become progressively smaller or even are absent in the posterior cells (BOLD 1951).

The eye-spot is said to be produced by the division of a pre-existing one or to arise *de novo*. In the *Volvocales*, the division of the protoplast may be accompanied by a division of the eye-spot or new eye-spots may be formed *de novo* in each of the protoplasts. In the filamentous green alga, *Ulva*, prior to gametogenesis the chloroplast becomes yellow-green. At the time of the first division, a tiny red region develops in the plastid. This region appears to grow more intensely red and to divide by repeated bipartitions each time the protoplast divides, so that each gamete swarmer ultimately receives an eye-spot derived genetically from the original red region of the plastid (SCHILLER 1923).

The eye-spot consists of a group of carotenoid droplets. In *Chlamydomonas reinhardi* (SAGER and PALADE 1954) the droplets are present in a small area within the chloroplast. The electron micrographs show that the eye-spot is made up of two parallel layers of dense uniform droplets packed hexagonally (Fig. 14 B). The droplets are probably formed in the interior of two adjacent disks. What factors of differentiation cause this region to accumulate the carotenoid droplets is unknown. In the yellow mutant plastid of *Chlamydomonas* which lacks lamellae, the eye-spot is present but not so well organized. One layer is made up of droplets of uniform size, but the second layer is represented by dense droplets of varying size which are scattered in the neighborhood of the eye-spot.

In *Euglena* (WOLKEN and PALADE 1953) the electron micrographs do not indicate whether the eye-spot is contained within the anterior chloroplast or lies outside of this chloroplast. The eye-spot is represented by a cluster of dense orange-red spheres 1000–3000 Å in diameter, embedded in a colorless matrix. The flagellar swelling in the cytopharynx lies close to the stigma region. The flagellar swelling is thought to be the organelle that is sensitive to light; and the function of the stigma is considered to be that of a light filter which, depending on the orientation, may prevent blue light from reaching the swelling (MAST 1941). In 1882 ENGELMANN had demonstrated that a shadow cast over the posterior portion of a *Euglena* did not affect its movement, but light was effective when it fell upon the region of the stigma. The action spectrum for a number of species of *Euglena* is identical with the absorption spectrum of astaxanthin (GOODWIN 1952). *Euglenaceae* that possess an eye-spot but lack chlorophyll are phototactic. The *Euglenaceae* which have no eye-spot as well as related colorless families, the *Astasaceae* and *Peranemaceae* which lack eye-spots are not phototactic. However, there must be an additional light sensitive reaction present, since members of these colorless families (*e.g. Peranema*) are sensitive to changes in light intensity. According to JAHN (1951) the eye-spot region divides during cell division and does not arise *de novo*. The colored granules, however, may disperse during prophase and then reaggregate during anaphase.

## 7. Summary.

Recent evidence from studies especially with the electron microscope suggest the following interpretation of the structure of the generalized mature chloroplast of higher plants (Fig. 8 A). Each chloroplast is surrounded by a semipermeable membrane about 100 Å thick. The chloroplast contains about 50 dense "grana" embedded in a colorless stroma or protein matrix.

A granum has the shape of a column or of a cylinder $\sim$4000–6000 Å in diameter and $\sim$5000–8000 Å in height. Each granum consists of a stack of 15 or more disks. Depending on the species, the diameter of the disks (4000–6000 Å) may be more or less constant or may be highly variable in the same chloroplast. A disk ($\sim$130 Å in thickness) consists of a dense upper and lower membrane (each about 35 Å thick) enclosing a space $\sim$65 Å thick. Attached to the outer surfaces of the dense disk membranes are fine membranes ($\sim$30 Å thick) which extend out into the stroma and connect one granum to another (STEINMANN and SJÖSTRAND 1955). In addition there is a space of about 65 Å which separates these fine membranes from each other (Fig. 3).

The chloroplast of an alga like *Euglena* would appear to represent a single granum which is surrounded by a chloroplast membrane. Such a granum contains lamellae or disks which extend the length and the width of the chloroplast; the lamellae are thicker and have thicker and denser membranes than those of higher plants.

Studies with the polarizing microscope indicate that the dense lamellae contain proteins and lipids. Chlorophyll is present in the grana and not in the stroma of the chloroplast. The chlorophyll content of a chloroplast is high. If chlorophyll is assumed to be concentrated in the lamellar membranes, the estimated concentration of chlorophyll would be 0.15–0.8 molar. Very little or no optic dichroism of chlorophyll can be detected in a chloroplast, a fact which has been interpreted as indicating that an ordered orientation of chlorophyll molecules may be absent. However, since studies of photosynthesis indicate that light energy may be transmitted through many chlorophyll molecules, it seems necessary to assume some ordered orientation of chlorophyll molecules.

A pyrenoid is a dense protein-containing region of the chloroplast around which starch may be deposited. In one type of pyrenoid the lamellae extend right through the pyrenoid; in another type, tubular elements may project into the pyrenoid from between the starch plates. Pyrenoids have been claimed to arise from pre-existing pyrenoids although other evidence suggests that they may arise *de novo*. The function of the pyrenoids is uncertain. Do they represent centers of higher enzyme activity and perhaps regions in which plastogenes are localized ?

The eye-spot consists of a group of carotenoid droplets. The eye-spot may arise from the division of a pre-existing one or may arise *de novo*. The function of the eye-spot in phototaxis is discussed.

## IV. Origin and development of chloroplasts.

The chloroplasts are of interest not only because they represent the primary seat of photosynthesis, but also because they represent a unit of cytoplasm which, like the nucleus, can be shown in favorable examples to be "self-reproducing". In general, the evidence for the self-reproducing ability of the chloroplasts is (a) In the lower plants the chloroplasts may be observed to divide, and their continuity can be established from one cell to the next by direct observation. (b) The transmission of the chloroplasts through the cytoplasm of only one parent demonstrates continuity of the chloroplasts and inheritance in a non-Mendelian fashion. If a plastid mutates, such a mutation is inherited in a non-Mendelian fashion (discussed in section V). (c) The chloroplasts, once lost from a cell, cannot be regenerated by that cell.

### 1. The division and continuity of the chloroplasts in the lower plants.

Division of the chloroplasts may be observed directly in the cells of the lower plants, especially in those plants which contain only one or a few chloroplasts

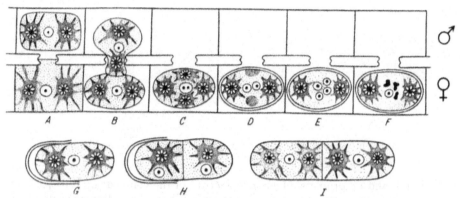

Fig. 18 A–I.  A–F Conjugation and zygospore formation in *Zygnema*. The male gamete passes through the conjugating tube and fuses with the female gamete. The two "male" chloroplasts degenerate. Thus, all subsequent chloroplasts arise by divisions of the "female" chloroplasts. In meiosis 4 nuclei are formed of which 3 degenerate. G–I Germination of zygospore and vegetative division of filament showing chloroplast division at H. (90) By permission from *Cryptogamic Botany*, G. M. Smith, McGraw-Hill Co.

per cell (Fig 17, Fig. 18 H, I). After cell and chloroplast division, the flat lamellae within the chloroplast grow or extend out as broad sheets while the chloroplast and cell enlarge. In the vegetative cell division of algae, the division of the chloroplast may occur at an earlier or somewhat later stage than the division of the nucleus. The pyrenoids may divide at the same time, but in many forms they disappear during cell division to reappear in each daughter cell. Primarily on the basis of direct observations of this kind the early workers, notably SCHIM-PER (1883) and MEYER (1883), proposed that plastids never originate *de novo*, but always arise from pre-existing plastids by division. Since, in the algae, a plastid is essentially a granum, the inference may be made that a granum never arises *de novo* but only from a pre-existing granum.

Not only may the chloroplasts of the lower plants be followed in vegetative division, but a number of cases have been reported where the plastid has been followed during formation of zygospores and gametes. These examples also demonstrate the fact that the plastid does not arise *de novo*, but from a pre-existing plastid. For example, in *Zygnema* each vegetative cell contains one nucleus and two plastids, all of which divide at each vegetative cell-division

(Fig. 18H, I). In sexual reproduction, the entire protoplast, with its nucleus and two plastids, passes through the conjugating tube as a "male" gamete and unites with a similar complete protoplast ("female" gamete) of another filament (Fig. 18A–F). The two nuclei fuse, forming the primary nucleus of the new individual (zygospore nucleus), while the two plastids contributed by the male gamete degenerate, leaving the two furnished by the "female" gamete as plastids of the new individual. A similar process occurs in *Spirogyra*.

A study now classic, of chloroplast inheritance in a species related to *Spirogyra*, was made by Chmielevsky (1890) (Fig. 19). In this alga, a *Rhynchonema* species, the vegetative cells contain one chloroplast per cell. At the time of comjugation tannins disappear from the cell, but starch and oil droplets increase.

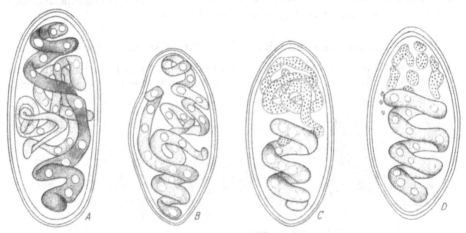

Fig. 19A–D. Degeneration of the "male" spiral chloroplast in the zygospore of *Rhynchonema*.
Chmielevsky (1890).

The starch and oil droplets which are plentiful in the conjugating cells pass into the young zygote and gradually disappear. When the cuticular wall of the zygote forms and becomes brown, the green spiral chloroplasts are still visible through the wall. The zygotes were fixed at a stage when the second and third coats of the zygote were formed. Fixation occurred in 1% $OsO_4$ for 5–10 seconds. The filaments containing the zygotes were then washed with water and placed in dilute glycerine; the glycerine was permitted to thicken by evaporation. The zygotes were then quite transparent. It was possible to observe by this technique that in the zygote the spiral chloroplast derived from the "male" gamete gradually became twisted, turned yellow, became thinner and broke up into particles that later united and were transferred to the vacuole. Thus, in the successive vegetative cell divisions, only the chloroplast of the "female" gamete was inherited.

Chloroplasts once lost from a cell cannot be regenerated by that cell. An interesting example of this fact and a further demonstration of the self-duplicating activity of the chloroplast is one described by Lwoff and Dusi (1935). In *Euglena mesnili* there are about 100 chloroplasts per cell and in this particular species chlorophyll can be formed in the dark. When the species is cultured in the dark, multiplication of the cell, although slow, is faster than the division of the chloroplasts. After some 15 months, the cells contain only one or two green chloroplasts. Finally, one may obtain individuals with no vestige of a chloroplast. These individuals are thenceforth incapable of giving rise to chloroplasts even when placed in the light.

In nature a large number of colorless algae exist which, on the basis of their similar morphology, utilization of similar organic substrates, and the storage of identical products, are recognized to be related to species of green algae. For example, species of the genus *Astasia* are undoubtedly related to species of *Euglena*. *Polytoma* species resemble *Chlamydomonas* species. *Prototheca* is probably a colorless *Chlorella* (PRINGSHEIM 1941).

A few other examples of chloroplast division and continuity in the higher cryptogams may be cited. In the liverwort, *Anthoceros*, the cells of the thallus contain a single plastid which divides with the nucleus at each cell division. The egg likewise contains a plastid, but the tiny spermatozoid has none. The zygote and the sporocyte cells which are later formed from the fertilized egg, are therefore characterized, like the cells of the gametophyte, by the presence of one plastid per cell, this plastid being derived from the division of the original egg cell plastid. Although it is difficult to demonstrate the plastid in the young sporogenous cells, every sporocyte contains one. As shown by DAVIS (1899), the sporocyte plastid divides twice during the prophase of the first division of the sporocyte nucleus, so that each spore of the resulting quartet receives one plastid. Upon germination, the spore produces a gametophyte with one plastid in each cell and the cycle is complete. The plastids, therefore, remain as morphological entities throughout the whole life cycle, multiplying exclusively by division. The studies of SCHERRER (1914) support the continuity of the chloroplasts through the life cycle of *Anthoceros*. Evidence for the continuity of chloroplasts in the liverworts and mosses has been obtained more recently by KAJA (1954).

In the ferns *Selaginella* and *Isoetes* the plastids have also been found to behave as morphological individuals, multiplying exclusively by division. In such cases, the plastids possess an individuality comparable to that of the nuclei, from which they differ conspicuously, however, in undergoing no fusion at the time of sexual reproduction. For example, in the sporophyte of *Isoetes*, STEWART (1948) found that in the meristematic cells of the root or leaf, there is present a single plastid per cell (as in the meristematic cells of many bryophytes and pteridophytes). The division of the single plastid precedes the division of the nucleus. Prior to cell division, the plastid of the meristematic cell becomes greatly elongated, the daughter plastids passing to the opposite poles of the spindle. At the interphase, a resting cell will contain only one plastid. As the meristematic cell matures and enlarges, the single plastid may become filled with numerous starch grains or the plastid may undergo several cell divisions, in which a variable number of plastids per cell are produced. The daughter plastids, thus formed, will become either starch-filled leucoplasts of storage cells or chloroplasts of leaf cells. During nuclear division, these plastids show no tendency to occupy the poles of the cells in contrast to the behavior of the plastids in a meristematic cell. The plastids of such maturing cells lose their intimate association with the nucleus and lie at random within the cell.

It is interesting to consider the hypothesis that the behavior of the chloroplasts in this higher fern may represent a transition to chloroplast behavior in the phanerogams. In *Isoetes* the young meristematic cell contains only a single chloroplast during cell division, but as the cell matures, it comes to have a number of chloroplasts per cell.

In gymnosperms and angiosperms, well defined plastids are not seen in meristematic cells. However, precursors of these plastids, called proplastids, are assumed to be present. The number of proplastids that occur in these meristematic cells is not known. If one or only a few proplastids were found to be present per meristematic cell of the phanerogams, this knowledge would be invaluable in the interpretation of variegations which suggest inheritance of mutated plastids in these higher plants.

Green plastids especially in younger leaves may be seen to divide by fission. REINHARD (1933) observed a division time of approximately 6 hours for the chloroplasts of the *Equisetum* prothallus; in moss leaves the time was approximately 8 hours.

## 2. Origin and development of the chloroplast of seed plants.

In the seed plants the continuity of the chloroplasts may be observed in the larger vegetative cells where the chloroplasts may be seen to divide by constriction or fission. In *Fuchsia* and *Sedum*, REINHARD (1933) observed that the division time of the chloroplast was 1–2 days. DANGEARD (1947) reports that chloroplasts multiply by division as can be readily seen in *Elodea canadensis* and in the moss *Mnium undulatum*. The fission can occur when starch grains are present in the chloroplasts. At times, the fission may be unequal. In the white lily, EMBERGER (1927) observed that during fission the plastids were cut in two by a transverse division; a clear line formed, then the line became more dense until it appeared as a trench between the two half portions.

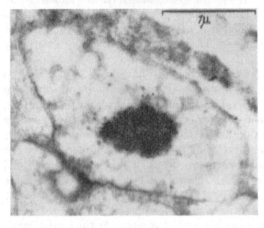

Fig. 20. Young chloroplast of *Chlorophytum* from pale-green leaves 1–3 cm. long that had been kept in the dark for $2^{1}/_{2}$ days. The granules of about 250 Å diameter appear to be organized in the arrangement of a crystalline lattice. HEITZ (1954).

In the embryo and in the meristematic cells at the growing points, no chloroplasts can be seen. The early proplastids may not be readily distinguishable from mitochondria with the light microscope (GUILLIERMOND 1941). However, at a later stage, when the elongated granules contain a starch grain at one end, they may be distinguished from mitochondria. The electron microscope should be of great aid in distinguishing between early proplastids and the mitochondria, since the latter are characterized by pocket-like projections or cristae within their structure (Fig. 12C and 8A).

In higher plants, HEITZ and MALY (1953) have examined the proplastids of living embryonic leaves of *Agapanthus umbellatus* and *Dracaena draco*. The young pale green plastids appear to fluoresce red in U.V. light; they are described as about 1–1.5 $\mu$ long and around 0.7 $\mu$ thick. The proplastids are amoeboid. They divide by fission. The youngest ones do not contain any dense granules. In the same cell there may also be proplastids that fluoresce bright red all over, but in addition they contain one or several tiny dots that fluoresce still more strongly. The fluorescent dots are considered to be grana. STRUGGER (1950, 1954) has examined the proplastids in living and fixed cells and reports that a primary granule is always present in a proplastid. The primary granule or "plastidogen complex", he believes, multiplies by self-reproduction. When

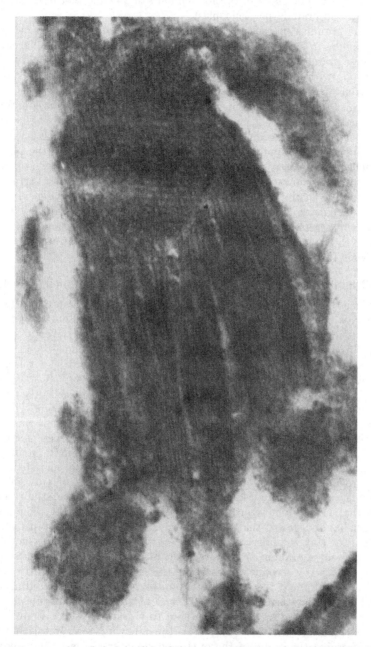

Fig. 21. Part of a young chloroplast of *Aspidistra elatior* from a section 1 mm. from the tip of the shoot. The groups of lamellae appear to be connected with the granules of the crystalline core. LEYON (1954c).

the granule divides to form two granules, the proplastid divides by fission, so that each proplastid contains one granule. On this hypothesis it is implied that the inheritance of the stroma is separate from the inheritance of the grana.

Perner (1954) has examined the leucoplasts of *Allium* and the stamen hairs of *Tradescantia*; in meristematic cells he always observed a granule within the amoeboid proplastids, even though a starch grain might also be present in the proplastid. The leucoplasts of the stamen hairs of *Tradescantia* were also reported to contain one granule per leucoplast.

Do the granules in these proplastids that develop in the absence of light, or the granules in leucoplasts, represent grana with a distinct lamellar structure? Or do these granules give rise to grana? These questions cannot be decided from observations with the light microscope.

Electron microscope studies by Heitz (1954) and by Leyon (1954 c) (Fig. 20, 21) have revealed that the proplastids in pale green meristematic tissue contain a three-dimensional array of tiny granules. In the proplastids of *Chlorophytum comosum* (Heitz 1954), the array of evenly spaced tiny granules resembles a crystal lattice. Each granule has a diameter of about 250 Å as estimated from the photograph. No lamellae are evident in this proplastid. In *Aspidistra elatior* (Leyon 1954) the photograph shows a granum which contains a crystalline array of tiny granules and from the sides of this crystallite extend the lamellae. In this case the tiny granules may be estimated to be 100–200 Å in diameter. The origin of these tiny granules, whether they divide, their composition, how they are related to the lamellar structure of the plastids of cryptogams and phanerogams, must await further studies.

## 3. Origin of the grana and the development of the lamellae or disks.

In algae the plastid does not arise *de novo* but only from a pre-existing plastid by a process of fission. Since the algal plastid represents essentially an enlarged granum, it may be inferred that a granum of an algal cell never arises *de novo* but only from a pre-existing granum by a process of fission.

In rapidly dividing cells of higher plants there appear to be relatively few and very small plastids. Thus, in liverworts and mosses. Kaja (1954) has observed with the light microscope that a meristematic cell like the apical cell may contain relatively few chloroplasts 1.5–2 $\mu$ in diameter. The number of grana in such chloroplasts is never less than 3–5 per chloroplast. As the cells mature and enlarge, the number of chloroplasts per cell increases; at the same time an individual chloroplast may increase to about 7 $\mu$ in diameter and may contain 60–80 grana per chloroplast. In the meristematic cells of phanerogams, the chloroplasts are reduced to the size of tiny proplastids. The number per cell is not known but is probably low. The studies of Heitz suggest that no granum is present in a very young proplastid. The studies of Strugger have suggested that there is always at least one granum per proplastid. The group of tiny granules arranged in a crystal lattice within the proplastid might represent a granum or the precursor of a granum. If these tiny granules have the ability to reproduce themselves, then it might be imagined that a very young proplastid might be made up of a limiting membrane surrounding a stroma material in which was embedded only one of these tiny granules. How the lamellae or discs arise in relation to these tiny granules is not known.

On the basis of the data available, the following interpretation might be made concerning the origin of the organized structures seen in the chloroplasts. In the *Aspidistra* chloroplast the fine membranes which extend out into the stroma and connect one granum with another (Fig. 3) are suggestive of the endoplasmic reticulum of Porter (1953). When *Chlamydomonas* is grown in the dark, it lacks chlorophyll and it also lacks the organized disk membranes; however, the plastid contains tubules with granules on the surface of the tubules. The granules on the tubules are suggestive of the granules of ribose nucleic acid (RNA) scattered on the surfaces of the endoplasmic reticulum as observed by PALADE (1955). On the basis of these homologies one might consider the granules of the tubules, if they are RNA, as self-duplicating units or plastogenes. (Might the granules of the crystalline array of HEITZ and of LEYON also represent RNA granules?) Growth of the endoplasmic reticulum and its secretions might provide the substances and mechanism for the organization of the disk membranes.

VON WETTSTEIN (1954) interprets the plastids of *Fucus* as being made up of continuous lamellae which are arranged like the scales of an onion. He suggests that the lamellae multiply by thickening and then separating. The separation of a lamella into two membranes would start in the middle of the plastid and proceed towards both ends.

## 4. Constancy of plastids per cell.

In the cells where the number of plastids is one to several, a marked constancy of plastids per cell is evident. For example, species of *Spirogyra* are known which contain only one or two or three spiral chloroplasts per cell. However, when the number of chloroplasts per cell becomes greater, the variation in number becomes greater. Thus, in *Spirogyra crassa*, the number may vary from 4–12 chloroplasts per cell.

The number of plastids per cell is maintained constant by several devices, even through the process of sexual fusion (SHARP 1934). For example, in *Zygnema*, each gamete provides 2 plastids, but in the zygote the male plastids degenerate. In subsequent vegetative divisions the two plastids divide at the time the nucleus divides to give rise to two new cells, each of which contains a nucleus and two plastids (Fig. 18).

In *Coleochaete*, during meiosis at one division of the zygote, the nuclei divide while the plastids do not divide, resulting in the production of one chloroplast per nucleus.

In *Anthoceros* the male gamete carries no plastid. Thus, the zygote has one plastid and at each division the plastid and nucleus divide simultaneously resulting in one plastid per cell.

In the meristematic cells of *Isoetes*, there is one chloroplast per cell, and for a time at each subsequent division only one chloroplast per cell is produced. Yet, as the cells enlarge, and in mature cells, several chloroplasts may be present.

What maintains the constancy of chloroplasts in these cells? One of the factors is obviously a mechanism correlated with the process of cell division. But other factors of differentiation also appear to be involved in the vegetative cells.

In the leaves of higher plants, the palisade cells contain more chloroplasts than those of the spongy mesophyll cells. "Mutant" chloroplasts of *Polypodium* may be of different sizes (KNUDSON 1940, MALY 1951). In general, the correlation is observed, that if a cell has large chloroplasts, the number of chloroplasts per cell are decreased. Since there is only one layer of plastids in the cytoplasm, the number of plastids per cell may be related to the available cell surface. Nutritional factors have been found to affect the number and size of chloroplasts per cell. TABENTSKII (1953) has observed that increase in nitrogen supplied to sugar beet leaf culture resulted in as much as 2.5 fold increase in number of the plastids per cell, but the plastids were smaller.

## 5. Maternal inheritance of cytoplasm and plastids.

In plants and animals at sexual fusion, the paternal cytoplasm of the male gamete appears to be discarded more or less completely. (Occasional deviations from this rule will be discussed below.) This important generalization has been known for a century, but as yet the reasons for discarding the male cytoplasm are not understood. In general, therefore, the mature plant or animal is a result of the interaction between male and female nucleus in the environment of female cytoplasm.

Early evidence which indicated that male cytoplasm was discarded at zygote formation came from direct observations on lower plants which contain one or a few chloroplasts. In general, the male gamete carries no chloroplast, or the chloroplasts derived from the male gamete disintegrate before fertilization occurs, or the "male" chloroplasts disintegrate in the zygote after fertilization. A few examples may be cited. In *Anthoceros*, the spermatozoid is believed to bear no remnant of a chloroplast, so no male chloroplast enters the egg; the chloroplast of the egg is then the chloroplast of the zygote. In *Zygnema* and *Spirogyra* the male chloroplast degenerates in the zygote as if the female were a foreign and toxic environment for the male plastid. In *Pinus*, disintegration of the cytoplasmic granules (mitochondria and proplastids) of the male also occurs in the oosphere. Mangenot (1938) observed that these granules of the male were larger than those of the female. The larger granules of the pollen tube could thus be followed during fertilization as the male cytoplasm penetrated into the oosphere. After fertilization, the larger granules could still be recognized. During the development of the embryo, the granules of male origin remained in that portion of the oosphere which did not contribute to the formation of the embryo and the "male" granules disintegrated later. There did not appear to be a mixing of the granules of male and female origin. These observations suggest that not only the proplastids, but also the mitochondria and other visible granular elements of the male are discarded on fertilization.

# 6. Summary.

Evidence is presented that plastids are derived only from pre-existing plastids. In the algae, plastids may be seen to divide in vegetative cell division. The division of the chloroplasts is by a fission which cuts across the extended lamellae within them. During enlargement of the young chloroplast and the cell, the lamellae enlarge as broad sheets. Under favorable conditions, as in *Rhynchonema*, the plastid may even be followed through meiosis and be seen to maintain its integrity. Plastids of higher cryptogams such as *Anthoceros*, *Selaginella* and *Isoetes* have also been followed through mitotic and meiotic divisions. Here also the plastids behave as morphological entities and multiply exclusively by division. *Isoetes* may represent a transition in plastid behavior between lower and higher plants. In the *Isoetes* sporophyte, meristematic cells contain only one plastid and at each division the plastid divides. However, as the cells enlarge a number of plastids come to be present per cell.

In the phanerogams, plastids can be seen to divide in older vegetative cells. However, in meristematic cells where plastids are very tiny, *i.e.*, "proplastids", they often cannot be distinguished by light microscopy, from other granules such as mitochondria. Segregation of proplastids in early embryonic tissues of chimeras suggest that relatively few proplastids may be present per cell at this stage. Since electron microscopy can distinguish mitochondria from other granules, it may be possible with this technique to obtain direct evidence on proplastid continuity in the meristematic cells and egg cells. Although some direct evidence is present in the phanerogams that proplastids are derived from pre-existing proplastids, indirect evidence is incontrovertible in support of this thesis. Plastids are generally inherited only from the cytoplasm of the egg cell, *i.e.*, maternal inheritance is the rule. If any of these plastids are abnormal, they will also be inherited in a non-Mendelian fashion and can be recognized in successive cell divisions. Thus the theory proposed toward the end of the last century by MEYER and SCHIMPER that plastids arise only from pre-existing plastids may be considered to be well established.

The stages in the development of the chloroplast of higher plants from proplastids are as yet poorly understood. The important finding has been made recently by HEITZ and by LEYON of a crystal lattice arrangement of tiny granules in the proplastid. How such a "granum" arises and how these tiny granules of this "granum" are related to the lamellar structure of larger grana awaits study. In the pale yellow proplastid, lamellar disk membranes form in the light only when chlorophyll is formed.

# V. External and internal heritable factors that influence chloroplast development.

The absence of chlorophyll in a growing plant, or in various tissues or regions of growing plants, is readily discernible. The lack of chlorophyll is a sensitive indicator that one or a number of abnormal factors, external and, or internal, are influencing the plastid so that chlorophyll is not formed.

## 1. External factors.

Some of the external factors that affect chlorophyll development may be mentioned. a) Light is essential in photosynthesis and therefore necessary indirectly to supply the energy and some materials for the enlargement and differentiation of the chloroplast. Light is also necessary for the reduction of protochlorophyll to chlorophyll, *i.e.*, for the formation of the last step in the biosynthetic chain of chlorophyll synthesis. b) Chlorophyll formation is sensitive to a lack of iron, perhaps because normal oxidative metabolism, via the cytochromes in the mitochondria and cytochrome *f* in the plastids, is required for normal plastid development. The lack of iron may lead to the formation of scattered pale green or white areas in the leaves. c) Virus diseases may interfere or redirect cell metabolism so that plastid growth and development is often inhibited; leaves affected by the viruses may often contain irregularly scattered areas that are splotched pale green or whitish. Electron microscope studies of cells infected with tobacco mosaic virus suggest that this virus may even develop within the chloroplasts (KAUSCHE and RUSKA 1940, BLACK *et al.* 1950, LEYON 1951).

## 2. Factors of tissue differentiation.

Four groups of internal or inheritable factors that affect the development of the chloroplast may be considered. One group of factors are connected with tissue differentiation. For example, the epidermis of mesophytic leaves develops so that the plastids in the cells are tiny and colorless except in the stomatal cells. In general, plastids remain undeveloped in roots, even if these organs are exposed to light. However, roots of some species such as those of *Cucurbita*, *Menyanthes*, and *Cazea* may become pale green in light. Little is known of these differentiation factors, and we shall not discuss them further. They may result from an interaction of the products of the other three groups of factors: the genom, plasmon, and plastidom.

## 3. Genom factors.

This group of inheritable factors resides in the chromosomes and segregates according to Mendelian laws: Those factors which influence the chloroplast may be due to genic changes or to chromosomal variations, and have been well studied.

According to DEMEREC (1935) the most common types of chlorophyll mutants are simple Mendelian recessives, indicating that the mutations are due to the action of single genes. For example in maize, the plant in which the most extensive genetic studies have been made, 13 genes are known, each of which independently produces an albino seedling. Some 20 genes are known which produce virescent seedlings, i.e., seedlings that are white in the early stages, but develop pigment later. Several genes are known which result in pale green seedlings where the amount of chlorophyll is reduced at certain stages of development. Some variegated seedlings have been found in which chlorophyll is lacking in spots (piebald) or in streaks (zebra). In the albinos, the plastid primordia are present and begin to develop in the seedling stage at the same rate as in the green seedlings, but no chlorophylls or carotenoids develop. Luteus seedlings which contain carotenoids result when, in addition to the albino gene, several other genes are present. The virescent types differ greatly from each other in the rate at which chlorophyll develops. At one extreme are the virescent plants which become green if grown under favorable conditions of temperature. Some virescent plants may become green at a rate which is too slow for survival. Among the pale-green types there is also considerable variation; certain types do not live beyond the seedling stage; other types are pale green as seedlings and as mature plants. Whether a seedling will survive does not merely depend on its chlorophyll content, but on various other factors involved in chloroplast growth and enzymic content. For example, Xantha-1 and Xantha-2 types are pale green viable plants that contain only 10 per cent of the normal content of chlorophyll although they contain a normal concentration of carotenoids. A pale green type which contains 75 per cent of the normal content of chlorophyll and a normal content of carotenoids dies at the same time as a pure albino seedling, i.e., at the three-leaf stage.

Other work on the plastid pigments of mutant maize is reported by SCHWARTZ (1949). SMITH and KOSKI (1947) found that one of the albino strains of corn did not turn green to an appreciable extent because light caused an excessive rate of destruction of chlorophyll. Several of their virescent corn mutants were found to be partially defective in the rate at which they made protochlorophyll, but they possessed the mechanism for the continued production and conservation of chlorophyll. HIGHKIN (1950) has studied the pigments of barley mutants. One of his mutants, Chlorina 2, was found to possess a normal content of chlorophyll a, but to be devoid of chlorophyll b.

Chromosomal variations may also affect the chloroplasts. For example, polyploidy may result in larger cells and larger chloroplasts.

If gene mutation or change in the number of chromosomes occurs in meristematic tissue of a very young bud, then the product of the bud, i.e., the branch, flower or flower cluster, may have a new genotype constitution and will exhibit an appearance often strikingly different from that of the other branches or flowers of the plant (JONES 1940). Such a change of genom may result in a *chimera* in which a distinct portion of the mature structure, commonly a sharply defined sector, differs constitutionally and in appearance from the other portions. For example, diploid *Datura* plants have been studied which may bear $2n+1$, $2n-1$ and $4n$ branches. The differing portions may at times also be arranged concentrically, *i.e.*, periclinally. In *Spinacia* the plerome and dermatogen may be $2n$ and the periblem may be $4n$ or $8n$. Such sectorial and periclinal types are known as "chromosomal chimeras".

Irregularities in the distribution of chromosomes or portions of chromosomes may also result in inheritable changes that affect chloroplasts. In maize, the studies of McClintock (1939) have shown that the mosaic patterns seen in the endosperm may be the result of an unstable chromosome which breaks and reforms with a high frequency. Other forms of unequal mitoses might arise by crossing-over or reciprocal translocation in somatic tissues (Jones 1940). Whether such phenomena will explain certain of the mosaic variegations that occur in the somatic tissues of plants is unknown. It may be noted that undoubted factor mutations occur in somatic cells of *Drosophila*, which result in mosaic individuals (Caspari 1948).

## 4. Plasmon factors or plasmogenes.

Another group of inheritable factors which may influence chloroplast development may reside in the cytoplasm outside of the chloroplast. Some 20 cases of cytoplasmic inheritance are discussed in a comprehensive review by Caspari (1948). Evidence from inter-species crosses (Sirks 1938) suggests that factors of the cytoplasm apart from those of the plastids, are independently inheritable. The extensive work of Michaelis (1954) on *Epilobium* has demonstrated clearly that inheritable factors in the cytoplasm influence the size of the plant, size differences of leaves, flowers, etc. Even in combination with a homozygous foreign nucleus these cytoplasmic factors may remain relatively unchanged.

Certain studies on variegated plants suggest that plasmon factors are active, but rigorous proof is still largely indirect. A classic case suggesting plasmon inheritance is that first described by Correns (1909). In the variegated plant *Mirabilis jalapa* var. *albomaculata* the leaves have white areas which are irregularly distributed over the leaf. Intergrades are also present, varying from green leaves with a few white spots to white leaves with a few green spots. The boundary between green and white tissues is not always sharp and the intensity of whitening is very unequal. The white regions contain colorless plastids; the green regions contain normal chloroplasts. Branches may form that are pure green and bear normal flowers; or branches may form that are colorless and bear colorless flowers and fruit. When flowers from a colorless branch are fertilized with pollen from a branch bearing green plastids, offspring with colorless plastids are produced which cannot photosynthesize and these plants die in the seedling stage. The reciprocal cross produces progeny with normal green plastids and the plant is viable.

Thus          ♀ green × ♂ white → green plants,

                 ♀ white × ♂ green → white plants.

These results are readily explained on the assumption that only the cytoplasm of the egg including its plastids are inherited, *i.e.*, maternal inheritance occurs. The male gamete only furnishes the nucleus. The cytoplasm and plastids of the male gamete either do not enter the egg, or if they enter, they degenerate. It cannot be decided from this experiment whether the absence of chlorophyll in the plastids of the colorless regions is a result of some inheritable factor in the plasmon or plastidom. The irregular splotching of the leaves appears to bear no relation to the developmental morphology of the leaves. The white areas may arise both early and late in leaf development. The assumption of a segregation of white and green plastids in the embryonic cells would not appear to explain satisfactorily the random distribution of the white leaf areas. Rhoades (1943, 1946) has suggested that there may be a ratio of normal to abnormal plasmagenes in the cytoplasm, *i.e.*, outside the plastid, which may affect the outcome of the cell plastid type. A biochemical interpretation may be that some enzyme system is limiting in concentration. For example, depending on environmental factors, certain leaf areas may not be entirely equivalent to other leaf areas, and the substance produced by this enzyme system may be too low in concentration to result in greening of the plastids in the white areas. Or differentiation factors may be at work, of the type that cause the epidermal cells to contain colorless plastids. The irregular splotching in this plant is the kind that might be expected of a virus disease but as yet there has appeared no evidence to implicate a virus.

*Humulus japonicus* var. *albomaculata* was found by Winge (1919) to produce variegated progeny from variegated parents irrespective of the type of pollen that had been used in the crosses. "Selfed" seeds of these mosaics never produced pure white or pure green branches or seedlings. *Winge* made reciprocal crosses of mosaic plants with normal green plants of this species of hop through four generations. The inheritance of green versus mosaic plants proved to be exclusively maternal, *i.e.*, no cytoplasm including plastids was contributed to the zygote from the pollen. The mosaic character in the hop has been thought by Winge to be due to some inhibitory factor carried in the cytoplasm, rather than to a factor in the plastids. In *Capsicum annuum* var. *albomaculata* the mosaic character was found by Ikeno (1917) to be inherited through both parents, suggesting that some cytoplasm of the pollen passed into the egg.

## 5. Plastidom factors.

These are a group of inheritable factors postulated to reside in the plastid. That the plastid has an individuality independent of the cytoplasm and nucleus may be inferred from an important chance observation of van Wisselingh (1920). In *Spirogyra triformis* he found an abnormal chloroplast together with normal chloroplasts in the same vegetative cell. This abnormal chloroplast lacked pyrenoids, and the starch grains instead of being formed as sheaths around the pyrenoids were scattered in the chloroplast. When the cell divided, this chloroplast as well as the normal ones divided and came to be present in the new cell. If a change had occurred in the nucleus or cytoplasm that would have been responsible for the abnormality in this chloroplast, it should have affected all the chloroplasts of the cell to the same degree. Since in the same cell an abnormal chloroplast was present together with the normal chloroplasts, and vegetative divisions maintained this condition, it seems necessary to conclude that chloroplasts possess mechanisms for their self-duplication and factors for the maintenance of their phenotypic individualities.

A classic example of plastid inheritance was discovered by Baur (1909, 1930) in *Pelargonium zonale* var. *albomarginata*. This plant has leaves with white margins. It often produces pure white and pure green branches which can bear flowers. If a flower from a green branch is pollinated with pollen from a white branch, the resulting progeny may occasionally be mosaic seedlings, *i.e.*, ♀ green × ♂ white → mosaic. Mosaic seedlings, in the course of growth, produce green, colorless and variegated leaves and branches. Likewise, the reciprocal cross of a flower from a colorless branch, when pollinated from a flower of a green branch, occasionally results in the same kind of mosaic seedlings, *i.e.*, ♀ white × ♂ green → mosaic. These crosses are of special significance since they help to distinguish between plasmon and plastidom inheritance. To explain the appearance of mosaic seedlings, it is assumed that cytoplasm and plastids from the male may sometimes enter the egg; in this way, for example "green" plastids would be brought into an egg which had "colorless" plastids. If the female plastids were colorless because of factors in the female cytoplasm, then the green male plastids entering into the female cytoplasm should also become colorless, i.e., colorless plants should result. Since mosaic plants are produced it is concluded that in this case the cytoplasms do not affect the greening or non-greening of the plastids. The factor or factors governing the greening must here reside in the plastids themselves. The plastids must be of two kinds. Either kind of plastid arises only from the division of previously existing similar plastids. The colorless or "mutated" plastid is colorless because it contains within itself some inheritable lesion.

The conclusion that a mutant plastid is present in this variety of *Pelargonium* is based on several assumptions which may be mentioned. First, it is assumed that plastids may be transferred from the pollen to the egg so that the fertilized egg will contain both kinds of plastids. Although the general rule is that only the nuclear material of the pollen is transmitted to the egg, there is some evidence that occasionally in some plants the cytoplasm and plastids do enter the egg. For example, ISHIKAWA (1918) in a study of the fertilization process of *Oenothera* found that starch grains in the pollen migrated with the cytoplasm through the pollen tube and entered the embryo sac. Presumably, the starch grains were contained in plastids.

A second assumption that is made in the interpretation of the above experiment is that the green and colorless proplastids contained in the embryo sac after fertilization segregate during cell divisions in the early embryo. Thus certain cells will only contain colorless plastids and it is these cells which will give rise to the colorless areas. If such a segregation is to occur, it seems necessary to make several additional assumptions. The idea of segregation might appear more plausible if it could be established that very few proplastids were present in the fertilized egg, and if the rate of division of the plastids in the meristematic cells was to be limited to the maintenance of a small number of plastids per cell. Another explanation of segregation may be that the male cytoplasm and plastids do not mix readily with the female cytoplasm and plastids so that the two different kinds of proplastids become segregated in the first few divisions of the fertilized egg. Relatively few instances of two types of plastids in the same cell have been reported. An assumption to explain the almost complete absence of both types of plastids in the same cell is that the colorless plastids may divide at a slower rate than the green ones, and eventually the colorless plastids would be lost. However, a number of workers have encountered green and pale plastids together in the same cell, in plants such as *Primula sinensis*, *Capsella bursa pastoris*, *Stellaria media albomaculata*, etc. KÜSTER (1935) observed three different kinds of plastids in mesophyll cells of the above species.

A further assumption which is necessary to explain the results observed in BAUR'S *Pelargonium* is that the green and colorless cells in the embryo will give rise to cell lineages and areas that are a consequence of the mechanics of the embryonic development of the plant. Embryological studies appear to support this assumption (NEILSON-JONES 1934).

An experiment, similar to that of BAUR'S, was carried out by GREGORY on *Primula sinensis albomaculata*. He grew pure yellow plants to flower and crossed these flowers with pollen from a normal green plant. The albomaculata character was transmitted only by the egg and not by the pollen. In the cells of quite young leaves he found simultaneously normal and chlorotic chloroplasts. The defect therefore was considered to lie in the chlorotic plastid rather than in the cytoplasm. For other literature on this subject compilations of Neilson-Jones (1934), Küster (1935), and Weier and Stocking (1952a) are recommended.

Added support for the concept of inheritable factors in the plastids is derived from the studies on interspecies crosses of *Oenothera* which have been carried out by Renner (1936) and by Schwemmle and his students (1938). The green plastids in one *Oenothera* species were found to differ in inheritable properties from the green plastids of another *Oenothera* species. For example, when reciprocal crosses between certain species were made, the cross A ♀ × B ♂ was found to produce green plants with healthy plastids, whereas B ♀ × A ♂ was found to produce plants with inhibited plastids. Here the chromosomes are identical in the two crosses, but the female parent provides the cytoplasm as well as the plastids. Thus:

Normal plant

A ♀ × B ♂ → | A chromosomes + B chromosomes + A cytoplasm + A plastids. |

Pale plant

B ♀ × A ♂ → | A chromosomes + B chromosomes + B cytoplasm + B plastids. |

To explain the difference in the appearance of the plastids, it is necessary to assume that the plastids of A were not inhibited by the chromosome complement of A + B but that the plastids of B were inhibited by the chromosome complement of A + B. It is not possible to conclude that the plastids of B were inhibited by the A chromosomes alone since the inhibition may be a result of the interaction of A + B chromosomes.

Did the A + B chromosomes produce some substance which affected the B plastids directly or did this substance act on the cytoplasm which in turn then caused the B plastids to be inhibited? The following type of analysis was used to test these possibilities. The analysis depended on finding some cases in which cytoplasm and plastids of the male were brought into the embryo sac, as occasionally happens. The zygote then would presumably have the following composition:

| A chromosomes + B chromosomes + A cytoplasm + B cytoplasm + A plastids + B plastids. |

If segregation of the plastids occurred, a variegated plant would arise. The green areas of the plant would have the composition:

| A chromosomes + B chromosomes + A cytoplasm + B cytoplasm + A plastids. |

The colorless areas of the plant would have the composition:

| A chromosomes + B chromosomes + A cytoplasm + B cytoplasm + B plastids. |

If the cytoplasms are the same in the green and in the colorless areas, then the substance produced by the A + B chromosomes must be inferred to inhibit the B plastids because the B plastids are different from the A plastids. The B plastids must differ in inheritable factors from the A plastids. In other words, since the genome and plasmone constituents are the same, the difference in response must lie in the plastidoms of A and B. Analysis indicated that this interpretation was highly probable.

A specific example may be cited. *Oenothera odorata* has two Renner complexes or two groups of chromosomes, *v* and *I*. During reduction division the diploid number of chromosomes do not segregate at random, but rather one group *v* goes to one pole to form one haploid nucleus, and the other group of chromosomes, *I*, goes to the other pole to form the other haploid nucleus. Thus the egg or sperm nucleus of *Oenothera odorata* will have either a Renner complex *v* or *I*. *Oenothera Berteriana* also has two Renner complexes, namely *B* and *l*. When reciprocal crosses are made between *O. Berteriana* and *O. odorata* the following data are obtained, as summarized by WEIER and STOCKING (1952a) in Table 2. Here the egg supplied the cytoplasm of the zygote.

*Table 2.*

| Cross<br>♀ *O. Berteriana* × ♂ *O. odorata* | | Cross<br>♀ *O. odorata* × ♂ *O. Berteriana* | |
|---|---|---|---|
| Appearance of hybrid with *Berteriana* cytoplasm including chloroplasts; the specific chromosome groups or Renner complexes are indicated. | | Appearance of hybrid with *odorata* cytoplasm including chloroplasts; the Renner complexes are indicated. | |
| Renner complex | Appearance | Renner complex | Appearance |
| $B+v$ | Not formed | $B+v$ | Normal |
| $B+I$ | Normal | $B+I$ | Weak and pale green |
| $l+v$ | Lower leaves sometimes yellow, otherwise normal | $l+v$ | Weak and pale green |
| $l+I$ | Lower leaves sometimes yellow, otherwise normal | $l+I$ | Non viable; dies in embryo stage |

It will be seen from the table that although the chromosomes, which constitute the nuclear complexes, may be the same in the diploid plants, yet the result is profoundly influenced by the composition of the cytoplasm + plastids. For example, in the diploid, with the Renner complexes (*i.e.*, chromosomes groups) $B+v$, if $B+v$ interacts with *Berteriana* cytoplasm + plastids, then the resulting plant dies. However, if $B+v$ interacts with *Odorata* cytoplasm + plastids, then the resulting plant is normal.

Next it is necessary to determine whether it is the plastids or the cytoplasm which is responsible for the above results. For such an analysis it is necessary that the cytoplasm and plastids of the pollen enter the egg. This happens occasionally, and can be detected when it does happen, under the following, conditions. Suppose a plant

$$\boxed{l + v \text{ chromosomes} + (\text{odorata cytoplasm} + \text{plastids})}$$

is crossed with pollen of *Berteriana*, *i.e.*, of

$$\boxed{B + l \text{ chromosomes} + (\text{Berteriana cytoplasm} + \text{plastids}).}$$

Then the following combinations will occur: $B+l$ which is lethal; $l+l$ which is lethal; $B+v$ which gives a normal plant; and $l+v$ which gives a weak, yellowish plant. However, if *Berteriana* cytoplasm + plastids of the pollen get into the egg, then a variegated white-green plant is obtained which can be readily distinguished from all of the other viable plants. This variegated plant was interpreted to have the following composition in the white areas:

$B + l$ chromosomes + odorata cytoplasm + odorata plastids
+ Berteriana cytoplasm.

In the green areas the composition was interpreted to be:

$B + l$ chromosomes + odorata cytoplasm + Berteriana cytoplasm
+ Berteriana plastids.

The above interpretation is based on assumptions of the kind that have been discussed for the case of *Pelargonium*. For example, if the cytoplasm of the *Berteriana* pollen tube enters the egg it is assumed that it will become thoroughly mixed with the *odorata* cytoplasm and no random distribution would be able to segregate the cytoplasms; whereas the plastids of *Berteriana* could be segregated from the plastids of *odorata*. The presence of *Berteriana* plastids has been tested for by crossing a flower from a green branch of the variegated plant with a standard *odorata* pollen. When this is done, the $F_1$ generation resembles that of the standard *O. Berteriana* $\times$ *O. odorata*, as seen in the left half of the table.

Thus it may be concluded that *odorata* plastids differ from *Berteriana* plastids in their response to the same chromosomes and that the difference is due to an inheritable mechanism within the plastids.

A curious and important observation that remains to be explained satisfactorily is the following: In the cross of *odorata* $\times$ *Berteriana* the weak pale green $B + I$ and $l + v$ plants are obtained which contain the *odorata* plastids. During the following generations these plants become normal green and healthy. The authors believe that this change from inhibited to normal plastids cannot be due to a progressive change in the cytoplasm or plastids, but that the genom has been gradually changed so that its products no longer act to inhibit the plastids.

Another result that is suggested from these studies is that the proplastids may have a vital function independent of developing into functional chloroplasts. Binder (1938) has studied the $l + I$ complex containing *odorata* plastids (Table 2). Fertilization occurs but the young embryo dies. Death is therefore not a result of the inability of plastids to develop to the stage where they carry on photosynthesis. Rather death may be due to the abnormal behavior of the *odorata* proplastids, perhaps to the inability of the *odorata* proplastids to supply some metabolite essential for the life of the young embryonic cell.

## 6. Summary.

Data have been presented which demonstrate that the chloroplast may be considered as a discrete entity with its own hereditary units, the plastidom. The nutrients which the plastid requires for its growth and division are highly complex. These nutrients are controlled by hereditary factors,—the plasmon residing in the cytoplasm outside of the plastid and the genom residing in the nucleus. A detailed analysis of these internal factors may become possible when tissue culture methods have been developed to grow isolated chloroplasts. In addition to these hereditary factors, external factors such as light, etc., will affect the plastid. Thus the size of the chloroplasts, their number, pigmentation, etc., *i.e.*, their phenotypic expression is a result of the interplay of the hereditary factors of the genom, plasmon, and plastidom together with external environmental factors.

# VI. Agents that bring about plastid "mutations".

The evidence has been presented above to show that plastids are "self-duplicating" bodies. Likewise, it has been shown that in the same internal environment of cytoplasm and nucleus, some plastids may respond differently than others. On the analogy with nuclear genes, it is logical to assume that plastid factors or plastogenes may be present in the plastids. Since specific substances like the carotenoids, chlorophylls, lipoproteins and certain enzymes are localized in the chloroplasts, it may be assumed that the syntheses of these compounds, at least in the end steps of their biosynthetic chains, as well as the organization of the lamellae, are controlled by not one but many plastogenes. Presumably, these plastogenes are made up in part of nucleic acid. Evidence is accumulating which suggests that nucleic acid is present in plastids (see Section VII).

It is difficult to observe plastid mutations because there may be a number of identical, *i.e.*, replicate, plastogenes per cell. For example, in a chloroplast there might be only one plastogene of each kind per chloroplast or there might be several replicates of these plastogenes per chloroplast. The number of replicate plastogenes per cell is increased if instead of the cell containing only one chloroplast, there are a number of chloroplasts per cell. The chances of finding a plastid with a mutated plastogene in a cell containing normal plastids would be very small, since, if such a mutated plastid happened to arise in a meristematic cell, the mutated plastid might tend to grow more slowly and would be lost in subsequent somatic cell generations.

From such considerations it would appear that in order to bring about an inheritable observable plastid change it may be necessary to change at the same time all replicates of a particular plastogene in all of the plastids of a cell. Several techniques for accomplishing this action may be considered, such as the action of a specific nuclear gene, or the action of streptomycin and possibly other chemicals, or the action of physical agents like heat, visible light, U.V. light, and X rays.

## 1. Action of nuclear gene in affecting plastogenes.

When a certain recessive nuclear gene is present in homozygous condition, a change will be brought about in the cell which will prevent the plastids from forming chlorophyll. This change is permanent. Even though the recessive gene is later replaced by its dominant allele, the plastids will remain colorless. Two cases of this kind have been studied in detail, one occurs in maize; the other in *Nepeta*.

In maize, a chlorophyll striping or variegation of the leaves and culm is produced when two recessive *ij* (iojap) genes are present in the cells (RHOADES 1943, 1946). The leaf pattern varies from fine white streaks to wide white bands which may occupy half or more of the leaf. Very few seedlings show the striping, since the stripes develop mostly on leaves that develop later. The iojap gene is situated in a known segment of maize chromosome 7. The presence of one recessive and one dominant iojap gene (ij + Ij) will not cause variegation. When plants bearing the double recessive dose (ij + ij) of this gene are used as a female parent and crossed with pollen from an (Ij + Ij) plant, there result white seedlings, variegated seedlings and green seedlings. Assuming that no cytoplasm and plastids passed from the pollen to the egg, the white seedlings may be interpreted to contain in their cells, the ij Ij genes and the maternal cytoplasm with its small colorless plastids. The variegated and green seedlings are assumed to have arisen when the cytoplasm and plastids from the pollen have been able to enter the egg. In the variegated plants, at the boundary between white and green areas, cells were found which contained both the green and colorless plastids in the same cell. In further crosses it was established that if the cells had the Ij + Ij gene constitution in the nucleus, and contained the cytoplasm and plastids derived from the ij + ij plant, then the plant would be colorless. From this analysis it may be concluded that when no dominant Ij gene is present in the nucleus, a damaging change will occur in the cytoplasm or plastids, which will be inherited; this change becomes evident as an inability of the plastids to form chlorophyll and the defect is thenceforth inherited independently of the nuclear genes.

These results suggest that, directly or indirectly, the action of the Ij gene may be involved in the multiplication of some hereditary units in the cytoplasm. The inhibition of multiplication of these units might be due to various incompatibility factors, on analogy with those factors which prevent a specific virus from multiplying in different tissues of the same organism, or in closely related species. One might even suppose that the Ij gene governed some step in nucleic acid biosynthesis in the cytoplasm. Then the lack of the dominant gene might result in a diminution in the rate of nucleic acid synthesis. As a consequence the hereditary particles in the cytoplasm might not multiply fast enough and would be diluted out in subsequent cell divisions. Thus cells would be formed which now lacked the hereditary particles and could not regain them even if the recessive genes were replaced by dominant ones.

In *Nepeta cataria* (WOODS and DU BUY 1951) a recessive and highly mutable gene, when present in homozygous condition, also appears to bring about inheritable changes in plastids, which are irreversible. The appearance of the plastid is thenceforth independent of the presence or absence of this recessive gene. In this case, some fifteen different plastid abnormalities have been observed, such as light green plastids, cream colored plastids, white vacuolate plastids, and also absence of recognizable plastids. These types appear to result from a number of factors that affect plastid development, *i.e.*, retention of plastid pigmentation, size of grana, plastid vacuolation, etc. There are also types where the tissue may contain several kinds of plastids, including normal ones, all in the same cell. The fact that several types of plastids may occur in the same cell along with normal green plastids might be considered as evidence that the abnormalities in the plastids are caused by some inheritable defects in the plastids per se and not in the cytoplasm.

## 2. Action of chemical and physical agents on chloroplast inheritance.

A chemical agent which enters a cell might be expected to act equally on replicates of a plastidom, *i.e.*, on the same factors which are present in all the plastids, and perhaps eliminate them simultaneously. Likewise, a physical agent like heat may also be expected to affect all the plastids similarly. Unique examples of such actions are found in *Euglena*. However, because this organism lacks a sexual cycle, it cannot be determined whether such agents affect the plastogenes directly or affect them indirectly through changes in the genom or plasmon.

*Euglena gracilis* var. *bacillaris* loses its color in the dark but becomes green when placed in the light (see p. 529). Streptomycin causes a loss of color in this organism. PROVASOLI, HUTNER, and SCHATZ (1948) found that once the organism becomes colorless by the action of this compound, it no longer will become green in the light even through successive generations. The concentration of streptomycin which causes "bleaching" (about 40 $\mu$g./ml.) does not affect its rate of growth in organic media. This concentration is 50,000 times less than the concentration which will kill. The bleaching appears to be most rapid when the cells are dividing rapidly, suggesting that the rate of division of the plastids may be diminished and that by cell division, individuals might arise that lack plastids (PROVASOLI *et al.* 1951). LWOFF and SCHAEFFER (1949) observed that if *Euglena* was grown in the dark on 500–600 $\mu$g./ml. of streptomycin for 5 days, they could become green again on a medium devoid of streptomycin. However, after 6 days in the dark on the streptomycin-containing medium, bleaching was found to be irreversible. Whether all remnants of the plastids disappear by streptomycin action has not been determined although such a determination may now be possible with the electron microscope.

The studies of ROBBINS, HERVEY, and STEBBINS (1953) on this strain of *Euglena* show that spontaneous mutations to colorless cells may occur with the high frequency of 1 in 25,000. On solid media containing streptomycin, colonies appeared that were pale green, yellow or white. Of special interest were the colonies which had a green center and were pale green or white along the edges; this result suggests that there was a progressive loss of chlorophyll with time (a diluting out of plastids?) as the cell divisions continued at the outer edges of the colony. Yellow cells when now grown in the absence of streptomycin, developed colonies which contained green patches or sectors. This latter phenomenon is also interesting since it signifies that streptomycin can act to bring about a change to a yellow plastid; this change is inheritable for many generations and then reverses spontaneously and at random to normal green. The frequency of change seems to be too high to be accounted for on the basis of spontaneous genic mutations.

Aureomycin (ROBBINS *et al.* 1953) was also found to be effective in producing chlorotic colonies of *Euglena*, but the differential between bleaching and killing was much lower than in the case of streptomycin.

Relatively high temperatures may convert some strains of *Euglena* to permanently colorless ones (PRINGSHEIM and PRINGSHEIM 1952). They may be bleached in 4–6 days at 34 to 35⁰ C. ROBBINS *et al.* (1953) found that green *Euglena* cells, when placed at 36⁰ in the dark for 22 days were all converted to colorless cells. Cell divisions were necessary if bleaching was to occur, which suggests that a diluting out of the plastids might have occurred, *i.e.*, the plastids did not divide while the cells divided so that soon most of the cells contained no chloroplasts.

Several observations on the eye-spot of *Euglena* should also be mentioned (PRINGSHEIM and PRINGSHEIM (1952). In strains of *Euglena* that are not completely bleached by heat or streptomycin, the eye-spot does not disappear. In strains which are bleached by heat or streptomycin, one of several results may be obtained: a) The eye-spot may disappear simultaneously with bleaching. b) The eye-spot may persist in the light even though the chloroplasts are bleached: however, once the bleached strain is placed in the dark, even six months after heat-bleaching, then the eye-spot is lost permanently. This latter result suggests that in the light a pattern is maintained which is lost in the absence of light; *i.e.*, the inheritable precursor or template which would normally start this pattern up again in the presence of light might here be absent. c) The eye-spot may persist even if the bleached strain is placed in the dark.

Other studies on higher plants with streptomycin indicate interference with chlorophyll production, but whether the mechanism involved is a temporary inhibitory effect on the cells or an inherited one cannot be determined from the available data. Seeds germinated on filter paper moistened with streptomycin solution greater than 2 mg./ml. developed colorless first leaves (VON EULER *et al.* 1948). The seedlings investigated were barley, lettuce, rye, spinach, and radish. With less concentrated streptomycin solutions, only the tips of the first leaves became green. Streptomycin appeared to retard or arrest chlorophyll formation in developing leaves, but chlorophyll was not affected if it was already present in the leaves before treatment. DE ROPP (1948) observed that streptomycin changed crown gall tumors from green to white and inhibited the growth of normal meristematic and tumerous tissues of sunflower. The action of streptomycin was studied on seedlings of *Pinus Jeffreyi* by BOGORAD (1950). At 0.2 per cent of streptomycin and in the dark, the cotyledons of almost all of the seedlings developed without chlorophyll, although these cotyledons would normally form chlorophyll in the dark. When the cotyledons were then removed from the megagametophyte tissue which surrounds them, and placed in a medium devoid of streptomycin, they became green only in the presence of light. This behavior would suggest that the enzyme system or its precursors which served to reduce, in the dark, protochlorophyll to chlorophyll in the pine cotyledons was damaged by streptomycin.

From the viewpoint of chloroplast inheritance it would be of interest to administer streptomycin to plants at the time of flower bud formation to see if a cytoplasmic or plastid effect could be induced that would be transmitted maternally. Several other effects of chemical and physical agents on the chloroplasts may be mentioned. D'Amato (1950) observed that treatment of barley seeds with acriflavine or acridine orange or 9-aminoacridine produced mutations that led to a decrease of chlorophyll in the plastids. Whether these chemical effects are a result of changes in the plastidom is not known.

Ultraviolet light has been found to inactivate the photochemical production of $O_2$ by chloroplasts (Holt et al. 1951). Chlorophyll is not destroyed by this treatment, but probably some enzyme system related to photosynthesis is adversely affected. No studies have been carried out to find whether this effect is inherited.

### 3. Summary.

The "conservative" character of the plastids, i.e., their apparent constancy in inheritance, is probably due to the fact that there are several or possibly many replicates of the plastid genes in the same cell. To observe changes in plastid genes, it is necessary to affect all replicates of one kind at the same time. Differential effects of chemical agents such as streptomycin or aureomycin, physical agents like heat, or certain specific nuclear genes conceivably may destroy or change simultaneously all the replicates of certain of these plastid genes.

## VII. Nucleic acids of the chloroplasts.

Since plastids are self-perpetuating bodies it has been assumed, on analogy with nuclei and viruses, that the plastids contained nucleic acids. Only recently has experimental evidence appeared to support the hypothesis of nucleic acid in plastids. Two techniques have been used to demonstrate nucleic acids; namely, the analytical and the histochemical techniques.

### 1. Analytical techniques and results.

In the analytical method, chloroplasts are isolated by differential centrifugation and then subjected to analysis. The success of this method requires that the isolated chloroplasts should not be contaminated with nuclear and other cytoplasmic materials. Weier and Stocking (1952b) have found that chloroplasts which were isolated from leaves that had been ground up in a blendor in 0.5 M sucrose and differentially centrifuged, were contaminated with nuclear material. Repeated washings did not serve to decrease the contamination of nuclear material adhering to the chloroplasts. However, by gently grinding the leaves, against sand covered with a sucrose solution, the chloroplasts and nuclei were observed to be released from the cells without appreciable disintegration. The chloroplasts could then be obtained relatively free of nuclear contamination by two centrifugation washings at 200 × gravity. Methyl green, when added to the suspension of material, rapidly stained the nuclei and nuclear fragments. Permanent slides could be made by fixing the material in osmic vapor and staining with 1 per cent toluidine blue for $1/_2$ hour; the nuclei, nuclear derivatives and disorganized chloroplasts were colored an intense blue while the intact chloroplasts appeared in a delicate shade of green. McClendon (1952) found by direct analysis that chloroplasts of tobacco, prepared with the use of the blendor, contained DNA; but this DNA, he concluded, was derived from adsorbed nuclear DNA as seen with an aceto-orcein stain or as seen in a phase microscope.

In preparing chloroplasts for analysis it is not only necessary to consider the possibility of adsorbed nucleic acids, but one must also consider the less likely possibility that nucleic acid might leach out of the chloroplasts during preparation. With these difficulties in mind, let us examine the available analytical data and the interpretations placed on the data.

McClendon (1952) used the method of Ogur and Rosen (1950) for the determination of RNA (ribose nucleic acid) and DNA (desoxypentose nucleic acid) after a preliminary extraction of lipides and nucleotides from the chloroplasts. RNA was extracted for 18 hours at $4^0$ C. with 1 N perchloric acid. (Some RNA, however, appeared to be as resistant to perchloric acid as did DNA in this treatment.) Then DNA was extracted with 1 N perchloric acid for 20 minutes at $70^0$. The RNA phosphorus of the chloroplasts was found to vary from 0.005 to 0.035 $\mu g/\mu g$ protein-N. The chloroplasts were estimated to contain 0.2 per cent RNA phosphorus or approx. 2 per cent RNA. The total phosphorus of chloroplast "granules" was 0.3 per cent according to Warburg (1949). Because of contamination with nuclear fragments it could not be decided whether DNA was present in chloroplasts. According to Jagendorf and Wildman (1954) the nucleic acid content of washed chloroplasts is very low. At most, only 1 per cent of total N or 10 per cent of total P was found to be due to the nucleic acid. It was estimated that on a dry weight basis the chloroplasts contained 0.3 to 0.7 per cent nucleic acid.

Menke (1938b) extracted fat soluble constituents with ether and alcohol-ether 3:1 (v/v). The material was then washed with water and extracted with 0.01 N NaOH and precipitated with acid. This fraction contained 0.7 per cent P; it gave a pentose test, and constituted 22 per cent of the lipid-free chloroplasts or approx. 0.1 per cent of the total original dried chloroplast. This P value appears to be very high. Sisakyan and Chernyak (1952) extracted isolated chloroplasts with 0.2 per cent NaOH and precipitated the extract with 20 per cent acetic acid. On the basis of color tests for pentoses they concluded that pentoses of the ribose type were the predominant ones; in addition traces of granules containing desoxyribose were also found.

## 2. Histochemical techniques and results.

The histochemical methods avoid the problems arising from contamination of chloroplasts with other cytoplasmic constituents if is proper fixation used. Studies of Kaufmann (1950, 1951) and of McDonald and Kaufmann (1954) reveal the pitfalls that may be encountered with histochemical methods as applied to nucleoproteins. The methods that have been used for detecting and differentiating RNA and DNA in the chloroplasts all depend on first extracting and decolorizing the chloroplasts with lipid solvents such as 95 per cent ethanol or ethanol: glycerine 4:1 (v/v). Other extractions or enzyme digestions may follow, but finally the nucleic acids are detected by two staining procedures, the Feulgen procedure and the methylgreen-pyronin procedure.

The Feulgen stain (1924) is used to detect DNA. It depends on the formation of an aldehyde group derived from the acid treatment of the desoxyribose in DNA. The aldehyde group can be detected with fuchsin-sulfurous acid. According to Metzner (1952) DNA was not detectable in the chloroplasts of *Agapanthus umbellatus*. He found that it was necessary to fix the plastids for 4 hours or longer in 96 per cent ethanol in order to dissolve out non-nucleic acid material that would otherwise give a positive Feulgen color after acid treatment. CHIBA (1951) reported that the Feulgen reaction was positive on chloroplasts of *Selaginella Savatieri*, *Tradescantia fluminensis* and *Rhoeo discolor*.

The methylgreen-pyronin mixture has been used to differentiate between DNA and RNA. However, it was found by KURNICK (1950) and KURNICK and MIRSKY (1950) that methyl green selectively stains highly polymerized nucleic acids and that pyronin selectively stains lower polymers of the nucleic acids. Usually, DNA is highly polymerized so that methyl green will stain it, and usually RNA, as fixed, is in a lower polymeric state and stains with pyronin.

METZNER (1952) found that methyl green stained only the "grana" of the plastids of *Agapanthus* and that pyronin stained the plastid stroma only diffusely. After treating the cells with 1 N HCl for 15 minutes at 60° C. in order to remove RNA, methyl green still stained the grana. He concluded that DNA was present in the grana and that RNA was present in the grana and stroma. CHIBA (1951) found that methyl green stained the chloroplasts blue-green; the chloroplasts were not stained after they had been treated with 0.3 M trichloracetic acid at 90° for 15 minutes which treatment would have removed all of the nucleic acids from the chloroplasts. CHIBA also found that with pyronin alone, the chloroplasts stained red-pink, but after ribonuclease treatment for 2 hours at 60° C. the chloroplasts still stained with pyronin although the stain was only light-pink. This result was interpreted to indicate that RNA was present in the plastids. When the plastids were doubly stained with methyl green and pyronin, the chloroplasts stained purple red, but after ribonuclease treatment the chloroplasts stained light purple.

## 3. Summary.

The data at present available by analytical techniques suggest that RNA is present in the chloroplasts. The histochemical techniques suggest that both DNA and RNA are present in the chloroplasts. Before these conclusions are accepted completely, further work is necessary. It will be important to determine analytically whether DNA is present in the chloroplasts. Also further work should be done with pure enzymes and solvents in connection with staining procedures. The large chloroplasts of algae or the colorless plastids, such as present in *Euglena* which has been grown in the dark, would seem better material for histological staining procedures than the chloroplasts of higher plants. Also studies on proplastids and young plastids would be of great interest in order to determine whether there is any directly segregating nucleic acid apparatus in these dividing bodies. In the algal chloroplast, the pyrenoid appears to be a center of high metabolic activity. The granules connected with the tubular elements of etiolated plastids might be homologized with the RNA granules of the endoplasmic reticulum of animal cells. What will staining techniques for RNA and DNA reveal in these bodies?

We desire to acknowledge our gratitude to Dr. ERNST CASPARI and to Dr. ARMIN C. BRAUN for their kind suggestions and criticisms of this review.

## Literature.

ALGERA, L., J. J. BEIJER, W. VAN ITERSON, W. K. H. KARSTENS and T. H. THUNG: Some data on the structure of the chloroplast, obtained by electron microscopy. Biochim. et Biophysica Acta 1, 517–526 (1947).

BAUR, E.: Das Wesen und die Erblichkeitsverhältnisse der „Varietates albomarginatae hort." von *Pelargonium zonale*. Z. Vererbungslehre 1, 330–351 (1909). — Einführung in die

Vererbungslehre, 2. Aufl. Berlin: Gebrüder Bornträger 1930. — BEAUVERIE, J.: La structure granulaire des grains de chlorophylle. Rev. Cytol. et Cytophysiol. végét. **3**, 80–109 (1938). — BINDER, M.: Die Eliminierung der 1 + I Embryonen mit Plastiden der *Oenothera odorata*. Z. Vererbungslehre **75**, 739–796 (1938). — BLACK, L. M., C. MORGAN and R. W. G. WYCKOFF: Visualization of tobacco mosaic virus within infected cells. Proc. Soc. Exper. Biol. a. Med. **73**, 119–122 (1950). — BOGORAD, L.: Effects of streptomycin on chlorophyll formation in dark-green seedlings. Amer. J. Bot. **37**, 676 (1950). — BOLD, H. C.: Cytology of algae. Edit. G. M. SMITH, Manual of phycology, p. 203–227. Waltham, Mass.: Chronica Botanica 1951. — BRANCH, G. E. K., and M. CALVIN: Theory of organic chemistry. New York: Prentice-Hall 1941.

CALVIN, M., and V. LYNCH: Grana-like structures of *Synechococcus cedorum*. Nature (Lond.) **169**, 455–456 (1952). — CASPARI, E.: Cytoplasmic inheritance. Adv. Genet. **2**, 1–66 (1948). — CHEN, S. L.: The action spectrum for the photochemical evolution of oxygen by isolated chloroplasts. Plant Physiol. **27**, 35–48 (1952). — CHIBA, Y.: Cytochemical studies on chloroplasts. I. Cytologic demonstration of nucleic acids in chloroplasts. Cytologia (Tokyo) **16**, 259–264 (1951). — CHMIELEVSKY, V.: Eine Notiz über das Verhalten der Chlorophyllbänder in den Zygoten der *Spirogyra*-Arten. Bot. Ztg **48**, 773–780 (1890). — CORRENS, C.: Vererbungsversuche mit blaß (gelb) grünen und buntblättrigen Sippen bei *Mirabilis Jalapa, Urtica pilulifera* und *Lunaria annua*. Z. Vererbungslehre **1**, 291–329 (1909). — CZURDA, V.: Morphologie und Physiologie des Algenstärkekornes. Beih. bot. Zbl., Abt. 1, **45**, 97–270 (1928).

D'AMATO, F.: Mutazioni clorofilliane nell'orzo indotte da derivati acridinici. Caryologia (Pisa) **3**, 211–220 (1950). — DANGEARD, P.: Cytologie végétale et cytologie générale. Paris: Le Chevalier 1947. — DAVIS, B. M.: The spore-mother-cell of *Anthoceros*. Bot. Gaz. **28**, 89 (1899). — DEKEN GRENSON, M. DE: Grana formation and synthesis of chloroplastic proteins induced by light in portions of etiolated leaves. Biochem. et Biophysica Acta **14**, 203–211 (1954). — DEMEREC, M.: Behavior of chlorophyll in inheritance. Cold Spring Harbor Symp. Quant. Biol. **3**, 80–86 (1935). — DOUTRELIGNE, J.: Note sur la structure des chloroplastes. Proc. Kon. Akad. Wetensch. Amsterdam, Sect. Sci. **38**, 886–896 (1935). — DUTTON, H. J., W. B. MANNING and B. M. DUGGAR: Chlorophyll fluorescence and energy transfer in the diatom *Nitzschia closterium*. J. Physic. Chem. **47**, 308–313 (1943). — DUYSENS, L. N. M.: Transfer of light energy within the pigment systems present in photosynthesizing cells. Nature (Lond.) **168**, 548–550 (1951).

EMBERGER, L.: Nouvelles recherches sur le chondriome de la cellule végétale. Rev. gén. Bot. **39**, 341–363 (1927). — ENGELMANN, T. W.: Die Erscheinungsweise der Sauerstoffausscheidung chromophyllhaltiger Zellen im Licht bei Anwendung der Bacterienmethode. Arch. ges. Physiol. **57**, 375–386 (1894). — EULER, H. v., M. BRACCO and L. HELLER: Les actions de la streptomycine sur les graines en germination des plantes vertes et sur les polynucléotides. C. r. Acad. Sci. Paris **227**, 16–18 (1948). — EYSTER, W. H.: Variation in size of plastids in genetic strains of Zea mays. Science (Lancaster, Pa.) **69**, 48 (1929).

FEULGEN, R., u. H. ROSSENBECK: Mikroskopisch-chemischer Nachweis einer Nucleinsäure vom Typus der Thymonucleinsäure und die darauf beruhende elektive Färbung von Zellkernen in mikroskopischen Präparaten. Z. physiol. Chem. **135**, 203 (1924). — FOGG, G. E.: Metabolism of algae. New York: Wiley 1953. — FRANK, S. R.: The effectiveness of the spectrum in chlorophyll formation. J. Gen. Physiol. **29**, 157–179 (1946). — FRENCH, C. S., and V. K. YOUNG: Fluorescence spectra of red algae and the transfer of energy from phycoerythrin to phycocyanin and chlorophyll. J. Gen. Physiol. **35**, 873–890 (1952). — FREY-WYSSLING, A.: Die Stoffausscheidung der höheren Pflanzen. Berlin: Springer 1935. — Submicroscopic morphology of protoplasm, 2. edit. Houston, Texas: Elsevier 1953. — FREY-WYSSLING, A., u. K. MÜHLETHALER: Über den Feinbau der Chlorophyllkörner. Vjschr. naturforsch. Ges. Zürich **94**, 179–183 (1949). — FREY-WYSSLING, A., u. E. STEINMANN: Die Schichtendoppelbrechung großer Chloroplasten. Biochim. et Biophysica Acta **2**, 254–259 (1948). — FREY-WYSSLING, A., u. K. WUHRMANN: Zur Optik der Chloroplasten. Helvet. chim. Acta **30**, 20–23 (1947).

GALSTON, A. W.: Phototropism. II. Bot. Review **16**, 361–378 (1950). — GARTON, G. A., T. W. GOODWIN and W. LIJINSKY: Studies in carotenogenesis. 1. General conditions governing β-carotene synthesis by the fungus *Phycomyces blakesleeanus* BURGEFF. Biochemic. J. **48**, 154–163 (1951). — GEITLER, L.: Über den Granabau der Plastiden. Planta (Berl.) **26**, 463–469 (1937). — GODDARD, D. R.: The respiration of cells and tissues. R. HÖBER, Physical chemistry of cells and tissues, p. 371–383. Philadelphia: Blakiston 1945. — GOODWIN, R. H.: The role of auxin in leaf development in *Solidago* species. Amer. J. Bot. **24**, 43–51 (1937). — GOODWIN, T. W.: Comparative biochemistry of the carotenoids. London: Chapman & Hall 1952. — GRANICK, S.: Quantitative isolation of chloroplasts from higher plants. Amer. J. Bot. **25**, 558–561 (1938). — Chloroplast nitrogen of some higher plants. Amer. J. Bot. **25**, 561–567 (1938). — GRANICK, S., and K. R. PORTER: The structure of the spinach chloroplast as interpreted with the electron microscope. Amer. J. Bot. **34**, 545–550 (1947). —

GUILLIERMOND, A.: The cytoplasm of the plant cell. Watham, Mass.: Chronica Botanica 1941.

HABERLANDT, G.: Physiological plant anatomy. New York: Macmillan 1914. — HÄRTEL, O., u. I. THALER: Die Proteinoplasten von Helleborus corsicus WILLD. Protoplasma 42, 417–426 (1953). — HASSID, W. Z.: Biosynthesis of complex saccharides. In D. M. GREENBERG, Chemical pathways of metabolism, Vol. I, p. 235–275. New York: Academic Press 1954. — HAXO, F.: Studies on the carotenoid pigments of Neurospora. I. Composition of the pigments. Arch. of Biochem. 20, 400–421 (1949). — HAXO, F. T., and L. R. BLINKS: Photosynthetic action spectra of marine algae. J. Gen. Physiol. 33, 389–422 (1950). — HEITZ, E.: Untersuchungen über den Bau der Plastiden. I. Die gerichteten Chlorophyllscheiben der Chloroplasten. Planta (Berl.) 26, 134–163 (1936). — Kristallgitterstruktur des Granum junger Chloroplasten von Chlorophytum. Exper. Cell Res. 7, 606–608 (1954). — HEITZ, E., u. R. MALY: Zur Frage der Herkunft der Grana. Z. Naturforsch. 8b, 243–249 (1953). — HIGHKIN, H. R.: Chlorophyll studies on barley mutants. Plant Physiol. 25, 294–306 (1950). — HILL, G. B., L. O. OVERHOLTS and H. W. POPP: Botany, 2. edit. New York: McGraw-Hill 1950. — HOLT, A. S., I. A. BROOKS and W. A. ARNOLD: Some effects of 2537 Å on green algae and chloroplast preparations. J. Gen. Physiol. 34, 627–645 (1951).

IKENO, S.: Studies on the hybrids of Capsicum annuum. Part II. On some variegated races. J. Genet. 6, 201–229 (1917). — ISHIKAWA, M.: Studies on the embryo sac and fertilization in Oenothera. Ann. of Bot. 32, 279–317 (1918).

JAGENDORF, A. T., and S. G. WILDMAN: The proteins of green leaves. VI. Centrifugal fractionation of tobacco leaf homogenates and some properties of isolated chloroplasts. Plant Physiol. 29, 270–279 (1954). — JAHN, T. L.: Euglenophyta. In G. M. SMITH, Manual of phycology, p. 69–81. Waltham, Mass.: Chronica Botanica 1951. — JONES, D. F.: Nuclear changes affecting growth. Amer. J. Bot. 27, 149–155 (1940). — JUNGERS, V., and J. DOUTRELIGNE: Sur la localisation de la chlorophylle dans les chloroplastes. Cellule 49, 407–417 (1943).

KAJA, H.: Untersuchungen über die Kontinuität der Granastruktur in den Plastiden der Moose. Ber. dtsch. bot. Ges. 67, 93–108 (1954). — KAUFMANN, B. P.: An evaluation of the applicability in cytochemical studies of methods involving enzymatic hydrolysis of cellular materials. Portugal. Acta Biol., Ser. A (R. B. GOLDSHMIDT) 1950. — KAUFMANN, B. P., M. R. McDONALD and H. GAY: The distribution and interpretation of nucleic acid in fixed cells as shown by enzymatic hydrolysis. J. Cellul. a. Comp. Physiol. 38, 71–101 (1951). — KAUSCHE, G. A., u. H. RUSKA: Über den Nachweis von Molekülen des Tabakmosaikvirus in den Chloroplasten viruskranker Pflanzen. Naturwiss. 28, 303 (1940). — KNUDSON, L.: Has the chloroplast a semipermeable membrane? Amer. J. Bot. 23, 694 (1936). — Permanent changes of chloroplasts induced by X rays in the gametophyte of Polypodium aureum. Bot. Gaz. 101, 721–758 (1940). — KÜSTER, E.: Anisotrope Plastiden. Ber. dtsch. bot. Ges. 51, 523–525 (1934). — Die Pflanzenzelle. Jena: Gustav Fischer 1935. — KURNICK, N. B.: Methyl green-pyronin. I. Basis of selective staining of nucleic acids. J. Gen. Physiol. 33, 243–264 (1950). — KURNICK, N. B., and A. E. MIRSKY: Methyl green-pyronin. II. Stoichiometry of reaction with nucleic acids. J. Gen. Physiol. 33, 265–274 (1950).

LEMBERG, R., and J. W. LEGGE: Hematin compounds and bile pigments. New York: Interscience Publishers 1949. — LEYON, H.: Sugar beet yellows virus: some electron microscopical observations. Ark. Kemi (Stockh.) 3, 105–109 (1951). — (1) The structure of chloroplasts: an electron microscopical investigation of sections. Exper. Cell Res. 4, 371–382 (1953). — (2) The structure of chloroplasts. III. A study of pyrenoids. Exper. Cell Res. 6, 497–505 (1954). — (3) The structure of chloroplasts. IV. The development and structure of the Aspidistra chloroplast. Exper. Cell Res. 7, 265–273 (1954). — (4) The structure of chloroplasts. VI. The origin of the chloroplast laminae. Exper. Cell Res. 7, 609–611 (1954). — LEYON, H., u. D. v. WETTSTEIN: Der Chromatophoren-Feinbau bei den Phaeophyceen. Z. Naturforsch. 9b, 471–475 (1954). — LIVERMAN, J. L., M. P. JOHNSON and L. STARR: Reversible photoreaction controlling expansion of etiolated bean-leaf disks. Science (Lancaster, Pa.) 121, 440–441 (1955). — LWOFF, A., et H. DUSI: La suppression expérimentale des chloroplastes chez Euglena mesnili. C. r. Soc. Biol. Paris 119, 1092–1095 (1935). — LWOFF, A., et P. SCHAEFFER: La décoloration d'Euglena gracilis par la streptomycine. C. r. Acad. Sci. Paris 228, 779–781 (1949).

MALY, R.: Cytomorphologische Studien an strahleninduzierten, konstant abweichenden Plastidenformen bei Farnprothallien. Z. Vererbungslehre 83, 447–478 (1951). — MANGENOT, G.: Notes sur la cytologie des laminaires. C. r. Soc. Biol. Paris 88, 522–523 (1923). — Sur les oosphères, les tubes polliniques et la fécondation chez le pin maritime. C. r. Acad. Sci. Paris 206, 364–366 (1938). — MAST, S. O.: Motor response in unicellular animals. CALKINS, G. N., and F. M. SUMMERS, eds.: Protozoa in biological research, p. 271–352. New York: Columbia University Press 1941. — McALLISTER, F.: The pyrenoid of Anthoceros. Amer. J. Bot. 1, 79–95 (1914). — The pyrenoids of Anthoceros and Notothylas with especial reference to their presence in spore mother cells. Amer. J. Bot. 14, 246–257 (1927). — McCLENDON,

J. H.: The intracellular localization of enzymes in tobacco leaves. I. Identification of components of the homogenate. Amer. J. Bot. **39**, 275–282 (1952). — McCLENDON, J. H., and L. R. BLINKS: Use of high molecular solutes in the study of isolated intracellular structures. Nature (Lond.) **170**, 577–578 (1952). — McCLINTOCK, B.: The behavior in successive nuclear divisions of a chromosome broken at meiosis. Proc. Nat. Acad. Sci. U.S.A. **25**, 405–416 (1939). — McDONALD, M. R., and B. P. KAUFMANN: The degradation by ribonuclease of substrates other than ribonucleic acid. J. Histochem. a. Cytochem. **2**, 387–394 (1954). — MENKE, W.: Chloroplasten-Studien. 2. Mitt. Protoplasma **22**, 56–62 (1934). — Über den Feinbau der Chloroplasten. Kolloid-Z. **85**, 256–259 (1938). — Untersuchungen über das Protoplasma grüner Pflanzenzellen. I. Isolierung von Chloroplasten aus Spinatblättern. Z. physiol. Chem. **257**, 43–48 (1938). — MENKE, W., u. E. KOYDL: Direkter Nachweis des lamellaren Feinbaues der Chloroplasten. Naturwiss. **27**, 29–30 (1939). — MENKE, W., u. H.-J. KÜSTER: Dichroismus und Doppelbrechung vergoldeter Chloroplasten. Protoplasma **30**, 283–290 (1938). — METZNER, H.: Cytochemische Untersuchungen über das Vorkommen von Nukleinsäuren in Chloroplasten. Biol. Zbl. **71**, 257–272 (1952). — MEYER, A.: Das Chlorophyllkorn in chemischer, morphologischer und biologischer Beziehung. Leipzig: A. Felix 1883. — Das ergastische Organeiweiß und die vitülogenen Substanzen der Palisadenzellen von Tropaeolum majus. Ber. dtsch. bot. Ges. **35**, 658–673 (1917). — Morphologische und physiologische Analyse der Zelle der Pflanzen und Tiere, Teil 1. Jena: Gustav Fischer 1920. — MICHAELIS, P.: Cytoplasmic inheritance in *Epilobium* and its theoretical significance. Adv. Genet. **6**, 287–401 (1954). — MILLER, E. C.: Plant physiology, 2. edit. New York: McGraw-Hill 1938. — MÖBIUS, M.: Über die Größe der Chloroplasten. Ber. dtsch. bot. Ges. **38**, 224–232 (1920). — Pigmentation in plants, exclusive of the algae. Bot. Review **3**, 351–363 (1937). — MOTTIER, D. M.: On certain plastids, with special reference to the protein bodies of *Zea*, *Ricinus*, and *Conopholis*. Ann. of Bot. **35**, 349–364 (1921). — MUSCHIK, M.: Untersuchungen zum Problem der Aleuronkornbildung. Protoplasma **42**, 43–57 (1953).

NEILSON-JONES, W.: Plant chimaeras and graft hybrids. London: Methuen 1934.

OGUR, M., and G. ROSEN: The nucleic acids of plant tissues. I. The extraction and estimation of desoxypentose nucleic acid and pentose nucleic acid. Arch. of Biochem. **25**, 262–276 (1950).

PALADE, G. E.: A study of fixation for electron microscopy. J. of Exper. Med. **95**, 285–298 (1952). — An electron microscope study of the mitochondrial structure. J. Histochem. a. Cytochem. **1**, 188–211 (1953). — A small particulare component of the cytoplasm. J. Biophys. a. Biochem. Cytology **1**, 59–68 (1955). — PARDEE, A. B., H. K. SCHACHMAN and R. Y. STANIER: Chromatophores of *Rhodospirillum rubrum*. Nature (Lond.) **169**, 282–283 (1952). — PERNER, E. S.: Zum mikroskopischen Nachweis des „primären Granums" in den Leukoplasten. Ber. dtsch. bot. Ges. **67**, 26–32 (1954). — PORTER, K. R.: Observations on a submicroscopic basophilic component of cytoplasm. J. of Exper. Med. **97**, 727–749 (1953). — PRINGSHEIM, E. G.: The interrelationships of pigmented and colorless flagellata. Biol. Rev. **16**, 191–204 (1941). — The loss of chromatophores in *Euglena gracilis*. New Phytologist **47**, 52–87 (1948). — PRINGSHEIM, E. G., and O. PRINGSHEIM: Experimental elimination of chromatophores and eye-spot in *Euglena gracilis*. New Phytologist **51**, 65–76 (1952). — PROVASOLI, L., S. H. HUTNER and I. J. PINTNER: Destruction of chloroplasts by streptomycin. Cold Spring Harbor Symp. Quant. Biol. **16**, 113–120 (1951). — PROVASOLI, L., S. H. HUTNER and A. SCHATZ: Streptomycin-induced chlorophyll-less races of *Euglena*. Proc. Soc. Exper. Biol. a. Med. **69**, 279–282 (1948).

RABINOWITCH, E. I.: Photosynthesis and related processes, Vol. 1. New York: Interscience Publ. 1945. — Photosynthesis and related processes, Vol. 2. New York: Interscience Publ. 1951. — REINHARD, H.: Über die Teilung der Chloroplasten. Protoplasma **19**, 541–564 (1933). — RENNER, O.: Zur Kenntnis der nichtmendelnden Buntheit der Laubblätter. Flora (Jena) **30**, 218–290 (1936). — REZENDE-PINTO, M. C. DE: Sur la structure hélicoïdale des chloroplastes. Portugal. Acta Biol. A **2**, 111–114 (1948). — RHOADES, M. M.: Genic induction of an inherited cytoplasmic difference. Proc. Nat. Acad. Sci. U.S.A. **29**, 327–329 (1943). — Plastid mutations. Cold Spring Harbor Symp. Quant. Biol. **11**, 202–207 (1946). — ROBBINS, W. J., A. HERVEY and M. E. STEBBINS: *Euglena* and vitamin B$_{12}$. Ann. New York Akad. Sci. **56**, 818–830 (1953). — ROPP, R. S. DE: Action of streptomycin on plant tumours. Nature (Lond.) **162**, 459–460 (1948).

SACHS, J.: Lectures on the physiology of plants. London: Oxford University Press 1887. — SAGER, R., and G. E. PALADE: Chloroplast structure in green and yellow strains of *Chlamydomonas*. Exper. Cell Res. **7**, 584–588 (1954). — SCARTH, G. W.: Colloidal changes associated with protoplasmic contraction. Quart. J. Exper. Physiol. **14**, 99–113 (1924). — SCHANDERL, H., u. W. KAEMPFERT: Über die Strahlungsdurchlässigkeit von Blättern und Blattgeweben. Planta (Berl.) **18**, 700–750 (1933). — SCHERRER, A.: Untersuchungen über Bau und Vermehrung der Chromatophoren und das Vorkommen von Chondriosomen bei *Anthoceros*. Flora (Jena) **107**, 1–56 (1914). — SCHILLER, J.: Beobachtungen über die Entwicklung des roten Augenflecks bei *Ulva Lactuca*. Österr. bot. Z. **72**, 236–241 (1923). — Schimper, A.: Über die Entwicklung der Chlorophyllkörner und Farbkörper. Bot. Ztg **41**, 105–112, 121–131, 137–146, 153–162 (1883). — Schmitz, F.: Beiträge zur Kenntnis der Chromatophoren. Jb. wiss. Bot. **15**, 2–177 (1884). — Schürhoff, P.: Die Plastiden. In K. Linsbauer, Handbuch der Pflanzenanatomie, Bd. 1, Liefg 10. Berlin: Gebrüder Bornträger 1924. — Schwartz, D.: The chlorophyll mutants of maize. Bot. Gaz. **111**, 123–130 (1949). — Schwemmle, J. u. a.: Genetische und zytologische Untersuchungen an Eu-Oenotheren; Teil I bis VI. Z. Vererbungslehre **75**, 358–800 (1938). — Senn, G.: Die Gestalts- und Lageveränderung der Pflanzenchromatophoren. Leipzig: Wilhelm Engelmann 1908. — Seybold, A.: Über die optischen Eigenschaften der Laubblätter. IV. Planta (Berl.) **21**, 251–265 (1933). — Sharp, L. W.: An introduction to cytology, 3. edit. New York: McGraw-Hill 1934. — Sirks, M. J.: Plasmatic inheritance. Bot. Review **4**, 113–131 (1938). — Sisakyan, N. M., and M. S. Chernyak: Nucleic acids of plastids. Dokl. Akad. Nauk SSSR. **87**, 469–470 (1952). — Skoog, F. (edit.): Plant growth substances. Madison, Wisc.: University of Wisconsin Press 1951. — Smith, G. M.: Cryptogamic botany, Vol. 2. New York: McGraw-Hill 1938. — Smith, J. H. C.: The relationship of plant pigments to photosynthesis. J. Chem. Educat. **26**, 631–638 (1949). — Smith, J. H. C., C. S. French and V. M. Koski: The Hill reaction: development of chloroplast activity during greening of etiolated barley leaves. Plant Physiol. **27**, 212–213 (1952). — Smith, J. H. C., and V. M. Koski: Chlorophyll formation. Carnegie Institution of Washington, Year book, No 47, p. 93–96, 1947/48. — Smith, J. H. C., V. M. Koski and C. S. French: The action spectrum for the formation of chlorophyll. Carnegie Institution of Washington, Year book, No 48, p. 91–92, 1948/49. — Steinmann, E.: An electron microscope study of the lamellar structure of chloroplasts. Exper. Cell Res. **3**, 367–372 (1952). — Steinmann, E., and F. S. Sjöstrand: Ultrastructure of chloroplasts. Exper. Cell Res. **8**, 15–23 (1955). — Stewart, W. N.: A study of the plastids in the cells of the nature sporophyte of *Isoetes*. Bot. Gaz. **110**, 281–300 (1948). — Strain, H. H.: Functions and properties of the chloroplast pigments. In J. Franck and W. L. Loomis, Photosynthesis in plants, p. 133–178. Ames, Iowa State College Press 1949. — The pigments of algae. In G. M. Smith (edit.), Manual of phycology, p. 243–262. Waltham, Mass.: Chronica Botanica 1951. — Straus, W.: Properties of isolated carrot chromoplasts. Exper. Cell Res. **6**, 392–402 (1954). — Strugger, S.: Die Vitalfärbung der Chloroplasten von *Helodea* mit Rhodaminen. Flora (Jena) **31**, 113–128 (1936). — Über den Bau der Proplastiden und Chloroplasten. Naturwiss. **37**, 166–167 (1950). — Über die Struktur der Proplastiden. Forschungsber. Wirtschafts- und Verkehrsministeriums Nordrhein-Westfalen, Nr 83, S. 1–27, 1954. — Szejnman, A.: Observations sur la formation des pyrénoides chez *Spirogyra*. Acta Soc. Bot. Poloniae **10**, 331–359 (1933).

Tabentskii, A. A.: Control of the processes of formation of green plastids. Izv. Akad. Nauk SSSR., Ser. Biol. **1**, 71–95 (1953). — Thomas, J. B., O. H. Blaauw and L. N. M. Duysens: On the relation between size and photochemical activity of fragments of spinach grana. Biochim. et Biophysica Acta **10**, 230–240 (1953). — Thomas, J. B., L. C. Post and N. Vertregt: Localisation of chlorophyll within the chloroplast. Biochim. et Biophysica Acta **13**, 20–30 (1954).

Voerkel, S. H.: Untersuchungen über die Phototaxis der Chloroplasten. Planta (Berl.) **21**, 156–205 (1933).

Warburg, O. H.: Heavy metal prosthetic groups and enzyme action. London: Oxford University Press 1949. — Warburg, O., u. G. Kruppahl: Über Photosynthese-Fermente. Angew. Chem. **66**, 493–496 (1954). — Weier, E.: (1) The structure of the chloroplast. Bot. Review **4**, 497–530 (1938). — (2) Viability of cells containing chloroplasts with an optically homogenous or granular structure. Protoplasma **31**, 346–350 (1938). — Weier, T. E., and C. R. Stocking: (1) The chloroplast: structure, inheritance, and enzymology. II. Bot. Review **18**, 14–75 (1952). — (2) A cytological analysis of leaf homogenates. I. Nuclear contamination and disorganized chloroplasts. Amer. J. Bot. **39**, 720–726 (1952). — Wenck, U.: Die Wirkung von Wuchs- und Hemmstoffen auf die Blattform. Z. Bot. **40**, 33–51 (1952). — Wettstein, D. v.: Formwechsel und Teilung der Chromatophoren von *Fucus vesiculosus*. Z. Naturforsch. **9b**, 476–481 (1954). — Willstätter, R., u. A. Stoll: Untersuchungen über Chlorophyll. Berlin: Springer 1913. — Winge, O.: On the non-mendelian inheritance in variegated plants. C. r. Trav. Labor. Carlsberg **14**, 1–21 (1919). — Wisselingh, C. v.: Über Variabilität und Erblichkeit. Z. Vererbungslehre **22**, 65–126 (1920). — Wolken, J. J., and G. E. Palade: An electron microscope study of two flagellates. Chloroplast structure and variation. Ann. New York Acad. Sci. **56**, 873–889 (1953). — Wolken, J. J., and F. A. Schwertz: Chlorophyll monolayers in chloroplasts. J. Gen. Physiol. **37**, 111–120 (1953). — Woods, M. W., and H. G. duBuy: Hereditary and pathogenic nature of mutant mitochondria in *Nepeta*. J. Nat. Canc. Inst. **11**, 1105–1151 (1951).

Zurzycki, J.: Arrangement of chloroplasts and light absorption in plant cell. Acta Soc. Bot. Poloniae **22**, 299–320 (1953).

# Die Farbstoffe.

Von

## K. Egle.

Mit 3 Abbildungen.

Die Plastiden können farblos sein und werden dann Leukoplasten genannt. Treten sie mehr oder weniger auffallend gefärbt in Erscheinung, dann faßt man sie unter dem Sammelnamen Chromatophoren zusammen und bezeichnet sie je nach ihrer Farbe — grün bis blaugrün, gelb bis orange, blaßrot bis carminrot — als Chloroplasten, Chromoplasten oder Rhodoplasten. Ihre Färbung beruht auf Pigmenten, die fast alle typisch für die pflanzliche Organisationsstufe innerhalb der Organismenwelt sind. Durch die Bindung der Pigmente an die Strukturelemente des Plastidenplasmas bezeichnet man sie auch als Plasmochrome oder plasmagebundene Farbstoffe, wodurch der Gegensatz zu den zahlreichen (hier nicht zu erörternden) im Zellsaft gelösten Pflanzenfarbstoffen, den Chymochromen, besser hervortritt (SEYBOLD 1942).

Die Chromatophoren enthalten niemals nur einen einzigen einheitlichen Farbstoff, sondern immer ein Gemisch aus mehreren und oft recht verschiedenartigen Pigmenten, unter denen jedoch einzelne Pigmentkomponenten mengenmäßig mitunter stark überwiegen können. Nach ihrer chemischen Konstitution teilt man die Plastidenpigmente in 3 Gruppen ein: die Chlorophylle, die Carotinoide und die Phycobiline (neuere zusammenfassende Darstellungen: RABINOWITCH, STRAIN 1949, 1951, BONNER).

Die *Chlorophylle*, die den Chloroplasten die typische Färbung verleihen, die aber auch in den Rhodoplasten durch andere Pigmente verdeckt vorkommen, sind Porphyrinverbindungen, bei denen 4 Pyrrolringe, durch 4 Methin-($-CH=$) Brücken ($\alpha$, $\beta$, $\gamma$, $\delta$) verknüpft, ein charakteristisch substituiertes Ringsystem bilden, das ein zentrales mit den Stickstoffatomen der 4 Pyrrolkerne in komplexer Bindung stehendes Magnesiumatom einschließt (Abb. 1). Zu den Substituenten des Tetrapyrrolringes gehören eine Äthyl- und eine Vinylgruppe, sowie ein Propionsäure- und ein Essigsäurerest. Das blaugrüne Chlorophyll a besitzt an jedem der 4 Pyrrolringe eine Methylgruppe, während das gelbgrüne Chlorophyll b am Pyrrolkern II an Stelle der Methylgruppe in Stellung 3 einen Formylrest aufweist. Der Essigsäurerest ist mit Methylalkohol, die Propionsäuregruppe mit dem in der Natur nur als Chlorophyllbestandteil vorkommenden ungesättigten Alkohol Phytol ($C_{20}H_{39}OH$) verestert. Die nicht veresterten freien Mg-Porphyrine, die neben den charakteristischen Substituenten auch den für die Chlorophylle typischen isocyclischen Pentanonring zwischen dem Pyrrolring III und der benachbarten Methinbrücke aufweisen, werden Chlorophyllide genannt. Die Chlorophylle a und b sind somit Phytyl-methyl-ester der entsprechenden Dicarbonsäuren, was zweckmäßig auch durch die Schreibweise ihrer Summenformeln zum Ausdruck gebracht wird:

$$MgN_4OH_{30}C_{32}\diagup\!\!\!\diagdown\begin{matrix}OOC_{20}H_{39}\\OOCH_3\end{matrix} \qquad\qquad MgN_4O_2H_{28}C_{32}\diagup\!\!\!\diagdown\begin{matrix}OOC_{20}H_{39}\\OOCH_3\end{matrix}$$

Chlorophyll a  Chlorophyll b

Die in Abb. 1 wiedergegebene Strukturformel für die Chlorophylle a und b kann heute als gesichert gelten, wenn auch die Anordnung der beiden „überzähligen" Wasserstoffatome (in Abb. 1 in 7,8-Stellung am Pyrrolkern IV eingezeichnet) noch nicht ganz geklärt ist (WILLSTÄTTER und STOLL, STOLL und WIEDEMANN 1938, 1952, FISCHER und ORTH, FISCHER und STRELL). Die Chlorophylle bilden mit organischen Solventien wie Petroläther, Alkohol, Benzol, Aceton echte Lösungen. In Wasser werden sie kolloidal gelöst. Sie besitzen ein charakteristisches Absorptionsspektrum mit mehreren Absorptionsbanden, von denen

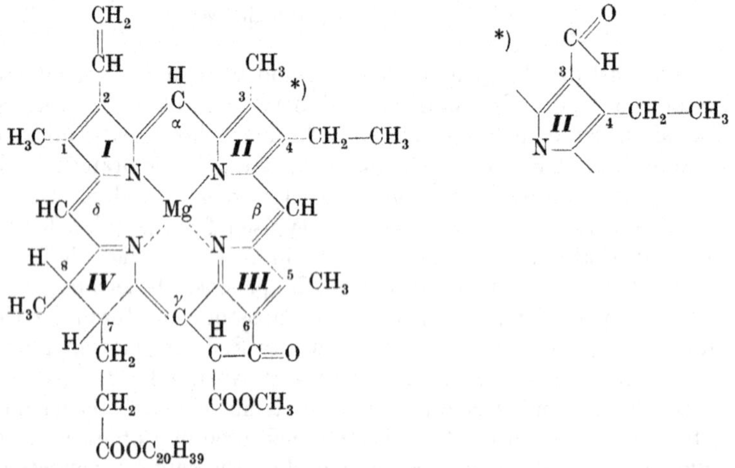

Abb. 1. Strukturformel von Chlorophyll a. Das Chlorophyll b besitzt am Pyrrolkern II (mit * markiert) an Stelle einer Methylgruppe einen Formylrest.

die im roten Teil des Spektrums liegende Hauptabsorptionsbande besonders scharf ausgeprägt ist. Ihre auffallende Rotfluorescenz kann zum Nachweis kleinster Chlorophyllmengen herangezogen werden. Der physikalisch-chemische Verteilungszustand der Chlorophylle in den Chromatophoren bzw. die Art ihrer Bindung an die Bestandteile der Grana ist noch nicht vollkommen aufgeklärt. Obwohl eine Bindung zwischen den Pigment- und Proteinmolekülen in den Grana der Plastiden sehr wahrscheinlich ist, für welche A. STOLL die Bezeichnung „Chloroplastin" eingeführt hat, konnte die chemische Analyse der Chloroplasten- bzw. „Grana"-Substanz noch keine gesicherten Ergebnisse über die reale Existenz eines Chlorophyllproteinkomplexes einheitlicher Zusammensetzung im Sinne eines konstanten stöchiometrischen Verhältnisses liefern. Trotz häufiger Versuche ist es bisher auch noch nicht gelungen, optisch homogene Chlorophyllpräparate aus Pflanzen zu isolieren, die in den beiden charakteristischen physikalischen Eigenschaften — der Bandenlage und der Fluorescenz — mit den Chloroplasten lebender Zellen genau übereinstimmen (SEYBOLD und EGLE 1940, RABINOWITCH; FREY-WYSSLING; ARONOFF; STOLL und WIEDEMANN 1952, EGLE 1953a).

Während in den Chloroplasten der Kohlenstoff-autotrophen Kormophyten, der Moose und mehrerer Algengruppen (Volvocales, Chlorophyceae, Conjugatae, Characeae) die Chlorophylle a und b enthalten sind, besitzen die Cyanophyceen, Diatomeen, Chrysophyceen, Heterokonten, Rhodophyceen und Phäophyceen kein Chlorophyll b (SEYBOLD und EGLE 1938). Auffallend ist das Fehlen von Chlorophyll b in der Grünalge *Vaucheria* und in der saprophytisch lebenden Orchidee *Neottia nidus-avis*. Da diejenigen Pflanzengruppen, deren Chromatophoren das Chlorophyll b fehlt, keine Assimilationsstärke bilden, hat SEYBOLD

(1941) die Theorie entwickelt, daß der b-Komponente eine bestimmte Funktion bei der Bildung der Stärke aus den primär anfallenden Photosyntheseprodukten zukommt.

Bei einigen Algengruppen, deren Chromatophoren kein Chlorophyll b besitzen, sind in neuerer Zeit neben Chlorophyll a als Hauptpigment verschiedene grüne Begleitfarbstoffe meist in sehr geringer Konzentration beobachtet worden. So ist in Pigmentextrakten aus Kiesel- und Braunalgen sowie aus Dinoflagellaten ein als Chlorophyll c bezeichneter Farbstoff gefunden worden, bei dem es sich wahrscheinlich um ein Magnesium-phäoporphyrin handelt, das den isocyclischen Pentanonring der beiden Chlorophylle a und b, jedoch kein Phytol enthält (STRAIN 1949, GRANICK). Chlorophyll d ist in geringer Menge bei zahlreichen Rotalgen beobachtet worden; in *Rhodochorton rothii* soll es bis zu 25% der Gesamt-chlorophyllmenge ausmachen. Ein als Chlorophyll e bezeichnetes Begleitpigment des Chlorophyll a ist bisher nur in Extrakten aus *Tribonema bombycinum* bekannt-geworden. Alle „Nebenchlorophylle" der Algen sind bisher hauptsächlich nur durch ihr Adsorptionsverhalten und durch die spektroskopischen Eigenschaften ihrer Lösungen näher charakterisiert worden. Das Gleiche gilt auch für die verschiedentlich beobachteten Chlorophyllisomeren. Die reale Existenz dieser Chlorophyllkomponenten in den Chromatophoren lebender Zellen ist noch nicht genügend gesichert (STRAIN 1949, 1951, RICHARDS).

Die Chromatophoren dunkel herangezogener Keimpflanzen enthalten geringe Mengen eines grünen Pigments, das als Protochlorophyll bezeichnet wird. Be-sonders reich an Protochlorophyll sind die dünnen Samenhäutchen vieler Kürbis-arten. Die chemische Erforschung dieses Pigments hat ergeben, daß es sich hierbei um einen Magnesium-vinyl-phäoporphyrin-$a_5$-phytylester handelt, der sich vom Chlorophyll a nur durch das Fehlen der beiden H-Atome am Pyrrolkern IV (vgl. Abb. 1) unterscheidet (NOACK und KIESSLING, STOLL und WIEDEMANN, FISCHER und OESTREICHER). Das aus Kürbissamenhäutchen extrahierte Proto-chlorophyll konnte chromatographisch in 2 Komponenten zerlegt werden, die nach ihren den beiden Chlorophyllen a und b sehr ähnlichen Adsorptionsver-halten sowie nach ihren den beiden Chlorophyllkomponenten entsprechenden spektroskopischen Eigenschaften als Protochlorophyll a und b bezeichnet worden sind (SEYBOLD 1937, SEYBOLD und EGLE 1938). Es ist bis jetzt jedoch noch nicht gelungen, das Protochlorophyll b eindeutig als eine Verbindung der „b-Reihe" festzulegen. Neuere Beobachtungen haben den Beweis erbracht, daß das Protochlorophyll in den Plastiden in Chlorophyll umgewandelt werden kann (SEYBOLD 1948; FRANK; KOSKI, FRENCH und SMITH).

Ein als Bacteriochlorophyll bezeichneter grüner Farbstoff, der die Purpur-bakterien (Thio- und Athiorhodaceen) zur Photosynthese befähigt, liegt in den Bakterienzellen offenbar nur als eine Komponente vor (SEYBOLD und EGLE 1939). In der chemischen Zusammensetzung unterscheidet er sich vom Chloro-phyll a durch den zusätzlichen Besitz der Elemente eines Wassermoleküls. Die Strukturaufklärung durch H. FISCHER und Mitarbeiter erbrachte den Beweis, daß es sich beim Bacteriochlorophyll um 2-Acetyl-3,4-dihydrochlorophyll a handelt (MITTENZWEI). Der in Chlorobakterien gefundene, dem Bacteriochloro-phyll ähnliche Farbstoff ist noch nicht näher bekannt.

Neuerdings ist die Aufklärung der Biosynthese der Chlorophyllfarbstoffe erfolgreich in Angriff genommen worden, wobei es sich gezeigt hat, daß der Aufbau dieser komplizierten Verbindungen in der lebenden Zelle aus sehr einfachen Bausteinen erfolgt (GRANICK; PIRSON; EGLE 1953b).

Die *Carotinoide*, welche die typische Färbung der Chromoplasten hervorrufen, aber auch in den Chloroplasten und Rhodoplasten, durch andere Farbstoffe verdeckt, enthalten sind, stellen gelb bis rot gefärbte Pigmente von aliphatischer oder alicyclischer Struktur dar, die in den meisten Fällen aus 8 Isopreneinheiten aufgebaut sind. Die Methylseitenketten in der Mitte der oft aus zwei vollkommen symmetrischen Hälften aufgebauten Moleküle stehen immer in 1,6-Stellung, während sich die übrigen Methylgruppen des aliphatischen Molekülteiles in 1,5-Stellung befinden (vgl. Abb. 2, I). Die alicyclischen Carotinoide besitzen einen oder zwei endständige sechsgliedrige Ringsysteme (Jononring). Die Strukturformeln der aliphatischen Carotinoide (Abb. 2, III) werden meist mit „offenem Ring" dargestellt, um die engen chemischen Beziehungen zu den alicyclischen Verbindungen besser hervorzuheben. Auf Grund ihres ungesättigten Charakters und durch das Vorhandensein zahlreicher konjugierter Doppelbindungen, die das chromophere System dieser Moleküle ausmachen, werden die Carotinoide als besondere Gruppe unter die Polyenfarbstoffe eingeordnet. Ihre charakteristischen Absorptionsspektren, die meist drei mehr oder weniger stark ausgeprägte Absorptionsbanden im kurzwelligen Teil des Spektrums aufweisen, sowie ihr differenziertes Verhalten bei der Absorption an verschiedenen Substanzen, hat wesentlich zur Erkennung, Isolierung und Konstitutionsaufklärung zahlreicher Carotinoide beigetragen (Zechmeister; Karrer und Jucker; Goodwin).

Nach ihrer chemischen Zusammensetzung und ihrem unterschiedlichen Verhalten gegenüber organischen Lösungsmitteln unterscheidet man 2 Gruppen von Carotinoiden:

Die *Carotine* sind reine Kohlenwasserstoffe, meist mit der Summenformel $C_{40}H_{56}$, die in aliphatischen und aromatischen Kohlenwasserstoffen, in Äther, Schwefelkohlenstoff und in anderen Lipoidlösungsmitteln leicht, jedoch in Alkohol nur schwer löslich sind.

Die *Xanthophylle* sind sauerstoffhaltige Kohlenwasserstoffe, die in 90%igem Alkohol meist noch recht gut, in Petroläther jedoch schwerer löslich sind, was sich die älteren Verfahren für ihre Abtrennung von den Carotinen zunutze gemacht haben. Man faßt die Xanthophylle als Derivate der Carotine auf, in denen der Sauerstoff in Form von Alkohol-, Keton-, Enol-, Aldehyd- oder Säuregruppen enthalten ist. In letzter Zeit hat Karrer (1948) natürlich vorkommende Xanthophylle beschrieben, die Epoxyde bereits bekannter Carotinoide darstellen, sowie auch solche, bei denen der Sauerstoff in einem Furanring enthalten ist (furanoide Oxyde; vgl. Abb. 2, V u. VI). Manche Xanthophylle kommen mit Fettsäuren verestert vor, andere bilden über ihre alkoholischen Hydroxylgruppen Äther.

Unter den bis heute bekanntgewordenen etwa 70 Carotinoiden, von denen hier nur auf die wichtigsten hingewiesen werden kann, gibt es solche, die in den Chromatophoren von Blättern, Sprossen, Wurzeln, Blüten, Früchten, Samen und Sporen der Kormophyten sowie in den Vegetations- und Reproduktionsorganen der kohlenstoffautotrophen Thallophyten weit verbreitet sind. Andere Carotinoide sind in ihrer Verbreitung auf gewisse systematische Gruppen und oft nur auf wenige nahe verwandte Arten beschränkt. Die am weitesten verbreiteten Chlorophyllbegleiter in den Chloroplasten und Rhodoplasten sind vor allem $\beta$-Carotin und Lutein (3,3'-Dioxy-$\alpha$-Carotin, das von Karrer ursprünglich als Xanthophyll bezeichnet worden ist). $\beta$-Carotin ist auch das Hauptcarotinoid der Chromoplasten in der Wurzel von *Daucus carota*, aus der es zuerst isoliert worden ist und nach der es auch seinen Namen erhalten hat. Dieses Carotinoid besitzt die stärkste Vitamin A-Wirkung, was aus der Tatsache verständlich wird, daß es bei seinem symmetrischen Bau durch Aufnahme von 2 Wassermolekülen in 2 Moleküle Vitamin A übergehen kann. $\alpha$-Carotin kommt in stark wechselnder

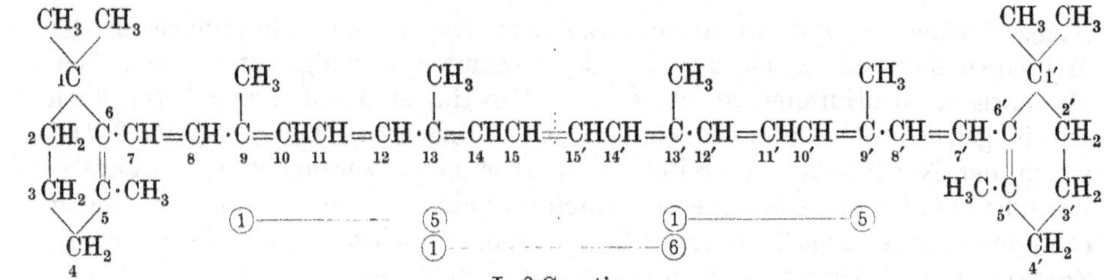

CH₃ CH₃ — structural formulas

$CH_3$ $CH_3$

$_1C$

$_2CH_2$ $C \cdot CH=CH \cdot \underset{7}{C}=\underset{8~9~10~11}{CHCH=CH} \cdot \underset{12}{C}=\underset{13~14~15}{CHCH} \dotplus \underset{15'~14'}{CHCH} = \underset{13'~12'}{C} \cdot \underset{11'~10'}{CH=CHCH} = \underset{9'~8'}{C} \cdot CH=CH \cdot \underset{7'}{C}$   $CH_2$

$_3CH_2$ $C \cdot CH_3$

CH₂

I. β-Carotin

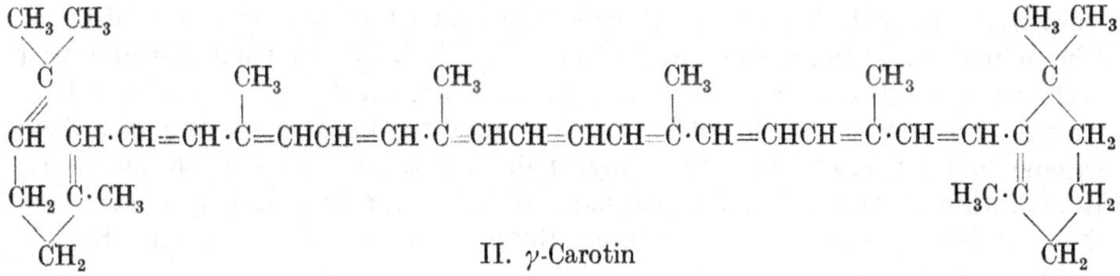

II. γ-Carotin

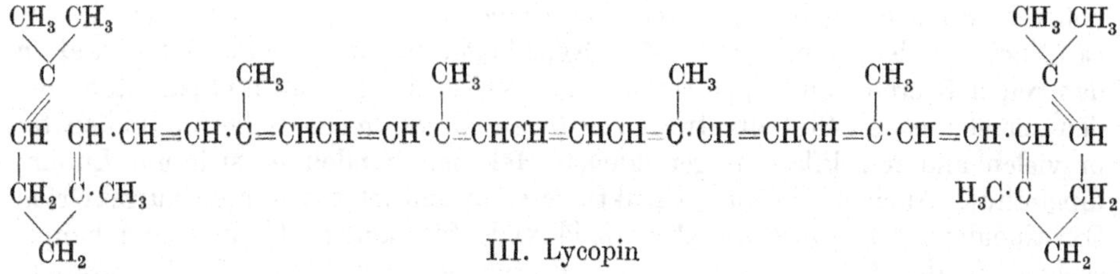

III. Lycopin

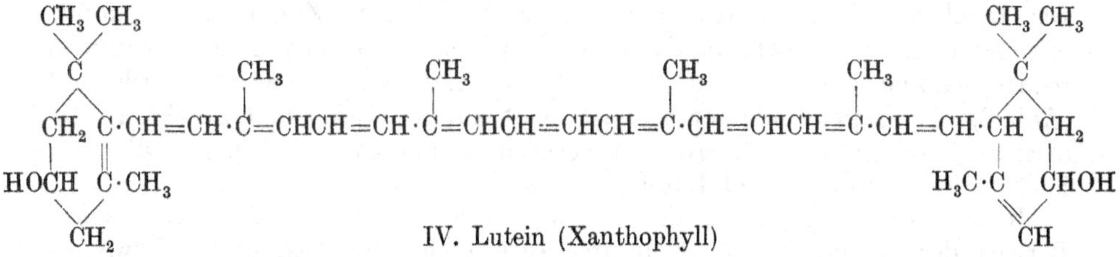

IV. Lutein (Xanthophyll)

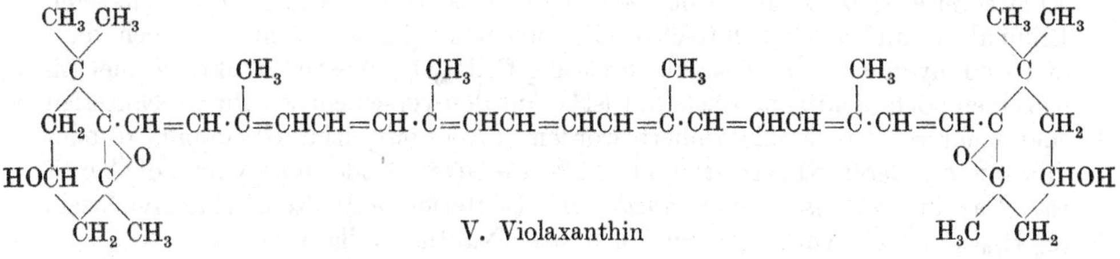

V. Violaxanthin

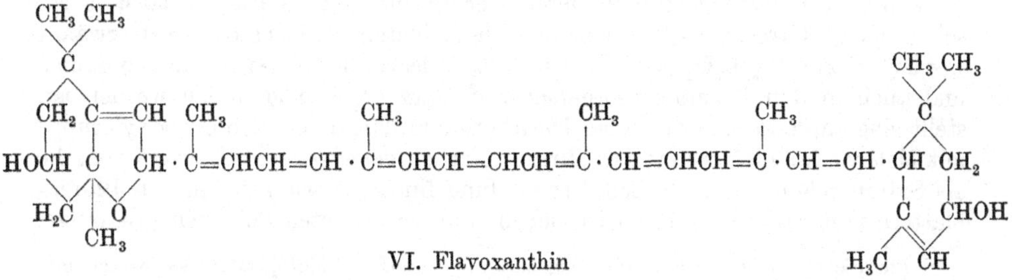

VI. Flavoxanthin

Abb. 2. Strukturformeln der häufigsten Carotinoide. Beim β-Carotin ist die Numerierung der C-Atome eingezeichnet (weitere Erläuterungen im Text).

Konzentration oft mit $\beta$-Carotin zusammen vor; in den Chromoplasten der Möhrenwurzel kann es bis zu 35% des Gesamtcarotinoidgehaltes ausmachen. Die übrigen Carotinisomeren ($\gamma$-, $\delta$-, $\varepsilon$-, $\zeta$-Carotin) sind selten und liegen auch nur in geringen Konzentrationen vor. Das aliphatische Lycopin (Abb. 2, III) ist in der Natur weit verbreitet, wird aber meist zusammen mit $\beta$-Carotin besonders häufig in vielen reifen Früchten gefunden. In hoher Konzentration ist es in den rotfrüchtigen Tomaten enthalten, woher auch sein Name abgeleitet ist. Zeaxanthin (3,3'-Dioxy-$\beta$-Carotin), ein mit Lutein isomeres Xanthophyll und mit diesem häufig zusammen vorkommend, ist zuerst aus Maiskörnern isoliert worden; es ist jedoch in den Chloroplasten und Chromoplasten von Blättern, Blüten und Früchten sowohl in freiem als auch in verestertem Zustand weit verbreitet. In Blütenorganen, Früchten und Samen treten neben den häufigeren Carotinoiden auch solche auf, die in ihrer Verbreitung nur auf diese Organe und oft sogar nur auf wenige nah verwandte Arten beschränkt sind. Rubixanthin (3-Oxy-$\gamma$-Carotin) ist hauptsächlich nur in den Hagebutten von *Rosa rubiginosa* und einigen anderen Rosenarten gefunden worden. Relativ häufig kommen veresterte Xanthophylle in den Reproduktionsorganen der höheren Pflanzen neben anderen Carotinoiden vor. Physalien, das Haupt-carotinoid in den gelben und roten Kelchblättern von *Physalis*-Arten und in den reifen Beeren von *Hippophaë rhamnoides* ist ein Zeaxanthindipalmitat. Bei den zuerst aus den Blütenblättern von *Helenium autumnale* isolierten und später in vielen anderen Pflanzen gefundenen Helenien handelt es sich um Lutein-dipalmitat. Auch die in ihrer Struktur erst in den letzten Jahren aufgeklärten Carotinoidepoxyde sowie die Carotinoidoxyde furanoider Struktur sind haupt-sächlich in den Chromatophoren von Blüten und Früchten gefunden worden. Das ziemlich weit verbreitete Violaxanthin hat sich als Zeaxanthin-di-epoxyd erwiesen (Abb. 2, V), während das in geringer Konzentration als Begleitcarotinoid relativ häufige Flavoxanthin als furanoides Luteinoxyd erkannt worden ist (Abb. 2, VI). Lutein-mono-epoxyd ist nicht nur in Blüten und Früchten, son-dern auch in grünen Blättern weit verbreitet, und zwar in Mengen, die den-jenigen des $\beta$-Carotins und Luteins oft gleichkommen.

In den Chromatophoren der Thallophyten kommen neben den aus dem Bereich der Kormophyten bekannten auch einige für bestimmte Verwandt-schaftskreise spezifische Carotinoide vor. Das Hauptcarotinoid der meisten Braunalgen und auch zahlreicher Diatomeen ist Fucoxanthin, ein hoch oxy-diertes Polyen von der Zusammensetzung $C_{40}H_{56}O_6$, dessen Strukturformel im einzelnen noch nicht genau bekannt ist[1]. Aus den verschiedenen Purpurbakterien sind mehrere Carotinoide isoliert worden (Rhodopurpurin, Rhodopin, Rhodo-vibrin u. a.), deren Konstitution nur teilweise bekannt oder noch ganz unbekannt ist. Das in *Thiocystis*- und *Rhodovibrio*-Bakterien enthaltene Rhodoviolascin ($C_{42}H_{60}O_2$) ist ein vom Lycopin ableitbarer Xanthophylläther.

Von den Polyenfarbstoffen, deren Verbreitung sehr stark spezialisiert ist, sollen noch 2 Carotinoid-Carbonsäuren mit kleinerem Kohlenstoffgerüst erwähnt werden. Bixin ($C_{25}H_{30}O_4$), der Farbstoff des Orleans, der bisher nur in den Samen und auch in den Vegetationsorganen von *Bixa orellana* gefunden worden ist, stellt eine aliphatische Carotinoid-Dicarbonsäure dar, deren eine Carboxylgruppe mit Methylalkohol verestert ist. In den Narben verschiedener *Crocus*-Arten, die als Safran schon seit alter Zeit Verwendung finden, sowie auch in den Blüten-blättern von *Verbascum*-Arten ist neben weit verbreiteten Polyenen ($\beta$-Carotin,

---

[1] Für die einen hohen Fucoxanthingehalt aufweisenden Chloroplasten der Kiesel- und Braunalgen eine besondere Bezeichnung („Phäoplasten") einzuführen, ist nicht gerecht-fertigt.

Lycopin und Zeaxanthin) Crocin enthalten, das als Di-gentiobiose-ester der Carotinoid-Dicarbonsäure Crocetin ($C_{20}H_{24}O_4$) erkannt worden ist. Die glykosidische Bindung führt hier zu hydrophilen Eigenschaften, womit die Möglichkeit gegeben ist, daß diese Carotinoidester auch im Zellsaft in Lösung gehen können.

Viele Carotinoide gehören zur ursprünglichen Pigmentausrüstung der Chloroplasten und werden daher als Primärcarotinoide bezeichnet. Alle Blumenblätter und Früchte besitzen in jungen Entwicklungsstadien Chloroplasten mit Primärcarotinoiden, zu denen jedoch vielfach im Verlauf der weiteren Entwicklung dieser Organe andere Polyene — teils auf Kosten der bereits vorhandenen, teils durch zusätzliche Bildung — hinzukommen, die als Sekundärcarotinoide bezeichnet werden (SEYBOLD 1943). Solche typischen Sekundärcarotinoide sind z. B. das Lycopin in den reifen Tomatenfrüchten, das Rhodoxanthin im Arillus von *Taxus baccata* und in den Blättern mehrerer Gymnospermen (vor allem im Winter), sowie die Xanthophyllester Capsanthin, Physalien und Helenien in Früchten, Kelch- und Blütenblättern mehrerer Arten. Auch in den herbstlich sich verfärbenden Laubblättern zahlreicher Pflanzen treten Sekundärcarotinoide meist in Form von Xanthophyllestern auf, die den normalgrünen Chloroplasten in den Blättern dieser Arten fehlen. Da die Plastiden des gelben Herbstlaubes sich in demselben physiologischen Zustand befinden wie diejenigen zahlreicher Blumen- und Fruchtblätter, wird die Entstehung der Chromoplasten aus Chloroplasten als Symptom einer tiefgreifenden Veränderung im Eiweiß- und Lipoidgefüge der Plastidensubstanz aufgefaßt (SEYBOLD 1943).

Seit einigen Jahren ist die Erforschung der Biosynthese der Carotinoide in Angriff genommen worden. Obwohl in letzter Zeit eine Reihe von Verbindungen als Bausteine für die Synthese der Carotinoide in der lebenden Zelle erkannt worden ist, kann man sich noch kein einheitliches Bild über den Syntheseweg machen (GOODWIN und Mitarbeiter, PORTER und LINCOLN; ARNAKI und STARY).

Die *Phycobiline* verleihen den Rhodoplasten die rote, purpurne und gelegentlich auch blauviolette Färbung und verdecken die gleichzeitig in diesen Chromatophoren enthaltenen Chlorophyll- und Carotinoidfarbstoffe oft vollständig. In ihrem Vorkommen im Pflanzenreich sind sie auf die Rhodoplasten der Rotalgen und auf das ebenfalls granaähnlichen Bau aufweisende Chromatoplasma der Blaualgen beschränkt (über das Auftreten ähnlicher Pigmente in den leghämoglobinhaltigen Wurzelknöllchen der Leguminosen kann hier nicht näher eingegangen werden). Ihr Farbstoffcharakter ist durch die chromophore Gruppe hochmolekularer Chromoproteide vom Molekulargewicht 263000 bis 291000 bedingt, die in zwei durch ihre Farbtönung sich deutlich unterscheidenden Gruppen vorliegen: den blauvioletten und gelegentlich auch grünblauen Phycocyaninen und den roten Phycoerythrinen. Diese Phycochromoproteide sind wasserlöslich und können durch Einlegen der Rot- und Blaualgen in destilliertes Wasser nach dem Abtöten der Zellen aus den Rhodoplasten bzw. dem Chromatoplasma — wenn auch nicht vollständig — herausgelöst werden (worauf auch die Vergrünung von Meeresrotalgen nach dem Einlegen in Süßwasser beruht). Über die Verteilung der Phycocyanine und Phycoerythrine in den Strukturelementen der Grana und über ihre Anordnung gegenüber den gleichzeitig vorhandenen Chlorophyll- und Carotinoidfarbstoffen im submikroskopischen Bereich ist noch wenig bekannt (BORESCH, KYLIN, RABINOWITCH, STRAIN 1949, 1951).

Die chemische Erforschung dieser Phycochromoproteide ergab enge Beziehungen ihrer chromophoren Gruppe zu den als Abbauprodukte des Hämoglobins bekannten Gallenfarbstoffen, besonders zu dem Biliverdin, weshalb auch LEMBERG die Bezeichnung Phycobiline für diese Algenpigmente eingeführt hat. Die Grundsubstanz der Farbkomponente der Phycobiline besteht aus einem

Tetrapyrrol mit offenem Ring (Abb. 3), das sich von dem Grundgerüst der
Chlorophylle, dem Porphyrinring, vor allem durch das Fehlen des Kohlenstoff-
atoms der α-Methinbrücke (vgl. Abb. 1) unterscheidet (FISCHER und SIEDEL).
Die proteinfreie Farbkomponente des Phycocyanins, das Phycocyanobilin,
konnte mit dem bereits früher bekannten Mesobiliviolin identifiziert werden.
Die chromophore Gruppe des Phycoerythrins, das Phycoerythrobilin, hat sich
mit Mesobilierythrin identisch erwiesen (LEMBERG und LEGGE). Die Bindung

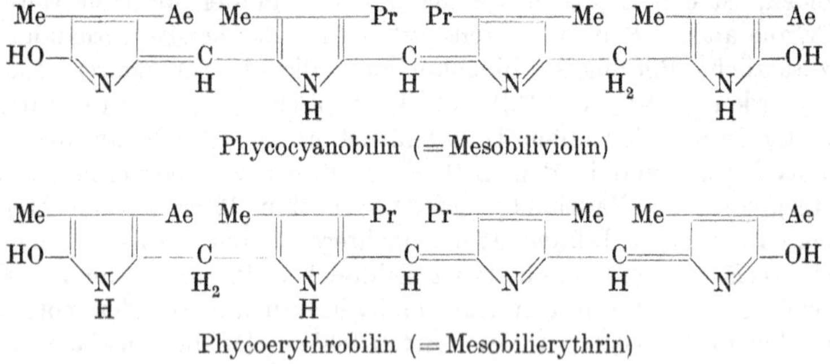

Phycocyanobilin (= Mesobiliviolin)

Phycoerythrobilin (= Mesobilierythrin)

Abb. 3. Strukturformeln der proteinfreien Farbkomponenten des Phycocyanins und des Phycoerythrins.
(Me = Methyl-, Ae = Äthyl-, Pr = Propylgruppe.)

der Farbkomponenten an dem noch nicht näher bekannten Eiweißkörper ist
bei den Phycobilinen außerordentlich fest; sie wird nicht einmal gelöst, wenn
der Eiweißanteil denaturiert wird. Hieraus wird geschlossen, daß die chromophore Gruppe durch eine echte chemische Bindung, wahrscheinlich in Form
einer Peptidbindung, mit dem Protein gekoppelt ist.

Die Phycobiline zeigen ein charakteristisches Absorptionsspektrum und eine
sehr stark in Erscheinung tretende gelbe und orangerote Fluorescenz. Die in
den verschiedenen Gattungen und Arten der Rot- und Blaualgen vorkommenden
Phycocyanine und Phycoerythrine scheinen nicht identisch zu sein, obwohl die
Unterschiede bisher nicht exakt erfaßt werden konnten. Man unterscheidet
4 Gruppen von Phycobilinen, die sich spektroskopisch und fluorescenzoptisch
unterscheiden lassen. Das Hauptphycobilin der Rhodophyceen ist das rote,
gelborange fluorescierende R-Phycoerythrin, das in zahlreichen Arten vom blauvioletten, rot fluorescierenden R-Phycocyanin als Nebenpigment begleitet wird.
Das grünblaue, rot fluorescierende C-Phycocyanin ist das Hauptphycobilin der
Cyanophyceen, zu dem in manchen olivgrünen Blaualgen noch das rote, orange-
farben fluorescierende C-Phycoerythrin in geringerer Konzentration hinzukommt.
Das Lichtfeld des Standortes übt auf die Zusammensetzung der Phycobiline aus
den einzelnen Komponenten bei manchen Rot- und Blaualgen einen deutlichen
Einfluß aus, eine Erscheinung, die als ,,chromatische Adaptation'' bezeichnet
wird (eine Diskussion über die zum Teil recht widerspruchsvollen Ergebnisse
gibt RABINOWITCH S. 421).

Wenn auch die im tierischen und menschlichen Organismus auftretenden
Gallenfarbstoffe einwandfrei als Abbauprodukte des Hämoglobins erkannt worden
sind, so ist die Annahme, daß auch zwischen den Chlorophyllen und den Phycobilinen eine ähnliche Beziehung bestehen möge, vorerst noch rein spekulativer
Natur (vgl. RABINOWITCH S. 478). Auch der Einbau der Phycobiline in das
Biosyntheseschema des Chlorophylls ist experimentell bis jetzt ebensowenig
gestützt wie die Einbeziehung dieser Pigmente in phylogenetische Diskussionen
(GRANICK, EGLE 1953b). Eine ausführliche Darstellung findet man in Bd. V,
von Carotinoiden in Bd. X.

# Literatur.

ARNAKI, M., u. Z. STARY: Untersuchungen über die Biosynthese der Carotinoide. Biochem. Z. **323**, 376—381 (1952). — ARONOFF, S.: Chlorophyll. Bot. Review **16**, 525—588 (1950).

BONNER, J.: Plant biochemistry. New York: Academic Press Inc. 1950. — BORESCH, K.: Algenfarbstoffe. In Handbuch der Pflanzenanalyse, Bd. 3, herausgeg. von G. KLEIN. Wien: Springer 1932.

EGLE, K.: Über den physikalisch-chemischen Verteilungszustand des Chlorophylls in lebenden Plastiden. Ber. dtsch. bot. Ges. **66**, 179—182 (1953a). — Die Biosynthese der Chlorophyllfarbstoffe. Naturwiss. **40**, 569—576 (1953b).

FISCHER, H., u. A. OESTREICHER: Über Protochlorophyll und Vinylporphyrine. Z. physiol. Chem. **262**, 243—269 (1940). — FISCHER, H., u. H. ORTH: Die Chemie des Pyrrols, Bd. II/2. Leipzig: Akademische Verlagsgesellschaft 1940. — FISCHER, H., u. W. SIEDEL: Naturfarbstoffe. II. Pyrrolsynthesen und Gallenfarbstoffe. In Naturforschung und Medizin in Deutschland 1939—1946, Bd. 39, S. 109—127. 1947. — FISCHER, H., u. M. STRELL: Naturfarbstoffe. IV. Chlorophyll. In Naturforschung und Medizin in Deutschland 1939 bis 1946, Bd. 39, S. 141—186. 1947. — FRANK, S.: Relation between carotenoid and chlorophyll pigments in Avena coleoptiles. Arch. of Biochem. **30**, 52—61 (1951). — FREY-WYSSLING, A.: Submicroscopic morphology of protoplasm and its derivatives. New York: Elsevier Publishing Comp. 1948.

GOODWIN, T. W.: The comparative biochemistry of the carotenoids. London: Chapman a. Hall Ltd. 1952. — GOODWIN, T. W., and M. JAMIKORN: Biosynthesis of carotenes in ripening tomatoes. Nature (Lond.) **170**, 104—105 (1952). — GOODWIN, T. W., W. LIJINSKY and J. S. WILLMER: Studies in Carotinogenesis 6. Biochemic. J. **53**, 208—212 (1953). — GRANICK, S.: Biosynthesis of chlorophyll and related pigments. Annual Rev. Plant Physiol. **2**, 115—144 (1951).

KARRER, P.: Carotinoid-expoxyde und furanoide Oxyde von Carotinoidfarbstoffen. Fortschr. Chem. organ. Naturstoffe **5**, 1—19 (1948). — KARRER, P., u. E. JUCKER: Carotinoide. Basel: Birkhäuser 1948. — KOSKI, V. M., C. S. FRENCH and J. H. C. SMITH: The action spectrum for the transformation of protochlorophyll to chlorophyll a in normal and albino corn seedlings. Arch. of Biochem. a. Biophysics **31**, 1—17 (1951). — KYLIN, H.: Über die Farbstoffe und die Farbe der Cyanophyceen. Kgl. fysiograf. Sällsk. Lund Förh. **7**, 131—158 (1937).

LEMBERG, R.: Pigmente der Rotalgen. Naturwiss. **17**, 541 (1929). — LEMBERG, R., and J. W. LEGGE: Hematin compounds and bile pigments. New York: Interscience Publishers, Inc. 1949.

MITTENZWEI, H.: Über Bacteriochlorophyll. Z. physiol. Chem. **275**, 93—121 (1942).

NOACK, K., u. W. KIESSLING: Zur Entstehung des Chlorophylls und seiner Beziehung zum Blutfarbstoff. Z. physiol. Chem. **193**, 97—137 (1930).

PIRSON, A.: Stoffwechsel organischer Verbindungen. I. Photosynthese. Fortschr. Bot. **14**, 289—333 (1953). — PORTER, J. W., and R. E. LINCOLN: Lycopersicon selections containing a high content of carotenes and colorless polyenes. The mechanism of carotene biosynthesis. Arch. of Biochem. **27**, 390—403 (1950).

RABINOWITCH, E. I.: Photosynthesis and related processes, Bd. 1. New York: Interscience Publishers, Inc. 1945. — RICHARDS, F. A.: The estimation and characterization of plankton populations by pigment analysis. I. J. Marine Res. **11**, 147—155 (1952).

SEYBOLD, A.: Zur Kenntnis des Protochlorophylls. Planta (Berl.) **26**, 712—718 (1937). — Über die physiologische Bedeutung der Chlorophyllkomponenten a und b. Bot. Archiv **42**, 254—288 (1941). — Pflanzenpigmente und Lichtfeld als physiologisches, geographisches und landwirtschaftlich-forstliches Problem. Ber. dtsch. bot. Ges. **60**, 64—85 (1942). — Zur Kenntnis der herbstlichen Laubblattfärbung. Bot. Archiv **44**, 551—568 (1943). — Zur Kenntnis des Protochlorophylls. III. Planta (Berl.) **36**, 371—388 (1948). — SEYBOLD, A., u. K. EGLE: Quantitative Untersuchungen über Chlorophyll und Carotinoide der Meeresalgen. Jb. wiss. Bot. **86**, 50—80 (1938a). — Zur Kenntnis des Protochlorophylls. II. Planta (Berl.) **29**, 119—128 (1938b). — Zur Kenntnis des Bacteriochlorophylls. Sitzgsber. Heidelberg. Akad. Wiss., Math.-naturwiss. Kl., 1. Abh. **1939**, 7—17. — Über den physikalischen Zustand des Chlorophylls in den Plastiden. Bot. Archiv **41**, 578—603 (1940). — STOLL, A., u. E. WIEDEMANN: Chlorophyll. Fortschr. Chem. organ. Naturstoffe **1**, 159—254 (1938). — Chlorophyll. Fortschr. chem. Forsch. **2**, 538—608 (1952).

WILLSTÄTTER, R., u. A. STOLL: Untersuchungen über Chlorophyll. Berlin: Springer 1913.

ZECHMEISTER, L.: Carotinoide. Berlin: Springer 1934.

# Chondriosomen und Mikrosomen (Sphärosomen).

Von

## Kurt Steffen.

Mit 7 Textabbildungen.

## Terminologie.

### Chondriosomen.

Die Chondriosomen sind Zellorganelle sui generis. Sie sind etwas stärker lichtbrechend als das Cytoplasma, in das sie eingebettet sind, und zeigen deswegen
grauen positiven Phasenkontrast. Im Hell- und Dunkelfeld erscheinen sie homogen, strukturlos; Dünnschnitte zeigen im Elektronenmikroskop eine charakteristische Feinstruktur. Die Chondriosomen sind mit Janusgrün B bei Sauerstoffgegenwart supravital färbbar. Sie kommen in sphärischen (0,5—1,2 $\mu$), stab-
(2—5 $\mu$) und fadenförmigen Modifikationen vor (vgl. GEITLER, Abb. 14 und 15).
Sie sind flexibel und stellen ein empfindliches osmotisches System dar. Durch
ihren Lipoidgehalt sind sie sehr fixierungslabil: durch alkohol- und essigsäure-
haltige Fixierungsmittel werden sie zerstört.

Die Chondriosomen wurden 1890 von ALTMANN als „Bioblasten" beschrieben und von
BENDA als „Mitochondrien" bezeichnet. Dieser Ausdruck ist zur Zeit hauptsächlich in
angelsächsischen Ländern im Gebrauch. Synonym wird hier der in der botanischen Literatur geläufige Ausdruck: „Chondriosomen (Summation: Chondriom)" verwendet. Unter
diesen Oberbegriff werden die einzelnen morphologischen Erscheinungsformen, wie sphärische (Mitochondrien der französischen Schule = Chondriosomen nach NEWCOMER 1951),
stab- und fadenförmige Organelle (Chondriokonten der französischen Schule = Mitosomen
nach NEWCOMER 1951) und durch Fragmentation bedingte Körnchenketten (Chondriomiten)
subsummiert. Bei Zweifel an der Chondriosomeneigenschaft der beobachteten Partikel wird
neuerdings von OPIE (1947) der neutrale Ausdruck „Cytochondrien" vorgeschlagen.

### Mikrosomen (Sphärosomen).

Die Mikrosomen sind 0,4—1,6 $\mu$ große sphärische, stark lichtbrechende Plasmaeinschlüsse. Sie zeigen im Gegensatz zu den Lipoidtropfen keine Tendenz
zusammenzufließen. Sie kommen in vielen Angiospermenzellen regelmäßig vor.
Sie bestehen aus einer Protein- und einer Lipoidkomponente. Ihre autonome
Vermehrung wird vermutet, ist jedoch bisher nicht nachgewiesen.

PERNER (1953) schlägt vor, diese Gebilde mit dem von P. A. DANGEARD (1919) geprägten
Ausdruck: „Sphärosomen (Summation: Sphärom)" zu bezeichnen. Der Ausdruck: „Mikrosomen" ist durch die Vorstellung belastet, daß diese Strukturen ergastische Gebilde von
lipoidem Charakter (granulations lipoidiques GUILLIERMOND 1924) sind. Diese Vorstellung
ist zum Teil schuld daran, daß so spät erst Methoden zur Unterscheidung der Mikrosomen
von Fett- und Lipoidtropfen entwickelt wurden. Es ist nicht zulässig, die Angaben älterer
Autoren über Mikrosomen ohne weiteres auf die Mikrosomen neuer Prägung (Sphärosomen)
zu übertragen, zumal in manchen Arbeiten in der weitesten Fassung des Begriffes alle mikroskopisch kleinen Plasmaeinschlüsse als Mikrosomen bezeichnet werden (historische Übersicht
bei PERNER 1953, dort ältere Literatur).

Nicht identisch mit den PERNERschen Mikrosomen sind die 50—200 m$\mu$
großen small granules der „Mikrosomenfraktion" CLAUDEs (1943), die einem
Vorschlag FREY-WYSSLINGs entsprechend als „Submikrosomen" zu bezeichnen
wären (vgl. auch LINDBERG und ERNSTER 1954).

# Differentialdiagnose.

Im lichtmikroskopischen Bereich besteht Verwechslungsmöglichkeit a) zwischen größengleichen sphärischen Chondriosomen, Mikrosomen (Sphärosomen) und Lipoidtropfen und b) zwischen stabförmigen Chondriosomen und Proplastiden.

a) Die Chondriosomen sind durch ihren Phasenkontrast, ihre Vitalfluorochromierung, ihre supravitale Färbbarkeit mit Janusgrün B und ihr osmotisches Verhalten hinreichend charakterisiert (vgl. Tabelle 1 und 2). Schwierig ist die Unterscheidung zwischen Mikrosomen (Sphärosomen) und Lipoidtropfen. Folgende Kriterien werden in der Literatur beschrieben, von ihnen scheinen die

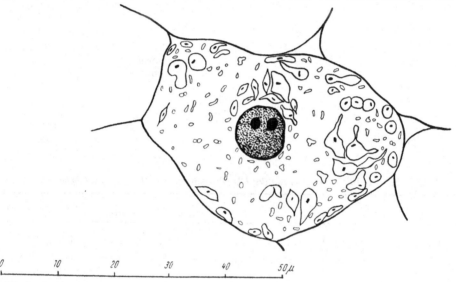

Abb. 1. *Agapanthus umbellatus.* Junge Mesophyllzelle aus der Streckungszone eines 10—20 cm langen Blattes mit Proplastiden und Chondriosomen (Mikrotomschnitt). Nach STRUGGER 1954.

unter 1, 2, 3 und 5 aufgeführten die zur Zeit sichersten: 1. auch bei schlechtem Ernährungszustand werden die Mikrosomen nicht wie die Reservelipoide resorbiert, 2. die Mikrosomen zeigen nicht wie die Lipoidtropfen Tendenz zum Zusammenfließen, 3. Mikrosomen geben erst nach Fixierung mit Fettfarbstoffen (Sudanfarbstoffe, Phosphin-3R, Rhodamin B u. a.) positive Fettreaktion, 4. Mikrosomen lassen sich mit Berberinsulfat elektiv fluorochromieren, sekundäre Fetttropfen nicht (PERNER 1952b), 5. fixierte Mikrosomen zeigen nach drei- bis vierstündiger Xylolbehandlung eine mit basischen Farbstoffen (Pyronin, Viktoriablau, Rosanilin) anfärbbare Proteinkomponente, während die Lipoidtropfen gelöst sind (PERNER 1952a), und 6. nach PERNER (1952a) sollen nur die Mikrosomen positive Nadi-Reaktion geben. Da aber das gebildete Indophenol lipophil ist, ist Sekundärspeicherung in Fetttropfen nicht ausgeschlossen. Ob die Indophenolspeicherung der Mikrosomen ein Primär- oder ein Sekundärvorgang ist, ist zur Zeit noch umstritten (zuletzt DRAWERT 1953).

b) Zur Unterscheidung von Proplastiden und Chondriosomen (vgl. Abb. 1) sind besonders wertvoll: 1. die elektive Färbung der Chondriosomen mit Janusgrün B (SOROKIN 1941), 2. die Amöboidie der Proplastiden und 3. das in den Proplastiden phako-optisch nachweisbare, doppelbrechende, feulgenpositive primäre Granum, das sich nach $OsO_4$-Fixierung mit Säurefuchsinfärbung nach ALTMANN nachweisen läßt, und das in den primären Grana bei *Dracaena*

Tabelle 1. *Aussehen, Form und Verhalten von Proplastiden, Chondriosomen, Mikrosomen und Fett- und Lipoidtropfen bei lichtmikroskopischer Betrachtung.*

| | Größe | Positiver Phasenkontrast | Dunkelfeld | Form-konstanz | Janus-grün B | Verhalten gegen Fixierungsmittel |
|---|---|---|---|---|---|---|
| 1. Pro-plastiden | 1.> 2. | grau mit dunklem primären Granum | dunkel mit mattgrau glänzender Grenzfläche | amöboid | — | labil |
| 2. Chondrio-somen | 2.< 1. | homogen grau | | flexibel | + | labil, durch alkohol- und essigsäure-haltige Fixierungs-mittel zerstört |
| 3. Mikro-somen (Sphäro-somen) | etwa größen-gleich | dunkel | silbrig hell | form-konstant | — | stabil |
| 4. Fett- und Lipoid-tropfen | | | | Tendenz zum Zusam-menfließen | — | durch lipoid-fällende Fixierung erhaltbar |

Tabelle 2. *Vitalfluorochromierung der Chondriosomen und Mikrosomen (Sphärosomen).*

| | Farbstoff | Autor | Sauerstoff-spannung | Fluorescenz-farbe | Filterkombination |
|---|---|---|---|---|---|
| Chondrio-somen | Rhodamin B | Strugger 1938 | unabhängig | goldgelb | BG 12, Sp 2,5 |
| | Berberin-sulfat | Drawert 1953 | Sauerstoff-gegenwart | goldgelb | UG 1, UV 2,5 |
| Mikrosomen | Janusgrün B | Drawert 1953 | Sauerstoff-mangel | grünlich-gelb bis weißgrün | UG 1, UV 2,5 |
| | Berberin-sulfat | Perner 1952b | Sauerstoff-gegenwart? | weißgrün | UG 1, UV 2,5 |
| | | Drawert 1953 | Sauerstoff-mangel | | |
| | Nilblausulfat frische Lösung | Drawert und Gutz 1953 | Sauerstoff-gegenwart | blendend-weiß | UG 1, UV 2,5 |
| | | | Sauerstoff-mangel | weißgelb | UG 1, UV 2,5 |
| | alte Lösung | Drawert 1953 | unabhängig | goldgelb | UG 1, Sp 2,5 |

*deremensis*, *Chlorophytum comosum* und *Agapanthus umbellatus* elektiv Chlorophyll-fluoreszenz zeigt (mündliche Mitteilung von Herrn Prof. Strugger). — Die Proplastiden sind zwar stets größer als die Chondriosomen, doch ist der Unter-schied im meristematischen Gewebe gering (z.B. *Chlorophytum comosum*, Chon-driosomen 1—1,4 × 0,8, Proplastid im postmeristematischen Gewebe 3 × 1,3, im meristematischen Gewebe 1,8 × 0,4 $\mu$ nach Strugger 1953).

Über weitere Unterscheidungsmöglichkeiten im lichtmikroskopischen Be-reich orientieren die Tabellen 1 und 2, im Elektronenmikroskop sind die Chon-driosomen bei ausreichender Dünne des Schnittes durch ihre charakteristische

Feinstruktur leicht zu erkennen. In den durch Differentialzentrifugierung gewonnenen Einzelfraktionen eines Homogenisats sind Chondriosomen durch ihren Phasenkontrast und ihr osmotisches Verhalten (Novikoff und Mitarbeiter 1953) zu identifizieren.

Da zur Diagnose von Chondriosomen in Homogenisaten zuweilen Janusgrün B herangezogen wird, und Chondriosomenäquivalente durch ihr Verhalten gegenüber Janusgrün B, Tetrazoliumsalzen und dem Nadi-Reagens charakterisiert werden (Muddsche Schule) scheint eine kritische Darstellung dieser Verfahren nötig.

## Supravitalfärbung mit Janusgrün B.

Janusgrün B wurde zuerst von Michaelis (1900) zum Nachweis von Chondriosomen verwendet. Da spätere Untersucher nicht das von ihm benutzte Diäthylderivat benutzten, geriet das Janusgrün wegen des Mißerfolges in Vergessenheit. Janusgrün G, bei dem die beiden Äthyl- durch Methylgruppen ersetzt sind, färbt wegen seiner geringen Reduktionsgeschwindigkeit (die Azo-Bindung wird nur schwer gesprengt) die Chondriosomen nicht mehr elektiv (Michaelis 1900, Cowdry 1918, Lazarow und Cooperstein 1953a und b, Cooperstein, Lazarow und Patterson 1953). Eine unterschiedliche Reduktionsgeschwindigkeit der handelsüblichen Janusgrün B-Präparate wurde neuerdings von Drawert (1953) gefunden. Janusgrün B wurde von Bensley (1910) erneut in die Cytologie eingeführt und in der Folgezeit (z. B. von Meves 1904, Sorokin 1938, 1941; weitere Literatur bei Newcomer 1951, Drawert 1953) auch häufig bei pflanzlichen Objekten benutzt (Konzentration 1:10000 oder 1:100000 in Leitungswasser oder 1:100000 in 10%iger Rohrzuckerlösung; Grenzkonzentration 1:500000).

Der Chemismus der elektiven Anfärbung der Chondriosomen wurde in neuerer Zeit durch Lazarow, Cooperstein und Patterson (1949 und 1953), Cooperstein und Lazarow (1950, 1953) und von Lazarow und Cooperstein (1950, 1953a und b) genauer untersucht.

Dies war besonders wichtig im Hinblick auf die Frage, ob das Janusgrün B auch in Homogenisaten die Mitochondrienfraktion selektiv färbt (Schneider 1947, Du Buy und Mitarbeiter 1949, Hogeboom, Schneider und Palade 1948, Swanson und Artom 1950). Die selektive Färbung der Chondriosomen innerhalb der Zelle beruht auf einer verminderten Reduktionsgeschwindigkeit und damit einem verminderten Reduktionsgrad. Die Reduktion des Janusgrün B (Diäthylsafranin-azo-dimethylanilin) erfolgt in vier Schritten: Janusgrün B (blau) ⇌ Leuko-Janusgrün B (violett) → Diäthylsafranin (rot) ⇌ Leukosafranin (farblos). Die Reduktion von Leuko-Janusgrün B zu Diäthylsafranin ist irreversibel (vgl. auch Makarow 1934, Lison und Fautrez 1939, Guilliermond und Gautheret 1940, Guilliermond 1949). Durch Reduktion von Cytochrom c wird mit Sicherheit Leukosafranin zu Diäthylsafranin und wahrscheinlich Leuko-Janusgrün B zu Janusgrün B (Lewis 1923, Lewis und Lewis 1924, Brenner 1949) oxydiert. Eine enzymatische Überprüfung der Reduktion konnte die Reduktionsprodukte spektrophotometrisch erfassen (Cooperstein und Lazarow 1953, Cooperstein, Lazarow und Patterson 1953) und nachweisen, daß jedes Enzymsystem, das Diphosphopyridinnucleotide (und über diese Flavoprotein) reduzieren kann, zu dieser Reduktion befähigt ist, so z. B. Milchsäure- und Glucosedehydrasen (Cooperstein und Lazarow 1953, vgl. auch Blakley 1951).

In der intakten Zelle werden zunächst außer den Chondriosomen auch andere Zellbestandteile durch Janusgrün B gefärbt, da aber die Reduktion in den Chondriosomen langsamer erfolgt, bleiben diese elektiv gefärbt (Lazarow und Cooper-stein 1953a und b). Der Reduktionsgrad und die Reduktionsgeschwindigkeit sind abhängig von der lokalen Konzentration von reduziertem Flavoprotein, das zunächst mit dem Farbstoff reagiert (Straub 1939). Die Konzentration des reduzierten Flavoproteins ist von seiner intracellulären Verteilung und von der Verteilung der Enzymsysteme abhängig, die das Flavoprotein reduzieren oder oxydieren. So vermindert z. B. die Cytochromoxydase die Konzentration an reduziertem Flavoprotein durch Oxydation (Theorell 1951). Da die Cyto-chromoxydase in den Chondriosomen lokalisiert ist (Schneider 1946, 1950, Schneider, Claude und Hogeboom 1948, Schneider und Hogeboom 1950), wird hier die Reduktionsgeschwindigkeit vermindert, außerdem reoxydiert das Cytochrom c-Cytochromoxydasesystem das Leuko-Janusgrün B zu Janusgrün B. Hemmt man das Cytochromoxydasesystem durch Sauerstoffabschluß (Bens-ley 1911, Bensley und Bensley 1938) oder durch Cyanidgaben (Lewis 1923, Lewis und Lewis 1924, Makarov 1934, Brenner 1949, Ritchie und Hazel-tine 1953), so kann die Reduktion durch das Dehydrasesystem (Banga, Laki und Szent-György 1933) erfolgen. Bei erneuter Sauerstoffzufuhr und Unter-brechung der Cyanidgabe wird durch Oxydation Janusgrün B zurückgebildet, vermutlich aus einem farblosen Leuko-Janusgrün B, das durch Reduktion des Azinringes vor erfolgter Sprengung des Azoringes gebildet wurde (Lazarow und Cooperstein 1953a und b). Damit dürfte sich der Einwand von Drawert (1953, S. 139) erübrigen, daß eine Oxydation vom Leukosafranin zum Janus-grün B chemisch nicht möglich sei.

Bei der elektiven Anfärbung der Chondriosomen spielt die Adsorption keine entscheidende Rolle, weil 1. die Adsorption nicht spezifisch ist: fein verteiltes Kollagen, hitze-denaturiertes Serumalbumin und Ultramikrosomen adsorbieren ebenfalls, und weil 2. auch Janusgrün G von den Chondriosomen stark adsorbiert wird, aber wegen seiner langsamen Reduktion trotzdem keine elektive Anfärbung hervorruft.

Nach Lazarow und Cooperstein (1953a und b) bildet Janusgrün B mit Aminosäuren (Alanin, Glycin, Cystein) und mit Proteinen (besonders mit un-löslichen) unlösliche Komplexe, worauf seine Giftigkeit für die Zelle zurück-geführt werden kann. Janusgrün B hemmt eine Anzahl enzymatischer Reak-tionen, so die Oxydation von Glucose, Lactat, Succinat und Formiat, die Tätig-keit von Fumarase, Cholinesterase, den RNS-Abbau und die oxydative Phos-phorylierung (Literatur bei Lazarow und Cooperstein 1953b). Nach Dra-wert (1953) kann die durch Leukosafranin bedingte elektive Mikrosomenfluoro-chromierung durch Speicherung in der lipoiden Phase der Mikrosomen oder durch Anreicherung von Fettsäuren in den Mikrosomen bedingt sein. Die An-reicherung von Fettsäuren könnte durch die Stoffwechseländerung ausgelöst sein.

Während die Janusgrün B-Färbung für die Chondriosomen der lebenden Zelle charakteristisch ist (vgl. auch Showacre 1953), dürfen in Homogenisaten nicht alle Partikel, die Janusgrün B aufnehmen, als Chondriosomen angespro-chen werden, da eine unspezifische Adsorption an Plasmaeiweißkörper erfolgt. Auch die Reduktion des Janusgrün B ist nicht auf die Chondriosomenfraktion beschränkt, da die Ultramikrosomenfraktion und die überstehende Lösung die gleiche Fähigkeit zur Reduktion zeigen (Stafford 1951, Lazarow und Cooper-stein 1953a und b). Die unterschiedliche Anfärbung von Chondriosomenfrak-tionen in Salz- oder Zuckerlösungen ist nicht durch unterschiedliche Farbstoff-adsorption, sondern durch die Quellung der Chondriosomen in isotonischer Salzlösung bedingt (Hogeboom, Schneider und Palade 1948, Perner und Pfefferkorn 1953, Lazarow und Cooperstein 1953a und b).

## Intraplasmatische Reduktion von Tetrazoliumsalzen.

Tetrazoliumsalze sind in neuerer Zeit häufig zur Darstellung von intraplasmatischen Reduktionsorten verwendet worden (z. B. von BIELIG, KAUSCHE und HAARDICK 1949, ANTOPOL, GLAUBACH und GOLDMAN 1950, NARAHARA und Mitarbeiter 1950, RUTENBERG, GOFSTEIN und SELIGMAN 1950, CALCUTT 1952, NORDMANN und GAUCHERY 1952; Literatur zur Technik bei ZIEGLER 1953a und b, SOMERSON und MORTON 1953, HARTMAN und Mitarbeiter 1953, PARKER 1953, SHAFFER, KUCERA und SPINK 1953). Sie gelten als spezifische Indicatoren für reduziertes Flavinenzym und darüber hinaus für eine das Flavinenzym reduzierende Reaktionskette (MATTSON, JENSEN und DUTCHER 1947, KUN und ABOOD 1949, KUN 1951, HAHN 1952, BRODIE und GOTS 1951, 1952a und b, SHELTON und SCHNEIDER 1952). Für den Cytomorphologen ist die Frage, ob die in der Zelle beobachteten Formazanspeicherorte auch die Reaktionsorte sind, entscheidend. Es ist wiederholt darauf hingewiesen worden, daß das gebildete Formazan fettlöslich ist, und daß die primär gebildeten Formazanpartikel zu größeren Grana zusammenfließen können (z. B. von BIELIG, KAUSCHE und HAARDICK 1952, WEIBULL 1953, DAVIS und Mitarbeitern 1953, ZIEGLER 1953a und b). Zudem ist das Formazan weiter zu farblosen Verbindungen reduzierbar (BIELIG, KAUSCHE und HAARDICK 1952). Diese Feststellungen mahnen zur mikroskopischen Kontrolle des Reduktionsvorganges. Die geringste Gefahr, sekundäre Speichereffekte zu bekommen, scheint noch bei Neotetrazolium zu bestehen, da es sehr schnell reduziert wird und kleine Kristalle bildet (SHELTON und SCHNEIDER 1952). Für die Identität von Reduktions- und Speicherungsort sprechen folgende Untersuchungen: WACHSTEIN (1949) beobachtete die Ablagerung des hydro- und lipophoben Tellurs und des hydrophoben, lipophilen Formazans am selben Ort; HÖLSCHER (1951) ließ das Reduktionsprodukt mit Kobaltionen einen unlöslichen und damit streng lokalisierten Komplex bilden. Der so charakterisierte Reduktionsort stimmte mit den sonst bei Formazanbildung beobachteten Zellorten überein.

In Homogenisaten ist besondere Vorsicht bei der Verwendung von Tetrazoliumsalzen geboten, sie lassen sich wohl verwenden, um die Enzymaktivität von Chondriosomenfraktionen zu testen (DIANZANI 1953a), aber nicht zur Identifizierung von Chondriosomen in der Fraktion. Nach WEIBULL (1953) kann auch die überstehende Flüssigkeit Tetrazoliumsalze reduzieren.

### Nadi-Reagens.

Auch bei Verwendung des Nadi-Reagens (Technik und Literatur bei PERNER 1952a) darf der Speicherort nicht ohne weiteres mit dem Reaktionsort gleichgesetzt werden (DRAWERT 1953, 1954), da das gebildete Indophenol lipophil ist.

# Chondriosomen.

1904 wurden zuerst von MEVES pflanzliche Chondriosomen beschrieben (Tapetumzellen von *Nymphaea*). Von den in rascher Folge erscheinenden Arbeiten der frühen, morphologischen Arbeitsperiode sind grundsätzlich wichtig die Arbeiten von DUESBERG und HOVEN (1910), die die vermutete Entstehung der Chondriosomen aus dem Kern widerlegen, und die Arbeiten von COWDRY (1917, 1920) und von MANGENOT und EMBERGER (1920), die die morphologische Gleichheit tierischer und pflanzlicher Chondriosomen beweisen (ausführliche Literatur GUILLIERMOND 1941, DANGEARD 1947, NEWCOMER 1940, 1951).

Zur Untersuchung pflanzlicher Chondriosomen kommen die cytomorphologische Analyse im Licht- und Elektronenmikroskop und die biochemische Untersuchung in den einzelnen Fraktionen der Differentialzentrifugierung in Frage.

# 1. Morphologie der Chondriosomen.

Für die Vitalbeobachtung der Chondriosomen sind besonders wasserlebende Pilze geeignet (z.B. *Allomyces* RITCHIE und HAZELTINE 1953, ältere Literatur bei GUILLIERMOND 1941) und außer der Zwiebelepidermis die Perianthblätter der weißen Tulpe und *Iris* (GUILLIERMOND 1941).

Als Fixierungsmittel werden gewöhnlich zur Erhaltung der Lipoidkomponente stark oxydierende Substanzen wie Kaliumbichromat (zusammen mit Formol im REGAUDschen Gemisch), Chromsäure und Osmiumtetroxyd (Gemisch von BENDA und MEVES) und neuerdings Pyrogallol in der Kombination mit Formol verwendet (MARENGO 1952, kritische Übersicht über die Fixierungsmittel bei NEWCOMER 1946, 1951 und ORTIZ-CAMPOS 1948). Auch neutrales Formol ist geeignet; optimal besonders für elektronenmikroskopische Untersuchungen ist gepuffertes $OsO_4$ (PALADE 1952a). Völlig ungeeignet zur Strukturerhaltung sind Gemische mit Alkohol und Essigsäure. Eine Schnellmethode (Fixierung mit Polyvinylalkohol und nachfolgende Gewebsmaceration durch Pektinase) wurde kürzlich von CHAYEN und MILES (1953) entwickelt (vgl. auch CHAYEN 1952). — Die Färbung des fixierten Materials kann unter anderem mit Säurefuchsin nach ALTMANN, mit Eisenhämatoxylin oder Kristallviolett erfolgen (weitere Angaben zur Färbetechnik NEWCOMER 1951, HARMAN 1950c, TANDLAR 1952. Simultane Fixierungs- und Färbemethode bei DEYSSON 1952 und MASCRÉ und DEYSSON 1953; neuere Vitalfarbstoffe: GUILLIERMOND 1949, BRENNER 1949, 1950, 1953).

## a) Vorkommen.

Gesichert ist das generelle Vorkommen von Chondriosomen bei Algen, Pilzen, Bryophyten, Pteridophyten und Spermatophyten, das Vorkommen in allen vegetativen Zellen der Spermatophyten und die Homologie der bei Tieren und Pflanzen gefundenen Chondriosomen. Bei den Schizophyten wurden in letzter Zeit besonders diskutiert: die Frage nach der Existenz der Mitochondrienäquivalente und deren räumliche und ontogenetische Beziehung zu Kernäquivalenten und zu den Metaphosphatgranula (Volutin).

### Bakterien.

MUDD (1953a und b) definiert die von ihm und seiner Schule beschriebenen Bakterienchondriosomen wie folgt: im Cytoplasma lokalisierte, ellipsoidische bis sphärische Granula, mit einer im Elektronenmikroskop deutlich sichtbaren Membran, elektronen- und phasenkontrastoptisch dicht erscheinend, phospholipoid- und in gewissen Fällen metaphosphathaltig (RUSKA und Mitarbeiter 1952). Mit koordiniertem System oxydierender und reduzierender Enzyme. Die größten ellipsoidischen bis sphärischen Formen grenzen größenordnungsmäßig an die kleinsten Säugetiermitochondrien an, während die kleinsten sphärischen unterhalb des Auflösungsvermögens des Lichtmikroskops liegen. Verlagerungen innerhalb der Zelle (KRÜGER-THIEMER und LEMBKE 1954) und funktionelle Unterschiede bei Bakterienchondriosomen wurden beobachtet.

Dieser Definition liegen folgende Tatsachen zugrunde: 1. die elektronenmikroskopische Beobachtung intraplasmatischer granulärer Strukturen bei Mycobakterien (LEMBKE und RUSKA 1940, WESSEL 1942, hier ältere Literatur, KNAYSI, HILLIER und FABRICANT 1950, MUDD und Mitarbeiter 1951a, RUSKA und Mitarbeiter 1952, MUDD und WINTERSCHEID 1953), Corynebakterien (RUSKA 1943, KÖNIG und WINKLER 1948) *Escherichia coli* und *Micrococcus cryophilus* (MUDD und Mitarbeiter 1951b, HARTMAN und Mitarbeiter 1953). Die Beobachtung von definierten „Reduktionsorten" mit Hilfe von Redox-Indicatoren, z. B. den Tetrazoliumsalzen (ANTOPOL, GLAUBACH und GOLDMAN 1948, BIELIG, KAUSCHE und HAARDICK 1949, 1952, MUDD und Mitarbeiter 1951a, WINTERSCHEID und MUDD 1953, DAVIS und Mitarbeiter 1953, SOROURI und MUDD 1953, MUDD 1953b, HARTMAN und Mitarbeiter 1953, dort weitere Literatur, MURRAY und TRUANT 1954). Das Nadi-Reagens (MUDD und Mitarbeiter 1951a und b, DAVIS und Mitarbeiter 1953, MUDD 1953a und b, dort weitere Literatur) und die Janusgrün B-Färbung (MUDD 1951a und b, 1953b, STEINBERG 1952, WINTERSCHEID und MUDD 1953, DAVIS und Mitarbeiter 1953, SOROURI und MUDD 1953) wurden ebenfalls zur Charakterisierung der Enzymorte herangezogen. Dabei zeigte sich, daß die MUDDschen Bakterienchondriosomen heterogen sind, denn nach den Untersuchungen von DAVIS und Mitarbeitern (1953) an drei *Salmonella typhosa*-Stämmen scheint die Reaktionsfähigkeit mit Tetrazoliumsalzen je nach Größe der Granula, Zellzustand und Entwicklungsbedingungen unterschiedlich zu sein, auch das Verhalten gegenüber dem Nadi-Reagens war verschieden (MUDD 1953a und b, dort ausführliche Diskussion). Bei *Escherichia coli* und *Azotobacter* konnte der mit Tetrazoliumsalzen nachweisbare „Reduktionsort" mit Berberinsulfat vital fluorochromiert werden (KRIEG 1954a und b). KRIEG hält es allerdings für möglich, daß am Reduktionsort durch Ablagerung von Metaphosphaten Granula sekundär entstehen (1954a).

Um die schwierige Frage nach der Existenz von Chondriosomenäquivalenten zu entscheiden, ist es nötig, die beiden Hauptuntersuchungsmethoden: die cytochemische und die elektronenmikroskopische, zu koordinieren. Für *Escherichia coli* (HARTMAN und Mitarbeiter 1953) und für *Bacillus cereus* (KELLENBERGER und HUBER 1953) konnte übereinstimmende Lokalisation der Formazanspeicherorte und der elektronenmikroskopisch sichtbaren Granula nachgewiesen werden. Auf Grund dieser und eigener Befunde wird von MUDD (1953b, dort weitere Literatur) angenommen, daß die Granula Reduktionsorte und den Chondriosomen homolog sind. Die Chondriosomenäquivalente („Chondrioide" nach KELLENBERGER und HUBER 1953) sind jedoch abweichend von den bei Tieren und höheren Pflanzen beobachteten Chondriosomen gebaut. Dünnschnitte durch *Bacillus cereus* zeigten die charakteristische Feinstruktur der Chondriosomen nicht (CHAPMAN und HILLIER 1953). Durch Doppelfärbungen (PREUNER und v. PRITTWITZ UND GAFFRON 1952, DAVIS und Mitarbeiter 1953, HARTMAN und Mitarbeiter 1953, WINTERSCHEID 1953, WINTERSCHEID und MUDD 1953, SOROURI und MUDD 1953, MUDD 1953b) konnte nachgewiesen werden, daß die Granula nicht mit den Kernäquivalenten zu identifizieren sind (vgl. auch BIELIG, KAUSCHE und HAARDICK 1952, HAHN 1952). Gegen Phagen sind die Granula resistent, während die Kernäquivalente durch sie zerstört werden (MUDD und Mitarbeiter 1952, 1953, PENSO, CASTELNUOVO und PRINCIVALLE 1952, HARTMAN und Mitarbeiter 1953 im Gegensatz zu BIELIG, KAUSCHE und HAARDICK 1949).

BISSET (1952a und b, 1953a und b, 1954, BISSET und HALE 1953) ist der Auffassung, daß die elektronenmikroskopisch sichtbaren Granula wenigstens zum Teil durch die Präparationstechnik bedingte Artefakte von Septen oder Zellwänden sind, die besonders bei Mycobakterien von einzelnen Autoren nicht berücksichtigt worden sind. Die auch von ihm (bei *Bacterium coli* 1953a) beobachteten Reduktionsorte werden für Wachstumspunkte gehalten.

In letzter Zeit wurde besonders die Beziehung der Chondriosomenäquivalente zu den Kernäquivalenten diskutiert. BRINGMANN (1950, 1952a) entwickelte die Vorstellung vom Komplexorganell (Karyoid) in dem Kern- und Chondriosomenäquivalente zum mindesten in isotoper Lage nebeneinander vorkommen (vgl. auch WALLHÄUSER 1954). Als Nucleoidstadium wird in Anlehnung an PIEKARSKI das Vorkommen von Kernäquivalenten im RNS-haltigen Cytoplasma bezeichnet, als „Paranucleoide" RNS-haltige Substanzanhäufungen, die neben den Nucleoiden bei einigen Schizomyceten vorkommen. Bei Coryne- und Mycobakterien und Aktinomyceten treten nebeneinander bzw. nacheinander während der Kulturentwicklung Nucleoid- und Karyoidstadien auf (BRINGMANN 1952b, dort weitere Literatur). Bei Coryne- und Mycobakterien erfolgt die Konzentration der RNS am Orte der Phosphatablagerungen im Laufe der ontogenetischen Entwicklung. Diese Phosphatzentren sollen gleichzeitig Fermentorte sein und werden wegen ihrer Fähigkeit zur RNS-Konzentration von Bringmann als Chondriosomenäquivalente angesehen. Bei den Coryne- und Mycobakterien stehen die Kern- und Chondriosomenäquivalente in regelmäßiger räumlicher Beziehung, ohne daß eine stoffliche Identität dieser beiden Organelle besteht (Bringmann 1953, 1954). Dieser sehr vorsichtigen Auslegung des Karyoidbegriffes steht die konkretere Vorstellung von Hoffmann (1951) gegenüber, der die Chondriosomenfunktion auf die äußere RNS-haltige Kugelschale eines Komplexorganells verlegen möchte. Bielig, Kausche und Haardick (1949, 1952) halten eine räumliche Beziehung zwischen Bakteriennucleoiden und auf Grund von Formazanbildung vermuteten mitochondrialen Reduktionsorten für gegeben, betonen aber, daß das Volumen des Reduktionsortes 15mal größer ist als das des Nucleoids.

Während Bringmann (1952b) der Auffassung ist, daß der Begriff „Volutin" sich erübrigt, da er mit dem Karyoidsystem identisch ist, sehen Krüger-Thiemer und Lembke (1954) die Polymetaphosphate (Volutin, vgl. Abschnitt 1 II DA) als granuläre Produkte der Chondriosomenäquivalente an. Diese sollen bei ausreichendem Nährstoff- und Phosphatangebot (z. B. bei den Mycobakterien) gebildet und wegen ihrer mit den Chondriosomen isotopen Lage mit diesen verwechselt werden (vgl. auch Krieg 1954a und b). Krieg (1954c) betont besonders, daß die Metaphosphatgranula bei den Mycobakterien den Kernäquivalenten aufgelagert und deswegen schlecht zu differenzieren sind.

### Cyanophyceen.

In der Zentralsubstanz der Cyanophyceen kommen DNS und RNS in isotoper Lage vor (Bringmann 1950, 1952a, v. Zastrow 1953). Diese Zentralsubstanz, in der außer den Nucleinsäuren noch Phosphate und Lipoide nachweisbar sind, wird von Bringmann (1952a) als ein Komplexorganell (Karyoid) aufgefaßt, in dem Kern- und Chondriosomenäquivalente vereinigt sind.

### b) Größe der Chondriosomen.

In der Zwiebelepidermis sind die sphärischen Chondriosomen 0,6—1,2 und die stabförmigen 2—4 $\mu$ groß (Perner und Pfefferkorn 1953). Meist wird in der Literatur eine Längenvariation im Verhältnis 1:10 bei konstant bleibendem Durchmesser angegeben. Extrem lange (25 $\mu$) fadenförmige Chondriosomen wurden von Ritchie und Hazeltine (1953) bei *Allomyces* gemessen. Größenveränderungen der Chondriosomen können durch Wachstum, Kontraktion, Aggregatbildung und Fragmentation hervorgerufen werden. Der durch Wachstum bedingte Übergang von sphärischen zu stab- und fadenförmigen Chondriosomen soll sich besonders gut bei *Saprolegnia* beobachten lassen (Guilliermond 1941), wo an der Hyphenspitze sphärische Formen vorherrschen, die allmählich in stab- und fadenförmige Formen übergehen (vgl. auch Fréderic und Chèvremont 1952). Reversible Verkürzung der stab- und fadenförmigen Formen unter Durchmesservergrößerung wird häufig beschrieben (z. B. von Fréderic und Chèvremont 1952). Die Chondriosomen können sich zu fadenförmigen Aggregaten zusammenlegen, wobei ihre Individualität erhalten bleiben kann, da sie ja nur aneinander gekoppelt zu sein brauchen (vgl. auch Chèvremont und Fréderic 1951a). Buvat (1953, dort weitere Literatur) vertritt die Auffassung, daß dieser Vorgang im normalen Entwicklungsablauf durch das Sistieren der Plasmaströmung bedingt ist. Bisher gebundenes Wasser soll zur Einwirkung auf die Chondriosomen frei werden und zunächst Stäbchenbildung, später Degradationsformen induzieren (vgl. auch PUYTORAC 1951). Kälte soll gleichsinnig wirken (GENÈVES 1951, 1952). Durch solche Aggregatbildung kann das Zahlenverhältnis der drei Chondriosomenformen verändert werden (PERNER 1952a, DRAWERT 1954). Es ist schwierig zu entscheiden, wo die pathologischen Veränderungen beginnen. Die unter anderen von BUVAT (1953), RITCHIE und HAZELTINE (1953) und DRAWERT (1954) beschriebenen Riesenformen dürften pathologisch verändert sein, wie schon PERNER (1952b) vermutet hat.

### c) Fragmentation und Teilung.

Eine Fragmentation der langen fadenförmigen Chondriosomen in kürzere stabförmige und der stabförmigen in sphärische (mit gleichem Durchmesser) ist oft beobachtet worden (z. B. von TIEGS 1928, WEATHERFORD 1933, DANNEEL und GÜTTES 1951). Sie scheint im normalen Lebenscyclus der Pflanze vorzukommen; so treten z. B. bei der Zoosporenbildung von *Saprolegnia*, *Leptomitus* und *Achlya* und bei der Gametenbildung von *Allomyces* stets sphärische Formen auf (HATCH 1935). Die Fragmentation erfolgt aber auch bei Schädigung durch Änderung des $p_H$-Wertes, des osmotischen Wertes der Umgebung und durch mechanische Einwirkung (Druck auf das Deckglas) (vgl. dazu GEITLER, Abb. 14c und d).

Die Teilung der Chondriosomen konnte durch den Mikrofilm von FRÉDERIC und CHÈVREMONT (1952, vgl. auch CHÈVREMONT und FRÉDERIC 1951a) nachgewiesen werden. Bereits früher war die Autoreproduktion auf Grund fixierter Teilungsstadien (Diplosomen) wahrscheinlich gemacht worden (z. B. MEVES 1904, DUESBERG 1911, FAURÉ-FREMIET 1910, DANNEEL und GÜTTES 1951, PAYNE 1952). Eine Längsverdoppelung, wie sie STEINER (1954) für möglich hält, ist unbewiesen und auf Grund der Feinstruktur auch unwahrscheinlich.

Die Vorstellung von der de novo-Entstehung der Chondriosomen ist jedoch auch in der modernen Literatur vertreten (vgl. auch LINDBERG und ERNSTER 1954, S. 114). Ihr liegt die nicht anzweifelbare, durch mikrokinematographische Aufnahmen gestützte Beobachtung zugrunde, daß Chondriosomen unter Verminderung der Lichtbrechung „verschwinden" können (FRÉDERIC und CHÈVREMONT 1952, dort Zusammenstellung der bisher hergestellten Filme). DANGEARD (1950a und b, 1951a) und DANGEARD und PARRIAUD (1953) behaupten, daß durch verdünnte Essigsäure oder Chloralhydrat das Chondriom aufgelöst werden kann, die chondriomfreien Zellen persistieren und aus den Auflösungsprodukten ein Chondriom de novo gebildet werden kann. Nach MARENGO (1949) sollen in den jungen Sporen von *Onoclea* alle plasmatischen Einschlüsse fehlen, so daß hier eine Lücke in der Kontinuität des Chondrioms bestünde. Von RITCHIE und HAZELTINE (1953) wird ein Fragmentieren in submikroskopische Partikel für möglich gehalten. Damit scheint auch ein Ansatzpunkt für die teilweise Erklärung dieser im übrigen noch nachzuprüfenden Phänomene gegeben. Wenn im mikroskopischen Bereich ein Wachstum für möglich gehalten wird, so dürfte es auch möglich sein, daß zunächst submikroskopische Formen durch Vergrößerung in den Bereich des Auflösungsvermögens des Lichtmikroskops kommen. Diese Meinung ist z. B. von OBERLING und Mitarbeitern (1950) für das von ihnen beschriebene „Ultrachondriom" (ausführliche Literatur hierzu bei DALTON 1953) vertreten worden. Diese Autoren nehmen an, daß die unter gewissen Umständen auftretenden Partikelchen von 40 $m\mu$ Größe Selbstreproduktion zeigen und durch Übergangsgrößen mit den Chondriosomen verbunden sind. Ähnliche Ansichten sind auch von ZOLLINGER (1948), CHÈVREMONT (1950), RÜTTIMANN (1951), EICHENBERGER (1953) und LINDBERG und ERNSTER (1954) geäußert worden. Die von ihnen als Mikrosomen bezeichneten Ausgangsstadien dürften mit den Pernerschen Mikrosomen (Sphärosomen) nicht identisch sein. Da in den elektronenmikroskopischen Arbeiten die Feinstruktur der Chondriosomen und der Übergangsstadien bisher nicht nachgewiesen werden konnte, fehlt der cytomorphologische Nachweis für die Entstehung von Chondriosomen aus Submikrosomen. Das Problem müßte mit anderer Technik weiter bearbeitet werden. In das Problem miteinbezogen werden sollte die Ansicht von Lenique, Hörstadius und Gustafson (1953), wonach es Hemmstoffe für die Mitochondrienentwicklung gibt (vgl. auch Hörstadius 1953 und Gustafsson 1953).

### d) Zahl.

Zwischen Chondriosomenzahl einerseits und Ernährungszustand, morphogenetischer Entwicklung und Aktivität der Zelle andererseits soll eine Korrelation bestehen (unter anderen Horning und Petrie 1928, Harman 1950a und b, Harman und Feigelson 1952a und b, Newcomer 1951, O'Brien 1951, Ito 1951, Perner 1952b, Paul und Sperling 1952, Lang 1953, vgl. auch Lindberg und Ernster 1954, S. 12). Von Lettré und Lettré (1953) wird sogar das Bestehen einer Kern-Plasma-Mitochondrienrelation für möglich gehalten. In diesem Sinne könnte folgende Beobachtung gedeutet werden: In den endomitotisch hochpolyploid gewordenen Mikropylar- und Chalazalhaustorien von *Veronica agrestis* ist die Zahl der Chondriosomen auf einen Durchschnittswert von $88,2 \pm 2,47$ gegenüber dem Durchschnittswert in den Endospermzellen $(57,2 \pm 1,53)$ gestiegen. Die Differenz ist mit einem p-Wert $< 0,1\%$ gesichert (Steffen 1954a).

Die Chondriosomenzahl ist aber höchstens ein Maß für die stoffwechselphysiologische Potenz der Zelle, sagt aber nichts über deren wirkliche Aktivität aus; so beobachtete Levitt (1952, 1954a und b), daß die Chondriosomen- und Mikrosomensubstanz, die durch Trockengewichtsbestimmung erfaßt wurde, in den Kartoffelknollen beim Übergang vom ruhenden zum aktiven Zustand abnahm.

### e) Bewegung.

Die Chondriosomen werden von der Plasmaströmung diskontinuierlich, jedoch langsamer als die Mikrosomen bewegt (Perner 1952a, Steffen 1953). Da sie flexibel (Geitler 1937) sind, werden die fadenförmigen Chondriosomen durch die Plasmaströmung durchgebogen, so daß eine wurm- (Ritchie und Hazeltine 1953) oder schlangenähnliche (Strangeways und Canti 1927) Bewegung resultiert (vgl. dazu Geitler, Abb. 14b). „Verzweigungen" der Chondriosomen können durch Auftreffen auf Hindernisse, durch Zusammenlagerung von Chondriosomen zu y-artigen Aggregaten und ohne erkennbare Ursache durch Formänderung erfolgen (Chèvremont und Fréderic 1951a, Fréderic und Chèvremont 1952). Den Chondriosomen scheint Eigenbeweglichkeit zuzukommen, wie Fréderic und Chèvremont auf Grund mikrokinematographischer Aufnahmen annehmen. Die Eigenbewegung soll durch Änderung der Oberflächenspannung zustande kommen.

### f) Beziehung zum Kern und zur Kernteilung.

Die Verteilung und Ausrichtung der Chondriosomen innerhalb der Zelle wird durch die Plasmaströmung (Ritchie und Hazeltine 1953) und nicht durch einen Diffusionsgradienten (Pollister 1941) bedingt. Steiner (1954) beschreibt für *Oospora* ein Anordnungsmuster, bei dem gleichartige Chondriosomen gepaart auftreten sollen. Die von Bram (1951) beschriebene, nach Einwirkung von Chemikalien auftretende Ballung der Chondriosomen um den Kern, scheint durch

Plasmasystrophe hervorgerufen. Jedoch sind auch Fälle polarisierter Anordnung bekannt geworden, so z. B. die von MARENGO (1949) beschriebene Anhäufung der Chondriosomen am Polarisationszentrum der Chromosomen im Pachytänbukett (für tierisches Gewebe vgl. LINDBERG und ERNSTER 1954, S. 14). Auf Grund mikrokinematographischer Aufnahmen wird eine zum Teil rhythmisch erfolgende Anlagerung an den Kern beschrieben (FRÉDERIC und CHÈVREMONT 1952, CHÈVREMONT und FRÉDERIC 1951a, b, c, 1952, FRÉDERIC 1951). Dabei soll ein Substanzaustausch zwischen Kern und Chondriosomen unter substantieller Beteiligung der Kernmembran stattfinden. Während der Kernteilung wird eine Verlangsamung der Chondriosomenbewegung beobachtet; LETTRÉ und LETTRÉ (1953) nehmen Inaktivierung während der Zellteilung an. Besonders in den letzten Stadien der Metaphase sieht man ein Dünnerwerden der Chondriosomen (ohne Längenveränderung) und das Auftreten von lokalen Anschwellungen innerhalb der Chondriosomen, so daß ein rosenkranzähnliches Aussehen resultiert. Diese Veränderungen betreffen die der Spindel benachbarten Chondriosomen, während die an der Zellperipherie liegenden keine Veränderung zeigen. Zum Teil tritt auch eine Verminderung der Lichtbrechung auf und die Chondriosomen „verschwinden" aus dem Gesichtsfeld. Eine Teilung der Chondriosomen während der Anaphase wurde nie beobachtet (CHÈVREMONT und FRÉDERIC 1951a). Die Verteilung der Chondriosomen auf die Tochterzellen erfolgt willkürlich, doch sollen sie sich etwa symmetrisch zur Teilungsebene anordnen (CHRISTIANSEN 1949). Das cytologische Bild wird durch die Zahl der in der Zelle befindlichen Chondriosomen bedingt; es kann so zu einer deutlichen Ringbildung um die Äquatorialplatte und zur Bildung einer dichten Platte zwischen den Tochterkernen kommen (zuletzt MARENGO 1949, dort und bei NEWCOMER 1940, 1951, weitere Literatur, vgl. auch DANNEEL und GÜTTES 1951).

### g) Verhalten beim Sexualakt und Beziehung zur plasmatischen Vererbung.

Beim Sexualakt der Angiospermen wäre theoretisch die Übertragung von Chondriosomen des männlichen Partners durch das Pollenschlauchcytoplasma oder durch die männlichen Gameten möglich. ANDERSON (1936) glaubt die Übertragung von Chondriosomen durch Pollenschlauchplasma in die Eizelle nachgewiesen zu haben. 1939 bildet er auch Chondriosomen in den Spermazellen ab, jedoch ist die Übertragung dieser Chondriosomen in die Eizelle nicht gesichert. Zwar gelang 1951 (STEFFEN) der cytologische Nachweis der Übertragung von männlichem Cytoplasma durch die Spermazellen in die Eizelle, jedoch mit Fixierungsmitteln, die nicht chondriosomenerhaltend waren.

Für *Pinus* soll nach MANGENOT (1938) der Übertritt der größeren Chondriosomen des männlichen Gameten gesichert sein, jedoch werden sie im oberen Teil des Proembryos durch die sog. Basalplatte abgeschlossen und haben keinen Teil an der Entwicklung des Embryos im eigentlichen Sinne. Weitere entwicklungsgeschichtliche Untersuchungen in dieser Richtung wären nötig, zumal kürzlich von MARQUARDT (1952) ausgehend von EPHRUSSIS Untersuchungen an der Hefe die Vorstellung entwickelt wurde, daß die reduplikationsfähigen Plasmaeinheiten (Summation Plasmon, vgl. IC) an die Chondriosomen oder „Grana" mit Mitochondrienfunktion gebunden sind. Dabei scheint aber nicht sicher, ob diese für die plasmatische Vererbung wichtigen Plasmaeinheiten mit den mikroskopischen Strukturen identisch oder submikroskopische Teile derselben sind. Die Plasmastrukturen sollen nach MARQUARDT (1952) Trägerstrukturen für lebensnotwendige Enzyme sein. Wird eine Sorte dieser Plasmaeinheiten experimentell in ihrer Zahl herabgesetzt, so kann es bei der Hefe nach mehreren Generationen durch willkürliche Verteilung zur manifesten Plasmaalteration kommen, die

biochemisch durch Ausfall der diesen Plasmaeinheiten eigenen Enzyme oder
Enzymsysteme faßbar wird. Michaelis (1953, 1954) hat gegen diese Vorstellung
mit dem Hinweis Einspruch erhoben, daß „die Erbkonstanz der Plastiden- oder
Mitochondrienfunktion nur zeigt, daß Plastiden oder Mitochondrien direkt oder
indirekt in die Reaktionsabläufe einbezogen sind, die von dem gesuchten Erbträger ausgehen".

Unter der Voraussetzung, daß z. B. die Hefezelle Chondriosomen unterschiedlicher Fermentgarnitur besitzt, könnte es bei Sprossung durch das Einwandern eines einzigen Chondriosoms in die Tochterzelle zu „vegetativen Mu-
tanten" kommen (Mundkur 1953).

### h) Verhalten gegenüber physikalischen und chemischen Einflüssen.

Gegenüber langsamer Temperaturerhöhung sind die Chondriosomen sehr
resistent (Perner und Pfefferkorn 1953, Ritchie und Hazeltine 1953).
Nach Ritchie und Hazeltine treten erst bei 75°, nach Dangeard bei 55—60°
(1951b) Degradationsformen auf. Widersprechende Angaben von Policard
und Mangenot (1922) wurden bereits von Guilliermond (1941) entkräftet.
Gegen radioaktive Strahlung (Milovidov 1929, 1930) sollen Chondriosomen
weniger, gegen Röntgenstrahlen jedoch weit empfindlicher als alle übrigen Zellbestandteile sein (Nadson und Rochlin 1933). Nach Scherer und Ringleb
(1953) hingegen sollen sich die Mitochondrien tierischer Zellen von der Röntgen-
schädigung schneller erholen als der Kern. Gegenüber UV-Strahlung sind die
nicht näher definierten Granula der Hefe verschieden resistent (Sarachek und
Townsend 1953).

Gegen Colchicineinwirkung scheinen Chondriosomen wenig empfindlich zu
sein (Garrigues 1940, Ryland 1948). Da die Chondriosomen ein osmotisches
System darstellen, reagieren sie leicht auf osmotische Änderungen der Umgebung,
jedoch zeigen die Chondriosomen aus verschiedenen Geweben ein unterschiedliches Verhalten. Während in gleichmolarer Zuckerlösung die Chondriosomen
der Meerschweinchenleber (Hogeboom, Schneider und Palade 1948) erhalten
bleiben, werden die Mitochondrien von Amphibien (Recknagel 1950) bereits
zerstört. (Weitere Literatur über die semipermeable Membran der Chondriosomen bei Duve und Mitarbeitern 1951, Berthet und Mitarbeitern 1951, Farrant, Robertson und Wilkins 1953 und Schneider 1953, Dianzani 1953b).

### i) Degradationsformen.

Als Degradationsformen treten besonders bei osmotischen Zustandsände-
rungen eine Quellung und Abkugelung (Farrant, Robertson und Wilkins
1953, Dianzani 1953b) der Chondriosomen und danach eine partielle Entmischung („Cavulation" nach Guilliermond) auf (Anitschkow 1923, Zollinger 1948, Opie 1948, Perner 1952a—c, Perner und Pfefferkorn 1953,
Ritchie und Hazeltine 1953, Fréderic und Chèvremont 1952, Buvat 1953).
Bereits die erste osmotische Schwellung ist vom Herausdiffundieren der Co-Enzyme begleitet (Raaflaub 1952, 1953). Da das Herausdiffundieren durch ATP
verhindert werden kann, erklärt sich aus dem unterschiedlichen physiologischen
Zustand der Zelle und der Chondriosomen der unterschiedliche Grad ihrer Schädigung. Reversible blasenförmige Schwellungen wurden bei Saprolegnia und
vor allem an tierischen Explantaten beobachtet (Fréderic und Chèvremont
1952, vgl. auch Puytorac 1951). Auffällig ist, daß Degradationsformen häufig
im Zusammenhang mit dem Auftreten von Vacuolen im Plasma beschrieben
werden, z. B. bei der Ascusbildung (Janssens, van de Putte, Heimsortel

1913, Varitschak 1931). Hierbei wäre an die Wirkung frei gewordenen Wassers zu denken (Buvat 1953). Daß Degradationsformen bei mangelhafter Fixierung auftreten, ist nicht verwunderlich, muß aber bei der kritischen Beurteilung von Studien an fixiertem Material berücksichtigt werden.

Bei Supravitalfärbung mit Janusgrün wurde in stabförmigen Chondriosomen tröpfchenförmige Farbstoffspeicherung von Drawert (1954) beobachtet. Während diese Tropfen bei Sauerstoffmangel Entfärbung zeigen, war der tropfenförmig gespeicherte Farbstoff in den wohl weiter fortgeschrittenen Degradationsformen von Perner (1952a) und Perner und Pfefferkorn (1953) nicht mehr reduzierbar (vgl. auch Opie 1947, 1948).

## k) Beziehung zwischen Chondriosomen und Plastiden.

Nur noch von historischem Interesse sind die von Pensa (1910) und Lewitzky (1910, 1911a und b) begründete Transformationstheorie, wonach Chondriosomen in Plastiden übergehen sollen, und die Dualitätstheorie von Guilliermond (1920a und b, Zusammenfassung 1941), wonach es zwei Chondriosomentypen gibt, in Plastiden verwandelbare und permanente (ausführliche Diskussion bei Guilliermond 1941, Newcomer 1940, 1946, 1951, Strugger 1953, 1954a u. b). Durch die Arbeiten Struggers (1950, 1953, 1954a u. b) und seiner Schule (Bartels 1954, Böing 1954, Grave 1954, Kaja 1954, Perner 1954) und durch die elektronenmikroskopischen Untersuchungen von Wolken und Palade (1953) wurde die Kontinuitätstheorie von Schimper-Meyer für die Plastiden erneut bestätigt. Die bisher schwierige Unterscheidung von Chondriosomen und Proplastiden (vgl. auch Rezende-Pinto 1952) wurde durch die Entdeckung des primären Granums in den Proplastiden erleichtert. Die Existenz dieses für die Differentialdiagnose wichtigen, mit kathodischen Farbstoffen färbbaren primären Granums kann für die Proplastiden der Gymnospermen und Angiospermen als gesichert gelten, ebenso das allgemeine und gesetzmäßige Vorkommen in den Proplastiden von *Agapanthus* und *Elodea* (Strugger 1953, 1954a u. b, Böing 1954, Grave 1954) und in den Leukoplasten der Wurzel von *Vicia faba* (Bartels 1954), der Zwiebelepidermis und der Staubfadenhaare von *Tradescantia* (Perner 1954). Strugger (1953, 1954a und b) führt die Feststellung von Heitz und Maly (1953), daß den Proplastiden primäre Grana als kontinuierliches System fehlen, auf ungeeignete Technik zurück (vgl. auch Perner 1954). Auf Grund elektronenmikroskopischer Aufnahmen kommt Leyon (1953) zu der Auffassung, daß den Proplastiden ein primäres Granum fehlt. Doch bleibt abzuwarten, ob die Deutung seiner Abbildungen richtig ist. Vergleichende lichtmikroskopische Untersuchungen von Leyon liegen nicht vor.

## 2. Chemische und biochemische Analyse mit Hilfe der Differentialzentrifugierung.

Die chemische und biochemische Analyse der Chondriosomen wird durch die Anwendung des Homogenisatverfahrens und der Differentialzentrifugierung ermöglicht. Da die mit dieser Methodik gewonnenen Ergebnisse rückläufig auf die lebende Zelle und ihre Funktionen übertragen werden, ist eine Besprechung dieses Verfahrens und der bei ihm auftretenden Mängel und Schwierigkeiten erforderlich. Dies ist auch nötig, weil vom Cytomorphologen die Diagnose der einzelnen Fraktionen erwartet wird und nur er wirksam vor Artefaktbildung warnen kann.

## a) Das Verfahren.

Die Differentialzentrifugierung von Homogenisaten wurde zuerst von Bensley und Gersch (1933) und Bensley und Hoerr (1934) zur biochemischen Analyse tierischer Objekte eingeführt, später von Claude (1946), Schneider (1946) und entscheidend von Hogeboom, Schneider und Palade (1948) verbessert (zusammenfassende Literatur über Darstellungsmethoden bei Schneider und Hogeboom 1951). Entsprechend ihrer Größe und Dichte sedimentieren nacheinander: Kerne, Mitochondrien, Mikrosomen, der cytoplasmatische Rest bleibt in der überstehenden Flüssigkeit zurück. In letzter Zeit ist wiederholt der Versuch unternommen worden, die heterogenen Mitochondrien- und Mikrosomenfraktionen weiterzuzerlegen (Chantrenne 1947, Barnum und Huseby 1948, Keller 1951, Novikoff und Mitarbeiter 1952, 1953, Paigen 1954, Kuff und Schneider 1954). Für die biochemische Arbeit scheint die Zerlegung in sechs Fraktionen am günstigsten zu sein, wobei die Mitochondrien- und Mikrosomenfraktion in eine gewaschene Mitochondrienfraktion, in zwei Mikrosomenfraktionen und in eine Mischfraktion, die kleine Mitochondrien und Mikrosomen enthält, zerfallen (Novikoff und Mitarbeiter 1953).

Für den Cytologen ist die Analyse und Identifizierung der einzelnen Fraktionen aus folgenden Gründen schwierig: 1. wegen der im Homogenisat oder in der Fraktion möglichen Strukturänderung (Artefakt- und Präzipitatbildung Stafford 1951) und 2. wegen der oft schwierigen sauberen Trennung der einzelnen Fraktionen (vgl. Hogeboom und Schneider 1950).

1. Der Erhaltungszustand im Homogenisat ist von der Art der Zellzertrümmerung, vom Resistenzgrad der Zellen, vom Medium, von der Einwirkung von ATP-Spaltprodukten, von Temperatur und Zeitdauer des Prozesses abhängig (Raaflaub 1953).

Da die Chondriosomen ein osmotisches System darstellen (Lehninger und Kennedy 1948, Potter 1946, Opie 1948, Dianzani 1953b), quellen sie in hypotonischen Lösungen unter Wasseraufnahme. Man verwendet deswegen als Homogenisationsmedium kalte (0—5° C) gepufferte ($p_H$ 6,8—7,2) isotonische KCl- oder schwach hypertonische Zuckerlösung. Zum Teil wird der $p_H$-Wert für Zuckerlösungen höher gewählt ($p_H$ 7,5—8) und β-Glycerophosphat dem Medium zugefügt. Die Bildung von oberflächlichen Blasen („blisters") soll nach Weber (1954) durch Zugabe von dem Dinatriumsalz des Äthylendiamintetraacetats zu KCl-Lösungen verzögert werden. Nach Lehmann und Wahli (1954) soll das genannte Komplexon Versene durch Blockierung der Ca- und Mg-Ionen strukturerhaltend wirken (vgl. auch Slater und Cleland 1952). Die Ursache dafür, daß pflanzliche Homogenisate wenig verwendet wurden, ist in ihrer großen Labilität zu suchen (Literatur über pflanzliche Objekte bei Millerd und Bonner 1953 und Perner 1953). Pflanzliche Homogenisate werden sekundär durch den in das Homogenisat eingehenden Zellsaft alteriert, wobei Ionenwirkungen und Beeinflussung durch frei werdende organische Stoffe, vor allem Säuren (vgl. Wildman und Cohen in diesem Handbuch) möglich sind. Doch sollen nach Perner und Pfefferkorn (1953) auch Homogenisate aus meristematischen zellsaftfreien Zellen morphologisch ähnliche Degradationsformen zeigen.

Die biochemische Leistung der Chondriosomen in vitro kann durch aus dem Plasma freigesetzte oder inaktivierte Enzyme, durch aus den Chondriosomen austretende Enzyme und durch Adsorption geändert werden. Die aus den Chondriosomen herausgelösten Enzyme können zudem frei ganz andere Eigenschaften als strukturgebunden haben (z. B. die Äpfelsäuredehydrase Huennekens 1951). Da die Co-Enzyme in den Chondriosomen verbleiben, bedürfen z. B. heraus-gelöste Dehydrasen nunmehr der Co-Enzyme. — Der durch Aufbewahrung bedingte Aktivitätsverlust der Homogenisate beruht zum großen Teil auf Aufspaltung der Co-Enzyme. (Die beiden wichtigsten pyridinnucleotidespaltenden Enzyme: DPN-Nucleosidase und Nucleotidphosphorylase kommen im Homogenisat in hoher Aktivität vor.) Mit Veränderung der biochemischen Leistung kann auch eine morphologische Änderung vor sich gehen (HARMAN 1950b, HARMAN und FEIGELSON 1952c), umgekehrt wird in einem morphologisch geordneten System der Reaktionsablauf durch Zerstörung dieser Ordnung und Desorientierung der Enzyme gestört (HARMANN und FEIGELSON 1952a, c; vgl. auch DANIELLI 1946).

Während PERNER und PFEFFERKORN (1953) und PERNER (1952c, 1953) die schon früher beschriebenen Degradationsformen der Chondriosomen mit irreversibler tropfenförmiger Janusgrün B-Speicherung beobachteten und der Auffassung sind, daß eine einwandfreie Isolierung aus pflanzlichen Dauerzellen unmöglich sei, konnte SCHNEIDER (1947) an tierischen Objekten guten Erhaltungszustand der isolierten Mitochondrien im Lichtmikroskop zeigen. PALADE (1953) konnte durch elektronenmikroskopische Untersuchung von Dünnschnitten durch isolierte Mitochondrien nachweisen, daß die Feinstruktur erhalten bleibt, wenn auch gewisse Schädigungen auftreten.

2. Homogenisate im Elektrolytmedium zeigen die Tendenz zu agglutinieren. Dadurch kann es bedingt sein, daß ein Teil der Chondriosomenfraktion bereits mit den Zellkernen sedimentiert. Das von GREEN (1951, dort Literatur) erarbeitete und Cyclophorasesystem genannte, gelartige Enzympräparat besteht z.B. nicht nur aus Mitochondrien, sondern auch aus Kernfragmenten (HARMAN 1950a, b, c), mit denen es verklebt ist. Verunreinigungen mit cytoplasmatischen Artefakten (Koacervatbildung) sind möglich (PERNER und PFEFFERKORN 1953, vgl. auch WILDMAN und COHEN in diesem Handbuch), jedoch zum Teil durch Waschen entfernbar, wie die Differentialmethoden von NOVIKOFF und Mitarbeitern (1953) und von JACKSON, WALKER und PACE (1953) zeigen.

## b) Identifizierung der gereinigten Fraktionen.

In älteren biochemischen Arbeiten erfolgt eine grobe Unterteilung nach Größe der Partikel: large granules mit Strömungsdoppelbrechung 0,5—3 $\mu$ (= Mitochondrien) und small particles: 50—200 m$\mu$ (= Mikrosomen der Biochemiker = Submikrosomen nach FREY-WYSSLING). Die Trennung ist willkürlich, ihr Ergebnis je nach angewandter Technik variabel. Beim Vergleich des biochemischen Verhaltens der Fraktionen verschiedener Autoren wird sich schon aus diesem Grunde eine Überschneidung der Werte nicht vermeiden lassen (vgl. Tabelle 3). Die Fraktionen können Mischfraktionen aus Chondriosomen und Submikrosomen sein oder die Hauptfraktionen sind in sich uneinheitlich.

Die Fähigkeit zur oxydativen Phosphorylierung ist ein gutes enzymatisches Kriterium für intakte Chondriosomen. Der Nachweis von Chondriosomen in den Fraktionen durch Janusgrün B oder Tetrazoliumsalze ist nicht möglich, da auch die überstehende Flüssigkeit reagiert (STAFFORD 1951). STAFFORD fand die Nadi-Reaktion auf größere Partikel einer Mischfraktion beschränkt. Da in der Fraktion der large granules größenordnungsmäßig auch die Sphärosomen PERNERs enthalten sein können, muß in Zukunft eine weitere Differenzierung dieser Fraktion versucht werden. Im lichtmikroskopischen Bereich ist eine Unterscheidung der Chondriosomen von den Sphärosomen durch den grauen positiven Phasenkontrast und das osmotische Verhalten der Chondriosomen möglich (z. B. NOVIKOFF und Mitarbeiter 1953). NOVIKOFF und Mitarbeitern gelang es, mit neuer Methode die

Chondriosomenfraktion in zwei Subfraktionen zu zerlegen, deren Partikel in der Größe und der biochemischen Aktivität (wenn auch nur gering) unterschieden sind (vgl. auch Paigen 1954, Kuff und Schneider 1954). Laird und Mitarbeiter (1952, 1953) wollen aus dem flockigen Sediment oberhalb der eigentlichen Chondriosomenfraktion eine zweite Chondriosomenform isoliert haben, die kleiner als die normale ist und sich durch hohen RNS-Gehalt (ähnliche Ergebnisse bei Vendrely-Randavel 1949) und hohe Succinoxydaseaktivität auszeichnet. Sie weisen den von Novikoff und Mitarbeitern (1953) auf Grund der biochemischen Ergebnisse geäußerten Verdacht, daß es sich um eine Mischfraktion handelt, durch Phako-Analyse zurück.

Novikoff und Mitarbeiter (1952, 1953) konnten in den von ihnen phakooptisch untersuchten Mikrosomenfraktionen Partikelchen unterscheiden, die nach optischer Dichte, Größe und Umrißform und in der Aktivität der sauren Phosphatase und Uricase unterschieden waren. Diese Partikel sind jedoch nicht mit den drei verschiedenen von Slautterback (1952, 1953) elektronenmikroskopisch dargestellten „Mikrosomen" identisch, da zwei der dargestellten Partikel unterhalb des Auflösungsvermögens des Lichtmikroskopes liegen. Eine Beziehung der von Novikoff und Mitarbeitern und von Slautterback gefundenen Partikel zu Zellorganellen läßt sich zur Zeit noch nicht herstellen.

In letzter Zeit sind Methoden entwickelt worden, die es gestatten, den Chondriosomengehalt der Fraktion zahlenmäßig auszuwerten (Allard und Mitarbeiter 1952, Shelton, Schneider und Striebich 1953a und b).

### c) Chemische und biochemische Analyse der Mitochondrien- und Mikrosomenfraktion.

Die wichtigsten analytischen Daten für die Mitochondrien- und Mikrosomenfraktion mögen aus den Tabellen 3 und 4 ersehen werden. Die Mikrosomenfraktion, die morphologisch in den meisten Fällen den Submikrosomen entspricht, ist reicher an RNS und Lipoiden, doch scheint es hiervon auch Ausnahmen zu geben, wie die Angaben von Levitt (1954a und b, dort weitere Literatur und Beispiele) in der Tabelle 3 zeigen. Auf jeden Fall geht aus den

Tabelle 3. *Analyse der Chondriosomen- und Mikrosomenfraktionen. Werte in Prozent des Trockengewichts.*

| | Pisum sativum STAFFORD 1951 | Solanum tuberosum LEVITT 1952, 1954a u. b | | Mäuseleber BARNUM und HUSEBY 1948 | |
|---|---|---|---|---|---|
| | Chondriosomen | Chondriosomen-fraktion | Mikrosomen-fraktion | Large granules | Mikrosomen |
| N | | 6,17 | 9,69 | 12,1 | 10,3 |
| P | | 0,985 | 1,54 | 1,11 | 1,87 |
| RNS | 0,5—1 | | | 3,7 | 9,1 |
| DNS | 0,7—0,9 | | | | |
| Proteine | 35—40 | | | | |
| Lipoide | 25—38 | 22 | 12 | 27,4 (56,6%) | 35,1 (62,7%) |

Die Fraktion der large granules von Barnum und Huseby dürfte etwa der Chondriosomenfraktion entsprechen, während die Mikrosomenfraktion eine Mischfraktion zweier elektronenmikroskopisch differenter Submikrosomen ist. ( ) = Phospholipoide in Prozent des Gesamtlipoidgehalts.

bisherigen chemischen Analysen hervor, daß die Submikrosomen keine Bruchstücke von Chondriosomen sind.

Die Chondriosomen bestehen hauptsächlich aus Lipoproteiden. Neuerdings wird von Chayen und Norris (1953) behauptet, daß die DNS in ihnen lokalisiert sei. Bei der FEULGEN-Reaktion werden die Chondriosomen zerstört, und die DNS soll in den Zellkern wandern (vgl. auch die Befunde von ŠAPOT und NEMZINSKAJA 1950 und STAFFORD 1951, die Chondriosomenfraktion von letzterer war allerdings durch Kernfragmente verunreinigt). Der RNS-Gehalt der Chondriosomen ist durch viele gleichlautende Angaben gesichert (z. B. CLAUDE 1950, dort weitere Literatur; HOERR 1943, LAZAROW 1943, DAVIDSON 1947, VENDRELY-RANDAVEL 1949, ZOLLINGER 1950; BROWN, FITTON-JACKSON und CHAYEN 1953 für pflanzliche Chondriosomen). Bei älteren Arbeiten dürfte der RNS-Gehalt durch Verunreinigung der Fraktion mit Mikrosomen zu hoch angegeben sein, so z. B. in den in Tabelle 3 zitierten Angaben von BARNUM und HUSEBY (1948). Nach den Untersuchungen von BROWN und Mitarbeitern (1953) an Bohnenwurzeln steigt mit abnehmender Teilchengröße der RNS-Gehalt. Nach HOGEBOOM, SCHNEIDER und PALADE (1948) beträgt der Gehalt an RNS-Phosphor je Milligramm des Gesamtstickstoffs in den Mitochondrien 11 $\mu$g, in den Submikrosomen jedoch 63 $\mu$g.

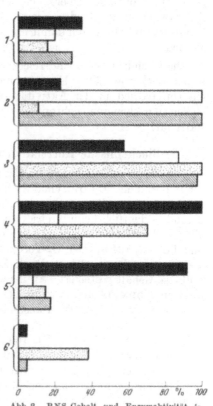

Abb. 2. RNS-Gehalt und Enzymaktivität je Gramm N in sechs Fraktionen eines Rattenleberhomogenisats nach NOVIKOFF und Mitarbeitern 1953 (in veränderter Darstellung). Die durchschnittliche Enzymaktivität jeder Fraktion wurde in Prozenten der höchsten, überhaupt in einer Fraktion gefundenen Aktivität (= 100%) angegeben. 1—6 Die einzelnen Fraktionen: 1 Kern-, 2 Mitochondrienfraktionen, 3 Mischfraktion aus Mitochondrien und Mikrosomen, 4 Mischfraktion aus zwei verschiedenen Mikrosomen, 5 Mikrosomenfraktion (einheitliche Fraktion von optisch weniger dichten Mikrosomen) und 6 überstehende Flüssigkeit. Die Kolonnen bedeuten: schwarz RNS, weiß Succinoxydase, punktiert Säurephosphatase und schraffiert ATP-ase.

Über die in den Mitochondrien- und Mikrosomenfraktionen bisher gefundenen Enzyme orientiert die Liste von LANG (1953). Um durch Adsorption bedingte Fehler zu vermeiden, sollte jedoch nach SCHNEIDER und HOGEBOOM (1951) nur dann ein Enzym als in der Fraktion vorkommend angesehen werden, wenn die Enzymkonzentration in der Fraktion größer als im Homogenisat ist. Auf Grund dieses Kriteriums verringert sich nach LINDBERG und ERNSTER (1954) die Zahl der in Mitochondrien vorkommenden Enzyme auf: Cytochrom-, Succin-, Oxaloacetat- und Octanoat-Oxydasen, DPN-Cytochrom-c-Reduktase und das bei der Synthese von p-Aminohippursäure beteiligte System.

Die Mitochondrien- und Mikrosomenfraktionen sind auch biochemisch keine Einheiten (NOVIKOFF und Mitarbeiter 1953, CHRISTIE und JUDAH 1953). NOVIKOFF und Mitarbeiter (1953) haben mit sehr sorgfältiger Waschung und

Tabelle 4. *Analyse der Lipoidkomponente der Mitochondrien der Rattenleber nach SWANSON und ARTOM (1950). Werte in Prozent des Gesamtlipoidgehalts.*

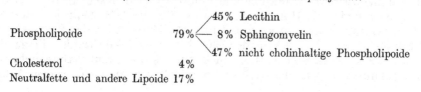

Phospholipoide          79% ⟨ 45% Lecithin
                              8% Sphingomyelin
                              47% nicht cholinhaltige Phospholipoide

Cholesterol             4%
Neutralfette und andere Lipoide 17%

Differentialzentrifugierung eine neunstufige Fraktionierung des Protoplasten erreicht. Von sieben untersuchten Enzymen fehlt keines in einer ihrer Fraktionen, es lassen sich nur quantitative Unterschiede in der Fermentaktivität feststellen (Abb. 2). Trotz aller Übergänge heben sich aber auch bei einer so vielstufigen Zentrifugierung die beiden Klassen der Chondriosomen und Mikrosomen scharf heraus und rehabilitieren damit nachträglich teilweise die gröberen summarischen Methoden.

Die Chondriosomen sind reicher an ATP-ase und Succinoxydase, die Mikrosomen reicher an RNS, Esterase und A 5′ P-ase. Innerhalb der Mikrosomen ist ein Unterschied in der Aktivität von saurer Phosphatase und Uricase festzustellen, die kleineren, lichtoptisch weniger dichten Mikrosomen zeigen geringere Aktivität.

### d) Biochemische Korrelation von Mikrosomen und Chondriosomen.

Die Frage nach der Enzymlokalisation darf in ihrer Bedeutung nicht überschätzt werden. Die Enzymaktivität wird durch das Substrat und dessen Konzentration geregelt. Die biochemische Korrelation aller Zellbestandteile wird besonders deutlich beim Krebscyclus, der seine volle Aktivität nur bei Gegenwart aller Zellbestandteile erreicht (HOGEBOOM und SCHNEIDER 1950, SCHNEIDER und POTTER 1949). Der $O_2$-Verbrauch isolierter Mitochondrien wird nach NIEMEYER und JALIL (1953) durch Zugabe der Mikrosomenfraktion mit hohem ATP-ase-Gehalt gesteigert. An der Oxydation von Isocitrat ist außer der Kernfraktion (JOHNSON und ACKERMANN 1953) auch die Mikrosomenfraktion (HOGEBOOM und SCHNEIDER 1950, PRESSMANN und LARDY 1952) beteiligt. 45% der Oxalessigsäureoxydase sind in den Chondriosomen lokalisiert, in den übrigen Fraktionen ist keine zu finden. Die 100%-Aktivität wird erst bei Zugabe der übrigen Fraktionen erreicht (SCHNEIDER 1953). Der Einbau von Alanin in ein Protein kann von einem Mikrosomenenzym nur in Gegenwart der Mitochondrienfraktion durchgeführt werden; von den Mitochondrien soll bei der aktiven Phosphorylierung ein löslicher Stoff (ATP-Derivat?) abgegeben werden (SIEKEVITZ 1952, SIEKEVITZ und POTTER 1953).

### e) Funktion der Chondriosomen.

Auf Grund der biochemischen Untersuchung von Mitochondrienfraktionen vor allem tierischer Objekte läßt sich zur Zeit folgende Übersicht über die Funktion der Chondriosomen geben (vgl. den ausführlichen Bericht in Teil 1, Bd. 2 dieses Handbuches). Die Übertragung der biochemischen Ergebnisse auf die lebende Zelle muß jedoch mit Vorsicht erfolgen, die Enzymleistungen in vitro sind eine oder zwei Zehnerpotenzen höher als die Leistungen der intakten tierischen Zelle bzw. deren Mitochondrien (LANG 1953).

### α) Energiegewinnung.

Während die Glykolyse im Cytoplasma der Zelle vor sich geht, finden die wesentlichen biologischen Oxydationsprozesse in den Chondriosomen statt (BRUMMOND und BURRIS 1953, zusammenfassende Darstellungen bei DOUNCE 1950, 1952, HOGEBOOM 1951, SCHNEIDER und HOGEBOOM 1951, KALCKAR 1952, GREEN 1951a und b und 1952, HOLTER 1952, LANG 1952, 1953, MILLERD und BONNER 1953, SCHNEIDER 1953, LINDBERG und ERNSTER 1954). Über das Enzymsystem des Citronensäurecyclus, das in den Chondriosomen lokalisiert ist, findet der Abbau von Essigsäureresten statt, die nicht nur aus Brenztraubensäure und Fettsäuren (LEHNINGER 1952, dort weitere Literatur), sondern auch aus dem Abbau von Aminosäuren stammen. Die pflanzlichen Chondriosomen scheinen sich von den tierischen durch den Besitz von Hexokinase (SALTMAN 1953, vgl. BONNER und MILLERD 1953) zu unterscheiden, so daß der Umsatz unvorbehandelter Zucker möglich zu sein scheint.

## β) Energieübertragung.

Die beim Elektronentransport frei werdende Energiemenge wird zum großen Teil durch einen enzymatischen Phosphorylierungsprozeß in eine nutzbare Form, in energiereiches Phosphat (ATP) überführt (Literaturübersicht bei MILLERD und Mitarbeiter 1951, COPENHAVER und LARDY 1952, LEHNINGER 1952, LINDBERG und ERNSTER 1952, 1954, MILLERD und BONNER 1953, LATIES 1953a und b). Jedoch scheint nach den bisherigen Untersuchungen (BONNER und MILLERD 1953) der thermodynamische Wirkungsgrad der oxydativen Phosphorylierung bei pflanzlichen Chondriosomen in vitro geringer als bei tierischen. Während bei tierischen Objekten drei bis vier Moleküle Phosphat je aufgenommenes Atom Sauerstoff durch Veresterung verschwinden, wird bei pflanzlichen Objekten *(Phaseolus aureus)* nur ein Phosphatmolekül verestert. Weitere Untersuchungen über diesen Energierückgewinnungsprozeß bei pflanzlichen Objekten sind abzuwarten. Die im ATP gespeicherte Energie wird zum Teil zur Strukturerhaltung der Chondriosomen benötigt (RAAFLAUB 1952, HARMAN und FEIGELSON 1952a, b, c).

## γ) Energienutzung.

Bei den energienutzenden Prozessen ist nicht nur an die in den Chondriosomen selbst ablaufenden Synthesen, sondern auch an die Prozesse zu denken, die außerhalb der Chondriosomen, aber durch Übertragung der in den Chondriosomen gespeicherten Energie (Aktivierungsprozeß) eingeleitet werden. Bei der Muskelkontraktion findet eine direkte Nutzung des in den spezialisierten Mitochondrien („Sarcosomen", LENIQUE 1953, LEVENBOOK 1953) gespeicherten ATP statt, nämlich eine Konversion von chemischer in mechanische Energie. Zu den in tierischen Mitochondrien ablaufenden Syntheseprozessen (ausführliche Darstellung bei LINDBERG und ERNSTER 1954, LANG 1953) gehören z. B.: die Bildung von Citrat und Acetoacetat, von Hippur- und p-Aminohippursäure, die Synthese von Phosphatiden aus vorhandenen Fettsäurekomponenten, die Bildung von Glutamin aus Glutaminsäure, die Knüpfung von Peptidbindungen und die Veresterung von Aneurin zu Cocarboxylase. Die Synthese von Proteinen und Ribonucleotiden scheint in nur geringem Maße zu erfolgen, während die Phosphatsynthese erheblich ist. Für pflanzliche Objekte ist nur ein Teil der genannten Syntheseprozesse gesichert (ausführliche Darstellung bei MILLERD und BONNER 1953): Peptidbindungen (Glutamin und Glutathion WEBSTER 1953), Phosphatbildung und innere Konversion von niederen Fettsäuren. An der Bildung von 1-Ascorbinsäure (MAPSON, ISHERWOOD und CHEN 1954) sind pflanzliche Chondriosomen zum mindesten beteiligt.

## δ) Physiologische Leistung normaler und spezialisierter Chondriosomen.

Durch biochemische Arbeiten kann stets nur die Leistung einer Mitochondrienpopulation erfaßt werden. Daß diese Populationen nicht einheitlich sind, haben die Arbeiten mit verfeinerten Fraktionierungsmethoden gezeigt. Auch in diesen nach Größe und Dichte sortierten Subfraktionen dürften noch Chondriosomen unterschiedlicher Leistung vorhanden sein. Aus den bisherigen biochemischen Ergebnissen läßt sich jedoch etwa die Standard-Enzymausrüstung der Mitochondrien erschließen. Von dieser Grundausrüstung gibt es durch Fehlen oder zusätzliches Auftreten von Enzymen Abweichungen.

Zur Grundausrüstung des Normalmitochondriums dürften die Enzyme gehören, die die Reaktionen des Citronensäurecyclus, die Koppelung zwischen Oxydation und Phosphorylierung und die ATP-Bildung katalysieren. Auch die Bildung gewisser Aminosäuren dürfte allen Mitochondrien möglich sein.

Im organisierten System des Chondriosoms erreicht die Respirationskapazität mit physiologischen Quantitäten von Enzymen und Co-Enzymen einen Höchstwert. Die physiologische Leistung der Chondriosomen erschöpft sich nicht in Energiegewinn und Synthese, darüber hinaus spielen die Chondriosomen eine wichtige Rolle in einem Regulationsmechanismus.

Da die Intermediärprodukte des Citronensäurecyclus teilweise zur Synthese verwendet werden, darf keine vollständige Oxydation des Substrates stattfinden. Verbrennung und Synthese wirken antagonistisch und müssen ausbalanciert werden. Das Gleichgewicht wird gestört, wenn zusätzliche (nicht die Intermediärprodukte des Citronensäurecyclus betreffende) ATP-verbrauchende Reaktionen ablaufen. Zur Bereitstellung der Energie muß erhöhte Oxydation einsetzen und damit eine Verminderung der für Syntheseprozesse verfügbaren Intermediärprodukte. Zu solchen ATP-verbrauchenden Reaktionen dürfte die Hexokinasereaktion gehören, die ihrerseits wieder hormonal kontrolliert wird. Grundsätzlich beeinflussen alle ATP-verbrauchenden Reaktionen die Syntheseprozesse, die Intermediärprodukte verwenden (LINDBERG und ERNSTER 1954).

Wenn man annimmt, daß die einzelnen in den Mitochondrien ablaufenden Prozesse miteinander konkurrieren, so muß jede Änderung in der Enzymaktivität oder jede Änderung in der qualitativen und quantitativen Zusammensetzung der erreichbaren Substrate die physiologische Leistung des Chondriosoms temporär verändern. Neben diesen nur temporär spezialisierten scheint es jedoch auch organspezifische Chondriosomen mit Sonderfunktionen zu geben, so ist z. B. die Harnstoffsynthese (zum mindesten in ihren ersten Stadien) auf die Lebermitochondrien der Säugetiere beschränkt (MÜLLER und LEUTHARDT 1949). Die Entgiftung aromatischer Substanzen durch Koppelung eines aromatischen Moleküls mit einem aktivierten Radikal findet ebenfalls nur in den Mitochondrien der Leber statt, hingegen fehlt den Lebermitochondrien das Enzymsystem zur Oxydation von Acetoacetat. Die Konversion von Citrullin zu Arginin kann nur in den Mitochondrien der Leber und Niere erfolgen (LEUTHARDT 1952). Es können aber auch im selben Gewebe Mitochondrien unterschiedlicher Leistung nebeneinander vorkommen, so z. B. im Muskel. Nach LENIQUE (1953) spielen die stabförmigen Mitochondrien eine Rolle beim Lipidstoffwechsel, die kleinen kugelförmigen („Sarcosome", WATANABE und WILLIAMS 1951) dienen als ATP-Quelle für die Muskelkontraktion. In diesem Fall ist mit der physiologischen Differenzierung auch eine morphologische verbunden, beide Formen sollen aus kleinen Mitochondrien des jungen Muskels hervorgegangen sein. Ob und wieweit physiologische Differenzierungen sich in Morphologie und Feinstruktur ausprägen, wird in Zukunft geprüft werden müssen. Durch WATSON (1953) sind für die Spermatogonien der Ratte bereits zwei unterschiedliche Mitochondrientypen beschrieben worden, deren Funktion allerdings noch unbekannt ist. Für pflanzliche Objekte liegen nur sehr wenige enzymologische Arbeiten und Studien über die Feinstruktur vor. Erst wenn ausreichendes Material vorhanden ist, wird sich die Frage klären lassen, ob die Sphärosomen PERNERs etwa Chondriosomen mit Spezialfunktion sind oder eine völlig andere Feinstruktur aufweisen und damit Zellorganelle besonderer Art sind. Ob pflanzliche Chondriosomen die Fähigkeit zur Sekretion haben, wie es die Schule von ZOLLINGER für die tierischen annimmt, muß ebenfalls noch geprüft werden. STEINER und HEINEMANN (1954a u. b) scheinen den Partikeln mit Mitochondrienfunktion bei *Oospora lactis* eine solche Funktion zuschreiben zu wollen.

# 3. Feinstruktur der Chondriosomen.

Für das elektronenmikroskopische Studium der Chondriosomenfeinstruktur sind Totalpräparate aus Zellkulturen (PORTER, CLAUDE und FULLAM 1945) und von isolierten Chondriosomen aus Homogenisaten (CLAUDE und FULLAM 1945, BUCHHOLZ 1947, DALTON und Mitarbeiter 1949, MÜHLETHALER, MÜLLER und ZOLLINGER 1950, PERNER 1952c, PERNER und PFEFFERKORN 1953, EICHENBERGER 1953, WEBER 1954) wegen zu großer Objektdicke und der dadurch bedingten Undurchstrahlbarkeit ungeeignet. Bei ersteren erscheinen die Chondriosomen durch Dichteunterschiede inhomogen, bei letzteren wird erst dann eine Membran deutlich, wenn die isolierten Chondriosomen längere Zeit der Extraktion durch wäßerige Lösungen unterlegen waren. Eine Darstellung der Chondriosomenmembran im Lichtmikroskop ist nicht möglich. Die Angaben von PERNER und PFEFFERKORN (1953, S. 101, 120, 127) beruhen auf der Mißdeutung eines im Dunkelfeld möglichen optischen Effekts (OETTLÉ 1950). Einblicke in die Feinstruktur der Chondriosomen wurden erst möglich durch die Verbesserung der Fixierungs- (BAKER und MODERN 1952, PALADE 1952a, PORTER und KALLMAN 1953), Einbettungs- (NEWMAN, BORYSKO und SWERDLOW 1949) und Schneideverfahren (LATTA und HARTMAN 1950, BIRBECK 1951, PALADE 1952a, SJÖSTRAND 1953a, PORTER und BLUM 1953, EAVES und FLEWETT 1954; weitere Literatur bei DALTON 1953; historische Übersicht bei BRETSCHNEIDER 1952). Die wesentlichen Untersuchungen wurden an verschiedenen Zelltypen des Säugetiergewebes (PALADE 1952b, hier weitere Literatur, PALADE 1953, SJÖSTRAND 1953b, SJÖSTRAND und RHODIN 1953, WATSON 1953) durchgeführt. Erweitert wurden sie durch vergleichende Untersuchungen an dem Gewebe von Vögeln, Amphibien, Mollusken und Anneliden (vgl. PALADE 1953) und Protozoen (*Euglena gracilis*, *Poteriochromonas stipitata* WOLKEN und PALADE 1953; *Tokophrya infusionum* RUDZINSKA und PORTER 1953). In neuerer Zeit (PALADE 1953) wurden die Untersuchungen auch mit Erfolg auf pflanzliches Gewebe (Blätter von „*Lemna viridis*" und *Nicotiana tabacum* [Abb. 5]) [PALADE 1953], Eizellen und Thallus von *Fucus vesiculosus* [LEYON und v. WETTSTEIN 1954, v. WETTSTEIN 1954]) ausgedehnt. Da nach PALADE (1953) bei allen bisher mit ausreichender Technik untersuchten Objekten Übereinstimmung besteht, erscheint es möglich, die hauptsächlich an tierischen Objekten gewonnenen Ergebnisse auf pflanzliche zu übertragen. Die von SPIRO (1953) beobachteten lamellenartigen und die von GLIMSTEDT und LAGERBERG (1953) gesehenen kabelähnlichen Binnenstrukturen können jedoch nicht in eindeutige Beziehung zu der von PALADE und SJÖSTRAND erarbeiteten Feinstruktur gesetzt werden (vgl. die Rekonstruktion bei LINDBERG und ERNSTER 1954, Fig. 6). Auch die von BEAMS und TAHMISIAN (1954) beobachtete Feinstruktur weicht vom Schema PALADEs ab und dürfte wohl einen Sonderfall darstellen. Die Ursache für die geringe Anzahl bisher untersuchter pflanzlicher Objekte dürfte in den Fixierungsschwierigkeiten liegen. Das sonst als optimales Fixierungsmittel für die Fixierung tierischer Objekte verwendete mit Veronal-Acetatpuffer gepufferte $OsO_4$ ($p_H$ 7,3—7,5 als Optimum, PALADE 1952a) dringt nur langsam ein (vgl. z. B. BRETSCHNEIDER 1950, ROZSA und WYCKOFF 1951). Bei älteren Arbeiten (z. B. der von BRETSCHNEIDER 1950, ROZSA und WYCKOFF 1951) war die Schnittdicke zu groß; anzustreben ist eine Schnittdicke unterhalb $0,1\,\mu$ (etwa $0,05\,\mu$, PALADE 1953). Nach PALADE (1953) zeigt das Chondriosom folgende Feinstruktur (vgl. auch Abb. 3, 4, 6 und 7): 1. Eine aus zwei Grenzlamellen bestehende Membran, deren innere Lamelle Einfaltungen, die Cristae mitochondriales (2.) bildet, und 3. eine Matrix genannte Binnensubstanz, die homogen erscheint und mit unseren heutigen Mitteln nicht weiter auflösbar ist. Bereits jetzt zeigt sich, daß das Verhältnis von Cristaezahl zu Matrixmasse in den Chondriosomen verschiedener Zelltypen variiert; so läßt sich zur Zeit folgende Reihe aufstellen: wenige Cristae, viel Matrix in parenchymatösen Leberzellen; vermehrte Zahl der Cristae, geringere Matrixsubstanz bei Drüsenzellen und viele Cristae mit wenig Matrixsubstanz im Nephronepithel in den gestreiften Muskelfasern, besonders in den Herzmuskelfasern (PALADE 1952b). In den Spermatogonien der Albino-Ratte wurden von WATSON (1953) zwei unterschiedliche Typen von Mitochondrien beobachtet.

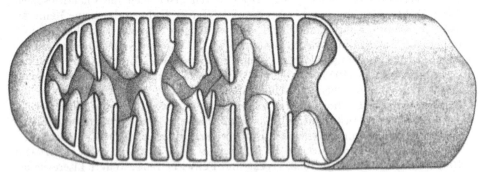

Abb. 3. Plastische Darstellung eines halben Chondriosoms, nach den Angaben von PALADE 1953 konstruiert Zur Demonstration der Binnenstruktur wurde das Chondriosom partiell angeschnitten.

### a) Die Chondriosomenmembran.

Die Dicke der Chondriosomenmembran variiert im elektronenmikroskopischen Bild von 7—25 mμ. Unterschiedliche Dicke bei Chondriosomen derselben Zelle ist bedingt durch die verschiedene Schnittrichtung, die geringste Dicke dürfte dabei dem echten Querschnitt entsprechen. Die im elektronenmikroskopischen Bild gemessene Membrandicke stimmt recht gut mit den mit anderer Methodik geschätzten Größen überein (MÜHLETHALER, MÜLLER und ZOLLINGER 1950: etwa 20 mμ; FARRANT, ROBERTSON und WILKINS 1953: 27 mμ). Die Membran erscheint dreikonturig, eine helle, homogene Mittelschicht wird beidseitig von einer Lamelle begrenzt. SJÖSTRAND und RHODIN (1953) möchten die 4,5 mμ dicken Begrenzungsmembranen für Proteinlamellen und die 7 mμ dicke Zwischenschicht für eine aus zwei bis vier Moleküllagen bestehende Lipoidschicht halten. Die Lipoidschicht hat durch den Einbettungsprozeß einen Teil ihres Lipoidgehaltes verloren und ist deswegen besser durchstrahlbar als die Begrenzungsmembran, deren Dichte durch das reduzierte Os bedingt ist. Falls Poren in der Membran vorkommen, müßten sie unterhalb des Auflösungsvermögens des Elektronenmikroskops liegen (vgl. dazu PALADE 1952b, 1953).

### b) Cristae mitochondriales.

Die Darstellung der Cristae mitochondriales ist ein Kontrastproblem. Sie sind im somatischen, tierischen Gewebe schlecht (da sie etwa die gleiche Dichte wie die Matrix haben), gut hingegen bei Spermatocyten und Spermatiden (PALADE 1953) und einigen Protozoen demonstrierbar, da sie hier dicker sind und der Kontrast zur hier helleren Matrix größer ist. PALADE 1953 (Abb. 6) konnte an einer quergeschnittenen Eindellung eines Chondriosoms (Epithelzelle des Nierenkanals der Ratte) nachweisen, daß die Cristae Einfaltungen der inneren Chondriosomemembran sind. Demgemäß erscheinen sie dreikonturig, ein hellerer, homogener Innenraum, der mit der Mittelschicht der Chondriosomenmembran in Verbindung steht, ist beidseitig von der gefalteten Innenmembran begrenzt

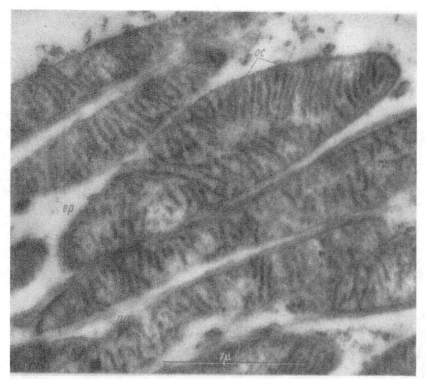

Abb. 4. Schnitte durch Mitochondrien des Epithels der proximalen, konvoluten Nierentubuli der Ratte. Das vollständig dargestellte Mitochondrium (ep) ist schräg geschnitten. In seiner Mitte ist der freie Kanal (fc) sichtbar. Die quergetroffenen Cristae sind deutlich dreikonturig (nc), schräggeschnittene (oc) sind an den Schnitträndern unscharf. In der Matrix einige dunklere Granula sichtbar. Vergrößerung: 38 500fach nach PALADE 1953.

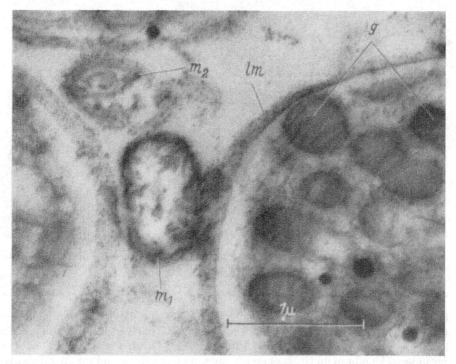

Abb. 5. *Nicotiana tabacum*. Ausschnitt aus einer Blattmesophyllzelle. Oben, links und rechts seitlich angeschnittene Chloroplasten. Im rechten Chloroplasten sind außer der Begrenzungsmembran (lm) die Grana (g) sichtbar. Die angeschnittenen Chondriosomen ($m_1$ und $m_2$) zeigen eine deutliche Begrenzungsmembran und Cristae. $m_1$ stellt einen Längsschnitt durch das Chondriosom dar. Vergrößerung: 36 800fach nach PALADE 1953.

(Abb. 3 und 7). Die Dicke der Falten beträgt im somatischen Gewebe 18—22 mμ, bei Spermatiden und Spermatocyten 70 mμ. Die gefaltete innere Membran ist 5, die lichte Zentralschicht 8 mμ dick (PALADE 1953, Epithel des Nierenkanals der

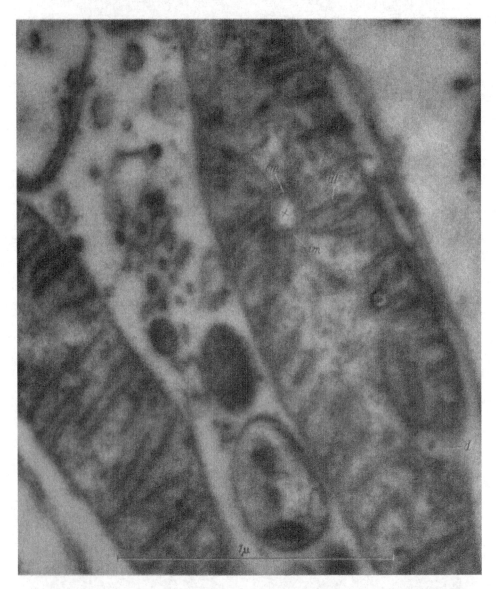

Abb. 6. Längsgeschnittenes Mitochondrium aus dem Epithel der proximalen, konvoluten Nierentubuli der Ratte mit zwei Eindellungen (*d* und *x*). Die Eindellung *d* ist längs, die Eindellung *x* quer geschnitten. An der quergetroffenen Eindellung wird deutlich, daß die innere Lamelle des Mitochondriums (*im*) in die Begrenzungsmembran der Cristae (*dl*) übergeht. Die äußere Membran (*em*) dürfte demnach eine kontinuierliche Lamelle darstellen, während die innere Membran (*im*) zu den Cristae eingefaltet ist. Vergrößerung: 74 500fach nach PALADE 1953.

Ratte), nach SJÖSTRAND und RHODIN (1953) beträgt die Dicke der Membran 4,5, die der Zwischenschicht 7 mμ. Die Werte sind gut vergleichbar, da gleiche Fixierung angewendet wurde (vgl. Abb. 6 und 7). Damit bewegt sich die Dicke der inneren Membran im Größenbereich eines monomolekularen Protein- oder eines bimolekularen Phospholipoidfilms. PALADE (1953) konnte an Hand von Serien-,

Quer- und medianen Längsschnitten nachweisen, daß die Falten keine Lamellen, sondern Septen sind, die verschieden weit in das Lumen des Chondriosoms

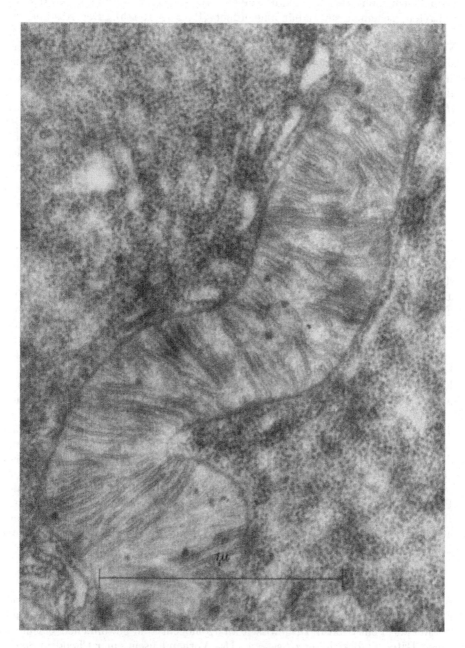

Abb. 7. Mitochondrien in einer exokrinen Zelle des Mauspankreas. Vergrößerung: 66 000fach. Nach SJÖSTRAND und HANZON 1954.

vorspringen, aber einen zentralen Kanal frei lassen (Abb. 3 und 4). Meist werden sie zu zweien oder vieren in einer Ebene des Chondriosoms gebildet, so daß sie also kein echtes Diaphragma darstellen. Sie entspringen mit breiter Basis und enden als Dreiecke oder Halbmonde (PALADE 1952 b, 1953). Meist sind die Cristae parallel

zueinander, quer zur Längsachse des Chondriosoms angeordnet. Parallelordnung
zur Längsachse wurde bei Neuriten und Dendriten beobachtet (Palade 1953). Bei
gequollenen, sphärischen Chondriosomen, wie sie bei Fixierung in saurem $p_H$
auftreten, scheinen sie radiär angeordnet (Palade 1952b). Der Abstand der
Cristae voneinander variiert von 17—47 m$\mu$ (Sjöstrand und Rhodin 1953).
Verzweigte Cristae werden beobachtet. Filamentartige Modifikationen der Cristae wurden bei Protozoen, und filament- und septenförmige Cristae nebeneinander bei Chondriosomen des Nierenkanalepithels gesehen (Palade 1953). Nach
Beams und Tahmisian (1954) zeigen die Mitochondrien der Samenzellen von
*Helix* eine von dem Paladeschen Schema abweichende Feinstruktur. Sie sollen
aus vier coaxial gestellten, ineinander geschachtelten Hohlzylindern bestehen. Der
äußerste Zylinder stellt die Mitochondrienmembran dar. Zwischen den Hohlzylindern und im Innern des kleinsten Hohlzylinders soll sich die Matrix befinden
(vgl. die plastische Darstellung bei Beams und Tahmisian). Die Dicke der
Hohlzylindermembran beträgt 400 Å.

### c) Die Matrix.

Der Raum innerhalb der Chondriosomenmembran, der durch die Einfaltungen gekammert ist, wird von einer strukturlos erscheinenden Substanz, der
Matrix, ausgefüllt (Palade 1953, Sjöstrand und Rhodin 1953). Sie erscheint
stets dichter als das Cytoplasma der Zelle. Die Dichte variiert mit dem Zelltyp
und ist abhängig vom $p_H$-Wert des Fixierungsmittels (Palade 1953) und der
Fixierungsdauer. So erscheint bei längerer Fixierungsdauer (24 Std) wahrscheinlich durch Extraktion die Matrix heller. Die Matrix ist im Leben als Flüssigkeit
aufzufassen, während den Chondriosomenmembranen etwa die Festigkeit eines
molekularen Filmes zukommt. Zuweilen (Chondriosomen der parenchymatösen
Leberzellen, Darm- und Nephronepithelzellen) sind in der Matrix dichtere
Körnchen mit dem Durchmesser von 20—30 m$\mu$ sichtbar (Palade 1952b, 1953).
Sjöstrand und Rhodin (1953) halten die von ihnen ebenfalls bis zu einem
Durchmesser von 70 m$\mu$ beobachteten Teilchen für Strukturelemente der Chon-
driosomen. Sie schließen dies aus dem ausschließlichen Vorkommen dieser Par-
tikelchen in den Chondriosomen und aus der durch sie verursachten Verbiegung
der Cristae. Die große Dichte dieser Teilchen wird von ihnen mit reduziertem
Os erklärt. Jedoch ist diese Angabe vorsichtig zu bewerten, da der Anteil des
reduzierten Os an der Dichte des Bildes nicht feststeht (vgl. die Diskussion des
Problems bei Palade 1952b). Auch Dalton (1953) erwähnt kleine kugelige
Partikel innerhalb der Chondriosomen; ähnliche osmiophile, runde Körper beobachteten Leyon und v. Wettstein (1954) bei Fucaceen. Ob eine Beziehung zu
Degradationsformen besteht, ist nicht geklärt.

### d) Beziehung der Feinstruktur zum Fermentsystem.

Die wichtige von Palade (1953) getroffene Feststellung, daß die Chondriosomen aus Homogenisaten zwar beschädigt, aber nicht ohne Feinstruktur sind,
erlaubt die Feinstruktur zu gewissen an Homogenisaten gewonnenen biochemischen Daten in Beziehung zu setzen. Das Vorhandensein einer Chondriosomenmembran war aus dem osmotischen Verhalten besonders der isolierten Chondriosomen (Claude 1946a und b, Hogeboom, Schneider und Palade 1948) und
aus der Tatsache gefolgert worden, daß der hohe Gehalt an löslichen Proteinen
(Hogeboom und Schneider 1950, 1951) und der Gehalt an löslichen Enzymen
(Palade 1951, de Duve und Mitarbeiter 1951, Kielley und Kielley 1951,
Schneider und Hogeboom 1952) bei Schädigung oder Zerstörung der Chondrio-

somen zurückgeht (weitere Diskussion bei PALADE 1952b). Die löslichen Enzyme dürften in der flüssigen Matrix lokalisiert sein und von der Chondriosomenmembran zurückgehalten werden. Es ist vorstellbar, daß die labileren Enzymsysteme, wie z. B. das Succinoxydasesystem der Chondriosomenoberfläche mehr genähert liegen als die stabileren (vgl. CLELAND und SLATER 1953). Die Membranfalten wären als Oberflächenvergrößerung aufzufassen, die die räumliche Trennung und koordinierte Anordnung der Enzyme eines Enzymsystems (SCHNEIDER und HOGEBOOM 1950) und damit die Stoffumsetzung in der richtigen Reihenfolge ermöglichen. Die Anordnung der Enzyme auf den Cristae wäre für die Teilung und Fragmentation der Chondriosomen günstig, da das Enzymsystem bei dieser Lokalisation weit weniger als bei Anordnung auf der Außenmembran geschädigt würde. Zur Stütze dieser Hypothese können zur Zeit folgende Punkte angeführt werden: 1. Für die Lokalisation der Enzyme in Membranen spricht die Tatsache, daß die Koordination der im Succinoxydasesystem beteiligten Enzyme verloren· geht, wenn die Chondriosomen mit Lecithinase A behandelt werden (NYGAARD und SUMNER 1953, NYGAARD, DIANZANI und BAHR 1954). Bei dieser Behandlung wird die Membranstruktur durch Hydrolyse der Phospholipoidmoleküle zerstört. Die am Enzymsystem beteiligten Einzelenzyme werden nicht beeinträchtigt (LANG 1953); jedoch wird der Ablauf einer längeren Reaktionskette unmöglich gemacht, weil die geordnete Bindung der Einzelenzyme verlorengegangen ist. Daß Grenzflächen eine Rolle spielen, dürfte aus der Umsatzerhöhung (bei Chondriosomen in vitro) nach elektrischer Reizung zu schließen sein (ABOOD, GERARD und OCHS 1952); lösliche Enzyme werden nicht beeinflußt. 2. Für die Lokalisation des Enzymsystems speziell in den Cristae sprechen folgende Punkte: a) das Auffinden von Partikelchen in der Größenordnung der Cristae in Suspensionen zerstörter Chondriosomen, die die meisten Enzyme des Succinoxydasesystems enthalten (HOGEBOOM und SCHNEIDER 1952); b) Experimente, aus denen hervorzugehen scheint, daß die Oxydation eines Stoffes erst nach seinem Eindringen in das Chondriosom eintritt (PETERS 1952, SCHNEIDER 1953).

## Mikrosomen (Sphärosomen).

Es ist nicht sicher, ob die PERNERschen Mikrosomen ein integrierender Bestandteil der pflanzlichen Zelle sind. In vielen Angiospermenzellen kommen sie regelmäßig vor. Für die Schizophyten fehlen Angaben; es muß allerdings darauf hingewiesen werden, daß die Bakteriengranula unterschiedliches Verhalten gegenüber Tetrazoliumsalzen und dem Nadi-Reagens zeigen (MUDD 1953a und b) und uneinheitlich zu sein scheinen. Die Angaben für Algen bedürfen der Überprüfung (vgl. PERNER 1953), bei Pilzen sind Hinweise für das Vorkommen von Mikrosomen vorhanden (GUILLIERMOND 1930, 1932 für *Saccharomycodes Ludwigii*). RITCHIE und HAZELTINE (1953) beschreiben für *Allomyces* nur Chondriosomen und Reservelipoide. Meistens werden die bei Hefe beobachteten Körper als Mitochondrien oder „Grana mit Mitochondrienfunktion" angesehen (BAUTZ und MARQUARDT 1953a und b, MARQUARDT und BAUTZ 1954a und b, SCHANDERL 1950, MEISSEL und Mitarbeiter 1950, HARTMAN und LIU 1953). STEINER und HEINEMANN (1954a und b) und STEINER (1954) halten bei *Oospora lactis* eine Metamorphose von Chondriosomen in Nadi-positive „Grana" für möglich, jedoch dürfen diese Partikel nicht ohne weitere Prüfung mit Mikrosomen gleichgesetzt werden. STEINER hält es für wahrscheinlich, daß die Nadi-positiven Partikel der „primäre Ort der Fettbildung" sind. BAUTZ und MARQUARDT (1953b) beobachteten, daß der Prozentsatz der Nadi-positiven Partikel gleichzeitig mit der Eigenschaftsänderung durch adaptive Enzyme verändert wird. HARTMAN und Liu (1953, dort weitere Literatur) sahen bei einer atmungsgehemmten Hefemutante „Grana", die keine positive Nadi-Reaktion, keine Reduktion von Janusgrün B und nur geringe Reaktion mit Tetrazoliumsalzen gaben. Demnach verhalten sich morphologisch gleiche Strukturen bei Enzymverlust oder Enzymadaptation den genannten Agentien gegenüber verschieden. Gleichzeitig zeigt sich die ganze Problematik einer auf solchen Reaktionen beruhenden Einteilung morphologischer Strukturen.

Nach Perner und Pfefferkorn (1953) variiert die Größe der bei *Allium cepa* beobachteten Mikrosomen von 0,5—0,8 $\mu$. Die abweichenden Angaben von Perner (1952a, S. 47) erklären sich wohl nur dadurch, daß die Messungen von 2—4 $\mu$ Durchmesser nach der Nadi-Reaktion und nach Indophenolanlagerung vorgenommen wurden. Diplostadien wurden beobachtet (Perner 1953), jedoch keine Durchteilung, damit ist die Autoreproduktion cytologisch bisher unbewiesen. Nach Perner (1953) sollen die Mikrosomen (Sphärosomen) eine Lipoid- und eine Proteinkomponente enthalten, Nucleoproteide werden vermutet. Da nur an fixiertem Material eine Fettreaktion mit Sudanfarbstoffen, Rhodamin B, Phosphin R, Nilblausulfat u. a. gelingt, nimmt Perner (1952a und 1953) an, daß im lebenden Zustand die Lipoidkomponente maskiert ist. Eine genauere Analyse dieser Lipoidkomponente wurde durch Perner (1953) versucht: durch Acetonbehandlung von fixiertem Material soll eine Fällung von Phosphatiden erfolgen. Extrahiert man die so behandelten Mikrosomen nachträglich mit Xylol, so färben sich nur die Grenzflächen mit Sudanschwarz und Rhodamin B an. Diese Grenzfläche wird für eine Phosphatidschicht gehalten, in derem Innern sich die löslichen Lipoide befinden sollen. Auch von Slautterback (1952 und 1953) wird auf Grund elektronen-optischer Untersuchungen das Vorhandensein einer Membran für die größten Partikelchen seiner Mikrosomenfraktion für möglich gehalten. Diese Partikelchen mit einer Durchschnittsgröße von 129 m$\mu$ reichen durch ihre Variationsbreite von 40—330 m$\mu$ etwa in den Bereich der Pernerschen Sphärosomen; Slautterback nimmt an, daß sie eine Proteinkomponente enthalten. Aus ihnen sollen nach Zerreißen der Membran die kleinsten in seiner Mikrosomenfraktion beobachteten Teilchen (Durchschnittsgröße 22,2 m$\mu$) ausgestoßen werden. Die von ihm beobachteten intermediären (Durchschnittsgröße 78,7 m$\mu$) und die genannten kleinsten Partikel dürften im Bereich der Submikrosomen (Frey-Wyssling) liegen, sie sollen lipoidhaltig sein oder zum mindesten ungesättigte Bindungen enthalten. Es müßte gesichert werden, daß diese Partikel reelle Strukturen und keine Artefakte sind. Novikoff und Mitarbeiter (1953) wollen durch Fraktionierung drei phako-optisch und biochemisch unterscheidbare Mikrosomengruppen analysiert haben (vgl. auch die Einwände von Schneider 1953). Zur größenmäßigen Begrenzung morphologischer Strukturen muß gesagt werden, daß durch das Lichtmikroskop jeweils nur ein willkürlicher Teilausschnitt der vorhandenen Strukturen erfaßt wird, ohne daß bekannt ist, ob diese Strukturen noch Vorläufer im lichtmikroskopisch nicht mehr sichtbaren Bereich haben. Auf Grund des positiven Ausfalls der Nadi-Reaktion kommt Perner (1952a) zu der Auffassung, daß die Cytochromoxydase in den Mikrosomen, genauer an deren Oberflächenstrukturen lokalisiert sei. Webster (1952) findet die Cytochromoxydase in nicht definierten Partikeln eines pflanzlichen Homogenisats, McClendon (1953) in Chondriosomen und anderen Partikeln des Homogenisats. Der Nachweis der Cytochromoxydase ist von Perner sicherlich geführt worden, jedoch muß man hinsichtlich deren Lokalisation vorsichtig sein, solange nicht die Bedenken, die wegen der lipophilen Speicherung des Indophenols bestehen (vgl. auch McClendon 1953), aus dem Wege geräumt sind. Ziegler (1953a und b), der in den Mikrosomen

Formazanspeicherung beobachtete, läßt die Frage, ob der Speicherungsort mit dem Reduktionsort isotop ist, offen. PERNER sieht eine weitere Stütze seiner Auffassung in der durch Berberin bedingten Atmungshemmung (MEISSEL und Mitarbeiter 1950, HILWIG und SCHMITZ 1951, SCHMITZ 1951 und PERNER 1953) und in der elektiven Speicherung von Berberinsulfat in den Mikrosomen. MAR-QUARDT und BAUTZ (1954b) beobachteten eine strenge Parallelität zwischen dem durch Atmungsgifte beeinflußten Atmungsverhalten und dem Ausfall der Nadi-Reaktion. Die Möglichkeit, daß Mikrosomen als Statolithen funktionieren könnten, wurde von ZIEGLER (1953a und b und in litteris 9. 4. 54) angedeutet, bedarf jedoch noch einer genaueren Untersuchung.

Abgeschlossen Juni 1954.

# Literatur.
## Zusammenfassende Darstellungen.
### Chondriosomen.

Allgemein: GUILLIERMOND 1932, NEWCOMER 1940, 1951.

Phasenkontrast-optische Untersuchung: ZOLLINGER 1948 u. 1950b, insbesondere Mikrofilm: CHÈVREMONT u. FRÉDERIC 1951a u. 1952, FRÉDERIC u. CHÈVREMONT 1952.

Janusgrün B-Färbung: COOPERSTEIN u. LAZAROW 1953, LAZAROW u. COOPERSTEIN 1953a, SHOWACRE 1953.

Nadi-Reagens: PERNER 1952a.

Tetrazoliumsalze: ANTOPOL, GLAUBACH u. GOLDMAN 1950, BRODIE u. GOTS 1951, 1952b, KUN 1951, PARKER 1953, SHELTON u. SCHNEIDER 1952, ZIEGLER 1953b.

Technik zur elektronenmikroskopischen Untersuchung: BIRBECK 1951, DALTON 1953, PALADE 1952a, insbesondere Gefriertrocknung: BELL 1952, GUSTAFSSON 1953, NEUMANN 1952, SJÖSTRAND 1952, WILLIAMS 1953, MOBERGER, LINDSTRÖM u. ANDERSSON 1954, Färbung: BAHR u. MOBERGER 1954.

Feinstruktur: PALADE 1952b, 1953, SJÖSTRAND 1953b.

Chondriosomenäquivalente der Bakterien: CHAPMAN u. HILLIER 1953, MUDD 1953a u. b, insbesondere Karyoidsystem: BRINGMANN 1952a u. b.

Homogenisatverfahren: HOGEBOOM, SCHNEIDER u. PALADE 1948, PERNER u. PFEFFER-KORN 1953, SCHNEIDER u. HOOGEBOOM 1951.

Funktion der Chondriosomen: CHESSIN 1951, CLAUDE 1950, DOUNCE 1950, GREEN 1951a u. b, KALCKAR 1942, MILLERD u. BONNER 1953 (speziell pflanzliche Chondriosomen), LANG 1952, 1953, SCHNEIDER 1953, insbesondere Energieübertragung: LEHNINGER 1952, LINDBERG u. ERNSTER 1952, 1954, BONNER u. MILLERD 1953.

### Mikrosomen (Sphärosomen).
DANGEARD 1947, PERNER 1953.

ABOOD, L. G., R. W. GERARD and S. OCHS: Electrical stimulation of metabolism of homogenates and particulates. Amer. J. Physiol. **171**, 134—139 (1952). — ALLARD, C., R. MATHIEU, G. DE LAMIRANDE and A. CANTERO: Mitochondrial population in mammalian cells. I. Description of a counting technic and preliminary results on rat liver in different physiological and pathological conditions. Cancer Res. **12**, 407—412 (1952). — ALTMANN, R.: Die Elementarorganismen und ihre Beziehungen zu den Zellen. Leipzig 1890. — ANDERSON, L. E.: Mitochondria in the life cycles of certain higher plants. Amer. J. Bot. **23**, 490—500 (1936). — Cytoplasmic inclusions in the male gametes of Lilium. Amer. J. Bot. **26**, 761—766 (1939). — ANITSCHKOW, N.: Über Quellungs- und Schrumpfungserscheinungen an Chondriosomen. Arch. mikrosk. Anat. **97**, 1—14 (1923). — ANTOPOL, W., S. GLAUBACH and L. GOLDMAN: The use of neotetrazolium as a tool in the study of active cell processes. Trans. New York Acad. Sci. **12**, 156—160 (1950).

BAHR, G. F., and G. MOBERGER: Methyl-mercury-chloride as a specific reagent for proteinbound sulfhydryl groups. Electron stains II. Exper. Cell. Res. **6**, 506—518 (1954). — BAKER, R. F., and F. W. S. MODERN: Controlled fixation with osmium tetroxide. Anat. Rec. **114**, 181—187 (1952). — BANGA, I., K. LAKI u. A. SZENT-GYÖRGYI: Über die Oxydation der Milchsäure und der Oxybuttersäure durch den Herzmuskel. Physiol. Chem. **217**, 43 bis 53 (1933). — BARNUM, C. P., and R. A. HUSEBY: Some quantitative analyses of the particulate fractions from mouse liver cell cytoplasm. Arch. of Biochem. **19**, 17—23 (1948). —

Bartels, F.: Studien an Leukoplasten der *Vicia Faba*-Wurzeln. Diss. Münster 1954. — Bartley, W., u. R. E. Davies: Secretory activity of mitochondria. Biochemic. J. **52**, XX—XXI (1952). — Bautz, E., u. H. Marquardt: Die Grana mit Mitochondrienfunktion in Hefezellen. Naturwiss. **40**, 531 (1953a). — Das Verhalten oxydierender Fermente in den Grana mit Mitochondrienfunktion der Hefezellen. Naturwiss. **40**, 531—532 (1953b). — Beams, H. W., and T. N. Tahmisian: Structure of the mitochondria in the male germ cells of *Helix* as revealed by the electron microscope. Exper. Cell Res. **6**, 87—93 (1954). — Bell, L. G. E.: The application of freezing and drying techniques in cytology. Internat. Rev. Cytol. **1**, 35—63 (1952). — Bensley, R. R.: On the so-called Altmann granules in normal and pathological tissues. Trans. Chicago Path. Soc. **8**, 78—83 (1910). — Studies on the pancreas of the guinea pig. Amer. J. Anat. **12**, 297—388 (1911). — Bensley, R. R., and S. H. Bensley: Handbook of histological and cytological technique. Chicago: Univ. Chicago Press 1938. — Bensley, R. R., and I. Gersh: Studies on cell structure by the freezing-drying method. Anat. Rec. **57**, 205—217 (1933). — Bensley, R. R., and N. L. Hoerr: Studies on cell structure by the freezing-drying method. VI. The preparation and properties of mitochondria. Anat. Rec. **60**, 449—455 (1934). — Berthet, J., L. Berthet, F. Appelmans and C. de Duve: Tissue fractionation studies. 2. The nature of the linkage between acid phosphatase and mitochondria in rat-liver tissue. Biochemic. J. **50**, 182—189 (1951). — Berthet, J., and C. de Duve: Tissue fractionation studies. 1. The existence of a mitochondria-linked, enzymically inactive form of acid phosphatase in rat-liver tissue. Biochemic. J. **50**, 174—181 (1951). — Bielig, H.-J., G. A. Kausche u. H. Haardick: Über den Nachweis von Reduktionsorten in Bakterien. Z. Naturforsch. **4b**, 80—91 (1949). — Über den Zusammenhang von TTC-Reduktionsorten und sog. Nucleoiden bei Bakterien. Naturwiss. **39**, 354 (1952). — Birbeck, M. S. C.: Histological techniques for the electron microscope. J. Roy. Microsc. Soc., Ser. III, **71**, 421—428 (1951). — Bisset, K. A.: The interpretation of appearances in the cytological staining of bacteria. Exper. Cell Res. **3**, 681—688 (1952a). — Bacterial cytology. Internat. Rev. Cytol. **1**, 93—106 (1952b). — Do bacteria have mitotic spindles, fusion tubes and mitochondria? J. Gen. Microbiol. **8**, 50—57 (1953a). — Bacterial cell envelopes. Symposium citologia batterica Rom 1953b, S. 9—17. — The cytology of *Micrococcus cryophilus*. J. Bacter. **67**, 41—44 (1954). — Bisset, K. A., and C. M. F. Hale: Complex cellular structure in bacteria. Exper. Cell Res. **5**, 449—454 (1953). — Blakley, P. L.: The metabolism and antiketogenic effects of sorbitol. Sorbitol dehydrogenase. Biochemic. J. **49**, 257—271 (1951). — Böing, J.: Beiträge zur Entwicklungsgeschichte der Chloroplasten. Diss. Münster 1954. — Bonner, J., and A. Millerd: Oxydative phosphorylation by plant mitochondria. Arch. of Biochem. a. Biophysics **42**, 135—148 (1953). — Bourne, G. H.: Cytology and cell physiology. Oxford: Clarendon Press 1951. — Brachet, J.: Biochemical and physiological interrelations between nucleus and cytoplasm during early development. Growth Symp. **11**, 309—324 (1947). — Bram, A.: Zum Verhalten der Mitochondrien bei Einwirkung verschiedener Pharmaka. Acta anat. (Basel) **13**, 385—401 (1951). — Brenner, S.: The demonstration by supravital dyes of oxydation-reduction systems on the mitochondria of the intact rat lymphocyte. S. Afric. J. Med. Sci. **14**, 13—19 (1949). — Supravital staining of mitochondria with amethyst violet. Stain Technol. **25**, 163—164 (1950). — Supravital staining of mitochondria with phenosafranin dyes. Biochim. et Biophysica Acta (Amsterdam) **11**, 480—486 (1953). — Bretschneider, L. H.: Elektronenmikroskopische Untersuchung der Pflanzenzellen. Proc., Kon. nederl. Akad. Wetensch. **53**, 1476—1489 (1950). — The electron-microscopic investigation of tissue sections. Internat. Rev. Cytol. **1**, 305—322 (1952). — Bringmann, G.: Vergleichende licht- und elektronenmikroskopische Untersuchungen an Oscillatorien. Planta (Berl.) **38**, 541—563 (1950). — Über Beziehungen der Kernäquivalente von Schizophyten zu den Mitochondrien höher organisierter Zellen. Planta (Berl.) **40**, 398—406 (1952a). — Die Organisation der Kernäquivalente der Spaltpflanzen unter Berücksichtigung elektronenmikroskopischer Befunde. Zbl. Bakter. II, **107**, 40—70 (1952b). — Erläuterungen zum Karyoidsystem von *Corynebacterium diphtheriae*. Zbl. Bakter. I Orig. **159**, 424—427 (1953). — Zur Interpretation elektronenmikroskopischer Bakterienbilder insbesondere der *Coli*-Gruppe. Z. Hyg. **139**, 155—159 (1954). — Brodie, A. F., and J. S. Gots: Effects of an isolated dehydrogenase enzyme and flavoprotein on the reduction of triphenyltetrazolium chloride. Science (Lancaster, Pa.) **114**, 40—41 (1951).— Bacterial DPN-cytochrome c reductase. Federat. Proc. **11**, 191 (1952a). — The reduction of tetrazolium salts by an isolated bacterial flavoprotein. Science (Lancaster, Pa.) **116**, 588—589 (1952b). — Brown, G. L., S. Fitton Jackson and J. Chayen: Cytoplasmic particles in bean root cells. Nature (Lond.) **171**, 1113—1114 (1953). — Brummond, D. O., and R. H. Burris: Transfer of $C^{14}$ by lupine mitochondria through reactions of the tricarboxylic acid cycle. Proc. Nat. Acad. Sci. U.S.A. **39**, 754—759 (1953). — Buchholz, J. T.: Methods in the preparation of chromosomes and other parts of cells for examination with an electron microscope. Amer. J. Bot. **34**, 445—454 (1947). — Buvat, R.: Die Ursache und Deutung der Chondriosomenumwandlungen. Endeavour **12**, 33—37 (1953).

CALCUTT, G.: The supravital staining of normal and malignant tissue with tetrazolium compounds. Brit. J. Canc. 6, 197—199 (1952). — CERUTI, A.: L'azione di alcuni cationi e dell'acqua sul condrioma isolato in vitro. Atti Accad. naz. Lincei, Rend. Cl. Sci. Fis., Mat. e Nat., Ser. 8 A, 5, 452—460 (1948). — Intensità e resistenza nel processo della contrazione del condrioma. Nuovo Giorn. bot. ital., N. s. 56, 251—253 (1949). — CHANTRENNE, H.: Hétérogénéité des granules cytoplasmiques du foie de souris. Biochim. et Biophysica Acta 1, 437—448 (1947). — CHAPMAN, G. B., and J. HILLIER: Electron microscopy of ultra-thin sections of bacteria. J. Bacter. 66, 362—373 (1953). — CHAYEN, J.: Pectinase technique for isolating plant cells. Nature (Lond.) 170, 1070—1072 (1952). — Ascorbic acid and its intracellular localization with special reference to plants. Internat. Rev. Cytol. 2, 78—131 (1953). —CHAYEN, J., and U. J. MILES: The preservation and investigation of plant mitochondria. Quart. J. Microsc. Sci. 94, 29—35 (1953). — CHAYEN, J., and K. P. NORRIS: Cytoplasmic localization of nucleic acids in plant cells. Nature (Lond.) 171, 472—473 (1953).— CHESSIN, R. W.: Die Isolierung der Cytoplasmagranula, ihr Bau und ihre Rolle im interzelligen Stoffwechsel. Sowjet. Wiss. 4, 319 (1951). — CHÈVREMONT, M.: Contribution à l'études des microsomes. Bull. Acad. roy. Méd. Belg. VI. s. 15, 29—37 (1950). — CHÈVREMONT, M., et J. FRÉDERIC: Contribution à l'étude du chondriome par la microscopie et la microcinématographie en contraste de phase. C. r. Assoc. Anat. 66, 268—277 (1951a). — Recherches sur les chondriosomes en culture de tissues par la microcinématographie en contraste de phase. C. r. Soc. Biol. Paris 145, 1243—1244 (1951b). — Recherches sur le comportement du chondriome pendant la mitose. C. r. Soc. Biol. Paris 145, 1245—1247 (1951c). — Évolution des chondriosomes lors de la mitose somatique étudiée dans les cellules vivantes cultivées in vitro par microscopie et microcinématographie en contraste de phase. Archives de Biol. 63, 259—277 (1952). — CHRISTIANSEN, E. G.: Orientation of the mitochondria during mitosis. Nature (Lond.) 163, 361 (1949). — CHRISTIE, G. S., and J. D. JUDAH: Intracellular distribution of enzymes. Proc. Roy. Soc. Lond., Ser. B 141, 420—433 (1953). — CLAUDE, A.: Distribution of nucleic acid in the cell and the morphological constitution of cytoplasm. Biol. Symp. 10, 111—129 (1943). — Fractionation of mammalian liver cell by differential centrifugation. I. Problems, methods and preparation of extraction. J. of Exper. Med. 84, 51—59 (1946a). — Fractionation of mammalian liver cell by differential centrifugation. II. Experimental procedures and results. J. of Exper. Med. 84, 61—89 (1946b). — Studies on cell morphology and functions: methods and results. Ann. New York Acad. Sci. 50, 854—860 (1950). — CLAUDE, A., and E. F. FULLAM: An electron microscope study of isolated mitochondria. J. of Exper. Med. 81, 51—62 (1945). — CLELAND, K. W., and E. C. R. SLATER: Respiratory granules of heart muscle. Biochemic. J. 53, 547—556 (1953). — CONSTANTINESCU, D.: Sur l'évolution du chondriome du sac embryonnaire de Digitalis purpurea L. C. Acad. Sci. Paris 216, 206—207 (1943). — COOPERSTEIN, S. J., and A. LAZAROW: Reduction of Janus green by isolated enzyme systems. Biol. Bull. 99, 321 (1950). — Studies on the mechanism of Janus green B staining of mitochondria. III. Reduction of Janus green by isolated enzyme systems. Exper. Cell Res. 5, 82—97 (1953). — COOPERSTEIN, S. J., A. LAZAROW and J. W. PATTERSON: Studies on the mechanism of Janus green B staining of mitochondria. II. Reactions and properties of Janus green B and its derivates. Exper. Cell Res. 5, 69—82 (1953). — COPENHAVER, J. H., and H. A. LARDY: Oxidative phosphorylation pathways and yield in mitochondrial preparations. J. of Biol. Chem. 195, 225—238 (1952). — COWDRY, E. V.: The mitochondrial constitution of protoplasm. Contrib. Embryol. 8, 39 (1918). — Historical background of research on mitochondria. J. Histochem. a. Cytochem. 1, 183—187 (1953). — COWDRY, N. H.: A comparison of mitochondria in plant and animal cells. Biol. Bull. 33, 196—228 (1917). — Experimental studies on mitochondria in plant cells. Biol. Bull. 39, 188—206 (1920).

DALTON, A. J.: Electron microscopy of tissue section. Internat. Rev. Cytol. 2, 403—417 (1953). — DALTON, A. J., H. KAHLER, M. G. KELLY, B. G. LLOYD and M. J. STRIEBICH: Some observations of the mitochondria of normal and neoplastic cells with the electron microscope. J. Nat. Canc. Inst. 9, 439—449 (1949). — DANGEARD, P. A.: Sur la distinction du chondriome des auteurs en vacuome, plastidome et sphérome. C. r. Acad. Sci. Paris 169, 1005—1010 (1919). — Cytologie végétale et cytologie générale. Paris: Paul Lechevalier 1947. — Sur la destruction du chondriome dans les méristèmes radiculaires et sur la possibilité de sa restauration. C. r. Acad. Sci. Paris 230, 27—29 (1950a). — Nouvelles observations sur la régénération du chondriome dans les radicules. C. r. Acad. Sci Paris 230, 496—498 (1950b). — Recherches expérimentales sur le chondriome dans les radicules des Phanérogames. Botaniste (Paris) 35, 35—81 (1951a). — Observations sur la destruction du chondriome par la chaleur. C. r. Acad. Sci. Paris 232, 1274—1276 (1951b). — DANGEARD, P. A., et H. PARRIAUD: Action des solutions de chloral sur le chondriome des méristèmes végétaux. C. r. Acad. Sci. Paris 236, 260—262 (1953). — DANIELLI, J. F.: Establishment of cytochemical techniques. Nature (Lond.) 157, 755—757 (1946). — DANNEEL, R., u. E. GÜTTES: Über das Verhalten der Mitochondrien bei der Mitose der Mesenchymzellen des Hühnerembryos.

Naturwiss. **38**, 117—118 (1951). — Davidson, J. N.: The distribution of nucleic acids in tissues. Symposia Soc. Exper. Biol. **1**, 77—85 (1947). — Davis, J. C., L. C. Winterscheid, P. E. Hartman and St. Mudd: A cytological investigation of the mitochondria of three strains of *Salmonella typhosa*. J. Histochem. a. Cytochem. **1**, 123—137 (1953). — Duve, Chr. de, J. Berthet, L. Berthet and F. Appelmans: Permeability of mitochondria. Nature (Lond.) **167**, 389—390 (1951). — Deysson, G.: Fixation et coloration extemporanée du chondriome à l'acide d'un réactif à l'orceine trichloracétique. C. r. Acad. Sci. Paris **235**, 1529—1531 (1952). — Dianzani, M. U.: Histochemical detection with ditetrazolium chloride of some enzymatic activities in isolated mitochondria. Nature (Lond.) **171**, 125—126 (1953a). — On the osmotic behaviour of mitochondria. Biochim. et Biophysica Acta (Amsterdam) **11**, 353—367 (1953b). — Dounce, A. L.: Cytochemical foundations of enzyme chemistry. In J. B. Sumner u. K. Myrback, The enzymes, Bd. 1, S. 188—266. New York: Academic Press 1950. — The interpretation of chemical analysis and enzyme determination on isolated cell components. J. Cellul. a. Comp. Physiol. **39**, Suppl. 2, 43—74 (1952). — Drawert, H.: Vitale Fluorochromierung der Mikrosomen mit Nilblausulfat. Ber. dtsch. bot. Ges. **65**, 263—271 (1952). — Vitale Fluorochromierung der Mikrosomen mit Janusgrün, Nilblausulfat und Berberidinsulfat. Ber. dtsch. bot. Ges. **66**, 134—150 (1953). — Vitalfärbung der Plastiden von *Allium cepa* mit Coelestinblau. Ber. dtsch. bot. Ges. **67**, 33—42 (1954). — Drawert, H., u. H. Gutz: Zur vitalen Fluorochromierung der Mikrosomen mit Nilblau. Naturwiss. **40**, 152 (1953). — DuBuy, H. G., M. W. Woods and M. D. Lackey: Enzymatic activities of isolated normal and mutant mitochondria and plastids of higher plants. Science (Lancaster, Pa.) **111**, 572—574 (1950). — Duesberg, J.: Plastosomen, „apparato reticolare interno" und Chromidialapparat. Erg. Anat. **20**, 567—916 (1911). — Duesberg, J., u. H. Hoven: Observations sur la structure du protoplasme des cellules végétales. Anat. Anz. **36**, 96—100 (1910).

    Eaves, G., and T. H. Flewett: Cutting of sections for electron microscopy with a modified "Cambridge" rocking microtome. Exper. Cell Res. **6**, 155—161 (1954). — Eichenberger, M.: Elektronenmikroskopische Beobachtungen über die Entstehung der Mitochondrien aus Mikrosomen. Exper. Cell Res. **4**, 275—282 (1953).

    Farrant, J. L., R. N. Robertson and M. J. Wilkins: The mitochondrial membrane. Nature (Lond.) **171**, 401—402 (1953). — Fauré-Fremiet, E.: Étude sur les mitochondries des protozoaires et des cellules sexuelles. Archives Anat. microsc. **11**, 457—648 (1910). — Fréderic, J.: Rapports de la membrane nucléaire avec les nucléoles et les chondriomes. Étude sur le vivant et en contraste de phase. C. r. Soc. Biol. Paris **145**, 1913—1916 (1951). — Fréderic, J., et M. Chèvremont: Recherches sur les chondriosomes de cellules vivantes par la microscopie et la microcinématographie en contraste de phase (1. partie). Archives de Biol. **63**, 109—131 (1952). — Frey-Wyssling, A.: Submicroscopic morphology of protoplasm. Amsterdam, Houston, London u. New York: Elsevier Press 1953.

    Garrigues, R.: Action de la colchicine et du chloral sur la racine de *Vicia faba*. Rev. Cytol. et Cytophysiol. végét. **4**, 261—301 (1940). — Geitler, L.: Chromatophor, Chondriosomen, Plasmabewegung und Kernbau von *Pinnularia nobilis* und einigen anderen Diatomeen nach Lebendbeobachtungen. Protoplasma (Wien) **27**, 534—543 (1937). — Genèves, L.: Altérations réversibles des cellules du tubercule du *Cichorium intybus* L. (Varieté-Endive) sous l'effect du froid et du réchauffement. C. r. Acad. Sci. Paris **232**, 1132—1134 (1951). — Effects de l'élévation de temperature sur le chondriome dans le méristème de racines d'*Allium cepa* cultivées a 0⁰ C. C. r. Acad. Sci. Paris **234**, 358—360 (1952). — Glimstedt, G., u. S. Lagerstedt: Observations on the ultrastructure of isolated mitochondria from normal rat liver. Kgl. Fysiogr. Sällsk. Hdl., N. F. **64**, Nr 3 (1953). — Grave, G.: Studien über die Entwicklung der Chloroplasten bei *Agapanthus umbellatus*. Diss. Münster 1954. — Green, D. E.: The cyclophorase system. In Enzyme and enzyme systems edited by John T. Edsall, S. 15—46. Cambridge. Mass.: Harvard Univ. Press 1951a. — The cyclophorase complex of enzymes. Biol. Rev. **26**, 410—455 (1951b). — Organized enzyme systems. J. Cellul. a. Comp. Physiol. **39**, Suppl. 2, 75—111 (1952). — Guilliermond, A.: Sur la coexistence dans la cellule végétale de deux variétés distinctes de mitochondries. C. r. Soc. Biol. Paris **83**, 408—411 (1920a). — Sur l'évolution du chondriome dans la cellule végétale. C. r. Acad. Sci. Paris **170**, 194—197 (1920b). — Recherches sur l'évolution du chondriome pendant le développement du sac embryonnaire et des cellules mères des grains de pollen dans les Liliacées et sur la signification des formations ergastoplasmiques. Ann. Sci. natur. Bot., Sér. X, **6**, 1—52 (1924). — Culture d'un *Saprolegnia* en milieux nutritifs additionées de colorants vitaux; valeur de la méthode des colorations vitales. Bull. Histol. appl. **7**, 97—110 (1930). — La structure de la cellule végétale: les inclusions du cytoplasme et en particulier les chondriosomes et les plastes (Sammelreferat). Protoplasma (Wien) **16**, 291—337 (1932). — The cytoplasm of the plant cell. Waltham, Mass.: Chronica Botanica Co. 1941 u. 1948. — La coloration vitale des chondriosomes. Bull. Histol. appl. **17**, 225—237 (1949). — Guilliermond, A., et R. Gautheret: Recherches sur la coloration vitale des cellules végétales. Rev. gén. Bot. **52**, 5—268 (1940). — GUILLIER-MOND, A., G. MANGENOT et L. PLANTEFOL: Traité de Cytologie végétale. Paris: Le François 1933. — GUSTAFSON, TR.: Sea-urchin development in the light of enzymic and mitochondrial studies. J. Embryol. a. Exper. Morphol. **1**, 251—255 (1953). — GUSTAFSSON, B. E.: Apparatuses and technique for freezing-drying of tissues. Kgl. Fysiogr. Sällsk. Hdl., N. F. **64**, 3—9 (1953).

HAHN, F. E.: Über Kernäquivalente und Reduktionsorte in Bakterien. Naturwiss. **39**, 527—528 (1952). — HARMAN, J. W.: Studies on mitochondria. I. The association of cyclophorase with mitochondria. Exper. Cell Res. **1**, 382—393 (1950a). — Studies on mitochondria. II. The structure of mitochondria in relation to enzymatic acitivity. Exper. Cell Res. **1**. 394—402 (1950b). — The selective staining of mitochondria. Stain Technol. **25**, 69—72 (1950c). — HARMAN, J. W., and M. FEIGELSON: Studies on mitochondria. III. The relationship of structure and function of mitochondria from heart muscle. Exper. Cell Res. **3**, 47—58 (1952a). — Studies on mitochondria. IV. The cytological localization of mitochondria in heart muscle. Exper. Cell Res. **3**, 58—64 (1952b). — Studies on mitochondria. V. The relationship of structure and oxidative phosphorylation in mitochondria of heart muscle. Exper. Cell Res. **3**, 509—525 (1952c). — HARTMAN, P. E., and CH. LIU: Comparative cytology of wild *Saccharomyces* and a respirationally deficient mutant. J. Bacter. **67**, 77—85 (1954). — HARTMAN, P. E., ST. MUDD, J. HILLIER and E. H. BEUTLER: Light and electron microscopic studies of *Escherichia coli*-coliphage interactions. III. Persistence of mitochondria and reductase activity during infection of *Escherichia coli* 3 with $T_2$ Phage. J. Bacter. **65**, 706—714 (1953). — HATCH, W. R.: Gametogenesis in *Allomyces arbuscula*. Ann. of Bot. **49**, 623—649 (1935). — HEITZ, E., u. R. MALY: Zur Frage der Herkunft der Grana. Z. Naturforsch. 8b, 243—249 (1953). — HILWIG, I., u. H. SCHMITZ: Über Beziehungen zwischen Verfettung und Stoffwechselhemmung an Gewebekulturen durch Zusatz von Berberin. Naturwiss. **38**, 336 (1951). — HÖLSCHER, H. A.: Über die Reduktionsorte von Tetrazoliumsalzen in Tumorzellen. Naturwiss. **38**, 116—117 (1951). — HOERR, N. L.: Methods of isolation of morphological constituents of the liver cell. Biol. Symp. **10**, 185—231 (1943). — HÖRSTADIUS, S.: Influence of implanted micromeres on reduction gradients and mitochondrial distribution in developing sea-urchin eggs. J. Embryol. a. Exper. Morphol. **1**, 257—259 (1953). — HOFFMANN, H.: The cytochemistry of bacterial nuclear structures. An examination of the Feulgen nucleal, Piekarski-Robinow, and the ribonuclease technics. J. Bacter. **62**, 561—570 (1951). — HOGEBOOM, G. H.: Separation and properties of cell components. Federat. Proc. **10**, 640—645 (1951). — HOGEBOOM, G. H., and W. C. SCHNEIDER: Sonic disintegration of isolated liver mitochondria. Nature (Lond.) **166**, 302—303 (1950). — Proteins of liver and hepatoma mitochondria. Science (Lancaster, Pa.) **113**, 355—358 (1951). — Cytochemical studies. IV. Physical state of certain respiratory enzymes of mitochondria. J. of Biol. Chem. **194**, 513—519 (1952). — HOGEBOOM, G. H., W. C. SCHNEIDER and G. E. PALADE: Cytochemical studies of mammalian tissues. I. Isolation of intact mitochondria from rat liver; some biochemical properties of mitochondria and submicroscopic particulate material. J. of Biol. Chem. **172**, 619—635 (1948). — HORNING, E. S., and A. H. K. PETRIE: The enzymatic function of mitochondria in the germination of cereals. Proc. Roy. Soc. Lond., Ser. B **102**, 188—206 (1928). — HUENNEKENS, F. M.: Studies on the cyclophorase system. XV. The malic oxidase. Exper. Cell Res. **2**, 115—125 (1951).

ITO, T.: Studies on the integument of the silkworm, *Bombyx mori*. III. Mitochondria and Golgi elements in the hypodermis. Bull. Sericult. Exper. Sta. (Japan) **13**, 613—628 (1951).

JACKSON, K. L., E. L. WALKER and N. PACE: Centrifugal preparation of rat liver mitochondria free of microsomes. Science **118**, 136—137 (1953). — JANSSENS, F. A., E. VAN DE PUTTE et J. HELMSORTEL: Le chondriosome dans les champignons. Cellule **28**, 445—452 (1913). — JOHNSON, R. B., and W. W. ACKERMANN: A role of nuclei in oxidative phosphorylation. J. of Biol. Chem. **200**, 263—269 (1953).

KAJA, H.: Untersuchungen über die Kontinuität der Granastruktur in den Plastiden der Moose. Ber. dtsch. bot. Ges. **67**, 93—107 (1954). — KALCKAR, H. M.: Metabolic enzymes of mitochondria. Acta med. scand. (Stockh.) **142**, Suppl. 226, 615—621 (1952). — KELLENBERGER, E., u. L. HUBER: Contribution à l'étude des équivalents des mitochondries dans les bactéries. Experientia (Basel) **9**, 289—291 (1953). — KELLER, E. B.: Turnover of proteins of cell fractions of adult rat liver in vitro. Federat. Proc. **10**, 206 (1951). — KIELLEY, W. W., and R. K. KIELLEY: Myokinase and adenosine triphosphatase in oxidative phosphorylation. J. of Biol. Chem. **191**, 485—500 (1951). — KING, R. L., and H. W. BEAMS: Ultracentrifugation and cytology of *Spirillum volutans*. J. Bacter. **44**, 597—609 (1942). — KNAYSI, G., J. HILLIER and C. FABRICANT: The cytology of an avian strain of *Mycobacterium tuberculosis* studied with the electron and light microscopes. J. Bacter. **60**, 423—447 (1950). — KÖNIG, H., u. A. WINKLER: Über Einschlüsse in Bakterien und ihre Veränderung im Elektronenmikroskop. Naturwiss. **35**, 136—144 (1948). — KRIEG, A.: Fluoreszenzmikroskopischer

Nachweis der Speicherung von Berberin in Chondriosomenäquivalenten von Bakterien in vivo. Naturwiss. **41**, 19—20 (1954a). — Mikroskopische Untersuchungen in vivo an *Azotobacter*-Zellen. Naturwiss. **41**, 147 (1954b). — Nachweis von Kernäquivalenten bei Bakterien in vivo. IV. Mitteilung: *Corynebacterium* und *Mycobacterium*. Z. Hyg. **139**, 64—68 (1954c). — Krüger-Thiemer, E., u. A. Lembke: Zur Definition der Mykobakteriengranula. Naturwiss. **41**, 146—147 (1954). — Kuff, E. L., and W. C. Schneider: Intracellular distribution of enzyms. XII. Biochemical heterogeneity of mitochondria. J. of Biol. Chem. **206**, 677 bis 685 (1954). — Kun, E.: Mechanism of enzymatic reduction of triphenyltetrazoliumchloride. Proc. Soc. Exper. Biol. a. Med. **78**, 195—197 (1951). — Kun, E., and L. G. Abood: Colorimetric estimation of succinic dehydrogenase by triphenyltetrazolium-chloride. Science (Lancaster, Pa.) **109**, 144—146 (1949).

Laird, A. K., O. Nygaard and H. Ris: Separation of rat liver mitochondria into two morphologically and biochemically distinct subfractions. Cancer Res. **12**, 276—277 (1952). — Laird, A. K., O. Nygaard, H. Ris and A. D. Barton: Separation of mitochondria into two morphologically and biochemically distinct types. Exper. Cell Res. **5**, 147—160 (1953). — Lang, K.: Lokalisation der Fermente und Stoffwechselprozesse in den einzelnen Zellbestandteilen und deren Trennung. 2. Kolloquium der Dtsch. Ges. für Physiol. Chemie 1952, S. 24 bis 42. — Das Cyclophorase-System. Angew. Chem. **65**, 409—415 (1953). — Laties, G. G.: The dual role of adenylate in the mitochondrial oxidations of higher plants. Physiol. Plantarum (Copenh.) **6**, 199—214 (1953a). — Transphosphorylating systems as a controlling factor in mitochondrial respiration. Physiol. Plantarum (Copenh.) **6**, 215—225 (1953b). — The physical environment and oxidative and phosphorylative capacities of higher plant mitochondria. Plant Physiol. **28**, 557—575 (1953c). — Latta, H., and F. J. Hartmann: Use of a glass edge in thin sectioning for electron microscopy. Proc. Soc. Exper. Biol. a. Med. **74**, 436—439 (1950). — Lazarow, A.: The chemical structure of cytoplasm as investigated in Professor Bensley's laboratory during the past ten years. Biol. Symp. **10**, 9 (1943). — Lazarow, A., and S. J. Cooperstein: The reduction of Janus green by liver cell constituents and a proposed mechanism for the supravital staining of mitochondria. Biol. Bull. **99**, 321 bis 322 (1950). — Studies on the enzymatic basis for Janus green B staining reaction. J. Histochem. a. Biochem. **1**, 234—241 (1953a). — Studies of the mechanism of Janus green B staining of mitochondria. I. Review of the literature. Exper. Cell Res. **5**, 56—69 (1953b). — Lazarow, A., S. J. Cooperstein, and J. W. Patterson: The chemistry of Janus green staining of mitochondria. Anat. Rec. **103**, 482 (1949). — Lehmann, F. E., u. H. R. Wahli: Histochemische und elektronenmikroskopische Unterschiede im Cytoplasma der beiden Somatoplasten des Tubifexkeimes. Z. Zellforsch. **39**, 618—629 (1954). — Lehninger, A. L.: Die Rolle der Mitochondrien bei Oxydations- und Phosphorylierungsprozessen. Z. Naturforsch. **7b**, 256—260 (1952). — Lehninger, A. L., and E. P. Kennedy: The requirements of the fatty acid oxidase of rat liver. J. of Biol. Chem. **173**, 753—771 (1948). — Lembke, A., u. H. Ruska: Vergleichende mikroskopische und übermikroskopische Beobachtungen an den Erregern der Tuberkulose. Klin. Wschr. **19**, 217—220 (1940). — Lenique, P.: Étude sur l'évolution du chondriome au cours de la genèse du tissu musculaire et de sa dégénérescence provoquée par la colchicine. Ark. Zool. (Stockh.) Andra Ser. **5**, 289—296 (1953). — Lenique, P., S. Hörstadius and T. Gustafson: Change of distribution of mitochondria in animal halves of sea urchin eggs by the action of micromeres. Exper. Cell Res. **5**, 400—403 (1953). — Lettré, H., u. R. Lettré: Kern-Plasma-Mitochondrien-Relation als Zellcharakteristikum. Naturwiss. **40**, 203 (1953). — Leuthardt, F.: Uréogenèse cycle tricarboxylique et mitochondries. Symposium sur le cycle tricarboxylique. IIᵉ Congrès International de Biochimie Paris 1952, S. 89—93. — Levenbook, L.: The mitochondria of insect flight muscle. J. Histochem. a. Cytochem. **1**, 242—247 (1953). — Levitt, J.: Two methods of fractionating potato tuber proteins and some preliminary results with dormant and active tubers. Physiol. Plantarum (Copenh.) **5**, 470—484 (1952). — Investigations of the cytoplasmic particulates and proteins of potato tubers. I. Bond water and lipid contents. Physiol. Plantarum (Copenh.) **7**, 109—116 (1954a). — Investigations of the cytoplasmic particulates and proteins of potato tubers. II. Nitrogen, phosphorus and carbohydrate contents. Physiol. Plantarum (Copenh.) **7**, 117 bis 123 (1954b). — Lewis, M. R.: The destruction of *Bacillus radicicola* by the connective tissue cells of the chick embryo in vitro. Bull. Hopkins Hosp. **34**, 223—226 (1923). — Lewis, W. H., and M. R. Lewis: Behavior of cells in tissue cultures. In E. V. Cowdry, General cytology, S. 385—447. Chicago: Chicago Press 1924. — Lewitzky, G.: Über die Chondriosomen in pflanzlichen Zellen. Ber. dtsch. bot. Ges. **28**, 538—546 (1910). — Vergleichende Untersuchungen über die Chondriosomen in lebenden und fixierten Pflanzenzellen. Ber. dtsch. bot. Ges. **29**, 685—696 (1911a). — Die Chloroplastenanlagen in lebenden und fixierten Zellen von *Elodea canadensis* Rich. Ber. dtsch. bot. Ges. **29**, 697—703 (1911b). — Leyon, H.: The structure of chloroplasts. II. The first differentiation of the chloroplast structure in *Vallota* and *Taraxacum* studied by means of electron microscopy. Exper. Cell Res. **5**, 520—529 (1953). — Leyon, H., u. D. v. Wettstein: Der Chromatophoren-Feinbau bei den

Literatur. 609

Phaeophyceen. Z. Naturforsch. **9** b, 471—476 (1954). — LINDBERG, O., and L. ERNSTER: On the mechanism of phosphorylative energy transfer in mitochondria. Exper. Cell Res. **3**, 209—239 (1952). — Chemistry and physiology of mitochondria and microsomes. Protoplasmatologia **3**, A4 (1954). — LISON, L., u. I. FAUTREZ: L'étude physicochimique des colorants dans ses applications biologiques. Étude critique. Protoplasma (Wien) **33**, 116—151 (1939).

MAKAROV, P.: Analyse der Wirkung des Kohlenoxyds und der Cyanide auf die Zelle mit Hilfe der Vitalfärbung. Protoplasma (Wien) **20**, 530—554 (1934). — MANGENOT, G.: Sur les oosphères, les tubes polliniques et la fécondation chez le pin maritime. C. r. Acad. Sci. Paris **206**, 364—366 (1938). — MANGENOT, G., et L. EMBERGER: Sur les mitochondries dans les cellules animales et végétales. C. r. Soc. Biol. Paris **83**, 418—420 (1920). — MAPSON, L. W., F. A. ISHERWOOD and Y. T. CHEN: Biological synthesis of L-ascorbic acid: the conversion of L-galactono-$\gamma$-lactone into L-ascorbic acid by plant mitochondria. Biochemic. J. **56**, 21—28 (1954). — MARENGO, N. P.: A study of the cytoplasmatic inclusions during sporogenesis in *Onoclea sensibilis*. Amer. J. Bot. **36**, 603—613 (1949). — Formalin-pyrogallol as a fixative for plant cytoplasm. Stain Technol. **27**, 209—211 (1952). — MARQUARDT, H.: Die Natur der Erbträger im Zytoplasma. Ber. dtsch. bot. Ges. **65**, 198—217 (1952). — MARQUARDT, H., u. E. BAUTZ: Sprunghafte Änderungen des Verhaltens der Mitochondrien von Hefezellen gegenüber dem NADI-Reagens. Naturwiss. **41**, 121—122 (1954a). — Die Wirkung einiger Atmungsgifte auf das Verhalten von Hefe-Mitochondrien gegenüber der NADI-Reaktion. Naturwiss. **41**, 361—362 (1954b). — MASCRÉ, M., et G. DEYSSON: Recherches sur la fixation et la coloration du chondriome. Bull. Soc. bot. France **100**, 11—14 (1953). — MATTSON, A. M., C. O. JENSEN and R. A. DUTCHER: Triphenyltetrazoliumchloride as a dye for vital tissues. Science (Lancaster, Pa.) **106**, 294—295 (1947). — McCLENDON, J. H.: The intracellular localization of enzymes in tobacco leaves. I. Identification of components of the homogenate. Amer. J. Bot. **39**, 275—282 (1952). — The intracellular localization of enzymes in tobacco leaves. II. Cytochrome oxidase, catalase and polyphenol oxidase. Amer. J. Bot. **40**, 260—266 (1953). — MEISSEL, M. N., N. A. POMOSCHTNIKOWA u. J. u. M. SCHAWLOWSKI: Die Unterdrückung der Atmungstätigkeit der Zelle bei selektiver Blockierung der Chondriosomen. Ber. Akad. Wiss. UdSSR., N. s. **70**, 1065 (1950). — MEVES, F.: Über das Vorkommen von Mitochondrien bzw. Chondriomiten der Pflanzenzellen. Ber. dtsch. bot. Ges. **22**, 284—286 (1904). — MEYER, J.: Dédifférenciation cellulaire et clivage des chondriocontes lors de l'évolution des cellules nourricières des galles de Diastrophus Rubi Htg. sur la Ronce. C. r. Acad. Sci. Paris **234**, 463—464 (1952). — MICHAELIS, L.: Die vitale Färbung, eine Darstellungsmethode der Zellgranula. Arch. mikrosk. Anat. **55**, 558—575 (1900). — MICHAELIS, P.: Der Nachweis der Plasmavererbung (das Prinzip und seine praktische Durchführung beim Weidenröschen, *Epilobium*). Acta biotheor. (Leiden) **11**, 1—26 (1953). — Wege und Möglichkeiten zur Analyse des plasmatischen Erbgutes. Biol. Zbl. **73**, 353—399 (1954). — MILLERD, A.: Respiratory oxidation of pyruvate by plant mitochondria. Arch. of Biochem. a. Biophysics **42**, 149—163 (1953). — MILLERD, A., and J. BONNER: The biology of plant mitochondria. J. Histochem. a. Cytochem. **1**, 254—264 (1953). — MILLERD, A., J. BONNER, B. AXELROD and R. S. BANDURSKI: Oxidative and phosphorylative activity of plant mitochondria. Proc. Nat. Acad. Sci. U.S.A. **37**, 855—862 (1951). — MILOVIDOV, P. F.: Influence du radium sur le chondriome des cellules végétales. C. r. Soc. Biol. Paris **101**, 676—678 (1929). — Sur l'influence du radium sur le chondriome des végétaux inférieurs. Protoplasma (Wien) **10**, 297—299 (1930). — MOBERGER, G., B. LINDSTRÖM and L. ANDERSSON: Freeze-drying with a modified GLICK-MALMSTRÖM apparatus. Exper. Cell Res. **6**, 228—237 (1954). — MUDD, ST.: The mitochondria of bacteria. J. Histochem. a. Cytochem. **1**, 248—253 (1953a). — The mitochondria of bacteria. Symposium Citologia Batterica Rom 1953b. — MUDD, ST., E. H. BEUTNER, J. HILLIER and P. E. HARTMAN: Nuclei and mitochondria in *Escherichia coli* cells infected with T2 bacteriophage. J. Nat. Canc. Inst. **13**, 241—243 (1952). — MUDD, ST., A. F. BRODIE, L. C. WINTERSCHEID, P. E. HARTMAN and E. H. BEUTNER: Further evidence of the existence of mitochondria in bacteria. J. Bacter. **62**, 729—739 (1951a). — MUDD, ST., J. HILLIER, E. H. BEUTNER u. P. E. HARTMAN: Light and electron microscopic studies of *Escherichia coli*-coliphage interactions. II. The electron microscopic cytology of the E. Coli B-T2 system. Biochim. et Biophysica Acta (Amsterdam) **10**, 153—179 (1953). — MUDD, ST., and L. C. WINTERSCHEID: A note concerning bacterial cell walls, mitochondria and nuclei. Exper. Cell Res. **5**, 251—254 (1953). — MUDD, ST., L. C. WINTERSCHEID, E. D. DELAMATER and H. J. HENDERSON: Evidence suggesting that the granules of mycobacteria are mitochondria. J. Bacter. **62**, 459—475 (1951b). — MÜHLETHALER, K., A. F. MÜLLER u. H. U. ZOLLINGER: Zur Morphologie der Mitochondrien (Elektronenmikroskopische Untersuchungen). Experientia (Basel) **6**, 16—17 (1950). — MÜLLER, A. F., u. F. LEUTHARDT: Oxydative Phosphorylierung und Citrullinsynthese in den Lebermitochondrien. Helvet. chim. Acta **32**, 2349—2356 (1949). — MUNDKUR, B. D.: Mitochondrial distribution in *Saccharomyces*. Nature (Lond.) **171**, 793—797 (1953). — MURRAY,

R. G. E., and J. P. Truant: The morphology, cell structure and taxonomic affinities of the moraxelle. J. Bacter. 67, 13—22 (1954).

Nadson, G. A., et E. J. Rochlin: L'effect des rayons X sur le protoplasme, le noyau et le chondriome de la cellule végétale d'après les observations sur le vivant. Protoplasma (Wien) 20, 31—41 (1933). — Nagai, S.: Experimental studies on the reduction of silver nitrate by plant cell. I. Dynamic process in reduction and precipitation. J. Osaka City Univ. Inst. Polytech., Ser. D Biol. 1, 33—44 (1950). — Narahara, H. T., H. Quittner, L. Goldman and W. Antopol: The use of neotetrazolium in the study of E. coli metabolism. Trans. New York Acad. Sci. 12, 160—161 (1950). — Neumann, K.: Grundriß der Gefriertrocknung. Göttingen: Musterschmidt 1952. — Newcomb, E. H., and P. K. Stumpf: Fatty acid synthesis and oxidation in peanut cotyledons. Phophorus Metabolism, Bd. II, S. 291. Baltimore: Johns Hopkins Press 1952. — Newcomer, E. H.: Mitochondria in plants. Bot. Rev. 6, 85—147 (1940). — Concerning the duality of the mitochondria and the validity of the osmiophilic platelets in plants. Amer. J. Bot. 33, 684—697 (1946). — Mitochondria in plants. II. Bot. Rev. 17, 53—89 (1951). — Newman, S. B., E. Borysko and M. Swerdlow: Ultra-microtomy by a new method. J. Res. Nat. Bur. Stand. 43, 183—199 (1949). — Niemeyer, H., u. J. Jalil: Increased oxygen uptake of rat liver mitochondria produced by adenosinetriphosphatases. Biochim. et Biophysica Acta (Amsterd.) 12, 492—494 (1953). — Nordmann, J., et R. et O. Gauchery: L'utilization du chlorure de 2,3,5-triphenyltetrazolium pour l'étude de l'activité déshydrogénasique des mitochondries. Experientia (Basel) 8, 22—23 (1952). — Novikoff, A. B., E. Podber, J. Ryan and E. Noe: Biochemical heterogeneity of the cytoplasmic particles isolated from rat liver homogenate. Federat. Proc. 11, 265—266 (1952). — Biochemical heterogeneity of the cytoplasmic particles isolated from rat liver homogenate. J. Histochem. a. Cytochem. 1, 27—46 (1953). — Nygaard, A. P., M. U. Dianzani and G. F. Bahr: The effect of lecithinase A on the morphology of isolated mitochondria. Exper. Cell Res. 6, 453—458 (1954). — Nygaard, A., and B. J. Sumner: The effect of lecithinase A on the succinoxidase system. J. of Biol. Chem. 200, 723—729 (1953).

Oberling, Ch., W. Bernhard, H. L. Febure, J. Harel et R. Klein: L'existence d'un ultra-chondriome dans les cellules normales et tumorales. C. r. Acad. Sci. Paris 231, 1260—1262 (1950). — O'Brien, J. A.: Plastid development in the scutellum of Triticum aestivum and Secale cereale. Amer. J. Bot. 38, 684—696 (1951). — Oettlé, A. G.: „Optical membranes": a common artefact. J. Roy. Microsc. Soc., Ser. III 70, 255—265 (1950). — Opie, E. L.: Cytochondria of normal cells, of tumor cells and of cells with various injuries. J. of Exper. Med. 86, 45—54 (1947). — An osmotic system within the cytoplasm of cells. J. of Exper. Med. 87, 425—444 (1948). — Ortiz-Campos, R.: On the fixation and staining of the vegetable chondriome. An. Estación exper. Aula Dei 1, 63—82 (1948).

Paigen, K.: The occurrence of several biochemically distinct types of mitochondria in rat liver. J. of Biol. Chem. 206, 945—957 (1954). — Palade, G. E.: Intracellular distribution of acid phosphatase in rat liver cells. Arch. of Biochem. 30, 144—158 (1951). — A study of fixation for electron microscopy. J. of Exper. Med. 95, 285—298 (1952a). — The fine structure of mitochondria. Anat. Rec. 114, 427—452 (1952b). — An electron microscope study of the mitochondrial structure. J. Histochem. a. Cytochem. 1, 188—211 (1953). — Panijel, J., et J. Pasteels: Analyse cytochimique de certains phénomènes de recharge en ribonucléoprotéins: Le cas de l'œuf de „Parascaris equorum" lors de la fécondation. Archives de Biol. 62, 353—370 (1951). — Parker, J.: Some applications and limitations of tetrazolium chloride. Science (Lancaster, Pa.) 118, 77—79 (1953). — Paul, M. H., and E. Sperling: Cyclophorase system. XXIII. Correlation of cyclophorase activity and mitochondrial density in striated muscle. Proc. Soc. Exper. Biol. a. Med. 79, 352—354 (1952). — Payne, F.: Do mitochondria divide ? J. of Morph. 91, 555—567 (1952). — Pensa, A.: Alcuni formazioni endocellulari dei vegetali. Anat. Anz. 37, 325—333 (1910). — Penso, G., G. Castelnuovo e M. Princivalle: Studi e ricerche sui microbatteri. IX. La demolizione fagica del Mycobacterium minetti. Rend. Ist. sup. Sanità 15, 549—554 (1952). — Perner, E. S.: Zellphysiologische und zytologische Untersuchungen über den Nachweis und die Lokalisation der Cytochrom-Oxydase in Allium-Epidermiszellen. Biol. Zbl. 71, 43—69 (1952a). — Die Vitalfärbung mit Berberinsulfat und ihre physiologische Wirkung auf Zellen höherer Pflanzen. Ber. dtsch. bot. Ges. 65, 51—59 (1952b). — Über die Veränderungen der Struktur und des Chemismus der Zelleinschlüsse bei der Homogenisation lebender Gewebe. Ber. dtsch. bot. Ges. 65, 235—238 (1952c). — Die Sphärosomen (Mikrosomen) pflanzlicher Zellen. Sammelreferat unter Berücksichtigung eigener Untersuchungen. Protoplasma (Wien) 42, 457—481 (1953). — Zum mikroskopischen Nachweis des „primären Granums" in den Leukoplasten. Ber. dtsch. bot. Ges. 67, 26—32 (1954). — Perner, E. S., u. G. Pfefferkorn: Pflanzliche Chondriosomen im Licht- und Elektronenmikroskop unter Berücksichtigung ihrer morphologischen Veränderungen bei der Isolierung. Flora (Jena) 140, 98—129 (1953). — Peters, R. A.: Lethal synthesis. Proc. Roy. Soc. Lond., Ser. B 139, 143—170

(1952). — POLICARD, A., et G. MANGENOT: Action de la température sur le chondriome cellulaire; un critérium physique des formations mitochondriales. C. r. Acad. Sci. Paris 174, 645—647 (1922). — POLLISTER, A. W.: Mitochondrial orientations and molecular patterns. Physiologic. Zool. 14, 268—280 (1941). — PORTER, K. R., and J. BLUM: A study in microtomy for electron microscopy. Anat. Rec. 117, 685—710 (1953). — PORTER, K. R., A. CLAUDE and E. F. FULLAM: A study of tissue culture cells by electron microscopy. J. of Exper. Med. 81, 233—244 (1945). — PORTER, K. R., and F. KALLMAN: The properties and effects of osmium tetroxide as a tissue fixative with special reference to its use for electron microscopy. Exper. Cell Res. 4, 127—141 (1953). — POTTER, V. R.: The essay of animal tissues for respiratory enzymes. IV. Cell structure in relation to fatty acid oxidation. J. of Biol. Chem. 163, 437—446 (1946). — PRESSMAN, B. C., and A. H. LARDY: Influence of potassium and other alkali cations on respiration of mitochondria. J. of Biol. Chem. 197, 547—556 (1952). — PREUNER, R., u. J. v. PRITTWITZ UND GAFFRON: Über den Nachweis von Reduktionsorten in Bazillen und Bakterien und ihren Zusammenhang mit den sog. Nucleoiden. Naturwiss. 39, 128—131 (1952). — PUYTORAC, P. DE: Recherches sur l'action de l'eau et de l'acide acétique sur le chondriome de certaines cellules végétales. Botaniste (Paris) 35, 125—163 (1951).

RAAFLAUB, J.: Die Korrelation zwischen Struktur und Aktivität von isolierten Leber-mitochondrien. Helvet. physiol. Acta 10, 22—24 (1952). — Die Schwellung isolierter Leberzellmitochondrien und ihre physikalisch-chemische Beeinflußbarkeit. Helvet. physiol. Acta 11, 142—156 (1953). — RECKNAGEL, R. O.: Localization of cytochrome oxidase on the mitochondria of the frog egg. J. Cellul. a. Comp. Physiol. 35, 111—130 (1950). — REZENDE-PINTO, M. C. DE: Über die Genese und die Struktur der Chloroplasten bei den höheren Pflanzen. Ergebnisse und Probleme. Protoplasma (Wien) 41, 336—342 (1952). — RITCHIE, D. D.: Reactions of fungus mitochondria to environmental changes. Trans. New York Acad. Sci., Ser. II, 15, 157—158 (1953). — RITCHIE, D. D., and P. HAZELTINE: Mitochondria in Allomyces under experimental conditions. Exper. Cell Res. 5, 261—274 (1953). — ROBERTS, H. S.: Changes in mitochondrial form. Anat. Rec. 104, 163—187 (1949). — ROBINSON, E., and R. BROWN: Cytoplasmic particles in bean root cells. Nature (Lond.) 171, 313 (1953). — ROSZA, G., and R. W. G. WYCKOFF: The electron microscopy of onion root tip cells. Exper. Cell Res. 2, 630—641 (1951). — RUDZINSKA, M. A., and K. R. PORTER: An electron microscope study of protozoan. Tokophrya infusionum. Anat. Rec. 115, 363—364 (1953). — RÜTTIMANN, A.: Über Aufbraucherscheinungen und Neubildung der Mitochondrien in den Nierenhauptstücken nach Speicherung. Schweiz. Z. Path. u. Bakter. 14, 373—387 (1951). — RUSKA, H.: Morphologische Befunde bei der bakteriophagen Lyse. Arch. Virusforsch. (Wien) 2, 345—387 (1943). — RUSKA, H., G. BRINGMANN, I. NECKEL u. G. SCHUSTER: Über die Entwicklung sowie den morphologischen und zytochemischen Aufbau von Mycobacterium avium (Chester). Z. wiss. Mikrosk. 60, 425—447 (1952). — RUTENBERG, A. M., R. GOFSTEIN and A. M. SELIGMAN: Preparation of a new tetrazolium salt which yields a blue pigment on reduction and its use in the demonstration of enzymes in normal and neoplastic tissues. Cancer Res. 10, 113—121 (1950). — RYLAND, A. G.: A cytological study of effects of colchicine, indole-3-acetic acid, potassium cyanide, and 2,4-D on plant cells. J. Elisha Mitch. Sci. Soc. 64, 117—125 (1948).

SALTMANN, P.: Hexokinase in higher plants. J. of Biol. Chem. 200, 145—154 (1953). — ŠAPOT, V. S., i V. L. NEMZINSKAJA: Über die wirkliche Natur der sog. Struktureiweißstoffe. Dokl. Akad. Nauk. SSSR. 70, 465—468 (1950). — SARACHEK, A., and G. F. TOWNSEND: The disruption of mitochondria of Saccharomyces by ultraviolet irradiation. Science (Lancaster, Pa.) 117, 31—33 (1953). — SCHANDERL, H.: Über das Studium der Chondriosomen pflanzlicher Zellen intra vitam. Züchter 20, 65—76 (1950). — SCHERER, E., u. D. RINGLEB: Beobachtungen an den Mitochondrien der Mäuseascites-Tumorzellen unter der Einwirkung von Röntgenstrahlen. Strahlenther. 90, 34—40 (1953). — SCHMITZ, H.: Über die Speicherung des Berberins in den Granula von Tumorzellen. Naturwiss. 38, 405 (1951). — SCHNEIDER, W. C.: Intracellular distribution of enzymes. I. The distribution of succinic dehydrogenase, cytochrome oxidase, adenosinetriphosphatase and phosphorus compounds in normal rat tissues. J. of Biol. Chem. 165, 585—593 (1946). — Nucleic acids in normal and neoplastic tissues. Cold Spring Harbor Symp. Quant. Biol. 12, 169—178 (1947). — Intracellular distribution of enzymes. VI. The distribution of succinoxidase and cytochrome oxidase activities in normal mouse liver and in mouse hepatoma. J. Nat. Canc. Inst. 10, 969—975 (1950). — Biochemical constitution of mammalian mitochondria. J. Histochem. a. Cytochem. 1, 212—233 (1953). — SCHNEIDER, W. C., A. CLAUDE and G. H. HOGEBOOM: The distribution of cytochrome c and succinoxidase activity in rat liver fractions. J. of Biol. Chem. 172, 451—458 (1948). — SCHNEIDER, W. C., and G. H. HOGEBOOM: Intracellular distribution of enzymes. V. Further studies on the distribution of cytochrome c in rat liver homogenates. J. of Biol. Chem. 183, 123—128 (1950). — Cytochemical studies of mammalian tissues: The isolation of cell components by differential centrifugation. A review. Cancer Res. 11, 1—22 (1951). — Intracellular distribution of enzymes. X. Desoxyribonuclease and ribonuclease.

J. of Biol. Chem. **198**, 155—163 (1952). — Schneider, W. C., and V. R. Potter: Intracellular distribution of enzymes. IV. The distribution of oxalacetic oxidase activity in rat liver and rat kidney fractions. J. of Biol. Chem. **177**, 893—903 (1949). — Shaffer, J. M., C. J. Kucera and W. W. Spink: The protection of intracellular brucella against therapeutic agents and the bactericidal action of serum. J. of Exper. Med. **97**, 77—90 (1953). — Shelton, E., and W. C. Schneider: On the usefulness of tetrazolium salts as histochemical indicators of dehydrogenase activity. Anat. Rec. **112**, 61—82 (1952). — Shelton, E., W. C. Schneider and M. J. Striebich: A method for counting mitochondria in tissue homogenates. Anat. Rec. **112**, 388 (1952). — A method for counting mitochondria in tissue homogenates. Exper. Cell Res. **4**, 32—41 (1953). — Showacre, J. L.: A critical study of Janus green B coloration as a tool for characterizing mitochondria. J. Nat. Canc. Inst. **13**, 829—845 (1953).— Siekevitz, P.: Uptake of radioactive alanine in vitro into the proteins of rat liver fractions. J. of Biol. Chem. **195**, 549—565 (1952). — Siekevitz, P., and V. R. Potter: The adenylate kinase of rat liver mitochondria. J. of Biol. Chem. **200**, 187—196 (1953). — Sjöstrand, F. S.: Freezing and drying. Published by the Institute of Biology, London 1952. — A new microtome for ultrathin sectioning for high resolution electron microscopy. Experientia (Basel) **9**, 114—115 (1953a). — Electron microscopy of mitochondria and cytoplasmic double membranes. Nature (Lond.) **171**, 30—32 (1953b). — Sjöstrand, F. S., and J. Rhodin: The ultrastructure of the proximal convoluted tubules of the mouse kidney as revealed by high resolution electron microscopy. Exper. Cell Res. **4**, 426—456 (1953). — Slater, F. C., and K. W. Cleland: Stabilization of oxidative phosphorylation in heart-muscle sarcosomes. Nature (Lond.) **170**, 118—119 (1952). — Slautterback, D. B.: Electron microscopic study of small cytoplasmic particles (microsomes). J. Appl. Physics **23**, 163 (1952). — Electron microscopic studies of small cytoplasmic particles (microsomes). Exper. Cell Res. **5**, 173—186 (1953). — Somerson, N. L., and H. E. Morton: Reduction of tetrazolium salts by pleuropneumonia like organism. J. Bacter. **65**, 245—251 (1953). — Sorokin, H.: Mitochondria and plastids in living cells of *Allium cepa*. Amer. J. Bot. **25**, 28—33 (1938). — The distinction between mitochondria and plastids in living epidermal cells. Amer. J. Bot. **28**, 476—485 (1941). — Sorouri, P., and S. Mudd: Evidence of the existence of mitochondria in *Proteus*. Indian J. Med. Res. **41**, 49 (1953). — Spector, W. G.: Electrolyte flux in isolated mitochondria. Proc. Roy. Soc. Lond., Ser. B **141**, 268 bis 278 (1953). — Spiro, D.: Ultrastructure of mitochondria from insect flight muscle. Federat. Proc. **12**, 136—137 (1953). — Stafford, H. A.: Intracellular localization of enzymes in pea seedlings. Physiol. Plantarum (Copenh.) **4**, 696—741 (1951). — Steffen, K.: Zur Kenntnis des Befruchtungsvorganges bei *Impatiens glanduligera* Lindl. Cytologische Studien am Embryosack der Balsamineen. Planta (Berl.) **39**, 175—244 (1951). — Zytologische Untersuchungen an Pollenkorn und -schlauch. 1. Phasenkontrast-optische Lebenduntersuchungen an Pollenschläuchen von *Galanthus nivalis*. Flora (Jena) **140**, 140—174 (1953). — Cytological observations on endosperm haustories. 8ᵉ Congrès International de Bot. Paris 1954, Sect. 8, 250. — Steinberg, B.: Les modifications de la structure interne de *E. coli* B sous l'action d'antibiotiques. Schweiz. Z. Path. u. Bakter. **15**, 432—443 (1952). — Steiner, M.: Hinweise auf eine Längsverdoppelung von Mitochondrien bei *Oospora lactis*. Naturwiss. **41**, 191 (1954). — Steiner, M., u. H. Heinemann: Grana mit positiver Nadi-Reaktion als Ort der primären Fettbildung in Pilzzellen. Naturwiss. **41**, 40—41 (1954a). — Über die Beziehungen zwischen den fettbildenden Grana und typischen Mitochondrien in den Zellen von *Oospora lactis*. Naturwiss. **41**, 90 (1954b). — Strangeways, T. S. P., and R. G. Canti: The living cell in vitro as shown by dark-ground illumination and the changes induced in such cells by fixing reagents.. Quart. J. Microsc. Sci. **71**, 1—14 (1927). — Straub, F. B.: Isolation and properties of a flavoprotein from heart muscle tissue. Biochemic. J. **33**, 787—792 (1939). — Strugger, S.: Die Vitalfärbung des Protoplasmas mit Rhodamin B und 6 G. Protoplasma (Wien) **30**, 85—100 (1938). — Über den Bau der Proplastiden und Chloroplasten. Vorl. Mitt. Naturwiss. **37**, 166—167 (1950). — Über die Struktur der Proplastiden. Ber. dtsch. bot. Ges. **66**, 439—453 (1953). — Die Proplastiden in den jungen Blättern von *Agapanthus umbellatus* L'Herit. Protoplasma (Wien) **43**, 120—173 (1954a). — Der fluoreszenzmikroskopische Nachweis des primären Granums in den Proplastiden. Naturwiss. **41**, 286 (1954b). — Swanson, M. A., and C. Artom: The lipide composition of the large granules (mitochondria) from rat liver. J. Biol. Chem. **187**, 281—287 (1950).

Tandlar, C. J.: Localización de los ácidos grasos en celulas vegetales y la presencia de un unevo organoide protoplástico. Ciencia e Invest. (Buenos Aires) **8**, 44—46 (1952). — Theorell. H.: In J. B. Sumner u. K. Myrback, The enzymes, S. 335—356. New York: Academic Press 1951. — Tiegs, O. W.: Surface tension and the theory of protoplasmic movement. Protoplasma (Wien) **4**, 88—135 (1928).

Varitschak. K.: Contribution à l'étude du développement des ascomycètes. Thèse Paris 1931. — Vendrely-Randavel, C.: Sur la présence d'acide ribonucléique au niveau du chondriome. Acta anat. (Basel) **7**, 225—233 (1949).

WACHSTEIN, M.: Reduction of potassium telluride by living tissues. Proc. Soc. Exper. Biol. a. Med. **72**, 175—178 (1949). — WALLHÄUSSER, K. H.: Die Darstellung der Zellteilungsvorgänge und des Wirkungsmechanismus antibiotischer Stoffe an vital gefärbten Bakterien. Arzneimittel-Forsch. **4**, 118—124 (1954). — WATANABE, M. I., and C. M. WILLIAMS: Mitochondria in the flight muscles of insects. I. Chemical composition and enzymatic content. J. Gen. Physiol. **34**, 675—689 (1951). — WATSON, M. L.: Spermatogenesis in the adult albino rat as revealed by tissue sections in the electron microscope. Health a. Biol. **185**, 1—104 (1953). — WEATHERFORD, H. L.: Chondriosomal changes in connective tissue cells in the initial stages of acute inflammation. Z. Zellforsch. **17**, 518—541 (1933). — WEBER, R.: Strukturveränderungen an isolierten Mitochondrien von *Xenopus*-Leber. Z. Zellforsch. **39**, 630—640 (1954). — WEBSTER, G. C.: The occurrence of a cytochrome oxidase in the tissues of higher plants. Amer. J. Bot. **39**, 739—745 (1952). — Enzymatic synthesis of glutamine in higher plants. Plant Physiol. **28**, 724—727 (1953). — WEIBULL, C.: Observations on the staining of Bacillus megaterium with triphenyltetrazolium. J. Bacter. **66**, 137—139 (1953). — WEIER, T. E.: The cytology of leaf homogenates. Protoplasma (Wien) **42**, 260—271 (1953). — WESSEL, E.: Übermikroskopische Beobachtungen an Tuberkelbazillen vom Typus humanus. Z. Tbk. **88**, 22—36 (1942). — WETTSTEIN, D. v.: Formwechsel und Teilung der Chromatophoren von *Fucus vesiculosus*. Z. Naturforsch. **9b**, 476—481 (1954). — WILLIAMS, R. C.: A method of freeze-drying for electron microscopy. Exper. Cell Res. **4**, 188—201 (1953). — WINTERSCHEID, L. C.: The cytology of mitochondria and nuclei in mycobacteria. Diss. Univ. Pennsylvania 1953. — WINTERSCHEID, L. C., and ST. MUDD: The cytology of the tubercle bacillus with reference to mitochondria and nuclei. Amer. Rev. Tbk. **67**, 59—73 (1953). — WOLKEN, J. J., and G. E. PALADE: An electron microscope study of two flagellates. Chloroplast structure and variation. Ann. New York Acad. Sci. **56**, 873—889 (1953). — WOODS, M. W., and H. G. DU BUY: Evidence for the evolution of phytopathogenic viruses from mitochondria and their derivates. I. Cytological and genetical evidence. Phytopathology **33**, 637—655 (1943).

ZASTROW, E. M. v.: Über die Organisation der Cyanophyceenzelle. Arch. Mikrobiol. **19**, 174—205 (1953). — ZIEGLER, H.: Über die Bildung und Lokalisierung des Formazans in der Pflanzenzelle. Naturwiss. **40**, 144 (1953a). — Über die Reduktion des Tetrazoliumchlorids in der Pflanzenzelle und über den Einfluß des Salzes auf Stoffwechsel und Wachstum. Z. Naturforsch. **8b**, 662—667 (1953b). — ZOLLINGER, H. U.: Cytologic studies with the phase microscope. II. The mitochondria and other cytoplasmic constituents under various experimental conditions. Amer. J. Path. **24**, 569—589 (1948). — Zum quantitativen Nucleoproteingehalt und zur Morphologie der Mitochondrien. Experientia (Basel) **6**, 14—16 (1950a). — Les mitochondries (Leur étude à l'aide du microscope à contraste de phases). Rev. d'Hématol. **5**, 696—745 (1950b).

# Chemie des Zellsaftes.

Von

## Arthur Pisek.

Mit 2 Abbildungen.

## I. Der Vacuoleninhalt und seine Bedeutung im allgemeinen.

Pflanzenzellen und -organe wachsen in der Regel nicht so sehr durch Volumszunahme der lebenden Teile als vielmehr infolge gewaltiger, unter Wasseraufnahme vor sich gehender Aufblähung der in der embryonalen Zelle nur der Anlage nach vorhandenen Vacuolen. Schließlich fließen diese zu einem meist einheitlichen, großen Saftraum zusammen, wofür alle Arten von Parenchym geläufige Beispiele liefern.

Die Vacuole ist Depot und Ablagerungsstätte für alle möglichen Substanzen des Stoffumsatzes. Hier werden *Ballast-Ionen* abgeladen, die die Pflanze mangels Ausschließungsvermögen mitsamt den lebensnotwendigen mineralischen Nährstoffen aufzunehmen gezwungen ist, aber nicht in den Stoffwechsel einbezieht, z. B. Na, Ca zum Teil, Cl, Kieselsäure (Rekretion, Frey-Wyssling 1935). Hierin werden Körper, die zwar im Lebensgetriebe erzeugt, für die weitere Verwendung jedoch irgendwie unbrauchbar geworden und erstarrt sind, als Endprodukte ausgeschieden (Terpene, Exkretion). Die Vacuole ist aber auch beliebter Stapel- und Umschlagplatz für Reservestoffe (Zucker usw.), ebenso wie für verschiedenste *Zwischenprodukte* des Stoffwechsels. Solche können sich gelegentlich vorübergehend anstauen, wenn auf irgendeiner Stufe einer Reaktionskette ein Engpaß entsteht, so daß die Erzeugung einer 'Zwischensubstanz deren gleichzeitigen Verbrauch übertrifft. In die Vacuole übergetretene Substanzen können, wenn sie reaktionsträge sind, liegenbleiben, andernfalls mit verschiedenen Zellsaftkomponenten sich umsetzen und stabilisieren. Soweit spielt die Vacuole auch die Rolle einer Nebenstätte des Stoffwechselgeschehens. Sie ist vom Hauptschauplatz durch den in vieler Hinsicht sehr resistenten Tonoplasten so wirksam abgeriegelt, daß darin selbst Plasmagifte wie die so verbreiteten Gerbstoffe, Saponine u. a. angehäuft werden können, ohne die lebenden Teile der Zelle zu schädigen.

So kann der Zellsaft ein gut Teil der Pflanzenstoffe überhaupt und wohl fast alle wasserlöslichen enthalten. Man findet mineralischen Inhalt, dazu primäre (z. B. Zucker, Aminosäuren) und viele der von diesen nicht immer scharf trennbaren, sog. sekundären Pflanzenstoffe, in deren Herstellung sich der autotrophe Organismus so grenzenlos erfinderisch und luxuriös betätigt (Paech 1950). Es sei nur an die Menge der Terpene, Glykoside, Gerb- und Farbstoffe, sowie Alkaloide erinnert. Manche Inhaltsstoffe sind noch kaum auf ihre Konstitution untersucht. Ein besonderes zentrales Übergangsgebiet stellen, wie Paech (1950) jüngst zusammenfassend herausgearbeitet hat, die gut wasserlöslichen, niederen ein- und mehrbasischen Fettsäuren vor, von welchen Beziehungen mindestens zum Kohlenhydrat- (Assimilation, Atmung) und Eiweiß-, wohl auch zum Fettstoffwechsel führen. Ihre Namen: Äpfel-, Citronen-, Weinsäure verraten, daß sie sich in manchen Früchten stark anreichern können, doch sind sie sehr allgemein und auch in anderen Organen verbreitet.

Das aus verschiedenen Kanälen gespeiste bunte Lösungsgemisch, worin je nach Art und Zustand bald diese, bald jene Stoffgruppen überwiegen, andere zurücktreten oder fehlen können, spielt nun wegen der osmotischen Eigenschaften vieler Komponenten eine wichtige Rolle für die Hydratur der Gewebe und Organe (WALTER 1931) und stellt die potentielle Energie zur Verfügung, auf der das Schaukelspiel von Saugkraft (diffusion pressure deficit, MEYER 1938) und Turgor hauptsächlich beruht. Viele der angedeuteten Stoffgruppen haben in dieser Hinsicht allerdings nur untergeordnete oder überhaupt keine Bedeutung, sei es, weil sie nicht wasserlöslich sind (z. B. Terpene), sei es, weil sie nur vereinzelt oder im allgemeinen in relativ geringen Mengen vorkommen, denen hohes Molekulargewicht gegenübersteht (Alkaloide u. a.), oder aus anderen Gründen osmotisch wenig wirksam sind (s. Gerbstoffe).

So gut feststeht, daß der Zellsaft eine Menge der verschiedensten Stoffe enthält, so wenig ist bisher darüber bekannt, was im konkreten Fall bestimmte Organe einer Art im einzelnen alles im Zellsaft führen. Es gibt keine annähernd erschöpfend ins einzelne gehende Generalanalyse. Schon wegen der Unmenge der Inhaltstoffe und der Schwierigkeit, die Komponenten in so komplexen Lösungsgemischen zu trennen, hat man sich begreiflicherweise stets damit begnügt je nach Gesichtspunkt solche Stoff*gruppen* zu erfassen, die in irgendeinem physiologischen oder ökologischen Zusammenhang von besonderer Bedeutung oder am osmotischen Wert hauptsächlich beteiligt sind; oder man ging überhaupt bloß auf bestimmte Einzelsubstanzen ein. Auch mag hier zu bedenken gegeben sein, daß nicht jeder — auch nicht jeder wasserlösliche — im Preßsaft, geschweige im Extrakt gefundene Stoff aus der Vacuole stammen muß.

## II. DE VRIES' grundlegende Analysen.

Die grundlegenden klassischen Arbeiten zur Kenntnis des Vacuoleninhaltes, deren wichtigste 1884 unter dem bezeichnenden Titel „Eine Methode zur Analyse der Turgorkraft" erschien, stammen vom vielseitigen HUGO DE VRIES. Mit Turgorkraft war gemeint, was wir heute als osmotischen Wert bezeichnen. Die Durchführung ist überholt, der Grundgedanke unvermindert aktuell geblieben. DE VRIES bestimmte grenzplasmolytisch den „Salpeterwert", d. h. den osmotischen Gesamtwert des Preßsaftes aus dem Mark von Blattstielen *(Gunnera, Heracleum, Rheum)* sowie von *Rochea*-Blättern, *Rosa*-Blütenblättern u. a.; andererseits ermittelte er quantitativ durch Reduktion mit FEHLINGscher Lösung die Zucker, durch Titration des Saftes die freien organischen Säuren und durch Titration der Asche die an Säuren gebundenen Alkalien und Erdalkalien, sowie Cl und $PO_4$. Aus den Analysenergebnissen berechnete er über die „isotonischen Koeffizienten", den Anteil der Komponenten am osmotischen Gesamtwert. Es ergab sich, daß Zucker, organische Säuren, sowie organische und anorganische Salze die wichtigsten osmotisch wirksamen Lösungsbestandteile vorstellen.

## III. Schwankungen in der Konzentration und Zusammensetzung des Zellsaftes.

Seit DE VRIES wurden die chemischen Methoden wesentlich verfeinert. An die Stelle der grenzplasmolytischen Bestimmung des osmotischen Wertes ist mehr und mehr die kryoskopische getreten (zusammenfassende Darstellung und Kritik bei WALTER 1931, und STEINER 1939). Um die den einzelnen Lösungsbestandteilen gemäß den jeweiligen Analysenergebnissen entsprechenden Partialdrucke zu ermitteln, werden empirisch-kryoskopisch Eichkurven festgelegt, soweit nicht schon Tabellen (WALTER 1931, 1936) hierfür zur Verfügung stehen. Dem wegen der Dissoziation und der verschiedenen Möglichkeiten der Ionenpaarung schwer berechenbaren Anteil der Elektrolyte am osmotischen Gesamtwert hat man dadurch beizukommen versucht, daß man gemäß den Analysendaten Modell-Lösungen reiner Substanzen herstellt und kryoskopiert (PITTIUS 1934, KNODEL 1938, PISEK 1950). Vorwiegend im Zusammenhang mit ökologischen Fragen, z. B. über die Beziehungen zwischen Frosthärte und Zuckergehalt, über die Rolle des Ca-Ions bei Besiedlung von kalkreichen und -armen Substraten, die Bedeutung des NaCl bei Halophyten, sowie auch vom praktischen Standpunkt sind in den letzten Jahren vielerlei Analysen durchgeführt worden, die unsere Kenntnisse über die Zusammensetzung des Zellsaftes, besonders auch im Hinblick auf den osmotischen Wert, erweitert und vertieft haben. Meist wurde der nach Töten der Proben

(Weckmethode, Heißluft, Chloroform) aus ihnen gewonnene Preßsaft geprüft; er kann im allgemeinen als nahezu identisch mit dem Zellsaft angesehen werden (STEINER 1939).

Dabei zeigte sich, daß nicht nur die Gesamtkonzentration des Zellsaftes in gewissen Grenzen spezifisch ist, so daß man mit WALTER (1931, 1950) von einem artcharakteristischen Optimum und Maximum (weniger Minimum) des osmotischen Wertes sprechen kann, sondern auch seine Zusammensetzung. *Begonia, Oxalis, Rheum* u. a. sind seit alters her als Oxalsäurepflanzen bekannt, *Sedum* reichert Äpfelsäure an, Salzpflanzen speichern oft in größter Menge Kochsalz, Ruderalpflanzen Nitrat u. dgl. mehr. Andererseits wechselt die Konzentration und Zusammensetzung selbst bei ein und derselben Pflanze in gewissen Grenzen je nach dem Entwicklungs- und Stoffwechselzustand und kann von Individuum zu Individuum mit den klimatischen Verhältnissen, vor allem aber mit der chemischen Beschaffenheit des Substrates schwanken. Der Zusammenhang mit letzterem erklärt sich daraus, daß die Pflanze bei der Stoffaufnahme wohl die einen Ionen zurückdrängen und andere bevorzugen kann (COLLANDER 1941); sie vermag aber kein Ion ganz auszuschließen (FREY-WYSSLING 1935) und nimmt in der Regel von einem bestimmten Ion um so mehr auf, je konzentrierter es in der Bodenlösung vorhanden ist, wenn auch deren Gehalt keineswegs proportional. (Vergleiche bezüglich Ca ILJIN 1932, bezüglich Na FREY-WYSSLING 1935 und SCHRATZ 1936).

Die Konzentration des Zellsaftes ist engst verknüpft mit dem Wasserhaushalt. Jede Unterbilanz macht den Zellsaft ungefähr linear mit dem Wasserverlust passiv eindicken (PISEK und CARTELLIERI 1931, STEINER 1939), wenn nicht ausnahmsweise „osmotische Regulation" (VOLK 1937) dem entgegenwirkt. Diese merkwürdige Ausnahme harrt ebenso noch der Klärung an Hand von Saftanalysen wie der umgekehrte Fall, daß nämlich im Laufe von Trockenperioden die Konzentration im wassergesättigten Zustand allmählich geringfügig zunimmt (MÜLLER-STOLL 1935). Durch Wasseraufnahme wird der Zellsaft entsprechend verdünnt. Demgegenüber läßt sich bei Blättern verschiedener

Tabelle 1a u. b. *Mengen und Partialdrucke der Preßsaftkomponenten (Blätter) im Vergleich zum kryoskopisch bestimmten osmotischen Gesamtwert (aus PITTIUS 1934). Qualitativ wurde nachgewiesen: K, Na, Ca, Mg; Cl, $PO_4$, $NO_3$ (in Spuren); Äpfelsäure, Glucose, Fructose, Saccharose, Gerbstoffe (osmotisch unwirksam).*

| Monat der Proben- entnahme | | Gesamt- zucker | Organische Säure (Äpfelsäure) | | $PO_4'''$ | $SO_4''$ | $Cl'$ | Summe der at 2—7 | Kryo- skopisch bestimmter osmotischer Wert at | Dif- ferenz 9—8 at |
|---|---|---|---|---|---|---|---|---|---|---|
| | | | frei | ge- bunden | | | | | | |
| | | 2 | 3 | 4 | 5 | 6 | 7 | 8 | 9 | |

a) Hedera Helix.

| Februar | mg/cm³ Saft | 55,0 | 8,0 | 22,0 | 2,2 | 5,0 | 1,9 | — | — | — |
| | at | 6,4 | 1,6 | 4,4 | 0,7 | 1,6 | 1,0 | 15,7 | 18,8 | 3,1 |
| Juli | mg/cm³ Saft | 13,0 | 7,0 | 20,0 | 4,3 | 5,0 | 1,8 | — | — | — |
| | at | 1,9 | 1,4 | 4,0 | 1,4 | 1,6 | 1,0 | 11,3 | 12,2 | 0,9 |

b) Ilex aquifolium.

| Februar | mg/cm³ Saft | 55,0 | 9,4 | 17,0 | 4,5 | 4,0 | 0,6 | — | — | — |
| | at | 6,5 | 1,9 | 3,4 | 1,4 | 1,3 | 0,3 | 19,0 | 21,6 | 2,6 |
| Juli | mg/cm³ Saft | 14,0 | 4,0 | 17,0 | 3,2 | 5,0 | 1,3 | — | — | — |
| | at | 1,9 | 0,8 | 3,4 | 1,0 | 1,6 | 0,7 | 11,9 | 12,8 | 0,9 |

Tabelle 2. *Aus* KNODEL *(1938)*.

| Monat der Proben-entnahme | Mono-saccha-ride | Saccha-rose | Gesamt-zucker | Salze | Ca als Chlorid | NH₄ als Chlorid | Gesamt-salze 4+5+6 | Organi-sche Salze | Chloride | Organi-sche Säuren | Osmoti-scher Wert 3+7+10 | Osmoti-scher Wert Kryo-skop | Dif-ferenz 12—11 |
|---|---|---|---|---|---|---|---|---|---|---|---|---|---|
| | 1 | 2 | 3 | 4 | 5 | 6 | 7 | 8 | 9 | 10 | 11 | 12 | 13 |
| 24. VIII. Mais | 2,16 | 0,94 | 3,10 | 3,50 | nicht best. | 0,22 | 3,72 | nicht best. | nicht best. | 1,34 | 8,2 | 10,4 | 2,2 |
| 1. VII. Roggen | 2,13 | 2,18 | 4,31 | 6,11 | nicht best. | 0,60 | 6,71 | 2,10 | 2,46 | 2,24 | 14,0 | 15,9 | 1,9 |
| 20. VIII. Futterrübe | 0,86 | 0,74 | 1,60 | 5,21 | 1,17 | 0,32 | 6,70 | 2,10 | nicht best. | 0,23 | 8,5 | 9,7 | 1,2 |
| 15. VII. Flieder | 3,90 | 0,00 | 3,90 | 8,42 | nicht best. | nicht best. | 8,42 | nicht best. | nicht best. | 1,80 | 14,1 | 17,8 | 3,7 |
| 15. III. Fichte | 9,40 | 0,55 | 9,95 | 2,42 | 1,70 | 0,54 | 4,66 | 2,07 | 1,01 | 2,65 | 17,3 | 18,6 | 1,3 |
| 15. III. Buxus | 5,27 | 3,31 | 8,58 | 7,30 | 1,90 | 0,65 | 9,85 | 4,37 | 2,38 | 1,34 | 19,8 | 22,7 | 2,9 |

Immergrüner regelmäßig eine jahresperiodische Schwankung in der Konzentration (Winter hoch, Sommer niedrig) auch bei gleichbleibendem Wassergehalt bzw. Reduktion auf konstanten Wassergehalt feststellen. Sie beruht darauf, daß im Herbst Hand in Hand mit der Frosthärtezunahme der Zuckerspiegel aktiv und reversibel erhöht wird (Sammelbericht bei STEINER 1939, neuerdings PISEK 1950). Bei manchen *Oleaceen*, aber auch beim Allerweltsunkraut *Veronica Tournefortii Mel.* tritt an Stelle der Zuckerschwankung eine solche im Mannitgehalt (ASAI 1932, 1937).

Im großen und ganzen wurde das erste orientierende Ergebnis von DE VRIES hinsichtlich der mengenmäßig wichtigsten und am osmotischen Wert hauptsächlich beteiligten Zellsaftkomponenten bestätigt: Zucker, Säuren und Salze machen zusammen durchschnittlich etwa 90% des osmotischen Wertes aus (Tabelle 1, 2; Abb. 1, Rhododendron).

# IV. Wichtigste Lösungskomponenten.

Der Enge des hier zur Verfügung stehenden Platzes steht die endlose Liste dessen gegenüber, was sich im Zellsaft alles vorfinden kann. Um einen Begriff davon zu bekommen, braucht man in einem beliebigen pflanzenchemischen Buch bloß die hochmolekularen Eiweißkörper, die Gerüstsubstanzen, Fette und den Großteil der gleich diesen ausnahmslos hydrophoben, wasserunlöslichen Terpene wegzulassen. Letztere sind Abkömmlinge der Isoprenreihe, typisch pflanzliche Produkte, deren niedere Glieder als Bestandteile ätherischer Öle bekannt, sich wie die Fette in besonderen Lückenräumen des Cytoplasmas sammeln, während Tetraterpene (Carotinoide) an Plastiden gebunden sind. Nur gewisse Triterpenabkömmlinge findet man — an Zucker geknüpft und hierdurch wasserlöslich geworden — als Saponine im Zellsaft gelöst in weiter Verbreitung. Die Polyterpene Guttapercha, Balata, und vor allem Kautschuk werden wahrscheinlich in Plastiden aufgebaut, treten dann aber in die Vacuole über und sind als kleine Partikel in sehr wechselnder Menge, oft neben Eiweiß und Alkaloiden (Papaver!) im Saft (Latex) der Milchzellen und -gefäße suspendiert, die eine ganze Reihe von Familien kennzeichnen[1].

Unsere Auswahl stellt die am osmotischen Wert nach bisheriger Erfahrung vornehmlich beteiligten Stoffgruppen in den Vordergrund.

---

[1] Betreffs Einzelheiten sei hier wie im allgemeinen auf die Werke von CZAPEK, FREY-WYSSLING, KLEIN und PAECH verwiesen, aus denen manches übernommen wurde, ohne sie stets ausdrücklich zu zitieren.

## 1. Zucker und Salze. Ihr Anteil am osmotischen Gesamtwert.

Von Kohlenhydraten finden sich im Zellsaft fast immer die Monosaccharide, Glucose und Fructose, von Disacchariden Saccharose in größerer Menge vor. Das wasserlösliche Fructosan Inulin ist selten, außer bei Angehörigen der

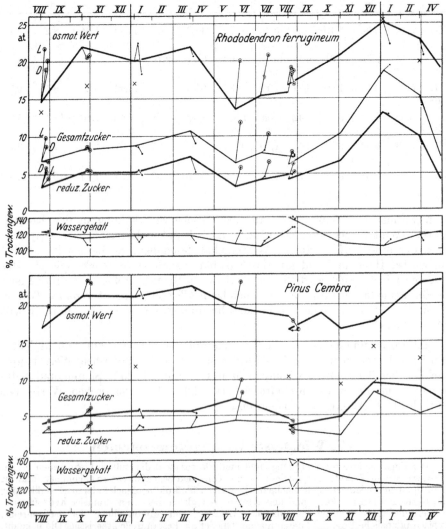

Abb. 1. Jahresgang von osmotischem Gesamtwert und Zuckerpartialdruck (at) des Preßsaftes der Blätter eines Zwergstrauches *(Rhododendron)* und Nadelbaumes *(Pinus)* der alpinen Waldgrenze auf dem Patscherkofel bei Innsbruck (1900 m, 1945 bis 1947). Jahresgang konform mit der Frosthärte. Die nach aufwärts (abwärts) den Kurven angehängten Punkte entsprechen den Werten, die nach Abhärtung (Verwöhnung) gefunden wurden. × Summe der Partialdrucke der Zucker und der analysierten An- und Kationen (Äpfel- und Citronensäure; [K, Ca, Mg). Diese Summe beträgt bei *Rhododendron* 80—90% des kryoskopisch bestimmten osmotischen Gesamtwertes, bei *Pinus Cembra* bestenfalls 65%. Aus Pisek 1950.

Compositen und Campanulaceen, in deren Speicherorganen es sich die Stärke vertretend stark anreichern kann. Glykogen ist nur bei Pilzen und Blaualgen verbreitet.

Für die Höhe des osmotischen Wertes sind in der Regel weniger die Zucker als die Salze maßgebend, selbst dann, wenn jene gewichtsmäßig weit vorherrschen, und obschon gewöhnlich die reduzierenden Zucker — Monosaccharide; Maltose, die jedoch selten vorkommen soll — die nichtreduzierenden (Saccharose) überwiegen. ILJINs (1932) extensive Analysen der Blattpreßsäfte von über 200 krautigen Arten verschiedener Standorte erbringen hierfür reichhaltiges Belegmaterial. Der osmotische Wert hängt ja ausschließlich von der Zahl der Teilchen in der Volumseinheit der Lösung ab. Diese Zahl ist, wenn gleiche Gewichtsmengen von Zuckern und Salzen in Lösung sind, bei letzterem, selbst abgesehen von der Dissoziation, schon wegen ihres gewöhnlich geringen Molekulargewichtes viel größer als bei den Zuckern. Bei 84% der von ILJIN untersuchten Arten überwiegt die Salzkonzentration (Mol), bei 63% ist sie mehr als doppelt, im Extrem vielmals größer als jene der Zucker, die gewöhnlich unter 0,1 mol bleiben, während Salze meist mehr als 0,2 mol verzeichnen. Man darf allerdings nicht übersehen, daß das Material fast nur krautige Sommergrüne umfaßt. Immergrüne Holzpflanzen (STEINER 1933), besonders die Ericaceen der alpinen Zwergstrauchheide wie Rhododendron ferrugineum L., Calluna, Loiseleuria, Arctostaphylos Uva ursi sind durchs ganze Jahr reich an Zucker, dessen Gesamtkonzentration bei letzteren selten unter 0,3 mol sinkt (ULMER 1937, PISEK 1950). Im Winter kann sie bei Rhododendron bloß zufolge aktiver Vermehrung der Zucker auf 0,7 mol steigen, wobei dann diese 70—75% des osmotischen Wertes und die gesamte Jahresschwankung desselben bestreiten (Abb. 1). Mediterrane Ericaceenhölzer (Arbutus Unedo L. und A. Andrachne L., Erica multiflora L.) sind hingegen nach Angaben bei GIROUX (1936) ausgesprochen zuckerarm.

## 2. Anorganische Kationen und Anionen.

Von allem was aus dem Substrat in den Pflanzenkörper eintritt, seien es Nähr- oder Ballastionen, findet sich immer etwas im Zellsaft, bald mehr, bald weniger je nach Selektions- oder Speichervermögen der Spezies und der Organe (HOAGLAND 1948, COOPER 1947).

K wird im großen und ganzen bevorzugt aufgenommen (COLLANDER 1941) und besonders in jungen Blättern in relativ beträchtlicher Konzentration gespeichert; in Hydrodictyon patenaeforme bis zum 4000fachen der Konzentration des Nährmediums (BLINKS und NIELSEN 1940). Die meisten von ILJIN untersuchten Arten enthalten zwischen 4 und 10 mg je Kubikzentimeter Saft, was 0,1—0,25 Gramm-Ionen entspricht.

Na hingegen ist oft bloß in Spuren, jedenfalls im allgemeinen wenig nachweisbar. Euhalophyten allerdings nehmen es mit Cl in bedeutenden Mengen aus dem kochsalzhaltigen Boden in die Vacuole auf (COLLANDER 1941). Partialdrucke von 10—20 at (Salzgehalt rund 15—30 mg je Kubikzentimeter Saft) sind hier Regel, bei Salicornia herbacea L. und Mangrovehölzern können es über 30 at sein (Abb. 2. STEINER 1933, WALTER und STEINER 1936). Zumindest zwei Drittel des osmotischen Gesamtwertes solcher Pflanzen gehen auf Rechnung des Kochsalzes; sein Partialdruck bildet gewissermaßen den Sockel, auf welchen der osmotische Wert der übrigen Zellsaftbestandteile gehoben wird, so daß das osmotische Gefälle Zellsaft/Boden im wesentlichen dasselbe ist wie bei Pflanzen salzfreier, sonst aber vergleichbarer Standorte (STEINER 1939).

Ca enthält der Boden wohl stets soviel, daß der Baustoffbedarf und der Antagonismus gegenüber den einwertigen Alkaliionen befriedigt werden kann; meist ist es überreichlich vorhanden und wird dann in der Regel in großem Überschuß aufgenommen. Besonders auf Kalkrohböden speichern einige Arten Ca in auffälligen Mengen (Alyssum saxatile, Anthyllis vulneraria, Centaurea scabiosa bis um 30 mg je Kubikzentimeter!). Im allgemeinen nimmt der Ca-Gehalt des Zellsaftes bei derselben Art mit dem Kalkreichtum des Bodens zu; auf demselben Boden schwankt er je nach Art, wohl weil die Wurzeln der einen

sich mit größerem, jene der anderen sich mit geringem Erfolg der Ca-Über-
schwemmung erwehren. Wenige Arten enthalten das Ca, mögen sie auch viel
davon speichern, großenteils im Zellsaft gelöst, daneben teilweise als Niederschlag
(„physiologisch calciophile", Iljin 1932, 1940). Dies ist nur möglich, wenn
keine oder wenig Oxalsäure in der Vacuole zugegen ist. Andernfalls bindet sich
das Ca mindestens zum Teil zum schwer aufschließbaren Oxalat. Auf letzterem
Wege (eingehende Darstellung der Ausscheidung von Oxalat, Carbonat, Kieselsäure bei Frey-Wyssling 1935) entledigen sich die meisten Pflanzen des Ca-
Ballastes oder halten seinen aktiven, gelösten Anteil wenigstens in Schranken

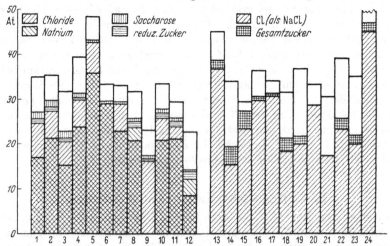

Abb. 2. Osmotischer Wert und Zusammensetzung des Zellsaftes nordamerikanischer Salzmarschpflanzen (1—12,
aus Steiner 1934, gekürzt; 1 *Spartina glabra*, 2 *Sp. patens*, 3 *Distichlis spicata*, 4 *Juncus Gerardi*. 5 *Salicornia
mucronata*, 6 *S. herbacea*, 7 *Plantago decipiens*, 8 *Atriplex patula hastata*, 9 *Aster subulatus*, 10 *Limonium caroli-
nianum*, 11 *Suaeda linearis*, 12 *Gerardia maritima*) und von ostafrikanischen Mangroven (13—24 nach Walter
und Steiner 1936, aus Steiner 1939, ebenfalls gekürzt; 13—15 *Sonneratia alba*, 16—18 *Rhizophora mucronata*.
19—21 *Ceriops Candolleana*, 22—24 *Avicennia marina*).

(„physiologisch calciophobe"[1], Tabelle 3). Dies setzt jedoch die Produktion und
Nachlieferung entsprechender Mengen von Säure als Fällungsmittel voraus. [Man
wird daher bei Erklärung der auffallenden Tatsache, daß manche Arten auf
kalkreiche, andere auf kalkarme Böden beschränkt sind, außer den physikalischen
Eigenschaften und der Reaktion ($p_H$) solcher Böden, künftig doch auch die
verschiedene Ca''-Konzentration der Bodenlösung mitbedenken müssen.

Fast stets enthält der Zellsaft auch Mg, das sich ähnlich wie Ca mit dem
Alter der Blätter anzureichern scheint (Iljin 1944). Junge Blätter führen nach
Michael (1941) annähernd gleichviel, während der Gehalt älterer je nach An-
gebot und Nachfrage stärker schwankt. Über mengenmäßige Verteilung und
Funktion vgl. unter anderem Zimmermann (1947).

In Buchenblättern fand Olsen (1948) K, Ca und Mg teilweise außerhalb
der Vacuole gebunden, vermutlich an Proteine; im Hochsommer kann bis $1/_3$ des
gesamten K-Gehaltes gebunden sein, während sich im Herbst vor dem Laubfall
alles in Lösung befindet.

---

[1] Es ist bezeichnend, daß an den von Iljin (1940) untersuchten Örtlichkeiten auf Kalk-
böden überwiegend calciophile, auf Silicat calciophobe Arten wachsen. Muster für letztere
sind: *Rumex acetosa* und *acetosella*, *Polygonum bistorta* und *persicaria*, *Silene vulgaris*, *Melandrium album* und *Oxalis acetosella*, die 90—100% der allerdings meist geringen Gesamtmenge
an Ca als Niederschlag enthalten. Physiologisch calciophile sind z. B. *Centaurea scabiosa*,
*Reseda lutea*, *Anthyllis Vulneraria* und *Alyssum saxatile*, die höchstens 35% ihres meist
hohen Ca-Gehaltes als Niederschlag führen (Iljin 1932).

Tabelle 3. *Verhältnis von Ca und Säuren bei Calciophilen (1—5) und Calciophoben (6—9)* (auszugsweise nach ILJIN 1940).

| | Ca mg/g Trockengewicht | | Konzentration in Grammäquivalenten | | | | |
|---|---|---|---|---|---|---|---|
| | im Nieder-schlag | gelöst | Ca | Oxal-säure | Citronen-säure | Äpfel-säure | Summe |
| | 1 | 2 | 3 | 4 | 5 | 6 | 4—6 |
| 1. Anthyllis vulneraria | 77 | 51 | 0,91 | 0 | 0,11 | 0,75 | 0,86 |
| 2. Robinia pseudacacia | — | 34,6 | 0,87 | 0 | 0,24 | 0,49 | 0,73 |
| 3. Centaurea scabiosa | 40,1 | 31,4 | 0,6 | 0 | 0,28 | 0,28 | 0,56 |
| 4. Reseda lutea | — | 40,4 | 0,53 | 0 | 0,16 | 0,31 | 0,47 |
| 5. Coronilla varia | 39,3 | 31,0 | 0,52 | 0 | 0,24 | 0,17 | 0,41 |
| 6. Melandryum album | 34,1 | 0 | 0 | 0,067 | 0 | 0 | 0,067 |
| 7. Rumex acetosa | 27,5 | 0 | 0 | 0,022 | 0 | 0 | 0,022 |
| 8. Silene vulgaris | 29,8 | 0 | 0 | 0,15 | 0,009 | : | 0,159 |
| 9. Ballota nigra | 32,0 | 0 | 0 | 0,098 | 0,016 | 0,012 | 0,126 |

Von mineralischen Anionen sind Sulfat und Phosphat in wechselnden, geringen Mengen regelmäßig nachweisbar. Nitrat und Chlorid sind wenigstens in Spuren sehr verbreitet. Wird dieses, wie erwähnt, von Euhalophyten besonders als Kochsalz im großen Ausmaß gestapelt, so findet sich ersteres im Zellsaft von Ruderalpflanzen, die nitratreiche Biotope besiedeln (*Urtica, Chenopodium, Capsella* usw.), relativ stark angereichert.

# 3. Organische Säuren.

Die häufigeren niederen Carbonsäuren sind als Zwischenprodukte des bei Tier und Pflanze im wesentlichen in denselben Bahnen laufenden aeroben Stoffwechsels erkannt oder doch wahrscheinlich. Sie mögen sich infolge von Stockungen im Umlauf der Reaktionskette gelegentlich stauen und dann in der Vacuole ansammeln. Zufolge der schwierigen Trennung und quantitativen Bestimmung sind ältere Angaben nur mit Reserve übernehmbar. Über neuere Bestimmungsmethoden vergleiche ILJIN (1939), VICKERY und PUCHER (1940) und PEYNAUD (1947).

Von den vielerlei Carbonsäuren, die in Pflanzen überhaupt vorkommen, kann je nach Art und Zustand bald diese bald jene vorherrschen. *Digitalis purpurea* L. z. B. reichert vorzüglich die zwar sehr verbreitete, aber mengenmäßig selten hervortretende Bernsteinsäure (ac. succinicum) an (KARRER und MATTER 1948); in den Blättern der Baumwollpflanze macht sie immerhin bis 17% der bisher nicht identifizierten Säuren aus, in Samen bis 29% (ERGLE und EATON 1951). *Bryophyllum calycinum* ist als ergiebigste Quelle der 1927 von NELSON erstmals in der Pflanze, nämlich in Brombeeren nachgewiesenen Isocitronensäure bekannt geworden (PUCHER 1942). In Maiskeimlingen kann die Milchsäure (ac. lacticum) bis zur Hälfte der Gesamtsäure ausmachen (SCHNEIDER 1939). Einige wenige solcher Säuren sind fast regelmäßig in größerer Menge anzutreffen, nicht nur in den saftigen Früchten, nach denen sie benannt wurden, sondern auch sonst im Obst und in den verschiedensten Organen wie Stengeln, Blättern und Speicherorganen, wo sie in wechselndem Mengenverhältnis in der Regel den Hauptanteil des Gesamtsäuregehaltes ausmachen. Es handelt sich vor allem um die zweibasische Oxal- und Äpfelsäure (ac. malicum); d(+)-Weinsäure (ac. tartaricum) scheint zwar verbreitet, auch bei Thallophyten, aber abgesehen von *Vitis* in Blättern nur wenig und dann meist als Niederschlag vorhanden zu sein (ILJIN 1938); dazu kommt die dreibasische Citronensäure.

Zusammen mit den anorganischen sorgen die Carbonsäuren unter anderem dafür, daß die Reaktion des Zellsaftes ($p_H$) durch eindringende Kationen nicht in den neutral-basischen Bereich gerät. Besonders die auch bei Bakterien und Pilzen verbreitete Oxalsäure verursacht, wenn sie frei oder als saures Salz in größerer Menge angehäuft wird, zufolge ihrer relativ sehr hohen Dissoziations-konstante ($25^0$ $6,5 \times 10^{-2}$) stark sauren Zellsaft: *Rheum, Rumex, Begonia, Oxalis* $p_H$ 3—2, wie Citrone. Für zahlreiche Arten steht fest, daß der Säuregehalt mit dem Ca-Gehalt des Bodens zunimmt. Dabei können sich Oxal-, Äpfel- und Citronensäure bis zu einem gewissen Grad mengenmäßig vertreten. Äpfelsäure erreicht besonders hohe Konzentration in Kalkpflanzen, die wenig oder keine Oxalsäure führen. Je mehr gelöstes Ca solche Pflanzen im Zellsaft enthalten, um so mehr auch Äpfelsäure, deren Ca-Salz am leichtesten löslich ist. Umgekehrt: Wo viel Oxalsäure abfällt und daher auch in Lösung bleibt, treten Äpfel- und Citronensäure in den Hintergrund (ILJIN 1936). Auch mit der Art der Stickstoff-ernährung schwankt der Säuregehalt — er ist bei $NO_3^-$-Gabe mehrmals größer als bei Versorgung mit $NH_4^+$, wohl weil die mit dem Nitrat aufgenommenen Kat-ionen die Säureproduktion anregen.

Von Aminosäuren gelten Leucin, Tyrosin, Arginin und Histidin als all-gemeiner verbreitet, vorzugsweise in Keimpflanzen. Von Amiden der Amino-säuren findet sich Asparagin und Glutamin in verschiedenen Organen bei ver-schiedenen Familien, ersteres besonders bei Leguminosen, letzteres bei Cruciferen.

## 4. Gerbstoffe.

Bei Moosen und Pteridophyten häufig, sind Gerbstoffe bei den Samenpflanzen ein ganz allgemein verbreiteter Vacuoleninhalt, besonders im Rindenparenchym, aber auch im Blattmesophyll und in Früchten, der sich mitunter auf eigene Zellen (Idioblasten) beschränkt. Am meisten werden sie nach uralter Erfahrung in Rinden, Borken und gewissen Gallen gespeichert (Eichenrinde enthält 9—12, Akazien- und Eucalyptusrinde 20—40%). Manchmal kann gerbstoffhaltiger Zell-saft sich mehr oder weniger verfestigen ("Inklusen" der Früchte und Blätter von *Tamarindus, Ceratonia, Rhamnus* u. a.). So leicht sich Gerbstoffe mit geläufigen Reaktionen, von denen allerdings keine eindeutig ist, pauschalmäßig nachweisen lassen, so wenig erfährt man damit über die genauere chemische Natur derartiger Substanzen, die sich auf mindestens zwei heterogene Gruppen verteilen. Die eine umfaßt hydrolisierbare Verbindungen, nämlich Verbindungen von Phenolcarbonsäuren unter sich oder mit ähnlichen Oxysäuren (Depside, sehr verbreitet z. B. Chlorogensäure) sowie Zuckerester (Tannine, z. B. in Eichen-gallen, Rinde und Blättern der Edelkastanie u. a.) und Glucosidverbindungen solcher. Die Angehörigen der zweiten Gruppe sind Flavanabkömmlinge, also Kondensationsprodukte (Catechingerbstoffe, z. B. in *Rheum*, Quebracho, Eichen-rinde). Von vielen natürlichen Gerbstoffen ist die Konstitution noch unbekannt. Meist sind es Gemische; ihre Entstehung im Organismus ist noch dunkel, die Meinungen über ihre Rolle im Stoffwechsel geteilt. Daß sie mit Polysacchariden und Eiweiß Komplexe bilden und sich auf solche Weise vorübergehend maskieren können, trägt zur Undurchsichtigkeit bei. Man plädierte für Reservestoffnatur (z. B. SPERLICH 1917), häufiger für Abbauprodukt (Exkret), das allenfalls sekun-där als Schutzstoff irgendwelche Bedeutung haben kann oder indifferent liegen-bleibt. Bei der Verschiedenartigkeit der Stoffe mag je nach Gemisch und Art von Fall zu Fall Verschiedenes zutreffen. Wir verweisen im übrigen auf CZAPEK (1922) und PAECH (1950).

Im hier gegebenen Zusammenhang interessiert, daß PITTIUS (1934) auf Grund kryoskopischer Untersuchung Tannin, Tannalbin und Gallussäure für osmotisch unwirksam erklärt; gerbstoffhaltiger Saft von *Ilex aquifolium* und *Hedera helix* ergab nach einstündigem Schütteln mit Hautpulver (zur Gerbstoffabsorption) dieselbe Gefrierpunktserniedrigung wie vorher. WALTER und STEINER (1936) fanden Tannin (Merck), das als relativ leicht löslich gilt, nicht ganz so belanglos. Sie geben für gesättigte, wäßrige Lösung einen osmotischen Wert von 2,2 at an. Wir erhielten nach Fällung mit Kaliumbichromat im Überschuß an gerbstoffhaltigen Fichtennadelpreßsaft unter Berücksichtigung der Verdünnung durch das Fällungsmittel einen um $0,13^0$ höheren Gefrierpunkt als vor der Fällung $(1,50^0)$, was ebenfalls für eine gewisse osmotische Bedeutung spricht. Sie wird wie bei allen derlei Substanzen je nach Löslichkeit, Menge und je nachdem, ob es sich um eine nieder- oder hochmolekulare Verbindung handelt, etwas schwanken, wenn auch anzunehmen ist, daß ihre osmotische Bedeutung im allgemeinen gering sein wird.

HÄRTEL (1951) hat gelöste „Phloroglucotannoide" bzw. „Catechine" als eine der Ursachen „voller" Zellsäfte nachgewiesen[1].

Gerbstoffe oder Trabanten solcher sind weiter mindestens in vielen Fällen Ursache einer sehr charakteristischen, matten Vacuolenprimärfluorescenz, die allgemein im Rindenparenchym von Holzpflanzen, aber auch in vielen Mesophyllzellen, besonders immergrüner Arten, beobachtbar ist. Die selteneren glykosidführenden Vacuolen, z. B. der Ericaceenblätter fluorescieren brillant und leuchtend in verschiedenen Farben. Bei geeigneten Objekten hat sich diese Primärfluorescenz als ebenso einfaches wie wertvolles Hilfsmittel bewährt, um an Schnitten durch teilweise geschädigte Organe rasch die lage- und mengenmäßige Verteilung lebender und toter Zellen zu überblicken (LARCHER 1954). Aus Zellen, deren Plasma nicht mehr semipermeabel ist, diffundiert der Fluorescenzträger ins umgebende Wasser. Die Vacuole erscheint dann dunkel.

Da die Lösungsbestandteile und die Konzentration einzelner Komponenten von Zelle zu Zelle teilweise wechseln, kann es nicht wundernehmen, daß die Vacuolen benachbarter Zellen und Zellagen (KASY 1951) oft in verschiedener Farbe und Intensität fluorescieren.

## 5. Glykoside, Farbstoffe.

Nicht nur einige Gerbsäuren, sondern auch verschiedenste andere Substanzen mit alkoholischen oder phenolischen OH-Gruppen, wie sie im „luxurierenden" Stoffwechsel der Autotrophen in bunter Fülle entstehen, können bei Gegenwart der spezifischen Fermente (Glykosidasen) ganz nach dem Muster der Disaccharidbildung mit Zuckern über deren Carbonylgruppe in Verbindung treten. Bei Hydrolyse spalten solche Glykoside demnach in eine Zuckerkomponente, als welche Monosen wie auch zusammengesetzte Saccharide figurieren, und einen zuckerfreien Bestandteil (Aglykon). Meist enthält dieses ein oder mehrere Benzolringe. Als einfachstes Beispiel für ein natürliches Glykosid sei Arbutin erwähnt, das aus Hydrochinon plus Glucose besteht. Nach dem Aglykon unterscheidet man Phenol-, Anthracen-, Indoxyl- usw. -Glykoside; im weiteren Sinne sind auch die weitverbreiteten Saponine, sowie die Senfölglykoside (Aglykon über eine S-Brücke an Zucker gebunden) hierher zu rechnen. Häufig ist das Aglykon in Glykosidbindung besser löslich als an sich, was die Ausscheidung in die Vacuole begünstigt. Doch ist die Löslichkeit recht verschieden — hängt auch von den Lösungsgenossen ab — und selbst schwer lösliche Glykoside können sich unter Umständen in der Vacuole anreichern. Manche sind in vielen nicht näher ver-

---

[1] HÖFLER (1949) meint damit Säfte, die irgendwelche Stoffe enthalten, welche in die Vacuole permeierende Vitalfarben wie Acridinorange unter grüner Fluorescenz weitgehend $p_H$-unabhängig zu binden vermögen.

wandten Familien zu finden (Hesperidin), andere auf nur wenige oder eine einzige Gattung beschränkt (Spiraein in *Spiraea, Filipendula*; Gein in *Geum*). Der Gehalt schwankt sehr stark von Art zu Art, aber auch je nach Standort und Entwicklungszustand.

Als Glykoside finden sich großteils auch die im Zellsaft — nicht bloß der Blütenhüllblätter — gelösten, meist $\pm$ blaßgelben Stoffe aus der Gruppe der Flavone und Flavonole (manchmal als Anthoxanthine zusammengefaßt) wie auch die Anthocyanidine (als Glykoside Anthocyane besser Anthocyanine genannt). Letztere (Sammelbericht auch bei BLANK 1947) sind einzeln und in Gemischen aus ihresgleichen Ursache der blauen und purpurroten wie violetten Farben, besonders in der Blütenregion; sie können auch leuchtend spektrales Rot zaubern; doch sind an diesem häufig plastidengebundene Carotinoide mitbeteiligt, auf die auch intensive Gelbfärbung meistens zurückzuführen ist. Flavone und Flavonole spielen bei Gelbfärbung eine bescheidenere Rolle. Die Intensität der Anthocyanfärbung wird abgesehen von der Mischung der Pigmente und ihrer fallweisen sehr schwankenden Konzentration (*Centaurea cyanus* unter 1%, dunkelblaue Violen mehr als 20% des Blütenblatt-Trockengewichts) auch dadurch beeinflußt, daß Anthocyane mit kolloiden Zellsaftbestandteilen und Tanninen, Flavonglykosiden und anderen „Copigmenten" Komplexe bilden können.

Wie den Catechingerbstoffen, den Flavonen und Flavonolen liegt auch dem Molekül der Anthocyanidine das Flavangerüst zugrunde. Je nach Oxydations-Reduktionszustand des mittelständigen Heterocyclus (Pyronringes) ergeben sich die angeführten Stofftypen. PAECH (1950) bringt sie in folgender Übersicht:

Flavangerüst:

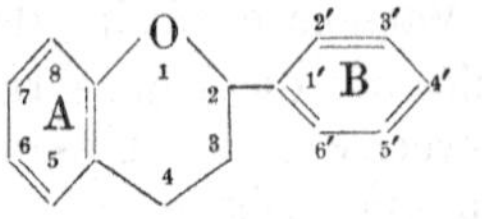

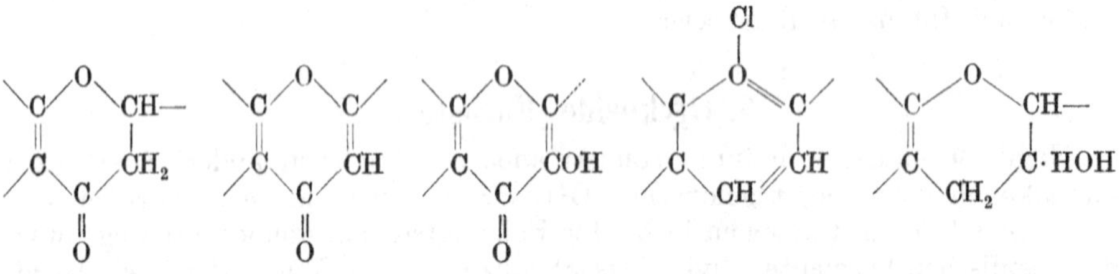

| | | | | |
|---|---|---|---|---|
| Flavanon | Flavon | Flavonol | Flavyliumsalz | Catechin |
| Beispiel: Hesperetin | Luteolin | Quercetin | Anthocyanidine | D, L-Epicatechin |

Durch Substitution von Hydroxyl- und Methoxylgruppen an diesen und jenen Stellen wird jeder Typus in sich weiter abgewandelt. Auch die natürlichen Anthocyanidine unterscheiden sich nur durch solche kleine Änderungen der Substituenten[1].

Aus der Verbindung mit bloß einem oder mit zwei Zuckermolekeln, die Mono- (meist Glucose allein oder mit Rhamnose) oder auch Disaccharide sein können, ergeben sich verschiedene Kombinationen zu Anthocyanen. Ihre *Aglyka* können in vitro durch Reduktion aus Flavonolen wie durch Dehydrierung aus Catechinen entstehen; auch die Pflanze dürfte derlei Umwandlungen beherrschen,

---

[1] Zum Beispiel ist Pelargonidin (Pelargonien-, Dahlienblüten) = 3,5,7,4'-Tetraoxyflavylium, Cyanidin (rote Rosen, *Centaurea cyanus* und viele andere Blüten, auch Laubblätter) = 3,5,7,3',4'-Penta-oxyflavylium, Delphinidin (*Delphinium-, Viola-, Gentiana*-Blüten) = 3,5,7,3',4',5'-Hexa-oxyflavylium.

und somit die erwähnten Flavanabkömmlinge ineinander überzuführen imstande sein. So würde verständlich, daß unter Umständen Anthocyane als Begleiterscheinung gewisser Stoffwechselvorgänge überall gebildet werden können. Bekannt ist das Erröten des Laubes verschiedener alpiner immergrüner im Herbst, wenn sie ohne Schneeschutz in der Kälte stehen; umgekehrt können manche Pflanzen in praller Sonne Anthocyan bilden (*Teucrium chamaedrys* und andere vgl. Müller-Stoll 1935).

## 6. Alkaloide.

Eine weitere Klasse ganz ausgesprochen pflanzlicher Stoffe von größter Mannigfaltigkeit bilden N-haltige Verbindungen von basischem Charakter, die den N in Ringbindung (Heterocyclus) enthalten. Sie mögen bei der Synthese von Aminosäuren auf Seitenwegen abgezweigt werden oder beim Abbau solcher als nicht weiter verwertbare Endprodukte abfallen. In ihre Synthese und Bedeutung in der Pflanze fängt erst in Betreff weniger, wie z. B. des Nicotins und Verwandten an, Licht zu kommen (Übersicht bei Mothes 1953). Man unterscheidet sie vorläufig nach der Art des heterocyclischen Grundgerüstes in Pyrrolidin-, Pyridin-, Purin- usw. Abkömmlinge. In Kryptogamen, Pteridophyten und Gymnospermen nur höchst vereinzeltes Vorkommnis (z. B. Taxin in *Taxus*), bei Monokotyledonen selten, sind sie bei den Dikotyledonen außerordentlich verbreitet und fehlen in gewissen Familien wie den *Ranunculaceen, Papaveraceen, Apocynacee, Gentianaceen, Loganiaceen, Solanaceen* kaum einem der Angehörigen. Vielfach sind ihrer mehrere in einer Art nebeneinander, wovon das mengenmäßig am stärksten vertretene dann als Hauptalkaloid von Nebenalkaloiden unterschieden wird.

Mit organischen und Mineralsäuren, wie mit Halogenwasserstoffsäuren bilden sie gut kristallisierbare Salze, die leicht wasserlöslich sind, so daß sie sich in dieser Form in großer Konzentration im Zellsaft ansammeln können, während die freien Basen im allgemeinen schwer oder gar nicht wasserlöslich sind.

Gerbstoffen, Glykosiden und Alkaloiden wird je nach Löslichkeit, Menge der gelösten Substanz und Molekulargewicht oft kein oder nur ein höchst bescheidener, manchmal aber vielleicht doch nicht zu vernachlässigender Anteil am osmotischen Wert zukommen. Er kann mithelfen, den gelegentlich, so besonders im Falle Pinus cembra (Abb. 1) auffallenden Fehlbetrag zwischen dem kryoskopisch bestimmten und dem aus immer lückenhaften Analysen ermittelten Summenwert zu überbrücken. Die Zahl der in dieser Hinsicht angestellten Vergleiche ist noch erstaunlich gering. de Vries' erster Versuch an bestimmten Beispielen die Hauptstoffgruppen des Vacuoleninhalts und ihren Anteil am osmotischen Wert quantitativ festzustellen, hat lange überhaupt keine und bis heute sehr wenige Nachfolger gefunden.

## Literatur.

Asai, T.: Untersuchungen über die Bedeutung des Mannits im Stoffwechsel höherer Pflanzen. I. und II. Jap. J. of Bot. **6**, 64 (1932); **8**, 343 (1937).

Blank, F.: The anthocyanin pigments in plants. Bot. Rev. **13**, 241 (1947). — Blinks, L. R., and J. P. Nielsen: The cell sap of Hydrodictyon. J. Gen. Physiol. **23**, 551 (1940).

Collander, R.: Selective absorption of cations by higher plants. Plant Physiol. **16**, 691 (1941). — Cooper, H. P., J. B. Mitchell and N. R. Page: The relation of the energy properties of soil nutrients to the chemical composition of the crop plants. Soil. Sci. Soc. Amer. Proc. **12**, 364 (1947). — Czapek, F.: Biochemie der Pflanze. Jena: Gustav Fischer 1922.

Ergle, D. R., and F. M. Eaton: Succinic acid content of the cotton plant. Plant Physiol. **26**, 186 (1951).

FREY-WYSSLING, A.: Die Stoffausscheidung der höheren Pflanze. Berlin: Springer 1935.
GEFFKEN, K.: Zur Bestimmungsmethodik biologisch wichtiger Kohlenhydrate im
Pflanzenmaterial. Bot. Archiv **36**, 345 (1934). — GIROUX, J.: Reserches biologiques sur
les Ericacees Languedociennes. Stat. Int. Mediter. et alpine Montpellier **47** (1936).
HÄRTEL, O.: Gerbstoffe als Ursache „voller" Zellsäfte. Protoplasma (Berl.) **40**, 338
(1951). — HOAGLAND, D. R.: Lectures on the inorganic nutrition of plants. Waltham, Mass.:
Chronica Bot. 1948. — HÖFLER, K.: Fluoreszenzmikroskopie und Zellphysiologie. Biol.
generalis (Wien) **19**, 90 (1949).
ILJIN, W. S.: Zusammensetzung der Salze in der Pflanze auf verschiedenen Standorten.
Kalkpflanzen. Beih. bot. Zbl. **50**, 95 (1932). — Zur Physiologie der kalkfeindlichen Pflanzen.
Beih. bot. Zbl. **54**, 569 (1936). — Calcium content in different plants an its influence on
production of organic acids. Bull. Assoc. Russe Res. Sci. **7**, 43 (1938). — Quantitativ micro-
analysis of salts and organic acid in plant. Bull. Assoc. Russe Res. Sci. **9**, 1 (1939). — Boden
und Pflanze. II. Physiologie und Biochemie der Kalk- und Kieselpflanzen. Abh. russ.
Forsch. Ges. Prag **10** (15), 75 (1940). — Salze und organische Säuren bei Kalkpflanzen. Flora
(Jena) **37**, 18 (1944).
KASY, R.: Untersuchungen über Verschiedenheiten der Gewebeschichten krautiger
Blütenpflanzen. Sitzgsber. Akad. Wiss. Wien, Math.-naturwiss. Kl. Abt. I, **160**, 509 (1951). —
KARRER, P., u. E. MATTER: Untersuchungen der sauren Bestandteile von Digitalis purpurea L
Helvet. chim. Acta **31**, 799 (1948). — KLEIN, G.: Handbuch der Pflanzenanalyse I.—IV.
Wien: Springer 1931—1933. — KNODEL, H.: Eine Methode zur Bestimmung der stofflichen
Grundlagen des osmotischen Wertes von Pflanzensäften. Planta (Berl.) **28**, 704 (1938).
LARCHER, W.: Schnellmethoden zur Unterscheidung lebender von toten Zellen mit Hilfe
der Eigenfluoreszenz pflanzlicher Zellsäfte. Mikroskopie (Wien) **8**, 299 (1954).
MEYER, B. S.: The water relations of plant cells. Bot. Review **4**, 531 (1938). — MOTHES,
K.: Über den gegenwärtigen Stand der botanischen Alkaloidforschung. Sci. pharmaceut.
**21**, 335 (1953). — MÜLLER-STOLL, W.: Ökologische Untersuchungen an Xerothermpflanzen
des Kraichgaues. Z. Bot. **29**, 161 (1936).
OLSEN, C.: Absorptively bound potassium in beech leaf cells. Physiol. Plantarum
(Copenh.) **1**, 136 (1948).
PAECH, K.: Biologie und Physiologie der sekundären Pflanzenstoffe. Berlin-Göttingen-
Heidelberg: Springer 1950. — PEYNAUD, E.: Contribution a l'etude biochemique de la
maturation du raisin et de la composition des vins. Ann. Chim. analyt., Paris **28**, 111 (1946). —
PISEK, A.: Frosthärte und Zusammensetzung des Zellsaftes bei Rhododendron ferrugineum,
Pinus Cembra und Picea excelsa. Protoplasma (Berl.) **39**, 129 (1950). — PISEK, A., u.
E. CARTELLIERI: Zur Kenntnis des Wasserhaushaltes der Pflanzen. I. Sonnenpflanzen.
Jb. wiss. Bot. **75**, 195 (1931). — PITTIUS, G.: Über die stofflichen Grundlagen des osmotischen
Druckes bei Hedera Helix und Ilex Aquifolium. Bot. Archiv **37**, 43 (1934). — PUCHER, G.:
The organic acids of the leaves of Bryophyllum calycinum. J. of Biol. Chem. **145**, 511 (1942).
SCHNEIDER, A.: Über das Auftreten von Milchsäure in höheren Pflanzen. Planta (Berl.)
**32**, 747 (1942). — STEINER, M.: Zum Chemismus der osmotischen Jahresschwankungen
einiger immergrüner Holzgewächse. Jb. wiss. Bot. **78**, 564 (1933). — Zur Ökologie der Salz-
marschen der nordöstlichen Vereinigten Staaten. Jb. wiss. Bot. **81**, 94 (1934). — Die Zusammensetzung des Zellsaftes bei höheren Pflanzen in ihrer ökologischen Bedeutung. Erg.
Biol. **17**, 151 (1939). — SCHRATZ, E.: Beitrag zur Biologie der Halophyten. III. Über Verteilung, Ausbildung und NaCl-Gehalt der Strandpflanzen in ihrer Abhängigkeit vom Salzgehalt eines Standortes. Jb. wiss. Bot. **83**, 133 (1936). — SPERLICH, A.: Jod, ein brauchbares
mikrochemisches Reagens für Gerbstoffe. Sitzgsber. Akad. Wiss. Wien, Math.-naturwiss.
Kl. Abt. I **126**, 1 (1917).
ULMER, W.: Über den Jahresgang der Frosthärte einiger immergrüner Arten der alpinen
Stufe, sowie der Zirbe und Fichte. Jb. wiss. Bot. **84**, 553 (1937).
VICKERY, H. B., and G. W. PUCHER: Organic acids of plants. Annual Rev. Biochem.
**9**, 529 (1940). — VOLK, O. H.: Untersuchungen über das Verhalten der osmotischen Werte
von Pflanzen aus steppenartigen Gesellschaften und lichten Wäldern des Mainfränkischen
Trockengebietes. Z. Bot. **32**, 65 (1937). — VRIES, H. DE: Eine Methode zur Analyse der
Turgorkraft. Jb. wiss. Bot. **14**, 427 (1884).
WALTER, H.: Die Hydratur der Pflanze. Jena: Gustav Fischer 1931. — Kryoskopische
Bestimmung des osmotischen Wertes bei Pflanzen. In ABDERHALDENs Handbuch der biologischen Arbeitsmethoden, Bd. XI/4, S. 353. 1931. — Tabelle zur Berechnung des osmotischen Wertes von Pflanzen-Preßsäften, Zuckerlösungen und einiger Salzlösungen. Dtsch.
bot. Ges. **54**, 328 (1936). — Grundlagen der Pflanzenverbreitung, Bd. III/1 Standortslehre.
Stuttgart: Eugen Ulmer 1949. — WALTER, H., u. M. STEINER: Die Ökologie der ostafrikanischen Mangroven. Z. Bot. **30**, 65 (1936).

# Der $p_H$-Wert des Zellsaftes.

Von

## H. Drawert.

Mit 3 Abbildungen.

## I. Einleitung.

Der Wasserstoffionenkonzentration ($c_H$) in den einzelnen Bestandteilen der Zelle kommt in verschiedener Hinsicht eine besondere Bedeutung zu. Es sei nur an die $c_H$-Abhängigkeit der Fermentaktivität, der elektrischen Ladung der Kolloide, der Quellung, der Viscosität, des Dissoziationsgrades der Elektrolyte usw. erinnert, die sich ihrerseits wieder auf osmotische und anosmotische Vorgänge, Intrabilität und Permeabilität, Speicherung, Spaltöffnungsbewegung, um nur einige Erscheinungen zu nennen, auswirken. Neben der $c_H$ des Plasmas hat man bisher der $c_H$ des Zellsaftes die größte Aufmerksamkeit geschenkt und mit den verschiedensten Methoden versucht, den $p_H$-Wert des Vacuoleninhaltes zu bestimmen.

Eine kurze Anleitung zur $p_H$-Bestimmung von Pflanzensäften mit den Indikatormethoden nach MICHAELIS und nach WULFF, sowie des Zellsaftes im lebenden Gewebe mit der „range indicator"-Methode nach SMALL finden wir noch im Pflanzenphysiologischen Praktikum von BRAUNER (1929). Wenn wir die oben kurz skizzierte Bedeutung der Acidität des Zellinnern für die Zellphysiologie und den Umfang der vor allem auch in methodischer Hinsicht vorhandenen Literatur über den $p_H$-Wert besonders des Zellsaftes berücksichtigen, erscheint es erstaunlich, daß wir aber in keinem der modernen pflanzenphysiologischen Praktika eine Anleitung zur Bestimmung etwa der uns hier besonders interessierenden Zellsaft-$c_H$ mehr finden. Das in dieser Hinsicht wohl zuständige Praktikum von STRUGGER (1949) erwähnt überhaupt nichts von der Reaktion des Zellsaftes der normalen Pflanzenzelle. Nur der bekannte Versuch zum Nachweis einer Änderung der Zellsaftreaktion durch Behandlung anthocyanhaltiger Zellen mit Säuren und Laugen unterschiedlichen Dissoziationsgrades wird aufgeführt.

Diese zunächst erstaunliche Tatsache hat aber ihren guten Grund. Wir verfügen bis heute über keine einwandfreien Methoden, in der lebenden Zelle den $p_H$-Wert der einzelnen Bestandteile zu bestimmen. Es ist daher eine wenig befriedigende Aufgabe, über den $p_H$-Wert des Zellsaftes berichten zu müssen. Die im folgenden geschilderten Methoden sind mit vielen Fehlerquellen behaftet und die damit erhaltenen Ergebnisse stellen bestenfalls Näherungswerte dar. Da die in der Literatur angeführten $p_H$-Werte mit einem mehr oder weniger großen Unsicherheitsfaktor belastet sind, ist davon abgesehen worden, die Literatur erschöpfend zu zitieren.

Auf folgende zusammenfassende Darstellungen vorwiegend methodischer Art, die aber zum Teil ausführliche Literaturverzeichnisse enthalten, sei hingewiesen: MICHAELIS (1922), ARRHENIUS (1924), REISS (1926), PFEIFFER (1927), MEVIUS (1927), CLARK (1928), SMALL (1929), LEUTHARD (1929), KEYSSNER (1931), KOLTHOFF (1932), KORDATZKI (1949), ENDER (1953), ESSELBORN (1953).

# II. Methodik und Ergebnisse.

## A. Kolorimetrische Methoden.

### 1. Zelleigene Indikatoren.

Die Indikatoreneigenschaften vieler Anthocyane legten es nahe, diese natürlichen, zelleigenen Farbstoffe für die $p_H$-Bestimmung des Zellsaftes anthocyanführender Zellen auszunutzen. Kraus, G. (1884) schreibt aber bereits: „Daß der Farbenwechsel anthocyanhaltiger Blüten mit einem wechselnden Säuregehalt des Zellsaftes zusammenhängt, ist vielfach vermutet oder behauptet worden, aber von niemandem bis jetzt bewiesen." Kraus, der zwar durch Titration nur die potentielle Acidität des Preßsaftes aus Borraginaceen-Blumenblättern bestimmt, weist darauf hin, daß rote Blumenblätter zwar eine höhere „Saftacidität" besitzen als blaue, aber davon abgesehen, die blauen auch — entgegen älteren Angaben — eine saure Gesamtreaktion aufweisen. Schwarz (1892) zieht ganz allgemein den Schluß, daß Zellen mit rotem Anthocyan einen sauren Zellsaft besitzen müssen und daß blaues Anthocyan auf alkalische Reaktion hindeutet. Die gegenteiligen Befunde von Kraus werden auf Fehler zurückgeführt. Schwarz weist darauf hin, daß der aus angeschnittenen und zerquetschten Pflanzenteilen austretende Saft zu Unrecht als „Zellsaft" bezeichnet wird, da er auch immer Stoffe aus dem Protoplasma enthält. Der allgemeinere Name „Pflanzensaft" wäre besser. Aus der weiten Verbreitung roter Blüten schließt der Autor, daß saure Zellsäfte vorherrschen, allerdings könnte die vorhandene Säuremenge nur gering sein, da schon Spuren von $NH_3$ einen Umschlag nach Blau bedingen.

Der Nachweis von Willstätter (1914), daß die rote Rose und die blaue Kornblume denselben Zellsaftfarbstoff besitzen, war eine große Stütze für die Auffassung, daß der Farbton des zelleigenen Anthocyans Aufschlüsse über die Zellsaftacidität gibt. Die Rose soll das rote Oxoniumsalz und die Kornblume das blaue Kaliumsalz des Cyanidins enthalten. Rohde (1917) bestimmte dann auch mit der Wasserstoffelektrode den $p_H$-Wert des Preßsaftes aus den Petalen der Rose mit $\sim 5{,}5$ und den aus den Kronenblättern der Kornblume mit $\sim 7{,}2$. Es ist deshalb nicht verwunderlich, wenn bis in die Gegenwart von vielen Autoren aus dem Farbton der zelleigenen Anthocyane auf den $p_H$-Wert des Zellsaftes geschlossen wird, so Cholnoky (1950, S. 117), Küster (1951, S. 492) und Bethe (1952, S. 127).

Nach eingehenden Untersuchungen von Haas (1916) hat besonders Smith (1933) versucht, mit Hilfe des zelleigenen Anthocyans genaue $p_H$-Werte für den Zellsaft verschiedener Blumenblätter anzugeben. Das Anthocyan wird durch Abkochen der Blumenblätter gewonnen und das Filtrat durch Pufferlösungen auf bestimmte $p_H$-Werte gebracht. Die so erhaltene Farbskala dient zum Vergleich, um den $p_H$-Wert des Zellsaftes zu bestimmen. Für verschiedene Farbvarietäten der *Cineraria* (vgl. auch Cholnoky) kommt Smith zu folgenden $p_H$-Werten: 6,2; 6,4; und 7,4. Bei diesen Untersuchungen können außerdem die Erfahrungen von Buxton und Darbishire (1929) bestätigt werden, daß eine ganze Reihe von Pflanzen Anthocyane führen, die selbst im stark alkalischen Bereich nicht nach Blau umschlagen.

Wenn Rohde (1917) unter den zelleigenen Farbstoffen mit guten Indikatoreneigenschaften z. B. das Anthocyan der roten Rübe aufführt, so ist das völlig unverständlich, da dieser Farbstoff ein Musterbeispiel für ein nicht umschlagendes Anthocyan ist.

Dies abweichende Verhalten vieler Anthocyane deutet schon darauf hin, daß es nicht möglich ist, ohne weiteres aus dem Farbton etwa eines Blumenblattes auf den $p_H$-Wert des Zellsaftes zu schließen. Auch die älteren Beobachtungen von FITTING (1912) an einigen *Erodium*-Arten über die Abhängigkeit des Farbtons der Blumenblätter von der Temperatur hätte bereits über die Zuverlässigkeit der in der Zelle befindlichen Anthocyane als Indikatoren für die Zellsaftreaktion Zweifel aufkommen lassen müssen.

Die Brauchbarkeit der zelleigenen Anthocyane als Indikatoren wird aber völlig fraglich durch die neueren Untersuchungen von ROBINSON. Der Farbton eines Blumenblattes wird danach nicht nur von der $c_H$ des Zellsaftes bestimmt, sondern von der Art und von der Konzentration des jeweils vorliegenden Anthocyans, von dem Verhältnis der Anthocyankonzentration zu der Konzentration der Co-Pigmente aus der Tannin- und Flavonol-Klasse und ferner durch die kolloidale Association wahrscheinlich mit Polysacchariden, die bei vielen Anthocyanen eine Farbtonänderung hervorruft. Entgegen den Angaben von ROHDE (1917) über die schwach alkalische Reaktion des Zellsaftes der blauen Kornblumenpetalen bedingt der Petalenpreßsaft eine Rotfärbung von blauem Lakmus und zeigt mit der Glaselektrode $p_H$ 4,9 (ROBINSON 1933, ROBINSON und ROBINSON 1939). Nach ROBINSON (1933) stimmen die Absorptionsspektren der blauen Cyaninlösung in vitro und die der blauen Kornblumenblätter überein, so daß es naheliegend war, anzunehmen, daß die Kornblume einen alkalischen Zellsaft besitzt. Diese allgemeine Auffassung läßt sich aber nicht aufrecht halten. Das Cyanin-Anion liegt in der Zelle wahrscheinlich in einer an kolloidalen Polysacchariden adsorbierten Form vor und wird dadurch stabilisiert, so daß auch bei saurer Zellsaftreaktion der blaue Farbton erhalten bleibt.

Wenn wir etwa in den Schiffchen der Blüten von *Vicia sepium* nebeneinander rote, blaue und violette Vacuolen antreffen (SCHORR 1938), so dürfen wir daraus nicht ohne weiteres auch auf unterschiedliche $p_H$-Werte der Zellsäfte in diesen Vacuolen schließen.

Vor ROBINSON hat bereits ATKINS (1922) darauf hingewiesen, daß das Anthocyan in den Blüten von *Hydrangea* kein Indikator ist, da der Zellsaft sowohl in den roten wie in den blauen Blüten $p_H$ 4,0—4,2 aufweist. Hier liegen wahrscheinlich Komplexsalze mit Eisen und Aluminium vor (ATKINS 1923). PFEIL (1936) erhielt allerdings $p_H$-Unterschiede bei Messungen von Preßsäften mit der Glaselektrode, und zwar ergaben die blauen Hortensienblüten $p_H$ 5,4, die roten dagegen $p_H$ 4,5. G. M. ROBINSON (1939) prüfte elektrometrisch den Preßsaft von Blumenblättern solcher Blüten, die während ihrer Entwicklung einen Farbwechsel von Rot nach Blau aufweisen. Der Preßsaft zeigte dabei kaum eine $p_H$-Änderung. Interessanterweise sind aber nur Blüten mit einer $c_H$ des Preßsaftes um $p_H$ 6 zu solchen Farbänderungen befähigt. Nahe verwandte Formen, die einen etwas sauren Preßsaft z. B. $p_H$ 5,6 besitzen, sind zu Farbtonänderungen während des Blühvorganges nicht mehr in der Lage.

Auch die aus dem Farbton verschiedener Spielarten von *Cineraria* geschätzten $p_H$-Werte des Zellsaftes (SMITH 1933, CHOLNOKY 1950) lassen sich nach DRAWERT (1954) nicht aufrecht halten. Entsprechend den Angaben von SMITH zeigt das Anthocyan roter *Cineraria*-Blüten auch in den Versuchen von DRAWERT nie einen Umschlag nach Blau, verhält sich also ganz anders als das blauer Blüten, das in vitro gute Indikatoreneigenschaften aufweist. SMITH schließt aus dem Farbton des zelleigenen Anthocyans für den Zellsaft roter Blüten auf $p_H$ 6,2 und für den blauer Blüten auf $p_H$ 7,4. Bestimmungen mit der Mikroglaselektrode am unverdünnten Preßsaft ergaben aber für beide Spielarten übereinstimmend $p_H$ 5,4 (DRAWERT 1954).

Gegen einen Vergleich der $p_H$-Werte von Preßsäften mit den aus dem Farbton des zelleigenen Anthocyans geschlossenen $p_H$-Werten, kann man natürlich einwenden, daß das Anthocyan vorwiegend nur in der Epidermis lokalisiert ist, während der Preßsaft ja auch die Zellen des Mesophylls mit enthält, deren Zellsaftacidität von der der Epidermis unter Umständen stark abweicht. Dieser ohne Zweifel berechtigte Einwand läßt sich aber dadurch entkräften, daß der Farbton des Preßsaftes im vorliegenden Fall keine Änderung zeigte, die auf eine stärkere Ansäuerung durch den Saft der Mesophyllzellen hingedeutet hätte. Über die Möglichkeit, aus Preßsäften auf die Zellsaftacidität zu schließen und die dagegen zu erhebenden Einwände vgl. S. 633ff.

Aus dieser kurzen Übersicht geht hervor, daß nicht nur KRAUS (1884) mit seiner Schlußfolgerung in bezug auf die potentielle Acidität über die saure Zellsaftreaktion auch der blauen Blüten Recht behalten hat, sondern daß dasselbe auch für die aktuelle Acidität zutrifft. — Es soll in diesem Zusammenhang nur kurz darauf hingewiesen werden, daß die potentielle Acidität meist nicht mit der aktuellen parallel läuft (GUSTAFSON 1924, 1925). — Wie aus dem dargelegten hervorgeht, kommen die zelleigenen Anthocyane für eine zuverlässige pH-Bestimmung des Zellsaftes nicht in Frage.

## 2. Vitalfärbung mit synthetischen Indikatoren.

Neben den zelleigenen Indikatoren sind auch die synthetischen Farbstoffe, soweit sie Indikatoreneigenschaften besitzen, dazu herangezogen worden, um durch eine Vitalfärbung den pH-Wert des Zellsaftes der von Natur aus farblosen Zellen zu bestimmen. Zu den Voraussetzungen für die Anwendung zellfremder Indikatoren gehört es aber, daß die Zellen für den Farbstoff nicht nur permeabel sind, sondern daß sie ihn auch im Zellsaft speichern können. Diese Voraussetzung trifft aber für viele Indikatoren nicht zu. Man hat deshalb versucht, diese Farbstoffe mit Hilfe des Mikromanipulators in die Zellen zu injizieren. Methodisch müssen wir demnach das Immersions- und das Injektionsverfahren unterscheiden.

### a) Immersionsverfahren.

Die meisten Versuche sind seit PFEFFER (1886) wohl mit dem Immersionsverfahren durchgeführt worden. Dazu eignen sich besonders die basischen Farbstoffe, weil sie durch den Zellsaft im allgemeinen stärker gespeichert werden als die sauren. Vor allem das Neutralrot ist für pH-Bestimmungen viel herangezogen worden (KÜSTER 1928). Von den sauren Farbstoffen hat nur Methylrot (SCHAEDE 1924) für das Immersionsverfahren eine größere Bedeutung erlangt. Die Farbstoffe kommen meist in einer 0,01%igen wäßrigen Lösung zur Anwendung. Leitungswasser ist auf Grund seiner vorwiegend schwach alkalischen Reaktion für die basischen Farbstoffe als Lösungsmittel besser geeignet als destilliertes Wasser. Bei größeren Gewebekomplexen empfiehlt es sich, mit Hilfe der Vakuum- (KELLER und GICKLHORN 1928) oder der Zentrifugenmethode (WEBER 1926) das Gewebe mit der Farblösung zu infiltrieren.

Nach den Erfahrungen mit dem zelleigenen Anthocyan muß es allerdings fraglich erscheinen, ob man mit den synthetischen Vitalfarbstoffen wirklich einwandfreie pH-Bestimmungen des Zellsaftes durchführen kann. Im Vergleich zum Anthocyan liegen bedeutend mehr Untersuchungen über mögliche Fehlerquellen bei der pH-Messung mit Indikatoren ganz allgemein vor. PFEIFFER (1927) führt folgende zu berücksichtigende Faktoren an: 1. Unterschiede in der Löslichkeit der Indikatoren; 2. störende Eigenfarbe der Zellbestandteile; 3. chemische Bindung oder Adsorption des Indikators an Strukturpartikeln des Plasmas oder des Zellsaftes. Hierher würde auch der sog. Proteinfehler zu rechnen sein. Nach GOLDACRE (1953) kann ein Indikator je nachdem, ob ein anwesendes Protein aufgerollte oder gestreckte Moleküle besitzt, bei gleichem $p_H$-Wert einen unterschiedlichen Farbton zeigen. Ein weiterer Faktor ist 4. der Salzfehler. Neutralsalze verschieben die Farbnuance eines sauren Indikators nach der alkalischen und die eines basischen Farbstoffes nach der sauren Seite. Für den amphoteren Farbstoff Prune pure gibt RUHLAND (1923) folgende Salzfehler an: Umschlag von Blau nach Violett in salzarmer Lösung bei $p_H$ 8, in Phosphatgemischen bei $p_H \sim 7$ und in Borat/Boraxgemischen ist der Umschlag bei $p_H$ 8,84 noch nicht erfolgt. 5. Auch die von reinem Wasser abweichende Dielektrizitätskonstante des Plasmas ist zu berücksichtigen, während Temperatur- und Alkoholfehler für biologische $p_H$-Messungen nur von untergeordneter Bedeutung sein dürften. Dazu käme mit BRENNER (1928) noch 6. der Konzentrationsfehler, auf den auch GOLDACRE (1953) besonders hinweist. Beim Anthocyan haben wir den Einfluß der Konzentration auf den Farbton bereits erwähnt. Nach BRENNER soll aber auch das Neutralrot, der gebräuchlichste Farbstoff für Zellsaft-$p_H$-Bestimmungen, auf Grund des Konzentrationseffektes als Indikator für Säuren völlig unbrauchbar sein. Fügt man zu einer 0,1%igen Neutralrotlösung ihr eigenes Volumen $^1/_{10}$ GM Salzsäure hinzu, so erhält man keinen Umschlag aber einige Tropfen GM KOH genügen bereits, um eine Farbtonänderung hervorzurufen. Erst eine 0,005%ige Neutralrotlösung zeigt nach einem Zusatz der gleichen Menge $^1/_{20}$ GM Salzsäure einen karminroten bis violetten Farbton. In relativ hohen Konzentrationen ist der Farbstoff also unempfindlich gegen Säuren. Nach GOLDACRE (1953) erscheint aber auch eine auf $p_H$ 8 gepufferte Neutralrotlösung bei hoher Konzentration rot und erst bei niedriger Konzentration gelb. Ein weiterer, bis heute noch nicht restlos geklärter Faktor, ist 7. die „Metachromasie", die bei vielen basischen Indikatoren zu beobachten ist. Als Metachromasie bezeichnen wir seit EHRLICH (1877) die Erscheinung, daß ein Farbstoff bestimmte Zellelemente in einem anderen Farbton anfärbt als er der benutzten Farblösung zukommt. Nach LISON (1935) kann der metachromatische Fehler bei $p_H$-Bestimmungen mit basischen Farbstoffen unter Umständen 6 $p_H$-Einheiten betragen. So ändert eine Spur Ca-chondroitinsulfat zu einer Brillantkresylblaulösung mit $p_H$ 4,0 zugesetzt den Farbton so, daß man auf $p_H$ 10,4 schließen würde, ohne daß sich der ursprüngliche $p_H$-Wert verschoben hat, oder bei einer Neutralrotlösung mit $p_H$ 2,2 deutet die Farbtonänderung auf $p_H$ 7,2—7,4 hin. Aus Färbungen mit Neutralrot und anderen basischen Farbstoffen schließt DANGEARD (1923), daß die alkalische oder neutrale Reaktion des Zellsaftes der Meristemzellen bei den Gymnospermen eine fast allgemeingültige Regel ist, und in den Pollenkörnern der Gymnospermen soll das Vacuolensystem der vegetativen Zelle neutral bis schwach sauer, das der generativen Zelle schwach alkalisch reagieren. DANGEARD nimmt an, daß die hier zu beobachtenden metachromatischen Farbtöne durch eine leichte Alkalinität des Zellsaftes bedingt sein sollen. Diese Schlußfolgerungen können aber nicht zutreffen, da der Autor auch mit Toluidinblau, dessen Umschlagspunkt über $p_H$ 9 liegt, und mit Methylenblau, das überhaupt keinen $p_H$-abhängigen Umschlagspunkt besitzt, die metachromatischen Zellsaftfärbungen erhält. Da hier nicht näher auf die Metachromasie eingegangen werden kann, sei auf die Zusammenfassungen von LISON und FAUTREZ (1939) und GILLISSEN (1953) verwiesen. Ein im allgemeinen nicht beachteter Fehler ist 8. die Verschiebung des Zellsaft-$p_H$-Wertes durch den aufgenommenen Indikator. Der den großen Zellen von *Nitella flexilis* entnommene Zellsaft zeigt nach IRWIN (1931) mit der Glaselektrode $p_H$ 5,36. Aus einer 0,004%igen Lösung von Brillantkresylblau mit $p_H$ 9,2 speichern die Zellen innerhalb von 15 min aber so viel Farbstoff (Farbbase), daß der $p_H$-Wert des Zellsaftes auf $p_H$ 6,66 ansteigt. Die wirklich vorliegenden Verhältnisse können auch noch dadurch unübersichtlich werden, daß der aufgenommene

Farbstoff im Zellsaft Entmischungen oder Niederschläge bedingt. So reagiert nach Schaede (1923) der Zellsaft in den Wurzelhaaren von *Hydrocharis morsus ranae* wohl neutral, die entstehenden Niederschläge sind aber sauer; „denn sie nehmen bei Methyl- und Gentianaviolett blauviolette, bei Safranin hellkarmin Farbe an".

Auf alle Fälle müssen wir auch bei den Vitalfarbstoffen wie beim Anthocyan in der Vacuole mit farbtonbestimmenden Adsorptionserscheinungen an Zellsaftkolloiden rechnen, die eine Bestimmung des reellen $p_H$-Wertes des Zellsaftes unmöglich machen. Dafür sprechen die Unstimmigkeiten der bei manchen Zellsorten mit verschiedenen Indikatoren erhaltenen $p_H$-Werte und die Diskrepanz zwischen kolorimetrischer und elektrometrischer $p_H$-Bestimmung. Nach Bailey und Zirkle (1931) kann man nach Neutralrotfärbung in den Cambiumzellen von Gymnospermen 2 Vacuolenarten unterscheiden, die eine Gruppe speichert den Farbstoff mit bläulich-magentarotem und die andere mit einem rotorange Farbton. Aus den Farbtönen würde man in dem einen Fall auf eine merklich saure und in dem anderen Fall auf eine schwach alkalische Reaktion des Zellsaftes schließen. Die „sauren" tanninhaltigen (!) Vacuolen ergaben mit 36 verschiedenen Indikatoren übereinstimmend $p_H \sim 4{,}4$. Die „schwach alkalischen" tanninfreien Vacuolen zeigten dagegen in Abhängigkeit von den benutzten 12 Indikatoren $p_H$-Werte von 7,2 (Neutralrot) bis 13 (Brillantkresylblau). Für das oben bereits angeführte Prune pure, das in salzarmer Lösung bei $p_H$ 8 von Blau nach Violett umschlägt, erwähnt Ruhland (1923), daß eine $^1/_2$%ige Hühnereiweißlösung diesen Umschlag verhindert. Nach demselben Autor läßt der Farbton einer mit Chrysoidin versetzten 0,2%igen Eiweißlösung auf $p_H$ 6,5 schließen, elektrometrisch gemessen beträgt ihr $p_H$-Wert aber 3,02.

Bei kolorimetrischen $p_H$-Wertbestimmungen in vitro ist es wohl möglich, Salz-, Protein- und Konzentrationsfehler zu berücksichtigen, dies ist aber nicht bei der lebenden Zelle durchführbar, da die Größenordnung dieser Fehler in vivo nicht bestimmbar ist (J. und D. Needham 1926). Alle durch Vitalfärbung mit Indikatoren erfolgten $p_H$ Bestimmungen des Zellsaftes müssen wir mit großer Skepsis betrachten. Das trifft auch in hohem Grade für die kolorimetrischen Bestimmungen der Zellsaftacidität von Schließzellen zu. So beträgt, aus dem Farbton gespeicherten Neutralrots geschlossen, nach Pekarek (1934) der $p_H$-Wert des Zellsaftes der Schließ- und Epidermiszellen von *Rumex acetosa* im Licht 5,3—6,0 bzw. 5,0—6,2, im Dunkeln dagegen 4,5 bzw. 7,6—8,0. Andererseits wissen wir aber, daß das Öffnen und Schließen der Stomata mit beträchtlichen kolloidalen Änderungen nicht nur des Plasmas, sondern auch des Zellsaftes der Schließzellen verbunden ist, so daß es sehr fraglich erscheinen muß, ob die von Pekarek beobachteten Farbtonänderungen nur auf eine $c_H$-Änderung des Zellsaftes zurückzuführen sind.

Derselbe Einwand ist auch gegen die Annahme von Drawert (1948) zu erheben, daß der Umschlagspunkt Membran-/Vacuolenfärbung außer von dem $p_H$-Wert der benutzten Farblösung noch von der Acidität des Zellsaftes abhängt, da diese aus dem Farbton des gespeicherten Neutralrots erschlossen wurde. Auch hier kann unter Umständen eine Änderung des Kolloidgehaltes des Zellsaftes sich einerseits auf den Farbton des Neutralrots und andererseits auf die Speicherfähigkeit des Zellsaftes und damit auf die Lage des Umschlagspunktes Membran-/Vacuolenfärbung auswirken (Drawert 1951). Jedenfalls kann sowohl eine Änderung der Zellsaftacidität wie auch eine Verschiebung des Kolloidgehaltes die Lage des Umschlagpunktes beeinflussen.

Schon aus diesen wenigen Beispielen geht hervor, daß wir auch durch Vitalfärbung mit dem Immersionsverfahren nicht in der Lage sind, einwandfreie Aussagen über den $p_H$-Wert des Zellsaftes zu machen.

## b) Injektionsverfahren.

Die meisten sauren Farbstoffe scheinen einigen der erwähnten Fehlerquellen nicht in dem Maße zu unterliegen wie die basischen. Sie haben aber den Nachteil, daß der Vacuoleninhalt der Pflanzenzellen — von wenigen Ausnahmen abgesehen— für sie keine Speicherfähigkeit besitzt. Man versuchte diesem Übelstand dadurch abzuhelfen, daß man die Indikatoren mit Hilfe des Mikromanipulators injizierte. Dabei muß natürlich die Verletzung des Plasmaschlauches in Kauf genommen werden, die sich auch in unbekannter Weise auf die Acidität des Zellsaftes aus-wirken kann. SCHMIDTMANN (1924) empfiehlt die Injektion kleiner Stückchen Indikatorsubstanz. Im allgemeinen werden aber wäßrige Indikatorlösungen inji-ziert. Es sei auf folgende Arbeiten verwiesen: RAPKINE und WURMSER (1926), PLOWE (1931), RIBBERT (1931), CHAMBERS und KERR (1932), KERR (1933), HOF-MEISTER (1940). Bei den injizierten Indikatoren handelt es sich vorwiegend um: Bromcresolpurpur, Bromthymolblau, Phenol- und Methylrot, aber auch um basi-sche Farbstoffe wie Neutralrot.

RIBBERT (1931) erhielt recht unterschiedliche Werte je nachdem, ob er Bromthymolblau in die Epidermiszellen von *Tradescantia virginica* injizierte oder die Zellen mit demselben Farbstoff nach der „range indicator"-Methode von SMALL anfärbte. In die Zellen injiziertes Bromthymolblau färbte den Zellsaft mit rein gelbem Farbton an, die nach der „range indicator"-Methode behandelten Zellen färbten sich dagegen gelbgrün, zeigten also eine mehr alkalische Reaktion. RIBBERT (1931) weist darauf hin, daß alle mit der „range indicator"-Methode erhaltenen Ergebnisse dringend einer Nachprüfung bedürfen.

## 3. Kolorimetrische Bestimmungen an Preßsäften.

Da das Hauptvolumen ausgewachsener Zellen von der Vacuole eingenommen wird, kann man den berechtigten Schluß ziehen, daß die Hauptmasse der durch Auspressen von Gewebekomplexen gewonnenen Flüssigkeitsmenge aus Zellsaft besteht. So ist häufig der Preßsaft zur Bestimmung des $p_H$-Wertes mit Indi-katoren benutzt worden. Diese Methode hat aber außer den für Indikatoren oben schon angeführten Fehlermöglichkeiten noch 3 nicht zu unterschätzende Fehlerquellen.

1. Man wird kaum wirklich reinen Zellsaft erhalten, sondern der Preßsaft enthält wohl stets Bestandteile des Plasmas, die sich auch auf den $p_H$-Wert in verschiedener Richtung auswirken können. Nach TAYLOR und WHITAKER (1927) erniedrigen bei elektrometrischen Messungen am Zellsaft von *Nitella* bereits ge-ringste Plasmamengen das Potential, wenn diese mit dem Zellsaft mechanisch gemischt werden. HOAGLAND und DAVIS (1923) zeigten ebenfalls an *Nitella*, daß der für sich isolierte Zellsaft eine ganz andere Salzkonzentration aufweist als der Preßsaft, wie aus Tabelle 1 hervorgeht.

Tabelle 1. *Die Zusammensetzung von isoliertem Zellsaft und von Preßsaft aus Nitella.* Nach HOAGLAND und DAVIS (1923).

| | Spezifischer Widerstand in Ohm | 1:1 000 000 | | | |
|---|---|---|---|---|---|
| | | K | Cl | Ca | Mg |
| Zellsaft . . . . . . . . . . | 81 | 2,200 | 3,200 | 600 | 430 |
| Preßsaft . . . . . . . . . . | 120 | 1,310 | 2,250 | 310 | 150 |
| Verdünnungsfaktor . . . . . | 1,48 | 1,68 | 1,43 | 1,94 | 2,82 |

Nach den Ergebnissen von HOAGLAND und DAVIS ist der Preßsaft verdünnter als der Zellsaft. Nach Meinung der Autoren ist diese Verdünnung wohl vorwiegend

auf das den Zellen noch anhaftende Wasser zurückzuführen. Sehr wahrscheinlich kommt aber auch dem zerstörten Plasma bei der Preßsaftgewinnung eine
Bedeutung als Adsorbens zu. Totes Plasma ist viel reaktionsfähiger als lebendes
(DRAWERT 1948), so daß hier das Plasma dadurch eine Art Filterwirkung ausübt
und auch als Ionenaustauscher von Bedeutung sein kann (vgl. auch MEVIUS
1927, S. 96/97). Diese Vorgänge müssen sich auch auf den $p_H$-Wert des Saftes
auswirken. HURD-KARRER (1939) berichtet allerdings, daß es für den $p_H$-Wert
des Preßsaftes aus Weizenpflanzen gleichgültig war, ob zur Gewinnung niedrige
oder hohe Drucke angewendet wurden. An sich wäre zu vermuten, daß mit
zunehmendem Druck der plasmatische Anteil am Preßsaft größer wird.

2. Durch die zur Preßsaftgewinnung unvermeidliche Verwundung oder Ab-
tötung der Gewebe werden die Struktur der Zellen und der ganze Stoffwechsel
so stark geändert, daß es verwunderlich wäre, wenn sich diese Eingriffe nicht
auch auf den $p_H$-Wert auswirken würden. Der Farbton mancher Zellen des
*Begonia*-Blattstiels ändert sich bei einem Aufenthalt in Bromkresolgrünlösung
bei Sauerstoffgegenwart bereits nach 15 min. so, daß man auf $p_H > 5$ schließen
muß, während die frischen Schnitte $p_H \sim 1,5$ vermuten lassen (ROSE und HURD-
KARRER 1927). Auch GUSTAFSON (1925) weist darauf hin, daß in den meisten
Pflanzenpreßsäften unter Umständen schon innerhalb von 30 min nach dem Auspressen bedeutende $c_H$-Änderungen auftreten können. Nur der Preßsaft von
*Bryophyllum calycinum* scheint davon eine Ausnahme zu machen, da er tagelang
seinen ursprünglichen $p_H$-Wert beibehält. Nach BAILEY und ZIRKLE (1931) kann
man bei *Nitella* aus dem Farbton des Zellsaftes nach einer Vitalfärbung mit
Neutralrot auf $p_H$ 7,0—7,8 schließen. Der ausgepreßte Saft ergibt aber sowohl
kolorimetrisch wie elektrometrisch $p_H$ 5,6. Andererseits soll nach HURD-KARRER
(1939) und auch nach HAAS (1941) ein Gefrieren des Gewebes vor der Preßsaftgewinnung kaum einen Einfluß auf den $p_H$-Wert haben.

3. Von den großzelligen Algen wie *Valonia* und *Nitella* abgesehen, wird es
besonders für elektrometrische $p_H$-Wertbestimmungen immer notwendig sein, zur
Preßsaftgewinnung ganze Zellkomplexe zu verwenden, um die benötigte Flüssigkeitsmenge zu erhalten. Damit erhält man aber nur einen Durchschnittswert
von einem Gewebe und nicht den $p_H$-Wert des Zellsaftes der einzelnen Zellen,
der sehr verschieden sein kann.

Die Feststellung der Acidität des Preßsaftes einzelner Zellen in einem ganzen
Gewebekomplex ist mit der kolorimetrischen Methode allerdings eher möglich,
als mit der elektrometrischen, besonders bei der Benutzung von Indikatorpapier.
Durch Ausdrücken eines dünnen Gewebeschnittes auf Filtrierpapier, das vorher
mit einer Indikatorlösung getränkt und wieder getrocknet worden ist, erhält man
eine topographische Übersicht der $p_H$-Wertverteilung. Mit diesem ,,Abdruck-
verfahren`` und Lakmuspapier hat bereits SACHS (1862) festgestellt, daß der Saft
der Siebröhren ebenso wie das in Teilung begriffene Gewebe der Wurzelspitze
alkalisch reagieren müßte, während Zellen der Streckungszone neutral und ausgewachsene Parenchymzellen sauer sind. Die alkalische Reaktion des Siebröhren-
saftes ist in der Folgezeit öfters bestätigt worden, so von RUHLAND (1912) bei
*Beta vulgaris*; und nach CRAFTS (1936) zeigt auch der Phloëmsaft vom Kürbis
mit der ,,spot-plate``-Methode $p_H \sim 7,3$. Die schwach alkalische Reaktion des
Phloëmsaftes scheint zwar weit verbreitet zu sein, man darf diese Erscheinung
aber nicht auf alle Pflanzenarten verallgemeinern, wie Tabelle 2 belegt.

Nach den Angaben von ROGERS und SHIVE (1932) besitzt der Siebteil von
*Oxalis repens* und *Rumex acetosella* sogar einen recht sauren Zellsaft. MARTIN
(1927) bestimmt den $p_H$-Wert des Siebteilgewebes von *Helianthus annuus* eben-

Tabelle 2. $p_H$-*Werte des Zellsaftes verschiedener Gewebe mit der „range-indicator"-Methode bestimmt. Nach* ROGERS *und* SHIVE (1932).

| Pflanze | Organ | pH des Zellsaftes | | | | | |
|---------|-------|-----------|-------|--------|-----------|------|----------|
| | | Epidermis | Rinde | Phloëm | Endodermis | Mark | Mesophyll |
| *Oxalis repens* | Stengel | 2,6—2,8 | 3,8—4,2 | 3,8—4,4 | 2,2—2,8 | 3,8—4,2 | — |
| *Rumex acetosella* | Blattstiel | — | 3,6—4,0 | 4,0—4,4 | — | 3,4—3,8 | — |
| *Zea mays* | Stengel Blatt | — | 5,6—6,0 | 5,8—6,2 | — | — | 5,6—6,0 |
| *Trifolium spec.* | Blattstiel | — | 6,4—6,8 | 6,8—7,2 | — | 6,4—6,8 | — |
| *Glycine soja* | Stengel | — | 6.6—7,0 | 6,8—7,4 | — | 6,6—7,2 | — |
| *Solanum tuberosum* | Stengel Blattstiel | — | 6,4—6,8 | 6,8—7,6 | — | 6,4—7,0 | — |

falls mit der „range-indicator"-Methode nach SMALL und findet $p_H$ 4,8—5,2 bei jüngeren und $p_H$ 5,8—6,0 bei älteren Pflanzen. ARRHENIUS (1924) gibt für die Zellen der verschiedenen Gewebearten einer Weizenpflanze die in Tabelle 3 zusammengefaßten kolorimetrisch gefundenen $p_H$-Werte an.

Tabelle 3. $p_H$-*Werte des Saftes verschiedener Zellen einer Weizenpflanze.* Nach ARRHENIUS (1924).

| | | | |
|---|---|---|---|
| Schließzellen des Blattes . . . | $p_H$ 4,1 | Wurzelzellen . . . . . . . . | $p_H$ 4,5 |
| Mittelnerv des Blattes . . . . | $p_H$ 4,1 | Epidermis des Stengels . . . | $p_H$ 6,1 |
| Epidermis der Blattbasis . . . | $p_H$ 6,0 | Gefäße des Stengels . . . . | $p_H$ 6,5 |
| Parenchym der Blattbasis . . | $p_H$ 6,5 | | |

In neuerer Zeit hat BOGEN (1938) im Abdruckverfahren mit der auf Filtrierpapier aufgetrockneten Indikatorserie nach SÖRENSEN die in Tabelle 4 zusammengefaßten $p_H$-Werte erhalten. Die mit dieser einfachen Methode erzielte Genauigkeit soll 0,3—0,5 $p_H$-Einheiten betragen.

Tabelle 4. *Mit dem Abdruckverfahren bestimmte* $p_H$-*Werte des Preßsaftes verschiedener Gewebe.* Nach BOGEN (1938).

| Art | Gewebe | pH-Wert |
|-----|--------|---------|
| *Pelargonium zonale* . . . . . . . . | Blattstiel, Mark | 1,5 |
| *Begonia angularis* . . . . . . . . | Blattstiel, Subepidermis | 1,9 |
| *Pelargonium zonale* . . . . . . . . | Blattstiel, Chlorenchym | 2,0 |
| *Oxalis Deppei* . . . . . . . . . . | Blattstiel, Chlorenchym | 2,0 |
| *Geranium macrorrhizon* . . . . . . | Blattstiel, Epidermis | 2,0—2,5 |
| *Polygonum cuspidatum* . . . . . . | Blattstiel, Mark | 2,5 |
| *Pelargonium zonale* . . . . . . . . | Blattstiel, Epidermis | 3,0 |
| *Polygonum cuspidatum* . . . . . . | Blattstiel, Epidermis | 3,0—3,3 |
| *Rumex scutatus* . . . . . . . . | Blattstiel, Epidermis | 3,0—3,5 |
| *Polygonum cuspidatum* . . . . . . | Blattstiel, Hypoderm | 3,0—3,5 |
| *Oxalis Deppei* . . . . . . . . . . | Blattstiel, Epidermis | 3,5—4,0 |
| *Gentiana frigida* . . . . . . . . | Stengel, Epidermis | 4,3 |
| *Trifolium pratense* . . . . . . . . | Stengel, Epidermis | 4,5—5,0 |
| *Taraxacum officinale* . . . . . . . | Blattmittelrippe, Epidermis | 5,0—5,5 |
| *Sempervivum* . . . . . . . . . . | Mesophyll | 5,3 |
| *Lupinus albus* . . . . . . . . . | Stengel, Hypoderm | 5,5—6,0 |
| *Caltha palustris* . . . . . . . . | Blattstiel, Subepidermis | 6,0 |
| *Sanchezia nobilis* . . . . . . . . | Blattrippen, Epidermis | 6,4 |
| *Lupinus albus* . . . . . . . . . | Blattstiel, Epidermis | 7,0 |

## B. Elektrometrische Methoden.

Die elektrometrischen Methoden der $p_H$-Wertbestimmung zeichnen sich verglichen mit den kolorimetrischen Verfahren durch größere Genauigkeit aus, so daß sie auch sehr bald in die Zellphysiologie Eingang fanden. Die gebräuchlichsten Elektroden sind die Chinhydron-, die Wasserstoff- und die Glaselektrode. In letzter Zeit ist die Glaselektrode beherrschend geworden. Von einigen Versuchen an großzelligen Algen abgesehen, ist es bisher aber noch nicht möglich gewesen, den $p_H$-Wert des Zellsaftes in der lebenden Zelle elektrometrisch zu messen, so daß man bei diesen Methoden auf Preßsaft angewiesen ist.

Mit eingestochener Wasserstoffelektrode haben Taylor und Whitacker (1927) den $p_H$-Wert des Zellsaftes von *Nitella* gemessen und $p_H$ 5,47—6,16 erhalten. Es ist von Interesse, dieses Ergebnis mit den von anderen Autoren an isoliertem Zellsaft gefundenen $p_H$-Werten zu vergleichen. Allerdings ist dabei zu bedenken, daß es sich nicht immer um dieselbe Art handelt und außerdem auch nichts über den physiologischen Zustand des jeweils benutzten Materials bekannt ist. Hoagland und Davis (1923) erhielten elektrometrisch $p_H$ 5,1, kolorimetrisch 5,2 und bei der kolorimetrischen Austestung von Einzelzellen ergaben sich individuelle Schwankungen zwischen $p_H$ 4,8—5,8. Nach Pearsall und Ewing (1924) beträgt der $p_H$-Wert kolorimetrisch gemessen 5,2 und nach Irwin (1926), ebenfalls kolorimetrisch bestimmt 5,6. Später benutzte Irwin (1931) eine Glaselektrode und erhielt bei *Nitella flexilis* $p_H$ 5,36. $p_H$ 5,6 geben Bailey und Zirkle (1931) sowohl nach elektro- wie kolorimetrischer Messung an und Albaum, Kaiser und Nestler (1937) fanden mit der Glaselektrode $p_H$ 5,2. Wenn man die vielen möglichen Fehlerquellen berücksichtigt, zeigen die Werte recht gute Übereinstimmung. Völlig aus der Reihe fällt nur der nach Vitalfärbung mit Neutralrot von Bailey und Zirkle (1931) gefundene Wert von $p_H$ 7,0—7,8, der wohl kaum zutreffen dürfte.

Recht beträchtliche Unterschiede zwischen der kolorimetrischen Methode nach Small und dem elektrometrischen Verfahren erhält Stanfield (1937) bei *Lychnis dioica*. Im blühenden Zustand zeigt diese Caryophyllaceae zwischen den weiblichen und den männlichen Pflanzen Differenzen im $p_H$-Wert der Säfte. An Querschnitten durch den Stengel können mit der „range-indicator"-Methode je nach Gewebeart folgende Werte bestimmt werden: weibliche Pflanze $p_H$ 3,8 bis 6,0 und männliche Pflanze $p_H$ 4,8—6,0. Die Preßsäfte ergeben aber in demselben Zustand der Pflanze elektrometrisch gemessen: weibliche Pflanzen $p_H$ 6,6 und männliche $p_H$ 5,5. Während also die weiblichen Pflanzen kolorimetrisch an Querschnitten bestimmt im Durchschnitt saurer reagieren als die männlichen, weist der Preßsaft derselben Pflanzenteile elektrometrisch gemessen bei den männlichen Pflanzen eine höhere Acidität auf.

Tabelle 5 gibt eine allgemeine Übersicht über einige an Preßsäften vorwiegend elektrometrisch erhaltenen Ergebnisse.

Weitere zusammenfassende Angaben über den $p_H$-Wert von Preßsäften sind bei Small (1929), Pfeiffer (1933) und Hurd-Karrer (1939) zu finden.

Aus der Übersicht geht hervor, daß nur wenige Pflanzen in einzelnen Organen einen sehr stark sauren Saft aufweisen und noch weniger einen schwach alkalischen. Die Hauptmasse der untersuchten Säfte liegt in dem Bereich von $p_H$ 5,0 bis 6,5. Auffallend ist die relativ geringe Acidität der Wurzelgewebe, allerdings liegen über die Wurzel auch nicht sehr zahlreiche Angaben vor. Über den Neutralpunkt hinaus gehen nur wenige Werte. Es sind dies unter anderem Blatt und Stengel von *Cucurbita pepo*. Besonders stark alkalisch reagieren nach Martin (1927) die Haare von *Helianthus annuus*, ihr $p_H$-Wert soll zwischen 9 und 10

*Tabelle 5.*

p$_H$-Werte von Preßsäften verschiedener Pflanzen. Die Pflanzen sind in der Reihe nach steigenden p$_H$-Werten aufgeführt. Enthalten in den Originalarbeiten die p$_H$-Angaben mehr als eine Stelle nach dem Komma, dann sind die hier wiedergegebenen Werte auf eine Stelle nach dem Komma auf- bzw. abgerundet worden. Ist ein p$_H$-Bereich aufgeführt, so ist damit die gefundene Variationsbreite in bezug auf innere oder äußere Faktoren wie Alter, Licht, Temperatur, Boden usw. umgrenzt. War aus den Angaben nicht klar ersichtlich, ob sich die Werte auf die Blätter oder den Stengel beziehen, so werden sie hier unter „Sproß" angeführt.

| Pflanze | Untersuchtes Organ | p$_H$-Wert | Benutzte Methode | Autor |
|---|---|---|---|---|
| *Begonia spec.* . . . . . . . | Blattstiel | 1,2—1,7 | kolorimetrisch | ROSE und HURD-KARRER (1939) |
| *Begonia spec.* . . . . . . . | Blatt | 1,3—1,6 | ? | RUHLAND und WETZEL (1926) |
| *Oxalis repens* . . . . . . . | Blatt | 1,9—2,2 | Wasserstoffelektrode | ROGERS und SHIVE (1932) |
| *Citrus medica* . . . . . . . | Frucht | 2,2 | kolorimetrisch | CLARK und LUBS (1917) |
| *Citrus medica* . . . . . . | Frucht | 2,3 | Glaselektrode | JØRGENSEN (1939) |
| *Citrus medica* . . . . . . . | Frucht | 2,3—2,5 | Wasserstoffelektrode | BARTHOLOMEW (1923) |
| *Citrus medica* . . . . . . . | Frucht | 5,3 ( ?) | Glaselektrode | HAAS (1941) |
| *Rumex acetosella* . . . . . | Blatt | 2,5—2,7 | Wasserstoffelektrode | ROGERS und SHIVE (1932) |
| *Vaccinium vitis-idaea* . . . . | Frucht | 2,7 | Glaselektrode | JØRGENSEN (1939) |
| *Rheum officinale* . . . . . . | Blattstiel | 3,2 | kolorimetrisch | DRAWERT (1952) |
| *Vitis vinifera* . . . . . . . | Frucht | 3,4 | Glaselektrode | JØRGENSEN (1939) |
| *Bryophyllum calycinum* . . . | Sproß | 3,5—6,0 | Wasserstoffelektrode | GUSTAFSON (1925) |
| *Bryophyllum calycinum* . . . | Blatt | 3,5—4,9 | Wasserstoffelektrode | INGALLS und SHIVE (1931) |
| *Bryophyllum tubiflorum* . . . | Blatt | 3,8—5,7 | kolorimetrisch | OVERBECK (unveröff.) |
| *Bryophyllum tubiflorum* . . | Blatt | 4,1—5,6 | Glaselektrode | OVERBECK (unveröff.) |
| *Bryophyllum daigremontianum* | Blatt | 4,0—5,3 | kolorimetrisch | DRAWERT (1952) |
| *Bryophyllum daigremontianum* | Sproßgipfel | 5,2 | Chinhydronelektrode | KEYSSNER (1931) |
| *Bryophyllum daigremontianum* | Blatt | 4,2—5,7 | kolorimetrisch | OVERBECK (unveröff.) |
| *Bryophyllum daigremontianum* | Blatt | 4,4—5,7 | Glaselektrode | OVERBECK (unveröff.) |
| *Rumex patientia* . . . . . . | Stengel | 4,1—4,5 | Wasserstoffelektrode | INGALLS und SHIVE (1931) |
| *Fagopyrum sagittatum* . . . | Stengel | 4,4—4,8 | Wasserstoffelektrode | INGALLS und SHIVE (1931) |
| *Fagopyrum sagittatum* . . . | Blatt | 4,9—5,4 | Wasserstoffelektrode | INGALLS und SHIVE (1931) |
| *Fagopyrum sagittatum* . . . | Blatt | 5,7—6,0 | Chinhydronelektrode | KEYSSNER (1931) |
| *Fagopyrum sagittatum* . . . | Wurzel | 6,2 | Wasserstoffelektrode | STOKLASA (1924) |
| *Sedum weinbergii* . . . . . | Blatt | 4,5 | Chinhydronelektrode | DRAWERT (1948) |
| *Sempervivum glaucum* . . . | Blatt | 4,6—6,1 | Chinhydronelektrode | KESSLER (1935) |
| *Impatiens parviflora* . . . . | Sproß | 4,7 | Wasserstoffelektrode | DITTRICH (1931) |
| *Allium cepa* . . . . . . . . | Zwiebel (Mesophyll) | 4,8—5,0 | kolorimetrisch | YAMAHA und ISHII (1933) |
| *Allium cepa* . . . . . . . | Zwiebel | 5,7 | Chinhydronelektrode | INGOLD (1930) |
| *Sedum reflexum* . . . . . . | Sproß | 4,9—5,4 | Wasserstoffelektrode | INGALLS und SHIVE (1931) |
| *Zea mays* . . . . . . . . . | Stengel | 4,9—5,7 | Wasserstoffelektrode | GUSTAFSON (1924) |
| *Zea mays* . . . . . . . . . | Stengel- nodien | 4,9—5,2 | Wasserstoffelektrode | PHILLIPS (1920) |
| *Zea mays* . . . . . . . . . | Blatt | 5,0—5,6 | Wasserstoffelektrode | GUSTAFSON (1924) |
| *Zea mays* . . . . . . . . . | Blatt | 5,3—5,9 | Chinhydronelektrode | KEYSSNER (1931) |
| *Zea mays* . . . . . . . . . | Wurzel | 6,4 | Wasserstoffelektrode | STOKLASA (1924) |
| *Sedum praealtum* . . . . . . | Blatt | 5,2 | Wasserstoffelektrode | DRAWERT (1948) |
| *Lycopersicum esculentum* . . | Stengel | 5,2—5,6 | Wasserstoffelektrode | INGALLS und SHIVE (1931) |
| *Lycopersicum esculentum* . . | Sproß | 5,4—5,8 | Wasserstoffelektrode | DUSTMAN (1925) |
| *Lycopersicum esculentum* . . | Blatt | 5,6—6,2 | Wasserstoffelektrode | INGALLS und SHIVE (1931) |
| *Lycopersicum esculentum* . . | Sproß | 5,8 | Wasserstoffelektrode | DITTRICH (1931) |
| *Sorghum saccharatum* . . . . | Blatt | 5,3—5,5 | Chinhydronelektrode | KEYSSNER (1931) |
| *Raphanus sativus* . . . . . | ganze Pflanze | 5,3—5,9 | Wasserstoffelektrode | PEARSALL und EWING (1929) |
| *Persea spec.* („Avocado") . . | Blatt | 5,4—6,0 | Glaselektrode | HAAS (1941) |
| *Sempervivum hausmannii* . . | Blatt | 5,4 | kolorimetrisch | WAGNER (1916) |
| *Senecio cruentus* (= *Cineraria hybrida*) . . | Blumen- blätter (blau u. rot) | 5,4 | Glaselektrode | DRAWERT (1954) |

*Tabelle 5.* (Fortsetzung.)

| Pflanze | Untersuchtes Organ | $p_H$-Wert | Benutzte Methode | Autor |
|---|---|---|---|---|
| *Phoenix dactylifera* . . . . . | unreife Früchte | 5,5—5,6 | Glaselektrode | Haas (1941) |
| *Sinapis alba* . . . . . . . | Blatt | 5,5 | kolorimetrisch | Wagner (1916) |
| *Asparagus officinalis* . . . . | Stengel | 5,5—5,7 | Wasserstoffelektrode | Ingalls und Shive (1931) |
| *Asparagus officinalis* . . . . | Phyllokladien | 5,7—6,1 | Wasserstoffelektrode | Ingalls und Shive (1931) |
| *Tradescantia zebrina* . . . . | Blatt | 5,5—5,9 | Chinhydronelektrode | Keyssner (1931) |
| *Sinapis arvensis* . . . . . . | Blatt | 5,5—6,1 | Chinhydronelektrode | Keyssner (1931) |
| *Triticum aestivum* . . . . . | Sproß | 5,5—6,7 | Chinhydronelektrode | Loehwing (1930) |
| *Triticum aestivum* . . . . . | Blatt | 5,9—6,4 | Wasserstoffelektrode | Hurd-Karrer (1939) |
| *Triticum aestivum* . . . . . | Blatt | 6,0—6,1 | Chinhydronelektrode | Keyssner (1931) |
| *Triticum aestivum* . . . . . | Blatt | 6,2 | Chinhydronelektrode | Drawert (1948) |
| *Triticum aestivum* . . . . . | Wurzel | 6,9 | Wasserstoffelektrode | Stoklasa (1924) |
| *Solanum tuberosum* . . . . . | Knolle | 5,5 | kolorimetrisch | Pearsall und Ewing (1924) |
| *Solanum tuberosum* . . . . . | Blatt | 5,5 | Chinhydronelektrode | Ingold (1930) |
| *Solanum tuberosum* . . . . . | Blatt | 5,8 | kolorimetrisch | Wagner (1916) |
| *Solanum tuberosum* . . . . . | Knolle | 5,8 | Wasserstoffelektrode | Dittrich (1931) |
| *Solanum tuberosum* . . . . . | Knolle | 5,8—5,9 | Glaselektrode | Pfeil (1936) |
| *Solanum tuberosum* . . . . . | Knolle | ~6,0 | Glaselektrode | Wartenberg (1939) |
| *Solanum tuberosum* . . . . . | Knolle | 6,0 | kolorimetrisch | Wagner (1916) |
| *Solanum tuberosum* . . . . . | Knolle | 6,0—6,2 | Wasserstoffelektrode | Reiss (1925) |
| *Solanum tuberosum* . . . . . | Blatt | 6,2—6,6 | Wasserstoffelektrode | Rogers und Shive (1932) |
| *Solanum tuberosum* . . . . . | Knolle | 6,2 | Chinhydronelektrode | Ingold (1930) |
| *Solanum tuberosum* . . . . . | Knolle | 6,3 | kolorimetrisch | Drawert (1952) |
| *Solanum tuberosum* . . . . . | Knolle | 6,5 | Wasserstoffelektrode | Weiss und Harvey (1921) |
| *Solanum tuberosum* . . . . . | Wurzel | 7,0 | Wasserstoffelektrode | Stoklasa (1924) |
| *Brassica oleracea var. sabauda* | Blatt | 5,6 | kolorimetrisch | Wagner (1916) |
| *Vicia faba* . . . . . . . . | Stengel | 5,6 | Glaselektrode | Reimers (unveröff.) |
| *Vicia faba* . . . . . . . . | Blatt | 5,8 | Glaselektrode | Reimers (unveröff.) |
| *Vicia faba* . . . . . . . . | Keimwurzel | 5,9—6,1 | Chinhydronelektrode | Gundel (1933) |
| *Prunus laurocerasus* . . . | Blatt | 5,6—6,0 | Chinhydronelektrode | Samuel (1927) |
| *Trifolium repens* . . . . . . | Stengel | 5,6—6,0 | Wasserstoffelektrode | Ingalls und Shive (1931) |
| *Trifolium repens* . . . . . . | Blatt | 5,8—6,4 | Wasserstoffelektrode | Ingalls und Shive (1931) |
| *Nicotiana rustica* . . . . . | Blatt | 5,6—5,8 | Wasserstoffelektrode | Ingalls und Shive (1931) |
| *Borrago officinalis* . . . . . | Sproß | 5,7 | Wasserstoffelektrode | Dittrich (1931) |
| *Hordeum vulgare* . . . . . . | Sproß | 5,7—6,0 | Wasserstoffelektrode | Hoagland (1919) |
| *Hordeum vulgare* . . . . . . | Wurzel | 6,1—7,1 | Wasserstoffelektrode | Hoagland (1919) |
| *Hordeum distichon* . . . . . | Blatt | 6,0—6,2 | Chinhydronelektrode | Keyssner (1931) |
| *Hordeum distichon* . . . . . | Wurzel | 6,9 | Wasserstoffelektrode | Stoklasa (1924) |
| *Glycine soja* . . . . . . . | Stengel | 5,7—5,8 | Wasserstoffelektrode | Ingalls und Shive (1931) |
| *Glycine soja* . . . . . . . | ganze Pflanze | 5,8—6,1 | Wasserstoffelektrode | Rogers und Shive (1932) |
| *Glycine soja* . . . . . . . | Blatt | 5,9—6,2 | Wasserstoffelektrode | Ingalls und Shive (1931) |
| *Glycine soja* . . . . . . . | Blatt | 6,0—6,4 | Chinhydronelektrode | Keyssner (1931) |
| *Glycine soja* . . . . . . . | Wurzel | 6,1—6,8 | Wasserstoffelektrode | Loehwing (1934) |
| *Glycine soja* . . . . . . . | Sproß | 6,2—6,3 | Wasserstoffelektrode | Loehwing (1934) |
| *Tradescantia viridis* . . . . | Stengel | 5,8 | Glaselektrode | Reimers (unveröff.) |
| *Tradescantia viridis* . . . . | Blatt | 6,0 | Glaselektrode | Reimers (unveröff.) |
| *Ricinus zanzibariensis* . . . | Blatt | 5,8 | Chinhydronelektrode | Keyssner (1931) |
| *Brassica rapa* . . . . . . . | Rübe | 5,8 | Wasserstoffelektrode | Dittrich (1931) |
| *Avena sativa* . . . . . . . | Halm | 5,8—6,1 | kolorimetrisch | Arland (1924) |
| *Avena sativa* . . . . . . . | Blatt | 5,9—6,5 | Chinhydronelektrode | Keyssner (1931) |
| *Avena sativa* . . . . . . . | Coleoptile | 6,0 | Chinhydronelektrode | Bonner (1934) |
| *Avena sativa* . . . . . . . | Coleoptile | 6,2 | Glaselektrode | Bonner (1934) |
| *Avena sativa* . . . . . . . | Blatt | 6,4—6,5 | kolorimetrisch | Arland (1924) |
| *Avena sativa* . . . . . . . | Wurzel | 6,6 | Wasserstoffelektrode | Stoklasa (1924) |
| *Helianthus annuus* . . . . . | Blatt | 5,8—6,4 | Chinhydronelektrode | Keyssner (1931) |
| *Helianthus annuus* . . . . . | Sproß | 6,0 | Wasserstoffelektrode | Dittrich (1931) |
| *Helianthus annuus* . . . . . | Stengel | 6,1 | Wasserstoffelektrode | Gustafson (1924) |

*Tabelle 5.* (Fortsetzung.)

| Pflanze | Untersuchtes Organ | $p_H$-Wert | Benutzte Methode | Autor |
|---|---|---|---|---|
| *Helianthus annuus* . . . . . | Blatt | 6,3—6,9 | Wasserstoffelektrode | GUSTAFSON (1924) |
| *Helianthus annuus* . . . . . | Sproß | 6,5—7,1 | Wasserstoffelektrode | LOEHWING (1934) |
| *Helianthus annuus* . . . . . | Wurzel | 5,3—5,8 | Wasserstoffelektrode | LOEHWING (1934) |
| *Cucurbita pepo* . . . . . . | Blatt | 5,8—6,1 | Chinhydronelektrode | KEYSSNER (1931) |
| *Cucurbita pepo* . . . . . . | Blatt | 6,8—7,5 | Wasserstoffelektrode | GUSTAFSON (1924) |
| *Cucurbita pepo* . . . . . . | Stengel | 6,8—7,6 | Wasserstoffelektrode | GUSTAFSON (1924) |
| *Beta vulgaris* . . . . . . . | Rübe | 5,9 | Wasserstoffelektrode | DITTRICH (1931) |
| *Beta vulgaris* . . . . . . . | Rübe | 6,0—6,5 | Glaselektrode | STAPP und PFEIL (1939) |
| *Beta vulgaris* . . . . . . . | Rübe | 6.4 | Wasserstoffelektrode | STOKLASA (1924) |
| *Phaseolus spec.* . . . . . . | Blatt | 5,9—6,1 | Wasserstoffelektrode | GUSTAFSON (1924) |
| *Phaseolus spec.* . . . . . . | Blatt | 6,0—6,1 | Chinhydronelektrode | KEYSSNER (1931) |
| *Gossypium spec.* . . . . . . | Blatt | 5,9 | ? | MASON und MASKELL (1928) |
| *Gossypium spec.* . . . . . . | Rinde | 6,1 | ? | MASON und MASKELL (1928) |
| *Daucus carota* . . . . . . . | Rübe | 6,0 | kolorimetrisch | PEARSALL und EWING (1924) |
| *Secale cereale* . . . . . . . | Blatt | 6,0—6,2 | Chinhydronelektrode | KEYSSNER (1931) |
| *Secale cereale* . . . . . . . | Wurzel | 6,8 | Wasserstoffelektrode | STOKLASA (1924) |
| *Pisum sativum* . . . . . . . | Blatt | 6,0—6,4 | Chinhydronelektrode | KEYSSNER (1931) |
| *Amaranthus retroflexus* . . . | Sproß | 6,1 | Wasserstoffelektrode | DITTRICH (1931) |
| *Lamium album* . . . . . . | Blatt | 6,1—6,5 | Chinhydronelektrode | KEYSSNER (1931) |
| *Spinacia oleracea* . . . . . . | Sproß | 6,2 | Glaselektrode | JØRGENSEN (1939) |
| *Cucurbita maxima* . . . . | Blatt | 6,4—6,7 | Wasserstoffelektrode | GUSTAFSON (1924) |
| *Chenopodium vulvaria* . . . | Blatt | 6,5—7,1 | Chinhydronelektrode | KEYSSNER (1931) |

liegen. Leider geht aus den meisten Mitteilungen über die mit der „range-indicator"-Methode gewonnenen Ergebnisse nicht klar hervor, ob sich die Werte auf den Zellsaft oder auf andere Zellbestandteile beziehen. Andererseits beobachtete aber auch DITTRICH (1931) mit dem Abdruckverfahren auf Indikatorpapier eine alkalische Reaktion des Zellsaftes von Chenopodiaceenhaaren, und zwar zeigten die Haare von *Chenopodium bonus henricus* $p_H \sim 7,3$ und die von *Atriplex spec.* $p_H \sim 7,6$. In diesem Zusammenhang ist es von Interesse, daß bereits PAYEN (1848) für die Blasenzellen von *Mesembrianthemum crystallinum* eine alkalische Zellsaftreaktion angibt.

## III. Abhängigkeit der Zellsaftreaktion von inneren und äußeren Faktoren.

### A. Innere Faktoren.

#### 1. Alter der Organe.

Vor allem an Blättern hat man eine Abhängigkeit des $p_H$-Wertes vom Alter beobachtet. Je nach der Insertionshöhe eines Blattes an der Sproßachse kann der Preßsaft einen unterschiedlichen Aciditätsgrad aufweisen. GUSTAFSON (1924) gibt nach Messungen mit der Wasserstoffelektrode folgende Unterschiede an: Bei *Zea mays* fällt die Acidität von den unteren zu den oberen Blättern von $p_H$ 5,0 bis auf 5,6, bei *Phaseolus spec.* von $p_H$ 5,9 bis auf 6,1 und bei *Cucurbita maxima* von $p_H$ 6,4 bis auf 6,7. Dieser Gradient kann bei anderen Arten aber auch in umgekehrter Richtung verlaufen; so sind bei *Cucurbita pepo* nach GUSTAFSON die apikalen Blätter schwach sauer, die basalen schwach alkalisch. Die gefundenen $p_H$-Werte liegen zwischen 7,5 und 6,8. Auch *Helianthus annuus* zeigt einen inversen Gradienten von $p_H$ 6,9—6,3.

Die Sproßachse kann sich den Blättern gerade entgegengesetzt verhalten. Das ist z. B. nach GUSTAFSON bei *Zea mays* der Fall. Die Sproßspitze besitzt hier $p_H$ 4,9 und die Basis $p_H$ 5,7. Blatt- und Stengelgradient können aber auch gleichsinnig verlaufen wie bei *Cucurbita pepo*, wo die Stengelspitze $p_H$ 6,8 und der basale Teil $p_H$ 7,7 aufweisen. Bei der Tomate zeigt der Preßsaft der beblätterten Sproßspitze $p_H$ 4,9 und der 7 Monate alte Stengel ohne Blätter $p_H$ 5,4 (DUSTMAN 1925). *Mercurialis perennis* besitzt in den Blättern einen longitudinalen Gradienten von $p_H$ 5,8—6,0 (Basis) bis $p_H$ 4,8—5,0 (Spitze) (MUKERJI 1927). Nach GUNDEL (1933) verschiebt sich sowohl in der Keimwurzel von *Vicia faba* wie im Hypokotyl von *Helianthus* und im Epikotyl von *Phaseolus multiflorus* mit zunehmendem Alter der Zellen der $p_H$-Wert des Zellsaftes nach der sauren Seite. Dieselbe Tendenz ist nach STANFIELD (1937) auch am Sproß von *Lychnis dioica* (Preßsaft) zu beobachten. BOGEN (1938) gibt für den Preßsaft der Stengelepidermiszellen von *Gentiana frigida* dagegen eine Verschiebung des $p_H$-Wertes von 4,3 auf 4,6 gegen den Schluß der Vegetationperiode an.

## 2. Aktivitätswechsel.

Der $p_H$-Wert des Preßsaftes eines Organes kann sich auch mit einem Aktivitätswechsel ganz beträchtlich ändern. Nach MITRA (1921) reagieren Wurzel, Blätter, Rinde und Sproßspitze von Apfelsämlingen in der Wachstumsperiode

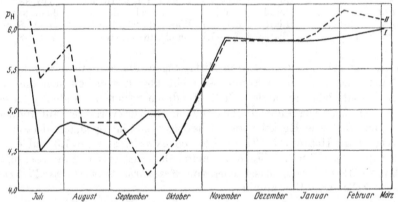

Abb. 1. Die jahreszeitliche $p_H$-Änderung des Preßsaftes aus den Sproßspitzen einjähriger Pfirsichsämlinge (Kurve I) und vierjähriger Apfelbäumchen (Kurve II) nach kolorimetrischen Bestimmungen. Ordinate: $p_H$-Werte, Abszisse: Monate (nach ABBOTT 1923).

deutlich sauer, während sie sich in der Ruhezeit in ihrer Reaktion dem Neutralpunkt nähern. ABBOTT (1923) untersuchte die Preßsäfte der Sproßspitzen einjähriger Pfirsichsämlinge und vierjähriger Apfelbäumchen kolorimetrisch ungefähr alle 10 Tage und erhielt die in Abb. 1 wiedergegebenen $p_H$-Werte. Wird die Aktivität des Gewebes durch Absinken der Temperatur herabgesetzt, so scheint damit parallel ganz allgemein auch ein Abfallen der Acidität des Preßsaftes einherzugehen. Für den Preßsaft von *Sempervivum glaucum* gibt KESSLER (1935) im Juli $p_H$ 4,58 an und im Dezember nach Frosteinwirkung $p_H$ 6,07. Nach BOGEN (1938) steigt in den Epidermiszellen des Blattstiels von *Pelargonium zonale* ebenfalls der $p_H$-Wert nach den ersten Frösten von $p_H$ 3,0 auf 3,5 an.

Ruhende Knollen von Knollenbegonien weisen nach RUHLAND und WETZEL (1926) $p_H$ 4,52 auf. Bei leichter Ankeimung fällt der $p_H$-Wert auf 3,6 und nach einem längeren Austreiben im Dunkeln sinkt er sogar auf $p_H$ 1,44. Diese starke Aciditätssteigerung ist auf eine Anreicherung von freier Oxalsäure zurückzuführen, die auf das 3000fache des ursprünglichen Gehaltes ansteigt. Bei der reifenden Zitrone steigt die Acidität des Preßsaftes von $p_H$ 4,46 auf $p_H$ 2,29 (BARTHOLOMEW 1923).

### 3. Vegetative und generative Phase.

Eine Aciditätssteigerung ist auch bei dem Übergang einer Pflanze von dem rein vegetativen in den generativen Zustand zu beobachten. So soll nach HURD-KARRER (1939) der $p_H$-Wert vom Preßsaft aus Weizenpflanzen in der reproduktiven Phase von $p_H$ 5,9—6,2 auf 5,3—5,5 fallen. Nach STANFIELD (1937) zeigt der elektrometrisch gemessene Preßsaft des Sprosses von *Lychnis dioica* im Rosettenstadium $p_H$ 6,8. Im blühenden Zustand ist der Aciditätsgrad angestiegen, und zwar bei den weiblichen Pflanzen auf $p_H$ 6,3 und bei den männlichen Pflanzen auf $p_H$ 5,0 (Preßsaft aus dem Sproß ohne Knospen und Blüten). Auch die Untersuchungsergebnisse von HOXMEIER (1953) an *Cannabis sativa* und *Spinacia oleracea* sprechen für eine Änderung der Preßsaftacidität bei dem Übergang der Pflanzen aus dem vegetativen in den generativen Zustand (vgl. Tabelle 6). Die geringste Acidität tritt kurz vor der Anlage der Blütenprimordien auf, dann erfolgt eine stärkere Ansäuerung (weitere Literatur s. HURD-KARRER 1939).

Tabelle 6. *Die $p_H$-Werte des Preßsaftes aus apikalen und basalen Teilen blühender Spinatpflanzen unter Lang- und Kurztagbedingungen. Nach* HOXMEIER (1953).

| Alter der Pflanzen in Tagen | p :-Wert | | | |
|---|---|---|---|---|
| | männlich | | weiblich | |
| | Spitze | Basis | Spitze | Basis |
| A. Langtag: 24 Std-Photoperiode | | | | |
| 57 | 6,20 | 6,35 | 6,04 | 6,15 |
| 86 | 6,03 | 5,83 | 5,94 | 5,76 |
| 92 | 6,32 | 6,11 | 6,59 | 6,40 |
| 99 | 5,96 | 5,75 | 6,07 | 5,67 |
| B. Kurztag: 10 Std-Photoperiode | | | | |
| 100 | 6,19 | 6,03 | 6,51 | 6,05 |

### 4. Diöcische Pflanzen.

Wie bereits im vorhergehenden Abschnitt und auf S. 636 erwähnt wurde, konnte STANFIELD (1937) im Preßsaft männlicher und weiblicher Pflanzen von *Lychnis dioica* einen Unterschied im $p_H$-Wert feststellen. Elektrometrisch gemessen reagieren die Säfte der männlichen Pflanzen saurer als die der weiblichen, der Unterschied beträgt über eine $p_H$-Einheit. Nach colorimetrischen Untersuchungen von SATINA und BLAKESLEE (1926) sollen aber zwischen den $p_H$-Werten der Säfte aus Blättern ♀- und ♂-Individuen einer Reihe von Samenpflanzen keine wesentlichen Unterschiede bestehen. In der von den Autoren gegebenen Übersicht weisen etwas größere Differenzen nur *Ailanthus glandulosa* (♀ $p_H$ 4,4 und ♂ $p_H$ 4,8) und *Rumex* (♀ $p_H$ 4,0 und ♂ $p_H$ 4,5) auf. Danach reagieren die Preßsäfte weiblicher Individuen saurer als die der männlichen. Dasselbe findet W. F. LOEHWING (1933) für *Cannabis* und *Spinacia*. HOXMEIER (1953) stellt dagegen in Übereinstimmung mit den Befunden von STANFIELD an *Lychnis* und TALLEY (1932) an *Cannabis* ganz allgemein bei *Cannabis sativa* und *Spinacia oleracea* elektrometrisch für die Preßsäfte aus ♂-Pflanzen eine höhere Acidität fest. Dies kann auch H. C. LOEHWING (1953) für *Cannabis* bestätigen. Die größten Unterschiede weisen die Blüten vom Hanf bei der Kultur unter Langtag auf: ♂-Blüten $=$ $p_H$ 6,09 und ♀-Blüten $=$ $p_H$ 6,62. In den verschiedenen Entwicklungsstadien können sich die Pflanzen aber auch unterschiedlich verhalten, wie Tabelle 6 belegt. Daraus erklären sich wohl die Unstimmigkeiten in den Ergebnissen der einzelnen Autoren. So wird von STANFIELD (1937) besonders darauf hingewiesen, daß bei *Lychnis* erst mit der Blühreife reproduzierbare Unterschiede im $p_H$-Wert beider Geschlechter zu beobachten sind. Im Rosettenstadium z. B. besteht keine Differenz im $p_H$-Wert.

## B. Äußere Faktoren.

### 1. Acidität des umgebenden Mediums.

Die Pflanzenzellen scheinen für freie H- und OH-Ionen weitgehend impermeabel zu sein oder der Zellsaft muß eine hohe Pufferkapazität besitzen, da der $p_H$-Wert des Zell- bzw. Preßsaftes nach den meisten Angaben gar nicht oder nur in geringem Maße von der Acidität des Außenmediums beeinflußt wird, solange die Zellen nicht geschädigt werden. Für *Nitella* liegen Angaben z. B. von Hoagland und Davis (1923) und Albaum, Kaiser und Nestler (1937) vor, nach denen der $p_H$-Wert des Zellsaftes bei einem Außenmedium mit $p_H$ 5—9 konstant $\sim$5,2 bleibt. Erst im stärker sauren Bereich steigt auch die Acidität des Zellsaftes an, parallel damit erfolgt aber auch eine Zellschädigung. Enthält das Außenmedium aber schwächer dissoziierte Salze, so kann es zu recht beträchtlichen Verschiebungen des Aciditätsgrades im Zellsaft kommen. In einer 0,005 mol $NH_4Cl$-Lösung mit $p_H$ 6,9 beginnt der $p_H$-Wert des Zellsaftes von *Nitella* bereits nach einem Aufenthalt in der Lösung von nur 5 min zu steigen und hat sich nach 30 min von $p_H$ 5,6 nach $p_H$ 5,94 verschoben, bleibt dann allerdings bis zum Tode der Zellen ungefähr konstant (Irwin 1926). Bekanntlich wird der Säuregrad des Zellsaftes leicht durch $CO_2$ und $NH_3$ geändert.

Bei Landpflanzen scheint der $p_H$-Wert des Preßsaftes — jedenfalls der oberirdischen Teile — weitgehend unabhängig zu sein von der Acidität des Nährmediums (Dustman 1925, Hurd-Karrer 1939, Haas 1941 u. v. a.). Bei einem Wachstum der Pflanzen in Nährlösungen mit verschiedenem $p_H$-Wert (3,7—7,5) konnte Keyssner (1931) allerdings zum Teil stärkere Unterschiede im $p_H$-Wert des Preßsaftes der Wurzelsysteme feststellen, die beobachteten Verschiebungen liegen zwischen 0,1 (Roggen) und 1,3 $p_H$-Einheiten *(Chenopodium)*.

## 2. Temperatur.

Über den Einfluß der Temperatur auf die Acidität des Preßsaftes liegen Untersuchungen von Hurd-Karrer (1929, 1939) an Weizenpflanzen vor. Danach ist der Aciditätsgrad um so höher, je wärmer ($> 20^0$ C) die Pflanzen aufgewachsen sind. Diese Aciditätssteigerung kann vielleicht auf eine Schädigung der Pflanzen durch höhere Temperaturen zurückgeführt werden, da sie mit der Empfindlichkeit der untersuchten Weizensorten gegenüber höheren Temperaturen zunimmt.

## 3. Licht.
### a) Tagesperiodizität.

Sehr stark hängt die Acidität des Preßsaftes von der Lichtintensität ab, so daß sich besonders ausgeprägt bei sukkulenten Formen und an sonnigen Tagen eine Tagesperiodizität im Aciditätsgrad ergibt. Eingehende Untersuchungen in dieser Hinsicht über die aktuelle und zum Teil auch die potentielle Acidität sind vor allem von Hempel (1917), Gustafson (1925) und Ingalls und Shive (1931) durchgeführt worden. Am Morgen reagieren die Preßsäfte saurer als am Nachmittag. Der $p_H$-Unterschied ist von der Lichtintensität am Tage abhängig. Für *Bryophyllum calycinum* fand Gustafson an sonnigen Tagen optimal morgens z. B. $p_H$ 3,5 und nachmittags $p_H$ 6,0. An einem bewölkten Tag wurden dagegen folgende Werte gemessen: morgens $p_H$ 3,8 und nachmittags $p_H$ 4,4. In Abb. 2 ist die Tagesperiodizität der aktuellen und der totalen (aktuellen + potentiellen) Acidität, an einem sonnigen Tage für *Bryophyllum calycinum* wiedergegeben und Abb. 3 zeigt die Änderung der aktuellen Acidität für eine nicht sukkulente Pflanze *(Trifolium repens)*. Das Gewebe von Weizenpflanzen weist bei sonnigem Wetter einen Tagescyclus mit dem maximalen Intervall von $p_H$ 5,5 am Morgen und $p_H$ 6,7 am Spätnachmittag auf (Loehwing 1930). Auch andere nicht succulente Pflanzen besitzen nach Loehwing (1953, dort auch weitere Literatur) einen Tagesrhythmus in der Zellsaftacidität mit einem Aciditätsmaximum am Morgen. Eine Ausnahme macht *Glycine soja* mit zwei entsprechenden Maxima am frühen Morgen und am frühen Nachmittag.

Kurz- und Langtag können sich ebenfalls auf den Säuregrad des Preßsaftes auswirken. Nach HURD-KARRER (1939) erhöht langer (17 Std) und erniedrigt kurzer (8 Std) Tag die Acidität bei Weizenpflanzen. Nach STANFIELD (1937) zeigt *Lychnis dioica* unter Kurz-

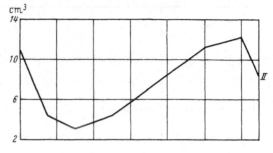

tagsbedingungen eine geringere Acidität als unter Langtagsbedingungen. Dieselbe Beobachtung macht HOXMEIER (1953) für *Spinacia oleracea* (vgl. in Tabelle 6 die gleichaltrigen Pflanzen unter Kurz- und Langtagsbedingungen). Der Einfluß der Tageslänge auf den $p_H$-Wert des Preßsaftes ist wohl mit der Abhängigkeit der Blütenbildung von der Tageslänge in Zusammenhang zu bringen. Wir haben ja bereits gesehen, daß zumindest in einigen Fällen der Preßsaft im vegetativen Zustand einer Pflanze einen höheren $p_H$-Wert besitzt als in der generativen Phase. Sollte dies allgemein zutreffen, dann müßten sich Kurz- und Langtagpflanzen unter gleichen Bedingungen verschieden verhalten. Bei den aufgeführten Arten handelt es sich nur um Langtagpflanzen. Über Kurztagpflanzen liegen bisher nur Angaben von HOXMEIER (1953) über *Cannabis sativa* vor. Den

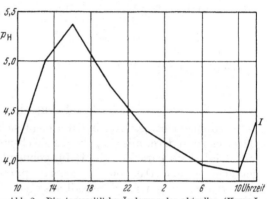

Abb. 2. Die tageszeitliche Änderung der aktuellen (Kurve I nach Messungen mit der Wasserstoffelektrode) und der totalen (Kurve II) Acidität des Preßsaftes ganzer Sproßspitzen von *Bryophyllum calycinum* an einem sonnigen Tage. Ordinate: $p_H$-Werte bei Kurve I und auf 10 cm³ Preßsaft bis zur Neutralisation verbrauchte 0,1365 nNaOH bei Kurve II. Abzisse: Uhrzeit. (Nach GUSTAFSON 1925).

Tabellen von HOXMEIER ist zu entnehmen, daß in der Tat die Preßsäfte aus vegetativen Organen unter Langtagsbedingungen gehaltener weiblicher Pflanzen etwas saurer reagieren als die unter Kurztagsbedingungen aufgewachsener. Für die männlichen Pflanzen ist allerdings aus den Zahlenwerten von HOXMEIER kein einigermaßen gesicherter Unterschied abzulesen.

### b) Etiolement.

Über Unterschiede im $p_H$-Wert des Preßsaftes grüner und etiolierter Pflanzen liegen auch Angaben vor, von denen einige in Tabelle 7 zusammengefaßt sind.

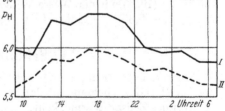

Abb. 3. Die tageszeitliche $p_H$-Änderung des Preßsaftes aus den Blättern (Kurve I) und dem Stengel (Kurve II) von *Trifolium repens* an einem sonnigen Tage nach Bestimmungen mit der Wasserstoffelektrode. Ordinate: $p_H$-Werte. Abzisse: Uhrzeit. (Nach INGALLS und SHIVE 1931).

Das unterschiedliche Verhalten von *Triticum*- und *Sedum*-Blatt könnte vielleicht auf das intensivere Wachstum des Grasblattes bei Lichtabschluß zurückzuführen sein. Nach neueren noch unveröffentlichten Messungen von REIMERS an unverdünnten Preßsäften grüner und etiolierter Pflanzen mit der Glaselektrode sind die Unterschiede allerdings bedeutend geringer als die sich aus Tabelle 7 ergebenden Werte.

Tabelle 7. *$p_H$-Werte des Preßsaftes aus grünen und etiolierten Pflanzen.*

| Pflanze | Untersuchtes Organ | $p_H$-Wert grün | $p_H$-Wert etioliert | Benutzte Methode | Autor |
|---|---|---|---|---|---|
| *Zea mays* | Keimlinge | 5,5 | 6,2 | Wasserstoff-elektrode | Haas (1920) |
| *Brassica oleracea var. capitata* | Blatt | 5,6—5,7 | 6,0—6,4 | Glas-elektrode | Chibnall und Grover (1926) |
| *Vicia faba* | Sproß | (5,6–5,8)[1] | 6,1—6,2 | elektro-metrisch | Phillips (1920) |
| *Helianthus annuus* | Hypokotyl | — | 5,9—6,2 | Chinhydron-elektrode | Gundel (1933) |
| *Sedum praealtum* | Blatt | 5,2 | 6,1 | Wasserstoff-elektrode | Drawert (1948) |
| *Sedum weinbergii* | Blatt | 4,5 | 5,3 | Chinhydron-elektrode | Drawert (1948) |
| *Bryophyllum daigremontianum* | Sproßgipfel | 5,2 | 5,7 | Chinhydron-elektrode | Drawert (1948) |
| *Triticum vulgare* | Blatt | 6,2 | 5,9 | Chinhydron-elektrode | Drawert (1948) |

## 4. Parasiten.

Nach einem parasitären Befall scheint sich die Preßsaftacidität des kranken Gewebes zu ändern, allerdings nicht gleichsinnig. Ob die Acidität ansteigt oder fällt, hängt vom Parasiten ab. Dafür können hier auch nur einige kurze Hinweise gebracht werden. Das von *Pseudomonas tumefaciens* erzeugte Tumorgewebe reagiert weniger sauer als der Preßsaft aus gesundem Nachbargewebe (Harvey 1920, Soru 1929, Klein und Keyssner 1929, Stapp und Pfeil 1939). Für die Zuckerrübe werden z.B. folgende Werte angegeben: gesundes Gewebe $p_H$ 5,8 und Tumorgewebe $p_H$ 6,35 (Harvey 1920, Wasserstoffelektrode). Messungen mit der Glaselektrode ergaben nach Stapp und Pfeil (1939) bei der Zuckerrübe $p_H$ 6,0—6,5 für gesunde Pflanzen und $p_H$ 6,0—7,6 für Tumorgewebe; bei Tomaten betragen die entsprechenden Werte $p_H$ 4,5—4,8 und $p_H$ 5,0—7,7.

Das von *Synchytrium endobioticum* befallene Gewebe der Knollen von *Solanum tuberosum* besitzt dagegen eine höhere $c_H$ des Preßsaftes als die gesunden Teile. Nach Weiss und Harvey (1921) fällt der $p_H$-Wert von 6,5 auf 6,0 und Messungen von Reiss (1925) mit der Wasserstoffelektrode ergaben $p_H$ 6,0—6,2 für gesundes und $p_H$ 5,6—5,85 für krankes Gewebe.

## IV. Schlußbetrachtung.

Wie in der Einleitung bereits betont, können wir von keiner der bis jetzt benutzten Methoden behaupten, daß sie uns reelle Aussagen über den absoluten $p_H$-Wert des Zellsaftes ermöglicht. Bestenfalls erhalten wir Näherungswerte. Man wird wohl immer, soweit es durchführbar ist, mehrere Methoden vergleichend anwenden müssen, um zu einigermaßen brauchbaren Resultaten zu kommen. Die zelleigenen Anthocyane scheinen nach den jüngeren Erfahrungen für Rückschlüsse auf die Zellsaft-$c_H$ völlig unbrauchbar zu sein. Wenn aber die Vitalfärbung mit zellfremden Indicatoren, das Abdruckverfahren mit Indicatorpapier und die elektrometrische Bestimmung des Preßsaftes einigermaßen übereinstimmende Werte ergeben, kann man wohl mit einiger Wahrscheinlichkeit annehmen, den

---

[1] Nach Reimers aus Tabelle 5 entnommen.

$p_H$-Wert des Zellsaftes erfaßt zu haben. Eine wirkliche Übereinstimmung der mit allen 3 Methoden erhaltenen Ergebnisse wird man aber wohl nur in den seltensten Fällen bekommen.

Ganz allgemein betrachtet ist selbstverständlich der Stoffwechsel der Zelle der H-Ionenlieferant für den Zellsaft. Verglichen mit dem Plasma scheint die Pufferung des Zellsaftes bedeutend schwächer zu sein, so daß schon geringe Stoffwechseländerungen eine Verschiebung der Zellsaft-$c_H$ zur Folge haben können. Die beobachteten Änderungen der Zellsaftacidität durch innere und äußere Faktoren sind letzten Endes auf eine Änderung des Stoffwechsels durch diese Faktoren zurückzuführen. Um so bedauerlicher sind die noch bestehenden rein methodischen Schwierigkeiten, den $p_H$-Wert des Zellsaftes der lebenden Zelle zu bestimmen und seine Verschiebung zu verfolgen; könnte doch diese $p_H$-Verschiebung für uns unter Umständen ein brauchbarer Indicator für eine Stoffwechseländerung in der Zelle sein.

# Literatur.

ABBOTT, O.: Chemical changes at beginning and ending of rest period in apple and peach. Bot. Gaz. 76, 167—184 (1923). — ALBAUM, H. G., I. KAISER and H. A. NESTLER: The relation of hydrogen-ion concentration to the penetration of 3-Indole acetic acid into *Nitella* cells. Amer. J. Bot. 24, 513—518 (1937). — ARLAND, A.: Der Hafer-Flugbrand, *Ustilago avenae* (Pers.) Jens. Biologische Untersuchungen mit besonderer Berücksichtigung der Infektions- und Anfälligkeitsfrage. Bot. Archiv 7, 70—111 (1924). — ARRHENIUS, O.: Hydrogenion, concentration, soilproperties and growth of higher plants. Ark. Bot. (Stockh.) 18, Nr 1, 1—54 (1924). — ATKINS, W. R. G.: The hydrogen ion concentration of plant cells. Sci. Proc. Roy. Dublin Soc., N.S. 16, 414 (1922). — The hydrogen ion concentration of the soil in relation to the flower colour of *Hydrangea hortensis* W. Sci. Proc. Roy. Dublin Soc., N.S. 17, 23 (1923).

BAILEY, I. W., and C. ZIRKLE: The cambium and its derivative tissues. VI. The effects of hydrogen ion concentration in vital staining. J. Gen. Physiol. 14, 363—383 (1931). — BARTHOLOMEW, E. T.: Internal decline of lemons. II. Growth rate, water content, and acidity of lemons at different stages of maturity. Amer. J. Bot. 10, 117—126 (1923). — BETHE, A.: Allgemeine Physiologie. Berlin-Göttingen-Heidelberg: Springer 1952. — BOGEN, H. J.: Untersuchungen zu den „spezifischen Permeabilitätsreihen" HÖFLERS. II. Harnstoff und Glycerin. Planta (Berl.) 28, 535—581 (1938). — BONNER, J.: The relation of hydrogen ions to the growth rate of the *Avena* Coleoptile. Protoplasma (Berl.) 21, 406—423 (1934). — BRAUNER, L.: Das kleine Pflanzenphysiologische Praktikum. I. Teil. Jena: Gustav Fischer 1929. — BRENNER, W.: Studien über die Empfindlichkeit und Permeabilität pflanzlicher Protoplasten für Säuren und Basen. Öfversigt Finska Vetensk. Soc. Förh. Afd. A 60, Nr 4 (1918). — BUXTON, B. H., and F. V. DARBISHIRE: Behavior of "anthocyanins" at varying hydrogen ion concentrations. J. Genet. 21, 71—79 (1929).

CHAMBERS, R., and T. KERR: Intracellular hydrion concentration studies. VIII. Cytoplasmic and vacuolar $p_H$ of *Limnobium* root hair cells. J. Cellul. a. Comp. Physiol. 2, 105—119 (1932). — CHIBNALL, A. C., and C. E. GROVER: A chemical study of leaf cell cytoplasm. I. The soluble proteins. Biochemic. J. 20, 108—118 (1926). — CHOLNOKY, B. J.: Beobachtungen über die Farbstoffaufnahme anthocyanführender Zellen. Mikroskopie (Wien) 5, 117—124 (1950). — CLARK, W. M.: The determination of hydrogen ions, 3. Aufl. Baltimore 1928. — CLARK, W. M., and H. A. LUBS: The colorimetric determination of hydrogen ion concentration and its application in bacteriology. J. Bacter. 2, 191—236 (1917). — CRAFTS, A. J.: Further studies on exudation in cucurbits. Plant Physiol. 11, 63—79 (1936).

DANGEARD, P.: Recherche sur l'appareil vacuolaire dans les végétaux. Botaniste 15, 1—262 (1923). — DITTRICH, W.: Zur Physiologie des Nitratumsatzes in höheren Pflanzen (unter besonderer Berücksichtigung der Nitratspeicherung). Planta (Berl.) 12, 69—119 (1931). — DRAWERT, H.: Zur Frage der Stoffaufnahme durch die lebende pflanzliche Zelle. IV. Der Einfluß der Wasserstoffionenkonzentration des Zellsaftes und des Außenmediums auf die Harnstoffaufnahme. Planta (Berl.) 35, 579—600 (1948). — Zur Frage der Stoffaufnahme durch die lebende pflanzliche Zelle. V. Zur Theorie der Aufnahme basischer Stoffe. Z. Naturforsch. 3b, 111—120 (1948). — Beiträge zur Vitalfärbung pflanzlicher Zellen. Protoplasma (Berl.) 40, 85—106 (1951). — Kritische Untersuchungen zur gravimetrischen Bestimmung der Wasserpermeabilität. Ein Beitrag zum Einfluß der $c_H$ auf die Wasseraufnahme und

-abgabe. Planta (Berl.) **41**, 65—82 (1952). — Über die Eignung zelleigener Anthocyane zur $p_H$-Wertbestimmung des Zellsaftes. Protoplasma (Berl.) **44**, 370—373 (1954). — Dustman, R. B.: Inherent factors related to absorption of mineral elements by plants. Bot. Gaz. **79**, 233—264 (1925).

Ehrlich, P.: Beiträge zur Kenntnis der Anilinfärbungen und ihrer Verwendung in der mikroskopischen Technik. Arch. mikrosk. Anat. **13**, 263 (1877). — Ender, F.: Wasserstoffionenkonzentration. In Hoppe-Seyler/Thierfelders Handbuch der physiologisch- und pathologisch-chemischen Analyse, Bd. I., S.527—616. Berlin-Göttingen-Heidelberg: Springer 1953. — Esselborn, W.: Colorimetrische Bestimmung der Wasserstoffionenkonzentration. In Hoppe-Seyler/Thierfelders Handbuch der physiologisch- und pathologisch-chemischen Analyse, Bd. I., S. 617—627. Berlin-Göttingen-Heidelberg: Springer 1953.

Fitting, H.: Über eigenartige Farbänderungen von Blüten und Blütenfarbstoffen. Z. Bot. **4**, 81—105 (1912).

Gillissen, G.: Die Färbungsmetachromasie (Sammelreferat). Protoplasma (Berl.) **42**, 448—457 (1953). — Goldacre, R. J.: Kommentar zu Ausführungen von D. M. Prescott. Nature (Lond.) **172**, 593—594 (1953). — Gundel, W.: Chemische und physikalisch-chemische Vorgänge bei geischer Induktion. Jb. wiss. Bot. **78**, 623—664 (1933). — Gustafson, F. G.: Hydrogen-ion concentration gradient in plants. Amer. J. Bot. **11**, 1—6 (1924). — Total acidity compared with actuel acidity of plant juices. Amer. J. Bot. **11**, 365—369 (1924). — Diurnal changes in the acidity of *Bryophyllum calycinum*. J. Gen. Physiol. **7**, 719—728 (1925).

Haas, A. R.: The acidity of plant cells as shown by natural indications. J. of Biol. Chem. **27**, 233—241 (1916). — $p_H$ determination in plant tissue. Plant Physiol. **16**, 405—409 (1941). — Harvey, R. B.: Relation of catalase, oxidase and $H^+$ concentration to the formation of overgrowths. Amer. J. Bot. **7**, 211—221 (1920). — Hempel, J.: Buffer processes in the metabolism of succulent plants. C. r. Labor. Carlsberg **13**, 1 (1917). Zit. nach Small 1929. — Hoagland, D. R.: Relation of nutrient solution to composition and reaction of cell sap of barley. Bot. Gaz. **68**, 297—304 (1919). — Hoagland, D. R., and A. R. Davis: The composition of cell sap in relation to the absorption of ions. J. Gen. Physiol. **5**, 629—646 (1923). — Hofmeister, L.: Studien über Mikroinjektion in Pflanzenzellen. Z. wiss. Mikrosk. **57**, 274—290 (1940). — Hoxmeier, Sr. M. C.: Buffer capacity and $p_H$ of press sap in relation to dioecism of phanerogams. Proc. Iowa Acad. Sci. **60**, 167—175 (1953). — Hurd-Karrer, A. M.: Relation of leaf acidity to vigor in wheat grown at different temperatures. J. Agricult. Res. **39**, 341—350 (1929). — Hydrogen-ion concentration of leaf juice in relation to environment and plant species. Amer. J. Bot. **26**, 834—846 (1939).

Ingalls, R. A., and I. W. Shive: Relation of H-ion concentration of tissue fluids to the distribution of iron in plants. Plant Physiol. **6**, 103—125 (1931). — Ingold, C. T.: Hydrogen-ion phenomena in plants. IV. Buffers of potato (tuber and leaf). Protoplasma (Berl.) **9**, 441—446 (1930). — Irwin, M.: Accumulation of brilliant cresyl blue in the sap of living cells of *Nitella* in the presence of $NH_3$. J. Gen. Physiol. **9**, 235—253 (1926). — Studies on penetration of dyes with glass electrode. IV. Penetration of brilliant cresyl blue into *Nitella flexilis*. J. Gen. Physiol. **14**, 1—17 (1931).

Jørgensen, H.: Ergibt das Chinhydronverfahren falsche Werte für das $p_H$ in Fruchtsäften, Ensilage usw.? Biochem. Z. **302**, 281—286 (1939).

Keller, R., u. J. Gicklhorn: Methoden der Bioelektrostatik. In Abderhaldens Handbuch der biologischen Arbeitsmethoden, Abt. V, Teil 2, S. 1189—1280. 1928. — Kerr, T.: The injection of certain salts into the protoplasm and vacuoles of the root hairs of *Limnobium spongia*. Protoplasma (Berl.) **18**, 420—440 (1933). — Kessler, W.: Über die inneren Ursachen der Kälteresistenz der Pflanzen. Planta (Berl.) **24**, 312—352 (1935). — Keyssner, E.: Die colorimetrische Bestimmung der Wasserstoffionenkonzentration. In G. Kleins Handbuch der Pflanzenanalyse, Bd. I, S.463—505. Wien: Springer 1931.— Der Einfluß der Wasserstoffionenkonzentration in der Nährlösung auf die Reaktion in der Pflanze. Planta (Berl.) **12**, 575—587 (1931). — Klein, G., u. E. Keyssner: Beiträge zum Chemismus pflanzlicher Tumoren. II. Über die Wasserstoffionenkonzentration in pflanzlichen Tumoren. Biochem. Z. **254**, 256—263 (1932). — Kolthoff, I. M.: Der Gebrauch von Farbenindikatoren, 4. Aufl. Berlin: Springer 1932. — Kordatzki, W.: Taschenbuch der praktischen $p_H$-Messung für wissenschaftliche Laboratorien und technische Betriebe, 4. Aufl. München: R. Müller u. Steinicke 1949. — Kraus, G.: Über die Wasserverteilung in der Pflanze. IV. Die Acidität des Zellsaftes. Sonderdruck S. 65 aus Abh. naturforsch. Ges. Halle **16** (1884). — Küster, E.: Vitalfärbungen. In T. Péterfi, Methodik der wissenschaftlichen Biologie, Bd. I. Berlin 1928. — Die Pflanzenzelle, 2. Aufl. Jena: Gustav Fischer 1951.

Leuthard, F.: Grundlagen und Grenzen biologischer $p_H$-Bestimmungen. Kolloidchem. Beih. **28**, 262—280 (1929). — Lison, L.: Sur la mesure du $p_H$ intracellulaire par la methode des colorations vitales. C. r. Soc. Biol. Paris **120**, 102—104 (1935). — Lison, L., et J. Fautrez: L'étude physicochimique des colorants dans ses applications biologiques. — Etude critique. Protoplasma (Berl.) **33**, 116—151 (1939). — Loehwing, H. C.: Diurnal changes in sap acidity of certain plants. Proc. Iowa Acad. Sci. **60**, 192—199 (1953). — Loehwing, W. F.: Effects of insolation and soil characteristics on tissue fluid reaction in wheat. Plant Physiol. **5**, 293—305 (1930). — Physico-chemical aspects of sex in plants. Proc. Soc. Exper. Biol. a. Med. **30**, 1215—1220 (1933). — Physiological aspects of the effect of continuous soil aeration on plant growth. Plant Physiol. **9**, 567—583 (1934).

Martin, S. H.: The hydrion concentration of plant tissues. III. The tissues of *Helianthus annuus*. Protoplasma (Berl.) **1**, 497—521 (1927). — Mason, T. G., and E. J. Maskell: Studies on the transport of carbohydrates in the cotton plant. II. The factors determining the rate and the direction of movement of sugars. Ann. of Bot. **42**, 571—636 (1928). — Mevius, W.: Reaktion des Bodens und Pflanzenwachstum. Freising u. München: Datterer & Cie. 1927. — Michaelis, L.: Wasserstoffionenkonzentration. I. 2. Aufl. Monographien aus dem Gesamtgebiet der Physiologie der Pflanzen und der Tiere, Bd. I. Berlin: Springer 1922. — Mitra, S.: Seasonal changes and translocation of carbohydrate materials in fruit spurs and two-year old seedlings of apple. Ohio J. Sci. **21**, 89—99 (1921). Zit. nach Abbott 1923. — Mukerji, S. K.: The biological relations of *Mercurialis perennis*. Proc. Linnean Soc. N.S. Wales **140**, 3—5 (1928). Zit. nach Small 1929.

Needham, J., et D. Needham: Observations sur le $p_H$ intérieure de la cellule. C. r. Soc. Biol. Paris **94**, 833—835 (1926).

Payen, A.: C. r. Acad. Sci. Paris **27**, 1 (1848). Zit. nach Sachs 1862. — Pearsall, W. H., and J. Ewing: The diffusion of ions from living plant tissues in relation to protein iso-electric points. New Phytologist **23**, 193—206 (1924). — The relation of nitrogen metabolism to plant succulence. Ann. of Bot. **43**, 27—34 (1929). — Pekarek, J.: Über die Aciditätsverhältnisse in den Epidermis- und Schließzellen bei *Rumex acetosa* im Licht und im Dunkeln. Planta (Berl.) **21**, 419—446 (1934). — Pfeffer, W.: Über Aufnahme von Anilinfarben in lebende Zellen. Unters. Bot. Inst. Tübingen **2**, 179—331 (1886). — Pfeiffer, H.: Der gegenwärtige Stand der kolorimetrischen Azidimetrie in der Gewebephysiologie. Protoplasma (Berl.) **1**, 434—465 (1927). — Bemerkungen zur Dittrichschen Regel über die Beziehung zwischen Nitratspeicherung und Preßsaftacidität pflanzlicher Gewebe. Protoplasma (Berl.) **17**, 301—316 (1933). — Pfeil, E.: Über Messungen mit der Glaselektrode. Angew. Chem. **49**, 57—59 (1936). — Phillips, T. G.: Chemical and physical changes during geotropic response. Bot. Gaz. **69**, 168—178 (1920). — Plowe, I. A.: Membranes in the plant cell. II. Localization of differential permeability in the plant protoplast. Protoplasma (Berl.) **12**, 221—240 (1931).

Rapkine, L., et R. Wurmser: Le potentiel de reduction des cellules vertes. C. r. Soc. Biol. Paris **94**, 1347—1349 (1926). — Reiss, P.: Données physicochimiques sur une tumeur végétale infectieuse. C. r. Soc. Biol. Paris **93**, 1371 (1925). — Le $p_H$ intérieur cellulaire. Paris: Les presses universitaires de France 1926. — Ribbert, A.: Beiträge zur Frage nach der Wirkung der Ammoniumsalze in Abhängigkeit von der Wasserstoffionenkonzentration. Planta (Berl.) **12**, 603—634 (1931). — Robinson, G. M.: Notes on variable colors of flower petals. J. Amer. Chem. Soc. **61**, 1606—1607 (1939). — Robinson, R.: Natural colouring matters and their analogues. Nature (Lond.) **132**, 625—628 (1933). — Robinson, R., and G. M. Robinson: The coloid chemistry of leaf and flower pigments and the precursors of the anthocyanins. J. Amer. Chem. Soc. **61**, 1605—1606 (1939). — Rogers, C. H., and I. W. Shive: Factors affecting the distribution of iron in plants. Plant Physiol. **7**, 227—252 (1932). — Rohde, K.: Untersuchungen über den Einfluß der freien H-Ionen im Innern lebender Zellen auf den Vorgang der vitalen Färbung. Pflügers Arch. **168**, 411—433 (1917). — Rose, D. H., and A. M. Hurd-Karrer: Differential staining of specialized cells in *Begonia* with indicators. Plant Physiol. **2**, 441—453 (1927). — Ruhland, W.: Untersuchungen über den Kohlenhydratstoffwechsel von *Beta vulgaris* (Zuckerrübe). Jb. wiss. Bot. **50**, 200—257 (1912). — Über die Verwendbarkeit vitaler Indikatoren zur Ermittlung der Plasmareaktion. Ber. dtsch. bot. Ges. **41**, 252—254 (1923). — Ruhland, W., u. K. Wetzel: Zur Physiologie der organischen Säuren in grünen Pflanzen. I. Wechselbeziehungen im Stickstoff- und Säurestoffwechsel von *Begonia semperflorens*. Planta (Berl.) **1**, 558—564 (1926).

Sachs, I.: Über saure, alkalische und neutrale Reaktion der Säfte lebender Pflanzenzellen. Bot. Ztg. **20**, 257—265 (1862). — Samuel, G.: On the shot-hole disease caused by *Clasterosporium carpophilum* and on the "shot-hole" effect. Ann. of Bot. **41**, 375—404 (1927). — Satina, S., and A. F. Blakeslee: Differences between sexes in green plants. Proc. Nat. Acad. Sci. **12**, 197—202 (1926). — Schaede, R.: Über das Verhalten von Pflanzenzellen gegenüber Anilinfarbstoffen. Jb. wiss. Bot. **62**, 65—91 (1923). — Über die Reaktion des lebenden Plasmas. Ber. dtsch. bot. Ges. **42**, 219—224 (1924). — Schmidtmann, M.: Über eine Methode zur Bestimmung der Wasserstoffzahl im Gewebe und in einzelnen Zellen. Biochem. Z. **150**, 253—255 (1924). — Schorr, L.: Über bunte Vakuolensysteme. Protoplasma

(Berl.) **31**, 292—297 (1938). — SCHWARZ, F.: Die morphologische und chemische Zusammensetzung des Protoplasmas. Beitr. Biol. Pflanz. **5**, 1—244 (1892). — SMALL, J.: Hydrogen-ion concentration in plant cells and tissues. Protoplasma-Monogr. **2** (1929). — SMITH, E. P.: The calibration of flower colour indicators. Protoplasma (Berl.) **18**, 112—125 (1933). — SORU, E.: Concentrations en ions H du tissu et des tumeurs de *Pelargonium zonale*. C. r. Soc. Biol. Paris **102**, 127 (1929). — STANFIELD, J. F.: Hydrogen ion concentration and sexual expression in *Lychnis dioica* L. Plant Physiol. **12**, 151—162 (1937). — STAPP, C., u. E. PFEIL: Der Pflanzenkrebs und sein Erreger *Pseudomonas tumefaciens*. Zbl. Bakter. II **101**, 261—286 (1939). — STOKLASA, J.: Über die Resorption der Ionen durch das Wurzelsystem der Pflanzen aus dem Boden. Ber. dtsch. bot. Ges. **42**, 183—191 (1924). — STRUGGER, S.: Praktikum der Zell- und Gewebephysiologie der Pflanze. Berlin-Göttingen-Heidelberg: Springer 1949.

TALLEY, P. T.: Some chemical differences in staminate and pistillate plants of hemp. Ph. D. Thesis, University of Wisconsin 1932. — TAYLOR, C. V., u. D. M. WHITAKER: Potentiometric determinations in the protoplasm and cell sap of *Nitella*. Protoplasma (Berl.) **3**, 1—6 (1927).

WAGNER, R. I.: Wasserstoffionenkonzentration und natürliche Immunität der Pflanzen. Zbl. Bakter. II **44**, 708—719 (1916). — WARTENBERG, H.: Studien über Redoxpotentiale der Gewebebreiaufschlämmungen und der Gewebepreßsäfte von Pflanzenteilen. Biochem. Z. **302**, 261—276 (1939). — WEBER, F.: Vitale Blattinfiltration (eine zellphysiologische Hilfsmethode). Protoplasma (Berl.) **1**, 581—588 (1927). — WEISS, F., and R. B. HARVEY: Catalase, hydrogen-ion concentration, and growth in the potato wart disease. J. Agricult Res. **21**, 589—592 (1921). — WILLSTÄTTER, R.: Über Pflanzenfarbstoffe. Ber. dtsch. chem. Ges. **47**, 2831—2874 (1914).

YAMAHA, G., u. T. ISHII: Über die Wasserstoffionenkonzentration und die isoelektrische Reaktion der pflanzlichen Protoplasten, insbesondere des Zellkernes und der Plastiden. Protoplasma (Berl.) **19**, 194—212 (1933).

# Physical chemistry of the vacuoles.

By

## Paul J. Kramer.

Vacuoles apparently occur in all plant cells, but they vary widely in number, size, shape, content, and even in color. One of the most characteristic structures of mature plant cells is a large central vacuole, occupying more than 50% of the cell volume and filled with liquid cell sap. On the other hand in meristematic tissue the vacuoles usually are small and variable in shape, ranging from spherical to thread-like or rod-shaped, though cambium cells sometimes contain relatively large vacuoles (BAILEY 1930). Their size and shape seems to depend in part on the condition of the cytoplasm, being more or less globular in quiet cytoplasm. but often slender and elongated in streaming cytoplasm (ZIRKLE 1937). Small vacuoles may coalesce to form large ones and large vacuoles may break up into numerous small ones. During seed maturation the vacuoles lose water and shrink until they are reduced to small structures superficially resembling mitochondria, but they swell again during germination (GUILLIERMOND 1941). SPONSLER and BATH (1942) suggested that numerous submicroscopic vacuoles exist in the framework of the protoplasm and that many biochemical reactions probably occur in these tiny chambers. It may be questioned whether these tiny spaces can be regarded as vacuoles, but it is difficult to define a vacuole very precisely. KÜSTER (1951) appears to regard only those droplets which are at least visible in the microscope as vacuoles.

Under certain conditions, not fully understood, spontaneous contraction of vacuoles may occur, the vacuoles rapidly shrinking in volume and their content becoming a more or less rigid gel. Apparently the liquid lost during contraction may either form a layer between the solidifying colloidal material of the vacuole and the cytoplasm or it may be absorbed by the cytoplasm which swells and occupies the space previously occupied by the vacuole. Vacuolar contraction has been most extensively studied in colored flower petals, especially those of the *Boraginaceae*, but it occurs in a variety of plant cells (GICKLHORN and WEBER 1926, KÜSTER 1940). KÜSTER (1950) and others have reported that accumulation of neutral red sometimes is followed by vacuolar contraction which may continue until the vacuolar contents solidify and the vacuoles appear to break up. Most investigators ascribe vacuolar contraction to syneresis (WEBER 1934), a process in which liquid spontaneously separates from a gel, but it appears that syneresis alone cannot explain all examples of it. BÜNNING (1949), KÜSTER (1950), and KENDA and WEBER (1952) have recently published accounts of this phenomenon and the latter believe changes in hydration of pectic substances in the cell sap are responsible for the rapid contraction which occurs in the *Boraginaceae*. Possibly in some instances the imbibitional capacity of the cytoplasm is increased, causing water to move out of the vacuole into the cytoplasm. Its rapid occurrence has led to suggestions that vacuolar contraction might play a part in certain turgor movements of plants.

The contraction of vacuoles just discussed is quite different from that of the contractile vacuoles which occur in protozoa, because during contraction of the latter type liquid is expelled from the cells. This type of contractile vacuole occurs in flagellates such as Euglena, the zoospores of various algae, and in some Myxomycetes (LLOYD 1928). According to LLOYD a special type of contractile vacuole occurs in those vegetative cells of Spirogyra which are developing into gametes. As the protoplasts of these cells contract and pass into the conjugation tubes contractile vacuoles form and empty the cell sap from the large central vacuole into the space outside of the protoplast. A discussion of contractile vacuoles also can be found in KÜSTER (1951).

**Origin of vacuoles.** Much discussion has occurred concerning the origin of vacuoles and the reader is referred to GUILLIERMOND (1941) and ZIRKLE (1937) for reviews of the earlier literature. It will also be discussed in a later chapter of this volume. Some of the earlier investigators believed that they originated de novo, but others thought that they originated from special bodies called tonoplasts by DE VRIES and hydroleucites by VAN TIEGHEM, and WENT claimed that they originated from preexisting vacuoles. DANGEARD and GUILLIERMOND believed for a time that they originated from mitochondria, but GUILLIERMOND and others have found that the small structures which enlarge into vacuoles are quite different chemically from mitochondria. BAILEY (1930) observed vacuoles of various sizes and shapes in cambial cells and ZIRKLE (1932) found the same situation in root and stem tips. They observed the division of preexisting vacuoles, but found no evidence that they were developing either de novo or from any cell inclusions. Because of the difficulty of observing the process, ZIRKLE (1937) admitted that failure to observe their formation is not proof that they never originate in this manner. GUILLIERMOND (1941) states that vacuoles have been observed to develop de novo in *Saccharomyces* and *Saprolegnia*. It seems quite possible that they originate both by division of preexisting vacuoles and also de novo by physicochemical processes which might normally be expected to operate in the cytoplasm.

GUILLIERMOND (1941) and FREY-WYSSLING (1953) discussed mechanisms of vacuole formation based on principles of colloidal chemistry which appear very plausible and at least indicate possible modes of origin. If cytoplasm is regarded as consisting of three principal phases, water, protein, and lipoid, in equilibrium with each other, it seems likely that a change in amount or condition of any of the components might lead to readjustment during which a portion of one or more components might be separated out. This "demixing" can be demonstrated experimentally in coacervate systems where water droplets can be caused to separate out and coalesce into large drops. BUNGENBERG DE JONG (1932) discussed its application to biological systems and the coacervate concept has been used by many other writers. GUILLIERMOND suggested that small particles of hydrophilic colloidal materials are deposited in the cytoplasm which have greater imbibitional forces than the other cytoplasmic colloids and as these particles become hydrated they swell, become fluid and form vacuoles. Such an origin is compatible with the observed structure of very small vacuoles and with the fact that they often contract. Because of the structure of the cytoplasmic framework such areas would at first be elongated, but as they enlarge they would push the framework aside and become more or less spherical (FREY-WYSSLING 1953). Although vacuole formation is doubtless brought about by the operation of physico-chemical principles it seems unwise to insist on too close an analogy to droplet formation in nonliving systems such as coacervates, because energy supplied by metabolic activity doubtless plays a part.

**The vacuolar membrane.** Vacuoles are inclosed in membranes known as tonoplasts, a name which was first used by DE VRIES for the vesicles which he thought gave rise to vacuoles. The membrane is a part and product of the cytoplasm although it is possible to separate vacuoles and their enclosing tonoplasts from the cytoplasm and keep them functioning actively for some time (BAILEY 1930, PLOWE 1931, EICHBERGER 1934). Some debate has occurred concerning the structure and composition of the tonoplast. It seems probable that as a vacuole enlarges a more concentrated layer of protein micelles is formed at the surface (FREY-WYSSLING 1953). Probably lipoidal material and other substances which reduce interfacial tension accumulate in this protein layer and the components become oriented in the interface, giving rise to a structure with semipermeability and other properties which are different from the interior of the cytoplasm. The tonoplast is believed by some workers to have a higher concentration of lipoidal material than the plasmalemma. Some investigators also believe that most of the differential permeability to solutes in a cell exists in the tonoplast rather than in the plasmalemma or the cytoplasm as a whole. The reader is referred to a recent paper by HOPE and ROBERTSON (1953) and to the chapter in this volume on Water Content and Water Turnover in Cells for further discussion of this problem.

It should be remembered that the maintenance of membranes such as the tonoplast depends both on the operation of physicochemical processes at interfaces and on the expenditure of energy released during respiration. Purely physical forces cannot as yet adequately explain their behaviour because they undergo marked changes in properties if the cell is killed or even if the rate of respiration is changed from normal.

**The content of vacuoles.** Vacuoles usually are regarded as cytoplasmic inclusions containing dilute solutions of organic and inorganic solutes, but vacuoles actually contain a wide variety of substances and are occasionally filled with a gel instead of a liquid. Some of our knowledge of the composition of the vacuolar sap was obtained by staining them with certain of the so-called vital dyes and some by the study of expressed sap. Both methods have their defects. Use of dyes is limited to those which will penetrate the vacuole and they probably produce changes from the normal condition, as when the accumulation of neutral red causes contraction of the vacuoles. Expression of the cell sap has been widely used because it yields large quantities of material, but the vacuolar sap obtained by this method is contaminated with cytoplasmic sap and at the same time some of its constituents probably are filtered out by the cytoplasmic membranes through which it is expressed. The most nearly representative samples of vacuolar sap are those obtained from the giant cells of the aquatic plant *Chara*, the freshwater alga *Nitella*, and the coenocytic marine alga Valonia. Small amounts of vacuolar sap have been obtained from cells of normal size by inserting tiny needles with a micromanipulator.

Among the substances found in vacuoles in addition to inorganic salts and sugars are enzymes, proteins, amino acids, amides and other nitrogenous compounds, lipids, mucilages, gums, resins, tannins, anthocyanins, flavones, organic acids, and crystals of protein and calcium oxalate. The salts and sugars usually are in true solution, but a number of the other substances are in the colloidal condition and as a result the viscosity of the vacuolar sap is often twice as great as that of water. In some instances the amount of colloidal material is so great that a gel is formed and solid vacuolar sap results (GUILLIERMOND 1941). The viscosity also decreases with increasing temperature to a critical point where, according to PEKAREK (1933), it begins to decrease again.

It is impossible to explain on a purely physical basis how all of these substances get into the vacuole. Most of the water and some of the other substances apparently enter by diffusion, but many ions, sugars, certain dyes, and some other materials accumulate in the vacuoles in much higher concentrations than occur outside the cell by processes which are not yet fully understood. Many of the organic compounds are synthesized in the cytoplasm and transferred into the vacuoles by what may be loosely termed excretion. Probably such movement is brought about by a combination of physiological and physico-chemical forces. Frey-Wyssling (1953) regards the vacuoles as excretory organelles in which substances incompatible with cytoplasmic structure accumulate. He states, "Vacuoles owe their existence to substances which are temporarily or definitively excluded from interaction with the framework of the cytoplasm". According to this view the vacuolar contents originate by the separation of water, possibly by syneresis, and other substances are then forced into the water phase by such physical factors as surface forces, differential solubility, and electrical potentials. In short, most of the substances present in the vacuole are there because their concentration exceeds that which can exist in the cytoplasm. For example, when the amount of lipoidal material in the cell exceeds that which can be bound by the cytoplasmic structure the excess accumulates as drops in the vacuole. While this view explains the accumulation of some materials it does not explain satisfactorily the accumulation of substances from the environment in higher concentrations than exist outside of the cells.

There has been some effort to distinguish between different types of vacuoles, chiefly on the basis of their staining behavior. Guilliermond (1941) summarized a number of observations which indicate that two types of vacuoles occur, often in the same cell. One type is low in $p_H$ and high in tannins, the other less acid and contains little or no tannin. The vacuoles high in tannin seem to correspond to the "A" type of Bailey and Zirkle (1931) and the "volle" cell sap of Höfler (1947, 1949) which according to Härtel (1951) stains heavily with certain dyes because they combine with tannins. The vacuoles containing little or no tannin probably correspond to the "B" type of Bailey and Zirkle and the "leere" type of Höfler. Probably the A-type vacuoles contain not only tannins, but also proteins, pectic substances, and lipoidal materials with which dyes combine readily (Drawert 1951, Härtel 1951, Scarth 1926). Although a typical vacuole is assumed to be filled with a dilute solution, many vacuoles contain large amounts of colloidal material. For example so much protein accumulates in some vacuoles that solid masses called aleurone grains develop. Bünning (1949) discussed the formation and enlargement of droplets or globules in the vacuoles of *Iris* leaves which he attributes to the swelling of gel particles, probably lecithin, which imbibe water from the cell sap. He also thinks that some drop formation occurs by syneresis. Among the vacuolar inclusions which have received considerable attention are the "Anthocyanophoren" or globules containing a high concentration of anthocyanin. According to Küster (1939a, 1951) they are formed by vacuole contraction and the material of which they are composed is of such a nature that anthocyanin tends to accumulate in them in preference to the remainder of the vacuole. Pectic compounds also occur in the vacuoles and sometimes are responsible for vacuolar contraction (Kenda and Weber 1952). In some cells small colorless vacuoles occur which appear to contain no colloidal material, but instead contain crystals. Guilliermond (1941) thinks these vacuoles arise in the same manner as droplets in coacervates. Küster (1951) also thinks that formation of droplets and globules in the cell sap (Entmischung phenomena) can be explained in terms of the coacervate theory of BUNGENBERG DE JONG (1932).

**Vital staining.** Many studies have been made of the accumulation of various dyes or stains by the different parts of living cells. The older literature has been summarized by SCARTH (1926) and by KÜSTER (1939b). The most common method of staining living cells is simply to immerse sections of tissue in the dye for various periods of time and observe it under the microscope. WEBER (1927) suggested that more uniform penetration can be obtained if the intercellular spaces are infiltrated with the solution of dye, but this is seldom done. Occasionally dyes are injected directly into the cytoplasm or the vacuole with a micro-injection apparatus (PLOWE 1931, CHAMBERS and KERR 1932, HOFMEISTER 1940). KÜSTER and others have also injected dyes into the xylem and allowed the transpiration stream to carry them into the cells of the leaves. Vital staining with indicator dyes has also been attempted frequently in efforts to determine the $p_H$ of various parts of the cell.

Vital stains have been selected partly for their supposedly low toxicity, but it seems probable that all of them are more or less toxic. Even though the cells are not killed or visibly injured at once, slow changes may occur and serious injury may eventually result from the presence of the dye. This is shown by changes in appearance and behavior of the cells. The penetration of the dye itself may cause changes in the cytoplasm and even in the vacuole, as demonstrated by the "Entmischung" or formation of droplets which often occurs in vacuoles after staining with methylene blue or neutral red (see SCARTH 1926, GICKLHORN 1929, BANK 1937). As mentioned earlier in this article the contents of these droplets sometimes solidify and even break up. Such behavior seems to be good evidence that staining has produced highly abnormal conditions in the cell.

DRAWERT (1951) has recently summarized work by himself and others on vital staining and on the factors which affect the absorption, accumulation, and distribution among the various cell constituents of vital stains in plant cells. Among the important characteristics which determine the behavior of vital stains are the size of the particles, the electric charge on the colored ion, the $p_H$ of the dye solution, and the relative solubility in water and hydrophobic solvents. Important cell characteristics are the pore size of the cytoplasm or cytoplasmic membranes, the lipoid content of the cytoplasm, the amount of colloidal material in the cell sap with which stains can combine (also see preceding section of this paper) and the $p_H$ difference between the vacuole and the external solution. In addition to these chemical and physical characteristics, the physiological condition of the cell is a very important factor because dye accumulation, like salt accumulation, is related to the metabolic activity of the cells.

**Functions of vacuolar contents.** Some of the substances which occur in the vacuoles are essential to the existence of the cell, others are reserve materials important to the plant, while still others seem to have no known function. The uses of reserves of water, sugars, salts, and nitrogenous compounds seems obvious, but the function of tannin is at least doubtful, and that of various alkaloids, gums, and resins is unknown. Many of these substances can be regarded as mere byproducts of plant metabolism which themselves have no known function in plants.

Although tannin is generally regarded as without any function, HAUSER (1935) thought the tannins might affect the degree of dispersion of the protoplasm and thereby modify its permeability and other physical properties. The function of the accumulated protein depends on the type formed. Accumulation of protein of low molecular weight results in a high concentration of colloidal material in the vacuoles and if dehydration occurs this may solidify and form aleurone grains or crystals. If the amount of protein of high molecular weight increases, protoplasmic fibers may form, giving rise to new protoplasm and conditions favorable to growth and cell division. The anthocyanins so conspicuous in many cells probably have no function, nor are the flavones known to have any.

The mineral salts and sugars present in the vacuole play a double role. They act as reserves which can be used in various metabolic processes and they also have important effects on the osmotic properties of the cells.

**Accumulation of ions in vacuoles.** The kinds and amounts of ions present in the vacuolar sap has been the subject of extensive study in connection with investigations of permeability and mineral uptake. The number of different elements found in plants is so great that a list of them has little significance except to indicate that plant cells are able to absorb almost any solute in their environment. Of much greater importance is the fact that ions can be accumulated in the vacuolar sap in concentrations much higher than those occurring outside the cells. Establishment of this fact by OSTERHOUT (1922), HOAGLAND and DAVIS (1923, 1929), COLLANDER (1936), and others formed a basis for modern studies of salt accumulation. Some results obtained in such studies are shown in Table 1.

Table 1. *Comparison of composition of vacuolar sap of Nitella with composition of the medium in which it was growing.* From HOAGLAND and DAVIS (1929).

| | Composition in Milliequivalents | | | | | | | |
|---|---|---|---|---|---|---|---|---|
| | Cl | NO$_3$ | SO$_4$ | H$_2$PO$_4$ | Ca | Mg | Na | K |
| Nitella sap . . . . . . . | 106.5 | | 20.2 | 2.4 | 19.3 | 17.2 | 78.8 | 58.2 |
| Pond water . . . . . . | 1.0 | | 0.67 | 0.0008 | 1.3 | 3.0 | 1.2 | 0.5 |
| Concentration factor . . . | 106.0 | | 30.0 | 300.0 | 15.0 | 5.7 | 65.0 | 116.0 |
| Nitella sap . . . . . . . | 84.6 | 3.3 | 8.3 | 3.5 | 16.9 | 7.9 | 25.9 | 56.4 |
| Culture solution . . . . . | 1.2 | 1.8 | 1.5 | 0.03 | 2.0 | 1.6 | 1.5 | 1.8 |
| Concentration factor . . . | 70.0 | 1.8 | 5.5 | 117.0 | 8.4 | 5.0 | 17.3 | 31.3 |

The mechanism of solute accumulation in vacuoles of plant cells is not fully explained. As a more extensive discussion will appear in volume 2 of this Hand Book, only a summary of possible mechanisms will be given here. Readers are also referred to OVERSTREET and JACOBSON (1952), and ROBERTSON (1951) for reviews on this topic. In the earlier studies emphasis was placed on the chemical, physical, and electrical properties of the cell membranes which determine their permeability to various kinds of solutes. Later great importance was attached to the GIBBS-DONNAN equilibrium as a mechanism for ion accumulation. When electrically charged micelles and ions occur on one side of a membrane impermeable to them, but permeable to other diffusible ions, an unequal distribution of the diffusible ions will occur on opposite sides of the membrane. This is known as the DONNAN equilibrium. Conditions such as occur in plant cells containing considerable colloidal material in the cell sap are favorable for the accumulation of ions in the cells by this mechanism (HOPE and ROBERTSON 1953). It is obvious, however, that ion accumulation cannot be explained completely in terms of physical processes because it is so closely related to the metabolic processes of cells that it must be either directly or indirectly dependent on energy supplied by respiration. For this reason interest is at present centered on physiological mechanisms, the role of enzymes and carrier molecules, and the relation of accumulation to respiration and cell metabolism in general. It seems that ion accumulation depends on both physical and physiological factors.

Considerable differences exist in the amounts of a given ion taken up by different plants and MILLER (1938) has summarized much of the data on this subject. An example is the difference in amount of sodium absorbed by different species. COLLANDER (1941) found that *Atriplex* absorbed over 40 times as much sodium from a nutrient solution as *Zea* and *Helianthus* growing in the same solution, but it absorbed only about half as much potassium and rubidium. Another example of differences between species of the same genus is the observation that *Astragalus bisulcatus* growing in soil containing only 2.1 p.p.m. of selenium accumulated 1250 p.p.m. while *A. missouriensis* growing in the same soil accumulated 3.1 p.p.m. (TRELEASE and MARTIN 1936). HARRIS, HOFFMAN, and LAWRENCE (1925) found that the sap expressed from the leaves of Egyptian cotton contained more chloride and less sulfate than sap from upland cotton growing in the same soil. This difference was inherited in crosses between the two types. No satisfactory explanations exist for such differences between species and varieties in respect to ion uptake, but they constitute further evidence that ion accumulation cannot be treated simply as a physical process.

Differences in amount of certain ions absorbed by cells can produce biochemical changes in plants. For example the succulence of certain halophytes has been attributed to the large amounts of chlorides which they absorb (VAN EIJK 1939) and it is claimed that an excess of chloride will result in thick tobacco leaves of poor quality (WILSON 1933). Truck growers in certain areas believe that fertilization with KCl produces succulent tomato fruit of poorer shipping quality than fertilization with $K_2SO_4$, while the reverse is said to be true of cabbage, in which succulence is desirable.

**Osmotic pressure.** The osmotic pressure of the cell sap has been studied extensively, partly because of its relation to osmotic phenomena and partly because of its supposed relation to cold and drought resistance. Many measurements of osmotic pressure have been made on pieces of plant tissue by plasmolytic methods and many others on expressed sap by cryoscopic methods. The results and significance of such studies have been summarized and discussed by WALTER (1931), CRAFTS, CURRIER and STOCKING (1949), and others. The latter also discuss the errors inherent in all measurements of osmotic pressure of plant sap. HARRIS and his coworkers made thousands of measurements of the osmotic pressure and electrical conductivity of the cell sap expressed from plants growing in a variety of habitats in an effort to correlate variations in the properties of the plant sap with variations in the environment in which it grew (HARRIS 1934).

Many factors in the plant and the environment affect the concentration of salt and sugar which largely determine the osmotic pressure of the sap. MILLER (1938, pp. 39–45) has summarized the data on this subject. Plants growing in moist soil usually have a lower osmotic pressure than those growing in dry habitats. The average osmotic pressure of leaf sap from the spring flora in a mesophytic region (Long Island, New York) was about 10 atmospheres while it was nearly 20 atmospheres in the Arizona desert. The highest osmotic pressures are found in desert plants, the highest known to the writer being 202.5 atmospheres which was found in the leaf sap of *Atriplex confertifolia*, a native of the alkali soils of the American desert. Osmotic pressures in excess of 50 atmospheres are uncommon, even in desert plants. Decreasing soil moisture results in an increase in osmotic pressure of the cell sap of most plants. MALLERY (1935), SIMONIZ (1936), STODDARD (1936), and KORSTIAN (1924), found the osmotic pressure to be a satisfactory method of evaluating the water stress to which plants were subjected in various habitats. Increase in osmotic pressure of the plant sap also occurs when the osmotic pressure of the soil solution is increased, although this is not invariably true (McCOOL and MILLAR, 1917). LEVITT'S (1951) summary of recent work shows that drought usually results in an increase in osmotic pressure.

In general trees and shrubs have a higher osmotic pressure than herbaceous plants and the osmotic pressure of trees and shrubs averages considerably higher than that of perennial herbs growing in the same environment. The osmotic pressure of leaves usually is higher than that of roots and several investigators found an increase in osmotic pressure from the lower to the upper leaves. In general older leaves appear to have higher osmotic pressures than young leaves. A winter increase in osmotic pressure of conifer needles also has been reported.

Table 2. *The $p_H$ of the sap of various species of plants. Selected from data compiled by HURD-KARRER (1939).*

| Plant | $p_H$ |
|---|---|
| Crop plants | |
| Alfalfa *(Medicago sativa)* | 5.6–6.4 |
| Bean *(Phaseolus, spp.)* | 5.6–6.2 |
| Beet *(Beta vulgaris)* | 5.4–6.6 |
| Buckwheat *(Fagopyrum esculentum)* | 4.4–6.0 |
| Cabbage *(Brassica oleracea capitata)* | 5.5–5.9 |
| Clover, red *(Trifolium pratense)* | 5.7–6.2 |
| Corn *(Zea mays)* | 5.0–5.7 |
| Pea, garden *(Pisum sativum)* | 5.7–6.4 |
| Potato *(Solanum tuberosum)* | 5.5–6.6 |
| Soybean *(Soya max)* | 5.6–6.8 |
| Squash *(Cucurbita maxima)* | 6.4–6.7 |
| Sunflower *(Helianthus annuus)* | 5.8–7.2 |
| Sweet clover *(Melilotus)* | 6.2–7.3 |
| Tobacco *(Nicotiana tabacum)* | 5.0–6.0 |
| Tomato *(Lycopersicum esculentum)* | 5.4–6.7 |
| Wheat *(Triticum)* | 5.3–6.0 |
| Plants with unusually acid sap | |
| Begonia spp. | 0.9–1.7 |
| Oxalis | 1.7–2.2 |
| Lemon (fruit) | 2.2–2.3 |
| Rumex spp. | 2.5–5.2 |
| Bryophyllum | 3.3–5.3 |
| Rheum rhaponticum | 3.1–4.6 |

Diurnal variations in osmotic pressure also occur, usually largely the result of decrease in water content resulting from heavy transpiration and possibly to some extent caused by increase in solutes produced by photosynthesis. The minimum osmotic pressure usually occurs in the morning and the maximum soon after midday (McCOOL and MILLAR 1917, KORSTIAN 1924, HERRICK 1933, STODDART 1935).

**Sap $p_H$.** The vacuolar sap is generally acid, but it appears to be less stable than that of the better buffered cytoplasm and varies over a somewhat wider range. Values varying from a low of 0.9 in Begonia to a high of $p_H$ 7 or 8 have been reported, but the average appears to be near $p_H$ 5.5 to 6.5 for expressed sap (HURD-KARRER 1939), while the average for protoplasm is nearer $p_H$ 6.8 to 7.0 (SEIFRIZ 1936). Some representative $p_H$ values are given in Table 2. Direct injection of dyes into the vacuole has been used to determine the $p_H$ of the sap and some basic dyes accumulate in the vacuole in sufficient concentration to allow estimation of the $p_H$ from their color. SMALL and his coworkers have made numerous measurements of the $p_H$ of plant material by the use of indicators and their results are summarized by SMALL (1929). These methods are subject to considerable errors because of the presence of salt and protein in the cell sap. Some plant cells contain pigments which change color with change in $p_H$ and therefore can be used as naturally occurring indicators. Examples of $p_H$ measurements made by this method are those of HAAS (1916) and SMITH (1933).

Diurnal variations in $p_H$ have been observed in many plants, especially in succulents which often are more acid at night than during the day because of accumulation of organic acids in darkness (MILLER 1938). HURD-KARRER (1939) found that unhealthy wheat plants became more acid than healthy, vigorously growing plants. High temperature and a long photoperiod checked growth and resulted in higher acidity while a short photoperiod was accompanied by vigorous growth and lower acidity. Liming the soil had no effect on the $p_H$ of wheat sap so long as it did not affect the rate of growth. HURD-KARRER's data suggest that environmental factors affect $p_H$ only to the extent that they modify the normal metabolism of the plant sufficiently to upset the organic acid content. The $p_H$ of the environment can usually be varied over a considerable range without much effect on the internal $p_H$ of the plant. HOAGLAND and DAVIS (1923), for example, found that the cell sap of *Nitella* remained at $p_H$ 5.2 while the $p_H$ of the surrounding medium was varied from 5 to 9.

The buffer action of plant saps is produced chiefly by weak acids and their salts. In some instances one system may predominate, as the citrate system is said to do in lemon (SINCLAIR and ENY 1946), but as a rule several organic acids are involved. Among those most common in plant cells are carbonic, citric, malic, oxalic, phosphoric, and tartaric. The amounts of the various acids present are extremely variable, depending on the species, metabolic activity, stage of development, and environmental factors such as light and temperature.

**Vacuolar sap in relation to cold and drought resistance.** Many studies have been made of the physicochemical properties of the vacuolar sap in relation to cold and drought resistance in plants. Most of the work has centered on changes in osmotic pressure, electrical conductivity, and bound water content of expressed sap. As mentioned earlier, the osmotic pressure usually increases during periods of drought and in the winter. The increase results in part from decrease in water content and in part from increase in soluble substances in the sap, as for example the conversion of starch to sugar. Changes in osmotic pressure are by no means proportional to the degree of cold or drought resistence, however. PISEK (1950), for example, found that although the killing temperatures of certain conifers ranged from $-5$ to $-43^0$ C. the osmotic pressures of their saps varied only from 17 to 23 atmospheres. He also found that *Rhododendron* could be caused to lose its cold hardiness in the winter by exposure to high temperature without any corresponding decrease in osmotic pressure of the cell sap. As would be expected, changes in environment also produce changes in electrolyte content of the cell sap which can be measured by changes in electrical conductivity of the plant sap. The numerous measurements made by HARRIS (1934) and others do not show any definite correlation between cold and drought resistance and the conductivity of the plant sap (see LEVITT 1941, pp. 118–120; also LEVITT 1951).

Hope was held at one time that cold and drought resistance could be explained in terms of bound water. NEWTON and MARTIN (1930) found a good correlation between the bound water content of the leaf sap of various wild and cultivated plants and their drought resistance and GREATHOUSE (1932) and others found that the bound water content of the leaf sap was correlated with cold resistance in various plants. LEVITT and SCARTH (1936) found by plasmometric methods that the bound water content of the vacuolar sap of *Catalpa* and *Liriodendron* phloem cells was higher in cold hardened than in unhardened tissue. More extensive research has shown, however, that there is no consistent relation between bound water content of the cell sap and either cold or drought resistance.

LEVITT (1951) thinks that the only bound water likely to have much effect on cold or drought resistance is that occurring in the protoplasm rather than that in the vacuoles. The role of bound water is discussed in more detail in another chapter of this volume.

ILJIN (1953) has recently summarized his studies on injury from dehydration and regards the size of the vacuoles as an important factor in drought resistance. He found cells containing large vacuoles less resistant to repeated plasmolysis and deplasmolysis than cells with small vacuoles, probably because the former undergo greater changes in volume and more injury to the protoplasm results than in cells with small vacuoles. The rate of dehydration and rehydration also is important, much less injury occurring when changes in water content are brought about slowly than when they occur rapidly. He also regards a high osmotic pressure as beneficial because it slows down the rate of water loss. It seems probable that high osmotic pressures and small vacuoles are usually found together, and although a high osmotic pressure may somewhat increase drought resistance it probably is more important as an indicator of the degree of moisture stress existing in the plant.

In general it seems that the causes of cold and drought resistance are not to be found in any properties of the vacuolar sap. Rather they are more closely related to the ability of the protoplasm to withstand dehydration without any permanent, irreversible injury to its colloidal structure.

**General remarks.** In reviewing the literature on vacuoles certain general trends seem to occur. Two decades ago great emphasis was placed on the explanation of cell structure and processes in terms of the operation of purely physicochemical principles. Vacuole development was explained in terms of droplet formation in coacervates and the presence of various substances in the vacuolar sap was ascribed to the fact that they are more soluble in water than in the cytoplasm. Formation of the vacuolar membrane was explained as resulting from surface forces operating at the interfaces between an aqueous and an nonaqueous phase. Cold and drought resistance were attributed to the osmotic pressure and the bound water content of the cell sap. While physicochemical principles undoubtedly govern the development and functioning of cells and cell vacuoles it is clear that cell structure and processes cannot be fully explained in terms of chemical and physical processes as we now understand them. The structure and permeability of cell membranes and the accumulation of substances in the vacuole are affected by cell metabolism, and cold and drought resistance are more dependent on complex properties of the protoplasm than on such simple physical factors as the osmotic pressure and conductivity of the cell sap. It seems that in general the more we learn about cell processes and conditions the more complex our explanations become.

## Literature.

BAILEY, I. W.: The cambium and its derivative tissues. V. A reconnaissance of the vacuome in living cells. Z. Zellforsch. 10, 651 (1930). — BAILEY, I. W., and C. ZIRKLE: VI. The effects of hydrogen ion concentration in vital staining. J. Gen. Physiol. 14, 363–384 (1931). — BANK, O.: Entmischung der gefärbten Vakuolen-Kolloide durch Farbstoffe. Protoplasma 27, 367–371 (1937). — BÜNNING, E.: Über quellbare Zellsaftkolloide in Irisblättern. Planta (Berl.) 37, 431–436 (1949). — BUNGENBERG DE JONG, H. G.: Die Koazervation und ihre Bedeutung für die Biologie. Protoplasma 15, 110 (1932).

CHAMBERS, R., and T. KERR: Intracellular hydrion concentration studies. J. Cellul. a. Comp. Physiol. 2, 105–119 (1932). — COLLANDER, R.: Der Zellsaft der Characeen. Protoplasma 25, 201–210 (1936). — Selective absorption of cations by higher plants. Plant Physiol. 16, 691–720 (1941). — CRAFTS, A. S., H. B. CURRIER and C. R. STOCKING: Water in the Physiology of the Plant. Waltham, Mass.: Chronica Botanica Co. 1949.

DRAWERT, H.: Beiträge zur Vitalfärbung pflanzlicher Zellen. Protoplasma **40**, 85–106 (1951).

EICHBERGER, R.: Über die "Lebensdauer" isolierter Tonoplasten. Protoplasma **20**, 606–632 (1934). — EIJK, M. VAN: Analyse der Wirkung des NaCl auf die Entwicklung, Sukkulenz und Transpiration bei *Salicornia herbacea* sowie Untersuchungen über den Einfluß der Salzaufnahme auf die Wurzelatmung bei Aster *Tripolium*. Rec. Trav. bot. néerl. **36**, 559–657 (1939).

FREY-WYSSLING, A.: Submicroscopic Morphology of Protoplasm, 2nd ed. Houston: Elsevier Publ. Co. 1953.

GICKLHORN, J.: Beobachtungen über vitale Farbstoffspeicherung. Kolloidchem. Beih. **28**, 367 (1929). — GICKLHORN, J., and F. WEBER: Über Vakuolenkontraktion und Plasmolyseform. Protoplasma **1**, 427–432 (1926). — GREATHOUSE, G. A.: Effects of the physical environment on the physico-chemical properties of plant saps, and the relation of these properties to leaf temperature. Plant Physiol. **7**, 349–390 (1932). — GUILLIERMOND, A.: The Cytoplasm of the Plant Cell. Waltham, Mass.: Chronica Botanica Co. 1941.

HAAS, A. R.: The acidity of plant cells as shown by natural indicators. J. of Biol. Chem. **27**, 233–241 (1916). — HÄRTEL, O.: Gerbstoffe als Ursache "voller" Zellsäfte. Protoplasma **40**, 338–347 (1951). — HARRIS, J. A.: The Physico-chemical Properties of Plant Saps in Relation to Phytogeography. Minneapolis: Univ. Minnesota Press 1934. — HARRIS, J. A., W. F. HOFFMAN and J. V. LAWRENCE: Differential absorption of anions by varieties of cotton. Proc. Soc. Exper. Biol. a. Med. **22**, 350–352 (1925). — HAUSER, W.: Zur Physiologie des Gerbstoffes in der Pflanzenzelle. Protoplasma **24**, 219–224 (1935). — HERRICK, E. M.: Seasonal and diurnal variations in the osmotic values and suction tension values in the aerial portions of *Abrosia trifida*. Amer. J. Bot. **20**, 18–34 (1933). — HOAGLAND, D. R., and A. R. DAVIS: The composition of the cell sap of the plant in relation to the absorption of ions. J. Gen. Physiol. **5**, 629–646 (1923). — The intake and accumulation of electrolytes by plant cells. Protoplasma **6**, 610–626 (1929). — HÖFLER, K.: Was lehrt die Fluoreszenzmikroskopie von der Plasmapermeabilität und Stoffspeicherung? Mikroskopie (Wien) **2**, 13 (1947). — Fluoreszenzmikroskopie und Zellphysiologie. Biol. generalis (Wien) **19**, 90 (1949). — HOFMEISTER, L.: Mikrurgische Studien an *Borraginoideen*-Zellen. II. Mikroinjektion und mikrochemische Untersuchung. Protoplasma **35**, 161–186 (1948). — HOPE, A. B., and R. N. ROBERTSON: Bioelectric experiments and the properties of plant protoplasm. Austral. J. Sci. **15**, 197–203 (1953). — HURD-KARRER, A. M.: Hydrogen-ion concentration of leaf juice in relation to environment and plant species. Amer. J. Bot. **26**, 834–846 (1939).

ILJIN, W. S.: Causes of death of plants as a consequence of loss of water: conservation of life in desiccated tissues. Bull. Torrey Bot. Club **80**, 166–177 (1953).

KENDA, G., and F. WEBER: Rasche Vakuolen-Kontraktion in *Cerinthe*-Blütenzellen. Protoplasma **41**, 458–466 (1952). — KORSTIAN, C. F.: Density of cell sap in relation to environmental conditions in the Wasatch Mountains of Utah. J. Agricult. Res. **28**, 845–907 (1924). — KÜSTER, E.: Über Vakuolenkontraktion und Anthocyanophoren bei Pulmonaria. Cytologia **10**, 44–50 (1939a). — Vital-staining of plant cells. Bot. Review **5**, 351–370 (1939b). — Neue Objekte für die Untersuchungen der Vakuolenkontraktion. Ber. dtsch. bot. Ges. **58**, 413–416 (1940). — Über Vakuolenkontraktion in gegerbten Zellen. Protoplasma **39**, 14–22 (1950). — Die Pflanzenzelle. Jena: Gustav Fischer 1951.

LEVITT, J.: Frost killing and hardiness of plants. Minneapolis: Burgess Publ. Co. 1941. — Frost, drought, and heat resistance. Ann. Rev. Plant Physiol. **2**, 245–268 (1951). — LEVITT, J., and G. W. SCARTH: Frost-hardening studies with living cells. II. Permeability in relation to frost resistance and the seasonal cycle. Canad. J. Res. C **14**, 285–305 (1936). — LLOYD, F. E.: The contractile vacuole. Biol. Rev. **3**, 329–358 (1928).

MALLERY, T. D.: Changes in the osmotic value of the expressed sap of leaves and small twigs of *Larrea tridentata* as influenced by environmental conditions. Ecol. Monogr. (Durham) **5**, 1–35 (1935). — McCOOL, M. M., and C. E. MILLAR: The water content of the soil and the composition and concentration of the soil solution as indicated by the freezing-point lowerings of the roots and tops of plants. Soil Sci. **3**, 113–138 (1917). — MILLER, E. C.: Plant Physiology, 2nd ed. New York: McGraw-Hill 1938.

NEWTON, R., and W. M. MARTIN: Physico-chemical studies of the nature of drought resistance in crop plants. Canad. J. Res. **3**, 336–427 (1930).

OSTERHOUT, W. J. V.: Some aspects of selective absorption. J. Gen. Physiol. **5**, 225–230 (1922). — OVERSTREET, R., and L. JACOBSON: Mechanisms of ion absorption by roots. Ann. Rev. Plant Physiol. **3**, 189–206 (1952).

PEKAREK, J.: VI. Der Einfluß der Temperatur auf die Zellsaftviskosität. Protoplasma **20**, 251–278 (1933). — PISEK, A.: Frosthärte und Zusammensetzung des Zellsaftes bei *Rhododendron ferrugineum*, *Pinus cembra* und *Picea excelsa*. Protoplasma **39**, 129–146 (1950). — PLOWE, J.: Membranes in the plant cells. Protoplasma **12**, 196–240 (1931).

ROBERTSON, R. N.: Mechanism of absorption and transport of inorganic nutrients in plants. Ann. Rev. Plant Physiol. **2**, 1–24 (1951).

SCARTH, G. W.: The mechanism of accumulation of dyes by living cells. Plant Physiol. **1**, 215–229 (1926). — SEIFRIZ, W.: Protoplasm. New York: McGraw-Hill Book Co. 1936. — SIMONIZ, W.: Untersuchungen über die Abhängigkeit des osmotischen Wertes vom Bodenwassergehalt bei Pflanzen verschiedener ökologischer Gruppen. Jb. wiss. Bot. **83**, 191–239 (1936). — SINCLAIR, W. B., and D. M. ENY: Stability of the buffer system of lemon juice. Plant Physiol. **21**, 522–532 (1946). — SMALL, J.: Hydrogen-ion concentration in Plant Cells and Tissues. Berlin: Gebrüder Borntraeger 1929. — SMITH, E. P.: The calibration of flower color indicators. Protoplasma **18**, 112–125 (1933). — SPONSLER, O. L., and JEAN D. BATH: Molecular structure in protoplasm in: The Structure of Protoplasm, pp. 41–79. Ames: Iowa State College Press 1942. — STODDART, L. A.: Osmotic pressure and water content of prairie plants. Plant Physiol. **10**, 661–680 (1935).

TRELEASE, S. F., and A. L. MARTIN: Plants made poisonous by selenium absorbed from the soil. Bot. Rev. **2**, 373–396 (1936).

WALTER, H.: Die Hydratur der Pflanze und ihre physiologisch-ökologische Bedeutung. Jena: Georg Fischer 1931. — WEBER, F.: Vitale Blattinfiltration: Eine zellphysiologische Hilfsmethode. Protoplasma **1**, 581–588 (1927). — Anthocyanophoren-freie *Enythraea*-Blütenzellen. Protoplasma **33**, 473–474 (1939). — Vakuolen-Kontraktion der *Borraginaceen*-Blütenzellen als Synärese. Protoplasma **22**, 4–16 (1934). — WILSON, L. B.: Effects of chlorine, bromine, and fluorine on the tobacco plant. J. Agricult. Res. **46**, 889–899 (1933).

ZIRKLE, C.: Vacuoles in primary meristems. Z. Zellforsch. **16**, 26–47 (1932). — The plant vacuole. Bot. Rev. **3**, 1–30 (1937).

# Die Entstehung der Vacuolen.

Von

## H. Drawert.

Mit 3 Abbildungen.

Das Hauptvolumen der ausgewachsenen Pflanzenzelle wird im allgemeinen von der Vacuole eingenommen. Um so verwunderlicher ist es, daß wir in der Frage nach der Entstehung der Vacuolen über Hypothesen noch nicht hinausgekommen sind. Diese Hypothesen lassen sich in zwei Gruppen einteilen. Das Charakteristische der einen Gruppe ist die Annahme, daß die Vacuolen die Folge einer Phasentrennung sind, also in dem Entwicklungsgang einer jeden Zelle vom meristematischen Zustand bis zur Dauerzelle immer wieder neu entstehen. Das Kennzeichnende der anderen Gruppe ist die Auffassung, daß die Vacuolen nie de novo sich bilden, sondern autonome, permanente Zellorganelle wie Kern und Plastiden darstellen, die nur aus ihresgleichen hervorgehen. Im Entwicklungsgang einer Zelle durchläuft die Vacuole nur verschiedene Hydratationsgrade, seien es die eines Plasten oder die einer angenommenen besonderen Vacuolensubstanz.

Die Darstellung der Entstehung der Vacuolen wird sich auf eine mehr oder weniger eingehende Schilderung der verschiedenen Hypothesen nach historischen Gesichtspunkten beschränken müssen und kann diese bestenfalls in ihrer Wertigkeit gegeneinander abwägen. Zusammenfassungen über die zur Diskussion stehenden Frage und eingehende Literaturhinweise finden wir bei GUILLIERMOND (1930, 1934, 1941), GUILLIERMOND, MANGENOT und PLANTEFOL (1933), ZIRKLE (1937), DANGEARD (1947), KÜSTER (1951).

Unter dem Begriff Zellsaft verstand man früher den ganzen Zellinhalt. So schreibt MEYEN (1837, S. 179): ,,Die Zellen sind bei ihrem frühesten Auftreten mit einer wasserhellen, durchsichtigen, farblosen oder gefärbten Flüssigkeit gefüllt, und man nennt dieselbe den Zellensaft. Erst in späteren Perioden ändert sich dieser Inhalt der Zellen und, nach Maßgabe des örtlichen Vorkommens und der Determination der Bildungsgesetze bildet sich derselbe in verschiedene anderweitige Stoffe um, . . .''. Nach der Entdeckung des Protoplasmas beschränkte man den Begriff Zellsaft auf den Inhalt der Vacuole und dachte sich die Entstehung der Vacuolen als einen Entmischungsvorgang. HOFMEISTER (1867, S. 5) schildert diesen Vorgang wie folgt: Das Protoplasma besitzt einen hohen Grad von Imbibitionsfähigkeit für Wasser. Sobald aber die Wasseraufnahme ein bestimmtes Maß überschreitet, ,,so wird wäßrige Flüssigkeit, eine Lösung der löslichsten Gemengteile des Protoplasma, im Innern der Protoplasmamasse in Tropfen ausgeschieden, welche als scharfbegrenzte sphäroidische Blasenräume, Vacuolen oder intracellulare Räume, innerhalb der zähen flüssigen Masse erscheinen.'' Derselben Auffassung ist auch SACHS (1874, S. 42): Die Bildung der Vacuole beruht offenbar darauf, ,,daß ein Theil des im Protoplasma vorhandenen Wassers sich an inneren Puncten sammelt und hier endlich Tropfen bildet, . . .''.

Die Vacuole spielt als Ausscheidungsprodukt des Plasmas in den Arbeiten der alten Pflanzenphysiologen nur eine ganz untergeordnete Rolle. In seinen ,,Vorlesungen über Pflanzen-Physiologie'' handelt SACHS (1887) als die ,,wesentlichen

Bestandteile der Zelle" Protoplasma, Zellkern und Zellwand ab, auf die Vacuole wird überhaupt nicht eingegangen. Nur an anderer Stelle erwähnt der Verf. einmal kurz (S. 89): „Indem die Zellen größer werden, vermehrt sich nicht in gleichem Maße das Protoplasma, sondern es bilden sich in ihm Hohlräume, welche mit Flüssigkeit, d.h. mit Zellsaft erfüllt sind, ...".

Diese Sachlage änderte sich erst mit der Erkenntnis von DE VRIES (1885), daß die Vacuole ein osmotisches System darstellt. Bringt man das Cytoplasma zum Absterben, so können die Vacuolen erhalten bleiben („Isolierung der Vacuole"). Das beweist, daß die Vacuole eine — schon von anderen Forschern postulierte — Haut, eine resistente, semipermeable Grenzschicht besitzt. Seit DE VRIES bezeichnen wir diese Vacuolenhaut als „Tonoplast". DE VRIES sah aber im Tonoplasten ein den Plastiden entsprechendes Zellorganell, das befähigt ist, in sein Inneres Wasser und alle in der Vacuole gelösten Substanzen auszuscheiden (Sekretionsorgan). Die Tonoplasten sollen nur aus ihresgleichen entstehen können, und die Bildung von Vacuolen de novo sei unmöglich. Dieselbe Auffassung wird auch von VAN TIEGHEM (1888) vertreten, er bezeichnet die Vacuolen als „Hydroleuciten". Eingehende Studien ganz im Sinne der Anschauungen von DE VRIES widmet F. A. F. C. WENT (1888—1890) der Vacuolenentstehung. WENT schreibt wörtlich (1888, S. 345): „Alle normalen Vacuolen einer Pflanze stammen durch fortwährende Theilung aus derjenigen der mütterlichen Eizelle", und „Die Tonoplasten sind als Organe des Protoplasmas den Kernen und Chromatophoren ebenbürtig." Nach WENT enthalten auch alle meristematischen Zellen kleine Vacuolen und eine sog. Neubildung kann nur auf einer Vergrößerung bereits vorhandener kleiner Vacuolen beruhen, falls nicht überhaupt ein pathologischer Vorgang vorliegt wie bei der Vacuolisation von Kernen oder Chromatophoren.

Zu ganz entgegengesetzten Vorstellungen kommt aber PFEFFER (1890) auf Grund seiner Untersuchungen an Plasmodien von *Chondrioderma*. Läßt man Plasmodien aus einer gesättigten Lösung unter anderem Asparaginkriställchen aufnehmen und bringt diese innerhalb des Plasmodiums in kurzer Zeit zur Lösung, dann entstehen künstlich Vacuolen, die in jeder Weise mit denjenigen übereinstimmen, die sich normal im Plasmodium finden. PFEFFER schließt aus seinen Ergebnissen, daß weder Hautschicht (Plasmalemma) noch Vacuolenhaut (Tonoplast) selbständige und sich fortpflanzende Organe sind wie Zellkern und Chromatophoren und daß Vacuolen im Cytoplasma neu entstehen können. NEMÉC (1900) gelang es dann, auch an umhäuteten Zellen Vacuolen künstlich im Cytoplasma zu erzeugen. Die alte Auffassung, daß meristematische Zellen vacuolenfrei sind und die Vacuolen erst im Laufe der Zellentwicklung sich neu bilden, wurde wieder allgemein anerkannt.

Die Frage nach der Herkunft der Vacuolen erhielt dann neue Impulse von seiten der Vitalfärbung. In diesem Rahmen ist besonders die Arbeit von DANGEARD (1923) zu erwähnen. Nach der Auffassung der französischen Schule ist der basische Farbstoff Neutralrot ganz spezifisch für die Vacuole. In meristematischen Zellen, wie auch in Pollenkörnern, färben sich mit Neutralrot viele kleine Granula, die nach der Auffassung von P. DANGEARD kleine Vacuolen darstellen, und zwar soll es sich um kondensierte Vacuolensubstanz (Metachromatin) handeln. Nach DANGEARD besitzen also auch die embryonalen Zellen Vacuolen mit hochkolloidalem Zellsaft. Während der Entwicklung der Zelle wird die Vacuolensubstanz hydratisiert, die zunächst runden Vacuolen werden fadenförmig, die Fäden fusionieren miteinander und bilden dann ein Netzwerk, dessen Teile allmählich zu einer großen Zentralvacuole zusammenfließen (Abb. 1 u. 2). Der Zellsaft wird bei diesem Prozeß immer flüssiger.

Ohne Vitalfärbung mit Neutralrot sind die Vacuolen in den embryonalen Zellen meist nicht sichtbar. In einigen anthocyanführenden Geweben kann man

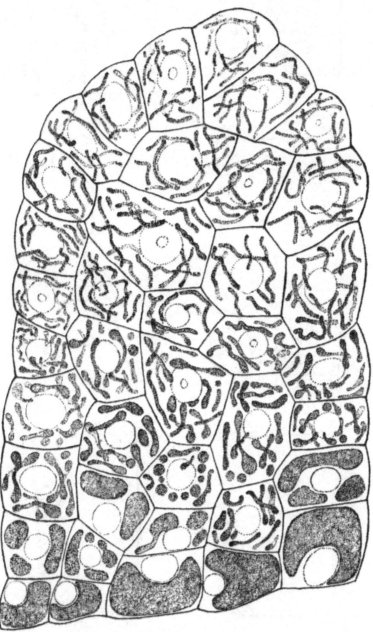

Abb. 1. Die Entwicklung anthocyanführender Vacuolen in dem Zähnchen eines jungen Rosenblättchens. (Nach GUILLIERMOND.)

allerdings auch die Vacuolen schon in den noch sehr jugendlichen Zellen erkennen. Die Blattzähne von Rosenblättern sind nach GUILLIERMOND hierfür ein gut geeignetes Untersuchungsobjekt (Abb. 1).

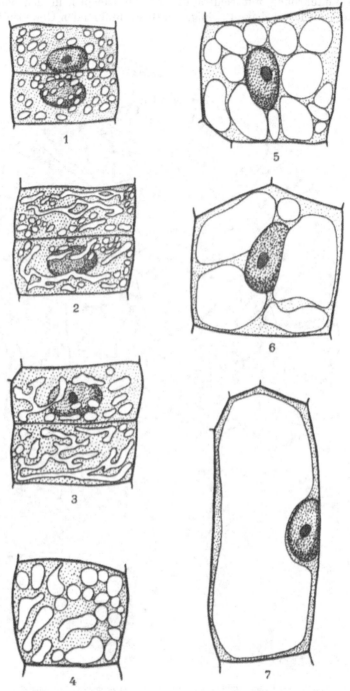

Abb. 2.  Die Entwicklung des Vacuolensystems in den Zellen einer Gerstenwurzel. Vacuolen durch Neutralrot vitalgefärbt. Fig. 1—3 meristematische Zellen, Fig. 4—6 Zellen aus der Differenzierungszone, Fig. 7 ausgewachsene Rindenparenchymzelle. (Nach GUILLIERMOND, umgezeichnet von THIELKE.)

Als Musterbeispiel für die Zustandsänderung des Zellsaftes im Entwicklungscyclus einer Zelle werden von den französischen Autoren die Aleuronkörner angeführt. In den Samen entstehen die Aleuronkörner in Vacuolen durch „Eindicken‘‘

des Zellsaftes. Bei der Samenkeimung gehen dann aus den Aleuronkörnern durch Wasseraufnahme wieder Vacuolen hervor. Entsprechend dem Hydratationszustand der Zelle kann also die Vacuole von dem Gel- in den Sol- und wieder in den Gel-Zustand übergehen.

Eine Änderung des spezifischen Gewichtes der Vacuolen während der Zellentwicklung konnte MILOVIDOV (1930) in Zentrifugierungsversuchen nachweisen. Um das Vacuolensystem in Keimwurzeln von *Hordeum* sichtbar zu machen, führte der Autor Vitalfärbung mit Neutralrot durch. Unter der Einwirkung der Zentrifugalkraft verschieben sich die Vacuolen in den der Wurzelspitze zunächstliegenden Zellen in zentrifugaler Richtung. In der anschließenden Zone enthalten die Zellen ein netzartiges Vacuolensystem, das nach dem Zentrifugieren kaum seine Stellung in der Zellmitte geändert hat. Daran grenzt ein Wurzelabschnitt mit großen und sich mit Neutralrot leicht färbenden Vacuolen, die sich in zentripetaler Richtung verschieben. Das spezifische Gewicht der Vacuolen nimmt entsprechend der Zunahme des Hydratationsgrades ab. An anthocyanführenden Vacuolen in den Trichomen der Nebenblätter von *Rosa* konnte dasselbe bestätigt werden.

Nach allen bisherigen Beobachtungen scheinen also im Plasma der embryonalen Zellen Granula vorhanden zu sein, die aus sehr hydrophilen Kolloiden bestehen müssen und die bei starker Hydratation des Plasmas unbegrenzt quellen und so die Vacuolen bilden. Eine andere Frage ist aber die, ob die nach Neutralrotfärbung sichtbaren Granula wirklich alle schon vorher vorhanden waren oder ob sie erst durch die Einwirkung des Farbstoffes — wenigstens zum Teil — durch Entmischungsvorgänge entstanden sind. Nach DANGEARD sollen keine Vacuolen sich de novo bilden, sondern nur aus ihresgleichen hervorgehen. Die Auffassung von DANGEARD deckt sich also weitgehend mit der von DE VRIES, nur mit dem Unterschied, daß DE VRIES sich teilende Plasten annahm, die den Zellsaft in ihr Inneres sezernieren und DANGEARD die Vacuolensubstanz (Metachromatinkörnchen) als ein sich teilendes Organell ansieht, das durch Hydratation zu Zellsaft wird.

Nach der Auffassung von DANGEARD müssen also die Metachromatinkörnchen Reduplikanten sein, damit ist man aber auch nicht weit davon entfernt, ihnen einen genetischen Wert anzuerkennen, zumal in Pollenkörnern sich ohne weiteres mit der Neutralrotfärbung nicht nur in der vegetativen, sondern auch in der generativen Zelle Metachromatinkörnchen nachweisen lassen. Und diese Auffassung konsequent weiter gedacht, könnte man neben Genom, Plastidom, Plasmon, Chondriom auch ,,Vakuom'' als genetischen Begriff aufstellen. Man dürfte dann den Begriff ,,Vakuom'' nicht mehr im morphologischen Sinne für die Gesamtheit der Vacuolen einer Zelle gebrauchen, sondern müßte ihn für die Genetik reservieren und für die morphologische Gesamtheit der Vacuolen mit GUILLIERMOND den Begriff ,,Vacuolensystem'' benutzen. Nach unseren bisherigen Kenntnissen ist es aber doch recht unwahrscheinlich, daß dem Vacuolensystem ein genetischer Wert zukommen sollte.

Eine weitere Frage ist die, ob aus allen mit Neutralrot nachweisbaren Granula Vacuolen werden können, ob allen Neutralrotgranula also Vacuolennatur zuerkannt werden muß. Bei einer Bejahung dieser Frage würden auch die meisten tierischen Zellen Vacuolen besitzen, nur das ihr Zellsaft sich vorwiegend im Gelzustand befindet. Eine Auffassung, die z.B. von PARAT (1928) vertreten wird. Interessanterweise sind ganz allgemein die für die Pflanzenzelle zur Vacuolenfärbung geeigneten Vitalfarbstoffe Granulafärber bei der tierischen Zelle (DRAWERT 1948).

GUILLIERMOND vertritt eine ähnliche Auffassung wie DANGEARD über die Entstehung der Vacuolen durch Hydratation bestimmter Plasmakolloide, nur sollen sich diese Kolloide nach GUILLIERMOND de novo im Plasma bilden, so daß keine entwicklungsgeschichtliche Kontinuität des Vacuolensystems besteht.

Zusammenfassend müssen wir also annehmen, daß das Cytoplasma der embryonalen Zellen Granula enthält, die — ähnlich den Asparaginkristallen in den Versuchen von PFEFFER — Ausgangsorte für Vacuolen werden können. Offen bleiben muß aber die Frage, ob diese Granula nur aus ihresgleichen durch Teilung hervorgehen oder ob sie — was wohl wahrscheinlicher ist — de novo im Cytoplasma als Entmischungsprodukte entstehen. Daneben können sich aber auch Vacuolen ohne besondere präformierte Ausgangspunkte einfach durch eine Phasentrennung bilden, ähnlich der unter bestimmten Bedingungen zu beobachtenden Vacuolisation von Kern und Plastiden. Dies wird man wohl ganz allgemein als den primären Prozeß ansehen müssen. Die hydrophilen Granula sind dann nur

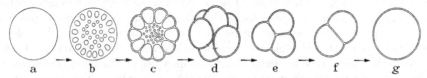

Abb. 3a—g. Der Ablauf des Vacuolisationsvorganges in einem aus Gelatine und Gummi arabicum bestehenden Coacervattropfen im Wasser. a Ursprünglicher Coacervattropfen, b primäre Vacuolisation, c Anschwellen der peripheren Vacuolen, d Schaumigwerden des ganzen Körpers, e und f Verschmelzen der einzelnen Vacuolen, g Coacervattropfen mit einer Zentralvacuole. (Nach BUNGENBERG DE JONG 1942).

die sichtbaren Sammelpunkte des ausgeschiedenen Wassers. Die Ausscheidung des Wassers könnte man sich analog der Vacuolisation von Coacervaten vorstellen, wie sie BUNGENBERG DE JONG (1932) beschreibt.

Wird in einem aus Gleichgewichtsflüssigkeit und suspendierten Coacervattröpfchen bestehenden System das Gleichgewicht in Richtung einer Dehydratation geändert, so erscheinen ganz allgemein in den Coacervaten Vacuolen. Diese Vacuolen werden von Gleichgewichtsflüssigkeit gebildet, die sich im Innern der Coacervattröpfchen durch Entmischung ansammelt. Durch Temperaturänderungen, Zufügen von Salzen oder anderen Kolloiden kann eine Dehydratation und damit eine Entmischung und Vacuolisation der Coacervattröpfchen herbeigeführt werden. So bedingen z. B. negativ geladene Coacervattropfen in Komplexcoacervaten aus positiv geladener Gelatine und negativ geladenem Gummi arabicum in Kontakt mit viel Wasser eine Vacuolenbildung (Abb. 3). Zuerst bilden sich kleine Vacuolen, dann erfolgt ein Anschwellen der direkt unter der Oberfläche liegenden Vacuolen. Dieser Vorgang greift auch auf das Innere der Coacervattröpfchen über, so daß das ganze Gebilde in einen blasigen Körper übergeht. Die die einzelnen Vacuolen noch trennenden Coacervatlamellen zerplatzen, und schließlich entsteht ein Coacervatkörper mit einer großen Zentralvacuole.

## Literatur.

BUNGENBERG DE JONG, H. G.: Die Koazervation und ihre Bedeutung für die Biologie. Protoplasma (Berl.) 15, 110—173 (1932). — Protoplasma — en celmodellen. Beteekenis der coacervatie voor de biologie. In KONINGSBERGER, V. J.: Leerboek der algemeene plantkunde. II. Amsterdam 1942.

DANGEARD, P.: Recherche sur l'appareil vacuolaire dans les végétaux. Botaniste 15, 1—267 (1923). — Cytologie végétale et cytologie générale. Paris 1947. — DRAWERT, H.: Zur Frage der Stoffaufnahme durch die lebende pflanzliche Zelle. V. Z. Naturforsch. 3b, 111—120 (1948).

GUILLIERMOND, A.: Le vacuome des cellules végétales. Protoplasma (Berl.) 9, 133—174 (1930). — Le système vacuolaire ou vacuome. Actual. scientif. et industr. Paris 171 (1934). — The cytoplasm of the plant cell. Waltham, Mass. 1941. — GUILLIERMOND, A., G. MANGENOT et L. PLANTEFOL: Traité de cytologie végétale. Paris 1933.

HOFMEISTER, W.: Die Lehre von der Pflanzenzelle. Leipzig 1867.

KÜSTER, E.: Die Pflanzenzelle, 2. Aufl. Jena 1951.

MEYEN, F. I. F.: Neues System der Pflanzen-Physiologie, Bd. I. Berlin 1837. — MILO-VIDOV, P. E.: Einfluß der Zentrifugierung auf das Vakuom. Protoplasma (Berl.) **10**, 452—470 (1930).

NEMEC, B.: Über experimentell erzielte Neubildung von Vakuolen in hautumkleideten Zellen. Sitzgsber. kgl. böhm. Ges. Wiss. Prag **1900**. Zit. nach P. DANGEARD 1923.

PARAT, M.: Contribution à l'étude morphologique et physiologique du cytoplasme, chondriome, vacuome (appareil de Golgi) enclaves, etc. $p_H$, oxydases, peroxydases, $r_H$ de la cellule animale. Archives Anat. microsc. **24**, 73—357 (1928). — PFEFFER, W.: Zur Kenntnis der Plasmahaut und der Vakuolen nebst Bemerkungen über den Aggregatzustand des Protoplasmas und über osmotische Vorgänge. Abh. math.-phys. Kl. sächs. Ges. Wiss. **16**, 185—343 (1890).

SACHS, J.: Lehrbuch der Botanik, 4. Aufl. Leipzig 1874. — Vorlesungen über Pflanzen-Physiologie, 2. Aufl. Leipzig 1887.

TIEGHEM, PH. VAN: Hydroleucites et grains d'aleurone. J. de Bot. **2**, 429—432 (1888).

VRIES, H. DE: Plasmolytische Studien über die Wand der Vakuolen. Jb. wiss. Bot. **16**, 465—598 (1885).

WENT, F. A. F. C.: Die Vermehrung der normalen Vacuolen durch Teilung. Jb. wiss. Bot. **19**, 295—356 (1888). — Die Entstehung der Vakuolen in den Fortpflanzungszellen. Jb. wiss. Bot. **21**, 299—366 (1890).

ZIRKLE, C.: The plant vacuole. Bot. Review **3**, 1—30 (1937).

# Die Chemie der Zellwand.

Von

## Erich Treiber.

Mit 30 Textabbildungen.

Die Zellwand gibt bekanntlich der Zelle ihre Form und begrenzt sie nach außen. (Unbedingt notwendig ist diese jedoch für die Zelle nicht, wie das Vorkommen nackter Zellen, z. B. Schwärmsporen und Plasmodien der Schleimpilze, beweist.) Der Aufbau mehrzelliger höherdifferenzierter Organismen ist jedoch stets an das Vorhandensein der Zellwände gebunden; diese übernehmen nicht allein mechanische Funktionen, vermöge der in ihnen enthaltenen Gerüstsubstanzen, sondern auch physiologische.

Im allgemeinen werden einmal gebildete Zellwände nicht mehr in den Stoffwechsel einbezogen, wenngleich es hier auch eine Reihe von Ausnahmen gibt. Physiologisch findet z. B. eine Resorption bei den Querwänden der Tracheen statt, ferner in lysigenen Intercellularen; in pathologischen Fällen findet sich Wandauflösung häufiger (Pilzinfektionen, Gummosis u. a.). Im Gegensatz zu Membranstoffen werden Reservestoffe wieder mobilisiert. Allerdings läßt sich eine Grenze zwischen Gerüst- und Reservestoffen, beides vorwiegend Ausscheidungsstoffe des lebenden Protoplasmas, nicht scharf ziehen. So besitzen z. B. die Hemicellulosen sowohl die Funktion von Stütz- als auch von Reservestoffen; erstgenannte Funktion wird besonders bei den Samen deutlich, letztere ist bei den Hemicellulosen mancher Hölzer, z. B. Weiden (Salix) beobachtbar. Ähnlich verhält es sich mit den Pektinstoffen, die in der Mittellamelle als selbständige Wandschicht in Erscheinung treten können; darüber hinaus spielen Pektine eine Rolle als Reservestoff, Haftelement und möglicherweise auch die eines Schutzkolloids.

Die Gerüst- oder Membranstoffe können nun primäre oder sekundäre Pflanzenstoffe sein. Zu den primären zählt das wichtigste Bauelement der Zellwand, die Cellulose; ferner gehören wohl hierher die übrigen Kohlenhydrate mit den verschiedenen Funktionen bis zu den eigentlichen Reservestoffen, die im Rahmen der Kohlenhydrate am Rande eine knappe Erwähnung finden werden. Unter den sekundären Pflanzenstoffen beanspruchen Lignin, Suberin und die Cuticularstoffe unser spezielles Interesse.

## I. Die Kohlenhydrate.

### 1. Einleitung.

#### Zucker und zuckerähnliche Polysaccharide.

Kohlenhydrate sind Stoffe, die die Elemente C, H und O enthalten, vorwiegend die Bruttoformel $C_n(H_2O)_m$ (eine Ausnahme bildet z. B. die Ribodesose) besitzen und vielfach als Polymere des Formaldehyds angesprochen werden können. Die Kohlenhydrate haben für alle Lebewesen eine wichtige Aufgabe. Der Energiebedarf wird vielfach zu einem erheblichen Teil direkt oder indirekt durch Umsetzung von Kohlenhydraten gedeckt. In den Pflanzen und bei den Insekten bilden Kohlenhydrate die chemische Grundlage der Gerüstsubstanzen.

Die Zahl der Kohlenhydrate ist recht groß; ihre Einteilung beruht zunächst darauf, daß manche durch Hydrolyse nicht, andere in eine kleinere oder größere Zahl von Spaltprodukten zerfallen. Es ergeben sich daraus zwanglos die Klassen der Monosaccharide, Oligosaccharide und Polysaccharide. Eine weitere Unterteilung richtet sich nach der Zahl der Sauerstoffatome (Biosen, Triosen, Tetrosen, Pentosen, Hexosen) und der chemischen Natur (Aldosen, Ketosen). Wenn man vom Glykolaldehyd und Dioxyaceton absieht, haben alle Monosaccharide — zugleich Bausteine der Polysaccharide — ein oder mehrere asymmetrische C-Atome, sind optisch aktiv und müssen daher als optische Antipoden auftreten. Die Mannigfaltigkeit wird noch durch Tautomerieerscheinungen vergrößert. Die natürlichen Monosaccharide besitzen, mit Ausnahme der Apiose, einiger Methylpentosen (z. B. Digitalose) und dem Zucker des Hamamelitannins eine normale, d. h. unverzweigte Kohlenstoffkette.

Abb. 1a. Raummodell der α-D-Glucose, konstruiert nach den Angaben von McDonald und Beevers (schwarz: O-Atom).

Traubenzucker und Fruchtzucker, wichtige Hexosevertreter, sind in der Natur im freien Zustand sehr verbreitet (z. B. süße Früchte); in weit größerem Maßstab beteiligen sie sich am Aufbau der Oligo- und Polysaccharide (Stärke, Cellulose, Inulin).

Das Bestehen von α- und β-Formen sowie eine Reihe weiterer chemischer und physikalisch-chemischer Tatsachen (z. B. auch das Fehlen einer Selektivabsorption im UV) hat zu der Annahme geführt, daß die Zucker normalerweise praktisch völlig in der Cyclohalbacetalform vorliegen (Tollens). Die offene Carbonylform der Glucose z. B. konnte nur in Form des Pentabenzoylderivats und des Pentaessigsäureesters isoliert werden. Die ringschließende Sauerstoffbrücke kann verschieden geschlagen werden, so daß z. B. die Kohlenstoffatome 1 und 4 oder

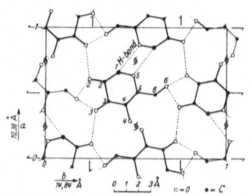

Abb. 1b. Projektion der Glucosestruktur (Fourierdiagramm) längs der c-Achse nach McDonald und Beevers (Wasserstoffbrücken gestrichelt eingezeichnet).

1 und 5 verbunden werden (Furanosen, Pyranosen). [Der gewöhnliche Traubenzucker scheint δ-oxydisch zu sein (Pyranose), zumindest in den meisten Derivaten.]

Das Vorliegen solcher „Ringstrukturen" geht auch aus röntgenoptischen Untersuchungen hervor; McDonald und Beevers (1950) konnten mittels Fourieranalyse zeigen, daß die α-D-Glucose einen sesselförmigen Pyranosering besitzt (Abb. 1a) und als weiteres wesentliches Ergebnis, daß alle OH-Gruppen an Wasserstoffbrücken beteiligt sind (Abb. 1b).

Durch Oxydation entstehen Aldehydcarbonsäuren, sog. Uronsäuren, die auch in der Natur weit verbreitet sind (z. B. Glucuronsäure, Galakturonsäure).

Die OH-Gruppen sind der Veresterung und Verätherung fähig; wichtig sind die Phosphorsäureester (z. B. Harden-Young-, Neuberg-, Embden- und Robison-Ester), Schwefelsäureester [Senfölglucoside, Chondroitin- und Mucoitinschwefelsäure (vgl. auch S. 696)] und die „Äther", die Glykoside. Durch analoge glykosidische Verknüpfung entstehen die Disaccharide (Maltose, Cellobiose, Saccharose), Trisaccharide (z. B. Raffinose, Gentianose, Pomose), Tetrasaccharide (Stachyose), Pentasaccharide (Verbascose) und die zuckerunähnlichen Polysaccharide.

Ersatz einer OH-Gruppe durch die Aminogruppe führt zu Aminozuckern (Glucosamin, Chondrosamin).

Während starke Laugen die Zucker zersetzen, bewirken schwache eine Epimerisation, die wahrscheinlich über eine Dienolform verläuft. Mineralsäuren wirken in der Hitze wasserentziehend unter Bildung von Furfurol (Pentosen) bzw. ω-Oxymethylfurfurol (Hexosen), welches sich weiter in Lävulinsäure umwandeln kann.

Mit den einfachen Zuckern bzw. Zuckeralkoholen haben die *Zyklite* eine gewisse Ähnlichkeit. Hierher gehören die Quercite (Eichelzucker) und Inosite [Muskelzucker (vgl. PLOETZ 1943)]. Mesoinosit scheint als Phosphorsäureester ein Phosphorsäurespeicher zu sein (Phytin der Pflanzen).

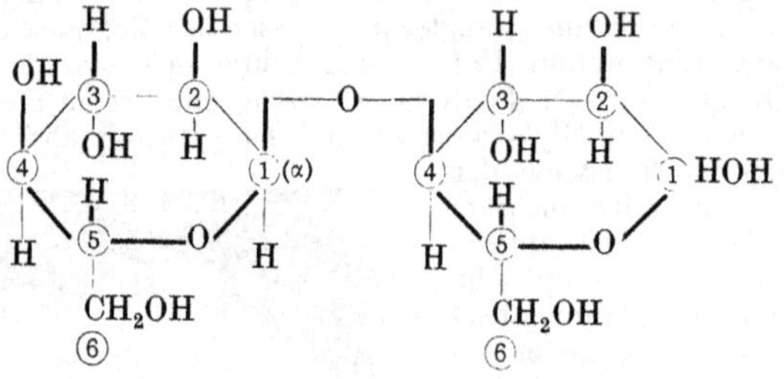

*Glucosidrest*　　　　　*Glucoserest*

Maltose

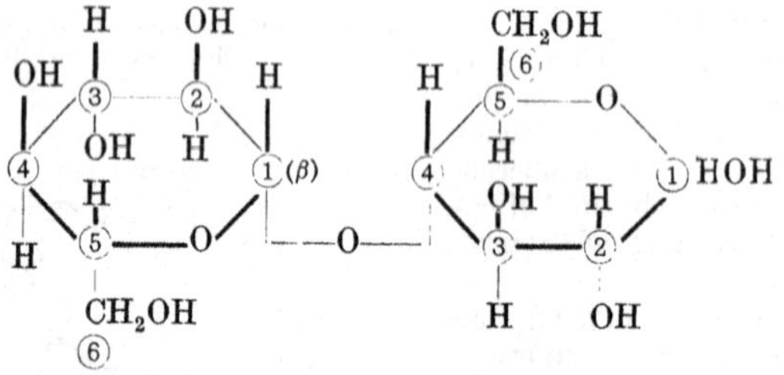

*Glucosidrest*　　　　　*Glucoserest*

Cellobiose

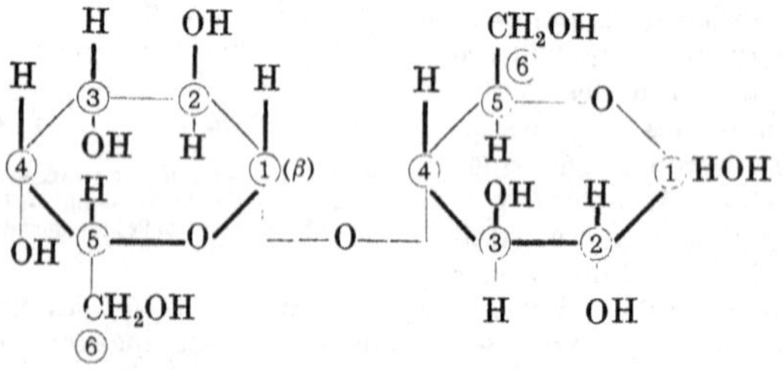

*Galaktosidrest*　　　　　*Glucoserest*

Lactose

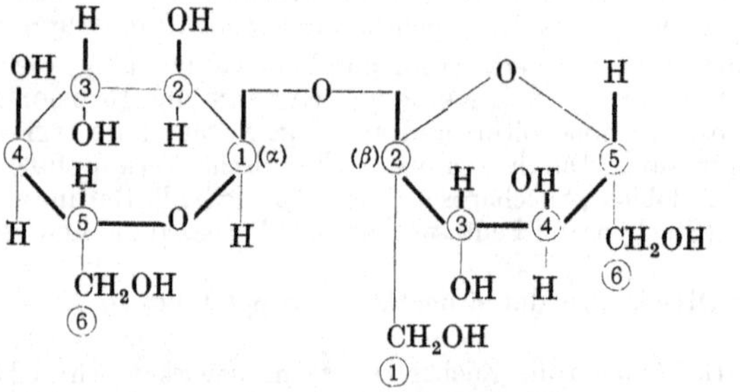

*Glucosidrest*　　　　　*Fructosidrest*

Saccharose

## Zuckerunähnliche Polysaccharide.

Diese überaus verbreiteten und in großen Mengen vorkommenden Verbindungen spielen entweder die Rolle von Reserve- oder von Gerüststoffen. Die verschiedenartigsten pflanzlichen Verbindungen, wie Stärke, Inulin, Lichenin, Gummiarten und Schleime, Pektinstoffe, Cellulose und Hemicellulosen gehören in diese Klasse.

Gemeinsame Kennzeichen ergeben sich aus dem Bauplan. Es besteht allgemein die Auffassung, daß die hochpolymeren Polysaccharide im wesentlichen aus glykosidisch verknüpften Monosaccharidresten bestehen.

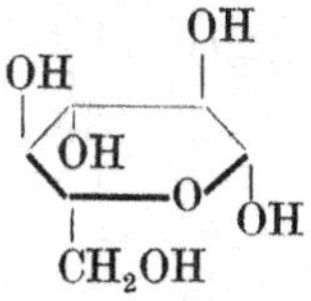

β-Glucose

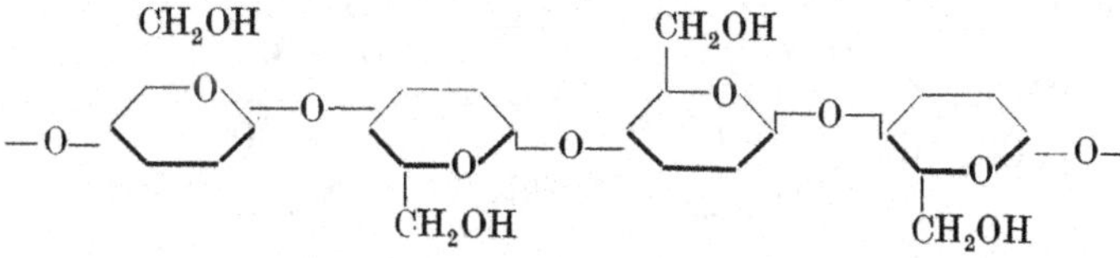

Cellulose

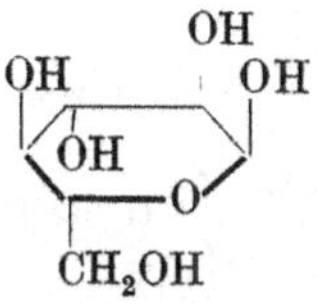

α-Glucose

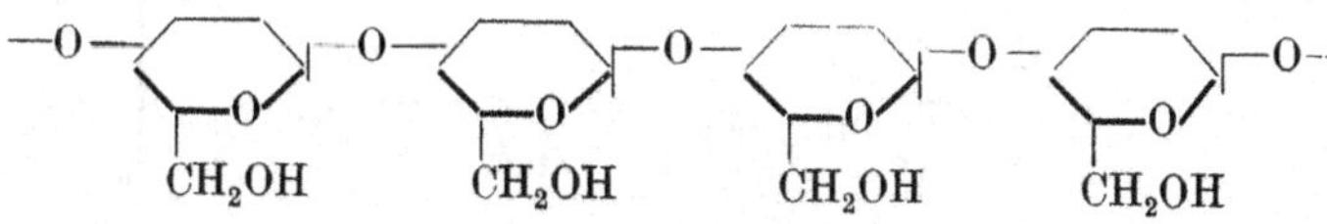

Stärke (Amylose)

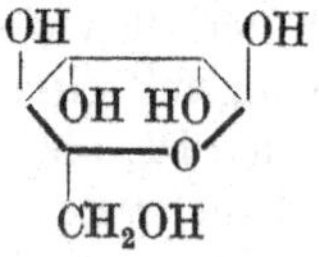

α-Mannose

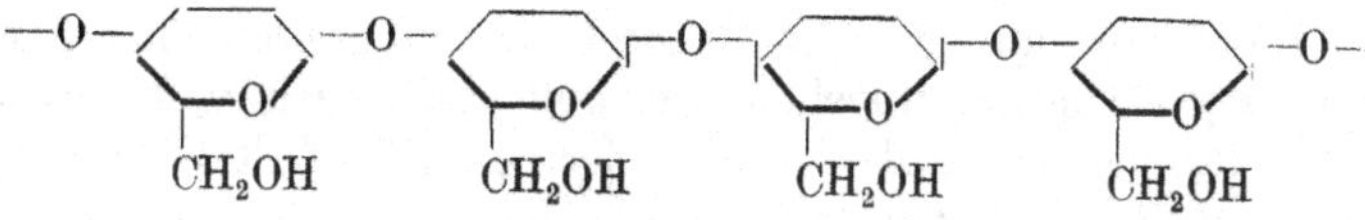

α-Mannan

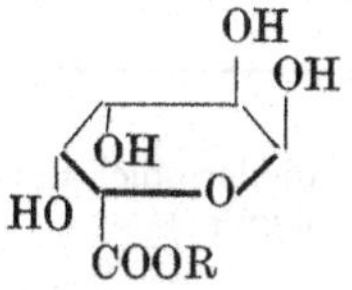

α-Galakturonsäure
(R = CH₃ oder H)

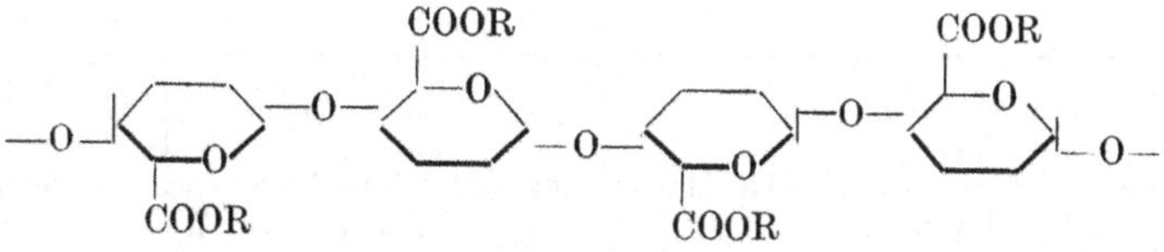

Polygalakturonsäure

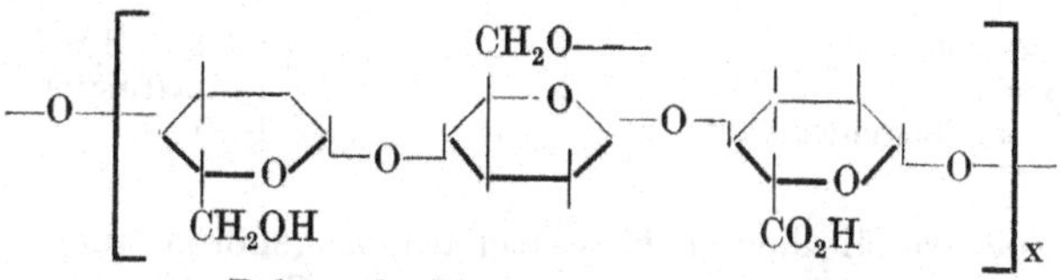

Polysaccharid von *Rhizobium radicicola*

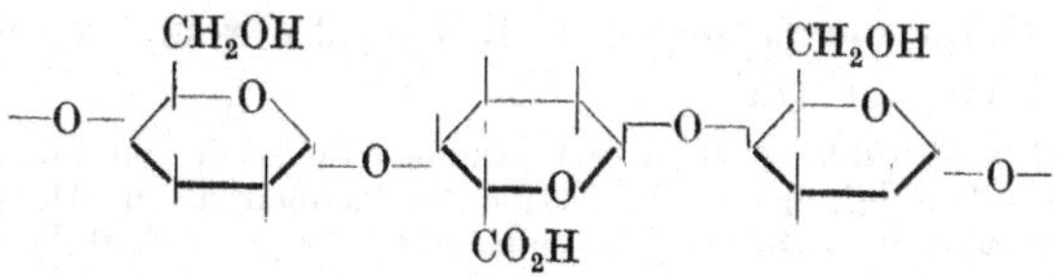

Oxydierte Cellulose

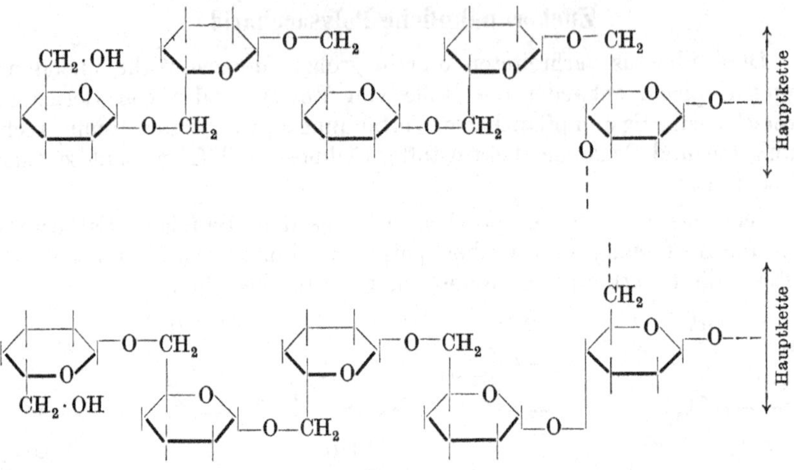

Dextran

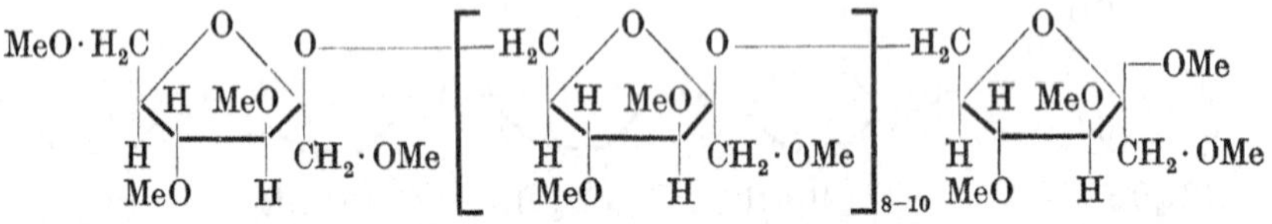

Lävan

Unter der hydrolysierenden Einwirkung von Mineralsäuren zerfallen sie in Monosen (am häufigsten D-Glucose). In den gebräuchlichen Lösungsmitteln sind die Polysaccharide unlöslich oder kolloidal löslich, wobei es fraglich ist, ob bzw. wieweit eine molekulardisperse Zerteilung auftritt. Aus den Ergebnissen der Röntgenspektroskopie darf geschlossen werden, daß die meisten Polysaccharide teilweise kristallinen Bau besitzen (vgl. S. 678).

Tabelle 1. *Übersicht über Polyglucosane bekannter Konstitution.*

| | | |
|---|---|---|
| Cellulose, Tunicin | $\beta$ 1—4 | |
| Laminarin | $\beta$ 1—3 | (relativ niedermolekular) |
| Polysaccharid aus der Gerüstsubstanz der Bäckerhefe | $\beta$ 1—3 | (hochmolekular) |
| Pustulin, Luteose | $\beta$ 1—6 | |
| Lichenin | $\beta$ 1—4 | und $\beta$ 1—3 (unverzweigt) |
| Bakterienlaevan | $\beta$ 1—2 | (?) |
| Bakteriendextran | $\alpha$ 1—6 | |
| Amylose | $\alpha$ 1—4 | |
| Amylopektin, Glykogen | $\alpha$ 1—4 | und $\alpha$ 1—6 (verzweigt) |
| Schardingerdextrin | $\alpha$ 1—4 | (ringförmig) |
| Polysaccharid aus Gerstenwurzeln | $\alpha$ 1—6 | |

Zu den zuckerunähnlichen Polysacchariden gehören auch — wie bereits erwähnt — die *Reservestoffe*, die teilweise auch die Funktion einer Gerüstsubstanz ausüben können und als solche im Rahmen der Hemicellulosen Erwähnung finden sollen. Unter die eigentlichen Reservestoffe sind vornehmlich Stärke, Glykogen und Inulin zu zählen.

*Stärke*[1]. Die Stärke findet sich in den Pflanzen in Form von Körnern, die sich leicht von den sie erzeugenden Zellteilen, den Plastiden, abtrennen lassen. Das durch Assimilation entstehende Kohlenhydrat wird als „Assimilations"- oder „autochthone" Stärke in den

---

[1] Vgl. M. SAMEC 1951, K. H. MEYER 1952 und J. A. RADLEY 1953.

grünen Plastiden (Chloroplasten) gebildet und vorübergehend abgelagert; auch die Reservestoffe (transitorische Stärke) in den Samen, Früchten usw. werden aus zugeführtem Zucker durch Plastiden gebildet. Alle als Reservestoffe geltenden Polysaccharide (Oligosaccharide, Stärke, Inulin, Hemicellulose), ja selbst die Cellulose als Gerüstsubstanz (z. B. im Endosperm vieler Palmensamen), können durch spezifische Fermente wieder mobilisiert werden. Spezifische Fermentsysteme gewährleisten die Reversibilität derartiger Umwandlungen. Die Form der Stärkekörner ist linsenförmig, sphärisch, oval usw.; die Größe liegt zwischen 0,002 und 0,17 mm. Sie sind aus Schichten länglicher, radial angeordneter Micellen (Abb. 2a) aufgebaut, die sich um den zentrisch oder exzentrisch liegenden Wachstumskern anlagern. Die Micellen geben Anlaß zu Kristallinterferenzen (vgl. S. 678), die beim völligen Entwässern verschwinden. Die Kristallite sind sehr klein (einige 100 Å); das Bestehen verschiedener Modifikationen (A, B, C) (Abb. 2b) ist feststellbar, jedoch ist derzeit eine Angabe über die Elementarzelle nicht möglich. Die Lage der Glucosereste dürfte der Abb. 2c entsprechen. Möglicherweise hängt die Schichtung des Stärkekorns mit dem Wechsel von Tag und Nacht zusammen (vgl. Wachstumsrhythmen des Baumwollhaares).

Abb. 2a. Schema der Anordnung der kristallinen Micelle (dick gezeichnet) und der sie verbindenden Hauptvalenzketten in einer Schicht des Stärkekorns.

Aus Hydrolyseversuchen mit Oxalsäure sowie enzymatischer Spaltung ist zunächst zu folgern, daß in der Stärke vornehmlich 1,4-α-glykosidisch verknüpfte Glucosereste vorliegen,

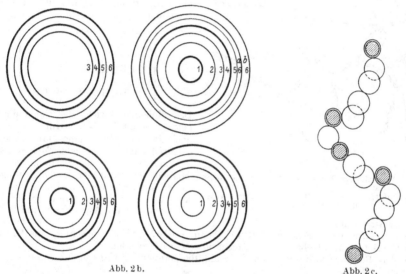

Abb. 2b.

Abb. 2c.

Abb. 2b. Röntgendiagramme der Stärke (die Intensität ist schematisch durch die Linienbreite dargestellt). 1 A-Spektrum; 2 B-Spektrum; 3 C-Spektrum mit mittelstarkem 1-Ring; 4 C-Spektrum mit schwachem 1-Ring.

Abb. 2c. Ausschnitt aus der Hauptvalenzkette der Stärke. Die Abbildung zeigt die Glucoseringe in seitlicher Betrachtung (weiß: C-Atome; schraffiert: O-Atome) (Nach K. H. MEYER.)

analog der Verknüpfung in der Maltose (vgl. die Formelbilder auf S. 670 und 671). Diese Folgerung wurde durch weitere Forschungsergebnisse gestützt.

Untersuchungen über die Leitfähigkeit führten zur Auffindung von Phosphorsäureresten (Robison-Ester) und mit dem Amylopektin gekoppelten Phosphatiden (POSTERNAK). Wir finden hier also einwandfrei Fremdgruppen in das Polysaccharid eingebaut; die Frage nach solchen Fremdgruppen spielt in der Erforschung der Feinstruktur der Polysaccharide eine große Rolle (z. B. Cellulose).

Die Stärke besteht aus 2 Kohlenhydraten verschiedener Konstitution, der *Amylose* (etwa 20%, die mit Jod eine blaue Einschlußverbindung liefert) und dem *Amylopektin*. Die native Amylose, die möglicherweise im Stärkekorninnern angereichert ist, besitzt gestreckte lange Ketten aus Glucose in 1,4-α-glykosidischer Bindung. Nach den neuesten Untersuchungen (HUSEMANN 1953) beträgt das Mindestmolekulargewicht eine Million. Eine ähnliche Größenordnung $(1,8 \cdot 10^6$ und darüber) ergab sich bei Messungen mit der Ultrazentrifuge durch LAMM und aus Lichtstreuungsmessungen von WITNAUER (bis $36 \cdot 10^6$). Stärke ist auch enzymatisch aus Glucose-1-phosphorsäure synthetisierbar. (Das verzweigte Amylopektin wird analog durch das Q-Enzym aufgebaut.)

Das Amylopektin, die Hauptsubstanz, ist ein Gemenge verzweigter Moleküle, wobei neben α-1,4- auch α-1,6-Bindungen vorkommen, durch die die Zweige angeheftet sind (Abb. 3). β-Amylase baut Amylopektin bis zu den Verzweigungsstellen ab und es hinterbleibt ein hochmolekulares Grenzdextrin (Erythrogranulose). (Beim Abbau der Stärke durch *Bac. macerans* erhielt SCHARDINGER kristalline Dextrine, die wahrscheinlich aus einer ringförmig geschlossenen Glucosekette mit Maltosebindung bestehen und im Zuge einer „Homologisierungsreaktion" [NORBERG] entstehen sollen.) Das Molekulargewicht des Amylopektins wird zu 40000—1000000 angegeben.

Von CHAMPBELL (1951) wurde Holzstärke näher untersucht. Der Amylosegehalt beträgt ∼ 20%. Holzstärke soll ein Vorläufer der Holzhemicellulosen sein, was seinerzeit auch von den Pektinstoffen des Holzes vermutet wurde. Der Holzstärkegehalt kann in Laubhölzern bis zu $5^{1}/_{2}$% betragen.

*Glykogen.* Pflanzliches Glykogen findet sich in Pilzen sowie im unreifen Samen von „golden Bantam". Es stellt chemisch ein amorphes, sehr stark verzweigtes Amylopektin dar und besteht offenbar aus einer niedermolekularen, wasserlöslichen und einer hochpolymeren Komponente. Für erstere sind Molekulargewichte von ein bis mehreren 100000 wahrscheinlich.

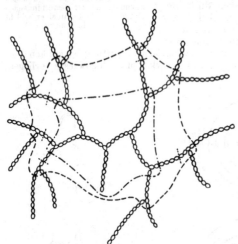

Abb. 3. Verzweigungsschema des Maisamylopektins: ooo Glucoserest; – – – Spaltung, welche zum Grenzdextrin I führt; — ·— ·— Spaltung, welche zum Grenzdextrin III führt.

*Inulin.* Inulin ist als Reservesubstanz in den Knollen der Compositen im Zellsaft gelöst. Es wird aus Fructofuranoseresten in 1,2-glykosidischer Bindung gebildet. HAWORTH schlägt folgendes Schema vor:

$$\left[ \begin{array}{c} -CH_2 \quad O- \\ >C< \\ CHOH \quad O \\ CHOH—CH\cdot CH_2OH \end{array} \right]_n$$

Das Molekulargewicht des Dahlieninulins beträgt ∼ 3000—5000. Kürzlich wurde noch die verwandte Artemose (aus *Artemisia vulgaris*) näher untersucht (MG ∼ 1947). Andere aus Fructose zusammengesetzte pflanzliche Polysaccharide, die vorwiegend in den unterirdischen Speicherorganen vorkommen, sind Irisin, Graminin, Triticin, Sinistrin, Secalin, Asparagosin und Phleïn.

Der Polymerisationsgrad ist zum Teil gering. Von SCHLUBACH wurden die Polyfructosane der Gräser untersucht (Festucin [DP 17], Poaïn [DP 47]). Ein gemischtes, aus Glucose- und Fructoseresten aufgebautes Polysaccharid ist das Asphodelin.

(Andere Polysaccharide, wie z. B. Lichenin, die Mannane, Xylane u. dgl. s. unter Hemicellulosen S. 691.)

## 2. Die Cellulose.

Das Cellulosemolekül wird durch eine kettenförmige Aneinanderlagerung von β-D-Glucose (Aldohexose) in 1,4-β-glykosidischer Bindung gebildet; sie stellt somit ein lineares Kettenpolymerisat von Glucose bzw. Cellobiose dar (Idealmodell; vgl. Abb. 4). Beweise für eine Kettenstrukturformel der Cellulose, aufgefaßt als Kette von Glucoseresten in Cellobiosebindung, liefert unter anderem die Isolierung von Cellotriose, -tetraose, -hexaose und neuerlich -heptaose (WOLFROM) aus den Abbauprodukten, die Untersuchung der optischen Drehung (FREUDENBERG), Hydrolyse (KUHN und FREUDENBERG) sowie die röntgenographische Untersuchung, die zum Celluloseraummodell führte (Abb. 6).

β-Glucose

Es erhebt sich nun noch die Frage nach der Konstellation der Cellulose, d. h. nach der Form, die das Molekül unter dem Einfluß freier Drehbarkeit einnehmen kann. Der Ring eines Glucosemoleküls ist gewellt (vgl. Abb. 1a) und kann 8 verschiedene, spannungsfreie Formen einnehmen, die ohne erheblichen Energieaufwand ineinander übergehen können. Die Ringe sind also keineswegs eben und in gewissem Sinne biegsam. Ferner sind die Achsen beiderseits des glucosidischen Verbindungs-O-Atoms drehbar. Allerdings kann durch diese Drehbarkeit das einzelne Fadenmolekül nicht alle erdenklichen Gestalten annehmen.

Die lange Kette ist in Wirklichkeit nur schwach geknäuelt; kurze Stücke bis zu 30 und mehr Kettengliedern sind nahezu gestreckt (NAKUSHIMA).

Von den zahlreichen Lagen, die 2 Kettenglieder in der Cellulosekette zueinander einnehmen können, seien die Formen C1 und 3B [Nomenklatur von REEVES (1949)] abgebildet (Abb. 5), die eine geradlinige Fortsetzung erlauben. Auf Grund der Längsperiodizität von 10,29 Å ist wahrscheinlich, daß die „Sesselform" C1, bei nicht völliger Streckung, vorliegt [vgl. auch SIPPEL (1950)].

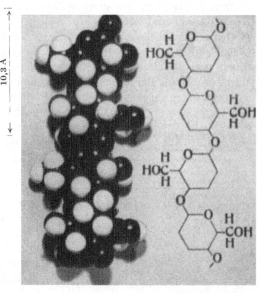

Abb. 4. Ausschnitt aus einer Cellulosekette. Links: *Stuart*modell, von vorne gesehen (Faserperiode eingezeichnet). (Nach FREUDENBERG.)

Als nächste Fragen nach der Klarstellung des chemischen Molekülbaues (Konstitution) erheben sich solche nach der Kettenlänge (Molekulargewicht) und Kettenlängenverteilung (Dispersität), Einheitlichkeit des Fadenmoleküls (Endgruppen, Fremdgruppen, Fehlerstellen; also Abweichungen vom Idealtyp) und Kristallstruktur (einschließlich Polymorphie), d. h. morphologische Betrachtungen der räumlichen Anordnung der Ketten zum Kristallit (Micell). Damit im Zusammenhang stehen dann weitere Fragen nach Größe, Form und Anordnung der Micellen und dem Kristallinitätsgrad, womit die Grundlagen gegeben sind für eine Diskussion des Feinbaues der Fibrillen und der Zellwände.

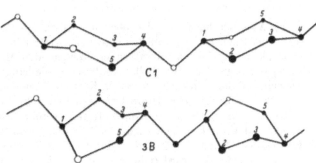

Abb. 5. (C₆ weggelassen!)

a) Molekülgröße. Über die Länge der Celluloseketten im ursprünglichen Zustand ist wenig Zuverlässiges bekannt; Durchschnittspolymerisationsgrade (DP-Werte) von ∼3000 Glucoseresten gelten heute schon als ungewollt abgebaut. Baumwolle in der ungeöffneten Samenkapsel besitzt einen relativ einheitlichen DP-Grad von über 4000, wobei die Polymerisationsgrade immerhin von ∼1450 (Cellulose der Primärwand?) bis über 4300 gehen. Auf Grund der viscosimetrischen Messungen wird man beim ungeschädigten nativen Material heute mit (Mindest-)-DP-Graden von etwa 5000—8000 rechnen müssen. (SCHULZ findet an unbelichteter Baumwolle 7700, Ramie 6800;

HESSLER an Baumwolle der Sekundärwand 10650). MEYERHOFF (1954) findet in der Ultrazentrifuge für Baumwolle einer gerade sich öffnenden Kapsel den Wert 8470. Wesentlich höhere Werte erhielt in der Ultrazentrifuge GRALÉN (Flachs 36000, Ramie 12400, Baumwolle 10800) und zu besonders hohen Werten sind IWANOW (15000) und GALOWA gekommen ($<$ 44500). (Zellstoffe besitzen infolge von Abbauvorgängen niederere DP-Werte, vorzugsweise solche zwischen 1000 und 3500[1].)

Die gelegentlich ausgesprochene Vermutung, daß native Cellulose weitgehend einheitlich ist, kann heute nicht recht vertreten werden. Bestimmt ist jedoch native Cellulose einheitlicher (d. h. sie besitzt eine ziemlich enge Verteilung im Molekulargewicht) als abgebaute Cellulose, bei der im Polydispersitätsspektrum gelegentlich mehrere Maxima gefunden werden. (Baumwolle der geöffneten Samenkapsel zeigt nach OTT Maxima bei 1800 und 4200 Resten.)

Tabelle 2. *Chemische Zusammensetzung der Baumwollfaser nach* ROLLINS.

| | Gesamte Faser % | Primärwand[2] % |
|---|---|---|
| Cellulose . . . . . . | 95,3 | 52 |
| Protein . . . . . . | 1,0 | 12 |
| Pectine . . . . . . | 1,0 | 12 |
| Wachs. . . . . . . | 0,8 | 7 |
| Asche . . . . . . | 0,9 | 3 |
| Cutin . . . . . . | — | 3 |
| Andere organische Verbindungen . . | 1,0 | 11 |

Es sei auch erwähnt, daß öfter die Frage diskutiert wurde, ob nicht jede Pflanze ihre eigene, d. h. eine im Feinbau differenzierte Cellulose aufbaut. Wenn auch die von KUBO angenommenen beiden Celluloseformen (Ramieform, Huflattichform) nicht existieren, so werden z. B. von SHARPLES an ägyptischer Baumwolle Lockerstellen etwas anderer Art als in Holzcellulose vermutet. SAMUELSON findet ebenfalls Unterschiede im Abbau zwischen Baumwolle und Holzzellstoff.

Auch die „Quellwiderstände", Kristallinitäten, ursprünglichen Polymerisationsgrade und die Polydispersität sollen unterschiedlich sein, so daß RÅNBY von der Gruppe: Baumwoll- und Ramiecellulosen und der Gruppe: Holz- und Strohcellulosen spricht.

Die reinste Cellulose liefert offenbar die Sekundärwand der Baumwolle. Die chemische Zusammensetzung der Baumwollfaser zeigt Tabelle 2.

**b) Abweichungen von der idealen Formel.** Es soll hier nicht auf die verschiedenartigen und verschieden engen Assoziationen mit den Cellulosebegleitern und Inkrusten, die von oberflächlich freiliegenden Molekülen ausgehen können, eingegangen werden. Wir wollen an dieser Stelle lediglich die Frage untersuchen, wieweit Cellulose wirklich ein reines Kettenpolymerisat von Anhydroglucose ist.

Aus den zahlreichen Hydrolyseversuchen geht hervor, daß reine Cellulose zu wenig mehr als 99% aus Glucoseresten besteht. Der Fehlbetrag dürfte Störstellen und Fremdgruppen zuzuschreiben sein. ABDEL-AKHER weist darauf hin, daß in der Cellulose 0,1—0,2% durch Perjodatoxydation nicht veränderbare „Glucose" gefunden wird (Glucose mit blockierter —OH-Gruppe). Unter Störstellen versteht man beispielsweise — vor allem nach PACSU — Acetal- und Halbacetalbindungen, die zu Vernetzungen, sog. Quervernähungen, Anlaß geben würden. Eine ähnliche Rolle würden Estergruppen spielen. Auf ältere Theorien sei hier nicht eingegangen.

Vor allem auf Grund von papierchromatographischen Untersuchungen nimmt man heute an, daß als Fremdgruppen auch andere Zucker (-Reste) in der Kette

---

[1] Bezüglich Meßmethoden usw. vgl. die Veröffentlichung der Vorträge, gehalten am Symposium of molecular weights of Cellulose 1953 [Ind. Engng. Chem. **45**, 2482—2532 (1953)], s. auch SCHULZ 1949, ferner GALOWA u. IWANOW 1953, SCHULZ u. MARX 1954.

[2] 1—5% der Gesamtfasermasse.

auftreten, wie z. B. Xylose, Mannose, Galaktose und Glucuronsäure — namentlich solche der „Glucosereihe" (D-Glucuronsäure und L-Xylose). Jedoch ist die Frage keineswegs leicht zu beantworten, ob die reine, von Begleitern befreite, ungeschädigte, genuine Cellulose alle oder einen Teil der genannten Reste primär eingebaut enthält. So gelang es VAN DER WYK und STUDER (1946) aus Baumwolle eine völlig carboxylgruppenfreie Cellulose herzustellen; diesem Befund nach müssen z. B. Carboxylgruppen lediglich den Polyosen entstammen [vgl. S. 692 sowie MEYER und MARK (1950)].

Die Tatsache, daß bei Abbaureaktionen häufig Bruchstücke übereinstimmender Kettenlänge (DP zwischen 100—1000) — bei niederer Temperatur vorzugsweise Spaltstücke vom DP-Grad $\sim$500 (bei höherer $\sim$250) — auftreten, hat zu der Annahme geführt, die oben erwähnten Störstellen und Fremdgruppen im Ausmaß von $\sim$0,2% seien gleichsam als *Lockerstellen* — vielfach mehr oder minder regelmäßig in bestimmten Abständen bzw. „Spaltebenen" — quer zur Faserachse angeordnet (vgl. Abb. 12). (NOBÉCOURT findet auch beim Zerreiben mit flüssiger Luft Spaltungen an bevorzugten Stellen.) Nach SCHULZ und HUSEMANN (1952) sollen solche — im allgemeinen 3000—5000mal schneller spaltende Bindungen — nach etwa je 465 $\pm$ 30 Glucoseeinheiten auftreten. PACSU ist, möglicherweise beeinflußt durch die „Regel" von BERGMANN und NIEMANN bei Proteinen, noch wesentlich weitergegangen und nimmt solche (nach seiner Auffassung vorzugsweise halbacetalische Vernetzungsstellen) nach $2^6$ und $2^7$ bzw. $2^8$ Glucoseresten längs der Kette an. Zu diesen Periodizitäten würde noch nach SCHULZ $2^9 =$ DP 512 hinzutreten. Wenn auch solche Modelle heute ziemlich ausnahmslos wieder abgelehnt werden, so ist doch die Existenz leichter spaltender Bindungen, zumindest in einigen Cellulosen, wahrscheinlich, und es ist durchaus möglich, daß gewisse Regelmäßigkeiten in der Anordnung bestehen. Die Regelmäßigkeit ist aber nicht streng, sondern gehorcht einer GAUSSschen Fehlerfunktion. Nach SHARPLES ist bei ägyptischer Baumwolle eine Lockerstelle unregelmäßig auf etwa 2900 Glucosereste verteilt; nach Alkalieinwirkung stellt sich ein neues Verhältnis 1:660 ein. Exaktere physikalisch-chemische Beweise für das Vorhandensein einer Überperiode sind noch nicht gefunden worden. Eine Deutung gelegentlich beobachteter Querstrukturen im Elektronenmikroskop (KINSINGER, HESS, WERGIN, HUSEMANN) von 150, 500 bzw. 2500 Å bzw. einer Überperiode von 610 Å (ZAŘDES), ferner für das Auftreten mikroskopischer Querspaltungen der Fasern (sowie Dermatosomenbildung), Verschiebungsfiguren usw. im Zusammenhang mit obigen Fragen ist noch offen.

Durch die jüngsten Untersuchungen von HESS wurde die in Vergessenheit geratene Beobachtung scharfer Röntgeninterferenzen (an Regeneratcellulose) von CLARK (1940) verifiziert und möglicherweise darf man darin sowie in den interessanten elektronenmikroskopischen Aufnahmen von HESS (1954) einen Hinweis auf die Existenz von Überperioden (650 bis 740 Å), in der Größenordnung der Micellänge erblicken (Abb. 14b) (vgl. auch KRATKY 1954).

Nebenbei sei bemerkt, daß auch noch andere Deutungen für das Zustandekommen periodischer Schwächungen der Bindungen existieren. So z. B. könnten einige Glucoseringe in einer anderen Konstellation sich befinden (BARTUNEK); es könnten $\alpha$-glykosidische Bindungen vereinzelt auftreten, mechanische Effekte eine Rolle spielen usw. Auch auf die Gedankengänge von FRANZ (1943) sei noch verwiesen.

Wenig Sicheres ist auch über die *Endgruppen* bekannt. Die beiden endständigen Glucosereste unterscheiden sich formelmäßig von den übrigen Kettengliedern durch den Besitz einer reduzierenden Halbacetal- bzw. einer zusätzlichen -OH-Gruppe; die Existenz letzterer darf vermutlich als nachgewiesen gelten. Der endständige Aldehyd in Halbacetalform wird offenbar sehr leicht in eine Carboxylgruppe umgewandelt. Weitere Veränderungen der Endgruppen durch Aufschluß-, Bleich- und Veredelungsprozesse treten natürlich an chemisch isolierter Cellulose auf, doch ist auch hierüber nichts Sicheres bekannt. Durch die Einwirkung von Oxydantien u. dgl. werden aber auch *in* der Kette sog. (sekundäre) Fehlerstellen erzeugt, die vornehmlich eine Dialdehyd- oder Dienolstruktur aufweisen (vgl. Kaverzniéva 1953). Auch mit der Bildung von Uronsäuren ist selbstverständlich zu rechnen. Solche Fehlerstellen sind ebenfalls prädistinierte Angriffsstellen einer Cellulosedegradation.

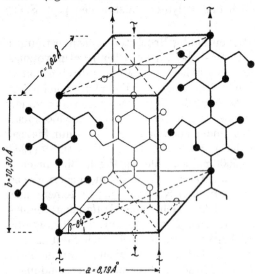

**c) Kristallstruktur.** Unsere Kenntnisse über die räumliche Anordnung der Ketten zum Kristallit verdanken wir röntgenoptischen Untersuchungen. Wie Nishikawa, Ono, Scherrer, Herzog, Jancke u. a. zeigen konnten, gibt Cellulose ein Faserdiagramm. Ausgedehnte Untersuchungen, vor allem an höher orientierten Präparaten, ergaben unter Einbeziehung chemischer Gesichtspunkte den in Abb. 6 dargestellten Elementarkörper.

Abb. 6. Elementarkörper der Cellulose I. (Nach Meyer und Misch.)

Die Kräfte, die die Gitterstruktur zusammenhalten, sind unter anderem eingehend von Mark und Kast diskutiert worden. In Richtung der b-Achse sind die Glucosereste durch eine Hauptvalenzbindung zusammengehalten. In der a-Achse (vgl. Abb. 7a u. b) sind die Glucoseringe nur durch einen Zwischenraum von etwa 2,5 Å getrennt, was bedingt, verglichen mit vielen organischen

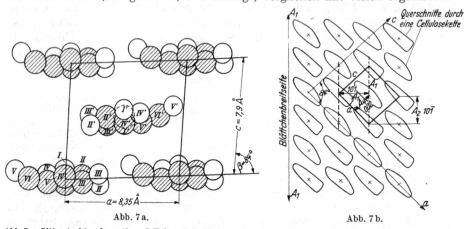

Abb. 7a.                              Abb. 7b.

Abb. 7a. Gitterstruktur der nativen Cellulose. Schnitt normal zur b-Achse (Faserachse). (Nach Meyer und Misch.)
Abb. 7b. Schnitt durch einen Kristallbereich von Cellulose I normal zur b-Achse (Faserachse). Elementarkörper, kristallographische Achsen und Spuren des Micells angedeutet.

Substanzen, insbesondere Polypeptiden, daß starke zwischenmolekulare Kräfte in Form von Wasserstoffbrücken zwischen 2 Sauerstoffatomen wirksam werden (zwischen den Hydroxylen 6 der einen und 2 bzw. 3 der Nachbarkette in der a-b-Ebene; vgl. auch die Abb. 8). Dies erklärt auch die dichtere Packung in

der a-b-Ebene und die Tatsache, daß Quellungsmittel diese Bindungen verändern oder partiell aufheben können. In der feuchten Faser der mercerisierten Cellulose sind z. B. die Wasserstoffbrücken zwischen den Ketten
durch Hydratisierung weitgehend gesprengt bzw. umgewandelt, so daß in
solchen „Inklusionscellulosen"[1] die nun freien Hydroxylgruppen z. B.
der Acetylierung zugänglich
sind. Ultrarotuntersuchun-
gen konnten diese Auffas-
sungen bestätigen.

In Richtung der c-Achse
ist der kleinste Abstand etwa
3,1 Å; die hier wirkenden
Kräfte werden daher von der
Natur der VAN DER WAALS-
schen Kräfte, im speziellen
wohl von der Art der Dis-
persionskräfte sein. Somit
wirken in drei kristallogra-
phisch verschiedenen Richtungen 3 Arten von An-
ziehungskräften. Die Festigkeit dieser Kräfte steht
auch einer Auflösung der
Cellulose in gewöhnlichen
Lösungsmitteln entgegen.

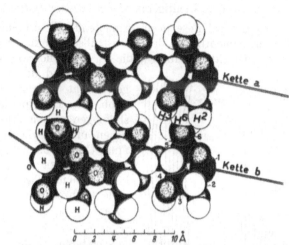

Abb. 8. Gegenseitige Lage zweier Celluloseketten (a und b) im
Elementarkörper und seitlicher Zusammenhalt durch Wasserstoffbrücken (-OH⁶ mit -OH³ bzw -OH³) (Kalottenmodell).

Nach MEYER und MARK bzw. MEYER und MISCH (1937) kommt der Cellulose I
eine monokline Elementarzelle [Raumgruppe offenbar $P2_1$ ($C_2^2$)] zu, wobei sich
durch den ganzen Kristallit Glucoseeinheiten in diagonaler Verschraubung (vgl.
auch Abb. 4) entlang
der Faserrichtung hin-
durchziehen (außerhalb
des Kristallverbandes
dürfte wohl eine belie-
bige Anordnung vor-
herrschen). Die „rönt-
genographische" Dichte
beträgt $1,63_3$.

Neben der hier be-
sprochenen Kristallform
(native Cellulose oder
Cellulose I), in der uns
die gewachsene Cellulose
(auch Tunicin) gegen-
übertritt, gibt es zufolge
der Polymorphie der

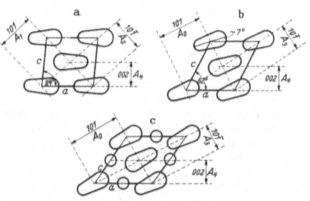

Abb. 9a—c. Schematisch dargestellter Schnitt normal zur b-Achse (Faserachse) durch die Elementarkörper von Cellulose I (a), Cellulose II (b)
und Wassercellulose (c) (Nach K. H. MEYER.)

Cellulose noch weitere Formen, von denen besonders die der mercerisierten bzw.
regenerierten Cellulose bedeutsam ist (Hydratcellulose oder Cellulose II), die
offenbar die stabile Form darstellt.

Von einigen Seiten wurde die Vermutung ausgesprochen, daß bei manchen
Algen (Halicystis) die Cellulose in der Form II (Hydratcellulose) vorliege. Genauere Untersuchungen (ROELOFSEN 1953) zeigten jedoch, daß die lamellierte

---

[1] Vgl. dazu z. B. STAUDINGER u. EICHER 1953.

Zellwand von *Halicystis Osterhoutii* als Hauptbestandteil ein Xyloglucan enthält neben einer weiteren alkaliresistenten unbekannten Wandsubstanz.

Der Vollständigkeit halber seien noch die Formen: Cellulosehydrat II (Wassercellulose), Cellulose III und Cellulose T (H.T.C.) dem Namen nach erwähnt. Den Unterschied gegenüber der im Prinzip ganz ähnlich gebauten Cellulose I erkennen wir am besten aus der Gegenüberstellung der schematisch dargestellten Schnitte für die beiden wichtigeren Cellulosegitter normal zur Faserachse (Abb. 9).

**d) Das Cellulosemicell.** Einleitend soll festgehalten werden, daß wir in der Natur, namentlich in den Gerüstsubstanzen des Pflanzen- und Tierreiches, zahlreiche Objekte vorfinden, die im röntgenographischen Sinne als Polykristalle anzusehen sind, die aber, wie ein genaueres Studium gezeigt hat, ein sog. „micellares System" darstellen (Abb. 10). Die Entwicklung unserer heutigen Vorstellungen vom micellaren System geht im wesentlichen auf die Arbeiten von Astbury, Frey-Wyssling, Gerngross, Hermans, Herrmann, Kratky, Mark, Meyer, Staudinger u. a. zurück. Wir können ein micellares System durch folgende 3 Merkmale kennzeichnen:

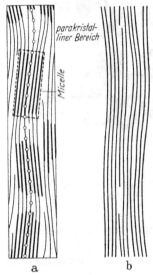

a    b

Abb. 10 a u. b. Anordnung der Celluloseketten im micellaren System der nativen Cellulose: a nach Kratky und Mark, b nach K. H. Meyer und van der Wyk.
(—o—o— durchlaufende Kette.)

1. Es ist aus kristallinen und amorphen Bereichen von kolloidalen Dimensionen aufgebaut. Die ersteren werden oft als Micelle, die letzteren gelegentlich als Fransenbereiche, heute öfter als parakristalline Bereiche bezeichnet.

2. Beide Arten von Bereichen bestehen aus mehr oder weniger langgestreckten Fadenmolekülen, doch ist die gittermäßige Parallelordnung nur in den kristallinen Bereichen gut, in den amorphen Bereichen schlechter. Die Übergänge können fließend sein, ebenso kann die Unordnung in den Fransenbereichen zwischen der eines wenig gestörten Kristalls (Parakristall) und völliger Regellosigkeit (amorph) schwanken.

3. Die Fadenmoleküle erstrecken sich meist durch mehrere Bereiche (kristalline und amorphe) hindurch, so daß wir kein gewöhnliches Gemisch aus beiden Bestandteilen vor uns haben, sondern eine Verknüpfung der beiden Typen von Bereichen mit verfließenden Übergängen vorliegt, die ohne Zerreißung der verbindenden Moleküle, also ohne weitgehenden chemischen Abbau nicht gelöst werden kann.

Solche micellare Systeme bildet nicht nur die Cellulose, sondern sie dürften auch zum Teil in vollsynthetischen Fasern und manchen Faserproteinen der tierischen Gerüstsubstanzen zu finden sein. Dabei liegt in der Regel eine histologisch bedingte „Wachstumsstruktur" vor, d. h. die stäbchen- oder blättchenförmigen, gittermäßig geordneten Bereiche liegen nicht wirr durcheinander, sondern weisen eine gesetzmäßige räumliche Richtungsverteilung (Textur) auf. Der bekannteste Fall ist die Fasertextur oder -struktur, worunter wir eine Parallelrichtung der Längsachsen der kristallinen Bereiche verstehen.

Wie auch der Mechanismus der Kristallisation im einzelnen verlaufen möge, Tatsache bleibt, daß die Wachstums- oder Kristallisationstendenz parallel zur 10Ī-Ebene größer ist als parallel zur Ebene 101 (die auch den größten Netzebenenabstand aufweist), so daß letztere am Micell die Blättchenbreitseite bildet

(Abb. 11). Diese Ebene $A_1$ ist auch die hydrophilste, da die Zahl der OH-Gruppen je Flächeneinheit hier größer ist als in der Ebene 002 ($\sim 65\%$), die mit den „Ringebenen" (e) zusammenfällt, oder der Netzebene 10$\bar{1}$ ($\sim 90\%$). Im Gegensatz zur Hydratcellulose ist aber die Blättchennatur der Micellen der Cellulose I scheinbar keineswegs so stark ausgebildet, so daß die meisten Rückschlüsse auf die Gestalt indirekte sind.

Eine höhere Orientierung, wie sie bei den offenbar ausgeprägten blättchenförmigen Hydratcellulosemicellen leicht erreicht werden kann (KRATKY), wird bei Cellulose I, wenn man von kleineren Effekten absieht, nicht erzwungen. Da auch das Tunicin, welches von vornherein eine Ringfaserstruktur besitzt (und dann beim Dehnen eine höhere Orientierung annimmt) und die natürliche selektive biaxiale (uniplanare) Orientierung der Cellulose in den Zellwänden der Algen *Valonia, Chaetomorpha, Cladophora* und *Fucus* nicht als eindeutiger Beweis für die Blättchengestalt anzusprechen ist, scheint ein solcher erst durch die Orientierungsversuche an Bakteriencellulose (SISSON) und insbesondere kürzlich am „Micellsol" von MUKHERJEE (1951) erbracht worden zu sein.

Die möglicherweise in den meisten Fällen von der Stäbchenform nicht stark abweichende Blättchen- oder besser Bändchengestalt (die Baum-

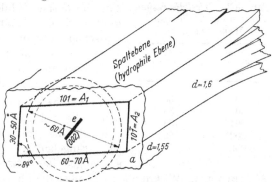

Abb. 11. Schematische Darstellung eines nativen Cellulosemicells. Dimensionen und Dichteangaben nach FREY-WYSSLING. (Netzebenen sind eingezeichnet; e Lage der Glucoseringebene; a parakristalline bzw. amorphe Rindenschicht.)

wolle soll eine solche Bändchengestalt ausgeprägter aufweisen) erlaubt auch kaum zuverlässige Aussagen hinsichtlich der seitlichen Dimensionen im einzelnen. Dabei darf auch nicht vergessen werden, daß die Micelle in bezug auf ihre Gestalt keineswegs exakt abgrenzbar ist, sondern die Flächen — wenn dieser bildhafte Ausdruck überhaupt hier erlaubt ist — in formlose, weitgehend gittergestörte Hüllen oder Grenzbezirke allmählich übergehen. An die Existenz von auch nur einigermaßen definierten Endflächen in der Längsrichtung ist überhaupt nicht zu denken. Daher ist es immerhin verwunderlich, daß die Micellen doch von relativ einheitlicher Größe sind, wie die Interpretation der Kleinwinkelreflexe von WEIDINGER und KRATKY lehrt. Aus diesen und anderen Gründen (z. B. Untersuchungen an „Micellpulver" von RÅNBY und RIBI) wird auch die Modifikation von MEYER und VAN DER WYK gegenüber der hier vorgetragenen Micellvorstellung eher für gewisse vollsynthetische Fasern zutreffen als für die native Cellulose [vgl. auch KAST (1953)]. Wenn die Abb. 10b den tatsächlichen Verhältnissen besser entspräche, so hätte es keinen Sinn eine Micellgröße bestimmen zu wollen, da es nach der MEYERschen Vorstellung völlig unklar ist, was man unter Micellgröße verstehen soll. Eine definierte übermolekulare Einheit könnte dann allenfalls erst die noch zu besprechende Mikrofibrille sein.

Soferne nähere Angaben über die einzelnen seitlichen Dimensionen gemacht werden, ergeben sich folgende Zahlenpaare: $45 \times 70$ Å (PACSU), $50 \times 150$ Å (MUKHERJEE), $30 \times 70$ bzw. $50 \times 60$ Å (FREY-WYSSLING), $53 \times 148$ Å (PRESTON). Im allgemeinen werden nur mittlere seitliche Dimensionen ($\varrho$) erschlossen (Abb. 11), die zwischen 44 und maximal 200 Å sich bewegen; vorzugsweise werden

Zahlen zwischen 55 und 80 Å genannt (s. auch Tabelle 3). Noch unsicherer ist unsere gegenwärtige Kenntnis von der Längserstreckung, wie wir auch kaum in der Lage sind, ein brauchbares Bild des Fibrillenlängsbaues heute zu geben. Sicher ist die Micellänge größer als 600 Å; es werden Werte zwischen 900 und 3000 Å genannt[1].

Besser ist die Frage nach der kristallinen Menge in der Faser beantwortet. Neben dem breiten Untersuchungsmaterial (nach der röntgenoptischen Methode) von HERMANS (vgl. PRESTON, HERMANS und WEIDINGER 1950) stehen noch weitere Ergebnisse, nach anderen Methoden (vor allem $D_2O$-Austausch) gewonnen, zum Vergleich zur Verfügung. Zum Teil auftretende Diskrepanzen haben vor allem ihren Grund in der Verschiedenheit, wie die einzelnen Methoden den Kristallit von der amorphen Umgebung abgrenzen.

Tabelle 3. *Micelldimension der nativen Cellulose nach* HENGSTENBERG.

| Vermessener Reflex hkl | Seitliche Dimension Å |
|---|---|
| 101 | 56 |
| 002 | 56 |
| 004 | 53 |
| 002 | 59 |
| 002 | 57 |

Für die native Cellulose ergeben sich Kristallinitätsgrade zwischen ~50 und 83%; nach der röntgenoptischen Methode vorzugsweise um ~71%. Niedere Werte findet man in der Primärwand (34—37%), die auch kleinere Micelldimensionen (WARDROP) aufweist und bei Bakteriencellulose (40%).

**e) Feinbau der Mikrofibrille.** Die nächste natürliche übermolekulare Einheit ist die *Mikrofibrille* (Abb. 15), die bereits im Elektronenmikroskop sichtbar ist. Eingehende elektronenmikroskopische Untersuchungen der letzten Jahre von FREY-WYSSLING, MÜHLETHALER, WYKOFF u. a. — über die im folgenden Abschnitt J/c (S. 731 ff.) näher referiert wird — mit einer sehr verfeinerten Präparationstechnik haben wahrscheinlich gemacht, daß ein solches natürliches, individualisiertes Strukturelement von möglicherweise recht konstantem Durchmesser (250—300 Å) existiert.

Während die bisher besprochene molekulare und micellare Struktur vorwiegend mit röntgenoptischen Methoden erschlossen werden konnte und die mikroskopische Struktur bis herab zur Mikrofibrille vornehmlich mit Hilfe des Elektronenmikroskops abbildbar wurde, fehlt im nun zu besprechenden Übergangsgebiet derzeit weitgehend noch die experimentelle Handhabe[2]. Es ist daher nicht verwunderlich, daß hier noch viele Unklarheiten herrschen und unsere gegenwärtigen Ansichten mehr als vorläufige Arbeitshypothesen aufzufassen sind.

Die Kernfrage ist zunächst die, ob die Mikrofibrille FREY-WYSSLINGs noch weiter unterteilbar ist oder nicht.

Ein Überblick über alle bisherigen, vor allem älteren Ergebnisse der Elektronenmikroskopie, die mehr oder weniger eindeutig die Tatsache eines Aufspleißens in submikroskopische Stränge mit einem Durchmesser von ~750 Å bis herab zu 60 Å zeigten, erweckt den Eindruck, daß das übermolekulare Gefüge zufolge seines komplizierten „Kohäsionsspektrums" auf verschiedene Zerteilungsmethoden, wie Zerquetschen nach Quellung oder chemische Hydrolyse, Angriff chemischer Agentien (vor allem $NO_2$), Mahlen (insbesonders in der Kugel- oder Schwingmühle) und schließlich Zerteilung durch Ultraschall in verschiedenster Weise anspricht. Diese zweifellos richtige Auffassung verträgt sich aber eher

---

[1] Eine zusammenfassende Darstellung über die Röntgenographie der Kolloide ist von KRATKY gegeben worden, und zwar: O. KRATKY, „Röntgenographie der Kolloide" im Kolloidchemischen Taschenbuch, 3. Aufl. Leipzig 1952 und „Der übermolekulare Aufbau der Cellulose" in R. PUMMERER, Chemische Textilfasern, Filme und Folien. Stuttgart 1951. Vgl. ferner die Ausführungen KRATKYs im 3. Band von H. A. STUART, Die Physik der Hochpolymeren (Berlin 1955 im Druck).

[2] Vgl. dazu auch die Übersichtsarbeit von ROLLINS 1954.

mit der Annahme, daß es ein natürliches submikroskopisches Strukturelement nicht gibt und eine Zeit hindurch wurde diese Auffassung auch von HERMANS, HUSEMANN und FREY-WYSSLING vertreten.

Die schon erwähnten neueren Arbeiten von FREY-WYSSLING u. a., denen geeignete und äußerst schonende Zerteilungsmethoden zugrunde liegen, zeigen aber nun ziemlich eindeutig, daß die mikroskopisch gerade noch sichtbare Fibrille leicht zu offenbar individualisierten Mikrofibrillen aufspaltet. Auch die Oberflächenabdrücke der Zellwand der *Valonia ventricosa* von PRESTON lassen in der intakten Zellwand Mikrofibrillen von ∼300 Å erkennen.

Wenngleich einige neuere Arbeiten vielleicht Zweifel an der von FREY-WYSSLING postulierten natürlichen Konstanz in der Dicke der Mikrofibrillen erwecken mögen, so darf doch die Existenz der Mikrofibrille als gesichert gelten. Bei energischerer Behandlung können aber — wie die gelegentlich beobachteten kleineren Dimensionen anzeigen — noch weitere Auffransungen über eine beschränkte Länge hinweg erfolgen, wobei Querdimensionen bis herab zur mittleren Micelldicke auftreten. Eine solche Aufspleißung würde jedoch anzeigen, daß die Mikrofibrille aus etwa 15—25 Micellsträngen besteht, also ein supermicellares Gebilde darstellt (vgl. Abb. 15 b). Der nichtindividualisierte Micellarstrang (Abb. 12 und 13) stellt im Querschnitt offenbar eine einzelne Micelle dar, ein Bündel von etwa 100 Hauptvalenzketten (vgl. Tabelle 4) in derem Innern dieselben gut geordnet sind, während die Ketten an den Rändern allmählich in Unordnung geraten und in eine amorphe oder parakristalline „Rindensubstanz" übergehen. Vom Kern aus gehen parakristalline „Cellulosefäden" zu benachbarten Micellsträngen, wodurch möglicherweise ein Längsschnitt etwa im Sinne der schematischen Abb. 14 a resultiert.

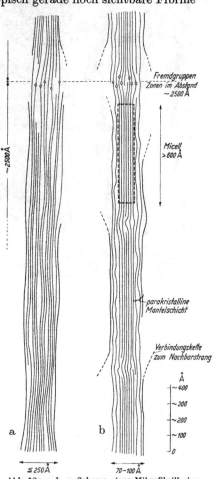

Abb. 12a u. b. a Schema einer Mikrofibrille im K. H. MEYERschen Sinne. b Schema eines Micellarstranges (seitliche Dimension etwa $2^{1}/_{2}$mal vergrößert).

Durch die nicht ganz ideale Packung der Micellarstränge — die vielleicht, ähnlich der Entstehung der Interfibrillarcapillaren teils mit der Wasserabscheidung bei der Cellulosekristallisation zusammenhängt, teils aber auch von Kristallisationsstörungen durch Fremdsubstanzen (Hemicellulose ?), Fremdgruppen usw. herrühren kann — entstehen intercellulare Spalten (Abb. 13 [i]), die z.B. für die chemische Reaktion der Cellulose von entscheidender Bedeutung sind.

Die intermicellaren Spalten i innerhalb der Mikrofibrille m sind von der Größenordnung $\geq 10$ Å. Feinere Spalten (bzw. kleinere Dichtewerte) dürften offenbar noch durch eine lamellare Aufspaltung zu den parakristallinen Bezirken resultieren, die wahrscheinlich bevorzugt parallel zu $A_1$ (vgl. Abb. 11) auftritt. Die Dichte der amorphen Cellulose wird von STEURER mit 1,482—1,489 angegeben (HERMANS 1,497; LAUER 1,35).

In der Sekundärfibrille, einem Bündel von etwa 250 Mikrofibrillen, befinden sich noch interfibrilläre Capillaren k. Diese sind gröber (Größenordnung 100 Å) und stellen Räume dar, in welche die Inkrusten der Zellwand eingelagert werden und in denen kolloidchemische Reaktionen vor sich gehen, wie z. B. die Adsorption von kolloiden Farbstoffen. Diese Spalten und gröberen Hohlräume, die offenbar miteinander kommunizieren und quasi ein Röhrensystem bilden, besitzen in biologischer und technischer Hinsicht eine besondere Bedeutung. Sie verdanken ihr Entstehen im wesentlichen wohl der Wasserabscheidung bei der Bildung der Cellulose, spielen später aber eine große Rolle für die Wasserführung der Zellwände. Die Größe dieser Hohlräume und ihre Form ist durch Edelmetalleinlagerungen erschlossen worden (FREY-WYSSLING, MARK, KRATKY, SCHOSSBERGER u. a.), wobei allerdings undiskutiert geblieben ist, wieweit durch eine eventuelle zwangsweise Aufweitung unter der Einwirkung der wachsenden Kristallite die erhaltenen Werte von 50—130 Å zu groß sind. FREY-WYSSLING (1938) konnte auch zeigen, daß die Kanäle, etwa 2500 Å im Durchschnitt lang, parallel zu den Faserachsen laufen, im übrigen jedoch irregulär verteilt sind. Größere Hohlräume, die möglicherweise einem feinen Röhrensystem angehören, konnte RUSKA im Querschnitt eines Baumwollhaares im Elektronenmikroskop sichtbar machen. Die von ihm gegebene Erklärung wird allerdings stark angezweifelt.

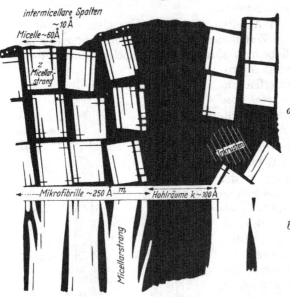

Abb. 13a. Ausschnitt aus einem schematischen Quer- (a) und Längsschnitt (b) durch eine Fibrille in Anlehnung an Modellvorstellungen von FREY-WYSSLING (schwarz: Spalten und Hohlräume; weiß: Cellulosekettenbündel, stark schematisch).

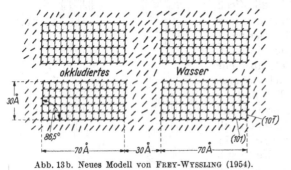

Abb. 13b. Neues Modell von FREY-WYSSLING (1954).

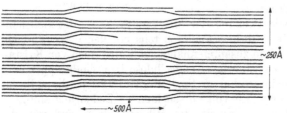

Abb. 14a. Schema einer supermicellaren Mikrofibrille. (Nach FREY-WYSSLING.) (Stark schematisiert.)

Zu praktisch demselben Bild, welches oben für den Micellarstrang geprägt wurde, kommen wir für die Mikrofibrille in ihrer Gesamtheit, wenn wir annehmen, daß vornehmlich eine einzige, höchstens jedoch einige wenige Micellen (mit über 150 Å Durchmesser) einen Mikrofibrillenkern oder die Kernpartie eines verklebten Mikrofibrillenbändchens bilden bzw. wenn die Mikrofibrillen einen variablen und vorzugsweise kleineren Querschnitt besitzen [nach PRESTON (1951) 530 × 380 bis 148 × 53 Å]. Der Mantel um den hochkristallinen Kern besteht wieder aus parakristalliner bzw. amorpher „Rindensubstanz". Eine weitere Variante ist denkbar durch die Annahme, daß die Hauptmasse der Mikrofibrille durch einen MEYER-WYKSchen Micellstrang gebildet wird (Abb. 12a), der, von eventuellen Diskontinuitäten im Bereich der Fremdgruppen abgesehen, selbst eine sehr verschwommene „Segmentierung" in *micellarer* Dimension, d. h. in DP-„Längen" von > 100 vermissen läßt, die eine lineare Aneinanderhängung von Micellen im KRATKYSchen Sinne bedingen würde. Im Gegensatz zu dieser letztgenannten Auffassung stehen auch die neuesten elektronenmikroskopischen Untersuchungen von HESS (1954), die anscheinend auf eine sehr regelmäßige Segmentierung hinweisen (vgl. auch S. 677 und Abb. 14b).

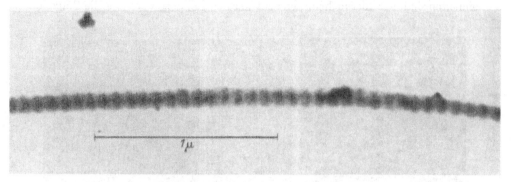

Abb. 14b. Native Cellulosefibrille (Zellstoff „Modocord X") nach Jodbehandlung. Größe der Periode etwa 720—740 Å (unveröffentlichte elektronenmikroskopische Aufnahme von K. HESS).

Eine Entscheidung für eine dieser Auffassungen ist derzeit nicht möglich. Es sei jedoch darauf hingewiesen, daß sich nun Befunde mehren, die dafür sprechen, daß die Mikrofibrillen keinen konstanten Durchmesser und darüber hinaus eine vielfach kleinere seitliche Dimension besitzen, als von FREY-WYSSLING vertreten. RÅNBY (1954) findet die Elementarfibrillen vorzugsweise in derselben Größenordnung (∼ 100 Å) wie die Micellstränge, die somit nur etwa 200 Celluloseketten umfassen können. Der hochkristalline Fibrillenkern wird in der Länge offenbar *nur* durch Zonen etwas *geringerer Kristallinität* „segmentiert"; die Micellstränge bestehen praktisch aus reiner Cellulose (Mannan, Xylan < 0,3%). An dem Aufbau der schlecht kristallisierten Rindenzone, die auch der Oxydation wesentlich zugänglicher ist, nehmen offenbar auch Hemicellulosen teil. Eine weitere Arbeit liegt von VOGEL (1953) an Ramie vor. Nach Auffassung des Verfassers ist die Mikrofibrille mit der Dimension 173—203 Å × 30 Å ein verklebtes Micellarstrangbändchen. Durch Hydrolyse erhält man Partikel von 80 bis 100 Å × 30 Å im Querschnitt und einer Länge von 300—1000 Å.

Vor allem in den dichten Sekundärwänden von Baumwolle, Ramie und Lein kommt es häufig zu Assoziationen zwischen den Mikrofibrillen, vor allem zu Bänderbildung; diese können in 2 Mikrofibrillen zerfallen. (Nach KLING und MAHL sollen auch Mikrofibrillenbündel als Übereinheiten existieren.) In der Primärwand und in Pflanzenschleimen wird keine seitliche Aggregation beobachtet; vermutlich werden hier die Oberflächenkräfte durch andere Polyosen usw. abgesättigt.

Etwa 200 Mikrofibrillen bilden die *Fibrillen* (Sekundärfibrillen oder Fila), die wieder zu weiteren Konglomeraten zusammentreten können (vgl. Tabelle 4).

Tabelle 4. *Bauelemente einer gewachsenen Cellulosefaser, in Anlehnung an* FREY-WYSSLING.

| | Querschnitt | Ungefähre Anzahl der Kettenmoleküle |
|---|---|---|
| Cellulosefadenmolekül . . . . . . | 8,3 × 3,9 Å | 1 |
| Micellarstrang . . . . . . . . . | 50 × 60 Å | 100 |
| Mikrofibrille . . . . . . . . . . | 250 × 250 Å | 2 000 |
| Mikrofibrillenbänder . . . . . . | (entsprechend 2 Mikrofibrillen) | 4 000 |
| Fibrillen . . . . . . . . . . . | 0,4 × 0,4 $\mu$ | 500 000 |
| Fibrillenbänder . . . . . . . . | variabel | — |
| Schichten (mittel) . . . . . . | ~ 12,6 $\mu$ | 30 000 000 |
| Faserzelle (Baumwollhaar) . . . | ~ 314 $\mu$ | 750 000 000 |

Die Frage, ob die Fibrille, insbesondere aber die weiteren Assoziate echte Struktureinheiten sind, soll hier nicht untersucht werden, genau so wenig wie die Frage, ob

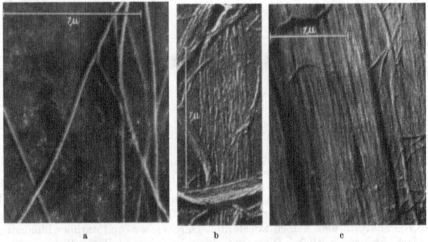

a           b           c

Abb. 15a—c. a Isolierte Mikrofibrille der Valonia nach PRESTON. b Gereinigte Baumwolle, ultrabeschallt, aufgespaltet in Fibrillenbänder und Micellbündel nach RÅNBY. c Ausschnitt aus einer Sekundärwand einer delignifizierten Tracheide von *Pseudotsuga taxifolia.* (Nach HODGE und WARDROP.)

eine Individualisierung durch eine Fremdhaut oder Kittsubstanz (Pektin, Hemicellulose), Dichteschwankung (Porosität), Ordnungszustand, Wassergehalt usw. hervorgerufen sein könnte und ob Querelemente existieren[1].

**f) Die Textur der gewachsenen Cellulose.** Die Micellen können in den pflanzlichen Geweben zu verschiedenartigen Texturen angeordnet sein. Die Art und der Grad der Orientierung können dabei außerordentlich wechseln.

Wenn die Längsachsen der Micelle parallel zur Faserachse liegen, spricht man von *Fasertextur.* Sie findet sich ausgeprägt im Wollgras und ziemlich vollkommen in den Bastfasern. Daß in diesen Fasern alle Hauptvalenzketten und Micellen parallel zur Faserachse liegen, geht sowohl aus dem Röntgendiagramm als auch aus dem mechanischen Verhalten hervor. Das spezifische Gewicht der trockenen Einzelfaser beträgt nach DAVIDSON 1,57 in Helium und ist damit innerhalb der Fehlergrenze fast ebenso groß wie die röntgenographisch bestimmte Dichte der Kristallite (1,63). Die Ketten müssen also fast durchwegs annähernd gleich dicht gepackt sein wie im gittermäßig geordneten Anteil. Daraus leitet K. H. MEYER den Schluß ab, daß in Bastfasern wirklich amorphe Partien

[1] Vgl. DOLMETSCH 1954.

mit regelloser Lagerung der Ketten etwa im Sinne der älteren Fransen-Micell-vorstellung nicht vorkommen.

Als Fasertextur im engeren Sinne bezeichnet man eine Anordnung, in der die Kristallite nur bezüglich der Längsachse zueinander parallel liegen (vgl. Abb. 15 c); die Lage der beiden anderen Achsen kann verschieden sein. Sind auch die anderen Achsen festgelegt, so daß eine Molekülanordnung ähnlich wie in einem Einkristall zustande kommt, spricht man von *höherer Orientierung*. Es darf darauf hingewiesen werden, daß wir zur Annahme berechtigt sind, daß eine solche in der Mikrofibrille vorherrscht und wahrscheinlich auch in den konzentrischen Schichten der Fasern (101-Ebene tangential zur Zylinder-wand) existiert (Abb. 16). In der Zellwand einiger Algen *(Valonia ventricosa, Chaeto-morpha sp., Cladophora prolifera)* ist auch eine solche einwandfrei nachgewiesen worden.

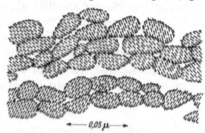

Abb. 16. Ausschnitt aus einem schematischen Querschnitt durch zwei aufeinanderfolgende Lamellen der Zellwand einer Cellulosefaser. Die Striche deuten die Lagerung der Glucoseringe an. (Der Maßstab bezieht sich auf die Mikrofibrillen, die Glucosereste sind etwa viermal so groß gezeichnet; es sollten somit mehr Glucosereste im Querschnitt sein. In den Zwischenräumen ist Wasser, Luft oder amorphe Kittsubstanz.) Abbildung nach K. H. MEYER.

Liegen die Micellen mit ihrer Längsachse in einer Ebene, innerhalb der sie regellos gestreut sind, so spricht man von *Folientextur*. Sie findet sich in den Wänden von Zellen, die keine ausgesprochene morphologische Achse besitzen.

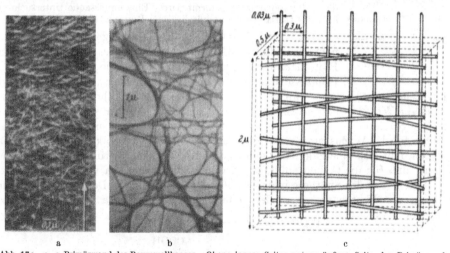

Abb. 17a—c. a Primärwand des Baumwollhaares. Oben: innere Seite; unten: äußere Seite der Primärwand. Der Pfeil kennzeichnet die Zellachse. (Aufnahme: ROELOFSEN.) (In den inneren Schichten ist die Orientierung mehr transversal, in den äußeren stärker axial („multi-net-growth").] b 8 Tage alte Cellulosemembran nach MÜHLETHALER. c Modell des Cellulosegerüsts einer lebenden Primär-Zellwand nach FREY-WYSSLING. Die Mikrofibrillen nehmen nur etwa 2,5% des Raumes ein.

Die damit sehr ähnliche *Streuungstextur* (FREY-WYSSLING) findet man in den Primärwänden (Abb. 17) und die der Folientextur ebenfalls ähnlich geartete *Ringfaserstruktur* (zwei Achsen in einer Ebene) im Tunicin, der Mantelsubstanz der Tunicaten.

Als *Schraubentextur* bezeichnet man eine Anordnung, in der die Micellen sich in einem Winkel zu der Längsachse um dieselbe herumschrauben (Abb. 18). Ein eindrucksvolles Beispiel ist hier die Kokosfaser mit dem Steigungswinkel ~45°. Sie ist dehnbar und wird als duktile Faser bezeichnet (vgl. auch die Lianen).

Manche Fasern, wie z. B. die Holzfasern der Coniferen, besitzen eine dünne Außenlamelle, in der die Fibrillen bzw. Micellarstränge praktisch tangential liegen und um die Faser herumgewickelt erscheinen (Abb. 18).

Der hier skizzierte Aufbau ist im allgemeinen durch die biologische Funktion bestimmt. Geeignete Quellfähigkeit, Quer-, Reiß- und Naßfestigkeit biologischer Faserstoffe wird durch Schuppenschichten (z. B. Wolle), Umwicklung von Fibrillenbündel mit Fibrillenbändern, seilförmige Anordnungen (Baumwolle), Querverbindungen und Kittsubstanzen (z. B. Lignin) erzielt.

**g) Bildung und Wachstum der Cellulose.** Es darf als bekannt vorausgesetzt werden, daß die pflanzliche Zellwand einen Cellulosemantel bildet und daß auch

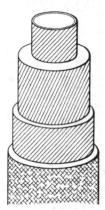

dieser Teil der Zelle, vornehmlich im Jugendstadium, ein lebendiges kompliziertes Gebilde darstellt, das von plasmatischen Elementen durchdrungen ist. So dürfte z. B. an der Spitze der Wurzelhaare das Fibrillengeflecht nach FREY-WYSSLING vollständig vom lebenden Plasma durchflutet sein.

In der lebenden Zellwand ist diese innen mit einem dicht anliegenden ununterbrochenen Plasmabelag ausgekleidet. Über die Struktur der Faserbegrenzung gegen das Lumen zu durch eine besondere (morphologisch individualisierte?) Membran [Innenschicht der Sekundärwand bzw. Tertiärlamelle (v. MOHL)] ist wenig bekannt. Vielfach wird angenommen, daß es sich bei der Tertiärlamelle um keine „tertiäre" Bildung handelt, daß sie substantiell differenziert ist (Anreicherung cellulosefremder Begleitstoffe) und ihr gegebenenfalls besondere Aufgaben zufallen. So vermeint FARR in der Tertiärlamelle den Ort erblicken zu dürfen, wo das protoplasmatische Material zu den Cellulosepartikeln geformt wird. Eine umfassende Untersuchung über die Tertiärlamelle ist kürzlich von BUCHER (1953) erschienen.

Abb. 18. Vereinfachtes Schema einer Holzfaser. Außen die Primärwand (Streuungstextur), innen Fibrillen mit Schraubentextur.

Auf Grund verschiedener Beobachtungen müssen wir annehmen, daß die Cellulose letzten Endes ein Produkt des Protoplasmas ist, gleichgültig, ob man die Bildung aus einer Primärsubstanz, aus Celluloseteilchen oder „Keimen" oder aus intercellularer Substanz, wandständiger Protoplasmaansammlung usw. annimmt. Auch die WIELERsche Theorie, nach der die Cellulose als chemischer Niederschlag zweier reagierender Lösungen aufgefaßt wird, sieht im Plasma den Ursprung. Selbst für das Wachstum der Sporenhäute, die nicht in Kontakt mit dem Plasma stehen, läßt sich eine Erklärung geben, die nicht im Widerspruch zur Auffassung steht, daß für die Cellulosebildung letztlich das Protoplasma verantwortlich ist.

Grundstoff der Synthese bilden zweifellos die Assimilate. Der wahrscheinliche Ablauf der Photosynthese in der ersten Phase dürfte etwa wie folgt wiederzugeben sein:

$$2 H_2O + 2 \text{ Ferment} \rightarrow 2 \text{ Ferment-}H_2 + O_2$$

$$CO_2 + C\diagup_{\diagdown H}^{O} \underset{CH_2OPO_3H_2}{\big|} + \text{ Ferment-}H_2 \rightarrow \overset{COOH}{\underset{CH_2OPO_3H_2}{\big| CHOH \big|}} + \text{ Ferment}$$

Jedenfalls finden sich unter den autoradiographisch festgestellten Primärprodukten Glycerinphosphorsäure und Hexosephosphate. Vielfach werden Primärprodukte der Photosynthese bereits als hochmolekular angesprochen (vgl. z. B. DOMAN 1952). Besonders RUBEN (1940) vertritt einen Bildungsmechanismus über Carbonsäuren direkt zu hochpolymeren Produkten. Einfache Zucker wären dann eher als Abbauprodukte aufzufassen[1], die enzymatisch aus Polysacchariden erhältlich sind. BALY konnte auch im Modellversuch an einem Nickelkatalysator

---

[1] Zur Cellulosesynthese in der Baumwollfaser kann nach KURSSANOW ausgenutzt werden: Glucose, Saccharose und teilweise Salicin, jedoch nicht Cellobiose.

Kohlendioxyd photosynthetisch direkt zu einem stärkeähnlichen Produkt polymerisieren.

Eine Polymerisation der Glucose kann auch experimentell durch Mineralsäuren hervorgerufen werden (Reversion), die in Konkurrenz steht mit der Säurehydrolyse. THOMPSON und Mitarbeiter (1954) konnten bei der Reversion folgende Zucker fassen: $\beta,\beta$-Trehalose, $\beta$-Sophorose, $\beta$-Maltose, $\alpha$- und $\beta$-Cellobiose, $\beta$-Isomaltose, $\alpha$- und $\beta$-Gentiobiose.

Anhaltspunkte über den Entstehungsmechanismus der Cellulose müßten Wachstumsbeobachtungen liefern[1]. Die Cellulosefibrillen bilden sich indessen so schnell aus, daß praktisch keine Zwischenstufen beobachtet werden können. Eine solche spontane Synthese wird auch beim Polysaccharid von *Oscillatoria princeps* beobachtet (FREDERICK). Die Existenz bisher angenommener Zwischenkörper ist sehr umstritten. ZIEGENSPECK will einen Zwischenkörper, das sog. Amyloid (vgl. auch S. 698) beobachtet haben, LÜDTKE spricht von Intercellulosen, und HESS und ENGEL bringen Primärpektin und Primärwachs, die zu Anfang der Wandentwicklung erscheinen, in engem Zusammenhang mit der weiteren Wandentwicklung. Schließlich wurden von FARR im Cytoplasma der Baumwolle Cellulosekeime (ellipsoidal particles) von der Größe $1 \times 1,5\ \mu$ beobachtet, die sich in einem späteren Stadium kettenförmig aneinanderlagern und schließlich zu Fibrillen verschmelzen sollen. Obgleich diese Beobachtungsergebnisse eine qualitative Ähnlichkeit mit der Dermatosomentheorie WIESNERs, mit den „Supermicells" von THIESSEN und mit den Celluloseteilchen von LÜDTKE, HESS, WERGIN und KERR besitzen, sind sie im Spiegel anderer Cellulose- und Zellwanddimensionen einer eindeutig ablehnenden Kritik begegnet.

Gewisse Einblicke gestattet die extracelluläre Entstehung der Bakteriencellulose. Auf Grund von übermikroskopischen Aufnahmen kommen FREY-WYSSLING und MÜHLETHALER (1948) zu dem Schluß, daß die Cellulosestränge aus einem elektronenoptisch amorphen, extracellulären Schleim im Zuge einer Kristallisation entstehen[2]. Von WALKER (1949) wurde als Zwischenstufe hierbei Glykolaldehyd ($CH_2OH \cdot CHO$) gefaßt. Bei Versuchen mit D-Glucose-$^{14}C^1$ fand man 82% $^{14}C$ in der 1-Stellung der Bakteriencellulose. Es wird angenommen, daß ein Teil der Hexoseeinheiten vor der Cellulosebildung gespalten wird (MINOR 1954). Zu ähnlichen Ergebnissen kommt auch BOURNE bei der Bildung von $^{14}C$-Cellulose durch *Acetobacter acetigenum*. Bei Zugabe von $^{14}C$-Milchsäure wird $^{14}C$ in der Cellulose zu 1—2% in $C^1$ bzw. $C^6$, 11—12% in $C^2$ und $C^5$ und 36—37% in $C^3$- und $C^4$-Stellung gefunden. Dieser Befund legt die Auffassung nahe, daß die Synthese aus 2 $C_3$-Fragmenten erfolgt. Hingegen wurde bei ähnlichen Versuchen an Baumwollpflanzen gefunden, daß von $^{14}C^1$-Glucose 44% in das Cellulosemolekül eingebaut wird, und zwar $^{14}C$ zu 99,97% in der $C^1$-Stellung. GREATHOUSE (1953) zieht den Schluß, daß $^{14}C^1$-Glucose durch ein Enzymsystem der Baumwollsamenkapsel direkt zu Cellulose polymerisiert wird. STACEY (1950) konnte aus Fructose mittels Zellbruchstücken aus *Acetobacter xylinum* Cellulose enzymatisch synthetisieren.

Auch die nunmehr befriedigende Aufklärung der enzymatischen Synthese von Amylose und Amylopektin, Glykogen, Dextran und Laevan läßt wohl

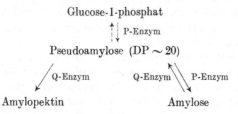

Glucose-1-phosphat

P-Enzym

Pseudoamylose (DP $\sim$ 20)

Q-Enzym      Q-Enzym    P-Enzym

Amylopektin             Amylose

---

[1] Vgl. auch FREY-WYSSLING und STECHER 1951, STECHER 1952; ferner WILLIAMS 1953.
[2] Wobei das Enzym an die Zellwand gebunden ist (BARCLAY 1954).

keinen Zweifel offen, daß die Cellulosesynthese enzymatisch analog aus irgend-
welchen Vorprodukten erfolgt, wobei anscheinend intermediär ein amorphes
Gelstadium durchlaufen wird. Solche flüssigen Vorprodukte scheinen z. B. im ,,boll
sap" von FARR bei der Baumwolle enthalten zu sein (vgl. auch KUSSANOW 1953).

Bei der Cellulosebildung in der Zellwand ist offenbar ein ganzer Ferment-
apparat beteiligt; es handelt sich voraussichtlich um ein ganzes System von
Katalysatoren und Aktivatoren. Nach WERGIN dürfte die Tätigkeit der Fermente
in diesem Falle an die Zellstruktur gebunden sein. Als solche käme die Zellwand
und vor allem die mit dieser eng verbundene Plasmagrenzschicht in Frage.

Bei der Teilung von Zellen (im Zellverband) entsteht zuerst die mit der
Primärwand genetisch eng verknüpfte isotrope Mittellamelle, ein Produkt der
Zellplatte, die primär vielleicht aus Eiweißkörpern (PRIESTLEY), im Frühstadium
aber wahrscheinlich aus pektinartigen Stoffen besteht. Im fortschreitenden Alter
tritt mehr und mehr eine Umwandlung in Lignin ein (vgl. S. 708). In der zweiten
Entwicklungsphase wird Cellulose neben anderen Wandsubstanzen (Pektine,
Wachse usw.) an die Mittellamelle abgelagert, die die Primärwand aufbaut. Die
Cellulose sehr junger Primärwände ist maskiert; im Röntgendiagramm z. B.
treten praktisch nur die Basisreflexe in Erscheinung (Kristallisationsbehinde-
rung?). Das Flechtwerk, welches die Primärwand darstellt (vgl. S. 687), kann
zustande kommen, wenn entweder alle Fibrillen im wandständigen Cytoplasma
gleichzeitig entstehen — wobei die Textur im Plasma vorbestimmt sein muß —
oder wenn die Fibrillen ein Spitzenwachstum aufweisen. Im letzten Falle könnten
sie als ,,Schuß" zwischen bereits vorhandenen, dem ,,Zettel" vergleichbaren
Fibrillen eingezogen werden.

Die folgende Zellstreckung oder das Streckungswachstum, jene Wachstumsperiode, während der die meristematischen Zellen ihre Längen in kurzer Zeit
vervielfachen, bevor sie in den Dauerzustand übergehen, geht nun keineswegs
auf eine passive Dehnung durch die Spannung des lebenden Inhalts durch Wasseraufnahme zurück. Ein elastisches Nachgeben ist nur als erster Schritt anzunehmen. Welche Leistungen die einzelnen Pflanzen während dieses Streckungs-
wachstums vollbringen, zeigen einige Zahlen: so wächst die Koleoptile von
*Avena sativa* um 3,7 cm je Tag und die Roggenfilamente um 2,5 mm in der
Minute. Es wird nach FREY-WYSSLING und Mitarbeiter (1948) vielmehr das
Flechtwerk lokal gelockert (MÜHLETHALER 1949) und neu entstandene Cellulose-
mikrofibrillen schieben sich ein. Die ,,Erweichung" der Wand, die Textur-
auflockerung, wird indirekt hormonal durch Wuchsstoffe geregelt. Nach der
Intussuszeption wird die Plastifizierung, zum Teil auch Auflösung gewisser
Wandpartien, rückgängig gemacht und die Membran verfestigt sich, worauf
sich neue, plastifizierte Felder ausbilden. Die Membran wächst also nicht als
Ganzes, sondern mosaikartig in die Fläche, wodurch sich bisher schwer erklär-
bare lokale Wachstumserscheinungen verstehen lassen. [Über das Wachstum
der Celluloseketten vgl. auch BAKER (1950).]

Die Sekundärwand, meist eine mächtige, lamellierte dichte Celluloseablage-
rung, entsteht beim nächsten Wachstumsschritt, beim Dickenwachstum. Die
Frage, ob das Dickenwachstum ein Appositions- oder Intussuszeptionswachstum
ist, scheint wohl zugunsten der ersten Auffassung, d. h. einer Ablagerung von der
Plasmaseite her, entschieden zu sein.

Über das Zustandekommen von Lamellen gibt VAN ITERSON (1937) ein mecha-
nisches Modell: Die submikroskopische Struktur wird durch die Strömungsrichtung des Protoplasmas bedingt, das Appositionsschichten gerichtet nieder-
legt. In der Tat kann man z. B. bei der Bildung von Gefäßen beobachten, wie
Plasmaströme kreisen und Ringe oder Spangen anlegen. Kreuzweise Schichtung,
wie z. B. in der Zellwand der Alge *Valonia* sei so zu erklären, daß das Protoplasma

nach Niederlage einer Schicht gezwungen werde, seine Strömungsrichtung um etwa 90° zu ändern. Ein interessanter Versuch der Deutung rhythmischer Ringstrukturen (Zonenbildung) bei biologischen Objekten als Quanteneffekt stammt von SCHAAFFS (1954).

## 3. Hemicellulosen.

Die natürlichen Cellulosefasern, insbesondere die Cellulosemembranen höherer Landpflanzen, enthalten außer Cellulose noch andere Kohlenhydrate, zu denen die Hemicellulosen (Holzpolyosen), Pektine, Gummi, Schleime und Polyuronsäuren zählen. Als Hemicellulosen[1] bezeichnet man nach SCHULZE eine Gruppe von Cellulosebegleitern, die ähnlich der Cellulose aus einer Kette glykosidisch verknüpfter Zuckerreste bestehen und die vielfach durch Säuren leichter hydrolysiert und im allgemeinen durch Alkalien leichter gelöst werden als Cellulose. Eine gewisse Ausnahme machen hierin die sog. resistenten Hemicellulosen. Zufolge des Fehlens streng spezifischer Lösemittel und sauberer Fraktioniermethoden ist die Abgrenzung des Gebietes der Hemicellulosen weder begrifflich noch analytisch scharf. Durch den Holzaufschluß sowie durch Bleich- und chemische Umsetzungsvorgänge entstehen in technischen Zellstoffen und Regeneratprodukten weitere Hemicelluloseanteile, die man vielleicht als sekundäre Hemicellulosen bezeichnen müßte.

Bis vor kurzem glaubte man in der Technik mit einer konventionellen Bestimmungsmethode auszukommen, die darauf basierte, daß man zwischen einer laugelöslichen und durch Essigsäure präcipitierbaren $\beta$- und einer durch Säuren nicht fällbaren $\gamma$-Fraktion unterschied. Die $\alpha$-Cellulose stellt dann die alkaliresistente, sehr hochmolekulare (DP $\gg$ 200) und praktisch reine Cellulose dar. Der Umstand jedoch, daß kalte Natronlauge zwischen 10 und 18% einerseits weder ein spezifisches, noch ein schonendes Lösungsmittel darstellt — erleiden doch die gelösten Polyosen verschiedene chemische Veränderungen und einen fortschreitenden Abbau (vom ursprünglichen Wert in der Gegend von DP$\sim$160 bis auf Werte zwischen 7 und 73)[2] — andererseits resistente Hemicellulosen wiederum nur sehr unvollkommen oder nicht herausgelöst werden[3], führte zu berechtigten Einwänden gegen ein solches Bestimmungsverfahren.

In den letzten Jahren sind auch Untersuchungen darüber angestellt worden, was man unter dem technischen Sammelbegriff $\beta$- und $\gamma$-Cellulose zu verstehen hat. $\beta$-Cellulose — welche sich nach WILSON in der Hauptsache erst sekundär beim Kochprozeß bildet — ist im wesentlichen stark abgebaute Cellulose (Glucosan), die teilweise anoxydiert ist (vgl. TREIBER 1953). Die sog. $\gamma$-Fraktion besteht zum Großteil aus anderen Polyosen (wahrscheinlich Xylane, Mannane und Glucosane mit niederem Carboxylgruppengehalt), die teils mit den genuinen Cellulosebegleitern identisch sind, teils diesen entstammen, soweit erstere nicht — meist nach partieller Hydrolyse — beim Aufschluß in die Ablauge gehen[4]. Man findet daher in der Sulfitablauge Mannose, Galaktose, Glucose, Xylose und

---

[1] Vgl. L. E. WISE 1949.

[2] Den alkalischen Abbau von Hemicellulose hat kürzlich PREY (1953) eingehend untersucht. Besonders starker Abbau tritt bei $p_H$-Werten über 12 bzw. Temperaturen über 100° C ein. Auch bei der meist notwendigen und vorausgehenden Delignifizierung (z. B. mit $ClO_2$) tritt ein Abbau ein.

[3] „Alkaliunlösliche" (resistente) Hemicellulosen — teils Xylane, teils Mannane — sollen mit Cellulose, teils auch mit Lignin assoziiert sein. Jedenfalls bestehen starke Nebenvalenzkräfte, und Adsorptionseffekte dürfen nicht außer Betracht bleiben (TREIBER 1953). Die DP-Werte dieser Linearpolymerisate liegen ebenfalls zwischen 150 und 160, wobei die Vermutung ausgesprochen wird, daß es sich um weitgehend einheitliche Kettenlängen handelt.

[4] In Reyon-Zellstoffen sind im wesentlichen nur Cellulosane (Glucosan, Mannan, Xylan) neben Spuren niedermolekularer Nichtkohlenhydrate (BUURMAN).

Arabinose. Eine scharfe Trennung in degradierte Cellulose bzw. primäre und sekundäre Glucosane und native Hemicellulose ist so *nicht* möglich (Jörgensen).

Auch in der Pflanzenanalyse geht man bis heute in der Isolierung ähnlich vor; man verwendet vorwiegend wäßrige oder alkoholische Alkali- und Alkalicarbonatlösungen (z. B. 5%ige Pottasche) verschiedener Konzentration. Gegenwärtig wird bevorzugt unter Sauerstoffausschluß, gelegentlich auch in Gegenwart reduzierender Substanzen gearbeitet. So extrahierte z. B. Wethern (1952) die Hemicellulose aus Schwarztanne mit 5%iger und anschließend 16%iger Kalilauge. Er fand hierbei im 1. Extrakt 7,7% Mannan, 21,1% Uronsäuren und 47,5% Pentosane mit dem Molekulargewicht von 6500—44000; im 2. Extrakt: 22,2% Mannan, 13,5% Uronsäuren und 38,9% Pentosane mit dem Molekulargewicht von 14000—30000.

Sowohl die eingangs skizzierte unscharfe Abgrenzung als auch die immer breitere Fassung des Begriffes Hemicellulose sowie die bisher beschrittenen Wege der Isolierung bringen es mit sich, daß wir letztlich recht verschwimmende Übergänge zu anderen Zellwandstoffen, wie den Pektinen, Pflanzengummi und Schleimen haben, so daß wir jene an dieser Stelle mitbesprechen wollen.

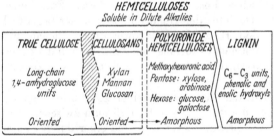

Die *Hemicellulosen* oder richtiger Polyosen, die etwa 15—50% des Pflanzenmaterials ausmachen, teilt man nach den Hydrolyseprodukten in *Pentosane* (Arabane, Xylane) und *Hexosane* (z. B. Mannane, Galaktane, Lichenin) ein bzw. nach ihrem Uronsäuregehalt, ihrer Löslichkeit sowie ihrer Beziehung und Bindung zur reinen Cellulose in (schwerer extrahierbare) Cellulosane und polyuronsäurereiche Hemicellulosen [Polyuronide; vgl. Norman (1937) und obenstehendes Diagramm]. Die beiden letztgenannten „Fraktionen" kommen in etwa gleichem Verhältnis im Pflanzenmaterial vor. Arm an Hemicellulosen sind Fasern hoher Festigkeit, wie Baumwolle, Flachs, Hanf und Ramie. Die Identifizierung der Hydrolyseprodukte ist nunmehr durch die Papierchromatographie wesentlich erleichtert worden.

Die „sauren" Hemicellulosen (Polyuronide Hemicelluloses) — die Uronsäuren und monomere Zucker als Hydrolyseprodukte geben — sind vorzugsweise Xylan-Glucuronsäurepolymerisate und finden sich vornehmlich in den Holzhemisubstanzen (z. B. White oak: Xylosemonomethylhexuronsäurekomplex); daneben findet sich Arabinose, Glucose und Galaktose. Sie sind verschieden im Aufbau und in der Löslichkeit. Einige können bereits mit Wasser extrahiert werden. Eine hierher gehörende Hemicellulose aus Weizenstroh wurde kürzlich näher beschrieben [Adams (1952)]; auch *Pseudotsuga taxifolia* besitzt ein wasserlösliches Polysaccharid vom Molekulargewicht $\approx$ 67000 (Galaktose, Arabinose, Xylose im Verhältnis 90:9:1 neben Spuren von Uronsäuren), ebenso das Holz von *Eucalyptus regnans*.

Die verschiedene Löslichkeit der diversen Hemicellulosen beruht auf einer Verschiedenheit in Kettenlänge, Kristallinität[1], Carboxylgruppengehalt, Assoziation oder Verknüpfung mit anderen Wandsubstanzen, wobei auch an eine

---

[1] Die Kristallinität der mit Säure gefällten $\beta$-Fraktion ist jedoch sehr hoch (50—62%) (Treiber).

mechanische Verflechtung zu denken ist (WHISTLER). Einen entscheidenden Einfluß übt auch der Kristallinitätsgrad, die Zahl der eingeschnappten Wasserstoffbrücken sowie die Menge inkludierter Fremdmolekeln (TREIBER 1953) aus. Die aus der alkalischen Lösung beim Ansäuern abscheidbaren Hemicellulosen bezeichnet man nach O'DWYER (1926) als A-Fraktion. Zugabe von Alkohol oder Aceton zum A-Filtrat ergibt schrittweise die B- und C-Fraktion. Weitere Fraktionierungen bzw. Reinigungen durch Fällung der Kupferkomplexe ergeben

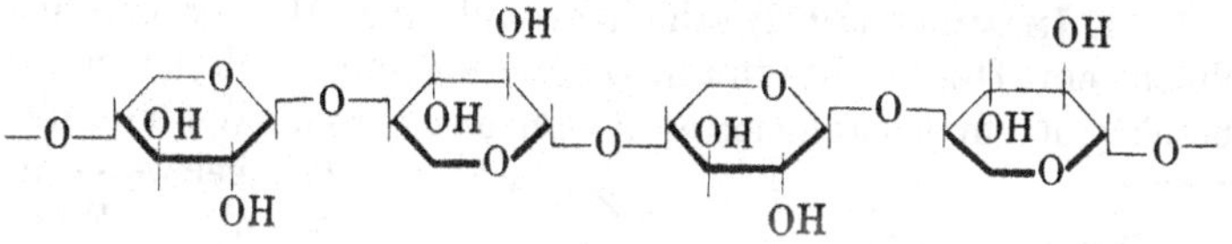

Xylan

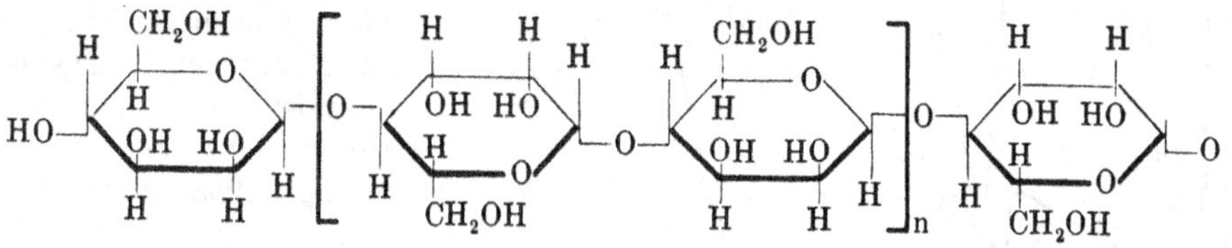

Mannan A

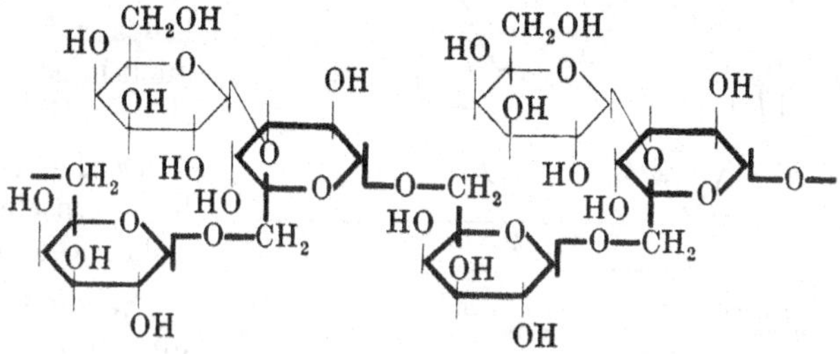

ε-Galactan

die entsprechenden Fraktionen $A_1$, $B_1$, $C_1$; durch anschließende Alkohol-Acetonfällung erhält man daraus die Fraktionen $B_2$ und $C_2$. Die Schärfe der „Fraktionierung" läßt, wie schon ausgeführt, sehr viel zu wünschen übrig. Je nach Konzentration des Alkali bzw. der Anwendungsform (Hydroxyd, Carbonat) unterscheidet man gelegentlich auch zwischen einer Hemicellulose I, II usw.

Die Cellulosane, die eigentlichen Hemicellulosen, sind im teilweisen Gegensatz zu den sauren Polyosen (Polyuroniden) reine Linearpolymerisate von Zuckern in pyranoider Form. Dadurch treten leicht Assoziationen durch Nebenvalenzkräfte mit Cellulose auf, die so fest sein können, daß diese im allgemeinen kurzen Ketten (DP 50—300) doch vielfach kaum quantitativ von der Cellulose abgetrennt werden können. Gymnospermen enthalten Glucosan, Mannan und etwas Xylan; Angiospermen Glucosan und Xylan. In Spuren finden sich noch Galaktose, Arabinose und Fructose.

Die Struktur einiger Pentosane und Hexosane ist heute einigermaßen bekannt. (Die Pentosane — vornehmlich aus Xylose und Arabinose — finden sich vorwiegend in den Gerüstsubstanzen, während Polyosen, die mehr den Charakter von Reservestoffen tragen, hexosereich sind.) Das aus fast allen verholzten Stellen — besonders der Harthölzer — extrahierbare *Xylan* ist ein 1,4-Linearpolymerisat aus D-Xylose. In geringen Mengen sind gegebenenfalls Uronsäuren

enthalten, so z. B. im Baumwollschalenxylan und im Birkenxylan, wo die Uronsäuren an den Enden der Kette sitzen dürften. Im Espartograsxylan finden
sich geringe Mengen von Arabinose. Assoziation mit α-Cellulose (DAS) und
Lignin (KAWAMURA) wird beobachtet. Den gegenwärtigen Erkenntnissen zufolge
dürften verschiedene Xylane existieren. Vielleicht sind so auch gewisse Unterschiede in den Versuchsergebnissen zu erklären.

So zeigt z. B. isoliertes Xylan nach HERZOG und GONELL Röntgeninterferenzen. Auch IMAMURA (1952) erhielt an auf einer Wasseroberfläche gewachsenen Filmen ein DEBYE-SCHERRER-Diagramm mit zahlreichen Ringen und
BISHOP beschreibt ein kristallisiertes Xylan aus Stroh. Wir konnten hingegen

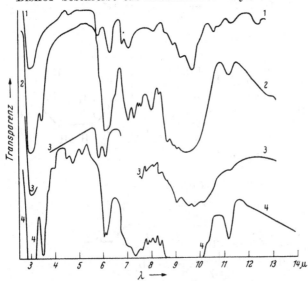

beispielsweise an einem etwa als $A_1$-Fraktion anzu-
sprechenden Buchenxylan
keine scharfen Interferenzen beobachten, was auf
einen amorphen Zustand
hinweist; nach parallelen
Untersuchungen von KA-
HOVEC (1953) handelt es
sich um eine Gelstruktur
ohne irgendwelche struk-
turelle Besonderheiten. Diskrepanzen ergaben sich
auch hinsichtlich der Xan-
thogenatbildung (MIGITA).

Der DP-Grad der Xylane
liegt zwischen 70 und 80,
sicher aber unter 150.

Abb. 19. Ultrarotaufnahmen von: Kurve 1 Alginsäure; 2 *Cydonia*-
Schleim; 3 Oxycellulose; 4 Hydratcellulose. (Nach KRATZL und
TREIBER.)

Neben Xylan finden sich
im Holz noch Arabane;
in der Eichenhemicellulose
kommt ein Xylan-Araban-Komplex vor. Ein Galaktose-Arabinosekomplex ist
in der Hülse von *Pisum sativum* und *Phaseolus vulgaris.*

Unter die *Hexosane* zählt man Glucosane, Mannane und Galaktane. Über
solche native Glucosane ist praktisch noch nichts bekannt. Untersuchungen
sind lediglich an Lichenin (Flechtenstärke) ausgeführt worden. *Lichenin* und
Isolichenin sind aus dem isländischen Moos und anderen Flechtenarten extrahierbar (HESS und FRIESE). Lichenin besteht aus Glucoseresten in β-1-4 und
β-1,3-Bindung (20%) in unbekannter, wahrscheinlich unregelmäßiger Reihenfolge. (Die Spaltung kann z. B. durch das Ferment Lichenase erfolgen.) Das
Molekulargewicht beträgt etwa 40000. (Bezüglich Holzstärke s. S. 674.)

Besser sind unsere Kenntnisse über Galaktane[1] und Mannane der Nadelhölzer
(*Picea canadensis* und andere Gymnospermen). Eingehend untersucht ist das
wasserlösliche ε-Galaktan (Galakto-araban) der Lärche *(Larix)*; das Polymerisationsschema der Galaktankomponente zeigt die Formel auf S. 693. Das
Molekulargewicht beträgt 7600—8900. (Galakto-araban findet sich auch in Erd-
nüssen und eine Polygalaktose enthält Agar.)

---

[1] Galaktosetypen finden sich auch in Rindengewebe, Holz der Coniferen, Blätter und
Früchten verschiedener Pflanzen (δ-Galaktan aus Luzerne, Bohnen und Gerstensamen,
γ-Galaktan aus Zuckerrübe). Im Galaktan von *Strychnos-nux-vomica*-Samen finden sich
Galaktose, Mannose, Xylose und Arabinose im Verhältnis ∼ 5:2:1:1 (ANDREWS).

Näher untersucht sind ferner die *Mannane* der Steinnuß (Phytelephasop.). Nach MEYER, MARK und KLAGES liegen lineare Ketten von $\beta$-1,4-glucosidisch verknüpften Mannoseresten vor; im Falle des Steinnußmannan A (vgl. Formel S. 693) etwa 70—86 Reste. Eine größere Kettenlänge besitzt Mannan B. Nadelholzmannan besitzt einen DP-Grad von 150—160. Weitere Kenntnisse besitzen wir von Konjakmannan, Cremastromannan und von Hemicellulosen der Weizenkleie, des Maiskolbens, der Stroh- und der Haferhülsen.

Abschließend kann festgestellt werden, daß die Zahl nichtcellulosischer Kohlenhydrate durch die verschiedenen Zuckerreste, die sich am Molekülaufbau beteiligen können, eine sehr große ist.

Jedoch ist bemerkenswert, daß im Gegensatz zu den Proteinen nur wenige verschiedene Kettenglieder in einem Makromolekül auftreten und daß in den ungeschädigten genuinen Polyosen sich vorwiegend nur gleichartige Grundreste vorfinden. Vergrößert wird die Zahl der so möglichen Kombinationen allerdings durch die verschiedenen Verknüpfungsmöglichkeiten, durch den Bau verzweigter Moleküle und möglicherweise durch Variation der Kettenlänge. Aus den bisherigen Arbeiten sind im allgemeinen DP-Werte zwischen 70 (CHANDA) und 160 (HUSEMANN) gefunden worden, wobei es noch ungeklärt scheint, ob vorwiegend DP-Werte zwischen 70 und 80 oder 130—170 auftreten. Auch die übermolekulare Struktur, über die noch äußerst wenig bekannt ist, scheint von Fall zu Fall verschieden zu sein. Es gibt offenbar alle Übergänge vom völlig amorphen bis zum hochkristallinen Zustand (vgl. Abb. 20).

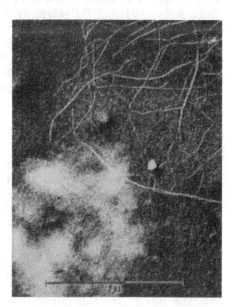

Abb. 20. Zerteilte Holocellulose von Tanne nach RÅNBY, die in der Aufnahme Micellstränge und wolkig strukturierte Hemicellulose zeigt.

Obwohl alle nun denkbaren Kombinationen nur zu einem kleinen Teil in der Natur verwirklicht sind, darf möglicherweise doch angenommen werden, daß jede Pflanzenart ihre eigenen Hemicellulosen besitzt. Ein solcher Umstand würde eine weitere Erschwerung in der Erforschung dieser bisher vernachlässigten Stoffklasse bedingen, die technisches und wissenschaftliches Interesse (Übergang zu den Bakterienpolysacchariden, Antigenen, tierischen Schleimsubstanzen) beanspruchen darf.

## 4. Pektine.

Pektinsubstanzen finden sich vornehmlich in jungen Pflanzen (Zellwände des jungen Meristemgewebes, Mittellamelle), Früchten und Wurzeln. Diese Kohlenhydrate zeichnen sich durch hohes Wasserbindevermögen aus und dürften auch deshalb eine besondere biologische Rolle spielen. Man findet lösliche (Pflanzen- und Fruchtsäfte) und unlösliche Pektine, sog. Protopektine (grüne Pflanzenteile). Calcium- und Calcium-Magnesiumsalze der Pektinsäure finden sich z. B. in der Mittellamelle.

Als Spaltprodukte der sauren Totalhydrolyse erhält man D-Galaktose, L-Arabinose und D-Galakturonsäure, Glieder einer „biologischen Reihe" analog zu: Glucose, Glucuronsäure → Xylose. Außerdem findet man Methylalkohol, der entsprechenden Galakturonsäureestern entstammt. Die Pektine sind offenbar lineare Makromoleküle; die Polygalakturonsäure, Hauptbestandteil der Pektinstoffe, ist als eine Kette $\alpha$-1,4-glucosidisch verknüpfter Galakturonsäurereste aufzufassen. Die Salze sind die Pektate. Pektin besteht aus teilweise methoxylierten Polygalakturonsäuren (H-Pektin: Veresterungsgrad $>50\%$, L-Pektin: $<50\%$), ihre Salze sind die Pektinate. Im Protopektin (Pektose) liegt eine salzartige Vernetzung von Pektinketten durch mehrwertige Metallionen vor. Pektinstoffe schließlich sind reich an Begleitstoffen (bis $\sim 30\%$), die vornehmlich dem Pektin beigemischt, mit diesem assoziiert, möglicherweise aber auch chemisch verknüpft sind. Die Konstitution dieser abtrennbaren Arabane (vor allem in der Erdnuß *Arachis*) und Galaktane (reichlich im Samen von *Lupinus albus*) hat HIRST näher untersucht.

Pektine bilden Filme und Fäden mit einer Faserperiode von 13—14 Å (3 Galakturonsäurereste). Die Molekulargewichte dürften mehrere 100000 betragen (Äpfel- und Citruspektin, HENGLEIN). ROELOFSEN (1951) findet in frischen Collenchymzellwänden *(Petasites vulgaris)* axial orientierte submikroskopische Fibrillen von axial orientierten Pektinkristalliten mit negativer Doppelbrechung.

## 5. Pflanzengummi und Schleime.

Auch diese Stoffe sind hochpolymere komplexe Polysaccharide, in denen hauptsächlich Hexose-, Pentose- und Uronsäurereste miteinander verknüpft sind (s. Tabelle 5). Gelegentlich finden sich seltene Zuckerarten, wie im Gummi des *Sterculia setigera*-Baumes (D-Tagatose). Durch das gelegentliche Auftreten acetylierter und methylierter Derivate wird die Struktur noch verwickelter — so kommt z. B. im Schleim von *Ulmus fulva* 3-Methyl-D-galaktose vor — und besonders schwierige Probleme ergeben sich bei Algenschleimen, die Schwefelsäureestergruppierungen (Natriumsalz) enthalten [Agar aus *Gelidium*, *Carraghen* (Schwefelsäurerest am C-Atom 4 des Galaktoserestes]; Sulfatfucose in Fucoidin usw.). Im Gummi sind noch vielfach Enzyme enthalten.

Tabelle 5. *Die Zusammensetzung einiger Pflanzengummis und -schleime.* Nach HIRST.

| Stoff | Vorhandene Zuckerreste |
|---|---|
| Gummi arabicum . . . . . . . . . . . . . . . | Glu, G, R, A |
| Kirschgummi . . . . . . . . . . . . . . . . | Glu, G, M, A, X |
| Damascenpflaumengummi . . . . . . . . . . | Glu, G, M, A, X |
| Eierpflaumengummi . . . . . . . . . . . . . | Glu, G, A, X |
| Mezquitegummi . . . . . . . . . . . . . . . | Glu, G, A und 4-Methylglucose |
| Traganthgummi . . . . . . . . . . . . . . . | Gal, G, F, A, X |
| *Sterculia setigera* Gummi . . . . . . . . . | Gal, G, R, und D-Tragatose |
| Leinölsamenschleim . . . . . . . . . . . . | Gal, G, R, X |
| Kressensamenschleim . . . . . . . . . . . . | Gal, G, R, X, A |
| *Plantago psyllium* und *P. fastigiata* Samenschleim | Gal, A, X |
| Samenschleim von *Plantago arenaria* . . . . . . | Gal, G, A, X |
| Schleim aus der Rinde von *Ulmus fulva* . . . . | Gal, G, R, und 3-Methyl-D-Galaktose |
| Carragheen . . . . . . . . . . . . . . . . | G und Sulfate |
| Fucoidin . . . . . . . . . . . . . . . . . | F und Sulfate |
| Agar . . . . . . . . . . . . . . . . . . . | G, L und Sulfate |
| Laminarin . . . . . . . . . . . . . . . . . | D-Glucose |
| Alginsäure . . . . . . . . . . . . . . . . | D-Mannuronsäure |

*Schlüssel.*

| | | |
|---|---|---|
| Glu = D-Glucuronsäure | M = D-Mannopyranose | A = L-Arabofuranose |
| Gal = D-Galakturonsäure | R = L-Rhamnopyranose | X = D-Xylopyranose |
| G = D-Galaktopyranose | F = L-Fucose | L = L-Galaktose |

Gummi arabicum ist ein Salz einer hochmolekularen Säure, der Arabinsäure. Das Molekül dieses Gummis ist sicher verzweigt gebaut und dürfte sehr kompliziert sein. Wahrscheinlich sind an einem resistenten Kern von Galaktose und Glucuronsäure Zweige aus L-Arabofuranose, L-Rhamnopyranose sowie 3-Galaktopyranosido-L-Arabofuranoseresten angeheftet. Nach DILLON ergibt sich wahrscheinlich folgendes Polymerisationsschema:

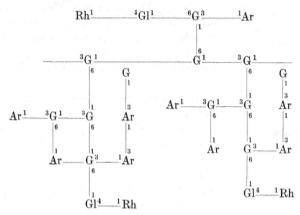

G D-Galaktose; Gl D-Glucuronsäure; Ar L-Arabofuranose; Rh L-Rhamnose.

Kirschgummi gibt bei der Hydrolyse Arabinose, Xylose, Galaktose, Mannose und Glucuronsäure. Gummi aus der Rinde damascenischer Pflaumenbäume ergibt bei der Hydrolyse Arabinose, Galaktose, Mannose, Xylose und Glucuronsäure; Eierpflaumengummi Galaktose und Glucuronsäure. Ähnlich komplizierte Verbindungen enthält Citronengummi. Traganth enthält etwa 50% Uronsäuren, daneben Arabinose.

Auch in den Pflanzenschleimen, die ebenfalls Zellwandbestandteile sind (z. B. Schleimmembranen), findet man Zucker und Uronsäuren als Spaltprodukte. So enthält Leinsamenschleim 2-D-Galakturonopyranosido-L-rhamnose neben L-Galaktose und D-Xylose. Der verzweigt gebaute Schleim aus Samen von *Plantago lanceolata* enthält Xylose, Methylpentose, Galaktose und Uronsäuren. Der Schleim aus Luzernensamen *(Medicago sativa)* ergibt bei der Hydrolyse D-Galaktose und D-Mannose. Sehr einfach scheint das „Salep"-Mannan gebaut zu sein. Aus *Hibicus esculentus*-Schleim konnte WHISTLER erstmals eine kristallisierte Galaktobiose isolieren.

Physikalisch-chemische Messungen wurden an Gummi traganth (GRALÉN) und Karaya-Gummi (KUBAL) durchgeführt. Die Molekulargewichte sind sehr hoch ($840000$ und $9 \cdot 10^6$), woraus sich Moleküldimensionen von weniger als $3000$—$4000$ Å in der Länge und mehr als $20$—$80$ Å im Durchmesser ergeben.

Interesse verdienen noch die aus Algen isolierbaren Alginsäuren, die als Polymannuronsäuren anzusprechen sind und sich technisch zu Fäden verspinnen lassen. Sie besitzen eine $\beta$-1-4-Verkettung; das Molekulargewicht dürfte in der Größenordnung von $150000$ liegen. Begleitet wird die Alginsäure von Laminarin (vgl. S. 672).

Es wird vermutet, daß einige Pflanzenschleime (Samen von *Cydonia vulgaris, Lepidium sativum, Brassica hirta* u. a.) auch abtrennbare Celluloseketten besitzen. Wir hätten hier wieder einen Fall von festen bzw. chemischen „Assoziaten", wie sie bei Cellulose-Hemicellulose, Cellulose-Polyuronsäure, Cellulose-Lignin, Hemicellulose-Lignin usw. vorliegen dürften. Vom Autor wurde kürzlich der von RENFREW (1932) näher beschriebene *Cydonia*schleim röntgenoptisch untersucht (TREIBER 1953); die Schleimfäden ergeben ein Faserdiagramm. Das Röntgendiagramm der abgetrennten „Cellulose" ist aber mit dem der normalen Cellulose nicht völlig identisch!

Die Löslichkeit der Schleime beruht wahrscheinlich auf der Anwesenheit von uronsauren Salzen. Elektronenmikroskopische Bilder pflanzlicher Schleime zeigen ein Netzwerk von Pektin- bzw. Cellulosefäden. Im gequollenen Zustand wird in den Maschen der Schleimsubstanz das Wasser gebunden.

Das Studium der Pflanzengummi und Schleime führt zu einigen fundamentalen Problemen der Chemie der Kohlenhydrate, nämlich der Untersuchung der Vorgänge, durch welche primäre Produkte der Photosynthese (offenbar

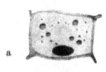

a

b

D-Glucose) in andere Hexosen übergeführt werden. Es scheint z. B. heute festzustehen, daß eine Umwandlung der D-Galaktose in die stereochemisch verwandte L-Arabinose nicht im Makromolekül stattfindet.

Die Bildung des Kirschgummis haben kürzlich CERUTI (1953) und Mitarbeiter studiert. Zuerst entsteht aus Körnchen polymerer Hexosen ein *Prägummi*, an dessen Randzone bald Pentosen und Uronsäuren erscheinen. Nach einem löslichen Zwischenstadium bildet sich der Gummi, der hierauf beginnt, die Zellwände zu imprägnieren (vgl. Abb. 21 a—c).

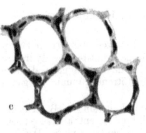

c

Abb. 21a—c. Einige Phasen der Entstehung des Kirschgummis nach CERUTI. a Körnchen von Prägummi in Parenchymzellen. b Großes Prägummitröpfchen. An der Oberfläche treten Uronsäuren und Pentosen auf. c Imprägnierung der Zellwand mit Gummi.

## 6. Besondere Zellwandstoffe — Chitin.

Besondere Zellwandstoffe werden in den Moosen gefunden, die möglicherweise an Cellulose gebunden sind, und zwar ein Celluloseäther (?) des phenolartigen Sphagnol und die Dicranumgolsäure. In den Wurzeln der Compositen finden sich Phytomelane, die neben H und O einen sehr hohen Gehalt an C besitzen. Cellulose in offenbar unbekannter Form enthalten einige Algen *(Spirogyra, Vaucheria)*. Auch über die Kallose[1] *(Mangin)* und das amorphe Karuban (Johannisbrot, *Ceratonia siliqua*) ist wenig bekannt.

Umstritten ist die Existenz des Amyloids, welches nach ZIEGENSPECK als Zwischenkörper der Cellulosebildung aufzufassen wäre, nach SCHLEIDEN, VOGEL und REISS zu den Hemicellulosen zählt, vornehmlich die Rolle eines Reservestoffes spielen soll und bei Hydrolyse in Dextrose, Galaktose und Xylose zerfällt. In der älteren Literatur (SCHWALBE) werden auch Präcipitate von Hydratcellulose aus Säurelösungen zufolge der blauen Jodreaktion als „Amyloid" bezeichnet.

Eine ähnliche Rolle wie die Cellulose spielt als Gerüstsubstanz das Chitin, welches sich im Tierreich und bei den Pilzen findet. Eine Reihe von Ähnlichkeiten mit der Cellulose führten MEYER und MARK (1928) zu der Auffassung, daß Glucosaminreste analog dem Bauplan der Cellulose in $\beta$-1,4-Bindung und diagonaler Verschraubung zusammengefügt sind. Die MEYER-MARKsche Auffassung ist in der Folgezeit durch BERGMANN, ZECHMEISTER und Mitarbeiter bestätigt worden. DIEHL und VAN ITERSON (1935) schließlich bewiesen, daß Pilzchitin identisch mit dem tierischen Chitin ist. (Chitin ist meist stark mit Eiweiß vergesellschaftet. In den letzten Jahren sind in vielen biologisch aktiven Substanzen auch gemischte Polysaccharide mit Hexosaminresten gefunden worden.)

---

[1] Kallose aus Cystolithen von *Ficus elastica* ist nach ESCHRICH (1954) ein unlösliches Glucosan.

Auch der Elementarkörper mit 8 Glucosaminresten ist nach MEYER und PANKOW ähnlich der der Cellulose, und zwar der metastabilen Cellulose III (Abb. 22) gebaut.

Orientierte Chitinketten finden sich in vielen Organteilen mit ausgeprägter Faserstruktur. Nach HEYN (1936) ist Chitin in den Wänden der Sporangien höher orientiert. In diesen Hohlzylindern liegen die Längsachsen der Chitinketten in der Faserrichtung, die c-Achse liegt radial und die a-Achse tangential; somit liegen auch die Ringebenen annähernd radial.

Nach FREY-WYSSLING (1950) bauen sich die *jungen* Sporangienträger von *Phycomyces Blakesleeanus* wie folgt auf: Unter einer Cuticula befindet sich eine Primärwand, die in der unteren Wachstumszone aus einer äußeren Lamelle, die ein unorientiertes Flechtwerk feinster Fibrillen darstellt, und einer inneren Lamelle mit gröberen Fibrillen (250—300 Å Durchmesser), die vorzugsweise quer zur Zellachse verlaufen, besteht. Möglicherweise existiert noch eine dritte Primärlamelle mit zwei sich unter einem Winkel von ~ 120° überkreuzenden Fibrillensystemen. In der Sekundärwand dürfte sich zunächst eine „Übergangslamelle" befinden, in der eine Fibrillenrichtung stark vorherrscht. Die übrige Sekundärwand zeigt eine sehr schöne Paralleltextur in der Wachstumsrichtung.

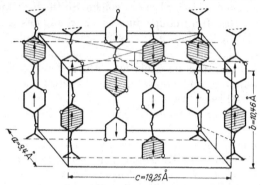

Abb. 22. Schema des Elementarkörpers des Chitins.
(Nach MEYER und PANKOW.)

# II. Die Chemie der übrigen Wandsubstanzen.

Während die Eiweißstoffe und die in diesem Abschnitt besprochenen Kohlenhydrate gewöhnlich als primäre Pflanzenstoffe bezeichnet werden, entstehen beim Stoffwechsel zahllose Zwischen- und Endprodukte verschiedenster chemischer Natur, die zum Großteil wieder besondere Aufgaben zu erfüllen haben. Im Gegensatz zu den allgemein verbreiteten primären Produkten können die sekundären Stoffwechselprodukte vielfach als systematische Merkmale herangezogen werden.

Aus der überwältigenden Fülle sekundärer Pflanzenstoffe können hier natürlich nur die wenigen, für die Zellwand bedeutsamen herausgegriffen werden. Im übrigen sei hier auf die betreffenden Kapitel der Bände 7, 8 und 10 verwiesen.

Während somit einige für die Pflanzenmembranen kaum bedeutsame Stoffe (Glucoside, Saponine, Alkaloide u. dgl.) nur flüchtig gestreift werden, sollen Kork, Wachs und Cutin sowie das Lignin eine etwas eingehendere Behandlung finden. Die genannten Membranstoffe sind wohl für die Eroberung des Landes durch die Pflanze und deren Höherentwicklung maßgebend gewesen; es sind dies jene Stoffe, die teils die Pflanzen vor übermäßiger Wasserabgabe schützen können, teils den Membranen bzw. Organen eine Stütze geben (Lignin), so daß die Pflanze ihr eigenes Gewicht tragen kann.

## 1. Einleitung.

Vor allem die absterbenden Gefäße können sich mit anorganischen Ablagerungen und harzähnlichen Stoffen u. dgl. füllen, die unter Umständen auch die Zellwand zum Teil imprägnieren können. In Sekretbehältern, welche durch die Auflösung von Teilen oder sogar der ganzen Membran entstehen, sind ätherische Öle enthalten, die den Terpenen zugehören. Zur selben Stoffgruppe gehören beispielsweise aber auch die farbigen Polyensäuren und Polyenalkohole. Erstere

(Chlorophyllin a und b), verestert mit farblosen Alkoholen (Phytol, Methanol), geben die Chlorophylle (verwandt damit ist auch der Farbstoff der Purpurbakterien), letztere, verestert mit farblosen Säuren, die Farbwachse (s. S. 707). Die nahe Verwandtschaft zwischen Wachsen und Fetten, Lecithinen und Sterinestern ergibt sich aus der Tatsache, daß es sich in all diesen Fällen um Ester farbloser Säuren mit farblosen Alkoholen handelt. Von physiologischer Bedeutung scheinen nach METZNER (1930) auch die Gerbstoffimprägnierungen zu sein, die eine starke UV-Absorption im langwelligen Gebiet besitzen. Diese Tatsachen machen es erforderlich, solche Stoffe hier kurz zu streifen.

Die oft auffällige dunklere Färbung des Kernholzes gegenüber dem funktionstüchtigen Splintholz sowie das Nachdunkeln geschnittenen Holzes rührt von Oxydationsprodukten der Gerbstoffe her und zeigt damit die Anwesenheit solcher auch im Holz an. Als gerbstoffreichstes Holz gilt das von *Schinopsis Balansae* und *Lorentzii*. Die Gerbstoffe der Hölzer können von denen der Rinde verschieden sein.

Unter *Gerbstoffen* verstand man früher amorphe Stoffe, die die Eigenschaft besitzen, Haut in Leder zu verwandeln. Sie fällen Alkaloide, Leim usw. und zeigen vielfach typische Eisenchloridreaktion, nach der man auch von eisenbläuenden (z. B. Tannengerbstoff) und eisengrünenden (z. B. Hemlockrindengerbstoff, *Tsuga canadensis*) Gerbstoffen spricht. Die (UV-) Lichtabsorption in den Spaltöffnungen soll z. B. nach METZNER (1930) von eisengrünenden Gerbstoffen, in Epidermis und Mesophyll häufig von eisenbläuenden Gerbstoffen herrühren.

Nach FREUDENBERG (1933) unterscheidet man:

a) eine esterartige Gruppe (Depside), die sich durch Hydrolyse zerlegen läßt und sich vorwiegend von der Gallussäure bzw. Digallussäure (I) (HERZIG) ableitet. Die Hydrolyse kann außer durch Mineralsäure auch durch das Ferment Tannase erfolgen. Die bestuntersuchten Vertreter sind hier das chinesische Tannin, türkische Tannin und Hamamelitannin (II). Es sind Gemische verschieden galloylierter Glucosen; vielfach ist noch Ellagsäure (III) enthalten (Ellagen-Gerbstoffe). Das einfachste Gallotannin ist das Glucogallin [Rhabarber *(Rheum officinale)*]. Komplizierter zusammengesetzt sind z. B. Tetrarin und Chebulinsäure. Ein Elagengerbstoff ist z. B. Corilagin, eine L-Galloyl-3,6-hexaoxydiphenyl-β-D-glucopyranose (SCHMIDT).

Zu dieser Gruppe gehören weiter die Depside [Ester aromatischer Oxycarbonsäuren mit Oxycarbonsäuren, wie z. B. m-Digallussäure (I), Lecanorsäure, Evernsäure, Chlorogensäure], die in den Flechten vorkommen (FISCHER 1919).

b) kondensierte Gerbstoffe (Catechin- oder Phlobatannine). Hierher gehören Catechine und andere nicht völlig aufgeklärte Substanzen, wie sie beispielsweise in der Eiche *(Quercus)*, Roßkastanie *(Aesculus)* und Quebracho *(Schinopsis Lorentzii)* gefunden werden. Die Catechine (IV), Muttersubstanzen vieler solcher Gerbstoffe, sowie Gerbstoffrote (Phlobaphene) sind als hydrierte Flavonole oder Anthocyanidine aufzufassen. Gewisse Parallelen zur Struktur des Lignins werden vermutet.

In manchen Fällen sind in den Zellwänden der Gefäßteile *Farbstoffe* enthalten, so daß man vielfach von Farbhölzern spricht. So enthält Blauholz *(Haematoxylon campechianum)* Hämatoxylin, Rotholz *(Erythroxylon, Caesalpinia)* Brasilin, das rote Sandelholz *(Santalum)* Santalin, Gelbholz *(Xanthoxylon)* Morin, Fisetholz *(Rhus Cotinus)* Fisetin usw. Meist sind die Farbstoffe — die vorwiegend Pyronderivate sind — als Glucoside enthalten (z. B. Quercitrin, Fustin). Im Gegensatz zum Sulfitaufschluß des Holzes gehen bei einer alkalischen Kochung diese Farbstoffe in die Ablauge.

Auch *Glykoside* treten in Zellwände über. Es sind Stoffe, die unter Wasseraufnahme in Zucker und ein Aglykon, meist eine aromatische oder fettartige Komponente aufgespalten werden können. Bekannt ist hier das Arbutin *(Ericaceen)*, Salicin (Weide), Aesculin (Roßkastanie), Fraxin (Esche), Primin, Daphnin, Indican, Purpureaglykosid usw. Interesse verdient das Rhapontin; Derivate seines Aglykons (z. B. Pinosylvin) sind offenbar als Schutzstoffe des Holzes gegen Pilze und Insekten aufzufassen (ERDTMANN). Eine gleichfalls fungicide Substanz, die sich im Holz vorfindet *(Thuja plicata)*, ist das siebenringige Thujaplicin. Bekannte Glykoside sind die Anthocyane.

Eine verwandte, gelegentlich auch in Zellwände übertretende Stoffgruppe stellen die *Saponine* dar, die besonders häufig in Nelkengewächsen vorkommen. Man unterscheidet

Genine oder Steroidsapogenine und Saponine, die Sapotalin als Hydrierungsprodukt liefern (z. B. Hederagenin, Oleanolsäure [1]) — ein Kohlenwasserstoff, der auch aus Triterpenverbindungen erhalten worden ist.

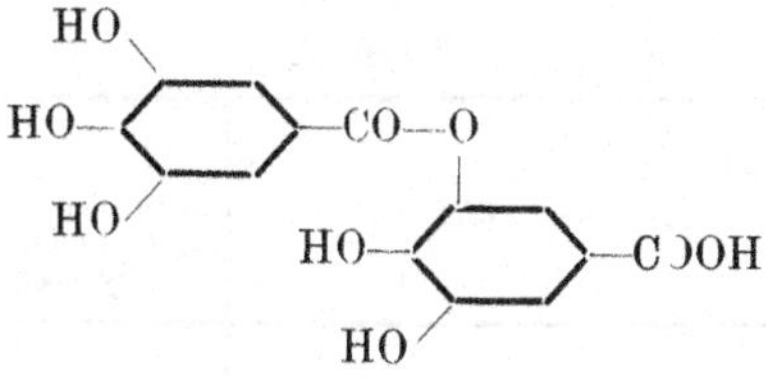

I

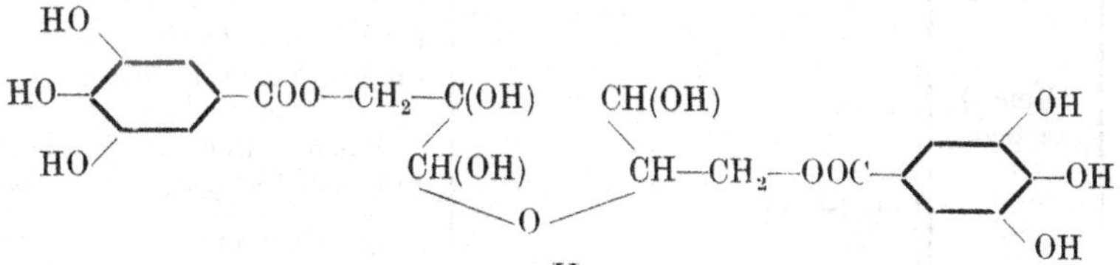

II

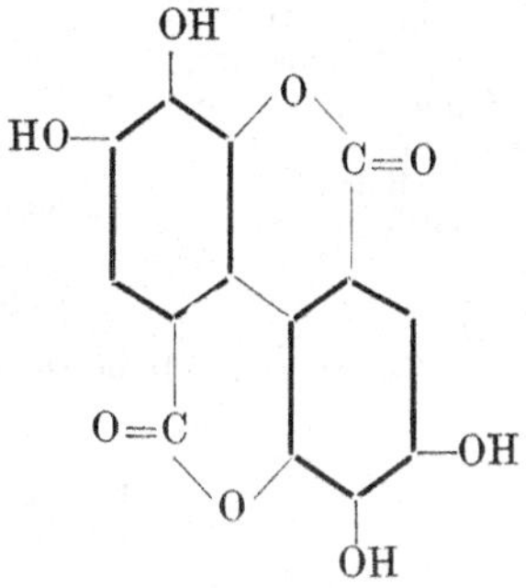

III

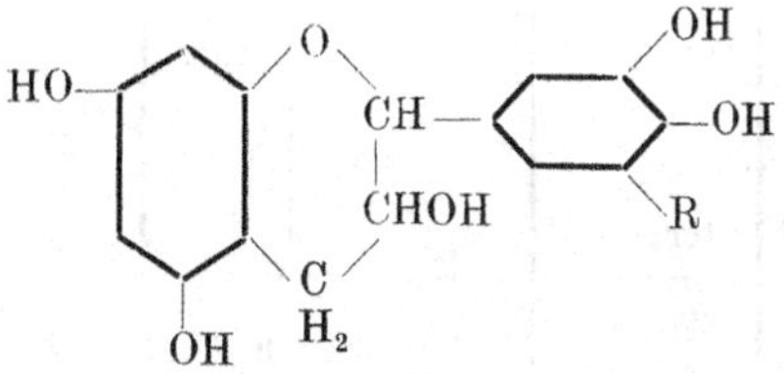

IV

R = H: D,L-Epicatechin
R = OH: Gallocatechin

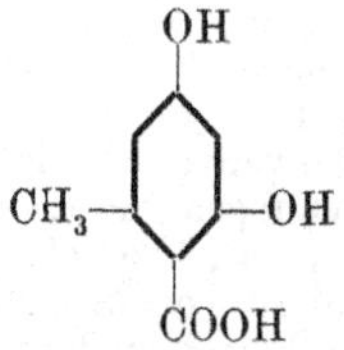

V

In höheren Pflanzen finden sich häufig *Alkaloide* (allerdings fraglich, ob in den Wandsubstanzen enthalten). Alkaloide im engeren Sinne sind Verbindungen mit heterocyclisch gebundenen Stickstoffatomen, mehr oder minder stark basischem Charakter, meist kompliziertem Molekülbau und vielfach ausgeprägter physiologischer Wirkung. Alkaloidähnliche Stoffe werden bei Pilzen gefunden (Ergotinin und Ergotoxin). Welche physiologische Rolle die Alkaloide in der Pflanze spielen, ist noch ungeklärt.

In engem Zusammenhang mit den Wandsubstanzen stehen aber zweifellos die *Terpene*, eine Stoffgruppe, zu der sehr viele sekundäre Pflanzenstoffe zu zählen sind. Hierher gehören beispielsweise die ätherischen Öle (vor allem die sog. „sauerstofffreien Öle"), Balsame, Harze, Campher und Kautschukarten (vgl. Tabelle 6). Aber auch die wesentliche Komponente mancher Farbstoffe (z. B. Safranfarbstoff Crocetin, Chlorophyll, Carotinoide), Farbwachse und anderer physiologisch interessanter Stoffe (Crocin, der Bewegungsstoff der Gameten der Grünalge, *Chlamydomonas*) hat Terpencharakter, und es ist nicht

---

[1] Zum Beispiel: Im *Morabukea*-Saponin (LAIDLAW).

unwahrscheinlich, daß auch Sterine, die vielfach im Aufbau eine Verwandtschaft mit Harzsäuren aufweisen, mit einfacheren Terpenverbindungen durch genetische Beziehungen verknüpft sind.

*Tabelle 6.*

| Formel des Grundkörpers | Terpenklasse | Art des Stoffes | Beispiele | |
|---|---|---|---|---|
| | | | eigentliche oder cyclische Terpene (Mono-, bi- und tricyclische Terpene) | acyclische oder olefinische Terpene |
| $C_5H_8$ | Hemiterpene | (kommt nicht frei vor) | — | (Isopren) |
| $C_{10}H_{16}$ | (Mono-) Terpene | ätherische Öle | (*Menthene, Menthadiene* und *bicyclische Terpene*) Pinen, Caren, Sabinen, Campher | Myrcen, Citral |
| $C_{15}H_{24}$ | Sesquiterpene [1] | | Bisabolen, Cadinen, Selinen, Copaen | Farnesol, Nerolidol |
| $C_{20}H_{32}$ | Diterpene | Harze / Balsame | Camphoren, Abietinsäure, Sapinsäure, Laevopimarsäure | Crocetin, Crocin [2] |
| $C_{38}H_{48}$ | Triterpene | | Betulin Oleanolsäure | Squalen |
| $C_{40}H_{64}$ | Tetraterpene | Carotinoide (Lipochrome) | Carotine, Phytoxanthine | Lycopin |
| $(C_5H_8)_n$ | Polyterpene | Kautschukarten (Latex) | — | *Hevea*-Kautschuk, Guttapercha, Balata |

*Ätherische Öle*, die sich in Sekretbehältern, die in Beziehung zur Membran stehen, ansammeln, sind vornehmlich leicht flüchtige Terpenkohlenwasserstoffe (Citronellol, Geraniol, Nerol, Farnesol; Menthan, Terpinen, Phellandren, Terpinolen, Limonen, Sylvestren; Sabinen, Caren, Pinen und Fenchon) und deren Derivate (z. B. Menthol, Menthon, Terpineol, Terpin, Cineol, Pulegon, Carvon, Jonon). Ihnen verdanken viele Pflanzen ihren Geruch. Die meisten ätherischen Öle sind komplizierte Mischungen vieler, oft ähnlich gebauter Terpensubstanzen mit anderen flüchtigen Verbindungen (Ester, Alkohole, Aldehyde, Phenole). Ihre Zusammensetzung unterliegt Schwankungen und ist vom Standort, Klima und Jahreszeit abhängig. Einige Riechstoffe sind nicht in freier Form, sondern als Derivate, speziell Glykoside, enthalten (z. B. Amygdalin, Sinigrin). Viele dieser Substanzen sind nach Kondensation für die Zellwandbildung bedeutsam, so in Harzen und Balsamen.

*Balsame* sind flüssige, *Harze* halbfeste organische Terpenverbindungen, die Stoffwechselprodukte darstellen (primäre oder physiologische Harze) oder die infolge von Verletzungen der Pflanzen als klebrige Substanz entstehen (sekundäre, pathologische Harze). Daneben finden sich noch aliphatische Harzkörper, Glykoside, Farbstoffe, ätherische Öle, Ester, Aldehyde, Enzyme, Bitterstoffe [3] usw.

Harzsubstanzen werden wie die übrigen Terpene im Protoplasma [4] gebildet und in den Saftraum abgeschieden oder durch die Zellwände in die Intercellularräume bzw. ins Freie befördert. Sie finden sich demnach im Zellinhalt, in den Zellwänden und den intercellulären Behältern, den sog. Harzgängen. Am harzreichsten ist bei Bäumen das Wurzelholz.

---

[1] Verwandt damit sind die Azulene.

[2] Beide den Carotinoiden verwandt. Den aliphatischen Diterpenen nahestehend ist das Phytol.

[3] Auch Komponenten der Bitterstoffe scheinen den Terpenverbindungen anzugehören. Nach Holzer und Zinke (1953) z. B. ist der Bitterstoff der Zichorie der Monoester der p-Oxyphenylessigsäure mit Lactucin, welches nach Wessely in die Klasse der Sesquiterpene, nach Tschesche allerdings in die der Steroide gehören soll.

[4] Die wichtige intermed. Zwischenstufe ist die Seneciosäure (β-Methylcrotonsäure) (Reichel).

Das Harz der Nadelhölzer z. B. besteht vorwiegend aus einem Gemisch von Harzsäuren (Di- und Triterpensäuren). (Die Alkalisalze bilden die Harzseifen. Ein Gemisch von Harz- und Fettseifen neben neutralen Stoffen [z. B. Alkohole] ist das Tallöl, welches sich als schaumige Masse nach längerer Zeit aus alkalischer Kocherlauge abscheidet.) Nach Aschan (1924) sind Abietinsäure sowie Isopimarsäure sekundäre Harzsäuren, während zu den natürlichen oder primären Harzsäuren Pimar- und Sapinsäure ($C_{20}H_{30}O_2$), Pinin- und Isopininsäure zählen sollen. Es existiert eine Einteilung in Harze vom Abietin- und Pimarsäuretyp (Harris).

Abietinsäure                         α-Sapinsäure

Von Wienhaus (1943) wurde Edeltannenterpentin untersucht; die moderne Arbeit soll als Beispiel kurz referiert sein. Das Terpentin enthält ~ 17% kristall. Stoffe (Abienolhydrat), 30% neutrale Stoffe (Pinen, Camphen, Pinolhydrat, Dipenten, Sesquiterpenalkohole, Aldehyde) und 50% Harzsäuren (Sylvinsäure, Laevopimarsäure u. a. Harzsäuren).

Neben den eigentlichen Harzbestandteilen findet man auch Fettstoffe, welche vorwiegend Triglyceride sowie Stearin-, Öl- und Palmitinsäure darstellen (vgl. S. 705); ferner *Campher*, die leicht flüchtige, feste Körper (bicyclische Terpene) sind. Unter *Kautschuk* werden die Koagulationsprodukte verschiedener Milchsäfte verstanden, die aber für den Zellwandbau keine Bedeutung zu haben scheinen.

Unter die extrahierbaren Holzkomponenten sind die phenolischen Resinole [Haworthsche Lignane (vgl. S. 712)] sowie 7-Ringverbindungen (Thujaplicin, Thujicsäure) noch zu zählen.

# 2. Rindenstoffe. — Das Suberin.

Bei der Umwandlung des Bastes in die Rinde bildet sich, neben einer einhergehenden Verringerung des Gehaltes an Cellulose, Hemicellulosen und Pektinstoffen (Scharkow) *Suberin*, der wesentliche Bestandteil der Zellmembranen in den Korkzellen.

Die Rinde, ein besonders im älteren Stadium kompliziertes Pflanzenorgan, besteht anfänglich aus einer Epidermis, dem primären Rindenparenchym und dem primären Bast. In den meisten Fällen nimmt mit der Zeit die Rinde an Dicke zu, während gleichzeitig eine Korkschicht entsteht, welche die äußeren absterbenden Gewebe als Borkenschicht abscheidet. Das eigentlich Bleibende an der Rinde ist der sekundäre Bast und die Korkschicht (Abb. 23a—c).

Die 4 Hauptgewebesysteme der Rinde — der Kork, das primäre Rindenparenchym, der primäre und sekundäre Bast — sind bei den verschiedenen Rinden unterschiedlich gebaut. In der Regel bildet der Kork nur dünne Schichten; eine Ausnahme findet sich bei der Korkeiche. Das primäre Rindenparenchym ist meist reich an Calciumoxalat, welches nach dorthin bevorzugt abgegeben wird [z. B. große Zwillingskristallbildungen in der Quillajarinde *(Quillaja saponaria)*]; außerdem können darin Harzgänge, Ölschläuche usw. vorkommen. Rinden sind vielfach reich an Gerbstoffen, wie z. B. Tannin [28,6% in Hemlocktanne *(Tsuga canadensis)*, 42,3% in *Ceriops Candolleana Arn.*]. So enthält z. B. nach Kurth die Rinde von *Abies grandis* 18—29% Extraktstoffe (Fette, Öle, Wachse, Tannin und Kohlenhydrate); der Rückstand besteht zu 40,4% aus Lignin und 59,5% aus Holocellulose. Auch spezifische Stoffe sind in den Rinden sehr häufig; man braucht nur an die Alkaloide der Chinarinden, an das Glycyrrhizin der Monesiarinde, das Betulin des Birkenkorkes, Alanin der Erlenrinde u. dgl. zu denken, ferner an bereits genannte Farbstoffe (Quercetin, Morin, Brasilin, Hämatoxylin, Fisetin, Lokao).

Während, wie bereits betont, unsere Holzgewächse unter der Epidermis nur eine dünne Korkschicht aus dem Phellogen entwickeln, die nur selten dicker wird (Scheinkork), wird bei der Korkeiche eine mächtige Schicht echten Korkes gebildet. Die Rinde der Korkeiche *(Quercus suber)* enthält bis zu 20% Gerbstoffe, daneben außer Suberin (31—33%): Wachse (Cerin; s. S. 705), Fett, Cellulose (29,8—33,5%), Lignin (27,7—31,9%) und Mineralstoffe (Manganreich!)

(ORANSKIĬ). Im Durchschnitt wird der Suberingehalt der Korke mit 28—36% (58% ?) angegeben (ZETZSCHE). In geringen Mengen kommt nach KÜGLER Vanillin vor, ferner Dakacrylsäure. Die Farbe stammt von Gerbstoffrot und

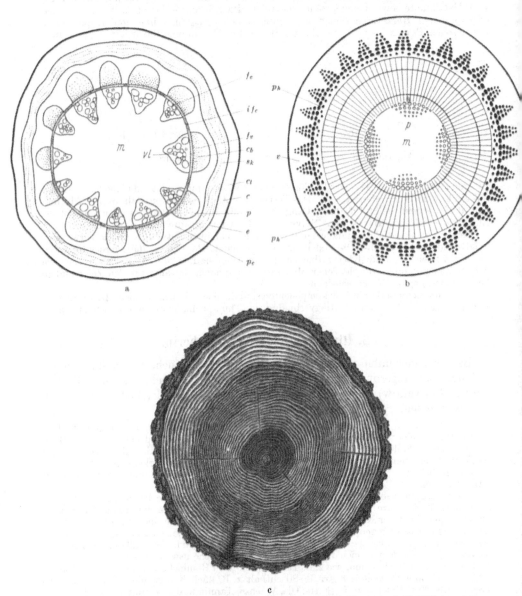

Abb. 23a—c. a Querschnitt durch einen 5 mm dicken Zweig eines Laubholzes *(Aristolochia Sipho)* nach STRAS-BURGER. *m* Mark; *fv* Gefäßbündel (*vl* Gefäßteil, *cb* Siebteil, *fc* Bündel-[Faszikular-]Cambium, *ifc* Markstrahl-cambium); *p* Außengrenze des Siebteils; *pc, e, c, cl* Regionen der Außenrinde, in dieser der Sklerenchymring *sk*. b Schematischer Querschnitt durch den dreijährigen Stammtrieb eines Laubholzes, dessen Holzkörper sich nicht mehr in einzelne Gefäßbündel, sondern nur noch in miteinander abwechselnde Holzstränge und (hier als zarte radiale Linien erscheinende) Markstrahlen gliedern läßt. *m* Mark; *p* innerste Teile der primären Gefäßbündel; *h* Jahresringe; *ph* Innenrinde (Bast); *v* Außenrinde (nach WIESNER). c Stammscheibe eines Nadelholzes mit deutlichen Jahresringen (Frühholzzone hell, Spätholzzone dunkel).

„gelbroten Säuren" (fettähnliche Farbstoffe ?). (Durch $H_2O_2$ werden letztere zerstört unter gleichzeitiger Bildung von Oxydosuberin. Die Reinigung des Suberins erfolgt nach ZETZSCHE mit Natriumsulfit).

Die Korkzellwand besteht aus 5 Schichten, von welcher die mittlere und die äußersten aus verholzter Cellulose, die beiden der Mittelschicht anliegenden Suberinlamellen aus Suberin und Cellulose bestehen. Die Cellulose ist mit dem Suberin offenbar nicht verknüpft. (Kork ist für kurzwelliges UV undurchlässig. Er zeigt hellblaue Fluorescenz.)

*Die Struktur des Suberins* ist noch nicht geklärt; sicher handelt es sich in der Hauptsache um einen Ester aus Stearin- bzw. Oxystearin-, *Phelon-* und Phloionsäure. Die aufbauenden Säuren besitzen zwei und mehr reaktionsfähige Gruppen, die zur Esterifikation und Ätherbrückenbildung befähigt sind. Es entsteht so ein räumliches Netzwerk aus Ester- und Ätherbrücken, ein hochpolymeres Produkt aus spezifischen höhermolekularen gesättigten und ungesättigten Oxyfettsäuren aufgebaut. Das Polymerisationsschema ist offenbar ähnlich den *Estoliden.* Der Polymerisationsgrad ist aber kleiner als beim Cutin. (Beim Jodabbau tritt Depolymerisation unter Jodierung ein.)

Als Zersetzungs- und Abbauprodukte von Suberin wurde erhalten (Lüscher 1936):

| | |
|---|---|
| Korksäure (Suberinsäure) | $COOH(CH_2)_6COOH$ |
| Phloionolsäure[1] (Trioxystearinsäure) | $CH_2OH(CH_2)_7(CHOH)_4(CH_2)_7COOH$ |
| Phloionsäure | $COOH(CH_2)_7(CHOH)_2(CH_2)_7COOH$ |
| *Phellonsäure* ($\alpha$-Oxybehensäure[2]) | $C_{21}H_{42}(OH)COOH$ |
| Eicosandicarbonsäure | $COOH(CH_2)_{20}COOH$ |

Suberinsäure ist keine primäre „Korksäure". Durch Erhitzen dieser im $CO_2$-Strom bei 140° tritt eine Rückpolymerisation zu einem suberinähnlichen Stoff ein. Von Jensen wurde noch eine olefinische Dioxymonocarbonsäure von der Bruttoformel $C_{18}H_{34}O_4$ isoliert. An anderen ungesättigten Säuren sind Suberol-, Corticinsäure und Oxyölsäure [letztere im Holunderkork (Scurti)] genannt worden.

Der Kork der Douglasfichtenrinde *(Pseudotsuga taxifolia)* wurde neulich von Hergert (1952) eingehend studiert. Etwa 40% der Korkzellsubstanz war extrahierbar und die Extrakte — vorwiegend Korkwachse — ergaben nach der Verseifung Lignocerinsäure und -alkohol, Ferulasäure, Hydroxypalmitinsäure, Dihydroquercetin und Glycerin. Daneben fanden sich Tannin, Phlobaphene, Zucker, Phytosterol und nichtidentifizierbare Säuren und phenolische Substanzen. Der nicht extrahierbare Teil war bis auf den hochmolekularen Kohlenhydratanteil in Alkali oder Salzsäure-Dioxan löslich. Neben einer phenolischen Säure unbekannter Konstitution fanden sich 11-Hydroxylaurinsäure und Undecansäure neben unidentifizierten phenolischen Säuren.

Pathologische korkige Ablagerungen an der Kartoffel sind von Phenolen begleitet (Hill 1939).

# 3. Wachse und Cutin.

Wachs und Cutin sind Ester. Solche sind bekanntlich in der Natur weit verbreitet; hierher gehören unter anderem die Fette, Wachse, Lecithine und Sterinester. Während die Fette und Öle Ester einbasischer Fettsäuren mit gerader Kohlenstoffanzahl (die $C_{18}$-Säuren sind am häufigsten) mit Glycerin sind, wird in den Wachsen das Glycerin durch höhere einwertige, selten zweiwertige Alkohole vertreten. Die gleichen Fettsäuren können auch in pflanzlichen Fetten auftreten.

*Pflanzenfette* und Öle sind ein- und mehrsäurige Triglyceride; daneben finden sich, offenbar durch Zersetzung, freie Fettsäuren. In sehr geringer Menge finden sich noch Phytosterine, Alkohole (z. B. Myricylalkohol und im Samen von *Brucea sumatrana* der cyclische Alkohol $C_{20}H_{39}OH$), Kohlenwasserstoffe [n-Eikosan (im Lorbeerfett), Octakosan, Spinacen] und Farbstoffe (Lipochrome).

---

[1] Auch im Cutin der *Agave americana.*

[2] Nach neueren Untersuchungen von Jensen $\omega$-*Hydro*xybehensäure.

Palmöl z. B. besteht zu $^1/_3$—$^1/_2$ aus Ölsäure, der Rest ist Palmitinsäure. Etwa 1% der gebundenen Fettsäuren verteilt sich auf Stearinsäure, Linolsäure u. a. Die Kakaobutter setzt sich zusammen aus: 24,9% Oleodistearin, 54,7% Palmitodiolein und 20,3% Oleo-palmitostearin. Daneben finden sich geringe Mengen von Palmitodi- und Tristearin, ferner Glyceride von Arachin- und Linolsäure. Cocosöl ist reich an Laurinsäure.

Eine häufig anzutreffende Fettsäure mit weniger als 18 C-Atome ist die Palmitinsäure; längerkettige finden sich unter anderem vornehmlich im Erdnußöl (Arachinsäure, Cerotin-säure, Lignocerinsäure). Die kurzkettigste ungesättigte Fettsäure ($C_{10}H_{18}O_2$) kommt im Pollen von *Ambrosia artemisifolia* vor; eine langkettige Olefincarbonsäure ist die Erucasäure des Senföls. Eine Oxyfettsäure kommt z. B. im Ricinusöl vor (Dioxy-stearinsäure), während eine cyclische Säure die Chaulmoograsäure ist (Samen von *Hydnocarpus*-Arten).

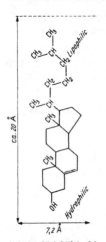

Abb. 24. Molekülstruktur von Cholesterin. Molekül-länge und -breite ist nach den Messungen von Bernal eingezeichnet. Die hydrophile Seite wird durch die OH-Gruppen gebildet.

Bei den Wachsen kann man nun zwischen einfachen aliphatischen Wachsen und solchen, die aliphatische Hydro-carbonsäuren, Ketonanteile oder ungesättigte Komponenten (z. B. n-Triacontanol) enthalten, unterscheiden.

Neben den eigentlichen Wachsestern finden sich im Wachs stets noch freie Säuren, freie Alkohole und häufig Kohlen-wasserstoffe unbekannter Konstitution.

Schließlich kann man noch von oft besonders gearteten Wachsen der Fruchtcuticula und der Samenschalen sprechen (hierin finden sich z. B. ungesättigte Säuren und Alkohole), ferner von Blattwachsen und den Wachsen, die in der Zelle dispergiert sind. Cutinisierte und verkorkte Membranen enthalten cyclische Wachse mit Cerin und Friedelin. Nach Lüscher (1936) u. a. handelt es sich dabei um Terpen-alkohole mit einem Ätherbrückensauerstoff.

Das Carnaubawachs, das Wachs der Fächerpalme (*Copernicia cerifera*) besteht z. B. in der Hauptsache aus Cerotinsäuremyricylester $C_{25}H_{51}COOC_{31}H_{63}$ (neben Estern der geradzahligen Säuren $C_{18}$—$C_{30}$), dem Carnaubasäure $C_{23}H_{47}COOH$, Cerotin-säure $C_{25}H_{51}COOH$, höhere Alkohole (Ceryl- und Myricylalkohol) und Kohlen-wasserstoffe beigemengt sind. Im Zuckerrohrwachs wurden unter anderem den Sterolen verwandte Substanzen identifiziert (Sitosterol, Stigmasterol u. a.). Baumwollwachs macht die Baumwollfaser wasserabstoßend; es ist trotz der hohen Verseifungszahl schwer verseifbar. Technische Bedeutung besitzen noch das Candelillawachs (*Pedilanthus Pavonis*) und Fibrawachs (Schilfwachs). Im Wachs der Buchenrinde fand Clotofski einen Ester eines $C_{20}$-Alkohols mit einer $C_{20}$-Säure. Ähnlich den Rindenwachsen ist auch das Korkwachs gebaut. Neben der Arachinsäure, Cerotinsäure, α-Oxyarachin- und α-Oxybehensäure fand Zetzsche Cerin, Friedelin und $C_{21}$- und $C_{24}$-Alkohole.

Die hydrophilen Gruppen im Wachs sind — ähnlich wie bei den Fetten — abgeschirmt bzw. maskiert; die Endgruppen sind ausgesprochen hydrophobe Gruppen, nicht reaktionsfähig und somit auch nicht zur Polymerisation befähigt. Moleküle mit hydrophilen und hydrophoben Endgruppen sind z. B. die Sterole Abb. 24), Proteine, Phosphatide und Cutin, die gegebenenfalls Bindeglieder zwischen hydrophilen und hydrophoben Wandsubstanzen bilden können.

Wachse als niedermolekulare Stoffe sind im allgemeinen gut kristallisiert und geben scharfe Röntgeninterferenzen, wobei vielfach auch große Netzebenen-abstände beobachtet werden können (z. B. 60 und 83 Å bei einigen Wachsen nach Hess; vgl. auch Tabelle 7 und Abb. 25).

Tabelle 7. *Röntgenkleinwinkelinterferenzen einiger Wachse nach* TREIBER *und* SEKORA
(unveröffentlicht).

Bienenwachs . . . . . . . . . . . . . . . . . . 72 ± 1 Å
Wachs von *Saccharum officinarum* . . . . . . . 78 ± 1 Å
Wachs von *Echeveria glauca × imbricata* . . . . . . 44 ± 1 Å
Wachs von *Picea Omorica* . . . . . . . . . . . 40 ± 1 Å

*Estolide.* BOUGAULT und BOURDIER (1908) fanden bei der Untersuchung bzw.
Verseifung von Fichtennadelwachs $\omega$-Hydroxyfettsäuren (z. B. Juniperinsäure,
die in Form des Oxyfettsäurelactons[1] zu den
vegetabilischen Moschusriechstoffen gehört), die
durch gegenseitige Veresterung Polyester (Estolide)
bilden.

*Farbwachse.* Auch die Farbwachse besitzen
eine nahe Verwandtschaft mit den Fetten, wie
sich durch Hydrierung zeigen läßt. Es sind Ester
von Polyenalkoholen (hydroxylhaltige Carotinoide)
mit Fettsäuren (KARRER 1948). Das Physaliën
der Judenkirsche *(Physalis Alkekengi)* und das
Farbwachs des Bocksdorns *(Lycium halimifolium)*
ist der Dipalmitinsäureester des Zeaxanthins
$(C_{15}H_{31}-COO-C_{40}H_{54}-OOC-C_{15}H_{31})$. Helenien
erweist sich als Luteindipalmitat und im Stief-
mütterchen *(Viola tricolor)* findet sich ein Viola-
xanthinester.

Näher untersucht ist das Farbwachs der Paprika-
frucht *(Capsicum annum)*. Man fand darin: Glycerin,
Wachsalkohole, Capsanthin, Myristinsäure, Palmitinsäure, Stearinsäure, Capsorubin, Zeaxanthin, Lutein,
Kryptoxanthin, Carnaubasäure und Ölsäure.

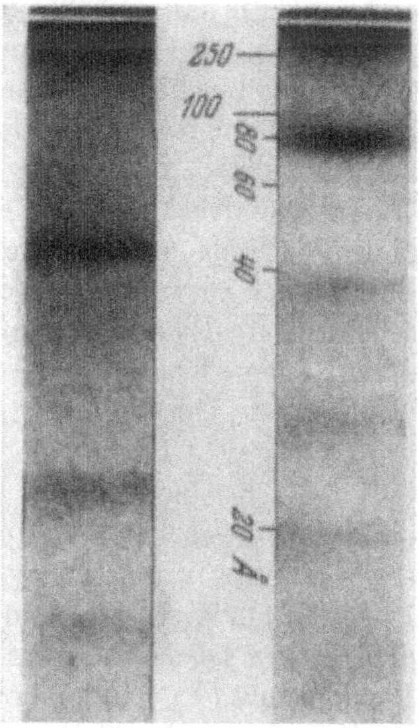

*Cutin.* Als Deckschicht der oberirdischen Organe besitzt das primäre Hautgewebe eine Cuticularschicht. Nach den optischen Untersuchungen
von M. MEYER (1938) (vgl. Abb. 26) scheinen die
Cuticular-Wandsubstanzen — Cellulose, Pektin,
Cuticularwachs und Cutin — vielfach mehr oder minder ausgebildete Zonen oder
Schichten zu bilden, die zusammenhängen, wobei nach einer älteren Auffassung

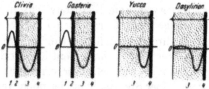

von FREY-WYSSLING, die neuerdings durch
die Arbeiten von HÄRTEL eine Stütze
findet, Cutin das Bindeglied zwischen
der hydrophilen Cellulose und dem hydrophoben Wachs spielen dürfte. Die Cuticularschicht besteht demnach aus einem
nichtschmelzbaren Gerüst von Cellulose
und Cutin und tangential orientierten
plättchenförmigen Cutinwachsmicellen,
die im cellulose- und pektinhältigen Teil

Abb. 25. Röntgenkleinwinkelaufnahme
von links: Wachs der *Echeveria glauca
imbricata*, rechts: Wachs von *Saccharum
officinarum*. (Reflexe höherer Ordnung
sind schwach erkennbar.)
(Nach TREIBER und SEKORA.)

Abb. 26. Aufbau der Cuticularschicht nach
M. MEYER. In das Schema ist die relative Stärke
der + - bzw. — -Doppelbrechung eingezeichnet.
*1* Celluloseschicht; *2* isotrope Pektinschicht;
*3* cutinisierte Wandschicht; *4* isotrope Cuticula.

die positive Doppelbrechung der Cellulose kompensieren können. Die äußerste
Schicht wird häufig durch einen Film submikroskopischer Dicke von reinem
Cutin gebildet, welches somit als unabhängige Wandsubstanz auftreten kann,
aber wahrscheinlich dann mit dem Cutin der Cuticularschicht nicht identisch

---

[1] $CH_2 \cdot (CH_2)_{14} \cdot CO.$
$\lfloor \underline{\quad} O \underline{\quad} \rfloor$

ist. In einigen Fällen kann auch die Cuticularschicht fehlen (*Bromeliaceen*). Eine isotrope Cuticula aus reinem Cutin soll z. B. an der *Agave americana* zu beobachten sein (LEGG und WHEELER), und kräftig ist beispielsweise die Cuticula bei *Myrtus pinnata* ausgebildet. Hingegen ist die Frage strittig, ob das Baumwollhaar eine individualisierte Cuticula besitzt. Nach den elektronenmikroskopischen Untersuchungen von KLING und MAHL (1951) besteht keine scharfe Trennung zwischen der Wachs-Pektinschicht — die für die Seidigkeit und Unbenetzbarkeit der Faser verantwortlich ist — und der Primärwand, die für einen Überzug aus reinem Cutin spräche.

Während die Cuticularwachse praktisch keine UV-Absorption aufweisen, absorbiert Cutin stark [2900—3000 Å (WUHRMANN-MEYER 1941), vgl. auch METZNER (1930), KÖHLER (1904)]. Jedoch ist die Absorption in der praktisch vorliegenden sehr dünnen Schicht keineswegs so groß, daß nach METZNER von einem Strahlungsschutzmittel gesprochen werden darf. Cutin fluoresciert intensiv goldgelb bis grüngelb.

Zufolge der Unlöslichkeit und optischen Isotropie wird für den Bau des Makromoleküls ein räumliches Netzwerk angenommen, wobei ähnlich wie im Suberin der Zusammenhalt durch Ester- und Ätherbrücken erfolgen dürfte. Cutin scheint saure Eigenschaften zu haben und somit ein hochpolymeres Gerüstanion darzustellen. Ferner dürfte es über freie Hydroxylgruppen verfügen, da es acetylierbar ist. Durch Kochen mit Glycerin erfolgt eine langsame Depolymerisation, während starke Laugen es zu Cutinfettsäuren verseifen. Der C-Gehalt liegt bei 69%. Nach einer älteren Anschauung, die gegenwärtig wohl nicht aufrechterhalten werden kann, sollte es sich beim Cutin um einen Nonylsäure-Cetylester sowie Caprinsäure-octadecylester handeln. Bisher wurden von ZETZSCHE und LÜSCHER nur gesättigte und ungesättigte Oxymono- und -dicarbonsäuren gefunden, ferner Oleocutin- und Stearocutinsäure. LEGG und WHEELER isolierten Phellonsäure ( ? ?), Phloionolsäure, Cutin-($C_{20}H_{50}O_6$) und Cutininsäure ($C_{26}H_{44}O_6$). Nach anderen Angaben wird Phellonsäure nie gefunden im Gegensatz zum Suberin. Nach GÉNAN DE LAMARLIÉRE sollen noch aldehydartige Stoffe anwesend sein.

Die Cuticula entsteht durch Erguß der Vorstufen nach außen, worauf, ähnlich wie beim Firnis, die Polymerisation erfolgt. Cutinartiger Natur scheint auch das Haftsekret der Haftballen zu sein, welches unter anderem wie Cutin gelbe Fluorescenz zeigt (HÄRTEL 1950).

Über der Cuticula kann es zur Ausbildung einer Wachsschicht aus den eingangs erwähnten Pflanzenwachsen kommen. In einigen Fällen entstehen Ölüberzüge *(Malus coronaria)*, Flavon-, Harz- und Firnisüberzüge.

Während zwischen Cutin und Suberin — von denen es möglicherweise verschiedene gibt — im wesentlichen offenbar nur graduelle Unterschiede bestehen, ist das noch viel schwerer verseifbare *Sporopollenin* stärker von beiden differenziert. Über Sporopollenin ist noch weniger bekannt (polyterpenartige Substanz ?).

## 4. Das Lignin.

Lignin ist mit der Cellulose des Holzes, Stroh usw. vergesellschaftet; der Ligningehalt steigt mit dem Grade der Verholzung und beträgt beim Holz durchschnittlich 19—30% (s. Tabelle 9). Nach GJOKIE (1895) und LINSBAUER (1899) wird Lignin bei Flechten, Moosen und Pilzen nicht beobachtet.

Als besonders ligninreich gilt die gealterte und von der Lignifizierung zuerst betroffene, praktisch cellulosefreie, isotrope und poröse Mittellamelle (Abb. 27), für die BAILEY (1936) ~71% Ligningehalt angibt. Nach JAYME, BAILEY, KERR,

Preston u. a. enthalten auch primäre und sekundäre Zellwände im Gegensatz zur Auffassung von Lüdtke (1931) Lignin; einen auffallend hohen Ligningehalt zeigen beispielsweise innere Wandschichten von Bambusfasern. Auf alle Fälle nimmt aber der Ligningehalt von außen nach innen ab und die Tertiärlamelle scheint stets frei von Lignin zu sein. Eine ähnliche Abnahme wurde auch in Richtung von der unteren Stammpartie zum Wipfel vermutet und Unterschiede im Ligningehalt zeigen sich zwischen Früh- und Spät- bzw. Zug- und Druckholz (Höpner). In zeitlicher Hinsicht werden bei neugebildeten Geweben zuerst die peripheren Randzonen, dann Parenchymzellen, Bastzellen usw. von der Lignifizierung erfaßt. Die Elektronenmikroskopie der Douglasfichte *(Pseudotsuga taxifolia)* (s. Abb. 15c) zeigte eindrucksvoll, wie das Lignin zwischen die

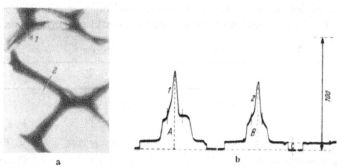

a               b

Abb. 27 a u. b. a Querschnitt von *Picea excelsa*, photographiert mit monochromatischem Licht von 2800 Å nach Lange (1950). Mitaufnahme eines Schwärzungskeils erlaubt eine quantitative Photometrierung. b Photometerkurven längs der Marken 1 bzw. 2, welche die Ligninverteilung in Mittellamelle und Zellwände erkennen lassen.

Fibrillen der Cellulose hineingelegt ist, so daß das System Verbundbaucharakter erhält. Im Sinne des Frey-Wysslingschen Modells stellen die Cellulosefibrillen die zugfesten Stäbe und das Lignin das druckfeste Füllmaterial dar. Das Lignin bleibt bei einer Herauslösung der Cellulose als kohärentes Netzwerk zurück, welches dann Stäbchendoppelbrechung zeigt. Elektronenmikroskopische Aufnahmen des Lignins von Rånby lassen bei den derzeitigen Auflösevermögen keine definierte morphologische Struktur erkennen. Zur erwähnten mechanischen Aufgabe dürfte auch noch eine Schutzwirkung vor chemischen und physikalischen Einflüssen hinzukommen. Die Cellulosefasern sind aber offenbar nicht nur mechanisch im Lignin eingebettet, sondern es dürften wohl auch lokale chemische Verknüpfungen zwischen den äußeren Celluloseketten [und Xylan (Rawamura) sowie Zuckern (Traynard)] und dem Lignin bestehen (vgl. Höpner 1941; Purves 1946, Richtzenhain u. a.). Nähere Kenntnisse sind hier von eminent technischem Interesse, weil diese Frage für den partiellen Holzaufschluß eine Rolle spielt und die Durchführbarkeit einer Trennung der Holocellulose vom Lignin davon beeinflußt wird. Eine Stütze dafür, daß irgendwelche der 9 theoretischen Möglichkeiten einer Lignin-Cellulosebindung realisiert sind, sind einerseits isolierte Cellulose-Ligninkomplexe[1] aus Sulfitablauge (Friese und Mitarbeiter 1937, im Gegensatz zu Lautsch), andererseits Ergebnisse röntgen- und lichtoptischer Untersuchungen [Mukherjee (an Jute), Lange] sowie Untersuchungen der Kinetik der Bisulfitreaktion (Hägglund 1936).

Der Nachweis der Verholzung — vom chemischen Standpunkt identisch mit dem Nachweis von Lignin — stützt sich auf Färbungen, Fluorescenz, Verdickungen

---

[1] Aaltio beschreibt jüngst einen Lignin-Pentosankomplex (im wesentlichen bestehend aus Xylose, Arabinose, Galactose, Glucose und Uronsäure) im Verhältnis L:P = 0,85 und 0,64 aus Aspenholz. Der Charakter der Lignine in beiden Anteilen ist verschieden; das Lignin des ersten Komplexes ist identisch mit dem der Mittellamelle.

und funktionelle Beschreibungen der untersuchten Zellwände, also vornehmlich anatomische Untersuchungen, und auf rein chemische Methoden, wie z. B. Abbaumethoden, sowie auf physikalisch-chemische Untersuchungen [UV- und UR-Messungen (vgl. Abb. 28)]. Zu den gebräuchlichsten Farbnachweisen [eine generelle Zusammenstellung findet sich im Buch von BRAUNS (1952) und HÄGGLUND (1951), wo auch eine eingehende Diskussion über die Ursachen der Färbung sich vorfindet] gehören die RUNGEsche Anilinsulfatreaktion und die WIESNERsche Phloroglucin-Salzsäurereaktion, Reaktionen, die auf die mengenmäßig sehr geringen Coniferylgruppen im Lignin zurückgehen (geringfügige Veränderungen am Lignin können die Farbreaktion zum Verschwinden bringen); es sind dies jene auch für den Chemiker wichtigen Holzreaktionen, bei denen eine bestimmte Komponente des Lignins mit einem an sich ungefärbten Reagens zu einem Farbstoffkomplex reagiert. Andere Holzfarbreaktionen, wie die MÄULEsche Reaktion, die Chlorbehandlung und die JAYMEsche Reaktion beruhen meist auf oxy-

Tabelle 8. *Überblick über sog. „Farbreaktionen des Holzes".*

| Reagens | Farbe | Autor |
|---|---|---|
| a) Organische Farbreagentien: | | |
| Anilin . . . . . . . . . . . . . . . . . | gelb | F. F. RUNGE |
| o-, m-, p-Chloranilin } | orangegelb | E. COVELLI |
| o-, p-Aminophenol } . . . . . . . . . | | |
| Dimethyl-p-phenylendiamin . . . . . . . . . | rot | C. WURSTER |
| Benzidin . . . . . . . . . . . . . . | rotgelb | S. DUKELSKY |
| Pyrrol . . . . . . . . . . . . . . . | rot | A. IHL |
| Phenol . . . . . . . . . . . . . . . | grünblau | F. F. RUNGE |
| Guajacol . . . . . . . . . . . . . . | gelbgrün | F. CZAPEK |
| Phloroglucin . . . . . . . . . . . . . . . | violettrot | J. v. WIESNER |
| Orcin . . . . . . . . . . . . . . . . . | dunkelrot | E. O. v. LIPPMANN |
| Barbitursäure . . . . . . . . . . . . . . | gelb | T. PAVOLINI |
| [ferner geben Aldehydreagentien wie Hydroxylamin und Semicarbazid (UNGAR) sowie Dimedon (ADLER) Farbreaktionen] | | |
| b) Anorganische Farbreagentien: | | |
| Eisenchlorid und Kaliumferricyanid . . . . . . | grünblau | C. F. CROSS und E. J. BEVAN |
| Chlor (in Gegenwart von Feuchtigkeit, . . . . . | gelb | A. PAYEN |
| nachbehandelt mit Ammoniak oder | | |
| Natriumsulfit) . . . . . . . . . . . . . . | rot | |
| Halogenwasserstoffsäuren . . . . . . . . . } | grün | E. UNGAR |
| Schwefelsäure >75% . . . . . . . . . . . } | | F. E. BRAUNS |
| Salzsäure in Methanol . . . . . . . . . . . | rot | I. H. ISENBERG und M. A. BUCHANAN |
| Völlig { Kobaltthiocyanat . . . . . . . . | blau | P. CASPARIS } |
| unspezifisch { Vanadiumpentoxyd in Phosphorsäure | gelbbraun | J. GRUESS } |
| { Zinkchlorid . . . . . . . . . . . | gelb | A. W. SCHORGER } |
| Stickstofftetroxyd . . . . . . . . . . . . . | gelb | G. JAYME } |
| nachbehandelt mit Triäthanolamin . . . . . . | braungelb | und M. HARDERS-STEINHÄUSER } |

C. MÄULE benutzt eine neutrale Kaliumpermanganatlösung und behandelt anschließend die Probe mit verdünnter Salzsäure und Ammoniak. Hartholz gibt eine rote Farbe, während Weichhölzer nur lichtbraun verfärbt werden.

R. COMBES behandelt die Probe zunächst mit Bleiacetatlösung und hierauf mit einer Zinkoxydsuspension — eine Vorbereitung, die nach ADLER und ELLMER unterbleiben kann —, worauf diese nach dem Waschen in Schwefelwasserstoffwasser eingebracht wird. Eine weitere Behandlung mit 75%iger Schwefelsäure ergibt eine rotviolette Farbe. (Vgl. auch die Reaktionen von SEIFERT sowie KÜRSCHNER u. SCHWEIZPACHER.)

dativen Veränderungen des Lignins. Die Farbreaktion entsteht dann durch sekundäre Kondensationsvorgänge. Eine Zusammenstellung der Farbreaktionen bringt Tabelle 8. Für den Nachweis der Lignineinlagerung, speziell der Primärverholzung, sind alle derzeitigen Methoden noch mit Unsicherheiten behaftet (Kratzl 1953).

Molekulargewicht und Konstitution der Ligninsubstanzen sind noch nicht restlos geklärt, während die Frage, ob das genuine Lignin aromatischer Natur ist, wohl als in diesem Sinne entschieden zu betrachten ist. Die oft geäußerte Vorstellung, daß Lignin aliphatischer Herkunft sei und erst bei der Aufbereitung durch chemische Mittel aus pektinartigen und den Zuckern nahestehenden Substanzen entstünde, ist heute nicht mehr aufrechtzuerhalten. Hingegen scheinen z. B. beim alkalischen Aufschluß weitere, alkoholunlösliche „Lignine" aus Hemicellulose zu entstehen (Prey 1953). Wenn man von älteren chemischen Untersuchungen, wie Druckhydrierung des Holzes (Goddard)

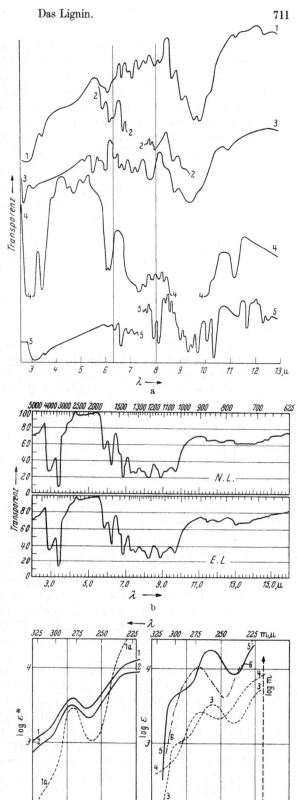

Abb. 28a—c. a Ultrarotaufnahmen von Kurve 1 4 μ-Radialschnitt von Tanne; Kurve 2 4 μ-Radialschnitt von Fichte; Kurve 3 4 μ-Radialschnitt von Buche; Kurve 4 Hydratcellulosefilm; Kurve 5 Coniferin. (Nach Kratzl und Tschamler.) b Ultrarotspektrum von Eichenlignin. *N.L.* Nativlignin; *E.L.* Nordlignin (enzymatisch isoliertes Lignin nach Kudzin und Nord). (Nach Nord.) c UV-Absorptionskurven von Kurve 1 Nativlignin von Tanne. (Nach Brauns.) Kurve 2 Ba-Lignosulfonat von Tanne. (Nach Brauns.) Kurve 1a Nativlignin von Western hemlock. (Nach Goldschmid.) Kurve 3 Fichtencambialsaft. (Nach Treiber.) Kurve 4 Fliedercambialsaft. (Nach Treiber.) Kurve 5 Coniferin. (Nach Hilmer.) Kurve 6 4-Hydroxy-3,5-dimethoxypropenylbenzol. (Nach Pearl.)

absieht, so konnte kürzlich Lange (1944) auf UV-spektrographischem Wege am mikroskopischen Schnitt mit der Caspersson-Apparatur den direkten Beweis erbringen, daß in der Mittellamelle ein Stoff vorhanden ist, der bei 2800 Å [$\sim 3570\,\nu'$ (mm$^{-1}$)] ein Absorptionsmaximum besitzt, welches — gleich wie die Messung des Brechungsexponenten des genuinen Lignins ($n_D = 1,6$) — auf das Vorliegen aromatischer Substanzen hinweist (vgl. Abb. 27). Auch Kratzl (1953) konnte an schwinggemahlenem Holz und Holzschnitten im UR aromatische C=C-Dehnfrequenzen beobachten (Abb. 28a). Die Existenz aromatischer Substanzen in der Pflanze ist ja keineswegs selten oder verwunderlich, sind doch

Tabelle 9. *Ligningehalte einiger Hölzer. Mittelwert nach verschiedenen Methoden, zum Teil auch verschiedenen Standorten nach* König *(1919).*

| | | | |
|---|---|---|---|
| Tanne *(Abies)* | . . . 28,8% | Pappel *(Populus)* | . 21,7% |
| Kiefer *(Pinus)* | . . . 29,9% | Esche *(Fraxinus)* | . . 24,6% |
| Birke *(Betula)* | . . . 24,6% | Erle *(Alnus)* | . . . . 24,2% |
| Buche *(Fagus)* | . . . 22,9% | Weide *(Salix)* | . . . 25,0% |

[Weizenstroh *(Triticum)* 16,1%]

auch andere Aromaten in großer Zahl vorhanden, und zwar außer Coniferin und Syringin Tannine und kondensierte Tannine, Catechine, Flavone und Anthocyane, Lignane (natürliche kristallisierte phenolische Substanzen [z. B. 1-Conidendrin] und Phenolharze, die dem Lignin nahestehen[1]) u. dgl. (vgl. auch S. 703).

Die Schwierigkeiten in der Erforschung des Lignins liegen darin, daß es bis jetzt nicht möglich ist, Lignin in unveränderter Form rein zu isolieren, daß eine glatte Hydrolysierbarkeit bzw. Aufspaltbarkeit in entsprechende Einheiten, wie beispielsweise bei den Proteinen, fehlt bzw. daß Abbaureaktionen sehr verschiedene Ligninbruchstücke und Bausteine liefern, so daß zunächst an einem geordneten Bauprinzip und der Existenz von Grundbausteinen gezweifelt werden muß — zumindest an einer kettenförmigen Struktur — (nach den älteren Vorstellungen von Cross und Bevan, Dorée und Barton und Fuchs sollte Lignin eine hochkondensierte Substanz mit Mehrkerngerüst sein) und daß Lignin sich letztlich nicht durch irgendwelche Eigenschaften ganz scharf charakterisieren läßt. Sicher gibt es zumindest 2 Gruppen von Ligninen, die Hartholz- und Weichholzlignine (letztere mit niederem Methoxylgehalt) bzw. den Fichten- und Buchentyp (Wedekind 1937) und möglicherweise sind im nativen Holz 2 sich verschieden verhaltende Anteile des Lignins vorhanden [Protolignin I und II nach Wacek (1953)] (vgl. Lignin A und B nach Erdtman). Auch von Ritter (1925, 1934) ist bereits eine Differenzierung in Lignin der Mittellamelle und amorphes Lignin der Zellwand vorgeschlagen worden und Freudenberg (1936) sprach von einem unlöslichen, geformten oder gebundenen „Lignin" und einem in organischen Solventien und Alkali löslichen „ungeformten Lignin".

Zur chemischen Kennzeichnung der Ligninpräparate dienen neben der Elementarzusammensetzung vornehmlich der Gehalt an Methyl-, Methoxyl- und Hydroxylgruppen. Die verschiedenen Ligninpräparate enthalten 61—66% C, 5—6% H und etwa 30% O; eine „Bruttoformel" eines Grundrestes würde im Mittel etwa lauten: $C_9H_{8,1}O_{2,4}(OCH_3)_{0,9}$, wobei vor allem das Wasserstoffdefizit bemerkenswert ist. Im einzelnen kommen auf ein $C_9$-Skelet (Fichtenholztyp) etwa 6,7 H-Atome, 0,6 phenolische Hydroxylgruppen, 0,9 aliphatische (vorwiegend primäre) OH-Gruppen, 0,9 Methoxylgruppen, 0,2 Carbonylsauerstoffe, 0,3 aromatische und 0,5 aliphatische Äthersauerstoffatome.

Über die Stabilität von Lignin ist noch wenig bekannt; sicher setzen mit der Zeit Veränderungen ein, offenbar auch mit zunehmendem Alter der Pflanze.

---

[1] Vgl. Haworth 1936.

Lignin erweist sich als optisch inaktiv. Die Dichte beträgt etwa 1,4, die Verbrennungswärme etwa 6,3 cal/g. Natives Lignin erweicht je nach den Versuchsbedingungen zwischen 80 und 120° und schmilzt unscharf zwischen 140—150°.

Die Mittel, die Lignin herauslösen, verändern es wahrscheinlich, wie schon aus gewissen Differenzierungen zwischen den einzelnen Präparaten hervorgeht. In manchen Fällen treten neue Gruppen in das Molekül ein; bei der Alkoholyse z.B. Alkohol, während beim Sulfitaufschluß eine Sulfurierung stattfindet. Man unterscheidet nach den Gewinnungsmethoden: *Säurelignin* (Schwefelsäurelignin nach KLASON, Salzsäure- oder WILLSTÄTTER-Lignin), *Cuproxamlignin* nach FREUDENBERG, *Alkalilignin* (LANGE), *Chloritlignin, Organosolvlignine* (FREUDENBERG, KLEINERT u. a.), zu denen auch das BRAUNS- und HOLMBERGsche Nativlignin zählen. Schließlich gibt es neben anderen Extraktionsmitteln [organische Säuren-(PAULY), Amine, Schwefelverbindungen] noch biologische Prozesse, die die Cellulose zerstören und Lignin zurücklassen (NORD-Lignin) [z. B. Pilz *Merulius lacrymans* (vgl. NORD 1952)].

In der Technik, wo der Herauslösung von Lignin bei der Zellstofferzeugung die größte Bedeutung zukommt, wird vorzugsweise mit Calciumbisulfit (zum Teil auch bereits mit Natriummonosulfit) oder Ätznatron und Ätznatron mit Natriumsulfid gearbeitet. Im ersten Falle, beim Sulfitprozeß, wird Lignin in Form der Lignosulfonsäure herausgelöst, beim Natron- und Sulfatverfahren als Alkalilignin bzw. Schwefellignin (HÄGGLUND 1953, ENKVIST). Für Bariumlignosulfonat wird die Bruttoformel $C_{40}H_{44}O_{17}S_2Ba$ bis $C_{43}H_{50}O_{18}S_2Ba$ angegeben. Die Sulfonierung erfolgt am $\alpha$-C-Atom der Seitenkette (KRATZL, ADLER).

$$-\langle = \rangle-CH-C-$$
$$\underset{SO_3H}{\big|}$$

Die Unlöslichkeit des Lignins — wenn nicht gewisse chemische Einwirkungen bzw. Abbauvorgänge vorausgehen — und die damit verbundenen Schwierigkeiten einer Isolierung sind die Ursache dafür, daß wir über die wahre Molekülgröße nichts wissen. Das ganze Verhalten spricht jedoch dafür, daß Lignin eine hochmolekulare Substanz darstellt, die möglicherweise dreidimensional vernetzt ist. Vielfach wird aber auch die Vermutung ausgesprochen, daß Lignin gar keine richtig hochmolekulare Substanz sei. Für die löslichen isolierten Lignine werden Molekulargewichte von 800—4000 gefunden; für die Bruchstücke werden Äquivalent- bzw. Molekulargewichte zwischen 356 und 945 angegeben. Nach HIBBERT und BRAUNS beträgt das Molekulargewicht der nativen Baueinheit etwa 840. GRALÉN (1946) findet an Thioglykolsäurelignin in der Ultrazentrifuge ein *mittleres* Molekulargewicht von etwa 7000; MIKAWA an Alkalilignin ein mittleres Molekulargewicht $\sim 500$. Das Gesamtmolekulargewicht des Pflanzenlignins liegt nach den verschiedenen divergierenden Angaben zwischen dem eines einzelnen Bausteins und dem Wert 11000—34000.

Wird das Lignin als ein Netzwerk von Hauptvalenzketten aufgefaßt — wobei in verschiedenen Pflanzen und in verschiedenen Altersstufen wahrscheinlich verschiedene Gruppen zusammengehängt sind —, das als Kittsubstanz für Cellulose dient und offenbar chemisch hie und da mit peripheren Celluloseketten verknüpft ist, so verliert die Frage nach der „Einheitlichkeit" ihren Sinn und man kann mit dem Begriff „Molekül" bzw. Molekulargewicht ebensowenig weiterkommen als etwa beim vulkanisierten Kautschuk.

Röntgenoptische Aufnahmen zeigten vorwiegend Bilder amorpher Substanzen (auch manche niedermolekulare Modellkörper sind amorph), sonst ergaben sich Aufnahmen ähnlich denen makromolekularer Stoffe.

*Ligninbausteine.* Vor mehr als 50 Jahren hat KLASON (1897, 1920) die Vermutung ausgesprochen, daß der Coniferylalkohol die Stammsubstanz des Lignins

sei und hat somit die erste Vorstellung über die chemische Struktur eines Lignin-
bausteines gegeben. Seiner Theorie zufolge entsteht Lignin durch Kondensation
von Coniferylalkohol, Coniferylaldehyd oder α-Hydroxyconiferylalkohol zu größeren Aggregaten. Die Spaltstücke des oxydativen und hydrierenden Abbaues
erwiesen sich nach dem Verfahren von FREUDENBERG als Veratrumsäure, Iso-
hemipinsäure, Dehydrodiveratrumsäure, nach der Methode von FREUDENBERG
und LAUTSCH (Oxydation mit Nitrobenzol) als Vanillin neben Phenolcarbonsäuren und nach der Druckhydrierung nach HARRIS und ADKINS als Derivate des
Phenylpropans. Auf Grund solcher Ergebnisse sind nach FREUDENBERG (1938)
die Einheiten des Fichtenlignins Derivate des 3,4-Dioxyphenylpropans.

Das Buchenlignin enthält noch eine weitere Komponente, einen aromatischen
symmetrischen Dimethyläther vom Typus des Sinapinalkohols bzw. der Syringa-
säure.

Ähnliche Vorstellungen entwickelten auch HIBBERT (Guajacylpropanderivate)
und ERDTMAN (Existenz auch dimerer Formen des Phenylpropanbausteins vom
Conidendrin- und Diisoeugenoltyp sowie des Phenylcumaransystems).

Die allgemeine Erkenntnis ist nun wohl die, daß Lignin ein polymeres Produkt
ist (offenbar 3-dimensional vernetzt), welches Propylbenzoleinheiten mit Hydroxyl-
und Methoxylgruppen enthält, vornehmlich in Guajacyl- oder Syringylanordnung.
Die offenbar ziemlich niedermolekularen löslichen „Lignine" sind wahrscheinlich
als Fragmente eines solchen 3-dimensionalen Gerüstes aufzufassen und ihre Hetero-
genität wird so verständlich.

Verknüpft mit der Frage nach den Ligninbausteinen ist selbstverständlich die
Frage nach der biogenen Synthese bzw. zunächst diejenige, wie die Pflanze aromatische Substanzen synthetisieren kann. Bekanntlich werden innerhalb 24 Std

aromatische Substanzen im Blatt gebildet (KRATZL). Nach KLASON (1942, 1945) entsteht Lignin direkt bei der Assimilation (z. B. aus Fructose). Nach KABSCH sind Cellulose, nach anderen Theorien Pentosan, Pektin, Zucker (Pentosen) die Vorstufen. Ein mögliches Schema wäre:

$$2\ \text{Pentose} \rightarrow 3,4\text{-Dihydroxycinnamylalkohol} + 5\ H_2O + CO_2,$$

welches teilweise auch von NORD (1952) vertreten wird (Kohlenhydrate über Acetaldehyd zu p-Hydroxyphenylpropaneinheiten). Einen leichten Übergang in

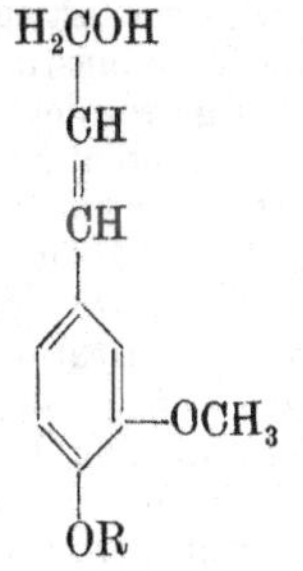

a) R = H Coniferylalkohol
b) R = $C_6H_{11}O_5$ Coniferin

I

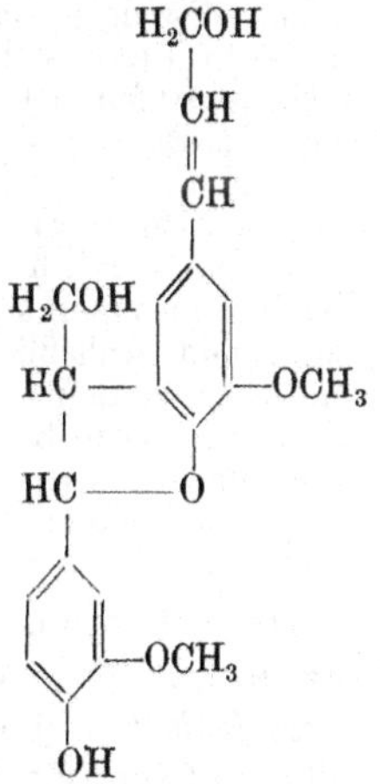

II

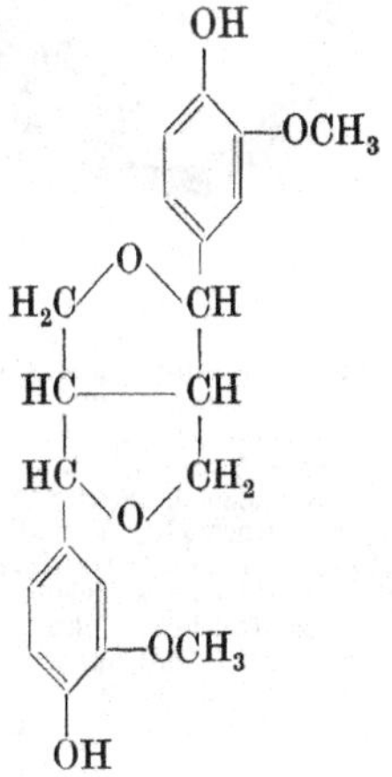

III

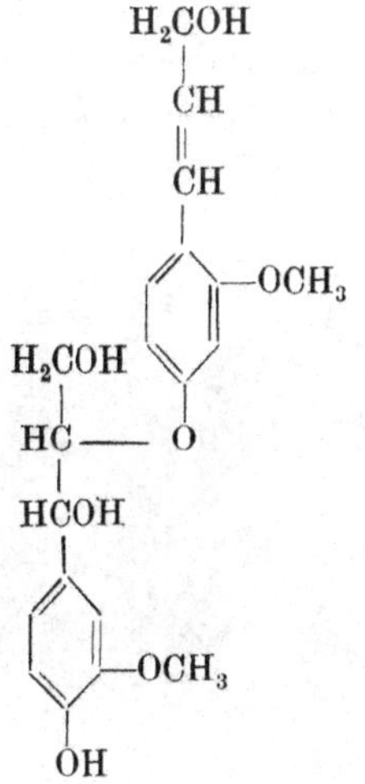

IV

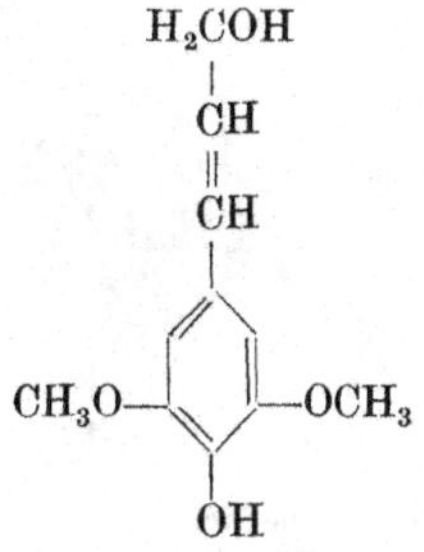

V

Benzolderivate zeigt auch die weit verbreitete Chinasäure und die Shikimisäure. Von hier wird auch unser Blick auf den leichten Übergang der Terpene in aromatische Kohlenwasserstoffe gelenkt. Besonders hingewiesen sei noch auf die Theorie von ENDERS (1943), der folgender Reaktionsablauf zugrunde liegt:

$$\text{Glycerin} \rightarrow \text{Triose} + \text{Glycerinaldehyd} \rightarrow \text{Ligninogen.}$$

PREY mißt dieser Theorie deswegen Bedeutung bei, weil Glycerinaldehydphosphorsäure tatsächlich ein Photosyntheseprodukt ist.

Neue Einblicke in die Biogenese des Lignins und zugleich eine Bestätigung der Theorien von KLASON und ERDTMAN erbrachten Arbeiten nach dem 2. Weltkrieg, vor allem die des Heidelberger Laboratoriums unter FREUDENBERG (1949). Man studierte die Wirkung wasserstoffentziehender Fermente auf Coniferylalkohol (I) und ihm verwandte chemische Substanzen, vor allem aus der Reihe der Hydroxyzimtalkohole. Die Fermente — die sich z. B. im Speisechampignon Psalliota campestris und im Cambium der Coniferen befinden

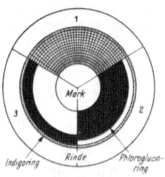

Abb. 29. Schema eines Querschnitts durch einen einjährigen Coniferensproß nach FREUDENBERG und Mitarbeiter (1951). Sektor 1: Schematische Darstellung des Holzkörpers und der Cambialzone. Sektor 2: Schematische Übersicht über die Topographie der Phloroglucinreaktion. Schwarz: verholztes Xylem mit positiver Farbreaktion. Sektor 3: Topographie der Indicanreaktion auf β-Glucosidase (Farbring im cambiumnahen Xylem).

wurden bei 20° C und $p_H$ 5,5—6 in Gegenwart von Luft auf äußerst verdünnte Coniferylalkohollösungen zur Einwirkung gebracht. Nach einigen Stunden beginnt die Ausflockung eines amorphen Materials — Kunstlignin genannt —, welches bis auf kleine Differenzierungen völlig mit dem BRAUNSschen Nativlignin übereinstimmt. Die Zwischenstufen bis zur Ausflockung wurden ebenfalls identifiziert. In etwa 40% Ausbeute entsteht intermediär Dehydrodiconiferylalkohol (II), ein Phenylcumaranderivat, zu ~ 15% wird D,L-Pinoresinol (III), zu 40—50% eine erst kürzlich identifizierte Substanz, der Guajacylglycerinconiferyläther, der die Konstitution IV besitzt, und schließlich entsteht zu ~5% ein Gemisch mit einer Zimtaldehydkomponente. Dieselben Komponenten in ähnlichen Konzentrationsverhältnissen finden sich auch im frischen Cambialsaft in sehr geringen Mengen (papierchromatographisch im Fichtencambialsaft nachgewiesen).

Der Vorgang der Ligninbildung wird von FREUDENBERG nun folgendermaßen gedeutet: *Grundsubstanz* bei den Coniferen ist Coniferylalkohol [Syringin bzw. Sinapinalkohol (V) kondensiert allein nicht, wird aber mit Coniferylalkohol zusammen mit einkondensiert; vielleicht werden auch einfachere, weniger substituierte aromatische Körper gebildet und mit eingebaut]. Durch Dehydrierung verwandelt sich I in optisch inaktive „Zweierstücke" (II—IV), die *sekundären Ligninbausteine*. Durch weiteren Wasserstoffentzug bzw. *selbsttätige* Umbildung (Polymerisation und

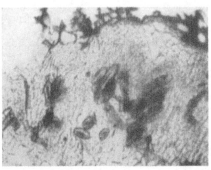 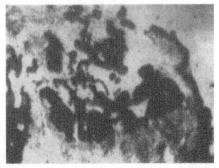

Abb. 30. Links: Karottengewebe auf coniferinfreiem Nährboden kultiviert. Rechts: Gewebe der gleichen Karotte auf coniferinhaltigem Nährboden (Kulturdauer 6 Wochen). [Nach VON WACEK und Mitarbeiter (1953).]

Kondensation) entsteht das Makromolekül Lignin. Über die letztliche Anordnung der Bausteine kann derzeit noch nichts ausgesagt werden; es steht nur fest, daß es sich um einen heteropolymeren Stoff handelt.

Der biologische Lignifizierungsprozeß verläuft wie folgt: Bekanntlich ist in der Cambialzone das Glucosid Coniferin vorhanden — TREIBER (1952) bestimmte spektrophotometrisch den Gehalt größenordnungsmäßig im Fichtenkambialsaft zu ~ $1/_2$% —, ferner finden sich dort, wie vorher erwähnt, wasserstoffentziehende Fermente, die die Wegnahme von H-Atomen an phenolischen Hydroxylen bewirken. Nicht können diese Fermente aber eine Spaltung von Coniferin zu Coniferylalkohol durchführen. Hier wirkt ein an der Stelle, wo die Verholzung im Gange ist, streng lokalisiertes anderes Ferment, eine Glucosidase, die Coniferin spaltet (FREUDENBERG 1952, Abb. 29).

Weitere Beweise, daß Coniferin für die Ligninbildung Grundstoff ist, wurde von FREUDENBERG (1953) selbst, sowie von SIEGEL und WACEK erbracht.

FREUDENBERG führte einem Fichtenbäumchen radioaktives Coniferin zu. Nach der Resorption wurden mehr als 90% der aktiven Substanz im Lignin gefunden. WACEK (1953) studierte den Einfluß von Coniferinzusatz auf die Verholzung von Karottengewebe, das im Nährmedium nach GAUTHERET gezüchtet wurde (Abb. 30). Bei allen Proben mit Coniferinzusatz ist der Verholzungsgrad *wesentlich* höher.

Die Ligninbildung in der Wachstumsspitze wird durch Wachstumshormone verhindert.

# 5. Anorganische Ausscheidungsstoffe der Pflanzen. Mineralisierung der Zellwand.

Bei den anorganischen Ausscheidungen beschränken wir uns hier lediglich auf feste anorganische Ablagerungen, die sich in Geweben höherer Pflanzen anhäufen. Speziell interessiert hier die Mineralisation der Zellwand, hingegen werden nicht erwähnt die verschiedenen wichtigen Spurenelemente, die auch lokale Anreicherungen erfahren können (z. B. Mangan in den Markstrahlen des Holzes) sowie die über der Cuticula entstehenden Salzkrusten der Tamarixarten. Auf das Festhalten von Metallen aus Salzlösungen durch pflanzliche Zellwände sei nur verwiesen. Die Eigenschaft als Kationenaustauscher zu wirken, kann z. B. bei der Alginsäure sogar analytisch ausgewertet werden (SPECKER 1953).

Im wesentlichen werden 2 Stoffgruppen abgeschieden, und zwar entweder als diffuse Ablagerung (Membraninkrustation) oder als diskrete Ablagerung (z. B. Cystolithen): Erstens Kalksalze, und zwar Calciumoxalat und -carbonat, zweitens Kieselsäureanhydride.

In den Membranen von *Acetabularia* findet sich z. B. in der Außenschicht Calciumcarbonat, in den inneren Schichten vorwiegend Oxalat. Selten kommt es zur Ausscheidung von Calciumphosphat (Globoiden), -sulfat (z. B. Zuckerrohr) und -tartrat (Weinrebe).

Von den Kieselablagerungen kann z. B. die Cuticula und besonders die Epidermis betroffen werden; letztere unter Umständen so stark, daß ein Kieselsäureskelett erhalten werden kann. Es scheint die Kieselsäure gegebenenfalls auch die Rolle einer Gerüstsubstanz spielen zu können; sie wird zu diesem Behufe auch in der Primärlamelle und den Sekundärschichten in Form eines zusammenhängenden Skelets eingelagert. Am stärksten aber sind die Epidermisanhänge der Verkieselung ausgesetzt, vor allem die Haare. So sind die Brennhaare von *Urtica* in der Spitze verkieselt, im unteren Teil mit Kalk inkrustiert. Große Kieselsäuremengen kommen vor im Lumen der Haare von *Morus* und *Broussonetia*; ferner können sie als Tabaschir in Bambusarten auftreten. Bekannt ist noch der Kieselsäurekörper in den Stegmata der Orchideen und Palmen.

Von LADENBURG (1872) wurde erstmals die Vermutung von der Existenz organischer Siliciumverbindungen ausgesprochen. Neuere Untersuchungen, vor allem über die Diffusion von $SiO_2$ durch Membranen sowie Extraktionsversuche an Geweben von HOLZAPFEL führten zur weitgehend gesicherten Annahme, daß die Kieselsäure mittels hydroxylhaltiger organischer Komponenten wasserlöslich gemacht und transportiert wird. So konnte ENGEL (1953) aus der Roggenhalmwand *(Secale)* Galaktose-Siliciumkomplexe isolieren; ähnliche Resultate zeitigten

Versuche mit Schachtelhalmen usw. HOLZAPFEL (1951) nimmt an, daß es sich um Esterbindungen handelt, etwa nach dem Schema I oder II und gegebenenfalls III:

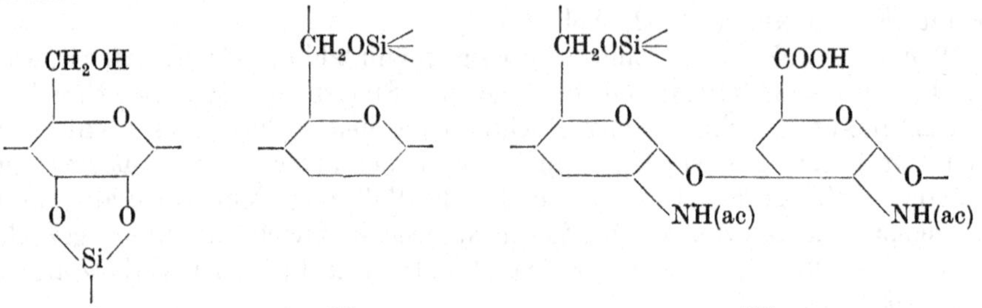

In sehr seltenen Fällen treten noch Ausscheidungen von Magnesium- und Aluminiumverbindungen auf. Bei *Hydrilla verticillata* und anderen Pflanzen (z. B. Phanerogamen) wird von einer Manganspeicherung berichtet, die zum Teil auch auf die Zellwände übergreift.

Eine spezialisiertere Darstellung der biogenen Kohlenhydrate findet man in Band V, von Lignin, Wachs, Kork und Cutin in Band X.

## Literatur.

### Zusammenfassende Darstellungen.

BONNER, J.: Plant biochemistry. New York: Acad. Press 1950.

FREY-WYSSLING, A.: Submicroscopic morphology of protoplasm and its derivatives. Amsterdam: Elsevier Publ. Co. 1948. — FREY-WYSSLING, A., u. K. MÜHLETHALER: The fine structure of cellulose in L. ZECHMEISTER, Fortschritte der Chemie organischer Naturstoffe, Bd. 8. Wien: Springer 1951.

HÄGGLUND, E.: Chemistry of wood. New York: Acad. Press 1951. — HESS, K.: Elektronenmikroskopie und Wandstruktur bei natürlichen Zellulosefasern. Kunstseide u. Zellwolle 28, 3—11 (1950). — HEUSER, E.: The chemistry of cellulose. New York: J. Wiley 1944.

KRATKY, O.: Der micellare Aufbau der Cellulose und ihrer Derivate. Angew. Chem. 53, 153—162 (1940). — Der übermolekulare Aufbau der Cellulose. In R. PUMMERER, Chemische Textilfasern, Filme und Folien. Stuttgart: Ferdinand Enke 1951.

MEYER, K. H.: Makromolekulare Chemie, 2. Aufl. Leipzig: Geest & Portig 1950.

OTT, E.: Cellulose and cellulose derivatives, 2. Aufl. New York: Intersciences Publ. 1946.

TREIBER, E.: Der übermolekulare Aufbau der Cellulose. Protoplasma (Wien) 40, 166—186, 367—396 (1951). — Ferner: Morphologische Strukturen bei natürlichen Fasern in H. A. STUART, Die Physik der Hochpolymeren, Bd. III. Berlin: Springer 1955 (im Druck).

\*

ADAMS, G. A., and A. E. CASTAGNE: Purification and composition of a polyuronide hemicellulose isolated from wheat straw. Canad. J. Chem. 30, 515—521, 698—710 (1952).

BAILEY, A. J.: Lignin in Douglas fir. Ind. Engng. Chem., Analyt. Edit. 8, 52—55 (1936). — BAKER, J. R.: Morphology and fine structure of organisms. Nature (Lond.) 165, 585—586 (1950). — BOUGAULT, J., et L. BOURDIER: Sur les cires des coniferes. Nouveau groupe de principes immediats naturels. C. r. Acad. Sci. Paris 147, 1311—1314 (1908). — BRAUNS, F. E.: The chemistry of lignin. New York: Acad. Press Inc. 1952. — BUCHER, H.: Die Tertiärlamelle von Holzfasern und ihre Erscheinungsformen bei Coniferen. Attisholz 1953.

CAMPBELL, W. C., J. L. FRAHN, E. L. HIRST, D. F. PACKMAN and E. G. V. PERCIVAL: Wood Starches. J. Chem. Soc. (Lond.) 1951, 3489—3498. — CERUTI, A., u. J. SCURTI: Sulla formazione delle gomme nel ciliegio I, II, Ann. Sperim. Agr. 1953. — CLARK, G. L.: Applied X-rays, 3. Aufl. New York: McGraw Hill 1940.

DIEHL, J. M., u. G. VAN ITERSON: Die Doppelbrechung von Chitinsehnen. Kolloid-Z. 73, 142—146 (1935). — DOLMETSCH, H.: Vorgebildete Spaltflächensysteme in nativen Zellulosefasern. Melliand Textilber. 35, 721—725 (1954). — DOMAN, N. G.: Die Natur der intermediären Produkte der Photosynthese. Dokl. Akad. Nauk. SSSR. 84, 1017—1020 (1952).

ENDERS, C. G.: Wie entsteht der Humus in der Natur? Chemie **56**, 281—285 (1943). — ENGEL, W.: Untersuchungen über die Kieselsäureverbindungen im Roggenhalm. Planta (Berl.) **41**, 358—390 (1953). — ESCHRICH, W.: Planta **44**, 532 (1954).

FRANZ, E.: Über einige Beziehungen zwischen morphologischem Aufbau und Güte von Faserstoffen. Angew. Chem. **56**, 113—120, 132—135 (1943). — FREUDENBERG, K.: Tannin, Cellulose, Lignin. Berlin: Springer 1933. Vgl. auch E. FISCHER, Untersuchungen über Depside und Gerbstoffe. Berlin: Springer 1919. — Neues über Lignin. Papierfabrikant **36**, 34—36 (1938). — Die Bildung ligninähnlicher Stoffe unter physiologischen Bedingungen. Sitzgsber. Heidelberger Akad. Wiss., math.-naturwiss. Kl. **1949**, Nr 5. Vgl. ferner K. FREUDEN-BERG, Zur Biogenese des Lignins, Holz als Roh- und Werkstoff **11**, 267—269 (1953). — FREUDENBERG, K., u. F. BITTNER: Versuche mit Coniferylalkohol, der radioaktiven Kohlenstoff enthält. Ber. dtsch. chem. Ges. **86**, 155—159 (1953). — FREUDENBERG, K., A. JANSON, E. KNOPF u. A. HAAG: Zur Kenntnis des Lignins. Ber. dtsch. chem. Ges. **69**, 1415—1425 (1936). — FREUDENBERG, K., H. REZNIK, H. BOESENBERG u. D. RASENACK: Das an der Verholzung beteiligte Fermentsystem. Ber. dtsch. chem. Ges. **85**, 641—647 (1952). — FREY-WYSSLING, A.: Micellarlehre erläutert an Beispielen des Faserfeinbaues. Kolloid-Z. **85**, 148—158 (1938). — Der submikroskopische Feinbau von Chitinzellwänden. Vjschr. naturforsch. Ges. Zürich **95**, 45 (1950). — Submicroscopic morphology of protoplasm and its derivatives, 2. Aufl. Amsterdam: Elsevier Publ. Co. 1953. Vgl. auch A. FREY-WYSS-LING, K. MÜHLETHALER u. R. W. G. WYCKOFF, Mikrofibrillenbau der pflanzlichen Zellwände. Experientia (Basel) **4**, 475 (1948). — FREY-WYSSLING, A., u. H. STECHER: Das Flächenwachstum der pflanzlichen Zellwände. Experientia (Basel) **7**, 420 (1951). — FRIESE, H., V. HÖGN u. H. WILLE: Zur Kenntnis der Sulfitablauge. Ber. dtsch. chem. Ges. **70**, 1072—1079 (1937). Vgl. ferner H. FRIESE u. E. CLOTOFSKI, Über die Sulfonierung des Lignins. Ber. dtsch. chem. Ges. **70**, 1986—1989 (1937) und F. SCHÜTZ, Bildung methoxyl- und ligninhaltiger Polysaccharide bei der Holzhydrolyse von Rotbuche bei 100—105°. Ber. dtsch. chem. Ges. **75**, 703—710 (1942).

GALOWA, O. P., u. W. J. IWANOW: Über das Molekulargewicht der Zellulose. Berlin: Akademie Verlag 1953. — GJOKIČ, G.: Über die chemische Beschaffenheit der Zellhäute bei den Moosen. Österr. bot. Ztg **45**, 330—334 (1895). — GRALÉN, N. J.: The molecular weight of lignin. Colloid Sci. **1**, 453—463 (1946). — GREATHOUSE, G. A.: Biosynthesis of specifically C14-labeled cottoncellulose. Science (Lancaster, Pa.) **117**, 553—554 (1953).

HÄGGLUND, E.: Fortschritte der Ligninchemie. Zellstoff u. Papier **16**, 570—574 (1936). Vgl. ferner E. HÄGGLUND, Holzchemie, 2. Aufl. Leipzig: Akademische Verlagsgesellschaft 1939 und E. HÄGGLUND, Untersuchungen über die Chemie der Sulfitzellstoffkochung. Sv. kem. Tidskr. **37**, 116—124 (1925); **38**, 177—192 (1926). — Über die chemischen Vorgänge beim Sulfatzellstoffprozeß. Holz als Roh- u. Werkstoff **11**, 251—257 (1953). — HÄRTEL, O.: Über die Haftorgane von Parthenocissus tricuspidata. Biol. generalis (Wien) **19**, 193—210 (1950). — HAWORTH, R. D.: Ann. Rpt. on Progress Chem. J. Chem. Soc. London **33**, 270 (1936). — HERGERT, H. L., u. E. F. KURTH: The chemical nature of the cork from Douglasfire bark. Tappi **35**, 59—66 (1952). — HESS, K., u. H. MAHL: Elektronenoptischer Nachweis großer Perioden bei Kunststoffen und Cellulosefasern. Naturwiss. **41**, 86 (1954). — HEUSER, E., u. L. JÖRGENSEN: Chain lenght distribution and chain-lenght distribution of wood cellulose as compared with cotton. Tappi **34**, 57—67 (1951). — HEYN, H. N. J.: Small angle X-Ray scattering in various cellulose fibers and its relation to the micellar structure. Textile Res. J. **19**, 163—172 (1949). Vgl. ferner H. N. J. HEYN, Small angle x-Rays Scattering of various cellulose fibers. J. Amer. Chem. Soc. **70**, 3138—3139 (1948). — HILL, L. M.: Untersuchung über Suberin und korkige Ablagerungen erkrankter Kartoffelknollen. Phytopathology **29**, 274—282 (1939). — HIRST, E. L.: Die Chemie der Pflanzengummi und -schleime. Endeavour **10**, 106—110 (1951). — HÖPNER, TH.: Gegenwartsprobleme der Holzchemie. Kolloid-Z. **93/94**, 98—106 (1941). — HOLZAPFEL, L.: Siliziumverbindungen in biologischen Systemen. Z. Elektrochem. **55**, 577—580 (1951). — HOLZER, K., u. A. ZINKE: Über die Bitterstoffe der Zichorie. Mh. Chem. **84**, 901—909 (1953). — HUSEMANN, E., u. H. BARTL: Über die Größe und Gestalt der Amylosemoleküle. Makromolekulare Chem. **10**, 183—184 (1953).

IMAMURA, R.: Colloidal properties of xylan in aqueous solutions. J. Soc. Text. Cell. Ind. (Japan) **8**, 445—448 (1952). — ITERSON jr., G. VAN: A few observations on the hairs of the stamens of tradescantia virginica. Protoplasma (Wien) **27**, 190—211 (1937).

KAHOVEC, L., G. POROD u. H. RUCK: Röntgenkleinwinkeluntersuchungen an dichtgepackten kolloiden Systemen. Kolloid-Z. **133**, 16—26 (1953). — KARRER, P., u. E. JUCKER: Carotinoide. Basel: Birkhäuser 1948. — KAST, W.: Die Teilcheneigenschaften der kristallinen Gebiete der Cellulosefasern. Z. Elektrochem. **57**, 525—530 (1953). — KAVERZNIÉVA, E. D.: Les transformations chimiques de la cellulose sous l'action des oxydants, in: Communications au XIII Congrès internationale de chemie pure et appliquée, Stockholm; Moskau 1953. — KLASON, P.: Beiträge zur Kenntnis des chemischen Baues des Tannenholzlignins. Ark. Kemi (Stockh.) **6**, 15—21 (1917). — Beitrag zur Kenntnis der Konstitution des Fichtenholzlignins. Ber. dtsch. chem. Ges. **53**, 1864—1873 (1920); **55**, 448—456 (1922). — Beiträge zur Konstitution des Fichtenholz-Lignins (Untersuchung des Nahrungssaftes der Fichte). Ber. dtsch. chem. Ges. **62**, 635—639 (1929). — Beiträge zur Konstitution des Fichtenholzlignins. Ber. dtsch. chem. Ges. **62**, 2523—2526 (1929). — Beiträge zur Konstitution des Lignins. Ber. dtsch. chem. Ges. **63**, 1548—1551 (1930). — Köhler, A.: Z. Mikrosk. **21**, 129, 273 (1904). — König, J., u. E. Becker: Die Bestandteile des Holzes und ihre wirtschaftliche Verwertung. Z. angew. Chem. **32**, 155—160 (1919). — Kratky, O., u. A. Sekora: Auffindung einer Längsperiodizität im Röntgenkleinwinkelbild einer jodierten Kunstseide. Z. Naturforsch. **9b**, 505—506 (1954). — Kratzl, K.: Über den qualitativen Nachweis der Verholzung. Holz als Roh- u. Werkstoff **11**, 269—276 (1953). Vgl. ferner K. Kratzl u. H. Tschamler, Ultrarotspektren von Holz und unlöslichen Ligninen. Mh. Chem. **83**, 786—791 (1952). — Kursanow, A. L., u. E. I. Vyskrebentseva: Wechsel in der Zusammensetzung der Baumwollfaser während der Celluloseentstehung. Biochimia **17**, 480—487 (1952). Vgl. auch Biochimia **18**, 448—451 (1953).

Ladenburg, A.: Über die Natur der in den Pflanzen vorkommenden Si-Verbindungen. Ber. dtsch. chem. Ges. **5**, 568—569 (1872). — Lange, P. W.: Nature and distribution of lignin in sprucewood. Sv. Papperstidn. **47**, 262—265 (1944). Vgl. ferner P. W. Lange, Ultraviolet absorption of solid lignin. Sv. Papperstidn. **48**, 241—245 (1945). — Some views on the lignin in the woody fiber during the sulfit cook. Sv. Papperstidn. **50**, 130—134 (1947). — Linsbauer, K.: Zur Verbreitung des Lignins bei Gefäßkryptogamen. Österr. bot. Ztg **49**, 317—323 (1899). — Lüdtke, M.: Untersuchungen über Aufbau und Bildung der pflanzlichen Zellmembran und ihrer stofflichen Komponenten. Biochem. Z. **233**, 1—57 (1931).

Maass, H.: Die Pektine. Braunschweig: Dr. Serger 1951. — McDonald, T. R. R., and C. A. Beevers: The crystal structure of $\alpha$-D-Glucose. Acta Crystallogr. **3**, 394—395 (1950). — Meyer, M.: Die submikroskopische Struktur der cutinisierten Zellmembranen. Protoplasma (Wien) **29**, 552—586 (1938). — Meyer, K. H.: Vergangenheit und Gegenwart der Stärkechemie. Experientia (Basel) **8**, 405—420 (1952). — Meyer, K. H., u. H. Mark: Über den Aufbau des Chitins. Ber. dtsch. chem. Ges. **61**, 1936—1939 (1928). — Makromolekulare Chemie, 2. Aufl. Leipzig: Geest & Portig 1950. — Meyer, K. H., u. L. Misch: Positions des atomes dans le nouveau modele spatial de la cellulose. Helvet. chim. acta **20**, 232—244 (1937). — Meyerhoff, G.: Molekulargewichtsbestimmungen an Cellulosenitraten in der Ultrazentrifuge. Naturwiss. **41**, 13 (1954). — Metzner, P.: Über das optische Verhalten der Pflanzengewebe im langwelligen ultravioletten Licht. Planta (Berl.) **10**, 281—313 (1930). — Michaelis, L.: Die Permeabilität von Membranen. Naturwiss. **14**, 33—42 (1926). Vgl. ferner K. H. Meyer u. J. F. Sievers, La permeabilite des membranes. Helvet. chim. Acta **19**, 987—995 (1936). — Minor, F. W., G. A. Greathouse, H. G. Shirk, A. M. Schwartz and M. Harris: Biosynthesis of $C^{14}$-specifically labeled cellulose by *Acetobacter xylinum*. J. Amer. Chem. Soc. **76**, 1658—1661 (1954). — Mukherjee, S. M., J. Sikorski and H. J. Woods: Micellar structure of native cellulose. Nature (Lond.) **167**, 821—823 (1951).

Nord, F. F., u. G. de Stevens: On the mechanism of lignification. Naturwiss. **39**, 479—480 (1952). — Norman, A. G.: The biochemistry of cellulose, the polyuronides. Lignin etc. Oxford: Clarendon Press 1937. Vgl. ferner A. G. Norman, Carbohydrates normally associated with cellulose in nature. In E. Ott, Cellulose and cellulose derivatives. 2. Aufl. New York: Interscience Publ. 1946.

O'Dwyer, M. H.: The hemicelluloses. Biochemic. J. **20**, 656—664 (1926). — Okamura, I.: Growth of cellulose. Teijin Times **22**, 3—4 (1952).

Ploetz, Th.: Inosite. Chemie **56**, 231—233 (1943). — Preston, R. D.: Fibrillar units in the structure of native cellulose. Discuss. Faraday Soc. **11**, 165—170 (1951). — Preston, R. D., P. H. Hermans and A. Weidinger: The crystalline-non-crystalline ratio in celluloses of biological interest. J. of Exper. Bot. **1**, 344—352 (1950). — Prey, V., E. Waldmann u. W. Krzandalsky: Über den alkalischen Abbau von Hemicellulosen. Mh. Chem. **84**, 888—900 (1953). — Prey, V., E. Waldmann u. F. Stiglbrunner: Zur Kenntnis des Alkalilignins. Mh. Chem. **84**, 824—827 (1953). — Purves, C. B.: Nature of the association between carbohydrate and lignin in wood. In E. Ott, Cellulose and cellulose derivatives, 2. Aufl. New York: Interscience Publ. 1946.

Radley, J. A.: Starch and its derivatives. London: Chapman Hall 1953. — Rånby, B. G.: Diskussionsbemerkung. Sv. Papperstidn. **57**, 9, 17 (1954). — Reeves, R. E.: Cuprammonium-Glycoside complexes. III. The conformation of the D-Glycopyranoside ring in solution. J. Amer. Chem. Soc. **71**, 215—217 (1949). — Renfrew, A. G., u. L. H. Cretcher: Quince seed mucilage. J. of Biol. Chem. **47**, 503 (1932). — Ritter, G. J.: Verteilung von Lignin in Holz. Ind. Engng. Chem. **17**, 1194—1197 (1925). Vgl. ferner Über die Zellwandstruktur von Holzfasern. Paper Ind. **16**, 178—183 (1934). — Roelofsen, P. A., V. Ch. Dalitz and C. F. Wijnman: Constitution, submicroscopic structure and degree of crystallinity of the cell wall of halicystis osterhoutii. Biochim. et Biophysica Acta 11, 344—352 (1953). — ROELOFSEN, P. A., and D. R. KREGER: The submicroscopic structure of pectin in collenchyma cell walls. J. of Exper. Bot. 2, 332—343 (1951). — ROLLINS, M. L.: Some aspects of microscopy in cellulose research. Analyt. Chem. 26, 718—724 (1954). — RUBEN, S., and M. D. KAMEN: Photosynthesis with radioaktive carbon; molecular weight of the intermediate products and a tentative theorie of photosynthesis. J. Amer. Chem. Soc. 62, 3451—3455 (1940). — RUBEN, S., M. D. KAMEN and W. Z. HASSID: Photosynthesis with radioactive carbon; Chemical properties of the intermediates. J. Amer. Chem. Soc. 62, 3443—3450 (1940). — RUBEN, S., M. D. KAMEN and L. H. PERRY: Photosynthesis with radioactive carbon; ultracentrifugation of intermediate products. J. Amer. Chem. Soc. 62, 3450—3451 (1940).

SAMEC, M.: Die Stärkeforschung im Rückblick und Ausblick. Kolloid-Z. 124, 135—141 (1951). — SCHAAFFS, W.: Kolloid-Z. 137, 121 (1954). — SCHULZ, G. V.: Über den heutigen Stand viskosimetrischer Molekulargewichtsbestimmung. Kolloid-Z. 115, 90—102 (1949). — SCHULZ, G. V., u. I. KÖMMERLING: Molekulargewichtsverteilung und Lockerstellen in Cellulosen nach Versuchen von EMERY und COHEN. Makromolekulare Chem. 9, 25—34 (1952). — SCHULZ, G. V., u. M. MARX: Über Molekulargewichte und Molekulargewichtsverteilungen nativer Cellulosen makromolek. Chemie 14, 52—95 (1954). — SIPPEL, A.: Die Verfeinerung des Wannenmodells der Zellulose. Kolloid-Z. 122, 20—23 (1951). — SPECKER, H., u. H. HART-KAMP: Abtrennung des Eisens von anderen Kationen durch Ionenaustausch an Alginsäure. Naturwiss. 40, 410—411 (1953). — STACEY, M.: Biological polymerisation with reference to Polysaccharide Synthesis. Chem. a. Ind. 1950, 727—729. — STAUDINGER, H., u. TH. EICHER: Über die Quellung resp. Inklusion der Zellulose mit niederen Fettsäuren. Makromolekulare Chem. 10, 254—260 (1953). — STECHER, H.: Über das Flächenwachstum der pflanzlichen Zellwände. Mikroskopie 7, 30 (1952). — STUDER, M.: Diss. Genf 1946.

THOMPSON, A., K. ANNO, M. L. WOLFROM and M. INATOME: Acid reversion products form D-Glucose. J. Amer. Chem. Soc. 76, 1309—1311 (1954). — TREIBER, E.: Optische Untersuchungen an Hemicellulosen. Österr. Papierztg 59, 23 (1953). — In Åke S. son STENIUS, The general discussion on cellulose held in Stockholm on 31. Juli 1953. Sv. Papperstidn. 57, 48 (1954). — TREIBER, E., W. LANG u. M. FLORIANTSCHITSCH: Die Ultraviolettabsorption einiger Cambialsäfte. Protoplasma (Wien) 41, 452—457 (1952). — TREIBER, E., G. POROD, W. GIERLINGER u. J. SCHURZ: Über die Adsorption von Makromolekülen an aktiven Oberflächen. Makromolekulare Chem. 9, 241—243 (1953). — TREIBER, E., H. TOP-LAK, M. u. H. RUCK: Physikalisch-chemische Untersuchungen an einigen Hemicellulosen, Holzforschung 9, 49—59 (1955). — TSCHIRCH, A.: Die Harze und die Harzbehälter, 2. Aufl. Leipzig: Gebrüder Bornträger 1906.

VOGEL, A.: Zur Feinstruktur von Ramie. Makromolekulare Chem. 11, 111—130 (1953). WACEK, A. v.: Ligninbausteine und Ligninbruchstücke. Österr. Chem. Ztg 54, 61—66 (1953). — WACEK, A. v., O. HÄRTEL u. S. MERALLA: Über den Einfluß von Coniferinzusatz auf die Verholzung von Karottengewebe bei Kultur in vitro. Holzforsch. 7, 58—62 (1953). — WALKER, T. K.: Gathways of acid formation on aspergillus niger and in related molds. Adv. Enzymol. 9, 579—584 (1949). — WALLACH, O.: Terpene und Campher. Leipzig: Veit & Co. 1909. — WEDEKIND, E.: Neuere Forschungen über die Lignine verschiedener Baumarten. Papierfabrikant 35, T 141—142 (1937). Vgl. ferner E. E. HARRIS, Some characteristics of wood lignins. J. Amer. Chem. Soc. 58, 894—896 (1936) und S. OGURI u. M. TAKEI, Die Absorptionsspektren der Lösungen von Bambusligninen. Waseda Appl. Chem. Soc. Bull. 14, 37—40 (1937). — WEHMER, C.: Die Pflanzenstoffe. Jena: Gustav Fischer 1929. — WETHERN, J. D.: Some molecular properties of the hemicelluloses of black spruce. Tappi 35, 267—271 (1952). — WIENHAUS, H., u. K. MÜCKE: Zur Chemie der Harze, IV. Mitt. Ber. dtsch. chem. Ges. 75, 1830—1840 (1943). — WILLIAMS, R. T.: Biological transformation of starch and cellulose. New York: Univ. Press 1953. — WISE, L. E.: The chemistry of the hemi-celluloses. Pulp Paper Mag. Canada 50, 179—186 (1947). — WUHRMANN-MEYER, K., u. M.: Untersuchungen über die Absorption ultravioletter Strahlen durch Cuticular- und Wachsschichten von Blättern. Planta (Berl.) 32, 43—50 (1941).

ZETZSCHE, F., u. G. ROSENTHAL: Untersuchungen über den Kork. Helvet. chim. Acta 10, 346—374 (1927). Ferner F. ZETZSCHE, CH. CHOLATNIKOW u. K. SCHERZ, Untersuchungen über den Kork. Helvet. chim. Acta 11, 272—276 (1928). — ZETZSCHE, F., u. G. SONDER-EGGER: Untersuchungen über den Kork. Helvet. chim. Acta 14, 632—641, 642—645 (1931). — ZETZSCHE, F. u. M. BÄHLER, Untersuchungen über den Kork. Helvet. chim. Acta 14, 846—849, 849—851, 852—856 (1931). — ZETZSCHE, F. u. E. LÜSCHER: Untersuchungen über den Kork. J. prakt. Chem. 150, 68—80, 140—144 (1938).

# Microscopic structures of plant cell walls.

By

## R. D. Preston.

With 5 figures.

## Cell size and shape.

When tissues of higher plants are treated with suitable reagents, they can be induced to fall apart into individual units—the cells. It follows, therefore, that each single cell may be considered to be surrounded by a wall which is individual to it. The material cementing cells together is often pectin in young tissue which may be transformed into calcium or magnesium pectate, or impregnated with lignin, as the tissue matures. Every cell originates, directly or indirectly, as an iso-diametric cell, in a meristem, which appears 5 or 6 sided in optical section and it has been a matter of some concern to determine the three-dimensional shape of these cells. It is now generally agreed that the shape is based on the 14-sided cubo-octahedron (Fig. 1) (VAN ITERSEN and MEEUSE 1941) almost in the form suggested by KIESER more than 100 years ago. The cubo-octahedron was originally suggested by LORD KELVIN as very nearly the shape assumed by soap bubbles when filling space completely in a froth, with the modification that the eight hexagonal faces are not plane but slightly curved, producing the "body of Thomson". Such curvatures have also been reported in isodiametric plant cells (VAN ITERSON and MEEUSE 1941).

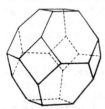

Fig. 1. Diagrammatic representation of the probable form of a cell of the apical meristem in a plant.

During differentiation, the dimensions of cells may change profoundly but, provided they continue to fill space completely, *i.e.* so long as no prominent intercellular spaces develop, they must still conform in shape to a derivative of the cubo-octahedron. The careful observations of LEWIS (1935), for instance, have fully confirmed this conception in the case of fibrous cells. Only in those cases where individual cells expand enormously and make new contacts to many other cells is this basic shape lost, and one such cell type would clearly be the vessel elements of ring-porous dicotyledonous trees.

The dimensional changes involved during differentiation from apical meristems are often very considerable. The size of mature cells of any type varies rather widely, not only between species but within individual plants so that the following figures may be taken only as rough guides. An undifferentiated meristematic cell will be of the order of 15–20 $\mu$ in diameter, with a volume of some $3–8 \times 10^{-4}$ cm³. In differentiating into a parenchyma cell, say in the cortex of the root (BROWN and BROADBENT 1950), there is an expansion of some $30 \times$ in volume, due almost entirely to water intake. This expansion occurs mainly in one direction so that the final parenchyma cell may be of the order of 30 $\mu$ broad and say, about 200 $\mu$ long. Differentiation into other more specialised cell types usually leads to dimensional changes of a much higher order. Vessel elements in xylem may range up to 300 $\mu$ in diameter, with a length, say, of 500 $\mu$ (PRESTON 1938).

Collenchyma cells, again, are of the order of 400–600 $\mu$ long and some 15 $\mu$ wide (MAJUMDAR and PRESTON 1941), while tracheids in conifer wood range in length from about 500 $\mu$ to many millimetres (PRESTON 1934). Fibres occur over a length range somewhat similar to that of tracheids, depending both on the species and on the tissue within individual plants.

Each cell type has its own particular characteristics by which it may be recognised. Thus a cell may be called a *vessel element* only if at least one end wall is perforated either completely or with bars traversing the perforation (scalariform perforation) (EAMES and MacDANIELS 1947). *Collenchyma cells* are typified by an uneven distribution of wall thickening and here three types may be recognised (a) *angular* collenchyma if the thickening is confined to cell corners, (b) *tubular* collenchyma if the thickening of the wall of each cell facing an intercellular space gives the appearance of a uniformly thickened intercellular space, (c) *plate* collenchyma when the thickening is spread over the walls lying tangentially in the stem. *Tracheids* are characteristically long, thread-like cells and may be distinguished from fibres in that they usually taper to blunt, chisel-shaped ends and their walls are prominently pitted. *Fibres*, on the other hand, are sharply pointed and have thicker cell walls in which pits are reduced, usually to slits, or are apparently absent. Intermediate types occur, the so-called *fibre tracheids*. The walls of all these cell types, with the exception of collenchyma cells, are usually heavily lignified. *Parenchyma cells*, on the other hand, are seldom lignified except in the wood; they are typically rather thin walled, are isodiametric, *i.e.* seldom more than a few times longer than broad, and usually have contents.

## The primary cell wall.

The walls of growing cells, which remain as the outermost layer in adult cells, are referred to as *primary* walls. In fresh preparations they may appear rather thick but the shrinkage to about $^1/_3$ of this thickness on dehydration shows that this is due to high water content. These walls have a low cellulose content; in *Avena* coleoptiles, for instance, about 40% (THIMANN and BONNER 1933) and in conifer cambium only 25% (ALLSOPP and MISRA 1940) of the dry weight, with a rather high content of pectin or hemicellulose. There is some evidence that these walls contain protein (PRESTON and WARDROP 1949). It has been estimated that in the primary walls of fresh *Avena* coleoptiles the cellulose occupies about 14% of the wall volume (FREY-WYSSLING 1936) and of conifer cambium only 8% (PRESTON and WARDROP 1949). The wall at this stage is commonly rather featureless though special staining techniques show the presence of plasmodesmata and, in certain cases at least, of a reticulate thickening structure.

Knowledge as to the origin, development and function of plasmodesmata is still very meagre (MEEUSE 1941). It was for a long time considered that plasmodesmata occurred only in walls produced at cell divisions, but there is now some evidence that "secondary" plasmodesmata arise, during the development for instance of elongating fibres, in walls which have come together during tissue displacements. Endosperm tissues of some seeds, *e.g. Diospyros*, form good material for the observation of plasmodesmata.

The main physiological feature of the wall at this stage is its capacity for continued increase in area, sometimes slowly, sometimes with astonishing rapidity. During this increase the wall does not become appreciably thinner as judged by microscopical observation so that new wall material must be synthesised during growth. Wall synthesis has, indeed, been demonstrated experimentally (Pre-ston and Clarke 1944) just as has protein synthesis (Blank and Frey-Wyss-ling 1944) and it is clear from such chemical analyses that the amount of wall material per unit wall area in fact decreases markedly as the wall area increases. It was for a long time believed that wall growth occurred through the inter-polation of new structural units among the old—the *intussusception* of Nägeli. Although this basic concept can still not be discarded, there is some evidence that some cells such as developing fibres grow only at their extremities (Schoch, Bodmer and Huber 1951) and this *tip growth* has also been claimed as operative even in parenchyma cells of oat coleoptiles (Frey-Wyssling 1952). Other workers, however, such as Stecher (1952) prefer to think in terms of *mosaic* growth, a local and random loosening of existing wall material followed by repair, which corresponds much more nearly with the older views on intussusception. These matters are, however, dealt with more fully in the next section.

When a growing cell divides, each daughter protoplast becomes eventually clothed in an individual wall, the two daughter cells being still retained in the old parent wall. This phenomenon was first conceived by Giltay (1882) and received support from the microscopical observations of Priestley and Scott (1938) and, later, Elliot (1951) and Wardrop (1952). The primary wall is thus structurally more complex than might be thought from microscopical examination of individual walls alone. The algae have long been recognised as examples of the complexity which can arise in this way (see *e.g.* refs. in Ast-bury and Preston 1940).

## The secondary wall.

About the time that a plant cell reaches its adult size the wall begins to thicken. This is supposed to occur by the addition of new thickening layers to the inside—the so-called *apposition* of von Mohl—forming collectively the secondary wall. Normally this has a much higher cellulose content than has its precursor, the primary wall, and is largely incapable of further growth. Micro-scopically, its thickness is widely variable among the various cell types. In parenchyma cells it is commonly thin. In vessels its thickness tends to vary, even within one vessel element, with the type of cell with which it is in contact, and can range up to $10\,\mu$ or more (Preston 1938). In tracheids it is usually of the order of $5\,\mu$ so that in these cells a wide central lumen is left. In fibres the wall is usually much thicker in relation to cell diameter so that in some types —*e.g.* bamboo fibres (Preston and Singh 1950)—the lumen becomes almost completely occluded. Secondary walls usually show microscopical detail of various kinds which are dealt with in the following pages.

### a) Layering.

In transverse section of almost all cells it can usually be demonstrated that the wall consists of several concentric layers (Fig. 2). Sometimes the layers differ from each other physically, when they can be distinguished in the unstained condition on account of a difference in refractive index. Sometimes the differ-ence is in chemical constitution, when the layers can be rendered more easily visible by suitable staining reagents. In other cases, the wall must be swollen in order to demonstrate the presence of layering. The differences between the layers can often be accentuated by observation under a polarising microscope (Fig. 3) or by photomicrography, using ultra violet or infra red radiation.

In transverse section of conifer tracheids there are commonly three layers, a narrow outer layer, a narrow inner layer, and a central layer which is broader in the tracheids of late wood. These layers are normally distinguishable under an ordinary light microscope, but the differences between them are far more striking under a polarising microscope (Fig. 3) since the outer and inner layers only are then bright. It should be noted that the whole wall of the conifer tracheid is thus 5-layered, the wall between two adjacent cells being therefore 9-layered. Each wall possesses three layers in the secondary wall and around this there remains the primary wall and the middle lamella. Most elongated cells show this type of structure, though in some fibres, *e.g.* bamboo (PRESTON and SINGH 1950) the layering is more complex, and some elements of wood show detail ranging down to the limits of microscopical visibility (BAILEY and KERR 1935). In all these cases the differences between the layers derives largely from differences in physical con-

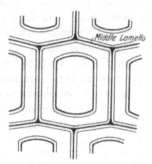

Fig. 2. Diagrammatic representation of a transverse section of tracheids in conifer wood, showing the wall layering.

stitution, though there are also less important chemical differences, which will be discussed later. The walls are commonly heavily lignified, and the presence of lignin can readily be demonstrated by staining in aniline chloride, with concentrated hydrochloric acid and phloroglucinol, or with silver following the method of COPPICK and FOWLER (1939). Maceration, *i.e.* separation of the cells, can be achieved only by removing the lignin by immersion for two or three days in 5% chromic acid, by alternate treatments with chlorine water and hot 3% sodium sulphite or by other lignin "solvent" methods.

Similar conditions occur in some vessels, though by no means all (PRESTON 1938). In *Fraxinus americana* for example, most vessel walls in transverse section appear quite structureless. Here and there, however, a few vessel elements in any section show the type of layering discussed here, and it is a peculiar fact that, if one element in a vessel shows this type of structure, then so do all the others in the same vessel.

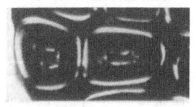

Fig. 3. Transverse section of wood fibres under a polarising microscope, nicols crossed. Note the bright inner and outer lamellae and the thick dark central layer.

In all these cell types, the lignin seems commonly to be not distributed uniformly throughout the wall but to be present in higher concentration towards the outside of each cell, so that the middle lamella is commonly the most highly lignified. The fibres of jute appear to be an exception.

The layering found in collenchyma cells is of an entirely different origin. Here the layers are visible only in the thickened parts of the wall. On dehydration these thickened parts shrink to less than one half of their normal thickness (PRESTON and DUCKWORTH 1946), so that it is commonly difficult to observe any layering in permanently mounted specimens. Even in fresh sections the layering is easily seen only after staining in pectin stains (methylene blue or Ruthenium red), or after the removal of pectin with suitable solvents or after staining in cellulose stains such as iodine and 70% sulphuric acid. It is indeed clear that, in the angular collenchyma of *Solanum lycopersicum* (ANDERSON 1927) and the strand collenchyma of *Heracleum sphondyllium* (MAJUMDAR and

Preston 1941), the appearance of the wall in sections is due to the presence of layers which are alternately cellulose-poor pectin-rich and cellulose-rich pectin-poor. In *Petasites vulgaris* (Preston and Duckworth 1946), however, the cellulose appears to be uniformly distributed through the wall though the pectic compounds are largely segregated into individual lamellae.

Cotton hairs show layering of yet a third type. The wall here can be relatively thin, and the hairs collapse on drying to a thin ribbon which is convoluted, *i.e.* twisted around its long axis. The wall is often clearly three-layered, an outer layer—the primary wall—enclothing two concentric layers, a narrow outer one and a thicker inner, both of which may be referred to as the secondary wall (Kerr 1946). In broad outline, therefore, these hairs correspond in microscopic structure to the other elongated cells already mentioned. On swelling in caustic soda, strong sulphuric acid or cuprammonium, however, the thicker inner wall layer shows numerous fine lamellations. These were first demonstrated by Balls and were suggested by him to be caused by the alternation of day and night during the growth of the hair. This suggestion was fully confirmed much later by Kerr (1937); there are therefore as many lamellae in this wall as there are days of growth. The wall consists of remarkably pure cellulose, as much as 90% or more of the wall dry weight being referred to this substance, so that there seems no possibility of the lamellation being of the collenchyma type. There seems general agreement on the view, based on the greater visibility of the layers in wet walls than in dry, that the lamellation is due to differences in porosity. The connection here between layering and the daily fluctuation in light intensity has been further confirmed by Anderson and Moore (1937), who have shown that the layering fails to develop in cotton plants grown in continuous illumination.

Although it is not the intention here to deal to any large extent with lower plants, it is worth noting that many algae show pronounced wall layering which seems often to be of the collenchyma type. The walls of the giant vesicles of *Valonia*, for instance, show as many as 40 microscopically visible layers (Preston and Astbury 1937). These layers are of the cellulose-rich, cellulose-poor type but it is not known whether the second component is entirely of a pectin nature. There is, in addition, a much finer lamellation which is not visible in the light microscope. This is dealt with elsewhere (p. 737). Layering of this type is also reported for *Chaetomorpha* (Nicolai and Frey-Wyssling 1938), *Cladophora* (Astbury and Preston 1940, Nicolai and Preston 1952) and for some other members of the *Cladophorales* (Nicolai and Preston 1952).

## b) Striations.

In surface view, the walls of many plant cells may be seen covered by a series of fine linear parallel markings known as striations. These are to be distinguished from the folds and other surface phenomena, which simulate striations proper, on account of their extreme fineness and of their clearer visibility in wet than in dry tissues.

Perhaps the outstanding example of striations, and certainly the one about which most is known, occurs in the walls of the vesicles of *Valonia*. Here, two sets of striations may be observed, one of which is commonly more obvious than the other, crossing at an angle of about 90°. These have been known for a long time (*e.g.* Correns 1892) and their interpretation has long been a matter of controversy. The fact that when the wall is torn, fine threads called *fibrils*, lying as a fringe along the torn surface, merge within the wall among the striations, suggested strongly that the striations represented something structural within the wall. There is no point here in entering into the polemics of an argument which is now merely historical; it has been established beyond reasonable doubt that, in *Valonia* and therefore presumably in other cells showing similar striations, the striations reflect accurately the underlying molecular architecture of the wall (PRESTON and ASTBURY 1937) (p. 737). Similar crossed striations occur also in the walls of *Cladophora* (ASTBURY and PRESTON 1940), some other members of the *Cladophorales* (NICOLAI and FREY-WYSSLING 1938, NICOLAI and PRESTON 1952) and one or two other members of the *Siphonales* (NICOLAI and PRESTON 1953).

Striations of this type are also visible in many of the elongated cells of higher plants. In fibres and tracheids (EAMES and MacDANIELS 1947) they commonly lie in a steep spiral[1] round the cell, the steepness of which varies from cell to cell (p. 739) (PRESTON 1948). Here, again, although this steeper spiral is always the more easily visible, a second flatter spiral can sometimes be observed (BAILY and KERR 1935). Undoubtedly these striations again reflect the underlying molecular architecture of the wall (p. 739). Similarly, some species of cotton hair show spirally wound striations, which reverse in sign periodically along the fibre in accord with the known facts of molecular structure.

## c) Pitting and other wall sculpturing.

Except in some sclerenchymatous cells, the secondary wall of the cells of higher plants is seldom entire. Here and there, small localised regions occur in which the secondary wall has failed to develop. In this way canals are formed passing from the interior of the cell, through the secondary wall up to the primary wall. These are the so-called *pits;* they may be parallel-sided, when the pit is referred to as *simple* (Fig. 4), or the secondary wall may overarch the original thin area, when the pit is referred to as bordered (Fig. 5). These pits occur almost invariably in pairs, *i.e.* wherever one cell has such a pit, the neighbouring cell in contact with it also has a pit, and the pits are then said to *correspond*. In the region of the pit, therefore, the two lumina (or the two protoplasts in living cells) are separated only by a thin membrane—the pit membrane—consisting of two primary walls and a middle lamella. The pit membrane is usually, in addition, perforated by a number of pores which range downwards from about $1 \mu$ in diameter. The size and nature of these has been studied by many workers and reference may be made especially to the work of STAMM (1946), who has subsequently made use of these and other relevant measurements in the cells of wood in an investigation of diffusion through wood. It seems probable that these pores correspond to some at least of the plasmodesmata whose presence in the primary walls of living cells can be demonstrated by special staining reactions. If these plasmodesmata are really permeated by cytoplasm, as they seem to be, then it is perhaps not surprising that the connection between wall and cytoplasm remains, in these localised regions, more intimate than elsewhere in the wall so that the development of a secondary wall, with its associated, if nowadays somewhat ill-defined, separation between wall and cytoplasm fails to occur.

The type of pit present in a wall depends in the first instance upon the type of cell. In general, parenchymatous cells have simple pits and most other cell types bordered pits. Careful examination of thin sections shows, however, that there are intermediates and it is sometimes difficult to be sure that a pit is simple.

[1] More properly, a helix, but the term spiral is now traditional in this context.

Simple pits, *par excellence*, may be seen in sections of the endosperm of some seeds, *e.g.* bluebell *(Scilla non-scripta)*. The wall is unusually thick as a result of a heavy deposition of what appears to be a hemicellulose and the pits are very prominent. Although the thickening layers tend to overarch the pit, the pit mouth (the end of the canal nearer the lumen) retains substantially the original diameter. Simple pits are prominent in many other parenchymatous tissues, *e.g.* ray parenchyma and wood parenchyma, but are sometimes so obscure that some swelling of the wall is necessary for distinct visibility, *e.g.* lateral walls of leaf epidermal cells. Pits are occasionally lacking, as they are said to be between the guard cells and the surrounding epidermal cells in leaves. They are said to be missing, or at least difficult to observe, in the union between scion and stock tissues.

In the strict sense, pit-pairs occur only in cells with secondary thickening, and they should therefore not be confused with *primary pit fields* typical of

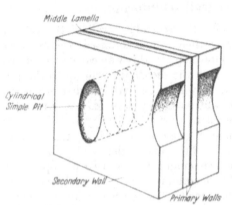

Fig. 4. Diagrammatic representation of a simple pit.

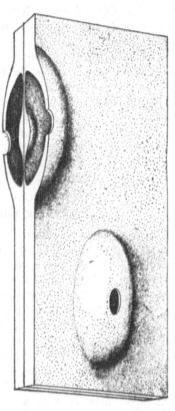

Fig. 5. Diagrammatic representation of a bordered pit.

meristems. The pits proper develop, within this primary pit field, only in the secondary wall. BAILEY and FAULL (1934) have called attention to the fact, for instance, that the cell walls of wood and ray parenchyma in *Sequoia sempervirens* are entirely primary in nature and conspicuous primary pit fields are present. These resemble only superficially the true pits found in wood and ray parenchyma of the *Abietoideae*. Bordered pits are present in almost all the elongated lignified elements of the wood. They are especially large in conifers, *e.g. Pinus sylvestris* which forms very suitable material for demonstration. The tracheids of conifers are cut off from the cambium in such a way that two opposite walls lie tangentially, and two radially, in the stem. The radial walls only are pitted except in the last few tracheids laid down each year where a few pits may be seen on tangential walls. The pits are large, usually uniseriate in the wall so that the original pit area, before the border began to form, occupies some two thirds of the breadth of the cell. The border is very prominent and the pore, *i.e.* the opening to the lumen, is circular or oval. In median sectional view, the overarching border is very prominent, and another feature, the *torus*, is conspicuous. This consists of a discoid thickened area of the pit membrane which is larger than the pore. It may lie medianly, or it may be pressed against one or other of the borders of the pit pair when the pit is said to be *occluded*. Bordered pits in other vascular plants vary very widely in form, structure, size and abundance and provide diagnostic features in the morphological study of woods (RECORD 1934, BROWN and PANSHIN 1940, EAMES and MACDANIELS 1947).

Taking the bordered pits of *Pinus*, in a loose sense, as the prototype, then the pits in other vascular elements are considerably modified in accordance both with the type of cell in which the pit occurs and with the type of adjacent cell containing the other member of the pit pair. In tracheids, fibre tracheids and fibres with very thick walls, the modification appears due largely to the excessive wall thickness. In these cases, the outer part of the pit, nearer the pit membrane, is a small-domed chamber somewhat resembling in shape the chamber of the *Pinus* pits but deeply embedded. The opening of this dome is circular; it forms the outer region of the pit canal which, on the inner face of the wall, has become a long slit as seen in face view, extending well beyond the limits of the pit cavity. The pit canal therefore resembles a flattened funnel (EAMES and MACDANIELS 1947, p. 47, Fig. 30). The longer axis of the slit mouth is usually tilted at an angle to cell length and the angle is fairly constant over one cell. A line drawn on the wall, parallel to the slit mouth would, indeed, trace out a spiral, and since the sign of the spiral is usually constant for any individual plant then it follows that the two slit mouths of an adjacent pit-pair are tilted at an angle equal to twice the angle of the spiral. The slit mouth direction follows rather closely the molecular structure of the secondary wall (p. 739).

The variation of pit type in a cell with the nature of the cell with which contact is made is well exemplified by the large vessels in the early wood of ring-porous dicotyledonous trees (PRESTON 1938). Here, one and the same vessel may make contact with tracheids, fibres, wood parenchyma, ray parenchyma and other vessels. In contact with tracheids and fibres the pit on the vessel side is a normal fully bordered pit. In contact with wood parenchyma the pit is less markedly bordered, and with ray parenchyma the (very large) pits are hardly bordered at all. In some species the pit membranes of these large pits in parenchyma contacts grow out or are blown into the vessel cavity and may block the vessel completely. These are known as *thyloses*.

### d) Other secondary and tertiary wall patterns.

In many trachiary elements the innermost wall layer consists only of a spirally arranged wall thickening. These are typical, for instance, of the elements of primary xylem. In the earliest elements (known collectively as the *protoxylem*) the thickening is annular, *i.e.* takes the form of separate transversely oriented rings. Later elements show the spiral type of thickening the pitch of the spiral become smaller the later the element is developed. Passing further outwards from the earliest protoxylem, the spirals become "buttressed" by bars joining adjacent turns of the spiral, forming a reticulate thickening with transversely elongated thinner regions, from which a transition to scalariform and finally normally pitted walls is relatively easy. These thickenings are laid down on a primary wall and are therefore themselves secondary. Similarly thickenings are sometimes laid down on a secondary wall and these thickenings are sometimes referred to as *tertiary*. There seems, however, no clear reason why these too should not be regarded morphologically as secondary.

## e) Sieve tubes and sieve plates.

The phloem of angiosperms is very frequently complex in structure but is invariably characterised by the presence of cells termed *sieve tube elements*. Ontogenetically these resemble the vessel elements of the xylem in being closely related to each other in longitudinal sequence to form *sieve* tubes. They are unique among living cells in being devoid of a nucleus on reaching maturity.

The walls of sieve tubes are interesting microscopically and, of course, physiologically owing to the development of *sieve plates* and *sieve areas*. These involve a structural modification of certain portions of the primary wall which. in face view, exhibit a reticulate appearance. It was first thought that the spaces between the network of wall substance were open pores connecting the vacuoles of two contiguous cells but later investigations have led to the idea that the pores are filled with strands of living protoplasm. There is, however, still much controversy concerning the nature of these openings (Esau 1939, Crafts 1939). Recent observations in the electron microscope confirm that the sieve pores are plugged by cytoplasm (Herton, Preston and Ripley 1955). It may be said in general that two types of sieve plate occur on the end walls of sieve tube elements in angiosperms. If the end wall is oblique, then a number of sieve areas may occur on it, separated by bars of wall substance. These are referred to as compound sieve areas (Esau 1948). When, however, the end wall is transverse or only slightly oblique, then the whole end wall becomes one sieve plate, the classical example occurring in the sieve tube elements of *Cucurbita*.

Comparable structures also occur on the lateral walls of sieve tube elements but are then often more delicate in appearance and the terms "sieve area" or "sieve field" is often applied to them.

The walls of sieve tubes must be termed primary. At a certain stage during differentiation, before the nucleus has disintegrated, the wall is often very thick in fresh undehydrated material. As the element matures and enlarges, however, the wall becomes progressively thinner.

# The submicroscopic structure of plant cell walls.

By

## R. D. Preston.

With 11 figures.

## The chemical components of cell wall.

The chemical substances in cell walls may, from the point of view of their physical and physiological attributes, be divided into two groups—the skeletal or structural components and the incrusting substances which are deposited arround them. The former are usually to some degree crystalline and the latter usually amorphous.

## Skeletal substances.

Among the skeletal substances, *cellulose* and *chitin* are outstanding. The former is characteristic of higher plants and some algae, though by no means all (NICOLAI and PRESTON 1952) and the latter is typical of fungi (though some fungi contain cellulose). In many algae the skeletal substance is unknown (NICOLAI and PRESTON 1952) but it is probably some polysaccharide or polysaccharide derivative. Among the incrusting substances, *hemicellulose* and *pectic compounds* (which may be referred to collectively as the *polyuronides*) are almost universal and *lignin* is prominent in woody tissues. The *cellulosans* such as mannan (in gymnosperms) and xylan (in angiosperms) may perhaps also be classed as incrusting substances though they are thought to be much more closely associated with cellulose than are the rest of these materials (ASTBURY, PRESTON and NORMAN 1935, PRESTON and ALLSOPP 1939). The following is a brief account of the structure, distribution and organisation of these substances in cell walls. For further details the reader is referred to other texts (NORMAN 1937, MEYER 1942, WISE 1944, HEUSER 1946, PRESTON 1952, FREY-WYSSLING 1953).

## Cellulose.

Cellulose may be regarded as a linear polymer of which the units are essentially $\beta$-glucose (Fig. 1). This contrasts strongly with the storage polysaccharide starch in which the units are $\alpha$-glucose. In cellulose some of the —OH groups may rarely have become carboxylated (MEYER 1942) and occasionally xylose residues may possibly replace glucose. The $\beta$-glucose units are united into long chains by 1:4 glucosidic bonds so that any two neighbouring residues form a cellobiose residue (HAWORTH 1929). The configuration of a section of such a chain may be imagined as depicted in Fig. 2, in which hydrogen atoms are omitted. In the crystalline regions of the cellulose, the chains are further arranged strictly parallel to each other and are spaced apart quite regularly (Fig. 3) forming arrays of parallel chains which are capable of giving rise to coherent "reflections" of a beam of X-radiation impinging on them. It is, of course, largely upon the interpretation of the resulting X-ray diagram that the structure of cellulose depends (MEYER 1942, PRESTON 1952).

The cellulose of plant cell walls is, however, not composed entirely of these highly crystalline regions. Since the time of NÄGELI it has been clear that there are periodically breaks in this regularity of structure, leading at first to the idea of discrete crystallites—the micelles of NÄGELI—separated by spaces containing only the amorphous non-cellulosic materials. The evidence for this type of discontinuity may be obtained in any one of a number of texts (*e.g.* PRESTON 1938, 1952, MEYER 1942, FREY-WYSSLING 1953). It is now known

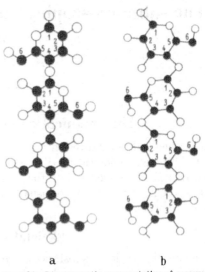

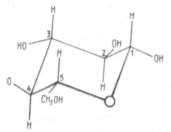

Fig. 1. Diagrammatic representation of β-glucose in the "armchair" modification found in cellulose.

a        b

Fig. 2a and b. Diagrammatic representation of a segment of a cellulose chain. (a) Following MEYER and MARK; (b) after STUART. Open circles, oxygen; full circles, carbon; hydrogen omitted.

that this conception went too far; general agreement has been reached that the molecular chains of cellulose run continuously from one "micelle" to the next,

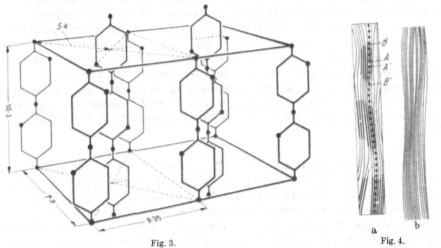

Fig. 3.        Fig. 4.

Fig. 3. The unit cell of native cellulose (cellulose I), showing the relative disposition of the chains in native cellulose.
Fig. 4a and b. Modern concepts of micellar organisation: (a) "fringe" micelles according to KRATKY and FREY-WYSSLING. (b) The continuous deformed structure proposed by MEYER (reproduced by courtesy of the late Professor K. H. MEYER).

the major difference between the so-called "micelles" and the "intermicellar spaces" being that in the latter the chains are either not parallel to each other or are not regularly spaced, or both (Fig. 4) (MEYER 1942, PRESTON 1952).

The occurrence of larger units has long been a matter of debate (see references in ASTBURY and PRESTON 1940) but the modern observations made in the electron microscope have so cleared up the situation that there is no longer any need to discuss the older literature in any detail. As first reported by PRESTON *et al.* (1948) for *Valonia*, a few weeks later by FREY-WYSSLING (1948) for cells of higher plants and later (see references in PRESTON 1952) for a wide variety of celluloses, natural cellulose occurs exclusively in the form of long "microfibrils" of the order of 100–300 Å in diameter (Fig. 5). At the moment it appears that the diameter is not constant throughout the Plant Kingdom but increases in the

Fig. 5. Electronmicrograph of the surface of the wall of a vesicle of *Valonia:* Pd-Au shadowed: Magnification 25,000 ×.

order: bacterial cellulose—higher plant cellulose (chiefly wood)—algal cellulose. It seems probable that the fine details of internal structure in these microfibrils is equally variable (PRESTON 1952). Even the arrangement of the microfibrils is again variable; in the algae *Valonia* (PRESTON *et al.* 1948, PRESTON and KUY-PER 1951, PRESTON, NICOLAI and KUYPER 1953), *Cladophora* and *Chaetomorpha* (NICOLAI and PRESTON 1952) the microfibrils lie beautifully parallel to each other; in the secondary walls of wood elements they tend to lie parallel but there is considerable dispersion from this preferred orientation (HODGE and WARDROP 1950, PRESTON and RIPLEY 1954, while in some primary walls it is difficult to detect any orientation whatever.

Returning for a moment to the internal organisation of the microfibrils, the only evidence available is concerned with the effects of acid hydrolysis. As first shown by RANBY (1949), if cellulose is boiled in sulphuric acid and washed in water, then at pH 3 an opalescent solution is obtained. A little of this dried down shows in the electron microscope innumerable rodlets 50–100 Å in diameter and some hundreds to some thousands of Å long, *i.e.* of much the same dimensions as the "micelles" said to preexist in cellulose by Hengstenberg and Mark (1928) using X-ray methods. These rodlets are flat and lie on their broader faces which correspond to certain molecular planes in the cellulose space lattice (the 101 planes spaced 6.1 Å apart, see Fig. 3) (Mukherjee and Woods 1953). This recalls the fact that the microfibrils themselves are flattish ribbons rather than rods and, further, that in the *Valonia* wall, and in the walls of the *Clado-phorales*, these same planes lie parallel to the wall surface. It is not, however, clear that the rodlets observed after acid hydrolysis are anything more than degradation products. They have not yet been observed with *Valonia* cellulose, perhaps on account of the severe treatment needed to hydrolyze this exceptionally highly crystalline cellulose.

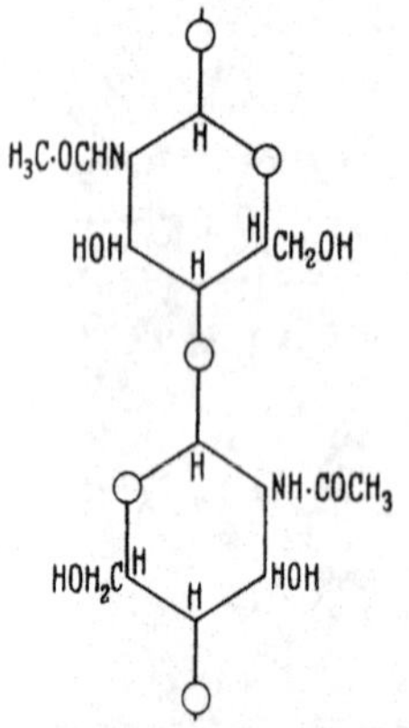

Fig. 6. Two adjacent acetyl-glucosamine residues in a molecular chain of chitin.

## Chitin.

Plant chitin, which is the typical skeletal substance of some, though not all, fungi, resembles animal chitin very closely. It differs from cellulose chiefly in that the unit of structure is not $\beta$-glucose but $\beta$-acetylglucose-amine (Fig. 6). This replacement of a hydroxyl by a larger radicle means that the chains of polyacetylglucos-amine cannot be so closely packed as can those of the polyglucose. Otherwise the structure of chitinous cell walls resembles closely that of cellulosic cell walls (Roe-lofsen 1949, Middlebrook and Preston 1952) and nothing further need be said about them.

## The incrusting substances.

The fact that the removal of the polyuronides including pectin and of lignin has no marked effect on the optical behaviour of the wall (Bailey and Kerr 1935, Freudenberg and Dürr 1932) on its cohesion (Bailey and Kerr 1935, Bonner 1936) or on the molecular structure of the cellulose (Astbury, Preston and Norman 1935, Preston and Allsopp 1939) means that these substances lie outside the crystalline regions of the cellulose. This undoubtedly explains why lignified tissue fails to give a positive staining reaction for cellulose until after the lignin is removed. There is some evidence that a fraction of the lignin is more closely associated with the cellulose in a cell wall. The cellulosans such as mannan in gymnosperma and xylan in angiosperms appear to associate with the cellulose in some way since extraction of these substances tends to increase the degree of crystallinity of the cellulose component (Astbury, Preston and Norman 1935).

As regards the distribution of these substances at the level of the light micro-scope, pectic compounds appear usually to be distributed rather uniformly in the wall. There are, however, exceptions in collenchymatous cells, in which this type of substance is to a large extent segregated into separate layers (Anderson 1927, Majumdar and Preston 1941, Preston and Duckworth 1946). Lignin tends to be present in larger amounts towards the outside of cells. This can readily be seen in transverse sections of lignified elements using the normal staining methods. The same phenomenon has also been demonstrated, and the "concentration" of lignin measured, using, instead of visible radiation, ultra-violet light to which lignin is opaque.

Like cellulose, the pectic compounds and the cellulosans are composed of long chain molecules in which the repeating units are, respectively, galacturonic acid or a derivative, and a pentose (xylose or mannose) instead of glucose. Pectin is properly regarded as the fully methylated polygalacturonic acid and pectic acid as the free acid, each associated in some way with an araban and a galactan. In mature tissue pectin has been transformed to the Ca salt or the Ca—Mg salt of pectic acid forming a hard cement in, for instance, the middle lamellae. Lignin, on the other hand, while presumably a high molecular weight compound, is not a linear polymer. Space will not allow a discussion of its structure here; reference must be made to other texts [1].

## The crystalline/non-crystalline ratio in cell walls.

Of late years, perhaps on account of the indefiniteness of the micellar concept, considerable attention has been paid to the determination of the relative amount of the crystalline component in cellulose-the so-called crystalline/non-crystalline

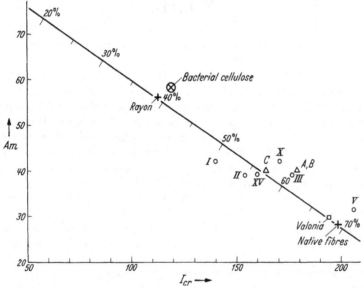

Fig. 7. Crystallinity nomograph plotted from the $I_{er}$ and $A_m$ values (representing the X-ray scattering from crystalline and non-crystalline regions respectively). Percentage of crystalline portion indicated along oblique line. O (V, X, XV *Pinus radiata*. CROSS and BEVAN celluloses; O (I, II, III) Bamboo; △ *Dendrocalamus strictus*.

ratio. This ratio is undoubtedly of some importance in fibre technology and cannot fail to hold some interest in the biological field.

The basic concept is that, since in some regions of the cellulose the chain molecules are arranged in a space—lattice whereas in other regions they are not, then the chains in these two regions should react differently both chemically and physically. This is due to the fact that only in the disordered regions are there many —OH groups lying more or less free. In consequence, a greater proportion of non-crystalline material will involve a greater water sorption, a greater ease of chemical attack, a lower density (in view of the less close packing of the chains in the non-crystalline regions) and different optical properties. All these have been worked up into methods for determining the value of the crystalline/non-crystalline ratio (MARSDEN 1949). Use has also been made of the different X-ray scattering in crystalline and non-crystalline regions (HERMANS and WEIDINGER 1948). Some of the results of this last method are given graphically in Fig. 7 (PRESTON, HERMANS and WEIDINGER 1950).

---

[1] Cf. WISE (1944).

It should be noted that since the criterion of crystallinity varies with the method used in determining the crystalline/non-crystalline ratio, the figures yielded by different methods on the same material are not necessarily by any means identical.

## Sub-microscopic structure of thick cell walls.

In dealing with the sub-microscopic structures of cell walls it is well to deal separately with primary (growing) walls and the secondary layers deposited upon them when the cell has reached maturity. In this section it is intended to deal with these secondary layers only; but since some attention must also be paid to the cell walls of algae where the terms primary and secondary are difficult to apply it has been thought preferable to refer here to *thick* cell walls.

## Methods.

In general, three distinct methods are now available for the structural investigation of cell walls and it will be as well to outline from the start the particular advantages and disadvantages of each.

The method of *X-ray analysis* allows, quite unequivocally, both the identification of the structural material of the wall (if this is a known substance such as cellulose), and of the preferred direction or directions of the constituent molecular chains, together with some measure of the degree of spread of the direction of individual chain bundles around this direction (the angular dispersion) and of the degree of crystallinity. These determinations refer, however, only to the crystalline regions; non-crystalline material lies outside the scope of this method. The material investigated must usually be at least air-dried, but it has been shown (PRESTON, NICOLAI and WARDROP 1948) that air drying has no effect on any observable details in the X-ray diagrams of cell walls. The specimen used in this method is commonly about 1 mm. thick so that, since the X-ray beam is normally 0.5 mm. diameter, the total volume of material used in obtaining one X-ray diagram is about 0.2 mm.$^3$. The use of low angle scattering also often enables the determination of the size of structural units such as the "micelles" and the "microfibrils".

Observation in the *electron microscope* of itself gives only the size and shape of the particles constituting the material examined. The resolving power of most electron microscopes at present is of the order of 30 Å so that the information yielded is well above the molecular level. In cellulose, for instance, the size and orientation of the microfibrils are evident, but the internal structure of the microfibrils is quite unresolved. The material is completely desiccated in the high vacuum of the microscope and bombardment by high speed electrons causes some heating. The material examined must be less than, say, 0.1 μ thick and with a magnification of 30,000× the area of wall in a single photograph may be about $3 \times 10^{-5}$ mm.$^2$. The total volume of wall examined in one photograph is therefore about $3 \times 10^{-9}$ mm.$^3$, enormously smaller than the volume corresponding to one X-ray photograph.

Under the *polarising microscope*, conditions are less stringent. The material may be living or dead, wet or dehydrated in alcohol etc., provided it is thin enough for penetration by visible light. The methods used give, for instance, the average direction of the cellulose chains in a wall (more or less accurately depending on the conditions) and some measure of the angular dispersion. If only one preferred direction is present, and this can sometimes be demonstrated in the polarising microscope itself, then this direction is thus readily obtained. Measurements of refractive indices can, in addition, give some idea of the degree of angular dispersion. The resulting data refer, of course, exclusively to material which is crystalline in the optical sense.

## Cell walls of algae.

Following the results both of X-ray analysis and of observations under the polarising microscope (NICOLAI and PRESTON 1952) as well as the more recent work in the electron microscope, the algae have provisionally been divided into three groups as far as wall structure is concerned.

The first group includes *Valonia* (PRESTON and ASTBURY 1937), *Chaetomorpha* (NICOLAI and FREY-WYSSLING 1938) and *Cladophora* (ASTBURY and PRESTON 1940) which had been worked previously. The more recent work (NICOLAI and PRESTON 1952) has extended the group to cover all species of the *Cladophorales*, except the *Spongomorpha* group of *Cladophora*, and a few members of the *Siphonales (e.g. Siphonocladus)*. The walls of this group of algae contain well-oriented, highly crystalline native cellulose (Cellulose I) as the skeletal material, organised into clearly defined microfibrils as seen in the electron microscope. The walls are finely lamellated and the constituent microfibrils lie, in the main, in two directions almost at right angles to each other and such that odd lamellae possess one direction and even lamellae the other. The thickness of the lamellae is probably of the order of the diameter of the microfibrils, *i.e.* about 200 Å. The mechanism which produces this periodic switch in microfibril direction is still obscure though an attack has recently been made upon the problem (NICOLAI and PRESTON 1953). In the aplanaspores of *Valonia* the microfibrils in the wall first laid down appear to be randomly oriented, but this changes rapidly to a lamella with the typical parallel arrangement (STEWARD and MUEHLETHALER 1953). This development of a well ordered from a disordered state may also occur in the new lamellae being deposited on the wall of an older vesicle (PRESTON and KUYPER 1952).

Occasionally a third orientation may be observed in the *Valonia* wall both in X-ray diagrams (PRESTON and ASTBURY 1937) and in electron micrographs (WILSON 1951, STEWARD and MUEHLETHALER 1953). Its appearance in adult vesicles is sporadic but, as far as the present evidence goes, it may be more frequent in younger vesicles. STEWARD and MUEHLETHALER (1953) have recently criticised the earlier work of PRESTON and ASTBURY (1937) and suggest a replacement of a two-lamella repeat by a three-lamella repeat with an angle of $60°$ between the constituent microfibrils. Both this and their criticism of the model of the whole vesicle proposed by PRESTON and ASTBURY (1937) are based as yet on rather insecure grounds and a decision must be postponed until more evidence is available as to the significance of the third orientation (PRESTON and ASTBURY 1954, PRESTON and RIPLEY 1954b).

A further intriguing structural feature of these algae has been already briefly mentioned. The X-ray diagram (Fig. 8) demonstrates that the molecular planes spaced 6.1 Å apart (Fig. 3) lie more or less parallel to the surface of the wall (PRESTON and ASTBURY 1937). This implies both that within the microfibrils these planes lie also more or less parallel to each other and that the microfibrils are selectively oriented about their own axes. It is not clear whether this last arrangement arises by virtue of the flat cross-section of the microfibrils or whether it has more to do with the high intensity of distribution of —OH groups in the 101 planes of cellulose.

In the second group of algae, to which most of the algae belong, the X-ray diagram of the wall suggests that the skeletal substance is hydrate cellulose (mercerised cellulose, cellulose II) (Nicolai and Preston 1952), though it may possibly be a polysaccharide derivative and not cellulose itself. Prominent among the group are e.g. *Halicystis, Spongomorpha* and *Ulothrix*. The X-ray diagram reveals that the material is poorly crystalline and usually without orientation. Electron micrographs nevertheless show the presence of micro-

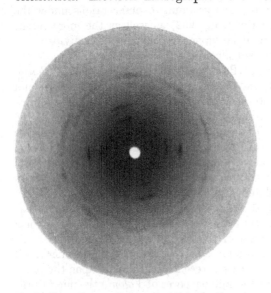

fibrils though these are much thinner than those typical of the native cellulose of the first group, of the order of 100 Å or less in diameter (Nicolai unpub.). They are randomly oriented, very contorted and embedded in an amorphous matrix which is much more abundant than in the algae of group I. In samples of *Spongomorpha* washed in distilled water when alive and then dried, the X-ray diagram of the poly-saccharide is masked by that of thetine—one of the few trans-methylating agents—which is apparently deposited in the wall (Nicolai and Preston 1953b). In some species this thetine may be washed away from the dried material by immersion in water.

Fig. 8.  X-ray diagram of the wall of *Valonia*. X-ray beam normal to wall surface. On each of the two rows of prominent arcs, the outer pair represent planes of 3.9 Å spacing and the nner pair 5.4 Å. The reflections corresponding to 6.1 Å are missing.

The third group contains as yet very few species, the out-standing being *Spirogyra, Vaucheria* and *Enteromorpha* (Nicolai and Preston 1952). The X-ray diagram corresponds to some unknown material with some orientation. In electron micrographs, the microfibrils appear very much as described for the second group.

## Cell walls of fungi.

Among the fungi, detailed study by modern structural methods has been devoted largely to the yeasts and the phycomycetes. In both these organisms, the chief skeletal substance is chitin (Heyn 1936, van Iterson, Meyer and Lotmar 1936, Middlebrook and Preston 1952, Roelofson 1949, Kreger 1954) associated with a variety of other substances, some of them unknown (Kreger 1954). In *Phycomyces* the molecular chains of chitin in the growth zone of the sporangiophores (immediately at the tip) have been shown to be arranged in a flat spiral (Middlebrook and Preston 1952, Roelofson 1949). In the secondary wall deposited lower down, however, they lie approximately parallel to the length of the sporangiophore (Middlebrook and Preston 1952). Electron micrographs reveal the presence of microfibrils embedded in amorphous material (Middlebrook and Preston 1952, Roelofson 1949). The chitin of plants is said to be identical with that of animals. Some species of fungus are said to contain cellulose, e.g. *Saprolegnia*.

# Cell walls of higher plants.

The cellulose in higher plants is identical chemically and physically with that of the algae of group I discussed above; the diameter of the constituent microfibrils is, however, somewhat less (HODGE and WARDROP 1950, RANBY 1949, PRESTON and RIPLEY 1954).

In most elongated cells of higher plants (*i.e.* fibres and tracheids) the cell wall is three to many-layered and the orientation of the microfibrils varies from one layer to the next much as it has been shown to do in *Valonia* (p. 737). In no case, however, are the microfibrils either so well oriented, or apparently so highly crystalline, as in the algae of Group I. In those species which have so far been examined (PRESTON 1952) the microfibrils within each layer lie in a spiral round the cell. It was first shown in conifer tracheids (PRESTON 1934, 1946) and later in other cell types (MISRA 1939, PRESTON and MIDDLEBROOK 1949, PRESTON and SINGH 1950) that, within the limits of dimensions of any one cell type, the angle ($\theta$) between the spiral winding and the direction of cell length decreases with cell length ($L$) in the form

$$L = A + B \cot \theta \qquad (1)$$

where $A$ and $B$ are constants. It is not clear that this relation has any fundamental significance other than stating the steepening of the spiral as cell length increases. The relation refers to the secondary layer only, *i.e.* the layer deposited after the bulk at least of cell expansion is over. It therefore implies not a pulling out of existing spirals, but an adjustment to cell length of the mechanism orienting the cellulose microfibrils. The adjustment is not, however, perfect since divergences from relation (1) are marked and real, *i.e.* beyond the limits of experimental error. Some of these divergences in conifer tracheids have been traced to an effect on secondary wall structure of the previous rate of growth of the primary wall (WARDROP and PRESTON 1950); for a tracheid of a given length the wall spiral is flatter the more rapidly the corresponding cambial initial reached its final length. There is, in addition, the question of the variability of cell width—the wider the cell the flatter the spiral.

The general effect of this connection between cell dimensions and wall structure can be seen in glancing over the variety of cell types in higher plants. Vessels in general, and particularly those of ring porous dicotyledons, tend to be short fat cells; with this goes a tendency for the cellulose chains to lie almost transversely. With tracheids, the wider range in cell length corresponds to a wide range in the angle of the structural spiral. Finally, phloem fibres are commonly so long that the spiral is usually very steep; in ramie, for instance, the cellulose chains lie almost longitudinally. These statements refer, of course, to the thickest layer in the wall, the layer which is the darker in transverse section between crossed nicols. They can readily be verified by determination of striation directions or of the run of the slit in the slit pits of fibres or, correspondingly, of the direction of the major axis in pits with oval mouths such as occur in some tracheids and in vessels. Indeed, in those vessels of *Fraxinus americana* in which the cellulose microfibril direction varies from layer to layer, the direction of the slit mouth changes accordingly giving the so-called spiral pit. It can further be seen that, since the structural spirals in, for instance, the fibres of any one plant are usually of the same sign (commonly left-handed in the English sense[1]), the slit mouths on neighbouring walls are crossed.

---

[1] *i.e.* a left-hand spiral is such that, on looking along the axis of the spiral the spiral winding turns anticlockwise away from the observer.

## Analytical methods for thick cell walls.

It therefore seems to be the rule that, whenever cellulose is found in the walls of plant cells with any considerable degree of orientation, there are at least two directions of orientation, and the mechanism of ordered deposition can switch from one to the other. The development of this "crossed fibrillar" structure constitutes one of the major problems in cell wall physics at the present time and it is worth while therefore to note that the demonstration of the crossed fibrillar structure depends on the use, not of one method of investigation, but of several. Each cell type has presented its own particular difficulties which have had to be met by appropriate modifications of the methods used in structural investigations. The following is a brief review of the methods used in the various cell types investigated, as a guide to future work on this and related problems.

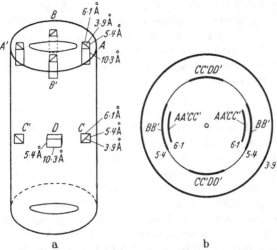

### a) Valonia.

Little need be added to what has already been said concerning this organism. The vesicle is so large that single pieces of wall can be removed for X-ray examination, and is so finely lamellated that single lamellae can be removed for examination in the electron microscope. The X-ray beam, directed normally to the wall surface indicates clearly the directions of the two (occasionally three) cellulose chain orientations. The major arcs on the diagram correspond to "reflections" from planes lying parallel to the chain direction; hence the chain direction lies at right angles to each row of prominent arcs. The electron micrographs (see Fig. 5) are in full confirmation; they add the further information that the individual directions are segregated into individual lamellae, a point which previously depended upon observation under the polarising microscope.

Fig. 9a and b. (a) Diagrammatic representation of a filament of *Cladophora*. The orientations of the unit cell of cellulose are indicated for longitudinally (upper) and transversely (lower) directed chains. (b) Diagrammatic representation of the X-ray diagramm of *Cladophora*, X-ray beam perpendicular to cell length. The letters refer to the corresponding positions in (a); outer ring, 3.9 Å; inner ring 5.4 Å; innermost arcs 6.1 Å.

### b) The Cladophorales.

The examination of filamentous algae presents the difficulty that, except in the larger species such as *Cladophora prolifera* which can be treated like *Valonia*, the experimental object is a hollow cylinder smaller in diameter than the X-ray beam. Using the X-ray method, a bundle of parallel filaments is arranged at right angles to the beam and yields the diagram illustrated diagrammatically in Fig. 9b. This naturally appears very different from the *Valonia*. diagram but nevertheless the interpretation is the same (cp. Fig. 9a and b). In the bulk of the cells of larger plants, and in the more apical cells of younger plants, one set of chains lies longitudinally and one transversely. In either case, both sets lie in spirals in the more basal parts of the filaments. As in *Valonia*, the chain direction is identical with striation direction. The variability of each direction has enabled the conclusion to be reached that in these algae the angle between the two sets of microfibrils tends to remain constant (ASTBURY and PRESTON 1940).

It may be noted that, in branching species such as *Cladophora*, the irregularly arranged branches do not interfere provided that a sufficient number of main filaments are arranged parallel.

Electron microscopy is also difficult with these small filamentous species. With care, however, lamellae can be stripped off with needles and the electron micrographs are then identical with those of *Valonia*. The fact that, under a polarising microscope, the central parts of the cells (where light is passing through an upper and a lower wall) are optically isotropic shows that the two sets of chain directions normally occur with about equal frequency.

## c) Elongated cells of higher plants.

**1. Sisal.** When, as in the leaf fibres of sisal, the fibres differentiate very slowly, the problem is easy of investigation. In these monocotyledonous leaves the fibres develop from the meristem at the leaf base and mature progressively up the leaf toward the tip. It is therefore a simple matter to pick out fibres with only the outer layer present and, further up the leaf, fibres with both layers present. These can then be investigated separately by the X-ray method. Measurement of the birefringence $(n_\gamma - n_\alpha)$ in transverse section makes it reasonably certain that the outer secondary layers of the younger cells are substantially the same as those of the mature cells. It is concluded that the cellulose chains in this outer layer make an angle of some 50° to cell length, the angle in the inner layer being about 20° (PRESTON and MIDDLEBROOK 1949).

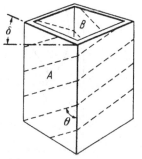

Fig. 10. Diagrammatic representation of a portion of a cell, the wall being built up of one molecular spiral whose tilt is indicated by the dotted lines.

**2. Conifer tracheids.** Next in order of difficulty come the elongated cells which are not readily available in the immature state but which, in the adult state, are regularly arranged. Consider a transverse section of one such cell (Fig. 10). The cell walls will be more or less birefringent according as the cellulose chains lie more or less nearly parallel to the transverse plane. If sections are cut at increasing angles, $\delta$, to the transverse plane then the birefringence will increase to a maximum at the point where the plane of section lies parallel to the cellulose chain direction. At this point $\delta = 90° - \theta$ where $\theta$ is the angle between the cellulose chain direction and cell length. Hence $\theta$ can be determined for any wall layer. The angles in the various layers turn out to be of the same order as those for sisal, depending of course on cell length (WARDROP and PRESTON 1950). For the application of this method it is essential that the cells should be regularly arranged. In conifer tracheids the tangential walls lie more or less parallel to each other and form therefore ready experimental material.

**3. Bamboo fibres.** The most difficult case arises when the only cells available are mature and irregularly arranged as in bamboo. Here, recourse has to be made to thoroughgoing optical analysis. In principle, the method is as follows (PRESTON and SINGH 1950).

The projection of the optical index ellipsoid on the surface of any layer of a wall can be set up as in Fig. 11, where $AB$ is the direction of the cellulose chains in the layer. From the equation for an ellipse it follows that

$$n_{\gamma}'' = \frac{n_{\gamma} n_{\alpha}}{n_{\alpha}^2 + (n_{\gamma}^2 - n_{\alpha}^2) \sin^2 \theta} \qquad (1)$$

$$n_{\gamma}^L = \frac{n_{\gamma} n_{\alpha}}{n_{\gamma}^2 - (n_{\gamma}^2 - n_{\alpha}^2) \sin^2 \theta} \qquad (2)$$

$n_{\gamma}''$, $n_{\gamma}^L$ and $n_{\alpha}$ can be measured. Hence $n_{\gamma}$ and $\theta$ may be calculated. Since the refractive indices can be measured only at the edges of a fibre, only the outer

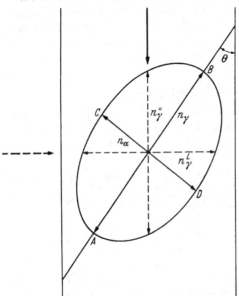

lamella can be treated in this way. This is the outermost layer which is bright in transverse section between crossed nicols. In bamboo there are several other similar bright layers inside the wall; for these it is necessary to assume the value for $n_{\gamma}$ determined for the outermost lamella, to measure the birefringence by a compensator $(n_{\gamma}^L - n_{\alpha})$ and to calculate $\theta$ from (2). The value of $\theta$ for the thicker dark layers may be obtained from the X-ray diagram of a bundle of fibres.

## The primary wall.

Primary walls are more difficult to deal with on account both of their tenuity and of their low cellulose content. The drying of the primary walls of conifer cambium for physical investigation, for instance, leads to a

Fig. 11. For explanation, see text.

shrinkage in wall thickness to about $1/_3$ of the thickness in the fresh condition. This involves, however, no serious modification of the X-ray diagram (PRESTON et al. 1948) so that data obtained from air-dried material can substantially be applied to considerations of fresh material.

Although some uncertainty was formerly expressed that primary walls contain cellulose, there can no longer be any doubt. The volume content of cellulose is, however, low, of the order of 14% in *Avena* coleoptiles, for instance, and only 8% in conifer cambium. It was early suggested, on rather inadequate evidence, that the molecular chains of cellulose run transversely to cell length; the fact has now been established that the chains lie in a flat spiral (PRESTON and WARDROP 1949) the spiral winding in conifer tracheids making an angle of some 16° to the transverse with very considerable angular dispersion (PRESTON 1947). Substantially the same condition is found in the growth zone of sporangiophores in *Phycomyces* (MIDDLEBROOK and PRESTON 1952) where, however, the substance involved is, of course, chitin and not cellulose. No great change in this orientation occurs as the cell grows.

Electronmicrographs reveal the presence of a meshwork of microfibrils which tend to be oriented, imbedded in a voluminous amorphous matrix which certainly contains pectic compounds and probably proteins (PRESTON and WARDROP 1949). It seems highly significant that in these walls the microfibrils appear to be twisted round each other. It is difficult to see how this twisting could be achieved if the cellulose is being deposited at a cytoplasm-wall interface. The inevitable conclusion seems to be that in these growing walls the cytoplasm interpenetrates the wall—that the wall is not a dead envelope but marks instead the outer limits of the living cytoplasm.

There is some further evidence for concluding that the wall is alive during cell growth. Thus, it is difficult to see how the almost transverse orientation could be maintained if a cell is expanding by intake of water and the wall is merely yielding under stress. The complete lack of any reorientation in growing walls has now been established over a wide field including the algae and it is abundantly clear that the extension in the wall can no longer be considered as a consequence of mechanical extension alone. Again, it is now reasonably certain that during cell extension neither the turgor pressure nor the extensibility of the wall increases. Finally, the observations firstly, that the removal of proteins from a primary wall causes changes in the X-ray diagram of the cellulose in the wall and that, secondly, the microfibrils in such a wall are twisted round each other imply that the cytoplasm and the wall interpenetrate. There is no question but that the primary wall is part of the living system and adjusts itself by growth to changes in cell dimensions.

It is still far from clear, however, how this adjustment is achieved. In general terms, the incorporation of new material in a wall involves the breaking of bonds, the insertion of the new material and the making of bonds. In some instances —*e.g.* many filamentous algae, fungi, and some fibrous cells in higher plants— this process is specifically localised at the tips of cells. The same localisation has been claimed to occur in parenchyma (FREY-WYSSLING 1952) but is denied by WARDROP (1955). In other cases it is said that the incorporation of new material can take place anywhere in a growing wall in the so-called *mosaic* growth (*e.g.* STECHER 1952).

## Literature.

ANDERSON, D. B.: Über die Struktur der Kollenchymzellwand auf Grund mikrochemischer Untersuchungen. Sitzgsber. Akad. Wiss. Wien, Math.-naturwiss. Kl., Abt. 1 136, 429 (1927). — ASTBURY, W. T., and R. D. PRESTON: The structure of the cell wall in some species of the filamentous green alga *Cladophora*. Proc. Roy. Soc. Lond., Ser. B 129, 54 (1940). — ASTBURY, W. T., R. D. PRESTON and A. G. NORMAN: X-ray examination of the effect of removing non-cellulose constituents from vegetable fibres. Nature (Lond.) 136, 391 (1935).

BAILEY, I. W., and T. KERR: The visible structure of the secondary wall and its significance in physical and chemical investigations of tracheary cells and fibres. J. Arnold Arbor. 16, 273 (1935). — BONNER, J.: Zum Mechanismus der Zellstreckung auf Grund der Micellarlehre. Jb. wiss. Bot. 82, 377 (1936).

FREUDENBERG, K., u. H. DÜRR: Kleins Handbuch der Pflanzenanalyse, Bd. 3, S. 142. 1932. — FREY-WYSSLING, A.: Growth of plant cell walls. Symposia Soc. Exper. Biol. 6, 320 (1952). — Submicroscopic morphology of protoplasm. Elsevier 1953. — FREY-WYSSLING, A., A. MÜHLETHALER u. R. W. G. WYCKOFF: Mikrofibrillenbau der pflanzlichen Zellwände. Experientia (Basel) 6, 12, 475 (1948).

HAWORTH, W. N.: The constitution of the sugars. London 1929. — HENGSTENBERG, J., and H. MARK: Über Form und Größe der Mi elle von Cellulose und Kautschuk. Z. Krystallogr. 69, 271 (1928). — HERMANS, P. H., and A. WEIDINGER: Quantitative x-ray investigation on the crystallinity of cellulose fibres. A background analysis. J. Appl. Physics 19, 491 (1948). — HEUSER, E.: Cellulose chemistry. New York: Wiley 1946. — HEYN, A. N. J.: Molecular structure of chitin in plant cell walls. Nature (Lond.) 137, 277 (1936). — HODGE, A. J., and A. B. WARDROP: An electromicroscopic investigation of the cell-wall organisation of conifer tracheids. Nature (Lond.) 165, 272 (1950).

ITERSON jr., G. VAN, K. H. MEYER and W. LOTMAR: Über den Feinbau des pflanzlichen Chitins. Rec. Trav. chim. Pays-Bas 55, 61 (1936).

KREGER, D.: Observations on cell walls of yeast and some other fungi by x-ray diffraction and solubility tests. Biochim. Biophys. Acta 13, 1 (1954).

MAJUMDAR, G. P., and R. D. PRESTON: The fine structure of collenchyma cells in *Heracleum sphondylium* L. Proc. Roy. Soc. B 130, 201 (1941). — MEYER, K. H.: High polymers,

Bd. IV. New York: Interscience 1942. — Middlebrook, M., and R. D. Preston: Spiral growth and spiral structure III. Biochim. et Biophysica Acta 9, 32 (1952). — Misra, P.: Spiral grain in *Pinus longifolia*. Forestry 13, 118 (1939). —Mukherjee, S. M., and H. J. Woods: X-ray and electronmicroscopic studies of the degradation of cellulose by sulphuric acid. Biochim. et Biophysica Acta 10, 499 (1953).

Nicolei, E., and A. Frey-Wyssling: Über den Feinbau der Zellwand von *Chaetomorpha*. Protoplasma (Wien) 30, 401 (1938). — Nicolai, E., and R. D. Preston: Cell wall studies in the Chlorophyceae I. Proc. Roy. Soc. Lond., Ser. B 140, 244 (1952). — Cell wall studies in the Chlorophyceae II. Proc. Roy. Soc. Lond., Ser. B 141, 407 (1953a). — Variability in the x-ray diagram of the cell walls of the marine alga *Spongomorpha*. Nature (Lond.) 171, 752 (1953b). — Norman, A. G.: Biochemistry of cellulose, polyuronides, lignin, etc. Oxford 1937.

Preston, R. D.: The organisation of the cell wall of the conifer tracheid. Philos. Trans. Roy. Soc. Lond. 224, 131 (1934). — The molecular chain structure of cellulose and its botanical significance. Biol. Rev. Cambridge Philos. Soc. 14, 281 (1939). — The fine structure of the wall of the conifer tracheid I. Proc. Roy. Soc. Lond., Ser. B 133, 327 (1946). — The fine structure of the wall of the conifer tracheid II. Proc. Roy. Soc. Lond., Ser. B 134, 202 (1947). — The molecular architecture of plant cell walls. Chapman a. Hall 1952. — Preston, R. D., and A. Allsopp: An x-ray examination of delignified and cellulosan-free cellulose and its significance for the problem of the structure of cell walls. Biodynamica 1939 No 53. — Preston, R. D., and W. T. Astbury: The structure of the wall of the green alga *Valonia ventricosa*. Proc. Roy. Soc. Lond., Ser. B 122, 76 (1937). — The structure of the cell wall of *Valonia*. Nature (Lond.) 173, 203 (1954). — Preston, R. D., and R. B. Duckworth: The fine structure of the walls of collenchyma cells in *Petasites vulgaris* L. Proc. Leeds Phil. Soc. 4, 343 (1946). — Preston, R. D., P. H. Hermans and A. Weidinger: The crystalline-non-crystalline ratio in celluloses of biological interest. J. of Exper. Bot. 1, 344 (1950). — Preston, R. D., and B. Kuyper: Electronmicroscopic investigations of the walls of green algae I. J. of Exper. Bot. 2, 247 (1951). — Preston, R. D., and M. Middlebrook: The fine structure of sisal fibres. J. Textile Inst. 40, 715 (1949). — Preston, R. D., E. Nicolai and B. Kuyper: Electronmicroscopic investigations of the walls of green algae II. J. of Exper. Bot. 41 40 (1953). — Preston, R. D., E. Nicolai, R. Reed and A. Millard: An electronmicroscope study of cellulose in the wall of *Valonia ventricosa*. Nature (Lond.) 162, 665 (1948). — Preston, R. D., E. Nicolai and A. B. Wardrop: Fine structure of cell walls in fresh plant tissues. Nature (Lond.) 162, 957 (1948). — Preston, R. D., and G. W. Ripley: Unpublished. 1954a. — Electron Diffractio diagrams of cellulose microfibrils in *Valonia*. Nature (Lond.) 174, 76 (1954b). — Preston, R. D., and K. Singh: The fine structure of bamboo fibres I. J. of Exper. Bot. 1, 214 (1950). — The fine structure of bamboo fibres II. J. of Exper. Bot. 3, 162 (1952). — Preston, R. D., and A. B. Wardrop: The submicroscopic organisation of the walls of conifer cambium. Biochim. et Biophysica Acta 3, 549 (1949).

Ranby, B. G.: Aqueous colloidal solutions of cellulosic micelles. Acta chem. scand. (København) 3, 649 (1949). — Roelofsen, P. A.: Note on the spiral growth and spiral cell wall structure of sporangiophores of *Phycomyces*. Biochim. et Biophysica Acta 3, 518 (1949).

Stecher, H.: Über das Flächenwachstum der pflanzlichen Zellwände. Mikroskopie (Wien) 7, 30 (1952). — Steward, F. C., and K. Muehlethaler: The structure and development of the cell wall in the *Valoniaceae* as revealed by the electronmicroscope. Ann. of Bot., N. S. 17, 295 (1953).

Wardrop, A. B.: Austral. J. Sci. Res. 1955. — Wardrop, A. B., and R. D. Preston: The fine structure of the wall of the conifer tracheid. V. Biochim. et Biophysica Acta 6, 36 (1950). — The submicroscopic organisation of the cell wall in conifer tracheids and wood fibres. J. of Exper. Bot. 2, 20 (1951). — Wilson, K.: Observations on the structure of the cell walls of *Valonia ventricosa* and of *Dictyosphaeria farulosa*. Ann. of Bot., N. S. 15, 279 (1951). — Wise, L. E.: Wood chemistry. New York: Reinhold 1944.

# Mechanical properties of the cell wall.

By

## R. D. Preston.

With 8 figures.

When a solid body is subjected to a set of balanced forces these forces are transmitted through all parts of the body and the body is said to be *stressed*. The *stress* at any point may be defined in terms of the force (in dynes, gm. weight or kgm. weight) per unit area at that point. A body under stress will usually change its dimensions or shape or both— it is said to be *strained*. In the case of change in dimensions only, the *strain* may be defined as the change in dimension divided by the original value of the dimen- sion *i.e.* the relative value of the change.

As the forces are increased from zero the strain at first varies directly with the stress (*a*, Fig. 1), and the ratio stress/ strain, the *modulus of elasticity*, is then a property of the material of which the body is made. Following this limited range over which the relation is linear there usually comes a stage (*b*) during which the strain increases more rapidly than the stress. If at any point on the

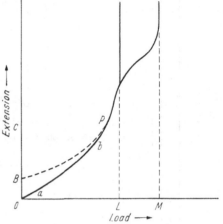

Fig. 1. For explanation, see text.

stress-strain curve such as *P*, Fig. 1, the stress is now gradually removed, the resulting stress-strain curve follows a line such as *PB* and the body no longer recovers its original dimensions at zero stress. The part *BC* of the strain *OC* at *P* is then referred to as *elastic*, *i.e.* recoverable and the part *OB* as *plastic* or non-recoverable. Beyond *P* the curve may follow either of two courses depend- ent on the nature of the deformed body. The body may suddenly collapse or *fail* at a stress *OL*; or, alternatively, the body may "*stiffen*" again and finally fail at a stress *OM*; *OL*, or *OM* is then called the *tensile strength* or the *ultimate strength*.

Stresses may be applied to a body in any or all of three mutually perpendicular directions and give rise to linear strains in the corresponding directions together with displacement or *shear* strains. Study of deformation under stress is there- fore, in the general case, exceedingly difficult especially with anisotropic bodies (see *e.g.* ALFREY 1948). In considering the cell wall we shall, however, concern ourselves chiefly with conditions which can be resolved into the simpler case of unidirectional stress.

## Simple unidirectional tensile stress.

Consider a rod *R*, Fig. 2, cross-sectional area *A*, loaded at one end with a weight *W*. Then the stress is $W/A$. If the rod elongates under this stress by an amount $\Delta L$ then the longitudinal strain is $\Delta L/L$. The ratio between these, the modulus of elasticity, is referred to as Young's Modulus, symbolised by $E$,

$$E = \frac{W\,L}{A\,\varDelta L} \qquad (1)$$

a measure of the resistance to extension. Under such a linear stress, the body also contracts laterally and the ratio between the lateral contraction and the longitudinal extension is called Poisson's ratio.

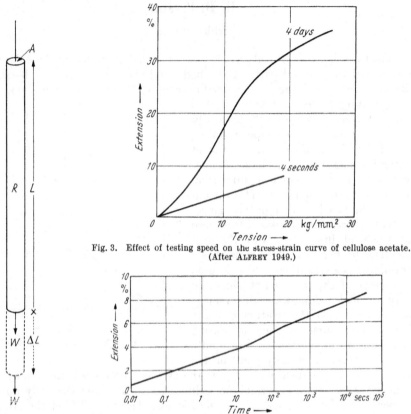

Fig. 3. Effect of testing speed on the stress-strain curve of cellulose acetate.
(After Alfrey 1949.)

Fig. 2. For explanation, see text.

Fig. 4. Creep curve for commercial viscose rayon at 70° F and 65% R.H. at a load of 0.67 gm./denier. (After Press, from Alfrey 1949.)

Equation (1) implies a linear relationship between stress and strain and applies therefore only within the region of $a$, Fig. 1, where the strain is purely elastic and thus recoverable. This is the Hooke's Law region. Beyond this region, part of the strain is plastic and equation (1) is no longer strictly applicable. It should be noted that if the rod, deformed past the elastic limit, is then released so that the load extension curve follows the path $OPB$ (Fig. 1) and then restressed, then the second stress-strain curve differs from the first. The earlier stress had deformed the rod permanently and the stress-strain relationship then depends on the previous history of the sample.

With high polymeric substances such as the cellulose of cell walls, other complications must be borne in mind. The most important are, in essence, three. Firstly, the course of the stress-strain curve depends on the rate at which the stress is applied or the strain is imposed (Fig. 3). Secondly, the stress-strain curve can be quite different if the specimen is wet from that observed in the dry condition. Thirdly, if the specimen is loaded with any weight, the concomitant dimensional change can continue over long periods—the material shows the phenomenom called *creep* (Fig. 4). Creep in plant fibres does not show the relatively simple behaviour of other plastoelastic bodies like rubber, the complexity being probably due to the initial crystallinity and orientation of the material and to the additional crystallisation imposed by the deformation (PRESS 1943). All cellulose materials can be considered somewhat arbitrarily as two-phase systems, one phase being crystalline and the other non-crystalline. The regions of crystallinity contribute most towards high moduli of elasticity and those of low crystallinity mostly towards distribution of stresses leading, among other things, to creep and towards long-rang elasticity. Most of the detailed analyses refer to regenerated celluloses and relatively little work has been done on native cellulose. Incrusting substances, such as lignin, in the latter may be expected to play a considerable role in determining some mechanical properties. The molecular factors involved in the extension of cotton and flax etc. have, however, been studied fairly extensively (*e.g.* HERMANS 1949). Most such fibres are built of cellulose microfibrils with a spiral orientation (see p. 741). It seems that when such a fibre is loaded in tension a tensile stress is developed parallel to the direction of preferred orientation; in addition, a compressive stress is set up between adjacent regions of the microfibril system and a shear stress between them. No attempt, will, however, be made here to assess the validity of these suggestions.

## Young's modulus and extensibility of cellulose fibres.

It has been realised for a long time that plant cells vary enormously in extensibility (see refr. in OPPENHEIMER 1930). Some collenchymatous tissues extend only by some 2.5% at break (*e.g.* collenchyma of *Heracleum sphondylium*) while others extend to 14% or more (*e.g. Rheum rhaponticum*). Similarly, the vascular bundles from leaves of *Plantago* ssp. are said to extend up to 20% although many fibre bundles are almost inextensible. Although these differences may undoubtedly be associated with anatomical features and with variations in chemical composition, there is no doubt that the configuration of the cellulose microfibrils is to a large extent responsible.

In a recent investigation of the relation between mechanical properties of cell walls and the orientation of the constituent cellulose microfibrils, SPARK and PRESTON (unpub.) have made use of the special features of sisal (PRESTON and MIDDLEBROOK 1949). Since sisal is a monocotyledonous plant with relatively slow differentiation of leaf fibres, fibres can be selected which probably differ in little but microfibril orientation forming, therefore, very suitable material for such a study. The results of this investigation are not yet fully analysed but nevertheless allow the following general statements to be made.

Sisal fibres the cellulose microfibrils of which make slow spirals round the walls show extensions when wet of up to 20% and this extension is fully recoverable. It is accompanied by changes in the X-ray diagram which prove a steepening of the constituent spirals. This high elastic extensibility coupled with changes in orientation of the cellulose component seems to constitute a new observation. The quantitative change in the angle of the spiral as the cell extends and contracts suggests that the spiral winding is extending in length at constant spiral girth. This in turn implies that the spiral winding is itself extending parallel to its length. It is of interest, therefore, to recall that the orientation of microfibrils in fibre walls is far from perfect; so that elongation could occur by a straightening out of sinuous microfibrils coupled with a re-orientation more nearly parallel to fibre length. Elastic recovery would then involve either a second wall component or, for instance, lateral adhesions between microfibrils. There is no evidence for any pronounced entropy factor. The full extension to break of individual fibres is considerably in excess of 20% reaching values of 60% or more. Some 40% of this extension is, of course, non-recoverable. The extension to break in sisal fibre with steep spirals, selected from positions nearer the leaf tip is, on the other hand, small of the order of only a few per cent. It seems fairly certain that this difference in extensibility is correlated specifically with orientation.

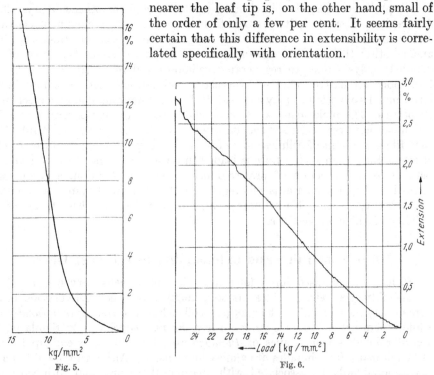

Fig. 5.                     Fig. 6.

Fig. 5. Load/extension curve of a single sisal fibre. The microfibrils initially made an angle of *ca* 50° to cell length.
Fig. 6. As in Fig. 5, except the spiral angle is 10°.

Two representative load-extension curves of single delignified fibres are given in Figs. 5 and 6. These were recorded automatically on a microextensometer specifically designed for studying single cells. The rate of loading in each case is of the order of 0.4 kgm. per squ. mm. of wall per second and the length of cell under test was about 2 mm. The fibre with a slow wall spiral (Fig. 5) has extended rapidly to 17% under a relatively small load of about 15 kgm/mm². The extension is fully elastic only over the first 1% extension and only within this range has Young's modulus any precise significance. The fibre with a very much steeper spiral (Fig. 6) has extended only 2.7% under a load of some 25 kgm/mm² so that the difference in extensibility with wall structure is very marked. The load-extension curve here is almost linear but even so, the fibre recovered only to an extension of 1.3% when the load was removed. Here again, therefore, only the initial Young's modulus has any real significance. Average values for some mechanical properties of these fibres are given in Table 1.

These figures are comparable with those reported for other fibres and for regenerated cellulose which refer usually, however, to fibre bundles. According to Meyer and Lotmar (1936), for instance, Young's modulus for dry ramie fibres lies in the range 5,000—7,000 kgm/mm² while highly oriented viscose

*Table 1.*

| Property | Slow wall spiral (50°) | Steep wall spiral (10°) |
|---|---|---|
| Extension at break . . . . . . . . . | 14.5% | 2% |
| Breaking strength · . . . . . . . . . | 8.3 kgm/mm² | 50 kgm/mm² |
| YOUNG's modulus | | |
| initial . . . . . . . . . . . . . | 300 kg/mm² | ca. 10,000 kgm/mm² |
| average over whole curve . . . . . | 55 kg/mm² | 2,500 kgm/mm² |

fibres reach values of 25,000–40,000 kgm/mm². Figures for disoriented cellulose are much lower (ramie decreasing to some 1500 kgm/mm²) and there is also a very considerable reduction on wetting the fibre (ramie going down to 1980 kgm per mm². Celluloses are therefore much more extensible wet than dry and when disoriented than oriented. MEREDITH (1946) has reported similar behaviour in cotton. The figures reported above for sisal are for the *first* stretching. Subsequent restretching leads to much lower figures, so that this first stretching causes a major disturbance in structure.

# Breaking strength.

The breaking strength of plant fibres, already quoted in Table 1 for sisal has been extensively investigated. We may perhaps note here only the most recent work of WARDROP (1951) as an example of the order of magnitude quoted in the literature. WARDROP's work is, however, of especial importance since he shows a clear relationship between the breaking strength under tension of the woods he investigated and the corresponding density and molecular architecture. From his published results we can compute roughly the tensile strength of the wall material (from the overall tensile strength and density, assuming the density of the wall substance to be 1.5) as shown in Table 2.

Table 2. *Tensile strength and spiral structure in wood.*

| Species | Tensile strength kgm/mm² | Spiral angle determined by the X-ray method ° | Species | Tensile strength kgm/mm² | Spiral angle determined by the X-ray method ° |
|---|---|---|---|---|---|
| *Pinus* . . . . | 13.4 | 25 | *Pinus* . . . . | 17.2 | 41 |
| *Radiata* . . . . | 14.0 | 19 | *Radiata* . . . . | 16.2 | 30 |
| Late . . . . . | 22.0 | 16 | Early. . . . . | 11.9 | 24 |
| Wood. . . . . | 21.4 | 14 | Wood. . . . . | 17.5 | 15 |
| Specimen . . . | 25.0 | 13 | Specimen . . . | 19.9 | 14 |
| | 20.8 | 12 | | 15.5 | 13 |
| | 26.5 | 11 | | 22.7 | 11 |
| | 33.6 | 10 | | | |
| | | | *Pinus* . . . . | 7.6 | 46 |
| | | | *Radiata* . . . . | 11.6 | 41 |
| | | | Late . . . . . | 15.9 | 40 |
| | | | Wood. . . . . | 8.9 | 37 |
| | | | Specimen 4 . . | 20.5 | 33 |
| | | | | 24.8 | 22 |

There can be no doubt but that there is a strong connection between breaking strength and wall structure of the same general type as that between YOUNG's modulus and structure.

The mechanism of failure under tension has been a subject of some controversy, but it is now generally agreed (see *e.g.* HERMANS 1949) that slipping between cellulose microfibrils plays only a minor role. It is probable that the microfibrils break across, fracturing primary valence bonds. The weakness in fibres with slow spirals would then lie possibly in the consequent lack of uniformity in loading.

The figures quoted above refer to air-dried material. If the water content is increased then the tensile strength of wood decreases. This contrasts markedly with the increase in tensile strength observed when single fibres (Hermans 1949) or single conifer tracheids (Wardrop 1951) are wetted. The increase in strength in this latter case, amounting to some 10% or more, is probably due to the action of water as a lubricant, allowing neighbouring cellulose units so to move that a more uniform distribution of internal stress is achieved on loading (Hermans 1949). The decrease in strength of whole tissues on wetting is possibly due to the effect of intercellular or intermicrofibrillar substances.

## Tensions in the walls of living cells.

While it is no part of the present purpose to discuss the interrelations of mechanical properties of cell walls and cell growth, it is instructive to compare the tensions expected to be present in living cells with the breaking strengths quoted above.

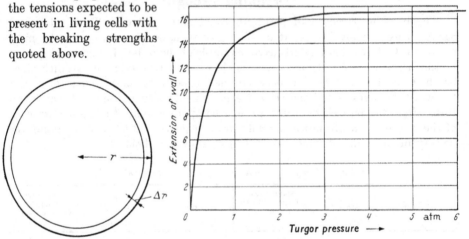

Fig. 7. For explanation, see text.

Fig. 8. The relation between the turgor pressure of a parenchyma cell and the percentage increase in cell dimensions over the plasmolysed dimensions (recalculated from Oppenheimer 1930).

Consider first a spherical cell, free from its neighbours (Fig. 7) $r$ mm. in diameter and with a wall $\Delta r$ mm. thick. Then if the internal pressure due to turgor is $P$ gm./mm² the wall will be stretched under a tension $T$ which may be calculated from the relation

$$P = \frac{2T \cdot \Delta r}{r}.$$

Taking $P$ as 200 gm/mm² on an average, $r$ as 30 $\mu$ and $\Delta r$ as 2 $\mu$, then $T$ turns out to be 1.6 kgm/mm². In a cylindrical cell the tension is anisotropic; the transverse tension is $\frac{2T \Delta r}{r}$ and the longitudinal tension $\frac{T \Delta r}{r}$. The transverse tension is thus about 3.2 kgm/mm². Remembering that the volume of the cellulose in parenchymatous walls may be as low as 10% of the wall volume, while it is of the order of 60% in the fibre walls discussed above, then, ignoring the effect of incrusting substances like pectin, these figures must be multiplied by a factor of about six for comparison with those for fibres. They then come dangerously close to some of the figures quoted in Table 2 and it is then clear that the extension of cells under turgor forces is often not fully elastic.

This may be confirmed by examining the studies which have been reported on the effects of changing turgor on cell extension. The relatively early results of OPPENHEIMER (1930) may be quoted as an example. OPPENHEIMER placed parenchyma cells of the pith of *Taraxacum officinale* successively in a series of solutions of sugar and determined the percentage extension of the cell over the length in the fully plasmolysed condition. The results, recalculated in such a way as to be analogous to a load-extension curve, are given graphically in Fig. 8. The extension of 16% and the curved nature of the graph suggest that the extension is not fully elastic. One peculiarity rests in the fact that the curve is concave to the load axis, not convex as in Fig. 5. This curve, it is to be remembered, corresponds to an unloading of the wall. If the reloading followed a similar course—which is probable though not certain—then the extension is easy at first but the wall stiffens later. This is perhaps to be explained by the nature of the wall. The microfibrils in parenchymatous walls are commonly sinuous. It may be that they straighten out before the load is fully borne by the extension of the microfibrils themselves.

## Literature.

ALFREY, jr., T.: Mechanical behaviour of high polymers. High Polymers Vol. VI. New York: Interscience 1948.

HERMANS, P. H.: The physics and chemistry of cellulose fibres. New York: Elsevier 1949.

MEREDITH, R.: J. Textile Inst. **37** (9), 205 (1946).

OPPENHEIMER, H. R.: Dehnbarkeit und Turgordehnung der Zellmembran. Ber. dtsch. bot. Ges. 48, 192 (1930).

PRESS, J. J.: Flow and recovery properties of viscose rayon yarn. J. Applied Phys. 14, 224 (1943). — PRESTON, R. D., and M. MIDDLEBROOK: The fine structure of sisal fibres. J. Textile Inst. 40, 715 (1949).

WARDROP, A. B.: Cell wall organisation and the properties of the xylem. I. Austral. J. Sci. Res. B 4, 391.

# Namenverzeichnis. — Author Index.

Die *kursiv* gesetzten Seitenzahlen beziehen sich auf die Literatur.
Page numbers in *italics* refer to the bibliography.

Baker, J. R. 251, *265*, 690, *718*.

Baker, R. F., u. F. W. S. Modern 595, *603*.

— s. D. C. Pease 462, *504*.

Bakke, A. L., u. N. L. Noecker 226, *240*.

— s. S. Dunn 235, *240*.

Balcar, J. O., W. D. Sansum u. R. T. Woodyatt 223, *240*.

Baldwin, E. *300*.

Ball, C. D. s. S. S. T. Djang *266*.

Ballentine, R., u. D. G. Stephens 249, *265*.

Baly 688.

Bandurski, R. S. s. J. Bonner 213, 214, 215, 217, *218*.

— s. A. Millerd 593, *609*.

Banga, I., K. Laki u. A. Szent-Györgyi 578, *603*.

— u. A. Szent-Györgyi 323, *337*.

Bank, O. 653, *658*.

Baptiste, E. C. D. 207, *218*.

Barber, H. N. 492, *500*.

Barbu, E. s. M. Joly *380*.

Barclay 689.

Barer, R. 418, *500*.

Barg, T. 407, *409*.

Barkas, W. W. 188, *191*.

Barker, J. W. s. T. A. Bennet-Clark 212, *218*.

Barnes u. Hampton 228.

Barnes, H. T., u. T. C. Barnes 190, *191*.

Barnes, T. C., u. E. J. Larson 190, *191*.

Barney, C. W. 207, *218*.

Barnum, C. P., u. R. A. Huseby 588, 590, 591, *603*.

Barratt, W. s. A. Coehn 376, *379*.

Barrnett, R. J., u. A. M. Seligman 252, *265*, 427, 432, *500*.

Barron, E. S. G. *300*.

Barry, V. C., T. G Halsall, E. L. Hirst u. J. K. V. Jones 404, *409*.

Bartels, F. 587, *604*.

Bartholomew, E. T. 637, 640, *645*.

Bartl, L. H. s. E. Husemann 673, *719*.

Bartlett, L. E. s. H. F. Rosene 207, 213, *221*.

Bartley, W., u. R. E. Davies *604*.

Barton 712.

Barton, A. D. s. A. K. Laird 590, *608*.

Barton, L. V. s. W. A. Crocker 195, *219*.

Bartunek 677.

Bary, de 336.

Bass, A. D. s. H. S. di Stefano *501*.

Bass, L. W. s. P. A. Levene 421, *503*.

Bateman, A. J. 91, *122*.

— u. K. Mather 91, *122*.

Bateman, J. B. s. R. Höber *502*.

Bateson, W. 8, *17*, 40.

— u. A. F. Gairdner *54*.

Bath, J. D. s. O. L. Sponsler 197, *221*, 224, 228, 241, 649, *660*.

Bauch, R. 480, *500*.

Bauer 216.

Baumgärtel, O. 404, 405, *409*.

Baur, E. 23, 24, 25, *54*, 550, 551, *560*.

Bautz, E., u. H. Marquardt 141, *164*, 601, 604.

— s. H. Marquardt 601, *603*, *609*.

Bawn, C. E. H., E. L. Hirst u. G. T. Young 187, *191*.

Bayliss, M., D. Glick u. R. A. Siem 254, *265*.

Bayliss, W. M. 353, *379*.

Beadle, G. W. 12, 17, 20, 28, *54*.

— u. E. L. Tatum 12, *17*.

— s. E. L. Tatum 12, *18*.

Beale, G. H. 29, 51, *54*.

Beams u. Evans 474.

Beams, H. W. 246, *265*.

— u. T. N. Tahmisian 595, 600, *604*.

— s. R. L. King *607*.

Beauchamp, P. de 145, *164*.

Beauverie, J. 519, *561*.

Becker, E. s. J. König *720*.

Becker, W. A. 156, *164*.

Bernhard, W. s. F. Hagenau 133, *165*.

Beevers, C. A. s. T. R. R. McDonald 669, *720*.

Behrens, M. 253, *265*, 417, *500*.

Beijer, J. J. s. L. Algeria 523, *560*.

Belar 129, 133, 145, 146, 148, 149, 150, 155, 159.

Belehrádek, J. *379*, 485, 486.

Bell 190, 603.

Bell, F. O. s. W. T. Astbury 421, 452, *500*.

Bell, L. G. s. W. G. P. Lamb *267*.

Bell, R. P. s. R. A. Robinson *193*.

Benda 580.

Bendich, A. 421, 434, *500*.

Bennet-Clark, T. A. 212, *218*.

— u. D. Bexon 212, 216, *218*.

— A. D. Greenwood u. J. W. Barker 212, *218*.

Bennett, H. S. 252, *265*.

Bensley, S. H. s. R. R. Bensley 578, *604*.

Bensley, R. R. 332, *337*, 577, 578, *604*.

— u. S. H. Bensley 578, *604*.

— u. I. Gersh 588, *604*.

— u. N. L. Hoerr 588, *604*.

Benton, J. G. s. A. W. Pollister *268*.

Berger, H. 379, *379*.

Berger, R. E. s. J. J. Biesele 487, *500*.

— s. R. C. Mellors *267*.

Bergmann, M. *337*, 698.

— u. C. Niemann 323, *337*, 677.

Bernal, J. D. 185, 188, *191*, 706.

— u. R. H. Fowler 173, 177, 179, 181, 185, *191*.

Bernard, Claude 390.

Bernhard, W., F. Haguenau u. Ch. Oberling 441, 457, 469, 470, *500*.

— s. Ch. Oberling 455, 469, *504*, 583, *610*.

Bernstein, M. H., u. D. Mazia *379*.

Bersa 353.

Berthet, J., L. Berthet, F. Appelmans u. C. de Duve 586, *604*.

Berthet, L. s. J. Berthet 586, *604*.

— s. Chr. de Duve 586, 600, *606*.

Best, R. J. 250, *265*.

Bethe, A. 627, *645*.

Beutler, E. H. s. P. E. Hartman 579, *607*.

Beutner, E. H. s. St. Mudd 580, 581, 601, 603, *609*.

Beutner, R. 365, 376, *379*.

Bevan, E. J. 692, 710, 712, 735.

Bexon, D. s. Bennet-Clark, T. A. 212, 216, *218*.

Beythien, A. s. H. v. Guttenberg 207, 209, 215, *219*.

Bhattacharjee, D., u. A. K. Sharma 254, *265*.

Biale, J. B. 202, *218*.

Biebl, R. 208, *218*.

Biegert, F. s. E. Bünning 134, 135, *164*.

Bielig, H.-J., G. A. Kausche u. H. Haardick 579, 580, 581, 582, *604*.

Biesele, J. J., R. E. Berger, M. Clarke u. L. Weiss 487, *500*.

Binder, M. 554, *561*.

— s. J. Schwemmle 44, *57*.

Bingham, E. C. 341, 343, 344, 347, *379*.

Binkley, F. 431, *500*.

Birbeck, M. S. C. 595, 603, *604*.

# Sachverzeichnis.

## (Deutsch-Englisch.)

Ä, Ö, Ü sind wie Ae, Oe, Ue eingereiht.

Bei gleicher Schreibweise in beiden Sprachen sind die Stichworte jeweils einfach aufgeführt.

Plättchen im Cytoplasma von Diatomeen, *platelets in cytoplasm of diatoms* 138.
plagiotroper Sproß, *plagiotropic shoot* 60.
Plasma s. a. Cytoplasma, Protoplasma.
Plasmaalteration, *cytoplasmic alteration* 585.
Plasmagene, *plasmagenes* 6, 140, 462.
Plasmalemma 651.
Plasmapartikel, *particles of cytoplasm* 133, 141.
Plasmasystrophe, *systrophe, protoplasmic* 585.
Plasmateilung (Cytokinese), *cytokinesis* 127
Plasmochrome, *plasmochromes* 565.
Plasmodesmen, *plasmodesmata* 334, 336, 723.
— in den Primärwänden, *in the primary walls* 723.
Plasmodien, *plasmodia* 126, 129.
Plasmolyse, *plasmolysis* 345, 383.
— in Blattzellen, *in leaf cells* 59.
plasmolytische Bestimmung, *plasmolytic measurement* 212.
Plasmon 37, 585.
—, Aufspaltung, *segregation* 48.
—, Eigenschaften, *properties* 48.
—, Epilobium 45.
—, Heterozygotie, *heterozygosity* 47.
—, Oenothera 43.
plasmonempfindliches Gen, *plasmon-sensitive gene* 46.
Plasmonsegregation 48.
Plasmosin 332.
Plasten *plasts* 137.
Plastid, Pigmente, *plastid, pigments* 278.
Plastiden, *plastids* 124, 133, 141.
— Konstanz, *constancy* 545.
—, Kontinuität, *continuity* 142.
—, Wassergehalt, *water content* 198.
Plastidenanlagen, *plastid primordia* 140.
Plastidenentmischung, *plastid segregation* 24.
Plastidenmutation, *plastid mutation* 23, 555.
Plastidenübertragung durch Pollen, *plastid, transmission of paternal* 47.
Plastidenveränderungen, Gen-induzierte, *plastid changes, gene-induced* 28.
Plastidom 44.
Plastin 434.
Plastizität des Eises, *plasticity of ice* 178.
— des Zellkerns, *of the nucleus* 474.
Pleiotropie (Polyphaenie), *pleiotropy* 2, 9.
— zwischen "white eye" und "vestigial wing" in Drosophila 9.
plektoneme Schraubenwindungen in Chromosomen, *plectonemic helices in chromosomes* 464.
Podophyllin 488.
polare Moleküle, *polar molecules* 373.
polarimetrische Methode, *polarimetric method* 227.
Polarisationsmikroskopie, *polarization microscopy* 461, 470.
Polarität in biologischen Objekten, *polarity in biological objects* 373.
— der Zelle, *of the cell* 132, 134ff.
Polkappen, *attraction plates* 441.

Pollen, Keimung nach Schichtung des Protoplasmas durch Zentrifugieren, *pollen, germination following centrifugal stratification of protoplasm* 246.
—, Übertragung durch Plastiden, *paternal plastid transmission* 47.
—, Unverträglichkeit, *incompatibility* 5.
Pollenkorn (Tradescantia), *pollen grain* 483.
Pollenmutterzelle (Lilium), *microsporocyte* 423.
—, Entwicklung (Lilium), *development* 422.
Pollenmutterzellen, *microsporocytes* 414.
Pollensterilität, *male sterility* 40, 41.
— beim Mais, *in maize* 41.
Polstrahlen, Viscosität, *astral rays, viscosity* 350.
Polyäthylen, *polyethylene* 327.
Polydispersitätsspektrum 676.
polyenergide Organismen, *polyenergid organisms* 126, 163.
Polygalakturonsäure, *polygalacturonic acid* 696, 735.
Polygalit, *polygalitol (1,5-D-sorbitan)* 285.
Polymerisation, DNS-, *polymerization, DNA* 483.
Polymetaphosphat, *polymetaphosphate* 405, 582.
Polymorphie, *polymorphism* 330.
— der Cellulose, *of cellulose* 679.
Polynucleotide, *polynucleotides* 465.
Polynucleotidketten, *polynucleotide chains* 453.
Polyosen, *polyoses* 691.
Polypeptide, *polypeptides* 299, 465.
Polypeptidketten, *polypeptide chains* 315, 317, 318, 452.
Polyphenol-oxydase, *polyphenoloxidase* 278.
Polyploidie, *polyploidy* 496.
Polyploidisierung, endomitotische, *polyploidization, endomitotic* 130.
Polysaccharid, *polysaccharide* 283, 287.
—, Anhäufung verursacht durch Dehydratation (Sukkulente), *accumulation caused by dehydration (succulentes)* 204.
— Bakterien-, *bacterial* 284.
Polysaccharide, Lokalisation in der Zelle, *polysaccharides, intracellular localization* 251.
— zuckerunähnliche, *non sugar-like* 671.
Polyuronide, *polyuronides* 692, 693.
— in den Membranen, *in cell walls* 731.
— und optisches Verhalten der Membran, *and optical behaviour of cell wall* 734.
Porphyrinverbindungen, *porphyrin compounds* 565.
„Position"-Effekte des Centromers und der Nucleolarzonen, *position effects of the centromere and the nucleolar zones* 466.
positive Doppelbrechung der Spindel, *positive birefringence of the spindle* 470.
Potential, kritisches, von Bakterien, *potential, critical, of bacteria* 377.
—, Strömungs-, *stream* 376.
Potentiale, Membran-, *potentials, "membrane"-* 376.

# Subject Index.

## (English-German.)

Ä, Ö, Ü are taken as Ae, Oe, Ue.

Where English and German spelling of a word is identical the italicised (German) entry is omitted.

CPSIA information can be obtained at www.ICGtesting.com
Printed in the USA
LVOW03s1948071214

417657LV00006B/140/P